FLORISTIC INVENTORY OF TROPICAL COUNTRIES:

The Status of Plant Systematics, Collections, and Vegetation, plus Recommendations for the Future

DAVID G. CAMPBELL AND H. DAVID HAMMOND, EDITORS

The New York Botanical Garden

A Joint Research Project of

THE NEW YORK BOTANICAL GARDEN
THE ARNOLD ARBORETUM
THE MISSOURI BOTANICAL GARDEN
THE WORLD WILDLIFE FUND

Published by

THE NEW YORK BOTANICAL GARDEN

Library of Congress Cataloging-in-Publication Data

Floristic inventory of tropical countries: the status of plant systematics, collections, and vegetation, plus recommendations for the future/David G. Campbell and H. David Hammond, editors; a joint research project of the New York Botanical Garden . . . [et al.].

   p. cm.
  Includes bibliographies and index.
  ISBN 0-89327-333-3
  1. Botany—Tropics. 2. Phytogeography—Tropics. 3. Botany—Tropics—Classification. 4. Botany—Tropics—Catalogs and collections. 5. Vegetation surveys—Tropics. I. Campbell, David G. II. Hammond, H. David, 1924–   . III. New York Botanical Garden.
  QK474.5.F56 1988   88-26058
  581.909′13—dc19   CIP

                            Scientific Publications Department
                                 New York Botanical Garden
                               Bronx, New York 10458, U.S.A.

    Copyright·© 1989 by The New York Botanical Garden. All rights reserved
                     Printed in the United States of America

Contents

FOREWORD   F. R. Fosberg ix

Section One: Introduction and Synthesis 1

INTRODUCTION   Ghillean T. Prance 3
THE IMPORTANCE OF FLORISTIC INVENTORY IN THE TROPICS   David G. Campbell 5

Section Two: Regional Reports 31

I. Asia

CHINA   Wang Huen-pu, Chen Sing-chi and Wang Si-yu 35
THE PHILIPPINES   Benito C. Tan and Justo P. Rojo 44
THAILAND   Tem Smitinand 63
VIETNAM, KAMPUCHEA AND LAOS   Maurice Schmid 83
SUNDALAND   Peter S. Ashton 91
JAVA   M. M. J. van Balgooy 100
SUMATRA   W. J. J. O. de Wilde 103
SULAWESI (CELEBES)   E. F. de Vogel 108
THE MOLUCCAS   E. F. de Vogel 113
NEW GUINEA   Peter F. Stevens 120
INDIA   Karolyn Kendrick 133
SRI LANKA   Karolyn Kendrick 141

II. Australia and Pacifica

AUSTRALIA   A. P. Kershaw and T. Whiffin 149
FIJI   J. Ash and S. Vodonaivalu 166
NEW CALEDONIA   Karolyn Kendrick 177
HAWAIIAN ISLANDS   D. Frame, W. L. Wagner, D. R. Herbst and S. H. Sohmer 181

III. Africa

WEST AFRICA (SIERRA LEONE TO NIGERIA)   F. Nigel Hepper 189
GABON   F. J. Breteler 198

EQUATORIAL AFRICA (CAMEROUN, PEOPLE'S REPUBLIC OF THE CONGO, CENTRAL
AFRICAN REPUBLIC, ANGOLA, ZAIRE, RWANDA AND BURUNDI)
Karolyn Kendrick 203
EAST AFRICA (KENYA, TANZANIA AND UGANDA)   R. M. Polhill 217
TANZANIA   J. C. Lovett 232
MADAGASCAR   (Laurence J. Dorr, Lisa C. Barnett and Armand
Rakotozafy 236

IV. Central America

OAXACA, MEXICO   David H. Lorence and Abisaí García Mendoza 253
TRANSISTHMIC MEXICO (CAMPECHE, CHIAPAS, QUINTANA ROO, TABASCO AND
YUCATAN)   Jerzy Rzedowski and Graciela Calderón de Rzedowski 270
GUATEMALA   Margaret Tebbs 281
BELIZE   Rachel J. Hampshire 286
HONDURAS   Cirilo Nelson 290
EL SALVADOR   Rachel J. Hampshire 295
NICARAGUA   Susan Y. Sutton 299
COSTA RICA   Luis D. Gómez 305
PANAMA   Rachel J. Hampshire 309

V. Caribbean

CUBA   Rene Pablo Capote López, Rosalina Berazaín Iturralde and Angela
Leiva Sánchez 315
HISPANIOLA   T. Zanoni 336
PUERTO RICO   José L. Vivaldi 341
THE LESSER ANTILLES   Richard A. Howard 347

VI. South America

COLOMBIA   Enrique Forero 353
VENEZUELA   Otto Huber and Dawn Frame 362
THE GUIANAS   J. C. Lindeman and S. A. Mori 375
NORTHWEST SOUTH AMERICA (COLOMBIA, ECUADOR AND PERU   Alwyn
Gentry... 392
BRAZILIAN AMAZON   Douglas C. Daly and Ghillean T. Prance 401
EASTERN, EXTRA-AMAZONIAN BRAZIL   Scott A. Mori 427
BOLIVIA   James C. Solomon 455

Section Three: Appendices—Collection Methods and Historical
Reviews.. 465

COLLECTION AND PREPARATION OF BARK AND WOOD SAMPLES
Ben J. H. ter Welle 467
COLLECTION AND PREPARATION OF KARYOLOGICAL SAMPLES   W. Morawetz 469

COLLECTING AND GERMINATING FERNSPORES M. C. Roos and G. P. Verduyn	471
COLLECTION AND PREPARATION OF POLLEN SAMPLES Annick Le Thomas	474
COLLECTING TROPICAL GERMPLASM Michael J. Balick	476
COLLECTION AND PREPARATION OF PALM SPECIMENS Michael J. Balick	482
BRYOLOGICAL STUDIES IN THE TROPICS William R. Buck and Barbara M. Thiers	484
REVIEW OF MYCOLOGICAL STUDIES IN THE NEOTROPICS Florence H. Nishida	494
QUANTITATIVE INVENTORY OF TROPICAL FORESTS David G. Campbell	523
Index	535

List of Authors

Julian Ash, School of Natural Resources, The University of the South Pacific, P.O. Box 1168, Suva, Fiji. Present Address: Botany Dept., Australian National University, P.O. Box 4, A.C.T., Australia

Peter S. Ashton, Arnold Arboretum, Harvard University, 22 Divinity Avenue, Cambridge, Massachusetts 02138, U.S.A.

M. M. J. van Balgooy, Rijksherbarium, Schelpenkade 6, P.O. Box 9514, 2300 RA Leiden, The Netherlands

Michael J. Balick, Institute of Economic Botany, New York Botanical Garden, Bronx, New York 10458, U.S.A.

Lisa C. Barnett, New York Botanical Garden, Bronx, New York 10458, U.S.A.

Rosalina Berazaín Iturralde, Jardín Botánico Nacional, Carretera del Rocío, Km 3½, Calabazar Boyeros, Ciudad Habana, Cuba

F. J. Breteler, Herbarium Vadense, Laboratory for Plant Taxonomy and Plant Geography, 37 Gen. Foulkesweg, P.O. Box 8010, 6700 ED Wageningen, The Netherlands

William R. Buck, Cryptogamic Herbarium, New York Botanical Garden, Bronx, New York, U.S.A.

Graciela Calderón de Rzedowski, Instituto de Ecología, Centro Regional del Bajío, Apartado Postal 386, 61.600 Pátzcuaro, Michoacán, Mexico

David G. Campbell, New York Botanical Garden, Bronx, New York 10458, U.S.A.

Rene Pablo Capote López, Herbario Academia de Ciencias, Instituto de Ecología y Sistemática, Calle 212, No. 17A09 esq. 19, Atabey, La Habana, Cuba

Chen Sing-chi, Institute of Botany, Academia Sinica, Beijing, People's Republic of China

Douglas C. Daly, New York Botanical Garden, Bronx, New York 10458, U.S.A.

Laurence J. Dorr, New York Botanical Garden, Bronx, New York 10458, U.S.A.

Enrique Forero, Instituto de Ciencias Naturales, Universidad Nacional, Apartado 7495, Bogotá, Colombia. Present Address: Missouri Botanical Garden, P.O. Box 299, St. Louis, Missouri 63166, U.S.A.

F. R. Fosberg, Dept. of Botany, National Museum of Natural History, Smithsonian Institution, Washington, D.C. 20560, U.S.A.

Dawn Frame, New York Botanical Garden, Bronx, New York 10458, U.S.A. Present Address: Dept. of Botany, University of California, Berkeley, California 94720, U.S.A.

Abisaí García Mendoza, Herbario Nacional, Instituto de Biología, Universidad Nacional Autónoma de México (U.N.A.M.), Apartado 70-233 Deleg., Coyoacán, C.U., 04150 México, D.F., Mexico

Alwyn Gentry, Missouri Botanical Garden, P.O. Box 299, St. Louis, Missouri 63166-0299, U.S.A.

Luis D. Gómez, P.O.B. 73, San Vito, Coto Brus, Costa Rica

Rachel J. Hampshire, Flora Mesoamericana Project, British Museum (Natural History), Cromwell Road, London SW7 5BD, United Kingdom

F. Nigel Hepper, Herbarium, Royal Botanic Gardens, Kew, Richmond, Surrey TW9 3AB, United Kingdom

D. R. Herbst, U.S. Fish & Wildlife Service, Environmental Services, Pacific Islands Office, P.O. Box 50167, Honolulu, Hawaii 96850, U.S.A.

Richard A. Howard, Arnold Arboretum, Harvard University, 22 Divinity Avenue, Cambridge, Massachusetts 02138, U.S.A.

Otto Huber, Apartado 80.405, Caracas 1080-A, Venezuela

Karolyn Kendrick, Free-lance Writer, 254 E. Seventh Street, New York City, New York 10009, U.S.A.

A. P. Kershaw, Department of Geography, Monash University, Clayton, Victoria 3168, Australia

Angela Leiva Sánchez, Directora, Jardín Botánico Nacional, Carretera del Bocío, Km 3½, Calabazar, Boyeros, Ciudad Habana, Cuba

J. C. Lindeman, Institute of Systematic Botany, State University of Utrecht, Heidelberglaan 2, P.O. Box 80.102, 3508 TC Utrecht, The Netherlands

David H. Lorence, Herbarium, Pacific Tropical Botanical Garden, P.O. Box 340, Lawai, Kauai, Hawaii 96765, U.S.A.

J. C. Lovett, c/o Brooke Bond, Box 40, Mufindi, Tanzania

Wilfried Morawetz, Institut für Botanik und Botanischer Garten der Universität Wien, Rennweg 14, A-1030 Vienna, Austria

Scott A. Mori, New York Botanical Garden, Bronx, New York 10458, U.S.A.

Cirilo Nelson, Departmento de Biología, Universidad Nacional, Tegucigalpa, Honduras

Florence H. Nishida, Natural History Museum of Los Angeles, 900 Exposition Blvd., Los Angeles, California 90007

R. M. Polhill, Herbarium, Royal Botanic Gardens, Kew, Richmond, Surrey TW9 3AB, United Kingdom

Ghillean T. Prance, New York Botanical Garden, Bronx, New York 10458, U.S.A. Present address: Director, Royal Botanic Garden, Kew, Richmond, Surrey TW9 3AB, United Kingdom

Armand Rakotozafy, Centre National de Recherche de Tsimbazaza (C.N.R.T.), Parc de Tsimbazaza, Antananarivo, Madagascar

Justo P. Rojo, Forest Products Research & Development Institute, College, Laguna, The Philippines

M. C. Roos, Institute of Systemic Botany, State University of Utrecht, Heidelberglaan 2, P.O. Box 80.102, 3508 TC Utrecht, The Netherlands

Jerzy Rzedowski, Instituto de Ecología, Centro Regional del Bajío, Apartado Postal 386, 61.600 Pátzcuaro, Michoacán, Mexico

Maurice Schmid, Laboratoire de Phanérogamie, Muséum National d'Histoire Naturelle, 16, Rue Buffon, F-75005 Paris, France

Tem Smitinand, The Forest Herbarium, Royal Forest Department, Bangkok 10900, Thailand

Seymour H. Sohmer, Department of Botany, Bishop Museum, P.O. Box 299, Honolulu, Hawaii 96817, U.S.A.

James C. Solomon, Missouri Botanical Garden, P.O. Box 299, St. Louis, Missouri 63166, U.S.A. Present Address: Casilla 20206, La Paz, Bolivia

Peter F. Stevens, Harvard University Herbaria, Harvard University, 22 Divinity Avenue, Cambridge, Massachusetts 02138, U.S.A.

Susan Y. Sutton, Flora Mesoamericana Project, Dept. of Botany, British Museum (Natural History), Cromwell Road, London SW7 5BD, United Kingdom

Benito C. Tan, Institute of Biological Sciences, University of The Philippines at Los Baños, The Philippines

Margaret Tebbs, Flora Mesoamericana Project, Dept. of Botany, British Museum (Natural History), Cromwell Road, London SW7 5BD, United Kingdom

Barbara M. Thiers, Cryptogamic Herbarium, New York Botanical Garden, Bronx, New York 10458, U.S.A.

Annick Le Thomas, Laboratoire de Phytomorphologie Générale et Comparée, École Pratique des Hautes Études, 16, rue Buffon, F-75005 Paris, France

G. P. Verduyn, Institute of Systematic Botany, State University of Utrecht, Heidelberglaan 2, P.O. Box 80.102, 3508 TC Utrecht, The Netherlands

José L. Vivaldi, Departamento de Recurso Naturales, P.O. Box 5887, Puerta de Tierra, Puerto Rico 00906

S. Vodonaivalu, Institute of Natural Resources, The University of the South Pacific, P.O. Box 1168, Suva, Fiji

E. F. de Vogel, Rijksherbarium, Schelpenkade 6, P.O. Box 9514, 2300 RA Leiden, The Netherlands

Warren L. Wagner, Department of Botany, Bishop Museum, P.O. Box 299, Honolulu, Hawaii 96817, U.S.A.

Ben J. H. ter Welle, Institute of Systematic Botany, State University of Utrecht, Heidelberglaan 2, P.O. Box 80.102, 3508 TC Utrecht, The Netherlands

Trevor Whiffin, Department of Botany, La Trobe University, Bundoora, Victoria 3083, Australia

W. J. J. O. de Wilde, Rijksherbarium, Schelpenkade 6, P.O. Box 9514, 2300 RA Leiden, The Netherlands

Wang Huen-pu, Institute of Botany, Academia Sinica, Beijing, People's Republic of China

Wang Si-yu, Institute of Botany, Academia Sinica, Beijing, People's Republic of China

Thomas Zanoni, Rafael M. Moscoso Herbarium, Santo Domingo, Dominican Republic

Problems of Complexity of the Plant World and a Floristic Inventory of Tropical Forests

F. R. Fosberg

Only after one has done serious work on one or more ecological problems and brought together what has been already learned can one appreciate the incredible complexity of a natural ecosystem. The size of the problem of understanding such a system, if appreciated in advance, would daunt most even vitally-interested people from undertaking the task. The number of kinds of plants (and animals), the web of environmental gradients, and the resource cycles, the food-webs, all interacting, create a complexity that becomes nearly intractable. The stochastic processes of evolution have gone on, starting with relatively simple primordial life in the Archeozoic seas which showed little diversity, through a time continuum in a diversifying environmental continuum. The continuum of living organisms responded to the increasing diversity of the world environment, changing in a myriad of directions, fitting its aspects into the intricacies of a three-dimensional maze of a million blind alleys and lines of descent, opening and closing through a fourth dimension of time. The possibilities were infinite and changing continuously through an estimated 3.5 billion years of time with an overall increase in diversity, the descent-lines proliferating, and limited only by extinction. After 3.5 billion years of such increasing diversity, we now have to deal with the millions of kinds of organisms existing in the present time-plane. In theory, these organisms are classified on the basis of phylogeny—descent from common ancestors—the lines having ramified and perhaps anastomosed for three billion years.

Since, by simple probability, most of the connections no longer survive, and fossils are rare and fragmentary, the creation of a logically sound and practicably workable classification of organisms even after 250 years of modern attempts has only partly been achieved.

The point of the above discussion is to bring home the seriousness of the problem of identification of organisms, and specifically for this inventory, the identification of plants that make up the tropical forest. Identification, when dealing with large numbers of kinds, depends on the creation and use of a classification. Both to devise and to use a classification of tropical plants requires a more than casual botanical ability and interest, as well as special training. Traditionally, people with the right kind of mind for taxonomic botany generally gravitated into this field, usually supporting themselves by teaching or museum work. The pleasure and satisfaction of doing this kind of thing made up for the relatively poor compensation compared to that of lawyers, physicians and other types of professionals. The present poor state of knowledge of even temperate, let alone tropical, floras, is a reflection of the inadequacy of the supply and support of such people. Ease of travel and improved collecting methods, causing a build-up of collections of specimens which, augmented by the realization that voucher specimens must be kept to support and document research of all kinds involving plants, has emphasized the inadequacy of numbers and training of taxonomists. Lack of adequate identifications is generally the most serious bottleneck in the way of ecological, economic, and many other branches of research on tropical ecosystems. And practically every good taxonomist of tropical plants is overworked by the demand for critical identifications.

One may ask what are the reasons for the current scarcity of good taxonomists of tropical plants. The best taxonomists are, as noted above, born, not made. Many such will certainly go ahead with careers in systematic botany, but some are not convinced that there are jobs enough in the field, and that salaries, at least beginning salaries, are too low to allow a proper standard of living. Consequently they, contrary to their basic desires, enter other fields that seem to promise more material rewards. There are many other students with good minds, in addition, who have some interest in science, but no compulsion in a particular direction. Most of us know at least a few successful taxonomists who were merely pushed in that direction by good teachers. Others are pushed or attracted in other directions, either by good teachers, by the glamour of fashionable or faddish fields, or by the fact that some of these fields are obviously much better supported than plant taxonomy.

Clearly then, one of the negative factors leading to the scarcity of taxonomists is inadequate support for taxonomic botany. Another is that relatively few institutions offer strong courses in plant systematics. Too often only a local flora course is available, and frequently it is a poor one. Active and often cutthroat competition for funds reduces the budget share for herbarium support, systematic teaching and research. This often is accompanied by propaganda that taxonomy, especially "alpha taxonomy" is old fashioned, or no longer necessary. The arrogance of many workers in the so-called advanced fields is well known, and many of these get into influential administrative positions. Taxonomy is then down-graded, and herbaria are given away or neglected. This attitude sometimes prevails even where the services of competent taxonomists are needed to support the work of its detractors.

Systematists are not only needed to identify the plants studied or collected in the course of other work, but, even more important, they are required to produce the basic monographic and floristic works that make possible accurate identification of plants.

The difficulty with the suggested network for plant identification is basically the scarcity of competent taxonomists compared with the size of the need for identifications, and, more important, the need for basic taxonomic research. The tropical rainforest flora is enormous and not well known.

The UNESCO International Committee for Humid Tropics Research, in its eight years of effort to promote ecological research in the tropical rainforest regions, was constantly faced with the problem of identification of the countless species of plants and animals in the ecosystems to be studied. It was obvious there were no shortcuts to the solution of this problem. In the reports of each of the eight annual meetings of this committee, the most urgent recommendation was for ample increased support for taxonomic research. Whether these pleas had any result is not obvious. There has been some increased attention to the need for systematics, but the emphasis has been more and more on intensive new approaches: biosystematic, numerical, chemotaxonomic, cladistic—all of which, however, deal very intensively with small numbers of species only. Phylogenetic understanding has doubtless advanced, but the problem of the identification of plants on a large scale remains.

A few cooperative efforts are being made to tackle the taxonomic problems of tropical regions, but these are meager compared to what needs to be done. It all helps, but the problem still remains.

It has been suggested that much of the time-consuming work of taxonomy, and of plant identification could be done by computers. If taxonomy were an exact, quantitative science, and if variation followed regular and predictable patterns, and, especially, if our data were more complete, there might be something to be said for this. However, the knowledge of the taxonomy of tropical plants is not of the degree of completeness, or of such a reliable nature, that computers are of much help, except for certain types of data storage and retrieval, and especially for lightening the bibliographic load that is a particular burden for the taxonomist. The sheer numbers of plant species, and especially of poorly-known ones, make the task of collection of the exact data required for the computerized identification staggering. Large numbers of species are known from single collections. Worse, these single collections are frequently fragmentary, and often not available where needed—there are hundreds of herbaria in all parts of the world. The magnitude of uncertainty in this field and the continua of variability make the flexibility and intuitive qualities of the mind of a good taxonomist an absolute requirement. The increasing demand for taxonomic assistance in ecologic, economic, and anthropologic fields, as well as in other branches of botany, make the present and immediately foreseeable pool of tropical taxonomists pitifully inadequate.

One can only suggest, perhaps better, demand, that the community of scientists of various disciplines that depend on the services of the taxonomic profession together throw their weight behind a major effort to accomplish the basic alpha taxonomy of tropical plants. Enterprises like the *Flora Neotropica*, the *Flora Malesiana*, the *Projeto Flora Amazônica*, the series of African floras being published by Kew, and many similar efforts represent substantial steps toward this end. However, unless far more substantial and consistent support is made available, it will be a long time before an adequate network for identification of plants will come about.

Section One: Introduction and Synthesis

Introduction

Ghillean T. Prance

The vast quantity of information that is presented in this volume is the result of a joint project of the World Wildlife Fund-US and the three major botanical institutions of the United States, the Arnold Arboretum, the Missouri Botanical Garden and the New York Botanical Garden. However, as can be seen by the multiplicity of authors involved, the book is the result of the work of a large number of people around the world, and we are extremely grateful to all these persons, who have so generously contributed from their vast pool of knowledge about the tropical rain forest habitat.

In order to coordinate this project it was set up as an activity of the Species Survival Commission (SSC) of the International Union for the Conservation of Nature and Natural Resources (IUCN). The project was advised by a special tropical forest working group of the SSC. The members of this group were: Ghillean T. Prance, The New York Botanical Garden (Chairman); Peter Ashton, Arnold Arboretum, Cambridge, USA; Edward Ayensu, Smithsonian Institutions, USA; Enrique Forero, Bogotá, Columbia; C. Gunatilleke, Sri Lanka; Alwyn Gentry, Missouri Botanical Garden, USA; the late Marius Jacobs, Leiden, Netherlands; T. N. Koshoo, Lucknow, India; João Murça Pires, Belém, Brazil; Peter H. Raven, Missouri Botanical Garden, USA; S. Sastrapradja, Indonesia; Seymour H. Sohmer, Bishop Museum, Hawaii, USA; Julian A. Steyermark, Caracas, Venezuela.

In discussions between the staff of the World Wildlife Fund-US and representatives of the botanical gardens mentioned above, it became apparent that our current inventory of the tropical rainforest habitat is far from adequate. Recent reports have indicated both the alarming rate at which the tropical rainforest is disappearing (e.g., Myers, 1980) and the need for a much greater research effort before it is too late (NAS, 1980). Given the need for an intensified inventory effort, as stressed in the latter report, it seemed appropriate to prepare a strategy document that would point to the specific needs for further tropical rainforest inventory. This publication attempts to fill that role.

The most important factors to be considered in planning for future inventory were (1) obtaining knowledge about specific areas of intact rainforest that are particularly threatened and likely to disappear within the next few years, and (2) the location of regions that are particularly uncollected in terms of botanical inventories. The various regional reports included here certainly point to places where these circumstances are true. Unfortunately, both conditions, pending destruction and inadequate inventory, often coincide in the same area.

A point that was stressed in the NAS (1980) report was the need for more than herbarium specimens as part of the inventory. These 'unusual collections' as they were termed, become even more useful and important as the forests are destroyed. For this reasson we have, in the Appendix, included details on some aspects of data collection beyond the usual herbarium collections, such as wood samples, karyology, pollen, fern spores and living germplasm material. In addition, the recent work of several investigators (e.g., Hubbell & Foster, 1983, Prance et al. 1976) has shown the importance of quantitative inventory carried out by botanists who can accurately identify the species in a plot of forest. Consequently, this report gives some emphasis to the status and the needs of quantitative forest inventory that uses sample plots and quadrats where every tree is identified and measured. These data are a considerable addition to those of specimen collection and identification. They are beginning to give us a better idea of the dynamics of the forest and the mechanisms of species diversity and are essential for any plans for conservation and utilization of the forest. The survey carried out for this report shows how little botanically accurate inventory exists. In addition, it becomes obvious that so many different methods and standards have been used that inventories from different regions are not easily compared. It is hoped that one of the results of this report will be a move towards greater standardization of methods.

The synthesis of this report, which follows, shows that there is an enormous and very urgent need for further inventory of the tropical rain forest biome. This report is in many ways a challenge to the systematic botany community of the world to obtain greater resources for the task that is before them. There is no doubt that rain forest plant species are being lost before they have been identified. However, it would be a pity to turn our increased effort into merely an effort to create an archaeological museum as a curiosity for future researchers. We must collect the type of data that will furnish information of value for conservation, data that provide the reasons for the preservation of the vast genetic diversity of the tropical rain forest. It is hoped that this report will encourage inventory, but inventory that is well planned to fill the gaps and to gather the data most useful for more prudent management of the tropical rain forests of the world.

The enormous effort of organizing the gathering of the data presented here and especially that of arranging its presentation in an orderly form from a vast range of different formats, and of providing a synthesis of the results is the work of David G. Campbell and H. David Hammond to whom I am most grateful.

Finally, this report would not have been possible without the funds provided by the World Wildlife Fund-US.

Literature Cited

Hubbell, S. P. & R. B. Foster. 1983. Diversity of canopy species in a neotropical forest and implications for conservation. *In* S. L. Sutton, T. C. Whitmore & A. C. Chadwick (eds.), Tropical rain forest. Ecology and management. Special Publication No. 2 of the British Ecological Society. Blackwell Scientific Publications, Oxford.

Myers, N. 1980 Conservation of tropical moist forest. National Research Council, National Academy of Sciences, Washington, D.C.

National Academy of Sciences. 1980. Research priorities in tropical biology. National Academy of Sciences, Washington, D.C.

Prance, G. T., W. A. Rodrigues & M. F. da Silva. 1976. Inventario florestal de um hectare de mata de terra firme km 30 da Estrada Manaus-Itacoatiara. Acta Amazonica **6(1)**: 9–35.

The Importance of Floristic Inventory in the Tropics

David G. Campbell

Contents

I. Introduction	6
II. What is Botanical Inventory?	6
III. When is Inventory Complete?	6
IV. Inventory and Extinction	13
V. Managing Data on Inventory	13
VI. The Importance of Inventory to Conservation	15
VII. Outline of Regional Chapters	15
A. Description of the Region	15
B. Description of the Tropical Forests and Vegetation Maps	15
C. Magnitude of Floristic Inventory	16
D. Completeness of Floristic Inventory	16
E. Endemism	16
F. New and Important Distributions	16
G. Extinction	16
H. Families and Genera Well- or Poorly-Collected	17
I. Scholars	17
J. Areas Threatened by Destruction or Conversion	17
K. Resources for Continued Floristic Inventory	17
L. Suitability of Floristic Inventory, Past, Present and Future	17
M. Additional Resources Required	17
VIII. Appendices on Unusual Collections and Ancillary Topics	17
IX. Summary Information: Priority Areas for Exploration	20
X. Cost of Future Inventory	20
XI. Literature Cited	29

I. Introduction

We live in an age when new life forms are being created in laboratories by man, but also an age when existing life forms—with lineages as ancient as protoplasm itself—are becoming extinct. Ours is a time when the complexities of ecosystems can finally be quantified and modelled, but yet a time when those most complex of ecosystems, the tropical forests, are being destroyed before they are fully known or understood. The daunting problems of habitat destruction and extinction that have faced conservationists during the waning decades of the twentieth century have exerted pressure on all disciplines of biology, but particularly on taxonomy, which is called upon to take stock—to make an inventory—of all the Earth's species before they disappear. The description of species has been going on since at least the time of Aristotle, yet today few people—including many biologists—appreciate that the discovery, description and inventory of species—basic taxonomy—is more important than ever, and that even in the space age taxonomy remains as vital and essential a science as it was during the times of Humboldt or Wallace.

As Wilson (1985) pointed out, 208 years after the death of Linnaeus, science does not know, even to the nearest order of magnitude, how many species there are in the world. It has been estimated that to date, approximately 1.7 million organisms have been formally described by taxonomists (Prance, 1977). Yet the total number of species on Earth remains wholly unknown, although estimates run as high as 30 million for insects alone (Erwin, 1982). Of these, approximately 428,500 to 558,500 are plants or fungi (250,000 flowering plants, 120,000 to 250,000 fungi, 12,000 ferns, 12,000 mosses, 11,000 hepatics, 17,000 algae and 16,500 lichens). Half or more of these species (272,000, not including algae and lichens) occur in the tropics (Ainsworth 1961; Jacobs 1977; Prance 1977; 1978; Raven, 1976), and in particular tropical forests, the focus of this book and this essay. Although we do not know the total number of plants in the tropics, what is certain is that, at present rates of scientific inquiry, there are not sufficient resources nor, more ominously, time, for the discovery and taxonomic description of the remaining underscribed majority of species on Earth. A system of priorities for botanical inventory is therefore necessary.

Much has been written about the rate of tropical deforestation (F.A.O./U.N.E.P. 1981a, 1981b, 1981c; Myers, 1980; U. S. Dept. of Energy, 1986), which is one of the gravest problems facing our generation and the generations that will follow. In the decades to come, Earth will undergo a pulse of extinction as great as any in its geological history, and most of these extinctions will occur in tropical forests. Mayo (1986), reporting to the IUCN Gran Canaria Conference on Botanic Gardens and the World Conservation Strategy, predicted that up to 60,000 plant species, mostly from the tropics, may become endangered during the next 40 years. This will result in the loss of primary productivity and therefore of sustainable yields of food, lumber, drugs and other products derived from tropical forest species (Qureshi et al., 1980). Furthermore, there is increasing evidence that worldwide destruction of tropical forests may disrupt atmospheric water and carbon cycling and result in global climatic change (Council on Environmental Quality, 1980; Salati, 1983).

Given the dire consequences of tropical deforestation, it has long been recognized that accelerated taxonomic and systematic research on tropical forest organisms, disciplines that are fundamental to understanding our biological inheritance, should be a priority both during the waning years of the twentieth century and on into the next. This goal has been called for by the World Conservation Strategy of the I.U.C.N., U.N.E.P., U.N.E.S.C.O. and F.A.O. (I.U.C.N., 1980). It is with this supremely important goal in mind that herbaria and museums all over the world are racing to complete the inventory of tropical forest species before they disappear and three major American herbaria and the World Wildlife Fund-US have collaborated to produce this volume, with the goals of reviewing progress to date in the floristic inventory of tropical areas and of setting priorities for future botanical exploration of these areas.

II. What is Botanical Inventory?

Botanical inventory is accomplished by means of systematic and taxonomic studies, the objectives of which are the identification of all species of plants in given geographical area. The species of plants must be vouchered by herbarium specimens, which are permanently conserved in herbaria and assigned unique collection numbers to enable future workers to locate them. Inventory is therefore requisite to all subsequent disciplines in biology, evolution and phytogeography. Wilson (1985) has called systematics "a fountainhead of discoveries and new ideas in biology." Floras are also socioeconomic entities, necessary for conservation and rational resource management. Accordingly, Toledo (1985) has termed herbaria and floras as "geopolitical" resources, concrete national or multinational treasures of economic importance as tangible as highways, buildings and other infrastructure.

III. When is Inventory Complete?

It is important to ask, at what level is biological inventory complete for a particular area? If we accept that species are discrete natural entities then, theoretically at least, the time must come when all of them are described in the flora or other taxonomic work for a particular region. Obviously, this enviable state will never be achieved in tropical forests, because extinction of many species will take place before they are ever described.

To answer the question of what is an adequate inventory, it is first necessary to consider that inventory has an essential geographical component; only by multiple collections of the same species over its distribution, can the variation and the range of a species be determined. Therefore, although a flora of a particular area may be complete in terms of representation all of the taxa found in that region, it will not provide sufficient information about species whose ranges exceed that particular area, and any monographic studies that result from these inventories, especially in the tropics, will be subject to revision as new collections are made.

In addition, Floras and monographic studies undergo a predictable evolutionary process in which, upon the basis of new information on variation and distribution, some of the described species are merged and the Flora contracts. In some cases hundreds of species have been contracted into one. The evolutionary stage of a particular Flora is often a function of the antiquity of botanical exploration. For example, certain components of the Floras of tropical Africa and Asia, of which the production began during the colonial period of the last century, have undergone contraction and consolidation, whereas the Floras of the neotropics, flooded by new discoveries and hindered by incomplete distributional data, are still rapidly expanding.

These considerations aside, the floral (as opposed to the distributional) component of a botanical inventory is complete when collecting activity ceases to yield new taxa. Are inventories of tropical forests still yielding new taxa? The answer is unquestionably "yes"; however, the data are few. Table I displays the rate of discovery of new species from two recent South American tropical forest inventory programs, *Projeto Flora Amazônica* and the Río Palenque Florula Project. The number of new taxa per number of determinations ranges from zero to 9.09%, with a mean of 1.01% for the Brazilian Amazon. These rates are typical for neotropical forest inventory. For example Prance & Campbell (1988) also showed a rate of one new taxon of Chrysobalanaceae to every 92 (1.1%) neotropical collections for the family. Clearly, the taxa in these neotropical forests are as yet far from being adequately described and the distributions of these taxa are even farther from being understood.

Table I also displays the percent of determinations per number of collections. This statistic, which ranges from seven to 63%, is an index of the efficiency of the inventory program in sampling the diversity of a particular area without repeating itself. Inefficient expeditions repeatedly collect common species. (However, it is interesting to note that the most efficient expedition had zero new taxa.) Accordingly, Kubitzki (1977), Prance (1984) and other authors, lamenting the congestion of herbaria with common species of little or no scientific value, have called for better-trained field botanists, able to distinguish between common and rare species.

Some authors (Prance, 1977; Vink, 1981 and others) have used a collection "density index" (no. of herbarium specimens/100 km^2) as being a measure of the completeness of the flora for a particular geographical area. The principal advantage of the density index is that it is quantitative. For example, one could arbitrarily declare a density index of 100 as being the minimum necessary to assure that a botanical inventory has sampled most of the taxa and gathered sufficient distributional information. Even this ostensibly low level is, however, well beyond the reach of most tropical nations, as shall be demonstrated later in this essay.

Although the number of herbarium specimens held in a particular region is an easy statistic to pin down (these data may be conveniently summarized from Holmgren et. al., 1981), the number of herbarium specimens from a particular region or forest type is an elusive statistic that is unfortunately essential for setting priorities for the inventory of tropical forests. The reason for this is that herbarium sheets frequently do not contain sufficient locality information to warrent classification by collection site or vegetation type, and even when they do, herbaria are usually not organized

Table I
Rates of discovery in Neotropical botanical inventory

Inventory Program	No. Collections	No. Determinations	Percent Determinations	No. New Species	Percent New Taxa
Projeto Flora Amazônica (Brazil):					
Expeditions 1 & 3	3,525	264	7.49	2	0.75
Expedition 2	1,921	919	47.84	16	1.74
Expedition 4	1,856	416	22.41	7	1.69
Expedition 5	5,342	1,161	21.73	10	0.86
Expedition 6	1,540	795	51.62	8	1.01
Expedition 7	514	323	62.84	0	0.00
Expedition 8	2,505	1,334	53.25	9	0.67
Expedition 9	4,915	1,358	27.63	8	0.59
Expedition 10	1,500	424	28.27	3	0.71
Expedition 11	2,739	734	26.80	7	0.95
Expedition 14	1,060	330	31.13	12	3.64
Total	27,417	8,058	29.39	81	1.01
Río Palenque (Ecuador)	3,000	1,100	36.67	100	9.09

Sources: Prance et al., 1984; Gentry, pers. comm.

in such a manner that information from the millions of specimens can be retrieved and sorted according to geographical region, let alone tropical forested parts of a particular region. Therefore, it is nearly impossible to get more than an estimate of the number of collections from tropical forests. We do not know, for example, how many collections there are in the world's herbaria from the critically endangered *restinga* forest of eastern Brazil, from the vanishing lowland dipterocarp forests of the Philippines, or from the biologically important Kakamega forest of Kenya. The collections in herbaria are simply not organized so that the data can be retreived geographically by vegetation type. Yet this type of information is crucial to the ascertainment of the adequacy of collections for specific regions of tropical forest.

In recognition of this limitation, the authors of this volume were requested to emphasize tropical forests in their regional chapters. Since it was usually impossible to do that, most authors relied on summary statistics on collections for all vegetation types in their region. The key question here is whether general information on inventory is an indicator for a particular region of the subset of vegetation types that is tropical forest, or is tropical forest inventory in some way quantitatively different from general inventory or, say, the inventory of savannas? Unfortunately, the answer is that the inventory of tropical forests *is* different. Forests, because they are tree-dominated and have many inaccessible epiphytes, are considerably harder to collect than other vegetation types. This is why one of the abiding recommendations for inventory of tropical forests is that future efforts need to emphasize trees and other neglected groups, such as epiphytes, lianas, bryophytes and lichens. The general collection data for a nation or a region can never be regarded as representative of the situation for tropical forests, which are only a small portion of that whole. The situation for tropical forests should be regarded as invariably more critical than for the region as a whole.

For tropical countries, particularly those with a long history of western colonization, many early—and valuable (e.g., type specimens)—herbarium specimens are held in foreign herbaria, usually European or American. Just as summary data regarding the number of herbarium specimens from a particular vegetation type cannot be derived from most herbaria, summary data regarding the number of specimens from a particular tropical nation, held in foreign herbaria, likewise cannot be easily ascertained.

Table II show the number of herbarium specimens held in tropical-forested countries, using data from the seven volumes of the *Index Herbariorum*, 1952–1981 (Lanjouw & Stafley, 1952, 1954, 1956, 1959, 1964; Holmgren & Keuken, 1974; Homlgren et al., 1981). In order to create Table II, the number of herbarium specimens held by every herbarium that has functioned within the borders of a tropical-forested country was tracked over the 29-year span, and the collection density indices for each tropical nation/region were calculated. Of course, these data are for collections from all plant communities, and not specifically for collections from tropical forests. Nor do these data provide information as to herbarium specimens held outside of the tropics, although they do provide an insight into the trends of collecting activity within each nation/region, because most tropical nations, particularly in the post-colonial years since 1952, have insisted that a set of all newly-collected herbarium specimens be deposited within their borders. Indeed, these statistics in the *Index Herbariorum* are the only worldwide data which objectively compare inventory activity on a world scale. It is important to note that this information is dated, and that since 1981 certain important collection programs–for instance, in Bolivia, Gabon, Cameroun, Madagascar and other tropical nations–have been initiated. As this essay goes to press, the latest edition of the *Index Herbariorum* is being compiled.

The status of collection depicted in Table II is highly variable; in 1981 the collection density index ranged from a high of 534 for Taiwan to a low of one, as seen in the Central African Republic, Gabon and Bolivia. Only in Taiwan, Malaysia & Singapore, Brunei, Sri Lanka, the Pacific Islands, El Salvador, Costa Rica and the West Indies did the density index exceed our arbitrary minimum of 100.

Figure 1 is derived from these data. Each region of the world is represented by two bars. The left bar is the mean acquisition rate of collections/year/100km^2 for the entire period from 1952–1981. The right bar is the same statistic for the period between the final two issues of the *Index*, 1974–1981. A nation/region with an accelerating inventory program will always have a right-hand bar that is higher than the left-hand bar. Unfortunately, this condition occurs only in continental Asia, West Africa, South America, Central America and the West Indies. Although increasing, the rate of collection for some of these regions still remains negligible: 0.7 specimens/year/100 km^2 for West Africa; 1.0 for South America. All other parts of the tropical world, including such critically important areas as Malesia, Central Africa, East Africa and Madagascar, have declining collection activity.

Figure 2 focuses on Asia and Australia. The state of collecting is particularly critical in the Malesian Archipelago, including Indonesia, parts of Malaysia, Brunei and Papua New Guinea, all of which have suffered declining collection rates. With the exception of Brunei, these are the same regions that are also suffering from the greatest deforestation.

Figures 3 and 4 depict African nations, where in 1981 the statistics were most dismal. The only African country with any sort of collecting vigor was Angola, with a collection rate of 4.5 specimens/year/100km^2. Uganda, because of civil war, had actually suffered a negative rate of acquisition. The collection rate for Madagascar, where deforestation is perhaps the worst on the planet, had by 1981 declined to near zero (although, due to the efforts of the Missouri Botanical Garden, collecting has recently resumed).

With the exceptions of Belize and the Guianas, the collection rates in Central and South America are all accelerating, generally reflecting the relative vigor of botanical exploration in the neotropics. Even so, collection rates in 1981 re-

Table II

**Number of herbarium specimens per tropical country/region
1952–1981**

Country	Collection Density Index[a]						
	1952	1954	1956	1959	1964	1974	1981
ASIA & AUSTRALIA							
China	27	27	27	33	33	(33)[b]	50
Taiwan	334	334	334	417	417	531	534
Philippines	9	13	20	26	30	35	42
Thailand	10	12	14	15	18	23	23
Malaysia & Singapore	87	88	141	141	148	204	219
Brunei	0	0	0	123	123	123	123
Indonesia	41	41	41	42	55	81	86
Papua New Guinea	1	1	3	5	10	44	46
Burma	0	0	2	2	5	5	5
India	?	51	53	55	(55)	107	107
Sri Lanka	130	130	130	130	130	130	152
Australia	35	35	(35)	(35)	(35)	53	56
Pacific Islands	152	152	152	182	193	483	529
AFRICA							
Sierra Leone	0	0	0	0	0	41	50
Liberia	0	0	0	0	3	6	6
Ivory Coast	0	1	1	1	1	7	7
Ghana	0	10	10	21	21	42	48
Benin & Togo	0	0	0	0	0	0	5
Nigeria	2	3	3	3	4	6	14
Cameroun	0	0	1	1	1	1	10
Gabon	0	0	0	0	0	0	1
Congo Democratic Republic	0	1	1	1	3	9	9
Zaire	0	0	4	4	4	5	5
Central African Republic	0	1	1	1	1	1	1
Angola	0	0	1	1	1	4	5
Uganda	3	15	15	15	17	30	7
Kenya	14	17	17	17	28	55	64
Tanzania	0	0	0	1	1	3	5
Madagascar	2	2	2	2	9	9	9
SOUTH AMERICA							
The Guianas	0	5	5	5	7	10	11
Venezuela	4	4	4	5	5	14	23
Colombia	5	7	8	8	13	17	30
Ecuador	13	13	13	13	13	13	21
Peru	2	5	5	6	7	21	28
Bolivia	0	0	0	0	0	0	1
Brazil	6	7	8	9	11	16	25
CENTRAL AMERICA							
Mexico	4	6	6	6	6	18	48
Belize	0	0	0	0	4	13	17
Guatemala	0	0	0	0	0	23	34
El Salvador	0	5	76	76	76	100	268
Honduras	0	0	0	0	80	85	136
Nicaragua	0	0	0	0	0	5	22
Costa Rica	?	49	68	68	68	194	236
Panama	0	0	0	0	0	13	44
WEST INDIES	12	41	46	76	(76)	196	261

Sources: Lanjouw & Stafleu, 1952; Lanjouw & Stafleu, 1954; Lanjouw & Stafleu, 1956; Lanjouw & Stafleu, 1959; Lanjouw & Stafleu, 1964; Holmgren & Keuken, 1974; Holmgren, Keuken & Schofield, 1981; The Times, 1981.

[a] Collection Density Index = number of collections/100 km^2 within the borders of the tropical-forested country.

[b] Parentheses indicate that the value is interpolated.

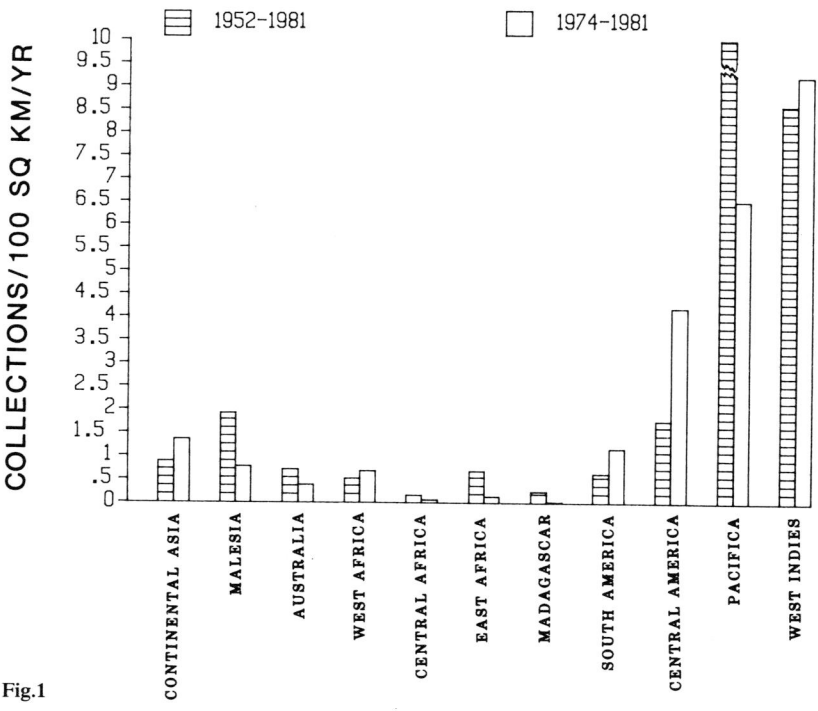

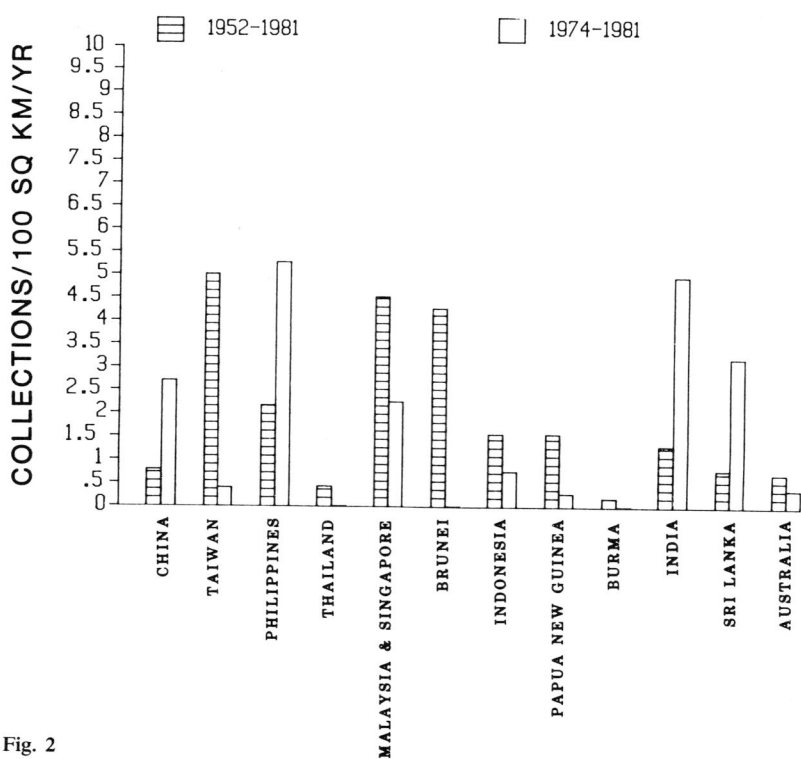

FIG. 1-6. Histograms of collection rates. Hatched bar is the number of collections per 100 km² per year for the interval from 1952–1956. The white bar is the same statistic for the interval from 1974–1981. Therefore if the white bar is taller than the hatched bar, then the pace of collecting is increasing. Data from same sources as Table II.

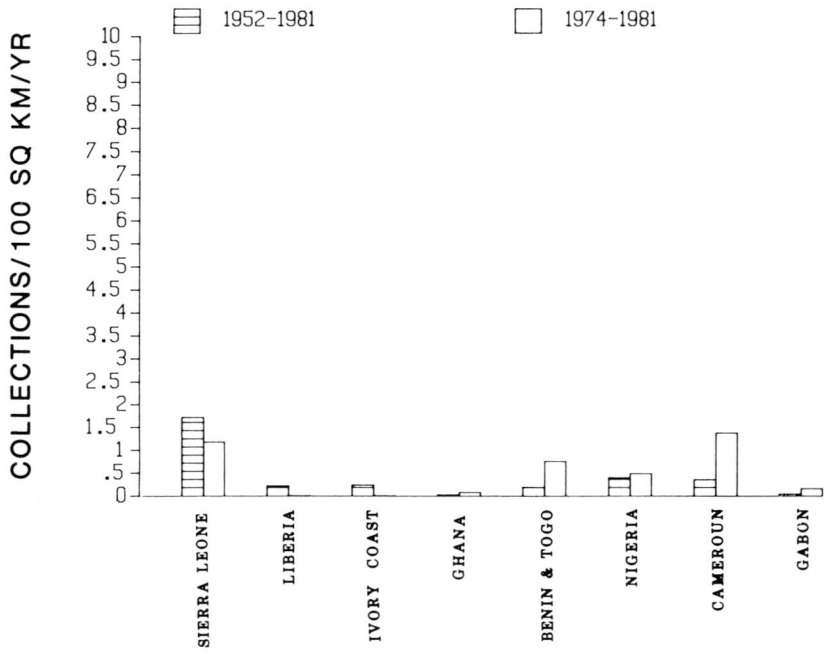

Fig. 3

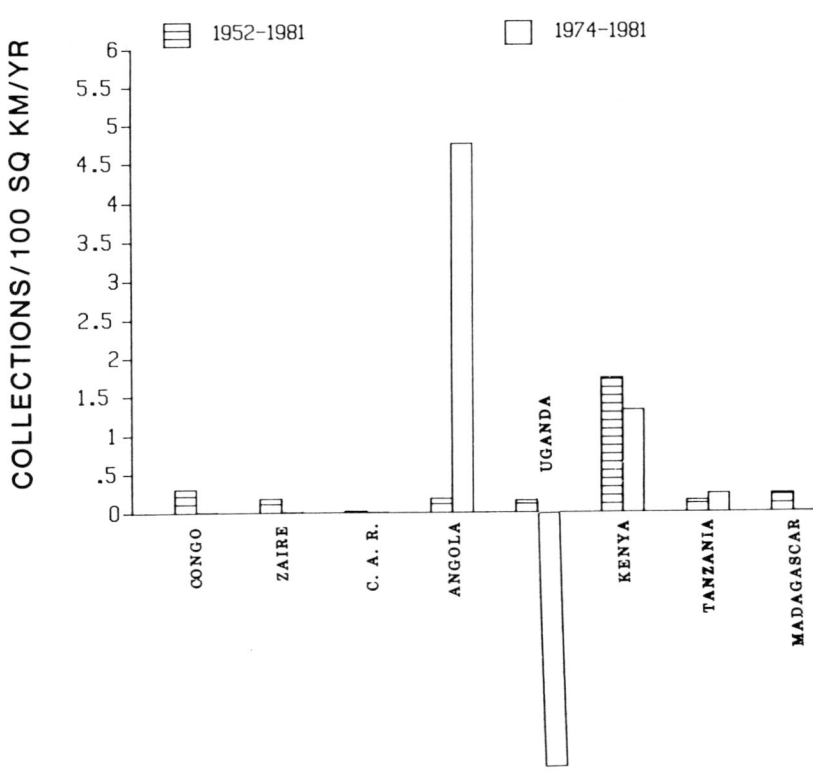

Fig. 4

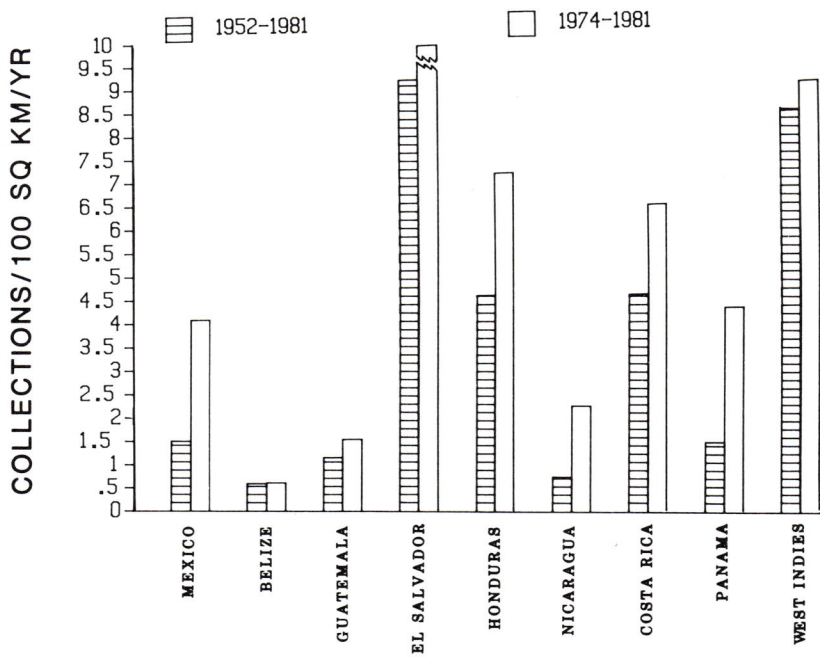

Fig. 5

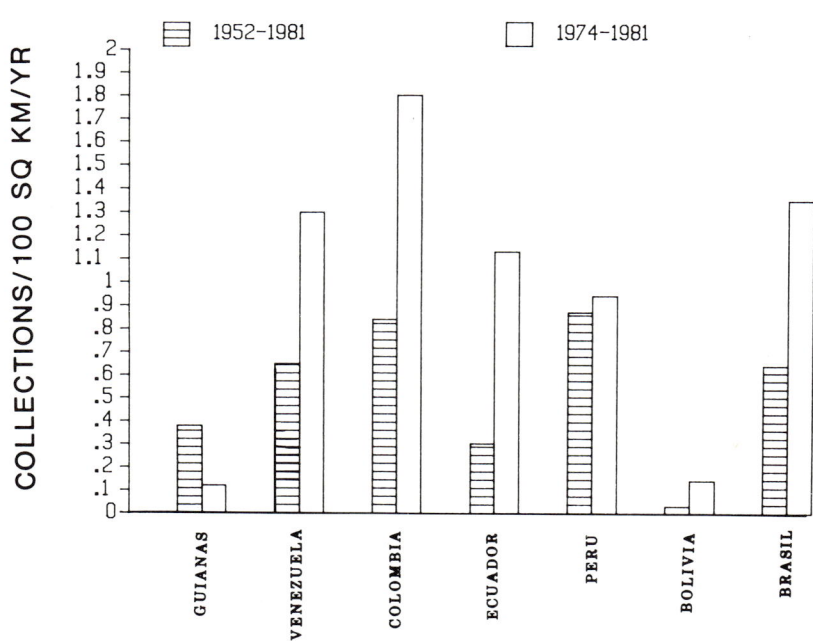

Fig. 6

mained very low, particularly in South America, where the maximum rate (Colombia) was still only 1.8 herbarium specimens/year/100km^2 (Figs. 5, 6).

Although the data derived from the seven volumes of the *Index Herbariorum* lack information on European and American collections, and therefore do not tell us the total collection density for each tropical-forested nation/region, we can accept that a minimal botanical inventory would require something like 100 herbarium specimens/100km^2 within the borders of the country, regardless of what is held abroad. Table III, projecting the collection rates from 1974–1981, predicts how long it will take each nation to achieve this minimal goal. Some areas, listed above, are already at or exceed 100 specimens/100km^2. On the other end of the spectrum, at 1981 collection rates, Thailand, Liberia, the Ivory Coast, the Congo Democratic Republic, Zaire, the Central African Republic, Uganda and Madagascar will *never* achieve the minimum. Others, such as Papua New Guinea, Burma, Benin & Togo, Nigeria, Gabon, Tanzania, the Guianas, Bolivia and Belize will take more than a hundred years. In 1981, in spite of repeated calls for the accelerated inventory of the tropics, the rate of botanical inventory in these nations was obviously wholly inadequate.

IV. Inventory and Extinction

In spite of the fact that thousands of species of plants and animals will inevitably become extinct during the coming decades as a result of the destruction of tropical forests, the conservation literature contains very few examples of the actual extinction of plant species in deforested parts of the tropics. The fundamental problem of monitoring extinction in tropical forest plants is that there are usually no baseline data on the species that existed before forest conversion, and rare species, which are the last ones to appear in the literature, are often the first ones to disappear. Given the inadequate inventory of tropical forests, it is inevitable that many species will become extinct in the tropics before they are ever discovered and formally described by taxonomists, a condition that I have termed "anonymous extinction" (Campbell, 1984). Indications are that the rate of anonymous extinction must be very high, but again, concrete examples are few. A notable exception is provided by Dodson and Gentry (1978) who, working in the Río Palenque Reserve, part of the critically endangered moist coastal forests of Ecuador, recorded a total of 1100 species of plants. Of these, approximately 100 were new to science and endemic to the coastal forest. Since the Río Palenque Reserve represents the last intact bit of Ecuadorian moist coastal forest, it is logical to assume that those 100 species would have gone extinct without ever having been described by taxonomists, if not for the efforts of conservationists and taxonomists working at Palenque.

More tragic is the example of Centinela Mountain, an isolated peak located about 12 km. from Río Palenque (Dodson, pers. comm.). Between 1976 and 1978, a floristic inventory of Centinela described 1000 species of plants, about 100 of which were endemic. Concomitant with the inventory, a new road was constructed to Centinela, resulting in the total deforestation of the mountain for purposes of agriculture and the extinction of the 100 endemics.

Because of the lack of such baseline information, not a single author in this volume was able to speculate on the rate of extinction for their region of expertise. Yet most could generally approach the problem. For example, where there is the concomitant occurrence of well-defined endemics and deforestation, the extinction rate, or at least the degree to which species are threatened with extinction, must be high. The tropical forests of eastern Brazil and Madagascar both have this unfortunate mix of circumstances.

V. Managing Data on Inventory

It is ironic that in order to set the priorities for floristic inventory contained in this strategy document, some of the difficult questions that were asked—such as the location of regions of high extinction rates, of imminent destruction, of high endemism, of areas that are yielding new and important distributions—have answers that could only be derived from inventory. This was not a tautological exercise; it is entirely logical that the feedback from current inventory should set the direction of future inventory.

Unfortunately, many of these questions could only be subjectively, not quantitatively, answered. The abiding problem of this survey has been the lack of systematic means for coordinated data retrieval. The world's herbaria are repositories of a vast amount of uncollated information—billions of bits of information—yet there have existed few means of systematically retrieving these data short of manually sifting through all of the herbarium labels. Herbarium labels often not only contain the answers to exactly those questions addressed by this survey, but also important information regarding ecology, phenology and ethnobotany. This was abundantly demonstrated by von Reis Altschul (1982), who extracted a wealth of ethnobotanical information stored on herbarium labels at the New York Botanical Garden.

The only practical manner to organize and retrieve information on herbarium labels is by computer. Indeed, the technology to do this has been widely applied to similar tasks for decades, such as telephone directories, scientific abstracts and the collections of libraries. For example, university libraries, which commonly hold millions of books in their collections, are now widely networked by computer, enabling retrieval of information according to title, author and subject. In an herbarium, computerized data storage should be structured according to taxa, locality of collection, date of creation of a new taxon or the elimination of an existing taxon, environmental information, phenology, ethnobotanical information and other subjects. Besides providing a modern and inexpensive housekeeping tool for herbarium management, a database such as this could be computer analysed to provide information as to strategies and priorities for future inventory research: the number of collections per geo-

Table III
Rates of botanical Inventory in tropical countries/regions

Country/Region	No. Collections /100 km² in 1981	Collections/100 km²/Year 1974–1981	No. Years Required to Reach Level of 100 Specimens/100 km² Within Country/Region
Asia			
China	50	2.39	21
Taiwan	534	0.40	0
Philippines	72	5.24	5
Thailand	23	0.00	∞
Malaysia & Singapore	219	2.24	0
Brunei	123	0.00	0
Indonesia	86	0.75	18
Papua New Guinea	46	0.28	193
Burma	5	0.02	676
India	108	0.00	0
Sri Lanka	152	3.27	0
Australia	56	0.38	117
Pacific Islands	529	6.52	0
Africa			
Sierra Leone	50	1.18	43
Liberia	6	0.00	∞
Ivory Coast	7	0.00	∞
Ghana	48	0.90	58
Benin & Togo	5	0.75	126
Nigeria	14	0.49	175
Cameroun	10	1.38	65
Gabon	1	0.16	618
Congo Democratic Republic	9	0.00	∞
Zaire	5	0.00	∞
Central African Republic	1	0.00	∞
Angola	5	0.50	90
Uganda	7	−3.27	∞
Kenya	64	1.32	27
Tanzania	5	0.24	398
Madagascar	9	0.00	∞
South America			
The Guianas	11	0.12	741
Venezuela	23	1.30	59
Colombia	30	1.79	39
Ecuador	21	1.13	70
Peru	28	0.94	77
Bolivia	1	0.14	707
Brazil	25	1.28	59
Central America			
Mexico	48	4.16	13
Belize	17	0.62	133
Guatemala	34	1.57	42
El Salvador	268	23.96	0
Honduras	136	7.26	0
Nicaragua	22	2.32	34
Costa Rica	236	6.46	0
Panama	44	4.49	12
West Indies	261	9.22	0

Sources: Lanjouw & Stafleu, 1952; Lanjouw & Stafleu, 1954; Lanjouw & Stafleu, 1956; Lanjouw & Stafleu, 1959; Lanjouw & Stafleu, 1964; Holmgren & Keuken, 1974; Holmgren, Keuken & Schofield, 1981; The Times, 1981.

graphic area, number of collections per taxon, range of taxa, rate of collections per taxon or geographic area over time and rate of new species described over time.

A modest and practical way for an herbarium to begin constructing a database would be to enter data from new collections or specimens returned after withdrawal and use. This would provide summary information for those specimens that are being actively used and would provide information on collection rates. Already, many herbaria are now using word processors to produce labels for new entries, a process that entails entering label data onto a magnetic storage system. It is a small extra step to enter this information into a computer database. Indeed, some herbaria, such as Bogotá and the Greene Herbarium have already started to computerize their collections. Others have started to computerize certain important collections (i.e., the Tropicos system at the Missouri Botanical Garden and the Projeto Flora Amazônica collections of the New York Botanical Garden and the Instituto Nacional de Pesquisas de Amazônia, in Manaus).

The information in the *Index Herbariorum* could likewise easily be databased, in order that botanists and conservationists can extract specific data as to collection density indices, collection activity and other strategic data from the summary statistics of herbaria.

VI. The Importance of Inventory to Conservation

Conservation cannot proceed without a thorough understanding of the components of the ecosystems that are being preserved. Therefore, inventory, which is nothing more than stock-taking, is requisite to conservation, particularly in tropical forests, where as yet many threatened species remain undescribed.

It is important to note that the priority areas for floristic inventory presented in this strategy document were selected using many of the same criteria that one would use to select areas for preservation: high level of endemism and imminence of destruction. Should, then, the list of priority areas for floristic inventory presented in this document be used as a wish list for parks and reserves? To a large extent, the answer is yes.

VII. Outline of Regional Chapters

The chapters of this volume are divided into two main categories: 1) regional status reports on particular geographical areas and 2) appendices regarding unusual collections and other topics ancillary to inventory. The regional chapters review the status of floristic inventory in most areas of the world where there are tropical forests (the single exception is Burma, where the authorities were unable to provide a chapter). Each author of a chapter covering a particular geographic region was provided with a standardized outline, drafted by the four sponsoring institutions, in order to insure that the information in the chapters would be comparable. The outline addressed questions ranging from the general to the specific. It is the general questions, such as A and B, below, dealing with the current demographic situation in the region and the kinds and extent of tropical forests, that may be of greatest interest to the general reader who is not a systematist, but who works in conservation, resource management and funding of research. These sections were designed to make the volume a valuable natural history reference in its own right, independent of the data provided for prioritization of research efforts.

The more specific questions C–M deal with prioritization. Although the outline was standardized for these items, the polyglot database in which the answers were sought was hardly uniform in specificity, resolution or quantitative detail. This unevenness of coverage was exacerbated by the unavailability of standardized summary data for most herbaria, the collections of which remain largely uncomputerized, as discussed above.

Regardless, there is a great wealth of information to be found in the pages that follow. The components of the outline presented to the authors, a discussion of the rationale behind them and the strengths and weaknesses of each statistic, are reviewed in the paragraphs that follow.

A. Description of the Region

1. Area (sq. km.), 2) Topography, 3) Foci of human population and 4) Human population density and expected demographic trends

This question sets the stage and provides baseline information for the region. Demographics are often the driving force behind deforestation and the trends in population predicit areas where deforestation will occur in the future. Examples are: the crowded eastern seaboard of Brazil (which not only places pressure on the highly endangered eastern forests, but also exports its population to Brazilian Amazônia) and the transmigration projects on Sumatra and Borneo. On the other hand, nations such as Gabon and Zaire, with low population densities, have relatively low rates of conversion of tropical forest, and those areas that are being converted are near urban areas.

B. Description of the Tropical Forests and Vegetation Maps

Before considering vegetation maps, it is necessary to define tropical forests for the purposes of this volume. Initially, the strategy outline adopted a broad-spectrum definition of tropical forests, classified according to the keys of U.N.E.S.C.O. (1973). Since it would have been impossible to include all of the tropical forest types in the U.N.E.S.C.O. scheme, an abbreviated list of 15 forest types were extracted from the UNESCO system for the purposes of this volume.

Not all authors were able to adhere to this scheme, however. Some authors (Dorr et al., this volume; Mori, this volume;) found the U.N.E.S.C.O. classification unsuited to their particular region. Other authors provided keys to translate the U.N.E.S.C.O. system into their own. Some found the state of knowledge regarding their particular region to be insufficient for the classification of forests. And still other authors (such as Smitinand and Wang) simply provided abbreviated species lists to characterize forest types. Therefore, although an attempt was made to use a standardized system of tropical forest classification, no single system proved practical for all regions covered by this volume.

Vegetation maps are one of the first products to result from botanical inventory and they were the first area for which the quality of information was highly variable. For example, it was not until 1984 that a modern vegetation map existed for Malesia (Whitmore, 1984), and given current rates of conversion in much of Malesia, even this map is doomed to rapid obsolecense. Most of the authors submitted some sort of map to accompany their chapters. However, the format and small scale of the pages of this volume constrained the amount of detail that could be depicted in the published versions of these maps, and consequently it proved impractical to include maps with more than five classifications of vegetation. In these cases, the author's map was simplified or not included. In other cases, a modern standardized vegetation map already existed for the wider region (i.e., U.N.E.S.C.O., 1981; White, 1983; Whitmore 1984), and a simplification of this map was drafted to conform with the demands of the individual chapter.

C. Magnitude of Floristic Inventory

1. The number of herbarium specimens held within the region and abroad and 2) The publications on the flora of the region

Herbarium specimens are the basic resource of inventory. This includes material within the borders of the region and well as stored in foreign herbaria, often within the borders of the former colonizer. The author was asked to enumerate both.

The author was also asked to review the status of published or projected monographs and floras. It is important to reiterate that published monographs and floras are not necessarily the end result of inventory, since they tend to shrink or expand as new information is assimilated and species are merged or created. In the matter of redefinitions of species, the process of inventory may continue long after collecting has ceased.

D. Completeness of Floristic Inventory

1. Identify areas that are: 1) Relatively well-collected (defined as more than 50 collections/ 100 km^2), 2) Poorly collected (defined as less than 50 collections/100 km^2)

The collection density index, as demonstrated above, despite its limitations, is a practical, standardized method to compare the completeness of floristic inventory among regions or countries. However, collection rates per geographical region smaller than a country area usually not data-based and therefore often cannot be extracted from herbarium statistics. Therefore, though some authors were able to respond quantitatively to this topic, most were not.

E. Endemism

1. Identification of the geographical areas that are highest in endemics

Endemism is a relative term, subject to the definition of the area in which a species is endemic. A species may be endemic to a county, state, country, or continent, and if it is endemic to the smallest geographical unit, then it is endemic to all the higher orders of organization. Endemism is further complicated by the merging of species that occurs as floras mature. By definition, a species collected in a single location is endemic to that location, but as more specimens are collected and its known range expands, the focality of its endemism is decreased.

Endemism is therefore difficult to define in homogeneous regions (i.e., the Amazon Valley); however, it is readily defined in plant communities that are isolated by geographical or climatic barriers, such as islands (i.e., the Moluccas; Madagascar), mountains (E. Africa) or coastal forests separated from similar habitats by a belt of savanna (E. Brazil). Indeed, it is exactly these habitats that tend to be high in endemics and under these circumstances, endemism is a powerful tool for setting of priorities for botanical exploration.

F. New and Important Distributions

1. Identification of the areas that are yielding new and important distributions

As floristic knowledge of a region matures, the ranges of species and the extent of plant communities become fine-tuned. Therefore, geographical areas that provide new information regarding ranges are priorities for inventory. Ecotonal areas, representing the shift from one vegetation type to another, are particularly important for these purposes.

G. Extinction

1. Identification of the areas that have high rates of extinction

As already mentioned, because of the lack of baseline information, none of the authors were able to calculate the rate of extinction. Yet most authors generally approach the problem. For example, where there is a mix of many well-defined endemics and deforestation, the extinction rate, or at least the degree to which species are threatened with extinction, must be high.

H. Families and Genera Well- or Poorly-Collected

This is a way of addressing the state of inventory in a manner that is not geographical. Certain plant families tend to be neglected by collectors, either because they are difficult to collect (i.e., the epiphytic Bromeliaceae and Orchidaceae) or just not glamorous (the Graminoids). Other families have not had a specialist working on them for a number of years. These neglected areas are therefore priorities for inventory.

I. Scholars

1. Identification of those working on each important family in the region

Most authors addressed this question with a table.

J. Areas Threatened by Destruction or Conversion

1. Identification of important threatened areas of tropical forest

These data are, unfortunately, well-defined. Although the species that are threatened by deforestation are not well known, the areas that are being destroyed are.

K. Resources for Continued Floristic Inventory

1. Academic institutions, 2. Government institutions, 3. Foreign programs and 4. Opportunities for collaboration with other nations and disciplines

The intention of this question was to determine the present day capabilities of the institutions, personnel and infrastructure in the region and abroad to pursue inventory and to describe the ongoing domestic and foreign programs in floristic inventory.

L. Suitability of Floristic Inventory, Past, Present and Future

1. How well these programs correlate with areas that are floristically poorly explored, 2. How well they correlate with areas of high species diversity, high endemism and high rates of extinction, and 3. How well they correlate with areas that are threatened with destruction or conversion

This is the synthesis. The question addresses the specificity and relevance of the inventory programs in dealing with the designated priority areas. The majority of inventory programs, unfortunately, have targeted their areas with less foresight than suggested herein, as functions of political expediency or financial opportunity. Some programs, such as *Projeto Flora Amazônica,* which was a joint program of the New York Botanical Garden and the Conselho Nacional de Desenvolvimento Cientifico e Tecnologico of Brazil, have focussed their activities on areas selected by means of the criteria discussed above. It is the purpose of this volume, of course, to set priorities derived, as far as is practical, from objective criteria.

M. Additional Resources Required

1. Estimation of requirements to insure an adequate floristic inventory of the tropical forests of the region

Given the present state of botanical inventory programs in the region and abroad and their degree of specificity in addressing priority areas, what more is needed? Of course, this question requires a definition of what is an "adequate" inventory for a particular region. It is important to note that most authors defined needs in terms of the enhancement of existing infrastructure, not the establishment of new programs. This included more educated personnel, herbarium equipment, vehicles, etc. A common requirement was simply the day-to-day financial resources to adequately curate existing herbarium collections.

VIII. Appendices on Unusual Collections and Ancillary Topics

Systematics and inventory embrace a variety of ancillary disciplines. For example, systematics includes quantitative ecological inventory, embracing the frequencies of species in a finite sample, the spatial relationships between these species, their life strategies and, over a period of time, the recruitment and mortality of species in a given area and the long-term changes in the plant community. Other aspects of systematics include specialized collections and analysis of 1) wood, 2) fernspores, 3) pollen, 4) germplasm, 5) palms, 6) specimens for chemotaxonomy, 7) bryophytes and 8) fungi. These specialized disciplines, each treated in an appendix at the end of this volume, are all components of the effort to botanically explore tropical forests. For each appendix, the author was instructed to review the status and techniques of his particular discipline and make recommendations for the standardization of methods.

Table IV
Priority areas in Asia for floristic inventory

Priority Area	Weight:	Low Coll. Density 2	High Endemism 1	New Distributions 1	Threatened 3	Total
I. ASIA						
China						
s.e. Guangdong, littoral hills					3	3
Qionglei						
seasonal rain forest in terraces & hills					3	3
Qiongnan Region				1	3	4
s.w. Guangxi limestone hills			1		3	4
s. Yunnan & s.w. Yunnan mtns. & valleys					3	3
s.e. Yunnan			1		3	4
Xishuan Banna seasonal forest			1		3	4
s.e. Tibet mtns. & river valleys				1		1
Taiwan			1			1
Philippines						
general						
beach type forest					3	3
dipterocarp forest			1		3	4
pine forest					3	3
Luzon						
Sierra Madre Mtns.		2				2
Mindanao I. mtn. divide between Bokidnon						
& Agusan Provs.		2				2
e. costal mtns.		2				2
Mindoro						
seasonal evergreen forest		2	1	1	3	7
Palawan		2	1	1	3	7
Thailand						
general						
lowland forests				1	3	4
peninsula						
peat swamp forests					3	3
Sakon Nakhon						
Tung Kula Ring Hai savanna forest					3	3
Central Palin						
dry deciduous dipterocarp forest					3	3
low tropical rainforest					3	3
n.w. & w. highlands						
upper tropical forest					3	3
hill evergreen forest					3	3
e. Khorat Plateau coniferous forest					3	3
Indochina						
Vietnam						
n.e. Vietnam		2	1	1	3	6
Danang-Hué Region		2	1	1	3	7
s. Laos, along Burmese & Thai borders		2				2
s.w. Cambodia		2				2
Borneo						
Pulau Laut, Meratus		2	1	1	3	6
n.w. Kalimantan		2	1	1	3	7
central Borneo mtn. system			1	1	3	5
Crocker Range of w. Sabah		2	1	1		4
Sumatra						
Anambas & Natunas		2		1	3	6
Barisian Range		2		1	3	6
Nort Aceh		2	1		3	6
s.e., especially Riau		2	1	1	3	7

(continued)

Table IV
Priority areas in Asia for floristic inventory (continued)

Priority Area	Weight:	Low Coll. Density 2	High Endemism 1	New Distributions 1	Threatened 3	Total
swamp forests		2			3	5
s. volcanoes		2		1	3	6
parts of Bangkalulu & Lampongs		2			3	5
Pagui Is.		2			3	5
Java						
general					3	3
Sulawesi						
n. arm (except Minahassa)		2			3	5
central Sulawesi		2			3	5
s.w. arm		2			3	5
Moluccas						
Halmahera						
e. part & s. arm		2			3	5
Morotai					3	3
Bacan Group (except Mt. Sibela)		2			3	5
Obi Group		2			3	5
Sula Group		2			3	5
Buru & Seram Group		2			3	5
Ambelon I, Vliassen Is.		2			3	5
Ambon					3	3
Leti, Babar, Tanimbam		2				2
Banda					3	3
Wetar Damar		2			3	5
New Guinea						
Papua New Guinea						
Great Papuan Plateau limestone area		2				2
Star Mtns.		2	1			3
Peg. Tamrau		2	1			3
Fly River lowland & low montane forests		2				2
Sepik, May & Horden Rivers headwaters		2		1		3
Papuan ultrabasic belt		2				2
Irian Jaya						
Peg. van Rees		2	1			3
Peg. Jayawijaya		2	1			3
Diogel River lowland & montane forests		2	1	1	3	7
Salawati, Batanta, Waigeo & Gebe Is.		2	1		3	6
New Britain						
limestone areas		2				2
Mt. Birinia		2	1			3
India						
n.w. India hilly regions		2				2
e. Himalayan range		2	1			3
Sikkim		2	1			3
Arunachal Pradesh		2				2
n. Bengal						
hills		2	1			3
Assam						
hills		2	1			3
Nagaland		2	1	1	3	7
Mizaram		2				2
s. & s.e. Madhya Pradesh		2				2
parts of Maharashtra Pradesh		2				2
parts of Andhra Pradesh		2				2
parts of Kerala		2				2
Sri Lanka						
lowland wet zone			1		3	4

IX. Summary Information: Priority Areas for Exploration

Tables IV to VIII are extended lists of the geographical areas designated by the authors in this volume as priorities for floristic inventory, with justification for their prioritization. The four criteria used for prioritization were derived from the above outline, and were weighted: imminence of destruction, because of its urgency, was given a weight of three; low level of floristic exploration a weight of two; high endemism and new and important distributions were both given a weight of one.

From these weighted criteria, a list of top priority areas, with a total weighted value of greater than or equal to five, are identified (Table IX); it is these areas that demand the immediate attention of taxonomic botanists.

X. Cost of Future Inventory

Inventory costs money, and it is the tropical-forested countries that can usually least afford to pay for it. This essay has demonstrated the merit of botanical exploration and the discovery and description of new taxa. But what does it cost to discover a new plant taxon in a tropical forest? Table X contrasts the cost of botanical inventory for the roving *Projeto Flora Amazônica*, in the Brazilian Amazon, and the fixed-base Río Palenque florula project. *Projeto Flora* was a logistically difficult expeditionary effort that spanned the entire Brazilian Amazon Valley. The Río Palenque florula focussed its activities on a small reserve, with minimal logistical problems. No consistent pattern emerges from the data. Although the cost of discovering a new taxon was higher for *Projeto Flora* than for Río Palenque ($3,756 vs $1,500), the cost per collection number and determination was significantly lower. (The high frequency of new species in Río Palenque is probably a reflection of the high level of endemic species.)

Let us for the sake of discussion accept the mean cost of $2,628 for the discovery of a new taxon of tropical forest plant (it is important to note that this is a conservative estimate that does not include the cost of the study, description and publication of the new taxon, nor does it consider the fact that as more new taxa are described, it will be increasingly hard to locate any remaining ones). Further, let us limit our discussion to flowering plants, (of which there are an estimated 125,000 in the tropics), excluding tropical fungi, algae and lichens, which are also components of herbaria, but for which the economics of discovery are probably different. For the sake of argument, we may assume that perhaps 5% of these flowering plants (that is 18,750) remain yet to be discovered (of course the true value is, by definition, unknowable). This gives us a cost, conservatively estimated, of approximately $49,275,000 to complete the botanical inventory of the tropics (if the number of undiscovered species is 10%,

Table V
Priority areas in Australia and Pacifica for floristic inventory

		Criteria for Prioritization:				
Priority Area	Weight:	Low Coll. Density 2	High Endemism 1	New Distributions 1	Threatened 3	Total
II. AUSTRALIA & PACIFICA						
Australia						
general						
all less accessible areas		2				2
Kimberly region		2				2
n.e. Queensland						
Cape Tribution & Russel River lowland rainforests			1			1
Bellendenker, Bartle Frere, Thornton Peak			1			1
Fiji						
general						
all extant forested areas		2	1		3	6
Hawaii						
Maui		2	1			3
Molokai		2	1			3
Hawaii		2	1			3
inaccessible areas on other islands		2	1			3
New Caledonia						
general						
inaccessible areas			1		3	4

Table VI
Priority areas in Africa for floristic inventory

Priority Area	Weight:	Criteria for Prioritization:				
		Low Coll. Density 2	High Endemism 1	New Distributions 1	Threatened 3	Total
III. AFRICA						
West African Forest Mosaic						
interior forests of Sierra Leone		2			3	5
Gola forest (Sierra Leone)		2			3	5
interior forests of Liberia		2	1	1	3	7
Sapo forest (Liberia)		2	1	1	3	7
Liberia-Ivory Coast border moist forests		2			3	5
Ivory Coast-Ghana border		2	1		3	6
Brong Ahafu, Ashanti (Ghana)					3	3
southwest Ghana		2	1		3	6
Bight of Benin Forest Mosaic						
Niger R. delta region (Nigeria)		2			3	5
north of Calabar (Nigeria)		2			3	5
Nigeria-Cameroun border moist forest		2	1	1	3	7
southwest Cameroun-Equatorial Guinea-Gabon frontiers		2		1	3	6
Guineo-Congolian Regional Mosaic						
vicinity of Franceville (Gabon)		2	1	1		4
northwest of Mayombe Escarpment (Zaire)		2				2
Luozi Territory (Zaire)		2				2
between Lekenie R. Toward Kindu (Zaire)		2				2
between Kibomo/Lomani R. & Opala & Kisingani (Zaire)		2				2
southwestern Shaba Province (Zaire)		2			3	5
vicinity of Manono (Zaire)		2				2
confluence of Luvua R. & Lubero (Zaire)		2				2
between Mambasa/Irumu/Wamba & Watse (Zaire)		2				2
between Bafwasende/Ponthierville/Lubutu & Lubero (Zaire)		2				2
north of Aru (Zaire)		2				2
Zaire-Central African Republic border		2				2
Lukaya Valley (Zaire)		2			3	5
Marungu Plateau, near Kasiki, Katanga Prov. (Zaire)		2			3	5
Margins of Likasi & Lufira Rivers (Zaire)		2			3	5
dry semievergreen forest n. of Basape (Zaire)		2			3	5
Niari Valley (Congo P. R.)		2			3	5
Bateke Plateau (Congo P. R.)		2			3	5
Sangha River Interval (Congo P. R.)		2			3	5
Zoka, Central Forest Reserve (Uganda)					3	3
Budongo & Siba, Central Forest Reserve (Uganda)					3	3
Semliki, Central Forest Reserve (Uganda)					3	3
Itawara, Central Forest Reserve (Uganda)					3	3
southern Kibale, Central Forest Reserve (Uganda)					3	3
Bwindi (impenetrable) Animal Sanctuary (Uganda)					3	3
Rift Lake Forests						
w. of L. Tanganyika (Zaire)		2				2
islands & margin of L. Victoria (Uganda)					3	3
Mabira, Central Forest Reserve (Uganda)					3	3
Kakamega, Forest Reserve (Kenya)		2	1	1	3	7
Mahale Mtns. (Tanzania)		2	1	1	3	7

(continued)

Table VI
Priority areas in Africa for floristic inventory (continued)

Priority Area	Criteria for Prioritization:				
Weight:	Low Coll. Density — 2	High Endemism — 1	New Distributions — 1	Threatened — 3	Total
Zanzibar-Inhambane Regional Mosaic					
Boni (Kenya)	2				2
Tana River. (Kenya)	2			3	5
Arabuko-Sokoke, Forest Reserve (Kenya)	2			3	5
Kayas (forest refuges of Mijikenda tribe) (Kenya)	2		1	3	6
Shimba Hills, National Reserve (Kenya)	2		1	3	6
Mrima Hill, Forest Reserve (Kenya)	2		1		3
Taveta (Kenya)	2		1		3
Rau, Forest Reserve (Tanzania)	2			3	5
Usumbara Mtns. (Tanzania)	2	1	1	3	7
Nguru Mtns. (Tanzania)	2	1			4
Uluguru Mtns. (Tanzania)	2	1		3	6
Uzungwa Mtns. (Tanzania)	2	1		3	6
Mahenge Plateau (Tanzania)	2	1		3	6
Pugu Hills (Tanzania)	2	1		3	6
Kichi Hills & Libangani (Tanzania)	2	1			3
Rondo Plateau (Tanzania)	2	1		3	6
Ngezi, Forest Reserve (Tanzania)	2	1			3
Somali-Masai Region					
Nairobi Forests (Kenya)				3	3
Afromontane Region					
Ruwenzori Mtns. (Uganda)	2	1		3	6
Virunga Mtns. (Zaire-Rwanda border)				3	3
Moroto (Uganda)	2		1	3	6
Suk Mtns. (Kenya)	2		1	3	6
Cherangani, Forest Reserve (Kenya)	2			3	5
Chyulu Hills (s. Tsavo Park, Kenya)	2			3	5
Teita Hills (Kenya)	2			3	5
Masai & Mbulu Mtns (Tanzania)	2			3	5
Kilimanjaro (Kenya-Tanzania border)	2			3	5
Pare Mtns. (Tanzania)	2			3	5
Island of Madagascar					
Montagne d'Ambre	2	1		3	6
Montagne St. Francis	2	1		3	6
Masoala Peninsula	2	1	1	3	7
forest of Ambohimanga du Sud	2	1		3	6
Antsingy region, near Antsalova	2	1		3	6
forest of Kasijy, n.w. of Maevatanana	2	1		3	6
Isalo region	2	1		3	6
calcareous plateau de l'Ankarana	2	1		3	6
Tsaratanana Peak		1			1
Marojejy Peak		1			1

Table VII

Priority areas in Central America and the Caribbean for floristic inventory

Priority Area	Weight:	Low Coll. Density 2	High Endemism 1	New Distributions 1	Threatened 3	Total
Mountains & Intermontane Valleys						
Mexico						
Oaxaca (Mixteca Alta & Baja)						
Sierra Madre del Sur		2			3	5
Sierra Madre de Oaxaca		2			3	5
Sierra de Juárez		2			3	5
Sierra Mixe		2	1	1	3	7
Sierra de Huautla		2			3	5
Sierra Villa Alta		2			3	5
Zempoaltepétl Massif		2	1		3	6
Chiapas						
Sierra Madre del Sur		2		1	3	6
vicinity of Pijijapan		2		1	3	6
Marqués de Camillas region		2		1	3	6
Belize						
Maya Mtns.		2				2
Cocksoorub Mtns.		2				2
Pine Ridge (ravines)		2				2
Victoria Peak		2				2
Branch River George		2		1		3
Guatemala						
Sierra de los Cuchumatanes		2		1	3	6
Baja Verapéz		2		1	3	6
Honduras						
Cordillera Merendon (Dept. of Cortes)		2			3	5
Cordillera Nombre de Dios (Dept. of Atlantica)		2			3	5
coastal cordillera (Dept. of Colón)		2			3	5
Sierra de Celaqua (Dept. of Lempira)		2		1		3
Cerro Santa Barbara		2			3	5
vicinity of San Esteban & Culmí (Dept. of Olancho)		2		1	3	6
Nicaragua						
Serranias de Yolzina		2			3	5
elfin summit forests		2	1	1		4
Cordillera Isabella		2		1	3	6
Cerro Baba		2			3	5
Cerro Yetuca		2			3	5
Costa Rica						
Cordillera de Talamanca		2		1	3	6
Panamá						
Serrania de Canzas		2		1	3	5
montane forests Chiriqui Prov.			1	1	3	5
Cerro Jefe cloud forest		2	1	1		4
Darién cloud forests		2	1	1		4
Caribbean Slope and Coastal Forests						
Mexico						
Tabasco			1	3		4
Campeche			1		3	4
Yucatan		2	1		3	6
northern lowlands of Chiapas		2		1	3	6
Belize						
Sarstoon River & s.w. Toledo Prov.		2				2
Columbia Forest Reserve		2		1		3

(continued)

Table VII

Priority areas in Central America and the Caribbean for floristic inventory (continued)

Priority Area	Weight:	Criteria for Prioritization:				
		Low Coll. Density 2	High Endemism 1	New Distributions 1	Threatened 3	Total
Guatemala						
s. Petén		2		1	3	6
Honduras						
Mosquitia Area						
Río Plátano & Rio Mocorbu		2			3	5
vicinity of Trujillo		2			3	5
Nicaragua						
Depts. of Río San Juan & Zelaya		2		1		3
Mosquitia Region		2		1		3
Costa Rica						
llanuras de Tortuguero		2		1	3	6
Río San Juan lowlands		2		1	3	6
Panama						
Provs. of Boca del Toro, Veraguas & Colón		2			3	5
s. Darién Province		2		1		2
Pacific Coastal Forests						
Mexico						
Isthmus of Tehuantepec		2	1		3	6
Costa Rica						
s.w. Osa Peninsula (Costa Rica)		2		1	3	6
Panama						
s.w. Azuero Peninsula		2				2
Dominican Republic						
Highlands						
Cordillera Central			1	1	3	5
Cordillera Septentrional			1		3	4
Sierra de Baoruco			1	1	3	5
Sierra Nieba			1			1
Lowlands						
Samaná Peninsula			1		3	4
Llanura Costera					3	3
Azua plains					3	3
Lesser Antilles						
Monserrat I.				1		1
Saba I.				1		1

Table VIII
Priority areas in South America for floristic inventory

Priority Area	Low Coll. Density (Weight: 2)	High Endemism (1)	New Distributions (1)	Threatened (3)	Total
Guayana Shield					
granite outcrops on shield border	2	1	1		4
all tepuis & inselbergs	2	1	1		4
la Macarena (tepui, Colombia)	2	1			3
Caribbean Coastal Forests					
Maracaibo Basin (Venezuela)				3	3
Extra-Amazonan River Valleys					
Orinoco Delta	2				2
Orinoco Basin					
States of Apure, Barinas & Portuguesa					
llanos (gallery forests)				3	3
lowland forest	2		1		3
Magdalena R. Valley (Colombia)	2		1	3	6
Barranca Bermeja swamps	2	1			3
Andean Forests					
Coastal Cordillera of Venezuela					
cloud forests	2	1		3	6
e. montane sections	2			3	5
lower Tuy Valley				3	3
s.e. Andes of Venezuela cloud forests					
(incl. Sierra San Luis)	2	1		3	6
Sierra de Perija (Venezuela)	2			3	5
Sierra Nevada de Santa Marta (Colombia)	2	1			3
Eastern Cordillera of Columbia					
eastern cloud forests (ceja de la montaña)		1		3	4
Santander	2				2
Central Cordillera of Colombia					
State of Antioquia (w. and c. cordilliera)				3	3
Huila	2				2
Western Cordillera of Colombia					
Cerro Tacurcuna (includes Chocó)	2	1	1	3	6
Ecuadorian Andes					
e. cloud forests		1			1
Peruvian Andes					
e. cloud forests	2			3	5
intermontane valleys	2	1			3
Pacific Coastal forests					
n. Chocó seasonal forest (Colombia)	2	1	1	3	6
Chocó Dept. rain forests (Colombia)	2	1	1	3	6
Valle (Columbia)	2			3	5
Cauca (Colombia)	2			3	5
Nariño (Colombia)	2			3	5
Alto de Dique (Colombia)	2	1	1	3	6
wet coastal lowlands of Ecuador		1	1	3	5
dry coastal lowlands of Ecuador		1		3	4
Amazonian Forest					
Brazilian Amazônia					
State of Maranhão (transition forest)	2		1	3	6
State of Amapá	2		1		3
Serra do Navío		1		3	4

(continued)

Table VIII
Priority areas in South America for floristic inventory (continued)

	Criteria for Prioritization:				
Priority Area Weight:	Low Coll. Density 2	High Endemism 1	New Distributions 1	Threatened 3	Total
State of Pará					
Belém region				3	3
R. Tapajós	2	1	1		4
Serra dos Carajás	2	1	1	3	7
Ilha do Marajó	2				2
Ilha do Bananal	2	1			3
Tucurui Dam Region (Rio Tocantins)				3	3
Paragominas Region				3	3
Jari (Amapá/Pará)				3	3
State of Amazonas					
Manaus regions				3	3
R. Trombetas	2	1	1		4
Tefé (center of endemism)		1	1	3	5
tepuis of Guiana Shield	2	1	1		4
upper Solimões	2		1		3
upper Rio Negro	2		1		3
Balbina Dam Region (Rio Uatumá)				3	3
Imeri	2	1	1		4
São Paulo de Olivença (center of endemism)	2	1	1		4
State of Mato Grosso (transition forest)	2		1	3	6
Rondônia-Aripuanã (center of endemism)	2	1	1	3	7
State of Acre	2	1	1		4
Peruvian Amazon					
Mayo Valley	2			3	5
Apurimac Valley	2			3	5
along Carretera Marginal				3	3
Chanchamayo	2			3	5
Huallaga	2			3	5
Bolivian Amazon					
Río Beni	2			3	5
Llanos de Mojos				3	3
lower Yungas				3	3
Colombian and Ecuadorian Amazon	2				2
llanos of e. Colombia (gallery forests)	2				2
Venezuelan Amazon					
State of Bolívar					
Lote Forestal de San Pedro				3	3
State of Guyana					
Gran Sabana				3	3
E. Brazilian Forests					
restinga forest		1		3	4
caatinga forest		1		3	4
campo rupestre				3	3
lowland moist forest	2	1	1	3	7

Table IX
Areas of highest priority for botanical exploration

I. Asia
  A. Continental Asia
    1. India
      A. Nagaland
    2. Thailand
      A. Central Plain: dry deciduous Dipterocarp forest & low tropical rainforest
      B. northwestern and western highlands: upper tropical forest & hill evergreen forest
    3. Indochina
      A. Vietnam: northeastern Vietnam; Danang-Hué Region
      B. southern Laos, along Thai and Burmese borders
      C. southwestern Cambodia
  B. Insular Asia
    1. Philippine Islands
      A. Luzon: Sierra Madre Mountains
      B. Samar-Leyte Islands: limestone forests
      C. Mindanao: the mountain divide between Bokidnon and Ausan Provinces and the eastern coastal mountains
      D. Mindoro & Palawan: seasonal evergreen forests
    2. Borneo
      A. northwestern Kalimantan
      B. western Sabah: Crocker Range
      C. Pulau Laut & Meratus
      D. central Borneo mountain system
    3. Sumatra
      A. Nort Aceh
      B. Anambas & Natunes
      C. Barisan Range
      D. eastern Sumatra: swamp forests
      E. southeast, especially Riau
      F. southern volcanos
      G. forested parts of Bangkalulu & Lampongs
      H. Pagui Islands
    4. Sulawesi
      A. northern arm (except Minahasa)
      B. central Sulawesi
      C. southwestern Sulawesi
    5. Moluccas
      A. eastern & southern Halmahera
      B. Morotai
      C. Bacan Group (except Mt. Sibela)
      D. Obi Group
      E. Sula Group
      F. Buru & Seram Group
      G. Ambelon Island & Vliassen Island
      H. Wetar Dama
    6. New Guinea & Adjacent Islands
      A. Papua New Guinea: lowland & low montane forests of Fly River Valley; headwaters of Sepik, May & Horden Rivers, Papual ultrabasic belt
      B. Irian Jaya: Peg. van Rees, Peg. Jayawijaya, lowland & montane forests of Diogel River
      C. Salawati, Batanta, Waigeo & Gebe Islands
      D. New Britain: limestone areas, Mt. Birinia
II. Africa
  A. West African Forest Mosaic
    1. Sierra Leone: interior forests, Gola Forest
    2. Liberia: interior forests, Sapo Forest, border with Ivory Coast
    3. Ivory Coast: borders with Liberia & Ghana
    4. Ghana: southwest
  B. Bight of Benin
    1. Nigeria: Niger River Delta, north of Calabar; border with Cameroun
    2. Cameroun: lowland coastal & montane forests

(continued)

Table IX
Areas of highest priority for botanical exploration (continued)

- C. Guineo-Congolean Regional Mosaic
  1. Zaire: southwestern Shaba Province, Lukaya Valley, Marungu Plateau in Katanga Province, margins of Likasi & Lufira Rivers, dry semievergreen forest north of Basape
  2. Congo People's Republic: Bateke Plateau, Sangha River Interval
- D. Rift Lake Forests
  1. Kenya: Kakamega Forest Reserve
  2. Tanzania: Mahale Mountains
- E. Zanzibar-Inhambane Regional Mosaic
  1. Kenya: Tana River, Arabuko-Sokoke Forest Reserve, Kayas, Shimba Hills
  2. Tanzania: Rau Forest Reserve, Usumbara Mountains, Uluguru Mountains, Uzungwa Mountains, Mahenge Plateau, Pugu Hills, Rondo Plateau
- F. Afromontane Region
  1. Kenya: Suk Mountains, Cherangani Forest Reserve, Chyulu Hills, Teita Hills
  2. Uganda: Ruwenzori Mountains, Moroto
  3. Tanzania: Masai & Mbulu Mountains, Kilimanjaro, Pare Mountains
- G. Madagascar

  Montagne d'Ambre, Montagne St. Francis, Masoala Peninsula, forests of Ambohimanga du Sud, Antsigny Region near Antsalova, forest of Kasijy northwest of Maevatanana, Isalo Region, Plateau de l'Ankarana

III. Central America
- A. Highlands & Valleys
  1. Mexico
     - A. Oaxaca: Mixteca Alta & Baja
     - B. Chiapas: Sierra Madre del Sur, vicinity of Pijijapan, Marqués de Camillas Region, northern lowlands
  2. Guatemala: Sierra de los Cuchumatanes, Baja Verapez
  3. Honduras: Cordillera Merendon, Cordillera Nombre de Dios, coastal cordillera of Department of Colón, Cerro Santa Barbara, vicinity of San Esteban & Culmí
  4. Nicaragua: Cordillera Isabella, Cerro Baba, Cerro Yetuca
  5. Costa Rica: Cordillera de Talamanca
  6. Panamá: Serrania de Canzas, montane forests of Chiriquí Province
- B. Lowlands & Coastal Slope
  1. Mexico: dry forests of Yucatan
  2. Guatemala: southern Petén
  3. Honduras: Río Platano & Río Mocorbu, vicinity of Trujillo
  4. Nicaragua: Serranias de Yolzina, Mosquitia Region
  5. Costa Rica: llanuras de Tortuguero, Río San Juan lowlands, southwestern Osa Peninsula
  6. Panama: Provinces of Boca del Toro, Veraguas & Colón
- C. Caribbean Islands
  1. Dominican Republic: Cordillera Central, Sierra de Baoruco

IV. South America
- A. Andean cloud forests (ceja de la montaña)
  1. Venezuela: eastern montane sections of the coastal Cordillera & the southeastern Andes
  2. Colombia: eastern cloud forests
  3. Ecuador: eastern cloud forests
  4. Perú: eastern cloud forests
- B. Highlands
  1. Venezuela: Sierra de Perija
  2. Colombia: Cerro Tacurcuna, Alto de Dique
- C. Pacific Coastal Forests
  1. Colombia: northern seasonal & rain forests of Chocó Department, moist forests of Valle, Cauca & Nariño Departments
  2. Ecuador: Pacific coastal wet and dry forests
- D. Magdalena River Valley (Colombia)
- E. Amazônia
  1. Brazil: State of Maranhão, Serra dos Carajás, State of Mato Grosso, State of Rondônia
  2. Perú: Chanchamayo, Huallaga, Mayo Valley, Apurimac Valley
  3. Bolivia: Río Beni
- F. Eastern Lowland Forests of Brazil

Table X

Costs of discovery in Neotropical botanical inventory

Inventory Program	Mean Cost in US$ Per:			
	Expedition	Collection No.	Determination	New Taxon
Projeto Flora Amazonica	25,358	11	38	3,756
Rio Palenque	150,000	50	136	1,500

Sources: Prance et al., 1984; Prance, pers. comm.; Gentry, pers. comm.

then the cost will be about $32,850,000; 5%, then about $16,425,000).

Is it worth it to spend all of this money to identify the last 5%, or even 15%, of the plant species on the planet? After all, an adequate inventory realistically need not be 100% complete; it need only have an acceptable standard of knowledge needed to facilitate sustainable land use, conservation, and science. The answer is, of course, yes, because the inventory of tropical forests will never be finished before deforestation overwhelms the effort. Species will go extinct, many anonymously, since tropical forests, whether their components are taxonomically described or not, will for the most part disappear. All of these unsavory reasons make the priorities set out in this book of timely and of inestimable value as a focus for the investment of our limited resources.

XI. Literature Cited

Ainsworth, C. D. 1961. Dictionary of fungi. Commonwealth Mycological Institute, Royal Botanic Garden, Kew.

Campbell, D. G. 1984. Presentation to the Seminar of Global Habitability, Columbia University New York, N.Y.

Council on Environmental Quality. 1980. Global future: Time to act, Report to the President on Global Resources, Environment and Population. U. S. Government Printing Office. Washington D.C.

Dodson, C. 1986. personal communication.

––––––– & A. Gentry. 1978. Flora of the Río Palenque Science Center. Selbyana. **4:** 1–628.

Erwin, T. 1982. Tropical forests: Their richness in Coleoptera and other arthropod species. Coleopterists Bull. **36(1):** 74–75.

F.A.O./U.N.E.P. 1981a. Tropical forest resources assessment project: Forest resources of tropical Africa, Part II: Country Briefs. F.A.O., Rome.

–––––––. 1981b. Tropical forest resources assessment project: Forest resources of tropical Asia. F.A.O., Rome.

–––––––. 1981c. Los Recursos Forestales de la America Tropical. Tropical Forest Research Assessment Project. F.A.O., Rome.

Flora Malesiana. 1974. Ser. 1,8 (1): 3.

Holmgren, P. K. & W. Keuken. 1974. Index herbariorum, Sixth edition. Oosthoek, Scheltema & Holkema. Urecht.

–––––––, ––––––– & E. K. Schofield. 1981. Index herbariorum, Seventh Edition. Dr. W. Junk B. V. Pub. Boston, Mass. The Hague.

Humbert, H. 1959. Origènes presumées et affinites de la flore de Madagascar. Pages 149–187 *in* Mem. Inst. Sci. Madagascar, Ser. B. Veg. 9.

I.U.C.N. 1980. World conservation strategy. I.U.C.N. Gland.

Jacobs, M. 1977. Editorial. Fl. Malesiana Bull. **30:** 2733.

Kubitzki, K. 1977. The problem of rare and infrequent species: The monographer's view. *In* G. T. Prance & T. S. Elias (eds) Extinction is forever. The N. Y. Bot. Gard. New York, N.Y.

Lanjouw, J. & F. A. Stafleu. 1952. Index herbariorum. Kemink en Zoon Utrecht.

––––––– + –––––––. 1954. Index herbariorum, Second edition. Kemink en Zoon. Utrecht.

––––––– + –––––––. 1956. Index herbariorum, Third edition. Kemink en Zoon. Utrecht.

––––––– + –––––––. 1959. Index herbariorum, Fourth edition. Kemink en Zoon. Utrecht.

––––––– & –––––––. 1964. Index herbariorum, Fifth edition. Kemink en Zoon. Utrecht.

Mayo, A. 1986. 60,000 plants may become endangered in the next 40 years. Species. No. **6:** 4–5.

Myers, N. 1980. Conversion of tropical moist forest conversion. Nat. Res. Council of the Nat. Acad. of Sci. Washington, D.C.

Prance, G.T. 1977. Floristic inventory of the tropics: Where do we stand? Ann. Missouri Bot. Gard. **64:** 659–684.

–––––––. 1978. Floristic inventory of the tropics: A correction. Ann. Missouri Bot. Gard. **65:** i–ii.

–––––––. 1984. Completing the inventory. *In* V. H. Heywood & D. G. Moore (ed.) Current Concepts in Plant Taxonomy. Academic Press. London.

––––––– & D. G. Campbell. 1988. The present state of tropical floristics. Symposium on Tropical Botany, Utrecht. Taxon. **37(3):** 519–548.

Qureshi, A. H., L. K. Hamilton, D. Meuller-Dombois, W. R. H. Perera & R. A. Carpenter. 1980. Assessing tropical forest lands: Their suitability for sustainable uses. East West Center. Honolulu, HA.

Raven, P. H. 1976. Ethics and attitudes. Pages 155–179 *in* J. Simmons, R. I. Beyer, P. E. Brandham, G. L. Lucas & V. T. H. Parry (Eds). Conservation of Threatened Plants. Plenum, New York, N.Y.

von Reis Altschul, S. 1982. New plant sources for drugs and foods from the New York Botanical Garden Herbarium. Harvard U. Press. Cambridge, Mass.

Salati, E. 1983. O clima actual depende da floresta. Pages 15–44 *in* E. Salati, H. O. R. Shubart, W. Junk & A. E. de Oliveira (Eds.). Amazônia, Desenvolvmento integração ecológia. CNPq. Brasilia.

The Times. 1981. Atlas of the World. Comprehensive Edition. Times Books. London.

Toledo, V. M. 1985. A critical evaluation of the floristic knowledge in Latin Amearica and the Caribbean. A report presented to the Nature Conservancy. Washington, D.C.

U.N.E.S.C.O. 1973. International classification and mapping of vegetation. U.N.E.S.C.O. Paris.

U.S. Dept. of Energy 1986. A comparison of tropical forest surveys. Office of Energy Res., Office of Basic Energy Sciences, Carbon Dioxide Res. Div., Wash. DC. DOE/NBB-0078.

Vink, W. 1981. Density indexes updated. Fl. Malesiana Bull. **34**: 3567–3568.

White, F. 1983. The Vegetation of Africa. UNESCO. Paris.

Whitmore, T. C. 1984. Vegetation of Malesia. Commonwealth Forestry Inst. Oxford.

Wilson, E. O. 1985. Time to revise systematics. Science. Vol. **230**: 4731.

Section Two: Regional Reports

Regional Reports

I. Asia

China

Wang Huen-pu, Chen Sing-chi and Wang Si-yu
Translated by Wang Si-yu

Contents

I. Description of the Region	36
A. Hills and Mountains of Southern Taiwan	36
B. Littoral Hills of Southeastern Guangdong	36
C. Qionglei Terraces and Hills	36
D. Qiongnan Hills and Mountains of Hainan Island	37
E. Limestone Hills and Mountains of Southwestern Guangxi	37
F. Valleys and Medium-high Mountains of Southern Yunnan	38
G. Basin and Range Landscape of Xishuan Banna	38
H. River Valleys and Mountains of Southwestern Yunnan	39
I. River Valleys and Mountains of Southeastern Tibet	39
J. Coral Islands of Dongsha, Zhongsha and Xisha	39
K. Coral Islands of Nansha Qundao	40
L. Summary	40
II. Vegetation Maps	40
III. Status of Floristic Inventory	40
A. Collections and Publications	40
B. Endemic Species	41
C. Extinction	41
IV. Future of Floristic Inventory	41
V. Literature Cited	43

I. Description of the Region

The tropical zone in China is situated at the northern margin of the Asian tropics. It is in a paleotropic floristic province where there are many relict species, some of which are important in the evolution of the modern flora. The flora in the eastern portion is composed mainly of elements derived from the eastern Malaysian flora, but it also has some special elements from the Sino-Japanese and Chinese subtropics. The flora of the western portion is composed mainly of Indo-Burmese elements as well as some special elements from the Sino-Tibetan and Chinese subtropics (Editorial Committee of China Physical Geography, Academia Sinica, 1983; Wu Cheng-yih, 1979).

We have divided the Chinese tropical zone into 11 regions (A–K) to be discussed in turn, starting from east to west and moving from the mainland to the islands.

A. Hills and Mountains of Southern Taiwan

This region (approximately 9800 km²) includes Tainan, Gaoxiong, Pingdong, Taidong and the nearby islands (Editorial Committee of Chinese Vegetation, 1980). Surrounded on three sides by the sea, this region is characterized by the warm and moist climatic features typical of tropical oceans. Annual precipitation averages 1500–2500 mm but may be as much as 3000–5000 mm in some places. Torrential rain, brought by typhoons in summer and autumn, is a distinctive climatic feature of this region. Topographically, the southern sections of Alishan and Zhongyang Shan are hilly terraces made of sandstone and shale, ranging in elevation from 200 to 300 m.

The primary forest is a seasonal rain forest dominated by *Artocarpus lanceolatus*, *Myristica cage*, and *Pterospermum niveum*. The region is mostly occupied by secondary forests. In dry places and where the seasonal rain forests have been ruined (due to man-caused deterioration of the environment), monsoon forests have developed. These are mainly composed of *Albizia procera*, *Bombax malabarica* and *Lagerstroemia subcostata* var. *hirtella*. Around bays, mangrove forests dominated by *Avicennia marina*, *Ceritops tagal* and *Rhizophora mucronata* may be found.

The northern part of the region is occupied by the northern section of Zhongyang Shan. Here the vegetation is well preserved and occurs in distinct horizontal belts. From 500 to 1300 m, there is evergreen broad-leaved forest, dominated by *Castanopsis borneensis*, *Lithocarpus ternaticupulus* and *Michelia kachirachirai*. From 1300 to 2000 m, the forest is principally composed of *Cinnamomum micranthum*, *Cyclobalanopsis longinux* and *Phoebe formosana*. From 2000 to 3000 m, the deciduous broad-leaved forest begins to give way to mixed evergreen forest. Some component plants include *Acer* spp. and *Cyclobalanopsis morrii* in association with *Chamaecyparis formosensis*, a native to Taiwan. From 3000 to 3600 m, the climax community is the subalpine needle-leaved forest represented by *Abies kawakamii*, *Juniperus morrisonicola* and *Picea morrisonicola*. Above 3600 m, alpine bush and alpine meadow appear. Any place within this region would be ideal for the establishment of nature reserves.

B. Littoral Hills of Southeastern Guangdong

This long and narrow coastal region (about 9800 km²) includes the littoral hills, terraces and islands in the southern parts of Huilai, Lufeng, Haifeng, Baoan, Zhuhai, Zhongshan, Doumen and Taishan counties. Facing the tropical South China Sea, the region is warm and has abundant water resources. The mean annual temperature is about 22°C; the yearly rainfall ranges from 1500 to 2000 mm. In summer and autumn there are occasional typhoons and violent thunderstorms. The hills are generally at elevations below 300 m, with a few peaks over 900 m.

The seasonal rain forest has been decimated here. The land it once occupied has been taken over by sparse thickets composed of *Baeckea frutescens*, *Pinus massoniana* and *Rhodomyrtus tomentosa*. The only remaining forest can be found in small patches near villages, where component species include: *Aporosa chinensis*, *Aquilaria sinensis*, *Bischofia javanica*, *Bridelia balansaei*, *Ficus microcarpa*, *Garcinia multiflora*, *Sterculia lanceolata* and *Syzygium odoratum* (Chen Shupei et al., 1983). Such natural forests should be protected, to control indiscriminate cutting of forest. In the littoral there are sandy and muddy beaches and, owing to crustal movement there are, in addition, funnel-shaped bays, drowned valleys, lagoons, littoral hilly terraces and islands existing side by side. The scenery is very appealing and makes this an ideal place for the establishment of seaside protected areas and for the well-managed development of tourism. Small clumps of mangrove forest may also be found; these are composed of dwarf trees such as *Aegiceras corniculatum*, *Acanthus ilicifolius*, *Avicennia marina* and *Kandelia candel*.

C. Qionglei Terraces and Hills

This region (approximately 58,800 km²) is located in the coastal area of Leizhou strait and Tonkin Gulf and includes northern Hainan Island, Leizhou Peninsula in southwestern Guangdong and southwestern Guangxi. Climatologically, this region is characterized by tropical monsoons. It has a mean annual temperature above 24°C; the yearly precipitation ranges from 1400 to 2000 mm, but up to 3000 mm may fall in some places, such as on the hills around Shiwan Dashan. The rain falls mostly in summer and autumn. The weather of the region is changeable because of the influence of the strong winds, especially typhoons. Leizhou Peninsula and Hainan Island were linked at one time; they became separated from the mainland of Guangdong during the Quaternary, when Qiongzhou Bay, a graben, sank. For this reason, close similarities exist between habitat and vegetation types of southern Leizhou Peninsula and northern Hainan Island. Terraces were formed over large areas with sediments from the Quaternary shallow sea, and from basalt and rhyolite derived from volcanoes. In the 1950's, the seasonal rain forest in this region was still well preserved in Hsuwen County (Chang Hongda et al., 1957; Chen Shupei et al., 1983). At present, however, very little remains and only small patches are left near Yugonglou. These are dominated by *Antiaris toxicaria, Canarium album, Carallia brachiata, Ficus altissima, Ficus microcarpa, Endospermum chinense, Schefflera octophylla* and *Syzygium hancei*. Early attempts to establish rubber (*Hevea brasiliensis*) plantations on the terraces failed. Nonetheless, recent attempts on squared terraces surrounded by windbreak forest have been successful (Ke Chichen, 1979). The trees chosen to provide windbreaks were introduced species of *Eucalyptus*, mainly *E. citriodora, E. exerta* and *E. rudis*, as well as *Casuarina equisetifolia* along the seashore. However, as a result of the planting of *Eucalyptus* and *Casuarina*, soil fertility is decreasing and concomitantly the growth rate of the windbreak species is slowing down. Henve, there is an urgent need to select indigenous, fast-growing species for planting in windbreaks. Obviously, it is imperative that careful management be coupled to intelligent species choice when designing, executing and maintaining windbreaks.

This is the only region in China where there is well preserved mangrove forest with tall trees. The principal species are: *Acanthus ilicifolius, Aegiceras corniculatum, Avicennia marina, Bruguiera sexangula, Ceriops tagal, Kandelia candel, Lumnitzera racemosa, Rhizophora apiculata* and *R. stylosa*. A nature reserve has been established in Dongda Gang, Qiongshan county, but effective regulations to insure protection are urgently required (Wang Huen-pu, 1982b). It is also necessary that the small patches of the mangrove forest distributed in the Leizhou Peninsula and Qiongzhou District, Guangxi, be brought under protection, in order to make full use of the littoral resources.

The granite and shale hills extending from the western sections of Luwan Dashan and Shiwan Dashan are below 500 m elevation. On these hills, most primary forest has been destroyed and is being replaced by *Pinus massoniana* forest or by shrubs, such as *Baeckea frutescens* and *Rhodomyrtus tomentosa* or by small patches of *Pinus latterii* forest. The relic seasonal forest, where present, is in small patches and is dominated by *Eberhardtia aurata, Hopea chinensis* and *Reevesia glaucophylla*. Forests of *Cinnamomum cassia* and *Illicium verum* are widespread.

D. Qiongnan Hills and Mountains of Hainan Island

This region (about 40,000 km²) includes the lands to the south of Wenchang, Qionghai, Dan Xian and Dongfang on Hainan Island. Mountains on Hainan Island, which range in height from 500 to 1870 m, are all located in this region. Around these mountains, terraces and plains occur down to the seashore. The annual mean temperature is 23–28°C; the annual average precipitation ranges from 1000 to 2000 mm, with most of the rain falling in the months from April to October. Nor is the rainfall geographically evenly distributed; the eastern mountains create a rain shadow on the leeward western part. Typhoons and thunderstorms are most prevalent in summer and autumn.

This region is the only one in China characterized by a vegetation typical of the tropics. In the southeastern part, the climax community found below 500 m is a rain forest dominated by *Heritiera parvifolia, Hopea hainanensis* and *Vatica astrotricha*. The seasonal rain forest which occupies most of the land on the hills is largely composed of *Albizia odoratissima, Bombax malabarica, Dillenia pentagyna, Ficus nervosa, Gironniera cuspidata, Kleinhovia hospita, Lannea grandis, Radermachera hainanensis, Spondias pinnata* and *Terminalia hainanensis*. In the mountains at 500–1500 m elevation, a montane rain forest is present, either *Podocarpus imbricatus* growing with *Adinandra hainanensis, Castanopsis borneensis* and *Syzygium chunianum*, or by *Dacrydium pierrei* growing with *Pinus kwangtungensis, Rhododendron simiarum* and *Ternstroemia gymnanthera* (Hou Kuanzhao et al., 1955; Hu Yujia, 1983; Yu Tong-quan, 1983). Recently, forestry organizations have been formed in Diaolo Shan, Jianfeng Ling, Bawang Ling and Limu Ling. On Hainan Island there are 11 nature reserves or protected sites, covering 25,400 ha.

In order to maintain the ecological equilibrium of Hainan Island, to develop its economy and to establish tourism there, a decision must be made soon to set aside more of these forest areas (including Wuzhi Shan and Lulian) as nature reserves. Protection plans must adhere closely to the requirements of a natural resource protected area: the area must be defined precisely, the natural forests must be under effective protection, regeneration of forests on the disturbed land and reforestation on the barren land must be encouraged, and scientific studies, teaching, and tourism must be undertaken in a planned way (Wang Huen-pu, 1980a, 1980b, 1981, 1982a, 1982b; Wang Xian-pu, 1984). The intelligent development of Hainan Island could serve as a model system for future planned utilization of China's precious natural resources. The original flourishing mangrove forest, in Quinglan Gang, and Wenchang Counties along the eastern coast, has almost entirely been cut down. Such forests may come back if they are protected from further destruction.

E. Limestone Hills and Mountains of Southwestern Guangxi

This region (about 19,600 km²) includes Zuojiang and Youjiang and extends to Nanning and Baise. The elevated portions in the northwest are at 1000 and 1300 m; in the southeast, about 500 m, with some peaks of Shiwan Dashan and Daqing Shan reaching more than 1000 m. In striking contrast, the river valley basins between Zuojiang, Youjiang and Yongjiang are relatively flat.

The region is surrounded on three sides by hills and mountains, with only one breach in the north. Therefore, although the annual mean temperature of this region is very close to that of the previous one, its absolute minimum temperature can occasionally drop to below 0°C. Most of the rainfall (1400 to 1500 mm) falls in summer and autumn (Li Zhiji, 1964; Li Zhiji et al., 1965).

On the limestone hills there are regions of extensive well preserved seasonal rain forest. This forest is composed of diverse species, including many Chinese endemics such as *Burretiodendron hsienmu* and *Parashorea chinensis* var. *kwangsiensis* (Li Zhiji et al., 1965; Hu Shun-shi et al., 1980; Forestry Department, Forestry Branch, Guangxi Agricultural College, 1978). The vegetation here is typical of a Sino-Vietnamese plant province, which is unusual since the area essentially is a part of the Indo-Malayan tropical floristic region. There are four preserves in this region. Longgang preserve in Longzhou County measures about 10,000 ha., Longrui preserve in Ningming measures 2,917 ha., Longmei preserve in Daxing measures 26,667 ha, and the area of Longhu Shan preserve in Longan is 1,250 ha. (Liang Choufen et al., 1981). It would be desirable to establish some preserves on the limestone mountains of Jingxi and Debao as well as on Daming Shan in Fusui. In addition, Napo would be a choice site for a preserve. It was here that two new species of Dipterocarpaceae, *Parashorea chinensis* var. *kwangsiensis* and *Vatica astrotricha*, were recently discovered (Chingtienshu Research Group, 1977). The composition of the seasonal rain forest growing on the sandstone and shale hills and mountains is quite different from that of the seasonal rain forest growing on the limestone hills and mountains. Nevertheless, both kinds of forest are dominated by *Eberhardtia aurata*, *Hopea chinensis* and *Saraca chinensis*. It is in this region that *Illicium verum* may be found (Li Zhiji, 1964). Unfortunately, most of the forests harboring *I. verum* have been damaged and only small patches remain in Shiwan Dashan and Daqing Shan (Wang Huen-pu et al., 1955; Li Shiying, 1956; Hu Shen-shi, 1979; Wang Huen-pu et al., 1982). Although forest farms and forestry management organizations have been established in the area, management of the remaining patches has not been optimal, owing to lack of experience.

F. Valleys and Medium-high Mountains of Southern Yunnan

Situated in southern Yunnan, this strip of the tropical zone (about 7840 km²) includes the lower reaches of Panlong Jiang, Nanxi He, Yuan Jiang, Tengtiao Jiang, Lixian Jiang and Xiaohei Jiang. The region is composed of mountains and valleys, the former may reach 2000 m and the latter range from 100 to 400 m. The mountains are granite, gneiss, phyllite and/or limestone. The annual mean temperature of Menglaba and Hekou in Jinping is 22–23°C and the annual rainfall is more than 2000 mm.

In the valleys there are small patches of rain forest dominated by *Crypteronia paniculata*, *Dipterocarpus yunnanensis* and *Hopea mollissima*. In the mountains, from 500 to 1500 m elevation, seasonal rain forest is common. The major tree species in this forest are: *Altingia yunnanensis*, *Antiaris toxicaria*, *Paramichelia baillonii*, *Semecarpus reticulata*, *Terminalia myriocarpa* and *Winchia calophylla*. On the drier slopes, especially the windward ones, a monsoon forest is present. Here, the dominant tree species include *Albizia chinensis*, *Bombax malabarica*, *Duabanga grandiflora*, *Erythrina stricta*, *Melia dubia* and *Stereospermum tetragonum*. From 1500 to 2200 m, the slopes are occupied by an evergreen broad-leaved forest dominated by *Castanopsis ceratacantha*, *Machilus* spp. and *Manglietia* spp. From 2200 m upwards, bush and elfin wood of *Rhododendron* spp. appear on mountain ridges and isolated peaks. Most of the forest in Southern Yunnan has been severely damaged and is being replaced by secondary scrub and grass land. The establishment of nature reserves and reforestation with rapid growing, economically valuable species, are urgently needed. Perhaps the best places for establishing reserves would be on the limestone mountains of Menglaba and Malipo in Jinping, where there are small patches of seasonal rain forest left. *Celtis cinnamomifolia* and *Garuga floribunda* are abundant in these forests.

G. Basin and Range Landscape of Xishuan Banna

Situated to the west of Ailao Shan, this region (approximately 31,360 km²) includes Xishuan Banna and southwestern Simao in Yunnan. The northern part of this region is considered the second stepping stone of the Yunnan Plateau. The average elevation is 900–1100 m, with the southern part of the region dissected by a broad river valley basin of 500–800 m elevation. There are some mountains of about 1500 m elevation, with the highest peaks reaching 2000 m. The climate of the region is characteristic of the tropics with an annual mean temperature range from 20 to 22°C. Yearly rainfall varies from 1200 to 1500 mm and mostly occurs in the summer months.

The seasonal rain forest growing on the terraces and hills below 800 m is predominantly composed of *Antiaris toxicaria*, *Aphanomyxis polystachya*, *Barringtonia longipes*, *Canarium album*, *Chukrasia tabularis*, *Dysoxylum glabra*, *Ficus altissima*, *Pometia tomentosa*, *Pterospermum lanciifolium*, *Terminalia myriocarpa* and *Pouteria grandifolia*. From 800 to 1000 m montane seasonal rain forest is present, the dominant species here are: *Actinodaphne henryi*, *Alstonia pachycarpa*, *Dysoxylum spicatum*, *Paramichelia baillonii*, *Phoebe nanmu* and *Semecarpus reticulata*. From 1000 to 1500 m, there is evergreen broad-leaved forest dominated by *Castanopsis hystrix*, *C. indica* and *Cinnamomum glanduliferum* (Teaching and Research Group of Ecology & Geobotany, Biology Department of Yunnan University, 1960; Xu Zaifu et al., 1983). Not long ago, small patches of *Parashorea chinensis* forest were discovered between Mengla and Fubang (Cooperation Group of *Parashorea chinensis*, 1977). The seasonal rain forest on the limestone mountains around Menglun has some unique species, one example is *Tetrameles nudiflora*. Small patches of *Pleomele cambodiana* forest also can be found along Menglain. Thirty years ago, the forest cover of this region used to be more than 60%, but now only half this amount is left.

There are four protected areas in this region, but forest destruction continues. The regulations regarding protection of natural area must be strictly enforced if forests are to remain in Yunnan. On slopes below 800 m many rubber trees (*Hevea brasiliensis*) have been planted, making this region the major base for the development of rubber plantations in Yunnan. In the future, clearing forest for rubber plantations will not be allowed. Hence, emphasis should be laid on management and improvement of existing plantations as well as on the development of new plantations on previously cleared lands.

H. RIVER VALLEYS AND MOUNTAINS OF SOUTHWESTERN YUNNAN

Situated in the southwestern parts of the Linchang and Dehong autonomous prefectures, this region (approximately 13,720 km²) borders Burma. It occupies the ends of the southern extensions of Lu Shan and Gaoligong Shan. Most of the mountains in this region are below 1500 m, but a few peaks reach 2500 m or more. The river valley basins along the reaches of Nanting He range from 400 to 500 m, while those in the Dehong prefecture vary from 750 to 800 m. The climate of this region is dominated by the monsoons coming from the southwest. High mountains in the north form a barrier blocking the cold air from reaching the region; consequently, the annual temperature variation is small (19 to 21°C). Annual rainfall ranges from 1200 to 1600 mm.

The rainy season begins in early May and continues until late October. The seasonal rain forest to the south of the lower reaches of Nanting He is composed of a large variety of species, mainly *Adina cordifolia*, *Dimocarpus yunnanensis*, *Duabanga grandiflora*, *Garuga yunnanensis*, *Pometia tomentosa* and *Terminalia myriocarpa*. On the limestone hills there can also be found small patches of seasonal rain forest containing *Tetrameles nudiflora*. Jin Zhenzhou (1982) has made a strong case for the urgent need to establish nature reserves in these areas and to strengthen management practices. The scattered forests, in the other parts of the region, are very small and are mostly seasonal rain forest dominated by *Bombax malabarica*, *Chukrasia tabularis*, *Eriolaena malvacea*, *Markhamia stipulata*, *Mesua ferrea*, *Sterculia villosa* and *Stereospermum tetragonum*. From 1000 to 1300 m, the evergreens *Chukrasia hystrix* and *C. indica* appear. At higher elevations, *Alnus nepalensis* and *Pinus yunnanensis* forests predominate and are reflective of the rapid vertical changes in climate on mountains at higher latitudes.

I. RIVER VALLEYS AND MOUNTAINS OF SOUTHEASTERN TIBET

Situated in southern Xizang, this region (approximately 90,160 km²) includes Zayu, Medog, the lower reaches of Yarlung Zangbo Jiang to the south of Dawang, Yadong and Yelamu. There are many high mountains and deep valleys in this region. The southern edge of the region slopes gently, and has many mountains below 1000 m. The bottom of the lowest valley is below 200 m. The climate is influenced by the monsoons originating in the Indian Ocean. Obstructed by the Himalayas, the warm wet air from the Indian Ocean converges in this region and brings to it a plenitude of warmth and water. For these reasons, there is the development of tropical vegetation at about 27°N latitude (Zhang Xin-shi, 1978). The annual mean temperature is approximately 20°C; annual precipitation, most of which occurs from June to November, ranges from 2000 to 3000 mm, and may exceed 5000 to 6000 mm in some places.

Since this is a sparsely populated area, there are large tracts of undisturbed natural forests. In the high mountains and deep valleys, the vertical distribution of vegetation types is evident. Nevertheless, plant species characteristic of different vegetation types are always found growing in close proximity. In the river valleys and foothills below 600 m, the rain forest is dominated by many members of the Dipterocarpaceae, e.g., *Dipterocarpus alatus*, *D. pilosus*, *D. turbinatus*, as well as *Canarium strictum* (Burseraceae) and *Tetrameles nudiflora* (Tetramelaceae). The seasonal rain forest on the hills is dominated by *Shorea assamica* and *S. robusta*, while that on the mountains (600 to 1300 m) is composed of *Altingia excelsa*, *Duabanga grandiflora*, *Dysoxylum binectariferum*, *D. gobara*, *Sphaeropteris brunoniana* and *Terminalia myriocarpa*. From 1000 to 2200 m, evergreen broad-leaved forest predominates, with, in the lower reaches, *Castanopsis hystrix*, *C. indica*, *C. tribuloides*, *Dasymaschalon rostratum* and *Machilus* spp. the dominant tree species, and in the upper reaches, *Cyclobalanopsis lamellosa*, *C. oxyodon* and *Lithocarpus xizhangensis* the dominant tree species. On the ridges and south-facing slopes there are *Pinus griffithii* and *P. yunnanensis* forests. Here, above 2200 m, subalpine needle-leaved forest is present. In the mountains from 2200 to 2800 m, *Tsuga dumosa* forest is present, as well as relic evergreen broad-leaved durisilvae. This latter vegetation type is found on the south-facing slopes, characterized by *Quercus pannosa* in the east, and *Pinus griffithii* and *Q. semecarpifolia* in the west. From 2800 to 4000 m. the primary forest is represented by *Abies delavayi* var. *motuonensis*, *A. georgei* var. *smithii*, *A. spectabilis*, *Picea likiangensis* and *Sabina tibetica*. The secondary forest is characterized by *Betula utilis* and *Populus davidiana*. From 4000 to 5000 m, there is alpine bush dominated by *Rhododendron* spp. Alpine meadow composed of *Kobresia* spp. may also be found. From 5000 to 6000 m sparse alpine vegetation occurs. Above 6000 m, the mountains are snow-capped all the year round (Chang King-wai et al., 1973; Chang King-wai et al., 1980; Zhang Xin-shi, 1978). Some natural protected areas have been established in this region.

J. CORAL ISLANDS OF DONGSHA, ZHONGSHA AND XISHA

This region (totaling less than 100 km²) includes all the coral islands of Dongsha Qundao, Zhongsha Qundao and

Xisha Quandao in the South China Sea. These islands are very small and low in elevation. Yonxing Dao, Dongdao and Zhongjian Dao, the larger islands in this region, have an area of about 1.5 km² each; the rest of the islands are less than 0.5 km². Intense solar radiation is a prominent feature of this region. The annual mean temperature is about 26°C; annual mean precipitation is about 1400 m, and mostly occurs in summer and autumn. The islands are often battered by strong winds and typhoons.

The vegetation on the coral islands is evergreen forest, which tends to form communities dominated by a few to a single plant species. *Guettarda speciosa*, *Pisonia grandis* and *Scaevola sericea* are among the more important tree species on these islands. More than 40 species of plants are under cultivation on the islands. Besides ornamental plants and trees for windbreak, most are cereal and vegetable crops. The guano on the islands is an important source of phosphate fertilizer (Editorial Committee of Chinese Vegetation, 1980).

K. Coral Islands of Nansha Qundao

Nansha Qundao consists of many small islands, sand banks, submerged reefs and hidden shoals, with a land area somewhat over 100 km². The terrain is very low, generally about 4–5 m and is strongly affected by tides. The annual mean temperature is above 28°C and the yearly rainfall is about 2000 mm.

The coral islands are essentially sedimentary and are composed of coral limestone, the remains of marine animals and deposits of guano phosphate rock. Beneath the forest the soil is a limestone-humus mix and along the shores there is a sandy soil.

The flora on these islands is rich in elements of tropical littoral and related floras; its closest affinities are with the Pacific tropical floras. Evergreen forests are common and widely distributed on Yonxing Dao, Dondao, Junyin Dao and Ganquan Dao of Xisha Qundao, as well as on the islands of Nansha Qundao.

The vegetation type is similar to the previous region; however, in addition, some islands have *Cordia subcordata* forest and littoral vegetation composed of *Euphorbia atoto* and *Thuarea involuta*. On Taiping Dai, there are many plants in cultivation (Editorial Committee of Chinese Vegetation, 1980).

L. Summary

In conclusion, the tropical regions in China do not lie primarily in the equatorial tropics but rather in a transitional zone between temperate and tropical climatic regimes. In spite of this, these regions are quite rich in natural resources. The development of agriculture, forestry, animal husbandry, and pisciculture on these lands is of overwhelming importance to China's economic future. The remaining tropical forests in China should be put to good use and not carelessly destroyed. With a little foresight, they may be intelligently integrated in to China's future economic development. An overall plan for development should be worked out for China that is compatible with both her ecological and her economic needs.

II. Vegetation Maps

The following maps covering the areas discussed above have been published:

Hou Hsioh-yu. 1979. Vegetation map of China (Scale 1:4,000,000). Cartographic Publishing House, Beijing.

Wu Cheng-yih et al. 1980. Vegetation of China, (vegetation maps at a scale of 1:10,000,000 included). Science press, Beijing.

South China Botanical Institute. 1976. Vegetation of Guangdong, (vegetation maps at a scale of 1:2,500,000 included). Science Press, Beijing.

III. Status of Floristic Inventory

A. Collections and Publications

There are approximately 1,600 genera and perhaps 12,000 species of vascular plants growing in the Chinese tropical zone. A total of 1,150,000 herbarium specimens are held in institutions within the region (Table I); and nearly as many (1,140,000) are held in institutions outside the region (Table II).

It is estimated that 80% of the lands (240,000 sq. km) in the tropical zone have been well collected. Taiwan has been best collected, followed by southern Guangdong, southern Fujian, Hainan, southeastern Yunnan and southern Guangxi.

Table I

Institutions within the Chinese tropical zone and their holdings of herbarium specimens.

Institution	Number of Specimens
South China Institute of Botany, Academia Sinica	550,000
National Taiwan University	160,000
Sun Yatsen University	150,000
Taiwan Museum	100,000
Yunnan Institute of Tropical Botany, Academia Sinica	50,000
Taiwan Forestry Research Institute	40,000
Guangxi Institute of Traditional Medical & Pharmaceutical Sciences	30,000
Yunnan Institute of Materia Medica	30,000
Guangdong Institute of Tropical Forestry	20,000
Guangxi Institute of Forestry	20,000
TOTAL	1,150,000

Table II

Institutions outside of the Chinese tropical zone with major holdings of tropical Chinese plants: The number of specimens.

Institution	Number of Specimens
Kunming Institute of Botany	400,000
Institute of Botany, Academia Sinica	300,000
European, American and Japanese Herbaria	200,000
Guangxi Institute of Botany	150,000
Fujian Teachers Training University	50,000
University of Amoy	30,000
Fujian Institute of Subtropical Botany	10,000
TOTAL	1,140,000

The lands which are poorly collected amount to approximately 60,000 sq. km, and are primarily in southeastern Tibet and southwestern Yunnan.

There have been numerous publications on the tropical Chinese flora. The most important are: *Flora Hongkongensis* (Behntham, 1861, 1872), *Flora of Hongkong and Kwantung, China* (Dunn & Tutcher, 1912), *Icones Plantarum Formosanarum* Vol. 1-10 (Hayata, 1911-1921), *Formosan Trees* (Kanehira, 1936), *Flora Kainantensis* (Masamune, 1943), *Flora of Guangzhou* (How et al., 1956), *Woody Flora of Taiwan* (Li, 1963), *Ligneous Plants of Taiwan* (Liu, 1972), *Flora Hainanica*, Vol. 1-4 (Guangdong Institute of Botany, 1964-1977), and *Flora of Taiwan*, Vol 1-6 (Editorial Committee of the Flora of Taiwan, 1975-1979).

B. ENDEMIC SPECIES

Of all the regions within the Chinese tropical zone, Taiwan has the highest number of endemic species. It is estimated that there are approximately 1,800 endemics there, 40% of a total of 4,300 species. Southwestern Guangxi and southeastern Yunnan are also high in endemics. Of 3,000 species, about 600 (20%) are endemic. Further collecting in southeastern Yunnan, coupled with relevant systematic research, may reveal new endemic species and/or genera. Interestingly, the Tongking Gulf Region (including southwestern Guangxi, southeastern Yunnan and northern Vietnam) also is rich in endemic genera. Some examples are: *Deutzianthus, Eberhardtia, Leptomischus, Lysidice, Pavieasia, Rhamnoneuron, Schizomussaenda, Siliquamomum* and *Zenia*. In contrast, typical tropical genera such as: *Alstonia, Antiaris, Barringtonia, Bauhinia, Canarium, Cleistanthus, Crypteronia, Dendropanax, Diospyros, Dracontomelon, Drypetes, Duabanga, Garuga, Horsfieldia, Knema, Mangifera, Myristica, Parkia, Pometia, Pouteria, Sapindus, Saraca, Schefflera, Semecarpus, Spondias, Sterculia, Terminalia,* and *Tetrameles*, are common in southern and southwestern Yunnan. It is in this region that many new distributions have been reported.

C. EXTINCTION

With the exception of remote or sparsely populated places (e.g. southeastern Tibet), there is rapid deforestation taking place in tropical China. It is likely that concomitant with this there is a high rate of species extinction. Although no accurate data exist, it is generally believed that there are many endangered species in Guangdong, Hainan, Guangxi and southwestern Yunnan.

A selective list of taxa, their current status (whether well or poorly collected) and the Chinese scientist(s) studying them, are presented in Table III.

As a result of the rapid population growth in China, much of the tropical forest is threatened by destruction or conversion. Although the volume of timber removed from many of the regions is great, the volume of fuelwood removed is vastly greater. In addition, large amounts of forest in Hainan and southern Yunnan have been converted to rubber plantations. The regions of tropical forest most threatened by destruction or conversion are in Hainan (especially in Jian Feng Ling, about 300 km^2) and southern Yunnan (especially in Xi Shuang Ban Na, about 1,500 km^2).

IV. Future of Floristic Inventory

At present, there are neither foreign programs nor cooperative programs with other institutions for floristic inventory in tropical China. Nevertheless, multidisciplinary studies in the fields of geobotany, ecology and zoology have been made several times. There are currently nine scientific institutions in tropical China concerned with tropical botany as well as four scientific institutions outside the region (Table IV). In addition, forestry bureaus may be found in almost all the prefectures and counties within the tropical zone.

Historically, botanical exploration in China has been erratic, although the Chinese tropical zone has been explored by modern botanists at least since the early nineteenth century. Perhaps the most significant plan for documenting the flora of China was drawn up in 1958. This is the Flora of China Project, which when completed will consist of 80 volumes covering some 30,000 species. To date, it is about two-thirds completed. This project has led to the continuous botanical exploration of China for the last twenty years. Nonetheless, as previously mentioned, some areas in China are still not well collected.

China is at the beginning of the long road to the creation of a national system of biological reserves. Over 30 nature sanctuaries have been set up in Guangdong, Hainan and southern Yunnan. China is becoming increasingly concerned about the amount of deforestation which has already taken place. Consequently, there is in China at the moment an atmosphere conducive to the protection and reclamation of tropical forest. The extent of China's commitment to conservation of natural resources will become apparent over the next few years.

A step towards evaluation of the tropical flora would be

Table III

List of Chinese taxa studied, their status and their specialists.

Taxon	Status[a]	Specialist(s)
Pteridophytes	P	Ren-chang Ching[†], Gung-hsia Shing, Wei-ming Chu, Chu-hao Wang
Gymnospermae	W	Li-kuo Fu
Angiospermae:		
Acanthaceae	P	Cheng-yih Wu
Anacardiaceae	W	Tien-lu Ming
Annonaceae	W	Ping-tao Li
Apocynaceae	W	Ping-tao Li
Araliaceae	W	Cheng-yih Wu
Araceae	W	Hen Li, Yen-Chen Tang
Asclepiadaceae	W	Ping-tao Li
Balsaminaceae	P	Yi-ling Chen
Begoniaceae	P	Te-tsun Yu[†]
Combretaceae	P	Ai-cheng Chao, Ting-chi Shu
Compositae	P	Yi-ling Chen, Chu Shi, Yung-chien Tseng, Yeou-ruen Ling
Cucurbitaceae	W	Cheng-yih Wu, An-ming Lu
Cyperaceae	P	Yen-chen Tang, Song-yun Liang, Pei-chun Li
Dioscoreaceae	W	Chih-tsun Ting
Dipterocarpaceae	P	Cheng-yih Wu
Ebenaceae	P	Shu-kang Lee
Elaeocarpaceae	P	Cheng-yih Wu
Ericaceae	W	Ren-chang Ching[†], Han-pi Yang, Au-luo Zhang, Tien-lu Ming
Euphorbiaceae	P	An-ren Li, Tsen-li Chin, Ping-tao Li
Fagaceae	P	Ching-chieu Huang, Yang-tian Chang
Flacourtiaceae	P	Shu-kun Lai
Gesneriaceae	P	Wen-tsai Wang
Gramineae	W	Yen-chen Tang, Liang Liu, Liang-chih Chia
Guttiferae	W	Yan-hui Li
Hamamelidaceae	P	Hung-ta Chang
Labiatae	W	Cheng-yih Wu, Hsi-wen Li
Lauraceae	P	Shu-kang Lee, Hsi-wen Li
Leguminosae	P	Te-chao Chen, Te-lin Wu, Pei-chun Li
Liliaceae	W	Sing-chi Chen, Yen-chen Tang, Song-zun Liang
Magnoliaceae	P	Yuh-wu Law
Meliaceae	W	Chieh Chen
Moraceae	P	Siu-shih Chang
Myristicaceae	P	Cheng-yih Wu
Myrsinaceae	W	Chieh Chen
Oleaceae	W	Ping-sheng Hsu
Orchidaceae	P	Sing-chi Chen, Zhan-huo Tsi, Kai-yung Lang
Palmae	P	Sheng-ji Pei
Piperaceae	W	Yung-chien Tseng
Proteaceae	W	Wen-tsai Wang
Ranunculaceae	W	Wen-tsai Wang
Rhamnaceae	W	Yi-ling Chen

(continued)

Table III (continued)

Taxon	Status[a]	Specialist(s)
Rhizophoraceae	W	Wan-cheung Ko
Rosaceae	W	Te-tsun Yu[†], Ling-ti Lu, Tsue-chih Ku
Rubiaceae	P	Wan-cheung Ko, Hsien-shui Lo
Rutaceae	W	Ching-chieu Huang
Sabiaceae	P	Yuh-wu Law
Salicaceae	P	Zhan Wang
Sapindaceae	W	Hsien-shui Lo
Sapotaceae	P	Shu-kang Lee
Simarubaceae	W	Pang-yu Chen
Solanaceae	W	An-ming Lu
Sterculiaceae	P	Hsiang-how Hsue
Symplocaceae	P	Young-fen Wu
Theaceae	P	Hung-ta Chang, Lai-kuan Ling
Verbenaceae	P	Shou-liang Chen
Zingiberaceae	P	Te-lin Wu

[a] P = poorly collected, W = well collected.
[†] Deceased.

to establish new special organizations (including teaching centers) in remote places of the tropical zone such as Hainan, southern Guangxi, Fujian, southeastern Yunnan and southeastern Tibet. It is estimated that at least 30 more botanists and 50 more technicians should be maintained in the tropical region, and that two million dollars per year extra should be provided for tropical research there.

Table IV

List of institutions within and outside the Chinese tropical zone with interests in tropical botany.

I. Within the tropical region:

1. South China Institute of Botany, Academia Sinica, Guangzhou, Guangdong.
2. Taiwan Museum, Taibei, Taiwan.
3. Yunnan Institute of Tropical Botany, Academia Sinica, Meng Long, Yunnan.
4. Taiwan Forestry Research Institute, Taibei, Taiwan.
5. Guangxi Institute of Traditional Medical & Pharmaceutical Sciences, Nanning, Guangxi.
6. Yunnan Institute of Materia Medica, Jing Hong, Yunnan.
7. Guangdong Institute of Tropical Forestry, Guangzhou, Guangdong.
8. Guangxi Institute of Forestry, Nanning, Guangxi.
9. Institute of Tropical Crops, Na Da, Hainan.

II. Outside the region:

1. Kunming Institute of Botany, Academia Sinica, Kunming, Yunnan.
2. Institute of Botany, Academia Sinica, Beijing.
3. Guangxi Institute of Botany, Guilin, Guangxi.
4. Fujian Institute of Tropical Botany, Amoy, Fujian.

V. Literature Cited

Bentham, G. 1861, 1872. Flora Hongkongensis. L. Reeve & Co., London.

Chang Honda et al. 1957. Vegetation of Leizhou Peninsula. Ser. Phytoecol. Geobot. 17. Science Press, Beijing.

Chang King-wai et al. 1973. A primary study of the vertical vegetation belt of Mt. Jolmo-Lungma (Everest) region and its relationship with horizontal zone. Acta Bot. Sin. **15(2)**: 221-236.

―――― et al. 1980. Latitudinal zonality of vegetation in Qinghai-Xizang plateau. Sci. Sin. **11**: 1091-1098.

Chen Shupei et al. 1983. Primary features of littoral vegetation between Zhenghai Bay and Yamen in Zhujiang Kou. Ecology **2**: 121-000.

Chingtienshu Research Group. 1977. A rare and valuable forest tree from Kwangsi—*Parashorea chinensis* var. *kwangsiensis*. Acta Phytotaxon. Sin. **15(2)**: 22-45.

Cooperation Group of *Parashorea chinensis*. 1977. A rare and valuable new tree discovered in Yunnan—*Parashorea chinensis*. Acta Phytotaxon. Sin. **15(2)**: 10-21.

Dunn, S. T. & W. J. Tutcher. 1912. Flora of Kwangtung and Hongkong. H. M. Stationary Office. London.

Editorial Committee of China Physical Geography, Academia Sinica. 1983. Physical geography of China. Vol. 1. Science Press. Beijing.

Editorial Committee of Chinese Vegetation. 1980. Vegetation of China. Science Press. Beijing.

Editorial Committee of the Flora of Taiwan. 1975-1979. Flora of Taiwan. Epoch Publishing Co. Taipei.

Forestry Department, Forestry Branch, Guangxi Agricultural College. 1978. Ecological and afforestation problems of *Burretiodendron hsienmu*. Phytoecol. Res. Bull. **1**: 1-96. Science Press. Beijing.

Guangdong Institute of Botany. 1964-1977. Flora Hainanica, Vol. 1-4.

Hayata, B. 1911-1921. Icones plantarum Formosanarum, Vol. 1-10. Supplementa by Y. Yamamoto, 1925-1932. Taihoku (Government of Formosa).

How et al. 1956. Flora of Guangzhou.

Hou Kuanzhao et al. 1955. A survey of the plants and vegetation of Hainan Island and the mainland of Guangdon. Ser. Phytoecol. Geobot. 4. Science Press. Beijing.

Hu Shun-shi. 1979. The phytocoenological features of evergreen broadleaf forest in Guangxi. Acta Bot. Sin. **21**: 362-370.

―――― et al. 1980. The phytocoenological features of limestone seasonal rain forest in Guangxi. J. Northeast. Forest. Inst. **4**: 11-26.

Hu Yujia. 1983. The phytocoenological features and types of *Dipterocarpus* forest in Hainan Island. Ecol. Sci. **2**: 16-24.

Jin Zhenzhou. 1982. Valuable tropical forests urgently requiring rescue and protection—typical types of rain forest and monsoon forest in the northern margin of Asian tropics. Bull. Stud. on Exploitation of Mountains and Hills in Tropics and Subtropics and Their Ecol. Equilib.: 118-122. Kexue Puji Printing House.

Kanehira, R. 1936. Formosan trees, Rev. ed. Taihoku Co., Formosa.

Ke Chichen. 1979. Successful practice of transformation of terrace grassland into rubber plantation in Qiongbei and tendency of exploitation of tropical natural sources in Hainan Island. Bull. Discussions on Trop. Sources and Their Exploitation and Utilization: 128-132.

Li, H.-L. 1963. Woody flora of Taiwan. Livingston Publishing Co., Narberth, Pennsylvania.

Li, Shiying. 1956. Plant communities in southwestern Lonjing and its adjacent areas. Ser. Phytoecol. Geobot. 4. Science Press. Beijing.

Li, Zhiji. 1964. Briefing the data for regionalization of tropics and subtropics in Guangxi. Ser. Phytoecol. Geobot. **2(2)**: 253-256.

―――― et al. 1965. Ecological and geographical distributions of main forest trees in Guangxi. Ser. Phytoecol. Geobot. **3(1)**: 1-50.

Liang Choufen et al. 1981. A natural protected area of karst forest, Longgang, in the northern margin of tropics in Guangxi. Guihaia **1(2)**: 1-6.

Liu, T.-J. 1972. Ligneous plants of Taiwan.

Masamune, G. 1943. Flora Kainantensis.

Teaching and Research Group of Ecology and Geobotany, Biology Department of Yunnan University. 1960. Vegetation of natural protected areas in Yunnan. J. Yunnan Univ. (Nat. Sci.) **1**: 50-53, 70-72.

Wang Huen-pu. 1980a. On the types and management of protected areas. J. Northeast. Forest. Inst. **2**: 1-6.

――――. 1980b. Nature conservation in China. The present situation. Parks **5(1)**: 1-10.

――――. 1981. On effective protection of protected areas. Wild Animals **1**: 17-19.

――――. 1982a. On establishment of natural protected areas. Sci. Technol. Environ. Protect. **3**: 5-9.

――――. 1982b. Status and tendency of natural protected areas in Hainan Island, Guangdong. Res. Nat. Sources **1**: 4-8.

―――― et al. 1955. Cultivation and ecological conditions of *Illicium verum*. Ser. Phytoecol. Geobot. 2. Science Press. Beijing.

―――― et al. 1982. The phytocoenological features of seasonal rain forest of acid soil region in Guangxi. Acta Bot. Boreal-Occid. Sin. **2(2)**: 69-86.

Wang Xian-pu. 1984. The relationship between the natural reserves and the development of tourism in Hainan Island, Guangdong. Guihaia **4**: 87-92.

Wu Cheng-yih. 1979. The regionalization of Chinese flora. Bull. Bot. Res. **1(1)**: 1-20.

Xu Zaifu et al. 1983. Studies on germplasm resources of tropical plants in Xishuan Banna. Bull. Trop. Pl. Res.: 7-15. Yunnan People's Daily Printing House.

Yu Tong-quan. 1983. The mountain rain forest of Hainan Island. Ecol. Sci. **2**: 25-33.

Zhang Xin-shi. 1978. The plateau zonality of vegetation in Xizang. Acta Bot. Sin. **20(2)**: 140-149.

The Philippines

Benito C. Tan and Justo P. Rojo

Contents

I. Introduction	46
II. Description of the Region	46
A. Geographical Extent and Area	46
B. Topography	46
C. Geology	46
D. Climate	46
E. Population	47
III. Vegetation Types	47
A. Forest Types	47
1. Mangrove	47
2. Beach	47
3. Molave	48
4. Dipterocarp	48
5. Pine	49
6. Mossy	49
IV. Vegetation Maps	49
V. Magnitude of Floristic Inventory	50
A. History of Philippine Botany	50
1. Pre-Linnean Spanish Colonial Period (1521–1753)	51
2. Post-Linnean Spanish Colonial Period (1753–1898)	51
3. The American Colonial/Commonwealth Period (1900–1942)	51
4. The Post-war Period (1946–Present)	52
B. Endemism	54
VI. Adequacy of Inventory for Major Groups	54
VII. Resources for Floristic Inventory	55
VIII. The Future of Floristic Inventory	55
IX. Acknowledgments	58
X. Addendum	58
XI. Literature Cited	58

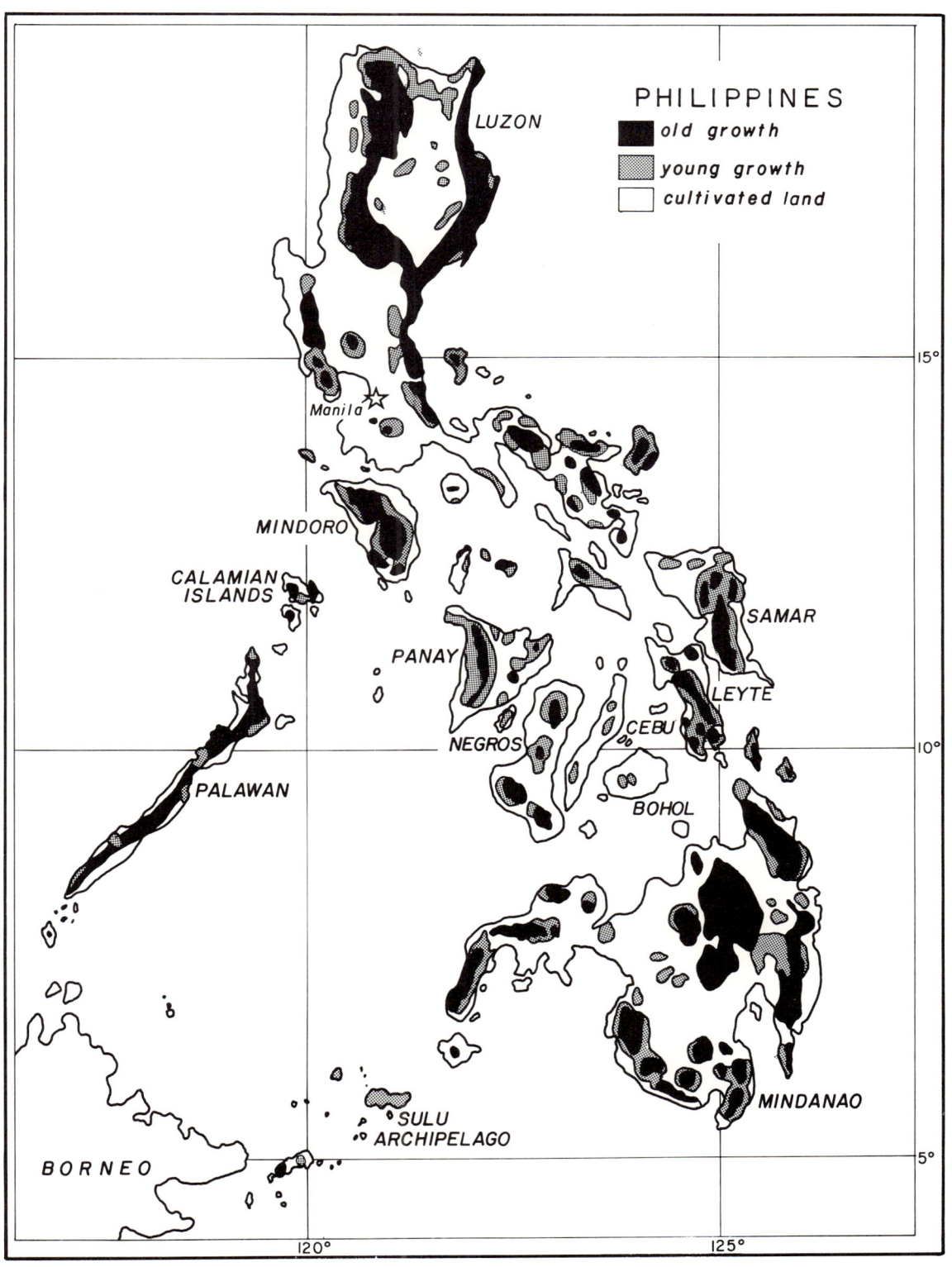

I. Introduction

Floristically, the Philippine Archipelago is a part of the plant geographical unit known as Malesia. It is grouped with Peninsular Malaysia, Sumatra and Borneo to form a subfloristic province known as the West Malesia (van Steenis, 1950; Jacobs, 1974). Accordingly, West Malesia is rich in endemics totalling 150 genera, 33 of which belong to the Philippines. The Philippine flora also represents the northernmost extension of the Malesian flora. Furthermore, the Philippines has an appreciably higher percentage of eastern Malesian floristic elements compared to that found in either Borneo, Sumatra or Peninsular Malaysia. This indicates that there has been a greater interchange of plants between the Philippines and East Malesia (Celebes, Moluccas and New Guinea) than the rest of West Malesia, and that the Philippines have been the stepping stones through which some eastern and western elements moved between the two subfloristic provinces.

II. Description of the Region

A. Geographical Extent and Area

The Philippines is an archipelago of some 7,107 islands closely scattered in a north-south orientation over some 1,295,000 km² of oceanic water. The islands form part of the great island arc off the southeastern coast of the Asian mainland and are located between 4°23′–21°30′N and 116°–126°E. In length, they stretch approximately 1,850 km between the southern tip of Taiwan and the northern parts of Borneo and Celebes. In breadth, they are about 965 km.

The total land area of the Philippines is about 300,000 km². Two-thirds of this total is represented by the two largest islands: Luzon, with an area of about 105,708 km², and Mindanao, about 95,587 km². The remaining area is represented by other large islands, 11 of which have areas between 10,000 and 13,000 km², and 24 with areas ranging from 180 to 970 km².

B. Topography

The Philippines is largely mountainous. There are 11 high mountain peaks ranging from 1,700 to above 2,900 m, of which the highest is Mt. Apo (2,929 m) in Mindanao, followed by Mt. Pulog (2,924 m) in Luzon. There are no high peaks on the medium-sized islands of Samar, Leyte, Masbate and Bohol.

The mountain ranges generally follow a north-south trend, except in Palawan where the ranges run to the northeast and southwest. The occurrence of the many mountain ranges and peaks has an important effect on the climate of the archipelago, particularly in regard to the prevailing winds and seasonal distribution of rainfall.

C. Geology

The geological history of the Philippines was reviewed by Merrill (1926), Smith (1924) and Dickerson (1928). More recent reviews of the complex geological history of the Philippines are those of Menard and Hamilton (1963), Hamilton (1979), and Fernandez (1982).

The Philippine archipelago may be considered structurally as a crumpled edge of the Asiatic continental platform. This is evidenced by the presence of the Mariana Trench close to the margin of the islands. It is in fact only 85 kms east of Mindanao, and is the deepest known part of the world. This deep is related to the one east of Taiwan and Japan. As a result of this shared feature, it might seem these three island groups have a similar relationship with mainland Asia. However, there is little evidence of a recent land connection between the Philippines and Taiwan. Archean rocks known from Taiwan are not found in the Philippines. Moreover, biological evidence confirms that the flora and fauna of Taiwan are allied to Asia rather than to the Philippines. Lesser deeps occur west of the Philippine islands, in what is presumed to be a sunken area known as a "graben" (Merrill, 1926).

There are today two long and narrow, interrupted, land bridges between Borneo and the Philippines, the Palawan and Sulu Archipelagoes. Also, geological evidence (see Taylor & Hayes, 1980; McCabe et al., 1982) as reviewed by Tan (1984) indicates that there may have been a direct land connection between the Palawan-Mindoro Islands and the Indochina-Hainan area during the Oligocene.

Regarding the relationship between the geologic history of the Philippine archipelago and its floristics, Merrill (1926) concluded that:

> *"The enormous coast line of the archipelago, the tropical location, the large land area, the diversified topography, the great altitudinal range (from sea level to 2,929 meters), the varying exposure to the shifting trade winds and typhoons, the peculiar seasonal distribution of the rainfall conditioned upon the topography, the winds, and the Kuro-Siwo or Japanese current which flows north along the east coast of the archipelago—all these, together with the study of the Philippine flora in the light of geologic history, give us the basis of an explanation of the enormously rich Philippine flora and fauna, the high percentage of specific endemism, and the strictly local occurrence of numerous species."*

D. Climate

The climate of the islands is dependent upon their topography. The numerous mountain ranges and high peaks have profound effects on prevailing wind patterns and the seasonal distribution of rainfall; these in turn exert a corresponding effect on vegetation type.

The maximum monthly mean temperature does not vary much during the year. In the warmer months (April to October) the mean monthly temperature ranges from 26.9° to 28.1°C; in the cooler months (November to March) it ranges from 25.4° to 26.5°C. May is the warmest month, January is the coldest (Merrill, 1926).

Although temperatures at low elevations are generally constant from north to south, annual average temperatures vary. The annual temperature range in Jolo is about 3°C while in Aparri it is 8°C. The wider range in temperature in the north (Aparri) is due mainly to the cooler "winter" months compared to the more even temperature in the south (Jolo). This almost negligible difference in the annual mean temperatures from north to south is due mainly to the warm equatorial waters which flow northward along the eastern coast of the Philippines—the Kuro-Siwo, or Japanese Current (Merrill, 1926). At higher elevations the temperature varies depending on altitudinal range. At about 1,500 m, in the vicinity of Baguio City, frost may occur in December and January. At a much higher elecation such as that of Mt. Pulog, ice of about a centimeter thick has been recorded in January.

As observed by Merrill (1926), temperature has little effect on north and south distribution of genera and species of plants. Most species found in the lowlands of Luzon are also found thriving in the lowlands in Mindanao and vice versa. It is considered, therefore, that the occurrence or non-occurence of plants growing exclusively at low elevations is likely the result of factors other than temperature, such as means of dispersal, physical barriers or rainfall distribution.

The Philippines, like many other tropical lands, has a heavy annual rainfall. Monthly distribution of rainfall differs markedly despite the country's narrow extension from east to west. The difference in climatic conditions is largely due to the relative position of the mountain ranges over which the prevailing winds blow.

There are two important periods of rainfall in the Philippines. One, brought by the northeast monsoon, occurs from December to March, causing heavy precipitation over the western part of the country. The other period of high precipitation, from July to November, is caused by typhoons. Generally the typhoons cross the country from east-southeast to west-northwest and exit over the northern end of the islands. Thus, these rains fall more abundantly in Luzon than in the rest of the country.

The local relative humidity is always high as would be expected in a country subjected to heavy rainfall and surrounded by seas. Regions experiencing a high rainfall with an absence of a dry season have the highest relative humidity.

E. POPULATION

The National Census and Statistics Office (1983) gave a total population of the Philippines at 52,259,000 inhabitants. Annual population growth is a little over three per thousand.

All the major islands are inhabited. Most people live in the large cities and towns or in smaller urban centers. Fewer reside in the provinces where their main livelihood is farming. Many families live as subsistence farmers working as tenants for those with large landholdings. Others eke out their living on public lands (mainly forest lands), practicing the age-old shifting method of agriculture. The clearing of forest for agricultural purposes has been ably documented by Myers (1980).

The encroachment of landless people onto the higher mountain slopes poses a serious threat not only to the montane forest cover but also to watershed areas vital for lowland irrigation and hydroelectric power generation. These people, in the so-called uplands or rainfed areas, are presently the target of government aid through agro-forestry schemes in which subsistence farmers are encouraged to plant food crops with forest trees.

With regard to energy generation to meet the demands of an increasing populace (which in one way or another affects the vegetation), the government has experimented with hydroelectric, geothermal, and most recently, dendrothermal systems. As a result of the latter, people are now encouraged to establish tree plantations to supply fuel for mini- or large-capacity dendrothermal plants. The success of dendrothermal plants will largely depend upon the success of growing trees in plantations at shorter rotation intervals and with higher yields.

III. Vegetation Types

A. FOREST TYPES

Philippine forest vegetation types as described by Whitford (1911) are still very much in use today although circumscriptions of some types have been altered through the years. There has been no recent attempt by local ecologists to update these descriptions or to equate them with modern schemes of classification such as the one proposed by UNESCO (1973). Below is a brief summary of forest types (based on Whitford, 1911) used by the Philippine forestry administration.

1. Mangrove

This is literally a forest of the sea. It is especially well developed on the mud flats at the mouths of the rivers entering the sea near heads of protected bays. When conditions are favorable, it occupies beaches washed by the tides. Descriptions of this type of vegetation have been reviewed by van Steenis (1958).

In the Philippines, as in other tropical countries where this type of forest occurs, the principal species found are members of the genera *Bruguiera*, *Ceriops* and *Rhizophora* (Rhizophoraceae). Other important species belong to Avicenniaceae and Sonneratiaceae. Representatives of Bombacaceae (*Camptostemon*), Euphorbiaceae (*Excoecaria*), Myrsinaceae (*Aegiceras*) and Rubiaceae (*Scyphiphora*) are also found.

2. Beach

This occupies sandy shores or beaches above high-tide

limits. Understandably, areas covered by this type of forest are easily settled by humans and consequently are at present nearly non-existent. In its original form, the frontal zone consists of a tangle of vegetation with pandans (*Pandanus* spp.), *Hibiscus tiliaceus*, *Pongamia pinnata* and *Thespesia populnea*. The principal trees are *Barringtonia asiatica*, *Calophyllum inophyllum*, *Erythrina orientalis*, *Hibiscus tiliaceus*, *Pongamia pinnata* and *Terminalia catappa*. In some places *Intsia bijuga*, *Mimusops parviflora* and *Pterocarpus indicus* occur. *Terminalia catappa* often occurs in patches or pure stands on rich river bottoms. On sandy flood plains, *Casuarina equisetifolia* may form pure stands.

3. Molave

The Molave type is so named because of the conspicuous presence of molave (*Vitex parviflora*) in these regions. There are also many leguminous species in this type of forest. Thus, it is known by some local botanists as the "leguminous species" type. It can also be rightly called a limestone forest type, because the species found are characteristic of limestone regions.

This type of forest occupies a topography similar to that of the yakal-lauan sub-type of the Dipterocarp forest (see below) except that in a majority of cases the underlying rock is usually limestone rather than volcanic in nature. Trees found in this type of forest are commercially valuable and are easily acessible for exploitation. They grow far apart and the spaces between them are filled with small trees and ofttimes with climbing or small erect bamboos. Moreover, most trees are short-boled with irregular spreading crowns. The forest has a decidedly deciduous foliage, especially when it is found in areas that have a pronounced dry season.

In some expressions of this type, the following trees are present: *Afzelia rhomboidea*, *Heritiera sylvatica*, *Intsia bijuga*, *Kingiodendron alternifolium*, *Lagerstroemia piriformis*, *Litchi chinensis* spp. *philippinensis*, *Mimusops parviflora*, *Parinari corymbosa*, *Pterocarpus indicus*, *Pterocymbium tinctorium*, *Sindora supa*, *Toona calantas*, *Vitex parviflora*, *Wallaceodendron celebicum*, *Wrightia pubescens* spp. *laniti* and *Ziziphus talanai*. Smaller trees may be found such as *Diospyros ferrea*, *D. philippensis*, *Mallotus floribundus*, *Pterospermum diversifolium*, and *Taxotrophis macrophyllus* may also be found here. It must not be supposed, however, that all these species occur at any one locality together.

Diospyros ferrea, *Kingiodendron alternifolium*, *Lagerstroemia piriformis*, *Parinari corymbosa* and *Vitex parviflora* are also found scattered throughout the open places of the dipterocarpaceous yakal-lauan sub-type. Even in some limestone areas, the slopes of valleys often contain clearly defined Dipterocarp type (see below) with trees of the Molave type scattered among thinner portions of the forest.

4. Dipterocarp

This is by far the most important forest type in the Philippines in terms of the timber it provides, not only for export but for local use. It comprises about 75% of the existing virgin forest and contains from 85 to 95% of the standing timber in the Islands (Weidelt & Banaag, 1982). It may be found in almost any type of topography, from immediately behind the frontal zone of the beach to an altitude of 1000 m. From the botanist's point of view, the composition of this forest is complex. On the other hand, from the standpoint of the lumberman, this type of forest is decidedly very simple. As the name implies, members of the Dipterocarpaceae are the dominant trees in these forests, and also represent the largest volume of standing timber. The remaining volume is represented by trees belonging to several other families such as the Anacardiaceae, Leguminosae, Meliaceae, Myrtaceae, and Sapotaceae.

With the exception of some members of *Hopea* and *Vatica*, all species of the Dipterocarpaceae are large trees. Most dipterocarps reach a height of 40 to 50 m and have diameters of 100 to 150 cm.

Dipterocarps in the Philippines are practically evergreen. A few of them, as well as species belonging to other families, are particularly deciduous, dropping some of their leaves during the dry season. In the Philippines, the Dipterocarp forest shows more or less distinct formations, making it possible to distinguish subtypes or associations. The subtypes are named after the familiar common names of the most numerous (in terms of individuals or volume per unit area) species found within them. These subtypes are:

a) Lauan or "Philippine mahogany." This subtype represents the most successful commercial forest in the Philippines and is confined to regions where the dry season is short or wanting. Its best development is found on the gentle slopes near the base of mountain masses, where it reaches altitudes of 300 to 400 m and then merges with the tangile-oak subtype. Sometimes, when the topography is rough, the stand is not dense. When near the sea, it may merge with the Yakal-lauan subtype or the Molave type of forest.

Species belonging to this subtype are mostly the so-called "Philippine mahogany" of commerce which include: *Parashorea malaanonan*, *Shorea almon*, *S. contorta*, *S. negrosensis*, *S. ovata* (=*S. agsaboensis*) and *S. palosapis*.

b) Hagakhak-lauan. Like the Lauan subtype, this subtype is confined to the regions where the dry season is short or wanting. However, unlike the lauan subtype, it is usually restricted to areas where the water table is near the surface. It achieves its most extensive development in narrow bottoms, especially on slightly raised river deltas.

The presence of hagakhak (*Dipterocarpus validus*) in this subtype also makes it different from the Lauan subtype. Other species associated with hagakhak are: *Cananga odorata*, *Canarium luzonicum*, *Celtis philippinensis*, *Dracontomelon dao*, *Koordersiodendron pinnatum*, various species of Meliaceae, *Octomeles sumatrana* and *Pometia pinnata*.

c) Yakal-lauan. This subtype reaches its best development on low coastal hills where the basal rock is volcanic and where the dry season is short. It often occurs on headlands projecting into the sea or on hills bordering large inner valleys.

The Yakal-lauan subtype is slightly deciduous during the driest time of the year. This subtype more nearly resembles the Hagakhak-lauan subtype than the Lauan. On ridges and exposed slopes, yakal (*Shorea astylosa*) and *Hopea* spp., which produce heavy yellowish-brown timbers, are abundant. In protected ravines and along streams *Parashorea malaanonan, Shorea contorta, S. palosapis, S. quiso,* and *Vatica mangachapoi* may be found. Other species similar to those occurring in the Hagakhak-lauan subtype also occur here.

d) Lauan-apitong. This is similar to the Lauan subtype in terms of altitude and topographic conditions but differs from it in that it occurs in regions with a pronounced dry season. During the dry months many species occurring here are deciduous. Except on particularly favorable sites, the forest cover tends to be open, allowing growth of climbing bamboos and lianas. The species composition is more complex than the Lauan subtype and resembles the hagakhak-lauan and the Yakal-lauan subtypes.

The dominant species included in this subtype are: *Anisoptera thurifera, Dipterocarpus gracilis, D. grandiflorus, Shorea contorta, S. polysperma* and *S. quiso*; and the subdominant species are: *Alstonia scholaris, Amoora* (= *Aglaia*) *aherniana, Artocarpus blancoi, Celtis philippinensis, Myristica philippinensis, Strombosia philippinensis* and *Vatica mangachapoi.*

e) Tangile-oak subtype. This subtype covers the area extending from the upper limits of the Lauan and Lauan-apitong subtypes to the lower limits of the Mossy type. It blankets the higher slopes of mountains. Its lower limits reach 400 m and its upper limits extend to 1000 m.

As the name connotes, the principal species are tangile (*Shorea polysperma*) and oaks (*Lithocarpus* spp.). Other species occurring are *Agathis philippinensis* (= *A. dammara,* sensu lato), *Cinnamomum mercadoi, Hopea acuminata, Syzygium* spp. and *Tristania decorticata.*

5. Pine

This vegetation type reaches its best development in the high plateaus of northern Luzon in the so-called "Mountain Provinces." The greatest proportion of this forest type grows at 900 to 1500 m in regions with a distinct dry season. Evidently in areas where pines now grow, extensive broad-leaved forests used to occur. As a result of anthropogenic activities, the broad-leaved forests were cut down and burnt, making the micro-climatic condition much drier, and thereby favoring the development of pine forest (Kowal, 1966). Land occupied by pine forest is increasing owing to the continued destruction of surrounding broad-leaved forest.

The principal pine species in the Philippines are *Pinus insularis* (= *P. kesiya*) and *P. merkusii.*

6. Mossy

It is estimated that eight percent of the land area of the Philippines is covered by this type of forest. Mossy forests are essentially protection forests. The Mossy forest gets its name from the abundant mosses and liverworts which grow there. The topography where this type of forest grows is rough and constantly changing due to frequent land slides brought about by heavy rains on steep slopes.

Generally, the climatic conditions are exceedingly moist because the peaks are usually shrouded by moisture-laden clouds. The temperature conditions are very much lower compared to the coastal regions.

The tree vegetation is not as complex as those found in lower elevations. On mountains above 1,200 m, the Mossy forest appears as its best. Species of *Dacrycarpus, Engelhardtia, Lithocarpus, Podocarpus, Syzygium, Ternstroemia* and *Tristania* represent some of the principal trees. Only a few species belonging to these genera are found at lower elevations.

IV. Vegetation Maps

To date, there is no published detailed vegetation map for the entire country. This may be due to the fact that the forests, which make up the bulk of the vegetation, are of exceedingly mixed types. Notable exceptions to this may be found in "Mountain provinces" (Luzon), and Zambales Province (Luzon) and along the west coast of Mindoro Island. Here pines, i.e., *Pinus insularis* (= *P. kesiya*) on the former, and *P. merkusii* on the latter two places, are in relatively pure stands. A few detailed vegetation maps for certain small areas such as a province or a watershed have, however, been prepared by the Natural Resources Management Center of the Ministry of Natural Resources in connection with local land utilization projects.

Merrill (1926) noted that the "average low-altitude primary forest in the Philippines is exceedingly complex, and not infrequently a hundred or more species of woody plants may be found in a single hectare." At present there are only about 1.5 million hectares of extensive lowland forest remaining in the Philippines, the rest has been destroyed by human settlements, agriculture and/or mechanical logging (which began on a large scale immediately after the Second World War).

While there have not been any vegetation maps published for the Philippines, the Bureau of Forestry (now renamed Bureau of Forest Development) published in 1964 a soil cover map with a five-color scheme in five sheets, showing the extent of commercial forest, non-commercial forest and brushland, cultivated land, *Imperata* and open lands, and marsh and swamp. This soil cover map was mentioned by Jacobs (1974). It is in fact a land use map which, among other things, was designed to indicate the extent of the remaining forests. This sort of map is essential for government policy makers and development planners. However, this map does not distinguish between vegetation types.

In connection with government land use policy, the Bureau of Forest Development annually publishes its Philippine Forestry Statistics. This publication contains information on the status of remaining forests. Table I presents data from this publication on the amount of forested areas (in ha) for four different years (1973, 1979, 1981, 1982).

Table I

Forests of the Philippines
(total in thousands of hectares).[a]

	1973	1979	1981	1982
Total Forests	15,669	12,661	12,252	11,204
Production Forest	13,885	10,901	10,493	9,010
Dipterocarp	13,389	10,461	10,061	8,610
Mangrove	284	245	239	211
Pine	212	195	192	189
Protection Forest	1,784	1,760	1,759	2,194(?)
Dipterocarp	1,438	1,422	1,421	–
Mossy	330	331	329	1,726(?)
Bamboo	16	7	–	–
Man-made plantation	–	–	–	468

[a] Source: 1982 Philippine Forestry Statistics, Bureau of Forest Development.

In 1982, the Bureau of Forest Development began using new category headings for forest land use in the Philippines. Added to the previous heading of "production forest" and "protection forests" were "agro-forest lands" and "pasture/range lands." Also, subheadings under production and protection forests were changed to (a) timber license areas, (b) industrial trees plantations and farms, and (c) others (cancelled timber license areas, open lands, etc.). For comparison between previous category headings and current ones, statistics from the 1982 Forest Land Use in the Philippines are presented in Table II.

Table II

1982 Forest land use status of the Philippines.[a]

Category	Thousand Hectares Area	%
Total Forest Lands	16,629,450	100.00
1. Agro-forest lands	4,300,750	25.86
a. agro-forestry lease-areas	28,240	–
b. land occupied by settlers—agri-plantation and crop lands	4,272,510	–
2. Pasture/range lands		
a. pasture lease and permit areas	535,000	3.22
3. Production forests	8,312,895	50.00
a. timber license areas	7,539,000	–
b. industrial tree plantation and tree farms	232,000	–
c. others (cancelled timber license areas, open lands, etc.)	565,395	–
4. Protection forests	3,479,805	20.92
a. wilderness	23,000	–
b. national parks	381,550	–
c. game refuge & wildlife sanctuaries	1,667,710	–
d. watershed forest reserves	342,160	–
e. reforestation projects	1,065,385	–

[a] Source: 1982 Philippine Forestry Statistics, Bureau of Forest Development.

The sources of data for Tables I & II came from aerial photographic surveys. Details of methods of reconnaissance in the interpretation of forest types are discussed in Bedard (1958). Separately, the Forestry Development Center of the College of Forestry, University of the Philippines at Los Baños and the Natural Resources Management Center in Manila are working on a project concerned with updating the soil cover map of the Philippines (1964) using Landsat imagery. One of the resulting maps (1984, UPLB College of Forestry, unpublished data) is reproduced here as Figure 1. Old growth, in Figure 1, represents virgin forests (either for production or protection purposes) while young growth includes logged-over areas together with forests of a secondary nature and/or of lesser-known tree species. The same map information had been given by the second author (J. Rojo) to Dr. T. Whitmore of Oxford University to be included in the latter's vegetation map of Malesia (Whitmore, 1984).

According to the 1978 Landsat data, the country has at most 38.2% forest cover (Myers, 1980), and of this, 74–77% represents a Dipterocarp type. Statistics also reveal that the total land cover of Dipterocarp forest was reduced by 38% from 1934 (See Fischer, 1934) to 1982. A great portion of the indigenous plant species thrive in this type of forest as well as in the mossy type. The present amount of several forest types and their distribution are shown in Table III. Today, 46.5% of the Dipterocarp forest is located in Mindanao, 32.6% in Luzon and 20.9% in the Visayan region.

V. Magnitude of Floristic Inventory

Organized botanical expeditions in the Philippines for serious floristic study and taxonomic revision started in earnest at the beginning of this century. To appreciate the difficulty and the extent of the work accomplished, a short review of the history of Philippine Botany seems in order.

A. History of Philippine Botany

The history of botanical work in the Philippines has been summarized by Robinson (1906), Merrill (1903, 1915a,

Table III

Distribution and area coverage of the different forest types in the Philippines.[a]

Forest Type	Luzon	Visayas	Mindanao	Total (thousand ha.)
Dipterocarp	2,805	1,799	4,006	8,610
Mangrove	17	87	107	211
Pine	189	–	–	189
Mossy	910	471	345	1,726
Man-made forest	257	119	92	468
TOTAL	4,178	2,476	4,550	11,204

[a] Source: 1982 Philippine Forestry Statistics, Bureau of Forest Development.

1926, 1953), van Steenis-Kruseman (1950), Quisumbing (1957, 1964) and recently by Madulid (1983), Madulid & Gutierrez (1981) and Santos (1984). These publications deal primarily with phanerogamic studies. For the botanical history of cryptogamic plant groups, one may consult Copeland (1958– 1960) and M. Price (1972a) for the ferns, Iwatsuki and Tan (1979) for the mosses, del Rosario (1971a) and Tan & Engel (1986) for the liverworts, Teodoro (1937) and Dogma (1975) for the fungi, Gruezo (1979) for lichens, and Velasquez (1962) and Cordero (1977) for the algae.

Indisputably, the study of Philippine plants commenced after the arrival of the Spanish colonizers in 1521 and reached a peak during the American colonial period. Before the advent of western colonization, knowledge of local plants and their uses was part of the rich indigenous folklore.

Briefly, the history can be divided into four periods, interrupted by the Philippine Revolution (1898-1900), and the Second World War (1942-1945).

1. Pre-Linnean Spanish Colonial Period (1521–1753)

This long period of more than two centuries is characterized by the preparation of a few amateurish theses on Philippine medicinal plants. Except for the historical value attached to these old publications, which indicated the approximate time for the introduction of some of the economic plants into the archipelago by the Spaniards, few botanists need to refer to them for nomenclatural and taxonomic purposes (Merrill, 1926). Furthermore, most of the specimens collected were kept for personal interest rather than for scientific purposes. Very few of these old botanical collections survive today except those made by G. Kamel (Madulid, 1984) which are preserved in the British Museum (Natural History) in London.

2. Post-Linnean Spanish Colonial Period (1753–1898)

This period saw the completion of a few major works on Philippine botany undertaken by the Spanish priests residing in the country. The most important of these is the four volume work *Flora de Filipinas* which was published in three editions (1837–1883). The work was written by M. Blanco, expanded later by A. Llanos and appended to by C. Fernandez-Villar and A. Naves (Merrill, 1903, 1915a). All together, there were 5871 species of plants credited to the country; most of these were flowering plants collected from settled areas in the archipelago. As a result of their working in isolation on the Philippine flora, with few contacts with other contemporary European botanists, errors made by the local Spanish priests in plant identification reached 61% (Merrill, 1915b).

Several foreign, mostly European, botanical expeditions reached the country during this period. The better known ones are the Malaspina Expedition (Madulid, 1983) and the Wilkes' United States Exploring Expedition (Merrill, 1903, 1908, 1926). The specimens collected were prepared in several sets of duplicates, which found their way to various herbaria in Europe and North America.

Before the end of the period, S. Vidal y Soler became the first Spanish botanist in the country to establish a sizeable botanical library and herbarium. These facilities, accidentally burned in 1897, formed the basis for his publications on the Philippine forest flora. Many of the Vidal collections can be found at the Arnold Arboretum, Leiden and Kew herbaria.

During this period many individuals who were either residents or visitors made substantial plant collections which later became types of many Philippine taxa described abroad. Some of the more important collections were made by H. Cuming, A. Loher and O. Warburg.

3. The American Colonial/Commonwealth Period (1900–1942)

This period marks the "golden age" of Philippine botany both in terms of quality and quantity of publications produced. The coming of the American colonizers saw the immediate reorganization of the Bureau of Science and Forestry Bureau and the initiation of botanical investigation under the able leadership of E. D. Merrill, who later was to reign over the Philippine botanical scene for more than half a century. The botanical activities of Merrill in the Philippines lasted from 1902–1923 and culminated in the publication of four volumes of *Enumeration of Philippine Flowering Plants* (1923–1926) and other important publications (see Merrill & Quisumbing, 1953). This monumental work by Merrill (1923–1926) still dominates Philippine flowering taxonomy today.

Other plant groups were likewise studied and revised by local workers, most of whom were either American or other foreign authorities. Much of the information generated was published in the "Philippine Journal of Science" and the "Leaflets of Philippine Botany." The latter journal was published between 1906–1920 and edited by A. D. E. Elmer. A nearly complete listing of pertinent taxonomic papers published on Philippine plants during the first three decades of American occupation can be found in volume four of Merrill's *Enumeration of Philippine Flowering Plants* (1926). Taxonomists who contributed much to Philippine botany during this period are listed in Table IV. For reasons not clear to us, many algal groups were left practically untouched.

In an effort to understand better the Philippine flora, botanical expeditions were extended to include Borneo, the Celebes, Indo-China and South China (Merrill, 1953). Manila, with its growing herbarium, library facility and a prestigious organ of publication, the Philippine Journal of Sciences (started in 1906), quickly became one of the centers of botanical research in Southeast Asia. The Philippine botanical literature was an important source for floristic investigations conducted in the neighboring countries.

According to Merrill (1926), the successful advances of Philippine botany were due to a centralization of all scientific work in one institute with full financial backing given by the government and also full cooperation among several units. Obviously, the dynamic and assiduous leadership of Merrill as the chief plant taxonomist in the country was another indispensable ingredient to the flourishing of Philippine botany during this time.

Table IV
Important contributors to knowledge of the Philippine flora from 1900–1942.

Specialist	Group(s)
Ames, O.	Orchidaceae
Baker, J.	Amaryllidaceae, Liliaceae
Beccari, O.	Palmae
Becker, W.	*Viola* (Violaceae)
Bennett, A. W.	*Potamogeton* (Potamogetonaceae)
Brown, W.	economic plants
Burkill, I. H.	*Gordonia* (Theaceae)
Candolle, de C.	Piperaceae
Clarke, C. B.	Acanthaceae, Cyperaceae
Dubard, M. M. M.	Sapotaceae
Elmer, A. D. E.	angiosperms
Foxworthy, F. W.	gymnosperms
Gamble, J. S.	Gramineae: Bambusoideae
Hackel, E.	Gramineae
Hooker, J. D.	*Impatiens* (Balsaminaceae)
King, G.	*Castanopsis* & *Quercus* (Fagaceae), *Ficus* (Moraceae)
Kränzlin, F. W. L.	Cannaceae, Cyrtandraceae
Martelli, U.	Pandanaceae
Merrill, E. D.	angiosperms
Pax, F. A.	Euphorbiaceae
Perkins, J. R.	Marantaceae, Meliaceae
Radlkofer, L.	Sapindaceae
Ridley, H. N.	*Liparis* (Orchidaceae), Zingiberaceae
Robinson, C. B.	Boraginaceae, Chloranthaceae, Myrtaceae, Urticaceae, mosses
Rolfe, R. A.	Orchidaceae
Schlechter, R.	Asclepiadaceae, Gesneriaceae
Schulz, O. E.	*Cardamine* (Cruciferae)
Schumann, K. M.	Marantaceae, Musaceae, Zingiberaceae
Alderwerelt van Rosenberg, C.	fern allies
Alston, A. H. G.	*Selaginella*
Bartram, E. B.	mosses
Brotherus, V. F.	mosses
Christ, K. H.	ferns
Copeland, E. B.	ferns
Evans, A. W.	liverworts
Graff, P. W.	fungi
Hieronymus, G. H.	*Selaginella*
Herter, W. G.	*Lycopodium*
Massee, G. E.	fungi
Mendoza, J.	fungi
Palo, M. A.	fungi
Rehm, H.	ascomycetes
Reinking, O. A.	fungi
Shaw, M. H.	algae
Stephani, F.	liverworts
Sydow, H.	fungi
Underwood, L. R.	ferns, liverworts
Wainio, E.	lichens
Warnstorf, C.	*Sphaghnum*
Williams, R.	mosses
Yates, H. S.	fungi
Zahlbruckner, A.	lichens

By 1926 the list of known Philippine plant species was greatly lengthened to include some 8,120 species of flowering plants and approximately 1,000 species of pteridophytes and nearly 3,000 species of non-vascular cryptogams. Many botanical explorations reached several distant and remote mountains as shown in a map prepared by Merrill (1915a).

Unfortunately in 1945, during the liberation of the city of Manila, the national herbarium was razed by fire. Some 305,367 mounted plant specimens accumulated in four decades were nearly completely destroyed (G. Price, 1976). When the war was over, hundreds of holotypes of Philippine vascular plant species were gone. Only a portion of the collections survived in foreign herbaria as replicate materials sent/deposited earlier by Merrill and his associates.

4. The Post-war Period (1946–present)

The task of restoring the national herbarium and an attendant botanical library was undertaken by E. Quisumbing soon after the war (Quisumbing, 1957). With the help of Merrill and Copeland, who now resided in the United States, Quisumbing was able to get some money and plant specimen driers (Quisumbing, 1957; Asis, 1975) to start with. In addition, requests for donation of replicates and return of pre-war loans of Philippine plant materials from foreign herbaria continued to receive positive responses. By 1953, the total mounted specimens deposited at the Philippine National Herbarium (PNH) reached 42,966 (Quisumbing, 1957).

Publications on the Philippine flora were resumed in the "Philippine Journal of Science" with Merrill, Copeland and Quisumbing, among others, initially taking the lead. However, compared to the pre-war period, monographs and taxonomic revisions of Philippine plant groups have become few. The retirement in 1961 of Quisumbing, who had proved himself to be the most productive Filipino plant taxonomist of the century, from his directorship of the Philippine National Museum, saw the decline of Philippine botanical study. Taxonomic research began to shift lopsidedly towards the economically important plant groups as witnessed by later publications (Aguilar-Santos & Doty, 1968; Allen, 1965; Galutira and Velasquez, 1963; Gutierrez, 1980–1982; Macabenta & Capina, 1984; Madulid, 1976; Monsalud et al., 1966; Pancho, 1961, 1962, 1976, 1983b; Pancho & Guantes, 1962; Pancho & Santos, 1963; Pancho & Hilario Jr., 1963; G. Price, 1974; Quimio and Puruganan, 1979; Quisumbing, 1951; Rojo, 1979; Sajise et al., 1974, 1979; Salvosa, 1963; Santos, 1973; Science Education Center, 1983; Uyenco & Cuevas, 1977; Velasquez, 1968; and Yen & Gutierrez, 1974). An important source of information for taxonomic or floristic literatures published during the period up to 1966 is the compilation annotated by Nemenzo (1967).

At present, the Philippine National Herbarium boasts a total collection of well over 150,000 mounted specimens of Philippine and foreign plants. The figure does not include their unaccessioned possessions. It also has a large backlog of unnamed and unmounted collections wrapped in hundreds of parcels. Yearly, at least three to five botanical expeditions are undertaken by the herbarium staff, at times in cooperation with foreign institutions such as the Rijksherbarium of the Netherlands, the Swedish Museum of Natural History (Stockholm), the Field Museum of Chicago in the United States, the Kyoto University and the Osaka City Museum of Japan.

Not unexpectedly, foreign botanists continue to make significant contributions to the post-war knowledge of Philippine flora, especially in groups where there is no local taxonomist working. Many of them have made personal collecting trips to remote areas through the arrangements and field assistance provided by local colleagues and friends. The results of these studies have been published in various foreign and local journals: Ashton (1982), Chandra (1977, 1980), Dransfield (1980), Hatusima (1966), Holttum (1973, 1974a, 1974b, 1975a, 1975b, 1976), Inoue (1965a, 1965b), K. Iwatsuki (1982), K. Iwatsuki and Price (1977), Z. Iwatsuki and Noguchi (1978), Z. Iwatsuki and Sharp (1966, 1974), Z. Iwatsuki and Tan (1980), Kobayashi (1980), Laubenfels (1978), Meijer (1954), Mizutani (1975, 1977), Moldenke (1978), Noguchi (1963), Reynolds (1966, 1967, 1981), Stone (1976), Tixier (1972, 1978), Touw (1962, 1971), van Zanten (1959, 1973), Whitmore (1980), and others.

No review of current progress of Philippine taxonomic botany is complete without mentioning the contributions of the Flora Malesiana Project, which is based at the Rijksherbarium, Netherlands. The country is fortunate to be located geographically within the Malesian region, where several families of pteridophytes and spermatophytes have been monographed. Jacobs (1974) made a comprehensive review of the progress of Malesian botany, including that of the Philippines, and a list of plant families revised and published in the *Flora Malesiana* series. The latest account of the taxonomic contributions of the Rijksherbarium to the Flora Malesiana Project was summarized in Blumea, vol. 25: 1–140 (1979).

Lately there have been attempts by local botanists to graduate from the traditional alpha level of taxonomy and to adopt a more biosystematic approach to the study of Philippine plants (Madulid, 1980c, 1980e; Mendoza & Zamora, 1980, 1981; Pancho & Hilario, 1963; Salgodo, 1982; Zamora & Vargas, 1973a, 1973b).

Perhaps the two most important post-war events in Philippines botany were the gradual takeover of authority on local plant groups from foreign workers by Filipino resident taxonomists, and the establishment of a handful of medium to small sized herbaria outside of Manila. Most of these newly constituted herbaria are attached to large universities where they serve a teaching function primarily and a research function secondarily. A few are registered in the 1981 edition of "Index Herbariorum." Four of these are herbaria located at the University of the Philippines at Los Baños (CAHP, CALP, LBC, CLP) while another one (PUH) is at the U.P. Diliman campus. The unregistered ones are the Herbaria of Silliman University at Dumaguete City, the University of San Carlos at Cebu City, Visayan College of Agriculture at Baybay, Leyte, Mindanao State University at Marawi City, Central Mindanao University at Musuan, Bukidnon, and St. Louis University at Baguio City. Of these, the herbaria of the University of the Philippines at Los Baños possess the largest collections, numbering some 40,000 mounted plant specimens, mostly of Philippine origin. The herbarium at the University of San Carlos ranks second with some 7,000 plant specimens (Madulid, 1984). All together, including the possessions of the Philippine National Herbarium, there are 265,000 mounted Philippine specimens deposited within the country at present (Madulid, 1984). Considering the richness and size of the flora, this number is small to support a floristic treatment of the Philippines plants.

In 1923 when Merrill resigned as Chief Botanist and Director of the Bureau of Science, he left some 275,000 specimens at the herbarium of the Bureau of Science (now the Philippine National Herbarium). Out of this figure, 180,090 numbers represented local plant materials. Using this figure, van Steenis (1950) calculated a collecting density index of $62/100$ km^2 for the country. This number is below the minimum density index of $100/100$ km^2 that van Steenis himself has set as an index for adequate knowledge of the flora of areas within Malesia.

Considering the latest count of mounted Philippine plant specimens kept at the Philippine National Herbarium (75,000) and adding to it the pre-war figure (180,090), the collecting density index works out to about 85 as of 1984. This is still below the 100 collecting number threshold. The data from the Philippine National Herbarium is not clear in terms of the regional concentration of plant collections. However, it is safe to state that Luzon Island is the most collected region in the country, with the Visayan Islands and Mindanao Island ranking second and third respectively. A correlation between relative numbers collected and relative distance away from Manila has been noted by some authors (see Touw, 1971; van Steenis, 1958).

Reviews of the localities visited and the frequency of past botanical collections from the time of Merrill (1915a) up to the present (Quisumbing, 1957, 1964; Madulid, & Gutierrez, 1981; our field data at UPLB, 1984) has made it apparent that botanical studies have not focused on the remote hinterlands away from the road systems. On the other hand, some areas are seemingly well collected, such as the Baguio-Mt. Data-Benguet district, the greater Manila district and the Mt. Makiling-Mt. Banahao forests.

Merrill (1915a) suggested "that botanical explorations from the country in all regions at high elevation, except for a few well explored mountains, no matter what type of vegetation, and likewise, most regions at medium elevation, will be found to contain species of much special interest, while that of virgin forests, no matter what its location, will always be found to yield a high percentage of novelties." Van Steenis (1950) and Jacobs (1974) have pointed out places in the Philippines where past botanical collections are few or wanting. These are the Sierra Madre mountain range of Luzon Island, and the interior mountains of Mindanao Island. To these should be added the limestone forest areas on the Samar-Leyte Islands and the swamp vegetation area on Mindanao Island. In setting priority for future botanical expeditions, the mountain divide between the Bukidnon and Agusan provinces and the eastern coastal mountains (Diwata Range) of Mindanao Island should get top consideration. To a lesser extent, the dry Dipterocarp mixed forests on Mindoro Island and the Palawan Island group need additional attention.

B. ENDEMISM

The Philippine flora appears rather poor in generic endemism: one to five percent for various plant groups (Merrill, 1926; Mitra, 1973; Tan, 1984a; Zamora and Co, 1980b). In terms of species, however, the percentage of endemism is relatively high and reaches more than 50% among certain flowering plant groups. A recent estimate credits some 12,000 species of plants to the country, 8000 of which are phanerogams (Madulid, 1982). About one-fourth of all the local plant species reported are endemic (Madulid, 1985). An exception to this observation is the precentage of Philippine endemic mosses, which scores only a low 10% (Tan, 1984a). At present, data for other cryptogamic plant groups are insufficient for conclusions on endemism. Herein lies the importance of having more floristic inventories of the country's rain forest in the near future.

VI. Adequacy of Inventory for Major Groups

Thus far, botanical collections in all parts of the country have been made mostly by general plant collectors and foresters who have concentrated on forest trees and other plants of economic importance. Trees, shrubs, lianas and herbs of unknown economic value and forest cryptogams have been neglected. In many areas which have been relatively well collected in the past, present collections undertaken by specialists continue to yield new plant records for both the locality and the country. This only underscores the desirability of having more intensive botanical explorations in all corners of the archipelago conducted by experienced specialists.

To stress the point further, our experience with the local moss flora can be cited to illustrate the incompleteness of the knowledge of this group of forest cryptogams. One of us (B. C. Tan) has added no less than 70 new records of moss taxa to the Philippine flora in a period of five years of extensive field work (1981–1985). Ten new synonyms were proposed in subsequent revisionary studies involving the examination of type specimens. Five other collections probably represent species new to science. Since World War II, the total number of known Philippine moss species has increased by nearly 30% (Iwatsuki & Tan, 1979). Even granting that seed plants are larger in size, more conspicuous in the field and already relatively well surveyed by American botanists formerly in residence, we conservatively estimate that the total number of phanerogamous species will still be increased by at least 10–15% through future botanical studies.

The incomplete state of knowledge about the Philippine flora, both the phanerogams and cryptogams, is borne out by the fact that only three floras of significance have been completed and published for the country. These are Merrill's *Flora of Manila* (1912), Bartram's *Mosses of the Philippines* (1939) and Copeland's *Fern Flora of the Philippines* (1958–1960). Their publication dates speak loudly of the dire need for modern revisions that will incorporate changing taxonomic concepts and nomenclatures.

Although there have been several checklists prepared for the various plant groups, many of these also require updating. The published ones are Merrill's list of flowering plants (1923–1926), Salvosa's list of Philippine trees (1963), Rojo's list of Philippine dipterocarps (1979), Iwatsuki and Tan's list of Philippine mosses (1979), the Velasquez et al.'s list of Philippine algae (1972), Reynolds' list of Philippine slime molds (1981) and Teodoro's list of Philippine fungi (1937). The last one has been updated by Dogma (1975) and by Quimio and Capilit (1981). Worthy of special mention is the critical review published by M. G. Price (1972a) updating the work of Copeland on Philippine ferns. Recently, Martinez (1984) published a checklist of Philippine blue-green algae to replace an earlier one authored by Velasquez and Soriano (1957).

In 1975 Dogma reviewed the progress of the study of Philippine fungi and concluded that little has been achieved in the past 40 years. It is our assessment that the same situation holds true for other plant groups. One of the problems that confronts the Philippine mycologists, according to Dogma (1975), is that there are exceedingly few local workers equipped with the training necessary to undertake an active profession. The situation is aggravated by the problem of having no ready access to the original collections of many Philippine species described in the last and the first part of this century.

A convincing case that demonstrated the importance of having professional workers to take charge of floristic studies is the recently concluded inventory of "endemic, endangered, rare, vanishing, and most economically important species" of Philippine flora and fauna. The project lasted for more than four years during the latter half of the 70's. It involved several inexperienced and self-taught taxonomists. The results of the project finally saw print in 1986 for certain groups of plants and animals. An examination of the final publication (*Guide to Philippine flora and fauna*, vols. I–IV, 1986) reveals the technical difficulties encountered by some authors in tackling the serious research problems of conducting floristic inventory under Philippine conditions. Owing to the obsolete literature consulted and the lack of reference to type specimens, uncertain identifications and outdated nomenclature abounds. Furthermore, a large portion of the information published can be traced to secondary sources. Because the new collections accumulated for the project were neither sufficient nor truly extensive, the intended goal of ascertaining the status of the many putatively endangered and endemic plant species remains unsatisfactorily accomplished. Local mountains with high concentrations of local endemics which would require full protection and conservation measures were not identified. Threatened ecosystems, if any, with high rates of destruction were not mentioned. By and large, the knowledge of the country's endangered flora has advanced only a little from where Merrill (1923–1926), Bartram (1939), Copeland (1958–1960) and Quisumbing (1967) left off. This lamentable situation has been partly addressed by an updated list of endangered Philippine plants published by Tan et al. (1986), who reassessed the doubtful status of many of the reported endangered species on the basis of their field observations. Altogether, they accepted only two phanerogams (*Rafflesia manillana* Teschem., *Tectona philippinensis* Benth. & Hook.), two pteridophytes (*Isoetes philippinensis* Merr. & Perry, *Podosorus angustatus* Holtt.) and one moss (*Buxbaumia javanica* C. Muell.) as truly or immediately threatened with extinction, with 62 others as rare or potentially threatened species.

VII. Resources for Floristic Inventory

Other problems besieging Philippine taxonomy are the meager financial support and the want of public recognition given by the national and local governments. As a result, there has been a rapid loss of workers in taxonomy to other areas of biology where larger amounts of research money are allocated. To date, many local plant taxonomists are fored by economic considerations to digress into applied plant science research. As shown in Table V, only a handful remain active in taxonomy today. Of these, a majority are concentrated in Manila and Los Baños, where the local flora is more or less well known. The situation is alarming, because of the accelerating rate of deforestation facing the country. Without a new generation of local plant taxonomists working full time, much of the flora may become extinct before it gets the chance to be studied and recorded.

Even the national center for plant taxonomic research, the Philippine National Herbarium, is not exempt from the perennial problem of a decreasing fund source. The herbarium is presently understaffed, with only 13 regular personnel, most of them non-technical in training. Because of this, efforts to catch up with the identification of the accumulated plant collections at the institution have been seriously hampered. Currently, the Philippine National Herbarium has no professional curators taking care of the pteridophytic and fungal collections.

It is because of the limited financial resources in many herbaria, the Philippine National Herbarium notwithstanding, that the minimum physical requirements for the proper storage of dried plant specimens under humid tropical conditions can not be met. Insect pest and fungal destruction of plant specimens have become rampant. This is tragic, when the specimens being destroyed are the valuable sets of replicates from past collections which are frequently referred to in earlier botanical literature.

It is not an exaggeration to claim that the Philippine plant taxonomists themselves are an endangered or vanishing breed. The profession has become the least rewarded discipline of all the biological sciences both in terms of salary and opportunities for academic growth. To government funding agencies, taxonomy is only a marginal discipline of the lowest priority, because it contributes nothing directly to the gross national product. In addition, funding agencies labor under the misconception that the flora "has been worked out already" by Merrill and his associates. The immediate ill consequence of this is that few funds are allocated to systematic studies, which results in the disappearance of professional taxonomists.

The situation has encouraged the appearance of the "overnight taxonomist" and "all-around taxonomist." These amateurs, unfamiliar with the rules of taxonomic nomenclature and lacking an appreciation of the need to examine type specimens, often content themselves with identification of plant specimens via photographs and illustrations. To them, taxonomy is a mere listing of arbitrarily determined plant names based on a set of dichotomous keys coupled to the personal interpretation of published descriptions. The effect is to further downgrade the quality of Filipino taxonomic research.

Confronted with all these problems, no Filipino taxonomists have been able to participate in revisionary or monographic work of families of flowering plants, gymnosperms and pteridophytes in a regional project like the *Flora Malesiana*. The final taxonomic works are done by foreign colleagues fortunate enough to be located in large centers of botanical studies, such as the Rijksherbarium (L). Other invited participants are taxonomists associated with the Kew (K), Bogor (BO), Singapore (SING), Kuala Lumpur (KLU) and Bangi (UKMB) herbaria. Table VI lists some of the major and important families belonging to the Philippine flora that have been worked out in full or in part by participants of the "Flora Malesiana" Project. The list is admittedly not exhaustive.

Other problems encountered by taxonomists have to do with the nature of the old botanical collections. In the past, too many Philippine plant species were described on the basis of a single collection which often contained scanty and/or incomplete plant material (Madulid, 1980d; Merrill, 1947; Rojo, 1973). Too often, accompanying these old collections were shoddily scribbled field labels. These shortcomings can be rectified, however, if modern collections are made from the type locality.

In summary, we paraphrase here the comments made by Prance (1977) in his excellent review paper published on the status and problems of floristic works in the tropics. He stated that in order to complete the inventory of tropical floras, it is necessary to stimulate more training of local resident botanists, to deposit properly labelled and identified plant materials in herbaria equipped to house them and to publish the inventory in local journals in the country where the inventory took place. When this is done, the results not only will improve greatly the knowledge about the local flora, raise the standard of botanical publications in the country, but also contribute significantly to conservation.

VIII. The Future of Floristic Inventory

Meijer (1980) estimated that only 30% of the original forest is left in the Philippines. Merrill (1915a) reported that during the time of Blanco, virgin forest occurred at or near the coast of Negros Island. Today this is a bygone memory for most of the island inhabitants. In fact, sometime previ-

Table V
Contemporary Philippine plant taxonomists.[a]

Plant Group	Taxonomist (Publications)	Institution
1. Grasses and other angiosperms	J. V. Santos† (1950, 1973, 1979)	Univ. of the Philippines at Diliman
2. Palms and other angiosperms	D. A. Madulid (1976, 1980 a–e)	Philippine National Herbarium, Manila
3. Dipterocarpaceae and medicinal plants	H. G. Gutierrez** (1968, 1973, 1974, 1975, 1976, 1980, 1981, 1982)	National Research Council of the Philippines, Metro-Manila
4. Angiosperms	D. R. Mendoza* (1964, 1967)	Araneta Univ., Metro-Manila
5. Palms and other forest trees	E. S. Fernando (1979, 1980, 1983)	Univ. of the Philippines at Los Baños
6. Leguminosae and forest trees	J. P. Rojo* (1969, 1972, 1973, 1976 a,b, 1977, 1978 a,b, 1979, 1981)	Univ. of the Philippines at Los Baños and Forest Products Research & Development Inst.
7. Orchidaceae	H. Valmayor* (1984)	Univ. of the Philippines at Los Baños
8. Orchidaceae and ornamental plants	G. R. Price** (1972, 1973 a,b, 1974)	Univ. of the Philippines at Los Baños
9. Weeds and other angiosperms	J. V. Pancho (1961, 1962, 1969, 1976, 1983a); Pancho & Santos (1963)	Univ. of the Philippines at Los Baños
10. Forage plants	N. O. Aguilar*	Univ. of the Philippines at Los Baños
11. Forest trees	L. L. Quimbo* (1979)	Univ. of the Philippines at Los Baños
12. Pteridophytes	P. M. Zamora* (1970, 1973 a,b, 1974, 1980, 1984); Zamora & Co (1980)	Univ. of the Philippines at Diliman
13. Pteridophytes	M. G. Price** (1972 a,b, 1973 a,b, 1974 a–c, 1977, 1982, 1983 a,b)	Univ. of the Philippines at Los Baños
14. *Selaginella* and bryophytes	B. C. Tan (1975, 1981, 1982, 1983 a,b, 1984)	Univ. of the Philippines at Los Baños
15. Pteridophytes	A. E. Salgado* (1982)	De La Salle Univ., Manila
16. Hepatics and other bryophytes	R. M. Del Rosario (1966, 1967 a,b, 1971 a,b, 1973, 1975, 1979a, 1980); del Rosario & van Zanten (1979)	Philippine National Herbarium, Manila
17. Phycomycetes and other fungi	I. J. Dogma Jr.* (1966, 1967, 1969, 1976, 1977, 1978, 1979, 1983)	Univ. of Santo Tomas, Manila
18. Fungi	T. Quimio (1972, 1976, 1977, 1979, 1981, 1983); Quimio & Lantican (1978); Quimio & Opina (1978); Quimio & Suayan (1976)	Univ. of the Philippines at Los Baños
19. Lichens, fungi and slime molds	F. R. Uyenco* (1973, 1977)	Univ. of the Philippines at Diliman
20. Lichens	W. S. M. Gruezo* (1977, 1979)	Univ. of the Philippines at Los Baños

[a] An asterisk (*) indicates currently part-time or semi-active worker. Double asterisk (**) indicates on leave from the institution concerned for foreign study or research assignment abroad. The publications cited here include only the more recent papers. For earlier workers and publications on the Philippine flora, see Nemenzo (1967).

† Died in 1987.

ous to 1837, according to Merrill (1915a), natural forest extended very close to the Angat area in Bulacan Province of Luzon, which is some 30 km north of Manila. Today the nearest forest (which is partially secondary) is on Mt. Makiling some 65 km south of Manila. Not even the Mt. Makiling forest reserve, a long established and a well known biological station since the days of American occupation (Pancho, 1974), has been spared from the destructive activities of man. Recently, parts of it were allowed to be logged and bulldozed for a geothermal energy exploration study.

Other long established national parks (see listing in Lansigan, 1939), which were legislated to conserve the locally representative forest ecosystems on the different islands, suffer a similar fate. Many are invaded by unscrupulous and illegal loggers, the slash-and-burn farmers ("kaingineros") and/or the politically influential landgrabbers, often with the connivance of local military personnel. Some are left unmanaged owing to the lack of funds for the adequate maintenance of parks. There are even isolated reports of the military using parks as training grounds for jungle warfare. The proposal submitted by Jacobs (1972) for the establishment of a natural reserve around the forested summits of Mt. Pulog-Mt. Panotoan-Mt. Tabayoc of northern Luzon highland, continues to receive no attention. This is unfortunate, since the flora represented there has a high number of endemics. Jacobs (1974) described the situation as one in which there is a richness of flora and a backward state of knowledge.

Local pleas for conservation have not been wanting (Madulid, 1982; Quisumbing, 1961, 1967; Reyes, 1980, 1983). Owing perhaps to the increasing pressure from local

Table VI

Some important Philippine plant families revised in connection with the Flora Malesiana Project (as of 1983).

Family	Researcher(s)
Angiosperms	
Aceraceae	M. Jacobs†
Anacardiaceae	Ding Hou
Araliaceae	W. R. Philipson
Betulaceae	C. G. G. J. van Steenis†
Burseraceae	P. W. Leenhouts, C. Kalkman
Celastraceae	Ding Hou
Combretaceae	A. W. Exell
Cyperaceae	J. H. Kern†
Dilleniaceae	R. Hoogland
Dipterocarpaceae	P. S. Ashton
Ericaceae	H. Sleumer
Fagaceae	E. Soepadmo
Flacourtiaceae	H. Sleumer
Juglandaceae	M. Jacobs†
Leguminosae	Knaap van Meeuwen (in part)
Malvaceae	J. van Borrsum Waalkes
Melastomataceae	F. Maxwell (in part)
Meliaceae	A. J. Kostermans, T. D. Pennington, D. J. Mabberley
Myristicaceae	J. Sinclair† (in part)
Ochnaceae	A. Kanis
Oxalidaceae	J. E. Veldkamp
Pandanaceae	B. C. Stone, H. St. John (in part)
Pittosporaceae	K. Bakker, C. G. G. J. van Steenis†
Rhizophoraceae	Ding Hou
Rubiaceae	C. Ridsdale (in part)
Salicaceae	M. Jacobs†
Sapotaceae	R. van Royen (in part)
Saurauiaceae	R. Hoogland
Sapindaceae	R. W. Leenhouts, M. Jacobs† (in part)
Simaroubaceae	H. P. Nooteboom
Staphyleaceae	B. L. van der Linden
Symplocaceae	H. P. Nooteboom
Thymelaeaceae	Ding Hou
Pteridophytes	
Cyatheaceae	R. E. Holttum
Gleicheniaceae	R. E. Holttum
Lindsaea group	K. U. Kramer
Schizaeaceae	R. E. Holttum
Thelypteridaceae	R. E. Holttum
Bolbitis, Platycerium	E. Hennipman
Drynarioideae	M. Roos
Pyrrosia	P. Hovenkamp

† = deceased

mission of an environmental impact assessment study is now required from private and governmental agencies when applying for licenses to utilize public lands and/or natural resources. What is lacking, however, is the strict implementation and enforcement of this conservation-related legislation.

Amidst all the frustrations, local conservationists and taxonomists fighting for a change in the local conservation scenario ssee hope with the recent agreement of the ASEAN member countries to bind themselves together with a legal document banning the export and import of all publicly acclaimed endangered plant and animal species from member countries (ASEAN Agreement on the Conservation of Nature and Natural Resources, mimeographed draft, 1982).

There is an immediate need for more young local botanists to be professionally trained abroad in plant taxonomy. This is a must, because the pertinent literature and the historical plant collections of the Philippine flora are not available within the country. Upon their return from schooling abroad, these botanists could take up the challenge of recording in a professional manner the species diversity of the local flora. In addition, they should be encouraged by the government and the university administration with enough material resources to become committed local plant taxonomists.

The Philippines should have at least three taxonomic centers of study located in the three political regions, namely the Luzon, Visayas and the Mindanao. Each of these categories should be fully staffed and funded for long term research programs on the regional flora. There should be one specialist for each of the major plant groups in all three centers of taxonomic study. This strategy is one of absolute necessity in order to win the race against time and to carry out an efficient floristic inventory program.

The grand projet of completing the country's flora, which will come in several volumes, needs the financial and moral support of the national government. Simultaneously, basic research in forest synecology and plant associations similar to those done by Brown (1919) on Mt. Makiling and Mt. Banahao, or those by Ashton (1964) in Brunei and Poore (1968) and Anderson (1963) in Malaya and North Borneo should be conducted. Future work, as suggested by Prance (1977), should also include studies of pollination mechanisms, phenology, plant-animal relationships, soil microbes, cytology, and breeding behavior of forest plants. Information derived from this sort of research is vital to a successful program of conservation and management of natural forest resources.

The Philippine Bureau of Forest Development (BFD) projected in 1980 that in 25 years time the country shall have totally exterminated its rain forests, if the present rate of forest destruction is not stopped. Others, like Raven (1977), who are more pessimistic, forecast a period of 5 to 10 years of survival for the Philippine rain forests. When this happens, the rich Philippine flora together with its vast economic and scientific value shall be lost forever to mankind.

With more outside help, in the form of research funding, training programs and public opinion, from international organizations such as the International Union for the Conser-

and international communities, more laws and several presidential decrees have been issued lately with the objective of conservation. This legislation has been compiled into two thick volumes for easy public perusal (NEPC, 1980). Sub-

vation of Nature and Natural Resources (IUCN), the Convention of International Trading in Endangered Species (CITES), the World Wildlife Fund and the UNESCO, together with an awakening citizenry inside the country, the battle for the conservation of Philippine flora and its unique rain forest ecosystems may not be a losing one after all. Quoting Raven (1981), it is well to remember that the vegetation of one country is a precious portion of the world biome which belongs to the whole of humanity.

IX. Acknowledgments

We are indebted to the College of Forestry, the Institute of Biological Sciences and the Museum of Natural History, of the University of the Philippines at Los Baños, also to the Director's Office of the Forest Products Research and Development Institute of the National Science and Technology Administration, for placing at our disposal the facilities, library and unpublished data on the Philippine forests. We thank Prof. J. Pancho, Drs. J. Santos, I. Dogma, Jr., T. Quimio, D. Madulid, and Messrs. E. Fernando, J. Regalado, V. Cuevas, E. Wijangco, B. Samson, F. Misa, N. Lacdan and M. Pantastico for the comments and criticism offered on our manuscript. The mesdames L. Atienza, R. Borromeo and Mr. A. Tandang helped type the manuscript.

X. Addendum

It has only recently come to our attention that the earliest map of Philippine vegetation was published in 1875 by the Spanish colonial government in Manila in a pamphlet entitled "Memoria-catalogo de la colección de productos florestales." The pamphlet was prepared in connection with the Philippine participation in the international trade exposition held in Philadelphia, USA, in 1876. The map shows three color patterns: area of unexplored forests, area of explored forests and cultivated area. As expected, about 4/5 of the archipelago were forested. Panay, Cebu and Bohol Islands are shown to be largely cultivated; as are the Central Plain, Mountain Provinces and Albay Province of the Luzon Island.

Today, however, the situation is different. The 1986 Philippine Forest Statistics have recently been published by the Bureau of Forest Development. The currently forested land area was admitted to be only 9.2 million hectares, a decrease of about 4 million hectares from the 11.2 million hectares reported in 1982 (see text). The percentage of original forests left in the country is thus a new low of 21%.

XI. Literature Cited

Aguilar-Santos, G. & M. S. Doty. 1968. *Caulerpa* as food in the Philippines. Philipp. Agric. **52**: 477–482.

Allen, P. H. 1965. Annotated index of Philippine Musaceae. Philipp. Agric. **48**: 320–411.

Anderson, J. A. R. 1963. The flora of the peat swamp forests of Sarawak and Brunei, including a catalogue of all recorded species of flowering plants, ferns and fern allies. Gard. Bull. Singapore **20**: 131–227.

Ashton, P. S. 1964. Ecological studies in the mixed dipterocarp forests of Brunei State. Oxford Forest. Mem. **25**: 1–75.

———. 1982. Dipterocarpaceae. Flora Males. ser. 1, **9**: 237–552.

Asis, C. V. 1975. Quisumbing and Friend. Nat. Appl. Sci. Bull. Univ. Philipp. **27**: 1–73.

Bartram, E. B. 1939. Masses of the Philippines. Philipp. J. Sci. **68**: 1–423.

Bedard, P. W. 1958. Reconnaissance, classification and mapping of Philippine forests. Pages 9–53 *in* Proc. Symposium on Humid Tropic's Vegetation, UNESCO.

Brown, W. H. 1919. The vegetation of the Philippine mountains. Bur. Sci. Publ. **13**: 1–434.

Chandra, S. 1977. Rhizome morphology of two lomariopsid ferns endemic to the Philippines. Kalikasan **6**: 55–62.

———. 1980. Sporangial morphology of dynarioid ferns. Kalikasan **9**: 31–42.

Copeland, E. B. 1958–1960. Fern Flora of the Philippines, vols. 1–3, NIST, Manila, vol. 1: 1–191; vol. 2. 193–376; vol. 3: 377–557.

Cordero Jr., P. A. 1977. Studies on Philippine marine red algae, Special Publ. Seto Marine Biol. Lab. ser. **IV**: 1–258.

Dickerson, R. E. 1928. Distribution of life in the Philippines. Bur. Printing, Manila. 322 p.

Dogma Jr., I. J. 1966. Philippine Clavariaceae. The Pteruloid series. Philipp. Agric. **49**: 844–861.

———. 1967. Addition to the genus *Clavulina*, Clavariaceae. Philipp. Agric. **50**: 771–778.

———. 1969. Observations on some cellulosic chytridiaceous fungi. Arch. Mikrobiol. **66**: 203–219.

———. 1975. Of Philippine mycology and lower fungi. Kalikasan **4**: 69–105.

———. 1976. *Diplophlyctis versiformis*, n.sp. and the *Siphonaria*-type of sexual reproduction. Kalikasan **5**: 121–142.

———. 1977. Philippine zoosporic fungi: *Olpidium sparrowii*, a new chytridiomycete parasite of rotifer eggs. Kalikasan **6**: 9–20.

———. 1978. Philippine zoosporic fungi: The algal parasite *Myzoctium megastomum* and its *Rozella* hyperparasite. Kalikasan **7**: 47–62.

———. 1979. Philippine zoosporic fungi: *Blyttiomyces verrucosus* n.sp. (Chytridiales) with a revision of the genus. Kalikasan **8**: 237–266.

———. 1983. Philippine zoosporic fungi: Three parasitic species of Chytriomyces (Chytridiales, Chytridiomycetes). Kalikasan **12**: 385–408.

Dransfield, J. 1980. On the identity of Sika in Palawan, Philippines. Kalikasan **9**: 43–48.

Fernandez, J. C. (ed.). 1982. Geology and mineral resources of the Philippines. Bur. Mines. and Geo-sciences, Manila. pp. 406.

Fernando, E. S. 1979. *Baccaurea* (Euphorbiaceae) in the Philippines. Kalikasan **8**: 301–312.

——— & J. V. Pancho. 1980. Mangrove trees of the Philippines. Sylvatrop **5**: 35–54.

———. 1983. A revision of the genus *Nenga*. Principes, J. Palm Soc. **27**: 55–70.

Fischer, A. F. 1934. Forest map of the Philippine Islands. Bur. of Forestry, Manila.

Galutira, E. and G. T. Velasquez. 1963. Taxonomy, distribution and seasonal occurrence of edible algae of Ilocos Norte, Philippines. Philipp. J. Sci. **92**: 483–522.

Gruezo, W. S. 1977. Contributions to the lichen flora of the Philippines. I. New and additional records. Kalikasan 6: 135–142.

———. 1979. Compendium of Philippine lichens. Kalikasan 8: 267–300.

Gutierrez, H. G. 1968. A revision of the genus *Hopea* Roxburgh of the Philippines. Acta Manilana 4: 1–86.

———. 1973. An archeological find: the genus *Psidium* (guava) in the Philippines. Philipp. J. Sci. 102: 143–145.

———. 1974. *Tricyrtis imeldae*, a new Philippine lily. Philipp. J. Sci. 103: 171–173.

———. 1975. *Hopea samarensis*, a new Philippine dipterocarp. Kalikasan 4: 235–329.

———. 1976. *Hopea dalingdingan*, another new Philippine dipterocarp. Kalikasan 5: 92–98.

———. 1980–1982. An Illustrated Manual of Philippine Materia Medica. vols 1–2. NRCP Publ. 1: 1–234; Publ. 2: 235–485. Manila.

Hamilton, W. 1979. Tectonics of the Indonesian Region, Geological Survey Professional Paper No. 1078, U.S. Govt. Printing Office, Washington, D.C. 345 p.

Hatushima, S. 1966. An enumeration of the plants of Batan Island, N. Philippines. Mem. Fac. Agric. Kagoshima Univ. 5: 13–70.

Holttum, R. E. 1973. Studies in the family Thelypteridaceae. VI. *Haplodictyum* and *Nannothelypteris*. Kalikasan 2: 58–68.

———. 1974a. New Philippine ferns. Kalikasan 3: 196–197.

———. 1974b. Studies in the family Thelypteridaceae. VII. The genus *Chingia*. Kalikasan 3: 13–28.

———. 1975a. Studies in the family Thelypteridaceae. IX. The genus *Sphaerostephanos* in the Philippines. Kalikasan 4: 47–68.

———. 1975b. The genus *Heterogonium* Presl. Kalikasan 4: 205–231.

———. 1976. New records of Thelypteridaceae from the Philippines. Kalikasan 5: 109–120.

Inoue, J. 1965a. A small collection of liverworts from Island Batan, Philippines. Hikobia 4: 272–276.

———. 1965b. A new species of *Balantiopsis* Mitt. from the Philippines. J. Jap. Bot. 40: 245–247.

Iwatsuki, K. 1982. Studies in the systematics of Filmy ferns. VI. The genus *Sphaerocionium* in Asia and Oceania. J. Faculty Sci. Univ. Tokyo (III) 13: 203–215.

——— & M. G. Price. 1977. The Pteridophytes of Mt. Burnay and vicinity, northern Luzon. South East Asian Stud. 14: 540–572.

Iwatsuki, Z. 1974. *Dendrocyathophorum paradoxum* and *Fissidens nymanii*, new to the moss flora of the Philippines. Misc. Bryol. Lichenol. 6: 1.

——— & A. Noguchi. 1978. Mosses of the Philippines collected by Dr. Mason E. Hale in 1964. J. Hattori Bot. Lab. 4: 195–200.

——— & A. J. Sharp. 1966 *Bryoxiphium norvegicum* subsp. *japonicum* (Berggr.) Love et Love new to the Philippines. Misc. Bryol. Lichenol. 4: 29–30.

——— & B. C. Tan. 1979. Checklist of Philippine mosses. Kalikasan 8: 179–210.

———, & B. C. Tan. 1980. Noteworthy Philippine mosses at the Hattori Botanical Laboratory (NICH), Japan. Kalikasan 9: 267–282.

Jacobs, M. 1972. The plant world on Luzon's highest mountains. Rijksherium, Leiden. 32 p.

———. 1974. Botanical panorama of the Malesiana archipelago (vascular plants). In: UNESCO (*ed.*), Natural resources of humid tropical Asia. Nat. Resources Res. 12: 263–294.

Kobayashi, T. 1980. Notes on the Philippine fungi parasitic to woody plants (3). Trans. Mycol. Soc., Japan 21: 311–320.

Kowal, N. E. 1966. Shifting cultivation, fire and pine forest in the Cordillera Central Luzon, Philippines. Ecol. Monogr. 36: 389–419.

Lansigan, N. P. 1939. The national park movement in the Philippines. Philipp. J. Forest. 2: 359–387.

Laubenfels, D. J., de. 1978. The taxonomy of Philippine Coniferae and Taxaceae. Kalikasan 7: 117–152.

Macabenta, J. P. C. and R. O. Capiña. 1984. The Marantaceae of the National Botanic Garden. Philipp. J. Sci. 113: 47–65.

Madulid, D. A. 1976. Soil and sand-binding grasses of the Philippines and its conservation value. Acta Manilana 24: 1–50.

———. 1980a. A monograph of the genus *Plectocomia*. Kalikasan 9: 111–210.

———. 1980b. Palynological studies in the genus *Plectocomia* Mart. ex Blume (Palmae: Lepidocaryoideae). Sylvatrop 5: 19–34.

———. 1980c. Chemotaxonomic studies in *Plectocomia* (Palmae: Lepidocaryoideae). Kalikasan 9: 69–80.

———. 1980d. Comments on the present state of taxonomic knowledge of the rattans in Southeast Asia. Canopy Int. 6: 10–11.

———. 1980e. Notes on the reproductive biology of rattans. I. Pollination in *Plectocomia*. Sylvatrop 5: 157–160.

———. 1982. Plants in Peril. Filipinas J. Sci. & Culture 3: 8–16.

———. 1983. The botanical results of the Malaspiña Expedition (1789–1794). Kalikasan 12: 1–14.

———. 1985. Status of plant systematic collections in the Philippines. Pages 71–75 *in* S. H. Sohmer (ed.), Forum on systematic resources in the Pacific. Bernice P. Bishop Mus. Special Publ. 74: 1–79.

——— & H. G. Gutierrez. 1981. Botanical expeditions in the Philippines (1953–1979). Natur. Res. Council Philipp. Res. Bull. 36: 78–90.

Martinez, M. R. 1984. A checklist of Blue-green Algae of the Philippines, BIOTECH, UPLB, Laguna, 96 p.

McCabe, R., J. Alamsco & W. Diegor. 1982. Geologic and paleomagnetic evidence for a possible Miocene collision in Western Panay, Central Philippines. Geology 10: 325–329.

Meijer, W. 1954. Notes on some Malayan species of *Anthoceros* L. (Hepaticae). Reinwardtia 2: 411–423.

———. 1980. A new look at the plight of tropical rain forests. Environ. Conserv. 7: 203–206.

Menard, H. W. & E. L. Hamilton. 1963. Paleogeography of the tropical Pacific. Pages 193–217 *in* J. L. Gressitt (ed.), Pacific Basin biogeography. Bishop Museum Press, Honolulu.

Mendoza, D. R. 1974. The genus *Xylosma* in the Philippines with one new species. Philipp. J. Sci. 93: 511–518.

——— & R. M. del Rosario. 1967. Philippine aquatic flowering plants and ferns. Philipp. Nat. Mus. Publ. no. 1: 1–55.

Mendoza, R. C. and C. V. Zamora. 1980. Studies on the *Selaginella* species in the Philippines. I. Leaf epidermal elements. Nat. Appl. Sci. Bull. Univ. Philipp. 32: 101–158.

——— and ———. 1981. Studies on the *Selaginella* species in the Philippines, II. Leaf epidermal elements. Nat. Appl. Sci. Bull. Univ. Philipp. 33: 175–220.

Merrill, E. D. 1903. Botanical work in the Philippines. Bull. Bur. Agric. 4: 1–53.

———. 1908. The Philippine plants collected by the Wilkes United States Exploring Expedition. Philipp. J. Sci. 3(c): 73–86.

———. 1912. A flora of Manila. Bur. Sci. Publ. 5: 1–490.

———. 1915a. The present status of botanical explorations of the Philippines. Philipp. J. Sci. 10(c): 159–167.

———. 1915b. Genera and species erroneously credited to the Philippine flora. Philipp. J. Sci. 10(c): 171–194.

———. 1923–1926. An enumeration of the Philippine flowering plants, vols. 4. Bur. of Sci., Manila.

———. 1928. Flora of the Philippines. Pages 130–167 *in* R. E. Dickerson (ed.). Distribution of life in the Philippines. Bur. of

Sci., Manila.

———. 1947. An amateur bryologist's observations on the Philippine moss flora. Bryologist 50: 4–13.

———. 1953. Autobiographical: Early years, the Philippines, California. Asa Gray Bull., new series 2: 335–370.

——— & E. Quisumbing. 1953. New Philippine plants II. Philipp. J. Sci. 82: 323–339.

Mitra, M. 1973. The Philippines in a botanist's eye. Bull. Bot. Soc. Bengal. 27: 67–70.

Mizutani, M. 1975. Epiphyllous species of Lejeuneaceae from the Philippines. J. Hattori Bot. Lab. 39: 255–262.

———. 1977. Lejeuneaceae from the Philippines. J. Hattori Bot. Lab. 43: 127–136.

Moldenke, H. N. 1978. Notes on new and noteworthy plants. CVII. Phytologia 38: 307–308.

Monsalud, M. R., A. Toñgacan, F. Lopez, & M. Lagrimas. 1966. Edible wild plants in Philippine forests. Philipp. J. Sci. 95: 431–561.

Myers, N. 1980. Conversion of tropical moist forests. National Research Council. U.S. Nat. Acad. Sci., Washington, D.C. 176 p.

National Environmental Protection Council (NEPC). 1980. Philippine Environmental Laws, Comments, and Materials, vols. 1–2. Ministry of Human Settlements, Q. C., vol. 1, 524 p.; vol. 2, 630 p.

National Census and Statistics Office. 1983. Monthly Bulletin of Statistics. October Issue. NCSO, National Economic and Development Authority, Manila. 33 p.

Nemenzo, C. A. 1967. "1969." The flora and fauna of the Philippines (1851–1966), an annotated bibliography. vol. 1—Flora. Nat. Appl. Sci. Bull. Univ. Philipp. 21: 1–307.

Noguchi, A. 1963. A collection of mosses from the Philippines. Sci. Reports Tohuku Imp. Univ. Series 4 (Biol.) 29: 145–151.

Pancho, J. V. 1969. Some rare endemic plants. The Philipp. Biota 4: 3–5.

———. 1974. Mt. Makiling as a station for biological research. Philipp. Agric. 57: 21–33.

———. 1976. Philippine Aquatic weeds. Kalikasan 5: 37–91.

———. 1983a. Vascular flora of Mt. Makiling and vicinity (Luzon: Philippines). Part I. Kalikasan Suppl. 1: 1–476.

———. 1983b. Plants poisonous to livestock in the Philippines. Kalikasan 12: 193–284.

———, E. Bardenas & J. Capinpin. 1961. Vegetative characters as an aid to the identification of common weed seedling in lowland rice fields. Philipp. Agric. 45: 73–87.

——— & M. M. Guantes. 1962. Seed identification of common weeds in lowland rice fields. Philipp. Agric. 46: 481–513.

——— & F. I. Hilario Jr. 1963. Chromosome numbers and intraspecific hybridization in Croton (*Codiaeum variegatum* Bl.). Philipp. Agric. 47: 104–112.

——— & F. I. Santos. 1963. Identification of common forage and pasture grasses by their vegetative characters. Philipp. Agric. 46: 733–757.

Poore, M. E. D. 1968. Studies in Malaysian rainforest, I. The forest on Triassic sediments in Jengka Forest Reserve. J. Ecol. 56: 143–196.

Prance, G. T. 1977. Floristic inventory of the tropics: Where do we stand? Ann. Missouri Bot. Gard. 64: 659–684.

Price, G. R. 1972. Philippine *Aerides*. Pollinia 1: 8–14.

———. 1973a. *Phalaenopsis* in the Philippines. Pollinia 2: 3–15.

———. 1973b. A review of Philippine species of *Paphiopedilum*. Orchid Dig. 37: 120–125.

———. 1974. Cultivated mussaendas in the Philippines. Kalikasan 3: 37–55.

———. 1976. Eduardo A. Quisumbing: portrait of the botanist as a Filipino. Kalikasan 5: 2–18.

Price, M. G. 1972a. A summary of our present knowledge of the ferns of the Philippines. Kalikasan 1: 17–53.

———. 1972b. Notes on Philippine Ferns. Brit. Fern Gaz. 10: 253–262.

———. 1973a. Overlooked names of Philippine ferns. Kalikasan 2: 109–113.

———. 1973b. A new species of *Grammitis* from Mt. Banahao, Luzon. Philipp. Agric. 57: 34–36.

———. 1974a. Nine new fern names. Kalikasan 3: 173–178.

———. 1974b. Three new ferns of the genus *Tectaria*. Kalikasan 3: 113–120.

———. 1974c. The pteridophytes described from Mt. Makiling, Luzon. Philipp. Agric. 57: 37–48.

———. 1977. Philippine *Dryopteris*. Garden's Bull. Singapore 30: 239–250.

———. 1982. The ferns of Steere and Harrington. Contr. Univ. Michigan Herb. 15: 197–204.

———. 1983a. Several unusual Malesian Diplazia (Athyriaceae). Gard. Bull. Singapore 36: 25–29.

———. 1983b. Two name changes for ferns of Mt. Makiling, Luzon, Philippines. Kalikasan 12: 155–156.

Quimbo, L. L. 1979. The taxonomic position of Philippine leguminous trees. Philipp. J. Sci. 106: 87–98.

Quimio, T. H. 1972. Agaricales of Mt. Makiling with emphasis on the edible ones and their culture potentials. Nat. Res. Council. Philipp. Bull. 69: 35–42.

———. 1976. Some discomycetes from Mt. Makiling (Philippines). Nova Hedwigia 28: 515–525.

———. 1977. Taxonomic consideration of *Auricularia* imported into the Philippines. Kalikasan 6: 69–72.

———. 1983. Some unreported Agaricales of Mt. Makiling (Philippines). Nova Hedwigia 38: 421–432.

——— & A. Capilit. 1981. Enumeration and bibliography of Philippine Fungi (1936–1977). BIOTECH-UPLB Publ. 79 p.

——— & M. T. Lantican 1978. Agaricales of Mt. Makiling: III. Genus *Lepiota*. Nat. Res. Council Philipp. Bull. 33: 1–11.

——— & N. L. Opina. 1978. Agaricales of Mt. Makiling (Philippines): II. Genus *Agaricus*. Nova Hedwigia 29: 847–858.

——— & F. G. Puruganan. 1979. Survey of *Pythium* associated with tobacco in the Ilocos Region. Philipp. Phytopathology 15: 127–136.

——— & Z. A. Suayan. 1976. Agaricales of Mt. Makiling (Philippines): I. Genus *Termitomyces*. Nova Hedwigia 28: 527–532.

Quisumbing, E. 1951. Medicinal plants of the Philippines. Bur. Printing, Manila, 1234 p.

———. 1957. Botanical expeditions in the Philippines (1940–1953). Proc. 8th Pacific Sci. Congress 4: 501–504.

———. 1961. The vanishing species of plants in the Philippines. Proc. 8th Pacific Sci. Congress 6: 191–197.

———. 1964. Botanical expeditions in the Philippines. Philipp. Geogr. J. 8: 21–38.

———. 1967. Philippine species of plants facing extinction. Araneta J. Agric. 14: 135–162.

Raven, P. H. 1977. Perspective in tropical botany. Ann. Missouri Bot. Gard. 64: 746–748.

———. 1981. Tropical rain forests: A global responsibility. Nat. Hist. 90: 28–32.

Reyes, M. R. 1980. Let us put more effort into forest protection. Canopy Int. 6: 1–2.

———. 1983. Conserving the Philippine dipterocarp forest. Canopy Int. 9: 14–15.

Reynolds, D. R. 1966. New records of Philippine fungi. Philipp. Agric. **50**: 784–790.

———. 1967. A key to known Philippine gasteromycetes. Philipp. Agric. **50**: 268–278.

———. 1981. Southeast Asian myxomycetes. II. Philippines. Kalikasan. **10**: 127–150.

Robinson, C. B. 1906. The history of botany in the Philippine Islands. J. New York Bot. Gard. **7**: 104–112.

Rojo, J. P. 1969. *Terminalia macrantha* (Combretaceae), a new Philippine species. Blumea **17**: 93–95.

———. 1972. *Pterocarpus* (Leguminosae-Papilionaceae) revised for the world. Phanerogamarum monographiae **5**: 1–119.

———. 1973. Recent changes in nomenclature of some Philippine trees, parts 1 & 2. Philipp. Lumberman **19**: 40–50 (Part 1); **19**: 37–45 (Part 2).

———. 1976a. Nomenclature notes on toog (*Petersianthus quadrialatus*) and kamatog (*Sympetalandra densiflora*). Pterocarpus **2**: 61–65.

———. 1976b. *Shorea virescens* Parijs, a hitherto unrecorded Philippine dipterocarp. Kalikasan **5**: 99–108.

———. 1977. Pantropic speciation of *Pterocarpus* and the Malesia-Pacific species. Pterocarpus **3**: 19–31.

———. 1978a. On the correct scientific name of the Benguet pine. Sylvatrop **3**: 31–36.

———. 1978b. On the taxonomy of *Shorea agsaboensis* Stern. Pterocarpus **3**: 63–70.

———. 1979. Updated enumeration of Philippine dipterocarps. Sylvatrop **4**: 123–145.

———. 1981. Philippine teak – an endangered tree. Canopy Int. **7**: 16.

Rosario, R. M., del. 1966. A noteworthy liverwort from the Mountain province, Luzon Is., Philippines. Philipp. J. Sci. **95**: 427–429.

———. 1967a. Philippine liverworts. Philipp. J. Sci. **95**: 427–429.

———. 1967b. Some liverworts from the Philippines. Bryologist **70**: 360–363.

———. 1971a. Herbertales of the Philippines. Ph.D. dissertation, University of Illinois at Urbana.

———. 1971b. New and noteworthy Philippine liverworts, II. Philipp. J. Sci. **100**: 227–242.

———. 1973. The Philippine species of the genus *Sphagnum*. Kalikasan **2**: 20–33.

———. 1975. Philippine liverworts, III. Calobryales and Herbertales of the Philippines. Philipp. J. Sci. **104**: 7–72.

———. 1979. Moss flora of the National Botanic Garden, Philippines. NIST, Manila. 133 p.

——— & **G. L. Penecilla.** 1980. Moss endemism in the Philippines. Nat. Mus. Bot. Papers **3**: 1–14.

——— & **B. O. van Zanten.** 1979. "1982". Mosses new to the Philippines. Philipp. J. Sci. **108**: 199–211.

Sajise, P. E., N. M. Orlido, L. C. Castillo & J. S. Lales. 1974. Studies on the genus *Themeda*. Kalikasan **3**: 71–82.

———, ———, ———, & ———. 1976. The ecology of Philippine grassland I. Floristic composition and community dynamics. Philipp. Agric. **59**: 317–334.

Salgado, A. E. 1982. Venation pattern in Philippine tactarioid ferns. De La Salle Monogr. **3**: 1–87.

Salvosa, F. M. 1963. Lexicon of Philippine trees. FPRI (FORPRIDECOM) Bull. **1**: 1–136.

Santos, J. V. 1950. A revision of the grass genus *Garnotia* Brongn. Nat. Appl. Sci. Bull. Univ. Philipp. **10**: 1–179.

———. 1973. Economic grasses of the Philippines. Philipp. BIOTA **8**: 14–28.

———. 1979. Observations on the dwarf bamboo of Mt. Pulog, Benguet, Philippines. Kalikasan **8**: 101–107.

———. 1984. Systematic botany in the Philippines: A perspective. ASBP Commun. **1**: 2–21.

Science Education Center. 1983. Guidebook to grassland plants. University of the Philippines, Diliman, Quezon City. 178 p.

Smith, W. D. 1924. Geology and mineral resources of the Philippines. Bur. Sci. Publ. **19**: 1–559.

Steenis, C. G. G. J., van. 1950. Flora Malesiana. ser. 1, vol. **1**: CVII–CXVI.

———. 1958. Ecology. *In*: Ding Hou, Rhizophoraceae, Fl. Mal. **4**: 429–495.

Steenis-Kruseman, M. J., van. 1950. Malaysian plant collection and collectors. Flora Malesiana ser. 1, **1**: XCII–SCIV.

Stone, B. C. 1976. Studies in Malesian Pandanaceae, XVI. Notes on Philippine taxa. Kalikasan **5**: 19–36.

Tan, B. C. 1975. Two *Selaginella* species new to the Philippines. Kalikasan **4**: 42–46.

———. 1981. *Orthorrhynchium elegans* (Hook. f. et Wils.) Reichdt. (Phyllogoniaceae), new record for the Philippine moss flora. Gard. Bull. Singapore **34**: 180–183.

———. 1982. Checklist of mosses of Mt. Makiling (Luzon Is., Philippines). Quart. J. Taiwan Mus. **35**: 135–148.

———. 1983a. Notes on SE Asiatic *Dicranodontium* (Dicranaceae). Ann. Bot. Fennici **20**: 233–235.

———. 1983b. The status of *Campylopus hemitrichius* (C. Muell.) Jaeg. Cryptogamie, Bryol. Lichenol. **4**: 357–361.

———. 1984. A reconsideration of the affinity of Philippine moss flora. J. Hattori Bot. Lab. **55**: 13–22.

——— & **A. R. Alvarez, Jr.** 1981. Additions to the "Moss Flora of the National Botanic Garden, Philippines." Kalikasan **10**: 165–176.

——— & **J. J. Engel.** 1986. An annotated checklist of Philippine Hepaticae. J. Hattori Bot. Lab. **60**: 283–355.

———, **E. Fernando & J. Rojo.** 1986. An updated list of endangered Philippine plants. Yushania **3**: 1–5.

——— & **A. C. Jermy.** 1981. Two new species of *Selaginella* from the Philippines. Fern Gaz. **12**: 169–173.

——— & **Z. Iwatsuki.** 1983. Nineteen new records of Philippine mosses. Kalikasan **12**: 328–350.

——— & **T. Koponen.** 1983. *Dicranoloma* (Musci, Dicranaceae) in Southeast Asia, with special reference to the Philippine taxa. Ann. Bot. Fennici **20**: 317–334.

——— & **A. Noguchi.** 1984. A small Philippine moss collection at Hiroshima University. Hikobia **9**: 43–50.

Taylor, B. & D. E. Hayes. 1980. The tectonic evolution of the South China basin. *In*: D. E. Hayes (ed.). The tectonic and geologic evolution of Southeast Asian seas and islands. Amer. Geophys. Union Geophys. Monogr. **23**: 89–104.

Teodoro, N. G. 1937. An enumeration of Philippine fungi. Dept. Agric. and Commerce Tech. Bull. **4**: 1–585.

Tixier, T. 1972. Mount Maquiling Bryoflora (Luzon). Gard. Bull. Singapore. **26**: 137–153.

———. 1978. Musci Igoroti (Mosses of the Mountain Province, Luzon, Philippines). Kalikasan **7**: 187–210.

Tuow, A. 1962. Revision of the moss genus *Neckeropsis* (Neckeraceae). I. Asiatic and Pacific species. Blumea **11**: 373–425.

———. 1971. A taxonomic revision of the Hypnodendraceae (Musci). Blumea **19**: 211–354.

UNESCO. 1973. International Classification and Mapping of Vegetation. Unesco, Paris.

Uyenco, F. R. 1973. Myxomycetes of the Philippines. Natur. Sci. Res. Center Tech. Report **32**: 1–30.

———— & V. C. Cuevas. 1977. Mycoflora of decomposing leaf litters. Nat. Appl. Sci. Bull. Univ. Philipp. **29**: 107–169.
Valmayor, H. L. 1984. Orchidiana Philippiniana, 2 vols. E. Lopez Foundation, Manila, 737 p.
Velasquez, G. T. 1962. The blue-green algae of the Philippines. Philipp. J. Sci. **91**: 267–375.
————. 1968. The edible seaweeds of the Philippines. Philipp. BIOTA **2**: 118–123.
———— & J. D. Soriano. 1957. The ecology of the Philippine Myxophyceae. Proc. 8th Pacific Sci. Congress **4**: 483–490.
————, G. C. Trono Jr., & M. S. Doty. 1972. Algal species reported from the Philippines. Philipp. J. Sci. **101**: 115–169.
Weidelt, H. J. & V. S. Banaag. 1982. Aspect of management and silviculture of Philippine dipterocarp forests. German Agency for Technical Cooperation, Frankfurt. 302 p.
Whitford, H. N. 1911. The forests of the Philippines. Bur. Forest. Bull. **10**: 1–113.
Whitmore, T. C. 1980. A monograph of *Agathis*. Plant Syst. Evol. **135**: 41–69.
————. 1984. A vegetation map of Malesia at scale 1:5,000,000. J. Biogeogr. **11**: 461–471.
Yen, E. D. and H. G. Gutierrez. 1974. The ethnobotany of the Tasaday: the useful plants. Philip. J. Sci. **103**: 97–139.
Zamora, P. M. 1970. Equisetaceae. Philipp. BIOTA **5**: 7–33.
————. 1974. A new species of *Cheilanthes*. Kalikasan **3**: 193–195.
————. 1984. Selaginellaceae of the University of the Philippines campus and vicinity, Quezon City. Philipp. BIOTA **17**: 85–94.
———— & L. L. Co. 1980a. New species of ferns (Filicopsida) from Palawan, Philippines. Nat. Appl. Sci. Bull. Univ. Philipp. **32**: 43–52.
————. 1980b. Guide to Philippine flora and fauna. V. Endemic fern genera. U.P. Nat. Sci. Res. Center, Diliman, Quezon City. 65 p. (mimeographed copy).
———— & M. G. Price. 1969. Twenty common Philippine ferns. Philipp. BIOTA **4**: 25–45.
———— & N. S. Vargas. 1973a. Notable variations in leaf forms of *Drynaria*. Philipp. Agric. **57**: 55–71.
————. 1973b. Nectary-costule association in Philippine drynariod ferns. Philipp. Agric. **57**: 72–88.
Zanten, B. O. van. 1959. Trachypodaceae, a critical review. Blumea **9**: 477–574.
————. 1973. A taxonomic revision of the genus *Dawsonia* R. Brown. Lindbergia **2**: 1–48.

Thailand

Tem Smitinand

Contents

I. Description of the Region 65
   A. Topography and Geology—General 65
      1. Northwest (Continental) Highlands 65
      2. Central Plain or Valley 65
      3. Northeast or Khorat Plateau 66
      4. The Southeast and Coast 66
      5. Southwest or Western Mountains 66
      6. Peninsula 66
   B. Climate 67
   C. Population 67

II. Vegetation 67
   A. Tropical Rain Forest 67
   B. Dry or Semi-evergreen Forest 68
   C. Hill Evergreen Forest 68
   D. Coniferous Forest 69
   E. Swamp Forest 69
      1. Fresh Water Swamp Forest 69
      2. Mangrove Swamp Forest 69
   F. Deciduous Forests 70
      1. Mixed Deciduous Forest 70
         a. Moist Upper Mixed Deciduous Forest 70
         b. Dry Upper Mixed Deciduous Forest 70
         c. Lower Mixed Deciduous Forest 70
      2. Dry Deciduous Dipterocarp Forest 70
   G. Savanna Forest 71

III. Vegetation Maps 71

IV. Magnitude of Floristic Inventory 71
   A. Specimens Collected 71

V. Publications on the Flora 73
   A. Pre-1920 73
   B. 1920–1950 73
   C. Post-1950 74

VI. Completeness of Floristic Inventory 74
   A. Factors Affecting 74
   B. Endemics and Extinction 74

VII. Current Collections and Work on Larger Families 75

VIII. Areas Particularly at Risk 75

IX. Inventory Resources 75
   A. Herbaria and Botanic Gardens 75
   B. Foreign Programs 77

X. Suitability of Floristic Inventory 77

XI. Selected Bibliography 77

I. Description of the Region

Phytogeographically, Thailand is central to South East Asian botany. Its flora is slightly different from the flora of neighboring countries, i.e., Burma, Laos, Kampuchea (Cambodia), and the Malay Peninsula, and a small amount of species endemism is apparent. At present it is considered to be rather under explored; recent explorations have resulted in quite a number of new records, and even new species.

A. Topography and Geology — General

The topography and geology of Thailand are rather complex, and the topography in particular is old. Faulting and folding movements in the past have caused a great deal of variation by forming highlands, plateaus, peneplanes, and depressions. The topography of Thailand is divided into six regions: the Northwest, or Continental Highlands, the Northeast or Khorat Plateau, the Central Plain (or Valley), the Southeast or Southeast Coast, the Southwest or Western Mountains, and the Peninsula. The following description of the six regions is mainly drawn on Pendleton (1963) and Smitinand (1958).

1. Northwest (Continental) Highlands

These consist of the mountainous area north and northwest of the Central Plain, continuous to the border of Burma and Laos. It is a region of parallel mountain ridges. The development of the relief of the mountains of North Thailand has taken place under the influence of two different bases of erosion. The Khong River forms the base level for the more northern part, while the catchment area of the Chao Phraya serves as the base in the south. Between the several rivers flowing to the Khong stand formidable mountain ranges with peaks often exceeding 1600 m above sea level; the highest one is the Pha Hom Pok (ca. 2350 m), west of the Fang River near the Burmese border. Through the greater part of the Northwest Highlands region, the valley systems with their intermontane basins are separated from one another by mountain ranges which run generally north to south. Between the Ping and its tributaries there is a large continuous range, of which highest peak, the Inthanon, northeast of Chiang Mai, rises to a height of about 2600 m; the highest of all Thailand. In the eastern drainage basins of the Ping, elevations are lower and continuous ranges of resistant granite are lacking. The highest peak, the Chiang Dao (ca 2250 m), being of permocarboniferous limestone forming characteristically steep peaks, seems to tower over the weak residual rocks of the ranges.

Also in North Thailand is part of the watershed of the Salween. This area is principally within Mae Hong Son province and includes the valleys of Pai, Yuam and Thanguin Rivers, all minor tributaries of the Salween. The Thanguin marks the Thailand-Burma border for some distance, as does the Salween itself. They are more or less isolated from the rest of North Thailand.

2. Central Plain or Valley

The alluvial plain of the Chao Phraya River, its tributaries and distributaries, and a surrounding piedmont belt, form the great Central Plain of Thailand, containing the most easily used soil, the main concentratoins of agriculture and population, and the best transportation network. It is clearly the heart of Thailand. The Central Plain extends approximately 400 km from the mountains of North Thailand to the Gulf of Thailand. From east to west is 120 to 200 km wide. Physically, the Central Plain may be divided into three subregions: the Krung Thep (Bangkok) Plain, the Upper Plain, and the Marginal Plains.

The Krung Thep Plain begins above Nakhon Sawan, where the water of the Ping and Nan River meet to form the Chao Phraya. Near Ayutthaya the Pasak River which drains the western slope of the Khorat marginal mountains enters the plain. It also receives the drainage of two other rivers the Mae Klong and the Prachin (Bang Pakong). This plain has an average width of 90 km and a length of some 230 km. The total area is approximately 20,000 square km. Inland, the plain rises very gradually, ranging from three m above sea level at Krung Thep about 20 km from the coast, to four m at Ayuthaya, and 25 m at Nakhon Sawan about 300 km from the coast. It is this very gradual slope, of 1:10,000, which allows flooding over the entire plain during high water in the rivers and streams. As far upstream as Ayutthaya, the levels of the rivers and their tributaries also rise and fall with the tide.

Borings have shown that the alluvium is made up of alternating silts, sands, and clays. At 110 m bedrock still is not reached. This plain is purely one of aggregation and has been built up in an apparently synclinal depression extending from the north to the south and onwards to the Thai Gulf.

On the east, west and north a belt of piedmont surrounds the Krung Thep Plain. The area is limited and very much interrupted by marginal outcrops of rock, low hills and mountains, which suggests that bedrock lies not far below the alluvium. The marginal region has been formed as a result of rising crustal movements of the earth. All the way to the Gulf in the Northern Peninsula, the Marginal Plains consist of a narrow strip along the west side of the Krung Thep Plain. This region extends for an average distance of about 50 km to the west of the Nakhon Chai Si and Mae Klong. In the east the Marginal Plains include much of the upper valleys of the Tha Chin and its tributaries.

The upper plain of the Central Valley includes the long narrow valley of the Pa Sak River in the north and northeast of the Krung Thep Plain. Separate mountains, or groups of mountains, stand out clearly from the plains along the western Marginal Plains and the western Upper Plain, as well as toward the interior of the Peninsula. On the basis of their morphological appearance these mountains resemble inselbergs, or island mountains, and are believed to be old islands. These sharp angular mountains are always of permocarboniferous limestone. Granitic mountains and those of quartzite have more gentle slopes; they are often scattered like inselbergs over the alluvial plain.

3. Northeast or Khorat Plateau

The saucer-shaped Northeast Plateau, is approximately 70,000 square km in extent. It is bordered on the north and east by the Khong River and mountains of Laos, on the west by the Phetchabun Mountains and Dong Phraya Yen and on the south by the San Kamphaeng and Dong Rek Scarps. These mountains and scarps are flat-topped sandstone formations, with higher elevations between 1000–1500 m above sea level. The highest peak of the Phetchabun Mountain, Phu Luang, reaches almost 1500 m, and that of the San Kampaeng Scarp, Khao Yai, reaches almost 1400 m. The interior of the region is undulating and dotted here and there with lower limestone hills and small shallow lakes. Large areas are flooded during the wet seasons, but the region suffers in the dry season for lack of water. The soils for the most part are thin and poor.

Almost the entire region is drained by a single river system, the Mun, which is a tributary of the Khong. Only on the northern and eastern edges are there a few small tributaries which drain directly into the Khong River. The rivers in this region generally flow through wide shallow valleys.

The high salt content of the soils of the northeast Plateau is due to the existing extensive salt pan in the substratum. Rising ground water in the summer rainy season dissolves salts and brings them up into the superficial layers. In the dry season an accumulation of salt develops at the surface of the soil through evaporation and by capillary rise of moisture from the deeper layers, forming crusts of salt on the surface in many localities.

4. The Southeast and Coast

This region is delimited on the north by the San Kam Phaeng Range, on the west by the Bang Pakong River and the east by a line of flat-topped hills, the Banthat Range, which marks the Thailand-Kampuchea border. The Southeast region is drained by numerous streams, most of which flow south. The principal rivers are the Bang Pakong, Chanthaburi, Pra Sae, Wen and Trat. There are a number of peaks in this region between 800 and 1800 m altitude.

The coast is fringed with rocky forested islands. The Sichang Island of Si Racha forms a natural anchorage for large ships to enter the Chao Phraya River. The largest island along the coast is Chang, between Chanthaburi and Trat, having an area about 90 square km and a peak that rises nearly 1000 m above sea level.

From southeast to northwest stretch the Cardamom mountains of Kampuchea, with peaks between 1100 and 1500 m. The Banthat Range, which marks the boundary between Thailand and Kampuchea, and the Chanthaburi Mountains, which reach an altitude of 1700 m, follow the same direction.

The arrangement and form of the separate mountains and ranges reflect the geological structure of the region. In Chanthaburi the higher peaks, such as Khao Soi Dao and Khao Sabap, are granitic batholiths. A low, anticlinal fold with a granitic core determines the morphology. In Rayong, separate mountains and ranges reach their greatest height in Khao Khieo, 800 m. Visible for a long distance, this peak is a landmark for navigators. This is a sinking coast, and numerous islands lying along it are peaks of a drowned landscape. All of them resemble the coastal areas.

5. Southwest or Western Mountains

This region consists of a sparsely inhabited strip of the Central Mountain range along the Burmese border, extending east to the plain of the Chao Phraya. It includes high rugged mountains in which streams have cut deep canyons and narrow valleys in the Thanon Thong Chai Range. The westernmost ridge of this range, the Tanao Si or Tenasserim, begins at Phra Chedi Sam Ong along the Thailand-Burma border and extends south along the western boundary into the Peninsula. A second ridge of bare limestone crags, the Mae Klong range, lies between Khwae Noi and Khwae Yai River Valley. The third ridge lies between the Khwae Yai and the Central Plain. A number of peaks rise above 1600 m, and a smaller number above 2000 m.

6. The Peninsula

Peninsular Thailand runs south about 800 km to the Thailand-Malaya border, varying in width from 15 to 200 km. Plains fringe the coastal areas, and highlands form the backbone of the Peninsula. Mountains extend from north to south and form short ridges. Physically and climatically, there is a distinct difference between the east coast and the west coast.

Along the western side of the Peninsula the Tanao Si (Tenasserim) Range continues to mark the Thailand-Burma border. South of Prachuap Khirikhan the Tanao Si levels off to the narrow coastal plain, and a further range begins southward, running from north-northeast to south-southwest. This range begins in a high mountain (about 1000 m) but is not continuous in elevation, being hardly 100 m in Chumphon. At the Pak Chan River it splits into two sections, a western range in Burma, and an eastern in Thailand. The eastern, known as the Phu Ket Range, extends south from the Kra Isthmus and follows the Indian Ocean into Phu Ket Island. There are numerous granite cores in the Phu Ket Range, many of them reaching more than 1100 m altitude. They are the source of the rich alluvial tin ores of the most important mining districts of Thailand.

An eastern range, the main range of the Peninsula, begins along the east coast at about 10°05′ north latitude on Tao Island. It continues through Phangan and Samui Island to the east coast mainland, east of Ban Don and parallels the coast all the way into Malaya. This is generally called the Nakhon Si Thammarat Range. Peaks in Phangan and Samui Islands rise to 685 m and 695 m respectively. Southeast of Ban Don, the peak reaches 1465 m, and majestic Khao Luang, west of Nakhon Si Thammarat, reaches 1750 m. Near its southern end, the peaks have an elevation of 650 to 750 m.

In general, the east coast of the Peninsula is smooth and regular, with few bays, but many long beaches, especially at

Nakhon Si Thammarat, Songkhla, and Pattani. A large inland sea, Thalesap, lies north of Songkhla. Off shores limestone islands are scattered.

The west coast shoreline is very irregular and much indented with estuaries. There are few beaches. The mountains extend down to the sea in many places. Mangrove swamps are numerous. The coast of Pak Chan River has the appearance of a drowned valley. Remains of buried mangrove trees have been exposed in the hydraulic mining many meters below the present sea levels and have been found along the shorelines of Takua Pa and Phu Ket.

B. Climate

Thailand falls into a subhumid tropical climate regime affected by the southwest and southeast monsoon, and an occasional typhoon from the South China Sea. The precipitation pattern of Thailand is varied. The northwestern Highlands, the Southwest and the Central Plain receive 1500 mm plus per year; the Northeast Plateau receives 1000 mm plus per year. The Southeast and the Peninsula receive two m plus per year, with the highest precipitation of four m plus in Trat and Ranong. Occasional storms are recorded, such as the typhoon which devastated forested lands in Chumphon during 19th century, and Surat Thani and Nakhon Si Thammarat in the present century.

The mean monthly temperature of most areas in Thailand ranges from 25°C for the coolest winter month (December or January) to 30°C for the warmest spring month (April or May). During the year the temperature seldom falls below 20°C and rarely rises above 35°C.

When the rainy season ends in October and November, the temperature declines considerably and shows a greater range in the North than in the South. The nights in the North may feel quite cool, although temperatures reach 25°–35°C during the daytime.

In February the cool season comes to an end, as air temperatures begin to rise with the increasing altitude of the sun. The real hot season commences in March; April and May bring the highest temperatures.

The temperature regime of the Northeast plateau is not very different from that of the Central Plain and the North, but in the cool season the overall temperatures are lower than those of the Central Plain. During the hot season they are somewhat higher.

In the Peninsula, temperature conditions are quite different from those of the rest of the Kingdom, owing to the short distances from the sea and the less distinct division of the year into dry and rainy seasons. Annual variation in temperature is very much lower. There is no cool season at all.

C. Population

The Thais are the major ethnic group in the country. The principal minorities are the Chinese, making up approximately 15 percent of the population, and the Thai-Malays of peninsular Thailand constitute about three per cent of the population.

According to the recent census, the population of Thailand is about 52,000,000 with a growth rate of three percent per year. The average density per square km is about 93 people. Areas of higher population density include Krung Thep, with about 130 people per square km. The population is predominantly rural (80%), with most people living in good-sized villages. Life expectancy was some 50–65 earlier this century, and will probably be close to 65–75 by the year 2000.

The influx of hill tribes along the northwestern and northeastern border of North Thailand, together with the Laotian and Cambodian refugees, also affects the population increase, even though the refugees are more or less temporary.

The population practices a conventional agriculture of rice, sugar cane, tapioca, maize and rubber cultivation; fruticulture is also very important, producing varieties of mangoes, durians, rambutans, longans, litchis, papayas and bananas. Variants of shifting cultivation are also widespread. Cash crops such as coconut, coffee, cocoa, oil palms and cotton are also grown. There are extensive mineral deposits being mined in various parts of the country, e.g., wolfram and fluorite in North Thailand, tin in the Peninsula and gem stones in the South East. Lignite in North and Peninsula Thailand, oil and natural gas in the Gulf of Thailand are currently exploited. This natural fuel is vital to the economic development. Logging, once a prosperous industry of extreme importance, is today declining owing to the opening up of vast areas of forested land for agriculture and other developments.

II. Vegetation

A. Tropical Rain Forest

This and other forest types are illustrated in the accompanying map. In the Chanthaburi or Southeastern and Peninsular Regions where contact with the monsoons is direct, the precipitation is very high (2500 mm up), and thus nearly the whole region is covered with this type of forest. After a careful study of these forests, two zones can be recognized, i.e. the Lower Tropical Rain Forest, and the Upper Tropical Rain Forest.

The Lower Tropical Rain Forests occupy the peneplains and hillslopes up to 600 m altitude. The forests are two-storied, with the upper story composed of gigantic trees, mostly Dipterocarps (*Dipterocarpus, Hopea, Shorea, Balanocarpus, Parashorea,* and *Anisoptera*), and others such as *Dyera, Endospermum, Horsfieldia, Melanorrhoea, Palaquium, Planchonella, Mangifera, Swintonia, Ailanthus, Cedrela, Artocarpus, Bischofia, Sandoricum, Tetrameles, Pterocymbium, Scaphium, Sterculia, Intsia, Mesua, Pterospermum, Schima, Cinnamomum, Calophyllum, Litsea, Alstonia, Ficus, Adenanthera, Koompassia, Lagerstroemia, Nephelium, Manglietia,* and *Podocarpus*.

The lower story is constituted of trees of medium height and girth of the genera *Vatica, Talauma, Baccaurea, Alchornia, Macaranga, Mallotus, Drypetes, Cleistanthus, Gloghidion, Croton, Cleidion, Sumbavia, Antidesma, Aporosa, Dichapetalum, Streblus, Syzygium, Phoebe, Alseodaphne, Aglaia, Garcinia, Memecylon, Polyathia, Mitrephora, Goniothalamus, Pseuduvaria, Orophea, Gluta, Semecarpus,* etc. Palms, such as *Orania, Oncosperma, Calamus, Korthalsia, Daemonorops,* and *Licuala,* are abundant.

Vines are also abundant such as *Bauhinia, Dalbergia, Milletia, Tetrastigma, Willughbeia, Aganosma, Poikiolospermum, Trachyspermum, Epigynum, Derris,* and *Entada.*

Bamboos are common in disturbed areas, belonging to the genera *Gigantochloa, Bambusa, Dinochloa, Schizostachys,* and *Dendrocalamus.*

The Upper Tropical Rain Forests occupy the slopes of 600–900 m. altitude, and are also two-storied. The upper story is represented by a great proportion of oaks and chestnuts (*Quercus, Lithocarpus,* and *Castanopsis*), interspersed with *Magnolia, Michelia, Syzygium, Pentace, Dipterocarpus costatus, D. grandiflorus, Myristica, Canarium* and *Podocarpus.* The lower story is composed of *Antidesma, Aglaia, Baccaurea, Glochidion.* Palms of the genera *Areca, Pinanga, Calamus,* and *Daemonorops* are abundant. A tree fern of the genus *Cyathea* is also present. Undergrowth is dense and mostly composed of Melastomataceae, Acanthaceae, and Zingiberaceae, with a great number of terrestrial ferns and orchids. Climbers are few and scattered, but epiphytes are abundant. Trees are heavily covered with mosses, ferns and orchids. This is a transitional area between the Hill Evergreen or Lower Montane Forest, which starts from the elevation of 1000 m upwards.

B. Dry or Semi-evergreen Forest

This type of forest is scattered all over the country along the depressions on the peneplain, along the valleys of low hill ranges of about 500 m. elevation, or forming galleries along streams and rivulets. The annual precipitation is between 1000–2000 mm.

Except the valleys of low hill ranges, the present dry evergreen forests are remnants of the once luxuriant and extensive forests covering the peneplains of the Central Region of the Chao Phraya Plain, and parts of the Korat Plateau or the Northeastern Region.

The forests are three-storied. The upper story consists of *Antisoptera costata, Dipterocarpus alatus, D. turbinatus, Hopea odorata, H. ferrea, Shorea thorelii, Alstonia scholaris, Pterocymbium tinctorium, Tetrameles nudiflora, Afzelia xylocarpa, Ailanthus triphysa, Ulmus lanceifolius, Antiaris toxicaria, Lagerstroemia ovalifolia,* and *Acrocarpus fraxinifolius.* The middle story is composed of *Cratoxylum maingayi, Chaetocarpus castanicarpus, Castanopsis nepheloides, Euphoria longana, Lithocarpus harmandii, Spondias pinnata, Cinnamomum iners, Irvingia malayana, Vatica cinerea, Sapium insigne* and *Diospyros* spp. The lower story is represented by tree species of smaller stature of the genera *Memecylon, Cleistanthus, Aporusa, Alchornia, Baccaurea, Macaranga, Mallotus, Knema, Melodorum, Mitrephora, Tarenna, Dillenia, Crataeva,* and *Niebuhria.*

Palms of the genera *Calamus, Areca, Livistona, Corypha,* and *Raphis* are scattered. Bamboos (*Gigantochloa, Bambusa, Dendrocalamus*) are sparsely present. Lianes, however, are abundant, belonging to the genera *Bauhinia, Dalbergia, Derris, Entada, Strychnos, Securidaca, Toddalia, Acacia, Hymenopyramis, Congea, Sphenodesme, Uncaria, Ventilago, Tetrastigma, Artabotrys, Desmos, Uvaria,* and *Pisonia.* Strangling figs are also frequent. Epiphytes, mainly orchids and ferns, are sporadic.

The undergrowth is dense and composed of members of the family Zingiberaceae belonging to *Curcuma, Boesenbergia, Alpinia, Catimbium, Cenolophon* and *Amomum*; others are *Tacca, Strobilanthes, Micromelum, Clausena, Barleria, Flemingia, Desmodium, Lourea, Capparis,* and ferns of the genera *Helminthostachys, Lygodium,* and *Thelypteris.*

C. Hill Evergreen Forest

Hill Evergreen Forest occurs in the upper elevations from 1000 m upwards, and appears discontinuously all over the country, with a larger percentage in the Northwestern Highland. This type of forest is also known as Temperate Evergreen Forest or Lower Montane Forest by some authors. The forest is two storied, and dominated by the oaks, chestnuts, laurels, magnolias, teas, and rhododendrons; gymnospermous elements are also present, such as *Podocarpus, Dacrydium, Cephalotaxus, Gnetum,* and *Cycas.* The soil is either red granitic, brown-black calcareous, or yellow-brown sandy. The humidity is very high, as explained by the moss-clad trees; the precipitation is 1500–2000 mm annually.

The upper story is represented by *Schima wallichii, Cinnamomum* spp., *Fraxinus excelsa, Dacrydium elatum, Podocarpus imbricatus, Cephalotaxus griffithii, Betula alnoides, Ulmus lancifolia, Cedrela toona, Nyssa javanica, Quercus, Lithocarpus, Castanopsis,* and *Calophyllum.* The second layer is composed of small trees of medium height and girth, such as *Gordonia, Camellia, Pyrenaria, Acer, Carya, Carpinus, Tristania, Sladenia celastrifolia, Symingtonia populifera, Notophoebe, Alseodaphne, Lindera, Phoebe, Helicia, Macaranga, Mallotus, Rhododendron, Symplocos,* and *Aquilaria.*

Shrubs are also abundant, belonging to: *Daphne, Wikstroemia, Melastoma, Osbeckia, Embelia, Maesa, Rapanea, Rhamnus, Cornus,* and *Osyris.* Herbaceous species form a rich ground flora and are represented by *Catimbium, Boesenbergia, Curcuma, Globba, Hedychium, Strobilanthes, Rungia, Asystasia, Calanthe, Phajus, Malaxis, Liparis, Habenaria, Anoectochilus, Anthogonium, Pollia, Forrestia, Streptolirion,* and *Ophiorrhiza.* Bamboos are *Teinostachys, Dinochloa, Gigantochloa,* and *Schizostachys.* Palms are relatively few (*Trachyspermum speciosum* and members of *Pinanga* and *Phoenix*).

Ferns are richly represented, including species of *Athyrium, Asplenium, Leptochilus, Polypodium, Phyma-*

todes, Thelypteris, Nephrolepis, Plagiogyria, Blechnum, Cyathea and *Osmunda*. Sphagnums are found in the boggy areas of high altitude. Epiphytes are abundant and found festooning trees; besides mosses and lichens, there are ferns of the genera *Drynaria, Asplenium, Humata, Davallia, Davallodes, Pyrrhosia* and epiphytic orchids of the genera *Dendrobium, Eria, Porpax, Bulbophyllum, Drymoda, Pleione, Coelogyne, Neogyne, Oberonia, Cymbidium, Gastrochilus,* and *Vanda*.

In the region where the summits and ridges are open and exposed, the vegetation is sparse, recalling a subalpine nature. Here are found *Primula, Kalanchoe, Sedum, Saxifraga, Circaea, Spiraea, Pedicularis, Gentiana, Parnassia, Viola, Sophora, Bupleurum, Seseli, Selinum, Heracleum, Cotoneaster, Geranium, Rhododendron, Iris, Lilium* and *Asparagus*.

D. Coniferous Forest

This type of forest is scattered in small pockets in the Northwest Highland and the Khorat Plateau of about 200–1300 m elevation, where poor acid soil occurs. The soil is either grayish sandy, or brownish gravelly and sometimes lateritic. The annual rainfall is about 1000–1500 mm.

The forest is 3-storied and is rather open in nature, in certain localities where forest fire is concurrent, the forest grades into a savanna. The upper story is composed of *Pinus kesiya* and *P. merkusii*, in certain localities where lateric soil is evident, *Dipterocarpus obtusifolius* and *D. tuberculatus* also come in, to form a *Pinus-Dipterocarpus* association.

The second story consists of saplings of upper story species, together with *Anneslea fragans, Quercus, Lithocarpus, Castanopsis, Styrax aprica, Myrica farquahariana* and *Tristania rufescens*; the lowest story is formed by small trees and tall shrubs such as *Adinandra, Embelia, Maesa, Phoenix humilis, Cycas pectinata, Vaccinium sprengelii, V. bracteatum, Rhododendron moulmeinense, R. lyi, Baeckia frutescens* and *Styrax rugosus*.

E. Swamp Forest

Along the depressions in the low-lying land, the estuaries and the muddy shores, a unique type of vegetation is developed. This type of vegetation is more or less subjected to occasional inundation, and is scattered in the wet region of the country, where the annual precipitation is high (2000 mm up).

The forest can be physiographically classified into two kinds: 1) the Fresh Water Swamp forest and 2) the Mangrove Swamp forest.

1. Fresh Water Swamp forest

This type of forest is usually found along depressions inland, the soil is either alluvial or sandy. If it is on alluvial soil the ground floor is muddy and fen-like. Whenever peat deposits occur, the ground develops into a domed bog and a sub-type vegetation top story is composed of trees of large dimension such as *Dyera costulata, Palaquium gutta, Koompassia malaccensis, Calophyllum teysmannii* and *Scaphium lychnophorum*. The second story is composed of *Nephelium lappaceum, Hydnocarpus sumatranus, Walsura trijuga, Hopea latifolia, H. pierrei, Cratoxylon arborescens, Heritiera littoralis, Ploiarium alternifolium,* and *Xanthophyllum glaucum*. The lowest story consists of *Aglaia, Gluta, Casearia, Melaleuca leucadendra* and *Alstonia spathulata*. The ground flora is very poor and represented by *Apostasia, Boesenbergia, Hanguana, Hedychium, Horntstedtia, Costus, Dipodium, Bromheadia, Forrestia, Donax, Schumanianthus,* and *Nepenthes*.

Palms are very common, consisting of many species of thorny climing rattans of the genera *Calamus, Korthalsia, Plectocomia,* and *Daemonorops*, together with *Areca triandra, Licuala, Nenga, Pinanga, Cyrtostachys lacca* and *Oncosperma tigillaria*. Epiphytes are rich in ferns and orchids.

Where the soil is sandy, the composition of the forest is quite different, more open and the trees rather stunted. *Fagraea fragrans* forms a majority in this type of forest with the bushy *Baeckia frutescens* and *Licuala* palms as associates; in some localities, *Melaleuca leucadendra* occurs in a pure stand of one-story formation.

The much disturbed area in this type of forest forms a peculiar secondary growth, called in Malaysia "Belukar." This secondary growth is composed of stunted trees and shrubs such as *Melaleuca leucadendra, Syzygium grande, Flagellaria volubilis, Derris elliptica* and *Spirolobium cambodianum*.

2. Mangrove Swamp forest

Mangrove swamps are to be found along the estuaries of rivers and the muddy coastlines, where the soils are deep alluvial deposits with a high saline content. The forest is periodically tidally inundated. In Thailand this type of forest is very extensive on the West coast from Satun to Ranong, and within the Gulf of Thailand from Samut Sakhon in the Southwest to Trat in the Southeast.

Species inhabiting this type of forest are semi-xerophytic, even though they are occasionally submerged by sea water which they cannot use. These plants store fresh water in their characteristic thick leather-like leaves.

The stand is uniform and forms a one-storied profile. At a close study the forest will reveal zonal formations. The most outlying region facing the sea or ocean can be called the *Avicennia-Sonneratia* zone, composed of *Avicennia officinalis, A. marina, Sonneratia griffithii, S. alba*, and *S. caseolaris*. After this outlying formation on the *Rhizophora* zone appears, mainly constituted by *Rhizophora mucronata* and *R. apiculata*. The inner zone can be called *Bruguiera-Kandelia-Ceriops* and is composed of *Bruguiera parviflora, B. caryophylloides, B. hainesii, Kandelia rheedii, Ceriops tagal* and *C. roxburghiana*. The inner-most region is a mixed formation of *Lumnitzera-Xylocarpus-Bruguiera* consisting of *Lumnitzera coccinea, L. racemosa, Bruguiera gymnorrhiza, B. eriopetala, Xylocarpus obovatus* and *X. granatum*. *Rhizophora apiculata* also occasionally occurs in this zone. Along the creeks *Heritiera littoralis* is also frequent.

Along the estuaries where there is shelter from winds and waves, the outer-most zone is fringed by *Rhizophora mucronata* and *R. apiculata*. In all these zones the ground flora is very poor and is represented by *Acanthus ilicifolius, A. ebracteatus, Derris trifoliata, Acrostichum aureum* and *A. speciosum*, together with *Aegiceras corniculata* and *Scyphiphora hydrophyllacea*.

F. The Deciduous Forests

Along the dry belt of the country, where precipitation is low (under 1000 mm) the climate is more seasonal, and the soil is either sandy or gravelly loam and sometimes lateritic. The vegetation here is classified as deciduous and tree species shed their leaves during the dry season. Trees growing in this forest type tend to develop growth or annual rings. The height of predominant trees is comparatively lower (20-25 m) than that of the evergreen forest. This forest is more or less subject to ground fire during the dry season.

Deciduous forests can be sub-divided into three main categories: Mixed Deciduous Forest, Dry Deciduous Dipterocarp Forest, and Savanna Forest.

1. Mixed Deciduous Forest

This type of forest is composed of all deciduous species in a good proportion but in certain localities a species such as teak (*Tectona grandis*) may become predominant and the area would generally be called a Teak forest for convenience.

The mixed deciduous forest condition also can be classified into three kinds, based on the terrain and climate: Moist Upper Mixed Deciduous, Dry Upper Mixed Deciduous, and Lower Mixed Deciduous.

a. Moist Upper Mixed Deciduous Forest. This type of forest occurs between the elevations of 300-600 m altitude, and is 3-storied in profile. The soil in this forest is usually loamy, either calcareous or granitic. The upper story consists of *Tectona grandis, Lagerstroemia tomentosa, Terminalia alata, T. calamansanai, T. bellirica, Afzelia xylocarpa, Xylia kerrii, Bombax insigne, Pterocarpus macrocarpus, Dalbergia cultrata, D. oliveri, Adina cordifloia, Gmelina arborea, Anogeissus acuminata, Millettia leucantha, Albizia lebbeck, A. procera, A. lebbekiodes, A. chinensis, Acacia leucophloea, Adenanthera pavonina* and *Dillenia pentagyna*. The second story consists of *Combretum quadrangulare, Careya arborea, Barringtonia racemosa, Millettia brandisiana, Albizia lucida, Dalbergia ovata, D. nigrescens, Peltophorum dasyrachis, Lagerstroemia floribunda, L. speciosa, L. macrocarpa, L. villosa, L. undulata, Diospyros mollis, D. montana, Syzygium cumini, S. leptanthum, Vitex peduncularis, V. canescens, V. pinnata,* and *Dillenia aurea*.

The lowest story is composed of *Cratoxylon formosum, Mallotus philippinensis, Gardenia coronaria, G. obtusifolia, Casearia grewiaefolia, Bauhinia racemosa, B. malabarica, Croton oblongifolius* and *C. hutchinsonianum*.

A small number of palms such as *Phoenix humilis* and some species of *Calamus* may be found scattered in this type of forest. Shrubs are represented by species of *Croton, Mallotus, Premna,* and *Randia, Harrisonia perforata, Bauhinia acuminata,* and many others. Lianes such as *Hymenopyramis brachiata, Congea tomentosa, Artabotrys siamensis, Desmos chinensis, Bauhinia bracteata, B. scandens, Butea superba, Spatholobus parviflorus,* and *Dalbergia rimosa* are also scattered in this forest.

The ground flora is composed of herbaceous species such as grasses of the genera *Capillipedium, Sporobolus, Themeda, Thysanolaena, Andropogon, Bothriochloa, Saccharum, Oryza, Eragrostis* and *Hyparrhenia*; others are *Kaempferia, Curcuma, Boesenbergia, Fimbristylis, Carex, Cyperus, Ceropegia, Aristolochia, Habenaria, Peristylus, Pecteilis,* and *Brachycorythis*.

b. Dry Upper Mixed Deciduous Forest. Along the ridges at the elevations of 300-500 m altitude, the vegetation becomes more open due to the evaporation, exposure, surface erosion and the leaching of organic components from the soil. The forest is also three storied. Some species occurring also in the Moist Upper Mixed Deciduous Forest are present but here they are rather stunted and crooked. The more pronouncedly deciduous species such as *Shorea obtusa, S. siamensis, Dipterocarpus tuberculatus, D. obtusifolius* and *D. intrincatus* are only lightly represented. The soil is either sandy loam or lateritic. The ground flora is frequently destroyed by fire. This type of forest, especially when constantly disturbed by human beings, will degrade into a bamboo sward which sometimes covers quite an extensive area. The main bamboo species are *Bambusa arundinacea* and *Thyrsostachys siamensis*.

c. Lower Mixed Deciduous Forest. This forest type occurs on low-lying country at 50-300 m altitude in the dry zone where the soil is either sandy loam or lateritic. The forest is three-storied. The absence of teak (*Tectona grandis*) from the upper story is a distinct characteristic, differentiating the Lower from the Upper Mixed Deciduous Forest.

Hopea odorata, H. ferrea and *Shorea roxburghii* are sometimes found in the upper story. Along waterways, semievergreen species such as *Syzygium cumini, Hopea odorata, Sapium insigne, Afzelia xylocarpa* and *Dipterocarpus alatus* form a narrow strip on both banks and are generally known as Gallery forest.

Because of its accessibility and the number of commercially valuable species it contains, this type of forest is second only to teak forest in economic importance. Unfortunately, its natural assets and the fact that it is usually found in areas suitable for agriculture, brings it close to the verge of total destruction.

2. Dry Deciduous Dipterocarp Forest

On the undulating peneplain and ridges, where the soil is either sandy or lateritic, and subjected to extreme leaching, erosion and annual burning, the vegetation is markedly changed into a subclimax type. The predominant species belong to the Dipterocarpaceae. The forest is rather open and can be considered as two-storied. The upper story consists of *Dipterocarpus obtusifolius, D. tuberculatus, Shorea obtusa, S. siamensis, Quercus kerrii,* and sometimes *Pterocarpus macrocarpus* and *Xylia kerrii.* Generally, the height of trees of the upper story is between 20–25 m but only 15–20 m in more arid areas. The second story is composed of low shrubby trees such as *Strychnos nux-vomica, S. nux-blanda, Dalbergia kerrii, Symplocos cochinchinensis, Diospyros ehretioides, Aporusa villosa, Phyllanthus emblica* and *Canarium subulatum.*

The ground flora consists of tuber and rootstock-bearing species, due to the effect of the fires, such as small bamboos of the species *Arundinaria pusilla, A. ciliata, Linostoma persimilis, Enkleia malaccensis, Phoenix acaulis, Pygmaeopremna herbacea* and numbers of the genera *Habenaria, Pecteilis, Hibiscus, Decaschistia, Kaempferia,* and *Curcuma. Dillenia hookeri* is common, forming clumps of low bushes.

Epiphytes are relatively abundant and mostly consist of ferns belonging to the genera *Drynaria, Platycerium* and *Pyrrosia,* and orchids of the genera *Aerides, Eria, Dendrobium, Bullophyllum, Cleisostoma* and *Ascocentrum. Dischidia rafflesiana, Dischidia minor, Hoya pachyclada,* and *Hoya kerrii,* are also common.

G. Savanna Forest

Savanna may be called the most extreme form of deciduous forest and is the result of fire. It is more frequent in the Northeastern region which has been cultivated from time immemorial. The soil is either sandy or lateritic. Precipitation is relatively low (50–500 mm). Small patches of savanna at different stages are scattered all over the region. The most extensive savanna, Thung Kula Rong Hai in Sakon Nakhon, has become a vast, desolate land.

Savanna forest is actually a grassland where trees of medium height grow, forming a very open stand. Besides trees, thorny shrubs such as *Feroniella lucida,* and *Carissa cochinchinensis* are interspersed with *Bambusa arundinacea.* In upper elevations, shrubs belonging to the genera *Aporusa, Ochna* and *Glochidion* are frequent. Tree species found in the savanna forest such as *Careya arborea, Mitragyna parvifolia, Acacia siamensis, A. catechu* and *Pterocarpus macrocarpus,* are fire resistant.

The grassland is composed of *Imperata* and *Vetiveria,* together with *Eulalia, Panicum, Sporobolus, Themeda, Eriochloa, Eremochloa* and *Sorghum.*

III. Vegetation Maps

The latest comprehensive vegetation map for Thailand in its entirety is that presented by the Royal Forest Department in 1982, based on Landsat Remote Sensing at a scale of 1:1,000,000 and showing considerable detail.

IV. Magnitude of Floristic Inventory

A. Specimens Collected

The collection of plant specimens in Thailand commenced in 1778 and since then a total of 116,000 specimens of seed plants has been collected for the country. Collection continues to increase as current botanical explorations are carried on by either Thai botanists or foreign ones. Duplicates of about half of the collections have been distributed to the main herbaria of the world, namely Royal Botanic Gardens, Kew (K); British Museum (Natural History), London (BM); Royal Botanic Gardens, Edinburgh (E); Botanical Museum of the University in Copenhagen (C); Rijksherbarium, Leiden (L); National Museum of Natural History, Paris (P); Kyoto University (KYO); Arnold Arboretum, Harvard (A); and U.S. National Museum, Washington (US).

Collections exceeding 400 specimens made prior to 1932 are described in the following paragraphs.

J. G. Koenig, a Danish botanist, collected between 1778–1779 in Ayutthaya, Prachin Buri, and Phu Ket. There is no record of the number since the collection was lost at sea, but he undoubtedly collected quite a number of herbarium specimens.

Arthur Keith, a British medical officer to the Gold Fields of Siam Limited, came to Thailand in 1889 and was stationed in Bang Saphan, Prachuap Khiri Khan. During the period of 1889–1891 he collected about 500 plants in Bang Saphan and its neighborhood. The collection is at the Singapore Herbarium (SING).

E. J. Schmidt, a Danish botanist later turned marine biologist, collected during 1899–1900 in Chanthaburi, Trat and particularly in Koh Chang, a group of islands offshore on the Thai-Kampuchea border. His collection of about 1300 numbers is being kept in C.

D. T. Gwynne-Vaughan, a British botanist joining the Skeat Expedition in 1899, collected in peninsular Thailand. His collection of about 450 numbers is deposited at K and Cambridge (CGE).

C. C. Hosseus, a German botanist, collected mostly in North Thailand between 1904–1905. His collection of about 830 numbers is deposited at Munich (M) (Germany), with duplicates in other European herbaria, namely K and E.

A. F. G. Kerr, originally a medical doctor in the service of the Royal Thai Government who later became the Government botanist, collected extensively (over 20,000 numbers) in Thailand between 1902–1932. The complete set is deposited at K and BM; also quite a number of duplicates are distributed to main herbaria of the world. The collection in the Department of Agriculture, Bangkok (BK) was intended to be the complete second set, but the early numbers were lost at sea during World War I.

H. B. G. Garrett, a British citizen who served as a forest officer under the Royal Thai Government, collected in North Thailand between 1899–1959. His collection consisted of

1,500 numbers, mainly kept in BM, K, and Bangkok (BKF), with a few duplicates in European herbaria.

Mrs. D. J. Collins, a British naturalist, collected chiefly at Chon Buri (Si Racha) and Chanthaburi 1902–1938. Her numbers ran to 2501 and were deposited in BM.

J. F. Rock, an American, made three collecting trips in North Thailand in 1919, 1920 and 1921. His collection of 1912 numbers was deposited at US.

H. S. H. Prince Lakshnakara Kasemsant, a Thai agricultural botanist, collected in central, Northeast, and peninsular Thailand during 1925–1932. His collection of about 1526 numbers was deposited at the Department of Agriculture (BK), with duplicates in BM and K.

A. Marcan, a British chemist serving under the Royal Thai Government as Chief Assayer to the Royal Mint and later as Director of Government Laboratory, collected during 1919–1931, running to 2814 numbers. The collection was distribtued to BM, SING and California (UC) (Grasses), with some duplicates in BK.

Put Phraisurind, a Thai assistant of Kerr, collected all over Thailand during 1920–1932. His collecting number ran to 4548 which were deposited at BK and K. He also collected for the Forest Herbarium (BKF) between 1933–1936, with a total number of about 500.

Mrs. Eryl Smith, a British medical doctor, collected mainly ferns in peninsular Thailand during 1922. Her collections, numbering 1948, were deposited in BM, with some duplicates in BKF and SING.

Phraya Vanprueck Phicharn (Vanpruk), a Thai forest officer, collected about 1200 numbers mainly in northern and peninsular Thailand during 1912–1920. The collection is deposited in BKF, with duplicates in K.

To conclude this period B. Hayata, a Japanese botanist collected in northern Thailand, with about 1000 numbers credited to his collection, which is kept at Tokyo (TI).

The collections made from 1932 onwards, with collecting numbers exceeding 400, are described in the following paragraphs.

Gunnar Seidenfaden, a Danish botanical student, collected in Chanthaburi and Surat Thani over a period of two months during 1934–1935, for a total of about 550. Later he became a diplomat and was posted the Danish Minister Plenipotentiary, and eventually became the Danish Ambassador residing in Bangkok. During his diplomatic career in Thailand he devoted his spare time to the study of the orchids of Thailand, with the collaboration of Tem Smitinand. A series of collecting trips were made during 1955–1973 in various parts of Thailand. The spirit collection and collection of living orchids, totalling over 9000 numbers, are being kept in Copenhagen.

Khid Suvarnasuddhi, a Thai forest officer, collected about 1000 numbers in North Thailand during 1936–1942. His collection is kept at BKF with some duplicates in K and Paris (P).

In 1946 the Kwae Noi Basin Expedition took place. The expedition was led by S. Bloembergen, with G. Den Hoed, and A. J. G. H. Kostermans, Dutch scientists, accompanied by Kasin Suvatabhandu, a Thai botanist. The collection of some 1200 numbers was made in Kanchanaburi and was deposited at Leiden (L), and BK.

Tem Smitinand, a Thai forest botanist, collected extensively from 1947 onwards with collections numbering more than 12,000. His collection is in BKF, with duplicates in main European and American herbaria. During 1947–1977 he also organized a group of Thai collectors to botanically explore various parts of Thailand, namely: Dee Bunpheng (ca. 1600 no.), Din Nakkan (ca. 1200 no.), Chit Nuphakdee (ca. 800 no.), Ploenchit Suvarnakoset (ca. 1800 no.), Bunnak Sangkhachand (ca. 3000 no.), Bunchu Nimanong (ca. 1000 no.), Sanan Phengnaren (ca. 1000 no.), Damrongsak Praphat (ca. 1200 no.), Sanoh Phengnaren (ca. 800 no.), Sanan Thaworn (ca. 1300 no.), S. Phusomsaeng (ca. 1000 no.), and Khanthachai Bunchai (ca. 2000 no.).

Besides these plant collectors, a number of Thai botanists attached to the Forest Herbarium who also actively collected there were Chamlong Phengkhlai (ca. 4500 no.), Thawatchai Santisuk (ca. 1600 no.), Chawalit Niyomdham (ca. 1500 no.), Leena Phuphatanapong (ca. 500 no.), and Wirachai Nanakhorn (ca. 1200 no.).

Llewelyn Williams, an American botanist, came to investigate the latex-bearing trees of Thailand in 1950. He made a three-month trip in northern, southeastern and peninsular Thailand, accompanied by Tem Smitinand. The collection of 500 numbers was deposited at BKF.

Ernest C. and Lucy Abbe, in their course of study on the Fagaceae of Southeast Asia, came to Thailand in 1956, and made extensive excursions in Thailand, accompanied by Tem Smitinand. The collections ran to 1000 numbers and was deposited in BKF and the Herbarium of the Missouri Botanical Gardens (MO).

Ingrid Alsterlund, a Swedish botanist, came to collect orchids for the University of Göteborg in 1967. She collected about 400 living orchids in Doi Inthanon and Doi Chiang Dao, Chiang Mai, northern Thailand. The collection is in Göteborg, Sweden (GB).

The Thai-Danish Botanical studies commenced in 1957, and two eminent Danish botanists, Professor Kai Gram and Dr. C. Syrach Larsen, came to investigate the possibility of conducting scientific research in Thailand and to prepare for the forthcoming expeditions. During the period of January–March, accompanied by Tem Smitinand, Gunnar Seidenfaden and Kasin Suvatabhandu, they travelled to northern and peninsular Thailand, collecting about 400 herbarium specimens, deposited in C.

The actual Thai-Danish botanical expeditions started from 1958 and ended in 1968. Danish botanists engaged in these expeditions were Thorvald Soerensen, Kai Larsen, Bertel Hansen, and E. Warncke. During the ten-year period they carried out eight expeditions in collaboration with Thai botanists from the Forest Herbarium. They made extensive excursions in Thailand with the total collecting number amounting to some 30,000.

In 1963 the UNESCO training expedition, led by T. Smitinand and H. Sleumer, collected about 500 numbers in

Chiang Mai, Loei, Saraburi and Surat Thani, mostly in limestone hills. The duplicates are deposited at BKF, L, and SING.

The Thai-Dutch botanical expeditions started in 1965 and ended in 1975. A. Touw and H. Hennipman of the Rijksherbarium, Leiden (L) accompanied by Chamlong Phengkhlai of the BKF collected in northern and northeastern Thailand. The collection, mostly mosses and ferns, amounted to 5000 numbers and were deposited in BKF and L.

After Touw and Hennipman the expeditions were carried on by H. P. Nooteboom (1968), F. van Beusekom (1968–1969) and R. Geesink (1971–1975). The collections of these successive expeditions amounted to 8700 numbers.

The Thai-Japanese botanical activity commenced in 1958 after the ninth Pacific Science Congress. T. Tuyama collected about 700 numbers in northern Thailand. A. Kira of the Osaka University led an ecological study team and with the collaboration of the Kasetsart University collected in northern, central and peninsular Thailand. The team collected some 3000 numbers of mostly sterile material. These collections are deposited in Tokyo (TI) and Osaka (OSA) respectively.

From 1965 onwards botanical activity was aimed at contributing to the *Flora of Thailand* project. The first expedition was led by M. Tagawa and K. Iwatsuki, of the University of Kyoto, in collaboration with the Forest Herbarium, Bangkok. The activity is still currently being carried on with fruitful results and the participation of other botanists, namely G. Murata, H. Koyama, T. Shimizu and N. Fukuoka. The collections currently amounted to about 120,000 numbers, mainly deposited in Kyoto (KYO and BKF).

It is worthwhile to mention that a number of Thai botanists attached to the Department of Agriculture, Bangkok also made some collections, namely Jarey Sadakorn (700), Chirayuphin Chandraprasong (2300), Prayat Sangkhachand (2000), Sakon Suthisorn (4500), Umphai Youngboonkird (500) and J. F. Maxwell (ca. 4000). The collections are kept in BK.

V. Publications on the Flora

Publications on the Flora of Thailand fall into three periods: pre 1920, 1920–1950, and post 1950.

A. Pre 1920

J. G. Koenig was the first botanist to make cursory observations on the Thai Flora. His *Chloris Siamensis*, a list of Thai plants, together with his diary, was translated into English and published more than 200 years after his death.

F. N. Williams (1904–1905) compiled and published a list of plants known from Thailand.

E. J. Schmidt edited the *Flora of Koh Chang* (1910–1916), which covered the whole plant kingdom, and was contributed to by a number of botanists of different nationalities.

A. F. G. Kerr published the result of his study, *Flora of Doi Suthep*, (1910), in the "Bulletin of Miscellaneous Information,: Kew ("Kew Bulletin").

C. C. Hosseus published the result of his study on Thai plants. These were *Die aus Siam bekannten Acanthaceen* (1907), *Beitrage zur Flora Siams* (1910), and *Die botanischen Ergebniss meiner Expedition nach Sian* (1911).

W. G. Craib, having studied Thai plants since 1907, published the result of his study, the List of Siamese plants with descriptions of new species (1911–1912) in the "Kew Bulletin," and later *Contributions to the Flora of Siam (Dicotyledones)* 1912, and *Contributions to the Flora of Siam (Monocotyledones)* 1913, in the "Aberdeen University Studies." These latter publications are the forerunners of a series of *Contributions to the Flora of Siam*. Additamenta were carried on by himself and his colleagues until 1938.

H. N. Ridley, after his botanical excursions in peninsular Thailand, published *An Account of a Botanical Expedition to Lower Siam* (1912); *The Plants of Koh Samui and Koh Pennan* (1915), and *On a Collection of Plants from Peninsular Siam* (1920).

B. 1920–1950

During this period the activity of the study on the Flora of Thailand was accelerated. The leading figure was W. G. Craib, who started to edit *Florae Siamensis Enumeratio*, which was carried on after his death in 1934 by A. F. G. Kerr. This preliminary enumeration consists of three volumes, the last issue, (number three of volume three) was published in 1962 and terminated in the family Gesneriaceae, edited by E. C. Barnett.

Among Craib's colleagues, H. R. Fletcher actively made studies on Apocynaceae, Sapotaceae, Sarcospermaceae, Styraceae, Myrsinaceae, Ebenaceae, Symplocaceae, Boraginaceae (With A. F. G. Kerr) and Verbenaceae; D. G. Downie studied the orchids and published her work in 1925, based on a manuscript left by Rolfe.

E. C. Barnett studied the Fagaceae of Thailand and partly published her thesis in 1942; E. T. Geddes studied the Rubiaceae and published her result in 1927; J. B. Imlay studied the Acanthaceae and partly published her thesis in 1939.

A. F. G. Kerr continued to study the Thai flora, and edited the *Flora Siamensis Enumeratio*, succeeding W. G. Craib. He did not live to complete this work. His manuscript of Volume Three was later edited by R. L. Pendleton (1951–1954) and E. C. Barnett (1962). He studied the Compositae, Ericaceae, Oleaceae, Asclepiadaceae, Loganiaceae, Hydrophyllaceae, Boraginaceae (with H. R. Fletcher), Convolvulaceae, Solanaceae and Scrophulariaceae. He also studied the orchids, keeping separate serial numbers of his collection, and published part of it in 1927.

Phya Vanaphruek Phichan published the *List of Common Trees and Shrubs of Siam* (1923).

H. R. Fletcher (1938) published "The Siamese Verbenaceae" in "Kew Bulletin" (1938) and published "Keys to the Siamese species of Myrsinaceae" in "Notes from the Royal Botanic Gardens, Edinburgh 20(48): 106–120. 1948.

The Royal Forest Department published the *Thai Plant Names, Botanical Names-Vernacular Names* (1948).

C. Post 1950

After World War II Thai botanical activity was revived and accelerated. The collections of Thai botanists began to accumulate and at the same time the rich and less known flora of Thailand attracted foreign botanists, resulting in many joint expeditions (e.g., Thai-Danish, Thai-Dutch and Thai-Japanese) as noted above.

The Thai-Danish botanical studies (1961–1969), by a number of botanists, edited by K. Larsen, were published in the "Dansk Botanisk Arkiv," covering various groups of plants. Gunnar Seidenfaden and T. Smitinand published *Orchids of Thailand*, a preliminary list (1959–1964). Seidenfaden then started the series of "Contributions to the Orchids of Thailand," published in "Dansk Botanisk Tiddsskrift" (1969–1977); and the series "Orchid Genera in Thailand," published in "Dansk Botanisk Arkiv" (1975–1989).

E. Nelmes published "The Genus *Carex* in Indo-China, including Thailand and Lower Burma" (1955) in "Memoire du Museum Nationale d'Histoire Naturelle" (Paris). M. Raymond published "Carices Indochinensis nec non Siamese" (1959) in "Memoire du Jardin Botanique," Montreal. J. Kern published "Cyperaceae of Thailand (Excl. *Carex*)" in "Reinwardtia" (1961–1962).

E. C. Barnett published "New species of the Gesneriaceae from Thailand" in the "Natural History Bulletin of Siam Society" (1961).

R. Heim published "Contribution a la flore mycologique de la Thailande I." in "Revue Mycologique" (1962).

C. Grey-Wilson published "New plants record from Thailand" in "Kew Bulletin" 26 (1971).

H. K. Airy-Shaw (1971) published "The Euphorbiaceae of Siam" in "Kew Bulletin" 26.

The *Flora of Thailand*, with T. Smitinand and K. Larsen as editors, commenced in 1970 and continues today. The *Flora* started with Volume Two, completed in 1984. Volume One, (the introduction), is still postponed. Volume Two is devoted to various families, contributed by foreign and Thai botanists.

Volume Three, devoted to pteridophytes, was prepared by the Japanese pteridologists, M. Tagawa and K. Iwatsuki. Part one (1979), comprises the Psilotaceae, Lycopodiaceae, Selaginellaceae, Isoetaceae, Equisetaceae, Ophioglossaceae, Marattiaceae, Plagiogyriaceae, Gleicheniaceae, Schizaeaceae, Hymenophyllaceae, Cyatheaceae, Dicksoniaceae and Dennstaedtiaceae. Part Two published: 1985.

Volume Four is devoted to the family Leguminosae, of which first part, (Caesalpinioideae), appeared in 1984, by K. Larsen, S. S. Larsen, and J. E. Vidal. Part Two (Mimosoideae) by I. Nielsen appeared in 1985.

Volume Five is devoted to various families. Part One published: 1987.

Since Number Six (1972) of the "Thai Forest Bulletin (Botany)," edited by T. Smitinand, a number of botanical papers have been contributed by Thai and foreign botanists. In 1980, T. Smitinand published *Thai Plant Names (Botanical Names-Vernacular Names)*.

VI. Completeness of Floristic Inventory

A. Factors Affecting

The completion of the floristic inventory of Thailand is faced by many complicating factors, the most important being the difficulty of travelling. Even though the development in the highway system after World War II permitted access to remote areas, travel is often impeded by insecurity along the borders and in certain areas due to insurrectionists.

Besides the aforesaid factor, others are worth discussing here.

1. The well-known collecting areas such as Doi Suthep, Doi Inthanon, Doi Chiang Dao in the north; Khao Yai in the central plain; Phu Luang, Phu Kradueng in the northeast; and Khao Luang in the peninsula, have been repeatedly collected following the same routes. Whenever different routes have been followed, new records and new taxa tend to show up in every collection.

2. When a monographic revision of certain groups of plants has been done on the collections from seemingly well-collected areas, it often turned out that the density of collecting was insufficient to sample the variation within the group.

3. The extent of local differentiation of the flora at the species level is still not clear. Adjacent mountains may not have the same species in a given genus. For example, *Rhododendron microphyton* is only found at Doi Inthanon and is absent from Doi Suthep; yet both areas have the same type of soil and vegetation.

4. The basic biology of most taxa is entirely unknown, and the infraspecific variation, both morphological and ecological, is poorly understood. To understand the variation and biology of the taxa, floristic inventory at a density of 50–100 collections per 100 sq. km is not sufficient.

According to Larsen (1979) Thailand is classified as an undercollected area as shown by the density index of 27 species per 100 sq. km. Therefore, Thailand still needs immediate, intense exploration, since the natural vegetation is rapidly disappearing. The areas along the borders of adjacent countries as well as the limestone formations are not well known, and thus need intensive collecting.

B. Endemics and Extinction

Areas of high elevation are particularly rich in endemics, but today endemics may happen to occur elsewhere either within Thailand or in neighboring countries.

At present the rates of extinction are still not known in

Thailand, except that some orchid taxa such as *Paphiopedilum sukhakulii, P. godefroyae, Pecteilis sagarikii, Vanda coerulea,* and *V. coerulescens,* have been decimated in their habitats.

VII. Current Collection and Work on Larger Families

Certain groups of plants have been collected and studied by specialists, i.e., mosses and hepatics, ferns, orchids, grasses and sedges, labiates, composites, ericads, dipterocarps, and other woody groups.

Some large families remain without active contributors. These are Araceae, Asclepiadaceae, Cucurbitaceae, Eriocaulaceae, Euphorbiaceae, Gramineae, Liliaceae, Moraceae, Myrtaceae, Sapindaceae, Sapotaceae, Sterculiaceae, Urticaceae, and Verbenaceae.

VIII. Areas Particularly at Risk

In addition to being heavily logged in the past decades, the lowland forests have been opened extensively for the economic development of the country. Only about 30% of the total forested land has been reserved under the National Parks Law and the Wildlife Conservation Law. The peat swamp forest is also threatened by the agricultural development in peninsular Thailand. The limestone formations are threatened by their exploitation for building material. Encroachment on the mangroves is increasing because of agro-fishery development.

IX. Inventory Resources

A. Herbaria and Botanic Gardens

Three government institutions are involved in floristic inventory. They are:

1. Royal Forest Department, Bangkok. The Forest Herbarium contains all groups of plants: Thallophytes, Gymnosperms, and Phanerogams, with seven active taxonomists.
2. Department of Agricultural Science, Bangkok. The herbarium contains Thallophytes, Gymnosperms and Phanerogams, with three active taxonomists.
3. Songklanakharin University, Songkhla. Their small herbarium contains mostly Phanerogams, with two active staff members.

Besides these institutions, five botanic gardens are under the supervision of the Forest Herbarium: the Maesa Botanic Garden at Chiang Mai, Phukhae Botanic Garden at Saraburi; Bhuddhamondhol Botanic Garden at Nakhon Pathom; Khao Hin Son Botanic Garden in Chachoengsao; and Khao Chong Botanic Garden at Trang.

Table I
Families of plants under study.

Acanthaceae	H. Terao, B. Hansen
Aceraceae	T. Santisuk
Actinidiaceae*	Hsuan Keng
Aegiceratiaceae	B. Nasongkhla
Aizoaceae	L. Phuphathanophong
Alismataceae	R. Haynes
Amaranthaceae	T. Myndel Pedersen
Anacardiaceae	S. Reynolds
Annonaceae	P. Kessler, Rogstad & H. Hennipman
Apocynaceae	Leuwenberg
Apostasiaceae*	Larsen & Vogel
Aquifoliaceae	T. Smitinand
Aquilariaceae	T. Smitinand
Aristolochiaceae	L. Phuphathanaphong
Balanitaceae	Sands
Balanophoraceae*	B. Hansen
Balsaminaceae	T. Shimizu
Basellaceae	K. Larsen
Begoniaceae	Sands & Thai Botanists
Berberidaceae	Chamberlain
Betulaceae	L. Phuphathanaphong
Bignoniaceae	T. Santisuk
Bombacaceae	Robyns
Bonnetiaceae*	Hsuan Keng
Boraginaceae	A. Ubolcholkhet
Buddleiaceae	J. E. Vidal
Burmanniaceae	B. Hansen
Butomataceae	R. Haynes
Caesalpiniaceae*	Larsen & Vidal
Campanulaceae	W. Eddie
Cannabinaceae	C. Phengklai
Capparidaceae	K. Chayamarit
Caprifoliaceae	N. Fukoka
Cardiopteridaceae	K. Larsen
Caryophyllaceae	K. Larsen
Casuarinaceae*	C. Phengklai
Centrolepidaceae*	K. Larsen
Cephalotaxaceae*	C. Phengklai
Chenopodiaceae	K. Larsen
Chloranthaceae	van Balgooy
Clethraceae	T. Smitinand
Combretaceae	W. Nanakhorn
Commelinaceae	R. Faden
Compositae	H. Koyama
Connaraceae*	J. E. Vidal
Convolvulaceae	Staples
Crassulaceae	H. Ohba
Cruciferae	I. Hedge
Crypteroniaceae	T. Santisuk
Cupressaceae*	C. Phengklai
Cycadaceae*	T. Smitinand
Cyperaceae	T. Koyoma
Daphniphyllaceae	Huang
Dilleniaceae*	R. Hoogland
Dipsacaceae	B. L. Burtt
Dipterocarpaceae	T. Smitinand

* An asterisk after a family name denotes treatment in the Flora of Thailand.

(continued)

Table I (continued)

Droseraceae	K. Larsen
Ebenaceae*	C. Phengklai
Elaeocarpaceae*	C. Phengklai
Elaeagnaceae	T. Santisuk
Epacridaceae	K. Larsen
Ericaceae	T. Smitinand
Escalloniaceae	Hiepko
Fagaceae	T. Smitinand & Soepadmo
Filicinae* 1–2	M. Tagawa & K. Iwatsuki
Filicinae 2–4	M. Tagawa & K. Iwatsuki
Flacourtiaceae	H. Sleumer
Flagellariaceae	K. Larsen
Gentianaceae	A. Ubolcholkhet
Geraniacae	P. Weo
Gesneriaceae	B. L. Burtt
Gnetaceae*	C. Phengklai
Goodeniaceae*	K. Larsen
Halorrhagaceae*	van der Meiden
Hamamelidaceae	C. Phengklai
Hanguanaceae*	K. Larsen
Hernandiaceae	B. Hansen
Hippocastanaceae*	C. Phengklai
Hydrocharitaceae	R. Haynes
Hypericaceae	Robson
Hypoxidaceae	B. L. Burtt
Icacinaceae*	H. Sleumer
Illiciaceae*	Hsuan Keng
Irvingiaceae*	C. Phengklai
Juglandaceae	L. Phuphathanaphong
Juncaceae*	K. Larsen
Labiatae	G. Murata
Lardizabalaceae	R. Geesink
Lauraceae	A. J. G. Kostermans
Lecythidaceae	van Payens
Loranthaceae	W. Nanakhorn
Lowiaceae*	K. Larsen
Lythraceae	T. Santisuk
Magnoliaceae*	Hsuan Keng
Malpighiaceae	Puangphen
Melastomataceae	C. Hansen
Meliaceae	Mabberley & Pennell
Memispermaceae	L. L. Forman
Menyanthaceae	A. Ubolcholkhet
Mimosaceae*	I. Nielsen
Monimiaceae	Philippson
Musaceae	Agent
Myristicaceae	de Wilde
Myrsinaceae	I. Nielsen
Najadaceae	R. Haynes
Nyssaceae*	C. Phengklai
Ochnaceae*	A. Kanis
Olacaceae	Hiep Ko
Oleaceae	P. Greens
Opiliaceae	Hiepko
Orchidaceae	G. Seidenfaden
Orobanchaceae	Yamazaki
Oxalidaceae*	J. F. Veldkamp
Palmae	J. Dransfield
Pandaceae	L. L. Forman
Pandanaceae	B. C. Stone
Papilionaceae	R. Geesink et al.
Parnassiaceae	T. Shimizu
Pentaphragmataceae	K. Larsen
Pinaceae*	C. Phengklai
Podoaceae	L. L. Forman
Podocarpaceae*	C. Phengklai
Podostemaceae	Cusset
Polygalaceae	van der Meiden
Polygonaceae	Kit Tan
Pontederiaceae	K. Boonkird
Portulacaceae*	R. Geesink
Posidoniaceae	R. Haynes
Potamogetonaceae	R. Haynes
Proteaceae	B. Nasongkla
Rafflesiaceae*	B. Hansen
Ranunculaceae	Tamura
Restionaceae*	K. Larsen
Rhamnaceae	T. Smitinand
Rhizophoraceae*	Ding Hou
Rhoipetalaceae	L. Phuphathanaphong
Rosaceae*	J. E. Vidal & Thuan
Rubiaceae	Fukuoka & Tirvengadam
Ruppiaceae	R. Haynes
Rutaceae	B. C. Stone, T. G. Hartley & B. Hansen
Sabiaceae	J. E. Vidal & Thai botanists
Salicaceae	K. Larsen
Santalaceae	K. Larsen & B. Hansen
Saurauiaceae*	Hsuan Keng
Saxifragaceae	T. Shimizu
Schisandraceae*	Hsuan Keng
Scrophulariaceae	Yamazaki
Simaroubaceae*	H. P. Nooteboom
Smilacaceae*	T. Koyama
Sonneratiaceae	T. Santisuk
Sphenocleaceae*	K. Larsen
Stenomeridaceae	K. Larsen
Strychnaceae	J. E. Vidal
Stylidiaceae*	K. Larsen
Symplocaceae*	H. P. Nooteboom
Taccaceae	C. Phengklai
Theaceae*	Hsuan Keng
Thismiaceae	K. Larsen
Thymelaeaceae	Bo Peterson
Tiliaceae	C. Phengklai
Trapaceae	J. E. Vidal & Thai botanist
Triuridaceae*	K. Larsen
Typhaceae	K. Larsen
Ulmaceae	L. Phuphathanaphong
Umbelliferae	I. Hedge
Vacciniaceae	T. Smitinand
Valerianaceae	K. Larsen
Violaceae	Moore
Xyridaceae	B. Hansen
Zannichelliaceae	R. Haynes
Zingiberaceae	K. Larsen
Zosteraceae	R. Haynes

* An asterisk after a family name denotes treatment in the Flora of Thailand.

B. Foreign Programs

Several foreign institutions are currently involved in the study of the Thailand flora on a regular basis. The Rijksherbarium, Leiden, has made five joint expeditions with the Forest Herbarium, Bangkok. The Botanical Museum in Copenhagen and the Botanical Institute, Aarhus, have made six joint expeditions with the Forest Herbarium. Recently the Royal Botanic Gardens, Kew, made a joint expedition in collaboration with the Danish to study the pollination of orchids. The University of Kyoto has made joint expeditions with the Forest Herbarium. The most recent one was the Kyoto-Shinshu expedition, in collaboration with the Forest Herbarium in 1984.

Collaboration with anthropologists, ecologists and others, both inside and outside the country, has been organized with successful results. The joint phytochemical survey carried out by the Forest Herbarium and Japanese pharmaceutical laboratories yielded very many interesting alkaloids, one of which may be useful in modern medicine.

X. Suitability of Floristic Inventory

Botanical expeditions in the past were hampered by limited funds, personnel, time, and accessibility to remote areas; the collections were thus cursorily made. A carefully coordinated and speedy solution is necessary.

In order to develop an effective program to inventory the flora, the collecting and identifying of plants has been incorporated into national goals. The work on a forester's manual is continuously carried out by the Forest Herbarium; so far three installments have been published in the Thai language. Guides to national parks and wildlife are also currently prepared by the same institution. One has been published in both Thai and English, but the work was relatively slow because of the shortage of manpower, funds and time.

The floristic inventories in Thailand are generally satisfactory, but the financial and personnel resources are insufficient to carry out these inventories.

The existing herbaria need an adequate and regular budget for maintenance, for air-conditioning, and for despatching the large amount of material requested on loan to and exchange from foreign botanists and leading herbaria abroad.

Thai botanists cannot work on taxonomy in Bangkok, where collections and references are insufficient. The cost of travelling, accommodation and living abroad is very high and financial support for this need is insufficient.

Unfortunately, the floristic inventory is not high on the list of priorities in Thailand at present.

XI. Selected Bibliography

Since exhaustive lists of Thai botanical literatures were compiled by E. H. Walker (1952) and B. Hansen (1973), the following account is intended to cite literatures of some taxonomic importance, and those that are excluded by the said authors. An asterisk before the reference indicates that it is cited in the text.

*Airy Shaw, H. K. 1971. The Euphorbiaceae of Siam. Kew Bull. **26(2)**: 191–363.

———. 1977. Additions and corrections of the Euphorbiaceae of Siam. Kew Bull. **32(1)**: 61–83.

Bakhuizen van den Brink Jr., R. C. 1975. A synoptical key to the genera of the Rubiaceae of Thailand. Thai For. Bull. (Bot.) **9**: 15–55.

Barnett, E. C. 1938. Contribution to Flora of Siam. Additamenta XLVII. Kew Bull. **1938**: 98–106.

*———. 1942. The Fagaceae of Thailand and their geographical distribution. Trans. Proc. Bot. Soc. Edinb. **33(3)**: 327–345.

*———. 1961. New species of the Gesneriacea from Thailand. Nat. Hist. Bull. Siam Soc. **20**: 9–25.

Boonkird, K. Pontederiaceae. Thai For. Bull. (Bot.) **9**: 12–14.

Bor, N. L. 1963–1968. Studies in the Flora of Thailand. 8, 26, 38, and 43. Gramineae. Dansk. Bot. Ark. **20**: 139–178, **23**: 141–168, 307 and 465–471.

Bremekamp, C. E. B. 1953. A new species of *Parasympagis* (Acanthaceae) from Thailand. Kew Bull. **7**: 565.

———. 1961–1969. Studies in the Flora of Thailand. 3, 32, 35, 57. Scrophulariaceae-Nelsonieae, Thunbergiaceae, Acanthaceae. Dansk. Bot. Ark. **20**: 55–88; **23**: 195–224, 273–279; **27**: 71–85.

Brotherus, V. F. 1901. Bryales. *In* J. Schmidt, Flora of Koh Chang 3. Bot. Tidsskr. **24**: 115–125.

Carroll, G. 1963. Studies in the Flora of Thailand. 24. Pyrenomycetes. Dansk Bot. Ark. **23**: 101–113.

Chaianan, C. 1972. Revision of *Germainia* Balansa & Poitrasson (Gramineae). Thai For. Bull. (Bot.) **6**: 29–47.

Charoenphol, C. C. 1973. Studies in the genus *Chlorophytum* of Thailand. Thai. For. Bull. (Bot.) **7**: 67–69.

———. 1974. Studies in Thai Liliaceae. Thai For. Bull. (Bot.) **8**: 88–94.

Christ, H. 1901. Pteridophyta excl. *Selaginella*. *In* J. Schmidt, Flora of Koh Chang 3. Bot. Tidsskr. **24**: 102–113.

Christensen, C. 1916. Filices. *In* J. Schmidt, Flora of Koh Chang 10. Bot. Tidsskr. **32**: 340–350.

———. 1931. Pteridophyta from Northern Siam. Asiatic Pteridophyta collected by J. F. Rock 1920–1924. Contr. U.S. Nat. Herb. **26**: 329–335.

Clarke, C. B. 1901. Cyperaceae. *In*. J. Schmidt, Flora of Koh Chang 3. Bot. Tidsskr **24**: 79–94.

———. 1902. Compositae, Umbelliferae. *In* J. Schmidt, Flora of Koh Chang 5. Bot Tidsskr. **24**: 241–248.

———. 1902. Lythraceae Melastomaceae, Scrophulariaceae, and Acanthaceae. *In* J. Schmidt, Flora of Koh Chang 6. Bot Tidsskr. **24**: 342–351.

———. 1904. Verbenaceae and Labiatae. *In* J. Schmidt, Flora of Koh Chang 8. Bot Tidsskr. **26**: 171–176.

———. 1916. Ochnaceae. *In* J. Schmidt, Flora of Koh Chang 10. Bot Tidsskr. **32**: 311.

Clayton, W. D. 1969. Studies in the Flora of Thailand. 56. A new species of *Parahyparrhenia*. Dansk Bot. Ark. **27**: 67–70.

Corner, E. J. H. 1963. Studies in the Flora of Thailand. 16. Moraceae. Dansk Bot. Ark **23**: 19–32.

*Craib, W. G. 1911–1912. List of Siamese plants with descriptions of new species. Kew Bull. 1911–1912.

*———. 1912. Contributions to the Flora of Siam (Dicotyledons). Aberdeen Univ. Studies **57**: 1–210.

*———. 1913. Contributions to the Flora of Siam

(Monocotyledons). Aberdeen Univ. Studies **61**: 1-39.

*_____. 1912-1933. Contribution to the Flora of Siam. Additamenta 1-15, 18-20, 23-28. Kew Bull. **1912-1925, 1926-1927, 1927-1933**.

*_____. (ed.) 1925-1931. Florae Siamensis Enumeratio. **1**: 1-809.

*_____ & A. F. G. Kerr (ed.). 1932-1939. Florae Siamensis Enumeratio **2**: 1-476.

*_____, _____, P. L. Pendleton, & E. C. Barnett (ed.). 1951-1962. Florae Siamensis Enumeratio **3**: 1-238.

Camberlege, P. F. & W. M. S. Cumerlege 1963. A preliminary list of orchids of Khao Yai National Park. Nat. Hist. Bull. Siam. Soc. **20**: 155-174.

Dammer, U. 1909. Palmae. *In* J. Schmidt, Flora of Koh Chang 9. Bot Tidsskr. **29**: 97-100.

Dissing, J. 1963. Studies in the Flora of Thailand. 25. Discomycetes and Gasteromycetes. Dansk Bot. Ark. **23**: 115-130.

*****Downie, D. G.** 1925. Contribution to the Flora of Siam. Additamenta 16-17. Kew Bull. **1925**: 367-394, 404-423.

Dransfield, J. 1983. *Kerridoxa*, A new coryphoid palm genus from Thailand. Principes 27: 3-11.

Engler, A. 1902. Araceae. *In* J. Schmidt, Flora of Koh Chang 5. Bot Tidsskr. **24**: 272-276.

Fletcher, H. R. 1937. Contribution to the Flora of Siam. Additamenta 41, 42, 44, & 45. Kew Bull. **1937**: 26-44, 71-75, 371-392, 505-510.

_____. 1938. Contribution to the Flora of Siam. Additamenta 46. Kew Bull **1938**: 199-209.

*_____. 1948. Keys to Siamese species of Myrsinaceae. Notes Roy. Bot. Gard. Edinb. **20**: 103-120.

Forman, L. L. 1954. A new genus from Thailand. Kew Bull. **7**: 555-564.

_____. 1962. *Aetheolirion*, A new genus of Commelinaceae from Thailand. Kew Bull. **16**: 209-221.

_____. 1963. Studies in the Flora of Thailand. 11. Menispermaceae. Dansk Bot. Ark. **20**: 187-190.

_____. 1981. A revision of *Tinospora* (Menispermaceae) in Asia to Australia and the Pacific. The Menispermaceae of Malesial and adjacent areas X. Kew Bull. **36**: 375-421.

Foslie, M. 1901. Corallinaceae. *In* J. Schmidt, Flora of Koh Chang 2. Bot Tidsskr. **24**: 15-22.

Fukuoka, N. 1967. *Sambucus* and *Viburnum* in Thailand. Acta Phytotax. Geobot. **22**: 163-174.

_____. 1970. Contribution to the flora of Southeast Asia III. *Hedyotis* (Rubiaceae) of Thailand. S. E. As. Stud. **8**: 305-336.

*****Geddes, E. T.** 1927. Contribution to the Flora of Siam. Additamenta 21. Kew Bull. **1927**: 164-174.

Geesink, R. 1975. Portulacaceae. *In* T. Smitinand & K. Larsen, Flora of Thailand **2**: 268-273.

Giesy, R. M. & P. W. Richards. 1959. A collection of Bryophytes from Thailand. Trans. Brit. Bryol. Soc. **3**: 575-581.

Gilg, E. 1916. Loganiaceae. *In* J. Schmidt, Flora of Koh Chang 10. Bot. Tidsskr **232**: 312-313.

Gomont, M. 1901. Myxophyceae hormogoneae. *In* J. Schmidt, Flora of Koh Chang. Bot. Tidsskr. **24**: 202-211.

*****Grey-Wilson, C.** 1971. New Plant Records from Thailand. Kew Bull. **26**: 141-151.

Hackel, E. 1901. Gramineae. *In* J. Schmidt, Flora of Koh Chang. 3. Bot. Tidsskr. **24**: 95-101.

Hallier, H. 1904. Convolvulaceae. *In* J. Schmidt, Flora of Koh Chang. Bot. Tidsskr. **26**: 170.

Hansen, B. 1961-1962. Studies in the Flora of Thailand 4. Sphagnaceae. Dansk Bot. Ark. **20**: 89-108, 204.

_____. 1966-1968. Studies in the Flora of Thailand. 36. Rutaceae 37.; Sphagnaceae: 42. Dioscoreaceae, Dansk Bot. Ark. **23**: 281-293; 295-300; 459-463.

_____. 1969 Studies in the Flora of Thailand. 51. Balanophoraceae; 52. A new species of Eriocaulon; 53. Xyridacee; Dansk Bot. Ark. **27**: 25-38.

_____. 1972. Balanophoraceae. *In* T. Smitinand & K. Larsen, Flora of Thailand **2**: 177-181.

_____. 1972. Rafflesiaceae, ibid., 182-184.

_____. 1973. Bibliography of Thai Botany. Nat. Hist. Bull. Siam Soc. **24**: 319-404.

_____ & K. Larsen. 1969. Studies in the Flora of Thailand. 49. Loranthaceae; 50. Santalaceae. Dansk Bot. Ark. **27**: 11-24.

Harms, H. 1902. Leguminosae. *In* J. Schmidt, Flora of Koh Chang 5. Bot. Tidsskr. **24**: 264-267.

Hattori, S. & M. Mizutani. 1969. Studies in the Flora of Thailand 59. Hepaticae. Dansk Bot. Ark. **27**: 91-98.

Heim, F. 1902. Dipterocarpaceae. *In* J. Schmidt, Flora of Koh Chang 7. Bot. Tidsskr. **25**: 42-47.

*****Heim, R.** 1962. Contribution a la flore mycoliogique de la Thailaned (1re Partie). Rev. Mycol. **27**: 123-158.

Hieromymus, C. 1901. Selaginellaceae. *In* J. Schmidt, Flora of Koh Chang 3. Bot. Tidsskr. **24**: 113-114.

Hill, A. W. 1940. A new *Strychnos* from Thailand. Kew Bull. **1940**: 199-200.

Hirano, M. 1967. Freshwater algae collected by the joint Thai-Japanese Biological Expedition to Southeast Asia 1961-1962. Nature and Life in Southeast Asia **5**: 1-71.

Hiroe, M. 1963. Studies in the Flora of Thailand 12. Umbelliferae. Dansk Bot. Ark. **20**: 191-195.

_____. 1967. Umbelliferae of Thailand, II. Acta Phytotax. Geobot. **22**: 141-144.

Hjelmquist, H. 1968. Studies in the Flora of Thailand. 44. Fagaceae, Betulaceae and Corylaceae. Dansk Bot. Ark. **23**: 473-516.

Holttum, R. E. 1961. Studies in the Flora of Thailand. 1. Filicinae, excl. of the Ophioglossaceae. Dansk Bot. Ark. **20**: 11-35.

_____. 1965. Studies in the Flora of Thailand. 33. Filicinae (second list). Dansk Bot. Ark. **23**: 225-244.

_____. 1966. Studies in the Flora of Thailand. 38. Filicinae. Some plants new or rare to Thailand with description of new species. Dansk Bot. Ark. **23**: 308-309.

Hoogland, R. D. 1964. Studies in the Flora of Thailand. 10. Dilleniaceae. Dansk Bot. Ark. **20**: 183-185.

_____. 1972. Dilleniaceae. *In* T. Smitinand & K. Larsen, Flora of Thailand **1**: 95-108.

*****Hosseus, C. C.** 1907. Die aus Siam bekanten Acanthaceae. Engler, Bot. Jahrb. **41**: 62-73.

*_____. 1910. Beitrage zur Flora Siams. Bot. Contralbl. Beih. **27(2)**: 455-507.

_____. 1911. Beitrage zur Flora van Wang Djao am Me Ping in Mittel-Siam. Engler, Bot. Jahrb. **45**: 366-374.

*_____. 1911. Die Botanischen Ergebnisse meiner Expedition nach Siam. Bot. Centralbl. Beih. **28(2)**: 357-457.

Hou, Ding. 1965. Studies in the Flora of Thailand. 30. Rhizophoraceae. Dansk Bot. Ark. **23**: 187-190.

_____. 1970. Rhizophoraceae. *In* T. Smitinand & K. Larsen, Flora of Thailand. **2**: 5-15.

Hu, S.-Y. 1968. Studies in the Flora of Thailand. 41. Araceae. Dansk Bot. Ark. **23**: 409-457; 48. The Ger.us *Barclaya* (Nymphaeaceae) ibid. 533-540.

*****Imlay, J. B.** 1939. Contribution to the Flora of Siam. Additamenta 51. Kew Bull. **1959**: 109-150.

Iwatsuki, K. 1963. Thelypteroid ferns of Thailand and Laos collected by Dr. T. Tuyama in 1957-58. J. Jap. Bot. **38**: 313-315.

———. 1972. Phytogeography of the pteridophytes in northern Thailand. Acta Phytotax. Geobot. **25**: 69-78.

Jacobs, M. 1962. Reliquiae Kerrianae. Blumea **11**: 428-493.

Kanis, A. 1970. Ochnaceae. *In* T. Smitinand & K. Larsen, Flora of Thailand. **2**: 24-30.

Keng, H. 1972. Saurauiaceae, Schizandraceae and Illiciaceae. *In* T. Smitinand & K. Larsen, Flora of Thailand **2**: 109-111, 112-114, 115-116.

———. 1972. Actinidiaceae, Theaceae and Bonnetiaceae. *In* T. Smitinand & K. Larsen, Flora of Thailand **2**: 139-141, 142-160.

———. 1975. Magnoliaceae, ibid., 251-267.

*Kern, J. H. 1961., Cyperaceae of Thailand (Excl. *Carex*). Reinwardtia **6**: 25-83.

*———. 1962. Cyperaceae of Thailand II. Reinwardtia **6**: 145-154.

Kerr, A. F. G. 1927. Contribution to the Flora of Siam. Additamenta 22. Kew Bull. **1927**: 212-220.

———. 1935. Contribution to the Flora of Siam. Additamenta 39. Kew Bull. **1935**: 326-335.

———. 1936. Contribution to the Flora of Siam. Additamenta 40. Kew Bull. **1936**: 34-47.

———. 1937. Contribution to the Flora of Siam. Additamenta 43. Kew Bull. **1937**: 87-94.

———. 1938. Contribution to the Flora of Siam. Additamenta 46. Kew Bull. **1938**: 24-32; 48, ibid. 127-133; 50, ibid. 445-454.

———. 1939. Contribution to the Flora of Siam. Additamenta 52. Kew Bull. **1939**: 456-465.

———. 1939. Early botanists in Thailand. J. Thailand Res. Soc. Nat. Hist. Suppl. **12**: 1-27.

———. 1940. Contribution to the Flora of Siam. Additamenta 53. Kew Bull. **1940**: 180-186.

———. 1941. Contribution to the Flora of Siam. Additamenta 54. Kew Bull. **1941**: 8-21.

Kitagawa, N. 1967. Studies on the Hepaticae of Thailand. 1. The genus *Bazzania*, with general information. J. Hattori Bot. Lab. **30**: 249-270.

———. 1968. Studies on the Hepaticae of Thailand. 3. The genus *Leucolejeunea*. S.E. As. Stud **6**: 608-613.

———. 1969. Studies on the Hepaticae of Thailand. 2. *Cephaloziea* and *Cephaloziella*. J. Hattori Bot. Lab. **32**: 290-306.

———. 1978. the Hepaticae of Thailand collected by Dr. A. Touw (I). Acta Phytax. Geobot. **29(1-5)**: 47-64.

———. 1979. The Hepaticae collected by Dr. A. Touw (II). Acta Phytotax. Geobot. **30(1-3)**: 31-40.

Koelpin Roun, F. 1902. Loranthaceae. *In* J. Schmidt, Flora of Koh Chang 5. Bot. Tidsskr. **24**: 6-13.

Kostermans, A. J. G. H. 1974. New species and combinations of Lauraceae from Thailand. Nat. Hist. Bull. Siam Soc. **25**: 229-244.

Koyama, H. 1981. Taxonomic studies in the Compositae of Thailand (I). Acta Phytotax. Geobot. **32(1-4)**: 56-67.

Koyama, T. 1963. Studies in the Flora of Thailand. 15. Smilacaceae. Dansk. Bot. Ark. **23**: 9-18.

———. 1975. Smilacaceae. *In* T. Smitinand & K. Larsen, Flora of Thailand **2**: 211-250.

———. 1979. Studies in the Cyperaceae of Thailand. II. Miscellaneous taxa of Fimbristylideae, Rhynchosporeae, Scirpeae and Sclerieae, Brittonia **31**: 284-93.

———. 1979. Studies in the Cyperaceae of Thailand III. New and critical species of the Cariceae. Bot. Mag. Tokyo **92**: 217-233.

Kränzlin, F. 1900. Orchidaceae and Apostasiaceae. *In* J. Schmidt, Flora of Koh Chang 1. Bot. Tidsskr. **24**: 6-13.

Larsen, K. 1961. Studies in the Flora of Thailand. 3. Liliaceae, Triuridaceae, Trilliaceae, Iridaceae, Polygonaceae. Dansk Bot. Ark. **20**: 37-54.

———. 1962. Studies in Zingiberaceae 1: The genus *Geostachys* in Thailand. Bot. Tidsskr. **58**: 43-49.

———. 1962. Studies in Zingiberaceae 3. On a new species of *Kaempferia* from Thailand and its relatives. Bot. Tidsskr. **58**: 198-203.

———. 1963. Studies in the Flora of Thailand. 20. Various families. Dansk. Bot. Ark. **23**: 57-75.

———. 1963. Studies in the Flora of Thailand. 6. Coniferae. Dansk. Bot. Ark. **20**: 123-128; 7. Various monocotyledon families. ibid. **20**: 129-138.

———. 1965. *Costus dhaniwatii*. A new species from S. E. Thailand. Siam Soc. Felic. Vol. S. E. As. Stud.: 149-152.

———. 1965. Studies in the Flora of Thailand 27. Thismiaceae. Dansk. Bot. Ark. **23**: 169-174; 28. Marantaceae. ibid. 175-182.

———. 1966. Two new Liliacae from the Khao Yai National Park. Bot. Nat. **119**: 198-200.

———. 1972. Centrolepidaceae, Flagellariaceae, Hanguanaceae, Juncaceae, Lowiaceae, Restionaceae, and Triuridaceae. *In* T. Smitinand & K. Larsen, Flora of Thailand **2**: 161-176.

———. 1975. Goodeniaceae and Sphenocleaceae. *In* T. Smitinand & K. Larsen, Flora of Thailand **2**: 278-280.

*———. 1979. Exploration of the Flora of Thailand. *In* K. Larsen & L. Holm-Nielsen, Tropical Botany: 125-133.

——— & S. S. Larsen. 1973. The genus *Bauhinia* in Thailand. Nat. Hist. Bull. Siam Soc. **25**: 1-22.

———, S. S. Larsen & J. E. Vidal. 1984. Leguminosae, Caesalpinioideae. *In* T. Smitinand & K. Larsen, Flora of Thailand **4(1)**: 1-129.

——— & E. F. de Vogel. 1972. Apostasiaceae. *In* T. Smitinand & K. Larsen, Flora of Thailand **2**: 131-138.

Markgraf, F. 1963. Studies in the Flora of Thailand. 5. Gnetaceae. Dansk. Bot. Ark. **20**: 117-122.

Meiden, R. van der. 1970. Haloragaceae. *In* T. Smitinand & K. Larsen, Flora of Thailand **2**: 1-4.

Mez, C. 1904. Myrsinaceae. *In* J. Schmidt, Flora of Koh Chang 8. Bot Tidsskr. **26**: 169.

Moldenke, H. N. 1963. Studies in the Flora of Thailand 22. Avicenniaceae, Symphoremaceae and Verbenaceae. Dansk Bot. Ark. **23**: 83-92; Verbenaceae *in* Studies in the Flora of Thailand. 28. Some plants new or rare to Thailand with descriptions of new species. ibid. 306.

Murata, Gen. 1970. New Labiatae from Thailand. Acta Phytotax. Geobot. **24**: 105-112.

———. 1971. Contributions to the Flora of Southeast Asia. IV. A list of Labiatae known from Thailand. SE As. Stud. **8**: 489-517.

———. 1976. Contributions to the Flora of Southeast Asia. VI. Additions and corrections to the knowledge of Labiatae in Thailand (I) S. E. As. Stud. **14**: 177-193.

NaSongkhla, B. 1973. Proteaceae. Thai For. Bull. (Bot.) **7**: 49-65.

Nayar, M. P. 1969. Studies in the Flora of Thailand. 55. Two new species of *Sonerilla*. Dansk Bot. Ark. **27**: 61-65.

* Nelmes, E. 1955. The genus *Carex* in Indo-China, including Thailand and Lower Burma. Mem. Mus. Nat. Hist. Nat. (n.s.)., Bot. **4**: 83-182.

Nielsen, I. 1980. Notes on Indo-chinese Mimosaceae, Adansonia Ser. 2, **19(3)**: 339-363.

Niyomdham, C. 1978. A revision of the genus *Crotalaria* Linn. (Papilionaceae) in Thailand. Thai For. Bull. (Bot.) **11**: 105-181.

———. 1980. Preliminary revision of tribe Sophoreae (Leguminosae-Faboideae) in Thailand: *Ormosia* Jacks. and *Sophora*

Linn. Thai For. Bull. (Bot.) **13**: 1–22.
Nooteboom, H. P. 1981. Simaroubaceae and Symplocaceae. *In* T. Smitinand & K. Larsen, Flora of Thailand **2**: 439–464.
Oestrup, E. 1902. Freshwater diatoms. *In* J. Schmidt, Flora of Koh Chang 7. Bot Tidsskr. **25**: 28–41.
_____. 1904. Marine diatoms. *In* J. Schmidt, Flora of Koh Chang 8. Bot Tidsskr. **26**: 115–161.
Ohwi, J. 1968. New Glumales from Thailand. Acta Phytotax. Geobot. **23**: 109.
Ostenfeld, C. H. 1902. Hydrocharitaceae, Lemnaceae, Pontederiaceae, Potamogetonaceae, Gentianaceae (*Limnanthemum*), Nymphaeaceae. *In* J. Schmidt, Flora of Koh Chang 5. Bot. Tidsskr. **24**: 260–263.
_____. 1902. Marine plankton diatoms. *In* J. Schmidt, Flora of Koh Chang 7. Bot. Tidsskr. **25**: 1–27.
_____. 1904. Cycadaceae, Taxaceae, Gnetaceae, Pandanaceae, Smilacaceae, Commelinaceae, Amaryllidaceae, Taccaceae, and Dioscoreaceae. *In* J. Schmidt, Flora of Koh Chang 8. Bot. Tidsskr. **26**: 162–166.
_____. 1905. A list of plants collected in the Raheng district, upper Siam by Mr. Lindhard. Bull. Herb. Boiss. **2(5)**: 709–724.
_____. 1909. Lentibulariaceae. *In* J. Schmidt, Flora of Koh Chang 9. Bot. Tidsskr. **29**: 101–103.
_____. 1916. Various families. *In* J. Schmidt, Flora of Koh Chang 10. Bot. Tidsskr. **32**: 319–320.
Paulsen, O. 1902. Fagaceae. *In* J. Schmidt, Flora of Koh Chang 5. Bot. Tidsskr. **22**: 255.
Pax, F. 1916. Euphorbiaceae. *In* J. Schmidt, Flora of Koh Chang 10. Bot. Tidsskr. **32**: 314.
Pendleton, R. L. 1963. Thailand: Aspects of landscape and life. Amer. Geogr. Soc. Handbook. New York.
Phanichapol, D. 1967. Check-list of fungi in the Forest Herbarium. Nat. Hist. Bull. Siam Soc. **22**: 263–269.
Phengkhlai, C. 1967. *Calocedrus macrolepis* Kurz, a conifer new to Thailand. Blumea **15**: 267.
_____. 1968. Studies in the Flora of Thailand. 46. Ebenaceae. Dansk. Bot. Ark. **23**: 521–525.
_____. 1968. A new species of *Diospyros* L. (Ebenaceae) from Thailand. Kew Bull. **23**: 267.
_____. 1972. Cephalotaxaceae, Cupressaceae, Podocarpaceae, and Gnetaceae. *In* T. Smitinand & K. Larsen, Flora of Thailand **2**: 195–210.
_____. 1980. Taccaceae. Thai For. Bull. (Bot.) **13**: 23–33.
_____. 1981. Ebenaceae, Cannabidaceae, Hippocastanaceae, Casuarinaceae, Nyssaceae, and Elaeocarpaceae. *In* T. Smitinand & K. Larsen, Flora of Thailand **2**: 281–438.
Phuphathanaphong, L. 1972. Revision of *Gironniera* Gaud. (Ulmaceae). Thai For. Bull. (Bot.) **6**: 49–59.
Poulsen, V. A. 1904. Eriocaulaceae. *In* J. Schmidt, Flora of Koh Chang 8. Bot. Tidsskr. **26**: 167.
Radkolfer, L. 1916. Sapindaceae. *In* J. Schmidt, Flora of Koh Chang 10. Bot. Tidsskr. **32**: 315–316.
*Raymond, M. 1959. Carices Indochinensis nec non Siamense. Mém. Jard. Bot. Montréal **53**: 1–125.
_____. 1965. Studies in the Flora of Thailand. 34. The genus *Carex*. Dansk Bot. Ark. **23**: 245–262.
_____. 1966. Studies in the Flora of Thailand. 39. Cyperaceae. Dansk Bot. Ark. **23**: 311–374.
Reihbold, Th. 1901. Marine algae (Chlorophyceae, Phaeophyceae, Dictyotales, Rhodophyceae). *In* J. Schmidt, Flora of Koh Chang 4. Bot. Tidsskr. **24**: 187–201.
*Ridley, H. N. 1912. An account of a botanical expedition to Lower Siam. J. Str. Br. Roy. Soc. **61**: 45–65.
*_____. 1915. The plants of Koh Samui and Koh Pennan. J. Fed. Malay States Mus. **5**: 158–168.
*_____. 1920. On a collection of plants from Peninsular Siam. J. Fed. Malay States Mus. **10**: 65–126.
Rostrup, E. & C. Massee. 1902. Fungi. *In* J. Schmidt, Flora of Koh Chang 6. Bot. Tidsskr. **24**: 355–367.
Royen, P. van. 1965. Studies in the Flora of Thailand. 29. Podostemaceae. Dansk Bot. Ark. **23**: 183–184.
_____. 1967. A new Podostemaceae from Thailand, *Polypleurella micrantha* van Royen. Blumea **8**: 522–524.
Ryan, F. D. & A. F. G. Kerr. 1911. Dipterocarpaceae of northern Siam. J. Siam Soc. **8**: 7.
Santisuk, T. 1973. Notes on Asiatic Bignoniaceae. Kew Bull. **28**: 175–185.
_____. 1974. Bignoniaceae. Thai For. Bull. (Bot.) **8**: 1–48.
_____. 1983. Taxonomy and distribution of terrestrial trees and shrubs in the mangrove formations in Thailand. Nat. Hist. Bull. Siam Soc. **31**: 63–91.
Schaller, R. 1944. Thai cardamon. Nat. Hist. Bull. Thai Res. Soc. **14**: 19–34.
Schlechter, R. 1916. Asclepiadaceae. *In* J. Schmidt, Flora of Koh Chang 10. Bot. Tidsskr. **32**: 317–318.
Schmidt, J. 1901. Peridinales. *In* J. Schmidt, Flora of Koh Chang 4. Bot. Tidsskr. **24**: 212–221.
_____. 1902. Rhizophoraceae. *In* J. Schmidt, Flora of Koh Chang 5. Bot. Tidsskr. **24**: 249–254.
Schoser, G. & K. Sehghas. 1965. *Paphiopedilum sukhakulii* ein unerwarteterfund aus Thailand. Orchidee (Hamburg) **16**: 109–110, 224–236.
Schumann, K. 1902. Scitamineae. *In* J. Schmidt, Flora of Koh Chang 5. Bot. Tidsskr. **24**: 268–271.
_____. 1902. Rubiaceae. *In* J. Schmidt, Flora of Koh Chang 6. Bot. Tidsskr. **25**: 329–341.
Seidenfaden, G. 1958. On a small collection of ferns from Thailand. Nat. Hist. Bull. Siam Soc. **19**: 84–87.
_____. 1965. *Bulbophyllum daminivatii*. A new orchid from Thailand. Flic. Vol. of S. E. As. Stud. **1**: 153–155.
_____. 1968. The genus *Oberonia* in mainland Asia. Dansk Bot. Ark. **25**: 1–125.
_____. 1969. The genus *Ione*. Bot. Tidsskr. **64**: 205–238; Contribution to the orchid flora of Thailand. ibid. 100–162.
_____. 1970. Contribution to the orchid flora of Thailand. Bot. Tidsskr. **65**: 100–162.
_____. 1970. Contributions to the orchid flora of Thailand II. Bot. Tidsskr. **65**: 313–370.
_____. 1971. Notes on the genus *Luisia*. Dansk Bot. Ark. **27**: 1–101.
_____. 1972. Contributions to the orchid flora of Thailand III. Bot. Tidsskr. **66**: 303–356.
_____. 1972. Contributions to the orchid flora of Thailand IV. Bot. Tidsskr. **67**: 76–127.
_____. 1973. Contributions to the orchid flora of Thailand V. Bot. Tidsskr. **68**: 41–95.
_____. 1975. Orchid genera in Thailand I. *Calanthe* R. Br. Dansk Bot. Ark. 29. **2**: 1–50.
_____. 1975. Orchid genera in Thailand II. *Cleisostoma* Bl. Dansk Bot. Ark. **29(3)**: 1–80.
_____. 1975. Orchid genera in Thailand III. *Coelogyne* Lindl. Dansk Bot. Ark. **29(4)**: 1–94.
_____. 1975. Contributions to the orchid flora of Thailand VI. Bot. Tidsskr. **70**: 64–97.
_____. 1976. Orchid genera in Thailand IV. *Liparis* L. C. Rich. Dansk. Bot. Ark. **31(1)**: 1–105.

_____. 1976. Contributions to the orchid flora of Thailand VII. Bot. Tidsskr. **71**: 1–30.

_____. 1977. Orchid genera in Thailand V. *Orchidoideae*. Dansk Bot. Ark. **31(3)**: 1–149.

*_____. 1977. Contribution to the orchid flora of Thailand VIII. Bot. Tidsskr. **72**: 1–14.

*_____. 1978. Orchid genera in Thailand VI. *Neottioideae* Lindl. Dansk Bot. Ark. **32(2)**: 1–195.

*_____. 1978. Orchid genera in Thailand VII. *Oberonia* Lindl. & *Malaxis* Sol. ex Sw. Dansk Bot. Tidsskr. **33(1)**: 1–94.

*_____. 1979. Orchid genera in Thailand VIII. *Bulbophyllum* Thou. Dansk Bot. Ark. **33(3)**: 1–228.

*_____. 1980. A new species of *Sunipia* (Orchidaceae) from Thailand. Nat. Hist. Bull. Siam Soc. **28**: 1–8.

*_____. 1980. Orchid genera in Thailand IX. *Flickingeria* Hawkes & *Epigeneium* Gagnep. Dansk Bot. Ark. **34(1)**: 1–104.

*_____. 1981. Contributions to the orchid flora of Thailand IX. Nord. J. Bot. **1**: 103–217.

*_____. 1982. Orchid genera in Thailand X. *Trichotosia* Bl. and *Eria* Lindl. Opera Bot. **62**: 1–157.

*_____. 1982. Contribution to the orchid flora of Thailand. X. Nord. J. Bot. **2**: 193–218.

*_____ & T. Smitinand. 1959–1965. The orchids of Thailand. A preliminary list. Bangkok.

Shimizu, T. 1970. Contribution to the Flora of Southeast Asia. II. *Impatiens* of Thailand and Malaya. S. E. As. Stud. **8**: 187–217.

Simmonds, N. W. 1965. Botanical results of the banana collecting expedition, 1954–5. Kew Bull. **11**: 463–489.

_____. Note on banana taxonomy. Kew Bull. **14**: 198–212.

Sinclair, J. 1953. Notes on Siamese Annonaceae. Gard. Bull. Sing. **14**: 40–44.

_____. 1958. A revision of the Malayan Myristicaceae. Gard. Bull. Sing. **16**: 205–472.

_____. 1959. Florae Maesianae Praecursores XXX. The genus *Knema* (Myristicaceae) in Malaysia and outside Malesia. Gard. Bull. Sing. **18**: 102–327.

_____. 1968. Florae Malesianae Precursores XLII. The genus *Myristica* in Malesia and outside Malesia. Gard. Bull. Sing. **33**: 1–540.

Skvortsov, A. K. 1963. Studies in the Flora of Thailand. 9. Salicaceae. Dansk Bot. Ark. **20**: 179–183.

Sleumer, H. 1958. The genus *Rhododendron* L. in Indo-china and Siam. Blumea, Suppl. **4**: 38–59.

_____. 1963. Studies in the Flora of Thailand. 21. Ericaceae (incl. Vacciniaceae). Dansk Bot. Ark. **23**: 77–81.

_____. 1966. Vacciniaceae, in Studies in the Flora of Thailand. 38. Some plants new or rare to Thailand with descriptions of new species. Dansk Bot. Ark. **23**: 303–305.

_____. 1965. Studies in the Flora of Thailand. 58. Some species of Flacourtiaceae and Ericaceae new to Thailand. Dansk Bot. Ark. **27**: 87–90.

_____. 1969. Materials towards the knowledge of the Icacinaceae of Asia, Malesia, and adjacent areas. Blumea **17**: 185–264.

_____. 1970. Icacinaceae and Cardiopteridaceae. *In* T. Smitinand & K. Larsen, Flora of Thailand **2**: 75–94.

Smitinand, T. 1958. The genus *Eria* Lindley (Orchidaceae) in Thailand. Nat. Hist. Bull. Siam Soc. **19**: 7–43.

_____. 1958. Identification keys to the genera and species of the Dipterocarpaceae of Thailand. Nat. Hist. Bull. Siam Soc. **19**: 57–83.

_____. 1961. Some noteworthy plants from Thailand (Siam). Nat. Hist. Bull. Siam Soc. **20**: 41–70.

_____. 1962. New records of plants from Thailand II. Nat. Hist. Bull. Siam Soc. **20**: 121–133.

_____. 1962. The genus *Actephila* Bl. (Euphorbiaceae) in Thailand. Nat. Hist. Bull. Siam Soc. **20**: 138–140.

_____. 1963. Studies in the Flora of Thailand. 19. Dipterocarpaceae & Lythraceae. Dansk Bot. Ark. **23**: 47–55.

_____. 1967. Orchids collected during the Japanese expeditions to Thailand in 1938 and 1955–1966. Nat. Hist. Bull. Siam Soc. **22**: 105–118.

_____. 1967. New records of Thai plants III. Nat. Hist. Bull. Siam Soc. **22**: 167–172.

_____. 1971. The genus *Cycas* L. (Cycadaceae) in Thailand Nat. Hist. Bull. Siam Soc. **24**: 163–1975.

_____. 1972. Cycadaceae. *In* T. Smitinand & K. Larsen, Flora of Thailand **2**: 155–194.

_____, T. Santisuk & C. Phengkhlai. 1980. The manual of Dipterocarpaceae of mainland South-East Asia. Thai For. Bull. (Bot.) **12**: 1–133.

Steenis, C. G. G. J. van. 1960. *Cladopus* in Thailand (Podostemaceae). Blumea **10**: 141.

_____. 1961. *Symplocos* (*Cordyloblaste*) *henschelii*. New for Thailand. Nat. Hist. Bull. Siam Soc. **20**: 83.

_____. 1963. Studies in the Flora of Thailand. 23. *Phyllanthodendron mirabile* (Euphorbiaceae). Dansk Bot. Ark. **23**: 93–99.

Stephani, F. 1902. Hepaticae. *In* J. Schmidt, Flora of Koh Chang 5. Bot Tidsskr. **24**: 277–280.

St. John, H. 1963. Revision of the genus *Panadanus* Stickman, Part 16: Species discovered in Thailand and Vietnam. Pac. Sci. **18**: 466–492.

Stone, B. C. 1970. Materials for a monograph of *Freycinetia* Gaud. (Pandanaceae) V. Singapore, Malaya, and Thailand. Gard. Bull. Sing. **25**: 189–207.

_____. 1971. A preliminary survey of the Pandanaceae of Thailand and Cambodia. Nat. Hist. Bull. Siam Soc. **24**: 4–32.

Tagawa, M. 1963. Polypodioid ferns collected by Prof. T. Tuyama in northern Thailand and adjoining Laos. J. Jap. Bot. **38**: 325–331.

_____ & K. Iwatsuki. 1965. On a small collection of Thailand ferns. S. E. Asia Stud. **3**: 70–89.

_____. 1967. Enumeration of Thai Pteridophytes collected during 1965–66. S. E. Asia Stud. **5**: 23–120.

_____. 1972. Families and genera of the pteridophytes known from Thailand. Mem. Fac. Sci. Kyoto Univ. Biol. **5**: 67–88.

_____. 1979. Pteridophytes: Psilotaceae-Dennstaedtiaceae. *In* T. Smitinand & K. Larsen, Flora of Thailand **3(1)**: 1–128.

Taylor, P. 1968. Studies in the Flora of Thailand. 47. Lentibulariaceae. Dansk Bot. Ark. **23**: 527–532.

Terao, H. 1980. Notes on some species of *Strobilanthes* (Acanthaceae) from Thailand (1). Acta Phytotax. Geobot. **31(1–3)**: 61–64.

_____. 1981. Notes on some species of *Strobilanthes* (Acanthaceae) from Thailand (2). Acta Phytotax. Geobot. **32**: 31–36.

Thuan, N. V. 1970. The genus *Rubus* in Vidal, Rosaceae. Fl. Thailand **2(1)**: 46–61.

Tixier, P. 1971. Bryophytae Indosinicae. Mousses de Thailande. Ann. Fac. Sci. **4**: 91–146.

_____. 1971–1972. Bryophytae Indosinicae. Mousses de Thailande: espèces nouvelles. Rev. Bryol Lichenol. **38**: 148–160.

_____ & T. Smitinand. 1966. Check-list of the moss collection in the Forest Herbarium, Royal Forest Department, Bangkok. Nat. Hist. Bull. Siam Soc. **21**: 161–195.

Toyokuni, H. 1981. Studies on the Gentianaceae of Thailand. I. Notes on *Exacum* (Gentianaceae-Exacinae). Acta. Phytotax. Geobot. **32**: 198–203.

Ubolcholket, A. 1983. Preliminary Study on Gentianaceae of Thailand. Thai For. Bull. (Bot.) **14**: 94–127.

Veldkamp, J. F. 1970. Oxalidaceae. *In* T. Smitinand & K. Larsen, Flora of Thailand **2(1)**: 16–23.

Vidal, J. 1963. Studies in the Flora of Thailand. 17. Sabiaceae & Connaraceae. Dansk Bot. Ark. **23**: 33–37.

———. 1970. Rosaceae. *In* T. Smitinand & K. Larsen, Flora of Thailand **2(1)**: 31–74.

———. Connaraceae. *In* T. Smitinand & K. Larsen, Flora of Thailand **2(2)**: 117–130.

———, **K. Larsen & S. S. Larsen.** 1984. Leguminosae-Caesalpinioideae. *In* Fl. Thailand **4(1)**: 1–129.

Wainio, E. A. 1909. Lichenes *In* Fl. Koh Chang 9. Bot. Tidsskr. **29**: 104–152.

Walker, E. H. 1952. Bibliography on Thai botany. Nat. Hist. Bull. Siam Soc. **15**: 27–88.

Warburg, O. 1902. Urticaceae. *In* Fl. Koh Chang 6. Bot. Tidsskr. **24**: 352–354.

———. 1916. Various families. *In* Fl. Koh Chang 10. Bot. Tidsskr. **32**: 321–325.

Warming, E. 1902. Podostemaceae. *In* Fl. Koh Chang 5. Bot. Tidsskr. **24**: 258–259.

West, W. & C. S. West. 1902. Fresh water Chlorophyceae. *In* Fl. Koh Chang 4. Bot. Tidsskr. **24**: 157–186.

Wilde, W. J. J. O. de. 1979. New account of the genus *Knema* (Myristicaceae). Blumea **25**: 321–478.

*****Williams, F. N.** 1904–1905. Liste des plantes connues du Siam. Bull. Herb. Boiss. **2(4–5)**.

Wood, D. 1974. A revision of *Chirita* (Gesneriaceae). Notes Roy. Bot. Gard. Edinb. **33**: 123–205.

Yahara, T. 1981. Taxonomic studies of the Urticaceae 1. The genus *Boehmeria* in Thailand. Acta Phytotax. Geobot. **32(1–4)**: 1–21.

Yamazaki, T. 1978. New or noteworthy plants of the Scrophulariaceae from Indochina (1) and (2). J. Jap. Bot. **53**: 1–11, 97–106.

———. 1979. New or noteworthy plants of Scrophulariaceae from Indo-china (3). J. Jap. Bot. **54**: 15–21.

———. 1980. New or noteworthy plants of Scrophulariaceae from Indo-china (4), (5) and (6). J. Jap. Bot. **55**: 1–13, 204–208, 328–336.

———. 1981. Revision of the Indo-Chinese species of *Lindernia* All. (Scrophulariaceae). J. Fac. Sci. Univ. Tokyo Sec. III. **13**: 1–64.

Vietnam, Kampuchea and Laos
Maurice Schmid

Contents

I. Description of the Region	86
A. Geographical Extent, Topography and Population	86
II. Vegetation Types	86
III. Vegetation Maps and Herbaria	87
IV. Present State of Inventory	87
V. Priorities for Inventory	88
VI. Selected Bibliography	89

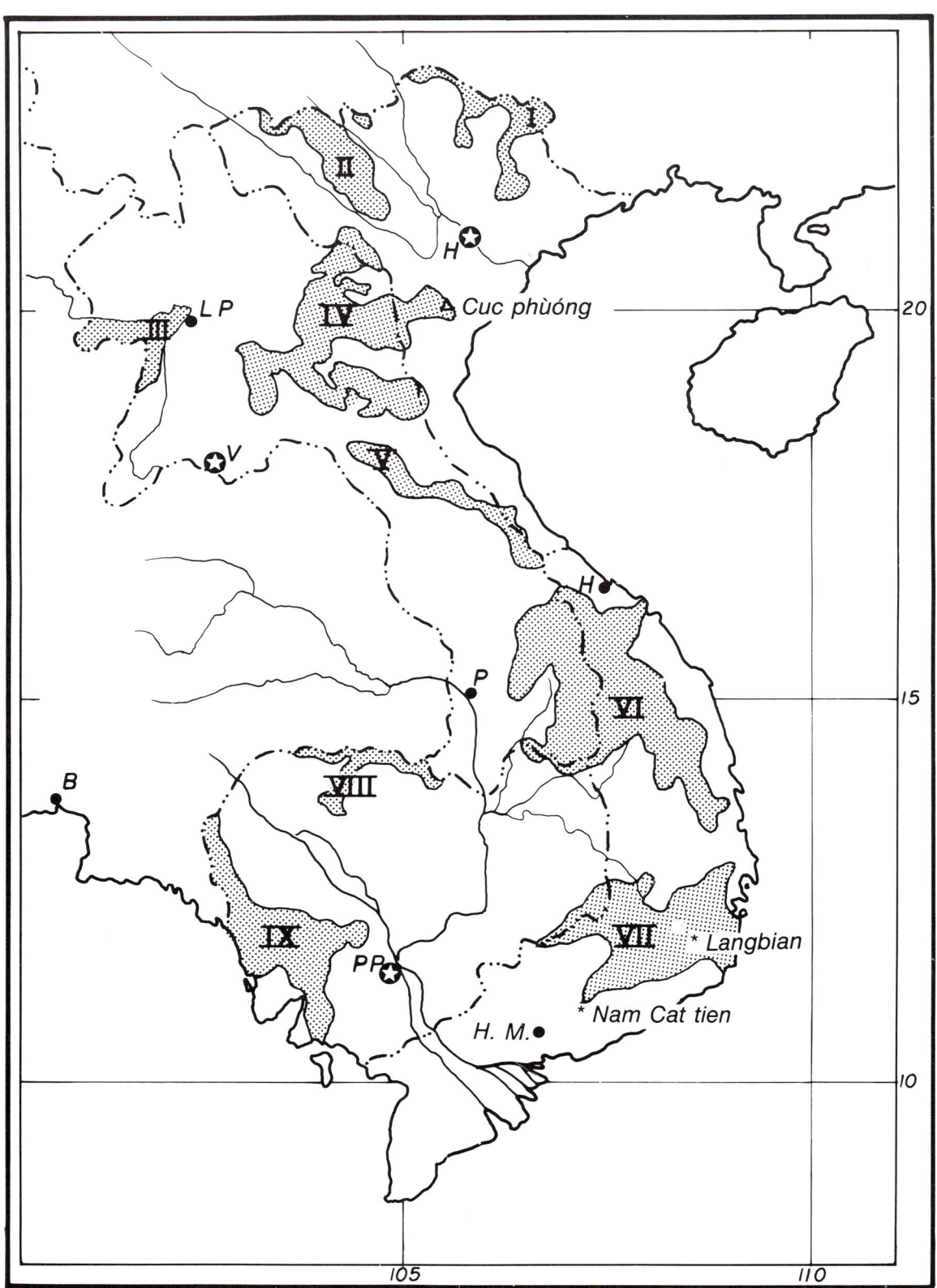

Map legend on p. 85.

Areas of special floristic interest (stippled)
Δ Cuc phuong: botanical reserve
* Protected areas (projects)

Main features of the delimited areas.

I. Topography: hilly
Climate: humid; winter relatively cold
Rocks: mainly limestones
Flora: some interesting species, but vegetation very disturbed

II. Topography: mountainous area (Fan si Pan: 3,140 m)
Climate: humid or perhumid; winter pretty cold
Rocks: granites or sandstones
Flora: of Sino-himalayan affinities: Gymnospermae: Pinaceae, Cupressaceae (*Fokienia*), Taxodiaceae (*Cunninghamia*), Fagaceae, Ericaceae (*Rhododendron, Vaccinium*), Magnoliales

III. Topography: hilly or mountainous (mountainous part not well known)
Climate: seasonal
Rocks: mainly sandstones
Flora: *Tectona grandis* area

IV. Topography: mainly mountainous (Phu Bia: 2,817 m)
Climate: humid; more seasonal in the west
Rocks: granites, sandstones, limestones (in the east)
Flora: in large part of Sino-himalayan affinities; Gymnospermae: Pinaceae (*Keteleeria*), Cupressaceae (*Fokienia, Calocedrus*), Taxodiaceae (*Cunninghamia*), Fagaceae (*Castanopsis, Lithocarpus*), Magnoliaceae, Hamamelidaceae (*Liquidambar, Altingia*), Lauraceae, Ericaceae (*Rhododendron, Vaccinium*), Platanus, etc.; at relatively low altitudes, Dipterocarpaceae (*Parashorea, Hopea*)

V. Topography: hilly (area not well known)
Climate: humid to seasonal
Rocks: mainly limestones
Flora: on limestones, *Dracaena, Cycas,* Araceae; on sandstones, *Pinus merkusii*

VI. Topography: for the largest part, mountainous area (Ngoc Pan: 2,600 m), still not well known
Climate: humid or perhumid towards the east, seasonal with a more or less severe dry season in the west and the southwest
Rocks: mainly granites and gneiss; basaltic tablelands in the west and the south
Flora: mainly of Sino-himalayan affinities towards the summits, of Indo-malesian affinities in the lower parts, especially in the west; Gymnospermae: Pinaceae (*Pinus khesiya*), Taxaceae (*Cephalotaxus, Amentotaxus*), Podocarpaceae; Fagaceae, Dipterocarpaceae (*Hopea*), etc.

VII. Topography: mountainous area (Chu yang Sinh: 2,400 m) surrounded by basaltic tablelands, at different levels, in the south and in the west
Climate: Dry along the coast, in the east, very humid on the slopes to the east, southeast and northeast (evergreen forest, mossy forest at about 1,200 m or 1,700 m, according to exposure); humid with a more or less severe dry season in the west and in the south at low altitude (deciduous forest)
Rocks: Granites and dacites for the mountains, schists and sandstones in the hilly parts, basalts for the tablelands
Flora: for a large part of Sino-himalayan affinities above 1,500 or 2,000 m, especially on granites, of Indo-malesian affinities at lower altitude, especially in the west; endemic species numerous, especially in the mountains and along the Pacific coast in dry conditions; Gymnospermae: Pinaceae; *Keteleeria, Pinus dalatensis, P. krempfii, P. khesiya, P. merkusii*), Taxaceae (*Taxus, Cephalotaxus*), Cupressaceae (*Calocedrus, Fokienia*), Taxodiaceae (*Glyptostrobus*), Podocarpaceae (*Dacrycarpus, Podocarpus fleuryi, Dacrydium*), Cycadaceae; Dicotyledones: Anacardiaceae (*Dracontomelum, Swintonia*), Dipterocarpaceae (mainly *Dipterocarpus* and *Hopea*), Ericaceae (*Agapetes, Rhododendron, Vaccinium*), Fagaceae (*Castanopsis, Lithocarpus, Quercus*), Juglandaceae (*Engelhardtia*), Lauraceae, Leguminosae (*Afzelia, Sindora, Pterocarpus*, etc.), Lythraceae (*Lagerstroemia*), Myrtaceae (*Syzygium*), Rubiaceae (Naucleae, etc.) etc.; Monocotolydones: Cyperaceae, Orchidaceae, Palmae, Poaceae, Zingiberaceae, *Musa*, etc.

VIII. Topography: hilly (highest point: about 800 m)
Climate: seasonal
Rocks: mainly sandstones
Flora: of Indo-malesian affinities; Dipterocarpaceae (*Dipterocarpus*, etc.), Guttiferae (*Mesua*, etc.)

IX. Topography: low mountains (highest point: 1,815 m, isolated in the northeast) with large piedmonts or valleys in the southeast and northwest
Climate: very humid in the south and southwest, seasonal with a more or less severe dry season in the north
Rocks: mainly sandstones with localized outcrops of limestones and basalts (especially in the northwest)
Flora: of Indo-malesian or Malesian affinities, poor in Gymnospermae (*Pinus merkusii,* Podocarpaceae), relatively rich in Dipterocarpaceae (*Dipterocarpus, Hopea, Shorea*), Fagaceae (*Lithocarpus*), Lauraceae, Myrtaceae (*Syzygium, Tristania*), Zingibera- ceae, etc.

I. Description of the Region

A. Geographical Extent, Topography and Population

Vietnam has an area of 330,000 km² and a population of about 60 million. Vietnam is mountainous in its central plateau and in the northwest. The mountains are not high (in the north: Fan Si Pan is 3,140 m; in the center, Ngoc Pan is ca. 2,600 m; towards the Dalat area, in the south, Chu Yang Sinh is 2,410 m). Vietnam has two large deltas, one in the south (Mekong), and one in the north (Song Hông).

Laos has an area of 240,000 km² and a relatively sparse population of 3.5 million. Like Vietnam, Laos has many low mountains. These may be found primarily in the northern regions and along the Vietnam border. The west central part is characterized by many plateaus and by the Mekong valley.

Kampuchea has an area of 180,000 km² and a population of six million. In Kampuchea, the Mekong Delta connects the central and southeastern parts of the country; plateaus (ranging to 950 m) dominate the landscape in the northeast; and the low Cardamomes Mountains (ranging to 1,815 m) are a characteristic feature of the south and southwest. The Cardamomes are an isolated mountain group separated from the Vietnamese highlands by the Mekong plain.

II. Vegetation Types

The main types of tropical forests represented within the area are the following:

There is evergreen tropical rainforest in southwest Kampuchea, as well as on the slopes exposed to the south and to the east, on the border of the Mekong Delta, in southern Vietnam and northeastern Kampuchea, and on the slopes exposed to the east in the central part of Vietnam.

Montane forest, often with Gymnosperms (Cupressaceae, Pinaceae, and Podocarpaceae), occurs in Vietnam and

Table I
Vietnamese institutions with significant holdings of herbarium specimens.

Institution	Location	Number of Specimens	Origin of Specimens
Institut d'Inventaire et d'aménagement forestier (Vien Dieu-Tra Quy-Hoach Rung)	Van dien Hanoi	40,000	Mainly North Vietnam
Université de Hanoi (HNU) (Truong Dai-Hoc Tong-Hop Ha-Noi)	Hanoi	31,000	North-Vietnam
Laboratoire de Botanique du C.N.R.S. (National Science Foundation) (HN) (Vien Khoa-Hoc Vietnam)	Tu Liem Hanoi	150,000	All over in Vietnam and Laos
Faculté de Pédagogie (Truong Dai-Hoc Su-Pham)	Hanoi	20,000	Mainly North Vietnam (a good collection of Mangrove species)
Faculté de Pharmacie (HNIP) (Truong Dai-Hoc Duoc Khoa)	Hanoi	10,000	Medicinal Plants
Ecole supérieure d'Agriculture (Truong Dai-Hoc Nong-Nghiep I)	Chau Quy Hanoi	20,000	Cultivated plants
Faculté de Sylviculture (Truong Dai-Hoc Lam-Nghiep)	Quang Ninh	10,000	Mainly from the North Vietnam (forest herbarium)
Faculté de Pédagogie (Truong Dai-Hoc Su-Pham)	Hué	5,000	From the central part of Vietnam
Herbier National (National Herbarium)	85 Tran Quoc Toan Ho Chi Minh City	150,000–200,000	All over (collected from 1834 to date)
Université de Ho Chi Minh Ville (Truong Dai-Hoc Tong-Hop) Tp Ho Chi Minh	Ho Chi Minh City	40,000	Mainly from South Vietnam
Faculté de Pédagogie (Truong Dai-Hoc Su-Pham)	Ho Chi Minh City	10,000	id.
Ecole Supérieure d'Agriculture (Truong Dai-Hoc Nong-Nghiep IV)	Thu Duc	10,000	id.

Laos, from the Dalat area northward. These forests are very localized in Kampuchea (Cardamomes), where they are poor in Gymnosperms (some Podocarpaceae and *Pinus merkusii*).

Dense deciduous forest (*Lagerstroemia*, Leguminosae) is especially important in central Kampuchea, and in the southern part of Laos and Vietnam, where it often grows on basalts.

Open Dipterocarp forest, with or without bamboo, occurs in northern Kampuchea, central Laos and central Vietnam, especially to the west.

In addition to these tropical forest types, semi-evergreen monsoon rainforest and open forest with bamboos occur widely throughout the three countries.

III. Vegetation Maps and Herbaria

There is a vegetation map of Vietnam on a scale of 1:2,000,000 and a map on a scale of 1:1,000,000 (Institut d'Inventaire et d'Aménagement des Forêts, Hanoi, 1980, unpublished), a vegetation map of South Vietnam (south of 17°) on a scale of 1:1,000,000 (Institut géographique, Nha Dia-Du Quoc-Gia, Dalat, 1969), and a map on a scale of 1:50,000 of the Ban Me Thuot area (Institute des Recherches Agronomiques, Saigon, 1955). For Kampuchea, there is a vegetation map on a scale of 1:1,000,000 (Institut de la Carte Internationale du Tapis Végétal, Toulouse, 1971). For Laos, there is only a map on a very small scale (Vidal, 1960).

According to Prof. Thai van Trung, Director of the National Herbarium, the institutions in Vietnam (Table I) hold about 400,000 herbarium specimens. The estimated numbers are still quite uncertain and probably too high. The Museum Herbarium (Laboratoire de Phanérogamie) in Paris holds about 150,000 specimens (see Table II). In addition, a few other herbaria have relatively small collections. Kew (mainly from the Hue area, in the central part of Vietnam), Arnold Arboretum (from the Cha Pa area in the mountainous part of North Vietnam), Berkeley (from the Cha Pa and Hue areas), The New York Botanical Garden, University of Tokyo (mainly from the Dalat area in the mountainous part of southern Vietnam), Komorov Institute in Leningrad (mainly recent collections from northern Vietnam).

The herbarium of the Kampuchea Forest Office (with perhaps 10,000 specimens), and the herbarium of the Phnom Penh University (with perhaps 5,000 specimens) were probably destroyed during the last troubles.

IV. Present State of Inventory

At the present time, there are about 10,000 species (vascular cryptogams and phanerogams) known within the three territories. The total number of native species might be between 12,000 and 15,000.

Concerning floristic evolution and speciation, the most interesting areas are probably (see Fig. 1): the Cha Pa area (region II) which is mountainous and has links to the Sino-

Table II

Holdings in the "Laboratoire de Phanérogamie" (Museum Nat. Hist. Nat., Paris) of herbarium specimens from Kampuchea, Laos and Vietnam — Main collections.

Collector	Dates	Number	Origin
\multicolumn{4}{c}{Old Collections}			
Béjaud, M.	1929–1935	1,100	Kampuchea
Bon, H. F.	1886–1895	6,000–8,000	Mainly North Vietnam
Chevalier, A.	1913–1919	15,000	All over
Eberhardt, P. A.	1906–1920	6,500	North Vietnam
Evrard, F.	1923–1929	3,500	South Vietnam (Dalat area)
Harmand, F. J.	1875–1906	3,500	South Vietnam
Lecomte, H. & Finet, A. E.	1911–1912	2,500	North Vietnam
Petelot, A.	1925–1955	9,000	Mainly North Vietnam
Pierre, J. B. L.	1877–1906	10,000	All over
Poilane, E.	1922–1961	40,000	All over
Thorel, C.	1875–1906	12,000	All over
Tsang, W. T.	1939–1940	4,000	Central and northern part of Vietnam
\multicolumn{4}{c}{Recent Collections (1940–1983)}			
Dournes, J.	1955–1965	1,000	Central part of Vietnam
Martin, A. M.	1965–1970	2,000	Kampuchea
Schmid, M.	1948–1964	5,000	Central part of Vietnam
Tixier, P.	1948–1970	1,000	All over
Vidal, J.	1939–1983	7,000	All over (mainly Laos)
Vu Van Cuong	1965–1970	1,250	South Vietnam (mainly hydrophytes)

Himalayan flora; the mountainous part of regions VI and VII; the coastal part of region VII (Phan Rang: a dry area which is pretty well known); the mountainous part and the low part in the southwest of region IX (with some links to the Malaysian flora). Many species (of which two are *Pinus*) are endemic to region VII, probably many species are peculiar to region IX and the contiguous part of southeast Thailand. There are probably endemic species restricted to limestone within region V, within region I (affinities with the limestone flora of South China), and in some very localized places in Kampuchea. Endemism at the genus level is very low in Laos and Kampuchea, and probably higher, although uncertain, in the central and mountainous southern part of Vietnam.

It is very difficult to estimate the rates of extinction. Probably the most interesting species are localized in the mountains of the central and southern part of Vietnam, where 20 years ago the vegetation was not disturbed too much.

Several species are exploited for their wood or bark and as a result are today very rare or threatened. These include: *Aquilaria, Calocedrus, Dysoxylum loureiri, Fokienia, Santalum* and some species of *Cinnamomum, Dalbergia* and *Diospyros*.

Among the well-collected families or floristic groups, may be mentioned: Cyperaceae, Dipterocarpaceae, Fagaceae (however, many species probably have not yet been collected), Gramineae (except bamboos), Gymnosperms, Lauraceae, Leguminosae, Orchidaceae (the same as for Fagaceae), and Rosaceae. Insufficiently collected families are the Musaceae, Myrtaceae, Palmae (especially lianas), Rubiaceae and Zingiberaceae.

The forests (or their remains) are threatened in the whole northern part of Vietnam, with the exception of the most inaccessible parts of the mountains, by overexploitation and conversion to agriculture. Large areas of forest reserves have been delimited, but clearing is authorized within some parts for shifting cultivation, which from a biological point of view defeats the object of the reserves.

On the plateaus of central Vietnam, forests are threatened by the expansion of cultivation, shifting cultivation by natives, or permanent cultivation, mainly by Vietnamese immigrants.

In southern Vietnam, the forests, which cover the slopes of the plateaus and the old alluvia on the northern part of the Donnai-Mekong plain, were extensive thirty years ago, but suffered very much during the war from cutting, bombing and defoliant spraying, and today are threatened by the extension of cultivation. Prof. Thai van Trung submitted to the government a project for the establishment of a reserve of about 30,000 ha in the northwestern part of this area, where the vegetation has not yet been too much disturbed and where the fauna—which may still include the rhinoceros—is particularly interesting.

There are no foreign programs concerned with floristic inventory currently implemented in the region. However, a revision of the "Flore Générale de l'Indochine" is being carried out at the Paris Museum (Laboratoire de Phanérogamie) with the collaboration of other research centers working at the same time on the Flora of Thailand. The last families revised were Caesalpiniaceae, Mimosaceae, Scrophulariaceae and Smilacaceae. The descriptions of new species have been principally published in "Adansonia" (Muséum National d'Histoire Naturelle, Paris). Links are maintained with Vietnamese botanists, particularly through the Vietnam National Herbarium (Prof. Thai van Trung).

Floristic research implemented by the Leningrad Komarov Institute in connection with the National Science Foundation (CNRS) in Hanoi must also be mentioned. Several families have been partly revised (Annonaceae, Apocynaceae, Araliaceae, Asteraceae, Cyperaceae, Dipterocarpaceae, Fabaceae, Lamiaceae, Magnoliaceae, Musaceae, Rosaceae, Rubiaceae), and new species have been described. Most works have been published in Botaníćeskij Żurnal or Nov. Syst. Pl. Vasc. (Leningrad) and Sinh-Hoc (Hanoi).

The main forest institutions where there are opportunities for collaboration in research on vegetation and flora are the Laboratoire de Phanérogamie (Prof. P. Morat, Museum) and ORSTOM, in Paris; the Leningrad Komarov Institute (Prof. A. Takhtajan); the University of Aarhus (Prof. K. Larsen) in Denmark; the Herbarium of the Royal Forest Department in Bangkok; and, for creation and maintenance of reserves, IUCN.

An inventory of the work published on the Indochinese flora (Kampuchea, Laos and Vietnam) is given in the *Bibliographie botanique de l'Indochine* (Pételot, 1955) and *Bibliographie botanique indochinoise* (Vidal, 1972). A complementary up-to-date bibliography has been published (1988) by J. & Y. Vidal and Pham Hoang Ho.

V. Priorities for Inventory

As noted above, floristic inventories may be sufficient for the northern and southern parts of Vietnam, but are quite insufficient for central Vietnam (recent surveys by Nat. Res. Council Botanists), the southern part of Laos and along the border of Burma and North Thailand, and in southwest Kampuchea. No inventory is currently planned in Kampuchea or Laos. A commission has been created in Vietnam for the inventory and conservation of the flora (Prof. Thai van Trung; Dr. Nguyen Tien Ban), but the possibilities for conducting inventories are very restricted (viz., no vehicles, not enough supplies) and, owing to the political situation, it is difficult to get permission to visit the most interesting areas (in some places, such as around Dalat, for lack of security).

The resumption of adequate floristic inventory in Kampuchea and Laos, and the creation in those countries of national herbaria, such as in Vietnam, is a matter of high priority.

Institutions and teaching centers concerned with floristic inventories are numerous in Vietnam, but these are concentrated in the north and in Ho Chi Minh City (Table III). There are no floristic centers in the regions where the flora is particularly rich and insufficiently known, and still not too much degraded, such as the Dalat and Kontum areas.

Table III

Botanists working in Vietnam on important families.

Location	Institution	Botanist	Family group
Hanoi	C.N.R.S. (National Science Foundation) (Vien Khoa-Hoc Vietnam)	Dr. Nguyen Tiên Ban	Annonaceae
		Dr. Ha Thi Dung	Araliaceae
		Dr. Le Kim Biên	Asteraceae
		Dr. Nguyen Khac Khoi	Cyperaceae
		Dr. Nguyen Dăng Khoi	Leguminosae
		Dr. Nguyen Tien Hiep	Rosaceae
		Dr. Tran Ngoc Ninh	Rubiaceae
		Dr. Vu Nguyen Tu	Pteridophyta
Hanoi	Université	Dr. Phan Ke Loc	General florisitic, useful plants
Hanoi	Faculté de Pharmacie	Dr. Tran Cong Khanh	Loganiaceae
Hanoi	Faculté de Pédagogie	Dr. Phan Nguyen Hong	Mangrove
Hanoi	Institut d'Inventaire et d'Aménagements forestiers	Dr. Vu Van Dung	Dipterocarpaceae
Ho Chi Minh City	Université	Dr. Le Thi To Quyen	Poaceae
		Dr. Le Cong Kiet	Pteridophyta, Gymnospermae
Ho Chi Minh City	Herbier National (National Herbarium) (Thao-Tap Quoc-Gia)	Dr. Tran Dinh Ly	Apocynaceae
		Pr. Dr. Thai Van Trung	Gymnospermae

There is a fairly adequate number of trained Vietnamese scientists and technicians. Some were schooled in France, the U.S.S.R., China or the U.S.A., and are competent. However, these latter are too few and additional foreign training of scientists would be desirable. Money is lacking for good maintenance of herbaria and for acquiring suitable equipment to make land surveys.

VI. Selected Bibliography

Taxonomy

Kampuchea, Laos, Vietnam

Laboratoire de Phanérogamie (Muséum, Paris) ed. 1905–1952. Flore Générale de l'Indochine, Masson, Paris, 7 vols. 1960. Flore du Cambodge, du Laos et du Vietnam. (22 fasc. publ., 55 families revised, including Anacardiaceae, Bignoniaceae, Combretaceae, Flacourtiaceae, Hamamelidaceae, Mimosoideae, Caesalpinioideae, Faboideae p.p., Loganiaceae, Ochnaceae, Pandanaceae, Rhizophoraceae, Rosaceae, Scrophulariaceae, Smilacaceae, Symplocaceae, Xyridaceae. In process: Dipterocarpaceae, Faboideae p.p., Thymelaeaceae, Proteaceae, Ericaceae, Juglandaceae, Primulaceae.)

Schmid, M. 1958. Flore agrostologique de l'Indochine. Agronomie Tropicale, Nogent sur Marne.

Vietnam

Grushvitsky, I. N., H. T. Skvortsova & Ha Thi Dung. 1985. Conspectus familiae Araliaceae Juss. Florae Vietnamii. Nov. Syst. Pl. Vasc. **22**: 153–191.

Le Kha Ke et al. 1969–1976. Common plants in Vietnam. 6 vols. Sciences & Techn., Hanoi.

Pham Hoang Ho. 1970–1972. An illustrated flora of South Vietnam. 2 vols. Bo Giao Duc, Saigon (Ho Chi Minh City).

Yakovlev, G. P., Nguyen Tien Ban & Nguyen Tien Hiep. 1982. On the scientific herbaria in the socialist republic of Vietnam. Bot. Zhurn. **67**, 9: 1306–1310.

Ecology, Phytogeography, Useful Plants

Champsoloix, R. 1954. Les forêts des Plateaux montagnards du Sud-Vietnam. Inst. Rech. Agr. et For., Saigon.

Do Tat Loi. 1981. Les plantes médicinales et les drogues médicinales du Vietnam. Sciences & Techn., Hanoi.

Dy Phon, P. 1981. Contribution à l'étude de la végétation du Combodge. Thèse Univ. Orsay.

Le Cong Kiet. 1962. La végétation psammophile de la presqu'île de Cam Ranh. Ann. Fac. Sc. Saigon: 367–434.

———. 1969. Le végétation des collines calcaires de la région de Hatien-Kienlu'o'ng. Thèse, Univ. Saigon.

Legris, P., R. Blasco et al. 1972. Notice de la carte végétation du Cambodge au 1:1,000,000. Inst. Fr. Pondichéry.

Maurand, P. 1943. L'indochine forestière. Inst. Rech. Agron. et Forest., Hanoi.

Nguyen Kha. 1966. Les forêts de *Pinus khesiya* et *P. merkusii* du Centre Vietnam. Ann. Sc. Forest. (Nancy) **23(2)**: 217–272.

Nguyen Ngoc Chinh & Vu Van Dung. 1971–1982. Essences forestières du Vietnam (Cay Go Rung Vietnam). 5 vols. Inst. Inventaire et Aménagement forest., Hanoi.

Nguyen Van Thuy. 1967. Les prairies marécageuses de la pénéplaine de Dalat. Ann. Fac. Sc. Saigon **2**: 373–450.

Pételot, A. 1952–1954. Les plantes médicinales du Cambodge, du Laos et du Vietnam. 4 vols. Centre Nat. Rech. Sc. Tech., Saigon.

———. 1955. Bibliographie botanique de l'Indochine. Centre Nat. Rech. Sc. Tech., Saigon.

Rollet, B. 1972. La végétation du Cambodge. Bois et For. Trop. (Nogent sur Marne): 144, 145, 146.

Schmid, M. 1969. Aperçu sur la végétation occupant les alluvions récentes de la partie méridionale de l'Indochine. UNESCO, Actes Coll. Dacca: 235–240.

———. 1974. Végétation du Vietnam: Le Massif Sud-Annamitique et les régions limitrophes. ORSTOM, Paris.

Thai Van Trung. 1970. The forest vegetation of Vietnam. Inst. Inventaire et aménagement forest, Hanoi.

Tixier, P. 1966. Les épiphytes du flanc méridional du Massif Sud-Annamitique. SEDES, Paris.

Vidal, J. 1956–1960. La végétation du Laos. 2 vols. Labor. forestier, Toulouse.

———. 1963. Les plantes utiles du Laos: Cryptogames, Gymnospermes, Monocotylédones. Muséum, Paris.

———. 1972. Bibliographie botanique indochinoise (from 1955 to 1969). B.S.E.I., Paris.

———. 1979. Outline of ecology and vegetation of the Indochinese Peninsula. Tropical Botany (Ac. Press): 109–123.

———. 1986. Problèmes et perspectives des études floristiques relative à la Péninsule indochinoise (Paris), **8/9:** 11–22.

———, **Y. Vidal & Pham Hoang Ho.** 1988. Bibliographies botanique indochinoise (1970–1985). Lab. Phanérogamie, Muséum, Paris.

Vu Van Cuong. 1964. Flore et végétation de la Mangrove de la région Saïgon-Cap St Jacques. Thèse, Univ. Paris.

Williams, L. 1965. Vegetation of South East Asia—Studies of forest types. Adv. Res. Proj. Agency, Washington, ord. 424, 1–9.

Sundaland

Peter S. Ashton

Contents

I. Description of the Region 93
   A. Geographical Extent and Area 93
   B. Topography, Geology and History 93
   C. Population 93
II. Vegetation Maps 94
III. Status of Floristic Inventory 94
   A. Patterns of Species Richness and Endemism 94
   B. Endemism 95
   C. Rates of Conversion 96
IV. Resources for Continued Floristic Inventory 97
   A. Regional Institutions 97
   B. Programs Based Outside the ASEAN Region 98
   C. Opportunities for International and Interdisciplinary Collaboration .. 98
   D. How Adequate is the Present and Currently Planned Floristic Inventory? 98
   E. The Minimum Resources Necessary to Ensure Adequate Floristic Inventory in the Coming Decade 98
V. Literature Cited 99

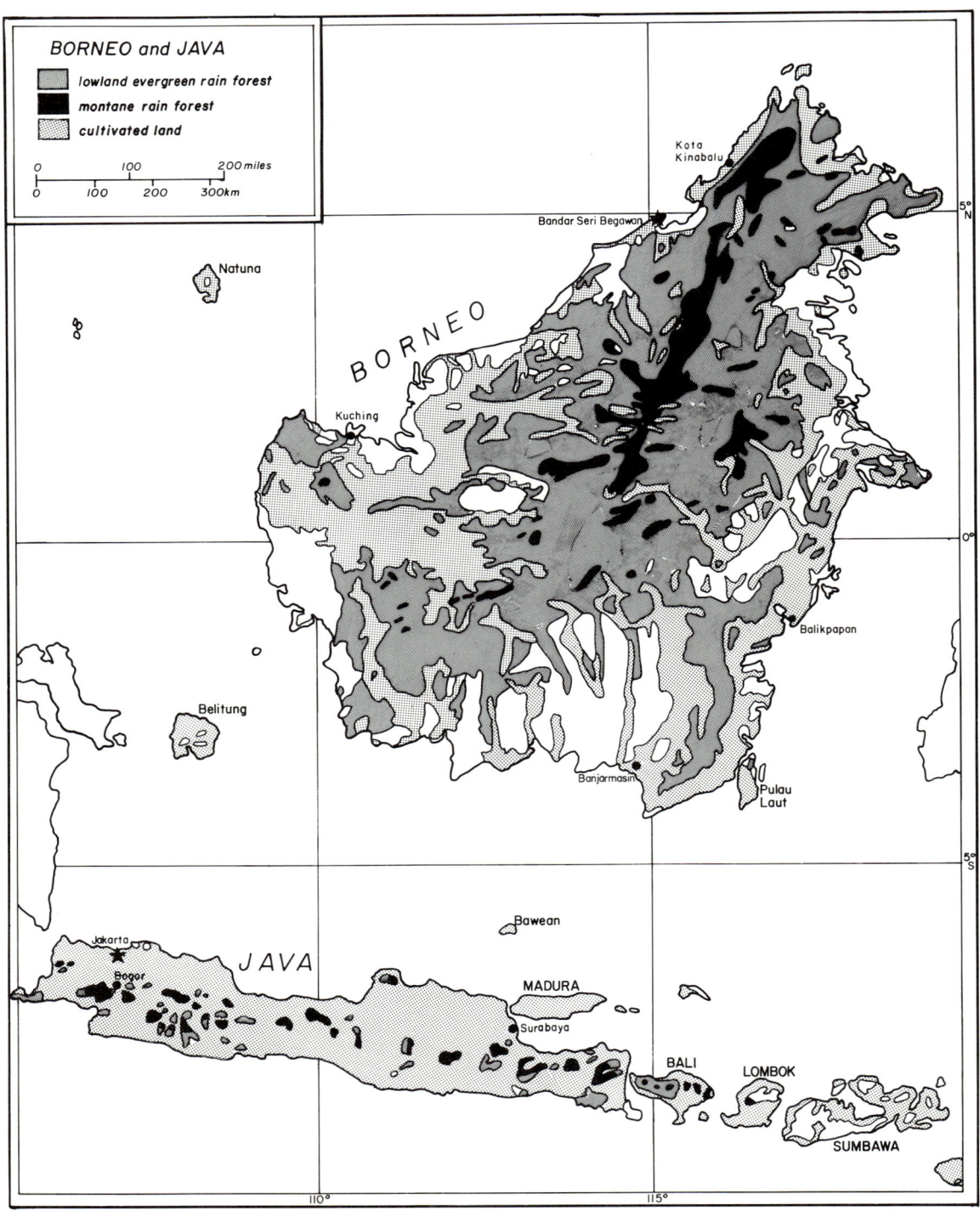

I. Description of the Region

A. GEOGRAPHICAL EXTENT AND AREA

The area under discussion, currently totaling 1,350,000 km², forms, with the Malaysian Peninsula, a geotectonic unit known as Sundaland. This is an area of large islands and shallow seas which, during most of post-Miocene times has been a single continental mass. Straddling the equator between the western edge of the Pacific Ocean and the eastern part of the Indian Ocean, the region is aseasonal and moist, with precipitation on average exceeding potential evaporation in all months and at all altitudes. Only at the margin of the area, in the rainshadows of Palawan and central-north Sumatra, and in northwestern Sumatra and central or east Java, where the dry trade winds penetrate, is there seasonality. The area is included therefore within that largest area of land in the world under tropical perhumid climates, which extends eastward through the Moluccas and New Guinea to the Pacific.

B. TOPOGRAPHY, GEOLOGY AND HISTORY

Lithologically and geomorphologically, however, there is great diversity. Briefly, Sundaland consists of a Laurasian continental plate, of which the Upper Paleozoic to Cretaceous crystalline core is exposed in southeast Sumatra, the main range of Peninsular Malaysia, the mountains of western Borneo, and the islands between.

A Gondwana plate, to the east of the Deccan plate and moving northward, continues to subduct what is now the southwest coast of Sundaland, having rotated the latter counterclockwise and pushed it northeast during the Tertiary. The Barisan Range, which forms a spine down the west coast of Sumatra, represents the uplifted margin of the Sunda plate, now mainly overlain by volcanic rocks, and leads southeast to Java's volcanic chain.

In Borneo, the continental core forms a triangle. One side follows the island's west coast. Another runs east-west and is bordered on the north by a Cretaceous to Recent geosyncline. The latter comprises an arc from the Kapuas Basin in the west to Palawan in the northeast. The third side, running west-southwest to east-northeast, abuts Quaternary and Recent sediments along most of its length, but towards the eastern end runs under the late Cretaceous and Neogene sediments of a lesser geosyncline along the southwest Bornean coast. In extreme southeastern Borneo, the Meratus Mountains and Pulau Laut, which consist of a complex of sedimentary, igneous and volcanic rocks, testify to Miocene and later folding consequent to the arrival from the south of the eastern, Australasian Sahul plate.

Whereas Peninsular Malaysia, eastern Sumatra, western and much of southern and eastern Borneo represent geomorphologically mature surfaces of undulating land and rounded hills, the Barisan Range, the Javanese mountains, the Meratus massif and the northwest Borneo neogeosyncline whose rim culminates in Mt. Kinabalu (4,175 m), represent steep to very steep young surfaces. In northwest Borneo characteristic parallel knife-edge cuestas, with sandstone crests and dip slopes, and shale or sandstone scarps, mandate a landscape of great lithological and edaphic diversity. Throughout the region, more or less isolated blocks and ranges of dramatic karst limestone, generally devoid of mineral soil but capped (where not burned) with raw humus, are a striking element in the landscape. Major areas occur in central Sumatra, northwest Peninsular Malaysia, northwest and east Borneo and northern Palawan. Down the east coast of Sumatra, in south Borneo, along the coast of Sarawak and Brunei in northwest Borneo, and in the basin and delta of the Mahakam are extensive plains of Quaternary sediments currently mantled with Holocene peats.

As yet, there is no unequivocal evidence for Quaternary climatic oscillations in the region. Arguments for or against therefore rely on biogeography and projections from the evidence of Quaternary seasurface faunas.

C. POPULATION

Until the 1960s, population centers were heavily concentrated in Java, where volcanic soils, extensive alluvial plains, and relatively seasonal climate (and hence controllable water supply and reliable harvesting seasons) supported ninety percent of the Indonesian population; and in the entrepôt of Singapore, which became a major city during the nineteenth century. The primary lowland vegetation of more humid and less seasonal west Java had become virtually confined to the poor soils of Bantam and parts of the southern coast by the beginning of this century. During the Hindu-Buddhist period, satellites of the Javanese empires flourished for a while in east Sumatra and the Padang Highlands of West Sumatra, with an outpost in northeast Peninsular Malaysia (Srivijaya), and on the lower Mahakam (Kutei), but Borneo and Sumatra remained overall one of the lowest populated regions in the habitable world. Plantation agriculture expanded in Sumatra, particularly around Medan in the northeast, from late in the nineteenth century. Population currently is rapidly increasing in the two islands and in Peninsular Malaysia, however, through immigration. Migrants from centers of unemployment, particularly in Java, are being settled on a very large scale in south Sumatra, and south and southeast Borneo. Present populations are approximately 95 million in Java, 22 million in Sumatra, and nine million in Borneo. Population increase in Indonesia is declining but still hovers around 2.1% per annum. Many of the early immigration schemes have proven unsuccessful, owing in part to poor agricultural conditions and to lack of familiarity with appropriate farming techniques by settlers. On this account deforestation and expansion of wastelands rapidly continues.

Peninsular Malaysia represents an exceptional case for the Tropics, where inherently infertile, forested lowlands have been converted, with careful planning and good management, to commodity crops, particularly rubber and oil palm. From the viewpoint of the aboriginal flora (though not the physical structure of the vegetation cover) the devastation has been at least as great as elsewhere, though the provision of good and reliable employment opportunities has all but eliminated slash-and-burn agriculture and thereby protected much of the hill forests.

II. Vegetation Maps

Few vegetation maps exist for the region. Until 1984, the only map for the whole area was compiled by C. G. G. J. van Steenis for UNESCO (1958). In 1984 T. C. Whitmore (Commonwealth Forestry Institute, Oxford) published a *Vegetation Map of Malesia* on a scale of 1:5,000,000. This map is particularly informative for plant explorers and conservationists because it depicts areas of cultivated land, secondary forest and scrub, forest plantations and natural grasslands. In Sumatra, an excellent new map has recently been published by L'Institut de la Carte Internationale du Tapis Végétal, Toulouse. For Sarawak, East Malaysia, a land-use map was published in 1956, indicating settled and shifting agriculture, mangrove, heath and the major categories of fresh water swamp forest. All other categories of putatively primary vegetation were dumped into a category of 'other forest.'

It is noteworthy that the region is by contrast rather well supplied with maps of surface lithology. Though they do not provide relevant information for mapping vegetation according to the 1973 UNESCO classification, they are very useful for finer classifications of primary forest, and thus in the identification of priority areas for inventory.

III. Status of Floristic Inventory

The region is excellently served by the monumentally conceived *Flora Malesiana*, edited by C. G. G. J. van Steenis, Rijksherbarium, Leiden. It contains comprehensive introductory chapters; a "Cyclopaedia of Collectors" (Steenis-Kruseman, 1950, supp. 1973, continuing in the annual "Flora Malesiana Bulletin"); chapters on phytography (1948–1954); an annotated bibliography (van Steenis, 1955–1958); collecting history and intensity (1950, 1973); and priorities for future collecting (1950, now partially out of date). The annual bulletin provides a comprehensive review, including progress with botanical collecting, expeditions, herbarium news and publications. The revisions in the Flora published so far, comprising about 20% of the angiosperm flora but less of the ferns, are semi-monographic in scope, and have almost without exception stood up well under use, although new taxa continue to appear. The species concept consistently adopted is broad, being based on qualitative differences and with quantitative local variation regarded as infraspecific. This almost certainly has led to underestimation of regional species richness in comparison to neotropical estimates, which are more often, even now, based on data from national and regional floras in which local quantitative variation is more recognized for distinction of species.

Table I
Collecting Intensities in Sundaland[a].

Region	Area (km²)	Approximate Total no. of Collection Numbers	Collecting Intensity (number 100 km⁻²)
Java & Madura	132,000	263,000	199
Sumatra & adj. Islands	473,630	103,000	22
Borneo: Total	736,695	255,000	35
Kalimantan			
(Indonesian Borneo)	539,665	65,000	12
Sarawak and Brunei	117,815	90,000	76
Sabah	79,235	100,000	126
Palawan	11,655	9,000	77
Whole region	1,356,337		46

[a] Peninsular Malaysia excluded.

The approximate collecting density, given on an island by island basis in Table I, gives an inadequate indication of collecting priorities, owing to the great regional variability in surface lithology and geomorphology, and hence the already mentioned variability in altitude and soils. Species richness of individual community types, as well as regional species diversity, is substantially correlated with these factors. Further, certain lithologies are particularly rich in species and/or endemics. I will therefore first address regional patterns of species richness and endemism.

A. Patterns of Species Richness and Endemism

Throughout Malesia, that is, from Sumatra and Peninsular Malaysia across to the eastern extremity of New Guinea and including the Philippines, New Guinea is the only region to vie with Borneo in the richness of its flora (van Steenis, 1950b). The spermatophyte flora of the Malaysian Peninsula is conservatively estimated at 7,500 species (Whitmore, 1973a), of which about 3,000 are trees. Although no recent estimates of that for Borneo are available, it is probably between 10,000–15,000 species. This estimate is based on comparisons of species richness between Borneo and Peninsular Malaysia of selected families revised in *Flora Malesiana*.

In Dipterocarpaceae, a well-known family in which only few unknown species are likely to turn up, there are 170% more species in Borneo than in Peninsular Malaysia; 58% out of the 267 are endemic to the island (Ashton, 1982). On the other hand, there are the same number of Euphorbiaceae in Borneo as in Peninsular Malaysia, only endemism in Borneo is 33% (Airy Shaw, 1975). Interestingly, in the euphorbiaceous genus *Claoxylon* there are 16 species, 13 of which are endemic to Borneo but only four in Peninsular Malaysia, whereas in *Baccaurea*, in which only four of the 25 species in Borneo are endemic, there are 20 species in Malaya (Airy Shaw, 1975; Whitmore, 1975). These figures suggest that part of the richness of the Bornean flora is attributable to the extraordinary habitat diversity on the island, particularly in its coastal northwestern and northeastern regions, into which endemics, especially taxa with poor seed dispersal such as *Claoxylon* and the dipterocarps, have differentiated ecotypically and regional species richness has reached its zenith.

Another reason for the relatively low score of *Euphorbiaceae* is probably undercollection. In 1921, the year of the foundation of the first forest department in Borneo by H. E. the Rajah's Government in Sarawak (systematic collecting beginning the following year), Merrill enumerated 102 species of dipterocarps known from Borneo. Since then, the combined efforts of the Sarawak, Brunei, Sabah and Indonesian forest departments have succeeded, for this preeminent hardwood family, to increase the species score 162%, while the economically unimportant, mainly weedy and understory Euphorbiaceae have increased only 106%, from 165 to 340. We may therefore reasonably expect the Bornean Euphorbiaceae to eventually exceed 430 species, whereas rather little increase can now be expected from the well-collected Malay Peninsula, with 371 species. These figures serve to emphasize that in most groups, but some more than others, the Bornean flora remains seriously undercollected.

B. Endemism

Although species endemism ranges as high as 60% in families with poor seed dispersal (although overall it is generally 40–50%), the number of endemic angiosperm genera has been set at 59 out of a total of approximately 1,500, while only one endemic family exists, the Scyphostegiaceae, consisting of a single species related to the Flacourtiaceae. These endemic genera include riparian or bottomland semi-weedy forms such as *Scyphostegia* and *Didesmandra* (Dilleniaceae), the endemicity of which defies easy explanation. Two patterns of endemicity stand out. One, the rarer, consists of an Australasian component, confined to low and medium altitudes on podsols and ultramafic rocks. An example is *Borneodendron* (Euphorbiaceae), of ultramafics in northeastern Borneo, whose nearest link appears to be *Cocconerion* of New Caledonia. The other pattern is a putative Gondwanan link, in Borneo confined to subcoastal hills on yellow leached soils in mixed dipterocarp forest, where *Didesmandra*, with affinities to *Schumacheria* of Sri Lanka, would appear to be an example. Others include *Upuna* (Dipterocarpaceae), possibly *Sarawakodendron* (Celastraceae), and many genera centered on Borneo but not strictly endemic to it, including *Axinandra* (Melastomataceae) of Borneo, the Malaysian Peninsula and Sri Lanka, and *Allantospermum* (Ixonanthaceae) of Borneo, Malaya and Madagascar. The explanation of this low-level generic but rather high species endemism seems to be that although northwest Borneo manifests exceptional habitat diversity, and may have escaped the climatic vicissitudes of the Pleistocene to a greater extent than other areas in western Malesia, the island as a whole has been connected to continental Asia and Sumatra for most of its history. The unusually high local species endemism of long-lived lowland rainforest plants there, especially trees, would attest to this. Also, Borneo, with Wallace's line lying to the east, is in some measure a meeting ground of eastern and western migrations, but nevertheless has experienced intermittent isolation from the west, and only intermittent (if any) land connection to the east by way of the Philippines.

An alternative explanation for the high level of endemism in northwestern Borneo may be the geomorphological youth and edaphic diversity of that area in comparison with the rest of the region. Not only are peats and podsols widespread, but the prevailing interbedded shales and sandstones are interrupted by isolated but significant beds of limestone, acidic as well as basic volcanic extrusions and, in the far west and at Kinabalu, granites. Local endemism is high on these exposures, while the leached coastal podsols and yellow podsolics support a distinct flora with high endemism in northwest Borneo, but also widespread though local elements which extend to the Riau archipelago, the coastal hills of Indragiri, east Sumatra, and east and south Borneo. Interestingly, the Holocene coastal peats of the region, though overall floristically poor and with few endemics, show the same regional pattern of species richness.

Within this northwest Bornean region, the Malaysian part is much better known than the Kapuas hinterland of western Indonesian Borneo, a long populated region whose fast-disappearing primary vegetation urgently needs further exploration. This region is rich in local taxa still insufficiently known for formal description.

Also still very poorly known are the basement granite areas in Indonesian western Borneo, an area which will certainly yield many range extensions and new species, as the 1982 Bogor-Leiden expedition to Gunung Raya in the Schwaner range has shown (Flora Malesiana Bulletin 36: 3892–3893, 1983).

The coastal hills of northeast Borneo, from Sandakan to the Sankulirang Peninsula, are also a center of endemism. Although now well known, this is an area where deforestation is particularly rapid. It therefore demands further exploration while the opportunity lasts. The ultramafic and limestone outcrops in this region and to the south of it are hardly known and deserve a very high priority.

The Meratus Mountains, and to a lesser extent Pulau Laut, consist mainly of basic, including ultramafic, rocks. The former shelter a rainshadow on their western slopes, where several endemics have been discovered. The collections from both are inadequate. The ultramafics of the Meratus remain promising for both endemics and range extensions.

The remaining parts of Borneo are apparently more uniform, and broadly poorer in endemics, although the mountains, and the basic extrusions and limestones, are certain to yield much new material. Southern Borneo has been little collected in this century, and the foothills of this province may well harbor unexpected species.

Palawan has a high collecting density, but the absolute number of collections is absurdly low and concentrated around the district capital of Puerto Princesa, in an area of sedimentaries. The island is lithologically and climatically very diverse. The central part is ultramafic and should prove of exceptional interest. The northern part is the oldest, contains much limestone, and is thought to represent a continental fragment from Asia. It remains little known botanically.

Sumatra, in spite of its size, consistently manifests a lower species endemism, not only lower than Borneo but also than the Malaysian Peninsula. The lower endemism than Borneo's is partly due to the substantial flora Sumatra shares exclusively with Peninsular Malaysia. But Sumatra remains poorly collected and, more particularly, the lack of resident botanical survey institutions has deprived us of adequate understanding of the ecology and biogeography of the island. Today the remaining lowland primary forest is so rapidly disappearing that few years remain for useful inventory. The following rationale for collecting priorities is, consequently, less substantially based than that for Borneo.

The Barisan Range provides Sumatra with the most extensive and continuous mountain region in Sundaland. With its rainshadow valleys, its present flora contains a significant continental element lacking from Borneo, including *Pinus* and elements such as *Altingia* (Hamamelidaceae) and *Primula* section *Candelabra*, which have spread east through Java via Sumatra. This flora is moderately well known from the Toba region and, increasingly, Gunung Leuser National Park, but most of the range, including the highest peak, G. Korinci, requires much more investigation. (See also the contribution on Sumatra by de Wilde in this volume.)

In the lowlands, the sedimentaries of the northwestern and eastern coastal hills as far south as Indragiri have distinct floras, with resemblances to both northwestern and east coastal Peninsular Malaysia, that remain poorly known and are now restricted to a few localities. Southern Lampong, now rapidly becoming totally deforested, is extensively volcanic and bears forest rich in Sapindaceae and Meliaceae. To the north, the eastern lowlands inland from the great peat swamps were formerly clothed with red meranti (Dipterocarpaceae) forest, apparently rather uniform and not rich in species. All but the peat swamps and the last-mentioned area urgently need further collecting. The coastal hills of Indragiri, in central-eastern Sumatra, and any remaining forest patches of the Riau Archipelago, remain inadequately collected, yet appear to hold great phytogeographic importance, owing to the presence of a distinct coastal Bornean element. They probably take top priority among these lowland areas.

Musala excepted, the floras of Sumatra's offshore Indian Ocean islands have not proven rich in endemics, although the herbaceous flora, including the epiphytes in particular, needs more collecting. In the South China and Java seas, Bangka and Belitung are now well known, but the Anambas and Natunas have been little visited by botanists and are expected to retain relicts from the former continental Sundaland flora.

Java, in strong contrast, is botanically the best known area in the Far East. Strongly influenced by volcanism, habitat diversity is principally climatic on account of the mountainous terrain and variability in seasonality. Aboriginal vegetation is now concentrated in the mountains; it has been described in some detail in a beautiful book by Hamzah, Toha and van Steenis (1972). Endemism is low. There has been less collecting in Java in the last two decades than in many other parts of Indonesia. Consequently van Steenis' (1950a) view that, although well collected, the lowlands of southwestern Bantam, south Priangan and south Java east to the Blambangan Peninsula still merit more work, holds. However, it is doubtful if these areas would yield more than comparable areas of Sumatra or Borneo. (See also the contribution on Java in this volume by van Balgooy.)

Thanks to "Flora Malesiana," a sound systematic base to the botany of the region is becoming available. Only phanerogams and ferns are being treated, however. Already, the revisions published indicate which taxa now require most further collecting.

As in other regions, the fungus, lichen and cryptogram floras are much less well known than that of the higher plants. Also, thanks to the work of the various forest departments, trees are generally better known than herbs and especially epiphytes and vines. However, the great speciose families, which often include a wealth of small trees, still require much further collecting: notably the Clusiaceae, Ebenaceae, Elaeocarpaceae, Ericaceae, Euphorbiaceae, Melastomataceae, Myrsinaceae, Myrtaceae, Palmae, Rubiaceae, and certain genera in Fabaceae, Meliaceae, Moraceae, Thymelaeaceae, Tiliaceae and Verbenaceae; also the important but difficult families Lauraceae and Sapotaceae.

Among herbaceous families, good pickled material is urgently needed for the large families Orchidaceae and Zingiberaceae.

C. Rates of Conversion

Practically all forest below 1,000 m in the region, which has not been included in national parks and nature reserves, has been given out for exploitation to logging concessionaires. Rates and intensities of exploitation within these concessions have varied, and many more inaccessible areas remain intact. At the time of writing (1987), the world economic recession has greatly reduced rates of exploitation, while increased export taxes and restrictions on export of unprocessed timber by Indonesia will prevent a return to the high, and certainly unsustainable, rates of the 1960s and early 1970s.

Logged forest can rarely, and only in the five years following exploitation, be distinguished from unlogged forest in satellite images. There is no means available, therefore, to estimate the area of primary forest remaining.

Logging causes variable damage and hence affects extinction of biota variously, according to the care exercised by the logger, the kind of equipment used and the nature of the terrain. Repeated creaming of prime timber within the regeneration time of the leading mature phase tree species, use of skidders and bulldozers (particularly when subcontracted on a piecework basis), and use of heavy equipment in steep terrain are the most damaging.

Whatever the intensity of damage from logging, its effect is to increase the area of the gap phase and decrease that of the mature phase. Inevitably, mature phase populations will decline in density, so that those species which require longer than the felling cycle to reach sexual maturity and those whose population densities are already critically low in nature, will inevitably disappear, even in those forests which are most carefully managed through natural regeneration. Strict nature preserves are therefore an indispensable part of rationally planned land use.

The region in question does not experience hurricanes. Volcanism is localized, the last major volcanic cataclysms being the formation of the Toba Caldera in north Sumatra at about 750,000 years B.P., and the eruption of Krakatau in 1883. As the region is mostly aseasonal, with rainfall exceeding ten cm in normal months, occasional fluctuations in rainfall distribution appear to be the major cause of natural widespread plant mortality. The effects of occasional prolonged droughts are greatly exacerbated by logging, while the presence of swidden agriculturists, who at the height of the dry periods burn felled fallow prior to planting, increase the likelihood of fire spreading into logged and even primary forest. Occasional reports of such fires have emanated from south Sumatra, Brunei and east Sabah and Kalimantan. During 1981–1983, a drought unprecedented since rainfall records began swept the Far East, affecting the regions mentioned, and also southeastern Peninsular Malaysia and almost certainly other eastfacing regions not reported. In Borneo alone, six, and perhaps as many as 12 million hectares of logged, and in some areas primary, forests were affected. Even in islands of forest which escaped fire, canopy trees were said to have died in east Kalimantan. Such occurrences have similar influences to those of logging. Where they periodically occur, the slow growing mature-phase flora is expected to be species poor, and there is some evidence that this is the case (Ashton, 1984). Also, epiphytes are particularly badly hit (J. Beaman, pers. comm. and pers. obs.) and the nature of the regional climate may therefore in part explain the apparent lower richness of the epiphyte flora, when compared with perhumid parts of the Neotropics (Gentry, 1984).

In summary, all lowland tropical forests of the region, excepting those in well managed national parks outside drought prone areas, are threatened with conversion within the next one to two decades, and eventually with destruction. The sanctity of preserves varies from island to island. It is high in Sarawak, increasing in Indonesia, declining in Sabah, and low in the Philippines. Forests above 1,000 m have greater chances of persisting, except in high population areas.

IV. Resources for Continued Floristic Inventory

A. REGIONAL INSTITUTIONS

The region is well provided with universities and herbaria. The problems are underfunding and understaffing, and past centralization of facilities which are no longer able to meet current inventory requirements, or are even defunct. The major regional botanical garden is Kebun Raya, the National Botanical Garden of Indonesia at Bogor. The botanical garden itself, with mountain and seasonal zone satellites, is one of the most extensive in the world and is relatively well maintained. The herbarium contains over 1,500,000 numbers from the whole of the Indonesian region. It includes duplicates of most numbers collected in the region particularly the Indonesian part, and includes duplicates of the nearby Forest Research Institute herbarium which houses about 45,000 numbers, almost entirely collected in Indonesia by its own staff. As part of the National Institute for Biological Research, Kebun Raya is an obligatory recipient of duplicates of all numbers currently collected within the country. Though it mounts modest collecting expeditions on its own in most years, Kebun Raya is inadequately funded to inventorize, even roughly, the forests undergoing conversion in Sumatra and Indonesian Borneo, let alone to rescue selected germplasm for introduction to the Garden.

Nor are scientific staff members in Indonesia adequate to pursue a significant program of floristic taxonomic revisions. Thanks to Flora Malesiana, the botany of the region is the subject of a comprehensive treatment that is among the best of its kind in the Tropics, and perhaps anywhere. However, because of the more monographic approach of this flora, field identification manuals for students and for foresters and agriculturists, are greatly needed. This is a task particularly well suited to Kebun Raya and its staff.

The leading herbarium in Peninsular Malaysia, with ca. 300,000 species, is at the Forest Research Institute, Kepong. The relatively numerous and well qualified staff is completing a flora of the peninsula. The collections predominantly represent woody plants, and concentrate on the peninsula, though there is good regional representation, particularly from the east of the country and from Brunei for which they probably have the best collection. The University of Malaya, at Petaling Jaya, has a good general herbarium, again concentrating on the peninsula. The Singapore Botanical Garden herbarium is of great botanical significance, but is now maintained by a skeleton staff.

In Sarawak, Sabah and Brunei, state herbaria are administered by the forest departments of Kuching, Sandakan and Bandar Seri Begawan, respectively. Although the botanists at the Forest Research Institute of Malaysia, Kepong, closely interact with those in east Malaysia, they do not contribute to eastern Malaysian programs. The herbaria at Kuching and Bandar Seri Begawan, which currently contain about 80,000 and 7,000 numbers, respectively, include rather comprehensive collections from their areas, although many early collections, notably those of Beccari, are unrepresented. The Sandakan herbarium was destroyed by fire in the 1960s but now once agains has more than 80,000 specimens. All three herbaria have active and relatively adequate collecting programs, although concentrating heavily on the tree flora.

No coordinated botanical inventory of Palawan has been carried out, and the Philippine National Herbarium has no funding for a consistent collecting program.

Although herbaria exist in the many universities of the Sundaland region, they mostly concentrate on local floras and do not, nor are likely to, constitute a significant resource in botanical inventory owing to the existence of clearly more appropriate agencies.

B. Programs Based Outside the ASEAN Region

The leading extraregional herbarium resources are at Leiden, where the materials for Indonesia are the most comprehensive of any; Kew, which has the most comprehensive collection for east Malaysia; and the Arnold Arboretum, which has good representation for Palawan, fair for the rest of the region. Important lesser collections are housed at the Smithsonian Institution, The New York Botanical Garden, the University of California (Berkeley), and the University of Michigan.

The only long-range foreign program of floristic inventory in the region is that of Foundation Flora Malesiana, with the Rijksherbarium as lead institution. In equal cosponsorship with the National Biological Institute of Indonesia, the Rijksherbarium continues to organize, at intervals of less than five years, a series of small to medium sized expeditions in Indonesia, somewhat over half of which are in the area under discussion. Occasionally, individual Rijksherbarium staff visit the region to collect, particularly for taxa in their field of interest.

Each year, some thirty other visiting foreign botanists independently collect in the region. The only other program with continuity is that organized by Dr. K. Iwatsuki of Tokyo University, which concentrates on herbaceous material, especially ferns.

C. Opportunities for International and Interdisciplinary Collaboration

Owing to the early progress of botanical inventory in the region, culminating (in Malaysia and Brunei) in the intensive collecting programs of the 1960s and 1970s, and owing to the existence of well developed regional centers with well qualified and experienced (albeit seriously insufficient) staff, the need is now for specialized collecting, and for general inventories in carefully selected areas, led by botanists experienced in the regional flora.

To be sure, some value always results from the collections of graduate students pursuing thesis work, and from the specialized collections resulting from interdisciplinary studies, but these do not address the dominant need: rapid, competent botanical inventory, including collections of live material, from areas undergoing conversion, and particularly those which are little known and suspected to be rich in endemics.

The regional institutions must be principal cosponsors of such initiatives. In general, they welcome collaboration with foreign groups that are led by experienced scientists who are known to and respected by them. Research visas are required to work in the countries concerned. In general, a sponsor within the counry in the region must first be identified. That sponsor and his/her institution would carefully evaluate the financial, scientific and political implications in proposals for collaborative ventures. This acts as a quality control and is a valuable additional insurance for outside funding agencies.

D. How Adequate is the Present and Currently Planned Floristic Inventory of the Region?

The statistics (Table I) indicate that while Malaysian Borneo and Java are now among the better collected areas of the humid tropics, collecting in Sumatra, Indonesian Borneo (Kalimantan), Palawan (where nearly all collections come from near the province capital), Sulawesi (Celebes), and New Guinea lag, and they are among the least adequately collected regions of the world. However, even in Java, areas remain to be explored. From the mid 1970s the rate of inventory of the regional flora has seriously declined to a point where it is now derisory; this at a time when notable increases have been achieved in the Neotropics.

The collecting programs in Sarawak and Sabah continue, but the herbaceous flora of these states, notably the epiphytes, fern and cryptograms, remain poorly known. The highest priorities are undoubtedly in Sumatra, Indonesian Borneo and Palawan, where opportunities are yearly declining owing to continuing rapid conversion and colonization.

The Leiden expeditions select areas according to criteria which closely match those used here, but the effort is totally inadequate in terms of the needs.

Areas of high endemicity where the last aboriginal vegetation is in threat, and which therefore require immediate inventory, include the following: the seasonal regions of North Sumatra, the Sapindaceae-Meliaceae forests of South Sumatra, the sedimentary coastal hills of East Sumatra and the adjunct Riouw Archipelago, the Anambas and Natunas archipelagos, Palawan, Indonesian west Borneo in general, the Crocker Range of western Sabah, Pulau Laut, the mid- and upper levels of the Meratus Mountains, limestone and ultramontane outcrops, and the main mountain systems of Borneo and Sumatra generally.

E. The Minimum Resources Necessary to Ensure Adequate Floristic Inventory in the Coming Decade

The herbaria of the region, and those worldwide, are institutionally adequate for long term needs, even though some regional herbaria are seriously underfunded. Bogor in particular can achieve little more than maintenance of existing holdings.

Van Steenis (1950c) has argued that a collections intensity of at least $100/100$ km^2 is necessary for the adequate documentation of the regional flora. This appears to be mostly unattainable. If, as a realistic but certainly barely satisfactory short term compromise, 50 plant collections/100 km² is regarded as an adequate collecting intensity, 340,000 further numbers need to be collected from Indonesian Borneo and Sumatra and their adjacent islands alone. This, however, is not an informed or realistic method of setting goals, because the areas where worthwhile collections can be got are already restricted and because institutional capacity worldwide is inadequate for such a task.

First, priority needs to be given for interim subsidy, over the next decade, to botanical inventory in Sumatra and Indonesian Borneo, which should be based on Kebun Raya, Bogor; and to that of Palawan, which should be based on the Philippine National Herbarium or the Forest Products Institute (FORPRIDECOM), Los Baños.

If a conservative target of 15,000 new numbers are to be obtained for Indonesian West Borneo, and 5,000 for each of the other regions identified in the preceding list, a total of 55,000 new collections needs to be made from those areas alone. This would conservatively require 11 man years actively in the field. When the needs of other parts of Indonesia are taken into account, it is seen that 15–20 man years for Indonesia alone are required. Even with the continuation of the Leiden program, nothing short of a project analogous to the Projeto Flora Amazonica would achieve this goal. The funding of a bilateral project between the United States and Indonesia analogous to the Projeto would fulfill most, though not all, of the Far Eastern regional needs using the present criteria. Based on the experience of the Projeto, and assuming comparable sharing between the two participating countries, costs would be approximately (1984) $200,000 p.a. over a ten year period for the U.S. contribution to the inventory alone. In addition, training grants for Indonesian participants, providing opportunities for research training at U.S. universitites, would be required, at an estimated total cost of $50,000 p.a.

V. Literature Cited

Airy Shaw, H. K. 1975. The Euphorbiaceae of Borneo. Kew Bull. Addit. Ser. IV. H. M. Stationery Office, London.

Ashton, P. S. 1982. Dipterocarpaceae. Flora Malesiana **9(2)**: 237–552.

_____. 1984. Biosystematics of tropical forest plants: A problem of rare species. Pages 497–578 *in* W. F. Grant (ed.), Plant biosystematics. Academic Press, Toronto.

Cobbett, D. J. & C. S. Hutchinson. 1973. Geology of the Malay Peninsula. Wiley-Interscience, New York.

Hamilton, W. 1979. Tectonics of the Indonesian region. U.S.G.S. Professional Paper 1078. U.S. Government Printing Office, Washington.

Hamzah, A., M. Toha & C. G. G. J. van Steenis. 1972. The mountain flora of Java. Brill, Leiden.

Liechti, P., F. W. Roe & N. S. Haile. 1960. The Geology of Sarawak, Brunei and the western part of North Borneo. Geological Survey Department, British Territories in Borneo, Bulletin 3 (2 vols.). Government Printer, Kuching.

Merrill, E. D. 1921. A bibliographic enumeration of Bornean plants. J. Straits Branch Roy. Asiat. Soc. Special Number. Fraser and Neave, Singapore.

Steenis-Kruseman, M. J. van. 1950. Alphabetical list of collectors. Flora Malesiana **1(1)**: 5–606; Supplements 1, 1958. ibid. **1(5)**: ccxxxvii–cccxliii, II, 1973, ibid. **1(8)**: i–cxv (subsequent additions in Flora Malesiana Bulletin).

Steenis, C. G. G. J. van. 1950a. The delimitation of Malaysia and its main plant geographic regions. Flora Malesiana **1(1)**: lxxvi–cvi.

_____. 1950b. Chronology of the collectors. Flora Malesiana **1(1)**: lxxvi–cvi.

_____. 1955–1958. Annotated selected bibliography. Flora Malesiana **1(5)**: l–cxliv.

Whitmore, T. C. 1973a. Tree Flora of Malaya. Vol. II. Longman, Malaysia.

_____. 1973b. A new tree flora of Malaya. Proceedings Precongress Conference. Pacific Science Association, Bogor.

Java

M. M. J. van Balgooy

Contents

I. Description of the Region	101
A. Geographical Extent and Area	101
B. Geology	101
C. Climate	101
D. Population	101
II. Vegetation Types	101
III. Vegetation Maps	102
IV. Magnitude of Floristic Inventory	102
V. Future of Floristic Inventory	102
VI. Literature Cited	102

I. Description of the Region

A. GEOGRAPHICAL EXTENT AND AREA

The region as here understood comprises the main island of Java, the large islands of Madura and Bawean, the Kangean and Karimunjawa islands and a number of smaller islands such as Panaitan, Nusa Kambangan, Nusa Barung and the islets in the Bay of Jakarta. The total area is about 132,000 km².

B. GEOLOGY

Java is a mountainous country. All the mountains are volcanic and most are of Tertiary age. Many are extinct, but some of the younger volcanoes are still active. The violent eruption of Krakatau in 1883, in particular, has received worldwide publicity. About half the island consists of volcanic deposits, the rest mainly of Tertiary marine deposits uplifted by tectonic movements. These are mostly limestone and are most widespread in the southern part of the island. There are some scattered, low Pretertiary outcrops especially in the western half. Most of the lowland is found in the eastern half of the island and is covered by alluvial deposits.

C. CLIMATE

The climate of most of west Java and the mountains in the central and eastern provinces is of an everwet type. Madura, the lowlands of east and central Java and the northwest part of Java experience dry spells lasting up to nine months. See particularly Schmidt and Ferguson (1951) and the more recent agroclimatic map by Oldeman (1975).

D. POPULATION

Java is one of the most heavily populated areas in the world, with 90–100 million people concentrated in the fertile alluvial lowlands and hills. The fertile volcanic soils of the island have allowed the growth of a large rural population, but the limit of population density has long been exceeded. The Indonesian government is making efforts to reduce the pressure on lands by transmigration to other islands and by promotion of birth control.

This high population and the long history of human occupation of the island has had a pronounced effect on the original plant cover, especially in the last century.

II. Vegetation Types

Java is floristically the poorest of the Greater Sunda Islands. This poverty is partly ascribable to recent destruction of the vegetation, although the proportion of actual species extinction is unknown. The areas with a strongly seasonal climate are particularly susceptible to human interference (fire). The original floristic poverty of Java is generally explained by the fact that Java is the most isolated and smallest of the Sunda Islands, by the high volcanic activity there, and because it actually consisted of several small islands in the past which only relatively recently have been connected to form the present island. The total number of native flowering plant and fern species is about 4,500 compared to an estimated 8,000 for the Malay Peninsula, which has approximately the same area. West Java is distinctly richer than central and east Java. Java is also poor in endemics. Several species formerly believed to be endemic have been found outside the islands. Some species occurring on Kangean and Karimunjawa are unknown from Java.

The largest concentrations of tidal forest (mangrove) are found in the coastal area east of Jakarta, near Surabaya, in the extreme southeast of the island and around Cilacap in the south. Members of the Rhizophoraceae (*Bruguiera* and *Rhizophora*) are dominant in this vegetation type, which is badly degraded in Java.

Coastal vegetation of sandy and rocky coasts include the 'pes-caprae formation' with herbaceous and shrubby species (*Ipomoea pes-caprae, Spinifex littoreus*) and the 'Barringtonia formation' with arboreous species (*Barringtonia asiatica, Calophyllum inophyllum*).

In the parts of the island with an everwet or feebly seasonal climate the climax vegetation in the lowland (up to 1,000 m a.s.l.) is evergreen rainforest (1.A.1). This is a tall forest (30–40 m) of mixed composition in which members of the Annonaceae, Euphorbiaceae, Guttiferae, Leguminosae, Meliaceae, Moraceae and Palmae prevail. Contrary to lowland forest in the other Greater Sunda Islands there is no dominance of Dipterocarps.

In the montane forest from 1,000 to 2,400 m (1.A.1b and 1.A.1c) the lowland families mentioned earlier gradually disappear and are replaced by other families such as Fagaceae, Lauraceae, Magnoliaceae, Myrtaceae, and Theaceaeae. This is also a tall forest up to 40 m with (in West Java) emergent trees of *Altingia excelsa* and *Dacrycarpus imbricatus*. It is somewhat poorer in species than the lowland forest but is richer in epiphytes, mosses and ferns.

Above 2,400 m the forest becomes increasingly stunted and poorer in species. Best represented among the arboreous species are the Ericaceae, Myrsinaceae and Myrtaceae, whereas the herbaceous and shrubby species include members of the Compositae, Graminae and Ranunculaceae.

Towards the east, Java is increasingly subject to a seasonal climate with dry periods lasting up to nine months. The natural climax vegetation is a seasonal evergreen (1.A.2) to deciduous monsoon forest (1.B.1). This forest is poorer in species than the everwet rainforest; epiphytes are scarce. Among the deciduous tree species teak (*Tectona grandis*) is the most common and is often also planted. Other deciduous species include *Aegle marmelos, Bombax malabarica* and *Garuga floribunda*. Common evergreen trees include *Azadirachta indica* and *Borassus flabellifer*. At altitudes between 1,500–3,000 m in east Java the forest often consists of almost pure stands of the fire-resistant *Casuarina junghuhniana*.

Most of the natural vegetation of Java is gone or heavily disturbed (logging, agriculture, cutting for firewood). Some small patches of lowland evergreen forest are left in parts of southwestern Java, Nusa Kambangan and some nature reserves. Small relics of monsoon forest are still to be found in southeast Java. The best preserved original vegetation types are the montane and subalpine forests. The montane flora is treated by van Steenis (1972).

III. Vegetation Maps

A map showing the extent of forest on Java is given in "Het Djatibosch bedrijf op Java" (1928). Vegetation maps were prepared by Hannibal (1950) and van Steenis (1958). They cover Indonesia and Malesia, respectively, at a scale of 1:5,000,000. A vegetation map of Java is given by van Steenis in Backer and Bakhuizen van den Brink (*Flora of Java*, 1966–1968), for which van Steenis wrote a valuable chapter "Concise plant geography of Java." A more recent map, again covering all of Indonesia, at a scale of 1:2,750,000 was issued by Direktorat Bina Program, Direktorat Jeneral Kehutanan Bogor in 1980. Java is also shown separately at a scale of 1:1,000,000. A new vegetation map, attempting to give a realistic idea of the remaining forest in Malesia based mainly on aerial photographs, has been prepared by Dr. T. C. Whitmore (Commonwealth Forestry Institute, Oxford) and appeared in 1984. A detailed vegetation map of the well-known Nature Reserve Ujung Kulon was prepared by Hommel (1987).

IV. Magnitude of Floristic Inventory

Botanically, Java is among the best known areas of Southeast Asia. Sri Lanka, Taiwan and Java are the only islands boasting a complete local flora. The total number of higher plants collected by 1972 was estimated by van Steenis-Kruseman (1974) to be about 260,000 numbers and is about 270,000 at present. Large scale collecting was started by the Dutch at the beginning of the last century. A large proportion of these collections is preserved in the Herbarium Bogoriense (BO). Another large herbarium in the same city is the one of the Forest Research Institute (BZF) (mainly trees) and the herbarium of BIOTROP. Other small herbaria, mostly housing only recent collections are found in Bandung (FIPIA), Yogya and Purwodadi. Outside Indonesia, by far the largest collection of Javanese plants is preserved in the Rijksherbarium, Leiden, Netherlands. Other foreign herbaria holding important collections of plants from Java are Kew, Singapore, Utrecht (mainly 19th century) and the British Museum.

The most important sources of information on the flora of Java are: Backer (1928–1934), Backer and van Slooten (1924), Backer and Posthumus (1939), Backer and Bakhuizen van Den Brink (1963–1968), Heyne (1952), Koorders (1911–1912), and Koorders and Valeton (1894–1914).

Various papers on Javanese plants have been published in "Bulletin du Jardin Botanique de Buitenzorg" (Bogor). "Reinwardtia" (Bogor) and "Blumea" (Leiden) as well as in other periodicals. Many families of Javanese plants have been treated in the unfinished *Flora Malesiana* (van Steenis, ed.).

V. Future of Floristic Inventory

A review of the density of collecting in Java illustrates that the island is generally well collected. Yet, there are still a few areas to be regarded as insufficiently known and which are worth visiting: the southwestern part of Java, the northern part of west Java around Indramayu (drought plants) and certain areas in the southeastern part of east Java.

The main local research institute is the Puslitbang Biologi (Biological Research and Development Centre) in Bogor, which includes Herbarium Bogoriense. Small scale collecting trips to gather both living and herbarium specimens are organized with foreign visitors. Active cooperation in this field exists, especially with the Netherlands (Leiden) and Japan, but is mainly concerned with areas outside Java, for which no coordinated exploration plan exists. See further the contribution by Dr. Ashton on Sundaland.

VI. Literature Cited

Backer, C. A. 1928–1934. Onkruidflora der Javasche Suikerrietgronden.

────── **& D. F. van Slooten.** 1924. Handboek der Javaanse Theeonkruiden.

────── **& O. Posthumus.** 1939. Varenflora voor Java.

────── **& R. C. Bakhuizen van den Brink.** 1963–1968. Flora of Java, 3 vol.

Heyne, E. 1952. De nuttige planten van Indonesie, ed. 3, 2 vol.

Hommel, P. W. F. M. 1987. Landscape-ecology of Ujung Kulon (West Java, Indonesia). Thesis, Wageningen.

Koorders, S. H. 1911–1912. Exkursionsflora von Java, 3 vol.

──────. 1913–1918. Atlas der Baumarten von Java, 4 vols.

──────. 1918–1923. Flora von Tjibodas.

────── **& Th. Valeton.** 1894–1914. Bijdragen tot de Kennis der Boomsoorten van Java, 13 vol.

Oldeman, L. R. 1975. An agro-climatic map of Java. Contr. Res. Inst. Agric. Bogor. 17.

Schmidt, F. H. & J. H. A. Ferguson. 1951. Rainfall types based on wet and dry period ratios for Indonesia with western New Guinea, Kem. Perh. Djaw. Meteor. Geof. Djakarta 42.

Steenis, C. G. G. J. van. 1965. Concise plant-geography of Java. *In* Backer & Bakhuizen van der Brink's Fl. of Java 2: 1–72.

──────. 1972. The mountain flora of Java. Brill, Leiden.

Steenis-Kruseman, M. J. van. 1974. Desiderata for future exploration. Fl. Mal. I, 8: iii.

Sumatra

W. J. J. O. de Wilde

Contents

I. Description of the Region	105
A. Geographical Extent	105
B. Topography and Geology	105
C. Population	105
II. Vegetation Types	105
A. Forest	105
III. Vegetation Maps	105
IV. Publications	106
V. Magnitude of Floristic Inventory	106
VI. Future of Floristic Inventory	106
VII. Literature Cited	107

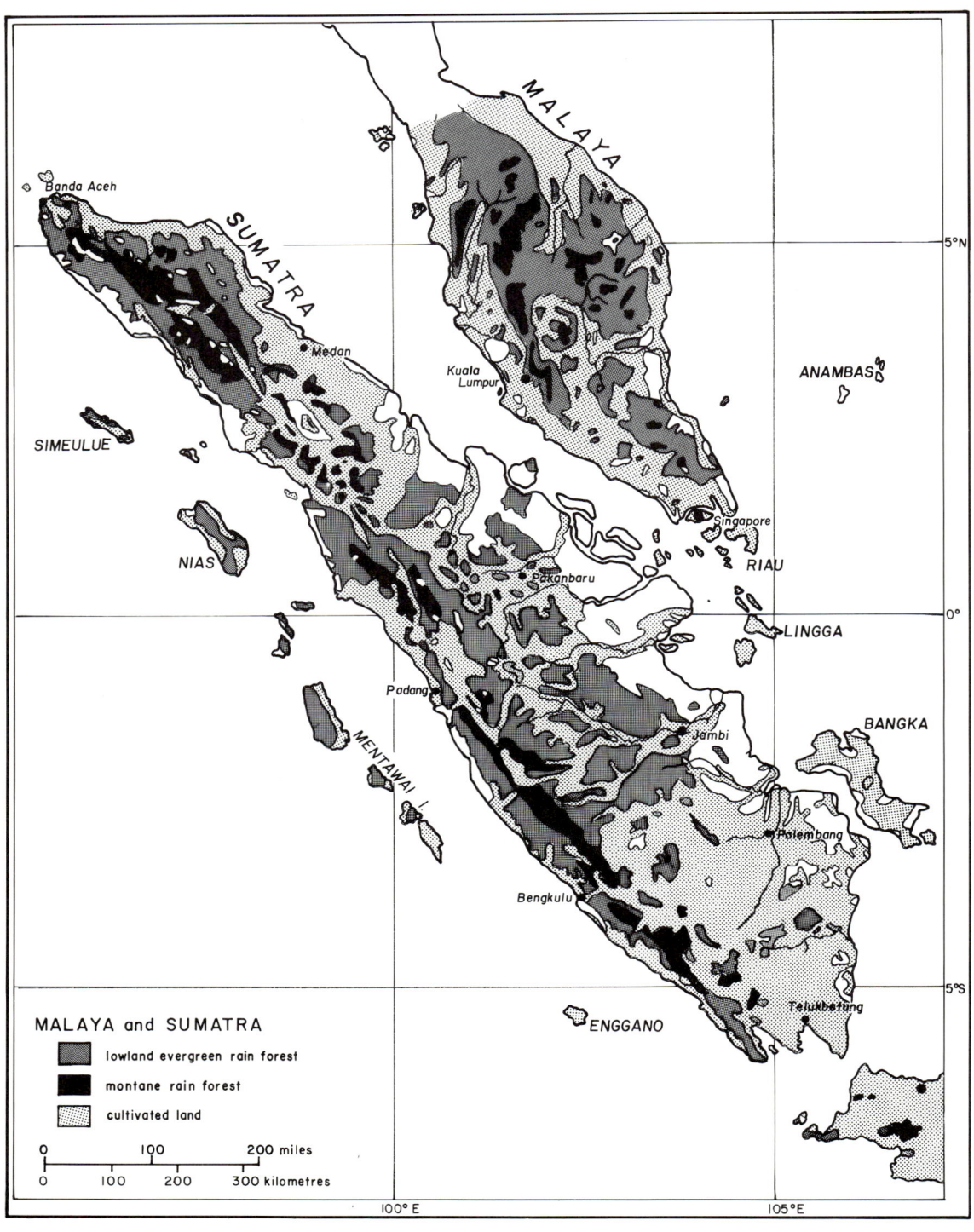

I. Description of the Region

A. Geographical Extent

Sumatra, including the Mentawai Islands and Simeuluë in the west, Bangka and Belitung and the Riau-Lingga Islands in the east, covers an area of about 470,000 km². Next to Borneo it is the largest of the Greater Sunda Islands on the Sunda Shelf.

B. Topography and Geology

Sumatra is elongate, about 1,650 km long, and nearly 350 km wide; it largely follows the "Circum Sunda Arc," to which belongs the Barisan Mountain chain that extends northwest to southeast along the west side of the whole island. This range has volcanic and nonvolcanic peaks reaching to over 3,000 m, for example, Gunung (Mount) Leuser in the north (ca. 3,400 m, volcanic). The Barisans are cleft lengthwise by a "graben" or main fault zone. To the east the mountain range flattens out to a broad sedimentary basin fringed by an alluvial plain of recent origin. The still active volcanoes are of Quaternary age. The Leuser area, situated in the north (mainly in Aceh Province) is mainly nonvolcanic. In places, Pretertiary limestone outcrops occur. As a consequence of the mainly everwet climate almost the whole area was originally covered by evergreen forest, mostly lowland and mountain forest of various types, and only in the Karo highlands (North Sumatra) and in the very north of Aceh are there dry pockets where a strong seasonal climate prevails (grass and scrub savanna after destruction of the original vegetation).

Whitten et al. (1984) in their book on the ecology of Sumatra give an overall review of the ecological status of the island, including much botany, and with extensive citation of literature.

C. Population

Medan, the capital of Sumatra, is the largest town; other larger towns are Banda Aceh, Padang and Palembang. Total population is about 24 million, about 51 inhabitants per km² (Java: about 700 per km²). The most densely populated areas are northern Aceh, the Medan-Karo Batak area (North Sumatra Province), and the Padang area (west Sumatra). Especially in Deli (Medan) and in the Karo Highlands soils are relatively fertile, owing to former volcanic activity. Population growth is fast. Locally there is extensive transmigration from Java, especially in south Sumatra and in southwest Aceh in the north.

II. Vegetation Types

A. Forest

As noted before, due to the mainly everwet climate, the original vegetation is largely evergreen rainforest (1.A). The Barisan Mountain chain, which receives the most precipitation (in many places much more than 3,000 m per year), carries various types of submontane and montane forest (1.A 1.b and 1.A 1.c) and cloud forest rich in epiphytes (1.A 1.e), all three to a large degree similar to that found, e.g., in nearby West Java. They are characterized by members of the families Lauraceae, Fagaceae (in North Sumatra including *Trigonobalanops verticillata*) and Conifers, in North Sumatra the low-stemmed, more open "elfin"-forest commonly with the beautiful *Rhodendron atjehensis*. In North and Central Sumatra (Mt. Kerinci area), the conifer *Pinus merkusii*, a species widely used for reafforestation of montane grassland, may be locally dominant on steep slopes (1.A 9.a/b). In the Kerinci area the conifer *Agathis dammara* and elsewhere other conifer species like *Podocarpus imbricata* may also be characteristic (1.A 9.b).

The lowland area west of the Barisans is generally wetter than the more extensively hilly and flat lowlands east of these mountains. According to the geographical position and the bioclimatological conditions, both for the dryland forests (1.A.1.a) as weel as for the inundated forests (1.A.1.g) various types of lowland rainforest can be distinguished (see Blasco et al., 1983; Laumonier, 1980; Laumonier et al., 1983, 1986). The dryland forest, especially, is usually of a very mixed character or with dominance of, e.g., Dipterocarpaceae, Euphorbiaceae, Annonaceae, etc. Locally, on better soils, there are several species of special interest like *Amorphophallus titanum* (Araceae), and the rare *Rafflesias* (like *R. arnoldii* and the local endemic *R. micropilorum*).

Some areas of lowland dryland forest (now largely exploited) are characterized by certain commercially important species, e.g., the dipterocarp *Dryobalanops aromatica* in the west; in the eastern plain some forests are characterized by *Eusideroxylon zwageri* (iron wood; Lauraceae).

The lowland swamp forest, mainly in the eastern part of Sumatra, consists mainly of tall alluvial forest (1.A.1.f) or, especially, various types of swamp forest or peat swamp forest (1.A.1.g). On the coast, which is accrescent on the east side due to the many large rivers debouching here, are broad fringes of Mangroves and Nipa-forest; both are types of swamp forest (1.A.1.g) determined by brackish water conditions.

III. Vegetation Maps

Sumatra was originally almost entirely covered by evergreen forest, including large proportions of all types of tropical ombrophilous forest. In Central Aceh and locally in the Northern Barisan Mountains there is some pine forest (*Pinus merkusii*), which is partly anthropogenous and fire resistant. The following vegetation maps of the area have been published:

1) van Steenis. 1939. Vegetation map of Indonesia, *in* Atlas van Tropisch Nederland. Scale 1:10,000,000. Outdated.
2) van Steenis. 1958. Vegetation map of Malesia,

UNESCO. Scale 1:5,000,000. Outdated, since cultivated areas and areas with destroyed forest and grass fields are nowadays very much more extensive.
3) Laumonier et al. 1982/1983, 1986. Vegetation maps of Sumatra. Scale 1:1,000,000. Published by SEAEO/BIOTROP; also published are South and Central Sumatra (North Sumatra scheduled). Contact Dr. F. Blasco, Université Sabatier, 39 Allées Guesde, 31400 Toulouse, France.

On these latter maps much detail is presented, and from these it can be estimated that only about 30% of the total land surface is presently covered by some sort of primary vegetation, e.g., forest or semi-primary forest. Of this, about half will be some type of mountain forest. Presently, most lowland forest commonly exists only in fragmentary forest.

The Forest Institute, Bogor, Indonesia, has rather detailed maps of all parts of Indonesia, including Sumatra, apparently mainly for private use.

IV. Publications

Except for minor publications on local areas, the flora of Sumatra is mainly treated taxonomically in "Flora Malesiana" (or in preliminary publications for the Flora), published by the Herbarium Bogoriense and the Rijksherbarium.

V. Magnititude of Floristic Inventory

The number of herbarium specimens held within Sumatra is very limited or none, all collections being kept in the state herbarium at Bogor (Herbarium Bogoriense), Java, or at the Rijksherbarium, Leiden, the Netherlands. Other foreign herbaria only hold limited collections, except for some recent medium-sized collections in Japan (e.g., in Kyoto University). The collections made by Americans in the 1920s and 30s, like Bangham's, Bartlett's, Rahmat's, and others, are mainly in American herbaria, e.g., the University of Michigan.

The total number of herbarium collections from Sumatra by 1972 (see van Steenis-Kruseman, 1974) is about 99,000; this is a density of 21, close to under-collected areas like West New Guinea (Irian Jaya) or Celebes (index at present ca. 24); the density of Java is about 200.

Almost all of Sumatra remains extremely poorly collected. Some important collections (mainly in L, BO) from limited areas are by: Lörzing (Sibolangit-Medan area); several pre-war American collectors (mainly in northern Sumatra: see Bartlett, 1935); van Steenis, de Wilde (G. Leuser area); W. Meijer (Mt. Sago-Pajakumbu); Jacobs, and others (G. Kerinci). Rather extensive pre-war collecting, mainly of trees, had been performed by the Forestry Service (Boswezen) in Sibolga, Jambi, and Palembang. Areas deserving special collecting efforts are: inter alia, north Aceh, the swamp forest in east Sumatra, certain parts of Bangkalulu (Bencoolen) and the Lampongs and the Pagai Islands.

The total number of phanerogamous species in Sumatra may be about 8,000–10,000. In general the vegetation links up floristically with those of lowland Malaya and Java (in Java the primary forest is very scarce now, and significantly poorer in species). In certain areas of Sumatra local endemism is relatively high, viz., the mountains of Aceh (van Steenis, 1938; de Wilde, in preparation) and likely places like the Indragiri foothills (see Ashton's paper on Sundaland) and, e.g., the Tigapuluh Mountains (Indragiri Province); the magnitude of endemism in other areas is little known. Special collecting in the named areas is urgently needed.

VI. Future of Floristic Inventory

In areas of rapid population expansion the rate of potential extinction accordingly will be high. Little is known about real extinction, although several species collected in the beginning of this century have not been recollected in recent times despite intensive exploration in north Sumatra (de Wilde) (e.g., *Passiflora sumatrana, Ulmus, Taxus,* and others).

In a vast under-collected area of diversified tropical rainforest, like Sumatra, no plant families are sufficiently collected.

Possibly all forest areas are of special importance, and almost all are threatened by destruction. Some areas are (or will be) indicated as (future) nature reserves (in Sumatra, e.g., the proposed G. Leuser National Park, G. Kerinci, and others).

There are no local academic institutions for carrying out a continuing floristic inventory in Sumatra.

All inventory is carried out by the national herbarium or the Forestry Department, both in Bogor, Indonesia. Occasionally there are local inventories in cooperation with the Rijksherbarium, sometimes also with Japanese or other foreign parties. Extensive foreign programs are nonexistent. Collaboration with other nations or other disciplines for plant collecting and botanical explorations is erratic and of minor importance. Formerly, too close cooperation with other disciplines was sometimes felt to hamper effective and extensive plant collecting for herbaria, because scientific plant collecting requires its own pace, skill and equipment. At present there is no correlation between exploration and potential areas for plant collecting, e.g., mainly the areas of timber exploitation, etc.; current botanical collecting activities are generally totally inadequate. An adequate floristic inventory, if ever possible, will require funds as soon as possible. In Indonesia all research should be done in cooperation with the government. It is regrettable that up to now so little scientific plant collecting has been performed in the many logging concessions, or in cooperation with the Forestry Service.

VII. Literature Cited

Bartlett, H. H. 1955. The Batak Lands of North Sumatra, from the standpoint of recent American botanical collections. Nat. Appl. Sci. Bull. Univ. Philipp. IV, **3**: 212–323.

Blasco, F., Y. Laumonier & Purnajaya. 1983. Tropical vegetation mapping: Sumatra. Biotrop. Bull. **22**: 23–43.

Laumonier, Y. 1980. Contribution a l'étude écologique et structurale des forêts de Sumatra. Thèse, Universitè Paul Sabatier, Toulouse.

———, **Purnajaya & Setiabudhi.** 1983. Vegetation map of South Sumatra; 1986: Vegetation map of Central Sumatra—Both published by I.C.I.T.V./Seameo-Biotrop (39, Allées J. Guesde, 31400 Toulouse, France).

Steenis, C. G. G. J. van. 1938. Exploraties in de Gajo-Landen (General results of the G. Losir expedition 1937)—Tijdschrift Kon. Ned. Aardr. Gen. **55**: 728–801, 32 photographs (in Dutch).

Steenis-Kruseman, M. J. van. 1974. Desiderata for future exploration. Flora Malesiana I, **8**: III.

Whitten, A. J., S. J. Damanik, J. Anwar, & N. Hisyam. 1984. The ecology of Sumatra. Gadjah Mada University Press, Yogyakarta, Indonesia.

Wilde, W. J. J. O. de. The vegetation of the Gunung Leuser Park (in preparation).

Sulawesi (Celebes)

E. F. de Vogel

Contents

I. Description of the Region	109
A. Geographical Extent	109
B. Topography and Geology	109
C. Soils	109
D. Climate	110
E. Population	110
II. Vegetation Types	110
III. Vegetation Maps	110
IV. Magnitude of Floristic Inventory	110
A. Important Collectors and Herbarium Collections	110
B. Publications	110
C. Collecting Density and Phytogeographic Relations	111
D. Endemism	111
E. Extinction	111
V. Future of Inventory	111
VI. Literature Cited	111
VII. Selected References	112

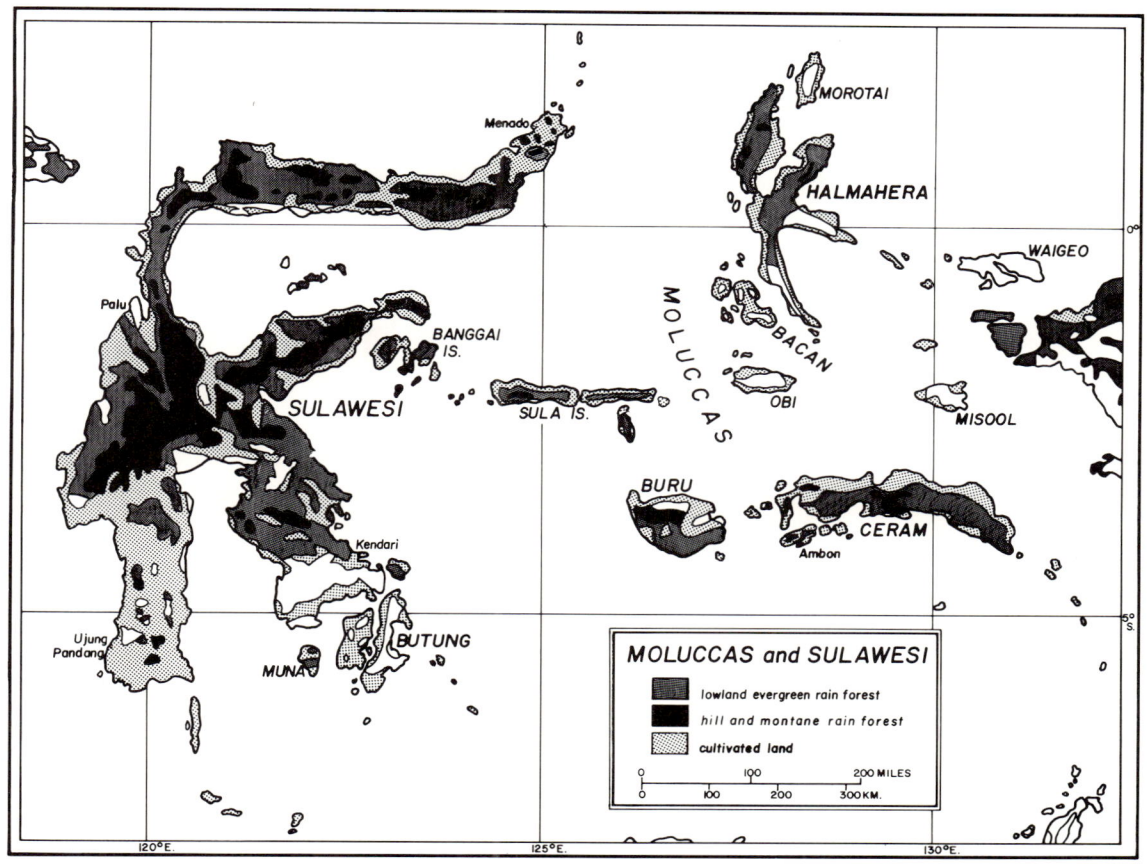

I. Description of the Region

A. Geographical Extent

Sulawesi (Celebes), including the surrounding islands, covers an area of about 189,000 km². Major islands and island groups are Kabaena, Muna, Butung (Buton), Banggai, Togian and Sangihe (Sangir)-Talaud. In size it is the fourth largest island in the Indonesian Archipelago.

B. Topography and Geology

Sulawesi is a geologically complex area. Its general shape can be described as two concentric arcs which are connected about halfway. The western arc merely consists of bodies of intermediate to basic volcanic rock and granite of Tertiary to Quaternary age, separated by areas with sediments of Quaternary age. On its north arm towards the Minahassa the volcanic deposits are more recent; in the latter area active volcanoes are still present. The arc thus formed continues in the Sangir (Sangihe) arc which bears active volcanoes as well. The Talaud Islands are a tectonic mélange, uplifted coral reefs are also present.

In the central part of Sulawesi where the two concentric arcs are connected, a north to south directed belt of blueschists and tectonic mélanges with limestone outcrops, amongst others, is present; in the southeast arm of the western arc this belt is continued. Towards the south this belt is increasingly covered with Quaternary alluvium; a similar situation is present on the islands Muna and Butung.

In the eastern arc large areas with ultrabasic rock, like ophiolites, peridotites and serpentinites are present in the central and northern parts, intercalated with Mesozoic and Tertiary sediments. In the east arm of the east arc rocks are partly covered with alluvium. In the extreme tip of the east arm an area with basalt is present. The ultra-basic rocks contain relatively large percentages of nickel, up to two percent, and a nickel mining company is present in the Soroako area. Kabaena Island consists of areas with ultrabasic rock and with tectonic mélanges.

The Banggai Islands consist of Palaeozoic rock; one island also has Jurassic outcrops, and on some islands Quaternary deposits are also present. The Togian Islands are entirely of Quaternary origin.

C. Soils

Soils in the northern arm are relatively fertile due to present volcanic activity. Fertile soils are also in the western arc;

very large lowland areas are deforested. Soil fertility of the ultrabasic areas is very low; here the vegetation is, to a certain extent, still present, although large areas have been logged over. Strip-mining in one area has caused the total loss of vegetation over a quite extensive surface. Soil fertility on the belt of tectonic melange is variable; the primary forest has been removed in many areas, even on several very steep mountain sides.

D. Climate

The climate is variable, with everwet conditions in the central highlands and parts of the north arm, and a highly seasonal climate in the arms of the island, especially both south arms. An extremely dry climate is present in the Palu Valley all year round; this is the driest part of Indonesia; average rainfall is less than 300 mm per year.

E. Population

The capital of Sulawesi Selatan, Ujung Pandang (Makassar), is the largest town of the island. Other large cities are Manado, Palu and Kendari. The total population is about 10.4 million, which makes an average of 54.7 per km² (data from 1980); the population is, however, very unevenly distributed. The most densely populated areas are the southwest arm and the tip of the northern arm, Minahassa. In these areas the natural vegetation has been largely destroyed by shifting cultivation.

Locally, population growth is fast; in other areas transmigration projects have been started, either by Javanese transmigrants or by farmers from densely populated areas on Sulawesi.

II. Vegetation Types

Originally Sulawesi must have been almost entirely covered with forest. Disturbances by settlers have taken place since prehistoric times. In the mountains, forests of UNESCO's types 1.A.1.b.–e. (evergreen, submontane, montane, and cloud forest) to 1.A.3.b. (semi-deciduous montane forest) are present in the central highlands and in the north arm. In the lowlands most forest is of types 1.A.2.a. (seasonal lowland forest) to 1.A.3.a. (semi-deciduous lowland forest). The dry Palu Valley is towards its southeast end covered with low shrubby, xeromorphic vegetation which presses towards the northwest into a savanna landscape with shrub vegetation in the gullies only; several parts around Palu are stone deserts without vegetation. Palms form a conspicuous part of the original vegetation in the west arc. *Livistona rotundifolia* (Woka) is common in the lowlands; its leaves are used for many purposes. *Pigafetta filaris* is a conspicuous element in the mountainous areas. After depletion of the forest these palms often remain. On the ultrabasics and tectonic mélanges palms are scarce.

III. Vegetation Maps

Sketch maps of the vegetation were published by C. G. G. J. van Steenis (1938) for Indonesia (scale 1:10,000,000) and Malesia (1958; 1:5,000,000). These are outdated, due to destruction of the forest by shifting cultivators, logging companies and strip-mining; secondary forests and grasslands are now much more extended. O. Hannibal, (1950) published a map of Indonesia, (outdated), on a scale of 1:2,500,000. T. C. Whitmore (1984) published a map of Malesia.

IV. Magnititude of Floristic Inventory

A. Important Collectors and Herbarium Collections

The number of herbarium specimens kept in Sulawesi is very limited. Small collections are presumably held at the universities in Manado and in Ujung Pandang. Major collections are kept in the state herbarium at Bogor (Herbarium Bogoriense), Java, and in the Rijksherbarium Leiden. Other foreign herbaria only hold limited collections. Important and extensive collections by R. Schlechter and the cousins Sarasin mostly from Central Celebes were lost in Berlin.

The total number of herbarium collections from Sulawesi by 1972 was ca. 34,000 (van Steenis-Kruseman, 1974); this represented a collection density of 19, the lowest one for Indonesia. Since 1972 extensive collections have been made on the island, including several hundred collections each by de Vogel (first sets L, BO), Dransfield (K), the British Museum (Natural History), Herbarium Bogoriense, W. Meyer (MO, BO, L), Johansson, Nybom and Riebe, and the Drake expedition (BO), over 6,000 collections by a Rijksherbarium-Herbarium Bogoriense expedition (first sets L, BO), and almost 1,000 collections by de Vogel and Vermeulen (first sets L, BO) during Project Wallace, bringing the collection density to almost 24 per 100 km².

B. Publications

Taxonomic revisions of plants from Sulawesi are mainly published in precursory publications for *Flora Malesiana* (in "Blumea" or "Reinwardtia"), or directly in *Flora Malesiana*. A bibliography up to 1954, listing the more important literature, was published in Flora Malesiana, I, 5, 1955 p. XXI and XXV–CXLIV. For more recent literature from that date onwards see the bibliographies in "Flora Malesiana Bulletin." See also the National Conservation Plan for Indonesia, Vol. VI (FAO, 1982).

C. Collecting Density and Phytogeographic Relations

The very low collecting density of Sulawesi indicates that the island is botanically very poorly known. Two areas, however, are relatively well known: Minahassa at the tip of the north arm, and the southwest peninsula. Large areas, for example, in the north arm, the central part, the east arm and the southeast arm, have never been visited by a botanist.

In Sulawesi, the area with the lowest collecting density of all parts of Indonesia, no plant family is well collected. Proof of this is the number of new species described after the Rijksherbarium-Herbarium Bogoriense expedition in 1979. The western arc has been much more intensely explored than the eastern, but from both arcs new species were described.

The total number of phanerogamous species in Sulawesi is not known; a fair estimate is 5,000 species. The island has a central position between the two major floristic areas of West Malesia and New Guinea. Already in 1945 Lam points to phytogeographical relations with West Malesia as well as East Malesia, although the relationship with West Malesia is more pronounced. Recent investigations by Van Balgooy, De Koning and Sosef (Van Balgooy, 1986), however, indicate a stronger relation of Sulawesi with the drier areas in the Malesian archipelago (Philippines, Java, Lesser Sunda Islands and the Moluccas), rather than with the moister areas (Borneo, New Guinea).

D. Endemism

High percentages of endemism at the genus level have been claimed for the island; in the Orchidaceae, for example, almost 80% (Schlechter, 1925), or over 71% (J. J. Smith quoted by Lam, 1945: 621), but recent revisions have shown that these figures are enormously exaggerated, and that endemism for most genera is much closer to zero. As far as I know, only one endemic genus of orchids is present on the island: *Bracisepalum* (Coelogyninae). The genus *Basigyne* from the same subtribe, formerly considered the only endemic genus, was later reduced by J. J. Smith to *Dendrochilum*. A similar situation is present in ferns, of which most species are widespread. Of the 27 species of *Rhododendron* 21 are endemic to Sulawesi. In other families endemism may be much higher. Local endemism may occur on isolated peaks, or may be confined to certain bedrock compositions. Both *Attelia mesenterium* (Hydrocharitaceae) and *Terminalia kjellbergii* (Combretaceae) occur only on ultrabasic rock. *Rhododendron bloembergenii* is endemic on high mountains. For most families endemism has not yet been estimated.

E. Extinction

About extinction little is known; quite a number of species were collected only once several decades ago, but recent collection trips have shown them to be not as rare as expected. Endemic species on ultrabasic soil may become extinct when more and more surface is cleared for strip-mining. Shifting cultivation and increasing numbers of transmigration projects likewise will cause increasing loss of species in the long run.

V. Future of Inventory

Areas deserving of special collecting efforts in Sulawesi are: the more western parts of the north arm; the central part; the east and the southeast arm, Butung and the Banggai Archipelago.

Forest cover of Sulawesi in 1982 was slightly over 50%, including all more or less depleted forest. Of this area 970,800 ha was included in existing and approved reserves. These reserves are somewhat less than 10% of the total surface of the island; however, in several reserves the forest is partly destroyed.

Important types of tropical forest most threatened by destruction or conversion are moist lowland forest on alluvium and on limestone, dry lowland forest on all sorts of soil and bedrock, and fresh water swamps.

Continuous floristic inventory depends on personal initiative of local university staff (and occasional interested amateurs). However, to my knowledge little collecting has been done by staff members of the universities of Ujung Pandang and Manado.

Occasional government expeditions are carried out by staff members of Puslitbang Biologi, Balitbang Botani (formerly Lembaga Biologi Nasional) (Bogor); the purpose of these trips is, in general, to obtain living plant collections, rather than herbarium specimens. Collection by local forestry officers is infrequent to nonexistent. Cooperation with foreign institutions is on an ad hoc basis; no permanent program exists for the inventory of Sulawesi, neither in Indonesia nor abroad. In 1985 a large-scale biological project (Project Wallace) organized by the Royal Entomological Society of London, worked in the Dumoga-Bone National Park, in which four botanists participated.

Plans to set up some sort of cooperation between local botanists and institutions abroad have had little result. The staff of local universities have occasionally accompanied foreign botanists, but all cooperation has been erratic.

VI. Literature Cited

Balgooy, M. M. J. van. 1987. A plant geographical analysis of Sulawesi. Pages 94–102 *in* T. C. Whitmore, Biogeographic evolution in the Malay Archipelago. Clarendon Press, Oxford University, New York and London.

FAO. 1982. Field Report 35. National conservation plan for Indonesia. Vol. **VI**: Sulawesi.

Hannibal, L. W. 1950. Vegetation map of Indonesia. Scale 1:2,500,000.

Lam, H. H. 1945. Notes on the historical phytogeography of Celebes. Blumea V: 600–640.

Schlechter, R. 1925. Die Orchidaceen der Insel Celebes. Fedde's Repert. **21**: 113–212.

Steenis, C. G. G. J. van. 1938. Vegetation map of Indonesia *in* Atlas van Tropisch Nederland. Scale 1:10,000,000.

———. 1958. Vegetation map of Malesia. UNESCO. Scale 1:5,000,000.

Steenis-Kruseman, M. J. van. 1974. Flora Malesiana. **I,8:III** pt. III.

Whitmore, T. C. 1984. Vegetation map of Malesia. Scale 1:5,000,000. J. Biogeogr. **11**: 461–471.

VII. Selected References

Balgooy, M. M. J. van 1987. The phytogeographical position of Sulawesi (Celebes). Pages 263–270 *in* P. Hovenkamp et al. (eds.), Systematics and evolution: A matter of diversity. Utrecht University, Utrecht.

——— & **I. G. M. Tantra.** 1986. The vegetation in two areas in Sulawesi. For Res. Bull., Spec. Ed.: 1–61.

Hamilton, W. 1979. Tectonics of the Indonesian region. Geol. Surv. Prof. Pap. **1078**: 159–190.

Meijer, W. 1984. Botanical explorations in Celebes and Bali. Natl. Geogr. Soc. Res. Repts. 1976 Projects. Pages 583–605.

Whitten, A. J., M. Mustafa & G. S. Henderson. 1987. The ecology of Sulawesi. Gadjah Mada University Press, Yogyakarta.

The Moluccas

E. F. de Vogel

Contents

I. Description of the Region	115
A. Geographical Extent and Topography	115
II. Vegetation Types	115
III. Vegetation Maps	115
IV. Status of Floristic Inventory	115
A. Extinction and Further Exploration	116
V. Regional Details	116
A. Halmahera, Morotai and the Bacan Group	116
B. The Obi Group	117
C. The Sula Group	117
D. Buru and the Seram Group	118
E. The Islands South of Buru and Seram	118
1. Ambelau, Ambon and the Uliasser Islands	118
F. The Outer Banda Arc Island Groups	119
1. Leti, Babar, Tanimbar and Kai	119
G. The Inner Banda Arc Island Groups	119
1. Wetar, Damar and Banda	119
H. The Aru Islands	119
VI. Literature Cited	119
VII. Selected References	119

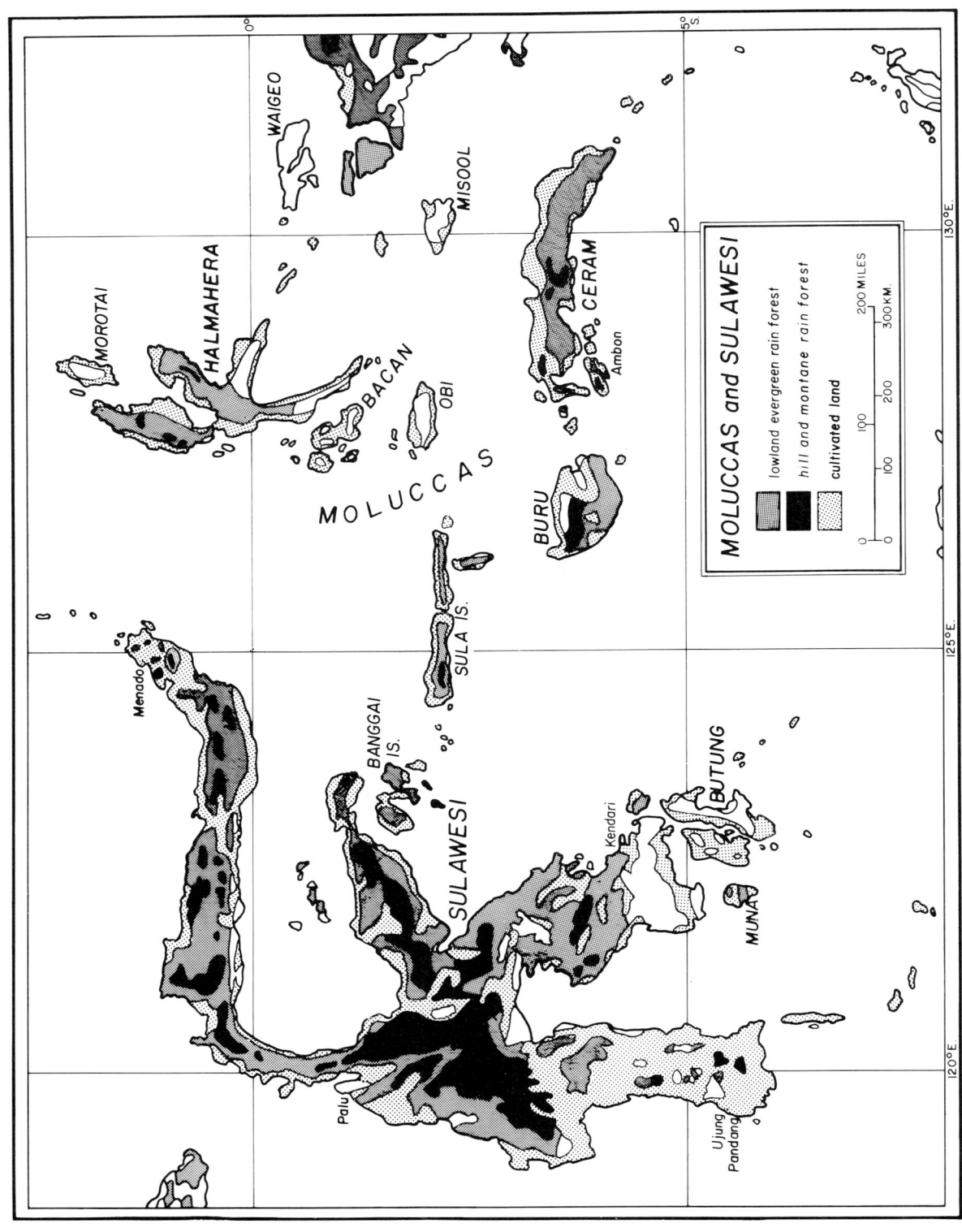

I. Description of the Region

A. Geographical Extent and Topography

The Moluccas are an island group scattered over a large area of sea between Sulawesi and New Guinea. Their land area is about 65,000 km². Major island groups are: (1) the Halmahera group, including Morotai and the Bacan islands; (2) the Obi Islands; (3) the Sula group; (4) Buru and the Seram group; (5) the islands south of Buru and Seram: Ambelau, Ambon and the Uliasser Islands; (6) the Outer Banda Arc island groups: Leti, Babar, Tanimbar, Kai; (7) the Inner Banda Arc island groups: Wetar, Damar, Banda; and (8) the Aru group. The islands vary considerably in size, topography and geology; the age of the bedrock is quite variable. Most island groups belong to at least three different geological subduction systems (arc systems), one group (the Sula Islands) is considered to be a disconnected piece of the Australian continental plate, and the Aru Islands, in fact, lie on the Australian continental shelf. The very complex geological composition of most islands fosters conditions such that even on small islands various forest types with very different compositions are present. For descriptions of the geology of the various island groups the author has relied heavily on the data given by Hamilton (1979) and Rutten (1927). The population of the Moluccas is ca. 1.4 million (1980) making an average of ca. 19 persons per km²; which by Indonesian standards is very low. The capital of the Moluccas is Ambon, a rather small city. Two large villages are present in the area: Ternate, and Soa-siu on Tidore; all other centers of population are more or less small villages. Although population increase is almost 3% per annum, pressure on the inland forests is still relatively low, because the interior of most islands is mountainous and of a rugged nature. Therefore, most people live in the neighborhood of the coast and on alluvial flats. The Aru Islands, on the contrary, consist entirely of lowlands, and here the population is more evenly distributed.

II. Vegetation Types

Sketch maps of the vegetation were published as listed below. Climate is variable with a severely dry period during May to October in the southern islands of both Banda Arcs. These bear tropical drought-deciduous lowland and montane forests of the types 1.B.1.a. and 1.B.1.b. (UNESCO, 1973). A moderate seasonal dry period occurs during September to November in the southern part of Halmahera, and in the eastern parts of Buru and Seram during July to November. Here, tropical semi-deciduous lowland forest is present (types 1.A.3.a. and 1.A.3.b.). All other islands have more or less everwet conditions, with tropical ombrophilous forest types (1.A.1.a.–g.) and tropical evergreen seasonal forest (1.A.2.a.–c.). In the mountains of the higher islands the conditions are everwet, with submontane and montane evergreen tropical ombrophilous forest (1.A.1.b.–c.). Since the population mainly lives along the coast, almost all forest on coastal plains has been cleared for agricultural purposes. Sago (*Metroxylon*, Palmae) forests are present in lowland swamp areas. Partly these are planted; sago is the staple food in the area. Swampy areas of broad leaved forest often contain high percentages of various genera of palms. In the deciduous and semi-deciduous forest *Melaleuca leucodendron* (Myrtaceae) and *Livistona rotundifolia* (Palmae) are conspicuous elements. Two species of *Shorea* (*S. selanica* and *S. montigena*, Dipterocarpaceae) are quite common in hill forest. On ultrabasic soil the genus *Siphokentia* (Palmae) can be locally very common. In montane forests *Agathis* (Araucariaceae) is locally common; this genus occurs also locally at low altitude. Most inland forests are in logging company concessions. These companies generally have permits to extract only one, two, or three kinds of trees (e.g., Dipterocarpaceae, *Agathis*, or *Intsia* (Leguminosae)). In many areas the forest is by now depleted to a more or less large extent, but much forest is still left. Entire destruction by settlers after logging operations occurs less frequently than in other parts of Indonesia. On the smaller islands of both Banda Arcs most lowland forest seems to have been converted into savanna.

III. Vegetation Maps

Vegetation maps of the area include the following: two maps by C. G. G. J. van Steenis—a 1938 map of Indonesia on a scale of 1:10,000,000 (now outdated) and a 1958 map of Malesia on a scale of 1:5,000,000 (outdated and, due to the scale and lack of precise data, not very accurate); a map of Indonesia by O. Hannibal (1950) on a scale of 1:2,500,000 (outdated); and a map of Malesia by T. C. Whitmore (1984).

IV. Status of Floristic Inventory

No herbarium is kept by any of the institutions in the region. Most collections made in the various areas are kept either in the state herbarium at Bogor, Java (Herbarium Bogoriense), or in the Forestry Research Institute at Bogor (mainly trees); in Tokyo, Japan (TI, KYO); and in the Rijksherbarium, Leiden, The Netherlands. Other herbaria hold only limited collections. The total number of herbarium collections for the Moluccas by 1950 was 27,525, by 1972 this amounted to 30,400 (van Steenis-Kruseman, 1950, 1974); this number is estimated to be still only about 48,500 in 1987. The first account by van Steenis-Kruseman gives the density of collections of the different island groups. It appears that the collecting density of the various island groups differs widely, and that it is low in most islands. After 1972 four important collecting trips were made in the Moluccas: one in the northern Moluccas by the present author (1,523 collections, first sets L, BO), and four in the southern Moluccas of which three were made by a group of Japanese collectors in Seram (almost 15,000 collections, first sets TI, KYO), and one by a Rijksherbarium team in Buru, (1,350 collections,

Table I

Number of collections and collecting densities for various island groups of the Moluccas.

Estimation	1950		1974		1987	
	Coll.	Density	Coll.	Density	Coll.	Density
Halmahera, Morotai and the Bacan Group						
Morotai	1600	76	Same		Same	
Halmahera	2400	13	3585		3585	19
Bacan Island	1000	29	1735		1735	51
The Obi Group						
Obi Islands	175	6	Same		550	19
The Sula Group						
Sula Islands	1375	28	Same		Same	
Buru and Seram						
Buru	1800	19	Same		3150	34
Isl. W. of Seram	50	14	Same		Same	
Seram	5350	31	Same		11,000	117
The Islands South of Buru and Seram						
Ambelau	few	low	Same		Same	
Ambon	7900	1039	Same		8000	1050
Uliasser Isls.	175	56	Same		Same	
The Outer Banda Arc Island Groups						
Leti Islands	50	7	Same		Same	
Babar Islands	50	6	Same		Same	
The Inner Banda Arc Island Groups						
Wetar	800	23	Same		Same	
Damar Islands	50	18	Same		Same	
Banda Islands	1250	1250	Same		Same	
The Aru Islands						
Aru Islands	2400	29	Same		Same	

first sets L, BO). For detailed data refer to the headings under the different island group later in this paper and Table I.

Best known in the botanical sense are Ambon, the Banda group, Geser Island and the Kai Islands. Most other islands and island groups have a collecting density lower than 50, and several are much lower. It must be kept in mind, however, that small collections on small islands results in high collecting densities. Quite extensive collections on large islands, however, many still result in a low collecting density, and the chance that all different habitats are visited and their representative species are collected is likewise low.

Taxonomic revisions of plants of the Moluccas are mainly published either in precursor publications for "Flora Malesiana" in "Blumea" and "Reinwardtia," or directly in "Flora Malesiana." A bibliography listing the more important literature up to 1954 was published by van Steenis (1955). For more recent literature from that date onwards, see the bibliographies in the "Flora Malesiana Bulletin"; see also the "National Conservation Plan for Indonesia," vol. VII, 1981 (FAO).

The total number of phanerogam species in the area has never been estimated; it may amount to ca. 3,000. Situated between the two major nuclei of tropical rainforest in the region, Sundaland and Sulawesi to the west, and New Guinea to the east, the Moluccas include elements of both.

A. Extinction and Further Exploration

Little is known about species extinction. Forests on ultrabasic soil are in danger in areas where nickel mining is planned because the strip-mining process denudes extensive areas of the soil. Lowland forests on alluvial flats are scarce on these islands, and most have already been converted to agricultural use. Logging companies are active on all larger islands in the area, see under the different island groups (data from 1972).

On island groups with a high collecting density (see the regional details) most plant families are probably well collected, on the others further exploration is urgently needed.

Logging goes on in all islands of the northern Moluccas, in the Sula Islands, in Buru and in Seram. The islands of the Banda Arc are not commercially logged.

No plan exists for continuous floristic inventory in the Moluccas. Expeditions to the region are few, and are mostly planned on an ad-hoc basis.

V. Regional Details

A. Halmahera, Morotai and the Bacan Group

This group of islands covers a land area of about 25,000 km². Like Sulawesi, Halmahera can be described as composed of two concentric arcs—belonging to one geological subduction system—which are connected halfway.

The western arc consists of active Quaternary volcanoes of which the northern ones lie on the mainland of the north arm of Halmahera, and the southern ones are situated offshore to the west of the island. Intermediate and basic volcanic rocks of Quaternary to Tertiary age, intercalated with clastic sediments and limestones, are present in the central western part of the island, continuing on the base, as well as on the northern part, of the north arm and on the island of Morotai.

The eastern arc is composed of the two eastwardly directed arms in the central part of the island as well as the southern arm. This arc consists of tectonic mélanges. Recent data on the geology are scarce and fragmentary, but the overall picture is that the rock types found are similar to those in Sulawesi, with basic rock outcrops intercalated with limestones and clastic sediments. Quaternary sediments overlay the tectonic mélanges in the north part of the south arm.

On alluvial flats near the coast the forest contains species like *Koordersiodendron pinnatum* (Anacardiaceae), several Myristicaceae such as *Horsfieldia sylvestris* and *Myristica* species, various species of *Diospyros* (Ebenaceae), and many palms of different genera.

On lowland terrain Ebenaceae (several species of *Diospyros*), Burseraceae (*Canarium* species), Alangiaceae (*Alangium* with two species) and Myristicaceae (*Myristica, Horsfieldia*) are conspicuous elements in the vegetation.

In hill country up to 500 m, Myristicaceae are still common with several species of *Myristica* and *Horsfieldia*. Other common genera are *Canarium*, *Alangium*, and Meliaceae (several species of *Dysoxylum* as well as *Aphanamixis grandiflora* and *Chisocheton sandoricarpus*). The rattan (*Daemonorops sarassinorum*, Palmae) is common in undisturbed forest; in more disturbed areas this species can form dense thickets.

The eastern part and south arm of Halmahera, especially, are in further need of botanical exploration. Several logging companies are active on the island and most forests are under concession. A very large portion of the south arm was given in concession to make a coconut plantation; it is not known by the present author if this is now in operation.

Bacan and neighboring islands consist of Quaternary deposits with much limestone and young volcanic deposits of Quaternary age. Bacan itself has, in addition, a large mountain of Palaeozoic age (Mt. Sibela) consisting of silicic rocks.

The vegetation of Bacan at altitudes up to 500 m is comparable to that of Halmahera. Common families here are Myristicaceae (*Myristica, Horsfieldia*), Alangiaceae (two species of *Alangium*), Burseraceae (several species of *Canarium*), Meliaceae (*Chisocheton* and *Dysoxylum*), Sapindaceae (*Pometia pinnata*), Anacardiaceae (*Dracontomelon dao* and *Buchanania arborescens*), and Ebenaceae (several *Diospyros* species). Palms are not very conspicuous in the vegetation.

At higher elevations, around 1,000 m on Mt. Sibela, Myristicaceae are comon (*Horsfieldia* and *Myristica*). Other genera which play a major role in the vegetation are *Lepiniopsis ternatensis* (Apocynaceae), *Neuburgia corynocarpa* (Loganiaceae) and *Neolitsea* and *Cryptocarya* (Lauraceae). Palms are not abundant but the rattan *Daemonorops* forms dense thickets in more open places, and the giant palm *Gulubia* is a common emergent up to 55 m high.

Of the Bacan group, only the Mt. Sibela area is well-known botanically, other areas urgently need further exploration. Most forests on the island are under logging concession.

Morotai is reasonably well known. All forests on the island are under logging concession.

B. The Obi Group

This group of islands covers a land area of about 2,750 km². The Obi islands are considered not to belong to the Halmahera group, but to represent the remains of a former subduction system. The northern part of Obi Island itself (about a third of its surface) consists of sedimentary rocks of mainly Tertiary age including limestone massifs and laterite areas. The southern part consists of tectonic mélange including large areas of ultrabasic rock; around 1970, a survey was made in the western part of Obi to investigate the possibilities for nickel mining.

On Obi itself only four small areas are thoroughly explored. A one-day trip by the present author to outcrops of ultrabasic rock on a hill in West Obi revealed several new and very rare species. Only two botanists have made extensive collections on the island.

The vegetation on laterite at ca. 500 m is quite different from that of Halmahera. Two *Shorea* species, *S. selanica* and *S. montigena* are quite abundant. Other important elements in the vegetation are *Crypteronia* (two species), *Myristica* and *Horsfieldia* with several species and *Knema tomentella* (Myristicaceae), *Gironniera celtidifolia* (Ulmaceae) and *Polyosma integrifolia* (Saxifragaceae).

On a limestone hill at 600 m palms are dominant, with the genera *Licuala, Siphokentia, Gronophyllum* and *Drymophloeus*.

The ultrabasic area in the western part of Obi is to a large extent devoid of forest and bears a low herbaceous vegetation. Small pockets of forest are mainly dominated by Myrtaceae (*Decaspermum, Metrosideros, Eugenia* and *Rhodamnia*). Other common plants are *Trichosperma kjellbergii* (Tiliaceae), *Vatica papuana* (Dipterocarpaceae) and *Palaquium* and *Planchonella* (Sapotaceae). Palms are very common, and include *Gronophyllum* and *Siphokentia beguinii*. The latter species is very variable in the leaf morphology, with the blade entire or compound, with few to 22 pairs of leaflets.

If strip-mining for nickel should start in the area—I have no information that it has actually started—most of the original vegetation will disappear. Most of Obi Island is under logging concessions. A logging company was active in the northern part of the island in 1974; they harvested only two species of dipterocarps, however.

The smaller islands of the Obi group are entirely under concession. They have never been visited by a botanist.

C. The Sula Group

This group of islands covers a land area of about 4,850 km². The Sula islands form, together with the Banggai group (see Sulawesi), a geological unit considered to be a piece of continental plate. It consists mainly of granites and metamorphic rocks which surface in two areas: the western part of Taliabu and the extreme east part of Mangole. Everywhere else on these two islands these rocks are covered with sediments of Mesozoic to Tertiary age, like arenites and carbonate-bearing sandstones and shales.

Sulabesi Island, while near Mangole, does not belong to this island group; in a geological sense it consists mainly of tectonic mélanges including shales.

All the islands of this group are undercollected and need more exploration. Only Bloembergen visited the area for a prolonged period. A typed report is present in the Forest Research Institute in Bogor; it was not available to the present author. On each of these islands a logging company is operating.

D. Buru and the Seram Group

This group of islands covers a land area of about 27,000 km². They constitute the northern part of the outer ridge of the Banda Arc.

Buru consists, in the south-southwest half, of an imbricated tectonic mélange of sedimentary rocks of young Mesozoic to Tertiary age, including large areas of limestone dating from the Mesozoic. Its north-northeast half consists of various types of crystalline basement rocks. Quaternary deposits are present in the north, but are otherwise scarce.

The northeast part of the island has a pronounced dry season with savannas dominated by *Melaleuca leucadendron*, of which large parts have been replaced by grassland. The *Melaleuca* is kept low, by burning and cutting, to keep the leaves within harvestable reach for the production of Kayu Putih oil.

The other parts of the islands, with an everwet climate, have a mixed evergreen forest dominated by *Shorea selanica* and *S. montigena*. The timber of these species is extracted by a number of logging companies which have a concession for these two species only. In the north *Casuarina* (Casuarinaceae) is dominant along the streams at ca. 350 m altitude. The terraces are dominated by *Neonauclea* (Rubiaceae); *Ficus* (Moraceae) is well represented. At ca. 500 m on very loose soil with stones and boulders *Shorea* is dominant, *Alstonia* (Apocynaceae), *Ficus* and *Octomeles* (Datiscaceae) are common.

In the northwest a limestone area is present mixed with other rock. *Casuarina* is again abundant on river terraces. *Neonauclea* and various species of *Eugenia* are dominant and *Albizia*, (Leguminosae), *Canarium*, *Elaeocarpus* (Elaeocarpaceae), *Terminalia* (Combretaceae), *Octomeles*, *Alstonia* and *Shorea* are common; the latter is however here not dominant. Above 600 m the vegetation changes with occurrence of *Dacrydium* (Podocarpaceae), *Gymnostoma* (Casuarinaceae), *Heritiera* (Sterculiaceae) and *Prumnopitys* (Podocarpaceae).

The forests at higher altitudes are said to be rich in *Agathis*, which does not form pure stands; logging of this species is prohibited.

Seram largely consists of tectonic mélanges, locally covered with coastal deposits of Quaternary age. The latter deposits are most extensive on the north coast from the constriction of the island in the western half towards the eastern end; they constitute a broad coastal belt of limestones and sandstones and conglomerates. A broad belt of imbricated and mélanged sedimentary rocks of Mesozoic to Tertiary age extends from the extreme east to the extreme northwest of Seram, consisting mainly of areas of sandstones, shales and limestones. From the center of the island eastwards this covers the entire width of the island. Mélanged metamorphosed sedimentary rocks, mainly schists, with intercalated basic and ultrabasic rocks are present in the southern half of West Seram. Crystalline basement rocks are present in several small outcrops all over the island; a larger area of these rocks is present on the tip of the peninsula opposite Ambon.

Lowland forests are 20–40 m tall and of mixed nature. They are dominated by *Canarium*, *Cananga odorata* (Annonaceae), *Celtis* (Ulmaceae), *Elaeocarpus* and others. At altitudes around 400–600 m *Shorea* is locally the main emergent.

Hill and lower montane forests are characterised by two Fagaceae: *Castanopsis buruana* and *Lithocarpus celebicus*. The latter species is found mainly on ridges. Other common species between 1,000–1,500 m are *Elmerillia sericea*, *Talauma celebica* and *T. oreadum* (Magnoliaceae), *Kadsura ultima* (Schisandraceae), *Galbulimima belgraviana* (Himantandraceae), as well as the fern *Cyathea biformis*.

Montane forest between 1,500–2,200 m contain large numbers of *Agathis* and *Dacrycarpus* (Podocarpaceae), Myrtaceae, *Castanopsis* and *Lithocarpus* (Fagaceae), *Tarrieta* (Sterculiaceae) and *Engelhardtia* (Juglandaceae).

Montane and elfin forests up to 20 m high are developed at 2,200–2,500 m altitude. They are characterised by *Phyllocladus* (Phyllocladaceae), *Quintinia* (Saxifragaceae), *Trimenia* (Trimeniaceae), *Dacrycarpus*, *Engelhardtia* and others.

Dwarf elfin forest, 5–10 m high at 2,500–2,700 m, consists of the tree fern *Cyathea*, *Dacrycarpus*, *Myrica* (Myricaceae), *Phyllocladus* (Phyllocladeaceae), *Rhamnus* (Rhamnaceae), and *Rhododendron* and *Vaccinium* (Ericaceae).

At the end of 1984 a Rijksherbarium expedition was made to Buru; ca. 1,350 collections were made. Only a few areas in the northwest were visited. Many more habitats, especially those on limestones and schists, and all mountain areas above 1,000 m need exploration urgently. According to the schedules all accessible areas will be logged over by 1989.

The islands west of Seram urgently need further exploration. Recently a Japanese team took almost 15,000 collections (first sets KYO, TI) in a limited area in the central part of Seram; they visited two of the highest peaks of the island. Many habitats on the island need further exploration. In several areas on Seram logging companies are active.

E. The Islands South of Buru and Seram

1. Ambelau, Ambon and the Uliasser Islands

This group of islands covers a land area of about 1,120 km². Ambelau (Amlau), Ambon and the Uliasser islands consist mainly of young volcanic soils and rocks. Along the coast elevated coral reefs are present locally. These islands are a continuation of the inner volcanic Banda Arc. The southern peninsula of Ambon consists of crystalline basement rock; this part is separated from the northern part by a stretch of Quaternary sediments.

All islands except Ambon need further exploration urgently; in Ambon hardly any undisturbed forest is left, on the other islands hardly any forest is left. The present author has had no access to reports on the vegetation, if any exist.

F. The Outer Banda Arc Island Groups

1. Leti, Babar, Tanimbar and Kai

This group of islands covers a land area of about 8,800 km^2: Leti 755, Babar 775, Tanimbar 5,820 and Kai 1,450 km^2. They form a continuation of the arc starting with Timor. They are mainly tectonic mélanges, consisting of a mixture of disrupted sediments and metamorphic rock. The Leti Islands all consist of a ring of young, elevated coral reefs surrounding a central area consisting of both ultrabasic rock and sedimentary rocks including sandstones, shales and some limestone. All these rock types are partly metamorphosed, partly unaltered.

The Babar Islands are mainly elevated coral reefs; the main island has a core of sandstones, shales, and ultrabasic outcrops.

Of the Tanimbar group, the western islands consist of elevated coral reefs; Yamdena Island also has, besides large areas of coral reefs, areas with tectonic mélanges containing Mesozoic sandstones, shales and limestones. A typed report on a field trip to Yamdena by Buwalola should be in the Forest Research Institute in Bogor; it was not available to the present author.

The western islands of the Kai group consist of young Tertiary to Quaternary clayey deposits in which crystalline and sedimentary rocks are present as fragments. The eastern island consists of tectonic mélanges.

The other islands of this arc towards Seram are small. They are mainly elevated coral reefs with a small central core of tectonic mélanges.

All islands except the Kai islands need further exploration urgently. Existing maps indicate large savanna areas on the smaller islands. No details on flora and vegetation were available to the present author.

G. The Inner Banda Arc Island Groups

1. Wetar, Damar and Banda

This group of islands covers a land area of about 4,200 km^2: Wetar, including Roma and Kisar, 3,850, Damar 285, and Banda 45 km^2. Together they form the inner Banda Arc, which is of volcanic origins. Both Wetar and Roma are extinct volcanoes of young Tertiary age. Along their north coast coral reefs are present which are uplifted to 800 m above sea level. The other islands of this arc are young, still partly active volcanoes; some bear elevated coral reefs.

Only a description of the vegetation of Wetar was available to the present author (Bloembergen, 1940). In the coastal area the forest is dominated by *Pterocarpus* (Leguminosae). Additional common species here are *Albizia*, also *Zizyphus* (Rhamnaceae). The inland area carries open *Eucalyptus* (Myrtaceae) forest dominated by four species of this genus, and open *Eucalyptus* savanna. Special mention was made of regular fires.

All islands except Banda need further exploration urgently; Banda itself is almost entirely deforested. On Wetar a logging company is active. Existing maps indicate large savanna areas on the smaller islands.

H. The Aru Islands

This group of islands covers a land area of about 2,255 km^2. The Aru islands belong geologically to the continental plate of New Guinea and Australia. They are elevated limestone reef plateaus of Quaternary to Plio-Pleistocene age. Little relief is present on these islands; nowhere does their altitude exceed 100 m.

The islands are uniform in topography and bedrock, and consequently the flora may be relatively uniform and poor. Although the collecting density is low, the flora is probably relatively well known. A typed report by Buwalda is present in the Forest Research Institute in Bogor; it was not available to the present author.

VI. Literature Cited

Bloembergen, S. 1940. Verslag van een exploratietocht naar de eilanden Timor en Wetar. Tectona **33**: 101–189.
FAO, Field Report 35. 1981. A national conservation plan for Indonesia, Vol. VII: Sulawesi.
Hamilton, W. 1979. Tectonics of the Indonesian Region. Geol. Surv. Prof. Paper **1078**: 114–159, 190–206.
Hannibal, L. W. 1950. Vegetation map of Indonesia. Scale 1:2,500,000.
Rutten, M. R. 1927. Voordrachten over de Geologie van Nederlandsch Oost-Indie. Pages 705–782.
UNESCO. 1973. International classification and mapping of vegetation. Paris.
Steenis, C. G. G. J. van 1938. Vegetation map of Indonesia. *In* Atlas van Tropisch Nederland. Scale 1:10,000,000.
———. 1955. Bibliography. Flora Malesiana **I,5**: XXIII, XXV–CXLIV.
———. 1958. Vegetation map of Malesia. UNESCO. Scale 1:5,000,000.
Steenis-Kruseman, M. J. van 1950. Alphabetical list of collectors. Flora Malesiana **I,1**: 5–606; Suppls. I (1958), **I,5**: ccxxxvii–cccxliii; II (1973), **I,8**: i–cxv. Subsequent additions in Flora Malesiana Bulletin.
———. 1974a. Flora Malesiana **I,1**: CXI.
———. 1974b. Flora Malesiana **I,8**: III.
Whitmore, T. C. 1984. Vegetation map of Malesia. J. Biogeogr. **11**: 461–471. Scale. 1:5,000,000.

VII. Selected Reference

Braak, C. 1929. Het klimaat van Nederlandsch-Indie, part 1, part 3, vol. 2. Verhandelingen Koninklijk Magnetisch en Meterologisch Observatorium Batavia 8.

New Guinea

Peter F. Stevens

Contents

- I. Introduction 122
- II. Description of the Region 122
    - A. Geographical Extent and Area 122
    - B. Topography and Geology 122
        - 1. The Southern Plain 122
        - 2. The Main Range 122
        - 3. Vogelkop and Islands 122
        - 4. Northern Basins and Ranges 122
        - 5. Northeastern Islands 123
        - 6. Southeastern Islands 123
    - C. Climate 123
    - D. Population 123
- III. Vegetation Maps 123
- IV. Magnitude of Floristic Inventory 124
    - A. Herbaria and Specimens Collected 124
    - B. Publications 124
        - 1. Prior to 1920 124
        - 2. 1920–1950 125
        - 3. 1950 to Present 125
- V. Completeness of Floristic Inventory 125
    - A. Factors Affecting Completeness 125
        - 1. Collectors' Predilections 125
        - 2. Suitability of Inventory 126
        - 3. Poor Understanding of Diversity in Taxa 126
        - 4. Unknown Extent of Local Differentiation 126
        - 5. Unknown Amount of Infraspecific Variation 127
        - 6. Conclusions 127
    - B. General Density of Collections 127
    - C. Areas Particularly High in Endemism 128
    - D. Extinction 128
- VI. Current Collections and Work on Larger Families 128
    - A. Collection of Particular Groups 128
    - B. Main Families and Their Taxonomists 128
    - C. Areas Particularly at Risk 128
- VII. Resources for Floristic Inventory 128
    - A. Academic Institutions 128
    - B. Government Institutions 128
    - C. Other Institutions 129
    - D. Foreign Programs 130
    - E. Collaborators 130
- VIII. Resources Needed for Future Inventory 130
    - A. Institutional Requirements 130
    - B. Funding 130
    - C. Goal Setting 131
- IX. Acknowledgments 131
- X. Literature Cited 131

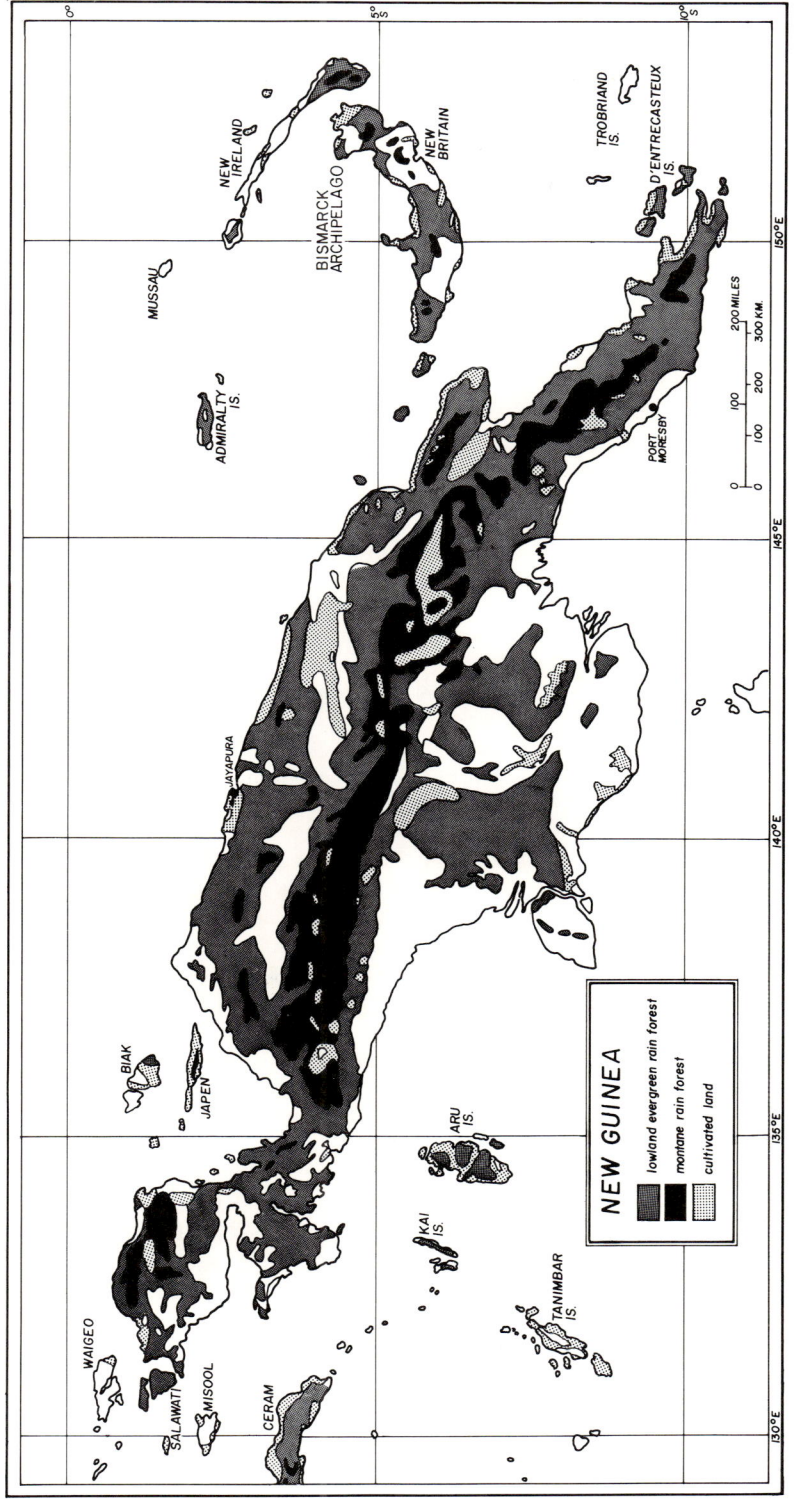

Fig. 1. Areas not otherwise indicated include subdivisions of lowland rain forest, e.g., peat forest, forest over limestone, etc.

I. Introduction

Phytogeographically New Guinea is a keystone for Pacific botany, as van Steenis (1950) and others have noted. It also shows substantial floristic differences from the western part of Malesia and is a major center of generic and specific endemism. Van Steenis considered that at least 50 years of coordinated expeditions would be necessary to evaluate the floristic richness of New Guinea. However, in the over 35 years since he made that statement, the pace of exploration has slowed and there has been much overlap in those areas that have been visited. Many of the areas that van Steenis notes as being in need of botanical exploration are in the same state today. We are seriously lagging behind his schedule.

II. Description of the Region

A. Geographical Extent and Area

The New Guinea region is taken to include the main island of New Guinea and surrounding islands, much as delimited in the Flora Malesiana (van Steenis, 1950, p. cxii & his map 2), but including Bougainville, the Solomon Islands and the Louisiade Archipelago and excluding the islands belonging to Australia in the Torres Strait. The total area is approximately 930,000 km² of which some 95% is accounted for by the island of New Guinea itself. All the country west of 141°E (with the exception of a small zone occupied by a westward bend of the Fly River) is Indonesian (Irian Jaya). The eastern part of the island of New Guinea and the islands east to Bougainville is the nation of Papua New Guinea. The Solomon Islands are made up of the islands between Bougainville and Choiseul Island southeast to the Santa Cruz islands.

B. Topography and Geology

The topography and geology of New Guinea are highly complex. The topography in particular is mostly extremely youthful and is actively changing. An outline of this complexity is necessary if plans for a floristic inventory of the region are to make sense. In the necessarily highly abbreviated account given below, I have drawn heavily on Hamilton (1979), Löffler (1974, 1977) and Pieters (1982). For the purposes of discussion I have divided the area up somewhat arbitrarily into the Southern Plain, the Main Range, the Vogelkop and Islands, the Northern Basins and Ranges, the Northeastern Islands, and the Southeastern Islands.

1. The Southern Plain

This area, which includes the islands in the Torres Strait and Kepulauan Aru, is an area of predominantly low relief through which flow two large rivers, the Digoel and the Fly, as well as a number of smaller rivers. During the last glaciation the Torres Strait was dry land, and the island of New Guinea (and Kepulauan Aru) were joined to Australia (see articles in Walker, 1972). In the northeastern part of this area in particular there is a large area of exposed limestone with karst topography (the Papuan Plateau), and there are fairly recent volcanoes. Current volcanic activity is practically absent, however, and the area is seismically stable, being underlain by the Australian shield.

2. The Main Range

The Main Range is a band of extreme relief that runs practically the whole length of the island. It only rarely drops below 2000 m in altitude for a distance of over 1600 km (13° latitude). An outlier in the extreme southeast comprises the Suckling/Simpson area (but see the Southeastern Islands). The Main Range is sharply demarcated from the Southern Plain in its western half, in places by cliffs rising over 1000 m, whilst to the north, as well as to the south in its eastern half, it is less sharply separated from other regions. Much of the Main Range is made up of rocks of the New Guinea Mobile Belt, although around 146°E, where it drops below 2000 m, the rocks are formed from sediments of the Aure Trough. Rocks of the Main Range are very diverse, many being more or less metamorphosed in the south east, and include granodiorite (Mt. Wilhelm, at 4510 m the highest point in Papua New Guinea) and limestone (Pegunungan Sudirman, of which Puncak (= Pk.) Jaya, or Mt. Carstenz, is, at 4880 m, the highest point in the whole of New Guinea). Between ca. 143° and 146°E there are numerous, mostly inactive, volcanoes, including the 4088 m Mt. Giluwe, and there are other volcanoes in the extreme southeast (E of 148°E) both to the N and S of the main range proper. Two of the higher mountains, G. Doorman and Mt. Suckling, have substantial ultramafic rock, the latter being part of the East Papuan Ultrabasic Belt. Traces of glaciation are widespread above 3500 m, and there are still glaciers on Pk. Jaya, Pk. Mandala and Pk. Trikara in Irian Jaya. The tree line may have been as low as 2000–2200 m in the last interglacial. There are extensive intermontane valleys in both Irian Jaya and Papua New Guinea.

3. Vogelkop and Islands

This is a heterogeneous area, including peaks over 3000 m in the Tamrau and Arfak ranges, but also several more or less isolated smaller ranges and extensive areas of low-lying, swampy land. It too is geologically diverse and, although part of the area is underlain by crystalline continental shield, it is seismically fairly active.

4. Northern Basins and Ranges

This area, which backs on to the Main Range, mainly consists of the basins of the Mamberamo and Sepik Rivers. These basins are enclosed on the seaward side by substantial ranges: the Peg. Van Rees, the Bewani-Torricelli mountains, and to the east the Adelbert-Saruwaged-Finisterre mountains (the last group represents an old island arc that has accreted onto the New Guinea mainland and then been uplifted). The Ramu and Markham rivers flow on the landward side of the latter mountains along a depression that is part of the same structural feature that forms the Sepik Basin. This area is seismically very active, and the Saruwaged-Finisterre mountains in particular are being actively uplifted and show correspondingly youthful topography.

5. Northeastern Islands

This area includes a series of island chains, the Bismarck Archipelago belonging originally to two island arcs. The area shows extreme seismic activity and there are numerous active volcanoes. On both New Britain and New Ireland, in particular, there are large areas of exposed marine limestone. The Solomon Islands, to the immediate southeast, continue this same general pattern of volcanic islands, raised reefs, and exposed rock of sea floor and mid-ocean ridge origin.

6. Southeastern Islands

This is a heterogeneous area. The mountainous islands of the D'Entrecasteaux and Louisiade Archipelagoes are a continuation of the Papuan Mobile Belt, while the Trobriand-Woodlark group consist of raised coral atolls. There is deep-seated seismic activity and some vulcanism, the latter in the D'Entrecasteaux group.

C. Climate

New Guinea as a whole experiences a humid tropical climatic regime, with southeasterly winds towards the middle of the year and northwesterlies (the Perturbation Belt) later. The Main Range is such a prominent feature that the intermontane valleys have largely independent local weather regimes. The central part of the south flank of the Main Range, the inland Gulf region, and the southern part of New Britain receive over 6000 mm to 8000 mm plus of rain per annum while extensive areas of the Southern Plains and the Sepik and Markham basins and smaller areas of the Vogelkop and the Mamberamo basin receive less than 2500 mm rain per annum. A few areas, such as part of the Markham and Bulolo Valleys and the south coast around Port Moresby, receive less than 1500 mm rain per annum. Major tropical storms are rare, but are sometimes recorded from the eastern part of the area. A severe cyclone affected the Solomons, particularly Guadacanal, in 1986.

D. Population

Accurate population figures for the whole area are still hard to come by, but by 1980 a population of some 3.4 million for Papua New Guinea was projected (van de Kaa in Ryan, 1972); this translates to a figure of about 4.5 million for Irian Jaya plus Papua New Guinea. This is equivalent to an average population density of about 5 people per km^2 (similar also in the Solomon Islands), but in some of the densely populated intermontane valleys there are some 150 people/km^2. Areas of higher population density include the Wissell Lakes and the Baliem valley in Irian Jaya, and the high valleys in the Western, Eastern and Southern Highlands and Chimbu Provinces, and around Wau (Morobe Province) in Papua New Guinea. Conversely, the central spine of New Britain and large areas of southern New Guinea are practically uninhabited. The population is predominantly rural, with most people living in small villages. Less than 5% of the population lives in the 15 larger towns of above 5000 inhabitants. Most of these towns are on the north coast and in the highlands, the extensive swampy hinterland of much of the south coast being unfavorable for population concentration. Life expectancy, estimated to be some 20-40 years during the earlier part of this century, is increasing rapidly, and will probably be close to 60 by 1990, especially if the often high levels of infant mortality are reduced. The rate of population increase was 2.6% per annum during the period 1966-1971, but it is expected to exceed 3% per annum, with concomitant doubling of the population between 1969 and 1991. The recent government-supported migration of Javanese and other Indonesians into Irian Jaya will probably affect the population increase. It is planned that about 136,000 families will be moved by the end of 1989.

Much of the population practices subsistence agriculture, with sago, yam, sweet potato, or taro as the main starch crop; bananas are of course very important. Variants of slash and burn agriculture are widespread. Cash crops, such as coconut, coffee, and oil palms (grown extensively only recently) are also grown. There are extensive mineral deposits being mined on the Ertsberg and on Bougainville, with other deposits in the western parts of both Irian Jaya and Papua New Guinea. There are also alluvial gold deposits, most notably in the Bulolo valley, and oil is extracted from the Vogelkop. The logging industry is of extreme importance throughout the area.

III. Vegetation Maps

More than a very generalized vegetation map of New Guinea cannot be given (Fig. 1). As will be apparent from the next section, the data from which a map could be drawn are very uneven, being excellent for Papua New Guinea and poor for the rest of the island. If an attempt is made to follow the UNESCO vegetation categories, the limits of the submontane and cloud forests will have to be adjusted to agree with the limits of the corresponding lower and upper montane forests in Grubb and Stevens (1985); the limits of the lower montane forest given by Paijmans (1975, 1976) are very broad. A further complicating factor is the mosaic nature of much of the vegetation, e.g., on the Southern Plains (see the excellent map in Paijmans, 1975). The Indonesian Forest Planning Bureau published a preliminary vegetation map of Indonesia in 1950 at a scale of 1:2,750,000.

As already indicated, the most comprehensive vegetation map for Papua New Guinea in its entirety is that presented by Paijmans (1975) at a scale of 1:1,000,000, allowing considerable detail to be shown. Limits of forest types above 1000 m for the whole of New Guinea are discussed in detail by Grubb and Stevens (1985). Johns (in Ryan, 1972) presents a generalized vegetation map for the whole of New Guinea.

IV. Magnitude of Floristic Inventory

A. Herbaria and Specimens Collected

On the order of 450,000 plant specimens (collection numbers) have been collected in New Guinea (figures taken from van Steenis-Kruseman, 1974, but modified: the total is probably higher if bryophytes are included). Duplicates of over half of these are to be found in institutions in Papua New Guinea and Indonesia, rather more scattered in numerous institutions elsewhere. For details, see the useful summary by Frodin and Gressitt (1982), and the literature cited by van Steenis (1950) and van Steenis-Kruseman (1974); as the latter notes, details of the activities of Japanese botanists are poorly known.

Collections made prior to 1945 are to be found predominantly in the country from which the collector came, or which financed the trip that he or she was on. Collections made by Gibbs, Boden Kloss, and Forbes are to be found in the British Museum (Natural History) (BM); there are very few duplicates of these collections, especially those made by the first two collectors. C. E. Carr, collecting in Papua in the 1930s (dying there of blackwater fever), made numerous duplicates; these are distributed widely, a fairly good set being at Lae (LAE). Collections made by other early Australian collectors, made mostly in what was then called Papua, are to be found in Melbourne (MEL). C. T. White, based in Brisbane (BRI), made fairly early collections in Papua. Collections made by Germans, of whom Lauterbach, Ledermann and Schlechter are especially well known, were housed in the herbarium at Berlin (B). Unfortunately, the latter was severely damaged in the Second World War, with only part of the monocotyledons and pteridophytes being saved; duplicates of dicotyledons may be found at any of a large number of herbaria, although material of many numbers has been entirely lost. Regrettably, no concerted attempt has been made to recollect in the areas visited by these early Germans, and this still seriously hampers work on the flora (see also Frodin & Gressitt, 1982). Collections made by M. S. Clemens occur in a few herbaria; there are large numbers at the University of Michigan (MICH), Zürich (Z), and at the Arnold Arboretum (A) (Cambridge, U.S.A.). Material collected by Beccari and d'Albertis is held in Florence (FI), and collections made by numerous Dutch collectors in what is now Irian Jaya are to be found predominantly in Leiden (L) and Utrecht (U) (Netherlands), and Bogor (BO) (Indonesia). The important Archbold expeditions began in the 1930s; duplicates from each expedition were distributed independently and may be found mainly at the Arnold Arboretum, Bogor, Leiden, and the New York Botanical Garden (NY). Major Japanese collections from western New Guinea made in the early 1940s are held in Fukuoa (FU) (Japan), Bogor, and the Arnold Arboretum, among other institutions.

Collections made after 1945 show a rather different pattern of distribution. Collections were made under the auspices of the two colonial governments, Dutch (Irian Jaya) and Australian (Papua New Guinea). Numerous duplicates of these collections were made and were distributed widely so that identifications made by foreign specialists would be transmitted to the home institutions. In Irian Jaya, collections were made mainly under the bb. (bossen buitengewesten) and BW (Boswezen Nieuw Guinea) series (the former begun before the war), although collecting largely stopped in the 1960s and has unfortunately not been resumed in any concerted way. The top set of the BW series is stored in the forestry department of Cenderawasih University (MAN) (Manokwari, Irian Jaya); duplicates are mostly in Bogor and Leiden; the bb. series is held mostly in Bogor, including the Forestry Research Institute, with duplicates at Leiden, the Arnold Arboretum and elsewhere. In Papua New Guinea the NGF (New Guinea Force, also incorrectly called New Guinea Forests and New Guinea Flora series) was started during the Second World War, and collecting in this series has continued since, although a "new" series, the LAE series, was started at number 50,001 (this series is currently in the 78,000's). Australian collectors, most working under the auspices of CSIRO, were very active in the eastern part of New Guinea until the mid 1970's; the first set of these collections are held at Canberra (CANB), duplicates are held at Lae and are widely distributed elsewhere. Numbers in the ANU (Australian National University) series, mostly from higher elevations and some, from the early 1970s, made from Irian Jaya, are similarly distributed. Top sets of other collections are to be found at Lae, Bogor, Paris (P), Leiden, Cambridge (Mass. (A) and England (CGE)), Kew (K), Tokyo (TI), Vienna (W), and many other institutions; lower sets of many of these collections are also in Lae. Finally, collections, made in the Solomon Islands in the BSIP series (some 20,000 numbers) add materially to our knowledge of the New Guinea flora. In addition to the top set, held in Honiara (BSIP), there are good sets in Lae, Kew, and Leiden in particular. Kew has the top set of a 1965 Royal Society, London, expedition that collected extensively in the Solomon Islands. Details of collections made since 1969 are to be found in the pages of the "Flora Malesiana Bulletin."

B. Publications

Publications on the flora of New Guinea are most conveniently discussed in three periods: prior to 1920, 1920–1950, and 1950 to present.

1. Prior to 1920

In this period Germans were very active, and C. A. G. Lauterbach edited a particularly important series of papers under the general title "Beiträge zur Flora von Papuasien" in "Engler's Botanische Jahrbücher" (22 parts published between 1912 and 1922), and there are almost flora-like compilations of parts of the erstwhile German New Guinea (Schumann & Lauterbach, 1901, 1905). Much important work appeared in articles in Dutch periodicals such as the "Bulletin of the Buitenzorg (Bogor) Botanic Garden" and in the botanical series of "Nova Guinea." Beccari's journal,

"Malesia," also includes descriptions of numerous taxa from New Guinea, while particularly important publications, enumerating the collections of individual expeditions, include those of Gibbs (1917) and Ridley (1916).

2. *1920–1950*

Perhaps the most important series of articles in this period appeared in the "Journal of the Arnold Arboretum" largely based on the enumeration of collections by L. J. Brass, C. E. Carr, M. S. Clemens, and S. F. Kajewski. L. M. Perry and E. D. Merrill together published almost 700 pages in a series of papers between 1939 and 1953, and there were many other contributors to the pages of this journal, including C. T. White. Kanehira and Hatusima published a series of papers based on their collections in the "Botanical Magazine, Tokyo," from 1938 to 1943. Numerous other contributions to the taxonomy of New Guinean plants appeared during this period, and also some noteworthy contributions to the knowledge of vegetation. These latter included Lam's classic *Fragmenta Papuana* (Lam, 1927–1929), as well as reports of the Archbold Expeditions which began appearing in the "Bulletin of the American Museum of Natural History" in 1935; L. J. Brass published interesting botanical accounts of these expeditions (e.g., Brass, 1938, 1941).

3. *1950 to Present*

With the gradual publication of the mammoth *Flora Malesiana* taxonomic work on the New Guinea flora has increased. Many families occurring in New Guinea have been treated in this flora, although early numbers tended to treat the very eastern part of New Guinea (and the Solomon Islands) rather perfunctorily. Many precursor papers to these treatments have appeared in the Dutch periodical "Blumea," although the "Kew Bulletin," "Reinwardtia," "Journal of the Arnold Arboretum," and, more recently, several Australian publications, have carried important articles. A series of *Handbooks to the flora of Papua New Guinea* started to appear in 1978, of which two volumes have been published to date. The title is somewhat of an understatement since New Guinea as a whole as well as the Solomon Islands are included. T. C. Whitmore (1966) provided a useful overview of the larger trees growing in the Solomon Islands. A four-volume *Alpine flora of New Guinea*, largely written by P. van Royen appeared between 1979 and 1982; this treats taxa growing at above 3,000 m elevation.

Important locally-written contributions include a treatment of the economically very important legumes by Verdcourt (1979), an illustrated account of weedy taxa (Henty & Pritchard, 1973), grasses (Henty, 1969), and the Combretaceae (Coode, 1969). The Anacardiaceae, Apocynaceae, Dipterocarpaceae, Eupomatiaceae, Himantandraceae, Magnoliaceae, and Sterculiaceae were also treated—somewhat more sketchily—in the series, *Manual of the forest trees of Papua and New Guinea*, and a conspectus of the Arecaceae has appeared (Essig, 1977). Other general accounts have also appeared recently (Johns, 1976; Johns & Bellamy, 1979). Preliminary checklists of several areas have been produced, including vascular plants growing above 9000 feet (2743 m) on Mt. Wilhelm (Johns & Stevens, 1971, now sadly in need of updating), the plants of Bougainville (Foreman, 1971) and most recently a list of some 1,789 phanerogams growing in the Upper Watut watershed (Streimann, 1983). There is a mimeographed series of keys in three volumes to the genera of flowering plants in New Guinea prepared by P. van Royen, and an unpublished *Illustrierte Flora des Bismarck-Archipels für Naturfreunde* by G. Peekel has also been made available in similar form (it was recently translated—see Peekel, 1984). Lists of taxa of smaller areas—individual mountains and the like, continue to be published (e.g. Stevens & Veldkamp, 1980; Hiepko & Schulze-Motel, 1981). Those in which taxonomic decisions are made without any real knowledge of the flora have rightly come in for severe criticism (see van Steenis, 1982). Percival and Womersley (1975) wrote an account of the mangrove vegetation of Papua New Guinea, and Leach and Osborne (1985) have prepared a survey of the aquatic vegetation of this area. D. G. Frodin is preparing a flora of the Port Moresby region.

V. Completeness of Floristic Inventory

A. Factors Affecting Completeness

Several factors complicate the evaluation of the completeness of the floristic inventory of New Guinea, as with most tropical countries where travel is difficult. Although these factors are not peculiar to New Guinea, they should be discussed here. I shall draw heavily on groups with which I am familiar in selecting my examples; another taxonomist would produce a different set of examples that would illustrate the same points.

1. *Collectors' predilections*

Collectors tend to follow similar routes when returning to areas that been collected before. That some 10,000 numbers have been taken from Mt. Wilhelm alone is no real indication of our knowledge of the flora of the mountains; most of these specimens have been taken from the Pindaunde Valley. Several taxa are known to have local distributions on the mountain; these include taxa in *Agapetes*, the Compositae, *Drimys*, and *Scleranthus* that do not even grow at particularly high altitudes. Several high altitude taxa of course have very restricted areas where they are able to grow; on Mt. Giluwe, *Senecio gnoma* van Royen grows only at the very top. T. G. Hartley in 1963 collected material from the frequently visited Mt. Kaindi that was the basis of a new genus in the Icacinaceae, *Hartleya*. Almost any substantial collection made from off the beaten track, even in "well collected" areas, will yield important records. Thus recent collections made from the Bulldog Trail, which starts (or finishes) in the Bulolo area, have yielded range extensions in taxa such as *Dimorphanthera* (Stevens, 1977), *Rhododendron*, etc.

2. Suitability of floristic inventory

Up to the 1960's it was simply not physically possible to get to all the places that seemed as though they might be interesting places from which to collect, and communications in parts of New Guinea are still not easy. Hence the course of inventory has in part been shaped by circumstance. The lure of the high mountains continues to divert needed attention from less glamorous but probably botanically more crucial lowland to montane areas. However, the Archbold expeditions in particular, as well as some of the other early large expeditions, such as those of Behrmann (Ledermann) and Carr, were noteworthy in that they a) focused on areas that looked floristically and faunistically interesting, b) emphasized making altitudinal transects whenever possible, and c) were highly collaborative ventures involving the then most modern forms of transport. Not surprisingly, they were both very expensive and involved considerable manpower. It will be noted that a number of the areas discussed below as being desirable targets for future collecting are certainly no easier to get to than these high mountain regions.

Particularly in Papua New Guinea, the inventory program effectively utilized the Department of Posts and Telegraphs as it established microwave communications in the island, and also as mapping sites were established in botanically interesting areas. However, more should be done to capitalize on the activities of logging companies; collecting in active logging operations can be very rewarding, albeit somewhat hazardous.

With the financial uncertainties in Indonesia, and especially Papua New Guinea, it is impossible to develop plans for inventorying the flora that will do more than scratch the surface of the problem. Current New Guinea-based general inventorying is minimal. Political decisions in Indonesia seem to account for much of the difficulty in collecting in Irian Jaya. Researchers from outside find it difficult to work there and little or no collecting is being done there at present. There are plans for collecting in Irian Jaya, emphasizing potential nature and forest reserves, but so far they have not been realized. Relationships between Provincial Governments and collectors in Papua New Guinea are sometimes not very satisfactory. Activities of collectors may sometimes be forbidden (e.g., East New Britain Province), or there may be disquiet expressed about the removal of dead or living plants from the province.

Although the collaborative ventures alluded to above have been to areas in need of collection, they are too few to do more than nibble at the problem of inventorying a fantastically diverse vegetation. A carefully coordinated and speedy solution is necessary.

3. Poor understanding of diversity in taxa

Even in areas that have been well collected, and in groups that have been recently studied taxonomically, it may still be difficult to understand just what the diversity of the group is, and hence how to plan further collecting in the group. Some examples will help. Vink (1970) analysed the variation in the montane genus *Drimys* with extreme care; he found that the density of collecting in many instances was insufficient for him to appreciate the pattern of variation in the genus. In addition, species concepts adopted in *Drimys* differ greatly from author to author. One can choose either to recognize a single species, *Drimys piperita* (Vink, 1970), or, following A. C. Smith (1969), group some 30 species in *Tasmannia*. When annotating specimens of genera like *Eurya, Rapanea* (=*Myrsine*) and *Saurauia*, following the accounts in van Royen (1982), it becomes clear that simple translation of taxa to biologically interesting entities can not be made. Sleumer (1986) revised *Rapanea* in New Guinea and reduced many of van Royen's species to synonymy. Mabberly (1979) not unreasonably opted to take a broad view of *Chisocheton lasiocarpus* in his monograph of the genus; an earlier local treatment (Stevens, 1975) recognized ten species in this complex. Whatever the rank at which these taxa should be recognized, some represent distinguishable entities, for example, *C. caroli*, which is found both N and S of the central part of the Main Range, an interesting and not uncommon distribution. Finally, in the *Calophyllum trachycaule* complex, the number of specimens assignable to the complex has more than tripled within a generation, and the number of taxa in the complex gives an imperfect representation of the variation within it.

Although it may seem that the decision whether to adopt narrow or broad species limits is a personal matter, it is clear that it will have a major effect on any inventory program and will determine both where and at what density we should collect (see Prance, 1977, for some comments on the species problem). It could be argued that if only one species of *Drimys* grows in New Guinea, then the genus has been adequately inventoried, at least at some level. However, new entities (or species!) are still being discovered, and it is clear that we have a less than satisfactory knowledge of the genus.

4. Unknown extent of local differentiation

The extent of local differentiation of the flora of New Guinea at the species level is still unclear, but is likely to be underestimated. Kalkman and Vink (1970) found that mountains that are close to one another nonetheless may not have the same species of Ericaceae on them. A comparison of the flora of Mt. Wilhelm and the adjacent Mt. Kerigomna also shows this; only on the latter mountain is found the distinctive *Eurya kerigomnica* while in *Drimys piperita*, entity "cordata," from fairly low altitudes on Mt. Wilhelm, has yet to be found east of that mountain while entity "heteromera" on the two mountains shows clear differentiation in petal number. The great topographical and geological heterogeneity of much of New Guinea must be borne in mind here. Even at very local scales, as around Kiunga, in the Western Province of Papua New Guinea, as well as in adjacent Irian Jaya, relief of less than 20 m nevertheless supports a great diversity of vegetation. Towards the south, there is a mixture of savannah and rainforest species, and of the latter, some belong to taxa that normally occur at much higher elevations. As one approaches the Main Range these "montane" elements increase. In addition, swamp forest interdigitates with the already heterogeneous ridge forest.

5. Unknown amount of infraspecific variation

The basic biology of most taxa is almost entirely unknown, moreover, and the infraspecific variation shown, both morphological (e.g., Stevens, 1974; Vink, 1970) and ecological (Coode & Stevens, 1972), is poorly understood. A comprehension of the correlation of this variation with environmental or other factors is almost non-existent. As an example of recent confusion, a recent report of pollination of *Drimys* on Mt. Kaindi implied apomixis (Thien, 1980), but the described pattern of variation on the mountain was in agreement neither with that described by Vink (1970) nor that which I have observed. Furthermore, Vink observed little regeneration by seed of *Drimys* and tentatively suggested that vegetative reproduction was common; I have commonly observed reproduction by seed of forest-dwelling entities. The broad species concept that tends to be adopted in a large flora such as *Flora Malesiana* will probably mask some biologically important local variation, a problem accentuated by the space constraints inevitable in these publications.

6. Conclusions

In terms of knowledge of the flora, inventorying at a density of 50–100 collections/100 km² ("relatively well collected") will not indicate satisfactorily the diversity of the area (how many kinds?). If, in addition, we wish to understand the variation and biology of the taxa that are there, we shall have to adopt a very different approach.

B. General Density of Collections

From what had just been said, it is clear that areas which have had more than 50 specimens taken from 100 km² can by no means be considered to be well collected. New Guinea will be found to be one of the worse collected areas in Malesia, only the Celebes and Sumatra having comparably low figures (and there have been subsequent expeditions to the Celebes), and it is more poorly collected than several of the Central American countries listed by Prance (1977).

Figure 2 shows in a general way areas (shaded) which could, by some stretch of the imagination, be called well collected. Note that many of the high mountains fall into this category, but the approaches to such mountains are often not well known; collectors are attracted to the subalpine and alpine areas and tend to ignore the montane forest.

Areas particularly in need of collecting include limestone areas like those on New Britain and Great Papuan Plateau. The colline and montane zones between 1000 and 2500 m would repay an extensive study, especially those of the Main Range in Irian Jaya, the little collected Peg. van Rees (north Irian Jaya), and the Baining Mountains, including Mt. Birinia, of the S. E. Gazelle Peninsula, on New Britain. Of the high mountains, the main range between Pk. Trikora and the Star Mountains of Papua New Guinea (between 139° and 141°E) need attention, as does Peg. Tamrau, which reaches 3000 m. The lowland to lower montane forests at the headwaters of the Digoel and Fly rivers, the eastern branch of the Mamberamo River, and the headwaters of the Sepik, May, and Horden rivers are of considerable diversity, but are little collected. Extensive collecting should be also carried out in the northern (and most extensive) outcrops of the Papuan Ultrabasic Belt, for instance east of of Garaina. Forests on the mainland west of 137°E need further collecting, in part because many taxa from Western Malesia reach their easternmost limits there (see also Grubb & Stevens, 1985). This area includes numerous isolated ranges that have been only poorly collected (with the partial exception of Peg. Arfak). Finally, the hilly islands just to the west of the Vogelkop, such as Salawati, Batanta, Waigeo, and Gebe, need further study; some are particularly interesting geologically.

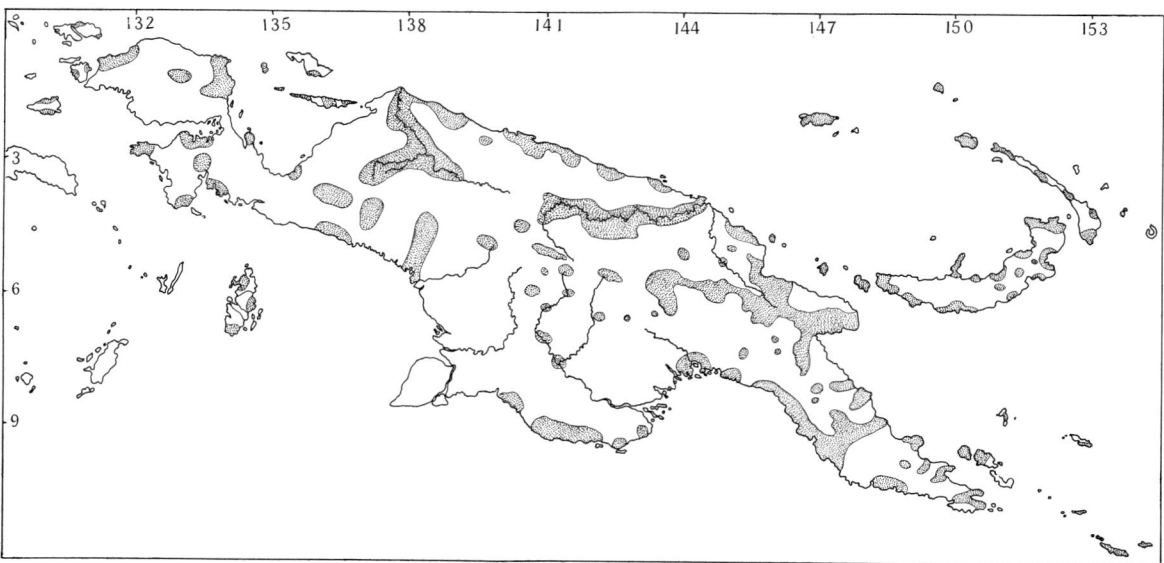

Fig. 2. "Well-collected" areas in Papuasia.

C. Areas Particularly High in Endemics

The term "endemic" is relative. Of the mountains that are fairly well known, many of the higher peaks such as Peg. Maoke, G. Doorman, Peg. Arfak, the Finisterres, and Mts. Wilhelm, Giluwe, Amungwiwa, Victoria, and Albert-Edward are particularly rich in endemics. It is difficult to pinpoint areas of endemics at lower altitudes because the distribution of taxa is so poorly known, but areas of particular richness include the headwaters of the Fly, Mamberamo and Sepik rivers; montane forest in the Aseki-Menyamya-Bulldog Road area; Lake Kutubu; Tagula, Rossel, and Misima Islands; and many places on the Vogelkop. Many of these areas are also yielding new and important distribution records.

D. Extinction

Nothing is known of rates of extinction in the region, except that a few orchid taxa, like *Paphiopedilum violascens*, have been decimated in certain areas because of the depredations of collectors. Sandalwood (*Santalum macgregorii*) have also been much collected.

VI. Current Collection and Work on Larger Families

A. Collection of Particular Groups

Generally speaking, montane to subalpine taxa with conspicuous flowers are fairly well collected, especially genera like *Rhododendron*, with its conspicuous flowers of horticultural interest. Herbaceous groups, including ferns and mosses, tend to be poorly collected because of the emphasis on the collection of woody groups by the local forestry departments in the period 1950–1970.

B. Main Families and Their Taxonomists

Table I includes only the bigger families of flowering plants, gymnosperms, and ferns of which large parts are, or recently have been, under study. Authors who have worked on a particular family fairly recently may no longer be particularly interested in it. Some large families are almost without contemporary active workers. These include the Annonaceae, Asclepiadaceae, Gesneriaceae, Lauraceae, Melastomataceae, Sapotaceae, and Urticaceae.

C. Areas Particularly at Risk

The island of New Guinea is peculiarly fortunate that a low population density in many places, combined with extreme ruggedness of much of the topography, has ensured that much of the forest is apparently still intact (but see White, 1976; Johns, 1986). Nevertheless, there are extensive areas in which human pressure has produced large areas of secondary vegetation; parts of the highlands, the Sepik basin, the northeast coast of New Guinea, and the Central Province spring to mind. The International Bank of Reconstruction and Development estimated that Papua New Guinea could support, on a permanent basis, a logging industry exporting 500,000,000 super feet of timber. This figure is only now being reached, and current estimates of forest resources have been substantially larger than earlier figures (e.g., R. J. B. Angus *in* Ryan, 1972). Half of New Guinea's forest is inaccessible (Davidson, 1976). However, the picture is not that rosy. Little land is set aside as permanent forest because of patterns of land ownership, and large-scale logging operations, especially along the north coast and the islands, are having extreme effects. In Irian Jaya many areas that are planned as forest reserves are under threat for logging, either as concessions or illegal cutting. Manus Island supports in places a unique lowland forest which has only a single species (*Calophyllum euryphyllum*) dominating. This forest is in severe jeopardy because it provides almost the only source of local income for the island. Over one quarter of New Britain has been classified as loggable forest and forests in the Solomon Islands represent the major natural resource of the country.

VII. Resources for Floristic Inventory

A useful outline of the herbaria in the New Guinea region is provided by Frodin (1985).

A. Academic Institutions

1. University of Technology, Lae.

The Leonard Brass herbarium is small, mostly containing ferns, with one active staff member.

2. University of Papua New Guinea, Port Moresby.

Its herbarium emphasizes the plants of the Central District. Damaged by fire in 1978, a new herbarium was opened in July, 1984. There are two taxonomists/ecologists and two technical staff here.

3. The College of Forestry of Cenderawasih University.

Although it houses the top set of the important BW series, it is not a center for taxonomic research.

B. Government Institutions

1. Herbarium Bogoriense and Forest Research Institute, Bogor (Java).

Although actually not in the region, it is the main taxo-

Table I
A List of the Plants Currently Under Study in New Guinea

Taxon	Specialist(s)
Anacardiaceae*	Ding Hou (L)
Apocynaceae	F. Markgraf (Z, deceased)
Araliaceae	D. G. Frodin (PH), W. R. Philipson (CANU)
Arecaceae	J. Dransfield (K), F. B. Essig (USF)
Asteraceae	J. Koster (L)
Burseraceae*	P. W. Leenhouts (L)
Celastraceae* (including Hippocrateaceae)	Ding Hou (L)
Clusiaceae (including Hypericaceae*)	N. K. B. Robson (BM), P. F. Stevens (A)
Combretaceae*	M. J. E. Coode (K), A. W. Exell (BM, deceased)
Cunoniaceae	R. Hoogland (L)
Cyatheaceae*	R. E. Holttum (K)
Cyperaceae*	H. P. Nooteboom (L, *Carex*), J. H. Kern (L, deceased)
Dilleniaceae*	R. Hoogland (L)
Elaeocarpaceae	M. M. J. van Balgooy (L), M. J. E. Coode (K), R. Weibel (G)
Ericaceae*	H. Sleumer (L), P. F. Stevens (A)
Euphorbiaceae	H. K. Airy Shaw (K, deceased)
Fagaceae*	E. Soepadmo (KLU)
Flacourtiaceae*	H. Sleumer (L)
Gleicheniaceae*	R. E. Holttum (K)
Icacinaceae*	H. Sleumer (L)
Lamiaceae*	H. Keng (SING)
Lauraceae	A. J. G. H. Kostermans (BO)
Leguminosae *sensu lato*	B. Verdcourt (K)
Lindsaeaceae*	K. U. Kramer (Z)

Table I (continued)

Taxon	Specialist(s)
Lomariopsidaceae	R. E. Holttum (K), E. Hennipman (U)
Meliaceae	D. H. Mabberley (OXF), C. M. Pannell (OXF)
Menispermaceae*	L. L. Forman (K)
Monimiaceae*	W. R. Philipson (CANU)
Moraceae	E. J. H. Corner (CGE), F. Jarret (K)
Myristicaceae*	D. B. Foreman (MEL), W. J. J. O. de Wilde (U)
Myrtaceae	T. G. Hartley (CANB *Syzygium*), etc.
Pandanaceae	B. C. Stone (PH), H. St. John (BISH)
Poaceae	E. E. Henty (LAE), J.-F. Veldkamp (L)
Podocarpaceae	D. J. Laubenfels
Polypodiaceae	E. Hennipman (U), etc.
Proteaceae*	H. Sleumer (L), D. B. Foreman (MEL)
Pteridaceae	K. U. Kramer (U)
Rhizophoraceae*	Ding Hou (L)
Rubiaceae	C. Ridsdale (L), S. H. Sohmer (BISH, *Psychotria*), S. Darwin (NO), O. Gideon (LAE, *Mussaenda*)
Rutaceae	T. G. Hartley (CANB)
Orchidaceae	E. de Vogel (L) *et al.*
Sapindaceae	P. W. Leenhouts (L)
Saurauiaceae	R. Hoogland (L)
Symplocaceae	H. P. Nooteboom (L)
Thelypteridaceae*	R. E. Holttum (K)
Winteraceae	W. Vink (L)
Zingiberaceae	R. M. Smith (E)

* There is an account of this family in the "Flora Malesiana."

nomic center in Indonesia. A major herbarium with an associated botanic garden, it is active and holds many plants collected in Irian Jaya.

2. Division of Botany, Lae.

A medium-sized herbarium, associated with a botanic garden, is here. It is in a modern, air-conditioned building (and so needs constant upkeep). An apparently unintentional failure to vote funds for the herbarium for the 1983 financial year seriously compromised even the basic maintenance of the collections. A *Lasioderma* infestation, since contained, built up in consequence. There are currently two botanists.

3. Papua New Guinea Forestry College, Bulolo.

A small herbarium emphasizing woody plants of the Bulolo Valley and environs is located here. There are two botany/ecology lecturers and two technical staff.

4. Honiara, Solomon Islands.

There is a small, currently inactive, herbarium emphasizing the woody plants of the Solomon Islands here.

C. OTHER INSTITUTIONS

A botanic preserve has recently been set aside near Goroka in the Mount Gahavisuka Provincial Park. This is the Lipizauga Botanical Sanctuary and is at an altitude of 2000–2500 m; its specialties are orchids and rhododendrons. On Mt. Kaindi, near Wau, is the Wau Ecological Institute, an offshoot of the Bishop Museum in Honolulu, Hawaii. This is a largely independent institute doing mostly entomological research, but there is a small herbarium there. There is currently one botanist in residence.

D. Foreign Programs

Several foreign institutions are currently involved in the field study of the New Guinea flora on a continuing, although especially latterly, rather intermittent basis. The Rijksherbarium, Leiden, has mounted joint expeditions with the Division of Botany, Lae, primarily to poorly collected mountainous regions of Papua New Guinea. The Bishop Museum, Honolulu, carried out extensive collecting in the high mountains in Papua New Guinea during the preparation of van Royen's Alpine Flora. The Kew herbarium also mounts occasional trips to poorly collected areas, the most recent (1975) being to the mountains in the southern part of New Ireland. In addition, several foreign researchers visit Papua New Guinea in the course of their work on particular groups. However, it should be remembered that such specialized collecting does not add appreciably to a general inventory program, rather, it leads to a much improved representation of individual groups (Prance, 1977, for example). Usually either a set of specimens is left behind in Java or Papua New Guinea, or an explicit agreement is reached that a set will be deposited there.

E. Collaborators

Generally, opportunities for collaboration are as good as they can reasonably be expected to be. Some recent collecting trips organized by the Division of Botany in Lae, have been accompanied by staff from the National Botanic Gardens, also in the Division, to collect material suitable for growing in the gardens. Collaboration with anthropologists, ecologists, and others both inside and outside the country, although generally worked out on a case-by-case basis, has generally been excellent, as numerous collaborative reports attest. As an example, Hartley et al. (1973) carried out a comprehensive survey of plants to see which contained alkaloids; the top set of specimens that Hartley collected is deposited at Lae.

IX. Resources Needed for Future Inventory

A. Institutional Requirements

The basic institutional background for floristic inventories throughout New Guinea is largely satisfactory, although the funding of these institutions is very precarious. Financial and personnel resources needed to carry out these inventories are unsatisfactory.

1. Herbarium maintenance

Herbaria need, especially in the tropics where there are high temperatures and humidity leading to molding, and very active insect pests to attack specimens, an appreciable and regular budget for their maintenance. Money is needed both for air-conditioning and the like (as at Lae), and for despatching the large amount of material that is requested on loan by botanists from all over the world. It would clearly be unwise to carry out a major inventory program, and then lose the specimens resulting from that program because of insufficient funds being alocated to store the specimens under the proper conditions.

2. Active collection programs

An active collection policy also requires considerable sums of money, not simply for the collections themselves but for their incorporation into the herbarium and distribution.

3. Personnel

Hardly surprisingly, there are few Papua New Guinean taxonomists, largely because such people have had the opportunity to move elsewhere in the civil service and universities. Although there are two Papua New Guinean botanists at Lae, it is unlikely that this number will increase in the near future; no suitable people have been identified, and there are no funds to pay them even if such people were to be found. However, their absence makes the long-term success of an inventory program more problematic.

4. Priority setting

For largely historical reasons floristic inventory is currently not high on the list of priorities for Papua New Guinea, and perhaps especially for Indonesia. Historical reasons have recently become reinforced by political and financial constraints. However, the need for such an inventory remains. Foreign institutions and scientists are largely problem-oriented when they work in New Guinea. This, too, is both understandable and desirable in the context of the studies that are being carried out (e.g., Prance, 1977). To oversimplify, there has to be a meeting of minds between those involved in the study of the New Guinea flora, but who are largely unable to carry out the necessary inventories, and those who could facilitate such inventories if suitably organized, funded, and directed.

B. Funding

It is possible to give a very tentative estimate of additonal resources needed to carry out a partial floristic inventory of New Guinea, focusing largely on the areas discussed above. Note that this earlier discussion focused on areas and quality of collecting, rather than overall amount, and did not attempt to ensure even coverage of the whole area. A figure of some 40 to 50 man-years for collecting alone is a reasonable one. Collectors will have to be experienced and knowledgable about the New Guinea flora; any program will be severely compromised if a succession of inexperienced collectors is used. Collectors should largely operate independently for maximum efficiency. Parties in the field are going to need field assistants and at times very considerable logistical support. Additional funds will be needed to produce field labels, and to process and send out specimens. Based on American salary scales, we are speaking of a figure of around

$1,700,000 current dollars for collectors' salaries (exclusive of benefits), the same for field and herbarium support, and about $250,000 for shipping, transport, etc.

C. GOAL SETTING

There is also a broader issue. To develop an effective program to inventory the flora, the very notion of collecting and identifying plants has to be incorporated into national goals. Thus it appears, with the advantage of hindsight, the connection of the Division of Botany with other government departments in Papua New Guinea was more distant than it needed to be, or should have been. There was no concerted effort to write forester's manuals, guides to identification of seedlings and adult plants, and the like. Not that some useful works along these lines were not produced, just that policy in this direction was not coherent. If a successful program to inventory the flora is to be developed that has governmental support, the issue of the place of botany in New Guinea life has to be addressed.

It goes without saying that there are two main goals for any inventory project. One is to make the country in which the project is being carried out aware of its resources, and so better able to utilize them and to make sensible decisions that affect the environment. In the course of the inventory program local expertise on the flora will be considerably increased. The other is to make the information on the flora available to the general scientific community. However, any inventory project that does not have the firm support of the country in which it is being carried out will be a failure. We do not want to know that a species was in a country once, but that it is there now.

IX. Acknowledgments

I am very grateful to K. Kartinawata (Bogor), E. E. Henty and J. Croft (Lae), and W. Vink and M. M. J. van Balgooy (Leiden) for useful comments on the manuscript. However, the opinions expressed are those of the author.

X. Literature Cited

Brass, L. J. 1938. Botanical results of the Archbold expeditions, IX. Notes of the vegetation of the Fly and Wassi Kussa Rivers, British New Guinea. J. Arnold Arbor. **19**: 174–190.

———. 1941. The 1938–39 expedition to the Snow Mountains, Netherlands New Guinea. J. Arnold Arbor. **22**: 271–342.

Coode, M. J. E. 1969. Manual of the forest trees of Papua and New Guinea. Part 1 (revised issue)—Combretaceae. Division of Botany, Department of Forests, Lae.

——— & P. F. Stevens. 1972. Notes on the flora of two Papuan mountains. Papua New Guinea Sci. Soc. Proc. (1971) **23**: 18–25.

Davidson, J. 1976. Interaction of production forestry with conservation objectives and multiple use of forest land in Papua New Guinea. Pages 49–71 *in* K. P. Lamb & J. L. Gressitt (eds.), Ecology and conservation in Papua New Guinea. Wau Ecology Institute Pamphlet 2.

Essig, F. B. 1977. The palm flora of New Guinea. Botany Bulletin 9, Office of Forests, Division of Botany, Lae.

Foreman, D. B. 1971. A checklist of the vascular plants of Bougainville. Botany Bulletin 5, Department of Forests, Division of Botany, Lae.

Frodin, D. G. 1985. Herbaria in Papua New Guinea and nearby areas. Pages 54–62 *in* S. H. Sohmer (ed.), Forum on systematic resources in the Pacific. Bernice P. Bishop Special Publication 74, Bishop Museum Press, Honolulu, Hawai'i.

——— & J. L. Gressitt. 1982. Biological exploration in New Guinea. Pages 87–130 *in* J. L. Gressitt (ed.), Biogeography and ecology of New Guinea. W. Junk, the Hague.

Gibbs, L. S. 1917. A contribution to the phytogeography and flora of the Arfak Mountains, etc. Taylor & Francis, London.

Grubb, P. J. & P. F. Stevens. 1985. The forests of Fatima Basin and Mt. Kerigomna and a review of montane and subalpine forests elsewhere in New Guinea. Australian National University, Research School of Pacific Studies, Department of Biogeography & Geomorphology, Publ. BG/8, Canberra.

Hamilton, W. 1979. Tectonics of the Indonesian region. United States Printing Office, Washington.

Hartley, T. G., E. A. Dunstone, J. S. Fitzgerald, S. R. Johns & J. A. Lamberton. 1973. A survey of New Guinea plants for alkaloids. Lloydia **36**: 217–319.

Henty, E. E. 1969. A manual of the grasses of New Guinea. Botany Bulletin 1, Division of Botany, Department of Forests, Lae.

——— & G. S. Pritchard. 1973. Weeds of New Guinea and their control. Botany Bulletin 7, Division of Botany, Department of Forests, Lae. (Ed. 2, 1975.)

Hiepko, P. & Schultze-Motel. 1981. Floristische und ethnobotanische Untersuchungen im Eipomek-Tal, Irian Jaya (West-Neuguinea). Reimer, Berlin. (Schriftenreihe Mensch, Kultur und Umwelt im zentralen Bergland von West-Neuguinea; Beitr. 7).

Johns, R. J. 1976. Common forest trees of Papua New Guinea. 12 parts. Forestry College, Bulolo.

———. 1986. The instability of the tropical ecosystem in New Guinea. Blumea. **31**: 341–371.

——— & A. Bellamy. 1979. Ferns and fern allies of Papua New Guinea. Lae.

——— & P. F. Stevens. 1972. Mount Wilhelm flora. Botany Bulletin 6, Division of Botany, Department of Forests, Lae.

Kalkman, C. & W. Vink. 1970. Botanical exploration in the Doma Peaks region, New Guinea. Blumea **18**: 87–135.

Lam, H. J. 1927–1929. Fragmenta Papuana. Natuurk. Tijdschr. Nederl.-Indie **87**: 110–180; **88**: 187–228, 252–324; **89**: 67–140, 291–388. (Transl. L. M. Perry, Sargentia **5**: 1–196, 1945).

Leach, G. J. & P. L. Osborne. 1985. Freshwater plants of Papua New Guinea. University of Papua New Guinea Press.

Löffler, E. 1974. Explanatory notes to the geomorphological map of Papua New Guinea. CSIRO Australia, Land Research Series 33, Canberra.

———. 1977. Geomorphology of Papua New Guinea. CSIRO Australia, and Australian National University, Canberra.

Mabberley, D. J. 1979. The species of *Chisocheton* (Meliaceae). Bull. Brit. Mus. Nat. Hist. Bot. **6**: 301–386.

Paijmans, K. 1975. Explanatory notes to the vegetation map of Papua New Guinea. CSIRO Australia, Land Research Series 35, Canberra.

——— (ed.). 1976. New Guinea vegetation. Elsevier, Amsterdam.

Peekel, P. G. 1984. Flora of the Bismark Archipelago for naturalists. Office of Forests, Division of Botany, Lae, Papua New Guinea.

Percival, M. & J. S. Womersley. 1975. Floristics and ecology of the mangrove vegetation of Papua New Guinea. Botany Bulletin 8, Division of Botany, Department of Forests, Lae.

Pieters, P. E. 1982. Geology of New Guinea. Pages 15–38 *in* L. J. Gressitt (ed.), Biogeography and ecology of New Guinea. W. Junk, the Hague.

Prance, G. T. 1977. Floristic inventory of the tropics: Where do we stand? Ann. Missouri Bot. Gard. **64**: 659–684.

Ridley, H. N. 1916. Report on the botany of the Wollaston Expedition to Dutch New Guinea, 1912–1913. Trans. Linn. Soc. London, ser. 2, Bot. **9**: 1–269.

Royen, P. van. 1982. The alpine flora of New Guinea. Vol. 2. J. Cramer, Vaduz.

Ryan, P. (ed.). 1972. Encyclopaedia of Papua New Guinea. 3 vols. Melbourne University Press, Melbourne.

Schumann, K. & K. Lauterbach. 1901. Die Flora der Deutschen Schutzgebeite in der Südsee. Borntraeger, Leipzig.

⎯⎯⎯ & ⎯⎯⎯. 1905. Ibid. Nachträge. Borntraeger, Leipzig.

Sleumer, H. 1986. A revision of the genus *Rapanea* Aubl. (Myrsinaceae) in New Guinea. Blumea **31**: 341–371.

Smith, A. C. 1969. A reconsideration of the genus *Tasmannia*. Taxon **18**: 286–290.

Steenis, C. G. G. J. van. 1950. Pages xi–clii, Introductory Part. Flora Malesiana I, 1. Noordhoff/Kolf, Djakarta.

⎯⎯⎯ (ed.). 1982. Critical notes of New Guinea plants described by A. Gilli. Blumea **28**: 165–169.

Steenis-Kruseman, M. J. van. 1974. Malesian plant collectors and collections. Flora Malesiana **1 (8)**: 1ff.

Stevens, P. F. 1974. The hybridisation and geographical variation of *Rhododendron atropurpureum* and *R. womersleyi*. Papua New Guinea Sci. Soc. Proc. 1973. **25**: 73–84.

⎯⎯⎯. 1975. Revision of *Chisocheton* in Pauasia. Contr. Herb. Australia **11**: 1–15.

⎯⎯⎯. 1977. Additional notes on *Dimorphanthera* (Ericaceae). J. Arnold Arbor. **58**: 437–444.

⎯⎯⎯ & J.-F. Veldkamp. 1980. Report on the Lae-Leiden Mt. Suckling expedition of 1972. Botany Bulletin 10, Division of Botany, Office of Forests, Lae.

Streimman, H. 1983. The plants of the Upper Watut watershed of Papua New Guinea. National Botanic Gardens, Canberra.

Thien, L. B. 1980. Patterns of pollination in primitive angiosperms. Biotropica **12**: 1–13.

Verdcourt, B. 1979. A manual of New Guinea legumes. Botany Bulletin 11, Office of Forests, Division of Botany, Lae.

Vink, W. 1970. The Winteraceae of the Old World. 1. *Pseudowintera* and *Drimys*. Blumea **18**: 295–354.

Walker, D. (ed.). 1972. Bridge and barrier: The natural and cultural history of Torres Strait. Australian National University, Research School of Pacific Studies, Dept. of Biogeography and Geomorphology, Publ. BG/5.

White, K. J. 1976. Ecology and conservation: Flora of Papua New Guinea. Pages 35–48 *in* K. P. Lamb & J. L. Gressitt (eds.), Ecology and conservation in Papua New Guinea. Wau Ecology Institute Pamphlet 2.

Whitmore, T. C. 1966. Guide to the forests of the British Solomon Islands. Oxford University Press, Oxford.

India

Karolyn Kendrick
Compiled from Information Provided by S. K. Jain and R. R. Rao, Botanical Survey of India

Contents

I. Introduction	134
II. Description of the Region	134
A. Geographical Extent, Topography, Climate and Vegetation	134
B. Population	135
III. Vegetation Maps	136
IV. Floristic Inventory	136
A. Collections, Herbaria and Botanical Gardens	136
B. Publications	137
1. Flora of India	137
2. Regional Floras	137
3. Botanical Survey of India	137
V. Completeness of Floristic Inventory	138
VI. Endemism, Distribution and Extinction	138
A. Endemism	138
B. Distributions	138
C. Extinctions and Endangered Species	138
VII. Resources for Continued Floristic Inventory	138
A. National and Regional Programs	138
B. Foreign Collaborations	139
C. Prospects for the Future	139
VIII. Literature Cited	139
IX. Selected References	140

I. Introduction

Although the vast area of India encompasses a broad range of phytogeographic areas, thousands of years of human habitation and severe population pressures, including the doubling of the population since 1947, have so altered the natural vegetation that primary forests are thought to exist only on small areas in hilly regions of northern, eastern and southern India and the sparsely populated but increasingly settled and exploited Andaman and Nicobar Islands in the Bay of Bengal. Only about one sixth of the subcontinent is now forested, and areas that have reattained their climax vegetation after shifting cultivation practices of the past are exceedingly rare and under great pressure, being mostly confined to isolated pockets of tropical evergreen and semi-evergreen forest in northeastern India and monsoon forest in the Western Ghats. Recent efforts by the reconstituted Botanical Survey of India to complete a new flora of India are hampered to some degree by the lack of trained personnel and the need for more regional research centers in poorly explored areas where endemism and species diversity run high, as does the threat of extinction.

II. Description of the Region

A. Geographical Extent, Topography, Climate, and Vegetation

The 3.27 million km² of subcontinental India harbor so great a variety of biomes that they can be only grossly defined by reference to the three major relief regions and the southwest monsoon of summer, when masses of hot, humid air push northeastward across the peninsula and dump their moisture. Where the relief blocks the southwest monsoon, regions in the rainshadow exhibit a tropical dry deciduous or thorn forest vegetation. Such is the case for the great Deccan Plateau, screened from the brunt of the monsoon rains by the Western Ghats. On the other hand, in northeastern India, lying athwart the air flow and at the junction of the Himalayas and the Eastern Hills that trend into Burma, the hill station of Cherrapunji on the southern slopes of the Khasi Hills records more than 11 m of rain a year, whilst Shillong, only 54 km away, receives only about 2.15 m of annual rainfall. Thus, the description of the region that follows can only hint at the great complexity and diversity of the subcontinent.

To the north and arcing southeastward, tower the young Himalayan peaks, formed in the upper Tertiary when geosynclinic deposits were compressed, folded and upraised by the northward movement of the Deccan block from the ancient continent of Gondwana. The eastern Himalayas rise from the Brahmaputra valley through a zone of swamp forest that gives way from about 200 m to 800–900 m to broad-leaved tropical evergreen rainforests, topped by subtropical grasslands and subtropical forests up to about 1800 m and succeeded by temperate forests to about 3500 m. In the humid Eastern Himalayas the timber line reaches 4570 m, but is only 3600 m in the drier western ranges. The Eastern Himalayan region has provided a gateway for the migration of plants and is considered a meeting ground of the Indo-Malaysian and Sino-Japanese floras. It is also claimed to be a "cradle of flowering plants" (Sahni, 1979) and a sanctuary for ancient angiosperms, sheltered from successive ice ages. The tallest trees in India grow here as do many botanical rarities that are subject to frequent over-collection. The Eastern Hills, running in a northeast-southwest direction, are a series of ranges, averaging 1500–1800 m, which divide India from Burma, and support tropical evergreen forests and bamboo thickets in previously cleared sections. The Brahmaputra, or Assam, Valley is the corridor that links Arunachal Pradesh to the rest of India. The river carries a heavy silt load and floods almost every year. Rice is the main crop in the valleys and tea on the slopes. The Assam Plateau, in a pocket between Bangladesh and Burma, is a fragment of the Deccan block and is broken by the Garo, Khasi and Jaintia Hills averaging 1200–1800 m in height. As mentioned above, rainfall on the south-facing slopes is among the heaviest in the world, and the humidity supports the growth of dense tropical evergreen forests. The dominant soil type is acrisol, very low in nutrients and subject to much erosion.

In the west, below the icy peaks and high plateaus of the Himalayas lies the heavily farmed intermontane basin, the Vale of Kashmir through which flows the Jhelum River. Himalayan conifers of tropical and subtropical types grow on the slopes of Kashmir, but the population density is very high, more than 200/km², and sheep and goats are driven up to graze the slopes in summer. However, many of the Himalayan ranges bordering Jammu and Kashmir as well as the dry deciduous forest of Pir Panjal Range to the west along the border of Pakistan are floristically underexplored.

Paralleling the mountain wall to the south are the lowlands of the alluvial Indo-Gangetic Plain, that sweeps more than 3,200 miles across the subcontinent from the Arabian Sea on the west through the deserts of Rajasthan across the floodplain bluffs and gullies around Delhi to the rice-bowl muds and clays of the Ganges-Brahmaputra delta and ends in the tidal mangrove forests of the Sundarbans where the Ganges River empties into the Bay of Bengal. The Indo-Gangetic divide, on which Delhi is situated, has several climatic and topographic zones. From the Pakistan border east to Delhi is a submontane strip of dry deciduous forest dropping off from the Siwalik Hills. The region becomes increasingly dry toward the south, but as the divide is a great crossroads of civilizations, the area has been long and densely populated. The divide gives way in a southeastward arc to the Upper Ganges plain, which receives between 64 cm and 100 cm of rainfall a year and is traversed by three major rivers, the Jumna, Ganges and Gagra, flowing southeastward in parallel. The plain is largely featureless except for the relief provided by the bluffs of older alluvial deposits, the *bhangar*, and the newer lowland deposits, the *khadar*—a relief largely absent from the floodplains of the Middle Ganges region, where rainfall increases to as much as 180 cm a year. The Ganges falls to the Bengal Plain (which averages less than 15 m above sea level, and is heavily cultivated and industrialized due to coal and iron ore in the Damodar Valley) and passes through Calcutta on the Hooghly distributary. Although Calcutta is situated about 160 km from the sea, the Hooghly is tidal for several kilometers beyond the city. East of Calcutta and into Bangladesh are the tidal mangrove forests of the Sundarbans at the old and new mouths of the Ganges.

Just north of the Tropic of Cancer the alluvial plains begin to rise through a complex topography of hill country, interrupted by the Vindhya Range escarpment, to the east-west Satpura Range, still forested and with heights of up to 1350 m, which marks the weld at the northern extent of peninsular India proper. Rainfall in the hill country ranges from more than 100 cm a year in the east, where dry deciduous forest dominates, to no more than 25 cm westward in the Aravalli Hills where thorn forest and scrub take over. The Narbada River valley cuts through the Dindhya Range, providing an east-west route across the plateau. Teak (*Tectona grandis*), sal (*Shorea robusta*), babul (*Acacia nilotica*) and other species of deciduous forest grow on the surrounding hills at 450–600 m. Lying across the Tropic of Cancer is the Malwa Plateau which is drained by the Chambal River, and is fairly fertile in the south where the soil is derived from the Deccan lavas.

Peninsular India is dominated by the great triangle of the eastward tilting Deccan Plateau, an Archaean shield some 39,000 km^2 in extent extending from about 22°N to 8°N and with an average elevation of 460–600 m. Although built fundamentally on the ancient crystalline rocks of Gondwana, the northwest Deccan in the State of Maharashtra is characterized by thick flows of basaltic lavas from the Cretaceous era that have formed clayey "black cotton soil," or *regur*, that retains the little moisture the region receives (about 100 cm a year, as it lies in the rainshadow of the Western Ghats). Just east of the Ghats is an area that gets no more than 50–80 cm of rain a year on average.

To the northeast, the topography is more varied with hills, plateaus, basins and valleys, and the rainfall is more certain and abundant. The Bastar-Orissa highland, at the great eastern bend of the peninsula, is still under monsoon forest cover, while to the south the silty Mahanadi basin and the rice-growing Godavari basin are under cultivation. The Eastern Ghats, fold ridges that peak in several points at more than 1,500 m are also still forested although the tribal groups inhabiting the uplands practice shifting agriculture. Separating the Mahanadi and Godavari basins on either side of 20°N is the Bastar Plateau.

In the south the plains of the plateau, sliced by outcroppings of granite or gneissic rock, rise through dry deciduous woodland, thorn forests and scrub to more than 900 m, where (in the state of Karnataka) the Eastern and Western Ghats converge. The Cauvery River valley separates the Deccan from a small plateau broken by the Nilgiri and Cardamom Hills, which themselves are separated by the Palghat Gap. The higher elevations of the Western Ghats and Nilgiri Hills, which reach up to 2500 m, catch the relief rains, recording up to 280 cm of rain during the monsoon months from June to October, and are wooded to about 1500 m with semi-evergreen and subtropical broadleaved hill forests, especially in Sholas. Tea, coffee and rubber plantations dominate the slopes of the Nilgiri and Cardamom Hills, while the plateau beneath them (the site of Coimbatore), in a rainshadow, is semiarid wasteland in parts and has a natural vegetation of dry deciduous woodland and scrub.

The eastern Coromandel Coast on the Bay of Bengal is divided into two regions roughly demarcated by Madras at 13°N. North of there are low deltalands where rice is grown without irrigation, while to the south are plains that receive only 100 cm or less of rainfall a year and most of that from the retreating monsoon of October to December. The western coastland, from the Gulf of Cambay to Cape Comorin some 1600 km south, forms a strip up to 50–65 km wide at the bottom of the steep Deccan escarpment. Below the transitional zone of Goa and Kanara, the Ghats retreat inland, giving way to the Malabar Coast that stretches from the slopes through low lateritic terraces and hill country to alluvial flats forming shallow lagoons and mudbanks for long stretches. The region above 16°N is narrow and broken by ravines, lying at the foot of the Ghat scarps. Between the Gulf of Cambay and the border of Pakistan is the peninsula of Saurashtra, built of Deccan lava. It is an area of sandy valleys and rocky ridges that supports scanty woodlands. Northward the semi-desert of Kutch is bounded north and east by extensive tidal marshy mudflats. Lying along the border of Pakistan is the Thar desert with about 25 cm of rainfall a year.

The Andaman and Nicobar Islands lie in a chain in the northeast Indian Ocean on the southward trend line of the Himalayan range. Sparsely populated but increasingly settled and exploited, the Andamans comprise 204 islands and are thought to contain India's main surviving primary forests—tropical evergreen rainforest dominated by *Dipterocarpus* and rich in *Pterocarpus* and *Mesua ferrea*. The tropical semi-evergreen and moist deciduous forests, with their commercially important teak and sal, are under severe pressure from logging and agriculture. The Nicobars are composed of 22 volcanic islands south of the Andamans, and the major upper canopy genera are *Hopea, Alstonia, Calophyllum*, and *Terminalia*. The low float islands of the northern Nicobars support scrub forest. Coastal areas in both island groups are bordered with mangrove forests and littoral vegetation.

The Laccadives are a group of 16 coral atolls totaling only 28 km^2 about 300 km off southern India's Malabar coast. Ten of the islands are inhabited, while three are small keys with no vascular plants. Only about 40 of the 130 or so recorded species are indigenous to the islands and none are endemics.

B. Population

The natural vegetation of India has been profoundly altered by more than 3,000 years of human activities. The population has doubled since independence to more than 750 million and grows by about 11 million people a year. Population pressures have thus led to the active colonization of sparsely populated and botanically rich sections of the country, such as the Himalayan hill regions of Uttar Pradesh and

Bihar, the Andaman Islands, and Assam, where refugees from Bangladesh were also partly settled.

To the human population of the subcontinent must be added some 200 million head of cattle, 50 million buffalo and more than 100 million sheep and goats and the consequent demands for fodder and forage.

The history of the massive deforestation of India goes back more than 200 years. By the 1860s teak was rare in central India due to British efforts to open up arable land, to build roads and railways, and to establish a logging industry that felled vast areas of Himalayan and Ghat hill forest. Many consider the Rajasthan desert to be at least partially manmade, and Stebbing (1921) thought the silting of the Krishna and Godavari deltas was due to the clearcutting of forests to the west. By the late nineteenth century the states of peninsular India were setting aside tracts of reserved forest. However, the management of these woodlands was stymied by illegal logging and trespassing for fuel, fodder and grazing. After independence, population pressures and the demands of nation-building led to ever more rapid deforestation. Several tracts of the Himalayan hills were clearcut and hill tops denuded, leading to vast erosion from the monsoon winds and rains. The rate of deforestation is estimated officially at 155,000 ha/year, but taking into account the unrecorded felling, many believe it to be closer to 1 million ha annually. Many of the areas denoted as reserve or administrative forests in the 1976 Atlas of Forest Resources of India are reported, in fact, to be denuded.

Vegetation in the densely populated, mostly dry areas like the Indo-Gangetic Plain and the central Deccan Plateau is stabilized at the lowest level of succession. In the humid evergreen rainforests of northeastern India and parts of the Western Ghats, shifting cultivation has opened up all but isolated pockets of the forest to invasion by deciduous species, and the oldest trees are about 300 years old. Immediately after cultivation, land in rainforest bioclimes supports savanna vegetation, which is followed by fire-resistant woody species characteristic of deciduous humid forests, such as *Careya arborea, Emblica officinalis, Macaranga peltata* and *Pterocarpus* as well as the gregarious sal and teak. Such semideciduous forests also occur naturally as climax vegetation in transition zones on the edge of humid evergreen forests, in riparian forests, or in humid climates with annual rainfall of 150–200 cm and a dry season of up to four months. When semi-deciduous forests are cultivated or logged, savannah grasses are followed by an open deciduous type of vegetation that is also the climax formation in subhumid regions with annual rainfall of 100–200 cm and a dry season of up to six months, or in those areas with more rainfall but a longer dry season. In these 20–25 year old bioclimes bamboo invades cleared areas, while sal and teak often dominate the upper canopy.

III. Vegetation Maps

The most recent overall vegetation map of India is the 1976 Atlas of Forest Resources of India, published by the National Atlas Organization in Calcutta (Das Gupta, 1976). The atlas bases forest classifications on the sixteen climatic types distinguished by Champion and Seth in their 1968 map (scale 1:15,000,000) published in *Forestry in India*. That in turn was a revision of Champion's 1936 classification and vegetation map of India scale 1:19,000,000), which thus remains the basic reference work. As noted above, some areas denoted as forested in the 1976 atlas have in fact been stripped. Other maps based on Champion's 1936 classification but with greater local detail are the series of regional vegetation maps (scale (1:1,000,000) by H. Gaussen published between 1957 and 1970, and those by Puri in 1958 and Qureshi in 1965. The 1957 *National Atlas of India* has a map by Chatterjee showing forest and land use on a 1:5,000,000 scale.

Among regional and state vegetation maps are a vegetation zone map of Bombay Province (1:4,100,000) published in 1950, a map by Chandrasekharan (1962) showing the forest types of Kerala State (1:590,000), a 1930 representation by Kingdon Ward of the forests of the Northeast Frontier (1:8,800,000), and a survey by Pandeya in 1962 of forest types in Madhya Pradesh (1:16,000,000). Among district maps are those of forest types in the Satpura Range (1958, 1:2,000,000), and of Tirunelveli district (1960, 1:1,000,000), Salem district (1959, 1:1,000,000), and Coimbatore district (1:1,000,000).

Debadghao published in 1960 a map of the grass cover of India (1:21,250,000), while Sidhu in 1963 showed the distribution of mangrove forests (1:39,500,000).

IV. Floristic Inventory

A. Collections, Herbaria and Botanical Gardens

More than 3.5 million specimens of plants are maintained at herbaria in India, of which some 2.5 million are tropical plants. The major herbaria are the Central National herbarium (CAL), Calcutta, where the 2.5 million specimens include the collections of Wallich, Hooker, Thomson and Clarke. The herbarium at the India Museum (BSIS) has about 50,000 specimens. In Shillong, the Eastern Circle Herbarium (ASSAM) maintains more than 110,000 specimens, including 30,000 algae, fungi, lichens, mosses, and ferns. Collections at the Western Circle Herbarium (BSI) in Poona and the Northern Circle Herbarium (BSD) in Dehra Dun total about 140,000 specimens. The herbarium of the Forest Research Institute in Dehra Dun (DD) houses the collections of Royle, Duthie, Gamble and Haines and has a total of about 300,000 specimens. The Blatter Herbarium (BLAT) in Bombay specializes in species from western India, while the relatively small herbarium of the French Institute (HIFP) in Pondicherry concentrates on the species of southern India. The Indian Agricultural Research Institute in New Delhi has a Mycological Herbarium (HCIO) of about 25,000 specimens. Other major collections are at the National Botanical Research Institute (LWG) in Lucknow and at the Presidency College Herbarium (PCM) in Madras.

About 400,000 specimens are held in herbaria outside India, mainly at the British Museum (BM) and the Royal Botanic Gardens at Kew (K) in Great Britain, at the Musèum National d'Histoire Naturelle (P) in Paris, and in Leningrad (LE).

In addition to their herbaria, the Botanical Survey of India administers eight botanic gardens in which a number of rare, endangered and endemic plants are cultivated. National orchidaria have been established at Shillong, where about 350 species are under cultivation, at Yercaud in Tamil Nadu, where about 170 species are grown, and at Howrah, to preserve some of the more than 900 species of Orhidaceae found in India. In addition, the National Botanical Research Institute in Lucknow is a member of the Botanic Gardens Conservation Coordinating Body. The Botanical Survey of India has also prepared a directory of botanic gardens in India that lists 55 gardens and institutions.

B. Publications

As early as 4000 B.C. the plants of India were under scientific scrutiny by ancient cultures. Sanskrit scriptures such as the "Charaka Samhita," "Susruta Samhita" and " Ashtangahridaya Samhita" are among the 700 or so works from the dawn of Indian botanical studies, when plants were studied mainly for their medicinal, agricultural and horticultural uses.

Europeans were introduced to the subcontinent's flora by the 1593 publication in Portuguese of botanical investigations by Garcia de Orta, followed in the seventeenth century by the appearance of *Hortus malabaricus* (Rheede tot Draakestein, 1678-1703). Linnaean nomenclature was introduced into India during the eighteenth century, setting the stage for modern floristic studies.

1. Flora of India

In 1787 what is now the Indian Botanical Garden was established, and the first half of the nineteenth century was a ferment of floristic exploration, collecting and publication by Europeans, such as Roxburgh, Wallich, Griffith, Royle and Wight. This classical era of botanical study culminated in the publication between 1872 and 1897 of J. D. Hooker's *Flora of British India* in seven volumes, in which he estimated the number of vascular plant species growing on the subcontinent at 13,000-15,000. There have been four additions to the *Flora*, with the most recent three relating only to modern India: C. C. Calder et al., List of species and genera of Indian phanerogams not included in Sir J. D. Hooker's Flora of British India, 1926; B. A. Razi's "Second list of species and genera . . ." published in 1959; a third addendum published in 1976 by M. P. Nayar and K. Ramamurthy; and a fourth in 1981 by M. P. Nayar and S. Karthikeyam. A number of general works have supplemented and updated Hooker's *Flora*, including general manuals for trees (Brandis, 1906), ferns (Beddome, 1972) and forest (Bor, 1953). In 1960 Bor published a modern treatment of the grasses, not including Bambuseae, listed in Hooker's Flora. In 1973 H. Santapau and A. N. Henry published an annotated dictionary of genera for India listing 2900 generic names with their range and an estimate of the number of species within each genus.

2. Regional floras

Throughout the late nineteenth and early twentieth centuries regional floras also rolled off the presses, many of which have recently been updated. The following are among the most noteworthy of the classical flora surveys: J. L. Stewart, *Punjab Plants*, 1869; R. H. Beddome's two-volume *Flora Sylvatica for Southern India,* 1869-1873; the two-volume *Forest Flora of Northeast and Central India,* 1874, by J. L. Stewart and D. Brandis; D. Prain, *Bengal Plants,* 1903; J. F. Duthie, *Flora of the Upper Gangetic Plain* (excluding Gramineae), 1903-1929; E. A. Partridge, *Forest Flora of HH, the Nizam's Dominions (Hyderabad/Deccan),* 1911; T. Cooke, The Flora of the Presidency of Bombay, *1903*-1908; J. S. Gamble and C. E. C. Fischer, *Flora of the Presidency of Madras,* 1915-1934; H. H. Haines, *The Botany of Bihar and Orissa,* 1921-1925; and Kanjilal et al., *Flora of Assam* 1939-1940. Other noteworthy works on expeditions, families of plants and regional distributions were published by King, Clarke, Watt, Prain, Burkill, Barber and Lawson.

3. Botanical survey of India

Many of these investigations had been spurred by the establishment in 1890 of the Botanical Survey of India at the Calcutta Botanic Garden. Scientific activities of the Botanical Survey were curtailed after World War I, and with the virtual abolition of the Survey in 1937 floristic and taxonomic studies were almost discontinued. Many universities reduced emphasis on plant taxonomy and systematics courses, and the important herbaria at Calcutta, Coimbatore, Poona, Shillong and Dehradun did not receive due attention. Despite the botanical lacuna, two works that remain standard references were published during this time: H. G. Champion's *A Preliminary survey of the forest types of India and Burma,* 1936, in which he set up the basic classification of forest types that is still used, and D. Chatterjee's *Studies on the endemic flora of India and Burma,* 1939, in which he surveyed 1831 genera of dicotyledons and concluded that 7.3% were endemic to the region as a whole and 4.9% to India alone. With the reorganization of the Botanical Survey of India in 1954, the slow rebuilding of taxonomic study and recruitment of new workers in the field was commenced. Four regional circles covering the whole of India were set up at Shillong, Dehradun, Coimbatore and Poona, with a laboratory and headquarters at Calcutta, and botanical exploration and collecting were reinitiated. University programs in taxonomy, however, remained seriously deficient, leading to the current paucity of Indian systematists. In 1978, the Botanical Survey began publication of a serial "Flora of India" in fascicles. To date, eighteen fascicles have been published dealing with various families, tribes and large genera. About 70 research programs on the flora of small regions are now under way in colleges and universities, with about 50 research fellows of the Botanical Survey of India also engaged in this work. The regional circles have been expanded to eight with the addition of Port Blair in the Andaman Islands, in addition to some regional field stations.

Since the reorganization of the Botanical Survey of India a number of regional floras and checklists have been published (see Selected References). The total number of publications on the flora of India is estimated at about 200 books and 3300 papers. A good source for learning of ongoing work is the "Bulletin of the Botanical Survey of India," commenced in 1959.

V. Completeness of Floristic Inventory

The following regions of India have been relatively well collected, with some 50–100 collections per 100 km^2: Himachal Pradesh; the plains of Punjab, Haryana and Uttar Pradesh; Rajasthan; Gujarat; Delhi; the plains of Madhya Pradesh; parts of Maharashtra; Karnataka; Tamil Nadu; Manipur; Meghalaya; the plains of Assam; the plains of Bihar; the plains of West Bengal; and the plains of Orissa.

Many hilly and isolated regions remain poorly collected or unexplored, including most of the northwestern and eastern Himalayan ranges. Also included in regions that have less than 50 collections per 100 km^2 are Sikkim, Arunachal Pradesh, the hilly north region of Bengal, the hilly regions of Assam, Nagaland, Mizaram, southern and southeastern Madhya Pradesh, parts of Maharashtra, parts of Andhra Pradesh, Kerala, and the Andaman and Nicobar Islands and other islands.

Families and genera that are believed to be well collected and for which recent revisions have been published or worked out include the Aceraceae, Annonaceae (Uvarieae), Apiaceae, Asclepiadaceae (*Ceropegia*), Balanitaceae, Convolvulaceae (*Ipomoea*), Coriariaceae, Cucurbitaceae, Cyperaceae (*Carex, Fuirena*), Dilleniaceae, Fabaceae (*Derris, Vigna, Phaseolus*) Ixonanthaceae, Liliaceae (Scilleae), Lamiaceae, Linaceae, Oleaceae, Orchidaceae (*Calanthe, Cymbidium, Coelogyne, Paphiopedilum*), Paeoniaceae, Pittosporaceae, Poaceae (Andropogoneae, Isachneae, Garnotieae), Ranunculaceae, Sapotaceae, Simaroubaceae, Violaceae, and Zingiberaceae (*Hedychium*).

Among the poorly-collected groups are aquatic plants, *Impatiens*, Podostemaceae, Lentibulariaceae, Zingiberaceae, Balanophoraceae and Stylidiaceae.

VI. Endemism, Distribution and Extinction

A. Endemism

India is said to harbor more endemic species of plants than any other region of the world except Australia. However, although endemism in the monocotyledons has been analyzed recently, figures for endemism among dicotyledons date back to Chatterjee's 1939 work. At that time he enumerated 11,124 species of dicots from 1831 genera and 173 families. For the Himalayan region, which includes areas now in Pakistan, Nepal, Tibet and Burma, he estimated specific endemism at 3169 plants. For peninsular India south of the Tropic of Cancer, he identified 2045 endemics.

India is thought to have about 14,000 species of wild flowering plants, with about 90% of them concentrated in the Himalayas, the Western Ghats, the Vindhya and Satpura ranges, the Eastern Ghats and the Khasi and Mizo Hills. The Northwestern Himalayas are estimated to harbor about 1000 endemics out of 3000 total species, while the Eastern Himalayas have about 1500 endemic species out of a total of 4000. About 2000 of the estimated 6000 species of peninsular India are endemics, and roughly 220 of the 2200 or so species on the Andaman and Nicobar Islands are thought to be endemic.

B. Distributions

Collection tours by the Botanical Survey of India have found new and important distributions of species in certain areas of the northwest Himalayas like Garhwal and Kashmir. In northeast India, spots in Arunachal Pradesh and Meghalaya are also producing data on new taxa and distributions, as are areas in the Western Ghats and on the Andaman and Nicobar Islands.

C. Extinction and Endangered Species

Data on rates of extinction are available only for small regions and for certain families or genera. About 100 species, mainly herbs, have been reported as not collected after their type collections. Areas most at risk are those under the most intense biotic pressure in northeast India, the Andaman and Nicobar Islands, the peninsular region and some spots in the northwest Himalayas and Western Ghats.

Although a Wildlife (Protection) Act was promulgated in 1972, not much has been done to preserve the more than 2,500 endangered species of angiosperms and gymnosperms in India. Many papers on threatened plants in various regions of India have been published, and the most comprehensive survey was issued by the Botanical Survey of India in 1983, listing about 900 rare and threatened taxa together with their distributions. The "Plant Conservation Bulletin," also published by the Botanical Survey of India, has printed many annotated lists of rare, endangered and endemic plants, and other journals and publications publish lists from time to time (see Selected References).

VII. Resources for Continued Floristic Inventory

A. National and Regional Programs

Only recently have efforts been made to redevelop curricula in floristics, plant geography and biosystematics at the university level, and the shortage of trained taxonomists is still acute. However, universities and colleges are now collaborating in the prepartion of district and state level floras that will form the background for the new Flora of India. The major governmental instrument for floristic inventory is the Botanical Survey of India. Other institutions interested in floristic activities include the National Botanical Research Institute in Lucknow, the Forest Research Institute in Dehra Dun and the National Bureau of Plant Genetic Resources in New Delhi. The Indian Council of Agricultural Research and the Council of Scientific and Industrial Research are also involved in some areas of floristics.

B. Foreign Collaborations

In terms of foreign programs, the IUCN is corresponding with the Department of Environment for developing some projects. A small U.S. Fish and Wildlife Services program has been operating in the Botanical Survey on Endangered Species, and the Smithsonian Institution helped in the study of a small region (the Hassan District) near Bangalore several years ago (Saldanha & Nicholson, 1976). International linkages are usually established through the Department of Environment. Internally, the Botanical Survey has collaborated with the Zoological Survey of India and the Anthropological Survey of India. The All-India Coordinated Research Project on Ethnobiology, which mounts some floristic studies in remote areas, functions jointly in about a dozen institutions.

C. Prospects for the Future

There is a dawning awareness in official institutions in India of the importance of a floristic inventory to lay the groundwork for the conservation, protection and utilization of the country's natural plant resources. It is hoped by the Botanical Survey that involving colleges and universities, and ultimately perhaps high schools, in the preparations of district floras will not only result in the production of trained taxonomists but also help instill in students a deeper interest in conservation. So far, colleges and universities are participating in about 70 research programs on the flora of small regions. A major objective of the program to create a new Flora of India is to investigate poorly explored areas, many of which also are areas of high species diversity and endemism and which are experiencing rapidly intensifying biotic pressure.

However, the resources for implementing the program of inventory, particularly in terms of trained workers, are estimated to be about 30–40% of what is required. Programs for the required training are being drawn up, including more emphasis on the teaching of taxonomy in the Botanical Survey and universities. It has even been suggested that a few universities make taxonomy their main discipline. Proposals for more research centers in floristically rich but poorly explored areas have also been made, and about 50 more taxonomists are needed to complete the survey. Direct costs of completing the new Flora of India would include the salaries and perquisites of the taxonomists as well as transportation, field and other infrastructual expenses.

There is thus evidence of renewed interest and activity in taxonomy and floristics in many research and educational institutions. The old herbaria are being strengthened and new ones established in different regions of the country.

VIII. Literature Cited

Anonymous. 1976. Atlas of forest resources of India. National Atlas Organisation. Dept. Science & Technology, Govt. of India. Calcutta.

Beddome, R. H. 1869–1873. The flora sylvatica for Southern India. Forester's manual of botany for South India 1, 2, 3.

———. 1876. Ferns of British India, Vols. 1 & 2. Reprint 1972. Oxford & IBH. New Delhi.

Bor, N. L. 1953. Manual of Indian forest botany. Oxford University Press. Oxford.

———. 1960. The grasses of Burma, Ceylon, India and Pakistan, excluding Bambuseae. Pergamon Press. New York.

Botanical Survey of India. 1978–. Fascicles of Flora of India. 18 published. BSI, Howrah.

———. 1983a. A directory of botanic gardens in India. BSI, Howrah.

———. 1983b. Materials for a green book of botanic gardens in India. BSI, Howrah.

Brandis, D. 1906. Indian trees. Archibald Constable & Co., London.

Calder, C. C., V. Narayanaswami & M. S. Ramaswami. 1926. List of species and genera of Indian phanerogams not included in Sir J. D. Hooker's Flora of British India. Rec. Bot. Surv. India 11(1): 1–157. (reprint, 1978. Dehra Dun).

Champion, H. G. 1936. A preliminary survey of the forest types of India and Burma. India Forest. Rec. 1.

——— & S. K. Seth. 1968. A revised survey of the forest types of India. GOI Press, Delhi.

Chatterjee, D. 1939. Studies on the endemic flora of India and Burma. J. Asiat. Soc. Bengal, Sci. 5: 19–67.

Cooke, T. 1903–1908. The flora of the Presidency of Bombay. 3 vols. Taylor & Francis, London. (Reprint, 1958, BSI, Calcutta).

Das Gupta, S. P. (ed.). 1976. Atlas of forest resources of India. National Atlas Organisation. Dept. Science & Technology, Govt. of India. Calcutta.

Debadghao, P. M. 1960. Types of grass covers of India and their management. Proc. 8th Int. Grasslands Congr.

Duthie, J. F. 1903–1929. Flora of the Upper Gangetic Plain. 3 vols. Supt. Gov't Printing, Calcutta.

Gamble, J. S. 1903. A preliminary list of the plants of the Andaman Islands. Chief Commissioner's Press, Port Blair.

——— & C. E. C. Fischer. 1915–1935. Flora of the Presidency of Madras. 3 vols. BSI, Calcutta. (New editions 1957 and 1967).

Gaussen, J. & Others. 1961–1968. Carte internationale du tapis végétal et de conditions écologiques. Institut Français, Pondicherry.

Haines, H. H. 1921–1925. The botany of Bihar and Orissa. 6 parts. Govt. of Bihar and Orissa.

Hooker, J. D. 1827–1897. Flora of British India. 7 vols. L. Reeve, London.

Kanjilal, U. N., P. C. Kanjilal & A. Das. 1934–1940. Flora of Assam. 4 vols. Govt. of Assam, Shillong.

Kingdon Ward, F. 1930. The forests of the North East frontier of India. Empire Forest J. 9: facing p. 12.

Nayar, M. P. & S. Karthikeyan. 1981. Fourth list of species and genera of Indian phanerogams not included in J. D. Hooker's The Flora of British India. Rec. Bot. Surv. India **21(2):** 129–152.

―――― & K. Ramamurthy. 1976. A third list of species and genera . . . Bull. Bot. Surv. India **15:** 204–234.

Orta, Garcia de. 1593. Aromaticum et simplicium aliquot medicamentum apud Indos nascentium historia. Antwerp.

Pandeya, S. C. 1962. Ecology as an aid to floristics. 2, Autoecology. Bull. Bot. Surv. India **4:** 151.

Partridge, E. A. 1911. Forest flora of HH, the Nizam's Dominions, Pillai, Hyderabad.

Prain, D. 1903. Bengal plants. 3 vols. BSI, Calcutta.

Razi, B. A. 1959. A second list of species and genera . . . Rec. Bot. Surv. India **18(1):** 1–56.

Rheede tot Draakestein, H. A. van. 1678–[1703]. Hortus indicus malabaricus. 12 vols. Fol. Joannis van Someren & Joannis van Dyck. Amsterdam.

Sahni, K. C. 1979. Endemic, relict, primitive and spectacular taxa in eastern Himalaya flora and strategies for their conservation. Indian J. Forest **2(2):** 181–190.

Saldanha, C. J. & D. H. Nicholson, eds. 1976. Flora of the Hassan District, Karnataka, India. Amerind Pub. Co., New Delhi. (Pub. for Smithsonian Inst. & NSF.)

Santapau, H. & A. N. Henry. 1973. A dictionary of flowering plants in India. CSIR of India, New Delhi.

Sidhu, S. S. 1963. Studies on the mangroves of India. I. E. Godavari region. Indian Forest 89.

Stebbing, E. P. 1921. The forests of India. 3 vols. John Lane the Bodley Head, London.

Stewart, J. L. 1869. Punjab plants. Govt. Press, Lahore. (Reprnt. 1977, Bishen Singh, Dehra Dun.)

―――― & D. Brandis. 1874. Forest flora of northwest and central India. Allen, London.

IX. Selected References

Abraham, Z. & B. N. Mehrotra. 1983. Some observations on endemic species and rare plants of the montane flora of the Nilgiris, South India. J. Econ. Taxon. Bot. **3(3):** 863–867.

Bahadur, K. N. & S. S. Jain. 1982. Rare bamboos in India. India J. Forest. **4(4):** 280–286.

Bhandari, M. M. 1978. Flora of the Indian Desert. Scientific Publishers, Jodhpur.

Blasco, F. 1970. Montagnes du Sud de l'Inde: savanes, forêts, écologie. Institut Français, Trav. Sect. Sci. Tech. 10. Pondicherry.

Bombay Province. 1950. Statistical atlas of Bombay State (Provincial part). Bureau of Economic and Statistics Secretariat, Bombay.

Cook, C. D. K. 1980. The status of some Indian endemic plants. Threatened Plants Comm. Newsletter. **6:** 17–18.

Dhar, U. & P. Kachroo. 1983. Alpine flora of Kashmir Himalaya. Scientific Publishers, Jodhpur.

Gamble, J. S. 1903. A preliminary list of the plants of the Andaman Islands. Chief Commissioner's Press, Port Blair.

Hajra, P. K. 1983. Annotated list of rare, threatened and endemic plants of the western Himalayas—monocotyledons. Pl. Conserv. Bull. **4:** 1–13.

Haridasan, K. & R. R. Rao. 1985. Forest flora of Meghalaya. 2 vols. Bishen Singh Mahenoron Pal Singh, Dehradun.

Henry, A. N., K. Vivekananthan & N. C. Nair. 1978. Rare and threatened flowering plants of south India. J. Bombay Nat. Hist. Soc. **75(3):** 684–697.

Jain, S. K. 1983. Flora and vegetation of India—An outline. BSI, Howrah.

―――― & K. L. Mehra (eds.). 1983. Conservation of tropical plant resources. BSI, Howrah.

―――― & A. R. K. Sastry. 1980. Threatened plants of India: A state-of-the-art report. BSI, Howrah.

―――― & ――――. 1983. Materials for a catalogue of threatened plants of India. Dept. of Environment, GOI, Calcutta.

Kataki, S. K. 1976. Indian orchids—a note on conservation. Amer. Orchid Soc. Bull. **46(2):** 117–121.

Koteswaram, P. 1974. Climate and meteorology of humid tropical Asia. Pages 27–86 *in* Natural resources of humid tropical Asia. UNESCO, Paris.

Kurz, S. 1870. Report on the vegetation of the Andaman Islands. Office of Govt. Printing, Calcutta.

Legris, P. 1974. Vegetation and floristic composition of humid tropical Asia. Pages 212–238 *in* Natural resources of humid tropical Asia. UNESCO, Paris.

Melville, R. 1970. Endangered plants and conservation in the islands of the Indian Ocean. Pages 103–107 *in* IUCN, 11th technical meeting papers and proceedings, 2, problems of threatened species. IUCN New Series 18, Switzerland.

Mukerjee, S. K. 1959–1962. Enumeration of Indian flowering plants. Bull. Bot. Surv. India **1:** 138–141; **2:** 99–107, 293–297; **3:** 99–101, 351–355; **4:** 39–47.

Patel, R. 1968. Forest flora of Melghat. Bishen Singh Mahendra Pal Singh, Dehra Dun.

Prain, D. 1893. Botany of the Laccadives. J. Bombay Nat. Hist. Soc. **7:** 460–486.

Parkinson, C. E. 1923. A forest flora of the Andaman Islands. Govt. Central Press, Simla.

Raghavan, R. S. & N. P. Singh. 1983. Endemic and threatened plants of western India. Pl. Conserv. Bull. **3:** 1–16.

Rao, R. R. & B. A. Razi. 1981. A synoptic flora of Mysore District. Intl. Bioscience Ser. 7. Today and Tomorrow Printers, New Delhi.

Rosayro, R. A. de. 1974. Vegetation of humid tropical Asia. Pages 179–196 *in* Natural resources of humid tropical Asia. UNESCO, Paris.

Saldanha, C. J. 1984. Flora of Karnataka. Vol. 1. Magnoliaceae to Fabaceae. Oxford & IBH, New Delhi. With the help of S. R. Ramesh et al.

Santapau, H. 1970. Endangered plant species and their habitats. Pages 83–88 *in* IUCN, 11th technical meeting papers and proceedings, 2, problems of threatened species. IUCN New Series 18, Switzerland.

Sivadas, P., B. Narayanan & K. Sivaprasad. 1983. An account of the vegetation of Kavaratti Island, Laccadives. Atoll Res. Bull. **266:** 1–9.

Vajravelu, E. 1983. Rare, threatened and endemic flowering plants of South India (Part 1). Pl. Conserv. Bull **4:** 14–30.

Varma, S. K. 1981. Flora of Bhagalpur: Dicotyledons. Today and Tomorrow Printers, New Delhi.

Willis, J. C. 1901. Notes on the flora of Minikoi. Ann. Roy. Bot. Gard. (Peradeniya) **1:** 39–43.

Sri Lanka

Karolyn Kendrick
(Compiled from information provided by
A. H. M. Jayasuriya, N. Gunatilleke
and C. V. S. Gunatilleke)

Contents

I. Description of the Region	142
A. Geographical Extent and Topography	142
B. Geology	142
C. Climate	142
D. Population	142
II. Vegetation Types	142
III. Vegetation Maps	143
IV. Floristic Inventory and Publications	143
V. Endemism and Extinction	143
VI. Resources for a Floristic Inventory	144
VII. Literature Cited	144
VIII. Selected References	144

I. Description of the Region

A. GEOGRAPHICAL EXTENT AND TOPOGRAPHY

Although the pear-shaped island of Sri Lanka, situated in the Indian Ocean at 5°55′N to 99°51′N lat. and 79°41′E long., occupies an area of only 65,584 km², it has four distinctive phytogeographic regions, ranging from montane humid evergreen and semi-evergreen in the rugged south-central highlands to small arid regions in the extreme northwest and southeast, where no more than about 12 species of flowering plants have been found. The island is 436 km long and, at its widest, 226 km wide.

B. GEOLOGY

In ancient geological time, Sri Lanka was part of the Indian subcontinent at the time the latter broke off from Gondwana, and the two did not separate until the Miocene. The central highlands were uplifted after the separation, in the Pliocene, Pleistocene, or even later. There is no evidence of any glaciation on the island. At various times after their separation Sri Lanka and peninsular India were probably connected for short periods by land bridges. The island is still linked to the tip of India by "Adam's Bridge," a series of coral reefs and islets strung across Palk Strait, which is about 32 km wide.

Archaean rocks cover most of the island and underlie the coastal plains. In the uplands, the Precambrian rocks form a complex schist-gneiss, surmounted by rocks comprised of a kondalite, or quartz, system. Under the Jaffna Peninsula, at the northern tip of the island, and extending along the northwest coast are soft limestone formations from the Miocene.

C. CLIMATE

Overall, the climate of the island can be categorized as equatorial and monsoonal, but relief modifies both these characteristics. The lowlands that occupy about three-quarters of the island's area are subdivided into wet and dry lowland zones. The "dry" zone of seasonal drought and variable rainfall comprises all of the island except the south-central highlands and the southwest coast facing the summer monsoon. In these subhumid regions, which generally receive 150–200 cm of annual rainfall, and have a dry season of five to six months, the natural vegetation consists of tropophilous savanna woodlands, scrub and thorn thickets. Within the dry zone are certain regional differences. The limestone region in the north and northwest is extensively cultivated in areas that can be irrigated, while coconuts fringe the northwest coast. The northeast and east coasts receive most of their rainfall during the winter northeast monsoon, as they lie in the rainshadow of the south-central hills. Sri Lanka's east coast, varying in width from 15 to 50 km, is lined with lagoons, dunes and marshland. Inland of the lagoon areas around Batticaloa, just below 8°N is the *talawa*, or savanna country. Sri Lanka has two small arid zones, one along the northwest coast above Chilaw, and another, drier region in the extreme southeast.

D. POPULATION

Sri Lanka has long been densely populated and cultivated. Both the wet and dry lowlands are densely dotted with remains of the Sinhalese civilization's irrigation reservoirs, or tanks. In the hill country, agriculture has opened up most, if not all of the forests. Holmes (1951) believed there to be no extant primary forest, but Abeywickrama (1965) cited pockets in the Adam's Peak wilderness.

The British introduced plantation agriculture, which remains the basis of the economy and accounts for about two thirds of the land under cultivation—tea, rubber and coconuts being the main crops. In all, about a quarter of the island's area is in permanent agricultural use, while shifting cultivation is still practiced in the dry zones. In the hill country tea plantations crowd the plateaus and valleys at around 900 m.

The 1981 census recorded a population of 14.8 million, but current UN estimates are as high as 15.6 million, with a growth rate of about 1.5% a year. The Jaffna Peninsula and the east coast around Batticaloa are densely populated, but the bulk of the population lives in the wet coastal plain and hill country. The northwest limestone region and the arid southeast are sparsely settled.

II. Vegetation Types

The wet lowlands girdle the central hills. These rolling lands are generally less than 185 m above sea level, but close to the mountains they climb to more than 300 m. The southwest coast, from about 8°N to 6°N, faces the southwest monsoon squarely, and receives more than 200 cm of rainfall each year, with no dry season. Thus, the climax vegetation in the southwest, up to about 900 m, is lowland evergreen forest, but less than 1600 km² of such forest remain, as the wet uplands are under plantation culture. Sinharaja, 88 km², is the largest remaining tract of these rain forests, now classified as primary. To the north, east and south of the hill massif, where the dry season may last up to four months, the climax vegetation is moist deciduous forest. Where the dry season is shorter—one to two months—there is mixed tropical evergreen and deciduous forest. Northwest of the hills around Kurunegala is a large area that receives more than 200 cm of rain a year, broken by a dry season of one to two months. Here the climax vegetation is dense evergreen forest, although the secondary growth is often deciduous; most of the forest is now cleared for cultivation and urbanization.

From the lowlands the land rolls up through hard quartz to a peneplain at about 600 m at the foot of the mountains. These have steep scarps and show young relief; they comprise about one fifth of the island's area. Although the highest peak, Pidurutalagala, in the center of the massif, is about 2,530 m in height, most of the hills average 1,200–1,800 m. The rugged country is divided naturally into three sections of eroded ridges by the basin of the Mahaweli Ganga River, between Knuckles Range and the central massif, and by the valleys of Kalu Ganga and Welawe Ganga in the south. During the summer monsoon months from June to October rain falls fairly constantly on the southwestern hills, which are cloaked in mist and cloud. The canopy above 1,500 m is dominated by Dipterocarpaceae (*Doona*), some of which grow more than 30 m high. Above 1,800 m the closed montane evergreen forest is often replaced by zones of wet patana grassland. Lower down from about 900 to 1,500 is semideciduous forest.

III. Vegetation Maps

The best general vegetation map is the 1965 *Carte International du Tapis Végétatal et des Conditions Ecologiques* by H. Gaussen et al., drawn on a 1:1,000,000 scale. This map, and the "Notice de la feuille Ceylon" that accompanies it, records the vegetation types and climatic regions. It is based on Andrews' 1:500,000 map (1961) which was drafted from aerial photos taken on a 1:40,000 scale. In 1983, the Sri Lanka Forest Department published a Forest Cover and Forest Land Management map on a scale of 1:250,000. Earlier vegetation maps include those by P. Legris et al. (1964) on a 1:1,000,000 scale, C. H. Holmes (1958) 1:900,800, and E. K. Cook (1931) 1:2,800,000. In 1958, R. A. de Rosayro published a regional climate and vegetation map of the Knuckles Range area (1:63,810).

IV. Floristic Inventory and Publications

The standard but now outdated flora for Sri Lanka is Trimen and Hooker's *Handbook of the Flora of Ceylon*, published in five volumes between 1893 and 1900. Nomenclature revisions and additions to Trimen's flora were made in 1911 by Willis and Smith and, more extensively, by Alston in 1931, whose work forms a sixth volume of the *Flora*.

In 1967 work got under way on a revised *Flora of Ceylon*. Parts of the first volume were published as fascicles in 1973. These were later incorporated in 1980 into the bound volumes constituting the *Revised Handbook of the Flora of Ceylon*, edited by M. D. Dassanayake and F. R. Fosberg. The fifth volume was published in 1985. Ten volumes in all are planned.

A number of regional flora and field guides for Sri Lanka have also been published (see Selected References), although there are a few studies of lower plants. Blatter's 1911 *A Bibliography of British India and Ceylon* was revised in 1952, 1958 and 1964 by H. Santapau. Other bibliographies containing references to floral research in Sri Lanka include a mimeographed *Bibliography of scientific publications relating to Sri Lanka* by N. A. Alwis (1978) and the three-volume *Bibliography of Ceylon* (1970–1976) by H. A. I. Gunatilleke. In addition, "Flora Malesiana Bulletin," published annually in Leiden, appraises many new publications on countries in this region.

The Sri Lankan National Herbarium (PDA) at the Peradeniya Botanic Garden maintains a collection of about 130,000 specimens. Voucher specimens, including numerous types, and originals of the drawings in Trimen's *Flora of Ceylon* are deposited there. About 325,000 specimens are outside the country, mainly in Great Britain at the Royal Botanic Gardens at Kew (K) and the British Museum (BM), and in India at the Indian National Herbarium (CAL) in Calcutta. Initial funding for the *Revised Flora of Ceylon* was provided by the Smithsonian Institution and specimens from those collecting expeditions are deposited at the U.S. National Herbarium (US) as well as at Peradeniya, and many elsewhere, as well. The personal collections of K. L. D. Amaratunga and the late T. B. Worthington have been drawn upon in revising the *Flora of Ceylon*.

Areas of Sri Lanka that are relatively well collected, with 50–100 collections per 100 km², include some of the forest reserves in the montane region, lowland rainforest and the dry lowlands. All other areas are poorly collected.

V. Endemism and Extinction

Although no family of plants is endemic to Sri Lanka, the richness of specific endemism has fascinated and attracted investigators. It is a trove of specific endemism. In all, about 3,368 species and from 1,294 genera and 192 families of phanerogams have been recorded (Abeywickrama, 1959), of which 830 species are considered endemic (Bandaranaike & Sultanbawa, 1969). Of the nonendemic species, Abeywickrama (1965) notes that 1,841 are found also in peninsular India and about one third of these are confined to India and Sri Lanka. Only four families of plants found in Sri Lanka are absent from peninsular India (Monimiaceae, Nepenthaceae, Cactaceae and Stylidiaceae). Most of the endemic species on the island appear to be relics, and some are confined to very small areas, such as a single mountain.

Alston (1931) considered 326 species to be endemic to the lowland wet zone—about 60–70% of the total number of species there—of which 195 were noted as "rare" or "very rare." Abeywickrama (1983) estimated 110–160 nonendemic species to be "rare" or "threatened." The rate of extinction is highest in the lowland wet zone due to population pressures and commercial exploitation. In 1981, the rate of deforestation of lowland closed deciduous forest was estimated by the FAO/UNEP at 580 km² a year, out of a total forest tract of 16,590 km². Overall, the annual rate of deforestation is 3.5%. In fact, almost all the natural forest left on the island are under threat of destruction of conversion. C. V. S. Gunatilleke (1978) estimated endemism among the tree flora at 60%. Of the 12 endemic genera, all but one are confined to the climax rainforests in the southwest, which are under the most severe pressure.

Of the 314 native pteridophytes identified by Sledge (1982), 57 are endemic to Sri Lanka, and of the nonendemics 28 are absent from India. Sledge also noted that 30 species of ferns had not been collected since the beginning of the century, although some may be discovered in poorly explored areas.

Sri Lanka has ten national parks and reserves totaling more than 172,000 ha. The national Fauna and Flora Protection Advisory Board, operating since 1938, was authorized by the 1937 Fauna and Flora Protection Ordinance. Crusz, in a review (1973) of conservation efforts in Sri Lanka, cites nine species, including six orchids, that are totally protected under the Fauna and Flora Protection Act. Two voluntary organizations are concerned with conservation: the Wildlife and Nature Protection Society of Sri Lanka and the March for Conservation, both in Colombo.

VI. Resources for a Floristic Inventory

The national herbarium and the departments of botany at the universities of Peradeniya and Colombo are active in the work on the *Revised Handbook of the Flora of Ceylon*. To complete the last one-third of the project and to ensure ongoing research and maintenance of collections, Sri Lanka will require additional resources. Financing from the Smithsonian Institution for the revised flora was discontinued in 1978 due to lack of funds, greatly slowing work on the project and the inventory of poorly explored areas. On a long-term basis, resources for teaching taxonomy in the universities need to be expanded, especially at the University of Peradeniya, where the revised flora is edited. About $500,000 is needed to complete the last third of the *Revised Handbook of the Flora of Ceylon*, as foreign and local plant taxonomists must be invited to revise the remaining families of plants.

Approximately $100,000 is required to renovate and improve facilities at the Perdeniya Herbarium. A break down of the estimated costs is as follows:

Air conditioning the herbarium buildings	$50,000
Books and journals for the library	$20,000
Herbarium equipment and furniture	$20,000
Office equipment	$10,000
Total	$100,000

VII. Literature Cited

Abeywickrama, B. A. 1959. A provisional check-list of the flowering plants of Ceylon. Ceylon J. Sci. (Biol. Sci.) **2**: 119–240.

———. 1965. The origin and affinities of the flora of Ceylon. Proc. of 11th Ann. Sess. of Ceylon Assoc. for Adv. Sci. Colombo Apothecaries' Co., Colombo.

———. 1983. Threatened or endangered plants of Sri Lanka and the status of their conservation measures. Pages 11–18 *in* S. K. Jain & K. L. Mehra (eds.), Conservation of tropical plant resources. Botanical Survey of India, Howrah.

Alston, A. H. G. 1931. A handbook to the flora of Ceylon 6. Dulau & Co., London.

Alwis, N. A. 1978. Bibliography of scientific publications relating to Sri Lanka. Mimeograph.

Andrews, J. R. T. 1961. A forest inventory of Ceylon. Government Press, Colombo.

Bandaranaike, W. M. & M. U. S. Sultanbawa. 1969. List of endemic plants of Ceylon. Mimeograph.

Blatter, E. 1911. A bibliography of the botany of British India and Ceylon. J. Bombay Nat. Hist. Soc. 20. Revised by H. Santapau in 1952, 1958, 1964, in same journal.

Cook, E. K. 1931. A geography of Ceylon. Macmillan, London.

Crusz, H. 1973. Nature conservation in Sri Lanka (Ceylon). Biol. Conserv. **5**: 199–208.

Dassanayake, M. D. & Fosberg, eds. 1980–1985. A revised handbook to the flora of Ceylon. 5 vols. to date. 1: Bakema, Rotterdam. 2–4: Amerind Publ. Co., New Delhi.

Gaussen, H. et al. 1965. Notice de la Feuille Ceylon. Carte Internationale du Tapis Végétal et des Conditions Ecologiques a 1:1,000,000. Hors série no. 5. Institut Français, Pondicherry, India.

Gunatilleke, C. V. S. 1978. Sinharaja today. Sri Lankan Forest. **13**: 57–64.

Gunatilleke, H. A. I. 1970–1976. A bibliography of Ceylon. 3 vols. Inter-Documentation Co., Zug, Switzerland.

Holmes, C. H. 1951. The grass, fern and savannah lands of Ceylon, their nature and ecological significance. Imperial Forestry Inst., Oxford.

———. 1958. The broad pattern of climate and vegetational distribution in Ceylon. Study of tropical vegetation. Proc. of Kandy Sump. UNESCO, Paris.

Legris, P. et al. 1964. International map of the vegetation and of environmental conditions—Ceylon. Institut Français, Pondicherry, India.

Rosayro, R. A. de. 1958. The climate and vegetation of the Knuckles region of Ceylon. Ceylon Forest. **3(1)**: 63, 810.

Sledge, W. A. 1982. An annotated check-list of the Pteridophyta of Ceylon. J. Linn. Soc., Bot. **84**: 1–30.

Trimen, H. & J. D. Hooker. 1893–1900. A handbook of the flora of Ceylon. 5 vols. Dulau, London.

Willis, J. C. & A. M. Smith. 1911. Additions and corrections to Trimen's flora of Ceylon. Ann. Roy. Bot. Gard. (Peradeniya) **5**: 175–214.

VIII. Selected References

Abeywickrama, B. A., ed. 1973. A revised handbook to the flora of Ceylon 1, parts 1, 2 (fascicles). University of Ceylon, Peradeniya.

Alston, A. H. G. 1937. Kandy flora. Government Press, Colombo.

Bond, T. E. T. 1953. Wild flowers of the Ceylon Hills. Oxford Univ. Press, London.

Fernando, D. 1954. Wild flowers of Ceylon. West, Mitcham.

Gunatilleke, C. V. S. & D. S. A. Wijesundara. 1982. Ex-situ conservation of woody plant species in Sri Lanka. Loris **16(2)**: 73–80.

Hooker, J. D. & T. Thomson. 1855. Flora Indica, with an introductory essay. W. Pamplin, London.

Legris, P. 1961. Bioclimes of S. India and Ceylon. Trav. Sect. Sci. Tech. **3**: 165–178. Institut Français, Pondicherry, India.

———. 1974. Vegetation and floristic composition of humid tropical continental Asia. Pages 217–238 *in* Natural resources of humid tropical Asia. UNESCO, Paris.

Thwaites, G. H. K. & J. D. Hooker. 1864. Enumeratio Plantarum Zeylaniae. Dulau, London.

Werner, W. L. 1981. From far and near . . . a plea for the conservation of three unique forests in Sri Lanka. Loris **15(6):** 331–332.

Worthington, T. B. 1959. Ceylon trees. Colombo Apothecaries Co., Colombo.

REGIONAL REPORTS

II. AUSTRALIA AND PACIFICA

Australia

A. P. Kershaw and T. Whiffin

Contents

I.	Introduction	150
II.	Description of the Region	151
	A. Geographical Extent and Area	151
	B. Topography and Associated Environmental Factors	151
	C. Population	151
III.	Vegetation	151
	A. History of the Australian Tropical Rain Forest Flora and Vegetation	151
	B. Mapping and the UNESCO Classification	152
IV.	Vegetation Maps	152
V.	Magnitude of Floristic Inventory	158
	A. Herbarium Holdings	158
	B. Publications	159
VI.	Completeness of Floristic Inventory	160
	A. Collection Density	160
	B. Areas of High Endemicity	160
VII.	Completeness of Taxonomic Knowledge	161
VIII.	Areas Under Threat	162
IX.	Present and Future Resources for Floristic Inventory	163
X.	Acknowledgments	164
XI.	Literature Cited	164

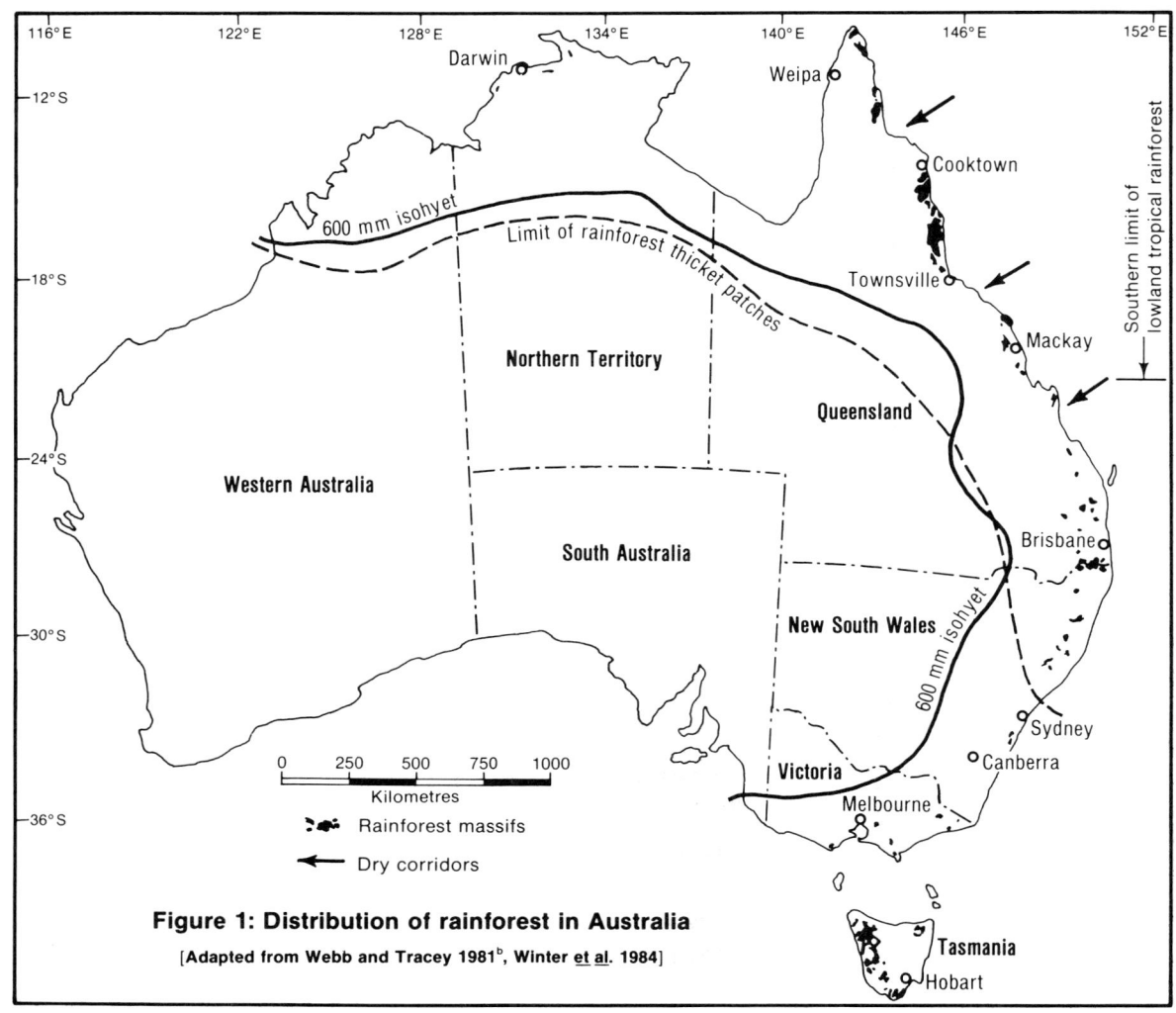

Figure 1: Distribution of rainforest in Australia
[Adapted from Webb and Tracey 1981[b], Winter et al. 1984]

I. Introduction

Australia, the dry continent, does not possess extensive areas of luxuriant tropical forest. However, around the eastern and northern margins of the country, conditions are suitable for the survival of a range of forest types which can be broadly classed as rain forest (or closed canopied forest) and open canopied forest. The open forests are dominated by sclerophyllous taxa such as *Eucalyptus* and *Acacia*, which form the wetter end of the continuum of sclerophyll vegetation that extends into the arid interior. Those sclerophyllous communities surround rain forest which exists as small, isolated patches.

This paper considers only rain forests which can be compared with tropical forests in other parts of the world. The sclerophyll forests, by contrast, are uniquely Australian. Additional reasons for emphasizing rain forests are their variety, their structural and floristic diversity and their vulnerability in the face of disturbance. The sclerophyll forests, which are considered to have largely derived from components of Australian rain forests, are much simpler and under less of a threat because they tend to be advantaged by disturbance, particularly fire. There is a certain amount of interspersion of the floras of the two formations but they are remarkably distinct considering their proximity to each other and their genetic relationships.

Australia is unusual also in possessing a broad latitudinal spread of rain forest patches. These extend down the east coast from the tropics, through the subtropics to warm and cool temperate latitudes. Although the species diversity and structural complexity of forests decrease towards higher latitudes, the extra-tropical forest types are not simply attenuations of the tropical forests: they possess unique features which point to different origins or long periods of isolation from the tropical types. They are relevant, however, to a consideration of forests within the tropics in that certain forest types which have their core areas outside tropical latitudes do extend into the tropics due to locally favorable environments and as a result of historical influences.

II. Description of the Region

A. GEOGRAPHICAL EXTENT AND AREA

The generalized distribution of rain forest is shown in Figure 1. Patches extend round the coastal parts of the continent from the Kimberleys of Western Australia, east to Cape York Peninsula, and then south to Tasmania. South Australia is the only State not possessing some rain forest. Concentrations of larger patches or 'massifs' occur in the relatively high rainfall areas of north-eastern Queensland, the Queensland/New South Wales border and in Tasmania. These provide the foci of high rainfall tropical, sub-tropical and cool temperate rain forest respectively. Very small patches exist to about the 600 mm isohyet in the north and northeastern parts of the continent but no such 'dry' rain forests occur in temperate latitudes.

Recent estimates of the extent of rain forest agree upon a figure of about 2 million hectares (or 20,000 km²) with some variation depending largely upon whether or not wet sclerophyll forests with well developed rain forest understories are included (Australian Forestry Council, 1974; Webb & Tracey, 1981a). This amounts to only about 0.3% of the area of the continent and less than a third of that estimated to have been present at the time of arrival of European man.

Within tropical Australia, the major rain forest massifs occur in clusters separated by dry corridors, effective in limiting genetic interchange. The most southerly corridor, south of Mackay, forms a convenient southern boundary for this study: it straddles the tropic of Capricorn and is at the southern limit of structurally defined lowland tropical rain forest, i.e., complex mesophyll vine forest (Webb, 1978). The area of rain forest in the tropics is estimated to be in excess of 11,000 km² or more than half the total for the continent (Winter et al., 1984). In contrast to the country as a whole though, only about 20% of the rain forest has been cleared (Winter et al., 1984). About 25% of the remainder is conserved within National Parks and other reserves.

B. TOPOGRAPHY AND ASSOCIATED ENVIRONMENTAL FACTORS

Within this continent of low relief, rain forest patches are generally associated with more rugged terrain, particularly along the eastern highlands which run almost the whole length of the eastern seaboard. They occur on the eastern slopes of the mountains and on the fronting coastal plains where precipitation amounting to at least 1300 mm per annum is derived from orographic uplift of the south-east trade winds. In Tasmania, however, the cool temperate forests are concentrated along the mountainous west coast and are watered by the prevailing westerly winds. The northern shores of the continent come under the influence of the summer north-west monsoons and rain forests are best developed along the coastal escarpments of the Kimberley Ranges in Western Australia and Arnhemland in the Northern Territory. Drier rain forest patches are often also related to broken terrain such as bouldery outcrops, coastal dunes and gullies because they afford protection from wildfires and are frequently locally moist. Away from these topographic features, drier forests are restricted to high nutrient status soils derived from basalt or alluvium.

C. POPULATION

Australia's population is concentrated along the better watered eastern coasts and poses a constant threat to surviving 'wetter' rain forest patches. Within the tropics population densities between 3 and 25 persons per km² are achieved only between Cairns and Mackay and are related to the intensive sugar cane industry. Sugar cane has replaced much lowland tropical rain forest and it is anticipated that the industry will continue to expand. The increasing popularity of the area for tourism ensures substantial future population growth.

The remainder of tropical Australia, outside minor settlements, has less than 1 person per km². The dominant land use is beef cattle grazing for which there is minimum intentional alteration to the native vegetation.

III. Vegetation

A. HISTORY OF THE AUSTRALIAN TROPICAL RAIN FOREST FLORA AND VEGETATION

There is little fossil information available from tropical latitudes. However, some reconstruction of the history of the tropical rain forest flora can be attempted from present biogeographic distributions, fossil floras from higher latitudes and a range of palaeogeographic data.

There has been a general belief, since the early work of Hooker (1860), that Australia's flora is composed of a number of invasive elements. The development of the theory of plate tectonics added weight to this view by providing pathways for entry—from other southern continents, when Australia was a part of Gondwanaland prior to 40 million years ago and through the southeast Asian region within the last 15 million years as the Australian continent collided with the Asian continental plate (Barlow, 1981; Raven & Axelrod, 1974).

It is becoming increasingly clear, though, from accumulating fossil evidence (Martin, 1978; Truswell et al., 1987) and from detailed analysis of present day floras (Webb & Tracey, 1981a) that a major component of the Australian rain forest flora has had a long history on the continent and that a large proportion of it must have evolved here (Webb et al., 1986; Werren & Sluiter, 1984). There still may have been a significant component derived from the Indo-Malaysian region, but recent geophysical data on the relative positions of the Australian and Asian plates combined with available fossil evidence suggests that entry to Australia may have been continuous but at low levels through the Late Cretaceous and Tertiary periods rather than occurring as a major invasion in the Late Cenozoic (Truswell et al., 1987).

The history of the rain forest flora in Australia has been reviewed recently by Kershaw et al. (1984). The modern rain forest flora developed gradually through the Late Cretaceous and Early Tertiary with rain forest vegetation achieving greatest diversity and maximum extent, at least in the southern part of the continent, by about 40 million years ago. After this time, progressively drier climatic conditions caused a gradual replacement of rain forest by more open-canopied communities from the center of the continent outwards. At the same time a steepening latitudinal temperature gradient brought about the extinction of warmth-demanding taxa in higher latitudes, but the northward movement of the continent probably allowed the survival of these taxa in the northern part of the continent.

There may have been a development also, or increase in importance, of vine thickets and other 'drier' rain forest communities in response to the deteriorating climatic conditions. The occurrence of these communities now only in lower latitudes could mean that they did not form under cool conditions. Alternatively they could have existed but failed to survive the more extreme climatic conditions, including a change from a summer to winter rainfall regime, considered to have been experienced here in the very late Tertiary (Bowler, 1982).

Further restriction and fragmentation of all types of rain forest would have occurred as a result of Pleistocene climatic fluctuations and increased burning by Aboriginal people in the Late Quaternary.

B. Mapping and the UNESCO Classification

As the UNESCO classification has not been used previously to describe Australian vegetation, an attempt has been made, with a great deal of assistance from Dr. L. J. Webb, to translate existing terms into UNESCO equivalents (Table I). The basic structural system of Webb (1978) indicates that Australia lacks a number of UNESCO tropical forest categories. These include evergreen seasonal submontane and montane forests because of a lack of mountains away from the wetter coastal areas; cloud forest, because no mountains reach this altitudinal zone; and any communities composed predominantly of needle-leaved conifers. It is also debatable whether true lowland ombrophilous forest exists or whether all evergreen lowland forests should be included in the evergreen seasonal forest category.

Conversely, Webb et al. (1984) consider that the UNESCO terminology for altitudinal gradients of evergreen forest is too restricted and that an extra category 'upper lowland or upland forest' between lowland and submontane forests is necessary to incorporate this additional type present in Australian forests. This forest type is structurally and floristically related to lowland subtropical rain forest which is extensively developed in Australia.

On the map (Fig. 2) only seven vegetation types are represented. Ombrophilous and evergreen seasonal lowland forests are grouped, as are ombrophilous and evergreen seasonal upper lowland or upland forests. Alluvial forest is included in the lowland group. Semi-evergreen vine thicket—which is more a subtropical than tropical vegetation type—is separated from the tropical semi-deciduous lowland forest category as tropico-subtropical semi-deciduous lowland forest; and swamp forest is too restricted in distribution to be mapped at this scale. The total distribution of rain forest is exaggerated in that mixtures and mosaics of rain forest and sclerophyll vegetation are represented as rain forest. Despite the inaccuracies, simplifications and generalizations, it is hoped that the map provides some indication of the complexity of vegetation and the relative importance of major rain forest types within the northeastern Queensland region.

IV. Vegetation Maps

A number of vegetation maps have been constructed for the whole continent of which the most widely used have been those of Williams (1955) and Carnahan (1976). Unfortunately the scales of these have prevented accurate portrayal of the distribution of rain forest or its subdivision into more than two or three major categories. Only relatively large scale maps of restricted areas provide any realistic indication of the distribution and nature of rainforest types. The major ones that exist for those parts of tropical Australia that contain rain forest are show in Figure 3.

The most comprehensive mapping program has been undertaken by Tracey and Webb (1975) who covered the important humid tropical region of North Queensland at a scale of 1:100,000. They recognized 17 subformations of rain forest as well as rain forest/sclerophyll mixtures. Those subformations have been used and further subdivided on 1:50,000 maps of two small areas within this region (Graham & Hopkins, 1980, 1983). Due to the rapid rate of forest clearance in the region and the availability of more advanced photogrammetrical techniques, the Queensland National Parks and Wildlife Service Branch in Townsville is presently updating the Tracey and Webb maps (John Winter, pers. comm.). The Webb and Tracey maps form the basis for the smaller scaled but areally larger map of the region by Winter et al. (1984).

Major rain forest patches in Cape York Peninsula were delineated by Pedley and Isbell (1971) and their 1:1,000,000 scale map has been updated for rain forest vegetation in that of Winter et al. (1984). This has utilized results from a number of studies undertaken in the intervening period.

There has been no detailed study of the distribution of rain forest south of Townsville, but some patches are portrayed on the 1:1,000,000 map of Isbell and Murtha (1972)

Table I
Relationships between tropical forest classifications.

UNESCO (1973) Formation Groups	WEBB (1978) Structural Types	NIX (1984) Structural Groups	WEBB et al. (1984) Provinces
1.A. Mainly evergreen forest			
1.A.1. tropical ombrophilous forest			
1.A.1.a lowland forest	Complex mesophyll vine forest – CMVF Mesophyll vine forest – MVF	Evergreen mesophyll vine forest – EMVF	B2
[a]upper lowland or upland	Complex notophyll vine forest – CNVF Simple notophyll evergreen vine forest – FNVF	Evergreen notophyll vine forest – ENVF Evergreen notophyll vine forest – ENVF	B2
1.A.1.b submontane forest	Simple notophyll evergreen vine forest – FNVF Microphyll vine/fern forest – MV/FF	Evergreen notophyll vine forest – ENVF Evergreen microphyll fern forest – EMFF	B2
1.A.1.c montane forest	Microphyll vine/fern thicket – MV/FT	–	B2
1.A.1.e cloud forest	–	–	
1.A.1.f alluvial forest	Complex mesophyll vine forest – CMVF	Evergreen mesophyll vine forest – EMVF	B2
1.A.1.g swamp forest	Mesophyll feather-palm vine forest – MFPVF Mesophyll fan-palm vine forest – MFAPVF	–	B2
1.A.2. tropical evergreen seasonal forest			
1.A.2.a lowland forest	Complex mesophyll vine forest – CMVF Mesophyll vine forest – MVF	Evergreen mesophyll vine forest – EMVF	B2
[a]upper lowland or upland	Complex notophyll vine forest – CNVF Notophyll vine forest – NVF Araucarian notophyll vine forest – ANVF	Evergreen notophyll vine forest – ENVF Evergreen notophyll/microphyll vine forest – EN/MVF[b]	C1+
1.A.2.b submontane forest	–	–	
1.A.2.c montane forest	–	–	
1.A.3. tropical semi-deciduous forest			
1.A.3.a lowland forest	Semi-deciduous mesophyll vine forest – SDMVF Semi-deciduous notophyll vine forest – SDNVF Semi-evergreen vine thicket – SEVT+	Semideciduous notophyll/mesophyll vine forest – SDN/MVF Semi-evergreen microphyll vine thicket – SEMVT[b]	B1 C2+
1.A.3.b montane or cloud forest	–	–	
1.A.9 tropical evergreen needle-leaved forest			
1.A.9.a lowland and submontane forest	–	–	
1.A.9.b montane and subalpine forest	–	–	
1.B. Mainly deciduous forest			
1.B.1. tropical drought-deciduous forest			
1.B.1.a lowland and submontane forest	Deciduous vine thicket – DVT	Deciduous microphyll vine thicket – DMVT	B3
1.B.1.b montane and cloud forest	–	–	

[a] New terms introduced by Webb et al. (1983).
[b] Essentially subtropical types extending into the tropics.

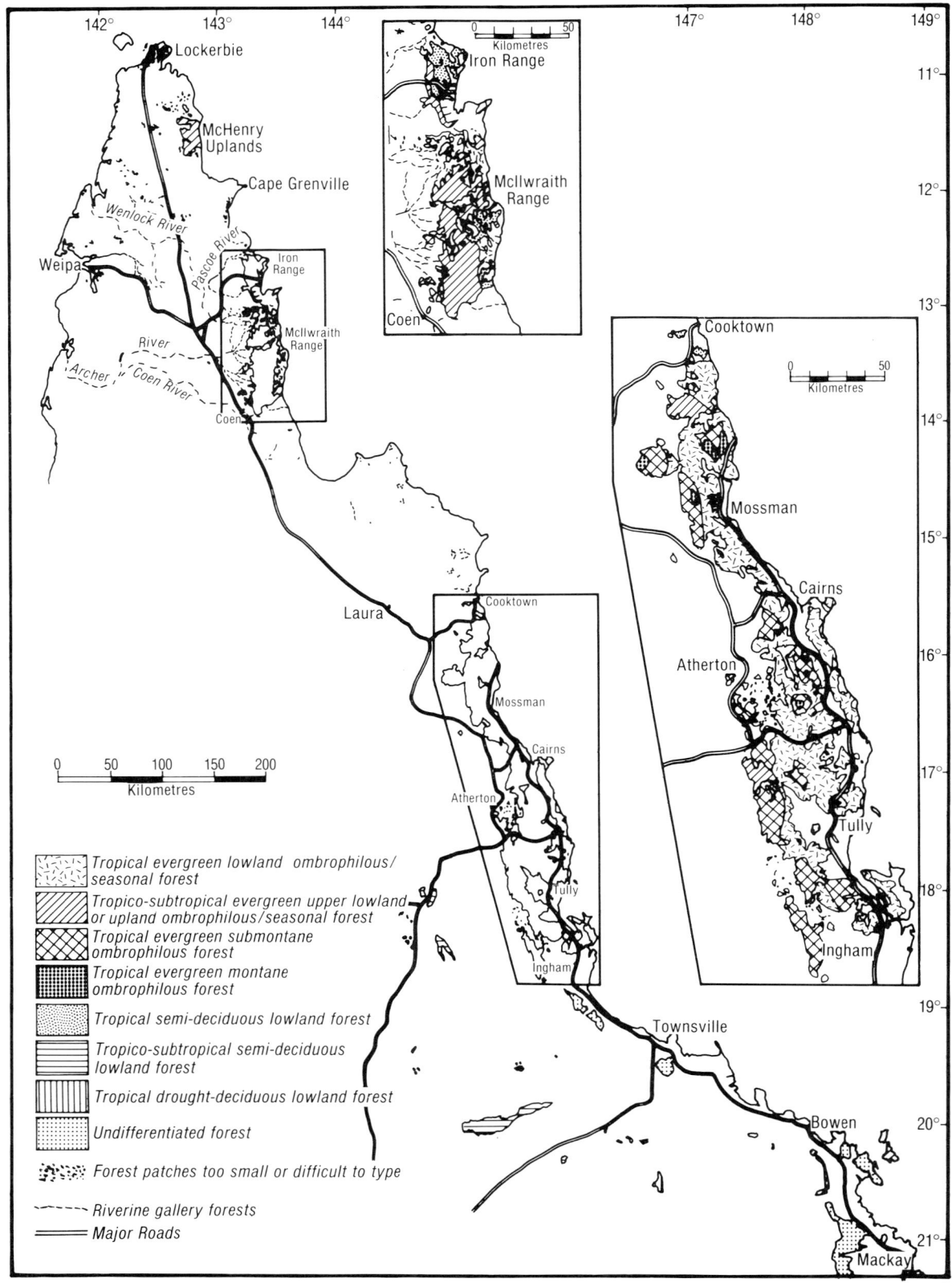

Fig. 2

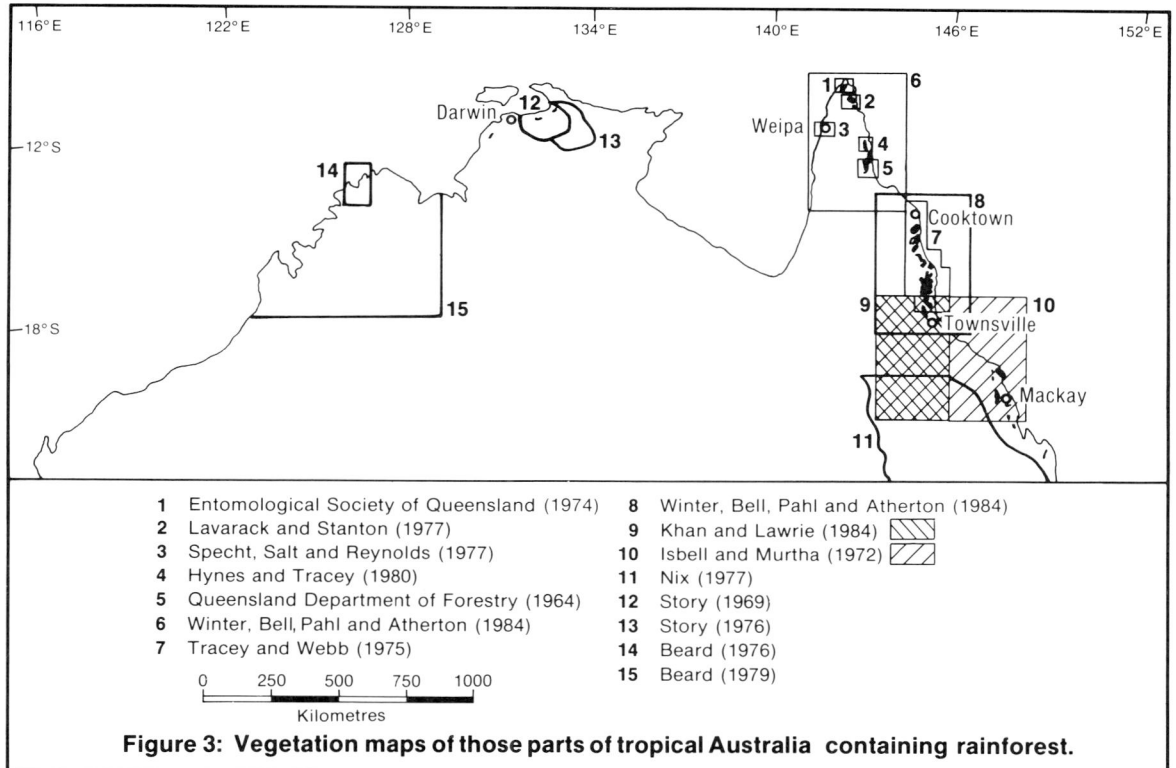

Figure 3: Vegetation maps of those parts of tropical Australia containing rainforest.

1 Entomological Society of Queensland (1974)
2 Lavarack and Stanton (1977)
3 Specht, Salt and Reynolds (1977)
4 Hynes and Tracey (1980)
5 Queensland Department of Forestry (1964)
6 Winter, Bell, Pahl and Atherton (1984)
7 Tracey and Webb (1975)
8 Winter, Bell, Pahl and Atherton (1984)
9 Khan and Lawrie (1984)
10 Isbell and Murtha (1972)
11 Nix (1977)
12 Story (1969)
13 Story (1976)
14 Beard (1976)
15 Beard (1979)

while larger inland deciduous and semideciduous forest patches are represented on the 1:1,500,000 map of Nix (1977) and the recent map of Khan and Lawrie (1984).

Attempts have been made to map the small patches of rain forest in parts of Arnhemland at a scale of 1:500,000 by the C.S.I.R.O. Division of Land Use Research. Story (1976) managed to include one type of forest (*Allosyncarpia* dominated rain forest) in the map from one area but in the adjacent area he could only schematically portray rain forest vegetation within the major open forest types in which it was known to occur (Story, 1969).

No rain forest patches were large enough to be represented on the 1:1,000,000 map of the Kimberley area of Western Australia (Beard, 1979), but some indication of the local distribution of rain forest, which achieves sizes only up to about 20 ha, is shown on a larger scale map of the Admiralty Gulf (Beard, 1976).

Although accurate mapping of rain forest in Australia at small scales is difficult, the recent development of site description data banks has allowed analysis of spatial patterns of rain forest vegetation—both on structural and floristic attributes. The bulk of rain forest site descriptions have been undertaken and organized by L. J. Webb and J. G. Tracey with further descriptions, from a variety of sources, added by H. A. Nix.

Nix (1984) uses 644 sites as a basis for the examination of the distribution of rain forest structural types in relation to major climatic variables. Climatic data for each site were generated by a C.S.I.R.O. devised system which allows average climatic conditions to be estimated for any point in Australia for which latitude, longitude and elevation is known. Using this information and the structural designation for each site, Nix calculated means and standard deviations for a variety of climatic parameters for each rain forest structural type. Figures 4a, b show the known locations of those rain forest types which are confined to or extend into the tropics, together with potential or predicted distributions derived from the climatic data. It is clear that known distributions do not cover potential climatic ranges, particularly with the drier semi-deciduous or drought deciduous types. This is partly because of incomplete sampling but it also reflects the importance of variables other than climate on rain forest distributions, particularly present day fire patterns and past climatic conditions and fire regimes.

Webb and Tracey (1981a) and Webb, Tracey and Williams (1984) have used a 561 site × 1316 rain forest tree species matrix as the basis for determination of ecofloristic regions and provinces. A classification of sites produced three regions for the whole of Australia, one being basically tropical, one subtropical and one largely temperate (see Fig. 5). Each could be subdivided with the tropical (B) and subtropical (C) regions separating basically in relation to rainfall amount and seasonality. Of the tropical types, provinces B2, B1 and B3 compare well with tropical evergreen lowland for-

156 Floristic Inventory of Tropical Countries

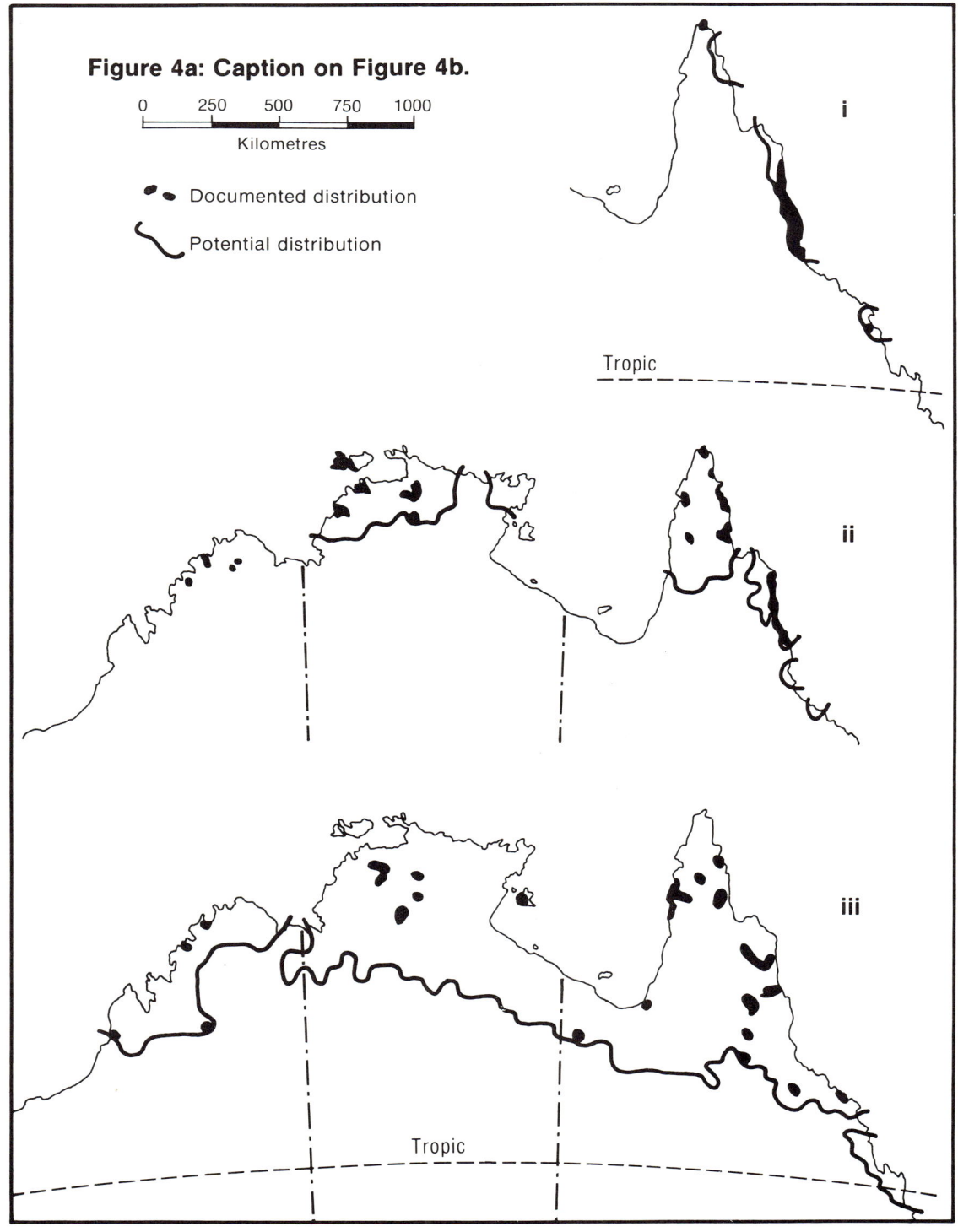

Figure 4a: Caption on Figure 4b.

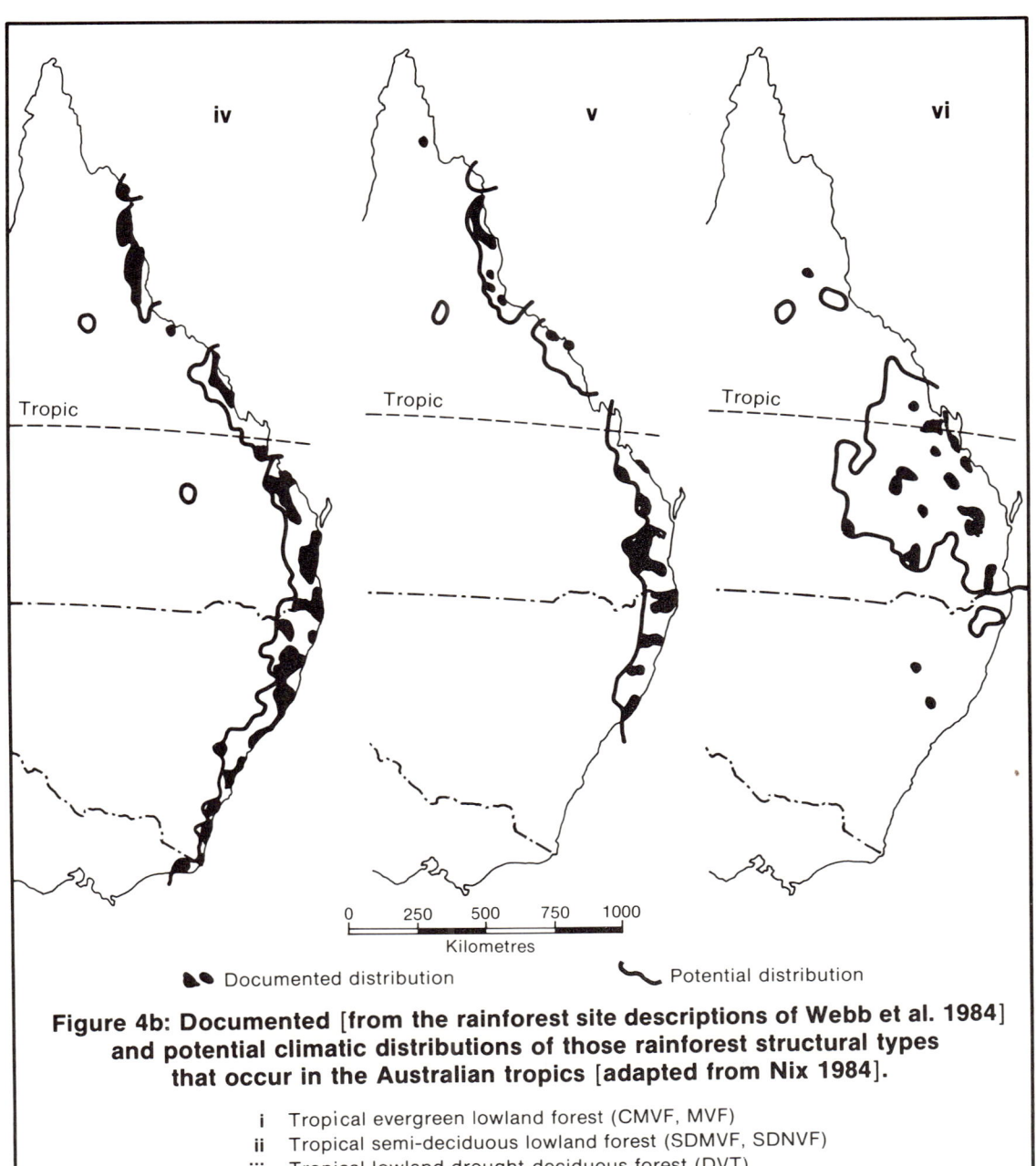

Figure 4b: Documented [from the rainforest site descriptions of Webb et al. 1984] and potential climatic distributions of those rainforest structural types that occur in the Australian tropics [adapted from Nix 1984].

- i   Tropical evergreen lowland forest (CMVF, MVF)
- ii Tropical semi-deciduous lowland forest (SDMVF, SDNVF)
- iii Tropical lowland drought-deciduous forest (DVT)
- iv Tropical-subtropical evergreen upper lowland or upland forest and submontane forest (CNVF, NVF, SNEVF)
- v   Araucarian forest (ANVF)
- vi Tropico-subtropical lowland semi-deciduous forest (SEVT)

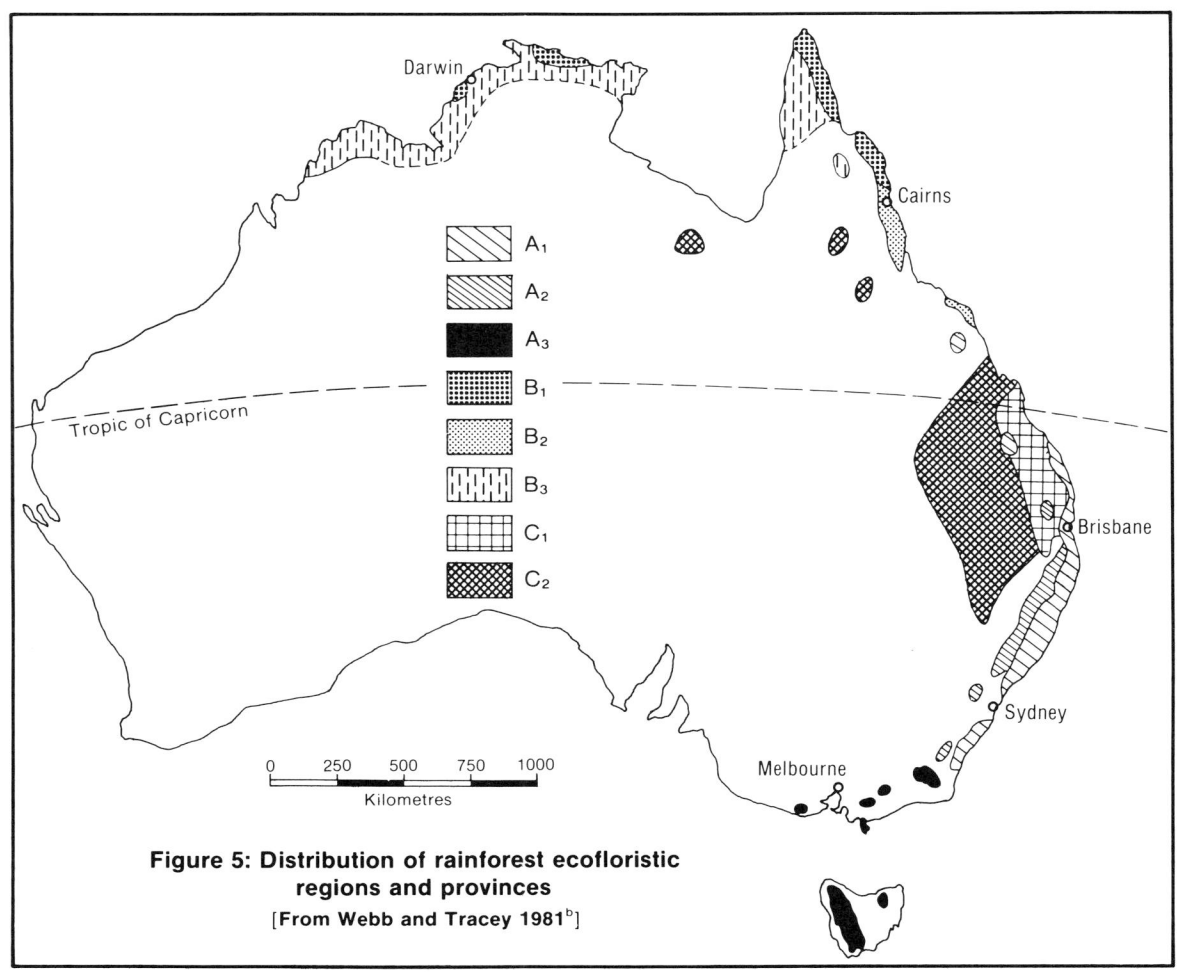

Figure 5: Distribution of rainforest ecofloristic regions and provinces
[From Webb and Tracey 1981[b]]

est, tropical semi-deciduous lowland forest and tropical lowland drought-deciduous forest respectively. Also with the subtropical types, C1 can be compared with araucarian forest and C2 with tropico-subtropical lowland semi-deciduous forest. Similarities extend to the intrusion of sub-tropical forest types as outliers well into tropical latitudes. This reinforces the importance of paleoclimatic conditions, particularly the cooler temperatures experienced during glacial periods (Kershaw, 1985).

V. Magnitude of Floristic Inventory

A. Herbarium Holdings

In terms of herbarium specimens, the most extensive sets of recent collections are held within Australia, but the older and historically more important collections are generally held overseas. An exception to this is that some historically important collections are held at the National Herbarium of Victoria (MEL) and, to a much lesser extent, at the Queensland Herbarium (BRI). Within Australia, the major centers of population are outside the tropics, and thus the older and generally larger herbaria are found away from the tropical forests. Only the herbaria at Darwin (DNA) and Atherton (QRS) are within the tropical forest area, while that at Brisbane (BRI) is within the subtropical area.

The main specialist collection of rain forest plants is held in the herbarium of the Queensland Regional Station, C.S.I.R.O., at Atherton (QRS). This is a recently established herbarium (1971), which has been able to undertake a vigorous program of collection of rain forest plants. The number of sheets currently held is about 80,000, the vast majority of which are rain forest plants. The main part of this collection comes from north-eastern Queensland and Cape York, but specimens are also held from other rain forest areas of Australia, notably from south-eastern Queensland and northern New South Wales. A small number of specimens from New Guinea and other parts of Malesia is also held.

The small rain forest patches of Arnhemland in the

Northern Territory are not well known, although survey work is continuing. The main collections from this area, a total of about 5000 sheets, are held at the Darwin herbarium of the Conservation Commission of the Northern Territory (DNA).

The rain forest areas of the Kimberley region of Western Australia have only recently been recognized, and floristic study is continuing. The main collections from this area are to be found in the Western Australian Herbarium (PERTH), and number approximately 15,000 sheets, many of which are represented as duplicates in other herbaria.

Within Australia, the other important herbaria would be in Brisbane (BRI) and Canberra (CANB). The Queensland Herbarium (BRI) would hold about 460,000 specimens, although not all of these are incorporated as yet. Of this total, 40–50% are from Queensland, 20–25% are from other parts of Australia, and 20–25% are from New Guinea. Extrapolating from the available information (Johnson 1983) indicates that around 60,000 specimens are from Australian tropical rain forest areas. Included in these are some early important collections by F. M. Bailey, and other important collections by C. T. White, S. T. Blake and L. S. Smith.

The Australian National Herbarium (CANB) has about 400,000 specimens, of which perhaps 120,000 are from tropical Australia. Less than half of these, perhaps 45,000, relate to Australian rain forests. Many of these are duplicates of collections found in other herbaria. Among the important collections are those from the surveys of the C.S.I.R.O. Division of Land Use Research, which were undertaken in many of the tropical parts of Australia and New Guinea. The herbarium also holds up to 40,000 collections from New Guinea.

The state herbaria in New South Wales (NSW), Victoria (MEL), and Tasmania (HO) hold collections of the subtropical, warm temperate and cool temperate rain forest floras to be found within their states. The National Herbarium of New South Wales (NSW) also holds collections of material from Australian tropical rain forests, but only a few of these are not duplicates represented elsewhere. The National Herbarium of Victoria (MEL) has few recent collections from tropical rain forest areas, but does hold historically important collections, principally those of F. Mueller and his collectors.

Herbarium specimens held abroad include most importantly the older collections, including many types. The most important herbaria in this respect would be at the Royal Botanic Gardens, Kew (K), and the British Museum (Natural History), London (BM). Also important are the herbaria at Geneva (G), at Harvard University (A and GH), and at Leiden (L). A number of other herbaria would have small collections of Australian rain forest material, but not as much as those mentioned.

B. Publications

Australian rain forests have in general been poorly served in terms of floras and keys. Studies on the ecology and biogeography of rain forest plants are only now beginning to make reasonable progress, and they continue to be hampered by a lack of taxonomic research.

The more important floras and keys are mentioned below. There are in addition a number of checklists from areas of varying sizes, but these in general do not contain keys or descriptions.

The first, and until recently the only, flora of Australia as a whole was *Flora Australiensis* by George Bentham (1863–1878). This was the standard flora for over a hundred years, and is still the only relevant flora for certain parts of the country. Its appearance stimulated the production of a number of state floras, including one for Queensland by F. M. Bailey (1899–1902) and one for the Northern Territory by Ewart and Davies (1917). In both cases, the State flora was based on *Flora Australiensis*, with the addition of subsequently described species. *Flora Australiensis* contains 8,125 species of vascular plants; as the Australian flora is now estimated to contain up to 25,000 species, the current limitations of *Flora Australiensis* can be easily seen.

The rain forest trees are covered in the book *Australian Rainforest Trees* by Francis (1970). The first edition, which was published in 1929, contained details only of the subtropical species. In the second edition, published in 1951, many of the more important of the species confined to the tropical parts of Australia were added, while the third edition (Francis, loc. cit.) involved basically an updating of the nomenclature. Many new species have been collected and described in recent times, and taxonomic studies undertaken. As a consequence, this book, while useful in its time, is now too out of date to be of practical use.

A more recent attempt to cover the rain forest trees is that by Hyland (1971, 1982). The *Revised Card Key to Rainforest Trees of North Queensland* (Hyland 1982) covers most of the known rain forest trees found between Townsville and Torres Strait, that is for north-eastern Queensland and Cape York. This consists of a polyclave key on computer cards, utilizing leaf and bark features. For each species there is in addition a short description of the distinctive features to aid in identification. Designed primarily for field use, this key is less useful for herbarium specimens, where many of the field characters are not available. A more complete key to Australian rain forest trees, incorporating all field and morphological characters (vegetative and reproductive), with full keys and descriptions, is currently being produced (Hyland & Whiffin, in prep.).

After deliberations over many years, the new *Flora of Australia* was started in 1979 (George, 1981). The work is planned to be completed in 30 years, with the vascular plants occupying 48 volumes and the nonvascular plants a number of subsequent volumes. The flora is not monographic, but in general the better known groups are being published first, so as to leave time for further studies on the less well known groups.

The new flora is arranged according to the classification system of Cronquist (1981), and is issued as a series of volumes. A preparation and publication schedule has been devised until 1990, with an approximate order for the volumes to appear after that date. So far (early 1988) the introductory volume and seven volumes of family treatments have appeared, with a number of other volumes at various stages of the preparation, editing and publication process. The flora is being coordinated by the Bureau of Flora and Fauna.

The *Flora of Australia* represents the current knowledge on Australian plants, in many cases (but not all) being preceded by a fuller taxonomic study of the families concerned. However, often the tropical rain forest plants are amongst the less well known groups, and so will not appear in the flora for a number of years (see later sections for more details).

The Queensland Herbarium (BRI) is currently writing a flora of Queensland in the form of four regional floras. The first to appear will be that for south-eastern Queensland, in three volumes. The first two volumes (Stanley & Ross 1983, 1986) have already appeared, and it is anticipated that the remaining volume will appear shortly. The floras for the other three regions will presumably then be undertaken in sequence.

The Western Australian Herbarium (Perth) is undertaking a flora of the Kimberley region over the next few years. At the moment there are no firm plans for a flora of the Northern Territory rain forests, although a computerized checklist is available at the Darwin herbarium (DNA).

VI. Completeness of Floristic Inventory

A. Collection Density

On a world scale, and in places also on an Australian scale, parts of the Australian tropical rain forest appear to be reasonably well collected, while other parts are still obviously poorly collected. The regions, in order of 'completeness' of collection would be: north-eastern Queensland, Cape York, central Queensland, Arnhemland in the Northern Territory, and the Kimberley region of Western Australia.

Data presented by Johnson (1983) show that the area of northeastern Queensland is well collected. Extrapolation from this data, with all the pitfalls inherent in such a procedure, indicates that the Queensland herbarium (BRI) alone has collections from this area at an average rate of about 200 collections per 100 square kilometers. As there will be additional, non-duplicate, material in other herbaria, especially at Atherton (QRS), then obviously this area appears to be well collected.

However, before such a conclusion can be reached, a number of points need to be considered. Firstly, the rain forest vegetation in this area shows high complexity and diversity. The data of Tracey (1982) and Webb and Tracey (1981a) for this area show a number of sites, mostly in complex mesophyll vine forest with over 150 tree species per hectare. Secondly, the collections already made indicate that there is a very pronounced local endemism in this area, with a relatively large number of species being known from only one or a few small areas. Finally, there are still many taxonomic problems to be resolved for which further collections are needed.

As an illustration of this, it can be seen that in the north Queensland key by Hyland (1982), there are 799 tree species listed, of which 130 need further taxonomic study, involving formal taxonomic description as new species in many cases. Hyland (pers. comm.) also estimates that in this area about 100 tree species still do not have good collections of flowering and fruiting material, thus preventing the preparation of a full description.

The rain forests of Cape York are found in a number of discrete areas, with many smaller scattered patches. Some of these areas have only been fairly recently recognized. The larger areas have in general been reasonably well collected, although even here more new species and range extensions are to be expected. The smaller scattered patches have not been as well collected.

The rain forests in the vicinity of Mackay, in central Queensland, are smaller, but are generally less well collected, being distant from the main botanical centers in Queensland. Although generally containing few endemics, the flora here may be more diverse than current collections would indicate. Range extensions in this area of very conspicuous rain forest trees have been discovered in the last few years, so more are to be expected when some of the less accessible areas are studied.

The rain forests of the Northern Territory are small in terms of area, and in terms of numbers of species. There are about 300 to 400 species of vascular plants here (Russell-Smith & Dunlop, 1984); it is estimated by Dunlop (pers. comm.) that the inventory of this area is perhaps 80% complete.

The rain forest areas of the Kimberley region of Western Australia have only recently been recognized, and many of the areas are inaccessible. The flora here would be smaller than in any of the other areas. There are currently in excess of 200 species of vascular plants known from these rain forests, but the floristic inventory must be considered very incomplete. Estimates by Kenneally (pers. comm.) indicate that the inventory is perhaps 75–80% complete, with the main gaps in the collections being from ephemeral and canopy flowering species where wet season collecting is required.

For all rain forest areas, three points need to be borne in mind as regards the floristic inventory. Firstly, many small isolated or less accessible areas have not been studied, and these may produce new species. Secondly, even in the relatively well collected areas of north-eastern Queensland, new species and even new genera are still being found; this is in part a reflection of the narrow endemism of many of these species. Finally, few areas have been systematically collected and, as a result of this, range extensions are continually being discovered, even for the larger and more conspicuous species.

B. Areas of High Endemicity

From the comments above, it is obvious that our knowledge of the distribution of many species is incomplete, as is our taxonomic knowledge. Nevertheless, enough is now known to enable us to make some reasonably general statements.

From a study of two of the larger and more important groups of rain forest trees, *Syzygium* and allied genera in the Myrtaceae, and *Cryptocarya* in the Lauraceae, Whiffin & Hyland (1986) found that there were two major centers of diversity and endemism (tropical north-eastern Queensland and subtropical south-eastern Queensland/northern New South Wales) and two minor centers (Cape York and Northern Territory). These centers probably represent centers of isolation and long term refuges for rain forest flora. Other areas, involving Western Australia and central Queensland in the tropical areas, and central and southern New South Wales in the subtropical and temperate regions, were seen as immigrant areas, receiving their flora from one or more of the centers of isolation.

North-eastern Queensland has long been recognized as the major center for Australian rain forest flora. Recent studies (e.g., Webb & Tracey, 1981a) have revealed two major features: a high degree of endemism, often very local endemism, and a relatively large number of primitive taxa often very restricted in their distribution. These two points are in major part explained by reference to the long term vegetation history of the area, which is gradually being elucidated (Kershaw, 1981; Webb and Tracey, 1981a).

In the distribution data of Webb and Tracey (1981a; unpubl.) there are records of 190 sites and 840 species of trees and shrubs for north-eastern Queensland. From these sites, 71 species (8%) are recorded from one site only and nowhere else within the entire data set, 41 species (5%) from two sites only, and 24 (3%) from three sites only. Although these numbers will vary according to sampling density and other effects, there is nevertheless a gradually emerging pattern of a significant number of species which are restricted to one or a few areas (i.e., local endemics).

Although there is a pattern of local endemism apparent for all rain forest areas of north-eastern Queensland, two vegetation types within this area appear to be of particular importance. These are the lowland tropical rain forests in areas of very high rainfall (exemplifed basically by the Cape Tribulation and the Russell River areas), and the montane rain forests of the higher mountains (such as Bellenden Ker, Bartle Frere, and Thornton Peak areas).

Many of the endemic species in north-eastern Queensland rain forests are primitive angiosperms, which are of importance for elucidating the origin and evolution of the angiosperms. Of the eighteen families in the more primitive orders Magnoliales and Laurales (Cronquist, 1981), ten are to be found in north-eastern Queensland. Two of these families, the Austrobaileyaceae and the Idiospermaceae, are in fact restricted to this area, and are very narrowly distributed, while the Eupomatiaceae are found in eastern Australia and New Guinea, and the Himantandraceae are found only from north-eastern Queensland through New Guinea to the Moluccas. In addition, at a lower level in the taxonomic hierarchy, representatives of some of the more primitive genera within a family are to be found in the north-eastern Queensland rain forests, the most notable examples being in the Proteaceae.

VII. Completeness of Taxonomic Knowledge

Many of the smaller genera are reasonably well known but some of the larger or more complex genera have not been studied recently. Often these genera are found to contain new species and numerous taxonomic problems. For example, in a recent study of *Syzygium* and allied genera in Australia (Hyland, 1983) 66 species were recognized, of which 21 were newly described; in addition there was the description of a new genus. In a similar way, studies in the Australian Lauraceae (Hyland, in prep.) show that there are about 115 species, of which at least 46 will need to be described, and in the Monimiaceae (Whiffin, in prep.) the corresponding figures are 25 and 9.

It is possible to make a number of statements concerning the completeness of our taxonomic knowledge using Australian rain forest trees for the basic data, as this information is currently being compiled (Whiffin and Hyland 1984, unpubl.). Within the Australian tropical rain forests, it is estimated that there are about 1100 species of rain forest trees in about 90 families. About 10% of these species are currently known but undescribed; however, as indicated above, detailed taxonomic study often shows there to be additional undescribed species.

Using this figure of the number of rain forest trees, it is possible to estimate the total vascular plant flora for Australian tropical rain forests by comparison of a number of complete site lists from different rain forest areas. Extrapolating from the proportion of tree species in these lists, the total vascular plant flora for Australian tropical rain forests is estimated to be about 2500 species. This includes only those species that are regular inhabitants of the rain forests, and not those sclerophyllous species or weedy species that may occasionally be found within the rain forest.

Information on the larger families of Australian rainforest trees is presented in Table II. Shown here are the number of tree species present in Australian tropical rain forests, some of the larger genera in the family, the volume of *Flora of Australia* in which the family will appear, and the actual or estimated date of appearance of that volume.

Taxonomists known to have specialist knowledge in, or to be currently working on, the Australian representatives of some of the more important families are listed in Table III. The main emphasis is on Australian taxonomists, as this information is less generally available. Information on overseas taxonomists, where they may be working on a genus or a family as a whole, is generally more readily available. Information on those taxonomists preparing flora treatments for the *Flora of Australia* is available from the Bureau of Flora and Fauna, and is also published at regular intervals in the "Newsletter" of the Australian Systematic Botany Society.

Table II

The larger families of Australian rainforest trees, showing the number of tree species present in the tropical rainforests, the volume of *Flora of Australia* in which the family will appear, the actual or estimated date of publication of that volume, and the more important genera in the family.

Family	Species	Volume	Date	Genera
Annonaceae	20	2	*	*Polyalthia*
Apocynaceae	14	28	*	*Alyxia, Alstonia, Cerbera*
Araliaceae	10	27	*	*Polyscias*
Caesalpiniaceae	10	12	1990	*Cassia*
Celastraceae	15	22	1984	*Maytenus*
Clusiaceae	15	6	*	*Garcinia, Calophyllum*
Combretaceae	14	18	1989	*Terminalia*
Cunoniaceae	14	10	*	*Ceratopetalum*
Ebenaceae	13	10	*	*Diospyros*
Elaeocarpaceae	34	7	*	*Elaeocarpus*
Euphorbiaceae	100+	23	*	*Croton, Glochidion, Mallotus*
Fabaceae	10	13,14,15	*	*Castanospermum*
Flacourtiaceae	12	8	1982	*Casearia*
Grossulariaceae	12	10	*	*Polyosma*
Lauraceae	90	2	*	*Cryptocarya, Endiandra*
Meliaceae	28	26	*	*Dysoxylum*
Mimosaceae	25	11,12	*	*Acacia, Albizia, Pithecellobium*
Monimiaceae	25	2	*	*Wilkiea*
Moraceae	34	3	*	*Ficus*
Myrtaceae	125	20,21	*	*Syzygium, Austromyrtus*
Proteaceae	44	16,17	1988	*Helicia, Macadamia*
Rhamnaceae	10	24	*	*Alphitonia*
Rubiaceae	40+	36	*	*Psychotria, Randia*
Rutaceae	60+	26	*	*Acronychia, Flindersia*
Sapindaceae	70+	25	1985	*Arytera, Cupaniopsis*
Sapotaceae	27	10	*	*Planchonella*
Sterculiaceae	20+	7	*	*Argyrodendron, Brachychiton*
Verbenaceae	12	30	*	*Callicarpa*

* to appear after 1990

Table III

Taxonomists known to have specialist knowledge of or currently working on the Australian representatives of some of the more important families.

Family	Specialist
Anacardiaceae	L. W. Jessup
Annonaceae	L. W. Jessup
Araliaceae	W. R. Philipson, D. G. Frodin (*Schefflera*)
Caesalpiniaceae	J. H. Ross, D. E. Symon (*Cassia*)
Celastraceae	L. W. Jessup
Combretaceae	N. B. Byrnes
Corynocarpaceae	G. P. Guymer
Cunoniaceae	R. D. Hoogland
Ebenaceae	L. W. Jessup, F. White
Elaeocarpaceae	G. P. Guymer, M. J. E. Coode
Euphorbiaceae	H. K. Airy-Shaw†, G. P. Guymer
Flacourtiaceae	L. W. Jessup
Icacinaceae	G. P. Guymer
Lauraceae	B. P. M. Hyland
Loranthaceae	B. A. Barlow
Malvaceae	A. S. Mitchell (*Hibiscus*)
Melastomataceae	T. Whiffin
Mimosaceae	A. Kanis†, L. Pedley
Monimiaceae	T. Whiffin
Moraceae	M. O. Rankin
Myrtaceae	B. P. M. Hyland (*Syzygium* and allied genera), P. G. Wilson (*Tristania, Xanthostemon*), B. A. Barlow (*Melaleuca*), G. P. Guymer (*Austromyrtus* and allied genera)
Proteaceae	L. A. S. Johnson and B. G. Briggs, B. P. M. Hyland, D. B. Foreman (*Helicia*)
Rubiaceae	C. Puttock (*Gardenia, Randia*), S. T. Reynolds (Coffeeae, Vanguerieae, Psychotrieae)
Rutaceae	T. G. Hartley
Sapindaceae	S. T. Reynolds
Sapotaceae	G. P. Guymer
Solanaceae	D. E. Symon
Sterculiaceae	G. P. Guymer
Symplocaceae	G. P. Guymer
Verbenaceae	A. A. Munir
Vitaceae	B. R. Jackes

In addition, a number of taxonomists have knowledge of certain rain forest types or areas. These include:

Rain forest trees, especially of Queensland	B. P. M. Hyland
Flora of Kimberley region, Western Australia	K. F. Kenneally
Flora of Northern Territory rain forests	C. Dunlop

† Deceased.

VIII. Areas Under Threat

As Australia's rain forests exist generally as small isolated, fire-intolerant patches surrounded by fire-promoting sclerophylls, they can all be seen as being under constant threat. This long term threat is increased in many areas by disturbances such as altered fire regimes, logging, trampling and associated activities of cattle, wild pigs and water buffalos, and clearing resulting from European man's activities.

Within the more accessible, complex, coastal forests, the major threats are logging and clearing for agriculture. It is the policy of the Queensland Department of Forestry to selectively log all accessible 'virgin' evergreen rain forest not preserved in National Parks and other reserves and subsequently log on a sustainable yield basis with logging cycles of 40–50 years (Queensland Department of Forestry, 1983).

The problems with this policy are that it is unlikely that the range of variation in rain forest is contained within conservation reserves and that many scientists feel substantial modification and degradation will result from attempted sustainable yield logging. Greatest pressure is on the most complex and floristically diverse lowland forests which are continuing to be cleared for agricultural production, particularly sugar cane (Winter et al., 1984). The only remaining significant area of undisturbed lowland forest lies between Cairns and Cooktown. It is feared that the recent construction of a road through it could be the first step to opening the area up for development. Some parts are already being sub-divided for small residential allotments (Fleetwood et al., 1984) and judging from a recent paper advertisement 'Agents/Sales People wanted to sell Daintree Rainforest' this threat is increasing.

The only complex rain forests that are not threatened in northeastern Queensland, under present levels of technology at least, are the higher submontane and montane forests and those of Cape York Peninsula.

Other rain forests for which there is great concern are the 'drier' semi-deciduous and deciduous thickets. These are largely unknown and poorly represented in conservation reserves because they are of low economic value and are aesthetically less pleasing than forests of the more humid areas (Gillison, 1984). Many are in topographically fireproof locations away from intensive developments and are probably quite secure, but others, in sensitive coastal dune systems and on desirable agricultural land are disappearing at a very rapid rate. Semi-deciduous lowland forest or semi-evergreen vine thicket associated with brigalow (*Acacia harpophylla*) lands which extend into the tropics in inland Queensland, have been almost destroyed within the last 15 years due to clearing for agricultural production (Gasteen, 1984).

IX. Present and Future Resources for Floristic Inventory

The Australian herbaria active in work on tropical rain forests were mentioned earlier. The more important centers for tropical research are at Atherton (QRS), Brisbane (BRI), and Darwin (DNA), with some workers also at Canberra (CANB), Sydney (NSW), and Perth (PERTH), with a few additional people in the other state herbaria and in the universities. It should be realized that all these herbaria are, by world standards, very small in terms of their scientific staff. In some of the herbaria mentioned, there is in effect only one scientific staff member, with some limited technical assistance. In no case are there more than two or three scientific staff members working on tropical groups. Add this to the rather low numbers of tropical taxonomists in the universities, or in other governmental instrumentalities such as the National Parks services, or various Divisions of C.S.I.R.O., and it will be seen that the total research effort is quite small.

Given this as a general background, there are two main problem areas that are currently facing future study of Australian tropical rain forests. The first is the low level of funding, and the second is the lack of overall coordination. In both of these there is an additional complicating factor. The various herbaria, government instrumentalities, and universities are for historical reasons under different administrative controls, some state government and some federal government. Thus, for example, they receive some or all of their funding from very different sources and, because of this, are often charged with very different primary functions.

The herbaria involved in tropical rain forest research within Australia are all secure, with their future existence guaranteed. However, in probably all cases, there are threats of reduced financial support due to financial stringency in the respective state or federal government. Thus, at a time when it might be expected that there would be an expansion in tropical research, there is in fact no guarantee that even the present level of funding will continue.

As mentioned earlier, the small number of scientists involved in the study of Australian rain forests is in fact scattered over quite a large number of different institutions. While this is to some extent inevitable, it must lead to a less coordinated approach to Australian rain forest research. There is little likelihood that this will change at the institutional level, as each institute or government concerned is generally unlikely to be willing to cede responsibility or operations to another. However, it is always possible to increase coordination of research activities through some more nebulous coordinating instrumentality. It must be said, however, that it would be necessary for this new coordinating body to have sufficient funds available as grants before it could achieve this.

Despite this gloom, there are some hopeful signs in recent federal government initiatives. The Australian Biological Resources Study (ABRS), first established in 1973, has promoted the collection and study of Australian biota through the provision of grants, and has produced reports on the longer term requirements for maintaining and increasing such activities. Included in the latter were firm recommendations on the production of the *Flora of Australia*. ABRS continues, through an Editorial Committee and an Advisory Committee, to oversee the production of the *Flora* and the provision of grants. The actual organization of these activities is carried on within the Bureau of Flora and Fauna, established in 1979, and now within the Department of Arts, Heritage and the Environment. Within this Bureau are, *inter alia*, the Executive Editor of the *Flora*, and two flora writers.

While the establishment of the Australian Biological Resources Study and the Bureau of Flora and Fauna have provided a very great impetus to taxonomic work within Australia, and have led to the production of a very fine flora, once again it should be noted that the level of funding is quite small. The funds available through the ABRS grants scheme are in general only sufficient to meet the more immediate needs of the production of the family treatments for the flora, with little money being available for longer term objectives. The main exception to this statement has been the provision of funds for the gradual computerization of the more important herbarium collections within Australia; when complete, this will prove of great benefit to all systematic and floristic studies.

A second recent initiative has been the setting up within the Department of Arts, Heritage and the Environment of a Working Group on Rainforest Conservation. This group undertook a broad study of rain forest conservation in Australia. Its report (Working Group on Rainforest Conservation, 1985) recommended the development of a national policy involving various aspects of survey, acquisition, management and research of Australian rain forest. A recent news release (June, 1986) by the Minister indicated that there would be an expenditure of about A$ 22.5 million over the next two years towards the establishment of a National Rainforest Conservation Programme. This will attempt to undertake the above aims, through joint venture projects with the States concerned.

X. Acknowledgments

We thank Dr. L. J. Webb for his advice on relationships between vegetation classification systems and the Directors and staffs of the various Australian herbaria for their willingness to provide information.

XI. Literature Cited

Australian Forestry Council. 1974. Forwood, Report of Panel 2, Forest Resources. Australian Gov't. Publishing Service, Canberra.
Bailey, F. M. 1899–1902. The Queensland flora. Diddams, Brisbane.
Barlow, B. A. 1981. The Australian flora: Its origin and evolution. Flora of Australia, 1: 25–75.
Beard, J. S. 1976. The monsoon forests of the Admiralty Gulf, Western Australia. Vegetatio 31: 177–192.
——. 1979. The vegetation of the Kimberley area. Explanatory notes to map sheet 1 of Vegetation Survey of Western Australia, 1:1,000,000 series: Kimberley. University of Western Australia, Nedlands.
Bentham, G. 1863–1878. Flora Australiensis: A description of the plants of the Australian Territory. Reeve, London.
Bowler, J. M. 1982. Aridity in the late Tertiary and Quaternary of Australia. Pages 35–45 in W. R. Barker & P. J. M. Greenslade (eds.), Evolution of the flora and fauna of arid Australia. Peacock Publications, Adelaide.
Carnahan, J. A. 1976. Natural vegetation. Pages 1–26 in Atlas of Australian Resources, 2nd. series. Dept. of Natural Resources, Canberra.
Cronquist, A. J. 1981. An integrated system of classification of flowering plants. Columbia University Press, New York.
Entomological Society of Queensland. 1974. Focus on Cape York. Entomological Society of Queensland, Brisbane.
Ewart, A. J. & O. B. Davies. 1917. The Flora of the Northern Territory. McCarron Bird, Melbourne.
Fleetwood, R., K. Means & M. Stannard. 1984. An attempt to resolve rain forest conservation and development conflicts north of Daintree. Pages 644–653 in G. L. Werren & A. P. Kershaw (eds.), Australian National Rainforest Study Report to the World Wildlife Fund (Australia), Vol. I. Proceeding of a workshop in the Past, Present and Future of Australian Rainforests. Geography Dept., Monash University, for the Australian Conservation Foundation.
Francis, W. D. 1970. Australian rainforest trees. 3rd Ed. Australian Govt. Publishing Service, Canberra.
Gasteen, W. J. 1984. The Brigalow lands of eastern Australia: agricultural impact and land use potential versus biological representation and stability. Pages 84–89 in G. L. Werren & A. P. Kershaw (eds.), Australian National Rainforest Study Report to the World Wildlife Fund (Australia), Vol. I. Proceedings of a workshop on the Past, Present and Future of Australian Rainforest. Geography Dept., Monash University, for the Australian Conservation Foundation.
George, A. S. 1981. The background to the Flora of Australia. Flora of Australia 1: 3–23.
Gillison, A. N. 1984. The 'dry' rainforests of *Terra Australis*. Pages 213–230 in G. L. Werren & A. P. Kershaw (eds.), Australian National Rainforest Study Report to the World Wildlife Fund (Australia), Vol. I. Proceedings of a workshop on the Past, Present and Future of Australian Rainforest. Geography Dept., Monash University, for the Australian Conservation Foundation.
Graham, A. W., & M. S. Hopkins. 1980. The forests and soils of the Jarra Creek area, North Queensland. CSIRO Australia, Division of Land Use Research Technical Memo 80/26.
—— & ——. 1983. The forests of the Liverpool Creek area, North Queensland. CSIRO Australia, Division of Water and Land Resources Technical Memo 83/7.
Hooker, J. D. 1860. Introductory essay. Botany of the Antarctic voyage of H.M. Discovery ships 'Erebus' and 'Terror' in the years 1839–1843, III. Flora Tasmaniae. Reeve, London.
Hyland, B. P. M. 1971. A card key to the rain forest trees of North Queensland. Queensland Dept. of Forestry, Brisbane.
——. 1982. A revised card key to rainforest trees of North Queensland. CSIRO, Melbourne.
——. 1983. A revision of *Syzygium* and allied genera (Myrtaceae) in Australia. Austral. J. Bot., Suppl. ser., 9: 1–164.
Hynes, R. A. & J. G. Tracey. 1980. Pages 11–30 in N. L. Stevens & A. Bailey (eds.), Contemporary Cape York Peninsula. Royal Society of Queensland, Brisbane.
Isbell, R. F. & G. G. Murtha. 1972. Burdekin-Townsville Region. Resources Series—Vegetation (Map at 1:1,000,000). Dept. of National Development, Canberra.
Johnson, R. W. 1983. The Queensland flora and its collection. Proc. Roy. Soc. Queensland 24: 1–18.
Kahn, T. P. & B. C. Lawrie. 1984. Vine thickets of the inland Townsville region. Pages 96–126 in G. L. Werren & A. P. Kershaw (eds.), Australian National Rainforest Study Report to the World Wildlife Fund (Australia), Vol. I. Proceedings of a Workshop on the Past, Present, and Future of Australian Rainforests. Geography Dept., Monash University, for the Australian Conservation Foundation.
Kershaw, A. P. 1981. Quaternary vegetation and environments. Pages 83–101 in A. Keast (ed.), Ecological biogeography in Australia. W. Junk, The Hague.
——. 1985. An extended late Quaternary record from northeastern Queensland and its implications for the seasonal tropics of Australia. Proc. Ecol. Soc. Australia 13: 179–189.
——, I. R. Sluiter, J. Dawson, B. E. Wagstaff & M. Whitelaw. 1984. The history of rainforest in Australia: Evidence from pollen. Pages 462–477 in G. L. Werren & A. P. Kershaw (eds.), Australian National Rainforest Study Report to the World Wild-

life Fund (Australia), Vol. I. Proceedings of a Workshop on the Past, Present and Future of Australian Rainforests. Geography Dept., Monash University, for the Australian Conservation Foundation.

Lavarack, P. S. & J. P. Stanton. 1977. Vegetation of the Jardine River catchment and adjacent coastal areas. Proc. Roy. Soc. Queensland 88: 39–48.

Martin, H. A. 1978. Evolution of the Australian flora and vegetation through the Tertiary: Evidence from pollen. Alcheringa 2: 181–202.

Nix, H. A. 1977. Land classification criteria: Vegetation. Pages 27–62 in Land units of the Fitzory region, Queensland. CSIRO Australian Land Research Series No. 39.

———. 1984. An environmental analysis of Australian rainforests. Pages 421–425 in G. L. Werren & A. P. Kershaw (eds.), Australian National Rainforest Study Report to the World Wildlife Fund (Australia), Vol. I. Proceedings of a Workshop on the Past, Present and Future of Australian Rainforests. Geography Dept., Monash University, for the Australian Conservation Foundation.

Pedley, L. & R. F. Isbell. 1971. Plant communities of Cape York Peninsula. Proc. Roy. Soc. Queensland 82: 51–74.

Queensland Department of Forestry. 1964. Coen (Map at 1:267,200 scale). Queensland Department of Forestry, Brisbane.

Queensland Department of Forestry. 1983. Rainforest research in North Queensland. Queensland Department of Forestry, Brisbane.

Raven, P. H. & D. I. Axelrod. 1974. Angiosperm biogeography and past continental movements. Ann. Missouri Bot. Gard. 61: 539–673.

Russell-Smith, J. & C. Dunlop. 1984. The status of monsoon vineforests in the Northern Territory: A perspective. Pages 155–196 in G. L. Werren & A. P. Kershaw (eds.), Australian National Rainforest Study Report to the World Wildlife Fund (Australia), Vol. I. Proceedings of a Workshop on the Past, Present and Future of Australian Rainforests. Geography Dept., Monash University, for the Australian Conservation Foundation.

Specht, R. L., R. B. Salt & S. T. Reynolds. 1977. Vegetation in the vicinity of Weipa, North Queensland. Proc. Roy. Soc. Queensland 88: 17–38.

Stanley, T. D. & E. M. Ross. 1983, 1986. Flora of South-eastern Queensland. Vols. 1, 2. Queensland Dept. of Primary Industries, Brisbane.

Story, R. 1969. Vegetation of the Adelaide-Alligator area. CSIRO Australian Land Research Series, No. 25: 114–130.

———. 1976. Vegetation of the Alligator Rivers area. CSIRO Australian Land Research Series, No. 38: 89–111.

Tracey, J. G. 1982. The vegetation of the humid tropical region of North Queensland. CSIRO, Melbourne.

——— & L. J. Webb. 1975. Vegetation of the humid tropical region of North Queensland (15 maps at 1:100,000 scale and key). CSIRO Australia. Long Pocket Labs., Indooroopilly, Queensland.

Truswell, E. M., A. P. Kershaw & I. R. Sluiter. 1988. The Australian/Southeast Asian connection: Evidence from the palaeobotanical record. Pages 32–49 in T. C. Whitmore (ed.), Biogeographic evolution of the Malay Archipelago. Oxford University Press, Oxford.

Webb, L. J. 1978. A general classification of Australian rainforests. Australian Plants 9: 349–363.

——— & J. G. Tracey. 1981a. Australian rainforests: Patterns and change. Pages 605–694 in A. Keast (ed.), Ecological biogeography of Australia. W. Junk, The Hague.

——— & ———. 1981b. The rainforests of northern Australia. Pages 67–101 in R. H. Groves (ed.), Australian vegetation. Cambridge University Press, Cambridge.

———, ——— & L. W. Jessup. 1986. Recent evidence for autochthony of Australian tropical and subtropical rainforest elements. Telopea 2: 575–589.

———, ——— & W. T. Williams. 1984. A floristic framework of Australian rainforests. Austral. J. Ecol. 9: 169–198.

Werren, G. L. & A. P. Kershaw (eds.). 1984. Australian National Rainforest Study Report to the World Wildlife Fund (Australia), Vol. I. Proceedings of a Workshop on the Past, Present and Future of Australian Rainforests. Geography Dept., Monash University for the Australian Conservation Foundation.

Werren, G. L. & I. R. Sluiter. 1984. Australian Rainforests—A reappraisal. Pages 488–500 in G. L. Werren & A. P. Kershaw (eds.), Australian National Rainforest Study Report to the World Wildlife Fund (Australia), Vol. I. Proceedings of a Workshop on the Past, Present, and Future of Australian Rainforests. Geography Dept., Monash University, for the Australian Conservation Foundation.

Whiffin, T. & B. P. M. Hyland. 1984. Current status of taxonomic knowledge of the Australian rainforest flora. Pages 231–250 in G. L. Werren & A. P. Kershaw (eds.), Australian National Rainforest Study Report to the World Wildlife Fund (Australia), Vol. I. Proceedings of a Workshop on the Past, Present and Future of Australian Rainforests. Geography Dept., Monash University, for the Australian Conservation Foundation.

——— & ———. 1986. Taxonomic and biogeographic evidence on the relationships of Australian rainforest plants. Telopea 2: 591–610.

Williams, R. J. 1955. Vegetation regions (map and notes) in Atlas of Australian Resources. Dept. of National Development, Canberra.

Winter, J. W., F. C. Bell, L. I. Pahl & R. G. Atherton. 1984. Specific habitats of selected northeastern Australian rainforest mammals. Unpublished Report to World Wildlife Fund (Australia).

Working Group on Rainforest Conservation. 1985. Rainforest conservation in Australia. A report to the Hon. Barry Cohen, M.P., Minister for Arts, Heritage and Environment.

Note added in proof: A number of references listed here are included in G. L. Werren & A. P. Kershaw (eds.), Australian National Rainforest Study Report to the World Wildlife Fund (Australia), Vol. 1. This report was published only in a limited quantity. However, these papers are being republished in three volumes as *The Rainforest Legacy, Australian National Rainforest Study*. This is being published by Australian Government Publishing Service, for the Australian Heritage Commission. These volumes should appear during 1988 and 1989.

Fiji

J. Ash and S. Vodonaivalu

Contents

I. Description of the Region	167
A. Location and Area	167
B. Topography and Geology	167
C. Human Population	167
II. Forest Types	167
III. Vegetation Maps: A Review	168
IV. Magnitude of Floristic Inventory	169
A. Herbarium Collections in Fiji	169
B. Herbarium Collections Overseas	169
C. Publications	169
V. Completeness of Floristic Inventory	170
A. General Pattern of Collecting	170
B. Intensity of Collecting	170
1. Endemic Species	170
2. Extinctions	170
VI. Adequacy of Inventory for Major Taxonomic Groups	170
VII. Areas of Forest Threatened by Destruction	171
A. Logging	171
B. Plantations	171
C. Conversion	171
VIII. Areas of Forest in Reserves	171
IX. Resources for Floristic Inventory in Fiji	172
A. Academic Institutions	172
B. Government Institutions	172
C. Foreign Programs	172
D. Opportunities for Collaboration in Floristic Research	172
X. The Effectiveness of Past and Present Floristic Inventories	172
A. Coverage of Entire Region	172
B. Coverage of Areas with Many Endemics or Extinctions	173
C. Coverage of Areas Threatened with Destruction	173
XI. Resources Required for Adequate Inventory of Fijian Forests	173
A. Institutions	173
B. Personnel	173
C. Costs	173
XII. Literature Cited	173
XIII. Appendix – References to the Flora of Fiji (1968–1988)	174

I. Description of the Region

A. Location and Area

Fiji is an archipelago in the South-west Pacific ocean with most of the islands 16°–19° south of the equator. The total land area is about 18,376 sq km, including the large islands of Viti Levu (10,388 sq. km), Vanua Levu (5,538 sq. km), Taveuni (435 sq. km), and Kadavu (409 sq. km). There are more than 300 islands with an area exceeding 2.6 sq. km and several hundred smaller rocky islets and sand cays.

B. Topography and Geology

All the islands are primarily of volcanic origin, the oldest eruptions being about 50 m.y. ago but many of the smaller and eastern islands are very much younger. Volcanic activity is recorded from Taveuni within the past few thousand years.

Taveuni is characterised by a basaltic volcanic landscape with cones, craters, lava flows and ash layers creating a very rugged and steep topography, with a mountain range rising to 1231 m above sea level. The landscapes of the other large islands are more complex, comprising deeply dissected old volcanic landforms, uplifted marine sediments and limestones, and alluvial coastal plains along parts of the coastline. On the largest island of Viti Levu there are over 30 peaks exceeding 900 m, with Mt. Tomaniivi (1,323 m) being the highest mountain in Fiji. On the other large islands the highest peaks are Mt. Batini (1,032 m) on Vanua Levu and Mt. Washington (839 m) on Kadavu.

The smaller islands mostly comprise a mountainous or hilly interior and very limited coastal alluvial deposits. In the Lau group there are many limestone islands with some karst features. Small islets are found on many of the extensive coral reefs around the islands. These comprise mounds of coral rubble and sand only a few meters above sea level. These islets are created, and may be destroyed, by wave action. Few of these islets have fresh ground-water, their flora is similar to mainland beaches.

Coastal landforms are influenced by the predominantly SE winds and waves, and the fringing reefs which reduce wave erosion. Despite the steep nature of the islands' topography there are few sea-eroded cliffs, and the coasts are mostly either rocky and sandy where they are exposed to wave action, or silty in less exposed bays. Coastal dunes are restricted to an area of 2 sq. km near the Sigatoka River. The deltas of the Rewa, Navua and Ba rivers on Viti Levu, the Labasa and Dreketi rivers on Vanua Levu, and other silty coastlines support 190 sq. km of mangrove.

Freshwater swamps are restricted to small areas of impeded drainage in some river valleys and parts of the coastal plains. In addition, several of the volcanic craters in Taveuni support swamp vegetation. The total area of swamp is about 8 sq. km.

With the exception of the beaches, wetter swamps, and steeper cliffs, it seems likely that all the Fijian islands supported forest vegetation prior to the arrival of man 3500 years ago.

C. Human Population

The population of Fiji represents several phases of immigration and population growth. Melanesians and Polynesians occupied at least 100 of the larger islands in prehistoric times and their villages were widespread in both coastal and inland areas. In the 19th century the population was estimated at 115,000–150,000. Forests were cleared for village sites and gardens. Fires probably created the extensive "talasiga" sedge-fern-grasslands on the hills in regions with a pronounced dry season. It is estimated that about 45% (8,300 sq. km) of Fiji still supports native forests, mostly in the wetter south-eastern parts of the larger islands. The forests which occupied the drier zones are now reduced to small pockets and many seem to be affected by occasional fires, grazing, and are colonized by introduced plants.

After Cession of Fiji to the United Kingdom in 1874, indentured laborers were brought from India to work on plantations. The plantations, mostly for sugar, were located on the alluvial coastal plains and foothills, especially on the north-eastern coast of Viti Levu and near Labasa in Vanua Levu. Towns developed in each of these areas to serve the large rural populations.

In 1986 the population of Fiji was 710,000 and was increasing at a rate of 1.8% each year. 80% of the population is on Viti Levu, mostly on the coastal floodplains and nearby hills. About 25% of the population is urban and 35% of the population lives within the environs of towns. The capital city, Suva, has a population of about 150,000 (in 1986) within its environs and is growing faster than the rural areas.

II. Forest Types

The structure of Fijian forests varies with the climate and it is possible to distinguish both upland and lowland zones, and regions with or without a marked dry season. Despite moderately high mean annual rainfalls, ranging from 1500–10000 mm, there is considerable variation from year to year and dry periods are experienced even in the wetter regions.

The lowland Fijian forests in wetter regions match most closely with the UNESCO vegetation class IAlc (Tropical evergreen ombrophilous montane forest). The upland Fijian forests in wetter regions match most closely the UNESCO vegetation class IAle (Tropical evergreen ombrophilous cloud forest). The forest in regions with a marked dry season seem to match the UNESCO vegetation classes 4Alc and 4Ale which are for subtropical evergreen montane and cloud forest regions with a dry season. The more logical tropical class 2Alc is excluded because of the presence of tree ferns in the forests and 2Ale is excluded because it is not recognized. In using the classes IAlc, IAle, 4Alc and 4Ale, it should be made clear that the lowland Fijian forests are not "montane" in the usually accepted sense of the word, nor are they "subtropical", but these physiognomic classes best describe the Fijian forests.

In addition class 5 (mangrove) forests are found in both

the wet and dry zones of the islands and these are indicated on the map. Small patches of alluvial forest (IA1f, 4A1f) and swamp forest (IA1g, 4A1g) occur but they are too small to be shown on the map. In general, patches of forest of less than 3 sq. km are not shown.

No attempt has been made to distinguish areas of virgin or logged forest on the map because it seems likely that, though selective logging changes the abundance and population structure of the forest trees, it does not necessarily cause exclusion of species. The forest edges adjacent to villages, cultivated and arable land, and to talasiga lands, often include patches of secondary forest.

III. Vegetation Maps: A Review

Aerial photographs covering the Fiji islands have been taken on several occasions, and these form the basis of all accurate vegetation maps that have been produced. The most recent complete coverage was taken in 1978 at a scale of 1:20000. Air photographs, land use maps, and topographic maps are available from the Ministry of Lands (Government Buildings, Suva, Fiji).

Using photographs taken in 1951 and 1954, at scales of 1:16000, & 1:40000, the Directorate of Overseas Surveys (1964, 1976) produced maps of Land Use in Viti Levu and nearby islands, and Vanua Levu and nearby islands at scales of 1:25,000 (D.W.S. (L.U.) 3022 FIJI 1:250,000). On these maps land use was classified as follows: forests, light forest, scrub, grass or reeds and fern, mangroves, coconuts, sugar cane or rice or maize and other crops, pasture (improved), urban and airfields.

Topographical and land use maps at a scale of 1:50,000 (Series 448) were produced in 1966–1972 for most of Fiji, except the Lau group, based upon the 1951–1954 aerial surveys with some additional information. On the 1:50,000 maps the land was classified as above but with the additional recognition of "congregations" of palm, bamboo, and marsh vegetation.

The Lau group, and the rest of Fiji, are covered by a 1:250,000 series of topographical maps (Series 648, Sheets 3 and 7, 1966–1978). On these maps vegetation is only classified as forest, mangroves, or cultivation and plantations.

A 1:500,000 topographical map (Fiji is - NZMS 242, 1980) shows all the islands of the group and distinguishes areas with "Native forest or coconut plantations".

Twyford & Wright (1965) described and mapped Fijian soils, land classes and land use capabilities at a scale of 1:126,720. These maps combine information on topography, soil fertility and climate, and suggest land use categories including:

1. Pastoral farming or some orchard crops or reafforestation or forest preservation for catchment protection, with either
    a). No improvement
    b). Minor application of fertilizer/soil conservation
    c). Major application of fertilizer/soil conservation
2. Afforestation for catchment protection
3. Timber production/shifting agriculture
4. Forest preservation for catchment preservation

Although these maps do not show vegetation types, they do include recommendations as to the most suitable vegetation in each area of Fiji. Certain forest types are indicated as suitable for timber production or shifting agriculture while other areas are recommended as protection forests.

Berry and Howard (1973) produced an inventory of Fijian forest resources comprising a survey, undertaken during 1966–1969, of the major areas of native forest on Viti Levu and Kadavu. Each major catchment area was described in terms of climate, geology, general vegetation, population, land tenure, land use, previous forest surveys, past and present exploitation, sawmills, plantations (mostly mahogany, *Sweitenia macrophylla* King, or pine, *Pinus caribaea* Morelot), and access routes. The forests were classified on the basis of field surveys and aerial photographs (mostly from 1954 & 1966 surveys) and mapped at a scale of 1:50000 into the following types.

Non-Commercial Forest (unsuitable for timber production, generally because of the low yield of merchantable timber ha^{-1})
Beach forest
Sapindaceae pole forest
Moderately stocked *Gymnostoma-Dacrydium* dry-zone forest
Low-stocked mixed forest on moderate to steep short slopes
Low-stocked, logged mixed forest on moderate to steep short slopes.
Logged-out forest
Low stocked fringe forest
Low stocked intermediate dry-zone fringe forest.
Low-stocked blocks of forest in grassland
Low-stocked old orchard *Pometia-Inocarpus* forest
Samanea woodland
Low stocked woodland
Gymnostoma woodland
Moderately stocked *Gymnostoma-Fagraea* forest
Low-stocked *Gymnostoma-Intsia* forest
Production Forest (suitable for timber production)
Well-stocked mixed forest on moderate to steep slopes
Moderately stocked, logged mixed forest on moderate to steep short slopes
Well-stocked mixed forest on flat to gentle slopes
Well-stocked *Agathis* forest on flat to gentle slopes
Well-stocked *Palaquium* forest
Moderately stocked *Agathis-Dacrydium* forest on undulating country
Well stocked *Agathis* forest at high elevation on steep slopes

Well stocked *Agathis* and *Decussocarpus* forest on steep slopes

Moderately stocked *Agathis* and *Decussocarpus* forest on gentle to moderate slopes

Moderately-stocked forest on flat to gentle slopes

Moderately-stocked mixed forest on moderate to steep short slopes

Moderately-stocked *Fagraea* forest

Moderately-stocked logged *Fagraea* forest

Moderately-stocked *Agathis* forest on steep slopes at high elevation

Moderately-stocked *Myristica-Gymnostoma* forest on moderate to steep short slopes

Moderately-stocked *Myristica-Gymnostoma* forest in river valleys

Moderately-stocked *Myristica-Gymnostoma* forest on hillsides & ridges

Moderately-stocked *Palaquium* forest

Moderately-stocked *Myristica* forest

Moderately-stocked *Endospermum-Parinari* forest

Moderately stocked *Endospermum-Gymnostoma* forest

Low stocked *Palaquium-Gymnostoma* forest

Moderately stocked *Intsia* forest

Protection Forest (to be protected from soil erosion: unsuitable for timber production)

Ridge thickets

Low-stocked forest on very steep ridges or precipitous slopes

Moderately-stocked forest on long, very steep slopes

Low-stocked dry-zone forest on long, very steep slopes

Low-stocked forest on short, very steep slopes

Low-stocked open canopy woodland on steep slopes.

The typical species composition, structure and timber volume of each forest type is outlined and a list of tree species collected during the field surveys is provided.

This survey indicated the common forest tree species and forest types over most of Fiji's native forests except for northeastern Viti Levu, Taveuni and smaller islands.

During the period since Berry and Howard's inventory was compiled, logging activity has increased considerably and significant areas of production forest would now require reclassification as low-stocked or logged-out forest. B. Thomerson of the Native Lands Trust Board [Suva] estimated some changes in forest area by sampling aerial photographs from a 1978 survey of Viti Levu and Vanua Levu.

Forest inventories have since been made of Taveuni and some areas of production forest by the Department of Forests [Suva]. These are primarily concerned with assessment of timber volume in the major commercial species.

Descriptions and maps of vegetation in Fiji include: the island of Vanua Balavu, Lau group (Garnock-Jones 1987), and Lakeba, Lau group (Garnock-Jones 1978, Latham 1979); sand dunes near Sigatoka, Viti Levu (Kirkpatrick & Hassall 1981); freshwater wetlands of Viti Levu (Ash & Ash 1984); and of Tagimaucia crater, Taveuni (Southern et al., 1986).

IV. Magnitude of Floristic Inventory

A. Herbarium Collections in Fiji

The only major general collection of plant material held in Fiji is the SUVA Herbarium [Institute of Natural Resources, University South Pacific, P.O. Box 1168, Suva, Fiji]. This herbarium, started in 1933, holds about 27,000 collections comprising about 41,000 sheets. The herbarium has specimens from 2,086 vascular plant species; approximately 78% of the species recorded as natives, established introductions and ornamentals in Fiji. In addition, the herbarium holds some collections from other Pacific islands and there is a small collection of bryophytes, algae and fungi.

The Department of Agriculture [Koronivia] has an herbarium collection of economically important plants, and various agricultural research sections maintain plots of desirable species.

The Ministry of Forests [Suva, Colo-i-Suva, Nasinu, and other centers] have collections from timber trees and also plantations of desirable species.

B. Herbarium Collections Overseas

Much of the material held by the SUVA herbarium is represented as duplicates in overseas herbarium. In addition, many of the earlier collections, which include much type material, are not represented at SUVA.

The overseas herbaria containing significant collections from Fiji are indicated by Smith (1979: 14–15, 37–88), according to the distribution of specimens from the various collectors, and the following list simply indicates which herbaria hold collections (consult *Index Herbariorium* for herbarium abbreviations).

Many specimens are held at: A, BISH, BM, GH, K, NY, P, UC, US.

Other specimens are held at: AK, AMES, B, BH, BO, BRI, C, CANU, CGE, CHR, CN, DAV, E, F, FH, FI, G, G-DC, GOET, HBG, HLA, IA, KIEL, KYO, L, LAE, LINN, LIV, MASS, MEL, MO, MW, PH, RSA, S, SBT, TDC, UC, UPS,W; WELTO, WRCS, Y, Z.

C. Publications

Seemann (1865–1873) produced the first major Fijian flora, including many descriptions of new species and 100 illustrations. Parham (1964, 1972) produced an annotated list of vascular plants recorded in Fiji, including 104 illustrations and a bibliography of 852 references to Fijian plants. Brownlie (1977) produced a flora for Fijian pteridophytes including keys and descriptions of all known species. Smith (1979, 1981, 1985) has produced a flora of the Gymnospermae and Monocotyledones, except the Orchidaceae, (1979), and of families 44–116 (1981), and 117–163 (1985) of the

Dicotyledones; a further volume, which should complete this flora, is anticipated.

In addition there are many other publications dealing with more restricted aspects of the flora, which are listed in the appendix.

V. Completeness of Floristic Inventory

A. General Pattern of Collecting

With the *Pteridophyte Flora* (Brownlie 1977) and the anticipated completion of the *Flora Vitiensis Nova* (Smith 1979, 1981, 1985) Fiji will possess up-to-date floras of the vascular plants. A number of species described in the floras are inadequately known, often lacking fertile material or field notes. Many species are known by a single specimen so there is no concept of their variation. In many more species there are only a few collections. It seems that at least 20% of the forest species are inadequately represented in herbarium collections, and few species have numerous collections. The bryophytes and algae are less well known and no floras have yet been produced.

It is probable that a number of species have not yet been collected or described, but it is difficult to predict how large this number might be. On the basis of the number of species known by only a single collection it seems probable that as many species may be unknown. It seems reasonable to expect a 5–10% increase in the number of species that are recognized, perhaps partially offset by changes in the status of existing species as their variation is better defined.

To improve our knowledge of the flora requires the collection of more specimens, and two different strategies are suggested. For species which are inadequately known, it is most sensible to begin by collecting around their original localities. For as yet undiscovered species, it is most sensible to collect in habitats or areas which have not previously received much attention: these areas are discussed below.

B. Intensity of Collecting

The intensity of collecting is the effort put into obtaining a complete collection of the species in a region, but this is difficult to assess. An easier index of collecting intensity is the number of collections that have been made in a given region. Overall the total of all Fijian herbarium collections, including those held overseas, is about 37,000. With a land area of 18,376 sq. km, this indicates a density of about 200 collections/100 sq. km, but this figure includes areas with numerous collections and other areas with none. The archipelagic nature of the Fiji islands, introduces a greater variation of species composition within the region than might be expected for a mainland region. The isolated island of Rotuma should probably be considered separately from the other Fijian islands, and separate floristic inventories of other island groups (e.g., Lau group) are desirable.

Collecting has been most intense in two types of localities; firstly, areas near Suva, and along major roads where access is easy. Secondly, remote localities such as particular mountains and some of the smaller islands, where collecting trips have been made.

The floristic diversity of these selected regions is greatly in excess of 100 species/sq. km and it is likely that at least 1000 herbarium collections/100 sq. km are required to obtain a reasonable estimate of the floristic composition of the area. On this basis there are few, if any, areas for which the species composition is known.

Most of the interior forested areas of Fiji, remote from roads until recently and not singled out for botanical investigations, have contributed few specimens to herbarium collections, and certainly less than 100 collections/100 sq. km. A few localities have been collected intensively, and extensive regions have been collected sporadically, while others have been scarcely collected. It is clear that the greater part of the forested areas and many smaller islands have been scarcely collected.

1. Endemic species

Endemic species are mostly restricted to forested areas, and few survive in habitats affected by cultivation, grazing or firing. Many of the endemics occur on several islands but most of the larger islands, and some of their mountain ranges, comprise the entire range of an endemic species. Virtually every forested region of Fiji bounded by a major valley or ocean is the entire range of one or more species.

There is one family (Degeneriaceae) endemic to Fiji, and several endemic genera: *Alsmithii, Sukunia, Degeneria, Goniocladus, Hedstromia, Neoveitchia, Gillespiea, Reodea, Squamellaria, Amaroria, Pimia*. The number of endemic species is dependent upon the species concept employed by the taxonomist. Virtually all the forest species are genetically isolated from overseas populations and many seem to have evolved distinct forms. It is estimated that there are 453 endemic species, 22% of the native flora.

2. Extinctions

Extinctions are difficult to document in the absence of written or fossil records and past, or future, extinctions can only be related to the loss of forest habitat. This is discussed more fully in section VII. The areas of forest threatened by destruction are generally fairly accessible and have been moderately well collected, but not sufficiently so to enable compilation of complete species lists. The palm *Neoveitchia storckii* is evidently threatened with extinction in its natural habitat through forest destruction (Gorman & Siwatibau 1975) and is representative of other species.

VI. Adequacy of Inventory for Major Taxonomic Groups

Most botanical collecting in Fiji has been of a general nature though some collectors have sought particular groups, c.f. Smith (1979).

Most taxa would benefit from further collection and this is especially true of some large families such as the Rubiaceae, Orchidaceae and Myrtaceae, and is generally true for bryophytes, algae and lichens. The recently published floras of Pteridophytes (Brownlie, 1977) and Spermatophytes (Smith 1979, 1981, 1985) indicate the adequacy of collections for particular species of vascular plants. Indications of the adequacy of knowledge of bryophytes are given by Whittier (1975) and Huerlimann (1978).

VII. Areas of Forest Threatened by Destruction

A. Logging

Virtually all logging in Fiji may be classified as "selective" removal of timber trees, and this generally involves the removal of large trees and associated destruction of smaller trees. Typically, 10–40% of the forest trees are killed and there is a slight alteration in the relative abundance of species which remain. It does not seem that selective logging of virgin forest is likely to cause immediate extinctions of forest and shrub species though the changes to their populations may be appreciable. Epiphytes, especially orchids but also ferns and bryophytes, may be more seriously affected since these species are often associated with the large trees removed for timber. The impact of logging on pollinating and seed dispersing animals is unknown.

Selective logging was recommended for about 30% of Fiji's forests (Berry & Howard 1973), and about 30% of the virgin forest was recommended as reserves for catchment protection. The remaining 40% was mostly either already logged or secondary forest. Most accessible native forests on Viti Levu and Vanua Levu will probably be selectively logged by 2010 A.D. Taveuni is the only island with extensive forests for which there have been no major logging operations but an assessment survey of these forests was made in 1984–1985 and selective logging is planned.

B. Plantations

After logging, the forest may be left to regenerate naturally or cut-over more intensively and planted with exotic trees such as Mahogany (*Swietenia macrophylla* King), *Cordia alliodora* (R. & P. Cham.), *Anthocephalus cadambra*, and the indigenous *Endospermum macrophyllum*. The preparation and tending of the forests for plantations causes the destruction of most existing trees and 60–80% of shrubs. The combined impact of silvicultural practises and competition from exotic trees is likely to eliminate or substantially reduce populations of many native species. The areas of plantations within logged forests are relatively small (16,000 ha in 1985) but the rate of planting is increasing and each year about 2000–5000 ha (15–40%) of the logged forest is converted to plantations. Extensive plantations of *Pinus caribaea* have been established in deforested areas.

C. Conversion

Conversion of native forest to non-forest land is mostly associated with gardens and small scale agriculture and has varied greatly in different forest types. The agricultural use of land is favored by proximity to an all-weather road, so that this conversion tends to follow the old road networks of logging companies and new Public Works Department roads.

Mangrove forests are threatened by land reclamation schemes in many areas of north-eastern Viti Levu and Vanua Levu where rice and sugarcane may be produced. In addition, mangrove areas near towns are infilled for buildings or used as rubbish dumps.

Sago palm forest, mostly in S.E. Viti Levu, is threatened by drainage schemes such as the Pacific Harbour development.

Beach forests were often converted to gardens by Fijian villagers and these areas were also favored for coconut plantations. Few areas of undisturbed beach forest remain.

Dry zone seasonal forests were probably once extensive but have now been greatly reduced in area by burning in prehistoric times. Patches of forest remain in localities which are protected from fires by reason of streams and rocky outcrops. Little of the dry forest remains.

Forests in wetter regions are less susceptible to burning, and after disturbance or cultivation ceases a secondary forest usually develops. The fringes of cultivated land in the wetter zones usually include patches of secondary forest and these may persist with a distinctive impoverished flora for 100–200 years. Overall, there is a gradual extension of agricultural land both around villages and associated with a few major agricultural schemes such as Seaqaqa and Batiri in Vanua Levu. Most of the alluvial lowland forests have been converted to agricultural land but lowland forests persist on steep slopes. In highland areas the forest have survived with little disturbance but this may change as the road system is extended.

VIII. Areas of Forest in Reserves

Several nature reserves were designated during 1958–1969 and there were detailed proposals by the National Trust of Fiji, in 1981, to set up a representative series of Nature reserves/National Parks but these have not yet been designated by parliament (1987). Designated nature reserves include:

Nadarivatu: 93 ha of upland forest near Nadarivatu, Viti Levu.
    Mt. Tomaniivi: 1322 ha of upland forest including Fiji's highest maintain near Nadarivatu, Viti Levu
Naqarabuluti: 279 ha of forest, near Nadarivatu, Viti Levu
Draunibota, Vuo & Labiko: 1.9 Ha, 1.2 ha & 0.26 ha; Limestone Islands near Suva, Viti Levu

Ravilevu: 4020 ha of lowland forest in Taveuni (future status is uncertain)
Calia: 427 ha of logged lowland forest near Navua, Viti Levu
Yaduataba: 87 ha Island reserve for iguanas, Bua

There are also a number of forest reserves controlled by the Ministry of Forests which are intended as production forests rather than nature reserves. The existing nature reserves are not adequate to preserve Fiji's forest flora and fauna and it is to be hoped that more forest areas will be preserved intact.

Forest reserves are most urgently needed in alluvial lowlands, coastal, and dryzones where human impact has converted much of the original forest to other uses. Exclusion of grazing animals and burning is desirable in these areas. Mountain ridge-top cloud forests and freshwater swamp forest are of very limited extent and for this reason require special attention even though they are not under immediate threat of damage.

IX. Resources for Floristic Inventory in Fiji

A. Academic Institutions

The only academic institution concerned with floristic studies is the University of the South Pacific [P.O. Box 1168, Suva, Fiji] which is based in Suva and operates in many of the island nations of the south-west Pacific:

i. The Institute of Natural Resources includes the SUVA Herbarium which has one technical officer. The institute conducts research on a contract basis and relies upon external funding for most of its activities. The institute is the usual base for scientists visiting the region and as such it is the obvious center for floristic studies and surveys both in Fiji and neighboring islands.

ii. The School of Pure and Applied Sciences includes biology, chemistry and physics departments and provides a basic science degree. Both botanical and phytochemical research is conducted.

B. Government Institutions

The Fiji Public Service includes several ministries and government funded organisations, all based in Suva, which are concerned to a varying degree with forests and/or botanical research:

i. The Ministry of Forests is concerned with exploitation of both natural forests and plantations. The focus of forest research is currently upon plantation forestry rather than native forests. There are, however, Forestry officials concerned with forest inventories and timber statements from the logging areas, and silvicultural studies of some native species are being undertaken.

ii. The Ministry of Agriculture & Fisheries is concerned with crop and weed plants, plant quarantine, etc. This department is not directly concerned with native forests except for clearance of forests for agricultural land and the significance of mangroves for fish production. In this latter role the Ministry does make inventories of mangrove forests.

The Ministry of Agriculture has a botany division and maintains a small herbarium of economically important plants.

iii. The Ministry of Lands & Mineral Resources is responsible for aerial surveys and general mapping and in this capacity is involved with any major surveys.

iv. The Ministry of Energy has an interest in wood as a fuel and fuelwood resources but this does not extend to floristic studies.

v. The Native Land Trust Board (N.L.T.B.) manages the land communally owned by Fijians. Forests are managed for the N.L.T.B. by the Ministry of Forests. The N.L.T.B. also leases land for forest plantations.

vi. The National Trust of Fiji is a government funded organization with the task of preserving the national heritage, both natural and cultural. This includes nature reserves and national parks. Although floristic inventories of these areas are of interest to the Trust, there are neither the staff nor the resources to support these studies.

vii. The Fiji Museum is primarily concerned with recording, collecting, and displaying Fiji's cultural heritage but there is also an interest in the native fauna, and to a lesser extent the flora.

C. Foreign Programs

There are no foreign programs of floristic studies in Fiji though there are foreign botanists working in Fiji and upon Fijian collections.

D. Opportunities for Collaboration in Floristic Research

There are no government restrictions on collaboration except for the conditions imposed by the Ministry of Home Affairs with regard to research visas and work permits. Plant specimens may be exported for research purposes subject to the deposition of appropriate duplicates at the SUVA Herbarium. Importing specimens, e.g., from neighboring island nations, is subject to quarantine procedures.

X. The Effectiveness of Past and Present Floristic Inventories

A. Coverage of Entire Region

Botanical collecting has centered on a few localities on the larger islands plus sporadic collections in other areas. It is estimated that 90–95% of the forest species have been collected and that the remainder may be found anywhere but most probably in more isolated upland regions and on some of the smaller islands. With the possible exception of the forests near Suva and a few mountain summits there are no forest areas for which reasonably complete species lists could be compiled.

Although the localities which are poorly collected may be readily defined, c.f. Figure 2, there are limited funds available to support collecting trips.

B. Coverage of Areas with Many Endemics or Extinctions

Endemic species occur throughout Fiji's forests and some are restricted to particular regions of the larger islands, such that virtually every region has endemics. Since the number of endemic species is very high, about 20% of the forest species, it is not particularly useful to distinguish them when planning collecting trips.

No extinctions have been documented and there is insufficient information to predict which species are at risk.

C. Coverage of Areas Threatened with Destruction

Forest destruction is mostly associated with agriculture and silviculture, and the forest types experiencing the greatest reduction in area are lowland, coastal, and dryzone forests. Surveys of beach forests, forest patches in the Talasiga areas, and forests on small islands which are being developed for agriculture, resorts, etc. should be given priority even though they are not likely to contain many unknown species.

XI. Resources Required for Adequate Inventory of Fijian Forests

A. Institutions

The existing institutions within Fiji, i.e., University of the South Pacific, combined with some overseas taxonomic expertise, are sufficient to carry out floristic inventories.

B. Personnel

The personnel available to conduct floristic inventories, excluding forest timber inventories, amounts to one technical officer at the SUVA Herbarium and interested staff at the University of the South Pacific. These are sufficient personnel to conduct collecting trips and process the specimens. Over a period of several years collecting trips could be made to the little known areas of Fiji and perhaps also to neighboring islands which have floristic affinities with Fiji.

C. Costs

The major factor limiting botanical collecting is the lack of financial support: the herbarium does not have sufficient funding to support major collecting trips and has to seek external funds. A series of collecting trips to the poorly known areas of Fiji (c.f. Fig. 2) would substantially improve knowledge of the flora.

XII. Literature Cited

———— & W. Ash. 1984. Freshwater wetland vegetation of Viti Levu, Fiji. New Zealand J. Bot. **22**: 377–391.
Ash, J. 1985. Growth rings and longevity of *Agathis vitiensis* in Fiji. Austral. J. Bot. **33**: 81–88.
Berry, M. J. & W. J. Howard. 1973. Fiji Forest Inventory: Land resources study 12, 3 vol. Overseas Development Authority, Surbiton, U.K. (including 36 maps).
Brownlie, G. 1977. The Pteridophyte flora of Fiji. Nova Hedwigia **55**: 1–397.
Garnock-Jones, P. J. 1978. Plant communites of Lakeba and Southern Vanua Balavu, Lau group, Fiji. Bull. Roy. Soc. New Zealand **17**: 95–117.
Gorman, M. L. & S. Siwatiban. 1975. The status of *Neoveitchia storckii*, a species of palm tree endemic to the Fijian Island of Viti Levu. Biol. Conserv. **8**: 73–76.
Hassall, D. C. & J. B. Kirkpatrick. 1985. The diagnostic value and host relationships of the dependent synusia in the forests of Mount Korobaba, Fiji. New Zealand J. Bot. **23**: 33–46.
Kirkpatrick, J. B. & D. C. Hassall. 1981. Vegetation of the Sigatoka sand dunes, Fiji. New Zealand J. Bot. **19**: 285–297.
———— & ————. 1985. The vegetation and flora along an altitudinal transect through tropical forest at Mount Korobaba, Fiji. New Zealand J. Bot. **23**: 33–46.
Latham, M. 1979. The natural environment of Lakaba. Pages 13–64 *in* H. C. Brookfield, (ed.) Man and the Biosphere programme, Project 7: Ecology and rational use of island ecosystems; UNESCO/UNFPA Population and environment project in the eastern islands of Fiji Island report 5: Lakaba: Environmental change, population dynamics and resources use.
Lal, P. N. 1984. Environmental implications of coastal development in Fiji. Ambio **13**: 316–210.
Leslie, D. M. & L. C. Blakemore. 1985. Properties and classification of selected soils from Vauna Balavu, Lau Group, Fiji. J. Roy. Soc. New Zealand **15(3)**: 313–328.
Munir, A. A. 1985. A taxonomic revision of the genus *Viticipremna* (Verbenaceae). J. Adelaide Bot. Gard. **7**: 181–200.
Parham, J. W. 1964, 1972. Plants of the Fiji Islands, 462 pp. Government Printer, Suva.
Seeman, B. 1865–1873. Flora Vitiensis, 453 pp. reprinted 1977 by J. Cramer, Vaduz.
Siwatibau, S. 1984. Traditional environmental practices in the South Pacific: A case study of Fiji. Ambio **13**: 365–368.
Smith, A. C. 1979. Flora Vitiensis Nova, Vol. 1, 495 pp., Pacific Tropical Botanic Garden, Lawai, Hawaii.
————. 1981. Flora Vitiensis Nova, Vol. 2, 810 pp., Pacific Tropical Botanic Garden, Lawai, Hawaii.
————. 1985. Flora Vitiensis Nova, Vol. 3, 758 pp., Pacific Tropical Botanic Garden, Lawai, Hawaii.
Southern, W., J. Ash, J. Brodie, & P. Ryan. 1986. The flora,

fauna, and water chemistry of Tagimaucia water, a tropical highland lake and swamp in Fiji. Freshwater Biol. **16**: 509–520.

Stone, B. C. 1985. On the 2nd Collection of *Pandanus halleorum*. Kew Bull. **40**: 287–290.

Twyford, I. T. & C. C. S. Wright. 1965. Soil resources of the Fiji Islands, 2 vol., Government Printer, Suva.

Whittier, H. O. 1975. A preliminary list of Fijian mosses. Florida Sci. **38**: 85–106.

XIII. Appendix I

References to the Flora of Fiji (1968–1988)

References are listed by year, and alphabetical order. References prior to 1971 are listed similarly in Parham (1972) *Plants of the Fiji Islands*, pages 414–454, and references are not duplicated in both lists, nor in the literature cited in the Fiji chapter of this monograph.

1968 **Tyler, D. B.** 1968. The Wilkes expedition, the 1st USA exploring expedition 1838–1842. Mem. Amer. Philos. Soc. **73**: 1–435.

1969 **Chapman, V. J.** 1969. Conservation of island ecosystems in the Southwest Pacific. Biol. Conserv. **1**: 159–165.

1970 **Roberts, O. T.** 1970. A review of pasture species in Fiji: 1 Grasses. Trop. Grassl. **4**: 129–137.

———. 1970. A review of pasture species in Fiji: 2 Legumes. Trop. Grassl. **4**: 213–222.

Smith, A. C. 1970. Research on the flora of Fiji. Fiji Agric. J. **32**: 213–222.

Weiner, M. A. 1970. Notes on some medicinal plants of Fiji. Econ. Bot. **24**: 279–283.

1971 **Campbell, E. O.** 1971. Liverworts collected in Fiji by A. C. Smith and W. Greenwood. J. Roy. Soc. New Zealand **1**: 7–30.

Chapman, V. J. 1971. The marine algae of Fiji. Rev. Algol. **10**: 164–171.

Hughes, H. R. 1971. Control of the water weed problem in the Rewa River. Fiji Agric. J. **33**: 67–72.

Schaeffer, J. 1971. Revision of the genus *Endospermum* (Euphorbiacae). Blumea **19**: 171–192.

Smith, A. C. 1971. Studies of Pacific Island plants XXII: New flowering plants from Fiji. Pacific. Sci. **25**: 491–501.

———. 1971. Studies of Pacific Island plants XXIII: The genus *Diospyros* (Ebenaceae) in Fiji, Samoa, & Tonga. J. Arnold Arbor. **52**: 369–403.

1972 **Bolza, E. & N. H. Kloot.** 1972. The mechanical properties of 56 Fijian timbers. Austral C.S.I.R.O. Div. For. Prod. Tech. Pap. **62**: 1–51.

Firman, I. D. 1972. A list of fungi and plant parasitic bacteria, viruses and nematodes in Fiji. Common. Mycol. Inst., Phytopathol. Pap. **15**: 1–35.

Parham, J. W. 1972. Plants of the Fiji Islands. 462 pp. Gov. Printer, Suva.

Smith, A. C. 1972. Studies of Pacific Island plants XXIV: The genus *Terminalia* (Combretaceae) in Fiji, Samoa & Tonga. Brittonia **23**: 394–412.

1973 **Brownlie, G.** 1973. The genus *Lindeaea* in Fiji. Amer. Fern J. **63**: 91–98.

Duarte, A. P. 1973. An attempt to explain the occurrence of 2 species of *Podocarpus* in Brazil. Arq. Jard. Bot. Rio de Janeiro **19**: 199–215.

Gillett, G. W. 1973. The genus *Cyrtandra* (Gesneriaceae) in Fiji. Taxon **22**: (Suppl.) 701.

Smith, A. C. 1973. Studies of Pacific Islands plants XXV. The Myrsinaceae of the Fijian region. J. Arnold Arbor. **54**: 1–41, 228–292.

1974 **Armstrong, J. E. & T. K. Wilson.** 1974. The morphology of the staminate flowers of *Horsfieldia* (Myristicaceae). Amer. J. Bot. **61**: (5 suppl.) 53 pp.

Miller, H. A. 1974. A bryological evaluation of the Polynesian subkingdom. Bull Soc. Bot. France **121**(spec. issue): 287–293.

Smith, A. C. 1974. Studies of Pacific Island plants XXVI: The genus *Gardenia* (Rubiaceae) in the Fijian region. Amer. J. Bot. **61**: 109–128.

———. 1974. Studies of Pacific Island Plants XXVII: The Guttiferae of the Fijian region. J. Arnold Arbor. **55**: 215–263.

1975 **Braithwaite, A. F.** 1975. Cyto-taxonomic observations on some hymenophyllaceae from the New Hebrides, Fiji and New-Caledonia. J. Linn. Soc., Bot. **71**: 167–189.

———. 1975. The phytogeographical relationships and origin of the New-Hebrides fern flora. Philos. Trans., Ser. B. **272**: 293–313.

Gillett, G. W. 1975. *Cyrtandra* (Gesneriaceae) in the Bismarck Archipelago and Solomon Islands. Kew Bull. **30**: 371–412.

Gorman, M. L. & S. Siwatibau. 1975. The status of *Neoveitchia storckii*, a species of palm tree endemic to the Fijian island of Viti Levu. Biol Conserv. **8**: 73–76.

Karan, B. 1975. Studies of Navua sedge *Cyperus aromaticus*, I: Review of the problem and study of morphology, seed output and germination. Fiji. Agric. J. **37**: 59–68.

Robinson, G. S. 1975. Macrolepidoptera of Fiji and Rotuma, a taxonomic and geographic study. E. W. Classey Ltd, Faringdon.

Smith, A. C. 1975. Studies of Pacific Island plants XXX: Notes on Fijian Apocynaceae and Asclepiadaceae. Brittonia **27**: 151–164.

———. 1975. Studies of Pacific Island plants XXXI: Notes on Fijian Sapotaceae. Brittonia **27**: 165–171.

——— & J.E. Haase. 1975. Studies of Pacific Island plants XXIX: *Bleasdalea* and related genera of Proteaceae. Amer. J. Bot. **62**: 133–147.

Smith, R. M. 1975. Note on gum turpentine of *Pinus caribaea* from Fiji. New Zealand J. Sci. **18**: 547–548.

——— & S. Siwatibau. 1975. Sesquiterpene hydrocarbons of Fijian Guavas. Phytochemistry **14**: 2013–2015.

St. John, H. 1975. Revision of the genus *Pandanus* part 38: *Pandanus* in Fiji 1st group except section *Pandanus*. Pacific Sci. **29**: 55–77.

———. 1975. Revision of the genus *Pandanus* 39: *Pandanus* of Rotuma Island, Pacific Ocean. Pacific Sci. **29**: 371–406.

Whittier, H. O. 1975. A preliminary list of Fijian mosses. Florida Sci. **38**: 85–106.

Winchester, R. V. 1975. Leaf amino-acids of *Psidium guajava*. New Zealand Sci. **18**: 239–242.

1976 **Firman, I. D.** 1976. Plant diseases in the area of the South Pacific Commission 3. Imi herbarium specimens of fungi on plant hosts. From Fiji S.P.C. Inf. Doc. (South Pac.

Comm.) 39, 14 pp.

St. John, H. 1976. Revision of the genus *Pandanus* 40: The Fijian species of the section *Pandanus*. Pacific Sci. 30: 249–315.

1977 Brownlie, G. 1977. The pteridophyte flora of Fiji. Nova Hedwigia 55: 1–397.

Darwin, S. P. 1977. The genus *Mastixiodendron* (Rubiaceae). J. Arnold Arbor. 58: 349–381.

Faden, R. B. 1977. The genus *Rhopalephora* (Commelinacae). Phytologia 37: 479–481.

Greenwood, W. 1977. The food plants or hosts of some Fijian insects, 5. Proc. Linn. Soc. New South Wales 101: 237–241.

MacEntee, F. J., Bold, H. C. & P. A. Archibald. 1977. Notes on some edaphic algae of the South Pacific and Malaysian areas with special reference to *Pseudotetraedron polymorphum*, new genus, new species. Soil. Sci. 124: 161–166.

St. John, H. 1977. The flora of Nivatoputapu Island, Tonga. Pacific Plant studies 32. Phytologia 36: 374–390.

1978 Burke, M. 1978. Meet Fiji's Rainforest. Dept. of Forestry, Suva, 33 pp.

Huerlimann, H. 1978. Hepaticae from the south Pacific Region, 6. Bauhinia 6: 293–414.

Smith, A. C. 1978. A precursor to a new flora of Fiji. Allertonia 1: 331–414.

Tomlinson, P. B. 1978. *Rhizophora* in Australasia, some clarification of taxonomy and distribution. J. Arnold Arbor. 59: 156–169.

1979 Norris, D. H. 1979. The systematic position of *Bryobrothera crenulata*. Bryologist 82: 305–309.

Smith, A. C. 1979. Flora Vitiensis Nova 1, 495 pp. Pac. Trop. Bot. Gard. Lawai, Hawaii.

Smith, R. M. & H. Kassim. 1979. The essential oil of *Piper aduncum* from Fiji. New Zealand J. Sci. 33: 127–128.

Tomlinson, P. B. 1979. Preliminary observations on floral biology in mangrove Rhizophoraceae. Biotropica 11: 256–277.

1980 Dhal, A. L. 1980. Regional ecosystems survey of the South Pacific area. South Pac. Comm. Tech. Paper 179, 99 pp.

Darwin, S. P. 1980. Notes on *Airosperma* (Rubiaceae) with a new species from Fiji. J. Arnold Arbor. 61: 95–106.

Kanis, A. 1980. The Malesian species of *Serianthes* Bentham (Fabaceae-Mimosoideae) (New taxa). Brunonia 2: 289–320.

Nowak, H. 1980. Revision of the moss genus *Mithyridium* for Oceania (Calymperaceae). Bryophytorum Bibliotheca 20: 1–236.

Page, C. N. 1980. Leaf micromorphology in *Agathis* and its taxonomic implications. Plant Syst. Evol. 135: 71–80.

Singh, S., I. Bola & J. Kumar. 1980. Diseases of plantation trees in Fiji islands:1. Brown rot of mahogany. Indian Forester. 106: 526–532.

Whistler, W. A. 1980. Coastal flowers of the tropical Pacific. Pac. Trop. Bot. Gard., Lawai, Hawaii.

1981 Airy Shaw, H. K. 1980. Notes on Asiatic, Malesian and Melanesian Euphorbiaceae (New taxa). Kew Bull. 36: 599–612.

Cown, D. J. 1981. Wood density of *Pinus caribaea* var. *hondurensis* grown in Fiji. New Zealand J. Forest. Sci. 11: 244–53.

Cox, P. A. 1981. Bisexuality in the Pandanaceae; new findings in the genus *Freycinetia*. Biotropica 13: 195–198.

Hannibal, L. S. 1981. The Hawaiian and Pacific *Crinums*. Plant LIfe 37: 104–106.

Molho, D., & B. Bodo. 1981. A chemically distinctive new *Ramalina* from Fiji. (*Ramalina luciae*, Lichens). Bryologist 84: 396–398.

Nicholson, S. A. 1981. Community and population level shifts in a young raw earth succession in Fiji. Trop. Ecol. 22: 116–122.

Peat, N. 1981. Focus on island forestry. Development, New Zealand 4: 3–9 Min. For. Affairs, Wellington, N.Z.

Singh, S. & I. Bola. 1981. Diseases of plantations trees in Fiji Islands.2. Indian J. Forest. 4: 86–91.

Smith, A. C. 1981. Flora Vitiensis Nova 2, 810 pp. Pacific Trop. Bot. Gard., Lawaii, Hawaii.

Smith, R. M., R. A. Marty & C. F. Peters. 1981. The diterpene acids in the bled resins of 3 Pacific Kauoi. Phytochemistry 20: 2205–2208.

———— & A. M. Robinson. 1981. The essential oil of ginger *Zingiber officinale* from Fiji. Phytochemistry 20: 203–206.

Stemmermann, L. 1981. A guide to Pacific wetland plants. 118 pp. U.S. Army Corps, Honolulu, Hawaii.

1982 Ando, H. 1982. *Hypnum* in Australia and the Southern Pacific (Musci). J. Hattori Bot. Lab 52: 93–106.

Cannon, J. R., K. D. Croft, Y. Matsuki, V. A. Patrick, R. F. Toia, & A. M. White. 1982. Crystal structure and absolute configuration of dextro-scadine hydrobromide. Austral. J. Chem. 35: 1655–1664.

Cribb, P. J. 1982. *Spathoglottis* (Orchidaceae) in Australia and the Pacific Islands (New taxa). Kew Bull. 36: 721–729.

Firman, I. D. 1982. Plant protection news, 90. 8 pp.

Hartley, T. G. 1982. A revision of the genus *Sarcomelicope* (Rutaceae). Austral. J. Bot. 30: 359–372.

Inoue, H. 1982. Speciation and distribution of *Plagiochila* in Australia and the Pacific (Hepaticae). J. Hattori Bot. Lab. 52: 45–56.

Koponen, A. 1982. The family Splachnaceae in Australia and the Pacific (Musci). J. Hattori Bot. Lab. 52: 87–91.

————. 1982. The family Mniaceae in Australia and the Pacific (Musci) J. Hattori Bot. Lab. 53: 87–91.

McMillan, C. & K. W. Bridges. 1982. Systematic implications of bullate leaves and isozymes for *halophila* from Fiji and Western Samoa. Aquat. Bot. 12: 173–188.

Scott, G. A. M. 1982. Bryofloristics in Australasia. Nova Hedwigia 71: 483–493.

Seppelt, R. D. 1982. *Ditrichum* and other genera of Ditrichaceae in Australia and the Pacific (Musci). J. Hattori Bot. Lab. 52: 107–112.

Singh, R. S. et al. 1982. The status of *Neochetina eichhorniae* Warner as biocontrol agent for water hyacinth *Eichhornia crassipes* (Mart.) Solms. (Fam.: Pontederiaceae) in Fiji. Pages 35–38 *in* Report of the regional workshop on Biological Control of Water Hyacinth, 3–5 May, 1982 NAL, Bangalore, India., Publ: Comm. Sci Council, London.

Stevens, G. N. 1982. *Ramalina leiodea*, a common maritime species in Oceania (Lichens). Lichenologist 14: 39–45.

1983 Cambie, R. C., R. E. Cox, K. K. Croft, & D. Sidwell. 1983. Phenolic diterpenoids of some podocarps. Phytochemistry 22: 1163–1166.

Handa, S. S. 1983. Plant anticancer agents XXV: Constituents of *Soulamea soulameoides*. J. Natural Products **46**: 359–364.

Jansen, M. E. & C. E. Ridsdale. 1983. A revision of the genus *Dolicholobium* (Rubiaceae). Blumea **29**: 251–311.

Smith, R. M. 1983. Kava lactones in *Piper methysticum* from Fiji. Phytochemistry **22**: 1055–1056.

1986   **Ash, J.** 1986. Growth rings, age and taxonomy of *Dacrydium* (Podocarpaceae) in Fiji. Austral. J. Bot. **34**: 197–205.

———. 1986. Demography and production of *Leptopteris wilkesiana* (Osmundaceae), a tropical tree-fern from Fiji. Austral. J. Bot. **34**: 207–215.

1987   ———. 1987. Demography and production of *Pandanus tectorius* (Pandanaceae) in Fiji. Austral. J. Bot. **35(3)**: 313–330.

Ash, J. 1987. Demography and production of *Cyathea hornei* (Cyatheaceae), a tropical tree-fern from Fiji. Austral. J. Bot. **35(3)**: 331–342.

1988   ———. 1988. Demography and production of *Balaka microcarpa* (Arecaceae), a tropical understorey palm in Fiji. Austral. J. Bot. **36**.

———. 1988. Stunted cloud-forest in Taveuni, Fiji. Pacific Sci. **41**: 191–199.

New Caledonia

Karolyn Kendrick

Contents

I.	Introduction	178
II.	Description of the Region	178
	A. Geographical Extent and Area	178
	B. Geology and Topography	178
	C. Population	178
III.	Vegetation Maps	178
IV.	Floristic Inventory	178
	A. Collections and Publications	178
	B. Endemism	179
	C. Extinction	179
V.	Resources for Continued Floristic Inventory	179
VI.	Selected References	179

I. Introduction

The island of New Caledonia, with its major dependencies the Loyalty Islands and the Isle of Pines, rich with archaic gymnospermas and angiosperms maintained by the islands' isolation, is renowned as a "botanical motherlode" of "missing links" in the evolutionary history of angiosperms (Thorne, 1963).

The uniqueness of New Caledonia stems from its geological history. According to current views (succinctly summarized by Morat, Veillon & MacKee, 1984), the origin of New Caledonia can be traced to an island archipelago lying off the Australian Plate of Gondwanaland during the Permian Epoch and Mesozoic Era. The island chain may have had at least partial land connection then and at various later times during the northward drift of the Australian plate, during which time the New Caledonia-New Zealand region was split off by the opening of the Tasman Sea about 130 million years ago. New Caledonia's volatile geological history of foldings, metamorphoses and partial submersions was culminated during the upper Eocene by the slow outflow of peridotites that eventually covered most of the island with layers up to 2000 meters thick. Volcanic activity ceased in the Oligocene, but much of the earliest flora was eliminated and the introduction (by long-range dispersal) of such modern families as the Compositae was precluded by the ultrabasic substrate. The isolation of the island by at least the end of the Cretaceous is invoked to explain the absence of native land mammals and the existence of the ancient flora, with the latter's survival and evolution a result of the consequent lack of competition.

II. Description of the Region

A. Geographical Extent and Area

A French territory, the island of New Caledonia lies diagonally across longitudes 163°E to 167°E from 19° to 23°S, comprising 16,750 km². The main island, La Grande Terre, is about 400 km long and averages 50 km wide. The Isle of Pines lies at the foot of the so-called mainland at 170°E, while about 100 km to the east (169°-170°E and 20°-23°S) are the uplifted coral atolls that comprise the Loyalty Islands. Other small, low islands are scattered to the west of La Grande Terre in the Coral Sea. The New Caledonia complex, marking the southern extent of the Melanesian archipelago, is situated about 1200 km east of Queensland, Australia, 1600 km north of New Zealand, 1800 km southeast of New Guinea, 1250 km west of Fiji, and 350 km southwest of the New Hebrides.

B. Geology and Topography

New Caledonia is surrounded by a barrier reef second only to that of Australia in size. The island is marked by great geological and edaphic diversity. From the coast, the land rises to a double chain of central mountains, with Mt. Panié (1650 m) being the highest peak. The east coast intercepts the prevailing trade winds, producing about 2000 mm of rainfall annually, whereas the westward facing slopes and west coast savannas receive less than 1000 mm of rain per year. The interior mountains get more than 3000 mm of annual rainfall. About 22% of the total surface area of the territory, mainly the central mountains and northeast coast, is covered with rain forest, comprising 250,000 to 300,000 ha on La Grande Terre and some 100,000 ha on the Loyalty Islands and the Isle of Pines. Ultrabasic substrates and their derivative lateritic soils now comprise about one-third of the island's surface and were probably once much more extensive.

C. Population

The islands have a rapidly expanding population of about 140,000, of whom some 65,000 live in the capital city of Nouméa. Rural Melanesians generally engage in subsistence agriculture on the 6% of the land that is cultivable. Nickel constitutes 95% of the island's exports, and nickel mining employs about 16% of the labor force. Mining companies hold mineral rights to some 466,000 hectares.

III. Vegetation Maps

The standard vegetation map of the islands is that of the Atlas de la Nouvelle-Calédonie (Map 15, Morat et al., 1981).

IV. Floristic Inventory

A. Collections and Publications

About 40,000 specimens of cryptogams and phanerogams are held at the ORSTOM herbarium in Nouméa, founded in 1964, and another 90,000 are distributed abroad, principally in Paris.

Many works have considered New Caledonia's flora in terms of its greatly mixed floristic affinities, its botanical relicts and endemic species, and in general the island complex's unique importance for probing the phylogeny of angiosperms.

Publication in fascicle form of *La Flore de la Nouvelle-Calédonie et Dépendences* by the Paris Museum of Natural History commenced in 1967. In Parts 1–11 (1982) the following families were covered: Acanthaceae, Amborellaceae, Apocynaceae, Atherospermataceae, Bignoniaceae, Boraginaceae, Chloranthaceae, Corynocarpaceae, Elaeocarpaceae, Epacridaceae, Flacourtiaceae, Icacinaceae, Lauraceae, Monimiaceae, Orchidaceae, Proteaceae, Sapotaceae, Solanaceae, Symplocaceae and Trimeniaceae. Four gymnosperm families have also been described. In all, revisions of about 65% of the rain forest flora have been published in the *Flore* and elsewhere.

R. Schlechter conducted early floristic studies of the island at the beginning of the century, and Rendle, Baker and Moore published in 1921 a systematic account of the angiosperm collections of R. H. Compton made in 1914. Much of the groundwork for the flora was laid by A. Guillaumin whose massive contributions were published between 1911 and 1974. In 1914 Guillaumin issued the first work in his long series "Matériaux pour la flore de la Nouvelle-Calédonie" consisting of family and generic revisions. Parts 1–85 of the "Matériaux" were indexed in 1946 in Bull. Soc. Bot. France 92. Guillaumin also authored a series of 130 Contributions consisting of collection lists and local floras. In 1948 he published the now outdated *Flore Analytique et Synoptique de la Nouvelle-Calédonie*. His report on the results of the 1950–1952 French-Swiss botanical expedition was published in five parts, the last posthumously, between 1957 and 1974.

Besides those who have contributed to the published volumes of the Flore, scholars currently working on various families include J. W. Dawson (WELYU), Myrtacae; T. G. Hartley (CANB), Rutaceae; Lowry (MO), Araliaceae; G. D. McPherson (MO), Euphorbiaceae; Pennington (K), Sapotaceae; and W. Vink (L), Winteraceae.

Almost all parts of the island complex are well collected, with 50–100 collections per 100 km^2. Most families and genera are well represented in collections, although more collecting is desirable for rare species and those of especial taxonomic interest. Efforts to explore the less-known and less accessible areas of the islands are continuing.

B. Endemism

The rich flora of New Caledonia is highly endemic, with specific endemism estimated at 76% (2474) of the 3256 native species so far described. Of the 787 native genera studied, 108 (14%) are endemic, and 16 of those have 25 or more species endemic to New Caledonia. Five of the 182 families of phanerogams are endemic: Oncothecaceae, Phellinaceae, Strasburgeriaceae, Amborellaceae and Paracryphiaceae. Of the fern flora, 42% is endemic. According to Morat et al. (1984), the rain forest contains 1511 species of seed plants in 365 genera and 108 families. All five endemic families are native to rain forest formations.

C. Extinction

New Caledonia's ultrabasic peridotites and serpentines are high in nickel, magnesium, chromium and manganese, consequently attracting extensive mining interests. Through the 1960s mining, prospecting and lumbering decimated large areas of rain forest and botanically unique locales, and threatened many of the remaining areas. The islands' rapidly growing population has also resulted in large-scale conversion of forest to agricultural and grazing uses, with attendant burning and overgrazing, especially on the western savanna. There is little undisturbed nonforest vegetation. However, much has been done to preserve what remains of New Caledonia's original flora through the establishment of conservation areas, and there is now little serious threat to botanically interesting areas.

V. Resources for Continued Floristic Inventory

In addition to maintaining the herbarium at Nouméa, the French Office de la Recherche Scientifique et Technique Outre-Mer (ORSTOM) supports its own collectors and expeditions, as does the Centre National de la Recherche Scientifique (CNRS). The Missouri Botanical Garden has maintained a collector in New Caledonia since 1980. Thus, the current institutions and personnel, if maintained, are sufficient to complete the floristic inventory of the islands.

VI. Selected References

An asterisk marks those works actually cited in the text.

Aubréville, A. 1965. Les reliques de la flore des Conifères tropicaux en Australi et en Nouvelle-Calédonie. Adansonia **5**: 481–492.

———. 1975. La flore Australo-papoue. Origine et distribution. Adansonia **15**: 159–170.

———. 1976. Centres tertiaires d'origine, radiations et migrations des flores angiospermiques tropicales. Adansonia **16**: 297–354.

——— et al., eds. 1967–. Flore de la Nouvelle-Calédonie et Dépendances. Fasc. 1–. Mus. Natl. de Histoire Naturelle; Paris.

Balgooy, M. M. J. van. 1960. Preliminary plant geographical analysis of the Pacific. Blumea **10(2)**: 385–430.

———. 1971. Plant geography of the Pacific, as based on a census of Phanerogam genera. Blumea, Suppl. **6**: 1–222.

Bernardi, L. 1979. The New Caledonian genera of Araliaceae and their relationships with those of Oceania and Indonesia. Pages 15–325 in K. Larsen & L. B. Holm-Nielsen (eds.). Tropical botany. Academic Press, London.

Centre Technique Forestier Tropical. 1975. Inventaire des resources forestières de la Nouvelle-Calédonie. CTFT Nogent-sur-Marne (227 pp., mimeo).

Däniker, A. U. 1932–1933. Ergebnisse der Reise von Dr. A. U. Däniker nach New-Caledonien und den Loyalty-Inseln (1924/6) Katalog der Pteridophyta und Embryophyta siphonogama. Vierteljahrsschr. Naturf. Ges. Zürich **77**: 1–235; **78**: 237–395, 397–507.

Good, R. 1964. The geography of the flowering plants. Longmans, London.

Griffith, J. R. 1975. New Zealand and the southwest Pacific margin of Gondwanaland. Pages 619–637 in K. S. W. Campbell (ed.), Gondwana Geology. Australian National University Press; Canberra.

Guillaumin, A. 1948. Flore analytique et synoptique de la Nouvelle-Calédonie. Phanérogames. Office de la Recherche Scientifique Coloniale, Paris.

———. Les caractères de la végétation néo-calédonienne. Compt. Rend. Sommaire Séances Soc. Biogéogr. **41**: 67–74.

———. Résultats scientifiques de la mission franco-suisse de botanique en Nouvelle-Calédonie (1950–1952). Mém. Mus. Natl. Hist. Nat. Sér. B, Bot. **8**, Part I (1957): 1–120; Part II (1962): 193–330; **15**, Part III (1964): 1–96; Part IV (1967): 97–132, **22**, Part V (1974): 1–36.

Halloway, J. D. 1979. A survey of the Lepidoptera, biogeography and ecology of New Caledonia. The Hague; W. Junk.

Jaffré, T. 1974. La végétation et la flore d'un massif de roches ultrabasiques de Nouvelle-Calédonie: Le Koniambo. Candollea 29: 427–456.

―――― & **M. Latham.** 1974. Contribution à l'étude des relations sol-végétation sur un massif des roches ultrabasiques de la côte ouest de la Nouvelle-Calédonie: Le Boulinda. Adansonia, ser. 2, **14**: 311–336.

*****Morat, P., J.-M. Veillon & H. S. MacKee.** 1984. Floristic relationships of New Caledonian rain forest phanerogams. Pages 71–96 in F. J. Radovsky, P. H. Raven & S. H. Sohmer (eds.), Biogeography of the Tropical Pacific. Assoc. of Systematics Collections and B. P. Bishop Museum, Lawrence, Kansas.

ORSTOM. 1981. Atlas de la Nouvelle-Caledonie. ORSTOM, Paris.

Raven, P. H. 1979. Plate tectonics and Southern Hemisphere biogeography. Page 1–24 in K. Larsen (ed.), Tropical botany. Academic Press, London.

*****Rendle, A. B., E. G. Baker & S. Moore.** 1921. A systematic account of the plants collected in New Caledonia and the Isle of Pines by Prof. R. H. Compton, M. A., in 1914. Part I. Flowering plants (Angiosperms). Proc. Linn. Soc., Bot. **45**: 245–417.

Sarlin, P. 1954. Bois et forêts de la Nouvelle-Calédonie. CYFT Nogent-sur-Marne (3 maps).

Schlechter, R. 1905. Pflanzengeograpische Gliederung der Insel Neu-Kaledonien. Bot. Jahrb., **36**: 1–41.

――――. 1907–08. Beiträge zur Kenntnis der Flora von Neu-Kaledonien. Bot. Jahrb. **39**: 1–274; **40**: 20–45.

Schmid, M. 1967. Note sur la végétation des Iles Loyauté. Nouméa: ORSTOM (70 pp., mimeo).

Steenis, C. G. G. J. van. 1955. Some notes on the flora of New Caledonia and reduction of *Nouhuysia* to *Sphenostemon*. Svensk Bot. Tidskr. **49**: 19–23.

――――. 1971. *Nothofagus*—Key genus of plant geography, in time and space, living and fossil, ecology and phylogeny. Blumea **19**: 65–98.

*****Thorne, R. F.** 1963. Floristic relationships of New Caledonia. Stud. Nat. Hist. Iowa Univ. **20(7)**: 1–64.

――――. 1969. Floristic relationships between New Caledonia and the Solomon Islands. Philos. Trans., Ser. B. **255**: 595–602.

――――. 1972. Major disjunctions in the geographic ranges of seed plants. Quart. Rev. Biol. **47**: 366–411.

Virot, R. La végétation canaque. Mém. Mus. Nat. Hist., N.S., Bot., **7**: 1–398.

Whitmore, T. C. 1984. Tropical rain forests of the Far East. Clarendon Press, Oxford.

Hawaiian Islands

D. Frame, W. L. Wagner, D. R. Herbst and S. H. Sohmer

Contents

I. Description of the Region	182
A. Geographical Extent and Area	182
B. Topography and Geology	182
C. Climate	182
D. Population	182
II. Vegetation Maps of the Region	182
III. Resources for Inventory	183
A. Herbaria	183
B. Publications	183
IV. State of Inventory	183
A. Endemics	184
B. Extinctions and Threatened Species	184
V. Future of Inventory	184
VI. Acknowledgments	185
VII. Literature Cited	186

I. Description of the Region

A. Geographical Extent and Area

The Hawaiian Archipelago is a complex of 132 volcanic islands, shoals, atolls and reefs that extend 2,451 kilometers southeast to northwest across the Tropic of Cancer. The eight main islands, Hawai'i, Maui, Kaho'olawe, Lana'i, Moloka'i, O'ahu, Kaua'i, and Ni'ihau, account for over 99% of the total land area of 16,634 sq. km. Small islands off the shores of the main islands and the Northwest Hawaiian Islands make up the remaining 15 sq. km of land (Armstrong, 1973).

Although relatively small in terms of total land area, the Hawaiian Archipelago comprises one of the longest mountain ranges on earth. The islands were created almost entirely by volcanic activity, and each one represents the top of a massive volcanic mountain.

B. Topography and Geology

The topography of the Hawaiian islands is a result of construction by volcanoes, living organisms, and sedimentary processes on the one hand, and of destruction by erosion on the other (Abbott et al., 1981). The major relief forms are those created by volcano building. Ni'ihau, the lowest of the main islands, is a volcanic mountain rising over 3,960 m above its ocean floor base. Intermediate relief elements are products of constructive and destructive forces. Cinder cones and tuff cones are formed by moderately explosive volcanic eruptions, and these topographic features may reach heights of nearly 300 m above their surroundings. Pu'u Makanaka, on Mauna Kea, is a 183 m high cinder cone; Diamond Head in Honolulu is a tuff cone nearly 245 m high. Valleys ranging in depth from several meters to greater than 600 m are formed by steam erosion. Hi'ilawe Falls, in Waipi'o Valley on the Island of Hawai'i has a vertical drop of about 300 m, making it one of the highest free falls in the world. Sea cliffs, carved by waves, range from a few to over a thousand meters. One of the highest sea cliffs in the world, 1,097 m high, is found on the north side of East Moloka'i.

C. Climate

Two patterns of rainfall are discernible for the Hawaiian Archipelago; the winter tropical storms occurring from November to April and the summer rains from May to October.

When the trade winds, which blow from the north or northeast, slacken during the winter months, storms originate from the south. These are known as "Kona" storms; kona is Hawaiian for 'leeward'. The Kona rains are generated by the typical pattern of a cold front meeting warm moist air and they provide a general rainfall. In sharp contrast, the summer rains are trade wind generated and exhibit a highly topical distribution.

During the summer, the large volcanoes of the Hawaiian Archipelago obstruct the flow of warm, moisture-laden air leading to orographic rain. In this sense, the Hawaiian Islands make their own weather. Where the mountains are enveloped by clouds, bogs and marshes develop, lush tropical forests grow below the cloud layer, and arid lands result in the leeward areas and where the mountains emerge above the clouds.

Summer rainfall distribution correlates well with topographic contours. Rain generated by the trade winds falls mostly on slopes and crests of mountains below 1,830 m. The highest recorded annual rainfall, averaging 11,938 mm a year, is known from Mt. Wai'ale'ale, which rises 1524 m above sea level. Over 1,830 m, on the slopes of the highest mountains, a lower annual rainfall is usually found. This is because an atmospheric inversion layer is created which effectively prevents clouds from rising higher; the clouds must split and move around the mountains rather than rise over the summits (Abbott et al., 1981).

Excellent maps depicting annual rainfall in the Hawaiian Islands may be found in Armstrong (1973). Nevertheless, the rainfall in uninhabited or inaccessible areas remains largely unknown.

Island temperatures reflect small seasonal variation in solar energy as tempered by the surrounding ocean. Temperatures generally range from 16–27°C throughout the year. Seasonal diurnal variation often is less than differences between day and night temperatures, which may differ by as much as 12°C, lending support to the statement that "nighttime is the winter of the tropics" (Blumenstock & Price, 1972).

D. Population

Given such an equable climate, it is not surprising that certain regions of the Hawaiian Islands are well populated. With the exception of Hilo on the island of Hawai'i, all major urban centers are on O'ahu. The basic focus of human population has centered on O'ahu because the unique geology of the area permits the accumulation of a great deal of underground fresh water.

Total resident population for the state is at about 900,000 and is expected to increase to 1.2 million by the end of the century, a modest increase by today's standards. However, because the population will be concentrated on the relatively small island of O'ahu, this will have considerable impact on local natural resources.

II. Vegetation Maps of the Region

Most of the literature on Hawaiian vegetation has been descriptions of rather broadbrush vegetation zones. Many of the published papers (e.g., Egler, 1939; Knapp, 1965; Krajina, 1963) do not include any detailed maps. The earliest, and still the only complete, mapping of Hawaiian vegetation is by Ripperton and Hosaka (1942) and it includes maps and brief descriptions of the vegetation zones. Although this work is widely utilized and quoted in many other scientific works and other sources such as the "Atlas of Hawaii" (Armstrong, 1973), it is widely admitted that the maps are too broadbrush and not really adequate for detailed description of Hawaiian vegetation. A number of other papers (e.g., Fosberg, 1972; Mueller-Dombois & Krajina, 1968) have rather detailed reports of specific areas. More recently, Mueller-Dombois and Fosberg (1974) have produced a series of detailed vegetation maps for Hawai'i Volcanoes National Park. Most recently, as an outgrowth of the U.S. Fish and Wildlife Service's "Forest Bird Survey," James Jacobi has been preparing detailed vegetation maps for the islands of Hawai'i, Moloka'i, Maui, and parts of Kaua'i. One map (Jacobi, 1978) has been published for the Kau Forest Reserve and adjacent lands on the island of Hawai'i. At the request of the Bishop Museum flora project, W. Gagné and L. Cuddihy have undertaken a compilation of vegetation types. Working in cooperation with the Heritage Program of the Nature Conservancy, Hawai'i, they have produced a draft classification of Hawaiian vegetation communities which will be published as a chapter in the Bishop Museum's "Manual of flowering plants of Hawaii" (Wagner, Herbst, & Sohmer, in prep.)

Table I
Summary of collections at Herbarium Pacificum.*

Collection	Estimated Contents as of 31 December 1985
Types	7,200
Hawaiian Phanerogams	87,300
Hawaiian Ferns	11,500
World Phanerogams	146,800
World Ferns	17,400
Bryophytes	37,100
Fungi	14,000
Lichens	4,800
Algae, pressed	65,900
Algae, wet	15,000
Algae, coralline	1,000
Vascular wet collections	2,100
Wood	5,700
Fruit	2,000
Oversize, *Pandanus*	5,000
Total	422,800

* S. H. Sohmer & P. O'Connor, pers. comm.

III. Resources for Inventory

A. Herbaria

The Herbarium Pacificum (BISH), at the Bernice P. Bishop Museum in Honolulu, has over 400,000 plant specimens making it the largest herbarium in Hawai'i. A breakdown of their plant collections is presented in Table I. The Department of Botany at the University of Hawaii (HAW) contains 20,000–30,000 specimens, the Lyon Arboretum of the same University (HLA) contains about 7,000 specimens, and the Pacific Tropical Botanical Garden (PTBG) houses about 3,000 specimens. Outside Hawai'i there are significant collections of Hawaiian plants at CU, F, GH, MASS, NY, POM, RSA and US. In addition, there is a substantial amount at the Herbarium of the Botany Department at the University of California, Berkeley (UC). Much of the Hawaiian material in these herbaria are duplicates of material at BISH. Abroad, BM, GB, K, LD, P, S, and W have historical collections from the Hawaiian Islands, those of BM, K, P, and W from activities of the 19th century exploring expeditions, and GB, LD and S because of the activities of Carl Skottsberg during the 1930's and 1940's. If there are now approximately 87,000 specimens of Hawaiian plants at BISH, it is likely that there are at least as many held in the herbaria mentioned. Most of these, as noted before, are duplicates of material at BISH.

B. Publications

There are very few bonafide floristic publications dealing with the Hawaiian flora, although much monographic material exists. The first, oldest, and to this day the best, is Hillebrand's *Flora of the Hawaiian Islands*, published in 1888. Recently reprinted, *The indigenous trees of the Hawaiian Islands* (Rock, 1913) is a useful sourcebook for arborescent plants. Degener's *Flora Hawaiiensis* (1933–1963) is incomplete and sometimes spotty in its treatment; it remains valuable because the information is founded upon Degener's extensive field experience, and coupled with detailed illustrations of nearly all treated species (see Mill et al., 1985). It represents the only endeavor, other than Hildebrand's flora, to treat the flora of these islands in a systematic way. St. John's *List and summary of the flowering plants in the Hawaiian Islands* (1973) is an extremely useful tool for obtaining access to the published names of the native and introduced plants in Hawaii. Marie C. Neal's *In Gardens of Hawaii* revised in 1965, has been, in lieu of other publications, one of the most important references to names and ethnobotanical uses of the indigenous flora (if they were cultivated at any time, the plants are in this book). As usual, there are a large number of "tourist" guides of a simplistic nature. The *Manual of Wayside Plants of Hawaii* by Willis T. Pope has its uses, but this is not a systematic treatment of even all the then known (1968) wayside (i.e., weed) plants.

IV. State of Inventory

In general, there are inaccessible areas on all the islands, particularly Kaua'i, Maui, Moloka'i and Hawai'i, that have been poorly collected. When estimating the completeness of floristic inventories in the Hawaiian Archipelago, typical calibrations of 50–100 collections/100 sq. km for relatively well collected areas and less than 50 collections/100 sq. km would suggest that the state has been very well collected. In point of fact, while one area of less than ten sq. km may be extremely well collected, over a ridge and down the other side an area may exist that is relatively poorly collected and which may contain many rare endemics.

The most commonly collected areas in Hawai'i are ridgetops and valley floors. These areas have well-used hiking trails that are accessible within a single day's hike. Lists of these hiking trails are available from the state and there are a number of popular guide books also available. Botanists in Hawai'i have for the most part preferred to collect in these areas because they are more accessible, and for this reason these areas have been returned to repeatedly and thus have been well collected (and sometimes over collected!). Areas less well collected can be placed in one of three categories: 1) the much less accessible, steep or nearly vertical slopes off of ridges, 2) the high plateau areas in the interior parts of some of the islands such as Kaua'i and Moloka'i, and 3) private lands where access is restricted; and entry into public lands, reached only via private, may be curbed (e.g., portions of floristically rich southwestern Kaua'i).

Much of the information in the following section is adapted from "Status of the flowering plants of the Hawaiian Islands" by Wagner, Herbst & Yee (1985).

A. Endemics

Considering the fact that Hawai'i is one of the most isolated island groups in the world, it is not surprising that approximately 90% of the flowering plant species are endemic. There are about 30 endemic genera as well as a few genera such as *Charpentiera* (Amaranthaceae) which have a relatively limited distribution elsewhere, resulting in approximately 16% endemism at the generic level. However, only 22% of the flowering plant families of the world are native to Hawai'i, and none are endemic. Those unexplored areas of Hawai'i that retain native vegetation may be expected to harbor many genera and species, some possibly as yet undescribed. It is equally likely that these potentially new taxa will be endemics.

B. Extinctions and Threatened Species

Fosberg and Herbst (1975) published a list of rare and endangered species; they estimated that 899 species could be so classed. Clearly the numbers will vary according to species concepts. As of 1980, about 150–200 native Hawaiian vascular plants were known or thought to have become extinct. This is roughly equivalent to 10–15% of the flora. Information from an ongoing but incomplete reevaluation suggests that, after minor variants are synonymized, about 80 species are presumably extinct, approximately 8% of the flowering plants (Wagner & Herbst, unpubl. data). An analysis of the endangered or threatened flowering plants published as a "notice of review" in the December, 1980 United States Federal Register, suggests that there is a definite pattern to their distributions both by island and by habitat (Wagner et al., 1985). O'ahu, the center of urban development, has the greatest percentages of such taxa (27%), followed by Hawai'i (18%), Maui (16%), and Kaua'i (14%). The small islands of Ni'ihau and Kaho'olawe have relatively low percentages each of about 1%. This belies a low habitat diversity and the fact that the native biota was greatly reduced prior to the earliest inventories. The ecological zones currently experiencing the severest degradation are the low to mid elevational ones such as mixed mesophytic forest and rain forest. Coastal and dryland vegetation experienced such an early degradation from human activity that the extent of change is a matter of conjecture.

The introduction of hooved animals in the latter part of the 18th century, followed by habitat elimination as a result of agricultural use and urbanization, has led to the progressive decline of the native Hawaiian ecosystems. Logging has not played an important role in the destruction of forests in Hawai'i as it has in other tropical regions; however, several places on the island of Hawai'i are currently threatened with logging. Perhaps the most serious problems with regard to destruction of Hawaiian tropical forest relate to the introduction of plant and animal pests. Conversion of forests for military and agricultural purposes also represents a threat.

V. Future of Inventory

The Bishop Museum is probably the major institution dedicated to a continuing floristic inventory. Other institutions and/or agencies that have performed inventory work in their jurisdiction include:
(1) U.S. Fish and Wildlife Service.
(2) Hawaii Department of Land and Natural Resources, Division of Forestry and Wildlife,
(3) The Nature Conservancy Hawaii Natural Heritage Program,
(4) University of Hawaii, Departments of Botany and Geography, and Lyon Arboretum, and
(5) Pacific Tropical Botanical Garden, Lawai, Kaua'i.

There is a collaborative spirit among individuals belonging to each of these organizations. BISH is the focal point, due to the availability of the collections, for floristic inventory work. Researchers at the Pacific Tropical Botanical Garden are primarily concerned with the collection and cultivation of rare or endangered indigenous plants; their aims frequently overlap with BISH and studies often involve researchers from both institutions.

A project nearing completion at the Bishop Museum, funded by the Irwin Charity Foundation of San Francisco and the National Science Foundation (NSF) is one to inventory the flowering plants of Hawai'i. The results will be published in a single volume, *Manual of the flowering plants of Hawaii* (Wagner, Herbst, & Sohmer, in prep.). The aim of the *Manual* project is to place the Hawaiian flora in a sound and consistent taxonomic framework; completion of the project will mark a major step towards a realistic floristic

Table II

Individuals contributing treatment to manual of the flowering plants of Hawai'i.

Person(s)	Plant group(s)
L. Albert de Escobar	Passifloraceae
F. Almeda	Melastomataceae
D. Austin	Convolvulaceae
D. Bates	Malvaceae
G. Carr	Asteraceae (*Argyroxiphium, Dubautia,* & *Wilkesia*)
S.-M. Chaw	Rubiaceae (*Bobea*)
G. Chippendale	Myrtaceae (*Eucalyptus*)
J. Coffey	Juncaceae
L. Constance & J. Affolter	Apiaceae
T. Croat	Araceae
S. Darwin	Rubiaceae (*Bobea*)
G. Davidse	Poaceae (*Dichanthelium* & *Panicum*)
J. Dawson & L. Stemmerman	Myrtaceae (Metrosideros)
T. Duncan	Ranunculaceae
F. Fosberg	Nyctaginaceae
F. Ganders & K. Nagata	Asteraceae (*Bidens*)
R. Geesink	Fabaceae (except *Canavalia, Chamaecrista, Crotalaria, Desmodium, Erythrina,* & *Lathyrus,* & *Senna*)
P. Goldblatt & J. Henrich	Iridaceae
J. Hayden	Euphorbiaceae (*Flueggea*)
R. Hobdy	Musaceae
M. Huft	Euphorbiaceae (introduced *Chamaesyce, Euphorbia* & *Mallotus*)
A. Jones	Asteraceae (*Aster*)
D. Koutnik	Euphorbiaceae (native *Chamaesyce*)
T. Koyama	Cyperaceae
J. Kress	Heliconiaceae
T. Lammers	Campanulaceae
T. Lowrey	Asteraceae (*Tetramolopium*)
P. Lowry	Araliaceae
S. Mill	Asteraceae (*Lagenifera*) & Zygophyllaceae
K. Nagata	Costaceae, Marantaceae & Zingiberaceae
D. Neill	Fabaceae (*Erythrina*)
P. O'Connor	Poaceae (except *Dichanthelium,* & *Panicum*)
H. Ohashi	Fabaceae (*Desmodium*)
R. Patterson	Goodeniaceae
T. D. Pennington	Sapotaceae
B. Peterson	Thymelaeaceae
R. Read & D. Hodel	Arecaceae
L. Schlutz	Asteraceae (*Artemisia*)
S. Skinner & D. Windler	Fabaceae (*Crotalaria*)
J. Solomon	Cactaceae
B. Stone	Pandanaceae & Rutaceae
D. Symon	Solanaceae
I. Telford	Cucurbitaceae
S. P. Vander Kloet	Ericaceae
S. Weller & A. Sakai	Lamiaceae (*Stenogyne*)
D. Windler & S. Skinner	Fabaceae (*Crotalaria*)
H. van der Werff	Lauraceae

evaluation of Hawaii. Botanists are contributing treatments on their areas of expertise. The individuals collaborating with the *Manual* project are listed in Table II. This manual, of which publication is projeccted for 1989, will be the best summary of the Hawaiian flora since the time of William Hillebrand (19th century). With keys, descriptions and illustrations, it will summarize what is known to species level, with less formal discussion of intraspecific taxa and with full bibliographic citations.

Nonetheless, at present there is only one systematic program to inventory poorly known (often inaccessible) areas. Not since the time of Charles Forbes, who worked at BISH from 1908–1920, has there been a systematic effort to send individuals to particular areas. The current project, funded by the Pacific Tropical Garden, is the systematic inventory of the Wai'anae and Ko'olau mountains, O'ahu, by S. Pearlman. In the future, the inventory work will be expanded to include floristically interesting places on other islands. Once the *Manual* is completed, scientists at the Bishop Museum intend to turn their attention to collaborative studies of poorly understood floristic regions, as well as in depth studies of particular genera.

The additional resources necessary to insure a floristic inventory of the sometimes inaccessible forests will consist of the following:

(1) Funds to get personnel into appropriate areas. This will often involve helicoptering people and supplies.
(2) Funds to hire two to three additional temporary staff to help collect, curate and distribute specimens.

The estimated travel and curation costs for this inventory are shown in Table III.

VI. Acknowledgments

We are greatly obliged to R. Barneby and D. Lorence for critical reading of the manuscript.

Table III

Travel and curation costs for a three-year inventory.

	Year 1	Year 2	Year 3
I. Personnel			
a. 2 Technicians	$30,000	$32,000	$37,950
b. Secretarial support	10,000	10,800	13,600
c. Fringe benefits (20%)	8,000	8,556	10,310
II. Equipment and supplies (including 5 herbarium cases each year)	10,000	10,000	10,000
III. Travel, per diem, vehicle and helicopter rental for 7–8 trips per year of 2–3 weeks duration each	17,500	17,500	20,125
Annual Total	$75,500	$78,856	$91,985
TOTAL = $246,341			

VII. Literature Cited

Abbott, A. T., E. A. Kay, C. H. Lamoureux & W. L. Theobald. 1981. Natural landmarks survey of the Hawaiian Islands. Manuscript prepared for the National Parks Service Natural Landmarks Program, Department of the Interior.

Armstrong, R. W. (ed.). 1973. Atlas of Hawaii. University Press of Hawaii, Honolulu.

Blumenstock, D. I. & S. Price. 1972. Climates of the states: Hawaii. *In* Kay, E. A. (ed.). A natural history of the Hawaiian Islands. University of Hawaii, Honolulu.

Degener, O. 1933–1963. Flora Hawaiiensis. 6 Vols. Honolulu. (Vol. 6 by O. Degener & I. Degener. Additional vols. in preparation.)

Egler, F. E. 1939. Vegetation zones on Oahu, Hawaii. Empire Forest. J. **18**: 44–57.

Fosberg, F. R. 1972. Guide to Excursion III. *In*: Tenth Pacific Sci. Congr., revised ed. University of Hawaii, Honolulu.

―――― **& D. Herbst.** 1975. Rare and endangered species of Hawaiian vascular plants. Allertonia **1**: 1–72.

Jacobi, J. D. 1978. Vegetation map of Kau Forest Reserce and Adjacent lands, Island of Hawaii. Pacific Southw. Forest Range Exp. Sta. Resource Bull. PSW-16. U.S.D.A. Forest Service.

Hillebrand, W. 1888. Flora of the Hawaiian Islands: A description of their phanerograms and vascular cryptogams. 1981 fascimile ed. Lubrecht & Cramer. Monticello, New York.

Knapp, R. 1965. Die vegetation von Nord- und Mittelamerika und der Hawaii-Inseln. Fisher, Stuttgart. (translated and published in Newslett. Hawaiian Bot. Soc. **14(5)**: 95–121. 1975.)

Krajina, V. J. 1963. Biogeoclimatic zones of the Hawaiian Islands. Newslett. Hawaiian Bot. Soc. **2**: 93–98.

Mill, S. W., W. L. Wagner & D. R. Herbst. 1985. Bibliography of Otto and Isa Degeners' Hawaiian Floras. Taxon **34**: 229–259.

Mueller-Dombois, D. & F. R. Fosberg. 1974. Vegetation map of Hawaii Volcanoes National Park. Cooperative Natl. Park Resources Studies Unit, Hawai'i. Techn. Rep. **4**: 1–44.

―――― **& V. J. Krajina.** 1968. Comparison of east-flank vegetation on Mauna Loa and Mauna Kea, Hawaii. *In* R. Misra & B. Gopal (eds). Recent advances in tropical ecology II.

Neal, M. C. 1965. In gardens of Hawaii. Bernice P. Bishop Mus. Spec. Publ. **50**: 1–924.

Pope, W. T. 1968. Manual of wayside plants of Hawaii. C. E. Tuttle Co. Rutland, Vermont.

Ripperton, J. C. & E. Y. Hosaka. 1942. Vegetation zones of Hawaii. Hawaii Agri. Exp. Sta. Bull. **89**: 1–60.

Rock, J. F. 1913. The indigenous trees of the Hawaiian Islands. Pacific Tropical Botanical Garden, Lawai, Kauai, Hawaii and Charles E. Tuttle Co., Rutland, Vt., and Tokyo, Japan. (Reprinted, 1974).

St. John, H. 1973. List and summary of the flowering plants in the Hawaiian Islands. Published by the Pacific Tropical Botanical Garden, Lawai, Kauai, Hawaii.

Wagner, W. L., D. R. Herbst & R. S. N. Yee. 1985. Status of the native flowering plants of the Hawaiian Islands. *In* C. Stone & J. M. Scott (eds.), Hawai'i's terrestrial ecosystems: Preservation and management. University of Hawaii Press, Honolulu.

Regional Reports

III. Africa

West Africa (Sierra Leone to Nigeria)

F. Nigel Hepper

Contents

I. Description of the Region	190
A. Geographical Extent and Topography	190
B. Forest Region	190
C. Population	190
II. Vegetation Maps	190
A. Regional Maps	191
B. Country Maps	191
III. Herbarium Resources from West Africa	191
A. Herbaria	191
B. Publications	192
IV. Completeness of Floristic Inventory	192
A. Areas Relatively Well Collected	193
B. Areas Poorly Collected	193
C. Areas High in Endemics	193
D. Areas with High Rates of Extinction	193
V. Tropical Forest Areas Threatened by Destruction or Conversion	193
VI. Academic and Governmental Institutions in the Region	193
VII. Foreign Programs	194
VIII. Literature Cited	194
IX. General References	195

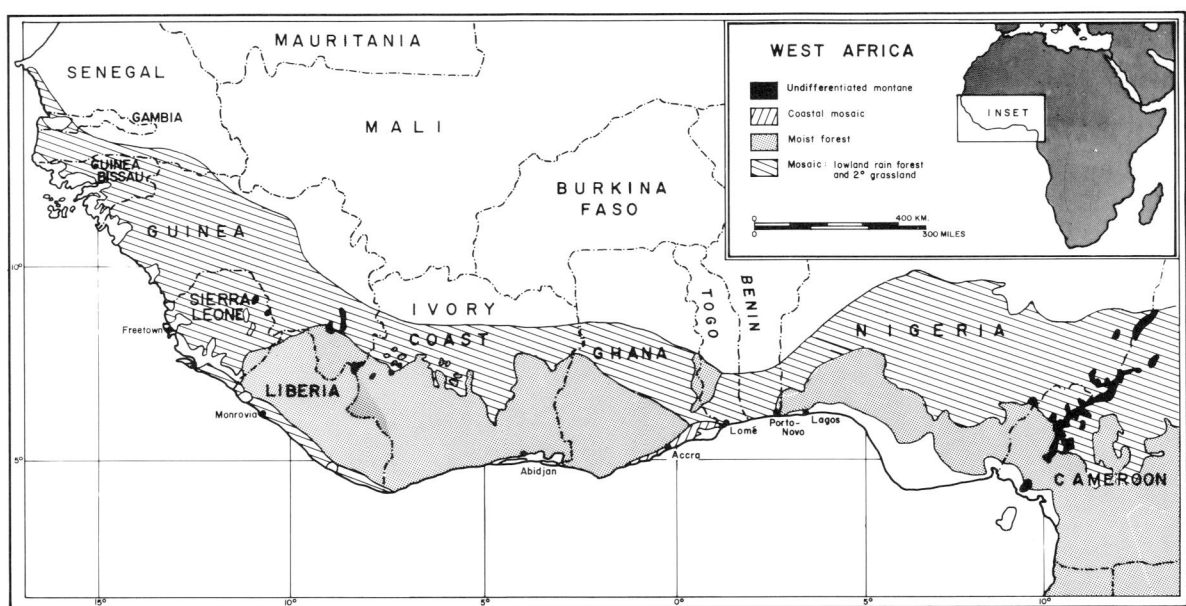

I. Description of the Region

A. Geographical Extent and Topography

West Africa lies generally between about 200 m and 500 m in altitude, and consists mainly of the worn, monotonous, and fairly level surfaces of the Pre-Cambrian rocks (Harrison Church, 1974). The forest zone extends from sea level northwards, still at a low altitude. Montane forest occurs as isolated patches on the scattered mountains (e.g., Guinée,: Fouta Djalon, Macenta, Nimba; Sierra Leone: Loma Mts.; Liberia: Nimba; Ivory Coast: Nimba, Mts. des Dans; Ghana: Togo Hills; Benin: Atacora; Nigeria: Jos Plateau, Obudu Plateau). Upland forest, intermediate between the lowland and montane forests, is distinguishable on many hills. Mangroves occupy areas along the coast, but they have diminished considerably.

B. Forest Region

This region extends from Senegal (Casamance) to the Nigeria/Cameroun frontier, an east to west distance of some 2700 km, and up to 400 km south to north from the Atlantic coast.

The present actual tropical forest in this region is a small fraction of what it was a century ago, indeed half a century or less, with deforestation continuing apace. 1980 FAO statistics, quoted by Leakey & Last (1983) for the above region, are shown in Table I.

It should be noted that these are overall figures for each country, and that over wide areas the forest has been entirely eliminated. Moist forest exists mostly in defined forest reserves of varying integrity. The fact that plantations appear to make up some of the loss is entirely irrelevant in the context of this volume, since the planted trees are mostly exotics, such as teak, *Gmelina* and pine, producing plantations with few indigenous species and lacking a natural forest structure.

C. Population

Cities are scattered along the coast as ports or fishing communities, but also elsewhere, for one reason or another. The second half of the twentieth century is seeing an urban population explosion. Agricultural villages occur throughout the forest zone and population density is highest in the eastern part of Nigeria. Settlement has affected the forest to such an extent that only remnant trees, or even just stumps, are all that indicate the former existence of forest. Only the presence of designated forest reserves has prevented a complete degradation. Forest reserves vary from country to country: Ghana has many, from small hilltop ones to extensive areas, while Ivory Coast has a few large reserves and Liberia is in the process of designating its first reserves (too late to save much).

II. Vegetation Maps

It is difficult to reconcile the UNESCO (1973) classification with that by White (1983a), also published by UNESCO. The latter has, therefore, been accepted, since the 1973

Table I
Statistics of deforestation in the west african region from Leakey & Last (1983).

	Total area of moist forest (ha)	Annual losses by deforestation (ha)	Total area of plantations (ha)
Sierra Leone	740,000	6,000	5,800
Guinea	(not recorded, insignificant)		
Liberia	2,000,000	41,000	6,300
Ivory Coast	4,458,000	310,000	44,900
Ghana	1,718,000	27,000	75,300
Togo	(not recorded, insignificant)		
Nigeria	5,950,000	285,000	163,300

classification is not widely used. White's text (1983b) to accompny the map should be consulted for detailed discussion.

Figure 1, derived from the relevant portions of White's map, shows the following categories. It must be stressed that nowhere is the forest continuous, as seemingly indicated by this map. (Underlined numbers are White's.)

1. Undifferentiated Afromontane Forest: *19a*.
2. West African Coastal Mosaic: *15* (little forest content).
3. Moist Forest: *1a, 2, 3*; Guineo-Congolian Rain Forest, both wet and dry types.
4. Mosaic: *11a, 12*; lowland rain forest and secondary grassland.

A. REGIONAL MAPS

White's UNESCO/AETFAT/UNSO Vegetation Map of Africa (1983a, 1983b) supersedes Keay's original AETFAT version (1958). White's map is on three sheets for the whole of Africa. West Africa, at a scale of 1:5,000,000, is on one sheet. A fourth sheet gives the explanations.

Distributions of many forest species appear on the loose-leaf folios *Distributiones plantarum africanarum*, published by the Jardin Botanique National de Belgique (Brussels, later Meise) (1969 onwards), and J.-P. Lebrun & A. L. Stork, *Index 1935–1976 des cartes de repartition des plantes vasculaires d'Afrique* (1977) and Hall & Swain (1981).

B. COUNTRY MAPS

1. Sierra Leone

Vegetation. No. 211/1951. 15 miles to the inch.
Vegetation map on p. 25 in Clarke (1966).

2. Liberia

Preliminary map. Scale 1:500,000. U.S. Economic Mission to Liberia, 1949, with major forest areas indicated.
Map of Liberia in Voorhoeve (1965), pages 11–12. Lists national forests with acreage.

3. Ivory Coast

Forêts Classées. Service des Eaux, Forêts et Chasses. Scale 20 mm = 30 km. 31 Dec 1956. In Aubréville, *Flore forestière de la Côte d'Ivoire, ed. 2*, 1:20 (1959).
Esquisse botanique de la Côte d'Ivoire, by G. Mangenot & J. Miège, ORSTOM. (1965). Scale 22 mm = 50 km.
Carte de la Végétaion de la Côte d'Ivoire, by J. L. Guillaumet & E. Adjanohoun. ORSTOM (1971). Scale 1:500,000.

4. Ghana

Vegetation Zones. Survey of Ghana 1959, in C. J. Taylor (1960). Scale 1:20,000,000.

5. Nigeria

Provisional map of vegetation zones. Survey Dept., 1953. Scale 1:3,000,000. Included in Keay (1953).
Vegetation and land use. Field studies 1977, Hunting Technical Services. Director of Forestry, PMB 12613, Lagos. Scale 1:250,000 (3 sheets).

III. Herbarium Resources from West Africa

A. HERBARIA

As a result of the long colonial period, important West African herbarium collections are held abroad: The major holdings are:

1. B (Berlin-Dahlem). This vast collection was largely destroyed in 1943. The remainder has now been increased by large scale collection in Togo and by exchanges received.
2. BM (British Museum, London). Important early collections, from the nineteenth century onwards, with type specimens, duplicates from elsewhere, and the first set of Talbot's plants from the Calabar area, Nigeria, are here.
3. BR (Brussels, now Meise). West African specimens are mostly duplicates from other herbaria. Its strength is in holdings from Zaire.
4. FHO (Oxford). Very important collections, mainly from forest, comprise perhaps one-third of the total number of sheets from West Africa (ca. 50,000 specimens).
5. K (Kew). The major collection for Anglophone West Africa, together with duplicates from other herbaria, it totals many thousand sheets.
6. MO (Missouri Botanical Garden). Although the main USA herbarium for African plants, West Africa representation is minimal, with duplicates from elsewhere, but new collections from Cameroun will alter this.
7. P (Paris). Francophone West Africa is, with the exception of the Ivory Coast, parts of Guinea and Senegal (Casamance), which are well represented, outside the forest zone.
8. WAG (Wageningen). This excellent modern herbarium has several thousand West African collections, mainly from Liberia and Ivory Coast, but also some from Ghana and Nigeria.

Table II

West African Herbarium Resources
(Number of herbarium specimens)

	Index Herbariorum (1981)	Estimation from other sources (1983)	Estimation of forest specimens (1983)
Sierra Leone			
Njala University College (SL)	21,000		8,000
Fourah Bay College (FBC)	12,000	6,000	2,000
Forestry Herbarium, Kenema (FHK)	2,857		2,000
Liberia			
Harley Herbarium, Monrovia (LIB)	7,000		5,000
Ivory Coast			
Laboratoire de Botanique, Abidjan (UCJ)	22,000	16,000	9,000
ORSTOM, Adiopodoumé (ABI)	50,000		3,000
Ghana			
Univ. of Science & Technology, Kumasi (KUU)	—	2,000	500
Univ. of Ghana, Legon (GC)	70,000	80,000	25,000
Forest Herbarium, Kumasi (KUM)	8,600	10,000	8,000
Forest Products Research Institute, Kumasi	—	500	100
Univ. of Cape Coast (CCG)	4,000	2,000	500
Nigeria			
Forest Herbarium, Ibadan (FHI)	100,000		40,000
Univ. of Ibadan (UCI)	17,000	(20,000 + sheets 15,000 gatherings of 3,300 species)	4,000
University of Ife (IFE)	25,000		15,000
Forestory Commission, Enugu	10,000		9,000

Families being worked on by specialist scholars are displayed in Table III.

B. Publications

Publications on the flora of West Africa are numerous. Important publications are listed in the References section, by region and by country.

IV. Completeness of Floristic Inventory

For an overview, see the attached map extracted from the second edition of the map of the extent of floristic exploration of Africa south of the Sahara (Hepper, Pages 157–162 *in* Kunkel, 1979).

Table III

Status of forest reserves in Ghana.

Forest Reserve	Reasons for Threat	Plants in Danger and other remarks
Atewa Range Area: 232 km^2	Area required for mining of bauxite; Government policy	Rare species of tree ferns; Sources of streams
Opon-Mansi Forest Reserve Area: 116 km^2	Area required for mining iron ore: Government policy	The tree cover; Sources of streams
Subri River Area: 580 km^2	Conversion to plantation of exotic tree species for producing wood for paper pulp: Government/FAO policy	Natural flora
Ankasa River Forest Reserve Area: 518 km^2	Part to be destroyed for a stone quarry for the construction of an international road linking Ivory Coast and Ghana	The center of a true rainforest type in the country (no month being rain free). Rich in species not found elsewhere in the country.

A. Areas Relatively Well Collected

With 50–100 collections/100 km², these areas are: 1) Coastal Sierra Leone; 2) Central Ivory Coast; 3) Around Kumasi in Ghana; 4) and certain forest reserves in Nigeria.

B. Areas Poorly Collected

Areas with less than 50 collections/100 km², and which potentially may yield new and important distributions, are: 5) Sierra Leone, Gola forest, adjoining the Liberian frontier; 6) most of liberia; 7) western forest of Ivory Coast, adjoining the Liberian frontier; 8) eastern forest of Ivory Coast, adjoining the Ghanaian frontier; 9) the highest rainfall forest of western Ghana, adjoining Ivory Coast; 10) parts of Nigeria, especially in the delta region and in the east, north of Calabar; and 11) the highest rainfall area of Cameroun, adjoining Nigeria.

C. Areas High in Endemics

Endemism is here defined as the number of endemics divided by the total number of species. These areas are: 1) parts of the Liberian forest (rich in Caesalpiniaceae); 2) the southeastern Ivory Coast; 3) southwestern Ghana; 4) southeastern Nigeria; 5) and southwestern Cameroun.

D. Areas with High Rates of Extinction

The Liberian forest recently has been severely depleted before adequate floristic exploration has been completed, with an unknown rate of extinction.

Forest clearance outside forest reserves has accelerated to such an extent that little remains in Ivory Coast, Ghana, and parts of Nigeria. This is exerting so much pressure on the forest reserves that the ecology is changing, with the increase of pioneer species and exotics crowding out native species. Extensive oil palm and coconut plantations in Ghana and elsewhere are annihilating the forest completely. Cocoa plantations are marginally less destructive, because shade trees are left.

V. Tropical Forest Areas Threatened by Destruction or Conversion

1) Guinée Republic: There is no recent information.

2) Sierra Leone: Almost all the remaining forest is threatened, according to 1983 information obtained from the Ministry of Agriculture and Forestry. Seriously threatened are: Tama Forest, Eastern Province; and Tonko Forest, Northern Province.

3) Liberia: Although no detailed information could be obtained specifically for this report, there is evidence from residents that wholesale clearance of the forests has taken place and little remains in many areas. The figure of forest area quoted from FAO (1980) in Table I is most probably too high. Instead of 2,000,000 ha, an area of 1,192,000 ha is cited by Bertrand (1983).

4) Ivory Coast: The forest depletion during the 1970s, principally owing to replacement by fruit, coffee, and cocoa plantations, was said to be accounting for some 400,000 ha of forest annually (Roche, 1979). This compares with 310,000 ha quoted above (Table I) for FAO 1980, and 300,000 ha mentioned by Bertrand (1983). The latter paper is probably the most recent and comprehensive evaluation of the state of forests in the Ivory Coast. Although Bertrand (1983) gives a lower figure for annual reduction, he cites an area of only 200,000 ha of intact, i.e., natural and unexploited forest, and 3,249,000 ha of exploited forest—a combined total of much less than the FAO figure of 4,458,000 ha.

5) Ghana: All forest outside designated forest reserves is at risk. In fact, there are only small relicts remaining, and pressure is now on the reserves and no unworked land is available (although there is much degraded deforested land uncultivated). The status of forest reserves is summarized in Table IV.

5a) Brong Ahafo, Ashanti area is threatened by farming and timber exploitation.

5b) Eastern Regional Forests: All are under great pressure.

6) Nigeria: The annual rate of forest destruction estimated by Ola-Adams & Iyambo (1977) was 260,000 ha, and it still seems to be on the increase, according to Allen (1981), which was confirmed by Okafor (1983, pers. comm.).

Since 1976 the export of timber has been banned, as all production is swallowed up by local markets. In fact, Nigeria imports wooden manufactured goods, such as house doors, from Ghana and elsewhere.

New oil palm plantations have replaced large areas of the species-rich forest north of Calabar, resulting in the loss of all indigenous species. Swamp forest and mangrove around the oil producing areas of the Niger delta are being cut and degraded.

The Federal Forestry Department has prepared an inventory of all high forest in Nigeria, in conjunction with FAO (FO: NIR/71/546 Technical Report 1979). It includes woody species above a certain size and excludes herbaceous ones. Further information may be obtained from Okali (1979).

The Nigerian Conservation Foundation (P.O. Box 467, Lagos) is initiating forest research and reserves at Okomu, Oban, Ogoja and elsewhere.

VI. Academic and Governmental Institutions in the Region

1) Sierra Leone: At the Kenema Herbarium (FHK), a forest species check-list is said to be in preparation.

2) Ivory Coast: Dr. L. Aké Assi, Laboratoire de Bota-

Table IV
Ghana academic institutions floristic programs.

Institution	Preparation of Floras	Preparation of Check-lists	Ecological & Taxonomic Investigations	Cooperation with Foreign Institutions
Botany Dept. Univ. of Ghana, Legon (GC)	yes	yes	yes	Royal Bot. Gardens, Kew
Forestry Dept.	–	yes	yes	ODA; Oxford Univ.
Forest Products Research Inst., Kumasi	–	yes	yes	–
Dept. of Game & Wildlife	–	–	yes	IUCN; WWF
Dept. of Bio. Sciences, Univ. of Science & Tech., Kumasi (KUU)	–	–	yes	–
Dept. of Botany, Univ. of Cape-Coast (CCG)	–	–	yes	–
Institute for Renewable Resources Univ. of Science & Tech., Kumasi	–	–	yes	UNEP; MAB; UNESCO
Environmental Protection Council	–	–	yes	
Cocoa Research Inst. Tafo	–	–	yes	–

nique de l'Universite, Abidjan (UCJ), is writing a flora of the whole of Ivory Coast, including the forest species. Some floristic work is also pursued by various botanists at the Centre ORSTOM d'Adiopodoumé (ABI).

3) Ghana: Table V lists the academic institutions engaged in floristic inventory within Ghana and their activities.

4) Nigeria: At Ibadan University (UCI), Dr. Joyce Lowe is preparing and editing several accounts for the *Flora of Nigeria*. The Forest Research Institute of Nigeria (P.O. Box 5054, Ibadan) maintains the largest herbarium (FHI) in the country, and study of the forests continue in several units. In addition, Dr. J. C. Okafor of the Forestry Commission (POMB 1028, Enugu) is studying forest fruit trees and preparing a taxonomic account of woody climbers for the *Flora of Nigeria*.

VII. Foreign Programs

There are several foreign programs engaged in floristic inventory in West Africa. They are:

1) West Germany: Botanischer Garten & Museum, Berlin-Dahlem. In June, 1985 they published the Togo project of a checklist of the flora, and interest is continuing.

2) England: A) Forest Herbarium (FHO), Oxford. They are working on taxonomy and vegetation of the area. B) Herbarium, Royal Botanic Gardens, Kew (K). They concluded the *Flora of West Tropical Africa* in 1972, but interest is continuing.

3) Scotland: Institute of Terrestrial Ecology, Bush Estate. Engaged in forestry research.

4) France: A) Laboratoire de Phanerogamie, Jardin des Plantes, Paris. There is a continuing interest in Francophone territories. B) ORSTOM, via a base at Adiopodoumé, Ivory Coast.

5) Belgium: Jardin Botanique National de Belgique at Meise (BR). There is a continuation of the Berhaut *Flore illustrée du Sénégal*, and interest in the vegetation of the Casamance (Senegal) continues.

6) Netherlands: Herbarium Vadense (WAG), Wageningen. There is work on the taxonomy of some groups well represented in West Africa.

Evaluation of the suitability of past, ongoing and planned floristic inventory of the tropical forest of West Africa must be considered in the context of colonial and independent Africa. During the "Colonial" period, a climax of publications appeared, based on European collections, with duplicates in regional institutions. Since then, economic and political difficulties, coupled with lack of suitable personnel, has slowed research and production as a general rule, although there are notable exceptions. International organizations have also become involved, bringing in capital and expertise.

VIII. Literature Cited

Allen, P. E. T. 1981. Land use in Nigeria. Forestry Management Seminar. Ibadan, Dept. of Forestry.

Aubréville, A. 1959. La flore forestière de la Côte d'Ivoire, 2nd. ed. Paris.

Bertrand, A. 1983. La déforestation en zone de forêt en Côte d'Ivoire. Bois et Forêts des Tropiques. **202**: 3–17.

Clarke, J. L. 1966. Sierra Leone in maps. London.

FAO: NIR/71/546 Technical Report. 1979. High forest development — Nigeria — the indicative inventory of reserved high forest in Southern Nigeria 1973-1977. (Report prepared for the Government of Nigeria by the Food and Agriculture Organisation of the United Nations as executing agency for the United Nations Development Program, based on work of Harold Sutter, Forest Inventory Officer). Published by UNDP/FAO, Ibadan.

Hall, J. B. & M. D. Swain. 1981. Distribution and ecology of vascular plants in a tropical rainforest. Forest vegetation in Ghana. The Hague.

Harrison Church, R. 1974. West Africa. Longman.

Hepper, F. N. in G. Kunkel. 1979. Taxonomic aspects of African economic botany. (A.E.T.F.A.T.). Gran Canaria & Royal Botanic Gardens, Kew.

Keay, R. W. J. 1953. An outline of Nigerian vegetation. Lagos.

Leakey, R. R. B. & F. T. Last. 1983. Past, present and future of west African hardwoods. Timber Grower 1983: 32. London.

Lebrun, J. P. & A. L. Stork. 1977. Index 1935-1976 des cartes de repartition des plantes vasculaires d'Afrique. Geneva.

Low, J. & D. P. Stanfield. 1970, 1974. Flora of Nigeria (Grasses, Sedges). Continuing, Ibadan University Press, Ibadan.

Okali, D. V. V. 1979. The Nigerian rainforest ecosystem. Publ. MAB, Ibadan.

Ola-Adams, B. A. & D. E. Iyambo. 1977. Conservation of natural vegetation in Nigeria. Environ. Cons. **4(3)**: 217-226.

Roche, L. 1979. Forestry and the conservation of plants and animals in the tropics. Forest Ecol. Managem. **2**: 103-122.

Taylor, C. J. 1960. Synecology and silviculture in Ghana. Accra & London.

UNESCO. 1973. International classification and mapping of vegetation. Paris.

Voorhoeve, A. G. 1965. Liberian high forest trees. Wageningen. Second edition, 1979.

White, F. 1983a. UNESCO/AETFAT/UNSO vegetation map of Africa. UNESCO, Paris.

_____. 1983b. The vegetation of Africa. UNESCO, Paris.

IX. General References

A. General References and Regional Floras

Adam, J.-G. 1969. Etude comparée de quelques forets ouest-africaines (Sierra Leone et Liberia). Bull. Inst. Fondam. Afr. Noire **31 sér. A**: 340-410.

Alexandre, D. Y. 1980. Caractère saisonnier de la fructification dans une foret hygrophile de Côte d'Ivoire. Terre et Vie **34**: 335-350.

_____. 1982. Etude de l'éclairement du sous-bois d'une forêt dense humide sempervirente (Taï, Côte d'Ivoire). Acta Oecol./Oecol. Gen. **3(4)**: 407-447.

Alston, A. G. H. 1959. The ferns and fern allies of West Tropical Africa. London. Supplement to FWTA revised edition.

Aubréville, A. 1950. La flore forestière Soudano-Guinéenne. Paris. (For woody species in drier areas, see Geerling, 1982).

_____. 1955. La typologie topographique forestière. Bois Forêts Trop. **41**: 3-7.

Bertrand, A. 1983. La déforestation en zone forêt en Côte d'Ivoire. Bois Forêts Trop. **202**: 3-17.

Chevalier, A. 1909. L'extension et la régression de la forêt vierge de l'Afrique tropicale. Compt. Rend. Hebd. Séances Acad. Sci. (Paris) **149**: 458-461.

_____. 1920. Exploration botanique de l'Afrique Occidentale Française.

Coombe, D. E. & W. Hadfield. 1962. An analysis of the growth of *Musanga cecropioides*. J. Ecol. **50**: 221-234.

Corner, E. J. H. 1954. The evolution of the tropical forest. Pages 34-46 in J. S. Huxley, A. C. Hardy & E. B. Ford (eds.), Evolution as a process. London.

Engler, A. 1925. Die Pflanzenwelt Afrikas. Pages 1-341 in A. Engler & C. G. O. Drude (eds.), Die Vegetation der Erde. IX. (Includes descriptions of West African vegetation.)

_____. 1900-1920. Botanische Jahrbücher für Systematik, Pflanzengeschichte und Pflanzengeographie. (Numerous contributions on African plants.)

Enti, A. A. 1968. Distribution and ecology of *Hildegardia barteri*. Bull. Inst. Fondam. Afrique Noire, Ser. A, Sci. Nat. **30**: 881-885.

Federov, A. A. 1966. The structure of the tropical rain-forest and speciation in the humid tropics. J. Ecol. **54**: 1-11.

Geerling, C. 1982. Guide de terrain des ligneux Sahéliens et Soudano-Guiéens. Wageningen. (Dry country woody species, supersedes Aubreville, 1950.)

Gledhill, D. 1972. West African trees. Longman, Harlow. (Popular treatment of a selection of trees, with colored illustrations.)

Grieg-Smith, P., M. P. Austin & T. C. Whitmore. 1967. The application of quantitative methods to vegetation survey. Part I. Association analysis and principal component ordination of rainforest. J. Ecol. **55**: 483-503.

Harrison Church, R. 1974. West Africa. Longman.

Hepper, F. N. (See Hutchinson & Dalziel.)

Hutchinson, J. & J. M. Dalziel. 1927-1936. Flora of West Africa, 1st ed. Keay, R. W. J. & F. N. Hepper. 1958-1972. 2nd ed. (The standard flora, with keys and brief descriptions, and distributions with cited herbarium specimens.)

Jeník, J. 1967. Root adaptations in West African trees. J. Linn. Soc., Bot. **60**: 25-29.

_____. 1969. The life-form of *Scaphopetalum amoenum* A. Chev. Preslia **41**: 109-112.

_____. 1970. The pneumatophores of *Voacanga thouarsii* Roem. & Schult. (Apocynaceae). Bull. Inst. Fondam. Afrique Noire, Ser. A, Sci. Nat. **32**: 986-994.

_____. 1970. Root systems in tropical trees. 4. The stilted peg-roots of *Xylopia staudtii* Engl. & Diels. Preslia **42**: 25-32.

_____. 1970. Root systems of tropical trees. 5. The peg-roots and pneumathodes of *Laguncularia racemosa* Gaertn. Preslia **42**: 105-113.

_____. 1971. Root structure and underground biomass in equatorial forests. Pages 323-331 in P. Duvigneaud (ed.), La productivité des écosystèmes forestiers. UNESCO, Paris.

_____. 1971. Root systems of tropical trees. 6. The aerial roots of *Entandrophragma angolense* (Welw.) C. DC. Preslia **43**: 1-4.

_____. 1971. Root systems of tropical trees. 7. The facultative peg-roots of *Anthocleista nobilis* G. Don. Preslia **43**: 97-104.

_____. 1973. Root systems of tropical trees. 8. Stilt-root and allied adaptations. Preslia **45**: 250-264.

_____ & A. A. Enti. 1969. Discontinuous distribution of *Allexis cauliflora*(Oliv.) L. Pierre in Equatorial Africa. Novit. Bot. Inst. Bot. Univ. Carol. Prag. **1968**:67-71.

_____ & B. J. Harris. 1969. Root-spines and spine-roots in dicotyledonous trees of tropical Africa. Österr. Bot. Z. **117**:128-138.

_____ & K. O. A. Mensah. 1967. Root systems of tropical trees. 1. Ectotrophic mycorrhizae of **Afzelia Africana** Sm. Preslia **39**:59-65.

Jones, E. W. 1963. Forest outliers in the Guinea zone of northern Nigeria. J. Ecol. **51**: 415–434.

Keay, R. W. J. (See Hutchinson & Dalziel.)

Kernan, H. S. 1980. Assistance in forestry development: Sierra Leone, summary report. FAOTCP/SIL/8907(1).

Kunkel, G. 1979. Taxonomic aspects of African economic botany. (A.E.T.F.A.T.) Gran Canaria & Royal Botanic Garden, Kew.

Leakey, R. R. B. & F. T. Last. 1983. Past, present and future of west African hardwoods. Timber Grower 1983: 32. London.

Letouzey, R. 1970. Manuel de botanique forestière, vols. 1 & 2. Nogent s/Marne. (Written for Cameroun and includes many West African species.)

Longman, K. A. & Jeník. 1974. Tropical forest and its environment. Longman, London.

Mensah, K. O. A. & J. Jeník. 1968. Root systems of tropical trees. 2. Features of the root system of iroko (*Chlorophora excelsa* Benth.). Preslia **40**: 21–27.

Myers, N. 1980. Conversion of tropical moist forest. National Academy of Sciences, Washington.

Neil, P. E. 1981. Problems and opportunities in tropical rainforest management. CFI Occasional paper No. 16. Oxford.

Namur, C. de. 1980. Etude des lianes en forêt nonperturbée. *In* Rapport de stage Août 78–Août 80. ORSTOM d'Adiopodoumé.

Okafor, J. C. 1988. Selection and improvement of tropical fruit trees.

Oliver, D. 1868–1937. Flora of Tropical Africa. London. (Later, Thiselton-Dyer, Prain & Hill, eds. Full keys and descriptions. Superseded by *F.W.T.A.*, but still useful.)

Richards, P. W. 1966. The tropical rain forest. Cambridge University Press, Cambridge.

Roche, L. 1979. Forestry and the conservation of plants and animals in the tropics. Forest Ecol. Managem. **2**: 103–122.

Saunders, H. N. 1958. A handbook of West African flowers. Oxford. (Descriptions and line drawings.)

Schnell, R. 1950. La forêt dense. Introduction à l'étude botaniques de la région forestière d'Afrique occidentale. Paris.

Stahel, J. 1971. Anatomical investigations on buttresses of *Khaya ivorensis* A. Chev. and *Piptadeniastrum africanum* (Hook.f) Brenan. Holz Roh-u. Werkstoffe **29**: 314–318.

Steentoft Nielsen, M. 1965. Introduction of the flowering plants of West Africa. University of London. (Textbook with much of relevance to forest flora.)

UNESCO. 1973. International classification and mapping of vegetation. Paris.

White, F. 1983a. UNESCO/AETFAT/UNSO vegetation map of Africa. UNESCO, Paris.

_____. 1983b. The vegetation of Africa. UNESCO, Paris.

B. Floras by Country

Senegal

Berhaut, J. 1967. Flore du Sénégal. 2nd ed. Dakar. (A continuous key.)

_____. 1971–1979. Flore illustrée du Sénégal. Dakar. (Outside moist forest zone, but includes some forest species; comprehensively illustrated.)

Gambia

Precival, D. A. 1968. The common trees and shrubs of the Gambia. Bathurst. (Mimeographed.)

Williams, F. N. 1907. Florula Gambica. Bull. Herb. Boissier, 2nd ser. **7**: 81–96, 194–208, 369–386. (No keys or descriptions.)

Guinea-Bissau

d'Orey, J. & M. C. Liberato. 1972–. Flora da Guiné-Bissau. (Formerly Guiné-Portuguesa). Lisbon.

Guinée

Adam, J. G. 1971–1981. Flore descriptive des Monts Nimba. 5 vols. Paris. (Comprehensive illustrated local flora with wider applications.)

Schnell, R. 1952. Végétation et flore de la région montagneuse du Nimba. Mem. Inst. Fondam. Afrique Noire 22.

_____. 1952. Vegetation et flore des Monts Nimba. Vegetatio **2**: 350–406.

Sierra Leone

Cole, N. H. A. 1968. The vegetation of Sierra Leone. Njala. (The only descriptive account.)

Deighton, F. C. 1957. Vernacular botanical vocabulary for Sierra Leone. London. (Useful cross references to scientific names.)

Saville, P. S. & J. E. D. Fox. 1967. Trees of Sierra Leone. Omagh (Mimeographed.)

Liberia

Harley, W. J. ca. 1955. Handbook of Liberian ferns. Ganta Mission. (Descriptions and line drawings.)

Kunkel, G. 1965. The trees of Liberia (Report No. 3 of the German Forestry Mission to Liberia). Munich. (See also Voorhoeve.)

Linder, D. H. 1930. Botanical report of Liberia. *In* R. P. Strong (ed.), The African Republic of Liberia and the Belgian Congo. Vol. 1. Harvard University Press, Cambridge.

Stapf, O. 1906. List of known plants of Liberia. *In* H. Johnston, Liberia. Appendix IV to Vol. 2.

Voorhoeve, A. G. 1965. Liberian high forest trees. Wageningen. (Keys and descriptions for field and herbarium; covers the same range as Kunkel, 1965, in more detail.) 2nd ed., 1979.

Ivory Coast

Anonymous. 1967. Resources forestières de la région Sud-ouest. Development and resources cooperation. Rapport au Gouvernement de la Republique du Côte d'Ivoire. (Not published.)

Aubréville, A. 1936. La flore forestière de la Côte d'Ivoire. 1st ed. 2nd. ed., 1959.

_____. 1957. A la recherche de la forêt en Côte d'Ivoire. Bois Forêts Trop. **57**: 17–32.

_____. 1958. A la recherche de la forêt en Côte d'Ivoire. Bois Forêts Trop. **58**: 12–27.

Aké Assi, L. 1963. Contribution a l'étude floristique de la Côte d'Ivoire et des territoires limitrophes. Paris. (Miscellaneous species.)

_____. 1984. Flore de la Côte d'Ivoire: étude descriptive et biogéographique. Vols. 1–3. Abidjan. (Mimeographed.)

Bousquet, B. 1978. Un parc de forêt dense en Afrique: le Parc National de Taï. Bois Forêts Trop. **179**: 27–46.

De Koning, J. 1983. La forêt du Banco. H. Veenman & Zonen BV, Wageningen.

Guillaumet, J. L. 1967. Recherches sur la végétation et la flore de la région du Bas-Cavally (Côte d'Ivoire). Paris. (Detailed study.)

_____ & E. Adjanohoun. 1971. La végétation de Côte d'Ivoire. Pages 157–263 *in* J. M. Avenard et al., Le milieu naturel de la Côte d'Ivoire. ORSTOM, Paris.

_____ et al. 1984. Recherche et aménagement en milieu forestière et humide: le Project Taï. UNESCO. Notes technique du MAB 15.

Huttel, C. 1975. Inventaire et structure de la végétation ligneuse. *In* Lemée et al., Recherches sur l'écosystème de la forêt subéquatoriale de la Basse Côte d'Ivoire. Terre et Vie **29**: 178–191.

———. 1977. Etude de quelques caractéristiques structurales de la végétation du bassin versant de l'Audrénisrou. Rapport Adiopodoumé.

Vooren, A. P. 1987(?). Analyse floristique et structure de la "voûte forestière." *In* Forêt de Taï, Côte d'Ivoire. ORSTOM. Adiopodoumé.

Ghana

Burtt Davy, J. & A. C. Hoyle (eds.). 1937. The first descriptive checklist of the Gold coast. Oxford. (Mimeographed.)

Hall, J. B. & M. D. Swaine. 1981. Distribution and ecology of vascular plants in a tropical rainforest: forest vegetation in Ghana. W. Junk, The Hague. (Includes distribution maps for each species in Ghana forest zone.)

Hawthorne, W. D. In press. Field guide to Ghana forest trees. Overseas Development Resources Institute, London.

Hepper, F. N. 1976. The West African herbaria of Schumacher & Thonning. Kew. (Specialist analysis of the specimens collected in late 18th century especially in Ghana)

Hossain, M. & J. B. Hall. 1969. A field-key to the trees of the Mole Game Reserve, Damong, Ghana. Legon, Institute of African Studies.

Irvine, F. R. 1961. Woody plants of Ghana, Oxford. (Provides information on uses and vernacular names, descriptions but no keys)

Lawson, G. W., K. O. Armstrong-Mensah & J. B. Hall. 1970. A catena in tropical moist semi-deciduous forest near Kade, Ghana. J. Ecol. **58**: 371–398.

Taylor, C. J. 1960. Synecology and silviculture in Ghana. Accra & London.

Nigeria

Allen, P. E. T. 1981. Land use in Nigeria. Forestry Management Seminar. Ibadan, Dept. of Forestry.

Clayton, W. D. 1958. Secondary vegetation and the transition to savanna near Ibadan, Nigeria. J. Ecol. **46**: 217.

———. 1961. Derived savanna in Kabba Province, Nigeria. J. Ecol. **49**: 595.

Evans, G. C. 1956. An area survey method of investigating the distribution of light intensity in woodlands with particular reference to sun-flecks, including an analysis of data from rain forest in Southern Nigeria. J. Ecol. **44**: 391–428.

FAO:NIR/71/546 Technical Report. 1979. High Forest Development—Nigeria—The Indicative Inventory of Reserved High Forest in Southern Nigeria 1973–1977. (Report prepared for the Government of Nigeria by the Food and Agriculture Organisation of the United Nations as executing agency for the United Nations Development Program, based on work of Harold Sutter, Forest Inventory Officer) Published by UNDP/FAO Ibadan.

Hopkins, B. 1962–1971. Vegetation of the Olokemeji Forest Reserve, Nigeria. Part I. J. Ecol. **50**: 559 (1962). Part II. **53**: 109 (1965). Part III. **53**: 125 (165). Part IV. **58**: 765 (1971). Part V. **58**: 765 (1971).

Jones E. W. 1955–1956. Ecological studies on the rain forest of Southern Nigeria. Part IV. J. Ecol. **43**: 564 (1955). Part V. **44**: 83 (1956).

Keay, R. W. J. 1947. Notes on the vegetation of Old Oyo Forest Reserve. Farm and Forest. Jan.–June: 36–46.

———. 1953. An outline of Nigerian vegetation. Lagos.

———. 1957. Wind dispersed seeds in a Nigerian forest. J. Ecol. **45**: 471–478.

———, Onochie, C. F. A. & D. P. Stanfield. 1960. Nigerian trees. Vol. 1. Vol. 2, 1964 Ibadan. (A field and herbarium flora with keys and descriptions.)

Kemp, R. H. 1963. Growth and regeneration of open savanna woodland in N. Nigeria. Commonw. Forest Rev. **42**: 113, 200–206.

Kennedy, J. D. 1936. Forest flora of Southern Nigeria. Lagos (Out-of-date, useful in its time.)

Low, J. & Stanfield, D. P. Flora of Nigeria (Grasses, 1970. Sedges, 1974; Continuing) Ibadan Univ. Press.

Neil, P. E. 1981. Problems and opportunities in tropical rainforest management. CFI Occasional paper No. 16. Oxford.

Njoku, E. 1963–1964. Seasonal periodicity in the growth and development of some forest trees in Nigeria. Part I.-Observations on mature trees. J. Ecol. **51**: 617–624 (1963). Part II. **52**: 19–26 (1964).

Okafor, J. C. 1978. Development of forest tree crops for food and supplies in Nigeria. Forest Ecol. Managem. **1**: 235–247.

———. 1980. Edible indigenous woody plants in the rural economy of the Nigerian forest zone. Forest Ecol. Managem. **3**: 45–55.

———. 1982. Horticulturally promising indigenous trees and shrubs of the Nigerian forest zone. Acta Hort. **123**.

Okali, D. U. U. 1979. The Nigerian rainforest ecosystem. Publ. MAB, Ibadan.

Ola-Adams, B. A. & D. E. Iyamabo. 1977. Conservation of natural vegetation in Nigeria. Environ. Conserv. **4(3)**: 217–226.

Rendle, A. B. et al. 1913. Catalogue of the plants collected by Mr. & Mrs. P. A. Talbot. London: BM (NH). (Mainly descriptions of new species.)

Richards, P. W. 1957. Ecological notes on West African vegetation. I.: The plant communities of the Idanre Hills, Nigeria. J. Ecol. **45**: 563–578.

Ross, R. 1954. Ecological studies on the rain forest of southern Nigeria. Part III. J. Ecol. **42**: 259–282.

Other Floras Outside Area

Aubréville, A. et al. 1969– . Flore du Cameroun. Paris. (Includes W. Cameroon which is also covered by FWTA and many species in common with West Africa proper.)

——— et al. 1969– . Flore du Gabon. Paris. (Includes many species in common with West Africa.)

Exell, A. W. et al. 1944. Catalogue of the vascular plants of S. Tomé. London. Supplements 1956, 1959, etc. (Includes Principe and Annobon.)

———. 1973. Angiosperms of the islands of the Gulf of Guinea. London; BM (NH). (Includes Fernando Po, Principe, S. Tomé and Annobon.)

Richards, P. W. 1963. Ecological notes on West African vegetation. III. The upland forests of Cameroons mountain. J. Ecol. **51**: 529–554.

Turrill, W. B. et al. (eds.). 1952– . Flora of tropical East Africa. London. (Outside West African area but important for the whole of Africa.)

Gabon

F. J. Breteler

Contents

I. Description of the Region	199
A. Geographical Extent and Topography	199
B. Population	199
II. Vegetation Maps	199
III. Herbarium Resources	199
IV. Publications	199
V. Present Status of Inventory	199
A. Completeness of Inventory	199
B. Endemism	199
C. Extinction	200
D. Threatened Areas	200
VI. Academic and Government Institutions	200
VII. Future of Inventory	200
A. Adequacy	200
B. Costs	201
VIII. Literature Cited	201

I. Description of the Region

A. Geographical Extent and Topography

Gabon covers an area of 267,667 km². It is quadrate in general outline, 600 km from west to east, 550 km from north to south.

With the exception of a rather flat coastal area, Gabon can be described as a country of hills and low mountains, mixed with some flatter parts—the plateaus of the interior. The mountains rarely exceed 1000 m in altitude. The Cristal Mountains in the NW, bordering Rio Muni (Equatorial Guinea), and the Chaillu Massif, in the central south, are the most obvious montane areas. The hills and mountains have steep slopes, which makes road-building and, hence, forest exploitation, difficult. The Ogooué is the main river. It drains about two-thirds of the country.

B. Population

Population statistics are most contradictory. From colonial times the following figures are given: 1906: 337,000; 1956: 338,000; 1960: 448,000. The official population figure for 1976 is 1,200,000. This would mean that the population had almost tripled in a period of 16 years (!), whereas the actual growth rate is more like 1% or slightly more. The actual population is calculated to be between 600,000 and 800,000 (i.e., between two and three inhabitants per km²). There are no signs that the population will increase more rapidly in the near future.

It is important to note, relevant to the purposes of this book, that the distribution of the population is changing very much. The two relatively large cities, Libreville, the capital, and Port Gentil, the main harbor, are growing very quickly. Port Gentil doubled its population in six years, from 88,000 in 1970, to 160,000 in 1976. This means that the interior has become less populated, at least as far as rural areas are concerned. Centers in the interior (of which some, like Franceville, are relatively large towns) are also growing faster in population than the average trend for the total population. In conclusion, it may be stated that rural areas are decreasing in population density.

II. Vegetation Maps

The tropical forests as a whole cover about 85% of the total area of Gabon. All these forests should be classified as tropical ombrophilous forests (UNESCO, 1973). In the north and northeast these forests tend slightly towards semideciduousness. In the central south, near the border with the Congo Republic, this semi-deciduous character becomes more obvious. Gabon's tropical ombrophilous forests have to be subdivided into lowland forest, submontane forest, and swamp forest.

The best vegetation map for the region is the AETFAT/UNESCO map by White (1983). Compared to the actual situation, this map does not show any major discrepancies.

III. Herbarium Resources

The Herbier National du Gabon, founded in 1983, holds between 8500 and 9000 specimens. The Institut de Recherche en Ecologie Tropicale at Makokou (IRET) holds 500 to 1000 specimens. This field station in NE Gabon is one of the research institutes belonging to CENAREST.

Foreign institutions holding significant numbers of specimens are:

1). Muséum National d'Histoire Naturelle, Laboratoire de Phanérogamie, Paris (P): ca. 20,000.

2). Department of Plant Taxonomy, Wageningen (WAG): ca. 10,000.

3). British Museum (Natural History), London (BM): 4500–5000, mainly duplicates of the Le Testu collection.

4). Jardin Botanique National de Belgique, Meise (BR): 3500–4500, mainly duplicates of the Le Testu collection.

Other institutions, such as K and LE have important duplicate sets of the Klaine collection.

IV. Publications

Of the *Flore du Gabon*, 29 fascicles have been published. Seventy families have been treated. See also Floret (1976), and for further literature on various regions see Pellegrin (1948), Saint Aubin (1963), Walker and Sillans (1961), Caballé (1978), Caballé and Fontes (1978). For the literature on local areas, see Hallé (1964, 1965), Hallé and le Thomas (1967, 1970), Hallé, le Thomas and Gazel (1967), Hladik and Hallé (1974), Florence and Hladik (1980), and Pellegrin (1938).

V. Present Status of Inventory

A. Completeness of Inventory

There are no areas in Gabon which may be classified as having been well collected, except perhaps for the surroundings of Libreville, where the number of collections probably reaches 50 per 100 km².

All families and genera are poorly collected in Gabon. The example given below illustrates this especially well, since the present author, the revisor of the Dichapetalaceae, has rather intensively collected this family in Gabon. The scholars working on different families are shown in Table I.

B. Endemism

Although the flora of Gabon is rich, it is but poorly col-

Table I

Scholars working on plant families for Gabon.

Name	Institution	Family
J. J. Bos	WAG	*Dracaena* (Liliaceae)
F. J. Breteler	WAG	Connaraceae, Dichapetalaceae
C. Farron	BAS	Ochnaceae
J. J. Floret	P	Rhizophoraceae, Lecythidaceae, Droseraceae Celastraceae
N. Hallé	P	Celastraceae
A. J. M. Leeuwenberg	WAG	Apocynaceae
J. J. F. E. de Wilde	WAG	Begoniaceae

lected. It is difficult, therefore, to identify areas that are high in endemics. The whole flora is high in endemics, judging from the relatively few collections that have been made. Further exploration will certainly yield many new records and new taxa. For example, collections made in Gabon in 1983 yielded one new record and two new taxa (one new species, one new variety) for the Dichapetalaceae. Table II illustrates the richness of the Gabonese flora. From these figures it may be concluded that Gabon is at least as rich as, and often much richer than, the whole of West Africa from Senegal in the west to Nigeria in the east, and that its endemism is much more evident.

C. Extinction

To estimate which species are threatened by extinction, the distribution of the species together with the rate of destruction of their natural habitat has to be known. Further exploration is needed to establish the ranges of plant species in Gabon. Habitat destruction occurs on a small scale only and almost only in relatively densely populated areas. How this may bring some species into a situation that they might be at risk of disappearance or have already disappeared is not known, but it is estimated that areas with a high rate of extinction are relatively few and small so far.

D. Threatened Areas

There are some small areas where the original forest is being converted into plantations of timber-producing species like Okoumé *(Aucoumea klaineana)*, or where the forest is completely removed in order to establish rubber and oil palm *(Elaeis guineensis)* plantations. Destruction or conversion of the tropical forest on an important scale has not yet occurred. However, forest exploitation is damaging the natural forest considerably. For the few trees per hectare that are being exploited, a relatively dense network of roads has been built, and this is done for hundreds of square kilometers of forest a year.

VI. Academic and Government Institutions

The only university in Libreville (Omar Bongo) has no botany department capable of helping with the floristic inventory of the region. The National Herbarium of Gabon has been working in close cooperation with the Herbarium Vadense, Wageningen, and the Paris Herbarium to accelerate botanical exploration. Since 1983 ca. 8000 numbers have been collected, mainly by botanists from Wageningen.

The National Herbarium of Gabon is quite willing to assist visiting botanists who want to contribute to the botanical exploration of the region. Letters should be addressed to: Herbier National du Gabon, Att.: Mr. A. Louis, CENAREST, B.P. 842, Libreville, Gabon. The same holds for scientists working in the fields of ecology and zoology, who are welcome at the Institut de Recherche en Ecologie Tropicale at Makokou (IRET), with letters to be addressed to the Director, c/o CENAREST, B.P. 842, Libreville, Gabon. Anthropologists should address themselves to the Centre Internationale de Recherches Medicales de Franceville, B.P. 769, Franceville, Gabon.

VII. Future of Inventory

A. Adequacy

As pointed out above, Gabon has a rich, though poorly explored flora, and it is difficult to define areas that are relatively well explored, or that have high rates of species diversity, endemism or extinction. Therefore, any collecting activity, wherever it may be done, is very useful, adding to our knowledge of the Gabonese flora. At present, collecting is done mostly in places where forest exploitation companies have penetrated vast forest areas, or where the forest is being converted or removed, for whatever purpose.

An adequate floristic inventory of Gabon is defined here as having a status of being "relatively well collected," i.e., 50–100 collections per km^2. For Gabon, having a surface of ca. 268,000 km^2, this means a total number of collections of between 135,000 and 267,000. For the following calculation, the average of these two figures (i.e., 200,000) has been taken as a starting point, because only 85% is covered with forest. Floret (1976) estimated the total number of collections made in Gabon at 26,000 at most. Since then not more than 10,000 collections have been added, bringing its total to about 35,000 at present, leaving a number of 165,000 to be collected. This figure corresponds with the result of a calculation based on the number of species (ca. 8000) present in the Gabonese forests and the average number of specimens per species (ca. 25) needed to make a reasonably good flora.

A very active program, that would yield 10,000 collections a year, would thus need 17 years to be accomplished, an enormous task when one takes into consideration several large areas of at least 10,000–20,000 km^2 which are poorly

Table II
Three examples to illustrate the richness of the Gabonese flora.

Taxon	Total number of species for continental Africa	Number of species in West Africa (Senegal–Nigeria)	Species in Gabon	Endemics in Gabon
Begonia	ca. 127	17	50	25 (50%)
Dichapetalaceae (see p. 200)	87	29	57	9 (16%)
Dracaena	ca. 50	ca. 25	24–30	5 (±20%)

known, or even unknown at present, and have no possibility of access by road.

The question should be posed, is a period of 17 years not too long to reach that goal? The answer is yes, much too long. Although the rate of destruction or conversion of the tropical moist forest is relatively low in Gabon, one should, however, take into consideration that plans for conservation, if they are to be accepted, need a sound floristic base, and that the implementation of a conservation strategy is a long-term process. There is no time to be lost.

An acceleration of the floristic inventory, for instance on the scale accomplished in the period 1983–1986, when ca. 8000 collections were made, would be a major step forwards. This would stimulate progress in the *Flore du Gabon*, which is rather slow at the moment, not only because of a lack of taxonomists working on it, but certainly also for lack of sufficient material.

There is no direct need for additional institutions, but it is of the utmost importance for the future if the academic training of Gabonese botanists could be started at the Omar Bongo University in Libreville. In the short term, this training would not be soon enough for the floristic inventory. Therefore, an acceleration of this inventory cannot be achieved without foreign aid.

For an adequate floristic inventory of the Gabonese tropical forests a good local botanical institute is a prime necessity. It is therefore urgent that the recently founded Herbier National du Gabon in Libreville be strengthened considerably. At the moment this institute has only the bare minimum needed for its functioning. The following is needed for the short term: 1) adequate housing for the institute, 2) a better salary for the only staff member, 3) a second, perhaps foreign, botanist mainly engaged in botanical exploration, and 4) two additional trained technicians.

B. Costs

A recent two month-long collecting trip to Gabon by two botanists from Wageningen yielded 1255 collections, with an average of ca. six duplicates, for a total of 7500 specimens. These specimens were collected at a cost of about one US dollar per specimen. Not included in this calculation are the cost of transport to and from Gabon, transportation and the housing facilities there—both offered by the Gabonese government—nor the salaries of the Dutch botanists. If the cost for all this had been included as well, it is calculated that the cost of a specimen would have amounted to about $3.00. Given our goal of 165,000 collections, with an average of six duplicates each, the total cost would be about $3,000,000. The price per collection could be lowered if fewer duplicates were to be collected, but the cost per specimen would increase.

The figures given above indicate what may be needed in terms of money for a good floristic inventory of Gabon. It is difficult to calculate the cost of an inventory program in more detail without knowing its major items. For instance, the total cost of a detailed foreign botanist is estimated at US$50,000 per annum. But if working facilities at the local institute in Gabon are not improved, the cost will be much higher. A joint effort of foreign and Gabonese institutions is needed to accomplish the task. It is expected that certain propositions from abroad will be matched by equal Gabonese efforts. Therefore, an expression of foreign interest from as many countries as possible, as shown by visiting botanists, for instance, is of major importance.

VIII. Literature Cited

Caballé, G. 1978. Essai sur la géographie forestière du Gabon. Adansonia, ser. 2. **17(4):** 424–440.

─── & J. Fontes. 1978. Les inventaires forestiers au Gabon: applications à la phytogéographie. Bois Forêts Trop. **177:** 15–33.

Florence, J. & A. Hladik. 1980. Catalogue des Phanérogames et des Ptéridophytes du Nord-Est du Gabon (sixième liste). Adansonia, ser. 2, **20(2):** 235–253.

Floret, J. J. 1976. Flore du Gabon. Boissiera **24:** 575–580.

Hallé, N. 1964. Première liste de Phanérogames et Ptéridophytes des environs de Makokou, Kemboma et Bélinga. Biol. gabon. **1:** 41–46.

───. 1965. Seconde liste de Phanérogames et Ptéridophytes du N.-E. Gabon (Makokou, Bélinga et Mekambo). Biol. gabon **1:** 337–344.

─── & A. le Thomas. 1967. Troisième liste de Phanérogames du N.-E. Gabon. Biol. gabon. **3(2):** 113–120.

─── & ───. 1970. Quatrième liste de Phanérogames et Ptéridophytes du N.-E. Gabon (Bassin de l'Ivindo). Biol. gabon. **6:** 131–138.

───, ─── & M. Gazel. 1967. Trois relevés botaniques dans

les forêts de Bélinga (N.-E. du Gabon). Biol. gabon. **3**: 3–16.

Hladik A. & N. Hallé. 1973. Catalogue des Phanérogames du N.-E. du Gabon (5e liste). Adansonia, ser. 2, **13(4)**: 527–544.

Pellegrin, F. 1938. La Flore du Mayombe, 1924–1938. Mém. Soc. Linn. Normandie, XXVI, 2: (1924); ibid., Nouv. Sér. Botanique I, 3, 1928; ibid., I, 4, 1938.

———. 1948. Les Legumineuses du Gabon. Mém. Inst. Études Centrafr. 1.

Saint Aubin, G. de. 1963. La Forêt du Gabon. Centre Technique For. Trop.

UNESCO. 1973. International classification and mapping of vegetation. UNESCO, Paris.

Walker, A. & R. Sillans. 1961. Les plantes utiles du Gabon. Paris.

White, F. 1983. Vegetation map of Africa. 4 sheets plus descriptive memoir. The vegetation of Africa. UNESCO, Paris.

Equatorial Africa (Cameroun, People's Republic of the Congo, Central African Republic, Angola, Zaire, Rwanda and Burundi)

Karolyn Kendrick

(With Additional Information from Paul Bamps and André Robyns)

Contents

I. Introduction	205
II. Description of the Region	205
A. Geographical Extent, Topography and Climate	205
1. General Topography	205
2. Cameroun	205
3. People's Republic of the Congo	206
4. Angola	206
5. Zaire	206
6. Rwanda and Burundi	207
B. Vegetation	207
C. Population	208
III. Vegetation Maps	209
IV. Floristic Inventory	210
A. Collections and Herbaria	210
B. Publications	211
V. Status of Inventory	212
A. Endemism and Extinction	212
B. Conservation and Destruction of Vegetation	212
VI. Literature Cited	213
VII. Selected References	214
VIII. Appendix – Soil and Vegetation Maps of Zaire, Rwanda and Burundi	215

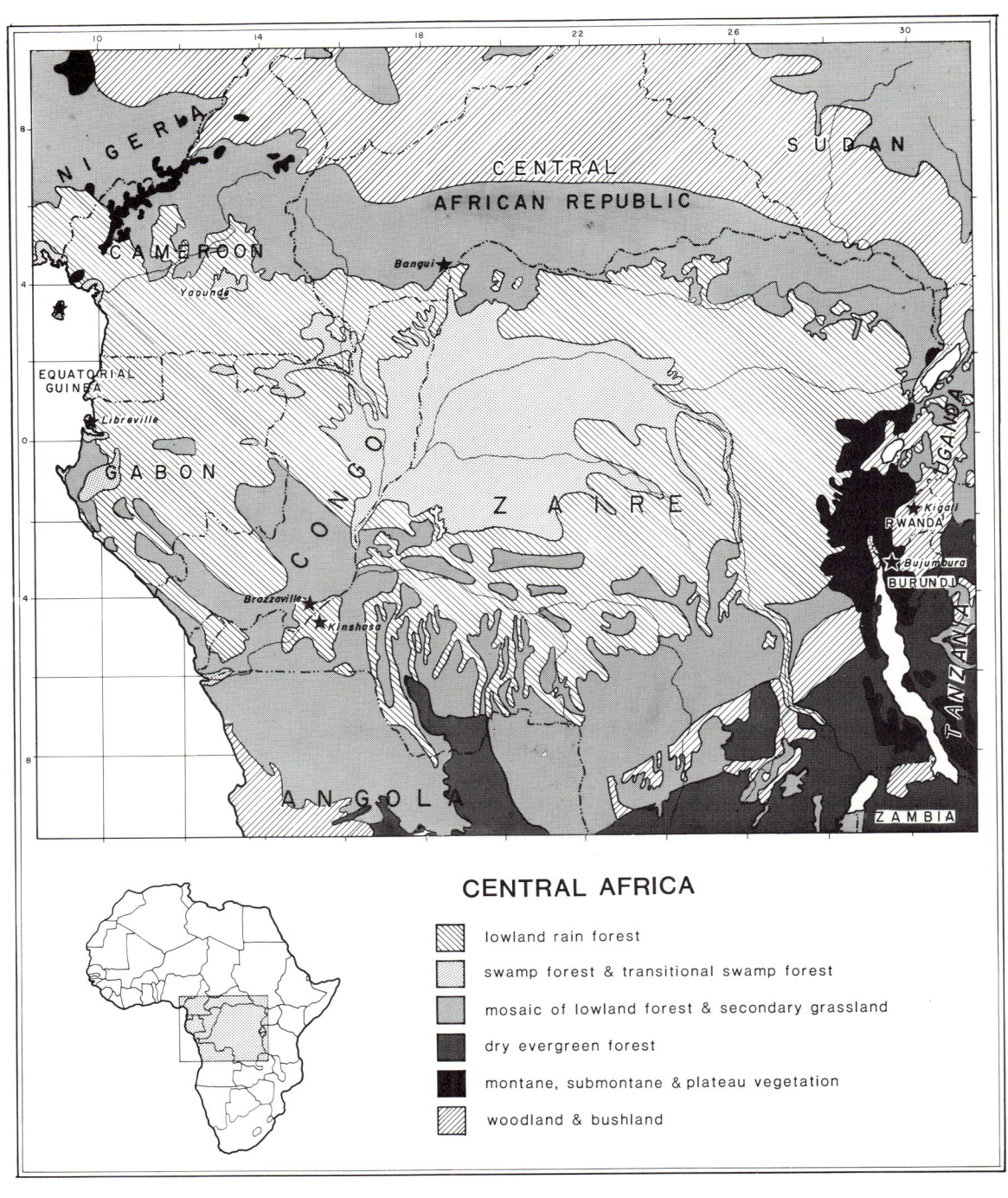

I. Introduction

Although the flora of tropical Africa as a whole is considered relatively well studied, floristic studies and inventories of tropical forests within the central and west equatorial African states of Angola, Burundi, Cameroun, Central African Republic, People's Republic of the Congo, Rwanda and Zaire are still incomplete, and the floras are known better physiognomically than floristically. In summarizing the state of floristic inventory in these countries, Brenan (1979) noted that by the end of the 1970s probably no more than 20–25% of Camerounian species and 40% of the species present in Angola and Zaire, Rwanda and Burundi had been described in the floras of the respective countries or regions. Although publication of the floras continues, as is the case for other areas of Africa, the pace of collecting and inventorying has slowed since the mid-1970s, in part because of financial, nationalistic and political constraints. In light of such problems, floristically interesting areas tend to fall victim to governments' efforts to generate export income from cash crops, timber and mining products, and to the rapid expansion in some areas of already dense rural populations, and the opening of previously forested areas to farming, grazing and commerce.

But, as Brenan also observed, less accessible areas tend to be under-collected. In noting the problems peculiar to inventorying the moist evergreen and semi-evergreen forests of the Guineo-Congolian region, Letouzey (1968) concurred with G. Mangenot's lament that species in a given tract are likely to be found "mixed together in a multitude of rare or very dispersed plants that are incompletely inventoried, variable from one point to another, and certain of which still are not well identified."

Nevertheless, the study of African floristics continues to benefit greatly from the cooperation of many international and national institutions, particularly the Association pour l'Étude Taxonomique de la Flore d'Afrique Tropicale (AETFAT), UNESCO and the FAO on the international level, the Institut d'Elevage et de Médecine vétérinaire des Pays Tropicaux (IEMVET) in Maisons Alfort, France, and the Office de la Recherche Scientifique et Technique d'Outre-Mer (ORSTOM) in Bondy, France. Moreover, the deposition of most of the collections in European herbaria has facilitated european research to such an extent that one authority on the Guineo-Congolian region has remarked that it is "easier for a specialist to examine material from all parts of Africa than from any comparable areas in the Northern Temperate Zone" (White, 1962).

II. Description of the Region

A. Geographical Extent, Topography and Climate

1. General topography

The peneplain of the Zaire River Basin is footed by pre-Cambrian rocks blanketed with ancient to recent sediments. A swampy alluvial plain dotted with the remnants of once extensive lakes or an inland sea occupies about 150,000 km^2, stretching from the basin's low ground of 338 m above sea level at Lake Mai-Ndombe northward into the Congo People's Republic. To reach the Atlantic Ocean the river cuts through the Monts de Cristal, descending in a series of cataracts and falls to a narrow, swampy coastal plain of mangrove thickets that mark its mouth. North of the river gorge is the Mayumbe hill country of Zaire and northeast Cabinda, the noncontiguous exclave of Angola. The Mayumbe escarpment, a 12,000 km^2 region of high endemism, is itself a southern spur of the plateaus of Cameroun and Gabon that separates the river basin from the coastal strip of hygrophilous rain forest. To the north and south of the river basin are broken plateaus and uplands of late pre-Cambrian sandstone, quartzite and schist. Straddling Zaire's northern border with the Central African Republic are the Ubangi-Uele plateaus that form the divide between the drainage basins of the Zaire and Nile Rivers. There, outliers of drier semi-evergreen forests follow rivers northward into the savanna country. South of the Zaire River basin are the Angolan plateaus of High Africa, which are generally above 1000 m. In the southeast the river basin is rimmed by the plateaus of Zaire's Shaba Region (Haut Katanga) arcing northeastward to the Lukuga River.

In the east the lowlands climb sharply to the Kivu Ridge (dorsale du Kivu), which is composed of highly metamorphic pre-Cambrian gneiss, amphibolite, quartzite and micaschist. From the northern end of Lake Tanganyika a system of mountains backs Kivu, rising through Burundi and Rwanda to the Ruwenzori Mountains astride the Zaire/Uganda frontier between Lake Albert and Lake Edward. These mountain chains of eastern Zaire, Rwanda and Burundi form the western rim of the Great Rift Valley and separate the Nile and Zaire River basins. Both Rwanda and Burundi lie, on average, above 1500 m, and rising from the Rwanda high plateaus are the Virunga Mountains, a system of active and extinct volcanoes, formed during the Pleistocene epoch, which marks the meeting place of Zaire, Rwanda and Uganda. To the east of the mountain chain, a high plateau of pre-Cambrian rock slopes down to the vast papyrus swamps surrounding the Kagera River. The many small rivers that dissect the plateau flow into the Kagera to form the farthest sources of the Nile. Southward, the plateau extends into Burundi, which ascends from Lake Tanganyika to the Zaire-Nile divide.

2. Cameroun

Cameroun, with its variety of bioclimes, embraces most of the vegetation types of West Africa. The triangular-shaped country embraces an area of 465,458 km^2, with its 750 km base oriented just above and below 2°N in latitude. A narrow coastal strip rounds the Bight of Biafra, widening as it proceeds southward past the volcanic rift that cuts northwestward from the island of Bioko (Fernando Pó) and pushed up the mountains on the country's northwestern coast. Mt. Cameroun, standing alone near the coast, is at 4,070 m the highest peak in West Africa and is an active volcano, having last erupted in 1959.

The southern lowland region lies south of the Sanaga River and runs from the coastal plains eastward to the country's borders with the Central African Republic and the Congo. The central region, bounded on the south by the Sanaga River and on the north by the Benoué River, rises to the Adamawa Plateau with elevations of 760–1360 m. The highlands, one of the few relief features in Low Africa more than 1,000 m high, are built of pre-Cambrian rocks thrown up in the late Cretaceous period.

Cutting across the country from west to east and dissecting the fault line is a watershed of granite gneiss and schist that divides the Niger River and Lake Chad drainage basins from the lowland rivers that drain into the Atlantic Ocean or the Zaire River. In the northern handle of the country savannas slope down from the Mandara Plateau, an area of distinctive mosaic vegetation, to the Lake Chad Basin.

That part of Cameroun falling roughly south of the latitudes of 5°N in the east and 7°N in the west lies in the Guineo-Congolian center of endemism, although outliers and islands or rainforest-type vegetation occur to the north, often on flanks of valleys in the Guineo-Congolia/Sudania transition zone.

3. People's Republic of the Congo

The People's Republic of the Congo covers an area of 342,000 km², of which about 60% is densely forested. A treeless coastal strip, no more than 60 km wide, edges the Atlantic Ocean. Spurs of the Mayombe Escarpment project on to the coast. The escarpment rises as a succession of forested plateaus. No relief features in the country are more than 800 m high. East of the Mayombe highlands is the Niari Valley, about 200 km wide, which rises to the Massif du Chaillu, with elevations of 2560–3680 m. on the Gabon border. In the south the land rises to the Plateau des Cataractes fronting the Zaire River gorge. East of the Niari Valley the land rolls through a series of plateaus steeply dissected by tributaries of the Zaire River. North of Brazzaville are the grassy Téké Plateaus, originally forested with drier semi-evergreen peripheral rain forest species, which slope into the peneplain of the Zaire River Basin. For 690 km the Zaire River constitutes the frontier with Zaire, succeeded at their swampy confluence by the Oubangui River, which meanders through a zone of swamp forest and thicket covering about 4 million hectares in the central eastern region of the country.

The climate varies somewhat with latitude: Ponte Noire on the Atlantic coast has a dry season of three to four months and about 1300 mm of rain annually. Inland, at Djambala on the Bateke Plateau, rainfall increases to more than 2000 mm per year, and the dry season diminishes to about two months, while in the heart of the Zaire Basin, around the confluence of the Zaire and Oubangui rivers, there is no dry season, but in February rainfall is minimal, about 60 mm.

4. Angola

Angola covers an area of 1.25 million km² and lies, except for the non-contiguous exclave of Cabinda north of the Zaire River, between 6°S and 18°S latitude and 12°E and 24°E longitude. There are three major relief features. A coastal strip of Secondary and Tertiary sediments, varying in width from about 50 km to 200 km, edges the Atlantic Ocean. The cold Benguela current and offshore winds extend the arid Kalahari-Highveld transition zone (UNESCO/AETFAT/UNSO Vegetation Map of Africa) along the coast as far north as 13°S. The southern coastal strip thus widens into the Moçamedes desert. Northward along the coast yearly rainfall rises to around 800 mm at the mouth of the Zaire River and the short dry season at midyear disappears, although the two- to three-month dry period of December–February remains. A mountain chain parallels the coastline. Moco Massif, the highest peak at 2620 m, is located inland from Lobito. In general, the hills range from 500–2000 m high, and are composed of ancient sedimentary rocks, probably from the Permian-Triassic, of folded limestone, shale and sandstones that show evidence of glaciation. They descend to a high peneplain, averaging 1000–1700 m in altitude, that tilts slightly to the east and covers about two thirds of the country. The plateau of eastern Angola is covered by a thick cloak of Kalahari Sand.

5. Zaire

Zaire encompasses 2.34 million km², stretching across the equator from 5°N to a tail at 13°S. The country has three well delimited relief features. Central Zaire is dominated by the Zaire River Basin, covering 900,000 km² in this country, at an average altitude of 300–500 m. Rainfall in the basin (Secteur Forestier Central) averages 1600–2000 mm per year, but it is unevenly distributed and near the equator may drop below 100 mm for one or two months. Mbandaka, on the equator at the swampy confluence of the Zaire and Ruki Rivers, averages 2066 mm of rainfall annually, but is one of the few localities in the basin to exceed 2000 mm of rain. Yangambi, a degree north of the equator at 25°E, averages 1964 mm of rain and has a dry season of a couple of weeks in January and a diminution of rainfall to about 70 mm at midyear. Bultot (1955, quoted by White, 1983) estimated that dry periods of a month or more occur there once every 12 years. However, researchers in general have remarked upon the poor correlation between species distribution in the basin and rainfall, moisture, or other obvious environmental factors, and many believe that soil type, geological history, and intervention by animals or humans may contribute heavily to the phytogeography of Zaire and the other countries surveyed here.

To the east of the river basin a rift slices through the lakes (Albert, Edward, Kivu and Tanganyika) that form Zaire's eastern border. Rising sharply from this *fosse* is the Kivu Ridge system, with maximum altitudes of 2000–5000 m. Savanna grasslands and remnants of transition rain forests occur on volcanic soils between 1000–1500 m, giving way at higher altitudes to Afromontane vegetation. The snow-covered peaks of the Massif du Ruwenzori, stretching along Zaire's border with Uganda, top out at 5109 m at Margherita Peak.

In the west, Zaire has a 40 km coastline on the Atlantic

Ocean at the mouth of the Zaire River. Mangrove thickets front the ocean and behind them is a 250 km² tract of swamp forest that has been reviewed by J. Lebrun and Gilbert (1954). The cold Benguela Current depresses temperatures to about 25°C on average, while rainfall drops to about 750 mm per year.

Northeast and south of the Zaire River Basin are plateaus. In the northeast, the Ubangi-Uele plateaus on the border with the Central African Republic are within the transition zone between the Guineo-Congolian and Sudanian phytochoria. There rainfall averages about 1500 mm a year and the dry season about two months. Most of the area is grassland, but drier outliers of the river basin rain forest extend far north in some of the river valleys. South of the *cuvette*, Zaire extends through a transition zone into the UNESCO/AETFAT/UNSO vegetation map's Zambezian center of endemism in High Africa. At about 1400 m on the Shaba plateau, on Zaire's border with Zambia, rises the Lualaba River, which is the major headstream of the Zaire River. Rainfall on the Shaba Plateau is of no more than 1000 mm per year, and the dry season lengthens to six months or more.

6. Rwanda and Burundi

Rwanda covers an area of 26,338 km² footed on pre-Cambrian rocks. In the west are the mountains, with peaks up to 3000 m, that part the Zaire and Nile river basins (Crête Zaire-Nil). Only in these mountains do expanses of primary forest still exist. East of the mountain chain a granite massif juts up north of Butare. Eastern Rwanda is a high plateau (averaging 1700 m altitude), sloping down in successive gradients to the marshy plains and papyrus swamps of the Kagera River on the eastern border. The northern part of the country is dominated by chains of extinct volcanoes, of which Karisimbi, 4507 m, is the tallest.

Burundi, which covers an area of 37,612 km², rises from 780 m on the shores of Lake Tanganyika to a high point of 2600 m. The country lies astride the Zaire-Nile divide on a high plateau—more than 1600 m in altitude—of pre-Cambrian rock, and is separated from Zaire by Lake Tanganyika and the Rusizi Valley, which lie in the trough of the West African Rift. To the east and southeast the plateau descends slowly to the lakes that feed the White Nile. The climate varies with altitude. At the crest of the Congo-Nile divide the temperature averages 14°-19°C and rainfall exceeds 1200 mm per year.

B. Vegetation

The rain forests of Cameroun, Central African Republic, People's Republic of the Congo, Zaire, Angola, Rwanda and Burundi belong to the phytochoria identified on the UNESCO/AETFAT/UNSO vegetation map of Africa as the Guineo-Congolian regional center of endemism—that is a phytochorion with more than 50% of its species confined to it and also more than 1000 endemic species—or to the transition or mosaic zones north, east and south of the Guineo-Congolian region.

Broadly speaking, there are two major types of rain forest within the Guineo-Congolian phytochorion. The mixed moist semi-evergreen forests of Central Africa are characteristic of the vast Zaire River Basin, which measures about 1300 km in diameter, sprawls over an area of some 3.46 million km² and extends into all the African nations surveyed here. The second major variant of rain forest—a hygrophilous evergreen rain forest of high endemism—is localized in these countries along the humid coast of Cameroun, arcing around the Bight of Biafra, where rainfall may average up to 3000 mm per year. On the peripheries of the two major rain forest types and in transition zones, a drier semi-evergreen forest extends even unto the Kalahari Sands of Angola and the Shaba Region of Zaire, an area of great plant diversity and endemism.

R. Letouzey's phytogeographical study of Cameroun provides an overview of the diverse vegetation groupings. The southeastern area of Congolese rain forest is considered poorly known, as are the highlands transecting the country north of 8°N. Hepper has summarized Letouzey's list of poorly known localities as follows: rock heaps on Mts. Mandara and Alantika; forest outliers on the Poli Mts.; grasslands on iron pans on the Adamawa Plateau; outliers and gallery forest in many parts of northern Cameroun; Mt. Oku and the south and west sides of the plateau; Massif de Ngor; the confluence of the Mbam and Sanaga; Massif du Mba Minkoum; chaine de Ndom; Inoubon forest and the mouths of the Sanaga and Nyong rivers; hill forests of Campo-Lolodorf-Nyabessan; swamp forests of Nyong; certain grasslands near Dja; and the *Gilbertiodendron* forest of southeast Cameroun. In addition, although the coastal rain forest is much degraded and, as Letouzey observes, disappears day by day, the forest is intact south and west of the Roumpi mountains, but there is little information about it.

The two major rain forest variants in Cameroun are demarcated by a transition zone of drier forest between them. Letouzey (1968) describes the hygrophilous coastal evergreen forest ('forêt biafréenne') as extending in a 100-150 km wide arc around the Bight of Biafra, widening in the south to 200-250 km, where altitude diminishes to no more than 300 m. In the north and northeast, the forest covers the slopes of the escarpment and Mt. Cameroun to a height of 1000-1200 m. In the east, and drifting to the southeast, the forest is found on slopes between 800-1000 m, which lead up to a plateau averaging 600-900 m in height. Everywhere in the arc rainfall averages around 3000 mm per year with no dry season, although around Campo in the south, rainfall diminishes in the summer, and the band of evergreen forest narrows. Atmospheric humidity is high throughout the year. Among the features distinguishing this forest type is the dominance of many endemic species of Caesalpinioideae, often in gregarious and pure stands.

A localized variant of the hygrophilous evergreen forest is a lowland zone of littoral vegetation, the existence of which cannot exactly be attributed to differences in physiography, soils or hydrology. The possibly secondary littoral forest, dominated by *Lophira alata* and *Sacoglottis gabonensis*, occupies a band 50 to 100 km wide around Douala and then narrowing along the coast southward to Kribi and Campo. A peculiarity of the *L. alata* forest is its total defoliation for one to two weeks when rainfall decreases to about 60 mm in December.

Mixed moist semi-evergreen forest with strong affinities to that of the northwest Zaire River Basin is found in the southeast, delimited in the north latitudinally between 3° and 4°N and east and west between 12° and 15°E. The northern peripheries of the forest follow roughly the courses of the Boumba and Dja rivers, while in the west drier peripheral forest grades gradually through agricultural zones into the Congolese rain forest. The terrain is a rolling plateau averaging 600–700 m in altitude in the west, falling to about 400 m in the southeast. Much of the area is poorly explored floristically, particularly the extreme southeastern region between the Boumba and Sangha rivers. The Congolese forest supports more lianes than does the coastal rain forest, and these are often found on firm ground, whereas on the coast they occur in water or waterlogged soils.

Drier peripheral semi-evergreen forest, characterized by Sterculiaceae and Ulmaceae, has a northern and eastern limit roughly demarcated by a sinuous line passing roughly through Yokadouma-Batouri-Bertoua-Nanga-Ebuko-Yaoundé-Bifia-Foumbot, on an undulating plateau averaging 600–900 m in altitude. However, vegetation characteristic of the moist evergreen and semi-evergreen forests extends into the zone, particularly on slopes of valleys and in the depths of secondary forest. Rainfall generally varies between 1450 and 1750 mm per year, with two to three months of less than 50 mm of rain and a summer spell of drier weather. The peripheral forest is much influenced by human activity, with abandoned fields giving rise to a diverse mosaic of vegetation. Northward of the peripheral forest are savannas and woodlands subject to seasonal fires.

Mt. Cameroun and the higher peaks of the Cameroun Highlans support Afromontane vegetation. Richards (1963) has described the undifferentiated Afromontane forest of Mt. Cameroun, where rainfall on the slopes can average 10,000 mm per year. The zone of transition between the moist lowland forests and the mountain forest is extremely degraded.

In the People's Republic of the Congo there are three major areas of mixed moist semi-evergreen rain forest concentration. Paralleling the coast, forest covers the Mayombe Escarpment and its foothills, as well as the plateau of Kouilou. Inland, along Gabon's southern border, is the Niari-Chaillu zone, encompassing about 3.5 million hectares of forest. The Zaire River Basin forest covers the entire northeast of the country. Its southern limits are delimited roughly by an arc cutting through Ewo, Etoumbi, Makoua and Mossaka, with a westward thrust along the Alima River.

On the Bateke Plateau, which averages 400–500 m in altitude, the forest cover has been largely stripped, and tracts characteristic of the original vegetation exist mainly on inaccessible sites. Makany (1976) notes that after several repeated clearings the forest disappears completely. Characteristic of the sandy clays on the higher reaches of the plateau, from about 700–750 m in altitude to about 400 m on its rim, and unique to it in the Congo, are forest dominated by *Parinari excelsa*, a species elsewhere widespread throughout tropical Africa, which here appears to compete with secondary savanna grasses. Another forest grouping, consisting of *Dialium corbisieri*, *Pentaclethra setveldeana*, and *Millettia laurentii*, is characteristic of the sandy soils at the edge of the plateau and on the hills of dry valleys.

As mentioned previously, the Sangha River interval, a "species-poor" zone, intervenes between the western forest and those of the Zaire River Basin, occupying a band at least 400 km wide between 14° and 18°E. The mean rainfall of 1600–1800 mm per year is less than that to the east or west, but a large part of the region falls within swamps dominated by "une forêt claire curieuse" of tall trees of inexplicable origin (Farron, 1968). On gentle slopes along waterways occur pure stands of *Gilbertiodendron dewevrei*, that cover about a third of the total area of the zone, while *Raphia* spp. dominate the lower waterlogged soils.

The northern half of Angola falls within the Guineo-Congolia/Zambezia transition zone, and large swaths of rain forest vegetation, sometimes several kilometers wide, penetrate the zone in the valleys of Zaire River tributaries in the northwest districts of Zaire, Uige and Cuanaza. Elsewhere on the plateau the vegetation forms a mosaic or intermingling of Guineo-Congolan and Zambezian species.

However, sprawling along the escarpment leading up to the plateau from about 7°50'S to 11°30'S, and about 100 km inland from the dry sea coast, is the Dembos "cloud forest," an exclave of rain forest vegetation found at altitudes of 350–1000 m. Rainfall on the escarpment ranges from 1100 to 1500 mm per year, but the forest owes its existence to condensation from the west sea breezes. Further south in the Zambezian regional center of endemism is the Seles exclave, a continuation of anomalous forest dominated by rain forest species.

Northeastern Cabinda falls within the Guineo-Congolian Region and is part of the Mayombe Escarpment region in its floristics, with many stands of *Gilbertiodendron dewevrei*.

In Burundi, the dense montane forest of the slopes has been cleared in most places and invaded by secondary species or savanna vegetation. Bamboo thickets were formerly widespread, but they too have been largely razed. The northeast plateau and Rusizi Valley experience two dry seasons—January–February and June–August—and the temperature is always more than 18°C. In Rwanda, in the mountains of the northwest above 2000 m, remnants of African montane forest still survive, but often reduced to small relic forests or succeeded by thick bamboo forest, especially on the volcanic slopes. Afroalpine vegetation occurs at elevations of 3800–4500 m.

C. Population

Some 140 tribes and ethnic groups inhabit Cameroun, which as a political unit unifies territories held by France and Britain after World War I. The northern savanna plateaus are inhabited by Sudanic and Arab pastoralists, who migrate seasonally, while the south is populated by Bantu agriculturalists. Pygmies live in the southeastern forests. By policy, the government has maintained agriculture as the keystone of the economy, even after the discovery of modest offshore oil deposits near the Nigerian border and after world prices for Cameroun's cash crops weakened. The country is virtually self-sufficient in food production. Major export crops are cocoa, coffee, cotton, rubber and timber. The government is actively supporting the introduction of cotton and rice farming, but the spread of forest farming is exerting great pressure on the rain forests.

The population of Cameroun numbers about 9.14 million (1983 UN estimate), and expands about 2% per year. Agriculture employs about 80% of the labor force. The littoral zone and Atlantic coastline are densely populated, and much of the original forest has been destroyed. Logging and plantation farming are rapidly opening up the interior forests. Cameroun has more than 20 million hectares of forestland, of which 17.4 million are considered exploitable. Although most industry is related to the primary sector, there are two aluminum plants at Edeah, using domestic and imported bauxite. Near the west coast oil palm plantations have been established north of Douala. Cocoa growing is concentrated around Mbanga and Kumba. Banana plantations are found toward Yabassi, Penja-Loum and Kumbai. Coffee groves are in the north. In central Cameroun oil palm and cocoa growing is concentrated around Edea, Eseka, Ebolowa and Ambam. The Congolese forest is less decimated, but shows the influence of centuries of shifting cultivation, introduction of taxa by the mosaic of cultures that have moved through the area, and invasion by species characteristic of drier forests. Farmers around Lomie and Zoulabot grow cocoa as a cash crop.

About 70% of the 1.65 million people (1983 UN estimate) of the People's Republic of the Congo live in the southern section of the country, and 40% are urban dwellers. Although the Niari Valley is intensively farmed and the country exports some coffee, cocoa and tobacco, timber is the major product of the primary sector. Logging is a big industry in the Mayombe district, where plantations of *Terminalia superba* abound, and the Niari-Chaillu zone is extensively harvested for *Aucoumea klaineana* (Okoumé) and Limba. The Bateke Plateau, as indicated above, is essentially deforested. The forests of the northeast and Zaire basin have yet to be exploited, although attempts are being made to build roads and upgrade river transportation into the interior.

Angola has a population of about 8.6 million, of whom about 20% live along the coastal belt, 40% on the central plateau and around Huambo, and most of the rest along the Luanda–Malanje railroad line. About 82% of the populace engages in agriculture, growing wheat, corn and cassava as subsistence crops; sisal and cotton as cash crops; and coffee, oil palms and sugar cane as export crops. Two species of wild coffee were part of the original shrubby stratum of the forests of Cuanza Norte and Cuanza Sul.

Zaire, with its population of more than 31 million, is the fourth most populous state in Africa, and the population growth rate is estimated at 3%. Kinshasa, the capital, has a population of about 3 million, and nine cities in all have more than 100,000 inhabitants. About 80–90% of the people live on the land in scattered villages, most densely in the southeast savanna regions of Kasai, Kwilu, the Bas-Zaire area around Kinshasa, and in Kivu Region, where the transition rain forests leading into Afromontane vegetation have been almost totally converted into heavily farmed grasslands. Although most of the population engages in subsistence agriculture, the plantation sector is fairly well developed and major cash crops include palm oil, rubber, coffee, fiber crops and sugar cane. Shaba Region is rich in minerals, and Zaire is the world's largest producer of cobalt and exporter of industrial diamonds, and the fourth-largest producer of copper.

Rwanda, with its population of more than 3 million, is the most densely populated country in Africa. Smallholding families—the average land holding is about one hectare—graze livestock and grow beans, peas, potatoes, cassava, sweet potatoes and corn as subsistence crops. Coffee, tea, pyrethrum and cinchona are the plantation and cash crops. The soil is thin and poor, except in some of the river valleys, and erosion and brush fires are major problems. This former mountain kingdom has more than 135 people per km^2, living mostly in family compounds scattered across the countryside, rather than in compact villages.

The plateau of central Burundi is densely populated and entirely under cultivation, the major cash crops being coffee, tea and tobacco. Northeast Burundi, more sparsely populated, has wooded savannas and forested alluvial valleys with a mixture of Zambezian and Guinean taxa. The Ruvubu River Valley transacts the plateau diagonally from the northeast. On the border with Rwanda the plateau drops off to marshy lake valleys dominated by vast swamps of papyrus. Reed marshes are also found in the southwest along the shores of Lake Tanganyika and the Rusizi River. Southern Burundi experiences a marked dry season.

III. Vegetation Maps

The FAO published (1980) a colored vegetation map of Cameroun on a scale of 1:1,000,000 and included that map in a pilot project, also incorporating Benin and Togo, for monitoring tropical forest cover. The FAO also sponsored the 1965 survey of the soils and ecology of West Cameroun made by Hawkins and Brunt. The first vegetation map of substantial parts of Cameroun was drafted by C. Ledermann in 1912 on a 1:1,000,000 scale covering "Deutsch-Adamaua."

Other sources include Letouzey's phytogeographical map (1958, 1:2,000,000 scale) in ORSTOM's *Atlas du Cameroun* and a geological map of Cameroun by Gazel et al. (1956, 1:1,000,000 scale). ORSTOM has published a pedological map of Cameroun (1961, 1:5,000,000), eastern Cameroun (Martin & Segalen, 1966, 1:1,000,000), and the Yagoua district of North Cameroun (Sieffermann & Vallerie, 1963, 1:1,000,000).

The best overall vegetation map of the Congo remains that from the *Atlas du Congo* (ORSTOM, 1:2,000,000) published in 1969. More localized maps include Rollet's map of vegetation types in the Zaire basin north of the Equator (1963, 1:1,000,000) and Koechlin's publication on savanna vegetation in the south of the Congo Republic (1961, 1:1,000,000).

Two phytogeographical maps have been published for Angola—the 1939 map of Gossweiler and Mendonça on a 1:2,000,000 scale, and Barbosa's 1970 map drafted on a 1:2,500,000 scale, which includes photographs and maps of some formations not included in the earlier work. Other vegetation and soil and vegetation maps have been published for various regions or park areas. A regional study including a vegetation map on a rain forest area is Monteiro's *Estudo da flora e da vegetação das florestas abertas do planalto do Bié* (1970, 1:500,000). Diniz' *Caracteristicas mesológicas de Angola* (1973) contains many small-scale vegetation maps.

The Zairean Institute National pour l'Étude et la Recherche Agronomique (INERA), formerly the Institut National pour l'Étude Agronomique du Congo Belge (INEAC), published a series of thirty-one maps between 1954 and 1970 relating vegetation to soils in Zaire, Rwanda and Burundi. The colored maps, begun as a series entitled *Carte des sols et de la végétation du Congo Belge du Ruanda-Urundi* and continued as *Carte des sols et de la végétation du Congo, du Rwanda et du Burundi*, are mostly drawn on scales of 1:50,000 or 1:100,000, but some range from 1:10,000 to 1:1,000,000. The maps, scales and authors are listed in an Appendix.

In 1936, J. Lebrun published two colored vegetation maps in *La forêt équatoriale congolaise*, on scale of 1:4,000,000 and 1:2,000,000, and in 1954 he and G. Gilbert devised an ecological classification of the forests, which was published with a small-scale vegetation map. R. Devred (1958) also published a small-scale map with his description of forest vegetation in Zaire, Rwanda and Burundi. Other researchers have included local vegetation maps in their reports on the floristics of different areas.

IV. Floristic Inventory

A. COLLECTIONS AND HERBARIA

In the late 19th and early 20th centuries, botanical collections from the rain forests of Cameroun were made by German colonizers, who also established a botanical garden at Victoria. Only descriptions of these explorations and taxa survive. Early surveyers of the hygrophilous coastal forest, whose major interest was in exploitable timber species, include Fr. Jentsch and M. Büsgen (1908-1909) and A. Bertin (1920). More scientific in nature were the sample plot surveys of J. Mildbraed (1930), and his various explorations and forest inventories remain a major source of floristic data for Cameroun. A summary of early collections of *sous-bois* species was made by A. Engler (1910-1925). Engler also included accounts of mountain vegetation in Cameroun in *Über die Hochgebirgsflora des tropischen Afrika* (1892). J. Mildbraed (1922) published a summary of his collections from around Kribi, and P. W. Richards (1963) studied the Bakunda Forest Reserve south of Kumba, where human action has introduced peripheral species into the coastal forest. Taxonomic work on many of these collections remains fragmentary. Important explorations of the Congolese forest were undertaken by the German Central African Expedition in 1910-1911 (Schultze, 1913; Mildbraed, 1922).

Important collections of Cameroun taxa are housed at the Museum National d'Histoire Naturelle, in Paris, at the Herbarium Vadense, Wageningen, and at Kew, England, with many duplicates at the Jardin Botanique National de Belgique, Meise. The Herbier National du Cameroun at Yaoundé, Cameroun, houses about 47,500 specimens, mostly duplicates.

In the People's Republic of the Congo, the Brazzaville herbarium, under the aegis of ORSTOM, has about 35,000 specimens. Duplicates of its collections are distributed among the herbaria of the National Museum of Natural History in Paris, Wageningen in The Netherlands, and the Jardin Botanique National de Belgique, Meise.

The Sangha River interval, except at its southern end, is one of the least-collected areas of Africa, and few collections have been made from the dense forest of the Zaire River Basin except along the Oubangui River. The southern part of the country is generally well collected, except for the district around Zanaga, near Gabon's southern border and much of the areas bordering Cabinda.

Although there are no plans for a systematic floristic inventory of the interior, forest surveys of various relatively well-known regions are being carried out by the Centre Technique Forestier Tropical (CTFT), which conducts research on forest and silviculture. Inventories already completed include those of Ouesso region, the Massif du Chaillu and the area around Dongou-Epena on the Oubangui River. Some of the more recent publications mentioned here are essentially updatings of Descoing's inventory (1961) of vascular plants to include collections made since then.

Three organizations sponsor work on the vegetation of the Congo: ORSTOM, Paris, CTFT, and the School of Sciences at the University of Brazzaville.

Botanists such as Tisserant and Le Testu were active in the region now the Central African Republic by the early 20th century.

Much of Angola, especially along the coast and extending well into the interior, is considered well collected. Pockets of poorly known areas occur inland from Ambrizete, in Benguela district, in parts of Lunda, and in much of Moxico and Cuando Cubango. Between 1937 and 1985 four volumes of the *Conspectus Florae Angolensis* had been published. The 2137 species described represented an estimated 40% of the country's taxa, but early volumes are now outdated, and publication is being delayed by the political situation. Angolan institutions involved in collecting and inventorying local vegetation include the Instituto de Investigação Agronomica de Angola, the Instituto de Investigação—Cientifica de Angola (LUAI), and the Botanical Laboratory of the University of Luanda (LUAU).

For what is now Zaire, the first collections from the Zaire River Basin were made in the 19th century, notably by G. Schweinfurth (1870-1871), Pogge (1875-1883), R. Büttner

(1885, 1886), F. Hens (1887–1888), F. Demeuse (1888–1892), J. Gillet (1893–1903), A. Dewevre (1895–1896), E. Verdick (1899–1900) and others. Some of these early collections were housed in Berlin and were destroyed during World War II, and many rare taxa are now known only by description. Duplicates of some of the Berlin pre-World War II collections are still intact in Hamburg, Kew, Meise and Paris.

The herbarium at the Jardin Botanique National de Belgique (BR), Meise, houses about 500,000 specimens from Zaire, Rwanda and Burundi, and is the repository for current collectin activities in Zaire, as well as for *Flore d'Afrique Centrale* source material. Other minor collections are at Kew and the British Museum, London, the Museum National d'Histoire Naturelle, Paris, Herbarium Vadense, Wageningen, and at the Missouri Botanical Garden. In Brussels, the Laboratoire de Botanique Systématique et d'Ecologie houses collections from Zaire (mainly Shaba, Kisangani) and Ruwenzori, while the Laboratorium voor Algemene Plantkunde en Natuurbeheer contains collections mainly from Shaba. Quite recent collections from Zaire, about 50,000 specimens, by S. Lisowski, a Polish botanist, have been sent to the herbarium at Poznan, Poland, with duplicates deposited in Meise. A history of botanical collecting in the Congo was published by W. Robyns in 1962.

In Zaire, duplicates of collections, totaling about 100,000 specimens, made between 1938 and 1960 were deposited at the Yangambi Herbarium near Kisangani, but the herbarium has no electricity and access to it is rather difficult. About 25,000 specimens, mostly from collections in Shaba region, are housed at Lubumbashi. In Kinshasa, a total of about 10,000 specimens are split between the University and the INERA herbaria. Another small collection of about 10,000 specimens is located at Iwiro, near Bukavu, and represents species found in Kivu and Rwanda.

The best-studied locality in the central forest region is around Yangambi, where Évrard estimates one collection per km^2. An inventory of all the species around Kisangani was accomplished by J. Lejoly, S. Lisowski and M. Ndejele (1983). For the Kivu region, collections average about one for every two km^2, and for the Tshuapa-Equateur region there is about one collection per ten km^2. Évrard (1968) inventoried from the 'Secteur Forestier Central' alone, 2593 species from 845 genera and 119 families, and considered that his list comprised about 85% of the known spermatophytes of the region. When he added on Germain's 1957 list for the Yangambi locality, he could cite 3130 species within the river basin rain forest. For the drier Tshuapa-Equateur region, Évrard enumerated 1726 species, representing 665 genera and 103 families. In Hepper's 1971 revision of Léonard's map (1965), on the extent of floristic exploration in Africa, P. Bamps identified the following areas as most in need of further exploration and collecting: northwest of the Mayombe escarpment; Luozi territory, which falls between Kwamouth-Kutu-Lukolela and the Zaire River; the area between the Lukenie River and 2°S from Kutu toward Kindu; the rectangle formed by Kibombo and the Lomani River in the south and Opala and Kisangani in the north; southeastern Shaba Region; the areas around Manono, where the Luvua River joins the Lualaba; the region west of Lake Tanganyika; the area between Mambas-Irumu-Wamba and Watsa; the region between Bafwasenda-Ubundu-Lubutu and Lubero; the territory north of Aru; and the northeast section of the country bordering the Central African Republic.

Little work is under way in advancing the floristic inventory of Zaire, and the prospects for progress are not encouraging. Few European botanists have been active in the country recently. S. Lisowski (University of Poznan, Poland), F. Malaisse (University of Lubumbashi), and J. J. Symoens (Flemish University of Brussels) have prospected intensively the high plateaus of Shaba.

For Rwanda, the flora is estimated at ca. 2000 species. The herbarium at Meise (BR) houses about 25,000 specimens from Rwanda. The National Herbarium of Rwanda at Butare was founded in 1973 and has about 12,000 specimens. P. Ndabaneze, a Burundian botanist, has inventoried all the grasses from Burundi. Canadian foresters, in collaboration with local botanists, are working on grasses found around Butare in Rwanda and Bujumbara in Burundi.

B. Publications

Publication of the *Flore du Cameroun* in fascicles commenced in 1963 and progresses slowly. The 27 volumes published through 1984 describe 97 families, in which were 472 genera and 1482 species, representing some 25–30% of the country's taxa. Among the few specialized works on Cameroun species are Tardieu-Blot's study of pteridophytes in the *Flore du Cameroun* (1964) and Russell's work on *Raphia* taxonomy (1965).

Specialized studies on the flora of the Congo include Blanchon's *Flore des Graminees du Congo (Republique du Congo-Brazzaville)*, Descoings' inventory of vascular plants, and the work by Bouquet on inventory of medicinal and toxic plants (1974).

The earliest checklist of the region was Durand and Durand's *Sylloge Florae Congolanae (Phanerogamae)* of 1909.

A flora of Mayombe was published between 1924 and 1938 by Pellegrin, based on Le Testu's collections. Recent works on the local flora include Sita's treatment of the vegetation of the island of M'Bamolu (Stanley Pool), Sauter's work on the Mbe Plateau, Makany's on the vegetation of the coastal region, and Descoings' mimeographed inventory of vascular plants deposited in the Brazzaville herbarium (1961).

In 1913 Chevalier published an enumeration of plants collected on a 1902–1904 expedition to the basins of the Oubangui and Chari rivers (the latter flowing northwestward to Lake Chad), in what is now the Central African Republic. Botanists such as Tisserant and Le Testu were active in the region by the early 20th Century.

The great expeditions of the late 19th century that opened Africa's interior to the imperial powers initiated botanical surveys as well. The first checklist of the vegetation of the Congo basin was Durand and Durand's *Sylloge Florae Congolanae (Phanerogamae)* (1909). Work on the *Flore du Congo Belge et du Ruanda-Urundi* commenced in 1942 and publication began in 1948. The name was changed in 1963 to *Flore du Congo du Rwanda et du Burundi.* Two volumes had been published by 1967 when the decision was made to continue the series in the form of fascicles. In 1971 the work was renamed again: the *Flore d'Afrique Centrale.* With the publication in 1984 of the first part of the treatment of orchids, more than 50% of Zaire's 10,000 or so species of spermatophytes have been described.

Localized floristic works include W. Robyn's *Flore des Spermatophytes du Parc National Albert* (3 vols., 1948–1955), covering what is now Virunga Volcanoes National Park, and G. Troupin's incomplete *Flore des Spermatophytes du Parc National de la Garamba* (1956). Three forester's manuals have also been published, the most recent of which, *Les Essences Forestières du Congo Belge et du Ruanda-Urundi* (1959) by J. Gillardin, is incomplete. In 1935 J. Lebrun published a forest manual on the eastern mountains and in 1931 came C. Vermoesen's work on the forests of Mayombe and the equatorial region. R. Pierlot (1966) published an important study on the forests of Kivu.

A team of Zairean botanists is beginning a work on the local vegetation at the University of Kisangani; three fascicles have already been published in the Annales de la Faculté de Kisangani. The government of Zaire, which is experiencing severe financial and social problems, can not sponsor botanical institutions.

By 1986 three volumes of the *Flore du Rwanda* had been published, treating 148 families of spermatophytes in all. Previous to publication of his flora, G. Troupin (1971) had issued a syllabus of the country's flora and later (1982) the *Flore des plantes ligneuses du Rwanda.* Bouxin (1974, 1975, 1976) has published several works on the vegetation of Akagera Park and on the Rugege forest in the highlands. P. Deuse (1963, 1968) has studied the marshes and bogs of both Rwanda and Burundi. Older works include Engler's on the plants of East Africa (1895). Canadian foresters, in collaboration with local botanists, are working on grasses found around Butare in Rwanda and Bujumbara in Burundi. P. Ndabaneze, a Burundian botanist, has inventoried all the grasses from Burundi.

There are few floristic works local to Burundi. In a study of a high valley on the Siguvyaye River near Lake Tanganyika, J. Lewalle (1975) found eight endemic species. Although the area was isolated and thinly populated, grazing has extended the savanna vegetation. Lewalle (1970) also prepared a checklist of the flora of eastern Burundi at various altitudes and has written a booklet (1971) on the trees of Burundi. Another local study was made by M. Reekmans (1980) in the Rusizi plain.

V. Status of Inventory

A. Endemism and Extinction

The Flora of the People's Republic of the Congo is too imperfectly known to assess its degree of endemism, although Brenan (1978) believes that it is probably comparable in specific endemism to Gabon, where 22% of the species are thought to be endemic. Three genera are known to be endemic to the Congo's rain forest areas.

For Angola Exell and Gonçalves (1973), in their analysis of extensive samples of the flora estimate specific endemism at around 38% for Angola proper and 12% for Cabinda.

Of the 3291 species described by 1980 in the *Flore du Congo Belge et du Ruanda-Urundi* and the *Flore d'Afrique Centrale,* 1280 (32.64%) were considered endemic to the three countries. Brenan (1978) cited an estimate for Zaire of 38.2% specific endemism made by Exell and Gonçalves (1973) based on a sample of 1218 species, among which were 12 endemic genera, almost all monotypic. In the tropical rain forests of Zaire, Brenan cited 19 endemic genera, representing 7.9% of total rain forest genera.

Brenan tabulated the distribution of the 1280 endemic species described in the floras and found 233 species native only to the central river basin rain forests. In the transitional region around Lakes Edward and Kivu were 62 endemic species. In the area of southern Zaire falling within the drier periphery of the moist, semi-evergreen rain forest and the Guineo-Congolia/Zambezia transition zone were 161 endemic species. Most surprising, however, were the 312 species endemic to the transitional Zambezian dry evergreen forest and savanna of High Katanga (Shaba), where there was, as well, a high degree of generic speciation, which was not the case for the central rain forest. Thus, this transition area represents a floristic enrichment rather than an impoverishment.

Rwanda and Burundi are areas of very low endemism, with fewer than 30 endemic species recognized. The original vegetation is much degraded and few tracts are intact.

Within the Guineo-Cogolian region Lebrun (1963) and others have recognized subcenters of endemism, with "species-poor" intervals in between. The coastal areas generally harbor a high number of endemic species, while Congolia has fewer, and the swampy Sangha River interval in between, extending through the Congo People's Republic, has few endemic species, lacking even one swamp species endemic to the region (White, 1979). In general, researchers have contrasted the lack of prolific speciation in the Guineo-Congolian region to the abundance of Generic endemism.

B. Conservation and the Destruction of Vegetation

Cameroun has several national parks and forest reserves: Campo Reserve on the southern Atlantic coast; Dja Reserve, incorporating Congolese and peripheral species; Faro Reserve in the highlands along the Nigerian border; Benoue National Park and Bouba Njidda National Park in the north-central highlands; and Waza National Park in the northern handle of the country.

At the 1966 meeting of AETFAT on conservation of vegetation in tropical Africa, C. Farron recommended that 16 rich and varied botanical regions be protected in what is now the People's Republic of the Congo. Six of the sites were in the southeast around Brazzaville, of which four were located on the Zaire River, including the Grand Bangou forest near Kindamba, an area rich in endemics and varied in biomes, and the Petit Bangou forest near Mayama, where many rare species were found; three tracts were located in the Niari Valley, including a primary forest near Aubeville; four sites were around Pointe-Noire, and three were in the northern forest massif. At the 1975 follow-up of that meeting, it was reported that all the areas had been largely devastated by exploitation of various kinds.

A Nature Protection Council was created in 1955 in Angola to oversee conservation efforts. Angola has three national parks—Quiçama, Cameia and Iona—two national integral reserves at Mupa and Luando, and 41 forest reserves. The 90,000 hectare Alto Maiombe forest reserve in Cabinda is representative of mixed moist semi-evergreen forest, while the Beu forest, covering 122,000 hectares, on the northern border near São Salvador, supports a mixed formation of drier peripheral rain forest and savanna species. The Golungo Alto forest reserve of 105,000 hectares on the escarpment in Cuanza Norte, is part of the Dembos forest exclave. Progress is reported in maintaining these reserves, and private interests have been expropriated in Quiçama National Park (Hedberg, 1976).

As far back as AETFAT's 1966 symposium on conservation in Africa, R. Germain had described two sites in Zaire as needing urgent protection, and both are still menaced. The forests of the Lukaya Valley near Kimuenza, threatened by the expansion of Kinshasa, is a site where several type specimens were collected in the early twentieth century. Some of these taxa are now known only by their descriptions, having been destroyed in Berlin during World War II. The second site is a small population of *Juniperus procera* near Kasiki on the Marungu Plateau in Shaba. However, the expansion of cobalt and copper mines in Shaba has already resulted in the disappearance of many species, including orchids, and dam construction on the Lufira River threatens many of the deciduous forest species. Other endangered sites are the Virunga Volcanoes National Park and the foothills of Ruwenzori, where cattle encroachments have already converted some areas of transitional forest to grassland. Other areas for which Germain urged protection are a dry semi-evergreen forest of *Khaya-Afzelia-Olea-Anogeissus* in the region north of Basape; *les ilôts relictes* of Triplochiton scleroxylon in Lebo; and certain areas under rapid development in the southern savannas and forests.

Zaire has a fairly large system of national parks and forest reserves, although poaching and trespassing are problems, as indicated above for Virunga. The first forest reserves were established in 1910, when about 600,000 hectares were set aside around the Sankuru, Kasai and Nepoko rivers. In 1925, Albert National Park was created in Kivu, gradually being extended to 800,000 ha. Other parks include Garamba (High Uele), established in 1939 and covering 1.173 million hectares. Kundelungu National Park, north of Lubumbashi, was created after 1960. Salonga National Park, comprised of two discontinuous tracts, Maiko National Park, and Kahuzi-Biega National Park were established since 1970 in the Zaire basin. There are also 146 forest reserves totaling 500,000 ha.

Two plant species have been fully protected since 1953: *Encephalartos laurentianus* De Wild, which occurs in the Kasai phytogeographical region, and *E. septentrionalis* Schweinf., occurring in the Uere valley and Bomu.

Priority in terms of floristic studies should be given to the continuation of the *Flore d'Afrique centrale* and the collaboration of Zairean botanists is highly to be desired. The most pressing need is money for the electrification, air conditioning, rehabilitation and upkeep of the Yangami herbarium. Such funds would almost certainly have to come from external sources.

In Rwanda there are two national parks. Akagera, in the northeast, on the borders with Tanzania and Uganda, has 179,000 protected hectares and an additional 71,000 hectares that are open to agriculture, but not hunting or fishing. The Parc National des Volcans, in the northwestern part of Rwanda, is mapped at 25,000 hectares, but the area of the park has been much reduced by encroachment of *Pyrethrum* plantations and the montane bamboo forest has been devastated by grazing. Part of the Zaire-Nile divide (Crête Zaire-Nil) is being set aside as reserve land. Plans to establish a reserve around the Nyabarongo Lakes have been stymied by the rapid settlement of the area.

Some progress in conserving threatened areas in Burundi was reported at the 8th AETFAT convention in 1975. The Burundi 'Departement des Eaux et Forêts' has taken action to protect threatened species from the expansion of tea plantations and was protecting more vigorously the Crête Zaire-Nile nature reserve. In addition, the creation of a national park in the Muyinga areas had been proposed, and the government had asked the United Nations for assistance in creating another park along the Ruvubu River.

VI. Literature Cited

Barbosa, L. A. Grandvaux. 1970. Carta fitogeográfica de Angola. Inst. Invest. Cient. Luanda, Angola

Bouquet, A. 1974. Inventaire des plantes médicinales et toxiques du Congo-Brazzaville. ORSTOM, Paris.

Bouxin, G. 1974. Étude phytogéographique des plantes vasculaires du marais Kamiranzovu (forêt de Rugege, Rwanda). Bull Jard. Bot. Nat. Belg. 44: 141–159.

———. 1975. Ordination and classification in the savanna vegetation of the Akagera Park (Rwanda, Central Africa). Vegetatio 29: 155–167.

———. 1976. Ordination and classification in the upland Rugege forest (Rwanda, Central Africa). Vegetatio 32: 97–115.

Brenan, J. P. M. 1978. Some aspects of the phytogeography of tropical Africa. Ann. Missouri Bot. Gard. 65: 437–478.

———. 1979. The flora and vegetation of tropical Africa today and tomorrow. *In* K. Larsen & L. B. Holm-Nielsen (eds.), Tropical botany. Academic Press, New York.

Chevalier, A. 1913. Etude sur la Flore de l'Afrique centrale. Enumération des espèces récoltées.

Descoings, B. 1961. Inventaire de vasculaires de la République du Congo, déposées dans l'Herbier de l'Institut d'Études Centre-Africaines a Brazzaville. ORSTOM, Brazzaville. Mimeo.

Deuse, P. 1963. Marais et tourbières au Rwanda et au Burundi. Publ. Univ. Elisabethville 6: 69–80.

_____. 1968. Rwanda. Pages 125–127 in I. Hedberg & O. Hedberg (eds.), Conservation of vegetation in Africa south of the Sahara. Acta Phytogeogr. Suec. **54**: 1–320.

Devred, R. 1958. La végétation forestière du Congo belge et du Ruanda-Urundi. Bull. Soc. Roy. Forest Belg. **65**: 409–468. With vegetation map.

Durand, T. & H. Durand. 1909. Sylloge Florae Congolanae (Phanerogamae). Albert de Boeck, Brussels.

Engler, A. 1892. Über die Hochgebirgsflora des Tropischen Afrika. Phys. Abh. K. Akad. Wiss. Berl. **2**: 1–461.

_____. 1910–1925. Die Pflanzenwelt Afrikas, insbesondere seiner tropischen Gebiete. In A. Engler & O. Drude (eds.), Die Vegetation der Erde. 9. Engelmann, Leipzig.

Évrard, C. 1968. Recherches écologiques sur le peuplement forestier des sols hydromorphes de la Cuvette central congolaise. Publ. Inst. Natl. Étude Agron. Congo Belge, Sér. Sci. **110**: 1–295.

Exell, A. W. & M. L. Gonçalves. 1973. A statistical analysis of a sample of the flora of Angola. Garcia de Orta. Sér. Bot. **1(1/2)**: 105–128.

FAO. 1980. Cameroun. Rome. With colored vegetation map, 1: 1,000,000. (UN 32/6. 1102-75-005.)

Farron, C. 1968. Congo-Brazzaville. Pages 112–115 in I. Hedberg & O. Hedberg (eds.), Conservation of vegetation in Africa south of the Sahara, Acta Phytogeogr. Suec. **54**: 1–320.

Germain, R. 1957.

Gillardin, J. 1959. Les Essences forestières du Congo Belge et du Ruanda-Urundi.

Gossweiler, J. & F. A. Mendonça. 1939. Carta fitogeográfica de Angola. Edição do Governo Geral de Angola, Lisbon.

Hedberg, I. 1976. Follow-up of the AETFAT meeting at Uppsala in 1966 on "Conservation of vegetation in Africa south of the Sahara." Boissiera **24**: 437–441.

Lebrun, J. 1935. Les essences forestières des regions montagneuses du Congo oriental. Publ. Inst. Natl. Etude Agron. Congo Belge Sér. Sci. **1**: 1–264.

_____. 1936. La forêt equatoriale congolaise. Bull Agric. Congo Belge **27**: 163–192.

_____ & G. Gilbert. 1954. Une classification écologique des forêts du Congo. Publ. Inst. Natl. Étude Agron. Congo Belge Sér. Sci. **63**: 1–89.

Léonard, J. 1965. Liste des flores africaines et malgaches recentes. Webbia **19**: 865–867.

Letouzey, R. 1968. Étude phytogéographique du Cameroun. Éditions Paul Lechevalier, Paris.

Lewalle, J. 1970. Liste floristique et répartition altitudinale de la flore du Burundi Occidental. Univ. Officielle de Bujumbura, Bujumbura. Mimeo.

_____. 1971. Arbres du Burundi. Ser. 1, Fasc. 1. Univ. Officielle de Bujumbura, Bujumbura. Mimeo.

_____. 1975. Endémisme dans une haute vallée du Burundi. Boissiera **24**: 85–89.

Makany, L. 1976. Végétation des Plateaux Téké (Congo). Trav. Univ. de Brazzaville **1**: 1–301.

Mildbraed, J. 1922. Wissenschaftliche Ergebnisse des Zweiten Deutschen Zentral-Afrika Expedition 1910–1911. 2. Botanik. Leipzig. Klinkhardt & Biermann.

_____. 1930. Probeflächen Aufnahem aus dem Kameruner Regenwald. Notizbl. Bot. Gard. **10(99)**: 951–976.

Monteiro, R. R. Romero. 1970. Estudo da flora e da vegetação das florestas abertas do planalto do Bié. Inst Invest. Cient. de Angola, Luanada.

Pellegrin, F. 1938. La Flore du Mayombe, 1924–1938. Mem. Soc. Linn. Normandie, XXVI, 2: (1924); ibid., Nouv. Ser. Botanique I, 3, 1928; ibid., I, 4, 1938.

Pierlot, R. 1966. Structure et composition de forêts denses d'Afrique centrale, specialement celles du Kivu. Acad. Roy. Sci. Outre-mer **16(4)**: 1–367.

Reekmans, M. 1980a. La flore vasculaire de l'Imbo (Burundi) et sa phénologie. Lejeunia, nov. ser. **100**: 1–53.

_____. 1980b. La végétation de la plaine de la Basse Rusizi (Burundi). Bull. Jard. Bot. Nat. Belg. **50**: 401–444.

Richards, P. W. 1963. Ecological notes on West African vegetation. II. Lowland forest of the Southern Bakundu Forest Reserve, and III. The upland forests Cameroons Mountain. J. Ecol. **51**: 123–149, 529–554.

Robyns, W. 1948–1955. Flore des spermatophytes du Parc National Albert. 3 vols. Inst. des Parcs Nationaux du Congo Belge, Brussels.

_____. 1962. La flore, livre blanc. Acad. Roy. Sci. Outre-mer **2**: 685–701.

Russell, T. A. 1965. The *Raphia* palms of West Africa. Kew Bull. **19**: 173–196.

Schultze, A. 1913. In Adolf Friedrich, Duke of Mecklenburg: From the Congo to the Niger and the Nile. (An account of the German Central African Expedition of 1910–1911. 2 vol.

Tardieu-Blot, M. L. 1964. Ptéridophytes. In Flor du Cameroun. No 3. 372 pages. Paris.

Troupin, G. 1956. Flore des spermatophytes du Parc National de la Garamba. Vol. 1. Inst. des Parcs Nationaux du Congo Belge, Brussels.

_____. 1971. Syllabus de la flore du Rwanda (Spermatophytes). Ann. Mus. Roy. Afrique Central Sér. -8°, Sci. Econ. 9. Butare, Rwanda/Tervuren, Belgium; Vol. 2, 1983; Vol. 3, 1986.

Vermoesen, C. 1931. Manuel des essences forestières de la Région Equatoriale et du Mayombe. Ministère des Colonies, Brussels.

_____. 1979. The Guineo-Congolian Region and its relationships to other phytochoria. Bull. Jard. Bot. Nat. Belg. **49**: 11–55.

_____. 1983. The vegetation of Africa: A descriptive memoir to accompany the UNESCO/AETFAT/UNSO Vegetation Map of Africa. UNESCO, Paris.

VII. Selected References

Airy Shaw, H. K. 1947. The vegetation of Angola. J. Ecol. **35(1/2)**: 23–48.

Aubréville, A. 1947. Les brousses secondaires en Afrique équatoriale. Bois Forêts Trop. **2**: 24–49.

_____. 1955. La disjonction africaine dans la flore forestière tropicale. Compt. Rend. Sommaire Séances Soc. Biogéogr. **278**: 42–49.

_____ et al. (eds.). 1963– . Flore du Cameroun Fasc. 1– . Museum National d'Histoire Naturelle, Paris.

Bamps, P. (ed.). 1967–1970. Flore du Congo, du Rwanda et du Burundi. From 1971– , Flore d'Afrique Centrale. Fascicles. Jardin Botanique National de Belgique, Meise.

Bernard, E. 1945. Le climat écologique de la Cuvette centrale congolaise. Publ. Inst. Natl. Etude Agron. Congo Belge., Coll. pp. 1–240.

Bouquet, A. 1976. État d'avancement des travaux sur la 'Flore du Congo-Brazzaville'. Boissiera **24**: 581.

Cahen, L. 1954. Géologie du Congo Belge. Vaillant-Carmanne, Liège.

Carrisso, L. et al. (eds.). 1937– . Conspectus florae angolensis.

Vos. 1- . Junta de Investigações Cientificas do Ultramar, Lisbon.
Deuse, P. 1966. Contribution à l'étude des tourbières du Rwanda et du Burundi. Inst. Nat. Rech. Sci. Butare, Rép. Rwandaise **4**: 53–115.
Diniz, A. Castanheira & F. Q. Barros Aguiar. 1969. Regiões naturais de Angola, 3rd ed. Inst. Invest. Agron. Angola, Sér. Cient. **7**: 1–6.
―――― & ――――. 1972. Os solos e a vegetação do Planalto occidental de Cela. Inst. Invest. Agron. Angola, Sér. Cient. **26**: 1–25.
―――― & ――――. 1973. Caracteristicas mesológicas de Angola. Nova Lisboa.
Duvigneaud, P. 1949. Voyage botanique au Congo Belge à travers le Bas-Congo, le Kwango, le Kasai et le Katanga. De Banbana à Kasenga. Bull. Soc. Roy. Bot. Belgique **81**: 105–110.
――――. 1952. La flore et la vegétation du Congo méridional. Lejeunia **16**: 95–124.
――――. 1955. Études écologiques de la végétation en Afrique tropicale. Les divisions écologiques du monde, pp. 131–148. (Colloques int. CNRS 59.) Also publ. in Année Biol. Sér. 3, **31**: 375–392.
――――. 1958. La végétation du Katanga et de ses sols métallifères. Bull. Soc. Roy. Bot. Belgique **90**: 127–286.
Farron, C. 1968. Congo-Brazzaville. Pages 112–115 *in* Hedberg & Hedberg (eds.), Conservation of vegetation in Africa south of the Sahara. Acta Phytogeogr. Suec. **54**: 1–320.
Frodin, D. G. 1964. Guide to standard floras of the world. Cambridge University Press, Cambridge.
Gérard, P. 1960. Étude écologique de la forêt dense à *Gilbertiondendron dewevrei* dans la région de l'Uele. Publ. Inst. Natl. Étude Agron. Congo Belge, Sér. Sci. **87**: 1–159.
Germain, R. 1968. Congo-Kinshasa. Pages 121–125 *in* Hedberg & Hedberg, eds., Conservation of vegetation in Africa south of the Sahara. Acta Phytogeogr. Suec. **54**: 1–320.
―――― & C. Évrard. 1956. Étude écologique et phytosociologique de la forêt à *Brachystegia laurentii*. Publ. Inst. Natl. Étude Agron. Congo Belge, Sér. Sci. **67**: 1–105.
Hedberg, I. & O. Hedberg (eds.). 1968. Conservation of vegetation in Africa south of the Sahara. Acta Phytogeogr. Suec. **54**: 1–320.
Lebrun, J. 1947. La végétation de la plaine alluviale au sud du lac Edouard. Exploration du Parc National Albert. Mission J. Lebrun (1937–1938), pt. 1 (2 vols). Inst Parcs Nat. Congo Belge, Brussels.
――――. 1955. Esquisse de la végétation du Parc National de la Kagera. Exploration du Parc National de la Kagera. Mission J. Lebrun (1937–1938), Vol. **2**: 1–89. Inst. Parcs Nat. Congo Belge, Brussels.
――――. 1956. La végétation et les territoires botaniques du Ruanda-Urundi. Naturalistes Belges **37**: 230–256.
――――. 1961. Les deux flores d'Afrique tropicale. Mém. Acad. Roy. Sci. Belgique, Cl. Sci., Mém. Coll. **32(6)**: 1–81.
―――― & A. L. Stork. 1977. Index des Cartes de Répartition: Plantes Vasculaires d'Afrique (1935–1976). Conservatoire et Jardin Botaniques, Geneva.
Lejoly, J. et al. 1983. Catalogue informatisé des plantes vasculaires des Sous-Régions de Kisangani et de la Tshopo (Haut-Zaire). Trav. Lab. Bot. Syst. Ecol. Univ. Libre Bruxelles. Brussels.
Léonard, J. 1951. Aperçu préliminaire des groupements végétaux pionniers dans la région de Yangambi. Vegetatio **3**: 279–297.
――――. 1953a. Les divers types de forêts du Congo belge. Lejeunia **16**: 81–93.
――――. 1953b. Les forêts du Congo belge. Naturalistes Belges **34**: 53–65.
Lewalle, J. 1968. Burundi. Pages 127–130. *In* I. Hedberg & O. Hedberg (eds.) conservation of vegetation in Africa south of the Sahara. Acta Phytogeogr. Suec. **54**: 1–320.
――――. 1972. Les étages de végétation du Burundi occidental. Bull. Jard. Bot. Nat. Belgique **42**: 1–247.
Louis, J. 1947. Contribution à l'étude des forêts équatoriales congolaise. Pages 902–915 *in* Publ. Inst. Natl. Étude Agron. Congo Belge Compt. Rend. Semaine Agric. de Yangambi.
Monteiro, R. F. Romero. 1962. Le massif forestier du Mayumbe angolais. Bois Forêts Trop. **82**: 3–17.
――――. 1970. Estudo da flora e da vegetação das florestas abertas do planalto do Bié. Inst. Invest. Cient. de Angola. Luanda.
Mullenders, W. 1953. Contribution a l'étude s groupements végétaux de la contrée de Goma-Kisenyi (Kivu-Ruanda). Vegetatio **4**: 73–83.
――――. 1954. La végétation de Kaniama (entre Lubishi-Lubilash, Congo belge). Publ. Inst. Natl. Étude Agron. Congo Belge Sér. Sci. **61**: 1–499.
――――. 1955. The phytogeographical elements and groups of the Kaniama District (High Lomami, Belgian Congo) and the analysis of the vegetation. Webbia **11**: 497–517.
Peeters, L. 1964. Les limites forêt-savane dans le nord du Congo en relation avec le milieu géographique. Rev. Belg. Géogr. **88**: 239–270.
Robyns, W. 1930. La flore et la végétation du Congo Belge. Rev. Quest. Sci. **97/98**: 261–299.
――――. 1950. Botanique. II. La flore. III. La végétation. IV. Les territoires phytogéographiques. Encycl. Congo belge **1**: 390–424.
――――. 1958. Flore du Congo Belge et du Ruanda-Urundi: Tableau Analytique des Familles. Inst. Natl. Étude Agron. Congo Belge, Brussels.
―――― et al. (eds.). 1948–1963. Flore du Congo Belge et du Ruanda-Urundi. Vols. 1–7, 8 part 1, 9–10. Inst. Natl. Étude Agron. Congo Belge, Brussels.
Schmitz, A. 1971. La végétation de la plaine de Lubumbashi (Haut-Katanga). Inst. Natl. Étude Agron. Congo Belge, Brussels.
Teixeira, J. B. 1968. Angola. *In* I. Hedberg & O. Hedberg (eds.), Conservation of vegetation in Africa south of the Sahara. Acta Phytogeogr. Suec. **54**: 193–197.
Thonner, F. 1915. The flowering plants of Africa. Dulaun, London. (Reprinted 1963, Cramer, Weinhein, Germany.)
Tisserant, C. 1950. Catalogue de la Flore de l'Oubangui-Chari. Mém. Inst. Études Centrafr. 2, Brazzaville.
Troupin, G. (ed.). 1978. Flore du Rwanda: Spermatophytes. Vol. 1. Ann. Mus. Roy. Afrique Centrale, Sér. in -8°. Sci. Econ. 9. Butare, Rwanda/Tervuren, Belgium. Vol. 2 (1983), Vol 3 (1986).
――――. 1982. Flore des plantes ligneuses du Rwanda. Ann. Mus. Roy. Afrique Centrale ser. in -8°, Sci. Econ. 12. Butare, Rwanda/Tervuren, Belgium.
White, F. 1981. The history of the Afromontane archipelago and the scientific need for its conservation. African J. Ecol. **19**: 33–54.
―――― & M. J. A. Werger. 1978. The Guineo-Congolian transition to southern Africa. *In* M. J. A. Werger (ed.), Biogeography and ecology of Southern Africa. The Hague, Junk. (Monogr. Biol 31.)

VIII. Appendix

Soil and Vegetation Maps of Zaire; Rwanda and Burundi published between 1954 and 1970 in the series 'Carte des

Sols et de la Végétation du Congo Belge et du Ruanda-Urundi' (later 'Carte . . . du Congo, du Rwanda et du Burundi') by the Belgian INEAC (later INERA).

Bourbeau, G. et al. 1955. Mosso (Urundi). Soils, vegetation 1:20,000.

Compère, P. 1970. Lower Zaire. Vegetation 1:250,000.

Denisoff, I. & R. Devred. 1954. Mvuazi, Bas-Congo. Soils, vegetation 1:50,000.

Devred, R. et al. 1958. Kwango. Soils, vegetation 1:1,000,000.

Focan, A. & W. Mullenders. 1955. Kaniama (Haut-Lomami). Soils 1:100,000, vegetation 1:50,000.

Frankart, R. 1960. Uele. Soils 1:40,000.

———. 1967. Paysannat Babua. Soils 1:40,000.

——— & **L. Liben.** 1956. Bugesera-Mayaga (Ruanda). Soils, vegetation 1:100,000.

Frankart, R. & G. Sottiaux. 1972. Planchette Muramvia. Soils 1:40,000.

Germain, R. et al. 1955. Vallée de la Ruzizi. Soils 1:100,000, vegetation 1:50,000.

Gilson, P. et al. 1956. Yangambi: Yangambi. Soils, vegetation 1:50,000.

——— et al. 1957. Yangambi: Lilanda. Soils, vegetation 1:50,000.

——— & **P. François.** 1969. Zone de la Haute-Lulua. Soils 1:50,000 and 1:200,000.

——— & **L. Liben.** 1960. Kasai. Soils, vegetation 1:200,000.

Holowaychuk, N. et al. 1954. Nioka (Ituri). Soils, vegetation 1:50,000.

ISABU. Cartes des sols et de la végétation du Burundi. (New name for INEAC in Burundi.)

Jamagne, M. 1965. Maniema. Soils 1:250,000.

Jongen, P. 1968. Ubangi. Soils 1:50,000 and 1:100,000.

——— et al. 1960 Ubangi. Sols 1:50,000, vegetation 1:100,000.

——— et al. 1970 Nord-Kivu et Région du lac Edouard. Soils 1:50,000 and 1:200,000.

——— & **M. Jamagne.** 1966. Région Tshuapa-Equateur. Soils 1:100,000 and 1:1,000,000.

Pahaut, P. et al. 1962. Bassin de la Karuzi (Burundi). Soils, vegetation 1:50,000.

Pecrot, A. & A. Léonard. 1960. Dorsale du Kivu. Soils, vegetation 1:500,000.

Sys, C. 1960. Congo belge et Ruanda-Urundi. Soils 1:5,000,000.

——— & **P. Hubert.** 1969. Mahagi. Soils 1:200,000.

——— & **A. Schmitz.** 1959. Région d'Elisabethville (Haut-Katanga). Soils, vegetation 1:60,000.

Van Wambeke, A. 1958. Bengamisa. Soils 1:100,000.

———. 1959. Région du lac Albert. Soils 1:100,000.

———. 1963. Rwanda et Burundi. Soils 1:1,000,000.

——— et al. 1960. Région de Yanonge-Yatolema. Soils 1:100,000.

——— & **C. Évrard.** 1954. Yangambi: Weko. Soils, vegetation 1:50,000.

——— & **L. Liben.** 1957. Yangambi: Yambaw. Soils, vegetation 1:25,000.

——— & **M. F. Van Oosten.** 1956. Vallée de la Lufira (Haut Katanga). Soils 1:100,000.

East Africa (Kenya, Tanzania, Uganda)

R. M. Polhill

Contents

I. Description of the Region	219
A. Geographical Extent and Topography	219
B. Forests	219
C. Forest Policy	219
D. Population	219
II. Vegetation Maps	219
A. Forest Classification	219
III. Vegetation Types	220
A. Guineo-Congolian Region and Lake Victoria	220
1. Uganda—Western Rift	220
2. Uganda—Victoria	221
3. Kenya	221
4. Tanzania	221
B. Zanzibar-Inhambane Regional Mosaic	221
1. Kenya	222
2. Tanzania	222
C. Somali-Masai Region	223
D. Afromontane Region	223
1. Uganda	224
2. Kenya	224
3. Tanzania	224
E. Mangrove Forests	225
IV. Floristic Inventory	225
A. Herbaria	225
B. Publications	225
V. Assessment	225
A. Status of Inventory	225
B. Protected Areas	226
C. Manpower and Facilities Needs	226
D. Research and Education Needs	227
VI. Specific Proposals	227
VII. Acknowledgments	228
VIII. Literature Cited	228

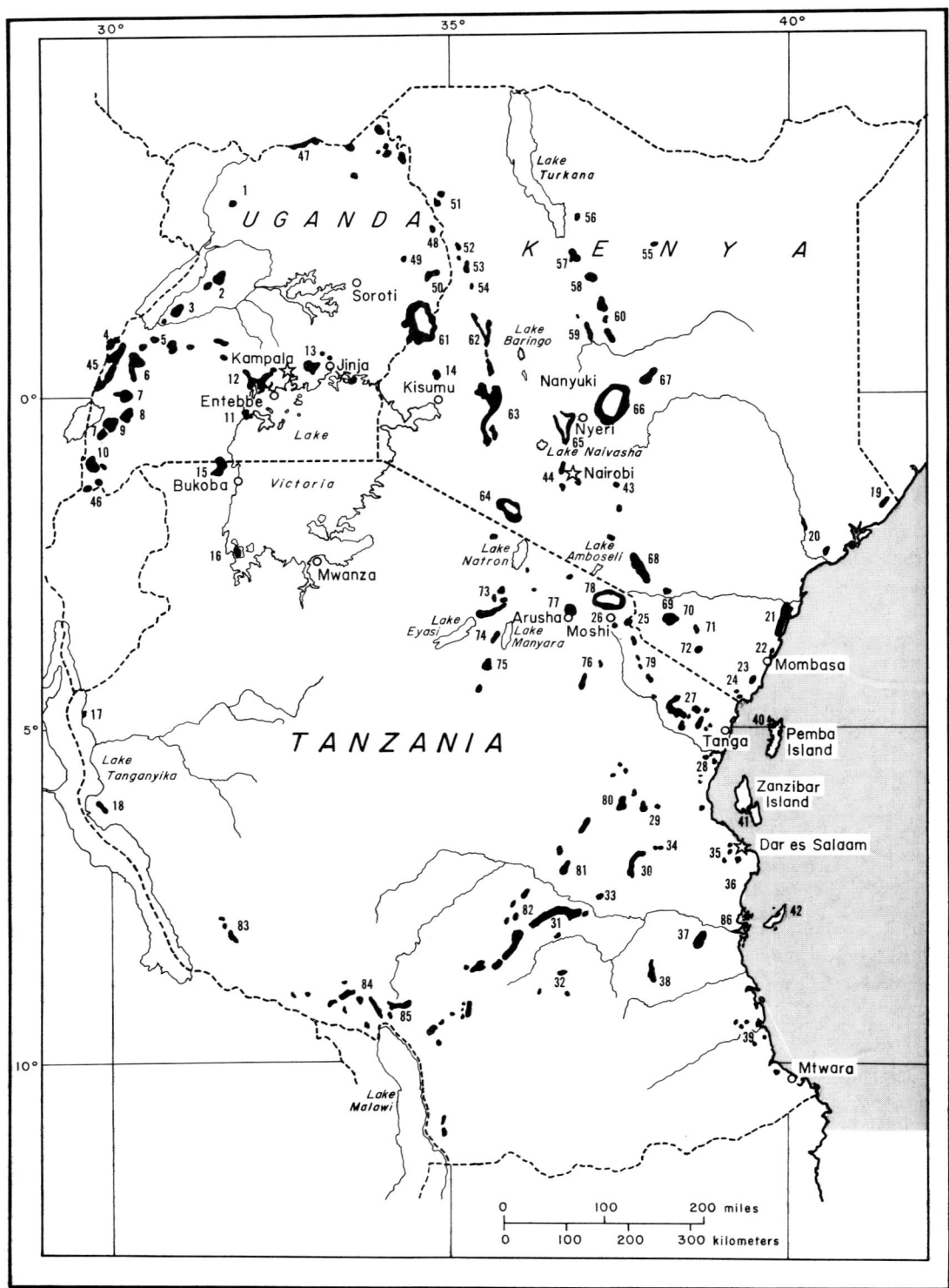

Map 1. Forests of East Africa, numbered as in text. Base map from Russell (1962).

I. Description of the Region

A. Geographical Extent and Topography

Snow-capped mountains on the equator epitomize the dramatic topography of East Africa. A scenic impression of Kenya, Tanzania and Uganda is now widely familiar from films, books and tourism. In an area of 1.7 million km² the land rises from a narrow coastal plain to an elevated plateau region bisected by the great Eastern Rift Valley, which is flanked by highlands and high volcanic mountains, notably Elgon, Aberdares, Mt. Kenya (5194 m), Meru, Kilimanjaro (5895 m) and the Mbulu Highlands (including Ngorongoro). Between the coast and the Rift there is a scatter of small hills and plateaux and a disrupted chain of old crystalline mountain blocks, notably the Usambara, Nguru, Uluguru and Uzungwa Mountains stretching to the Southern Highlands. To the west of the Eastern Rift the vast plateau in Tanzania, including the Serengeti Plains, dips to Lake Victoria and the many swampy valleys of southern Uganda, beyond which the country rises again to the flanks of the Western Rift Valley with Ruwenzori, the eastern Virunga Mountains and dissected highlands along Lake Tanganyika and north of Lake Malawi.

B. Forests

Forests cover only 2–3% of the land surface (map 1) and continually diminish. They exist principally in the Highlands, on hill tops, along the drainage systems and in patches along the coastal belt. From the beginning of this century governments have set aside reserved areas for water conservation, forestry, game preservation and more recently some relatively unspoilt tracts as National Parks. Considerable areas have since been withdrawn for farming, especially in the Kenya Highlands, and for planting exotics. The residual areas, especially those not in National Parks, suffer increasing encroachment. The costs of policing and its political acceptability present considerable problems for the governments and were exacerbated in Uganda by the collapse of economic and political structure in the 1970's.

C. Forest Policy

Forest policy has long differed among the three countries. The Kenyan forests were early regarded as nonproductive and replacement with exotic softwoods was favored from the mid-20's. In Tanzania and Uganda the natural forest estate was conserved and exploited more carefully and still provides a significant economic resource. Many species previously considered uneconomic to exploit are now included in lists of classified commercial timbers. Outside government reserves, small patches of forest survive on private estates, in sacred groves, on isolated hills, in gorges, and more extensively in areas of exceptionally low population. Despite the tiny size of most of these forests many are of exceptional biological interest and have been adopted in a few places by universities, museums and conservation bodies.

Reasonable security of forest ecosystems is, however, assured most effectively in the National Parks. These are famous worldwide, and tourist revenue has become a major contribution to the economy, especially of Kenya, and has great potential throughout the area. The popular mountain parks in Kenya and northern Tanzania have fine forest stands, but forests elsewhere, with very different structure and biota, are now the main target for conservation plans. The centralized growth of the tourist industry in Nairobi in the 1960–70's caused considerable resentment both locally and internationally, and the need for a more equitable reward to local district councils and service organizations has been appreciated. There is a growing local appreciation of the Parks, notably fostered by wildlife clubs.

D. Population

Population is concentrated in the high rainfall areas around Lake Victoria, in the highlands and along the coast. There is an annual population increase of 3–4%. Kenya has one of the highest birth rates in the world and the population, 14.7 million in 1978, will exceed 30 million by the turn of the century at the present rate of growth (Hamilton, 1982). Massive moves to urban areas will strain the provision of employment and social services, aggravated by a rise in material expectations. The economy is largely based on agriculture and there will be pressure on land for export crops to meet the higher per capita costs of an increasingly urban population. Apart from Forest Reserves and parts of the Southern Highlands of Tanzania, there are few areas of high rainfall that are not already extensively managed for food production. Wood-based products provide more than 80% of the total energy used and improved agricultural practices are not easy to implement. Owino (1983) recommends a fourfold increase in plantations for fuel requirements in Kenya. There is no doubt about the pressures that will be put on residual, already fragmented, forest areas. The success of the governments and conservation lobbies in protecting what they presently have is truly commendable. Nonetheless, the depletion of forests and other plant communities is so serious that priorities and strategies for the inventory, assessment and improved management or conservation of the remaining forests must be done soon.

II. Vegetation Maps

A. Forest Classification

The traditional classification of East African forests is that of Greenway (1943, 1973), based on physiognomy, elevation and precipitation. The main categories may be loosely equated with those of the International Classification and Mapping of Vegetation (UNESCO, 1973) as follows:

Greenway (1973)   UNESCO (1973)

Lowland Rain Forest	I,A,1,a Tropical ombrophilous lowland forest
Upland Rain Forest	I,A,1,c Tropical ombrophilous montane forest
Lowland Dry Evergreen Forest	I,A,2,a Tropical evergreen seasonal lowland forest
Upland Dry Evergreen Forest	I,A,2,c Tropical evergreen seasonal montane forest
Ground Water Forest Riverine	I,A,1,f Tropical ombrophilous alluvial forest
Fresh-water Swamp Forest	
Saline-water Swamp Forest	I,A,1,g Tropical ombrophilous swamp forest

The UNESCO/AETFAT/UNSO Vegetation Map of Africa (White, 1983) retains a physiognomic basis, but the accompanying descriptive memoir, *The Vegetation of Africa*, treats the formations on a floristic basis. That system suits conservation purposes and is used for map 1. The more notable forests are numbered and considered briefly below in terms of their status, area, biological interest and documentation.

Map 1 is adapted from Russell (1962). Modifications have been made from subsequent literature and comments from correspondents listed in the Acknowledgments. Maps in the recent report of the Kenya Rangeland Ecological Monitoring Unit (KREMU), Nairobi (Doute et al., 1983), using remote sensing techniques, have been consulted for Kenya. A comparable survey for Tanzania has been done recently by the Institute of Resource Assessment, University of Dar es Salaam (Rodgers, Mziray & Shishira, in ed.). A continuation of these monitoring techniques, supported by further groundwork, is much needed. Reports on the individual forest areas throughout the region should lead also to a more refined classification of East African forests, which form a remarkable series of relict fragments, with the precarious survival of characteristic and unique communities of plants and animals.

III. Vegetation Types

Further general introductions to East African vegetation and conservation are provided by the following: Chapman & White, 1970; FAO–UNEP, 1981; Hamilton, 1982; Hedberg, I. & O., 1968; Hedberg, O., 1979; IUCN, 1987; Lamprey, 1977; Langdale-Brown et al., 1964; Lind & Morrison, 1974; Morgan, 1968; Russell, 1962; Trapnell, 1966–1986.

The forests are considered under the following headings:
Guineo-Congolian Region and Lake Victoria Regional Mosaic. Areas 1–18 on Map 1.
Zanzibar-Inhambane Regional Mosaic. Areas 19–42.
Somali-Masai Region. Areas 43, 44.
Afromontane Region. Areas 45–85.
Mangrove Forests. Area 86.
Cross references are used for conservation areas that come under two headings, e.g., certain mountains and scarps in eastern Tanzania that extend from Zanzibar-Inhambane to Afromontane, or occur in two countries, e.g., Elgon.

A. GUINEO-CONGOLIAN REGION AND LAKE VICTORIA REGIONAL MOSAIC

The Guineo-Congolian forests as a whole comprise about 8,000 species of vascular plants with more than 80% endemic. The levels of diversity and endemism are highest around the Gulf of Guinea from southern Nigeria to Angola. The forests of East Africa are notably drier, poorer in species and more fragmented. Levels of endemism rise a little along the Western Rift Valley in a transition zone between the lowland and Afromontane forests in eastern Zaire. The forests that remain in Uganda all have slightly different floristic compositions with considerable variation in elevation, rainfall, and groundwater conditions. The Kakamega Forest in western Kenya is strikingly different from forests further east and thus nationally important as a conservation area. The Lake Victoria Regional Mosaic is recognized by White to express the heterogeneity of floristic elements mixed in this relatively recent land-form. It has few endemics and the Guineo-Congolian element, though remaining fairly discrete, occurs in patches within the complex. General reference: White (1979).

1. Uganda—Western Rift

1. Zoka. Central Forest Reserve; an isolated northern outlier, 50 sq. km², it is threatened by fire and elephant damage. References: Lind & Morrison, 1974; also see Stuart-Smith, 1962.

2. Budongo and Siba. Central Forest Reserve, of 350 km² on scarp above lake at about 1100 m; it is immature forest with *Cynometra alexandri* dominant, also swamp forest; about 90% is exploited for hardwoods, but sawmilling is in abeyance and with only limited pitsawing at present. References: Eggeling, 1947; Osmaston, 1956; Philip, 1964; Synnott, 1985.

3. Bugoma. Central Forest Reserve, about 250 km²; it is similar to Budongo and relatively untouched. Reference: Osmaston, 1959a.

4. Semliki. Central Forest Reserve. It is partly on poor clay soils, with a scrubby *Cynometra* forest. There is a richer mixed forest towards Ruwenzori which is considerably cleared; much further encroachment has occurred since 1978.

5. Itwara. Central Forest Reserve, about 77 km². A sawmill is operating part of the time and there is some pitsawing. While presently 50–70% unexploited, surrounding tea estates are being rehabilitated and demand on the forest may increase.

6. Kibale. Central Forest Reserve, 550 km,² at 1050–1600 m. There is a 60 km² Nature Reserve, including lowland and intermediate forests. Fairly untouched in the north, it is badly encroached by settlers in the south (about 100 km² disturbed between 1971 and 1982), but control is planned by the government. Conservation interest is centered on the occurrence of 11 primate species. The Kibale Forest Project is directed by Dr. T. Struhsaker, New York Zoological Society. There has been IUCN/WWF funding for protection and monitoring effects of logging outside Nature Reserves. References: IUCN, 1987; Osmaston, 1959b; Struhsaker & Leland, 1979.

7. Maramagambo. In Ruwenzori National Park, 500 km². It is intact.

8. Kasyoha-Kitomi. The Central Forest Reserve, of perhaps 120 km², is more or less intact apart from some pitsawing. Reference: Leggat, 1957.

9. Kalinzu. Central Forest Reserve, 150 km² is intact. One sawmill is operating on low production and there is some pitsawing. Some illegal cultivation occurs (with many fires seen in 1979). Reference: Osmaston, 1960.

10. Bwindi (Impenetrable). Animal Sanctuary, of 80 km², at an elevation of 1,400–2,400 m, with transition from lowland to montane forest. There is a rich flora and fauna, including one of the few remaining potentially viable populations of mountain gorilla. It is an important water catchment area, but there is a considerable amount of pitsawing and charcoal burning and some encroachment by settlers. A WWF/New York Zoological Society/African Wildlife Leadership Foundation survey has been devised for 1983–1986, with National Park status planned. References: Hamilton, 1969; Harcourt, 1981; Leggat & Osmaston, 1961.

2. Uganda—Victoria

11. Lakeside and Islands. Central Forest Reserves include considerable areas of forest on the Sese Islands and along the lake edge, but all are subject to encroachment and clearance, except on small remote islands and swampy places. References: Ball, 1965; Jackson & Gartlan, 1965.

12. Mpanga and West Mengo. Central Forest Reserve. Mpanga, a very small area, is intact, but some illegal pitsawing and charcoal burning (serious but not irretrievable) occurs. It was formerly used as an experimental area for forest regeneration, and plots are intact. West Mengo Forests are reasonably preserved, but some pitsawing and charcoal burning occurs. Reference: Dawkins & Philip, 1962.

13. Mabira. Central Forest Reserve. There is serious encroachment, especially near the west side and along roads, with possibly 20% lost to agriculture in the last decade, but not as depleted as recent reports have suggested. Some illegal felling and charcoal burning occurs near settlements. It was formerly the most extensive of the Lake Victoria Basin forests and renowned for its rich flora. Reference: Webster (1961).

3. Kenya

14. Kakamega. A Forest Reserve, mostly much disturbed, but three Nature Reserves, Kisere (480 ha), Yala (470 ha) and Kakamega Forest Station (210 ha) were gazetted in 1967. The Yala Reserve is completely destroyed; Kakamega National Park, a small area S. of the Yala River, gazetted in 1983, has not yet been specifically investigated for flora and fauna. Maps and photographs can be found in Doute et al. (1983). It is famous as the easternmost extension of Guineo-Congolian flora and fauna. Another nature reserve is nearby in the N. Nandi Hills, but is much disturbed. References: Dawkins, 1960; Diamond, 1979; Faden, 1970; IUCN, 1987; Lucas, 1968; Rowell, 1982.

4. Tanzania

15. Minziro. Forest Reserve, about 250 km². It is an unusual form of semi-swamp forest dominated by *Podocarpus*. Reference: Lind & Morrison, 1974.

16. Rubondo Island. A National Park established in 1977, of 220 km², of which 180 km² is forest. It is probably the best example of lake island forests in Tanzania, and is used by the Frankfurt Zoological Society as a refuge for threatened species. References: IUCN, 1987; Rodgers et al., 1977.

17. Gombe. A protected area from the early part of the century. The National Park was established in 1968 as a research center for chimpanzee studies. A little forest exists along rivers and mid-slope of scarp, which has provided a number of new plant records for Tanzania. References: Clutton-Brock & Gillett, 1979; IUCN 1987.

18. Mahale Mountains. A National Park has been proposed, but not yet gazetted. Forest patches in remote mountainous area of about 250 km², which have been studied recently, mostly in connection with chimpanzee research. They are floristically interesting, with a considerable element not yet recorded further east. References: IUCN, 1987; Nishida & Uehara, 1981, 1983; Stevens, 1962.

B. Zanzibar-Inhambane Regional Mosaic

The Zanzibar-Inhambane Regional Mosaic forms a narrow belt 50–200 km wide along the East African coast from southern Somalia to the Limpopo River. The region is highly populated and only fragments of forest remain in a complex mosaic of bushland, woodland, thicket and wooded grassland. The wettest forests are developed on the windward side of the mountains or as groundwater forest. These are richest in species, notably the Shimba Hills in Kenya, the East Usambara and Uluguru Mountains in E Tanzania, but further inland the foot of the Uzungwa Scarp is proving comparable. Nearly 50% of the tree species appear to be endemic or nearly so, with strong affinities to the Guineo-Congolian Region. Higher on the mountains there is a transition to the Afromontane Region. Isolation and habitat diversity on these ancient mountain blocks, which are contiguous to a rich lowland flora and subject to climatic changes, have contributed to considerable local speciation. Lovett (in ed.), combining intermediate and montane forests, calculates that 25% of forest species are endemic to Tanzania, 30% if a small part of SE Kenya is also included in the area.

The drier forests have a simpler physiognomy and fewer species, but are remarkably diverse, with notable variation even within small patches depending on aspect, soils and local water conditions. There are numerous regional endemics, often seemingly very local, and new records or novelties are still commonplace. The affinities are diverse but include the Somali-Masai Region, Madagascar and Asia as well as West Africa. General references: Hawthorne, 1984; Moll & White, 1978.

1. Kenya

For general references, see Dale, 1939; Hawthorne, 1984; Lucas, 1968; and Moomaw, 1960.

19. Boni. National Reserves, with Groundwater forests at Boni (134,000 ha) and Dodori (88,000 ha) near the Somali border have hardly been explored and warrant a high priority for investigation.

20. Tana River. Patches of forest along river and delta exist (see maps in Doute et al., 1983); several large settlement schemes may increase pressure on the trees. The Tana River Primate Reserve was established in 1976, of 170 km^2; The Utwani Forest is in the delta area. References: Andrews et al., 1975, Marsh, 1976, 1985; Hughes 1986; IUCN, 1987.

21. Arabuko-Sokoke. Forest reserve, of 400 km^2. It is a dry coastal forest, now largely coppiced, with continual pressure for building materials, and with felling continuing in the Nature Reserve. It was surveyed by the University of East Anglia Expedition, in 1983. A species list by the enumeration section of the Kenya Forest Department was filed at the Herbarium, National Museums (EA) Nairobi. References: Kelsey and Langton, 1984; IUCN, 1987.

22. Kayas. Fragmentary forest refuges of the Mijikenda people are scattered between Malindi and the southern border, with considerable plant diversity, often on limestone or coral rag. Some are still considered sacred by the older people, but some are being cleared (e.g., at Kaloleni) and all are threatened by increased agricultural pressure. Tiny but interesting forests also survive near the coast—Muhaka, Gedi, Gogoni, Buda, Shimoni and Diani (see map in Doute et al., 1983; IUCN, 1987). Sadly, Diani has been virtually destroyed by tourist development since this paper was drafted. A survey of the Kayas is reported by Hawthorne et al. (1982), Hawthorne (1984), and the work is being extended by Mrs. Ann Robertson (1984) in association with the East African Herbarium, National Museums, Nairobi. A guide and nature trails were devised at the Gedi Nature Reserve by Miss K. Gerhardt and Miss M. Steiner, supported by SIDA in 1985, as an extension of earlier work by Dr. and Mrs. R. B. Faden, and a Botanical Garden is planned. Reference: Gerhardt & Steiner, 1986.

23. Shimba Hills. This National Reserve, run by the National Parks, consists of small of patches of varied lowland forest in an area of 200 km^2, with many noteworthy plant records over the last twenty years. Logging and forestry are spoiling the natural scenery and the conflict in policy between the Forest Department and National Parks remains unresolved. Elephant migration is impeded by settlement inland. References: Glover, 1968; IUCN, 1987; Kates et al., 1968; Lucas, 1968; Ross, 1982.

24. Mrima Hill. This Forest Reserve is a few km^2 of wetter coastal forest, much disturbed by mining operations in the last two decades, but now abandoned and recovering. It is floristically rich and peculiar. Jombo Hill, a little further inland, is drier and not quite so rich. Reference: Lucas, 1968.

25. Taveta. A groundwater forest, with some large trees and interesting species on the border with Tanzania; it is in good condition. It is under the control of the Ministry of Environment and Natural Resources.

2. Tanzania

26. Rau. A Forest Reserve near Moshi township, it is groundwater forest similar to Taveta, greatly reduced and disturbed, but with about 20 ha of good natural forest, with stands of the rare *Oxystigma msoo*. Reference: Rodgers, 1983.

27. Usambara Mountains. This region is densely settled with a number of Forest Reserves in a general area of over 2,000 km^2. There are lowland and intermediate wet (rain) forests on the E Usambaras, intermediate to Afromontane forest (including relictual fragments of a drier type) on the W Usambaras. Recently reviewed by Rodgers & Homewood (1982b), this was ranked by them as a prime forest area in tropical Africa in terms of biological diversity and endemics. The Amani Institute, the East African botanical center for half a century from 1902, is here. The Forest Department has its herbarium and silvicultural research center at Lushoto (TFD); Sokoine University, Morogoro, has a reserve at Mazumbai. Floristic and ecological surveys are in progress by the consortium WWF/IUCN/SAREC/Universities of Dar es Salaam, Sokoine and Uppsala.. References: Backeus, 1982; Borhidi et al., 1984, 1985; Hall, 1982a; Hedberg et al., 1982; IUCN, 1987; Iversen, 1986, 1987; Lovett et al., 1983a; Rodgers & Homewood, 1982b; Rodgers, et al., 1982; Willigen & Lovett, 1979.

28. Tanga-Pangani. A number of dry evergreen Forest Reserves between Tanga and Bagamoyo are of great interest (Hawthorne, 1984), but their current status is uncertain. Recent ground reports indicate much depletion, but satellite imagery suggests that there is a good preservation in more inaccessible places.

29. Nguru Mountains. Forest Reserves exist on higher slopes in general, with an area of about 250 km^2, some of which are still in good condition. These are dry evergreen and Afromontane forest and probably still some fragments of wetter lowland forest. The mountains are a part of a chain of mountains between the Usambaras, Ulugurus and Southern Highlands (see also Ukaguru (80) and Rubeho (81), which have some endemics but which are little surveyed). Reference: Lovett et al., 1983b.

30. Uluguru Mountains. A Forest Reserve is on the higher parts, of less than 100 km^2. The slopes are already cleared to inadvisedly high levels and the residual forest is of vital water catchment value for Dar es Salaam. The remaining parts are mostly Afromontane but with numerous endemics with Zanzibar-Inhambane affinities. The Kimboza Forest Reserve, a fragmentary patch of 4 km^2 on limestone at the eastern foot, has nearly 400 species with numerous endemics or near endemics (Rodgers et al., 1983). The Forest Division of Sokoine University, with Pocs' Herbarium, is sited outside Morogoro at the northern foot. An ecological survey of the northern peaks is in progress by J. B. Hall and J. C. Lovett. References: Pocs, 1976a, 1976b; Rodgers, et al., 1983.

31. Uzungwa Mountains. Several Forest Reserves, 450 km^2, from 300–2,800 m, exist, with continuity of lowland to Afromontane forest. A groundwater forest at Magombera immediately to the east, is seriously disturbed but part is now assigned to the Selous Game Reserve. The Uzungwa scarp is currently under survey by the Universities of Dar es Salaam and Sokoine, and the WWF/IUCN for National Park proposals. Initial work has revealed some remarkable novelties and records of species previously only known from the foothills of Usambaras and Ulugurus (Rodgers & Homewood, 1982a), and a more detailed survey, undertaken by Mr. J. C. Lovett in 1983–1985, combined with a multidisciplinary expedition supported by the National Geographic Society in 1984, will be continued by the Missouri Botanical Garden and WWF-US (J. C. Lovett, 1985, in press). Reference: IUCN, 1987.

32. Mahenge. A Plateau with forest-topped hills and outcrops of limestone. It is densely populated, but forest fragments are botanically rich and relatively unexplored. Reference: Haerdi, 1964.

33. Malundwe Hill. This is a few km^2 of forest within Mikumi National Park, at 800–1200 m, and is a closed evergreen forest with locally high rainfall. In composition it is comparable to the Uzungwa scarp and foothills of the Uluguru and Usambara Mountains. Reference: Hall, 1982b.

34. Kitulanghalo and Dindili Hills. Forest Reserves on isolated hilltops, with dry forest characteristic of the plains of Tanzania's eastern plateau. Reference: Kielland-Lund, 1982.

35. Pugu Hills. These are the best known of the fragmentary forest patches all along the coast. The Forest Reserve, 22 km^2 and close to the capital, shows much encroachment and disturbance, with part used as a military camp and part as a kaolin and brick factory; only about 3 km^2 is left with natural forest. Nevertheless, remarkable floristic diversity and numerous endemics or near endemics exist. References: Hawthorne, 1984; Howell, 1981; Wingfield, 1967.

36. Kisiju. A Coastal forest patch of 2 km^2 safe from cultivation in the past by tidal creeks. Dense stands of gum copal trees (*Hymenaea verrucosa*), exist, probably typical of much of the wetter coastal area in the past. It will be designated a Forest Reserve. References: Rodgers, Mwasumbi & Hall, 1985.

37–38. Kichi Hills and Libangani. 37. The Kichi Hills are virtually uninhabited at present, with no protective status. Comprised of extensive areas of dry coastal semi-deciduous forest on ridge-tops, they are botanically little known and potentially a promising area for research and conservation. 38. The Libangani Hills inside the Selous Game Reserve have similar but less extensive stands; interesting patches of riverine and groundwater forest also occur in the Selous. References: Rodgers & Ludanga, 1973; Vollesen, 1980.

39. Rondo Plateau. A Forest Reserve in part, exploited for timber mid-century and with considerable population pressure, but forest remnants survive here and in other parts of the Mtwara-Lindi region. Botanically rich, it is relatively unexplored. References: Bennett et al., 1979; Griffith, 1951.

40. Ngezi. A Forest Reserve, about 15 km^2 of swamp forest on NW Pemba. It is divided into compartments, which are well maintained, but management has had an impact throughout the forest, with commercial trees extracted and rigorous weeding of the understory. There is a notable Madagascan element, with endemics (e.g., *Chrysalidocarpus* palm, wild banana) and anomalous, typically high montane genera (*Philippia, Polyscias*) doing well. A report has been prepared by the University of Dar es Salam (Rodgers et al., 1986). Ras Kyuyu, a coral rag forest, still has good stands of tall trees, but its understory is damaged by cattle. Msitu Mkuu, on shallow soil over coral, has an interesting mesic forest free from maritime influences; it is somewhat disturbed.

41. Jozani. A small semi-swamp forest, surrounded by semi-deciduous dry forest on coral rag, in the middle of Zanzibar. Recently surveyed by the University of Dar es Salaam, it is the major habitat for the Zanzibar Red Colobus monkey. Reference: Salvadori, 1985.

42. Mafia Island. An extensive area of coastal semi-deciduous forest on coral rag along eastern seaboard. References: Rodgers & Wingfield, in ed.

C. SOMALI-MASAI REGION

Upland dry evergreen forest of relatively simple physiognomic structure occurs as relic stands around the edges of the Eastern Rift of Kenya and northern Tanzania, and also on isolated hills to the east, progressively grading into Zanzibar-Inhambane dry lowland forest. The flora is reasonably well known, but the ecology is little studied and the ecosystem deserves better protection. General reference: White, 1983.

43. Ol Doinyo Sabuk. A National Park, 18 km^2; also there is a very small but interesting forest in the Chania River gorge at Thika a little further west, with notable endemics and disjunct populations. Reference: IUCN, 1987.

44. Nairobi Forests. These consist of a Forest Reserve at Karura (just outside Nairobi), considerably degraded fragments in the Nairobi National Park and vicinity, the Ngong Hills, a fragment at the Agricultural and Forestry Research Institute at Muguga (much degraded), with larger stands in the Forest Reserve along the Kedong scarp. These are everywhere prone to encroachment and disturbance. One small sample at Ololua, near Nairobi, is protected by the National Museums of Kenya. Plans are in hand to survey a number of relict forests SE of Nairobi, under the auspices of the National Museums. See map in Doute et al., 1983. Reference: Verdcourt, 1962.

D. AFROMONTANE REGION

The Afromontane Region has been subjected to greater floristic study than the other forest types of East Africa. It has a relatively simple physiognomic structure and floristic coherence. Reasonably good examples of both the wetter and drier types exist in the mountain National Parks, and elsewhere often form water catchments of recognized importance. Nonetheless, the ecology and detailed biological and floristic relationships are virtually unexplored. Better protection of some of the forest in the Southern Highlands of Tanzania is also recommended. General references: Hall, 1984b; Lovett, in ed.; Synnott, 1979a; White, 1981.

1. Uganda

45. Ruwenzori Mountains. This Forest Reserve, about 880 km², is recognized as a vital water catchment area, but protection is difficult to enforce along the boundary with Zaire. There is perhaps 5% encroachment on the lower section of the reserve, but a considerable amount of forest and all the altimontane flora is intact.

46. Virunga Mountains. The Kigezi Gorilla Game Reserve, 30 km², on the Zaire-Rwanda border, is threatened but reportedly fairly intact and gorillas are coming back from Zaire. Reference: IUCN, 1987.

47-50. Karamoja. (47. Imatong Mountains; 48. Moroto; 49. Napak; 50. Kadam). These are mostly drier types, and relictual. It is more varied and wetter on Imatongs (largely in Sudan), but the Ugandan part has not been surveyed recently. Moroto is badly overgrazed and cut on the lower slopes in more accessible places. It is not yet under government control. Napak and Kadam are fairly intact, with less population pressure than the other areas. References: Jackson, 1956; Wilson, 1962.

[**61. Elgon.** — see Kenya — probably 800 km² in Uganda, with some encroachment on the lower slopes of the reserve (at least 80 km² lost since 1979), but some efforts to reassert government control have been made).]

2. Kenya

51-54. Suk Mountains. (51. Loima Hills on Muruanisigar, 100 km²; 52. Lorosuk Hills on Kachagalau; 53. Sepich Hill (Cheberua) and 54. Mtelo Mountain (Sekerr), 180 km²). Similar to the Karamoja mountains (47–50), they are mostly drier forest types, all little explored. Fine stands exist on Loima. They are prone to fire and dry season grazing.

55. Marsabit. A National Reserve, 200 ha, with good olive forest on upper slopes; isolated and biogeographically interesting. Reference: IUCN, 1987.

56-60. E. Turkana Mountains. (56. Kulal; 57. Nyiru; 58. Ndoto; 59. Lerochi; 60. Mathews Range). These are Forest Reserves maintained for water catchment, with relictual, mostly dry, montane forest. Kulal is a Biosphere Reserve (700,000 ha) with a center for the Integrated Project in Arid Lands (IPAL); Nyiru was recently reported in excellent shape, with paths largely overgrown. Ndotos was similarly in good shape, perhaps due to reduced elephant pressure. References: Hepper et al., 1981; Herlocker, 1979; IUCN, 1987; Kokwaro & Herlocker, 1982; Synott, 1979a, 1979b, 1979c.

61. Elgon. A good stand of montane forest in the National Park (17,000 ha) on the wetter south-western side of the mountain. This is a superb park but off the current tourist route. References: Dale, 1940; Hamilton & Perrot, 1980; Tweedie, 1976; IUCN, 1987,

62. Cherangani; 63. Mau; 64. Nguruman. These Forest Reserves, markedly reduced in recent decades by settlement, logging for *Juniperus*, replacement with softwoods (especially on Mau), and cattle grazing and fires, are still important water catchment areas. Cherangani still has some fine forest, but is being depleted rapidly. The SW Mau Nature Reserve, 430 km², and Nguruman are still in good shape. See maps in Doute et al. (1983). References: Fayad & Fayad, 1950; Glover & Trump, 1970; Kerfoot, 1964; Mabberley, 1975.

65. Aberdares. This National Park (76,000 ha) on the northern part, includes the main peaks, with good stands of both wet and dry montane forest at higher altitudes. Reference: IUCN, 1987.

66. Mt. Kenya. A Forest Reserve (140,000 ha) of which the only highest parts are in the National Park (59,000 ha). The altimontane communities over 2500 m are still relatively intact, but the montane forest is likely to come under increasing pressure in the near future. A management policy is needed to sustain forestry yield, reduce encroachment and poaching, and to provide alternative firewood sources. References: Abraham, 1958; F. White, 1950; IUCN, 1987.

67. Nyambeni Hills. Forest reserve, 53 km²; with some pressure from tea estates. It is moist montane forest. See map 53 in Doute et al. (1983).

68. Chyulu Hills. Incorporated in Tsavo National Park, its 45 km² are but rarely patrolled; the best wetter montane forest is in the south, with patches on hills further north. Of volcanic origin, the lack of permanent water has helped to preserve the forests from settlement pressure, but the Masai graze their cattle all along the south side and even to the tops of some hills; fires have also been detrimental. A small forest at Kibwezi, along the main road to the east, along the river and on lava flows, is now being encroached. Reference: IUCN, 1987.

69-72. Teita Hills. There are remnants on peaks and gullies of 70. Teita Hills, which are in a heavily settled area, but more extensive on 69. Mbololo Hill, a little further north. A floristic affinity with the W. Usambara Mountains is seen and some endemics are found in the transition from the Zanzibar-Inhambane Region. 71. The Sagala Hills, 72. Kasigau, and other hills standing up from the dry plains towards the coast have small patches of botanically interesting if little studied forest. The Teita Hills were recently surveyed by National Museums (Beentje et al., 1987).

3. Tanzania

73-76. Masai and Mbulu Mountains. Moist and dry montane forest are found on calderas and wall of the rift, some in Forest Reserves, but are mostly unprotected except 73. Ngorongoro; 74. Marang Forest is being included in the Manyara National Park; 75. Hanang, see Greenway (1955); 76. Lelatema, a dry forest, is remote and unstudied. References: Herlocker & Dirschl, 1972.; IUCN, 1987.

77. Meru. Moist and dry montane forests, about 100

km², in a Forest Reserve and partly within Arusha National Park. Unpublished reports by Vesey-FitzGerald, National Parks, Tanzania; IUCN, 1987.

78. Kilimanjaro. Moist and dry montane forests, in a Forest Reserve which is rather heavily exploited and with dense settlement below. The National Park, (75,000 ha) mostly above the Forest Reserve, has six corridors down through the original natural forest, and is partly unspoilt, as are some more inaccessible parts of the Forest Reserve. About 2000 species of flowering plants are listed. References: Bigger, 1968; Gilbert, 1970; Greenway, 1965; IUCN, 1987; Mwesaga, 1984.

79. Pare Mountains. These are largely denuded except on the north and south slopes. They have not been surveyed recently.

80. Ukaguru; 81. Rubeho; 82. Image. These are Forest Reserves, with moist montane forest on the upper parts, and relatively remote. The Ukaguru Mountains have a few known endemics, but there are extensive softwood plantations. These form links between the eastern and southern mountains. See also areas 27, 19–31.

83. Mbisi. A Forest Reserve, 30 km², with moist montane forest. It forms a link between the western and southern forests. Reference: Rodgers et al., 1984.

84. Rungwe and Poroto Mountains. Forest Reserves, about 250 km², with moist montane forest. There are some softwood plantations and some felling, but large scenically beautiful tracts with relatively intact forest still exist. Reference: Leedal, undated.

85. Ndumbi. This Forest Reserve at the SW tip of Kipengere Range, 25 km², has dry montane forest. It is the southernmost locality for *Juniperus*. On ancient rocks, (whereas other forests in 100 km radius are on volcanic soils) it is prone to fire. It could form part of a more extensive National Park on the Kipengere Range. Reference: Leedal, undated.

E. Mangrove Forests

Most extensive in estuaries of the larger rivers, it is much cut for building poles and charcoal; the best stands are now probably in 86. Rufiji Delta. Kenya stands are shown in maps in Doute et al., (1983). References: Graham, 1929; Grant, 1938; Kokwaro, 1982, 1985; Kuyper, 1982; Van Leeuwen, 1982; Walter & Steiner, 1936.

IV. Floristic Inventory

A. Herbaria

Holdings of specimens in the main local herbaria are as follows:

East African Herbarium, National Museums, Nairobi (EA)	600,000
Nairobi University Herbarium (NAI)	80,000
KREMU Herbarium, Nairobi	20,000
Dar es Salaam University (DSM)	80,000
Lushoto Forest Herbarium, Tanzania (TFD)	20,000
National Herbarium, Tropical Products Research Institute, Arusha, Tanzania (NHT)	4,000
College of African Wildlife Management, Mweka, Tanzania	2,000
Makerere University, Kampala (MHU)	40,000
Forest Herbarium, Entebbe, Uganda (ENT)	4,000
Kawanda Herbarium, Uganda (KAW)	4,000
Kingupira, Selous Game Reserve	5,000

Smaller herbaria exist at the Division of Forestry, Sokoine University, Morogoro, and in some National Parks and colleges.

The proportion of forest specimens varies with the primary research interest of the various institutes, perhaps generally 10%, but it is notably higher in forest herbaria.

The main East African collections elsewhere are at Kew, with half a million specimens at a guess, with many duplicates principally at BR, FI, LISC, P, PRE, and SRGH in the main expansion period from 1947–1970, with a subsequent increase and exchange with C, MO, UPS and WAG. Significant earlier collections are at BM and FHO. The enormous collections from Tanzania at Berlin-Dahlem were mostly lost in the Second World War, but parts remain (Hiepko, 1978) and duplicates are widely scattered, notably at BM, BR, EA, HBG, and P. Apart from EA and K, the fastest growing herbarium is now the Missouri Botanical Garden, which has a good representation of the general flora, but needs augmentation especially of the forest element. To effect this a resident collector is now stationed in Tanzania (see Lovett, this volume).

B. Publications

Milne-Redhead et al. (1952–) "Flora of Tropical East Africa", covers three-fifths of the flowering plants (rather more if unpublished parts are included); most of the major forest families are now available in some stage of preparation, with the exception of some Gamopetalae, notably Acanthaceae, and the Malvales. Other principal floras are: Agnew, 1974; Brenan, 1949; Dale and Greenway, 1961 (revised edition in preparation); Eggeling and Dale; 1952; Hamilton, 1981. Gillett and McDonald (1970) is a useful checklist. Blundell (1982, 1987) has provided popular, well illustrated introductions. Dictionaries include Glover, 1967 and 1969; Haerdi, 1964; Hora and Greenway, 1940; Kokwaro, 1972, 1976; Kokwaro and Herlocker, 1982; Nishida and Uehara, 1981; Sangai, 1963.

V. Assessment

A. Status of Inventory

Lists of collections are mostly compiled under serial numbers of collectors or institutes, floras give distributions in only a rather generalized way and check-lists rarely indicate the intensity of collecting, so it is not easy to assess in many cases what has been collected from a particular forest. Nonetheless, it is probable that at least one specimen has been collected per km^2 in all notable forests except the most inaccessible, principally in W and SE Tanzania and NE Uganda. By the criteria of this volume they are relatively well collected. Despite this it is amazing that almost any sizeable collection of plants from forests of the Zanzibar-Inhambane Region, and to a lesser extent the Guineo-Congolian Region, still produces novelties and new records. Furthermore, each of the fragmentary forests seems to have a slightly different composition and considerable internal variation. Moreover, a dearth of detailed ecological studies, or understanding of even the rudiments of pollination and dispersal biology, compared to the rates at which the forests are dwindling or being disturbed, makes the situation considerably less encouraging. Certainly the floristic inventory has not reached a stage to begin to monitor rates of extinction, though by inference these must have been quite high, especially in the last thirty years.

Nor is it possible to identify families and genera that are particularly well or poorly collected, though it is notable how many rubiaceous undershrubs are collected in young fruit, how difficult it is to collect forest giants, aerial epiphytes and parasites, and how many forest plants rarely bear fertile parts. A list of contributors to the "Flora of Tropical East Africa" and other specialists with East African interests is always available from Kew or the East African Herbarium, National Museums, Nairobi. Specialist studies, preferably on a continental basis, are vital to the next generation of taxonomic revisions as modern systematics increasingly demands a biological explanation of the variation seen in herbarium collections. The more pressing need, however, is to arouse local interest and field expertise, channelled into check-lists, local floras, detailed ecology and a study of plant-animal relationships throughout the ecosystems, on a long-term basis and with adequate field study resources.

B. Protected Areas

By tropical standards East Africa has an outstanding network of protected areas. Thorsell (1983) warns against a regimen of benign neglect, however. There is a real need for action to improve their management and a firm commitment to implement policy long-term. Numerous projects are set up on two to three year plans, which are politically expedient, but are, however, extraordinarily wasteful of expertise and training. New buildings and facilities are frequently provided by external aid with an appalling disregard for maintenance costs and scientific staffing. A high proportion of local scientists are now engaged in full-time administration and routine enquiry services, while overseas training, widely acclaimed as a panacea, has produced virtually no increase in field botanists and is very often quite inappropriate. In the present world situation, the number of experts who visit for useful periods has dwindled and again rarely extends either to significant involvement in field work or effective field training. The proportion of expenditure by international organizations on administration, meetings and remote documentation compared to field work needs severe scrutiny. The resources for biological research, even in the present economic situation, are substantial, but considerably frittered away in unintentional or thoughtless ways.

C. Manpower and Facilities Needs

All too often specific requirements are reduced to pleas for further facilities in cities and more overseas tours. Buildings and botanic gardens are really needed, and promising young scientists deserve the best training available. It seems short-sighted, however, to end the list there. A plea is also needed for extending research and education into the forests themselves.

Reasonable security of ecosystems in East Africa has been achieved against the tide of development pressures by the creation of National Parks or Nature Reserves with a resident staff provided by the National Parks, universities, museums or international organisations such as IPAL. The area has been extremely fortunate in having a College of African Wildlife Management at Mweka in Tanzania for twenty years now, which has provided 1000 graduates trained in wildlife park management for East Africa and numerous other countries. Training is also provided by forest schools at Moi University, Eldoret, Kenya, Egerton College, Njoro, Kenya, at Olmotonyi, Arusha, Tanzania, and at Makerere University, Kampala, Uganda. In a majority of cases the justification for a National Park has been based on the attraction of larger mammals, in the case of forests principally primates, though the avifauna, scenic beauty, water conservation and low settlement pressure may have been contributory. An exceptional diversity of primates predictably coincides with a rich ecosystem for which the plants form the base. Essentially, what is good for monkeys is good for botanists. Certainly there are forest fragments which have already lost their larger mammals, but a floristic inventory will arguably have less effect than provision of a forest guard and sympathetic involvement with the local residents. The Gombe Stream Reserve, the Kigezi Gorilla Reserve, the Kibale Forest Project, and the Parks of northern Tanzania are associated in everyone's mind with exceptional scientists and administrators who have devoted considerable parts of their lives to staying and working in these forests, sometimes against odds that proved insuperable (especially on international borders). There are legendary game wardens, museum directors, university teachers and filmmakers who have shaped the conservation successes of East Africa and provided the inspiration and logistical support for research, policing, tourism and funding.

In those respects the botanist is overshadowed. There is no doubt that tropical forests touch a deep chord in the human psyche that may relate to our origins, but much of human endeavor has been directed to creating for ourselves a more comfortable environment. In time there is the highest potential for forest parks, but they must have tourist facilities, access to the surroundings with canopy walkways, sympathetic guidance and information on the more subtle lives of forest plants and animals. Field centers must be comfortable (but inexpensive) and local scientists remunerated to compensate for the still apparent sociological disadvantages.

D. RESEARCH AND EDUCATION NEEDS

It is vital to extend plans beyond collecting herbarium specimens and making check-lists, however worthy these preliminaries may be, to serious interdisciplinary ecological and biological research. The sheer number of plant species daunts any tropical flora writer or ecologist. All but the most recent local floras are out of print but, since this was written, a new edition of *Kenya Trees and Shrubs* is half completed and a simply designed and well-illustrated *Forest Flora of Tanzania*, with background information on the vegetation, phytogeography and local plant uses has been planned (see Lovett, this volume). From there further localization and simplification could introduce children, students and casual visitors to an insight into the plant world. The proposal by Verdcourt (1968) to appoint a field forest botanist in each county deserves renewed consideration. His further proposal for educational establishments to adopt local forests for introducing teachers and students to a storehouse of innumerable biological problems has been partly implemented by the universities in Kenya and Tanzania (though, interestingly, for distant sites), but also warrants further consideration.

In Kenya there is a renewed interest in forest research and education, and this is being supported in part by several donor organizations. A workshop on 'The Plant Communities of Kenya' was organised by the Environmental Training and Management in Africa (ETMA) at Nairobi in March 1984 (Bassan, 1984) to assess the position and identify the needs for action to arrest the depletion of natural resources and explore ways that they can be conserved and used more effectively. The Chief Conservators of Forests have been very aware of the need for greater emphasis on the evaluation and management of natural forests (Mburu, 1983). The monitoring of forest cover changes by the Kenya Rangeland Ecological Monitoring Unit (KREMU) provides an excellent basis for future groundwork (Doute et al., 1983). Owino (1983) outlines the urgent need for a more dynamic and coordinated forest policy, stressing both the need for the survey and conservation of natural forests as well as an effective plantation programme. A useful assessment of the role of agroforestry is given in the proceedings of the 1980 ICRAF seminar (Buck, 1981).

Local interest and action to survey and protect forests could be stimulated through the District Environmental Assessment Project of the National Environment Secretariat (NES) and through its Division of Environmental Education, and also through clubs such as the Wildlife Clubs of Kenya and Kenya Energy Non-Governmental Organizations (KENGO). Extended facilities at the East African Herbarium, National Museums, Nairobi, could be used for data gathering. An atlas of the rare trees of Kenya is being prepared by the Herbarium itself. Especial emphasis is being given to the coastal forests, which are botanically quite outstanding. By a miracle, representative samples are in good order, and deserve the highest conservation priority among the forests of Kenya. Mrs. Ann Robertson is now preparing a floristic inventory at the East African Herbarium, National Museums, Nairobi. Special emphasis is also being given to the Taita Hills (Beentje et al., 1987).

The current programmes for forest research and conservation in Tanzania have been effected by imagination, flair, political skill and sheer hard work in the University of Dar es Salaam. It is hoped that the input of international support will continue well beyond initial floristic surveys. In practice it is possible that emphasis on areas of high conservation potential will overspill into surveys and make a better case for protection of at least some of the smaller forests. It should also lead to a better general education in the management of productive forest, the appreciation of water catchment areas and the exploration of medicinal plant values. All of these are well understood by the government, the university, the Forest Department and private enterprise, but the options of quick or sustained profit and the cost and political acceptability of policing and providing alternative wood supplies do present long-term problems that have to be faced. The priorities are set out by Rodgers (1982) for the WWF/IUCN Tropical Forests and Primates Programme, with emphasis first on the Usambara and Uzungwa Mountains, then on the Ulugurus and other intervening mountains, then on the Mahale Mountains of western Tanzania, and, one would hope also one or two good sites in the Southern Highlands. Hall (1984a) emphasizes the need to collate basic knowledge on ecology and taxonomy as a basis for determining natural resource utilization.

In Uganda the reconstruction of the economy is becoming apparent and there are hopes that the primate-oriented research in the Western Rift forests will signal the way forward for conservation generally in the country.

VI. Specific Proposals

1. Extension of the East African Herbarium, National Museums, Nairobi.

A building was erected during 1984–1985, but furnishings and equipment are needed. Information is available from the Director, National Museums, Box 40658, Nairobi.

2. Extension of the Herbarium, Botany Department, University of Dar es Salaam.

Plans and estimates are available from the Head of the Botany Department, University of Dar es Salaam, Box 35060, Dar es Salaam. The National Herbarium at Arusha should also be considered.

3. Postgraduate training.

There is a need for four taxonomic botanists and two or three ecologists in Kenya, and also for some technical training in overseas herbaria.

4. Botanic Gardens in Kenya.

The Ololua Forest, Langata, Nairobi, has been designated by the National Museums of Kenya as a plant awareness area. A nature trail has been established and a checklist of the plants was published in 1987, and the program should be extended. The possibility of a botanic garden on the Kenya coast was considered at a training workshop on Coastal Zone management Planning, held at Kilifi in July 1984. The municipal garden outside Mombasa might be improved, and the Nature Reserve at Gedi is scheduled for more resources from the National Museums to make a Botanic Garden. Plans are available from the Director, National Museums, Kenya.

5. Botanic Gardens in Tanzania.

New Botanic Gardens have been proposed for the University of Dar es Salaam, Sokoine University, the National Herbarium, Arusha, as well as re-establishing the collections at Amani, but plans need to be coordinated. A regional conference on tropical African botanic gardens is planned for 1989 under the auspices of the IUCN Botanic Gardens Conservation Secretariate to discuss such problems.

6. Appoint a field forest botanist in each country attached to the principal herbarium.

7. National Park Uzungwa, Tanzania.

The cost of this would be US $80,000, plus the cost of a small forest field research center with a director (which could be combined with number 5). See WWF/IUCN (1982).

8. Kenya Trees, Shrubs and Lianes.

This project, initiated in 1984 by Dr. H. Beentje, East African Herbarium, Nairobi, is now half completed. US $75,000 is still needed for publication.

9. Forest Flora of Tanzania.

The cost of this would be US $80,000. Project plans are obtainable from Mr. J. C. Lovett, Missouri Botanical Garden.

10. Urgent help for Ugandan herbaria.

The Herbarium at Makerere University has been well maintained, but warrants revitalization. The Forest Herbarium at Entebbe lost cupboards in 1979 and although the specimens were saved and have now gone to Makerere University, they need remounting and rehousing. The Botanical Gardens at Entebbe are still maintained. The Herbarium at Kawanda Research Station survived the 1981–1982 conflict but needs attention. A new management plan for the herbaria and facilities for staff training, especially for ecologists, is much needed.

VII. Acknowledgments

I am most grateful to the following people who have commented on the first draft and supplied much additional information—Mr. M. G. Gilbert, Flora of Ethiopia Project, Kew; Dr. V. C. Gilbert, formerly ETMA, Nairobi; Professor J. B. Hall, College of N. Wales, Bangor University of Dar es Salaam; Dr. A. C. Hamilton, New University of Ulster; Dr. I. and Professor O. Hedberg, University of Uppsala; Miss C. H. S. Kabuye, East African Herbarium, Nairobi; Professor J. O. Kokwaro, University of Nairobi; Dr. M. F. Lamprey, IPAL, Nairobi; Mr. J. C. Lovett, Missouri Botanical Garden; Mr. J. Mabonga-Mwisaka, EIAP, Nairobi; Mr. R. A. Plumptre, University of Oxford; Mrs. Ann Robertson, Malindi, Kenya; Dr. W. A. Rodgers, Wildlife Institute of India, Dehra Dun; Dr. G. R. Cunningham van Someren, National Museums, Nairobi; Dr. B. Verdcourt, Kew; and Mr. K. Vollesen, Kew.

VIII. Literature Cited

Abraham, M. F. H. 1958. The East African Camphor forests of Mt. Kenya. E. African Agric. J., Kenya. **24:** 139–141.

Agnew, A. D. Q. 1974. Upland Kenya wild flowers. 827 pp. Oxford University Press.

Andrews, P., C. P. Groves & J. F. M. Horne. 1975. Ecology of the lower Tana River flood plain (Kenya). J. E. African Nat. Hist. Soc. & Nat. Mus. **151:** 1–31.

Backeus, I. 1982. Report on a study tour of the indigenous forests of the West Usambara Mts., Tanzania, with special reference to regeneration. Meddel. Växtbiol. Inst. Uppsala.

Ball, J. B. 1965. Working plan for Lake Forests, Masaka District. 35 pp. Uganda, Forest Department. mimeo.

Bassan, E. (ed.). 1984. Endangered Resources for Development. Strategy conference for the management and protection of Kenya's plant communities. 21 pp. National Environment and Human Settlements Secretariat, Nairobi.

Beentje, H. J., Ndegwa Ndiangui & J. Mutangah. 1987. Forest islands in the mist. Swara **10:** 20–21.

Bennett, J. G., L. C. Brown, A. M. W. Geddes, C. R. C. Hendy, A. M. Lavelle, L. G. Swell, & R. Rose Innes. 1979. Mtwara/Lindi Regional Integrated Development Programme, Vol. 1. The physical environment. Ministry of Overseas Development internal report, mimeo.

Bigger, M. 1968. A check list of the flora of Kilimanjaro. 54 pp. College of Wildlife, Mweka, mimeo.

Blundell, M. 1982. The wild flowers of Kenya. 160 pp. Collins, London.

———. 1987. Collins guide to the wild flowers of East Africa. Collins, London.

Borhidi, A., Sebsebe Demissew, B. C. M. Hedrén, S. T. Iversen, W. R. Mziray & T. Pòcs. 1984. Preliminary report of the field expedition in February-March, 1984, within the frame of the Usambara rain forest project. (7 pp., unpubl. report).

———, **S. T. Iversen, W. R. Mziray, L. Peregovits, T. Pòcs & R. P. C. Temu.** 1985. Preliminary report of field expedition in January-February, 1985, within the frame of the integrated Usambara rain forest project. (4 pp., unpubl. report).

Brenan, J. P. M. 1949. Check-lists of the forest trees and shrubs of the British Empire, No. 5: Tanganyika Territory, Part II. 653 pp. Imperial Forestry Institute, Oxford.

Buck, L. (ed.). 1981. Proceedings of the Kenya National Seminar on Agroforestry. 638 pp. ICRAF, Nairobi.

Chapman, J. & F. White. 1970. The evergreen forests of Malawi. 190 pp. Commonwealth Forestry Institute, Oxford.

Clutton-Brock, T. H. & J. B. Gillett. 1979. A survey of forest composition in the Gombe National Park, Tanzania. African J. Ecol. **17:** 131–158.

Dale, I. R. 1939. The woody vegetation of the Coast Province of Kenya. Inst. Pap. Imp. Forest Inst. **18**: 1–28.

———. 1940. The forest types of Mt. Elgon. J. E. African Nat. Hist. Soc. **15**: 74–82.

——— & P. J. Greenway. 1961. Kenya trees and shrubs. 654 pp. Buchanan's Kenya Estates, Ltd., Kenya.

Dawkins, H. C. 1960. Observations on regeneration of the Kakamega and Malaba Forest. 5 pp. Uganda Forest Department mimeo.

——— & M. S. Philip. 1962. Working plan for Mapanga Forest, 1960–1965. Forest Department mimeo.

Diamond, T. 1979. Kakamega, is there a way to stop the rot? Swara **2(1)**: 25–26.

Doute, R., N. Ochanda, & H. Epp. 1983. Forest cover mapping in Kenya using remote sensing techniques. KREMU, Nairobi.

Eggeling, W. J. 1947. Observations on the ecology of the Budongo rain forest, Uganda. J. Ecol. **34**: 20–87.

——— & I. R. Dale. 1952. The indigenous trees of the Uganda Protectorate, ed. 2. 491 pp. Government Printer, Entebbe.

Faden, R. B. 1970. A preliminary report on the Kakamega forest in Kenya. 15 pp. East African Herbarium, Nairobi, mimeo.

FAO-UNEP. 1981. Tropical Forest Resources Assessment Project—Forest Resources of Tropical Africa, Part II: Country Briefs.

Fayad, V. C. & C. Fayad. 1950. An ecological survey of the Nguruman Forest, Kenya. 33 pp. + maps, figures. Mimeographed. Nairobi (no publisher).

Gerhardt, J. & M. Steiner. 1986. An inventory of a coastal forest in Kenya; at Gedi National Monument, including a checklist and a nature trail. Working Paper 36, Swedish University of Agricultural Sciences, Uppsala.

Gilbert, V. C. 1970. Plants of Mt. Kilimanjaro. 117 pp. U.S. National Park Service, Washington, D.C.

Gillett, J. B. & P. G. McDonald. 1970. A numbered check-list of trees, shrubs, and noteworthy lianas indigenous to Kenya. 67 pp. Government Printer, Nairobi.

Glover, P. E. 1967. A Botanical-Kipsigis Glossary. E.A.A.F.R.O., Nairobi, mimeo.

———. 1968. Report on an ecological survey of the proposed Shimba Hills National Reserve. Soil Survey Unit, Department of Agriculture.

——— et al. 1969. A Digo-Botanical Glossary. Kenya National Parks, Nairobi, mimeo.

——— & E. C. Trump. 1970. An ecological survey of the Narok District of Kenya Masailand, Part 2. Vegetation. 157 pp. + map. Kenya National Parks, Nairobi.

Graham, R. M. 1929. Notes on the mangrove swamps of Kenya. J. East African Nat. Hist. Soc. **36**: 157–164.

Grant, D. K. S. 1938. Mangrove woods of Tanganyika Territory, their structure and dependent industries. Tanganyika Notes and Records, April 1938.

Greenway, P. J. 1943. Second draft report on vegetation classification for the approval of the Vegetation Committee Pasture Research conference, Nairobi. Mimeo.

———. 1955. Ecological observations on an extinct East African volcanic mountain. J. Ecol. **43**: 544–563.

———. 1965. The vegetation and flora of Mt. Kilimanjaro. Tanganyika Notes and Records, **64**: 97–107.

———. 1973. A classification of the vegetation of East Africa. Kirkia **9**: 1–68.

Griffith, A. L. 1951. East African Enumerations. I. The Rondo Plateau, South Tanganyika, Empire Forest Rev. **31**: 146–149.

Haerdi, F. 1964. Die Eingeborenen-Heilpflanzen des Ulanga-Distriktes Tanganjikas (Ostafrika). 278 pp. Verlag für Recht und Gesellschaft AG, Basel.

Hall, J. B. 1982a. Current research in Tanzanian montane rain forest. USDM/IDRC Forest Research Training Course, mimeo.

———. 1982b. Visit to Malundwe Hill, Mikumi National Park, from March 20–22, 1982. Division of Forestry, University of Dar es Salaam. 2 pp., mimeo.

———. 1984a. Ecological and taxonomic knowledge and natural resource utilization in Tanzania: a survey of position, problems and prospects. *In* J. Middleton, A. M. Nchundiwe, S. F. Banyikwa, & J. R. Mainoya. (eds.), Proceedings of the Symposium on the Role of Biology in Development, Dar es Salaam, September 1983: 22–29.

———. 1984b. *Juniperus excelsa* in Africa: A biogeographical study of an Afromontane tree. J. Biogeogr. **11**: 47–61.

Hamilton A. 1969. The vegetation of SW. Kigezi. Uganda J. **33**: 175–199.

———. 1981. A field guide to Uganda forest trees. Makerere. Available from author, University, Coleraine, Londonderry, N. Ireland.

———. 1982. Environmental history of East Africa. 328 pp. Academic Press, London.

——— & R. A. Perrott. 1980. The vegetation of Mt. Elgon, East Africa, Mimeographed. Research report for the Government of Kenya.

Harcourt, A. H. 1981. Can Uganda's gorillas survive?—A survey of the Bwindi Forest Reserve. Biol. Conserv. **19**: 269–282.

Hawthorne, W. D. 1984. Ecological and biogeographical patterns in the coastal forests of East Africa. Unpublished D. Phil. Thesis, University of Oxford.

———, K. Hunt & A. Russell. 1982. Kaya: An ethnobotanical perspective. Oxford University, mimeo.

Hedberg, I. & O. Hedberg (eds.) 1968. Conservation of vegetation in Africa south of the Sahara. Acta Phytogeogr. Suec. **54**: 1–320.

Hedberg, O. 1979. Systematic botany, plant utilization and biosphere conservation. Almquist and Wiksell International, Stockholm.

———, A. Borhidi, T. Poćs & J. B. Hall. 1982. Report on factfinding mission on conservation of Tanzanian rain forests (9 pp. unpubl. report).

Hepper, F. N., P. M. L. Jaeger, J. B. Gillett & M. G. Gilbert. 1981. Annotated check-list of the plants of Mt. Kulal, Kenya. 123 pp. IPAL Technical Report D-3.

Heriz-Smith, S. 1976. The wild flowers of the Nairobi Royal National Park. Hawkins, Nairobi.

Herlocker, D. J. 1979. Vegetation of South-Western Marsabit District, Kenya. 68 pp. and map. IPAL Technical D-1.

——— & H. J. Dirschl. 1972. Vegetation of the Ngorongoro Conservation Area, Tanzania. 37 pp. + map. Canadian Wildlife Service, Ottawa.

Hiepko, P. 1978. De Herbario Berolinensi Notulae. Willdenowia **8**: 389–400.

Holmgren, P. K., K. Keuken, & E. K. Schofield. 1981. Index Herbariorum, ed. 7; Regnum Vegetabile 106.

Hora, F. B. & P. J. Greenway. 1940. Check-lists of the forest trees and shrubs of the British Empire, No. 5, Par I. 312 pp. Imperial Forestry Institute, Oxford.

Howell, K. M. 1981. Pugu Forest Reserve; biological values and development. African J. Ecol. **19**: 73–81.

Hughes, F. M. R. 1986. The Tana River flood plain forest, Kenya; ecology and the impact of development. Ph. D. thesis, University of Cambridge, England.

IUCN. 1987. Directory of Afrotropical Protected Areas. 1034 pp.

IUCN, Gland.
Iversen, S. T. 1986. Integrated Usamabara rain forest research project: Preliminary report of the third field expedition., 11 October–20 November, 1986. (13 pp., unpubl. report).
———. 1987. Integrated Usambara rain forest research project: Preliminary report of the fourth field expedition, 27 April–20 May, 1987. (11 pp., unpubl. report).
Jackson, G. & J. S. Gartlan. 1965. The flora and fauna of Lolui Island, Lake Victoria. J. Ecol. **53**: 573–597.
Jackson, J. K. 1956. The vegetation of the Imatong mountains. J. Ecol. **44**: 341–374.
Kates, Peat, Marwick & Co. 1968. A land use strategy for the Shimba Hills. Canadian International Development Agency Report.
Kelsey, G. M. & T. E. S. Langton. 1984. The conservation of the Arabuko-Sokoke Forest, Kenya. ICBP/University of East Anglia. Cambridge, England.
Kerfoot, O. 1964. The vegetation of the South-west Mau Forest. E. African Agric. Forest J. **29**: 295–318.
Kielland-Lund, J. 1982. Structure and morphology of four forest and woodland communities of the Morogoro area, Tanzania. Pages 69–93 in H. Diertschke (ed.), Struktur und Dynamik von Waldern. Kramer, Vaduz.
Kokwaro, J. O. 1972. Luo-English Botanical Dictionary. E. African Publishing House, Nairobi.
———. 1976. Medicinal plants of East Africa. 384 pp. East African Literature Bureau, Nairobi.
———. 1982. Economic importance and local use of Kenyan mangroves. Pages 377–386 in Proceedings of the Kenya National Seminar on Agroforestry. ICRAF, Nairobi.
———. 1985. The distribution and importance of the mangrove forests of Kenya. J. E. African Nat. Hist. Soc. & Nat. Mus. **188**: 1–12.
——— & D. Herlocker. 1982. A check-list of botanical, Samburu and Rendille names of plants of the IPAL Study Area, Marsabit District, Kenya. 164 pp. IPAL/UNESCO Technical Report D-4, Nairobi.
Kuyper, J. B. J. M. 1982. The human influences on the vegetation in a part of the Kilifi area. A landscape guided approach. Preliminary Report No. 4.
Lamprey, H. F. 1977. A preliminary review of the conservation status of East African habitats. A review paper of the Scientific and Technical Committee of the East African Wildlife Society.
Langdale-Brown, I., H. A. Osmaston & J. G. Wilson. 1964. The vegetation of Uganda. 159 pp. + maps. Government Printer, Entebbe.
Leedal, P. H. (undated). Ecology of the Mbeya Region. 99 pp., mimeo.
Leggat, G. J. 1957. Working plan for the Kitomi-Kasyoha Central Forest Reserve, 1957–1961, extended to 1967. Forest Department mimeo.
——— & H. A. Osmaston. 1961. Working plan for the Impenetrable Forest Reserve, Kigezi District, 1961–1971. Forest Department mimeo.
Lind, E. M. & M. E. S. Morrison. 1974. East African vegetation. 257 pp. Longman, London.
Lovett, J. C. 1985. Results of the First International Uzungwa Expedition 1984. Variable area large tree survey. 19 pp. 2. Collection Number List. 26 pp. Mimeo. Missouri Botanical Garden.
———. 1988. Endemism and affinities of the Tanzanian montane forest flora. Monogr. Syst. Bot. Missouri Bot. Gard. **25**: 591–598.
——— & J. M. Lovett. 1985. Preliminary Report of the First International Uzungwa Expedition. 21 pp. mimeo. WWF.

———, J. M. Lovett, & R. M. Polhill. 1983a. Report of a visit to the West Usambara Mountains, Tanzania, 9–20 July 1983. 27 pp. mimeo. WWF, Kew.
———, ———, & ———. 1983b. Kanga Mountain Forest Visit, 4–5 July 1983. 4 pp. mimeo. WWF, Kew.
Lucas, G. Ll. 1968 Kenya. In I. Hedberg & O. Hedberg (eds.), Conservation of vegetation in Africa south of the Sahara. Acta Phytogeogr. Suec. **54**: 156–163.
Mabberley, D. J. 1975. Notes on the vegetation of the Cherangani Hills, NW. Kenya. J. E. African Nat. Hist. Soc. **150**: 1–11.
Marsh, C. 1976. A management plan for the Tana River Game Reserve. New York Zoological Society Report.
———. 1985. A re-survey of Tana River primates and their forest habitat. Report to National Museums of Kenya.
Mburu, O. 1983 Forest and forestry research in Kenya. Workshop on Strengthening Forest Research in Kenya. 9 pp., mimeo.
Milne-Redhead, E. et al. (eds.) 1952-. Flora of Tropical East Africa. Balkema, Rotterdam.
Moll, E. J. & F. White. 1978. The Indian Ocean Coastal Belt. Pages 561–598 in M. J. A. Werger (ed.), Biogeography and ecology of southern Africa. Junk, Hague.
Moomaw, J. C. 1960. A study of plant ecology of the Coast Region of Kenya Colony. 54 pp. with map. Government Printer, Nairobi.
Morgan, W. T. W. 1968. East Africa: Its people and resources. Oxford University Press, Nairobi.
Mungai, G. M., J. B. Gillett & C. F. Eagle. (undated). Plant species in Kenya: Survival or extinction. 6 pp. Wildlife Clubs of Kenya.
Mwesaga, B. C. 1984. M. Sc. (For.) thesis. University of Dar es Salaam.
Nishida, T. & S. Uehara. 1981. Kitongwe name of plants: A preliminary listing. African Study Monogr. **1**: 109–131.
——— & ———. 1983. Natural diet of chimpanzees (*Pan troglodytes schweinfurthii*): Long-term record from the Mahale Mountains, Tanzania. African Study Monogr. **3**: 109–130.
Osmaston, H. A. 1956. Working plan for the Budongo, Siba and Kitogo Forest Reserve, 2nd revision, 1955–1964. Forest Department mimeo.
———. 1959a. Working plan for the Bugoma Forest, 1960–1970. 33 pp. Forest Department mimeo.
———. 1959b. Working plan for the Kibale and Itwara Forests, 1959–1965. Forest Department mimeo.
———. 1960. Working plan for the Kalinzu Forest, 1960–1970. 25 pp. Forest Department mimeo.
Owino, F. 1983. Analysis of current forestry research programmes in Kenya. Workshop on Strengthening Forest Research in Kenya. 11 pp. mimeo.
Philip, M. S. 1964. Working plan for Budongo Central Forest Reserve, 3rd revision. 130 pp. Forest Department mimeo.
Pòcs, T. 1976a. Vegetation mapping in the Uluguru Mountains (Tanzania, Africa). Boissiera. **24**: 477–498, with map.
———. 1976b. Bioclimatic studies in the Uluguru Mountains (Tanzania, East Africa). Acta Bot. Acad. Sci. Hung. **22**: 163–183.
Robertson, A. 1984. The status of Kaya forest in Endangered Resources for Development. National Museums of Kenya, Nairobi.
Rodgers, W. A. 1982. WWF/IUCN Tropical Forests and Primates Programme, Tanzania. 24 pp.
———. 1983. A note on the distribution and conservation of *Oxystigma msoo* Harms (Caesalpiniaceae). Bull. Jard. Nat. Bot. Belg. **53**: 161–164.
———. 1984. Wildlife and habitat conservation. Pages 30–37 in Middleton, J., Nchundiwe, A. M., Banyikwa, S. F. & Mainoya,

J. R. (eds.), Proceedings of the symposium on the role of biology in development. Dar es Salaam, September 1983.

———, T. T. Struhsaker & C. C. West. 1984. Observations on the red colobus (*Colobus badius tephrosceles*) of Mbisi Forest, south-west Tanzania. Afr. J. Ecol. **22**: 187–194.

———, J. B. Hall & L. Mwasumbi. 1983. The conservation values and status of Kimboza. 80 pp. mimeo. University of Dar es Salaam.

———, ———, ——— & K. Vollesen. 1986. The conservation status and values of Ngezi Reserve, Pemba Island, Tanzania. Mimeo., 38 pp. University of Dar es Salaam.

——— & K. M. Homewood. 1982a. Biological values and conservation prospects for the forests and primate populations of the Uzungwa Mountains. Biol. Conserv. **24**: 285–304.

——— & ———. 1982b. Species richness and endemism in the Usambara mountain forests, Tanzania. J. Linn. Soc., Biol. **18**: 197–242.

——— & R. I. Ludanga. 1973. The vegetation of the Eastern Selous Game Reserve. 67 pp. Miombo Research Center, Dar es Salaam, mimeo.

———, ——— & H. P. DeSuzo. 1977. Biharamulo, Burigi and Rubondo Island Game Reserves. Tanzania Notes and Records. **88-82**: 99–124.

———, K. Mtomowema & J. B. Hall. 1982. Conservation of tropical rain forests: deliberations of the Tanga workshop/seminar 16–18 February 1982. Unpublished report.

———, L. B. Mwasumbi, & J. B. Hall. 1985. The floristics of three coastal forests near Dar es Salaam. Tanzania Forest Working Group. Typescript, 24 pp.

———, W. Mziray & W. Shishira. In Press. The extent of forest cover in Tanzania using satellite imagery. Ined.

——— & R. G. Wingfield. In Press. The vegetation of Mafia Island [exact title not known]. Kirkia.

Ross, K. 1982. Ecology of large herbivores in Shimba Hills, and its significance to their management. Ann. Rep. WWF.

Rowell, T. 1982. Kakamega Forest. Swara **5(2)**: 8–9.

Russell, E.W. (ed.) 1962. The natural resources of East Africa. 144 pp. + maps. Hawkins, Nairobi.

Salvadori, C. 1985. Zanzibar: What is left of its wildlife? Swara **8**: 10–13.

Sangai, G. W. 1963. Bondei, Shambaa and Zigua botanical dictionary. E.A.A.F.R.O., Nairobi, mimeo.

Stevens, T. E. 1962. Oxford University Tanganyika expeditions 1958–1959. Tanganyika Notes and Records **58–59**: 110–115.

Struhsaker, T. T. & L. Leland. 1979. Kibale, an inheritance still preserved. Swara **2(1)**: 18–24.

Stuart-Smith, A. M. 1962. Working plan for Zoka Central Forest Reserve 1962–1972. 11 pp. Forest Department mimeo.

Synnott, T. J. 1979a. A report on the status, importance and protection of montane forest. 57 pp. IPAL, Technical Report No. D-2a.

———. 1979b. A report on prospects, problems, and proposals for tree planting. 41 pp. IPAL Technical Report D-2b.

———. 1979c. Implementing forestry programmes for local community development, South-Western Marsabit District, Kenya. 10 pp. IPAL Technical Report D-2c.

———. 1985. A checklist of the flora of Budongo Forest Reserve, Uganda, with notes on ecology and phenology. Commonwealth Forestry Institute Occasional Papers No. 27. 99 pp. University of Oxford.

Thorsell, J. W. 1983. Some observations on management planning for protected areas in eastern Africa. 22nd Working Session of IUCN Commission on National Parks and Protected Areas, Zimbabwe, May 1983, **4**: 5–8.

Trapnell, C. G. et al. 1966–1986. Kenya Vegetation, sheets 1–4 (maps 1:250,000). Directorate of Overseas Surveys, Tolworth.

Tweedie, E. M. 1976. Habitats and check-lists of plants on the Kenya side of Mount Elgon. Kew Bulletin **31**: 227–257.

UNESCO. 1973. International classification and mapping of vegetation. 99 pp. UNESCO, Paris.

Van Leeuwen, M. W. N. 1982. Vegetation and landuse map (scale 1:100,000) of the Kilifi area. A landscape guided Approach. Preliminary Report No. 3.

Verdcourt, B. 1962. The vegetation of the Nairobi Royal National Park. Pages 38–56, with map, *in* S. Heriz-Smith, The wild flowers of the Nairobi Royal National Park. Hawkins, Nairobi.

———. 1968. East Africa: Regional synthesis. Pages 186–192 *in* I. Hedberg & O. Hedberg (eds.), Conservation of vegetation in Africa south of the Sahara. Acta Phytogeogr. Suec. 54.

Vollesen, K. 1980. Annotated check-list of the vascular plants of the Selous Game Reserve, Tanzania. Opera Bot. **59**: 1–117.

Walter, H. & M. Steiner. 1936. Die Ökologie der Öst-Afrikanischen Mangroven. Z. Bot. **30**: 65–193.

Webster, G. 1961. Working plan for South Mengo forest, 1st revision, 1961–1971. 43 pp. Forest Department mimeo.

White, F. 1950. The forests of Mt. Kenya. Forest Soc. J. **5**: 1–7.

———. 1979. The Guineo-Congolian Region and its relationship to other phytochoria. Bull. Jard. Bot. Nat. Belg. **49**: 11–55.

———. 1981. The history of the Afromontane archipelago and the scientific neeeds for its conservation. African J. Ecol. **19**: 33–54.

———. 1983. Vegetation Map of Africa. 4 sheets plus descriptive memoir, The Vegetation of Africa. 356 pp. UNESCO.

Willigen, T. A. van & J. Lovett. 1979. Report of the Oxford Expedition to Tanzania, 1979. 95 pp., mimeo.

Wilson, J. G. 1962. The vegetation of Karamoja District, Northern Province of Uganda. Mem. Res. Div., ser. 2, **5**: 1–163. Department of Agriculture, Uganda, Mimeo.

Wingfield, R. G. 1967. A description of Pugu Hills Forest Reserve. University of Dar es Salaam, unpublished typescript.

WWF/IUCN. 1982. The WWF Tropical Forests and Primates Campaign. Forest Pack n.5 - Tanzania. 11 pp. cylostyled.

Tanzania

J. C. Lovett

Contents

I. Introduction	233
II. Description of the Region	233
A. Geographical Extent and Area	233
B. Topography	233
C. Population	233
III. Vegetation Maps of Tanzania	233
IV. Publications on the Flora of the Region	233
V. Importance of the Forests	234
A. Endemism	234
B. Comparisons to Forests of Other Countries	234
VI. Status of Floristic Inventory	234
VII. Acknowledgments	234
VIII. Literature Cited	234

I. Introduction

Tanzania has already been discussed in this volume by Roger Polhill, but a more detailed analysis of the Tanzanian forest flora is necessitated by its high level of endemism and uniqueness in continental Africa. The East African flora is relatively well known, but this cannot be regarded with complacency in a world strategy for tropical forest floristic inventory. Rather it should be regarded as a springboard of knowledge from which to prepare a much-needed regional flora so that decisions on conservation can be made upon a sound scientific basis.

II. Description of the Region

A. Geographical Extent and Area

Tanzania, a large country with an area of 939,400 km², is extremely dry, with nearly half the country receiving less than 760 mm of rain a year, and 96% of the country receiving less than 1270 mm of rain a year (Griffiths, 1972). Consequently only a very small percentage of the country is, or was, covered by tropical rain forest. Currently the best estimate for the area of tropical rain forest (TRF) in Tanzania is 9308 km², which is the area occupied by permanently closed high forest reserves, a mere 1% of the country area.

B. Topography

In view of the small amount of tropical rain forest, I have limited my description of the topography of the region to those areas with sufficiently high rainfall to support tropical rain forest. A more general description of the country has been given by Berry (1972).

The forest grows in five main topographic areas, three of which are mountainous. These high altitude regions intercept rain-bearing winds, permitting TRF to develop. The other two areas are the coastal forests, which have a patchy, and rapidly diminishing, distribution on the hills and islands of the east coast; and the forest in the west of the Lake Victoria basin, which is really an extension of the Ugandan flora. The other three areas are disjunct mountains, and can be likened to forested islands in a sea of dry savanna. It is this isolation which has allowed evolution to proceed independently of the other forest areas, resulting in high degrees of endemism. Of these areas, those with the poorest forest flora are recently active volcanics: Meru, Kilimanjaro, Hanang and Poroto (Rodgers & Homewood, 1982). Richer forests are found on the rift highlands of Mahali and Mbizi in the west of the country, and the drier forests of Mbulu which are in the central north. But by far the most interesting and diverse flora is found on the block mountains in the east of the country, which I will define here as the eastern arc mountains, which from north to south are the Upare, Usambara, Nguru, Ukaguru, Uluguru, Rubeho, Malundwe, Uzungwa, Mahenge, and the southern highlands (excluding the volcanics). These areas have an extraordinarily high degree of endemism, and despite the fact that they are very small when compared to the main African forest blocks, they must rank as some of the most biologically interesting areas in the world.

C. Population

However, the rain which supports the growth of the TRF also brings a high agricultural potential to the land. Tanzania relies heavily on agriculture, which provides a livelihood for 90% of the country's population, and 75% of the total value of exports (1976 figures, from Anonymous, 1978). As the country is predominantly dry, then the main centers of rural population are those once occupied by forest. For example, population densities of over 300 people/km² occur in the Usambara Mountains, which are among the highest densities in the world for agriculture-based economies (Tanzania et al., 1976). The population is also rapidly expanding; it doubled between 1958 and 1978. These people need food. To grow that food they need water, which occurs mainly in the high rainfall areas of the mountains, where increasing pressure is brought to bear on the dwindling forests (Lovett, 1988a).

III. Vegetation Maps of Tanzania

Apart from the UNESCO/AETFAT/UNSO Vegetation Map of Africa (White, 1983), the classic vegetation map of Tanzania is by Gillman (1949) and this is still superior to many more recent maps. Alan Rodgers of the University of Dar es Salaam is currently preparing a forest map of Tanzania from satellite photographs, backed up by ground control which will provide an accurate description of the distribution of the forest in the country. Other maps of a more regional basis are: Gilchrist (1952) of the southern highlands and Uzungwa Mountains; Pocs (1976) of the Uluguru Mountains, and TIRDEP (Egger & Glaser, 1975) of the Usambara Mountains. A criticism of the TIRDEP map with regard to representation of forest can be found in van der Willigen and Lovett (1979).

IV. Publications on the Flora of the Region

Apart from the definitive *Flora of Tropical East Africa* (Milne-Redhead et al., 1952–), and the classic work by Brenan (1949), there are a number of other works which have check-lists of limited parts of the forest flora. These are: Greenway and Vesey-FitzGerald (1972) for Lake Manyara National Park ground-water forests; Clutton-Brock and Gillett (1979) for Gombe National Park riverine forests; Steele (1966) for the montane trees and shrubs of south Kilimanjaro; Hall (in manuscript) for the montane forest of the University Forest Reserve, Mazumbai, in the West

Usambara Mountains; Hawthorne (1984) for the coastal forests north of the Rufiji River; Rodgers, Homewood and Hall (1979) for the Magombero ground-water forest; Rodgers, Hall, Mwasumbi and Vollensen (1984) for the Kimbosa ground-water forests; Mwasumbi (in manuscript) for the coastal forests in the Pugu Hills; Vesey-FitzGerald (unpublished) for the flora of Mount Meru; Vollesen (1980) for the lowland, riverine and ground-water forests of the Selous Game Reserve; and Lovett and Thomas (1986) for Mwanihana Forest Reserve, northern Uzurgwa Mountains. There are few resident taxonomists in Tanzania, and much of the taxonomic work falls on untrained but skilled technicians.

V. Importance of the Forests

A. Endemism

The degree of endemism for the whole of the forest flora of Tanzania has not yet been calculated, but a recent estimate of the endemism in the montane forest flora is available (Lovett & Polhill, 1984; Lovett, 1988b). In this study, the montane forests were defined in the broad sense to include lowland and ground-water forest associated with the mountains. It excludes the coastal forest flora, which in itself is rich in endemics, the forest flora of Lake Victoria, and the mangrove forest flora. The study was based on published parts and manuscripts of the *Flora of Tropical East Africa*, which comprise 53% of the families occurring in the flora region (Kenya, Tanzania and Uganda). To obtain and estimate for the whole flora, the figures obtained from the study were multiplied by 1.88 to give the equivalent of 100% analysis. This gives a figure of 2085 species in 801 genera with 509 forest species endemic to Tanzania. The degree of endemism is thus 24.4% of the total forest flora. Of the endemics, 93% occur in the forests of the eastern arc mountains, testifying to their biological importance.

B. Comparisons to Forests of Other Countries

However, to really demonstrate the biological interest of the Tanzanian forests, they must be compared to those of other countries. This can be done by reference to an excellent paper by Brenan (1978). The total tropical African flora is estimated at 30,000 species and 2497 genera in 20,000,000 km^2. The montane forest flora of Tanzania therefore contains 7% of these species, and 32% of the genera, in less than 0.05% of the areas. Comparison with neighboring countries shows that Uganda has an estimated 0.13 endemic species/1000 km^2 Kenya has 0.45, while Tanzania has 1.19, with the montane forests containing 71.4% of Tanzania's estimated 1122 endemic species, or 50 endemic species/1000 km^2. Gabon can be regarded as having the highest level of endemism of any continental African country, the endemism being estimated at 22.4% of the country's total flora. However, this degree of endemism in a flora that is principally forest is less than the estimated 24.4% endemism of the Tanzanian montane forests. By this criterion the Tanzanian montane forests are the most important forests on continental Africa.

VI. Status of Floristic Inventory

Work has been proceeding on the definitive *Flora of Tropical East Africa* (Milne-Redhead et al., 1952–) since 1951, a period of over 30 years. This provides an excellent basis for writing a regional flora of the Tanzanian forests. Indeed, such a regional flora will enable the earlier parts of the Flora to be updated in the light of recent name changes and the discovery of new species. It is estimated that such a flora will take three years to produce (Lovett & Polhill, 1984), and cost 80,000 US dollars. The flora would be composed of three sections: ecology and phytogeography; conservation practice and prospects; and systematics of vascular plants (with a later supplement for lower plants). Further fieldwork would also be necessary, for despite over 100 years of botanical work in Tanzania (Gillett, 1962), there are many areas of forest which are poorly known. The eastern scarp of the Uzungwa Mountains has recently been yielding many new species and a new family for East Africa (Vollesen, 1982), and there are many areas, such as the southern Ulugurus and Mahenge which remain little known.

In view of the immense biological importance of the Tanzania Forests, and the current pressures on them, the forest flora would not only be of great academic interest, but would also greatly assist in the effort to rationally conserve as much of the forest as possible.

VII. Acknowledgments

I am grateful to the Bentham-Moxon Trust and The World Wildlife Fund for financial support during the preparation of this paper; the Tanzanian Scientific Research Council for permission to conduct fieldwork in Tanzania; and Dr. R. M. Polhill and Prof. J. B. Hall for continued advice and encouragement. I owe a special debt to my wife, Jill Lovett, for assistance in the field and for help with editing and typing.

VIII. Literature Cited

Anonymous. 1978. Towards self-reliance. Development, employment and equity issues in Tanzania. International Labour Office, Jobs and Skills Programme for Africa, Addis Ababa.

Berry, L. 1972. Physical features. *In* W. T. W. Morgan (ed.), East Africa, its people and resources. Oxford.

Brenan, J. P. M. 1949. Check-lists of the forest trees and shrubs of the British Empire, No. 5: Tanganyika Territory, Part II. Imperial Forestry Institute, Oxford.

———. 1978. Some aspects of the phytogeography of tropical Africa. Ann. Missouri Bot. Gard. **65**: 437–478.

Clutton-Brock, T. H. & J. B. Gillett. 1979. A survey of forest com-

position in Gombe National Park, Tanzania. African J. Ecol. **17**: 131–158.
Egger, A. & T. Glaser. 1975. Usambara ecological development project proposal to TIRDEP: Tanga Region, Tanzania.
Gilchrist, B. 1952. Vegetation. Pages 57–67 *in* Report of Central African rail link development survey 1–2. Overseas Consultants Inc. and Sir Alexander Gibb and Partners. Colonial Office, London.
Gillman, C. 1949. A vegetation-types map of Tanganyika Territory. Geogr. Rev. **39**: 7–37.
Gillett, J. B. 1962. The history of the botanical exploration of the area of "The Flora of Tropical East Africa" (Uganda, Kenya, Tanganyika, and Zanzibar). Pages 205–229 *in* A. Fernandes (ed.), Compt. Rend. IV Réunion Plènière de l'Association pour l'Étude Taxonomique de la flore d'Afrique Tropicale. Lisbon.
Greenway, P. J. & D. F. Vesey-FitzGerald. 1972. Annotated checklist of plants occurring in Lake Manyara National Park, J. East African Nat. Hist. Soc. Natl. Mus. **28**: 1–29
Griffiths, J. F. 1972. Climate. *In* W. T. W. Morgan (ed.), East Africa, its people and resources. Oxford.
Hall, J. B. (in manuscript). Check-list of the University Forest Reserve, Mazumbai, West Usambara Mountains, Tanzania.
Hawthorne, W. D. 1984. Ecological and biogeographical patterns in the coastal forest of East Africa. Unpublished Ph. D. Thesis. University of Oxford.
Lovett, J. C. 1988a. Practical aspects of moist forest conservation in Tanzania. Pages 491–496 *in* P. Goldblatt & P. Lowry II, (eds.), Modern systematic studies in African botany. Monogr. in Systematic Botany, Missouri Botanical Garden 25. St. Louis.
———. 1988b. Endemism and affinities of the Tanzanian montane forest flora. Pages 591–598 *in* P. Goldblatt & P. Lowry II, (eds.), Modern systematic studies in African botany. Monogr. in Systematic Botany, Missouri Botanical Garden 25. St. Louis.

——— . **R. M. Polhill.** 1984. Flora size and endemism. Part of a feasibility study for the Montane Flora of Tanzania. Mimeograph, Kew.
——— **& D. W. Thomas.** 1986. The ecology of pteridophytes in the Mwanihana Forest Reserve, Tanzania. Fern Gaz. **13**(2): 103–107.
Milne-Redhead, E. et al. (eds.). 1952–. Flora of Tropical East Africa. Balkema, Rotterdam.
Pocs, T. 1976. Vegetation mapping in the Uluguru Mountains (Tanzania, Africa). Boissiera **24**: 477–498, with map.
Rodgers, W. A., J. B. Hall, L. Mwasumbi & K. Vollesen. 1984. Check-list of the Kimboza Forest Reserve, Tanzania. Mimeograph, University of Dar es Salaam.
——— **& K. Homewood.** 1982. Species richness and endemism in the Usambara mountain forests, Tanzania. Biol. J. Linn. Soc. **18**: 197–242.
———, ——— **& J. B. Hall.** 1979. An ecological survey of Magombero Forest Reserve, Kilombero district, Tanzania. Mimeograph, University of Dar es Salaam.
Steele, R. C. 1966. A check-list of the trees and shrubs of the South Kilimanjaro Forests. Part 1. Tanzania Notes Rec. **65**: 97–102.
Tanzania, Government of, and the Federal Republic of Germany. 1976. Tanga Water Master Plan, 7 vols.
Vesey-FitzGerald, D. F. (unpublished manuscript). Meru Vegetation.
Vollesen, K. 1982. A new species of *Seychellaria* (Triuridaceae) from Tanzania. Kew Bull. **36**: 733–736.
———. 1980. Annotated check-list of the vascular plants of the Selous Game Reserve, Tanzania. Opera Bot. **59**: 1–117.
van de Willigan, T. & J. C. Lovett. 1979. Report of the Oxford Expedition to Tanzania, 1979. Mimeograph, Oxford.
White, F. 1983. Vegetation map of Tanzania. Four sheets plus descriptive memoir. The vegetation of Africa. UNESCO.

Madagascar

Laurence J. Dorr, Lisa C. Barnett and Armand Rakotozafy

Contents

I. Introduction	238
II. Description of the Region	238
A. Topography and Climate	238
1. East Malagasy Region	238
2. West Malagasy Region	239
B. Population	240
III. Vegetation Map of Madagascar	240
IV. Magnitude of Floristic Inventory	241
A. Herbarium Specimens in Malagasy Institutions	241
B. Herbarium Specimens in Non-Malagasy Institutions	241
V. Publications on the Flora	241
VI. Completeness of Floristic Inventory	243
A. Relatively Well-Collected Regions	243
B. Poorly Collected Regions	243
C. Regions Rich in Endemics	244
D. Regions Yielding New and Important Distributions	244
E. Extinction	244
VII. Current Collections and Work on Large or Important Families	244
A. Collection of Particular Groups	244
B. Important Families	245
VIII. Important Areas of Forest Threatened by Destruction	245
IX. Inventory Resources	246
A. Academic Institutions	246
B. Government Institutions	246
C. Non-government Institutions	246
D. Foreign Programs	246
E. Collaborative Programs	247
X. Suitability of Floristic Inventory	247
XI. Resources Needed for Continued Inventory	247
A. Institutions	247
B. Personnel	248
C. Cost	248
XII. Acknowledgments	248
XIII. Literature Cited	248

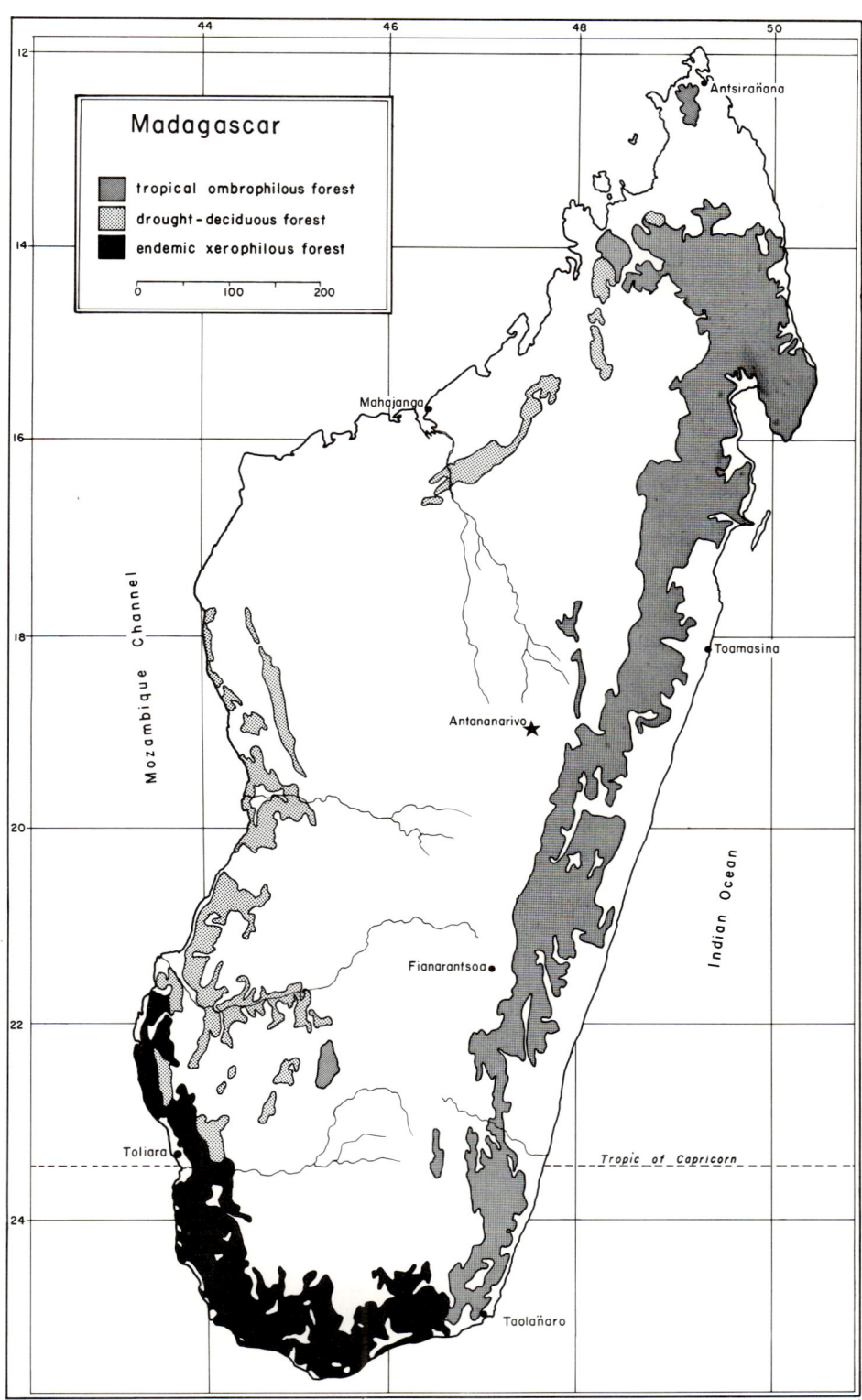

FIG. 1. Existing tropical forest in Madagascar, according to Humbert and Cours Darne (1965). Forest classification units follow UNESCO (1973); 1A1—Tropical ombrophilous forest; IBI—Tropical drought-deciduous forest; and *—Endemic xerophilous forest.

I. Introduction

Madagascar is noted for the richness of its flora in terms of the number of species, especially when compared to the flora of continental Africa. Approximately 8500 species of vascular plants are known (White, 1983). However, higher estimates of the size of the flora range from 10,000 species (Humbert, 1959; Koechlin, 1972; Koechlin et al., 1974) to 12,000 (Dejardin et al., 1973; Guillaumet, 1984). The flora is diverse, with slightly more than 200 vascular plant families represented on the island.

Madagascar is also remarkable for the level of endemism in its flora. Endemism at the species level has been calculated as 81% (Humbert, 1959), although further study may reveal this estimate to be high, especially with regard to herbaceous taxa. The flora is outstanding in terms of supraspecific endemism. About 20% of Madagascar genera and seven angiosperm families (Asteropeiaceae, Didiereaceae, Didymelaceae, Diegodendraceae, Humbertiaceae, Sarcolaenaceae, and Sphaerosepalaceae) are restricted to the island (White, 1983). Geosiridaceae should not be included among the endemic families. The sole genus, *Geosiris*, can be accommodated in the Iridaceae, subfamily Nivenioideae, according to Goldblatt et al., 1987.

The identifiable affinities of the Malagasy flora are strongest with Africa, particularly with eastern-coastal and southern Africa. Perrier de la Bâthie (1936) estimated that the African element in the flora comprised 27%, the Asian element 7%, the Austral element 3%, and the pantropical element 42%. A recent element accounted for 15% and the true Madagascar endemic element was estimated to be 6%. More contemporary phytogeographic studies point out the dearth of basic taxonomic data on which to base analyses (Dejardin et al., 1973; Leroy, 1978). A widely-held hypothesis contends that the island houses a large number of relictual taxa (Leroy, 1978). The recent discovery in South Africa of fossil pollen of Sarcolaenaceae (Coetzee & Muller, 1984), a family today confined to Madagascar, supports this hypothesis.

Unfortunately, Madagascar is also noteworthy in terms of the destruction of its natural vegetation. According to Rauh (1979), the total surface area of Madagascar is about 580,000 km^2 (a low figure, see below), of which natural vegetation covers only 50,000 km^2. This includes 25,000 km^2 of tropical rain forest, 15,000 km^2 of southern xerophytic bush and 10,000 km^2 of mountain sclerophyllous and deciduous forest of the plateau and west. Similarly, Battistini and Verin (1972) estimate that only 21.1% of the island is covered by undisturbed forest. In light of Chauvet's (1972) calculation that native forest is cleared at the rate of 100,000 to 200,000 hectares (1000 to 2000 km^2) per year, it is difficult to remain optimistic that much primary forest will survive to be inventoried.

II. Description of the Region

Madagascar, the fourth largest island in the world, is located 400 km east of Mozambique and the African continent. The island is approximately 590,000 km^2 in area. The northernmost point of the island, Tanjon'I Bobaomby (Cap d'Ambre), and the southernmost, Tanjon'I Vohimena (Cap Sainte Marie), are separated by 1600 km. Almost 600 km separate the east and west coasts at the broadest part of the island. The Tropic of Capricorn passes throught the south in the vicinity of Toliara (Tuléar).

A. Topography and Climate

The following account has been organized according to Humbert's (1955a) phytogeographic regions and domains in order to place topographic and climatic information in a floristic context. This discussion is based largely on the work of Donque (1972), Humbert (1965), Koechlin (1972), and White (1983). The most recent comprehensive review (Koechlin et al., 1974) of the flora and vegetation of Madagascar departs slightly from Humbert (1955a) with respect to terminology and definition of phytogeographic units, but Humbert's (1955a) system is still widely utilized.

Humbert (1955a) recognized two major phytogeographic regions, the East Malagasy Region and the West Malagasy Region, on the basis of their fundamentally different vegetation. The East Malagasy Region consists essentially of evergreen rain forest or evergreen forest of lower stature and the West Malagasy Region consists of dry deciduous forest and deciduous thicket. These regions are in turn subdivided into domains based on features of the vegetation and climate.

1. East Malagasy Region

This region was subdivided by Humbert (1955a) into four domains: Eastern Domain (Domaine de l'Est); Sambirano (Domaine du Sambirano); Central Domain (Domaine du Centre); and High Mountain Domain (Domaine des Hautes Montagnes).

a. Eastern Domain. The Eastern Domain is that belt extending the length of the eastern coast from the basin of the Loky River north of Vohemar south to Tolanaro (Fort Dauphin). Elevations range from sea level to 800–900 m inland. This narrow coastal band (rarely more than 30 km wide) is clearly delimited on the west by the steep escarpment rising to the Central Plateau. Most of the coastal plain is composed of recent alluvium.

Quantity and regularity of rainfall are the distinguishing climatic features of the region. Windward of the Central Plateau, the Eastern Domain receives an average of more than 2000 mm of rain each year and is not subject to an appreciable dry period.

Humbert (1965) has characterized the climax vegetation of the Eastern Domain as dense, ombrophilous, evergreen forest; taxa do not exhibit a high degree of seasonality. Well-represented families include Euphorbiaceae, Rubiaceae, Araliaceae, Orchidaceae, and Melastomataceae. However, most eastern vegetation exists today as secondary vegetation (*savoka*) or savanna.

b. Sambirano. The northwestern corner of the island, known as the Sambirano, comprising the basin of the Sambirano River between the Helodranon 'Ambaro (Baie d'Ambaro) and the Helodranon 'Sahamalaza (Baie de Sahamalaza) and including Nosy-Bé is an extension of the Eastern Domain and is recognized as such principally on aspects of climate and vegetation.

Annual rainfall is high, averaging 2200 mm, and more or less regularly distributed throughout the year. Converging easterly currents in the lee of the Massif de Tsaratanana are responsible for the high and constant levels of precipitation. The mean annual temperature, 26°C, is also high because of latitude.

Floristically, the remaining Sambirano forests are dense and ombrophilous, resembling those of the east in terms of their very high species diversity, but with the actual component species being different from those of the east. A number of western taxa are common in the Sambirano at lower altitudes.

c. Central Domain. The Central Domain includes those regions averaging 800 to 2000 m elevation. Included are isolated mountains separated from the main part of the domain: Montagne d'Ambre in the north and the Isalo Plateau in the southwest. Often misleadingly referred to as the "Hauts Plateaux," the Central Domain is not level but highly irregular in profile, being composed principally of intensely metamorphosed Precambrian bedrock that is blanketed extensively by layers of lateritic clays.

The Central Domain in general experiences a cooler climate (17–20°C) than the rest of the island (except for the high mountains). Although the region is subject to seasonal changes in temperature, in all areas the diurnal temperature ranges are greater than the annual range. Above 1300 m elevation temperature reduction becomes an important ecological factor, declining an average of 0.6°C for every 100 m. Annual rainfall averages about 1500 mm; depending on elevation and latitude, between 81 and 94% of this precipitation falls during the austral summer.

The botanical landscape of the Central Domain consists mainly of grassy pseudo-steppe punctuated by profound erosion scars (*lavaka*). Rain forests exist mainly as relictual pockets of vegetation in isolated valleys. These forests are rich, being especially diverse in understory species. The arborescent genera *Tambourissa* (Monimiaceae) and *Weinmannia* (Cunoniaceae) are often characteristic canopy components. The so-called *tapia* woodland, another type of climax vegetation, is usually localized on rocky outcrops of the western slopes, and consists principally of *Uapaca bojeri* (Euphorbiaceae) plus diverse members of the Sarcolaenaceae.

d. High Mountain Domain. This domain is discontinuous, composed of those massifs and isolated mountains exceeding 2000 m elevation. (Koechlin, 1972, questions the validity of recognizing a phytogeographical domain restricted to a single elevation.) Maromokotro, part of the Massif de Tsaratanana in the extreme north, is the island's highest summit at 2876 m; the Massif de Marojejy, also in the north, reaches 2134 m; Mont Tsiafajavona at 2643 m lies in the Massif d'Ankaratra in the center of the island; and Pic Boby in the Massif de l'Andringitra stands at 2658 in the southeast. These mountain regions are of diverse geologic origin; those in the north and center are essentially volcanic in origin, and the Andringitra granitic.

Continuous climatological data are not available for these isolated peaks although it is assumed that the high mountains receive large amounts of precipitation, including occasional snow and frost in the austral winter, and that the vegetation is exposed to wide fluctuations in temperature, insolation, and humidity.

The climax vegetation of the peaks is variable: dense, ombrophilous, evergreen forest; dense, sclerophyllous forest of short stature (10–12 m in height) with very little canopy stratification; or montane thicket dominated by *Erica* spp. (Ericaceae) and other bush ericoid in appearance.

2. *West Malagasy Region*

This Region was subdivided by Humbert (1955a) into two domains: Western Domain (Domaine de l'Ouest) and Southern Domain (Domaine du Sud).

a. Western Domain. The Western Domain is discontinuous, being subdivided by the Sambirano in the northwest (see Fig. 1). The region is composed of numerous sedimentary plateaus that slope gradually toward the Mozambique Channel; elevations range from sea level to about 800 m.

Climatically, the western part of the island is notably drier, with annual rainfall averaging 1500 mm in the north, tapering to 500–1000 mm toward the south. Precipitation is seasonally restricted as well, with approximately 90% of the rains falling during the wet season from November to March.

Climax vegetation consists of dry deciduous forest in which trees remain dormant during the 7–8 month arid period. Soil composition, either laterite, sand, limestone, or alluvium, determines the nature of the climax vegetation. Lateritic clays, principally in the north and northwest of the Western Domain, support forest 12–15 m in height with occasional taller trees including *Dalbergia* spp. (Leguminosae), *Stereospermum euphorioides* DC. (Bignoniaceae), *Givotia madagascariensis* Baill. (Euphorbiaceae) and others. Forest on sandy soil differs only slightly from that on laterite; *Tamarindus indica* L. (Leguminosae) is found frequently on humid soils, while *Euphorbia enterophora* Drake predominates in drier areas. Calcareous plateaus support dry deciduous forest of shorter stature, with a tendency toward more spiny and pachycaulous growth forms; canopy taxa include *Albizia* spp. (Leguminosae) and *Protorhus* spp. (Anacardiaceae), with *Adansonia za* Baill. and *A. fony* Baill. (Bombacaceae) as emergents. Forests on alluvial soils have disappeared to a large extent; relictual patches consist of tall forest, 25–30 m or more in height, containing deciduous as well as persistant-leaved dominants. Savanna and secondary grassland also cover an extensive portion of the Western Domain.

b. Southern Domain. This region extends along the southern coast south of Morombe almost as far east as Tolanaro (Fort Dauphin). The Southern Domain consists generally of plains and plateaus between 200 and 400 m elevation. However, small mountains in the east (Androy) reach 800 m elevation.

The south is characterized by an arid climate throughout the year. Infrequent and localized rains average 300–500 mm annually, although relative humidity is high (Humbert, 1965). The mean annual temperature of 26°C is subject to little seasonal fluctuation.

The flora of the Southern Domain is xerophilous, being characterized by the presence of the endemic family Didiereaceae, as well as the genus *Euphorbia*. Arborescent species are not predominant features, but include *Adansonia* spp., *Tetrapterocarpon geayi* Humbert (Leguminosae), *Dicoma* spp. (Asteraceae), and *Acacia* spp. (Leguminosae).

B. Population

Recent census data for Madagascar do not exist, but a current (mid 1986) population estimate is 10.3 million (Kent & Haub, 1986). An average population density for the island is about 17.5 people per km^2, but this figure is misleading because the population is unevenly distributed. Approximately 22% of the population is urban (Kent & Haub, 1986). The largest city, Antananarivo, is also the capital and may have as many as 1 million inhabitants. The next largest city, Toamasina (Tamatave), is the principal port and is located on the east coast. It has about 60,000 residents. Only four other cities have populations that exceed 30,000 people. These are Fianarantsoa and Antsirabe, both on the central plateau south of Antananarivo, and Mahajanga (Majunga) and Antseranana (Diégo-Suarez), ports on the west and north coasts, respectively. The eastern portion of the central plateau from Antananarivo to Fianarantsoa is the region of highest population density, averaging more than 50 people per square kilometer (see Gourou, 1966). The eastern coastal plain also has a high population density, especially near Toamasina and from Mananjary south to Farafangana. Other areas of higher population density include Sambava and Antalaha in the north, Tolanaro (Fort-Dauphin) in the south, and two major rice growing regions, one near Marovoay and the other near Lac Alaotra. The western two-thirds of the country is sparsely inhabited as is the Massif de Tsaratanana in the north.

Life expectancy at birth is now 50 years. The rate of population growth (birth rate minus death rate) is at present calculated to be 2.8% per annum (Kent & Haub, 1986). This figure is roughly the average annual rate of population growth for African nations and is somewhat greater than that of Asian nations. At current rates the population "doubling time" is 25 years and the population projected to the years 2000 and 2100 is 15.6 million and 52.2 million respectively (Kent & Haub, 1986).

The great majority of the population is rural and the principal occupation is agriculture. Rice is the most important crop and the dietary staple. In the west and in parts of the central plateau cultivation is permanent. However, in the escarpment forest and along the east coast slash-and-burn (*tavy*) techniques are used. Temperate crops are grown on the central plateau while tropical fruits are abundant along the coast. Important export crops are vanilla, cloves, coffee, cocoa, black pepper, raphia, and sisal. Cattle are important culturally, both as food and as a symbol of wealth. There are an estimated 10 million head of cattle on the island and portions of the dry south have been seriously degraded by overgrazing. There is some small scale commercial mining in Madagascar, but no large or heavy industry exists. Logging is important locally in the east, but export markets have not been developed significantly. Oil exploration is now underway in the west and north.

III. Vegetation Map of Madagascar

A generalized vegetation map of Madagascar is presented in Fig. 1. Delimitation of forested areas follows the vegetation map of Humbert and Cours Darne (1965) that, inasmuch as it is 20 years old, offers an optimistic estimate of the extent of remaining forest. The map is, however, the most recent and probably the most accurate, comprehensive analysis available.

The UNESCO/AETFAT/UNSO Vegetation Map of Africa (White, 1983) provides an articulate characterization and depiction of forest types of Madagascar. We therefore offer here a more generalized presentation of the three major vegetational regimes: tropical ombrophilous forest; tropical drought-deciduous forest; and endemic xerophilous forest characterized by *Didierea* (Didiereaceae) and other unique spiny and succulent taxa. The phytogeographical areas outlined in Fig. 1 are essentially the vegetational domains described by Humbert and Cours Darne (1965). White (1983) offers a recent and concise discussion of these vegetational domains, although his treatment also essentially follows that of Humbert and Cours Darne (1965).

The most detailed vegetation map of Madagascar is that prepared by Humbert and Cours Darne (1965) at a scale of 1:1,000,000. Aymonin (1970) surveyed vegetation maps that covered Madagascar in whole or in part published from 1894 to 1969 and she provided further updates on the progress of vegetation mapping to 1971 (Keraudren-Aymonin, 1971) and 1973 (Keraudren-Aymonin, 1976). Subsequently, three additional vegetation maps at a scale of 1:250,000 have been prepared for regions in the north (Rossi, 1977a, 1977b) and south (Saloman, 1978) of the island. Madagascar was included in the recently published UNESCO/AETFAT/UNSO Vegetation Map of Africa (White, 1983) that was prepared at a scale of 1:5,000,000.

There is an urgent need to update the comprehensive vegetation map prepared by Humbert and Cours Darne (1965) since much of the forest that existed then has undoubtedly been degraded or cleared in the past 20 years. Humbert and Cours Darne (1965) relied on aerial photographs to determine vegetation types and their localization and boundaries on the island. The aerial photographs used date from 1950 for the eastern and 1955 for the western portion of the island (FAO, 1981) and thus the information on boundaries of vegetation types may be 30–40 years old. A new vegetation map for Madagascar could be prepared rapidly using satellite imagery (SPOT; LANDSAT Multispectral Scanner, MSS; or NOAA Advanced Very High Resolution Radiometer, AVHRR) provided that appropriate funding and administration could be arranged. Although such a project would be ineffective in determining precise vegetation types (satellite imagery would have to be coupled with field verification) those areas of the island still forested could be determined rapidly. Comparison of successive chronological satellite images would also reveal those areas under the greatest pressure of deforestation and thus be an invaluable tool for conservation planning.

IV. Magnitude of Floristic Inventory

A. Herbarium Specimens in Malagasy Institutions

The two principal Malagasy herbaria (TAN and TEF) are discussed in section IX, B. Combined, the herbaria contain 60–70,000 specimens. However, there is much overlap and duplication of specimens both between and within the two herbaria and there may be no more than 35–40,000 unique collection numbers available for study on the island. The majority of specimens in these herbaria are represented also in Paris (P).

B. Herbarium Specimens in Non-Malagasy Institutions

Humbert (1962) estimated that there were 250,000 specimens in the Madagascar and Mascarene Herbarium at Paris (P). Nonetheless, it is difficult to estimate reliably the number of Madagascar specimens that exist now in all foreign institutions. If we assume conservatively that Humbert's (1962) estimate included at least 200,000 specimens from Madagascar and we add 40,000 for collections made by French and Malagasy botanists in the 1960s and 1970s, 15,000 for the number of collections in European herbaria (e.g., BM, G, K, L, etc.) that are not otherwise represented at Paris, and 10,000 for the recent efforts of American botanists, then a rough estimate is that there are 260,000 specimens (collection numbers) available to inventory the Malagasy flora.

The earliest collections from Madagascar are those made by Flacourt in the 17th century. Other early collections are those of Commerson, du Petit-Thouars, Chapelier, and Bréon from the end of the 18th century. All these early collections are preseved in Paris (P). However, with the exception of the Flacourt specimens, duplicates exist outside Paris, although not in great numbers. This first phase of botanical exploration involved non-resident collectors who were restricted to the coastal areas of Madagascar. Flacourt is an exception in that he was a resident at Fort Dauphin (now Tolanaro).

Bojer and Hilsenberg were the first plant collectors to penetrate into the interior and collect on the Central Plateau. The first set of Bojer specimens is found at Vienna (W) and subsequent sets were widely distributed. The collections of Hilsenberg may be found in numerous herbaria. There was continued French interest in Madagascar through the early part of the 19th century and as with specimens prepared by earlier French explorers, the first set of collections by Bernier, Boivin, Grandidier, Humblot, and Catat were sent to Paris (P). The latter part of the 19th century saw intensive collecting on the part of English missionaries and travellers; the first set of Baron, Parker, Deans Cowan, Forsyth Major, and Scott Elliot collections can be found at Kew (K) with duplicates elsewhere. Rutenberg, a German botanist, also collected in Madagascar in the late 19th century. His collections are at BREM. Those of another German botanist of the period, Hildebrandt, are much more widely distributed. The Swiss botanists Goudot, Mocquerys, and Rusillon who visited Madagascar in the late 19th century sent their collections to Geneva (G).

Collection in the 20th century has been on a larger scale. Perrier de la Bâthie, Humbert, Decary, Capuron, and Bosser each collected on the order of 20,000 numbers and the principal sets are now at Paris (P). Extensive duplicates of Capuron and Bosser collections were left in Madagascar and are deposited in TEF and TAN respectively. Note should be made also of the Réserves Naturelles (RN) and Service Forestier (SF) series of collections. (Many of Capuron's collections were distributed in this latter series). Both are well represented in Paris (P) with top sets of duplicates at TEF and TAN.

The few significant exceptions to the French domination of collecting in the 20th century include collections by Lam and Meeuse at Leiden (L), L. Bernardi at Geneva (G), H.-J. Schlieben at Berlin (B), and collections by various staff members of the Missouri Botanical Garden. The latter collections can be found in St. Louis (MO), with top sets also in Antananarivo (TAN), Paris (P), and London (K). Unfortunately plant collecting virtually came to a standstill in the mid 1970s and only now is it beginning again in a serious fashion.

V. Publications on the Flora

The first significant taxonomic contributions to the flora of Madagascar are those of du Petit-Thouars. His *Histoire des végétaux recueillis sur les îles de France, la Réunion, et Madagascar* (Paris, 1804), *Mélanges de Botanique et des voyages* (Paris, 1811) and *Histoire particulière des plantes Orchidées recueillies sur les trois îles australes d'Afriques, de France, de Bourbon, et de Madagascar* (Paris, 1822) contain numerous descriptions of genera and species that du Petit-Thouars collected in Madagascar and the Mascarenes.

The first European voyagers, including du Petit-Thouars, were restricted to coastal areas of Madagascar, but early in the 19th century botanists, including Bojer and Hilsenberg, were allowed to visit the Central Plateau. The collections of Bojer and Hilsenberg were widely distributed, and those at Geneva (G), at least, were frequently cited in early volumes of de Candolle's *Prodromus* (Paris, 1823–1873). Bojer published *Hortus mauritianus* (Mauritius, 1837), with many of the plants thought also to occur on Madagascar, but because he published names without descriptions his book has been a source of taxonomic problems.

The description of the flora of Madagascar began in earnest in the last quarter of the 19th century. The collections of Baron and other members of the English missionary community served as the basis for numerous papers by J. G. Baker in the "Journal of the Linnaean Society. Botany" and the "Journal of Botany". Over 1,000 new species were described in English journals during this period. Baron published a *Compendium des Plantes Malgaches* (1900–1906) that appeared in serial form in "Notes, Reconnaissances et Explorations" and in the "Revue de Madagascar" (published in Tananarive, now Antananarivo and Paris, respectively). Baron's *Compendium* included 4,700 species and varieties in 970 genera and represented the first attempt to treat the flora in one work (Dorr, 1987). Baron (1890) also published the first phytogeographical synthesis of the island. Unfortunately, the synthesis is incomplete because of his lack of experience with the vegetation of the arid south.

The end of the 19th century also saw increased interest in Madagascar on the part of the French. Grandidier published his monumental *Histoire physique, naturelle et politique de Madagascar* (Paris, 1875–1901). The botanical portion, prepared by Baillon and later by Drake del Castillo, was published between 1886 and 1903. Prior to contributing to Grandidier's *Histoire*, Baillon had published numerous papers on Madagascar plants, many of them in the "Bulletin mensuel de la société linnéenne de Paris".

M(adagascar became a French colony in 1895 and for the next three-quarters of a century botanical exploration and publication was dominated by the French. An exception, however, was Palacky's publication of a *Catalogus plantarum madagascariensium* that appeared in five fascicles between 1906 and 1907 and was published in Prague. The preeminent collector and author in the early 20th century was Perrier de la Bâthie. His very important contributions include "La végétation malgache" (1921) that appeared in the "Annales du musée colonial de Marseille"; *Biogéographie des plantes de Madagascar* (Paris, 1936); and *Catalogue des plantes de Madagascar* (1930–1940) published in fascicles by the Académie Malgache (Tananarive, now Antananarivo). (The last work included contributions from other authors). Important journals in this period were "Archives de botanique, bulletin mensuel", Caen, and "Archives de botanique, mémoires", Caen. The latter contained revisions of large genera (e.g., *Impatiens*, Balsaminaceae and *Philippia*, Ericaceae) that have yet to be treated elsewhere.

The foundation of the Académie Malgache in 1902 was followed by the publication of "Bulletin trimestrial de l'Académie Malgache" and later by "Mémoires de l'Académie Malgache", both of which contain botanical articles. One of the more significant essays in the latter serial is Humbert's (1927) "Destruction d'une flore insulaire par le feu. Principaux aspects de la végétation à Madagascar."

Humbert was the principal figure behind the organization of the *Flore de Madagascar et des Comores*, the first fascicle of which appeared in 1936. The *Flore* clearly provided a stimulus for renewed and concerted investigation of the flora. Many of the treatments that appeared in the *Flore* had precursors in the "Bulletin de la Société botanique de France"; "Notulae Systematicae", Paris; "Mémoires du Muséum National d'Histoire Naturelle"; and "Compte rendu hebdomadaires des séances de l'académie des sciences", Paris.

For obvious reasons botanical activity slowed during World War II, but in 1946 the institut de Recherche Scientifique de Madagascar (I.R.S.M.) was founded in Antananarivo and activity was resumed. The institute published two periodicals that contained botanical articles, "Le Naturaliste Malgache" and "Mémoires de l'Institut Scientifique de Madagascar, sér. B, Biologie Végétale." The latter contained many revisions that led to treatments in the *Flore*. The Institut had a short life, being abolished administratively in 1957. Many of the activities of the Institut were taken over by the Office de la Recherche Scientifique et Technique Outre-Mer (O.R.S.T.O.M.) and articles on taxonomy and ecology were published in the "Mémoires O.R.S.T.O.M.". The demise of I.R.S.M. also saw the publication of treatments of Malagasy taxa shift to "Adansonia, sér. 2", a publication of the Muséum National d'Histoire Naturelle in Paris.

One of the most active botanists in the 1950s, 1960s, and early 1970s was Capuron. His mimeographed *Essai d'Introduction à l'Étude de la Flore Forestière de Madagascar* (1957) is one of the few works that contains keys to families of Madagascar plants. Thonner's (1908) key to families is not generally useful in Madagascar, since it was written in German and translated into English (Thonner, 1915; Geesink et al., 1981), but not French. Capuron's contribution to an understanding of the taxonomy of woody plants was significant (see Leroy, 1972, for a list of Capuron's publications and mimeographed reports).

A popular handbook for the flora of Madagascar and the Mascarenes, *Végétation et groupements végétaux de Madagascar et des Mascareignes*, was published in four volumes by Cabanis et al. (1969–1970). Synthetic treatments were also provided by various authors in Battistini's (1972) review of the biogeography and biology of Madagascar and later by Rauh (1973). The *Flore et Végétation de Madagascar* by Koechlin et al. (1974) was a more ambitious effort marking a change from a descriptive to an analytical phase in the understanding of the flora.

The late 1970s and early 1980s have seen work in Madagascar broaden to include a wider spectrum of Malagasy, European, and American contributors. Accordingly, articles on the taxonomy and ecology of Madagascar plants have appeared recently in the " Annals of the Missouri Botanical Garden"; "Blumea"; "Bulletin du Jardin Botanique National de Belgique", Meise; "Candollea"; Kew Bulletin"; "Mededeelingen Landbouwhogeschool", Wageningen; "Webbia"; and other journals. A new French periodical, "Bulletin du Muséum National d'Histoire Naturelle, 4e sér., Section B, Adansonia, Botanique. Phytochimie", incorporating "Adansonia, sér. 2", has appeared recently. Useful guides to the more dispersed literature of this period are the publications of the Association Pour l'Étude Taxonomique de la Flore Tropicale d'Afrique (A.E.T.F.A.T.). Also, "APARSMAD", published by the Laboratoire de Phanérogamie, Muséum National d'Histoire Naturelle in Paris, reviews current scientific literature pertaining to Madagascar.

The *Flore de Madagascar et des Comores* seems to have dominated floristic work on Madagascar to the exclusion of smaller, less ambitious projects. Surprisingly there are no local floras for any area in Madagascar, including the Réserves Naturelles Intégrales and other protected areas. (There are reports of expeditions that have been published, but they do not contain complete checklists of species). Similarly, with the exception of Capuron's (1957) *Flore forestière* there are no guides to forest trees or keys to families of Madagascar flowering plants. Clearly, much basic and valuable work remains to be done.

VI. Completeness of Floristic Inventory

A. Relatively Well-Collected Regions

Floristic inventory of any area of primary vegetation in Madagascar is far from being at a satisfactory level of completion. Collecting at a density of 50–100 specimens per 100 km² (suggested criteria of the World Strategy for the Floristic Inventory of Tropical Forests), is insufficient for even the most basic inventory purposes (such as the creation of a species list), for reasons outlined by Stevens, (this volume) and Prance (1977). In addition, simply amassing herbarium specimens is an inadequate approach to attaining a biological knowledge of the flora, as discussed by Heywood (1979), Janzen (1977), and Tomlinson (1977).

Certain areas of Madagascar can be said to be "well-collected" in terms of the number of collections per unit area, especially when compared to areas of South America and New Guinea. However, a relatively high number of collections per unit area in Madagascar is to some extent a deceptive indication of how well the flora has been surveyed. Historically, there have been strong biases toward collecting in specific limited areas. In general, Réserves Naturelles Intégrales and forestry stations have tended to receive a disproportionate amount of collecting attention, at least since 1927 when most of the former were officially created. There are, however, exceptions, since the Réserves were chosen for their importance in representing forest types and not for their accessibility. Collections from Réserves and forestry stations are well represented in the Réserves Naturelles and Service Forestier collecting series (see section IV,B).

The southern *Didierea* forest has also received much collecting attention. The bias is an understandable consequence of the originality of this vegetation. The vicinity of the capital, Antananarivo, is well known floristically and this can be attributed to the relative ease of access to the surrounding countryside, the comparative poverty of the highly disturbed plateau vegetation, and the aggregation of botanical, forestry, and university offices. In the east, the coastal plain from Fenerive south to Mananjary (including Nosy Baraha, formerly Isle Ste. Marie) is well known, having been collected since the early 18th century, but there remain interesting and inadequately-known forests such as Ambila-Lemaitso and Analalava, to name only two. In the west, only the forestry station at Ampijoroa can be said to be well-collected. Fortunately, Ampijoroa is representative of a much larger forest that extends from the Betsiboka River near Ambato-Boeni northeast to the Mahajamba River.

B. Poorly Collected Regions

Using the admittedly rough estimate of 260,000 collection numbers from Madagascar, we calculate that there are no more than 45 collections/100 km² for the island as a whole. This would rank Madagascar as "poorly collected" according to criteria of the World Strategy for the Floristic inventory of Tropical Forests (less than 50 collections/100 km²). In light of these statistics, Koechlin's (1972) statement that "the inventory of the flora is largely completed today" appears to be ungrounded. Léonard (1965) has classified Madagascar as being "moderately well known" in terms of floristic exploration, his criterion being that the number of collections from the island equals one to three times the number of species believed to occur on the island.

Underemphasized in terms of collecting are the more inaccessible regions of Madagascar. Heavy rains, especially during the austral summer, make travel in the north and in the Sambirano region impossible for most of the year. Consequently the northern part of Madagascar is botanically the least-explored region of the island, even though the mountains of that region have yielded some of the most intriguing Malagasy endemics such as *Takhtajania* (Winteraceae), *Ascarinopsis* (Chloranthaceae), and *Diegodendron* (Diegodendraceae). Specific areas in the north that need intensive floristic work are Montagne d'Ambre and Montagne des Français, both near Antseranana (Diégo-Suarez).

The French, beginning with Perrier de la Bâthie and Humbert, made a concerted effort to sample the vegetation of all the high mountains of Madagascar. Nonetheless the high mountains are still poorly known. A cursory examination of the botanical literature pertaining to Tsaratanana (Perrier de la Bâthie, 1927), Marojejy (Guillaumet et al., 1975; Humbert, 1955), Pic Boby and the Andringitra (Paulian et al., 1971; Perrier de la Bâthie, 1927), Itremo and Ibity (Guillaumet et al., 1975), and the Anosyenne Chain (Humbert, 1941; Paulian et al., 1973) will convince one that these peaks are inadequately collected. Furthermore it is unreasonable to expect adequate floristic inventory of these mountains when they have been collected only once or twice and not necessarily in the best seasons for flowering and fruiting.

In the east the Masoala Peninsula must be considered as very poorly known principally because it is inaccessible. The road to Maroantsetra, the largest village near the peninsula, has not been open for several years and now the only access to the edge of the peninsula is by airplane or boat. The forest of Ambohimanga du Sud, east of Ambositra is another area in the east that has not been adequately collected.

The Antsingy region of the west near Antsalova is poorly known despite the efforts of Leandri (1936). The remaining forests are restricted to limestone and exploration will undoubtedly yield interesting additions to the flora. A smaller, but nevertheless equally poorly-known and potentially-interesting forest is that of Kasijy, northwest of Maevatanana.

Finally, the Isalo region, which represents an outlier of the vegetation of the Central Plateau, can not be considered to be sufficiently explored. This unusual sandstone formation has yielded numerous narrowly endemic species and to date has recieved only cursory exploration, principally along its southern edge.

C. Regions Rich in Endemics

Madagascar in general is outstanding in terms of its endemics, especially at the supraspecific level. This originality alone is one of the foremost justifications for extensive floristic study. According to Koechlin (1972), the greatest concentration of endemics is found in the south (48% generic/95% specific endemism); followed by the west (41%/90%); the east (37%/90%); the Sambirano region (23%/89%); and finally the center (21%/89%). Regarding the level of endemism in specific habitats, Koechlin (1972) estimated endemism of forest and shrubby formations at 89%; rocky outcrops at 82%; marsh areas at 56%; and littoral regions at 21%.

Specific areas that seem to be rich in endemics are the previously mentioned Isalo (section VI,B) and the calcareous Plateau de l'Ankarana in the north of the island. The two highest mountain peaks, Tsaratanana and Marojejy, seem to share a large number of species, but otherwise each of the high mountains appears to have a strong endemic element.

D. Regions Yielding New and Important Distributions

Since exploration has not been especially active in recent years few areas can be said now to be yielding new and important distributions. The examination of palm collections from the Masoala Peninsula has yielded, however, an interesting new genus and several new species (Dransfield & Uhl, 1984a, 1984b). Further work in the areas mentioned in section VI B will undoubtedly reveal new species and genera.

E. Extinction

There is no documentation of areas that have high rates of extinction of species. Although habitat destruction (see section VIII) will lead to species extinctions most of the identifiable threats to species appear to arise from the selective exploitation of taxa for economic gain. Thus the worldwide horticultural trade has placed undue pressure on populations of orchids, *Pachypodium* (Apocynaceae), and *Aponogeton* (Aponogetonaceae). Indigenous uses of certain plants are also excessive (Andriamampianina *in* Hedberg, 1979; Decary, 1946; Humbert, 1927; Perrier de la Bâthie, 1933b). Conspicuous examples of endangered Malagasy plants are the tree ferns (Cyatheaceae), overexploited because of their use as vases for potting plants (Andriamampianina *in* Hedberg, 1979). Species sought for their latex, but especially *Euphorbia intisy* Drake (Euphorbiaceae), are apparently becoming rare (Decary, 1946; Humbert, 1927; Koechlin et al., 1974). Other species exploited for their wood, such as Malagasy ebony (*Diospyros perrieri* Jum. and *D. microrhombus* Hiern, Ebenaceae) (Decary, 1946; Perrier de la Bâthie, 1933b) and "palissandre" (*Dalbergia* spp., Leguminosae) (Perrier de la Bâthie, 1933b) are on the verge of extinction. *Baudouinia rouxevillei* Perrier (Leguminosae) commonly known as "manjakabetany" or "bois sacré" is restricted to a small area northeast of Toliara (Tuléar); because the wood is highly prized for its aesthetic and mystical properties it will almost certainly become extinct within the next few decades (Capuron, 1968; Keraudren, 1968; Perrier de la Bâthie, 1933a). At least two species of baobab face threats. *Adansonia grandidieri* Baillon, which has edible seeds and is also used for making rope, is not regenerating because of overconsumption of seeds (Perrier de la Bâthie, 1933b; Perrier de la Bâthie & Hochreutiner, 1955). In the arid south *A. za* is frequently cut and debarked to provide water for grazing goats (Miège, 1974). The list of endangered or threatened species could continue, but we think the point is well made; economic exploitation, whether legitimate or not, has begun to take its toll on the flora of Madagascar.

VII. Current Collections and Work on Large or Important Families

A. Collection of Particular Groups

No group of Malagasy taxa can be said to be well collected. There are numerous instances in the *Flore de Madagascar et des Comores* in which the fruit of a given species is unknown or in which a taxon is known only from the type collection. Woody taxa in general tend to be under-collected, despite the efforts of the Service des Eaux et Forêts. Certain families that are notoriously difficult to collect need to be collected critically. These include the palms (Arecaceae) and the screw-pines (Pandanaceae).

The diversity of the palm flora of Madagascar when compared to that of neighboring Africa is greatly out of proportion to the size of the island. Jumelle (1945) recognized 18 genera (12 endemic), but a new, endemic genus has been described recently (Dransfield & Uhl, 1984b) and it seems certain that collections by palm specialists will lead to further discoveries of new species and possibly even genera.

Similarly, Madagascar is a secondary center of diversity for screw-pines, with over 100 species of *Pandanus* known to occur on the island (Stone, 1983). In the late 1960s and early 1970s a concerted effort was made to collect *Pandanus*

Table I

Plant families of Madagascar that have been worked on recently. The dates in parentheses are when a treatment was published in the *Flore de Madagascar et des Comores* for that family. The names are of specialists studying that family.[a]

Acanthaceae (1967, p.p.)	H. Heine
Apocynaceae (1976)	A. J. M. Leeuwenberg
Araceae (1975)	J. Bogner
Araliaceae	L. Bernardi, P. P. Lowry II
Arecaceae (Palmae) (1945)	J. Dransfield; N. W. Uhl
Aristolochiaceae	O. Poncy
Asclepiadaceae	B. Descoings
Balanophoraceae (1984)	B. Hansen
Balsaminaceae	C. Grey-Wilson
Bignoniaceae (1938)	A. Gentry
Chloranthaceae	J. Jérémie
Clusiaceae (incl. Guttiferae & Hypericaceae) (1951, 1951)	H. K. Robson; P. F. Stevens
Commelinaceae (1937)	R. B. Faden
Dichapetalaceae (1961)	F. J. Breteler
Didiereaceae (1963)	W. Rauh
Didymelaceae	J. -F. Leroy
Diegodendraceae	J. -F. Leroy
Elaeocarpaceae (1985)	C. Tirel
Equisetaceae	F. Badré
Ericaceae (inc. Vacciniaceae)	L. J. Dorr (*Erica, Vaccinium*); W. Judd (*Agarista*)
Euphorbiaceae (1958, p.p.)	W. S. Armbruster (*Dalechampia*); G. Cremers (*Euphorbia*)
Flacourtiaceae (1946)	A. J. G. H. Kostermans
Gentianaceae	J. Klackenberg
Hamamelidaceae	E. Rakotobe
Iridaceae (1945)	P. Goldblatt
Lauraceae (1950)	A. J. G. H. Kostermans; H. van der Werff
Leguminosae	R. M. Polhill (*Crotalaria*); R. Rabevohitra (*Dalbergia*); B. Rakouth; C. Stirton (Sophoreae); J. -F. Villiers (Mimosoideae)
Lentibulariaceae (1955)	P. Taylor
Loganiaceae (1984)	A. J. M. Leeuwenberg; E. M. Norman (*Buddleja*)
Lythraceae (1945)	S. Graham
Malvaceae (1955)	O. J. Blanchard, Jr. (*Kosteletzkya*); L. J. Dorr
Melastomataceae (1951)	H. Jacques-Félix
Meliaceae	J. -F. Leroy
Monimiaceae (1959)	D. H. Lorence
Moringaceae (1982)	B. Verdcourt
Onagraceae (incl. Oenotheraceae) (1950)	P. H. Raven
Orchidaceae (1939)	J. Bosser; P. Cribb
Oxalidaceae	A. Lourteig
Pandanaceae	B. C. Stone
Poaceae	J. Bosser
Psilotaceae	F. Badré
Rubiaceae	J. -F. Leroy; C. Puff
Selaginellaceae	F. Badré
Solanaceae	W. G. D'Arcy; A. Rakotozafy
Sterculiaceae (1959)	L. C. Barnett
Tiliaceae	C. Tirel
Winteraceae	J. -F. Leroy

[a] Clearly most families in the *Flore* are not being studied now by experts and it is unlikely that the critical need for comprehensive treatments of the Acanthaceae, Euphorbiaceae, Leguminosae, and Rubiaceae will be met soon.

(St. John, 1968; Stone, 1970a, 1970b; Stone & Guillaumet, 1970, 1972), but this needs to be resumed. Finally, families with many succulent species, especially Asclepiadaceae, Crassulaceae, and Euphorbiaceae, also need to be collected critically with selected material being preserved in liquid.

B. IMPORTANT FAMILIES

The *Flore de Madagascar et des Comores* is projected to include treatments of 209 families (222 if one subdivides the Polypodiaceae, s.l.). Currently 52 families lack treatments and revisions of two, Acanthaceae and Euphorbiaceae, have been produced only in part. Table I includes those families that have been revised recently or are currently under study. Following the family name the date of the treatment for the *Flore*, if one exists, is given in parenthesis. The fact that a specialist is studying a family that has not been treated in the *Flore* does not necessarily imply that that person will prepare a treatment appropriate for the *Flore*.

VIII. Important Areas of Forest Threatened By Destruction

Almost every forested area in Madagascar is threatened by destruction. According to Battistini and Verin (1972), only 21.1% of the original vegetation remains undisturbed. Most discussions of the extent of deforestation in Madagascar are based on Perrier de la Bâthie (1921) and Humbert's (1927) hypothesis that the entire surface of the island was dominated more or less (though not exclusively, see Humbert, 1955a) by forest formations. Recently, Battistini and Verin (1972) and Koechlin (1972) discussed evidence that the island was almost completely forested in the past. However, MacPhee et al. (1985), utilizing new palynological and paleontological evidence, argue that central Madagascar, at least, may have been a mosaic of woodlands, bushlands, and savanna prior to man's arrival. Whatever the pre-settlement vegetational cover of the island was, there is no question that man today has had an enormous impact on the environment.

Data recently compiled by the FAO (1981) suggest that forest cover has diminished at an average rate of 1,650 km² per year between 1976 and 1980. The same report projected that an average of 1,500 km² of forest would be lost each year between 1981 and 1985. Most of the destruction is a result of slash-and-burn (*tavy*) agriculture practiced by the indigenous people, who must rely on this land for their own survival (FAO, 1981; Jolly & Jolly, 1984).

The eastern forest has been singled out as the region under highest risk (FAO, 1981), or at least the region least resilient to damage (Koechlin, 1972). Also at high risk are those forested regions close to cities; the forests around cities such as Toliara (Tuléar), Tolanaro (Fort Dauphin), and Antananarivo must provide a constant source of wood and charcoal for cooking and heating.

As a response to growing pressure on the remaining native biota, eleven (out of an original twelve) Réserves Naturelles Intégrales have been maintained. These Réserves cover a combined area of approximately 5,500 km² (Bernardi, 1974; Andriamampianina, 1981). Each Réserve was selected to preserve a specific forest type. In addition, over 20 Réserves Speciales have been established to complement the range of biotypes represented in the Réserves Naturelles Intégrales. Millot (1972), Bernardi (1974), and Rauh (1979) have been frank about the difficulty of maintaining Réserves intact, especially as land and forest within Réserves become more valuable as resources to the local populations.

IX. Inventory Resources

A. Academic Institutions

1. Université de Madagascar, Antananarivo

The Département de Biologie Végétale has a small teaching herbarium and a small library. The botany faculty includes specialists in plant anatomy and cytology.

2. Université de Madagascar, Toliara (Tuléar).

Although the academic program is oriented towards teacher training there is one faculty member interested in plant ecology. There is also a marine station affiliated with the Université.

3. Académie Malgache, Antananarivo

The Académie maintains a rich library and frequently sponsors talks by resident and visiting scientists.

B. Government Institutions

1. Service des Eaux et Forêts, Antananarivo

This agency is responsible for maintaining all natural and protected areas in Madagascar. It also issues collecting permits and grants permission for visitors to enter national parks and reserves. The Service maintains an herbarium (TEF) of 20–30,000 specimens and a wood collection (TEFw) of approximately 2,500 specimens (Stern, 1967).

2. Centre National de Recherche de Tsimbazaza (C.N.R.T.), Parc de Tsimbazaza, Antananarivo

The Parc comprises botanical and zoological gardens, a natural history museum, and a library. The herbarium (TAN) contains 40,000 mounted sheets representing some 35,000 collection numbers and there are 5–6000 unmounted collections waiting to be processed (1984 estimate). The library is fairly rich in systematic literature, but most journal subscriptions stopped in 1975. The very rich Grandidier Collection is part of the library of the Parc, as is the Poisson Collection.

C. Non-Government Institutions

1. World Wildlife Fund (WWF)-International.

This organization maintains an office in Antananarivo and is involved in programs jointly with several of the Ministries of the Malagasy Government. These programs promote conservation and are designed to extend and strengthen the system of reserves. At present WWF-International funds projects involving the Parc National de la Montagne d'Ambre, Réserve Speciale Botanique d'Ambohitantely, Réserve de Périnet-Analamazoatra, Réserve Naturelle Intégrale No. 6 de Lokobe, Nosy Mangabé, and Station Forestière d'Ampijoroa and Réserve Naturelle Intégrale No. 7 d'Ankarafantsika. Some of these projects include provisions for floristic inventory.

D. Foreign Programs

Until recently all research by foreigners in Madagascar came under the jurisdiction of the Direction de la Recherche Scientifique (DRS) in the Ministère de l'Enseignement Superieur (MESup). This ministry now is responsible for all basic research, including floristic and faunistic surveys. Applications for research in Madagascar are expected to be relevant to nature conservation and, at the request of MESup, proposals were screened by an International Advisory Group of Scientists (IAGS). However, the IAGS term recently expired (Feb 1986) and it is not clear whether this external review committee will be continued. A second scientific Ministry, the Ministère de la Recherche Scientifique et Technologique pour le Développement (MRSTD), was established in 1985 to handle applied research.

E. Collaborative Programs

1. The Missouri Botanical Garden, St. Louis, Missouri

Missouri and the Centre National de Recherche de Tsimbazaza (C.N.R.T.), Parc de Tsimbazaza currently have an agreement that provides for botanical inventory.

2. The Duke University Primate Center, Durham, North Carolina

The Centre National de Recherche de Tsimbazaza (C.N.R.T.), Parc de Tsimbazaza and the Service des Eaux et Forêts are cooperating with Duke's Primate Center in research on lemurs. Studies of lemur diet, a portion of this collaboration, could be integrated with vegetation studies.

3. The University of Strasbourg, Strasbourg, France

The Université de Madagascar, Antananarivo; Hôpital Ravoanangy, Andrianavalona-Antananarivo and the Centre National de Recherche de Tsimbazaza (C.N.R.T.), Parc de Tsimbazaza collaborate in research on lemurs with Strasbourg.

4. Beza-Mahafaly Reserve

The Université de Madagascar, Antananarivo, Yale University, New Haven, Connecticut, Washington University, St. Louis, Missouri, and WWF-U.S. have created a natural reserve southeast of Toliara (Tuléar). Their continuing program includes vegetation sampling and vegetation mapping.

5. Analabé Reserve.

The Malagasy Government, WWF-International, the Musée National d'Histoire Naturelle, Paris and the Foundation de Haulme signed an agreement to establish a 40 km² reserve on the de Haulme property north of Morondava.

The listing of collaborative research that might concern vegetation inventory is not exhaustive. Indirectly, programs sponsored by the United Nations to combat deforestation and the Swiss Government to promote reforestation will have an impact on floristic inventory.

X. Suitability of Floristic Inventory

Several important contributions have been made already toward floristic knowledge of Madagascar. Perrier de la Bâthie and Humbert did excellent jobs as individuals in terms of collecting plants in some of the more remote regions. (Their itineraries are sketched in Perrier de la Bâthie, 1936, and Keraudren-Aymonin and Aymonin, 1969, respectively). Similarly, Capuron, working with the assistance of the Département (now Service) des Eaux et Forêts, made remarkable progress in terms of collecting in remote and diverse areas.

The scientific institutions I.R.S.M. and O.R.S.T.O.M. made great headway in terms of floristic inventory in large part because they had the support to organize the type of large-scale expedition necessary to attain the most remote territories. The French Centre National de la Recherche Scientifique (RCP) assisted by the civil and military authorities of Madagascar and O.R.S.T.O.M. made a concerted effort to prospect the highest peaks on the island (Guillaumet et al., 1975; Paulian et al., 1971, 1973). These organizations made a remarkable contribution toward floristic knowledge although there is still a critical need to follow up on their initial surveys.

Unfortunately, in practice, exploration is contingent less upon floristic needs than upon actual physical circumstances. A large obstacle to past and present-day exploration is the severely-deteriorated road system (Jolly & Jolly, 1984). During the rainy season, major national highways (roads from Antananarivo to Mahajanga, Toamasina, Toliara, and Tolanaro) may be impassable even with the aid of a 4-wheel drive vehicle, and the road to Antsiranana is closed to all traffic from December to May. A new road between Antananarivo and Toamasina has been completed, which should expedite research to some degree on the species-rich eastern forest. To a large extent, the state of the road system will determine whether a given expedition is physically possible or not.

XI. Resources Needed For Continued Inventory

A. Institutions

The institutional structure for continuing floristic inventory in Madagascar exists and much credit must be given to the Malagasy Government for recognizing and maintaining two herbaria, a Service des Eaux et Forêts, botanical and zoological gardens, and programs in botany and forestry at the university level. What is lacking, as is true for other tropical countries (see Stevens, this volume) faced with difficult decisions concerning the allocation of scarce monetary resources, are substantial and predictable operating funds.

If large-scale floristic inventory of Madagascar is undertaken, considerable attention will have to be devoted to upgrading herbarium facilities. This would include minimally the purchase of additional herbarium cases, mounting paper, mounting supplies, and fumigants. Few buildings in Madagascar are air-conditioned and therefore arrangements would have to be made at least to dehumidify these herbaria.

The existing collections also, to be generally useful, require much curation. Without reliably-identified material the identification of new collections is hopeless or at the very least filled with problems. Although expert determinations are critical for many groups, neither herbarium now is able to loan specimens.

B. Personnel

The two herbaria are run by competent botanists, but their duties include large amounts of administration. There are at present no taxonomists with graduate level degrees in Madagascar. University programs will need to encourage students who show a particular aptitude for systematic botany and perhaps collaborative programs with foreign institutions could provide not only the advanced education of students, but also the training of exceptional technicians.

C. Cost

We cannot estimate now the cost of continued floristic inventory. Suffice it to say that the cost will have to be borne by the Malagasy Government and foreign institutions together. Several of the agreements with foreign institutions that presently exist may well serve as models or may themselves give rise to collaborative programs that will provide adequate floristic inventory of the tropical forests of Madagascar. As was emphasized by Stevens (this volume) in order for inventory to succeed it will have to be firmly supported by the host country. The Malagasy Government, in funding the programs that it now funds, has already indicated that it considers knowledge of the Madagascar flora important.

XII. Acknowledgments

We would like to thank the Ministère de L'Enseignement Superieur (MESup) without whose generous cooperation field work in Madagascar would not have been possible. The views put forth in this paper are our own and do not necessarily reflect the official views of the Malagasy Government.

XIII. Literature Cited

Andriamampianina, J. 1981. Les réserves naturelles et la protection de la nature à Madagascar. Pages 105–111 *in* P. Oberlé (ed.) Madagascar, Un Sanctuaire de la Nature. Librairie de Madagascar, Antananarivo.

Aymonin, M. 1970. Madagascar. Pages 154–175 *in* A. W. Küchler, International bibliography of vegetation maps, Vol. 4. University of Kansas Libraries, Lawrence, Kansas.

Baron, R. 1890. The flora of Madagascar. J. Linn. Soc., Bot. **25**: 246–294.

Battistini, R. 1972. Madagascar relief and main types of landscape. Pages 2–5, *in* R. Battistini & G. Richard-Vindard (eds.). Biogeography and ecology in Madagascar. Dr. W. Junk B. V., Publishers, The Hague.

———— & **P. Verin.** 1972. Man and the environment in Madagascar. Past problems and problems of today. Pages 311–337 *in* R. Battistini & G. Richard-Vindard (eds.). Biogeography and ecology in Madagascar. Dr. W. Junk B. V., Publishers, The Hague.

Bernardi, L. 1974. Problèmes de conservation de la nature dans les îles de l'Océan Indien I. Méditation à propos de Madagascar. Saussurea **5**: 37–47.

Cabanis, Y., L. Chabouis, & F. Chabouis. 1969–1970. Végétaux et groupements végétaux de Madagascar et des Mascareignes. Vols. I–IV. Bureau pour le Développement de la Production Agricole, Tananarive.

Capuron, R. 1968. Contributions à l'étude de la flore forestière de Madagascar. A.-Notes sur quelques Cassiées malgaches (2e partie). Adansonia, sér. 2, **8**: 199–217.

Chauvet, B. 1972. The forest of Madagascar. Pages 191–199, *in* R. Battistini & G. Richard-Vindard (eds.). Biogeography and ecology in Madagascar. Dr. W. Junk B. V., Publishers, The Hague.

Coetzee, J. A. & J. Muller. 1984. The phytogeographic significance of some extinct Gondwana pollen types from the Tertiary of the southwestern Cape (South Africa). Ann. Missouri Bot. Gard. **71**: 1088–1099.

Decary, R. 1946. Quelques plantes malgaches rares ou en voie d'extinction. Bull. Mus. Hist. Nat., Paris, sér. 2, **18**: 495–499.

Dejardin, J., J. -L. Guillaumet & G. Mangenot. 1973. Contribution à la connaissance de l'élément non endémique de la flore malgache (végétaux vasculaires). Candollea **28**: 325–391.

Donque, G. 1972. The climatology of Madagascar. Pages 87–144 *in* R. Battistini & G. Richard-Vindard (eds.). Biogeography and ecology in Madagascar. Dr. W. Junk B. V., Publishers, The Hague.

Dorr, L. J. 1987. Rev. Richard Baron's *Compendium des Plantes Malgaches*. Taxon **36(1)**: 39–46.

Dransfield, J. & N. W. Uhl. 1984a. A magnificent new palm from Madagascar. Principes **28**: 151–154.

————. 1984b. *Halmoorea*, a new genus from Madagascar, with notes on *Sindroa* and *Orania*. Principes **28**: 163–167.

FAO. 1981. Madagascar. Pages 281–302, *in* Forest resources of Tropical Africa Part II: Country briefs. Food and Agricultural Organization of the United Nations, Rome.

Geesink, R., A. J. M. Leeuwenberg, C. E. Ridsdale & J. F. Veldkamp. 1981. Thonner's Analytical Key to the Families of Flowering Plants. Leiden University Press, The Hague.

Goldblatt, P., P. Rudall, V. I. Cheadle, L. J. Dorr & C. A. Williams. 1987. Affinities of the Madagascan endemic *Geosiris*, Iridaceae or Geosiridaceae. Bull. Mus. Nat. Hist. Nat. Paris, 4e sér., 9, 1987, section B, Adansonia, **3**: 239–248.

Gourou, P. 1966. Madagascar. Localisation de la Population (1:1,000,000); Madagascar. Densité de la Population (1:2,000,000). Institut Geographique Militaire, Bruxelles. (map).

Guillaumet, J.-L. 1984. The vegetation: an extraordinary diversity. Pages 27–54 *in* A. Jolly, P. Oberlé & R. Albignac. (eds.). Madagascar. Pergamon Press, Oxford.

————, **J.-M. Betsch, C. Blanc, P. Morat & A. Peyrieras.** 1975. Étude des écosystèmes montagnards dans la région malgache. III. Le Marojejy. IV. L'Itremo et l'Ibity. Géomorphologie, climatologie, faune et flore (Campagne RCP 225, 1972–1973). Bull. Mus. Hist. Nat., Paris, sér. 3, n° 309, Écologie générale **25**: 29–57.

Hedberg, I. 1979. Possibilities and needs for conservation of plant species and vegetation in Africa. Pages 83–104, *in* I. Hedberg (ed.). Systematic botany, plant utilization and biosphere conservation. Almqvist and Wiksell International, Stockholm.

Heywood, V. H. 1979. The future of island floras. Pages 431–441, *in* D. Bramwell (ed.). Plants and islands. Academic Press, London.

Humbert, H. 1927. La destruction d'une flore insulaire par le feu. Principaux aspects de la végétation à Madagascar. Mém. Acad. Malgache **5**: 1–79, pls. I–XLI.

————. 1941. Le massif de l'Andohahelo et ses dépendances (Madagascar, Réserve naturelle n° XI). Compt. Rend. Sommaire Séances Soc. Biogéogr. **18**: 31–37.

———. 1946. La protection de la nature à Madagascar. J. Arnold Arbor. **27**: 470–479, pls. I–V.

———. 1955a. Les territoires phytogéographiques de Madagascar. Leur cartographie. Ann. Biol. (Paris) **31**: 439–448, map.

———. 1955b. Une merveille de la nature à Madagascar. Première exploration botanique du Massif du Marojejy et de ses satellites. Mém. Inst. Sci. Madagascar, sér. B, Biol. Vég. **6**: 1–210.

———. 1959. Origines présumées et affinités de la flore de Madagascar. Mém. Inst. Sci. Madagascar, sér. B, Biol. Vég. **9**: 149–187.

———. 1962. Histoire de l'exploration botanique à Madagascar. Compt. Rend. IVᵉ Réunion Plén. A.E.T.F.A.T.: 127–144.

———. 1965. Description des types de végétation. Pages 46–78 in H. Humbert & G. Cours Darne. Carte internationale du tapis végétal et des conditions écologiques à 1/1.000.000. Notice de la carte. Madagascar. L'Institut Français de Pondichéry, Pondichéry, India.

——— & G. Cours Darne. 1965. Carte internationale du tapis végétal et des conditions écologiques à 1/1.000.000. Notice de la carte. Madagascar. L'Institut Français de Pondichéry, Pondichéry, India.

Janzen, D. H. 1977. Promising directions of study in tropical animal-plant interactions. Ann. Missouri Bot. Gard. **64**: 706–736.

Jolly, A. & R. Jolly. 1984. Malagasy economics and conservation: a tragedy without villains. Pages 211–217 in A. Jolly, P. Oberlé, & R. Albignac (eds.). Madagascar. Pergamon Press, Oxford.

Jumelle, H. 1945. 30ᵉ Famille. Palmiers (Palmae). Flore de Madagscar et des Comores. Imprimerie Officielle, Tananarive.

Kent, M. M. & C. Haub. 1986. 1986 World Population Data Sheet. Population Reference Bureau, Inc., Washington, D.C.

Keraudren, M. 1968. Madagascar. Pages 260–265 in I. Hedberg & O. Hedberg (eds.). Conservation of vegetation in Africa south of the Sahara. Acta Phytogeogr. Suec. **54**: 1–320.

Keraudren-Aymonin, M. 1971. Les publications phytogeographiques concernant Madagascar. Mitt. Bot. Staatssamml. München **10**: 484–486.

———. 1976. Progrès accomplis dans la cartographie de la flore de Madagascar et des Comores. Boissiera **24**: 653.

——— & G. Aymonin. 1969. Les explorations et les collections botaniques du Professor Henri Humbert. Mus. Hist. Nat. (Paris), Travaux Lab. "La Jaysinia," **3**: 11–33.

Koechlin, J. 1972. Flora and vegetation of Madagascar. Pages 145–190 in R. Battistini & G. Richard-Vindard (eds.). Biogeography and ecology in Madagascar. Dr. W. Junk B. V., Publishers, The Hague.

———, J.-L. Guillaumet & P. Morat. 1974. Flore et Végétation de Madagascar. J. Cramer, Vaduz.

Leandri, J. 1936. Reconnaissance botanique de la partie médiane de l'Ouest malgache. La région de l'Antsingy et le Menabe septentrional. Bull. Trimestriel Acad. Malgache, n.s., **19**: 1–37.

Léonard, J. 1965. Map of the extent of floristic exploration in Africa south of the Sahara. Webbia **19**: 911–914, map.

Leroy, J.-F. 1972. René Capuron (1921–1971). Foundateur de la botanique forestière de Madagascar. Adansonia, sér. 2, **12**: 13–38.

———. 1978. Composition, origin and affinities of the Madagascan vascular flora. Ann. Missouri Bot. Gard. **65**: 535–589.

MacPhee, R. D. E., D. Burney & N. A. Wells. 1985. Early Holocene chronology and environment of Ampasambazimba, a Madagascar subfossil lemur site. Int. J. Primatol. **6(5)**: 463–489.

Miège, J. 1974. Étude du genre *Adansonia* L. II. Caryologie et blastogenèse. Candollea **29**: 457–475.

Millot, J. 1972. In conclusion. Pages 741–756 in R. Battistini & G. Richard-Vindard (eds.). Biogeography and ecology in Madagascar. Dr. W. Junk, B. V., Publishers, The Hague.

Paulian, R., J.-M. Betsch, J.-L. Guillaumet, C. Blanc & P. Griveaud. 1971. Étude des écosystèmes montagnards dans la région malgache. I. Le Massif de l'Andringitra, 1970–1971. Géomorphologie, climatologie et groupements végétaux. Bull. Soc. Écol. **2(2–3)**: 189–266.

———, C. Blanc, J.-L. Guillaumet, J.-M Betsch, P. Grieveaud & A. Peyrieras. 1973. Étude des écosystèmes montagnards dans la région malgache. II. Les chaînes Anosyennes. Géomorphologie, climatologie et groupements végétaux (Compagne RCP 225, 1971–1972). Bull. Mus. Hist. Nat., Paris, sér. 3, nº 118, Écol. gén. **1**: 1–40.

Perrier de la Bâthie, H. 1921. La végétation malgache. Ann. Inst. Bot.-Geol. Colon. Marseille, sér. 3, **9**: 1–268.

———. 1927. Le Tsaratanana, l'Ankaratra et l'Andringitra. Mém. Acad. Malgache **3**: 1–71.

———. 1933a. Le Manjakabetany de Tuléar (*Baudouinia Rouxevillei* sp. n.). Bull. Trimestriel Acad. Malgache, n.s., **15**: 7–8.

———. 1933b. Les plantes introduites à Madagascar. Henri Basuyau and Cie. Toulouse.

———. 1936. Biogéographie des plantes de Madagascar. Société d'éditions Géographiques, Paris.

———. 1952. 165ᵉ Famille. Ébénacées (Ebenaceae). Flore de Madagascar et des Comores. Firmin-Didot et Cie, Paris.

——— & B. P. G. Hochreutiner. 1955. 130ᵉ Famille. Bombacacées (Bombacaceae). Flore de Madagascar et des Comores. Firmin-Didot et Cie, Paris.

Prance, G. T. 1977. Floristic inventory of the tropics: where do we stand? Ann. Missouri Bot. Gard. **64**: 659–684.

Rauh, W. 1973. Über die Zonierung und Differenzierung der Vegetation Madagaskars. Tropische und subtropische Pflanzenwelt, 1. Mainz, Akad. Wiss. Lit. 146 p.

———. 1979. Problems of biological conservation in Madagascar. Pages 405–421 in D. Bramwell (ed.). Plants and islands. Academic Press, London.

Rossi, G. 1973–1974 (1977a). Carte de tapis végétal, Ambanja (1:250,000). Foiben-Taosarintanin'I Madagasikara, Antananarivo, Madagascar. (map).

———. 1971–1973 (1977b). Carte de tapis végétal, Diégo-Suarez (1:250,000). Foiben-Taosarintanin'I Madagasikara, Antananarivo, Madagascar. (map).

St. John, H. 1968. Revision of the genus *Pandanus* Stickman, Part 27. *Pandanus* novelties from Madagascar. Pacific Sci. **22**: 104–137.

Saloman, J. N. 1976–1977 (1978). Carte du tapis végétal, Tulear (1:250,000). Foiben-Taosarintanin'I Madagasikara, Antananarivo, Madagascar. (map).

Stern, W. L. 1967. Index Xylariorum. Reg. Vegetabile **49**: 1–36.

Stevens, P. F. 1988. New Guinea. This volume.

Stone, B. C. 1970a. Observations on the genus *Pandanus* in Madagascar. J. Linn. Soc. Bot. **63**: 97–131.

———. 1970b. New and critical species of *Pandanus* from Madagascar. Webbia **24**: 579–618.

———. 1983. A guide to collecting Pandanaceae (*Pandanus, Freycinetia*, and *Sararanga*). Ann. Missouri Bot. Gard. **70**: 137–145.

——— & J.-L. Guillaumet. 1970. Une nouvelle et remarquable espèce de *Pandanus* de Madagascar. Adansonia, sér. 2, **10**: 127–134.

———. 1972. Un nouveau *Pandanus* (Pandanacées) sub-aquatique de Madagascar. Adansonia, sér. 2, **12**: 525–530.

Thonner, F. 1908. Die Blütenpflanzen Afrikas. R. Freidländer & Sohn, Berlin.

———. 1915. The Flowering Plants of Africa. Dulau & Co., Ltd. London.

Tomlinson, P. B. 1977. Plant morphology and anatomy in the tropics—the need for integrated approaches. Ann. Missouri Bot. Gard. **64**: 685–693.

White, F. 1983. The Vegetation of Africa. Natural Resources Research Series, No. XX. The UNESCO Press, Paris.

Regional Reports

IV. Central America

Oaxaca, Mexico

David H. Lorence and Abisaí García Mendoza

Contents

I. Description of the Region	254
A. Topography	254
B. Geology	254
C. Soils	255
D. Climate	255
E. Population	255
II. Vegetation Maps and Systems of Classification	255
III. Vegetation Types Recognized for Oaxaca	256
A. Tropical Ombrophilous Forests (IA1)	256
1. Lowland (IA1a), Submontane (IA1b), and Riparian (IA1f(1))	256
2. Cloud (Broad-leaved) (IA1c), Montane (IA1e(1))	256
B. Tropical (Subtropical) Evergreen Forest (IA2)	257
1. Seasonal Lowland (IA2a), Submontane (Broad-leaved) (IA2b(1))	257
C. Tropical (Subtropical) Semideciduous Forest (IA3)	257
1. Lowland (IA3a)	257
D. Evergreen (Nongiant) Conifer Forest with Rounded Crowns (IA9b)	258
E. Evergreen (Nongiant) Conifer Forest with Conical Crowns (IA9c)	258
F. Drought Deciduous Lowland (and Submontane) Forest (IB1a)	259
G. Thorn Forest (IC2)	259
1. Mixed Deciduous-Evergreen Thorn (IC2a), Purely Deciduous Thorn (IC2b)	259
H. Evergreen Coniferous Woodland (IIA2)	260
1. Evergreen Coniferous Woodland with Rounded Crowns and Sclerophyllous Understory (IIA2a(1)), Evergreen Coniferous Woodland with Prevailingly Conical Crowns (IIA2b)	260
I. Drought Deciduous Woodland (IIB1)	260
J. Deciduous Subdesert Scrubland with Succulents (IIIC2(b))	261
K. Mangrove Forest (IA5)	261
IV. Floristic Inventory in Oaxaca	261
A. Important Collectors	262
B. Important Herbaria	262
V. Publications on the Oaxacan Flora	262
VI. Completeness of the Floristic Inventory	263
A. Areas High in Endemism or Yielding New and Important Distributions	263
B. Extinction	264
C. Taxa Well Collected	264
D. Tropical Forest Areas Threatened by Destruction or Conversion	264
VII. Resources for Continued Floristic Inventory in Oaxaca	265
VIII. Suitability of Past, Ongoing and Proposed Floristic Inventory	265
IX. Additional Resources Required	266
X. Acknowledgments	266
XI. Literature Cited	266
XII. Appendices	268

I. Description of the Region

Oaxaca, situated in the southern portion of the Mexican Republic, covers 95,364 km² and is bordered on the north by Veracruz and Puebla, on the east by Chiapas, on the west by Guerrero, and on the south by the Pacific Ocean. Mexico's fifth largest state, Oaxaca is subdivided politically into 30 districts and 570 municipalities.

A. Topography

Oaxaca has an extremely accentuated relief, suggestive of a piece of crumpled paper that has been spread out. The state's complex topography is determined by two major mountain ranges, the Sierra Madre del Sur and the Sierra Madre de Oaxaca. The Sierra Madre del Sur enters Oaxaca from the west and runs parallel to the Pacific Ocean in a NW–SE direction to the Isthmus of Tehuantepec. With maximum altitudes ranging between 2000 and 2500 meters, this range is known locally as the Sierra de Miahuatlán. The Southwest Coastal Plain lies between the Sierra and the Pacific, extending to the Río Tehuantepec. It has an average width of about 25 km and is in turn composed of smaller plains, all derived from erosion of the Sierra.

The Sierra Madre de Oaxaca is a continuation of the Sierra Madre Oriental that extends from just south of the Pico de Orizaba (Veracruz) in the north to the Isthmus of Tehuantepec in the south, where it then joins the Sierra Madre del Sur. The Sierra Madre de Oaxaca runs in a NW–SE direction and measures about 300 km long by 75 km wide, with average altitudes exceeding 2500 m and summits often exceeding 3000 m (3500 m at Cerro Zempoaltépetl). Portions of this range are known locally as the Sierra de Huautla, Sierra de San Juan del Estado, Sierra de Juárez, Sierra de Cuajimoloyas, Sierra de Villa Alta, Sierra Mixe, and Zempoaltépetl Massif. Along its Gulf slopes it delimits the broad alluvial Gulf Coast Plain which, in Oaxaca, encompasses the southern part of the Río Papaloapan system and its affluents. This almost flat area known as the Tuxtepec region is shared with Veracruz. The western part of the Sierra Madre borders the deep Cañon de Tomellín. This is the southernmost extension of the Tehuacán-Cuicatlán Valley system, the bottom of which lies at about 500 m altitude.

Orographic contact between these two major mountain systems has resulted in a series of relatively low mountains in between that run SW–NE. Although not constituting a single uniform structural entity, these ranges cover much of the state and are variously known as the Oaxacan Complex, the Mixtecan Shield (Tamayo, 1962) or the Oaxacan Highland Province (Ferrusquía-Villafranca, 1976).

In the middle of the Isthmus of Tehuantepec, a cordillera known as the Sierra Atravesada represents the NW terminus of the Sierra Madre de Chiapas. Although mostly under 1000 m, its highest peaks reach ca 2300 m. A plain averaging 10–40 km wide called the Isthmus-Chiapas Coastal Plain runs between the Pacific Ocean and the Sierra Atravesada-Sierra Madre de Chiapas, delimited in the west by the Río Tehuantepec. Finally, a small portion of NW Oaxaca is situated in the Balsas Depression with average altitudes of about 1000 m. For more details of the state's topography, consult the following references: Alvarez, 1961; Departamento Cartográfico Militar de la República Mexicana 1959–1972; García & Falcón, 1977; Secretaría de Programación y Presupuesto, 1981; Tamayo, 1962.

Five major river systems occur in the state. On the Gulf slope lie tributaries of the Río Papaloapan that arise in the Sierra de Juárez and run NW through the Cañon de Tomellín, eventually unite with the Río Salado from Tehuacán, and then cross the Sierra Madre de Oaxaca in a NE direction. These tributaries finally join the Papaloapan proper, which empties into the Gulf of Mexico. The Papaloapan is fed primarily by runoff from Oaxaca, the area contributing on average approximately 39,176 million cubic meters annually of a total 41,135 million cubic meters at the estuary (Tamayo, 1962).

The headwaters of the Río Coatzacoalcos, principally at the Río del Corte and Río Uxpanapa, lie on the northern slopes of the Sierra Atravesada in the Isthmus of Tehuantepec and follow a northerly direction. The Coatzacoalcos also empties into the Gulf of Mexico and carries an estimated annual runoff of 22,394 million cubic meters at its estuary (García & Falcón, 1977).

On the Pacific slope in northwestern Oaxaca the Río Mixteco, an affluent of the great Río Balsas system, drains the western watershed of the Sierra Mixtecas and northern portion of the Sierra Madre del Sur.

The Río Verde system lies completely within the state. It originates in the Valley of Oaxaca, gathers runoff from the southern slopes of the Sierra de Juárez and eastern slopes of the Mixteca, crosses the Sierra Madre del Sur and empties at its estuary in the eastern part of the Chacahua Bay.

The Río Tehuantepec, also situated entirely within the state, has its origins in the region between the Sierra de Miahuatlán and Sierra Mixe, follows an E–SE direction parallel to them and empties into the La Ventosa Bay in the Gulf of Tehuantepec.

B. Geology

There are essentially three geologically recognizable regions in Oaxaca: 1) the Tlaxiaco Province in the center and north; 2) the Sierra Madre del Sur and Oaxacan High Plains Province in the center and southeast; 3) the Sierra de Juárez Subprovince in the north and northeast.

The Tlaxiaco Province, situated in the Mixtecan physiographical complex, is largely composed of metamorphic rocks. The Sierra Madre del Sur-Oaxacan High Plains Province is predominantly composed of volcanic and metamorphic rocks. The Sierra de Juárez Subprovince in the north and northeast is composed of folded sedimentary rocks with series of younger granitic intrusions whose ages range from Paleozoic to Cenozoic, the majority being Mesozoic. Finally, the Sierra Atravesada is considered to represent a segregate geological entity because it is granitic and belongs structurally to the Sierra Madre de Chiapas. For further geological details, see López-Ramos (1974, 1981), and Secretaría de Programación y Presupuesto (1981).

C. SOILS

The major soil types found in Oaxaca comprise Andosols, Lithosols and Acrisols (Secretaría de Programación y Presupuesto, 1981), although the state's intricate topography and geology embrace a much wider range of subcategories. Maps and descriptions of all soil types based on the FAO-UNESCO classification scheme, with modifications for Mexico, are presented in Flores (1974), García & Falcón (1977), Secretaría de Agricultura y Recursos Hidráulicos (1979), and Dirección de Estudios del Territorio Nacional (1979).

D. CLIMATE

The climatic regimes of Oaxaca have been classified by García (1981). Based on modifications of the Köppen system they are comprised of three major categories: A, C, and B. Hot and humid climates (Af and Am) occur in the Gulf slope lowlands of the Choapan and Tuxtepec Districts, and in the Juchitán District of the Chimalapa region. Hot subhumid climates of the Aw type are localized in the drier southern lowland regions, primarily in the Isthmus of Tehuantepec along the Pacific coast. Temperate subhumid climates of the Cw type occur for the most part in mountainous zones above 1000 m. Finally, hot dry climates of the Bs type are restricted to the central upland valleys and intermediate parts of the Río Mixteco, Río Tehuantepec and Río Papaloapan Basins.

The most recent and comprehensive climatological map is that of the Instituto de Geografía, U.N.A.M. (1970), which contains isohyets and isotherms based on data from most of the state's meteorological stations together with corresponding climatograms. In addition, maps with maximum and minimum temperatures, hottest months, annual temperature variation, precipitation regimes, and other like data, can be found in Soto and Jaureguí (1965) and Mosiño (1974).

E. POPULATION

Of the 31 Mexican states (excluding the Federal District), Oaxaca is ninth in terms of population density, with an estimated 2,860,600 inhabitants, or an average of 21.13 persons per km^2, as of 1976 (García & Falcón, 1977). In terms of population growth, Oaxaca is in last place, with an increase of 16.68% between 1960 and 1970 (García & Falcón, 1977).

Population centers with more than 10,000 inhabitants as of 1976 (García & Falcón, 1977) include the capital, Oaxaca, with over 150,000 (1983 estimate), Tuxtepec (80,000; road sign, 1984), Matías Romero (36,000; road sign, 1984), Juchitán de Zaragoza (30,000), Loma Bonita (23,000), Salina Cruz (22,000), Pinotepa Nacional and Tehuantepec (16,000), Ixtepec, Huajuapan de León and Tlacolula (all 14,000), Jamiltepec (11,000), and Tlaxiaco (10,000). The balance of the population mostly lives in small towns and villages scattered throughout the state.

Of the total population, 42% is economically active, of which 72% are engaged in traditional activities such as agriculture, livestock husbandry, hunting and fishing, 9% are employed by industries, 6% work in offices, and the remaining 13% work in commerce and diverse activities.

II. Vegetation Maps and Systems of Classification

Vegetation studies in Oaxaca essentially began with Conzatti (1918, 1922, 1929). He produced a map of the state's phytogeographical regions, but he did not designate vegetation types. Smith and Johnston (1945) published a small and generalized vegetation map of Latin America including Mexico, in which they recognized six vegetation types.

Recent attempts have been made to classify the vegetation of Mexico, including Oaxaca, by the following authors: Leopold (1950), Miranda and Hernández X. (1963), Sarukhán K. (1968), Flores et al. (1972), Rzedowski (1978) and the Comisión Tecnico Consultiva para la Determinación Regional de los Coeficientes de Agostadero (COTECOCA, 1980). Each of these has used physiognomic and floristic criteria to define communities or formations and, in addition, characteristics of climate, topography and soil have been included as part of the overall vegetation descriptions.

Leopold (1950) was the first to publish a credible vegetation map of Mexico where 12 vegetation types (= formations) were recognized, at a scale of 1:10,000,000 (later reproduced at 1:8,000,000). A highly detailed and precise classification system recognizing 32 vegetation types for Mexico was elaborated by Miranda and Hernández X. (1963), but unfortunately a map was not included with this scheme. Rzedowski (1978) presented a schematic vegetation map of Mexico which, although accurate, is too general to be useful for Oaxaca. The first detailed vegetation map to include Oaxaca was elaborated by Flores et al. (1972) for the Mexican Republic at a scale of 1:2,000,000. In this work, 11 original vegetation types are recognized for Oaxaca and the map clearly shows the complex, mosaical nature of the state's vegetation. In this work, however, the distribution of most of Oaxaca's vegetation types was extrapolated from topographical and ecological data and considered in light of the scant literature, rather than being founded upon field studies. As a result, at least for Oaxaca, this map contains may inaccuracies and generalizations.

Fine-grained vegetation maps depicting small areas have been elaborated for three areas in Oaxaca utilizing aerial photos, extensive field studies and floristic inventories. Two of these include upland vegetation such as pine-oak forest, deciduous forest and xerophytic scrub in the Mixteca Alta, one at Tepelmeme de Morelos (Cruz C. & Rzedowski, 1980), and the other in the Sierra de Tamazulapan at Teposcolula (García M., 1983). The third includes lowland wet forest in the Tuxtepec District of northeastern Oaxaca (Gómez-Pompa et al., 1964).

More recently, COTECOCA (1980) of the Mexican Secretaría de Agricultura y Recursos Hidráulicos (SARH) has elaborated a detailed and highly accurate vegetation map of Oaxaca based on extensive field observations and aerial photos. Here 20 original vegetation types are recognized for the state. Unfortunately, the map and accompanying text detailing soil types, climate, localities and component species have not been published. Although the work is extremely useful, some inaccuracies were noted in lists of component species.

Our classification, in which 12 original vegetation types are recognized, is based primarily on COTECOCA (1980) and modified to fit the UNESCO (1974) system. A number of problems were encountered during the elaboration of our system, due in part to the different vegetational criteria employed by different authors. Furthermore, vegetation types are difficult to define and delimit in Mexico. In particular, because of Oaxaca's complex topography and resulting mosaic of microclimates and microhabitats, coupled with its subtropical location, it boasts a myriad of plant communities, some of which defy classification.

First, difficulty was encountered in accommodating commonly used concepts of Mexican vegetation types into the UNESCO scheme even though the UNESCO key was followed. Particularly problematic was our varied *Pinus* and/or *Quercus* formations which fall into at least seven UNESCO categories, both tropical (subtropical) and temperate (i.e., IA1e(1), IA9b, IA9c, IB1a, IIA2a(1), IIA2b, and IIB1). On the one hand, distinctions could not be maintained between two categories of evergreen coniferous woodland, i.e., those with rounded crowns and sclerophyllous understory (IIA2a(1)) and those with conical crowns (IIA2b), because they intergrate both physiognomically and floristically. On the other hand, some of the areas designated as evergreen coniferous woodland actually represent or integrate with forests having large trees (*Pinus* spp.) 20 or more meters tall and closed canopies.

Secondly, Oaxaca's highly accentuated relief at times imposed constraints on the number of ultimate subcategories we recognized, particularly in terms of altitudinal breakdown. Nor do Oaxaca's communities always lend themselves to this sort of breakdown. Therefore, Lowland, Submontane and Riparian Tropical Ombrophilous Forest (IA1a, IA1b and IA1f(1)) were grouped together. Similarly, distinctions were not made between Tropical Ombrophilous Montane and Broad-leaved Cloud Forest (IA1c, IA1e(1)), nor between Tropical (Subtropical) Evergreen Seasonal Lowland and Submontane Broad-leaved Forest (IA2a, IA2b(1)), nor between purely Deciduous and Mixed Deciduous-Evergreen Thorn Forest (IC2a, IC2b).

The UNESCO (1974) system is purely physiognomic and we have adapted it to frame our vegetational concepts. Each of the following sections is headed by the UNESCO vegetation type(s) which we conceive to represent a single vegetational unit. This is followed by general remarks regarding previous authors' concepts and/or equivalents. Then, brief descriptions of the formations and where they occur in Oaxaca are provided as well as information on climate, soils, physiognomy and component species.

III. Vegetation Types Recognized for Oaxaca

A. Tropical Ombrophilous Forests (IA1)

1. Lowland (IA1a), Submontane (IA1b), and Riparian (IA1f(1))

These three categories of Tropical Ombrophilous Forest correspond with communities described as Selva Alta Perennifolia (Miranda & Hernández X., 1963; Sarukhán K., 1968), and Bosque Tropical Perennifolio (in part) of Rzedowski (1978). As they do not differ greatly either physiognomically or floristically in Oaxaca, they are considered here as one vegetation type.

This community is restricted to the northern edge of the state in the middle of the Río Papaloapan Basin, extending from the northern tip at the Presa Miguel Alemán southeastward through the lower parts of the Tuxtepec and Choapan Districts. In addition, two small areas occur in the Río Coatzacoalcos and Río Grijalva Basins in the middle of the Isthmus of Tehuantepec (Selva del Ocote) at the boundaries with Veracruz and Chiapas.

In Oaxaca, this forest develops at altitudes ranging from 50 to 800(–1000) m in hot, humid zones with mean annual temperatures between 22 and 26°C, and mean annual precipitation exceeding 2000 mm. The soils are in general well drained, deep, lateritic and associated with a limestone substrate, as for example in the Río Uxpanapa zone.

This physiognomically dense community is dominated by tall evergreen trees generally exceeding 30 m that branch in the extreme upper parts of the trunks. In addition to the three arboreal strata, a shrubby understory, an herbaceous ground cover, and abundant lianas and epiphytes are also present.

Gómez-Pompa et al. (1964) list the following species as characteristic and dominant: *Acosmium panamense, Andira galeottiana, Calophyllum brasiliense, Dialium guianense, Robinsonella mirandae, Scheelea liebmannii, Terminalia amazonia* and *Vochysia hondurensis. Acacia tomentosa, Dussia mexicana, Brosimum alicastrum, Spondias radlkoferi, Pouteria sapota* and *Pterocarpus rohrii* are also characteristic in many areas. In riparian or periodically inundated areas Sousa S. (1964) cites *Brosimum alicastrum, Lonchocarpus cruentus* and *Spondias mombin*, as well as *Andira galeottiana, Calophyllum brasiliense, Pachira aquatica* and *Tabebuia rosea*.

Many areas previously supporting this community have been converted into cattle pasture and agricultural land, and most of the remaining areas are threatened.

2. Cloud (Broad-leaved)(IA1c), Montane (IA1e(1))

The Bosque Caducifolio of Miranda & Hernández X. (1963), Sarukhán K. (1968) and COTECOCA (1980), and

Bosque Mesófilo de Montaña of Rzedowski (1978) and their associations most closely correspond to UNESCO's Tropical Ombrophilous Montane and Broad-leaved Cloud forest formations. The latter are considered jointly here because altitudinal separation is not feasible, at least in Oaxaca.

This formation forms a band between (1000–)1400–2250 m altitude along the northern and eastern slopes of the Sierra de Juárez and Sierra Mixe, as well as on the summits of small mountains emerging from the lowland and submontane zones. In addition, on the Pacific slopes it occupies intermittent areas on the western flanks of the Sierra Madre del Sur up to the Guerrero boundary.

Characterized by a humid montane climate (Cf, also Af or Am), the mean annual temperature ranges from 14 to 20°C and the mean annual precipitation exceeds 2000 mm, probably reaching 6000 mm in places such as Vista Hermosa (Rzedowski & Palacios C., 1977). In Oaxaca, this forest occurs in situations that are more humid than those occupied by Pine-Oak forest, and warmer than those occupied by *Abies* forests.

Physiognomically, the dominant trees average 20 to 30 or even 40 m tall and have soft, medium-sized leaves. Both evergreen and deciduous species occur in combination, together with palms, tree ferns, epiphytes, vines and hygrophilous herbs. Floristically, this formation is a mixture of both neotropical and holarctic elements. Rzedowski & Palacios C. (1977) studied this community at Vista Hermosa in the Sierra de Juárez, where they report the following species as dominant: *Oreomunnea mexicana*, *Weinmannia pinnata* and *Liquidambar styraciflua*. Additional characteristic taxa are *Alchornea*, *Clethra*, *Clusia*, *Hedyosmum mexicanum*, *Ilex*, Lauraceae (including *Persea*), *Meliosma dentata*, *Nyssa sylvatica*, *Oreopanax*, *Prunus*, *Podocarpus*, *Quercus*, *Saurauia* and *Styrax*. At another location near Putla, Sarukhán K. (1968) reports *Brunellia comocladifolia*, *Hymenaea courbaril*, *Liquidambar styraciflua*, *Viburnum* and *Xylosma*, among others.

Lying between this community and the Lowland and Submontane Ombrophilous Forest, i.e., below 1000 m where the termperatures are warmer (20 to 24°C) and the precipitation lower (ca 1500 mm) because it is below the cloud zone, there are ecotones dominated by *Quercus magnoliaefolia*, *Q. skinnerii* and *Q. sororia* mixed with lowland tropical elements such as *Brosimum alicastrum*, *Manilkara zapota*, *Scheelea liebmannii*, *Tabebuia chrysantha* and *Terminalia amazonia*. These have been designated by COTECOCA (1980) as the Bosque Esclerófilo Perennifolio. This formation is included here because of its physiognomy, although it does not correspond very well floristically or otherwise. Finally, in the Chimalapa region *Quercus-Liquidambar* forests mixed with tropical elements occur as low as 200–300 m elevation in some places.

Many areas supporting Tropical Ombrophilous Montane and Broad-leaved Cloud Forest are rapidly being converted for coffee culture or other agricultural purposes.

B. Tropical (Subtropical) Evergreen Forest (IA2)

1. Seasonal Lowland (IA2a), Submontane (Broad-leaved) (IA2b(I))

These two categories correspond to the Selva Mediana Subperennifolia of various authors (Miranda & Hernández X., 1963; Saurkhán K., 1968; COTECOCA, 1980). One area supporting this formation is located in northern Oaxaca just north of the Presa Miguel Alemán, where it is surrounded by Tropical Ombrophilous Lowland Forest (IA1a). In addition, this community may be found on karst outcrops and/or rocky hills along a mountain front, e.g., the hills just south of Tuxtepec and in the Uxpanapa area (T. Wendt, pers. comm.; Pennington & Sarukhán, 1968). Two other zones supporting this type of vegetation are on the Pacific slope. One, near sea level, surrounds the Lagunas de Chacahua and de Pastoría; the other is north of Pochutla near Candelaria Loxicha and Pluma Hidalgo, at altitudes of 600 to 900 m.

In the north, this formation occurs on karstic limestone hills with steep slopes and with a highly dissected topography, whereas in Chacahua the soil is sandy. The climate is generally similar to that of the Lowland and Submontane Ombrophilous Forest, but with a well defined, albeit short, dry period. This same dry period is generally present in Oaxacan Tropical Ombrophilous Forest (IA1a, IA1b, IA1f(1)), but its effect is less accentuated due to the deep soils. Mean annual temperatures range from 20 to 24°C, and the mean annual precipitation exceeds 1500 mm.

The canopy trees of this community average 25 to 30 m tall; palms are generally abundant in the lower strata. A distinguishing feature of this formation is that 25 to 50% of the trees lose their leaves during the dry season. *Brosimum alicastrum* is frequently the dominant tree species of this community. In the north, other characteristic species include *Bursera simaruba*, *Dendropanax arboreus*, *Malmea depressa*, *Manilkara zapota*, *Ouratea mexicana*, *Pouteria sapota*, *Robinsonella mirandae*, *Ruprechtia costata*, *Simira rhodoclada*, *Spondias radlkoferi*, *Ternstroemia tepezapote* and *Vatairea lundellii*. On the Pacific slope and coast the dominants include *Acosmium panamense*, *Bravaisia integerrima*, *Calophyllym brasiliense*, *Calycophyllum candidissimum*, *Chrysophyllum cainito*, *C. mexicanum*, *Coccoloba barbadensis*, *Crataeva tapia*, *Enterolobium cyclocarpum*, *Manilkara zapota* and *Poulsenia armata*.

Where this formation occurs on the Pacific slope and coast it has been subject to human disturbance. Human impact seems to have been less in the north because the karstic limestone substrate is unsuitable for agricultural purposes.

C. Tropical (Subtropical) Semideciduous Forest (IA3)

1. Lowland (IA3a)

This formation corresponds with the Selva Alta and Mediana Subcaducifolia of Miranda & Hernández X. (1963),

Sarukhán K. (1968) and COTECOCA (1980), and to the Bosque Tropical Subcaducifolio of Rzedowski (1978). A salient feature of this forest is that 50 to 75% of the dominant species lose their leaves during the dry season. The Semideciduous Lowland Forest is the dominant vegetation type over large areas of the Pacific Coastal Plain to the Guerrero boundary. A much smaller band also occurs in the Isthmus of Tehuantepec.

The climate is hot and subhumid (Aw and Am) with a well defined dry season of about four months. The mean annual temperatures generally do not exceed 22°C and the mean annual precipitation is below 1600 mm. Apparently this forest is not restricted to any particular soil type.

In this community, the trees are less densely stocked than in the former lowland forest types and have a more open canopy. The trees range in height from 20 to 30 m. Most species possess small to medium sized leaves that are soft or occasionally coriaceous in texture, and the maximum flowering period usually coincides with the leafless period.

Floristically, most species have neotropical affinities. Dominants include *Albizia caribaea*, *A. guachapele*, *A. tomentosa*, *Andira inermis*, *Brosimum alicastrum*, *Caesalpinia coriacea*, *C. velutina*, *Calycophyllum candidissimum*, *Enterolobium cyclocarpum*, *Ficus tecolutensis*, *Hymenaea courbaril*, *Lafoensia punicaefolia*, *Orbignya guacuyule* and *Pterocarpus acapulcensis*.

In Oaxaca, this formation occurs in regions supporting large indigenous populations. Consequently, many areas have been cut or disturbed.

D. Evergreen (Nongiant) Conifer Forest with Rounded Crowns (IA9b)

This formation is equivalent to the Bosque Aciculifolio and Bosque Escuamifolio (in part) of COTECOCA (1980), the Pinar (in part) and Bosque de Escuamifolios of Miranda & Hernández X. (1963), and part of Rzedowski's (1978) Bosque de Coníferas, i.e., Bosque de *Pinus* and Bosque o Matorral de *Juniperus*. The coniferous forests that characterize vast areas of the Mexican landscape are surprisingly diverse both ecologically and floristically. Although some appear to represent secondary communities due to human disturbance, many pine forests constitute edaphic and climatic climaxes (Rzedowski, 1978).

In Oaxaca, the majority of coniferous forests with rounded crowns are dominated by *Pinus* with 15, or nearly half, of Mexico's 35 species recorded for the state (COTECOCA, 1980; Martínez, 1945). Essentially pure stands of *Juniperus flaccida* (the Bosque Escuamifolio) occur in two relatively small areas in the state. This community is probably not climax and has been considered as a transition zone (Miranda & Hernández X., 1963) or as secondary in nature (Rzedowski, 1978). In addition, small stands of *Juniperus flaccida* may occur scattered within or between pine-oak forests, pastureland, scrubland or deciduous forest. For this reason, we have placed it together with the communities to which it corresponds most closely, i.e., evergreen coniferous forest (here) and evergreen coniferous woodland (below). Finally, it should be noted that *Pinus* and *Quercus* communities often have similar ecological requirements and as a result may occupy similar niches, developing side by side or intermixed and forming mosaics and/or other complex associations. For this reason, pine and oak forests have been considered as a single vegetation type by many authors (see Rzedowski, 1978).

Evergreen Coniferous Forests with Rounded Crowns occur in the humid mountainous regions of Oaxaca mostly between 1500 and 3000 m altitude, primarily in the Sierra Madre del Sur, Sierra Madre de Oaxaca and the Mixtecan ranges. They occupy a zone above the Tropical Ombrophilous Montane and Cloud Forest or Evergreen Coniferous Woodland zones and below the Evergreen Coniferous Forest with Conical Crowns. The climate here is characteristically temperate-humid (Cw) with mean annual temperatures ranging between 18 and 22°C. The precipitation ranges from 1500 to 4000 mm annually and is distributed over seven to 11 months of the year. These pine forests prefer, but are not restricted to, mildly acidic soils derived from igneous rocks.

The form and disposition of the needle-like pine leaves imparts a distinctive physiognomy to this formation. The trees are evergreen, 25 to 40 m tall, and have monopodial trunks branching near the top to form more or less rounded crowns. There communities are dominated by one or more of the following species of *Pinus*: *P. ayacahuite*, *P. cornuta*, *P. douglasiana*, *P. lawsonii*, *P. macrocarpa*, *P. michoacana*, *P. montezumae*, *P. oaxacana*, *P. oocarpa*, *P. patula*, *P. rudis* and *P. teocote*.

Although considerable expanses of pine forest still occur in Oaxaca, these areas are being constantly reduced through exploitation by paper and lumber companies for wood, pulp and cellulose. Little resin exploitation is carried out in the state. Where soil conditions permit, limited areas are cleared for cultivation of maize, beans and other grains by the indigenous people.

E. Evergreen (Nongiant) Conifer Forest with Conical Crowns (IA9c)

Corresponding to this formation are the Bosque de Abetos y Oyameles (Miranda & Hernández X., 1963), the Bosque de *Abies* (Rzedowski, 1978), and the Bosque Aciculinearifolio (COTECOCA, 1980).

This community is dominated by various trees belonging to the genera *Abies*, *Pinus* and *Quercus*, and is restricted to the cold mountainous regions above 2750 m in the Sierra de Juárez, Sierra de Miahuatlán and Zempoaltépetl Massif. The Cw climate is characterized by mean annual temperatures of 7 to 15°C and a mean annual precipitation above 1000 mm. Nocturnal frosts occur throughout the year.

Physiognomically, this is a dense formation with various strata of trees and shrubs, many with coriaceous leaves.

Vines and epiphytes (mostly bryophytes and lichens) are abundant. The dominant trees reach 20 to 30 m and include *Abies guatemalensis, A. hickeli, A. oaxacana, Cupressus lindleyi, Pinus hartwegii, P. patula, P. rudis, P. strobus* var. *chiapensis,* and *Quercus laurina.* Other component species include sclerophyllous trees and shrubs such as *Arbutus xalapensis, Arctostaphylos polifolia, A. pungens* and *Litsea glaucescens.*

Although this community occupies only limited areas at the highest altitudes, it is threatened by exploitation for wood and pulp.

F. Drought Deciduous Lowland (and Submontane) Forest (Ib1a)

The Drought Deciduous Lowland (and Submontane) Forest is equivalent to the formations known as Selva Baja Caducifolia (COTECOCA, 1980; Miranda & Hernández X., 1963; Sarukhán K., 1968 (in part)) and the Bosque Tropical Caducifolio (Rzedowski, 1978).

Numerous floristic variants of this forest cover large portions of Mexico's semiarid and arid zones. It is characterized by short trees 4 to 10(–15) m tall, the majority of which lack spines and lose their leaves during the dry season. This community occurs in Oaxaca from near sea level to about 1700(–1900) m in areas where the climate is hot and dry (Aw or occasionally Bs), with mean annual temperatures ranging from 19 to 29°C. The mean annual precipitation varies from about 600 to 900 mm. Most importantly, there is a marked dry period lasting from five to eight months, with most of the precipitation falling during the summer. The Drought Deciduous Forest develops in a variety of topographical situations having a wide range of soil types, mostly well drained. In Oaxaca, the largest expanses occur in the northwest, north central, south central and southeast parts between the Sierra Madre del Sur and the Sierra Madre de Oaxaca and on the southern side of the Sierra Atravesada.

In terms of physiognomy, this community has a fairly dense stocking of trees, although the canopy is often sparse or open. The trees form a single, more or less uniform stratum. The trunks are short and frequently twisted, often branching low down or near the base. Few species are armed in sharp contrast to the next category, the thorn forest. Perhaps the most striking feature of this formation is that the majority of species lose their leaves during five to eight months of the year. The leaves are frequently compound with small leaflets. A shrubby stratum is generally present, often including cacti and agaves, but the herbaceous stratum is usually poorly developed or even lacking. Vines and epiphytes such as *Tillandsia* are usually scarce, but may be locally abundant in favorable situations.

Although floristic composition varies greatly at different locations and altitudes, some characteristic arboreal species include: *Acacia macilenta, A. tenuifolia, Amphipterygium adstringens, Apoplanesia paniculata, Bauhinia pauletia, Bucida macrostachya, B. wigginsiana, Bursera morelense, B. simaruba, Bursera* spp., *Caesalpinia coriaria, C. sclerocarpa, Capparis incana, Casaeria nitida, Ceiba aesculifolia, C. parvifolia, Celtis iguanea, Cercidium praecox, Comocladia engleriana, Conzattia multiflora, Cyrtocarpa procera, Euphorbia schlechtendalii, Fouqueria formosa, Gyrocarpus americanus, Haematoxylon brasiletto, Leucaena esculenta, Lonchocarpus emarginatus, L. obovatus, Lysoloma acapulcensis, L. divaricata, Parthenium tomentosum, Pithecellobium mangense, Plumeria rubra, Prosopis juliflora, Pseudosmodingium multifolium, Senna atomaria, Thevetia peruviana, Yucca* spp. and *Zizyphus amole.* The Leguminosae are particularly diverse in this floristically rich formation.

Human impact has generally been less here than in many of Oaxaca's other vegetation types, primarily because the shallow, rocky soils and arid climate are unsuitable for agriculture. However, goats and burros are allowed to graze in some areas. In addition, although the trees are unsuitable for commercial exploitation, they are used locally for construction and fuel.

G. Thorn Forest (IC2)

1. Mixed Deciduous-Evergreen Thorn (IC2a), Purely Deciduous Thorn (IC2b)

These formations best correspond with the Selva Baja Caducifolia Espinosa (COTECOCA, 1980), the Selva Baja Espinosa Caducifolia (Miranda & Hernández X., 1963; Sarukhán K., 1968), and the Bosque Espinoso (Rzedowski, 1978).

In Oaxaca, these forests are not clearly separable and will therefore be considered jointly. Furthermore, although this community can usually be distinguished from the Drought Deciduous Forest (Selva Baja Caducifolia) because it contains a predominance (60–80%) of spiny tree species, it is not always well delimited and often intergrades with the former and with the semideciduous forest.

The Thorn Forest is restricted to the lowland plains and hills along the southern part of the Isthmus of Tehuantepec where the climate is hot and subhumid (Aw). Here the average annual temperatures range from 25 to 28°C and the mean annual precipitation is between 900 and 1200 mm. The dry season may be from six to seven months long.

Physiognomically, this formation shares many features with the Drought Deciduous Forests of which it is a variant. The trees are low, from four to 15 m, with short stout trunks often branched near the base. The canopies are sparse and open, and the leaves are mostly compound and/or pubescent and deciduous. Most species are armed with spines or thorns. The shrubby stratum may be poor or well developed; agaves and cacti are often present, appearing conspicuous during the dry season. Epiphytes are scarce or lacking, although lianas may be abundant locally. This community is rich in Leguminosae and characteristic species include: *Brongniartia parviflora, Caesalpinia eriostachys, C. exostemma, Cienfuegosia rosei, Cercidium praecox, Coccoloba liebmannii, Crescentia alata, Esenbeckia berlandieri, Fouquieria formosa, Haematoxylon brasiletto,*

Mimosa eurycarpa, M. platyloba, M. tenuiflora, Pachycereus pecten-aboriginum, Parkinsonia aculeata, Pereskia lychnidiflora, Pithecellobium pallens, P. seleri, Prosopis juliflora, Randia thurberi and *Senna wislizeni* var. *pringlei*.

Where this formation occurs on plains in the Isthmus of Tehuantepec the soils are often deep and relatively rich. As a result, where irrigation is possible (i.e. around the Río Tehuantepec), this land is being cleared for agriculture.

H. Evergreen Coniferous Woodland (IIa2)

1. Evergreen Coniferous Woodland with Rounded Crowns and Sclerophyllous Understory (IIA2a(1)), Evergreen Coniferous Woodland with Prevailingly Conical Crowns (IIA2b)

These two communities are best accommodated under the categories of Bosque Aciculiesclerofilo and Bosque Escuamifolio (in part) of COTECOCA (1980), and also correspond (in part) to the Bosque de Pino-Encino, Pinar, and Encinar of various authors. As here construed, these evergreen woodlands are dominated by *Pinus* and/or *Quercus*, as well as *Juniperus flaccida* in some cases. They are communities sharing ecological affinities with, or are transitional to, the Evergreen Coniferous Forests mentioned earlier. They differ, however, by having a less dense tree stocking.

Evergreen Coniferous Woodland covers vast areas in the Mixteca Alta and Sierra Madre del Sur in southern and southwestern Oaxaca at altitudes of (500–)1000 to 2500 m. Characterized by temperate subhumid or humid climates (Cw and BS), the mean annual temperature ranges from 18 to 24°C with a mean annual precipitation ranging from (1000–)2500 to 4000 mm, falling mostly during the summer months.

As previously mentioned, certain areas (primarily in the north and northeast) designated as belonging to this vegetation type, actually constitute forests (i.e., with trees 20 m or more tall and with a relatively dense stocking) rather than woodlands (see below). These occur primarily in more mesic areas of the Sierra Madre de Oaxaca and the Isthmus of Tehuantepec, lying above both the Tropical Ombrophilous Lowland and Submontane Forest zones, and below the Tropical Ombrophilous Montane and Broad-leaved Cloud Forest zones. Occasionally this formation may be found below the Evergreen Forest with Rounded Crowns. It is not possible to accommodate this vegetation type satisfactorily in existing UNESCO categories such as Tropical Ombrophilous Needle-leaved Cloud Forest (IA3e(2)), because lianas and epiphytes are scarce and it lies below the cloud zone. Nor does it represent Temperate Evergreen Forest (IA9) because of the subtropical climate, frequent presence of deciduous elements such as *Liquidambar*, and the abundance of tropical understory elements. For these reasons, and because this formation often intergrades with Evergreen Coniferous Woodland, we have chosen to include it here. Pure stands of *Pinus oocarpa* in the Isthmus exemplify this formation type. The forest grows down to 250 m elevation, and often interdigitates in a complex manner with tropical oak forest and/or Tropical Ombrophilous Forest (T. Wendt, pers. comm.).

Woodlands (i.e., open stands of trees with the crowns not touching) occur in drier regions of the south and southwest. Here the trees are generally low, from four to 30 m tall, with straight trunks (*Pinus*) or branched trunks (*Quercus*), and densely branched crowns.

Characteristic component species of these pine-oak woodlands include: *Juniperus flaccida, Pinus ayacahuite, P. leiophylla, P. michoacana, P. oaxacana, P. oocarpa, P. patula, P. pseudostrobus, P. rudis, P. teocote, P. tenuifolia, Quercus acutifolia, Q. castanea, Q. crassifolia, Q. laurina, Q. liebmannii, Q. macrophylla, Q. magnoliaefolia* and *Q. rugosa*. A sclerophyllous understory sometimes occurs and may consist of ericaceous shrubs and low trees such as *Arctostaphylos polifolia, A. pungens, Arbutus glandulosa*, and *A. xalapensis*.

Pine-oak woodlands have long been easy targets for exploitation by man. Besides local use of wood for construction and fuel, commercial exploitation for lumber and pulp has begun. Once the land has been cleared, it is used for raising crops and livestock.

I. Drought Deciduous Woodland (IIB1)

The Drought Deciduous Woodland corresponds best with the Bosque Esclerófilo Caducifolio (COTECOCA, 1980), Bosque de *Quercus* (in part) of Rzedowski (1978), and Encinar (in part) of Miranda & Hernández X. (1963).

Formations dominated by oaks (*Quercus* spp.) comprise some of the most characteristic plant communities of the mountainous Oaxacan uplands. Floristically, physiognomically and ecologically diverse, they vary from low, dry scrublands to open woodlands or to dense, humid cloud forests. Although oaks are dominant, species of *Pinus, Juniperus* and other trees are often intermixed, and many of the understory shrubs are sclerophyllous.

Oak formations best fitting the Drought Deciduous Woodland category are restricted to parts of the Mixteca Alta of northwestern Oaxaca at altitudes of 2000 to 2600 m. The climate here is temperate-subhumid (Cw) with an average of four to five dry months per year and a mean annual precipitation of 650 to 800 mm falling mostly during the summer. Although nocturnal temperatures below 0°C are common during the coldest months, the mean annual temperature is 16 to 18°C. The limestone bedrock is covered by shallow lithosols that are neutral to slightly basic (García M., 1983).

In this formation the trees range from four to 30 m tall and are highly branched, usually with dense crowns. The leaves are sclerophyllous, small to large, often narrow and frequently pubescent below, and mostly deciduous during the dry season. Constituent species in Oaxaca include *Quercus acutifolia, Q. castanea, Q. crassifolia, Q. impressa, Q. lanigera, Q. macdougalii, Q. mexicana, Q. microphylla, Q. peduncularis, Q. sebifera, Q. segoviensis, Q. urbanii, Pi-*

nus lawsonii, P. oaxacana, P. patula, P. rudis and *Juniperus flaccida*. A well developed understory of sclerophyllous small trees and shrubs is found here and may be composed of *Amelanchier denticulata, Arbutus xalapensis, Arctostaphylos polifolia, A. pungens, Ceanothus coeruleus, C. greggii, Cerocarpus fothergilloides, C. pringlei, Garrya ovata, Lindleyella mespiloides, Rhus chondroloma, R. oaxacana, R. standleyi, R. virens* and *Vauquelinia australis*. Epiphytes such as *Tillandsia* and parasitic Loranthaceae are often common locally, and a rich herbaceous ground cover generally develops during the rainy season.

Oak wood is used extensively for construction, firewood and charcoal by the local people, and the bark and leaf galls for tanning. More often than not, the vegetation is disturbed or secondary because of past clearing for agricultural purposes.

J. Deciduous Subdesert Scrubland with Succulents (IIIC2(b))

This formation is best equated with the Matorral Mediano Subinerme and Matorral Oligocilindrocaule Afilo of COTECOCA (1980), and the Matorral Xerófilo of Rzedowski (1978).

The desertic and subdesertic scrublands that characterize many of Mexico's arid zones are so diverse floristically and physiognomically that a number of categories have been recognized by various authors (Rzedowski, 1978). However, essentially all are shrubby formations with xerophytic adaptations that include both succulent and deciduous elements, some of which are armed with spines.

The climate (Bs) is characterized by a pronounced diurnal variation in temperature with annual means of 20 to 24°C, a relatively low precipitation ranging from 500 to 650 mm annually, and high insolation with low atmospheric humidity. The number of dry months varies from five to seven. The soil is characteristically pale and grayish, porous, poor in organic matter and with high levels of calcium.

In Oaxaca, two types of Subdesert Scrubland are recognized by COTECOCA (1980) which we combine here for convenience. The first type (Matorral Oligocilindrocaule Afilo) is restricted to a narrow zone in the northwest along the Puebla border at ca 1000 m altitude and represents the southern limits of the Tehuacán Desert flora. It is characterized by large columnar or candelabriform cacti such as: *Cephalocereus chrysacanthus, C. hoppenstedtii, Escontria chiotilla, Heliabravoa chende, Myrtillocactus geometrizans, Neobuxbaumia macrocephala, N. mezcalensis, N. tetetzo, Pachycereus hollianus, Stenocereus pruinosus, S. stellatus* and *S. weberi*. These cacti impart a characteristic physiognomy to the formation and are most conspicuous during the leafless dry season. Intermixed are low deciduous trees and shrubs, for example: *Acacia acatlensis, A. coulteri, Amphipterygium adstringens, Brongniartia oligosperma, Bursera* spp., *Ceiba parvifolia, Capparis incana, Cercidium praecox, Ipomoea intrapilosa, Mimosa lacerata, Plumeria rubra* and *Senna unijugata*. The palm *Brahea dulcis* and rosette succulents such as *Agave* spp., *Dasylirion* spp. and *Yucca periculosa* often form dense, local populations.

The second type of Deciduous Subdesert Scrubland (Matorral Mediano Subinerme) occurs sporadically in and around the valley of Oaxaca at altitudes of 1500 to 1800 m. It consists of mostly deciduous shrubs, one to two meters tall, intermixed with a few small trees and cacti such as *Opuntia* spp. and *Stenocereus stellatus*. The shrubs are unarmed and have small, coriaceous leaves. Component woody species include: *Acacia farnesiana, Amelanchier denticulata, Bursera galeottiana, Calliandra* spp., *Croton ciliatoglandulosus, Dodonaea viscosa, Euphorbia rossiana, Fouquieria formosa, Ipomoea arborea, I. muricoides, Karwinskia humboldtiana, Malpighia mexicana, Mimosa monancistra, Neopringlea viscosa, Rhus virens, Senna skinneri* and *Wimmeria persicifolia*.

Although formerly predominant in the Valley of Oaxaca and vicinity, much of this scrubland has been cleared for agriculture or is grazed by goats. Moreover, as the region has long been inhabited by humans, it is possible that this vegetation type is partly secondary in nature (see Rzedowski, 1978), and presumably derived from Drought Deciduous Forest.

K. Mangrove Forest (IA5)

The Mangrove Forest or "Manglar" occurs intermittently along Oaxaca's Pacific coast, principally on the shores of coastal lagoons, sheltered bays, and river estuaries where the muddy soil is fine and deep, always within the tidal range. The largest expanses are located in the Gulf of Tehuantepec and along the western part of the coast. The floristically poor community is composed almost exclusively of evergreen trees ranging in height from four to 15 m, with broad, succulent or sclerophyllous leaves. The community's most striking physiognomic feature is the presence of either stilt roots or pneumatophores on most of the arboreal species.

In Oaxaca, the only four component species are *Rhizophora mangle* (the most frequent), *Laguncularia racemosa, Avicennia germinans* and *Conocarpus erectus*. Other life forms such as epiphytes and vines are scarce or absent. The herbaceous ground stratum is limited to stands of *Acrostichum danaefolium* and *Batis maritima*. Additional references pertaining to this well known formation are available in Rzedowski (1978).

Although lands occupied by the Mangrove Forest are unsuitable for agriculture, the wood of *Rhizophora* is frequently exploited to make charcoal.

IV. Floristic Inventory in Oaxaca

Oaxaca is only moderately well collected in relation to the other Mexican states. Although a relatively large number of collectors have botanized in Oaxaca (we record over 100 from the literature and exsiccata), by far the majority of collections have been made along or near existing highways, roads (especially National Routes 125, 131, 175, 195, and 200), railroads, and near archeological sites. Many were made by botanical "transients", and zones that have been intensively and systematically collected are few indeed. Numerous inaccessible regions, poorly known botanically, exist as a result of the state's extremely accentuated relief. Considering that Oaxaca is probably Mexico's second richest state botanically (after Chiapas), with perhaps alomst 8,000 vascular plant species, much more collecting remains to be done.

A. Important Collectors

Botanical exploration of Oaxaca began in the 1790's when J. M. Mociño of the Sessé and Mociño expedition crossed the state on his way to Guatemala (McVaugh, 1977, 1980). Botanical activity increased substantially during the 19th century, beginning with W. F. von Karwinski (Hemsley, 1887), followed by F. Andrieux (Sousa S., 1979), K. A. Ehrenberg, H. Galeotti (McVaugh, 1978), C. Jürgensen, and T. Hartweg (McVaugh, 1970). Other 19th century collectors include: F. M. Liebmann, A. Ghiesbreght (Hemsley, 1887), C. C. Deam, W. Hough, J. N. Rose, C. G. Pringle, C. L. Smith (Rico A., 1980), and E. W. Nelson (Morton & Schultes, 1942; Goldman, 1951).

By the turn of the 20th century collecting in Oaxaca had become sporadic. Some visits were made by E. Seler (Loesenser, 1923), and C. A. Purpus (Sousa S., 1961); however, it was C. Conzatti who carried out most of the botanizing at this time. It is noteworthy that although the Mexican National Herbarium dates back to 1788 (Germán R. & Sousa S., 1980), Conzatti was apparently the first collector of Oaxacan plants to leave a complete set in Mexico (in his personal herbarium and at MEXU*). During the 1930's important collections were made by Y. Mexia (Bracelin, 1938), R. E. Schultes and B. P. Reko. From the 1940's to the 1960's significant and often extensive gatherings were made by G. Martínez C., T. MacDougall (a dedicated amateur), and by brigades hired by the Comisión de Dioscoreas of the Instituto Nacional de Investigaciones Forestales (INIF). INIF inventoried the northern lowlands around Tuxtepec. Collecting activity in Oaxaca flourished during the 1960's in response to increased botanical studies undertaken by both Mexican and non-Mexican scholars. A number of these include long term projects still in progress, such as the Leguminosae of Oaxaca (under the direction of M. Sousa S. at MEXU), the Cactaceae of Mexico (H. Bravo-Hollis (1978), MEXU), and the Pteridophytes of Oaxaca (J. Mickel, NY). These projects are generating thousands of specimens. Extensive collections have also been made by B. Hallberg, a long time Oaxaca resident, and by at least 20 other collectors active in Oaxaca during this decade.

Botanical activity in the state continued to increase during the 1970's. Gatherings were made by over 30 collectors during this decade, more than half by Mexican nationals or residents, including A. Delgado S., J. Rzedowski, D. Zizumbo V. and P. Colunga G. M. Collecting efforts by the Herbario Nacional (MEXU) in Oaxaca were coordinated by M. Sousa S. until 1980, when D. Lorence joined the staff and assumed the responsibility (until 1987). In 1981, MEXU received a three year Biological Collections grant from Consejo Nacional de Ciencia y Tecnología (CONACYT) which enabled a collecting brigade of two persons equipped with a new field vehicle to collect key areas for two weeks per month. Consequently, from 1981–1984 over 6,000 numbers (some 30,000 specimens) had been collected in Oaxaca with the eventual aim of producing a state flora. The most active collectors during the 1980's have been R. Cedillo T., M. Sousa S., O. Téllez V. and R. Torres C. In addition, intensive collecting of the following zones is in progress: the Tehuacán-Cuicatlán Valley (which includes the Cañon de Tomellín in NE Oaxaca) by F. Chiang C. (MEXU), the Sierra de Juárez by D. Lorence, the Río Uxpanapa and adjacent Chimalapas by T. Wendt (CHAPA), and parts of the Sierra de Tamazulapan in the Mixteca Alta by A. García M. (MEXU).

B. Important Herbaria

Some 32 herbaria currently exist in Mexico (Germán P. & Sousa S., 1980), the largest being the Herbario Nacional (MEXU) at the Instituto de Biología of the Universidad Nacional Autónoma de México, which houses approximately 400,000 collections, over 40% of the country's total. Our estimate of the vascular plant holdings from Oaxaca at MEXU is ca. 37,000. The exact figure will become available when the data are finally incorporated into a computerized data bank, a process now half complete. Mexico's second largest herbarium, the Escuela Nacional de Ciencias Biológicas at the Institute Politecnico (ENCB) with over 200,000 collections, has about 2500 specimens from Oaxaca (Rzedowski, 1976, and pers. comm.). Other Oaxacan collections may be found in the herbarium of the Escuela Nacional de Agricultura (CHAP, with ca. 1000), the herbarium of the Colegio de Postgraduados (CHAPA, with ca. 3,000), and in the herbarium of the Institute Nacional de Investigaciones Forestales (INIF, with ca. 2000), bringing the total number of Oaxacan collections in Mexico to approximately 45,500.

Significant holdings of Oaxacan plants outside of Mexico are distributed in various major world herbaria: AMES, BM, BR, C, F, GH, K, LE, MICH, MO, NY, P, UC and WI (see Lanjouw 1945 and Rzedowski 1976 for more details). Our rough estimate of the number of Oaxacan specimens held in institutions outside of Mexico is over 100,000 and probably about 130,000.

V. Publications on the Oaxacan Flora

A considerable amount of literature exists dealing with various facets of the botany of Oaxaca, varying in both qual-

---

* Unless otherwise stated, all acronyms for Herbaria may be found in Index Herbariorum (Holmgren et al., 1981)

ity and detail. Some publications represent no more than general observations, mere lists of species encountered and vegetation types seen on field trips or excursions (e.g., MacDougall, 1948; Sánchez M., 1958); nonetheless, some of these are quite important since they represent the only published accounts of the vegetation of poorly known areas (e.g., MacDougall, 1971, for the Chimalapa region). Others are somewhat more comprehensive, dealing with specific areas such as the northeast (Mullerried, 1948), Cuicatlán (Martínez, 1948), the Sierra de Juárez (Paray, 1951), or even the entire state (Bravo H., 1960; Conzatti, 1929).

More recently, a number of publications comprising much more detailed floristic or ecological studies of clearly defined areas, sometimes carried out in conjunction with extensive collecting, have appeared. Such studies exist for parts of Oaxaca lying within river systems such as the Río Balsas (Miranda, 1947) and Río Papaloapan (Miranda, 1948a, 1948b; Gómez-Pompa et al., 1964; Barreto V. & Hernández X., 1970), or completely within the state such as the Mixteca Alta (Cruz C. & Rzedowski, 1980; García M., 1983), the Sierra de Juárez (Ortiz C., 1970; Rzedowski & Palacios C., 1977), and the south coast (Conzatti, 1918, 1922; Fuentes A. et al., 1980). Ecological studies on secondary succession in northern Oaxaca near Tuxtepec were carried out by Sarukhán K. (1964) and Sousa S. (1964). A number of ethnobotanical studies exist for selected areas (Schmieder, 1938; Martínez-Alfaro, 1970; Cervantes S., 1979; Zizumbo V. & Colunga G. M., 1980; González O., 1982). Still other studies encompass distinct taxonomic groups such as the Pteridophytes (Mickel & Beitel, 1988), the Cactaceae (Bravo H., 1931; MacDougall, 1959), or life forms such as trees (Williams, 1938, 1939; Attolini, 1948). Finally, the excellent bibliography of Kaplan Langman (1964) collates most of the known literature published on the Mexican vegetation and angiosperm flora up to 1962, and should be consulted for further details.

VI. Completeness of the Floristic Inventory

As discussed previously in the section dealing with the magnitude of the floristic inventory, by far the greatest part of Oaxaca remains poorly collected, i.e., with less than 50 collections per 100 km^2. Briefly, these areas embrace most of the Mixteca Alta and Mixteca Baja including the Sierra Madre del Sur in southwestern and southern Oaxaca and the portion of the state adjacent to Guerrero, a large part of the Sierra Madre de Oaxaca including the state's northern tip and vast areas of the Sierra de Juárez, Sierra Mixe, Sierra de Huautla, Sierra de Villa Alta and Zempoaltépetl Massif, and most of the Isthmus of Tehuantepec.

Many botanists, both early and recent, followed existing routes and trails i.e., from Tehuacán (Puebla) south to Oaxaca City via Huajaupan de León and the Mixteca Alta; through the Mixteca Alta from Cuicatlán or Teposcolula southwestward to Putla de Guerrero or even to Pinotepa Nacional near the Pacific coast; from Oaxaca City southward to Puerto Angel on the coast via Miahuatlán; from Oaxaca City northward and eastward through the Sierra Madre de Oaxaca to the lowlands of the Chinantla (Choapan District) near the Veracruz border. Many areas bordering these routes, as well as the classical botanical and archeological sites mentioned before, can be considered relatively well collected, i.e., with 50–100 collections per 100 km^2.

Other relatively well collected regions include the Cañon de Tomellín in northeastern Oaxaca (Teotitlán District); parts of the Sierra de Juárez in the Districts of Ixtlán; Tuxtepec and Villa Alta; parts of the Mixteca Alta in the Teposcolula and Coixtlahuaca Districts; the vicinity of Cafetal Concordia north of Pochutla in the Pochutla District; the Valley of Oaxaca and vicinity; and the peninsula including San Mateo del Mar.

A. Areas High in Endemism or Yielding New and Important Distributions

Due to the incomplete state of knowledge regarding Oaxaca's floristic inventory, it is currently impossible to estimate the number of endemics occurring there. However, for the Leguminosae, one of the state's largest and most widely distributed families prevailing in many vegetation types, there are five major centers of endemism (M. Sousa S., pers. comm.): Salina Cruz and vicinity; the Sierra Madre del Sur; western Oaxaca around Putla and the Guerrero border, a part of the Río Balsas Basin shared with Guerrero and Michoacán; the Sierra de Juárez; the Cañon de Tomellín, an extension of the Tehuacán-Cuicatlán Desert flora. It is estimated that 6–7% of the Leguminosae species are endemic to Oaxaca (M. Sousa S., pers. comm.).

Another area known to harbor significant numbers of endemics is the Sierra de Juárez, one of the state's wettest and floristically richest areas. Our estimates of species known to be endemic to Oaxaca for three families occurring in this Sierra are: Asteraceae 0% (0 of 96, although some have narrow distributions including either Chiapas or Puebla); Rubiaceae 18% (10 of 55); Monimiaceae 40% (2 of 5). Most of these endemics are also restricted to the Sierra de Juárez or vicinity, although further collecting might extend their ranges. Bearing in mind the small sample size presented, these data would suggest that endemism varies between families.

Additional areas apparently high in endemism include other parts of the Sierra Madre de Oaxaca such as the wet Sierra Mixe and Zempoaltépetl Massif, the wet lowlands and mountains of the Chimalapas and Chinantla, the dry Cañon de Tomellín, and the arid areas in the Isthmus of Tehuantepec.

Data on endemism for a portion of the Sierra de Tamazulapan in the Mixteca Alta are presented in García M. (1983). This study encompasses 11 different vegetation types, many widespread in Oaxaca. Since the vegetational diversity of the study area is great, the resultant data may provide insights into the nature of endemism for the entire state. Of the nearly 500 vascular plant species occurring in the study area (27.4 km²), 25 (or 5%) are endemic to Oaxaca. More accurate figures on endemism must await the more thorough collection of Oaxaca and adjacent states.

Among the areas yielding new and important distributions in Oaxaca are the lowlands of the Río Uxpanapa zone, including the Chimalapas. Extensions of many lowland Mesoamerican genera, species and even families are being found there (T. Wendt, pers. comm.). The northernmost limits of many Mesoamerican montane taxa lie in the Sierra de Juárez, and it is thought to have been a refuge during the Pleistocene (Toledo, 1982). Moreover, it is likely that the Sierra Madre de Oaxaca in general was a Pleistocene refuge.

B. Extinction

Given the limited state of knowledge concerning plant populations in Oaxaca, it is difficult to estimate extinction rates. Destruction of large portions of a species' habitat may not necessarily result in its extinction if lands still remain where breeding populations can maintain themselves. Thus, large scale decimation of populations cannot be considered as synonymous with extinction of a species, although the genetic implications for such small and fragmented populations are obvious. In addition to population size, a multitude of other factors such as type of habitat, breeding systems, pollination and dispersal syndromes must be assessed and taken into account before generalizations can be made regarding extinction rates.

C. Taxa Well Collected in Oaxaca

Few genera or families have been collected extensively in Oaxaca. In fact, current efforts to produce a floristic treatment of the Leguminosae of Oaxaca by Mario Sousa S. and collaborators at MEXU seem to be unique as far as Angiosperms are concerned. The family as a whole has been well collected in Oaxaca, and revisions of several genera already exist in the form of student theses (e.g., *Acacia*, Rico A., 1980; *Leucaena*, Zárate P., 1982). Treatments for other genera are in preparation (e.g., *Astralagus, Crotalaria, Dalea, Desmodium, Erythrina, Lonchocarpus, Marina, Mimosa, Tephrosia*), and the overall Leguminosae treatment will be published in the near future.

John Mickel (NY) has collected pteridophytes extensively in Oaxaca and his floristic treatment of this group (with Joseph Beitel) for the state is in press.

Other families and genera that have been comparatively well collected in Oaxaca are listed in Appendix I. A number of these represent particular genera or infrageneric groups that have been recently revised or monographed (e.g., *Agave, Fuchsia, Quercus*); others constitute existing or planned family treatments for Mexico as a whole (e.g., Cactaceae, Malvaceae) or could be logical extensions of treatments for Flora Mesoamericana, Flora of Chiapas or Flora de Veracruz (e.g., Monimiaceae, Rubiaceae). In most cases, at least some field work and collecting was carried out by the scholars listed in Appendix I. The majority of remaining families and genera can be considered either not well or poorly collected.

D. Tropical Forest Areas in Oaxaca Threatened by Destruction or Conversion

In pre-Columbian times land use by the indigenous Mexican populations was both limited and rational. Man survived without large scale or total destruction of the plant communities harboring the numerous floral and faunal species upon which he depended for his survival. Since the Conquest, and particularly during the last century or so, forest destruction has increased dramatically. This is only partially attributable to local population growth, as the rate for Oaxaca is the lowest in Mexico. Much of the responsibility for extracting wood, clearing and cultivating the land lies in the hands of private and corporate interests. Sadly, unnecessary waste and destruction often result because the latest technology is either not used or is not properly applied.

Regions currently suffering high rates of deforestation and conversion include virtually all the lowland hot-humid zones formerly supporting Tropical Ombrophilous Lowland, Submontane and Riparian Forest. Most of these are being or have recently been converted into cattle pastureland, with occasional plantings of sugarcane, citrus or pineapple as well as subsistence agriculture. Remnants of intact forest still exist in the north around the Miguel Alemán Reservoir in the Tuxtepec District and in inaccessible areas of the northeast in the Choapan and Mixe Districts; more extensive areas also remain in the east around the Chimalapas in the northern Juchitán District.

Over the centuries, parts of the Tropical Ombrophilous Montane and Cloud Forest zones have been exploited for shifting agriculture, by the indigenous populations. Subsequent to the Conquest, coffee was introduced for cultivation. As a result, these zones together with those supporting Subtropical and Tropical Seasonal Evergreen Submontane Forest have largely been converted for coffee culture in the north, south and east. Conversion to coffee is total and unequivocal, although sometimes native canopy trees may be left to provide shade.

Lands supporting Evergreen Seasonal Lowland Forest, Evergreen Coniferous Forest, Coniferous Woodland, Drought Deciduous Lowland and Submontane Forest, and Drought Deciduous Woodland, have seemingly been subjected to shifting maize culture and mixed agriculture for centuries by indigenous people. A slash and burn technique is employed. Oftentimes fires intended to clear land burn out of control, and so can alter the dynamics of the pine-oak forest. After several years of planting, the fields may be left fallow or abandoned, permitting limited regeneration and imparting a patchwork appearance to the landscape. Thus, many of the aforementioned formations have probably suffered some disturbance during the past or are secondary in nature. As maize is often planted on extremely steep slopes, erosion is commonplace. So too, overgrazing, primarily by goats and sheep, can be a severe problem. Finally, zones in many of the valleys with permanent rivers have been cleared for irrigated agriculture.

Portions of the dry hills and lowlands of the Isthmus around Juchitán and Tehuantepec formerly supporting Drought Deciduous Forest or Thorn Forest have also been cleared for irrigated agriculture. Further clearing is being carried out in anticipation of a proposed reservoir planned for the wet Chimalapa lowlands. Planning for the construction of this reservoir seems to be proceeding after initial delays.

The only proclaimed nature reserves in Oaxaca are the Lagunas de Chacahua National Park of the western Pacific coast (Fuentes et al., 1980) and the Benito Juárez National Park in the southern Sierra de Juárez near Guelatao de Juárez. The former includes areas of Tropical Semideciduous Forest, Palm Savannah, Drought Deciduous Forest and Mangrove Forest. However, even here a village is present, and the infrastructure and means to maintain the reserve are scarcely adequate. The latter is much smaller and encompasses a small pine-oak forest and some Tropical Deciduous Forest plus a small lagoon.

Continued destruction of primary forest can therefore be expected for the wet lowland regions in particular, with the resulting elimination of large populations of numerous plant and animal species. Ultimately, forests are likely to survive only on lands unsuitable for agriculture or grazing such as on karstic limestone or steep slopes. Most alarming of all, a number of privately owned paper companies are deforesting vast areas throughout the Republic to satisfy Mexico's rapidly increasing demand for paper, cellulosic products and wood. These companies apparently practice uncontrolled clear-cutting with no subsequent replanting programs. Such practices pose the greatest threat to Oaxaca's remaining forests. Furthermore, attempts by some communities to develop more rational community-controlled forestry programs are hampered by the political influence of the large companies.

VII. Resources for Continued Floristic Inventory in Oaxaca

The most significant collecting effort in Oaxaca currently being undertaken by any academic institution is clearly that of the Herbario Nacional (MEXU) of the U.N.A.M. In 1981, the Instituto de Biología (U.N.A.M.) received a three year Biological Collections grant from CONACYT, providing MEXU funding for four field vehicles with collecting brigades. One of these was assigned to Oaxaca (as noted in sect. IV.A. above) and hence, from 1981–1984 over 6,000 numbers comprising about 30,000 specimens were collected there. Recently, a CONACYT grant enabling the Herbario Nacional to collect critical areas of Oaxaca for two years was approved. These funds were used to support several full-time local collectors for two years (1985–1986).

Systematic inventory of the Río Uxpanapa zone of Veracruz and Oaxaca in the Isthmus of Tehuantepec by Tom Wendt at the Escuela Nacional de Agricultura, Colegio de Postgraduados, Chapingo (CHAPA) constitutes another important program. This effort is aimed at producing a woody flora for the region. Intensive collecting of this poorly known area lying in the hot wet lowlands has been in progress since 1980 and includes the northern portion of Oaxaca's Juchitán District. This region also lies within the limits of Flora Mesoamericana, and Dr. Wendt recently obtained funding from CONACYT and CHAPA to collect the Chimalapa lowlands and mountains in Oaxaca and Veracruz for a three and one-half year period beginning in 1984. In addition to collecting there himself for eight months per year, this funding provides for three full time local collectors.

Other programs include that coordinated by Dr. Fernando Chiang C. (MEXU) aimed at inventorying and producing a flora of the Tehuacán-Cuicatlán Valley, a region which includes a small portion of northwestern Oaxaca, the Cañon de Tomellín. This effort, funded by the U.N.A.M., has been in progress for four years and will continue until the zone has been well inventoried.

A literature search of Oaxacan plants was begun in 1981 by Dr. Raúl Ibarra O. at the Instituto Tecnológico Regional de Oaxaca. This effort, however, has subsequently stagnated.

Apart from several localized ethnobotanical collections made by the Culturas Populares section of the Secretaría de Educación Pública de México (SEP), and a small reference herbarium maintained by the COTECOCA (SARH), efforts by non-academic Mexican governmental institutions towards a floristic inventory of Oaxaca are nil.

In terms of foreign programs, of which we are aware, there is an ethnobotanical project in the Sierra de Juárez being undertaken as the basis for a doctoral dissertation at the University of California, Berkeley by Mr. Gary Martin (pers. comm.). Since 1980, Mr. Martin has collected about 700 numbers, comprising some 600 species. Our botanical inventory of the Sierra de Juárez will of course be cooperative with and complementary to his, and will include an interchange of data, material and determinations.

Unlike some Latin American countries, Mexico has a well established system for promotion and execution of scientific research. Public institutions such as CONACYT, Subsecretaría de Desarrollo Urbano y Ecología (SEDUE), SARH, and SEP, coupled with the many academic institutions provide a sound framework for continued research in Mexico. Ample opportunities exist for bilateral cooperation between Mexican and foreign institutions, both in floristics (e.g., Flora Mesoamericana) as well as in other disciplines.

VIII. Suitability of Past, Ongoing and Proposed Floristic Inventory in Oaxaca

Floristic inventory in Oaxaca prior to the 1970's was generally random if not sporadic. At best, collecting focused upon specific taxonomic groups, or else comprised ecological or ethnobotanical studies of small areas.

Ongoing efforts at MEXU and CHAPA are aimed at investigating and inventorying critical areas, i.e., those poorly explored floristically, those high in species diversity and endemism, and those threatened by destruction. These efforts are both appropriate and suitable and, it is hoped, will continue until their completion.

IX. Additional Resources Required

Mexico undoubtedly possesses the necessary academic institutions, including herbaria and personnel, to insure an adequate floristic inventory of Oaxaca's forests. In addition, the Herbario Nacional and other institutions are training additional personnel both locally and abroad, as well as expanding existing facilities. For example, a new museum building is planned at the U.N.A.M. to supplement the already overcrowded Herbario Nacional.

Not surprisingly, the primary limiting factor towards early completion of an inventory is funding. As a result of Mexico's recent high inflation rate, with prices often doubling every year, budget constraints on field work can be severe.

The cost of carrying out an adequate floristic inventory in a region as large and diverse as Oaxaca is considerable. Apart from an initial investment of approx. $US 14,000 for a vehicle, it costs approx. $US 16,600 to equip and maintain a team of two collectors for a year, including salaries, gasoline, food and lodging, and label typing (Appendix II). This is a considerable sum for a Mexican university budget. In addition, this cost in pesos nearly doubles each year due to inflation. Although it is much less expensive to maintain resident collecting brigades, there are drawbacks to this. Resident collectors may lack botanical expertise and incentive, and therefore the resultant collections may tend to become repetitive. In our opinion, it is most advantageous to maintain mobile collecting brigades that can enter key areas, establish a base there, and collect the area thoroughly for several weeks or a month. An additional three year collecting program with two such brigades concentrating on critical areas should result in an inventory sufficiently adequate to initiate work on a flora of Oaxaca.

X. Acknowledgments

We sincerely thank Dawn Frame, Tom Wendt, Mario Sousa S., Fernando Chiang, Lourdes Rico A., and Oswaldo Téllez V. for critically reviewing the manuscript; and Dr. J. Rzedowski for furnishing data on Oaxacan collections at ENCB.

XI. Literature Cited

Alvarez, Jr., M. 1961. Provincias fisiográficas de la República Mexicana. Bol. Soc. Geol. Méx. **24(2):** 1–15.

Attolini, J. 1948. Las especies forestales de la cuenca del Papaloapan. Rev. Econ. **11:** 29–39.

Barreto V., F. & A. H. Hernández X. 1970. Relación suelo-vegetación en la región de Tuxtepec, Oax. In: Contribuciones al estudio ecológico de las zonas cálido-húmedas de México. Bol. Esp. Inst. Nal. Invest. For. México. **6:** 63–113.

Bracelin, N. F. 1938. Ynes Mexia. Madroño **4:** 273–275.

Bravo H., H. 1931. Cactáceas del Valle de Oaxaca. Anales Inst. Biol. Univ. Nac. México. **2:** 117–126.

―――. 1954. Iconografía de las Cactáceas Mexicanas (tercera serie). Cactáceas de las Mixtecas Atlas. Anales Inst. Biol. Univ. Nac. México. **25:** 473–552.

―――. 1960. Algunos datos acerca de la vegetación del Estado de Oaxaca. Revista Méx. Estud. Antropol. **16:** 31–47.

―――. 1978. Las Cactáceas de México, Vol. 1. U.N.A.M. México, D.F. 743 pp.

Cervantes S., L. M. 1979. Plantas medicinales del Distrito de Octlán en la región de los valles centrales de Oaxaca. Bachelor's Thesis, Facultad de ciencias. U.N.A.M. México, D.F. 301 pp.

Conzatti, C. 1918. Exploración botánica por la costa meridional de Oaxaca. Bol. Dirección Est. Biól. México. **2(3): 309**–325.

―――. 1922. Una expedición botánica a la costa oaxaqueña del suroeste. Imprenta del Gobierno del Estado de Oaxaca, Oaxaca. 33 pp.

―――. 1929. Las regiones botánico-geográficas del estado de Oaxaca. In: Proc. Internat. Congr. Pl. Sci. (Ithaca) **1:** 525–539.

COTECOCA. 1980. Oaxaca. Impreso por las memorias de COTECOCA-SARH. V. 1, 2, with map of vegetation, scale of 1:500 000. 295 pp. Unpublished.

Cruz C., R. & J. Rzedowski. 1980. Vegetación de la cuenca del río Tepelmeme, Alta Mixteca, Estado de Oaxaca (México). Anales Esc. Nac. Cienc. Biól. Méx. **22:** 19–84.

Departmento Cartográfico Militar de la República Mexicana. 1959.1972. (For Oaxaca there are 37 topographic maps with legend, at a scale of 1:100 000. Sheets: 14P-c(1–3), 14Q-c(1–3), 14Q-i(10–12), 14Q-k(3,6,9,12), 14Q-l(1–12), 15Q-g(1,4,6,10), 15Q-j(7–12).

Dirección de Estudios del Territorio Nacional. 1979. Descripción de la leyenda de la Carta Edafológica DETENAL. Secretaría de Programación y Presupuesto. México, D.F. 104 pp.

Ferrusquía-Villafranca, I. 1976. Estudios geológico paleontológicos en la región Mixteca. I. Geología del área Tamazulapan-Teposcolula-Yanhuitlán, Mixteca Alta, Estado de Oaxaca, México. Bol. Inst. Geol. **97:** 1–160.

Flores, D. A. 1974. Los suelos de la República Mexicana. In: El escenario geográfico; Recursos Naturales. Inst. Nal. Antropol. Hist. México. Pages 9–108. México: panorama histórico y cultural II.

Flores M., G., J. Jiménez L., X. Madrigal S., F. Moncayo R. & F. Takaki T. 1972. Mapa y descripciones de los tipos de vegetación de la República Mexicana. Secretaría de Agricultura y Recursos Hidráulicos. México, D.F.

Fuentes A., L. et al. 1980. Estudio interdiciplinario sobre la conservación y el aprovechamiento de un Parque Nacional (Lagunas de Chacahua, Oaxaca) con una población humana establecida. Comisión de Biologías de Campo. Facultad de Ciencias, U.N.A.M. México, D.F. 465 pp. Unpublished.

García, E. 1981. Modificaciones al Sistema de Clasificación Climática de Köppen (para adaptarlo a las condiciones de la República Mexicana). 3rd edition. Offset Larios, México, D.F. 252 pp.

García, E. & Z. Falcón de Gyves. 1977. Nuevo Atlas Porrúa de la República Mexicana. 3rd edition. Editorial Porrúa, México, D.F. 197 pp.

García M., A. J. 1983. Estudio ecológico-florístico de una porción de la Sierra de Tamazulapan, Distrito de Teposolula, Oaxaca, México. Bachelor's Thesis, Facultad de Ciencias, U.N.A.M. México. 112 pp.

Germán R., M. T. & M. Sousa S. 1980. Herbario Nacional de México-MEXU. Instituto de Biología, U.N.A.M. México, D.F. 49 pp.

Goldman, E. W. 1951. Biological investigations in México. Smithsonian Misc. Collect. 115: 1–476.

Gómez-Pompa, A., L. Hernández P., & M. Sousa S. 1964. Estudio fitoecológico de la cuenca intermedia del río Papaloapan. Bol. Esp. Inst. Nal. Invest. For. México. 3: 37–90.

González O., S. 1982. Contribución a la Etnobotánica de la costa de Oaxaca entre los puertos de Salina Cruz y Puerto Angel. Bachelor's Thesis, Facultad de Ciencias, U.N.A.M. México, D.F. 76 pp.

Hemsley, W. B. 1887. A sketch of the history of the botanical exploration of Mexico and Central America. *In* Goldman & Salvin (eds.), Biologia Centrali Americana 4: 117–137.

Holmgren, P.K., W. Keuken & E. K. Schofield. 1981. Index Herbariorum. Part I. 7th edition. Bohn, Scheltema & Holkema, Utrecht/Antwerpen; Dr. W. Junk B. V., The Hague/Boston.

Instituto de Geografía, U.N.A.M. 1970. Hojas climáticas. Comisión de Estudios del Territorio Nacional. México, D.F. (the following 1:500 000 scale sheets include Oaxaca: 14P-II, San Pedro Pochutla; 14Q-VIII, Oaxaca; 15Q-VII, Tuxtla Gutiérrez; 14Q-VI, Veracruz; 15Q-V, Coatzacoalcos).

Kaplan, Langman, I. 1964. A selected guide to the literature on the flowering plants of Mexico. University of Pennsylvania Press. Philadelphia, Pa., 1015 pp.

Lanjouw, J. 1945. On the location of botanical collections from Central and South America. Pages 228, 229 *in* Plants and plant science in Latin America. Chronica Botanica Co., Waltham, Mass.

Leopold, A. 1950. Vegetation zones of Mexico. Ecology 31: 507–518.

Loesener T. 1923. Edward Seler: Nachruf. Verh. Bot. Vereins Prov. Brandenburg 65: 78–83.

López-Ramos, E. 1974. Carta Geológica del Estado de Oaxaca. 2nd. edition. Comité de la Carta Geológica de la República Mexicana. Scale 1:500 000.

———. 1981. Geología de México. Vol. 2, 3. Edición Escolar, México, D.F. 549 pp.

MacDougall, T. B. 1948. Afoot in Mexico. J. New York Bot. Gard. 49: 153–163.

———. 1959. Lugares nativos de algunas Cactáceas y suculentas de Oaxaca y Chiapas. Cact. Suc. Méx. 4(1): 12–16.

———. 1971. The Chima wilderness. Explorer 49(2): 86–103.

Martínez, M. 1945. Las Pináceas Mexicanas. Anales Inst. Biol. Univ. Nac. México. 1: 1–302.

———. 1948. Algunas observaciones relativas a la flora de Cuicatlán, Oaxaca. Anales Inst. Biol. Univ. Nac. México 19(2): 365–391.

Martínez-Alfaro, M. A. 1970. Ecología humana del ejido B. Juárez o Sebastopol, Tuxtepec, Oax. Bol. Esp. Inst. Nal. Invest. For. México. 7: 1–157.

McVaugh, R. 1970. Introduction to the facsimile reprint of George Bentham's "Plantae Hartwegianae". Lehre. J. Cramer. Pages 1–102.

———. 1977. The botanical results of the Sessé and Mociño expedition (1787–1803). I. Summary of excursions and travels. Contr. Univ. Michigan Herb. 11: 97–195.

———. 1978. Galeotti's botanical work in Mexico: the numbering of his collections and a brief itinerary. Contr. Univ. Michigan Herb. 11: 292–297.

———. 1980. The botanical results of the Sessé and Mociño expedition(1787–1803). II. The Icones Florae Mexicanae. Contr. Univ. Michigan Herb. 14: 99–140.

Mickel, J. T. & J. M. Beitel. 1988. Pteridophyte flora of Oaxaca, Mexico. Mem. New York Bot. Gard. 46: 1–568.

Miranda, F. 1947. Estudios sobre la vegetación de México. V. Rasgos de la vegetación en la cuenca del Río de las Balsas. Revista Soc. Mex Hist. Nat. 8: 95–127.

———. 1948a. Observaciones botánicas en la región de Tuxtepec, Oax. con notas sobre plantas útiles. Anales Inst. Biol. Univ. Nac. México. 19: 105–136.

———. 1948b. Datos sobre la vegetación de la cuenca alta del Papaloapan. Anales Inst. Biol. Univ. Nac. México 19: 333–364.

——— & E. Hernández X. 1963. Los tipos de vegetación de México y su clasificación. Bol. Soc. Bot. México. 28: 29–179.

Morton, C. V. & R. E. Schultes. 1942. Localidades visitadas y rutas recorridas por E. W. Nelson en el estado de Oaxaca. Anales Inst. Biol. Univ. Nac. México 13: 47–51.

Mosiño, P. A. 1974. Los climas de la Repúblic Mexicana. Pages 57–172 *in* El escenario geográfico; Introducción ecológica. Inst. Nal. Antropol. Hist. México. México: Panorama histórico y cultural I.

Mullleried, F. K. G. 1948. La naturaleza en el noreste de Oaxaca. Revista Soc. Mex. Hist. Nat. 9: 137–143.

Ortíz C., D. 1970. Contribución al conocimiento de la flora de la Sierra de Juárez. Bachelor's Thesis, Facultad de Ciencias, U.N.A.M. México, D.F. 33 pp., Illus.

Paray, L. 1951. Exploraciones en la Sierra de Juárez. Bol. Soc. Bot. México 13: 4–10.

Pennington, T. D. & J. Sarukhán. 1968. Arboles Tropicales de México. I.N.I.F., S.A.G., México, D.F.

Rico A., M. de L. 1980. El género Acacia (Leguminosae) en Oaxaca. Bachelor's Thesis, Facultad de Ciencias, U.N.A.M. México, D.F. 116 pp., Maps., Illus.

Rzedowski, J. 1976. Catálogo de los herbarios institucionales mexicanos. Sociedad Botánica de México, México, D.F. 74 pp.

———. 1978. Vegetación de México. Editorial Limusa. México, D.F. 423 pp.

——— & R. Palacios C. 1977. El bosque de *Engelhardtia* (*Oreomunnea*) *mexicana* en la región de la Chinantla (Oaxaca, México); una reliquia del cenozoico. Bol. Soc. Bot. México 36: 93–123.

Sanchez M., H. 1958. Relación de una excursión a Oaxaca. Cact. Suc. Mex. 3(2): 36–39.

Sarukhán K., J. 1964. Estudio sucesional de un área talada en Tuxtepec, Oax. Bot. Esp. Inst. Nal. Invest. For. México 3: 107–172.

———. 1968. Los tipos de vegetación arbórea de la zona cálidohúmeda de México. Pages 3–46 *in* T. D. Pennington & J. Sarukhán K. Manual para la identificación de los principales árboles tropicales de México. INIF & FAO. México, D.F.

Schmieder, O. 1930. The settlements of the Zapotec and Mije Indians, state of Oaxaca, México. Univ. Cal. Publ. Geog. 4: 1–184.

Secretaría de Agricultura y Recursos Hidráulicos (SARH). 1979. Cartografía Sinóptica: Uso actual del suelo. Estado de Oaxaca. (two sheets, scale 1: 500 000).

Secretaría de Programación y Presupuesto (SPP). 1981. Atlas Nacional del Medio Físico. Talleres gráficos de la Dirección General de Geografía del Territorio Nacional. México, D.F. 224 pp.

Smith, A. C. & I. M. Johnston. 1945. A phytogeographical sketch of Latin America. In: Plants and Plant Science in Latin America, pp. 11–18. Chronica Botanica Co., Waltham, Mass.

Soto, M. C. & E. Jáuregui O. 1965. Isotermas extremas e índice de aridez en la República Mexicana. Instituto de Geografía, U.N.A.M. México, D.F. 120 pp.

Sousa S., M. 1961. Las colecciones botánicas de C. A. Purpus en México, período 1898–1925. Univ. Cal. Publ. Bot. **51**: 1–36.

———. 1964. Estudio de la vegetación secundaria en la región de Tuxtepec, Oaxaca. Bot. Esp. Inst. Nal. Invest. For. México **3**: 91–150.

———. 1979. Itinerario botánico de G. Andrieux en México. Taxon **28**: 97–102.

Tamayo L., J. 1962. Geografía General de México. Vol. 1. 2nd edition. Instituto Mexicano de Investigaciones Económicas. México, D. F. 562 pp.

Toledo M., V. M. 1982. Pleistocene changes of vegetation in tropical Mexico. Pages 9–111 *in* G. T. Prance (ed.), Biological diversification in the tropics. Columbia Univ. Press, New York.

UNESCO. 1974. Tentative physiognomic-ecological classification of plant formations of the earth (revised from Ellenberg and Mueller-Dombois). Pages 466–488 *in* Mueller-Dombois, D. and Ellenberg, H. (eds.), Aims and methods of vegetation ecology. John Wiley & Sons, New York.

Williams, Ll. 1938. Forest trees of the Isthmus of Tehuantepec, Mexico. Trop. Woods. **53**: 1–11.

———. 1939. Arboles y arbustos del Istmo de Tehuantepec, México. Lilloa 4: 137–171.

Zárate P., S. 1982. Las especies de *Leucaena* Benth. de Oaxaca con notas sobre la sistemática del género para México. Bachelor's Thesis, Facultad de Ciencias, U.N.A.M. México, D. F. 167 pp, Illus.

Zizumbo V., D. & M. P. Colunga G. 1980. La utilización de los recursos naturales entre los Huaves de San Mateo del Mar, Oaxaca. Bachelor's Thesis, Facultad de Ciencias, U.N.A.M. México, D. F. 367 pp. Illus.

XII. Appendices

Appendix I

List of families and genera relatively well collected in Oaxaca, scholars working on them, and herbaria with most important exsiccata. Scholar is usually but not always a principal collector of the taxon. *signifies deceased

Family	Genus	Scholar and Institution	Important Exsiccata
Agavaceae	*Agave*	H. S. Gentry (DES)	DES, MEXU
	Beschorneria	A. García M. (MEXU)	MEXU
Burseraceae	*Bursera*	J. Rzedowski et al. (ENCB)	ENCB, MEXU
Cactaceae	all genera	H. Bravo (MEXU)	MEXU
Compositae	various genera	A. Cronquist (NY)	NY
		J. Villaseñor (MEXU)	MEXU
		B. L. Turner et al. (TEX)	TEX
Dioscoreaceae	*Dioscorea*	Comisión de Dioscoreas	INIF
		B. Schubert (GH)	GH
		O. Téllez V. (MEXU)	MEXU
Ericaceae	*Gaultheria*	C. M. Corcoran (WI)	EI, MEXU
Fagaceae	*Quercus*	C. H. Muller (TEX)	MEXU
		M. Martínez* (MEXU)	MEXU
Iridaceae	*Tigridia*	E. Molseed* (UC)	UC, MEXU
Labiatae	*Salvia*	T. P. Ramamoorthy (MEXU)	MEXU
Leguminosae	all genera	M. Sousa S. et al. (MEXU)	MEXU
Liliaceae	*Schoenocaulon*	D. Frame (NY)	MEXU, NY, GH
Malvaceae	all genera	P. Fryxell (TAMU)	MEXU, TAMU
Monimiaceae	*Mollinedia, Siparuna*	D. Lorence (PTBG)	MEXU
Onagraceae	*Fuschsia*	D. Breedlove (CAS)	CAS, MEXU
		P. Raven (MO)	MEXU, MO
Pinaceae	*Abies, Cupressus Juniperus, Pinus*	M. Martínez* (MEXU)	MEXU
		T. Zanoni (NY)	MEXU
Polygalaceae	various	T. Wendt (CHAPA)	CHAPA, MEXU
Rubiaceae	various	D. Lorence (PTBG)	MEXU, MO
Pteridophyta	various	J. Mickel (NY)	MEXU, NY

Appendix II

Cost of equipping and maintaining a mobile plant collecting brigade of two persons in Mexico for one year at a conversion rate of 1 US dollar = 1200 pesos (as of June 1987) and assuming 200 days field collecting.

Item	Cost Mexican Pesos	US Dollars
Salary 2 collectors for a year	8,074,800	6,729
Living allowance (200 days)	3,364,800	2,804
Gasoline (average 10,000 pesos per day × 200)	2,000,000	1,667
Label typing	1,752,000	1,460
Supplies and expendables	1,122,000	935
Field equipment	1,682,400	1,402
Field vehicle (Jeep Wagoneer with 4 wheel drive)	16,822,800	14,019
TOTAL	34,818,800	29,016

Appendix III

Vegetation types of Oaxaca; using Classification System of UNESCO (1974)
(Adapted from COTECOCA, 1980)

Code	Vegetation Type (original)
IA1a	Tropical Ombrophilous Forest, Lowland
IA1b	Tropical Ombrophilous Forest, Submontane
IA1f(1)	Tropical Ombrophilous Forest, Riparian
IA1c	Tropical Ombrophilous Forest, Montane
IA1e(1)	Tropical Ombrophilous Cloud Forest (Broad-leaved)
IA2a	Tropical (Subtropical) Evergreen Seasonal Lowland Forest
IA2b(1)	Tropical (Subtropical) Evergreen Submontane Forest (Broad-leaved)
IA3a	Tropical (Subtropical) Semideciduous Lowland Forest
IA5	Mangrove Forest
IA9b	Evergreen (Nongiant) Conifer Forest with Rounded Crowns
IA9c	Evergreen (Nongiant) Conifer Forest with Conical Crowns
IB1a	Drought Deciduous Lowland (and Submontane) Forest
IC2a	Mixed Deciduous-Evergreen Thorn Forest
IC2b	Purely Deciduous Thorn Forest
IIA2a(1)	Evergreen Coniferous Woodland with Rounded Crowns and Sclerophyllous Understory
IIA2b	Evergreen Coniferous Woodland with Conical Crowns Prevailing
IIB1	Drought Deciduous Woodland
IIIC2(b)	Deciduous Subdesert Scrubland with Succulents

Transisthmic Mexico (Campeche, Chiapas, Quintana Roo, Tabasco and Yucatán)

Jerzy Rzedowski and
Graciela Calderón de Rzedowski

Contents

- I. Description of the Region 271
  - A. Introduction 271
  - B. Geographical Extent and Topography 271
  - C. Climate 271
  - D. Population 271
- II. Vegetation and Vegetation Maps 272
- III. A Synthetic Review of the Vegetation 273
  - A. Tropical Evergreen Forest (IA1a, IA1b, IA1f, IA1g(1)) 273
  - B. Tropical Semi-evergreen Forest (IA2a) 273
  - C. Tropical Subdeciduous Forest (IA3a, IA3b) 273
  - D. Tropical Deciduous Forest (IB1a(1), IB1a(2)) 273
  - E. Thorn Forest (IC2a, IC2b) 273
  - F. Palm Forest 273
  - G. Montane Mesophyllous Forest (IA1c, IA1e, IA2b) 273
  - H. Riparian Forest 274
  - I. Oak Forest 274
  - J. Coniferous Forest (IA9b, IA9c) 274
  - K. Mangrove Forest (IA5) 274
  - L. Non-forest Vegetation 274
- IV. Floristic Inventory 274
  - A. Important Collectors 274
  - B. Important Herbaria 275
  - C. Best Explored and Collected Areas 275
  - D. Publications 275
  - E. Endemism 276
  - F. Extinction 276
- V. Academic and Governmental Institutions Involved in Floristic Inventory 277
  - A. Local Institutions 277
  - B. Other Mexican Institutions 277
  - C. Main Foreign Institutions 277
- VI. Future of Inventory 277
- VII. Acknowledgments 277
- VIII. Literature Cited 278

I. Description of the Region

A. Introduction

Tropical forests are characteristic of about a third of the territory of Mexico, reaching along the coastal lowlands of the northern states of Baja California Sur, Sonora, Chihuahua, and Tamaulipas. However, their best representation in terms of area coverage, as well as in structural and floristic complexity and diversity, can be found in the southern part of the country, especially in the portion of Mexico situated east of the Isthmus of Tehuantepec, considered by many authors to be a part of Central America.

Transisthmic Mexico lies entirely within the intertropical belt and is (or rather was) very largely a densely forested region. Nevertheless, due to the presence of high mountains in the area, not all the forests are truly tropical in nature. More than a third of the State of Chiapas is covered with woods dominated by conifers and oaks. Similarly, along many rivers the riparian or "gallery forest," composed of willows, bold cypresses, sycamores, ashes and poplars, often descends into the hot lowlands. These forests and woodlands, although not identical with analogous types from higher latitudes, often lack most of the "tropical" features.

As is discussed further on, several such communities do not fit adequately into any of the categories of Mueller-Dombois and Ellenberg's (1974) vegetation classification, which was recommended to be followed in this book. They are all included, however, in the following synthesis.

B. Geographical Extent and Topography

The joint surface of the Mexican states of Chiapas, Tabasco, Campeche, Quintana Roo and Yucatan amounts to about 240,000 km². The area includes the westernmost segment of the Central American core, merging with the Isthmus of Tehuantepec, plus the greater part of a long and voluminous protrusion toward the north—the Yucatan Peninsula.

From a geomorphological point of view, three very different portions can be distinguished: (1) the almost arrheic and mostly flat and shallow-soiled, low-leveled limestone plateau of the Peninsula; (2) the alluvial, flat, often ill-drained Tabascan coastal plain, where the Grijalva and Usumacinta rivers form a complex fluvial system, with a large number of lakes, meanders and extensive marshes; (3) the Chiapas complex, a prevailingly mountainous region, in which several physiographic subdivisions have to be recognized.

Two main, parallel mountain chains cross the state of Chiapas from the northwest to the southeast; the northernmost one, the Central Plateau, mostly 1,200–2,200(2,900) m high, is largely formed by limestone; the southern one, the Sierra Madre, mostly 1,500–3,000(4,050) m high, is chiefly composed of metamorphic and granitic rocks, but with a short volcanic segment near the Guatemalan border. An elongated, deep (altitudes 400–800 m) valley, called the Central Depression, separates these two chains, and toward the Pacific Ocean a narrow (20–35 km wide) strip of the Pacific Coastal Plain is situated. North and east of the Central Plateau, lower (400–1500 m) but often abrupt, mostly limestone and sandstone mountain areas extend toward the Tabascan Plain and toward the Guatemalan Petén flats; these are the Northern and Eastern Highlands.

C. Climate

The diversity of climates can be schematically outlined in terms of Koeppen's classification. A very wet area (Af) extends along a belt through much of Tabasco and extreme northern and northeastern Chiapas. Most of the Yucatan Peninsula shows an Aw climate, but a narrow northern coastal strip is of BS type. In Chiapas the Am and Aw categories prevail, but the higher mountains register Cw (mainly in the Central Plateau) and Cf (mainly in the Sierra Madre) types. The drier parts of Chiapas are the Central Depression and the Pacific Coastal Plain.

D. Population

The transisthmic part of Mexico was the site of the development of important Pre-columbian civilizations. The Mayan culture flourished for about 20 centuries (between 1000 B.C. and 1000 A.D.) and some authors (Morley, 1953) estimated that the population of the Peninsula may have reached as many as 50 million inhabitants in some epochs. This is probably a much exaggerated figure, but doubtless the population density in many areas was much higher than at present.

The reasons for the decline of the Mayan empire are unknown; several hypotheses, however, emphasize the exhaustion of natural resources as a possible cause.

At the beginning of the twentieth century the area had poor communications with the rest of Mexico and was scarcely populated, the only important centers being the northern half of the Yucatan Peninsula, central Tabasco and central Chiapas. In the last 30 years a strong effort has been made to integrate and incorporate the area with the rest of the country economically. Much new colonization has been promoted, especially in Campeche and Quintana Roo, but also in some areas of Tabasco and Chiapas. The older pattern of semi-nomadic or permanent agriculture and forest exploitation is now being substituted by a grazing and petroleum oriented economy, with the growing importance of tourist centers.

Table I shows some figures from the last population census. As can be seen, between 1950 and 1980, the number of people doubled in Chiapas and Yucatan, tripled in Tabasco, increased 3.5 times in Campeche and 8.5 times in Quintana Roo. Tabasco and Chiapas retain a large percentage of rural population, and the population density in Tabasco is considerably higher than elsewhere. Important Indian-inhabited areas still exist in Chiapas and Yucatan.

Table I
Some population data for the area.

States	Total Population 1950	Total Population 1980	Rural Population 1980	Population density/km² 1980	Non-Spanish Speaking Population 1980
Campeche	122,087	420,553	198,203	8.25	7,453
Chiapas	904,588	2,084,217	1,661,076	28.21	211,429
Quintana Roo	26,975	225,985	116,959	4.47	12,819
Tabasco	362,195	1,062,961	772,450	41.95	5,860
Yucatan	518,798	1,063,733	490,662	27.62	71,511
	1,934,643	4,857,449	3,239,350	20.26	309,072

II. Vegetation and Vegetation Maps

Although Rovirosa (1909: 6–33) in his *Pteridografia del sur de Mexico* outlined some important features of the plant cover of Tabasco and adjacent Chiapas, the first true vegetation study in the area was done by Millspaugh (1916) for the islands of the Alacran Reef, off the coast of the Yucatan Peninsula. Later, Bequaert (1933), Lundell (1934, 1938), Bartlett (1935), as well as Hernández Corzo (1950) and Bravo (1955), provided some descriptions of the mainland vegetation of the peninsular area. Alvarez del Villar (1952) presented a geobotanical sketch of the state of Chiapas.

Doubtless, most of the actual knowledge on the vegetation of transisthmic Mexico is due to the research of Miranda, who published in 1952–1953 a global account for Chiapas and in 1958 for the Yucatan Peninsula. The same author published additional information in 1942, 1957 and 1961. Miranda recognized and described 12 vegetation types for Chiapas. His subdivisions for the Peninsula, however, are much finer—26 communities are distinguished.

West (1956) wrote on the plant cover of the Tabascan lowlands. Puig (1972) devoted his study to the savannas of the Huimanguillo region in the western part of the same state. Somewhat later, Quiroz (1977) investigated the soil-vegetation relationships in the Balancán-Tenosique area of eastern Tabasco. Finally, López Mendoza (1980) described the vegetation of the entire state. Tabascan mangroves and other halophytic communities were studied by Thom (1967) and by López Portillo (1982).

Vázquez Soto (1963) refined the information on the vegetation of Campeche, while Sarukhán (1968, see also Sarukhán & Hernández Xolocotzi, 1968) essayed a synthetic evaluation of the lowland *Terminalia amazonia* evergreen forest in southeastern Mexico.

Breedlove (1973, 1981) offered a revised classification of the vegetation of Chiapas, recognizing 19 phytogeographic areas. Pérez and Sarukhán (1970) published on the plant cover of the Pichucalco region in the northern part of the state, while Zuill & Lathrop (1975) studied the montane mesophyllous forest in the same general area. Somewhat later Calzada and Valdivia (1979), as well as Meave (1983), contributed to the knowledge of the tropical evergreen forest of the "Selva Lacandona" of northeastern Chiapas.

Bonet and Rzedowski (1962) and Flores (1984a) reported on the changes of plant cover of the Alacran Reef. Sauer (1967) and Espejel (1984, 1986a, 1986b) studied the vegetation of the coastal dunes of the Peninsula's shores, while Rico Gray (1982) worked on the coastal zone of northwestern Campeche. Finally, Flores (1984b) described the vegetation of all the islands surrounding the Yucatan Peninsula.

Chavelas Pólito (1968) described the vegetation of the Forest Research Station in Escárcega, Campeche.

Concerning maps, the first thorough regional vegetation charts were published by Miranda (1952–1953, 1958) for Chiapas (scale 1:1,200,000) and the Yucatan Peninsula (scale 1:4,000,000).

Vázquez Soto's (1963) paper on the vegetation of Campeche includes a map on a 1:1,000,000 scale.

Flores Mata et al. (1971) prepared a vegetation map of the Republic of Mexico on a 1:2,000,000 scale.

López Mendoza's (1980) vegetation map (scale 1:600,000) includes Tabasco and some adjoining parts of Chiapas.

Forest inventory maps of variable scales and qualities were published for the entire area (Anonymous, 1976a, 1976b, 1985a, 1985b, 1985c).

The Instituto Nacional de Estadística, Geografía e Informática (formerly Comisión Nacional de Estudios del Territorio Nacional and Dirección General de Estudios del Territorio Nacional) has been publishing detailed vegetation maps of Mexico on different scales, most of them under the name of "carta de uso del suelo." Transisthmic Mexico is covered completely by maps on a 1:1,000,000 scale (Anonymous, 1981) and almost completely by maps on a 1:250,000 scale (Anonymous, 1982–1985).

The Secretaría de Agricultura y Recursos Hidráulicos published its *Cartografía sinóptica—uso actual del suelo* (actually vegetation maps, chiefly based on satellite images, mainly on a 1:500,000 scale) for all the states of Mexico. The five transisthmic states are covered by Anonymous (1976–1981).

III. A Synthetic Review of the Vegetation

The following is a very brief synthesis of the information contained in the contributions enumerated in the preceding section. In most instances reference is made only to original or potential vegetation. A very large percentage of this vegetation does not exist any more in the area. Whenever possible, reference is made to Mueller-Dombois and Ellenberg's (1974) classification, indicating their symbols in parentheses; in many cases the assigned categories are approximate only. The following important plant communities can be distinguished in the areas:

A. TROPICAL EVERGREEN FOREST (IA1a, IA1b, IA1f, IA1g(1))

This forest once covered about a fourth of the state of Chiapas, especially toward the north, and a very large part of Tabasco, extending into the southwestern corner of Campeche and ranging from sea level to approximately 1,400 m in altitude. Many hundreds of tree species participate in this community, some of the more common ones of the upper canopy are: *Alchornea latifolia, Aspidosperma megalocarpon, Bernoullia flammea, Blepharidium mexicanum, Bravaisia integerrima, Brosimum alicastrum, Calophyllum brasiliense, Dialium guianense, Dussia cuscatlantica, Erblichia xylocarpa, Licania platypus, Pithecellobium arboreum, Poulsenia armata, Pouteria campechiana, Pseudomedia oxyphyllaria, Quararibea funebris, Sloanea ampla, Sterculia apetala, S. mexicana, Swietenia macrophylla, Talauma mexicana, Terminalia amazonia, T. obovata, Ulmus mexicana, Vatairea lundellii, Vochysia hondurensis, Zinowiewia integerrima.*

B. TROPICAL SEMI-EVERGREEN FOREST (IA2a)

Forests of this type covered much of Quintana Roo and a very large part of Campeche and also existed in some parts of Tabasco and Chiapas; mainly at altitudes below 500 m. The most abundant tall trees are: *Brosimum alicastrum* and *Manilkara zapota*. Other frequent arboreal elements are: *Alseis yucatanensis, Bucida buceras, Cedrela odorata, Cryosophila argentea, Dendropanax arboreus, Exothea diphylla, Metopium brownei, Swartzia cubensis, Swietenia macrophylla.*

C. TROPICAL SUBDECIDUOUS FOREST (IA3a, IA3b)

Mainly represented in eastern and southern Yucatan and in adjoining areas of Campeche and Quintana Roo, also in the Central Depression and on the Pacific lowlands of Chiapas, ranging from sea level to 1,250 m in altitude. Some of the common trees are: *Astronium graveolens, Brosimum alicastrum, Bumelia persimilis, Bursera simaruba, Calycophyllum candidissimum, Ceiba pentandra, Dalbergia granadillo, Ficus cotinifolia, Mastichodendron gaumeri, Enterolobium cyclocarpum, Piscidia piscipula, Platymiscium dimorphandrum, Vitex gaumeri.*

D. TROPICAL DECIDUOUS FOREST (IB1a(1), IB1a(2))

This forest covered most of the State of Yucatan with a few extensions into Campeche and Quintana Roo, also well represented in Chiapas in the Central Depression and on the Pacific lowlands, at altitudes between 0 and 1,250 m. Among the common trees, the following can be mentioned: *Alvaradoa amorphoides, Bumelia celastrina, Bursera bipinnata, B. simaruba, Euphorbia pseudofulva, Fraxinus purpusii, Gymnopodium antigonoides, Hampea trilobata, Heliocarpus reticulatus, Lonchocarpus longipedicellatus, Lysiloma desmostachys, Pistacia mexicana, Plumeria obtusa, Pterocereus gaumeri, Swietenia humilis, Zuelania guidonia.*

E. THORN FOREST (IC2a, IC2b)

This community occupies the arid narrow strip of coastal northwestern Yucatan and also some areas of poorly drained soils in central and southern Campeche and in southern Quintana Roo, mostly not much above sea level. Some important trees are: *Achatocarpus nigricans, Bucida buceras, Bumelia retusa, Caesalpinia vesicaria, Cameraria latifolia, Ceiba aesculifolia, Cephalocereus gaumeri, Diospyros cuneata, Eugenia lundellii, Guaiacum sanctum, Haematoxylon campechianum, Podopterus mexicanus, Pterocereus gaumeri, Stenocereus griseus.*

F. PALM FOREST

Mueller-Dombois and Ellenberg (1974) offer two possibilities for palm-dominated forest (IA1g(2) and IA1h(2)), but none of these correspond to the majority of Mexican palm forests. In the region discussed here, tall palms often dominate along river courses, also in some seasonally flooded areas and also on limestone hillsides. These communities occupy rather restricted and scattered areas in Campeche, Quintana Roo, Tabasco and Chiapas. The most important species are: *Brahea prominens, Orbignya cohune, Sabal mexicana, S. morrisiana, Scheelea liebmannii, S. lundellii, S. preussii.*

G. MONTANE MESOPHYLLOUS FOREST (IA1c, IA1e, IA2b)

This assemblage of forest communities once covered important areas of the wetter slopes of the mountains of Chiapas, both in the northern and in the southern parts, at altitudes between 1,200 and 2,300 m. Some of the more common trees are: *Brunellia mexicana, Carpinus caro-*

liniana, Clethra matudae, C. suaveolens, Cornus disciflora, Liquidambar styraciflua, Matudaea trinervia, Nyssa sylvatica, Oecopetalum mexicanum, Olmediella betchleriana, Oreopanax sanderiana, Pinus strobus var. *chiapensis, Persea schiedeana, Phoebe helicterifolia, Podocarpus guatemalensis, Quercus acatenangensis, Q. candicans, Q. skinneri, Saurauia villosa, Symplocos pycnantha, Turpinia paniculata, Ulmus mexicana, Weinmannia pinnata.*

H. Riparian Forest

Along many rivers in Chiapas and Tabasco narrow strips of a distinctive, often monospecific, and mostly deciduous forest can be observed. The following are the most outstanding tree components: *Fraxinus uhdei, Platanus mexicana, Populus mexicana, Salix chilensis, Taxodium mucronatum.*

I. Oak Forest

The diverse south-Mexican and Central American oak forests can be placed in several types of Mueller-Dombois and Ellenberg's (1974) classification, but really do not fit well in any of them. Some are evergreen and some totally or partially deciduous, some are dense and some open, many but not all the dominant oaks have hard leaves, some are rich in epiphytes, some are only scrubs. These forests cover extensive areas in the highlands of Chiapas, between 700 and 3,000 m, but a few descend almost to sea level, penetrating into Tabasco and even into southernmost Campeche. Some of the more common oaks are: *Quercus acatenangensis, Q. brachystachys, Q. candicans, Q. conspersa, Q. corrugata, Q. oleoides, Q. oocarpa, Q. peduncularis, Q. polymorpha, Q. skinneri.* Additional frequent tree components are *Arbutus glandulosa, Prunus serotina* ssp. *capuli* as well as species of *Pinus.*

J. Coniferous Forest (IA9b, IA9c)

Extensive areas of the Chiapas highlands are covered by pines and other conifers, usually at altitudes between 1,000 and 4,000 m, but sometimes at lower elevations. The main tree components of these forests are: *Abies quatemalensis, Cupressus lindleyi, Juniperus comitana, J. gamboana, Pinus ayacahuite, P. hartwegii, P. oocarpa, P. pseudostrobus, P. rudis, P. tenuifolia, P. teocote,* as well as species of *Quercus.*

K. Mangrove Forest (IA5)

This vegetation type is mostly confined to littoral saline aquatic or subaquatic environments. It is very plentiful along the coasts of Yucatan, but also frequent in other states. In most places mangroves assume only the form of scrub, but in some areas of Campeche and Tabasco they may attain a height of more than 20 m. The most important elements are: *Avicennia germinans, Conocarpus erecta, Laguncularia racemosa* and *Rhizophora mangle.*

L. Non-forest Vegetation

The following vegetation types dominated by herbaceous or shrubby species are reported from the area: savanna, high-mountain grassland, as well as a large variety of aquatic, subaquatic and halophytic communities.

IV. Floristic Inventory

A. Important Collectors

Botanical collecting in the area began in the first half of the 18th century with W. Houstoun's visit to Campeche. The first intensive exploration was done by J. M. Mociño who explored Chiapas in 1796 and 1798. Between 1830 and 1840 A. B. Ghiesbreght, J. J. Linden and T. Hartweg traveled mainly in Chiapas and Tabasco. A. Schott and C. T. Millspaugh made important collections in Yucatan in the second half of the 19th century, but the most significant plant gatherings from the Peninsula are those of G. T. Gaumer who collected more than 5,400 numbers between 1885 and 1923. J. N. Rovirosa was active in Tabasco and northern Chiapas from 1875 to 1900. G. E. Seler and C. Seler collected plants in Yucatan, Campeche and Chiapas between 1896 and 1911.

E. W. Nelson and A. E. Goldman traveled and collected all over Mexico, visiting Campeche in 1900 and 1901, Chiapas in 1895, 1896, 1900 and 1904, Quintana Roo in 1901, Tabasco in 1895 and 1900 and Yucatan in 1901. C. A. Purpus made important collections in southern and western Chiapas between 1913 and 1927.

After 1930 a large number of collectors visited the region. The following deserve special mention: E. Matuda, who explored southeastern Chiapas and eastern Tabasco; C. L. Lundell, mainly active in the Yucatan Peninsula; and F. Miranda, who explored in connection with his studies of the vegetation of Chiapas and, later, of the Yucatan Peninsula.

Several modern floristic projects are under way in the area. In 1964, D. E. Breedlove initiated an intensive collecting activity with the aim of preparing a Flora of Chiapas. Several collectors have participated in this effort and so far some 42,000 numbers have been gathered. The first set is deposited at CAS, but duplicates are being distributed to many other herbaria. Besides Breedlove, the most active collectors of this project are A. S. Ton and R. M. Laughlin.

In 1978 C. P. Cowan started a collecting activity preparatory to a Flora of Tabasco. In 1981 S. Zamudio joined the project. The first set of their materials is conserved at CSAT, but many duplicates have been distributed to other herbaria (especially MEXU, ENCB and CHAPA). Altogether, Cowan, Zamudio and collaborators collected about 6,000 numbers. In 1985 the sponsoring institution, Colegio Superior de Agricultura Tropical, was dissolved and the herbarium activities, as well as the collecting program, are at least temporarily discontinued. However, the facility is being transferred to the Colegio de Postgraduados and there are high hopes that the program will be reactivated.

Since 1980 the Instituto Nacional de Investigación sobre Recursos Bióticas created a botanical center in Mérida and its main goal is the elaboration of a Flora of the Yucatan Peninsula. So far, more than 25,000 numbers have been collected, the main collectors being C. Chan Vermont, J. S. Flores and E. Ucán Ek. The first set is deposited at XAL.

Between 1980 and 1983 the Instituto de Biología of the Universidad Nacional Autónoma de México in collaboration with the Centro de Investigaciones de Quintana Roo organized intensive explorations of the state of Quintana Roo, the flora of which was heretofore very little known. More than 8,000 numbers were collected, mainly by E. F. Cabrera and O. Téllez. The first set is deposited at MEXU and many duplicates were distributed elsewhere.

In 1980 an international project called Flora Mesoamericana was started with the aim of preparing a manual of the flora of Central America and the transisthmic part of Mexico. Five Mexican herbaria (CHAPA, CSAT, ENCB, MEXU and UAMIZ) have been participating in the collecting activities sponsored by the project. Till 1986 altogether about 40,000 numbers were obtained. The most outstanding collectors are: E. Cabrera, E. López, E. Martínez, M. Sousa, A. S. Ton, E. Ventura, F. Ventura, and S. Zamudio.

B. Important Herbaria

At present, six institutional herbaria exist in the area: one in Chiapas, one in Quintana Roo, two in Tabasco and two in Yucatan. Almost all of them were founded in the last 10 years and their joint regional holdings amount to about 30,000 sheets. Approximately 120,000 sheets from the area are kept in other Mexican herbaria (mainly MEXU, ENCB, XAL and INIF). Some 300,000 sheets are probably held in foreign herbaria, the richest ones being CAS, F, GH, LL, MICH, MO and US. It must be kept in mind, however, that the total amount of numbers collected in the area scarcely exceeds 150,000, giving a general average of about 60 collections per 100 km^2.

C. Best Explored and Collected Areas

The best-explored areas (with more than 100 collections per 100 km^2) are: (1) the surroundings of larger towns and important archeological sites, such as Mérida, Progreso, Uxmal, Chichén Itzá, Campeche, Ciudad del Carmen, Escárcega, Chetumal, Villahermosa, Cárdenas, Huimanguillo, Teapa, Pichucalco, Palenque, San Cristóbal, Comitán, Tuxtla Gutiérrez, Tapachula, Huixtla, Tonalá, and a few others; (2) areas bordering the main highways; (3) islands surrounding the Yucatan Peninsula; (4) some areas particularly attractive on account of their natural beauty, such as the Cañón del Sumidero, the Montebello Lakes, etc.; (5) much of the higher mountain areas of Chiapas, especially of the Central Plateau; (6) the greater part of the eastern half of the State of Yucatan.

Much exploration still remains to be done in most of Campeche, in large portions of Tabasco, in Quintana Roo and in adjacent Yucatan, as well as in northeastern and northwestern Chiapas.

Among the better-collected plant groups are: Compositae, Cyperaceae, Gramineae, Labiatae, Malvaceae, Melastomataceae, Onagraceae, Rubiaceae, and pteridophytes.

Much more collecting is required for: Cactaceae, Convolvulaceae, Cucurbitaceae, Iridaceae, Loranthaceae, and in general, tall trees, aquatics and epiphytes.

The published preliminary lists reveal the following numbers of species of vascular plants (Breedlove, 1986; Cowan, 1983; Sosa et al., 1985; Sousa & Cabrera, 1983):

Yucatan	1,120
Campeche	938
Quintana Roo	1,292
Mexican part of the Yucatan Peninsula	1,936
Tabasco	2,147
Chiapas	7,657

On the basis of these data, the vascular flora of the entire area can be estimated to contain some 10,000–11,000 species, more than 50% of which are woody plants, including at least 2,000 species of trees.

D. Publications

No complete true modern flora or floristic manual exists at present for any part of transisthmic Mexico, but a large amount of literature has accumulated in the last 100 years towards this goal. Only the most significant contributions will be reviewed here. Much additional information can be found in Langman's (1964) book.

The first botanical study published on the area was apparently the book of Dondé and Dondé (1873), in which 65 species of Yucatan plants are treated. More complete lists were published a few years later by Dondé (1878) for the surroundings of Mérida and by Rovirosa (1892) for the surroundings of Villahermosa (formerly San Juan Bautista).

Following a visit to the peninsula, Millspaugh (1895, 1896, 1898) published a series of contributions to the Flora of Yucatan. A few years later the same author decided to write a formal Flora with descriptions, keys and illustrations, but only the treatment of five families (Compositae, Cyperaceae, Gramineae, Polypodiaceae, and Schizaeaceae) appeared in print (Millspaugh & Chase, 1903–1904).

In 1909 Rovirosa's finely illustrated book on the pteridophytes of southeastern Mexico was posthumously published.

The large collection of Yucatan plants accumulated in the herbarium of the Field Museum of Chicago was thoroughly studied by Standley, who published his *Flora of Yucatan* in 1930. It includes 1,263 species, 1,068 of which are believed to be native. No keys and illustrations are provided and the descriptions are mostly very short, but much information is given concerning vernacular names and uses of the plants. A somewhat modified Spanish version of this *Flora* was published later (Standley, 1945).

Between 1931 and 1938 H. H. Bartlett and C. L. Lundell, of the University of Michigan, organized several botanical expeditions to the Yucatan Peninsula. In connection with these expeditions several written contributions were produced. Some of the outstanding ones were published in the *Botany of the Maya area*, a two-volume book, in which the following taxonomic groups are treated: fungi (Mains, 1935), lichens (Hedrick, 1935), marine algae (Taylor, 1935), Malpighiaceae (Morton, 1936a), Gramineae (Swallen, 1936, but see also Swallen, 1934), Acanthaceae (Leonard, 1936), Dioscoreaceae (Morton, 1936b), Passifloraceae (Killip, 1936), Apocynaceae (Woodson, 1940), Bromeliaceae (Smith & Lundell, 1940), Eriocaulaceae and Verbenaceae (Moldenke, 1940), Labiatae (Epling, 1940), Cyperaceae (O'Neill, 1940), Melastomataceae (Gleason, 1940) and Bignoniaceae (Seibert, 1940). Separately, Steere (1935) published on mosses. In 1945 Souza Novelo's *Apuntes a la flora de Yucatan* appeared.

As the result of his own explorations in southeastern Chiapas, Matuda (1950a, 1950b) published two floristic lists enumerating 777 species for Mt. Ovando and 2,628 species for the districts of Soconusco and Mariscal. Among other smaller contributions to the flora of Chiapas, four (Matuda, 1948, 1950c, 1952, 1953) dealing with the Bromeliaceae, Burmanniaceae, Marantaceae and Meliaceae deserve special mention. An account of about 700 of Matuda's collections from eastern Tabasco and adjacent areas was published by Lundell (1942).

Miranda (1952–1953), in his book on the vegetation of Chiapas, devotes 556 pages to the enumeration, description, ecological affinities, vernacular names, illustrations, properties and uses of more than 1,000 species of the state flora. In a later paper (Miranda, 1961) an important list of trees observed in the *Selva Lacandona* of northeastern Chiapas is included and local names are provided.

Pennington and Sarukhán (1968) wrote a manual for identification of the more common tropical trees of Mexico. With the exception of seven, the 150 described and illustrated species inhabit transisthmic Mexico. An illustrated compendium of common names of a large number of plants was produced by Chavelas Pólito (1982).

The Flora of Chiapas project has so far produced three important publications: an introductory account (Breedlove, 1981), a treatment of the pteridophytes (Smith, 1981), and a preliminary floristic list (Breedlove, 1986).

Cowan (1983) published a similar list for the flora of Tabasco, Sousa and Cabrera (1983) for Quintana Roo and Lundell and Lundell (1983) for northern Yucatan and the Cobá area of Quintana Roo. Finally, Sosa et al. (1985) enumerated the species of vascular plants and their known local names for the entire Mexican part of the Yucatan Peninsula.

A profusely illustrated manual of 116 species of the flora of Quintana Roo was published by Téllez, Sousa et al. (1982).

For purposes of general identification of plant specimens from the reviewed areas the most useful books are still the *Flora of Guatemala* (Standley et al., 1946–1977; Ames & Correll, 1952–1953; Bartram, 1949) and the *Trees and Shrubs of Mexico* (Standley, 1920–1926).

Among the contributions concretely devoted to useful plants and ethnobotany the following are outstanding. Manuals and enumerations of Yucatan medicinal plants were written by Ossado (1834), Cuevas (1894, 1913), Lavadores (1969) and Mendieta and del Amo (1981). Moscoso (1981) published on Chiapas highlands medicinal plants. Souza Novelo (1940, 1950) wrote on food and seasoning plants as well as on polliniferous and melliferous species of Yucatan. Mills' (1957) book deals with timber trees of northern Chiapas. The most important ethnobotanical contributions are those of Roys (1931), Steggerda (1941, 1943) and Barrera Marín et al. (1977) for Yucatan, and that of Berlin et al. (1974) for the highlands of Chiapas.

E. Endemism

Due to its peculiar configuration and soil conditions, the Yucatan Peninsula and especially its northern portion has an outstanding wealth of endemic taxa. According to Standley (1930: 165), 17% of the species and three genera of flowering plants are restricted in their distribution to the Peninsula.

No similar data are available for the rest of the area, but it can be easily seen that the flora of the Tabascan lowlands is extremely similar to that of Guatemala on one side and to that of Veracruz on the other. Only a very limited number of endemics exists (mainly on limestone hillsides situated at its southern extreme), although a fair number of new distributional records is still expected.

A different situation prevails in the mountains of Chiapas where at least 5% of the species are of restricted distribution, some extending their area to adjacent regions of Guatemala. According to Breedlove (1981: 1–3), the areas richest in endemism are the Sierra Madre, the Central Depression, the Central Plateau and the Northern Highlands.

F. Extinction

So far the study of endangered plant species and communities in Mexico has received little attention and almost no concrete information is available for the southeastern part of the country. Nevertheless, the following general considerations are in order.

About 80 to 90% of the forests of transisthmic Mexico have been converted into farmland, cattle pastureland and secondary vegetation. The intensity of the conversion has been rapidly increasing in the last 20 years in connection with the existence of better means of communication, colonization of new territories, demographic increment and technological advances.

Little is left of the pristine plant cover of the State of Yucatán, where tropical deciduous forest once prevailed and the same is true for Tabasco, originally covered mostly by tropical evergreen forest and aquatic or semi-aquatic vegetation. Not as drastic at present is the situation in some parts of Campeche and Quintana Roo, but most of the existing forest there is in immediate danger of disappearance.

In many parts of Chiapas the rough topography slows

down the destruction of the primary vegetation, but a large percentage of very steep slopes has been already cleared, especially in the wet areas where often only some portions completely unsuitable for agriculture or grazing remain forested.

In consequence, it can be stated that the most threatened plant species in our area are those exclusively related to the primary tropical evergreen lowland forest (including trees, epiphytes, parasites, lianas, sciophilous understory elements, etc.) and to the semi-aquatic and aquatic communities of Tabasco.

V. Academic and Governmental Institutions Involved in Floristic Inventory

A. Local Institutions

1. Campeche

a) The Mexican forest research system established a research unit in Escárcega. In past years some floristic studies have been performed there.

2. Chiapas

a) The Instituto de Historia Natural (formerly Instituto de Botánica) de Chiapas in Tuxtla Gutiérrez keeps a botanical garden and a small herbarium. At present research on medicinal plants is carried out.

b) The Centro de Investigaciones Ecológicas del Sureste in San Cristóbal has a small group of plant ecologists studying the forest of Chiapas highlands.

3. Quintana Roo

a) The Centro de Investigaciones de Quintana Roo in Puerto Morelos sponsors floristic, ecological and ethnobotanical research. A small herbarium is kept there.

b) The Mexican forest research system keeps a research unit in Bacalar.

4. Tabasco

a) The Colegio de Postgraduados recently took over the research facilities of the former Colegio Superior de Agricultura Tropical. These include the largest regional herbarium, in which intensive floristic, ecological and ethnobotanical research has been performed.

b) The Instituto de Biologia of the Universidad Juárez Autónoma de Tabasco keeps a small herbarium and is active in floristic research.

c) The Universidad Autónoma Chapingo keeps a field unit in Puyacatengo, near Teapa, in which studies on the vegetation of Tabasco have been done.

5. Yucatán

a) The Instituto Nacional de Investigaciones sobre Recursos Bióticos in Xalapa established a branch in Mérida with the aim of studying the plant resources of the Peninsula. An important local herbarium is kept there. This is at present the most active botanical center of southeastern Mexico.

B. Other Mexican Institutions

a) The Instituto de Biología of the Universidad Nacional Autónoma de México in Mexico City co-sponsors the Flora Mesoamericana project and maintains an active collecting program.

b) The Escuela Nacional de Ciencias Biológicas of the Instituto Politécnico Nacional in Mexico City also participates in the Flora Mesoamericana project and maintains an active collecting program.

c) The Departamento de Biología of the Universidad Autónoma Metropolitana, Unidad Iztapalapa in Mexico City participates in the Flora Mesoamericana project.

d) The Colegio de Postgraduados in Chapingo participates in the Flora Mesoamericana project and performs ethnobotanical research in Yucatan.

C. Main Foreign Institutions

a) The California Academy of Sciences in San Francisco sponsors the Flora of Chiapas project.

b) The Missouri Botanical Garden in St. Louis co-sponsors the Flora Mesoamericana project.

c) The British Museum in London co-sponsors the Flora Mesoamericana project.

d) The Middle American Research Institute of Tulane University in New Orleans sponsors some botanical investigation in Yucatan.

VI. Future of Inventory

In view of the rapid disappearance of Mexican tropical forests, it is urgent to complete as soon as possible their thorough floristic and ecological inventory. As stated before, many local and foreign institutions are deeply interested and involved in this task, but due to the scarcity of human and material resources, the progress is slow and, given the actual speed of work, the first faithful floristic approximation for our area will not be ready before the year 2000. By that time, however, many plant species and communities will be gone or irreversibly impoverished.

It is therefore very important to speed up and make more efficient the efforts of exploring, studying and protecting the flora and the vegetation of the area. It can be estimated that an additional investment of an equivalent of US $2,000,000 per year is necessary for this purpose for about ten years.

VII. Acknowledgments

The assistance of Armando Butanda, Irene García, Mario

Sousa and especially of Sergio Zamudio in the preparation of the manuscript, is gratefully acknowledged.

VIII. Literature Cited

Alvarez del Villar, J. 1952. Esquema geobotánico de Chiapas. Bol. Soc. Mex. Geogr. Estad. **73:** 96-124.

Ames, O. & D. S. Correll. 1952-1953. Orchids of Guatemala. Fieldiana, Bot. **26:** 1-727.

Anonymous. 1976a. Inventario forestal del estado de Quintana Roo. Dirección General del Inventario Nacional Forestal, México, D.F.

Anonymous. 1976b. Inventario forestal del estado de Chiapas. Dirección General del Inventario Nacional Forestal, México, D.F.

Anonymous. 1976-1981. Cartografía sinóptica—uso actual del suelo. (Scale 1:500,000). Sheets: Campeche, Chiapas, Quintana Roo, Tabasco & Yucatan. Secretaría de Agricultura y Recursos Hidráulicos, México, D.F.

Anonymous. 1981. Carta de uso del suelo. (Scale 1:1,000,000). Pages 184-187 *in* Atlas nacional del medio físico. Secretaría de Programación y Presupuesto, México, D.F.

Anonymous. 1982-1985. Carta de uso del suelo y vegetación. (Scale 1:250,000). Sheets D15-2,5; E15-3,5,6,9; E16-1,2,4,5,7; F15-9,12; F16-7,8,10. Instituto Nacional de Estadística, Geografía e Informática, México, D.F.

Anonymous. 1985a. Inventario forestal del estado de Tabasco. Publicación Especial No. 54. Secretaría de Agricultura y Recursos Hidráulicos, México, D.F.

Anonymous. 1985b. Inventario forestal del estado de Yucatan. Publicación Especial No. 55. Secretaría de Agricultura y Recursos Hidráulicos, México, D.F.

Anonymous. 1985c. Inventario forestal del estado de Campeche. Publicación Especial No. 56. Secretaría de Agricultura y Recursos Hidráulicos, México, D.F.

Barrera Marín, A., A. Barrera Vázquez & R. M. López Franco. 1976. Nomenclatura etnobotánica maya: una interpretación taxonómica. Instituto Nacional de Antropoligía e Historia. Colección Científica 36. México, D.F.

Bartlett, H. H. 1935. A method of procedure for field work in tropical American phytogeography upon a botanical reconnaissance in parts of British Honduras and the Peten forests of Guatemala. Publ. Carnegie Inst. Wash. **461:** 1-25.

Bartram, E. B. 1949. Mosses of Guatemala. Fieldiana, Bot. **25:** 1-442.

Bequaert, J. C. 1933. Botanical notes from Yutacan. Publ. Carnegie Inst. Wash. **431:** 505-524.

Berlin, B., D. E. Breedlove & P. H. Raven. 1974. Principles of Tzeltal plant classification. An introduction to the botanical ethnography of a Maya-speaking people of highland Chiapas. Academic Press, New York.

Bonet, F. & J. Rzedowski. 1962. La vegetación del Arrecife Alacranes, Yucatán, México. Anales Esc. Nac. Ci. Biol. **11:** 15-50.

Bravo, H. 1955. Algunas observaciones acerca de la vegetación de la región de Escárcega, Campeche, y zonas cercanas. Anales Inst. Biol. Univ. Nac. México **26:** 283-301.

Breedlove, D. E. 1973. The phytogeography and vegetation of Chiapas (Mexico). Pages 149-165 *in* A. Graham (ed.), Vegetation and vegetational history of northern Latin America. Elsevier Scientific Publishing Company, Amsterdam.

———. 1981. Introduction to the flora of Chiapas. Flora of Chiapas. Part I. California Academy of Sciences, San Francisco.

———. 1986. Listados florísticos de México. IV. Flora de Chiapas. Instituto de Biología. Universidad Nacional Autónoma de México, México, D.F.

Calzada, I. & P. Valdivia. 1979. Introducción al estudio de dos zonas de la Selva Lacandona, Chis., México. Biotica **4:** 149-162.

Chavelas Pólito, J. 1968. Estudio florístico-sinecológico del campo experimental forestal "El Tormento," Escárcega, Campeche. Thesis. Facultad de Ciencias. Universidad Nacional Autónoma de México, México, D.F.

———. 1982. Catálogo de nombres comunes de plantas recogidos por la Comisión de Estudios sobre la Ecología de Dioscóreas. Instituto Nacional de Investigaciones Forestales. Catálogo No. 8. México, D.F.

Cowan, C. P. 1983. Listados florísticos de México. I. Flora de Tabasco. Instituto de Biología. Universidad Nacional Autónoma de México, México, D.F.

Cuevas, B. 1894. Ensayo botánico. Mérida.

———. 1913. Plantas medicinales de Yucatán y guía médica práctica doméstica. Mérida.

Dondé I., J. & J. Dondé R. 1873. Apuntes sobre las plantas de Yucatán. Mérida.

Dondé, R. J. 1878. Calendario botánico de Mérida y sus alrededores. Emulación **3:** 152, 166, 184, 224, 236, 260, 288.

Epling, C. 1940. The Labiatae of the Yucatan Peninsula. Publ. Carnegie Inst. Wash. **522:** 225-245.

Espejel, I. 1984. La vegetación de las dunas costeras de la Península de Yucatán. I. Análisis florístico del estado de Yucatán. Biotica **9:** 183-210.

———. 1986a. Studies in coastal sand dune vegetation of the Yucatan Peninsula. Thesis. Uppsala University, Uppsala.

———. 1986b. La vegetación de las dunas costeras de la Península de Yucatán. II. Reserva de la biósfera Sian Kaán, Quintana Roo, México. Biotica **11:** 7-24.

Flores, J. S. 1984a. Dinámica de emersión del suelo y sucesión de la vegetación en el Arrecife Alacranes del Canal de Yucatán. Biotica **9:** 41-63.

———. 1984b. Vegetación insular de la Península de Yucatán. Bol. Soc. Bot. México **45:** 23-37.

Flores Mata, G. et al. 1971. Memoria del mapa de tipos de vegetación de la República Mexicana. Secretaría de Recursos Hidráulicos, México, D.F.

Gleason, H. A. 1940. The Melastomaceae of the Yucatan Peninsula. Publ. Carnegie Inst. Wash. **522:** 323-371.

Hedrick, J. 1935. Lichens from the Yucatan Peninsula. Publ. Carnegie Inst. Wash. **461:** 107-114.

Hernández Corzo, A. 1950. Estudio geobotánico, agrícola y forestal de Yucatán. Bol. Soc. Mex. Geogr. Estad. **69:** 163-201.

Killip, E. P. 1936. Passifloraceae of the Maya region. Publ. Carnegie Inst. Wash. **461:** 301-328.

Langman, I. K. 1964. A selected guide to the literature on the flowering plants of Mexico. University of Pennsylvania Press, Philadelphia.

Lavadores, V. G. 1969. Las 119 plantas medicinales más conocidas en Yucatán, Méx. Mérida.

Leonard, E. C. 1936. The Acanthaceae of the Yucatan Peninsula. Publ. Carnegie Inst. Wash. **461:** 141-238.

López Mendoza, R. 1980. Tipos de vegetación y su distribución en el estado de Tabasco y norte de Chiapas. Universidad Autónoma Chapingo, Chapingo.

López Portillo G., J. A. 1982. Ecología de manglares y de otras comunidades de halófitas en las costa de la Laguna de Mecoacán, Tabasco. Thesis. Facultad de Ciencias. Universidad Nacional Autónoma de México, México, D.F.

Lundell, C. L. 1934. Preliminary sketch of the phytogeography of the Yucatan Peninsula. Publ. Carnegie Inst. Wash. **136**: 255–321.

———. 1938. The 1938 botanical expedition to Yucatan and Quintana Roo. Carnegie Inst. Wash. Year Book **37**: 143–147.

———. 1942. Flora of eastern Tabasco and adjacent Mexican areas. Cont. Univ. Michigan Herb. **8**: 1–74.

——— & A. A. Lundell. 1983. The flora of northern Yucatan and the Coba area of Quintana Roo. Wrightia **7**: 170–228.

Mains, E. B. 1935. Rusts and smuts of the Yucatan Peninsula. Publ. Carnegie Inst. Wash. **461**: 93–106.

Matuda, E. 1948. Meliáceas de Chiapas. Anales Inst. Biol. Univ. Nac. México **19**: 407–425.

———. 1950a. A contribution to our knowledge of the wild flora of Mt. Ovando, Chiapas. Amer. Midl. Naturalist **43**: 195–223.

———. 1950b. A contribution to our knowledge of the wild and cultivated flora of Chiapas. Districts of Soconusco and Mariscal. Amer. Midl. Naturalist **44**: 513–616.

———. 1950c. Estudio de las plantas de Chiapas. VIII. Marantáceas de Chiapas. Anales Inst. Biol. Univ. Nac. México **21**: 319–343.

———. 1952. Las bromeliáceas de Chiapas. Anales Inst. Biol. Univ. Nac. México **23**: 85–153.

———. 1953. Burmanniáceas de Chiapas. Bol. Soc. Bot. México **15**: 19–23.

Meave, J. A. 1983. Estructura y composición de la selva alta perennifolia en los alrededores de Bonampak, Chiapas. Thesis. Facultad de Ciencias. Universidad Nacional Autónoma de México, México, D.F.

Mendieta, R. M. & S. del Amo. 1981. Plantas medicinales del estado de Yucatán. Instituto Nacional de Investigaciones sobre Recursos Bióticos y Compañía Editorial Continental, México, D.F.

Mills, T. H. 1957. Timber trees of northern Chiapas. Imprenta "Cosmos," México, D.F.

Millspaugh, C. F. 1895. Contribution to the flora of Yucatan. Publ. Field Mus. Nat. Hist., Bot. Ser. **1**: 1–56.

———. 1896. Contribution II to the coastal and plain flora of Yucatan. Publ. Field Mus. Nat. Hist., Bot. Ser. **1**: 281–339.

———. 1898. Contribution III to the coastal and plain flora of Yucatan. Publ. Field Mus. Nat. Hist., Bot. Ser. **1**: 345–410.

———. 1916. The vegetation of the Alacran Reef. Publ. Field Mus. Nat. Hist., Bot. Ser. **2**: 421–431.

——— & A. Chase. 1903–1904. Plantae yucatanae. Plants of the insular, coastal and plain regions of the Peninsula of Yucatan, Mexico. Fascicles I & II. Publ. Field Mus. Nat. Hist., Bot. Ser. **3**: 1–151.

Miranda, F. 1942. Estudios sobre la vegetación de México. II. Observaciones preliminares sobre la vegetación de la región de Tapachula, Chiapas. Anales Inst. Biol. Univ. Nac. México **13**: 53–70.

———. 1952–1953. La vegetación de Chiapas. Ediciones del Gobierno del Estado de Chiapas, Tuxtla Gutiérrez. 2 vols.

———. 1957. Vegetación de la vertiente del Pacífico de la Sierra Madre de Chiapas (México) y sus relaciones florísticas. Proc. 8th Pacif. Sci. Congr. Pacif. Sci. Assoc. Quezon City. Vol **4**: 438–453.

———. 1958. Estudios acerca de la vegetación. Pages 215–271 in Los recursos naturales del sureste y su aprovechamiento. II parte, tomo 2. Instituto Mexicano de Rucursos Naturales Renovables, México, D.F.

———. 1961. Tres estudios botánicos en la Selva Lacandona, Chiapas. Bol. Soc. Bot. México **26**: 133–176.

Moldenke, H. N. 1940. The Eriocaulaceae, Verbenaceae and Avicenniaceae of the Yucatan Peninsula. Publ. Carnegie Inst. Wash. **522**: 137–245.

Morley, S. G. 1953. La civilización maya. 2nd ed. Fondo de Cultura Económica, México, D.F.

Morton, C. V. 1936a. Enumeration of the Malpighiaceae of the Yucatan Peninsula. Publ. Carnegie Inst. Wash. **461**: 125–140.

———. 1936b. Notes on *Dioscorea*, with special reference to the species of the Yucatan Peninsula. Publ. Carnegie Inst. Wash. **461**: 239–253.

Moscoso, P. 1981. La medicina tradicional de los Altos de Chiapas. Editorial Tradición, México, D.F.

Mueller-Dombois, D. & H. Ellenberg. 1974. Aims and methods of vegetation ecology. Wiley & Sons, New York.

O'Neill, H. T. 1940. The sedges of the Yucatan Peninsula. Publ. Carnegie Inst. Wash. **522**: 247–322.

Ossado, R. 1834. Medicina doméstica. Mérida.

Pennington, T. D. & J. Sarukhán. 1968. Manual para la identificación de los principales árboles tropicales de México. Instituto Nacional de Investigaciones Forestales y FAO, México, D.F.

Pérez J., A. & J. Sarukhán K. 1970. La vegetación de la región de Pichucalco, Chiapas. Instituto Nacional de Investigaciones Forestales. Publicación Especial No. 5. México, D.F.

Puig, H. 1972. La sabana de Huimanguillo, Tabasco, México. Pages 389–411 in Memorias del I Congreso Latinoamericano y V Mexicano de Botánica. México, D.F.

Quiroz, A. J. 1977. Estudio preliminar de la relación suelo-vegetación en la zona Balancán-Tenosique, Tabasco. Thesis. Facultad de Ciencias. Universidad Nacional Autónoma de México, México, D.F.

Rico Gray, V. 1982. Estudio de la vegetación de la zona costera inundable del noroeste de Campeche, México: Los Petenes. Biotica **7**: 171–190.

Rovirosa, J. N. 1892. Calendario botánico de San Juan Bautista y alrededores. Naturaleza (Mexico City) II, **2**: 106–126.

———. 1909. Pteridografía del sur de México. Imprenta de I. Escalante, México, D.F.

Roys, R. L. 1931. The ethno-botany of the Maya. Middle American Research Series. Publ. 2. Tulane University, New Orleans.

Sarukhán, J. 1968. Análisis sinecológico de las selvas de *Terminalia amazonia* en la planicie costera del Golfo de México. Thesis. Colegio de Postgraduados. Escuela Nacional de Agricultura, Chapingo.

——— & E. Hernández Xolocotzi. 1968. Sinecología de las selvas de *Terminalia amazonia* en la vertiente del Golfo de México. Agrociencia **3**: 1–17.

Sauer, J. 1967. Geographic reconnaissance of seashore vegetation along the Mexican Gulf coast. Coastal Studies Institute. Techn. Rep. 56. Louisiana State University, Baton Rouge.

Seibert, R. J. 1940. The Bignoniaceae of the Maya area. Publ. Carnegie Inst. Wash. **522**: 375–434.

Smith, A. 1981. Pteridophytes. Flora of Chiapas. Part II. California Academy of Sciences, San Francisco.

Smith, L. B. & C. L. Lundell. 1940. The Bromeliaceae of the Yucatan Peninsula. Publ. Carnegie Inst. Wash. **522**: 103–136.

Sosa, V. et al. 1985. Etnoflora yucatanense. Lista florística y sinonimia maya. Instituto Nacional de Investigaciones sobre Recursos Bióticos, Xalapa.

Sousa, M. & E. F. Cabrera. 1983. Listados florísticos de México. II. Flora de Quintana Roo. Instituto de Biología. Universidad Nacional Autónoma de México, México, D.F.

Souza Novelo, N. 1940. Plantas melíferas y poliníferas que viven en Yucatán. Mérida.

———. 1945. Apuntes relativos a la flora de Yucatán. Inst. Tecn. Agric. Henequenero. Mérida.

_____. 1950. Plantas alimenticias y plantas de condimento que viven en Yucatán. Mérida.
Standley, P. C. 1920–1926. Trees and shrubs of Mexico. Contr. U.S. Natl. Herb. **23**: 1–1721.
_____. 1930. Flora of Yucatan. Publ. Field Mus. Nat. Hist., Bot. Ser. **3**: 157–492.
_____. 1945. La flora. Pages 273–523 in Enciclopedia yucatanense. Tomo I. Edición Oficial del Gobierno de Yucatán. Mérida.
_____, J. A. Steyermark, L. O. Williams et al. 1946–1977. Flora of Guatemala. Fieldiana, Bot. **24**, parts 1–12.
Steere, W. C. 1935. The mosses of Yucatan. Amer. J. Bot. **22**: 395–408.
Steggerda, M. 1941. Maya Indians of Yucatan. Publ. Carnegie Inst. Wash. **531**: 1–280.
_____. 1943. Some ethnological data concerning one hundred Yucatan plants. Smithsonian Institution. Bureau of American Ethnology Bulletin 136, Anthropological Papers, **29**: 193–226.
Swallen, J. R. 1934. The grasses of the Yucatan Peninsula. Publ. Carnegie Inst. Wash. **436**: 325–355.
_____. 1936. The grasses of British Honduras and the Peten, Guatemala. Publ. Carnegie Inst. Wash. **461**: 141–189.
Taylor, R. W. 1935. Marine algae from the Yucatan Peninsula. Publ. Carnegie Inst. Wash. **461**: 115–124.
Téllez V., O., M. Sousa S. et al. 1982. Imágenes de la flora quintanarroense. Centro de Investigaciones de Quintana Roo, Puerto Morelos.
Thom, B. G. 1967. Mangrove ecology and deltaic geomorphology: Tabasco, Mexico. J. Ecol. **55**: 301–343.
Vazquez Soto, J. 1963. Clasificación de las masas forestales de Campeche. Instituto Nacional de Investigaciones Forestales. Boletín Técnico No. 10. México, D.F.
West, R. C. 1956. The natural vegetation of the Tabascan lowlands, Mexico. Rev. Geogr. (Brasil) **64**: 109–122.
Woodson, R. E., Jr. 1940. The Apocynaceous flora of the Yucatan Peninsula. Publ. Carnegie Inst. Wash. **522**: 59–162.
Zuill, H. A. & E. W. Lathrop. 1975. The structure and climate of a tropical montane rain forest and an associated temperate pine-oak-*Liquidambar* forest in the northern highlands of Chiapas, Mexico. Anales Inst. Biol. Univ. Nac. México, Ser. Bot. **46**: 73–118.

Guatemala

Margaret Tebbs

Contents

I. Description of the Region	283
A. Geographical Extent and Topography	283
B. Population	283
II. Vegetation Maps	283
III. Herbarium Resources	283
IV. Present State of Inventory	283
V. Academic and Governmental Institutions	283
VI. Future of Inventory	284
VII. Literature Cited	284

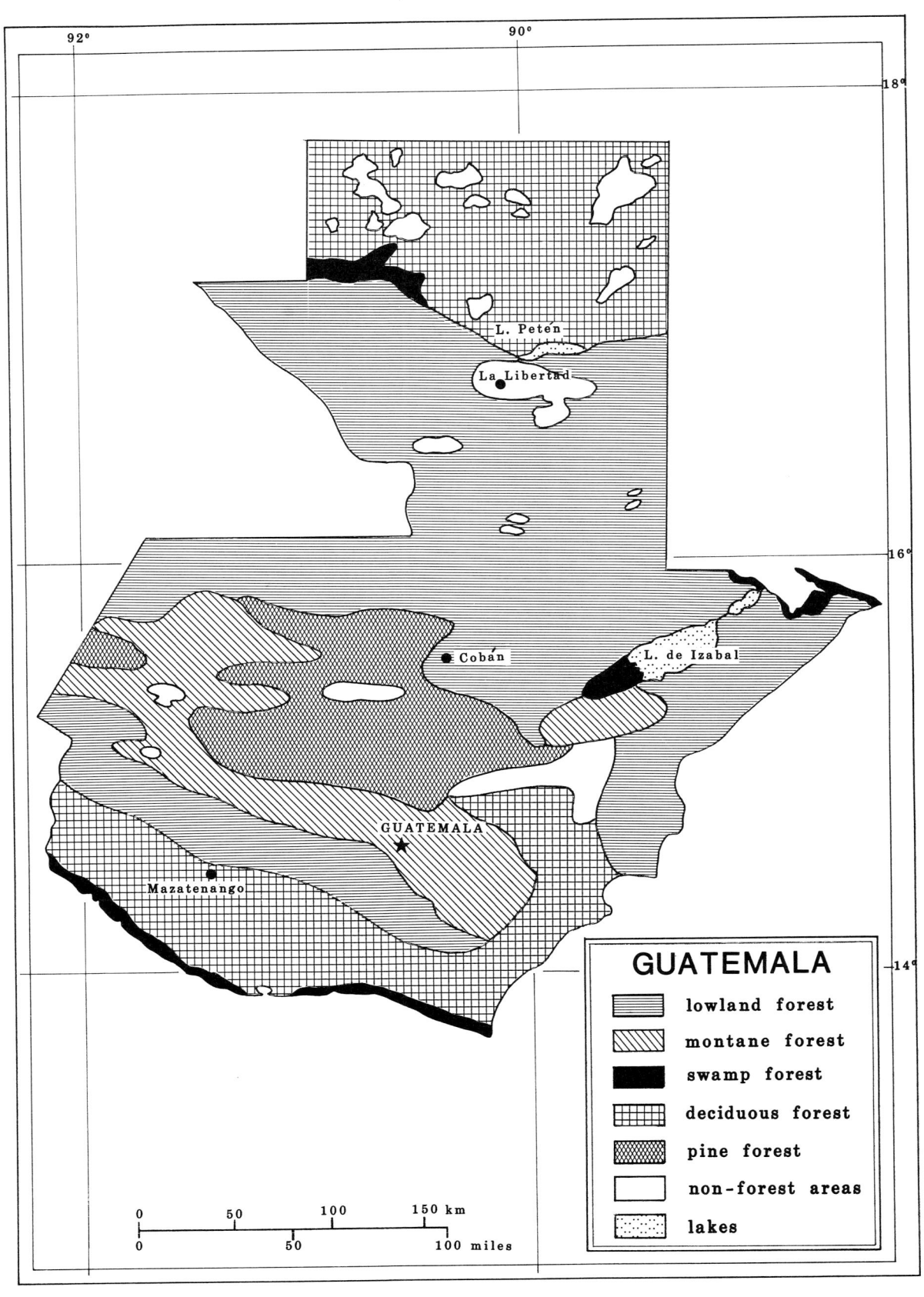

I. Description of the Region

A. GEOGRAPHICAL EXTENT AND TOPOGRAPHY

Guatemala has an area of 108,780 sq. km. High mountain ranges to over 4000 m. cross the center of the country from east to west. There are some 33 volcanoes, some still active. In the north are the plains of Petén. North Petén consists mainly of tropical dry forest; south Petén of tropical moist forest. Small patches of pluvial forest remain in the west and northwest of the country, near the Mexican border. There is also a small amount of paramo in the highest mountains (Fig. 1).

B. POPULATION

The population of Guatemala is 7.2 million (1982), a population density of 66 people per km². The population of Guatemala City is 1,180,000 (1982) and that of Coban is 11,000 (1982). The Central Highlands, extending east to west, support the largest part of the population. The state of Petén has less than 100,000 inhabitants, although it includes more than one-third of the land mass.

II. Vegetation Maps

The series prepared by the Instituto Geografico Nacional (IGN) de Guatemala, has general vegetation maps. These maps, published in 1966 and 1973, are marked with mangrove, marsh, woods and plantation symbols. The Atlas Nacional de Guatemala (1972) and the Directory of Military Survey (1977) both have vegetation maps.

III. Herbarium Resources

The Escuela de Biologia, Museo de Historia Natural y Jardin Botánico, Universidad de San Carlos de Guatemala (USCG), Guatemala City, holds 31,253 specimens.

The herbarium of "Ulises Rojas", Museo Nacional de Historia Natural (GUAT), Guatemala City, holds 5,000 specimens.

The herbarium, Universidad del Valle de Guatemala (UVAL), Guatemala City holds 5,000 specimens. Therefore the total number of herbarium specimens for Guatemala held in the country is 41,253. There is a small collection in Herbario Paul Standley, (EAP), Tegucigalpa. There are 80,000 specimens held in herbaria abroad, most in the Field Museum, Chicago.

IV. Present State of Inventory

Relatively well collected areas are northern and parts of

Table I

List of plant families and their specialists for Guatemala.

Family	Specialist and Home Institution
Orchidaceae	Margaret Dix (UVAL),[a] Elfriede de Pol (USCG)
Pines	Karen Lind (UVAL)
Commelinaceae	Laura Schuster (UVAL)
Piperaceae	Christopher Humphries (BM), Margaret Tebbs (BM)
Ferns	Mario Dary (USCG), Robert Stolze (F)
Conifers	Jose Maria Aguilar (USCG)
Solanaceae, New World Tropics	Michael Nee (NY)
Araceae	Thomas Croat (MO)
Gramineae	Gerritt Davidse (MO)
Rubiaceae	John Dwyer (MO)

[a] Index Herbariorum, Ed. 7 (1981) acronyms.

southern Petén, Alta Verapaz, and the highlands stretching from Izabal on the Caribbean coast to Huehuetenango and Quezaltenango near the Mexican border, where roads make these areas more accessible. The lowlands on the Pacific coast are also well collected.

Areas relatively poorly collected include the Sierra de los Cuchumatanes in the northwest of the country (an area of special interest), the more inaccessible parts of Petén (assuming that there is still forest in that region), and Baja Verapaz, on the north side of the mountains.

It is estimated that there are 8,000 plant species in Guatemala. Well collected groups include the Orchidaceae with 542 species (1965) and the pteridophytes, with 652 species (1983). Refer to the *Flora of Guatemala* for a comprehensive review of families.

Areas at risk of destruction include the lowland tropical dry forests of Petén, as indicated by clearings marked on the Tactical Pilotage Charts.

V. Academic and Governmental Institutions

Institutions active in floristic inventory in the region include the Escuela de Biologia, Museo de Historia Natural y Jardin Botánico, Universidad de San Carlos de Guatemala (USCG), the Museo Nacional de Historia Natural (GUAT), the Departamento de Parques Nacionales y Vida Sylvestre, Instituto Nacional Forestal (INAFOR), Guatemala City, the Escuela de Biologia y Farmácia, Universidad de San Carlos, Guatemala City, and the Instituto Guatemalteco de Turismo, Guatemala.

The protected area of Guatemala is 57,600 ha, excluding disturbed sites. Article 108 of Constitution of the Republic establishes a basis for the protection of the environment.

Forest Law 1925 and Agrarian Law 1936 are designed to

maintain the necessary natural cover for conserving sources of water.

Protected areas are:

1) Tikal National Park, with 57,600 ha, which includes ruins of Maya City, and dry tropical and subtropical forest.

2) Lake Atitlán NP, Dept. of Solola, 70 km from Guatemala City, with 13,000 ha. There is no information about vegetation. The area has experienced much disturbance, resulting from the construction of homes and hotels.

3) Rio Dulce NP, with 24,200 ha. There is some virgin forest, humid tropical and sub-tropical forest, and low montane wet forest. However, several small settlements are in the region and there is much deforestation and illegal hunting. Nickel mining is planned. There is no functioning management or protection.

4) El Rosario NP, Dept. of Petén, has 1,030 ha. There is no information about vegetation or disturbance.

5) Pacaya Volcano Natural Monument has 2,000 ha of broadleaf and coniferous species. There is cultivation and pastureland on slopes, and a program of reforestation on areas destroyed by forest fires caused by volcanic activity. There is no management or protection.

6) University Biotope for the Conservation of the Quetzal, Dept. of Baja Verapaz, in Sierra de las Minas has 900 ha of cloud forest. It is well protected with no disturbance.

VI. Future of Inventory

The flora of Guatemala was more or less complete in 1977. What is needed is an enforced government conservation program to protect the natural forests of the country from the insidious and relentless invasions by logging combines, slash and burn farming and general industrial pollution. Countries with influence in Guatemala (i.e., that have money invested there) would no doubt be able to exert pressure on the government.

Obviously teaching centers should be available, especially to educate the young, to show that conservation is vitally important to all. These centres should cover all important aspects of education—e.g., alternative sources of power, population control, etc.

VII. Literature Cited

Ames, O. & D. S. Correll. 1952–1953. Orchids of Guatemala, 1 & 2. Fieldiana Bot. **26(1–2):** 1–727.

Anonymous. 1905. Specimens gathered by Professors Kellerman and Hine in Guatemala. Ohio State Lantern, School of Journalism, Ohio State University. 30 pp.

Atlas Nacional de Guatemala. 1972.

Bartlett, H. H. 1935. A method of procedure for field work in parts of British Honduras and Peten Forest in Guatemala. Publ. Carnegie Inst. Wash. **461:** 1–26.

Beard, J. S. 1953. The savanna vegetation of northern tropical America. Ecol. Monogr. **23(2):** 149–215.

———. 1955. The classification of tropical American vegetation types.

Correll, D. S. 1965. Supplement to the orchids of Guatemala and British Honduras. Fieldiana Bot. **31(7):** 177–221.

Directory of Military Survey. 1977. Tactical pilotage charts. Ministry of Defense. London.

Gomez-Pompa, A. 1967. Some problems of tropical plant ecology. J. Arnold Arbor. **48(2):** 104–121.

Graham, A. 1973. Vegetation and vegetation history of northern Latin America. Elsevier.

Holdridge, L. R., F. B. Lamb & B. Mason. 1950. Los bosques de Guatemala. Instituto Interamericana de Ciencias Agricolas. 174 pp.

Holmgren, P. K., W. Keuken & E. K. Schofield. 1981. Index Herbariorum. 7th edition, Ultrecht.

Inst. Geogr. Nac. (IGN) de Guatemala. 1973. Vegetation maps.

Iltis, H. H. 1978. Extinction or preservation: What biological future for the South American tropics. University of Wisconsin.

Inst. Nac. Forest (INAFOR). 1981. Estudio sobre exportaciones de fauna y flora silvestre de Guatemala de Enero/78 a Diciembre/80. Dept. de Parques Nacionales y Vida Silvestre. 24 pp.

IUCN. 1982. Directory of neotropical protected areas.

Kellerman, W. A. 1908. A vacation trip through Guatemala. Tropical and subtropical America **1:** 113–117.

Koopowitz, H. & H. Kaye. 1983. Plant extinction: A global crisis. Stonewall Press. Washington.

Lizama, C. 1981. Orchids of Guatemala. In J. Stewart & C. N van der Merwe (eds.), Proceedings of the Tenth World Conference. Durban, South Africa.

Luer, C. A. 1975. *Pleurothallis* of Mexico and Central America. Selbyana **1(2):** 196–212.

Lundell, C. L. 1937. Vegetation of Petén. Publ. Carnegie Inst. Wash. **478:** 1–244.

———. 1962 Plant exploration of s. lowlands of Petén. Amer. Philos. Yearbook, 308–311.

Morton, F. 1931. Guatemala. Vegetationsbilder **22(1):** 15pp, 6 plates.

Padilla, S. A. 1928. El arbol nacional de Guatemala. An. Soc. Geogr. Hist. Guatemala **4(3):** 236–239.

Perry, D. R. 1984. The canopy of the tropical rain forest. Sci. Amer. **251(5):** 114–122.

Popenoe, J. 1941. Plant resources of Guatemala. Chron. Bot. (Waltham, Mass.) **6:** 16–19.

Prance, G. T. & T. S. Elias. 1978. Extinction is forever. New York Bot. Garden, New York. 437 pp.

Record, S. J. 1927. Trees of Gualan, Guatemala. Trop. Woods **11:** 10–18.

———. 1928. Trees of Santa Inez, Guatemala. Trop. Woods **14:** 25–29.

——— & H. Kuylen. 1926. Trees of lower Rio Motagua Valley, Guatemala. Trop. Woods **7:** 10–29.

Richards, P. W. 1952. The tropical rain forest. University Press, Cambridge.

———. 1973. The tropical rain forest. Sci. Amer. **229(6):** 58–67.

———. 1977. Tropical forests and woodlands: An overview. Agro-Ecosystems **3:** 225.

Schwerdtfeger, F. 1953. Los Pinos de Guatemala. Informe FAO/ETAP **202:** 57, ff. 139.

Standley, P. C. 1941. The forests of Guatemala. Trop. Woods **67:** 1–18.

———. 1945. Notes of some Guatemalan trees. Trop. Woods **84:** 1–18.

———, J. Steyermark & L. O. Williams. 1946–1977. Flora of Guatemala. Fieldiana, Bot. 24. 12 volumes including index.

Stolze, R. S. 1976, 1981, 1983. Ferns and fern allies of Guatemala. Fieldiana, Bot. **39**: 1–30; Fieldiana, Bot., N. S. **6**: 1–522 (Pt. 2—*Polypodiaceae*); Fieldiana, Bot. **12**: 1–91 (Pt. 3—*Marsileaceae, Salviniaceae* and the Fern Allies).

Williams, L. O. 1970. Geographical locations in Guatemala. Taxon **19**: 486–487.

Belize

Rachel J. Hampshire

Contents

I. Description of the Region	288
A. Geographical Extent and Topography	288
B. Population	288
II. Vegetation Maps	288
III. Herbarium Resources	288
IV. Present State of Inventory	288
V. Academic and Governmental Institutions	288
VI. Future of Inventory	289
VII. Literature Cited	289
VIII. Selected References	289

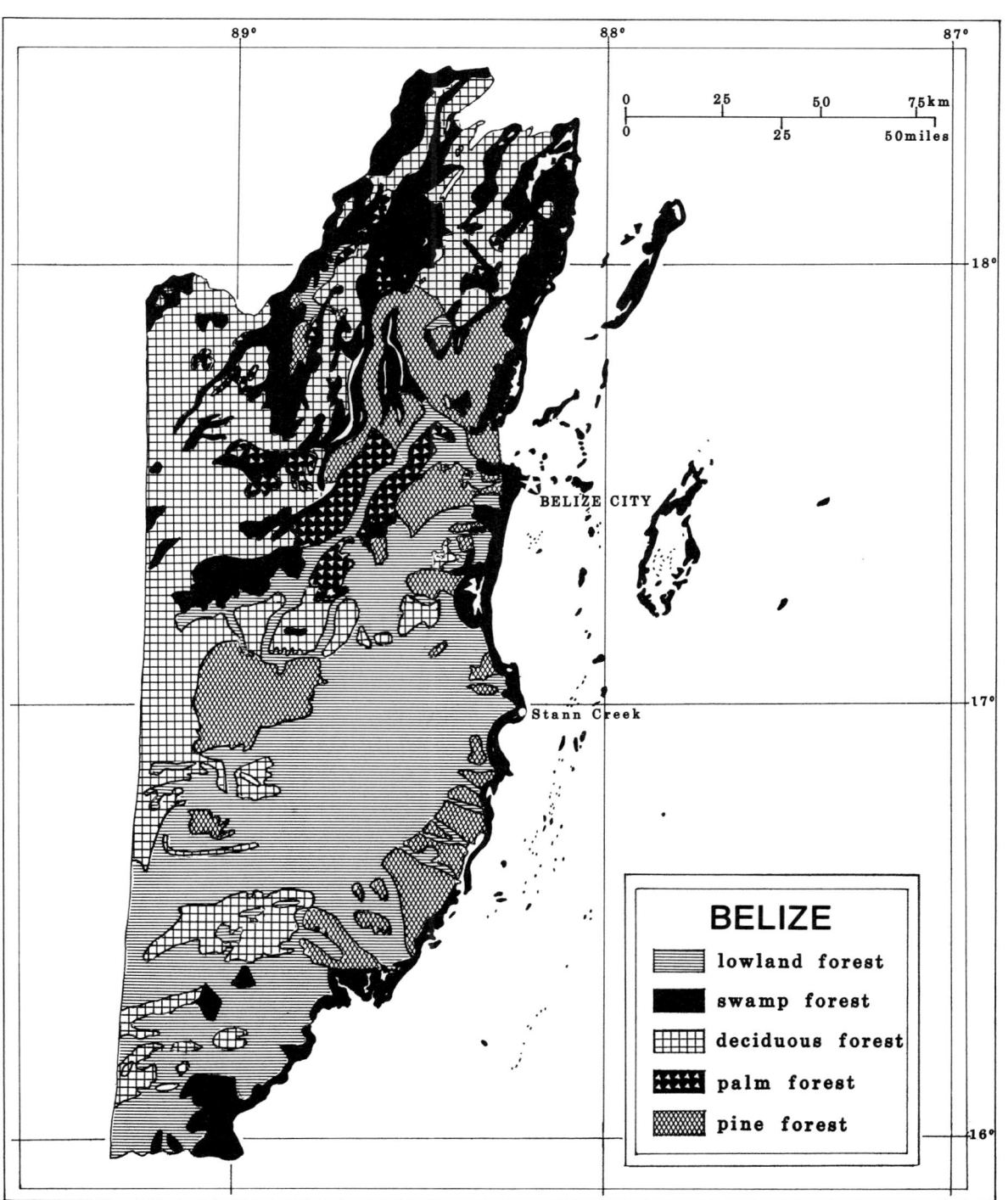

I. Description of the Region

A. Geographical Extent and Topography

Belize has an area of 22,965 km², including cays. Of this, 12,570 km² are forested (Langley, 1981).

The coastlands of Belize are flat and swampy. The land rises gradually towards the interior, although the northern half of the country only reaches 120 m above sea level. The land here is hilly in parts, with areas of low tableland. The Maya Mountains, reaching 1,100 m, form the backbone of the southern half of Belize.

There are also a number of offshore islets (cays), the smallest being sandy and barren, the larger ones sometimes supporting mangroves or coconut plantations. (Fig. 1).

B. Population

The rate of population expansion is approximately 3.7% per annum.

II. Vegetation Maps

The important vegetation maps for Belize are Wright et al. (1958), Survey Dept.—Belize (1945, 1952) and Directorate of Overseas Surveys (1973). However, it should be noted that the forests of Belize are being felled at such a rate that most vegetation maps are out of date.

III. Herbarium Resources

Approximately 4,000 herbarium specimens from Belize are held in the herbarium of the Forestry Department, Ministry of Natural Resources, Belmopan (BRH). An additional 21,000 collections are held abroad, principally at BM, F, LL, and MO.

IV. Present State of Inventory

In total, about 25,000 specimens have been collected from Belize. This gives an overall figure of more than 100 collections per 100 km². However, this figure is misleading, since some areas have been overcollected, while less accessible parts are poorly collected.

The well-collected regions of Belize include: the Turneffe Islands, areas near main roads, (e.g., Mile 30, Western Highway), Mountain Pine Ridge, and the areas around Orange Walk Town.

Poorly-collected regions of Belize include: the Maya Mountains (Dwyer & Spellman, 1981; Gentry, 1978), Sarstoon River and southwest Toledo Province (Gentry, 1978;

Table I
The population of Belize in 1980.

District	Population Total	Rural	Density/km²
Corozal	22,902	16,003	12.3
Orange Walk	22,870	14,431	4.8
Belize	50,801	11,030	12.1
Cayo	22,837	14,286	4.3
Stann Creek	14,181	7,520	6.5
Toledo	11,762	9,366	2.5
Grand total	145,353	72,636	Av. 6.3

and Dwyer & Spellman, 1981), the Cockscomb Mountains, the gorge around "1000 Foot Falls," ravines around Pine Ridge, and the top of Victoria Peak.

Standley and Record (1936) estimated that Belize had 138 endemic species; IUCN (1982) estimate over 150 endemic species. The total number of species of vascular plants was estimated at 2,500–3,000 by Gentry (1978) and 2,500 dicotyledons by Dwyer and Spellman (1981). The percentage of endemic species is therefore 4.6–6%. The latest IUCN statistics (Davis et al., 1986) show Belize has seven threatened endemic plant species.

New distributions are being revealed from some cays, some gorges, including the Branch River Gorge and the Columbia Forest Reserve.

Families which have been well collected in Belize include Melastomataceae, Leguminosae, Rubiaceae and Piperaceae. Poorly-collected groups include those difficult to reach or press, such as tall trees and the Bromeliaceae. Authors of "Flora Mesoamericana" are working on all families known to occur in Belize.

The estimated rate of deforestation of broadleaved forest in Belize is 90 km² annually. Destruction of forest occurs mainly in the more densely populated north. Many forested areas of Belize are threatened, partly by refugees fleeing from El Salvador and Guatemala. Little primary forest remains in Belize (D'Arcy, 1977: 97).

V. Academic and Governmental Institutions

Academic institutions which could become involved in floristic inventory work include: St. John's College, Belize City; Belize College of Art, Science and Technology, Belize City; Fletcher College, Corazal; University of the West Indies Extra-Mural Department, Belize City; and Wesley College, Belize City.

The only government institution engaged in inventory is the Herbarium (BRH), Forest Department, Belmopan. In 1982 the herbarium was small and had severe insect infestations.

The main foreign program is the "Flora Mesoamericana" project.

Belize has 4,169.8 ha under protection. Protected areas include the Guanacaste Park (Bird Sanctuary) and eight cays consisting of coconut groves and mangroves. There are also ten forest reserves, totaling 418,994 ha.

VI. Future of Inventory

The undercollected areas of Belize have been highlighted by, for example, Gentry (1978) and Dwyer and Spellman (1981). Future collecting programs should therefore correlate with areas recognized as currently undercollected. To date, plant collecting programs have not correlated well with known areas of endemism. The Maya Mountains might support the richest flora of Belize and, like many other botanically interesting parts of the country, they are among the least accessible.

An herbarium where specimens would be stored in stable, insect-free conditions is a priority. The herbarium would cost about $60,000. A botanical garden (currently lacking), with a teaching center, would be useful in showing people the importance of the local flora and developing a public interest in it. The garden and herbarium could be located in Belmopan. For the herbarium, two scientists and two technical officers would be needed. This could be reduced to one scientist and one technical officer after five years. Additional staff would be needed to develop a botanic garden. Personnel for these institutions would cost $40,000 per annum initially, reducing to $20,000 after five years (at current salaries), (i.e, two scientists at $14,000 and two technicians at $6,000). Materials and transport for the institutions would cost $20,000 in the first year, reducing to $8,000 in subsequent years.

VII. Literature Cited

D'Arcy, W. G. 1977. Endangered landscapes in Panama and Central America: the threat to plant species. Pages 89–104 *in* G. T. Prance & T. S. Elias (eds.), Extinction is forever. The New York Botanical Garden, New York.

Davis, S. D., S. J. M. Droop, P. Gregerson, L. Henson, C. J. Leon, J. L. Villa-Lobos, H. Synge & J. Zantovska. 1986. Plants in danger: What do we know? IUCN, Gland, Morges, Switzerland.

Directorate of Overseas Surveys. 1973. Land use and vegetation. British Honduras. Maps 4 and 5 (Belize Valley) 1:100,000. British Government's Overseas Development Administration (Land Resources Division).

Dwyer, J. D. & D. L. Spellman. 1981. A list of the Dicotyledoneae of Belize. Rhodora 83: 161–236.

Gentry, A. H. 1978. Floristic knowledge and needs in Pacific Tropical America. Brittonia 30: 134–153.

IUCN. 1982. Directory of neotropical protected areas. Morges, Switzerland.

Langley, J. P. (ed.). 1981. Tropical Forest Resources Project Part Latin America. FAO/UNEP, Rome.

Standley, P. C. & S. J. Record. 1936. The forests and flora of British Honduras. Field Museum Nat. Hist. Bot. 12: 1–432.

Stoddart, D. R., F. R. Fosberg & D. L. Spellman. Cays of the Belize Barrier Reef and Lagoon. Atoll Res. Bull. 256: 1–76.

Survey Department—Belize. 1945. Vegetation Map. Scale 1:50,000. Sheets 1–31. D.C.S. Misc. 8.

———. 1952. Vegetation Map. Scale 1:50,000. Sheets 1–31. D.C.S. Misc. 8.

Wright, A. C. S., D. H. Romney, R. H. Arbuckle, R. C. Carcel & A. C. Williamson. 1958. British Honduras Natural Vegetation Map, Sheets 1 and 2. Scale 1:250,000. British Honduras Survey, Forestry and Geological Departments.

Various authors. 1936 and 1940. Botany of the Maya Area. Miscellaneous Papers. I–XXI. Publ. Carnegie Inst. Wash. Nos. 461 and 522.

VIII. Selected References

Fosberg, F. R., D. R. Stoddart, M.-H. Sachet & D. L. Spellman. 1982. Plants of the Belize Cays. Atoll Res. Bull. 258: 1–77.

Lundell, C. L. 1942. The vegetation and natural resources of British Honduras. Chron. Bot. 7(4): 169–171.

Nelson Smith, J. H. 1945a. Forest associations of British Honduras. II. Caribbean Forest. 6: 45–70.

———. 1945b. Forest associations of British Honduras. III. Caribbean Forest. 6: 131–158.

Spellman, D. L., J. D. Dwyer & G. Davidse. 1975. A list of the Monocotyledonae of Belize including a historical introduction to plant collecting in Belize. Rhodora 77: 105–140.

Standley, P.C., J. A. Steyermark & L. O. Williams. 1946–1977. Flora of Guatemala. Fieldiana, Bot. 24: 1–13.

Stoddart, D-R, F. R. Fosberg & M.-H. Sachet. 1982. Ten years of change on the Glover's Reef Cays. Atoll Res. Bull. 257: 1–41.

The majority of publications about the flora of Belize were written during the period 1920–1945, and give proposals for the development of forestry and the timber trade.

Honduras

Cirilo Nelson

Contents

I. Description of the Region	291
A. Geographical Extent and Topography	291
B. Population	291
II. Vegetation Maps	291
III. Herbarium Resources	291
IV. Present State of Inventory	292
V. Academic and Governmental Institutions	292
VI. Future of Inventory	292
VII. Literature Cited	292
VIII. Selected References	293

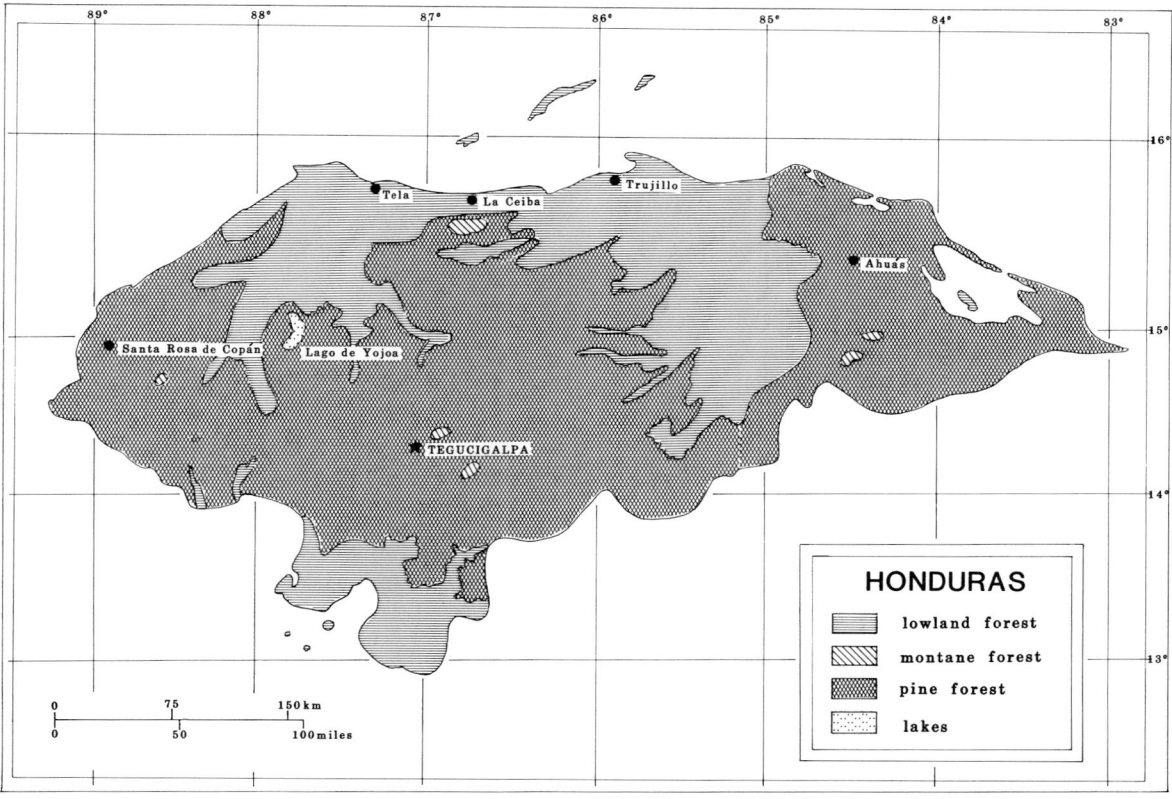

I. Description of the Region

A. Geographical Extent and Topography

Honduras has an area of 112,008 km². The Honduran relief is 63 percent highlands, with a mean altitude of 1,000 m, and the rest is plains. Honduras has three physiographic regions: the Pacific Lowland, which includes the Motagua Plain, the Ulúa-Chamelecón Plain, the Nombre de Dios Piedmont, the Aguán-Negro Plain, and the Mosquito Coast, and the Serranía region, which includes the Northern Cordillera and the Southern Cordillera (Wilson & Meyer, 1982) (Fig. 1).

B. Population

Honduras is a rapidly growing country with a 3.5 percent annual population growth rate. If this trend continues, by the year 2000 the Honduran population will reach 7 million. The population density is about 24 inhabitants per km², the western part being the most dense. The population of Honduras is about 70 percent rural (Campanella et al., 1982).

II. Vegetation Maps

There is an ecological map of Honduras by Leslie R. Holdridge, issued by the Organization of American States in 1962. Wilson and Meyer (1982), based on their personal observations, have presented some variants of it.

III. Herbarium Resources

The Escuela Agrícola Panamericana has an estimated 55,000 Honduran specimens. The Universidad Nacional Autónoma de Honduras has an estimated 20,000 Honduran specimens. La Escuela Nacional de Ciencias Forestales at Siguatepeque, Honduras, has an estimated 5,000 Honduran specimens. The United Brands at La Lima, Honduras, has an estimated 1,000 specimens.

It is unknown how many Honduran specimens are held in institutions abroad, but the U. S. National Herbarium (US), Field Museum (F), the New York Botanical Garden (NY), and the Missouri Botanical Garden (MO) are the foreign institutions that hold the largest numbers.

IV. Present State of Inventory

Areas relatively well collected include the Departments of Francisco Morazán, Comayagua, Atlántida, and Gracias a Dios.

Families and genera well collected in Honduras are listed in Table I, along with their principal collectors.

Areas poorly collected include the remainder of the 14 departments.

The number of endemics and extinctions are poorly known.

The Departments of Atlántida, Olancho, Gracias a Dios, and Lempira are yielding new and important distributions (Hazlett, 1977a,b; Molina, 1977; Nelson, 1976b, 1977, 1978a, 1979; Proctor, 1983).

Important areas threatened by destruction or conversion include Cerro Santa Bárbara, tropical forests around Trujillo, tropical forests past Culmí in eastern Olancho, tropical forests around San Esteban in Olancho.

Table I

Families and genera well collected in Honduras.

Polypodiaceae: *Adiantum, Asplenium, Blechnum, Diplazium, Dryopteris, Elaphoglossum, Polypodium, Thelypteris.*
  Notable collectors: Standley, Molina, Allen, Clewell, Hazlett, Nelson.
Pinaceae: *Pinus.*
  Notable collectors: Standley, Molina, Allen, Nelson, Styles, Hughes.
Gramineae: *Andropogon, Axonopus, Bouteloua, Digitaria, Eragrostis, Panicum, Paspalum.*
  Notable collectors: Swallen, Standley, Molina, Pohl, Davidse, Nelson.
Cyperaceae: *Bulbostylis, Cyperus, Eleocharis, Rhynchospora, Scleria.*
  Notable collectors: Standley, Molina, Davidse, Nelson.
Araceae: *Anthurium, Philodendron.*
  Notable collectors: Standley, Molina, Croat.
Bromeliaceae: *Catopsis, Tillandsia.*
  Notable collectors: Allen, Yuncker, Williams, Standley, Molina.
Musaceae: *Heliconia.*
  Notable collectors: Standley, Molina, Williams.
Orchidaceae: *Epidendrum, Maxillaria, Pleurothallis.*
  Notable collectors: Edwards, Standley, Molina, Clewell, Nelson, Williams.
Piperaceae: *Peperomia, Piper.*
  Notable collectors: Yuncker, Standley, Molina, Nelson.
Fagaceae: *Quercus.*
  Notable collectors: Standley, Molina, Nelson.
Moraceae: *Ficus.*
  Notable collectors: Standley, Molina, Nelson.
Loranthaceae: *Phoradendron.*
  Notable collectors: Standley, Molina, Nelson.
Asteraceae: *Eupatorium, Vernonia.*
  Notable collectors: Standley, Molina, Nelson, Williams.

V. Academic and Governmental Institutions

At present, only the Universidad Nacional Autónoma de Honduras (TEFH) is doing a floristic inventory of the region. The activity of the Escuela Agrícola Panamericana (EAP) has dropped considerably from what it was in the past.

The Universidad Nacional had an estimated budget of $5,000 per year, but it also has dropped considerably in the last 3 years. This does not allow the institution to hire additional personnel or to increase the field trips. There are only two persons doing a floristic inventory at the Universidad Nacional. They bring in, at present, around 2,000 numbers per year. More people are needed for more intensive field trips, gluing specimens, and dispatching duplicates for exchange.

There is no government institution engaged in the compilation of a floristic inventory.

Opportunities for collaboration with institutions in other nations and with other disciplines are open, but financing in advance would be desirable for field trips and shipping costs.

VI. Future of Inventory

In order to insure an adequate floristic inventory of the tropical forests of the region, additional personnel, such as technicians and scientists, would be needed to ensure prompt identifications, and more intensive collecting. The approximate cost would be US $20,000 per year (author's personal calculations).

VII. Literature Cited

Campanella, P. et al. 1982. Honduras. Perfil Ambiental. AID contract No. AID/SOD/PDC-C-0247. JRB Associates. McLean, Virginia (USA). 201 p.

Hazlett, D. 1979a. Arboles maderables y otros árboles desconocidos de la cordillera Nombre de Dios. Ceiba 23: 76–84.

———. 1979b. A first report on the vegetation of Celaque. Ceiba 23: 114–128.

Molina, A. 1977. Nuevas contribuciones a la flora de Honduras. Ceiba 21(2): 61–66.

Nelson, C. 1976b. Plantas nuevas para la flora de Honduras. Ceiba 20(2): 58–68.

———. 1977. Plantas nuevas para la flora de Honduras. Ceiba 21(1): 51–55.

———. 1978a. Contribuciones a la flora de La Mosquitia, Honduras. Ceiba 22(1): 41–64.

———. 1979. Plantas nuevas para la flora de Honduras, III. Ceiba 23: 85–92.

Proctor, G. R. 1983. New plant records from the Mosquitia region of Honduras. Moscosoa 2(1): 19–22.

Wilson, L. D. & J. R. Meyer. 1982. The snakes of Honduras. Milwaukee Publ. Museum. Wisconsin. 159 p.

VIII. Selected References

Aguilar, J. 1938. Flora tradicional. Tip. Nac. San Salvador. 75 p.

———. 1945. Contribución al estudio de nuestra flora tradicional. Revista Arch. Bibl. Nac. (Tegucigalpa) 24(5/6): 244–254.

———. 1946. Contribución al estudio de nuestra flora tradicional. Revista Arch. Bibl. Nac. (Tegucigalpa) 24(6/7): 336–342; 25(2/5): 168–171; 24(9/10): 453–458.

Allen, P. H. 1953. Two new fan palms from Central America. Ceiba 3: 173–178.

———. 1955. The conquest of Cerro Santa Bárbara, Honduras. Ceiba 4: 253–270.

Alvarado, M. C. 1986. Clasificación e inventario de árboles y arbustos del Cerro Juana Laínez, Tegucigalpa, Honduras. B. S. Monograph. Univ. Nac. Aut. Honduras. Tegucigalpa.

Ames, O. 1932. A new *Epidendrum* from Spanish Honduras. Bot. Mus. Leafl. 1: 1–3.

———. 1933. A new *Ostomeria* from Spanish Honduras. Bot. Mus. Leafl. 4: 1–3.

———. 1934. *Epidendrum cystosum*, a new species from the Republic of Honduras. Bot. Mus. Leafl. 2: 105–111.

Bartram, E. S. 1929. Honduran mosses collected by Paul C. Standley. Field Mus. Nat. Hist., Bot. Ser. 4(9): 349–364.

Blake, S. F. 1922. Native names and uses of some plants of eastern Guatemala and Honduras. Contr. U. S. Natl. Herb. 24: 87–100. pl. 29–33.

Carr, A. F., Jr. 1950. Outline for a classification of animal habitats in Honduras. Bull. Amer. Mus. Nat. Hist. 94: 563–594.

Castro, H. 1979. Contribución al estudio de la flora medicinal de Honduras. Doctoral Thesis. Fac. Chem. Pharm. Natl. Univ. Honduras.

Clewell, A. F. 1973. Floristic composition of a stand of *Pinus oocarpa* Schiede in Honduras. Biotropica 5(3): 175–182.

———. 1975. Las Compuestas de Honduras. Ceiba 19: 119–244.

———. 1986. Observations on the vegetation of the Mosquitia in Honduras. Sida 11(3): 258–270.

Conzemius, E. 1927. Economic plants of the Bay Islands (Honduras). Gard. Chron. III. 81: 50–51, 69–70, 81, 117–118, 133, 180–181, 217, 270, 305, 413–414, fig. 42.

Cruz, F. 1901. Botica del pueblo. Ed. 3. Impr. Enrique Rojas. Madrid.

Cruz, G. & M. Erazo. 1977. Análisis de la vegetación del bosque nebuloso "La Tigra" (Reserva forestal de San Juancito). Ceiba 21(2): 19–60.

——— et al. 1978. Cuenca del río Plátano. Dirección General Recursos Naturales Renovables. Tegucigalpa. 133 p.

Chable, A. C. 1967. Reforestation in the Republic of Honduras, Central America. Ceiba 13(2): 1–56.

Eoff, G. 1976. Lowland orchids of the Mosquitia, Honduras. The Florida Orch. 19(1): 13–17.

FAO Report. 1968. Survey of the pine forests of Honduras. Final Report U. N. Dev. Program. FAO/SF. 26-HON 50. 80 p.

Fiallos, E. 1919. Apuntes de la flora de Honduras. Impr. Nac. Tegucigalpa. 174 p.

García, M. 1979. Vegetación de la quebrada La Orejona. B.S. Monograph. Biol. Dept. Natl. Univ. Honduras. Tegucigalpa.

Gilmartin, A. J. 1965. Las bromeliáceas de Honduras. Ceiba 11(2): 1–80.

Gould, F. W. 1961. *Opizia stolonifera* Presl in Honduras. Ceiba 9(2): 59–60.

Hazlett, D. 1975. El aprovechamiento del *Swietenia macrophylla* y la vegetación en asociación con lo mismo en un bosque lluvioso olanchano. Revista Acad. Hond. Geogr. Hist. (Tegucigalpa) 59(10): 60–63.

———. No date. Arboles de la reserva forestal de Lancetilla. Mimeographed work. 2 p.

Helbig, K. M. 1965. Areas y paisajes del nordeste de Honduras. Banco Central Honduras. Tegucigalpa. 287 p.

Hemsley, W. B. 1879–1888. Biologia Centrali-Americana; Botany. London. 5 vols.

Johannessen, C. L. 1963. Savannas of interior Honduras. Ibero-Americana: 46. University of California Press. Berkely and Los Angeles. 160 p.

King, R. & H. Robinson 1979. Studies in the Eupatorieae (Asteraceae). CLXXVI. The relationship of *Eupatorium cyrillinelsonii*. Phytologia 44(2): 84–88.

Landa, L. 1940. Botánica. Tip. Nac. Ariston. Tegucigalpa. 149 p.

Leonard, E. 1950. Five new species of Acanthaceae from Honduras. Ceiba 1(2): 103–115.

Little, E. L. 1977. *Pinus hartwegii* in Honduras. Ceiba 21(1): 103–115.

Molina, A. 1951. Nuevas especies de plantas de la república de Honduras. Ceiba 1: 255–263.

———. 1952. Nuevas plantas de Nicaragua y Honduras. Ceiba 3: 91–97.

———. 1953a. Cinco nuevas plantas leñosas de Honduras. Ceiba 3: 165–171.

———. 1953b. Revisión de las especies de *Cephaëlis* de México, Centro América y las Antillas. Ceiba 4: 1–38.

———. 1964. Coníferas de Honduras. Ceiba 10(1): 5–21.

———. 1970. Cuatro nuevas compuestas de Honduras. Ceiba 16(2): 51–56.

———. 1974. Vegetación del valle de Comayagua. Ceiba 18(1–2): 47–69.

———. 1975. Enumeración de las plantas de Honduras. Ceiba 19(1): 1–118.

———. 1978. Un nuevo *Eupatorium* de Honduras. Ceiba 22(1): 39–40.

Nelson, C. 1976a. Algunas plantas del departamento de Ocotpeque, Honduras. Ceiba 20(1): 27–41.

———. 1978b. Correción a la ortografía de un *Eupatorium* nuevo. Ceiba 22(2): 162.

———. 1981. A new *Psychotria* (Rubiaceae) from Nicaragua and Honduras. Phytologia 50(1): 1–2.

———. 1982a. A new *Robinsonella* (Malvaceae) from Honduras. Phytologia 51(6): 381–383.

———. 1982b. Nociones de taxonomía vegetal. Editorial Univ. Tegucigalpa. 223 p.

———. 1984. Una *Ocotea* (Lauraceae), una *Salvia* (Labiatae) y un *Eupatorium* (Compositae) nuevos de Honduras. Ceiba 25: 173–176.

———. 1986. Plantas comunes de Honduras. 2 Vols. Editorial Univ. Tegucigalpa. 922 p.

Pérez, A. & T. Pérez. 1935. Catálogo de la Quinta Pérez Estrada. Inglés-Español. Impr. Pérez Estrada. San Pedro Sula. 31 p.

Pfeifer, Howard Wm. 1960. Vascular plants of Mount Uyuca. Ceiba 8(2): 102–142.

Record, S. J. 1927. Trees of Honduras. Trop. Woods 10: 10–47.

Reina, J. M. 1942. Don Francisco Cruz y la botica del pueblo. Tip. Nac. Tegucigalpa. 229 p.

———. 1944. Treinta tisanas populares en Honduras. Tip. Nac. Tegucigalpa. 22 p.

———. 1958. Monografía botánica. Soc. Geogr. Hist. Honduras. Tegucigalpa. 24 p.

Rosa, J. M. T. 1918. Flora y fauna santabarbarense. Tip. Nac. Tegucigalpa. 73 p.

———. 1970. Cualidades de algunas plantas. Revista Ariel III. **12(226):** 22.

Roys, R. L. 1931. The ethnobotany of the Maya. Publ. Middle Amer. Res. Ser. Tulane Univ. Louisiana No. 2. 359 p.

Salguero, D. R. S. de. 1986. Estudio etnobotánico sobre: *Argemone mexicana, Plantago major, Sechium edule* y *Tagetes lucida* en diferentes municipios de Honduras. B. S. Monograph. Biol. Dept. Univ. Nac. Aut. Honduras. Tegucigalpa.

Smith, J. D. 1889–1907. Enumeratio plantarum guatemalensium necnon salvadorensium hondurensium nicaraguensium costaricensium. 8 vols. Oquawkae.

Standley, P. C. 1930a. The woody plants of Siguatepeque, Honduras. J Arnold Arb. **11:** 15–46.

———. 1930b. A second list of trees of Honduras. Trop. Woods **21:** 9–41.

———. 1931. Flora of the Lancetilla Valley. Field Mus. Nat. Hist., Bot. Ser. **10:** 1–418.

———. 1934. Additions to the trees of Honduras. Trop. Woods **37:** 27–39.

———. 1950. Teosinte in Honduras. Ceiba **1(1):** 58–61.

———. 1952. *Begonia fonsecae*, especie nueva hondureña. Ceiba **3:** 149–151.

——— & **L. O. Williams.** 1950a. *Dioon mejiae*, a new cycad from Honduras. Ceiba **1:** 36–38.

——— & ———. 1950b. Plantas nuevas hondureñas y nicaragüenses. Ceiba **1:** 74–96.

——— & ———. 1952. *Synandrina*, género nuevo hondureño de las flacurciáceas. Ceiba **3:** 74–75.

——— & ———. 1954. A new *Senecio* from Honduras. Ceiba **4:** 190–191.

Steyermark, J. A. 1950. A new *Utricularia* from Honduras. Ceiba **1:** 125–126.

———. 1952. A new *Carex* from Guatemala and Honduras. Ceiba **3:** 23–24.

United Nations. 1968. Survey of pine forests, Honduras, final report. Food & Agric. Organization, FAO/SF: 26-HON 50. 79 p + maps.

Villeda, M. E. 1979. Distribución geográfica de la familia Bignoniaceae en Honduras. B. S. Monograph. Biology Dept. Univ. Nac. Aut. Honduras. Tegucigalpa.

Vogel, F. H. 1954. Los bosques de Honduras. Ceiba **4(2):** 85–121.

Williams, L. O. 1954. A new *Odontoglossum* from Honduras. Ceiba **4(4):** 225–226.

Yuncker, T. G. 1938. A contribution to the flora of Honduras. Field Mus. Nat. Hist., Bot. Ser. **17(4):** 236–407.

———. 1940. Flora of the Aguán Valley and the coastal regions near La Ceiba, Honduras. Field Mus. Nat. Hist., Bot. Ser. **9(4):** 245–346.

El Salvador

Rachel J. Hampshire

Contents

I.	Description of the Region	296
	A. Geographical Extent and Topography	296
	B. Population	296
II.	Vegetation Maps	296
III.	Herbarium Resources	296
IV.	Academic and Governmental Institutions	297
V.	Present State of Inventory	297
VI.	Future of Inventory	297
VII.	Literature Cited	297
VIII.	Selected References	298

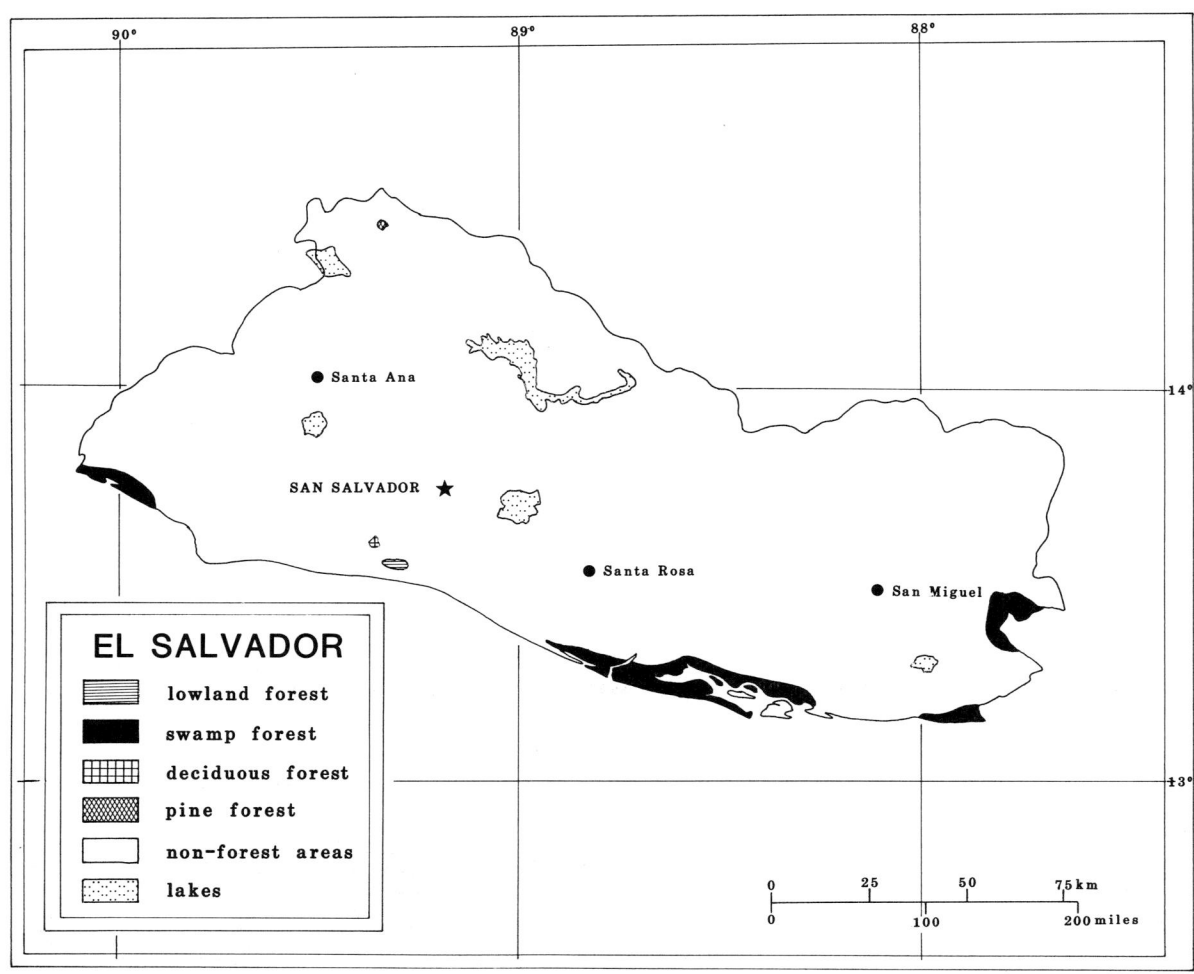

I. Description of the Region

A. Geographical Extent and Topography

El Salvador is the smallest country in Central America, occupying an area of 21,040 km².

Along the south coast of El Salvador is a narrow coastal plain, occupying about ten percent of the country. A volcanic mountain chain runs behind the coastal plain across El Salvador from east to west. The chain consists of more than 20 volcanoes in the central plain at 200–600 m. The plain is divided by the valleys of the Río Lempa and Río Grande de San Miguel. North of this is a mountain chain formed by the Metapan and Chalatenango Mountains. The boundary with Honduras is disputed (Fig. 1).

B. Population

El Salvador is the most densely populated country in Central America. The 1983 census was 5,229,096, or 249 people per km². Statistics for July 1981 show that for a total population of 4,873,385, the urban population was 1,902,339 and the rural population 2,971,046. The rate of population increase is approximately 3.5% per annum. Sadly, there is no sign of a reduction in this rate.

II. Vegetation Maps

The vegetation maps for El Salvador include Holdridge (1959) and Lauer (1954).

III. Herbarium Resources

El Salvador has been well collected. Gentry (1978) estimated that a total of 8,000 specimens have been collected from the country. In 1974, the herbarium of the Departamento de Biologia, Universidad de El Salvador (ITIC) held

6,000 specimens mainly from El Salvador. By 1981 the institution held about 30,000 specimens, with emphasis on El Salvador and Central America. More than 6,000 specimens from El Salvador are now held in the country.

Approximately 2,000 specimens may be held in institutions abroad. US holds important collections, especially the collections of Standley, and Renson's specimens sent for identification and subsequently described by Pittier. The Renson specimens retained in El Salvador were unfortunately destroyed. P. Bernhardt (then in the Peace Corps) sent many specimens to MO and UC in the mid-seventies. Important collections made in El Salvador are briefly described in the introduction to Standley and Calderon (1925).

IV. Academic and Governmental Institutions

Local institutions engaged in biological research include the Department of Biology of the Universidad de El Salvador, Ciudad Universitaria, San Salvador, the Universidad Centro-americana "Jose Simeon Canas," San Salvador; and the Parque Zoológico Nacional y Jardín Botánico, San Salvador. The principal effort to coordinate floristic inventory is currently by the "Flora Mesoamericana Project" based at the British Museum (Natural History), the Missouri Botanical Garden, and UNAM (Mexico City).

V. Present State of Inventory

The total number of species of flowering plants known from El Salvador is 2,070 (Standley & Calderon, 1926). Standley and Calderon had predicted that when the country had been fully explored botanically, the number of species would be 3,000–4,000. Gentry (1978) rejects this estimate as being too high. To allow for species published since 1926, an estimate of 2,500 species will be used to calculate the rate of endemics. El Salvador has 17 endemics (Davis et al., 1986). This represents 0.68% of the total. IUCN statistics show that of the 17 endemic species in El Salvador, 12 are vulnerable, rare or indeterminate. As there is virtually no forest remaining in El Salvador it is unlikely that new or important distributions will be found within the indigenous flora. However, Rohweder (1956) mentions that the coffee plantations hide an enormous weed flora and study of the weeds might reveal new distributions for introduced species.

In writing this report, the author has found no references to families occurring in El Salvador which are either well or poorly collected. Authors of the "Flora Mesoamerica" are currently working on families from the region.

The population density has resulted in destruction of most of the forest in the country and its conversion to agricultural land. Now, virtually all available land is used for crop production and El Salvador has had to develop industrially to support its growing population. By 1945 (Kovar, 1945), most of the forests had been destroyed, and man's effect on the vegetation has been such that it is virtually impossible to speculate on the natural structure of the vegetation of El Salvador. For example, Rohweder (1956) states that it is difficult to say whether grass savanna is a natural or man-induced phenomenon in El Salvador. The deforestation has had serious consequences, such as the destruction of habitat and the local extermination of species of animals, for example, white-lipped peccarys, spider monkeys, howler monkeys, cottontail rabbits and white-tailed deer (Daugherty, 1972). Many of the exterminated animals had provided most of the animal protein in the human subsistence diet and, despite commercial livestock production in coastal regions, this has not been replaced.

A further consequence of deforestation has been severe soil erosion. For example, Daugherty (1973) attributed the accelerated rate of siltation of the Cinco de Noviembre Dam to erosion, after destruction of forests in the Montecristo region. It was estimated in 1963 that because of the siltation, the dam, which is El Salvador's major producer of hydroelectric power, would lose 40% of its storage capacity within 30 years.

The pine and oak forest described by Lotschert (1956) from a remnant at Los Esemiles has now been destroyed (Daugherty, 1973). In 1981 the estimated rate of deforestation for closed broadleaf forest was 40 km² per annum, from a total of only 1010 km² (Langley, 1981).

VI. Future of Inventory

In 1982 El Salvador had no legally established national parks. (IUCN, 1982) and the IUCN list (1985) still excludes El Salvador. However, strong protective measures have been taken in the following areas. At Montecristo, 19.9 km² of pine and oak forest have been proposed for protection. Other protected areas are Cerro Verde (8 km²) and Deininger (2 km²). Sadly, subsistence farming in Montecristo has eliminated much of the pine/oak association that existed below 400 m prior to protection. The "national parks" are not adequately policed and, although illegal, pilfering is frequent (D'Arcy, 1977). The development of teaching centers would be important in El Salvador. This might deter the theft from the national parks.

While there is little forest remaining in El Salvador, trees supporting epiphytes may remain uncut on farm or plantation boundaries. Inaccessible areas, such as ravines, may also support remnants of forest.

VII. Literature Cited

D'Arcy, W. G. 1977. Endangered landscapes in Panama and Central America: The threat to plant species. Pages 89–104 *in* G. T. Prance & T. S. Elias (eds.), Extinction is forever. The New York Botanical Garden.

Daugherty, H. E. 1972. The impact of man on the zoogeography of El Salvador. Bio. Conserv. **4(4)**: 273–278.

——. 1973. The Montecristo cloud forest of El Salvador—A

chance for protection. Biol Conserv. **5(3):** 227–230.
Davis, S. D. et al. 1986. Plants in danger: What do we Know? IUCN, Gland, Morges, Switzerland.
Gentry, A. H. 1978. Floristic knowledge and needs in Pacific Tropical America. Brittonia **30(2):** 134–151.
International Union for the Conservation of Nature. 1982. Directory of neotropical protected areas. Tycooly International Publishing, Ltd. Dublin, Ireland.
_____. 1985. United Nations list of national parks and protected areas. IUCN, Gland, Switzerland.
Kovar, P. A. 1945. Idea general de la vegetación de El Salvador. Pages 56, 57 *in* F. Verdoorn (ed.), Plants & plant science in Latin America. Chronica Botanica Company, Waltham, Massachusetts.
Langley, J. P. (ed.). 1981. Tropical forest resources assessment project. Part Latin America. FAO/UNEP, Rome.
Lotschert, W. 1956. La vegetación de El Salvador. Comun. Inst. Trop. Invest. Ci. Univ. El Salvador **4(3/4):** 65–79.
Rohweder, O. 1956. Die Farinosae in der Vegetation von El Salvador. Cram, de Gruyter & Co., Hamburg.
Standley, P. C. & S. Calderon. 1925. Lista preliminar de las plantas de El Salvador. La Union, Dutriz Hermanos, San Salvador.

VIII. Selected References

Armitage, K. B. & N. C. Fassett. 1971. Aquatic plants of El Salvador. Arch.Hydrobiol. **62(2):** 234–255.
Bernhardt, P. & E. A. Montalvo. 1978. Selected collecting sites in El Salvador. I. Private property. Bull. Torrey Bot. Club **105:** 9–13.
Calderon, S. 1925. Maderas preciosas de El Salvador. Revista Agric. Trop. **3(2):** 89–94.
Carlson, M. C. 1948. Additional plants of El Salvador. Bull. Torrey Bot. Club **75:** 272–281.
Guzman, D. G. 1909. 100 arboles maderables del Salvador. An. Mus. Nac. Repúb. Salvador **4(26):** 56–58.
_____. 1950. Especies útiles de la flora Salvadorena. 2nd ed. Imprenta Nacional, San Salvador.
Hamer, F. 1974. Las Orquideas de El Salvador. Vols. I & II. Ministerio de Educación, Dirección de Publicaciones, San Salvador.
_____. 1981. Las Orquideas de El Salvador. III. Suplemento. The Marie Selby Bot. Gards., Florida.
Holdridge, L. R. 1975. Zonas de vida ecológicas de El Salvador. FAO, Proyecto Desarrollo, Forestal y Ordenación de Cuencas Hidrográficas.
Lauer, W. 1954. Las Formas de la vegetación de El Salvador. Comun. Inst. Trop. Invest. Ci. Univ. San Salvador **3(1):** 41–45.
Lotschert, W. 1953a. Sobre la ecología de vegetación de los barrancos de El Salvador. Comun. Inst. Trop. Invest. Ci. Univ. El Salvador **(2):** 47–53.
_____. 1953b. La sabana de morros de El Salvador. Comun. Inst. Trop. Invest. Ci. Univ. El Salvador **2(5/6):** 122–128.
_____. 1959. Vegetation und standortdilima in El Salvador. Heft 10, Botanische Studien. VEB Gustav Fischer Verlag Jena.
Tucker, J. M. & C. H. Muller. 1945. Additions to the oak flora of El Salvador. Madroño **8(4):** 111–117.
Weber, H. 1958. Contribuciones al conocimiento de plantas Salvadorenas. Comun. Inst. Trop. Invest. Ci. Univ. El Salvador **7(1/2):** 23–32.
Witsberger, D., D. Current & E. Archer. 1982. Arboles del Parque Deininger. Ministerio de Educación, Dirección de Publicaciones, San Salvador.

Nicaragua

Susan Y. Sutton

Contents

I. Description of the Region	301
A. Geographical Extent and Topography	301
B. Climate	301
C. Population	301
II. Vegetation Maps	301
III. Herbarium Resources	301
IV. Publications	302
V. Present State of Inventory	302
A. Endemics and New Species	302
B. Threatened Areas	302
VI. Academic and Governmental Institutions	302
A. Native Programs	302
B. Foreign Programs	303
VII. Future of Inventory	303
VIII. Resources Needed	303
IX. Literature Cited	303

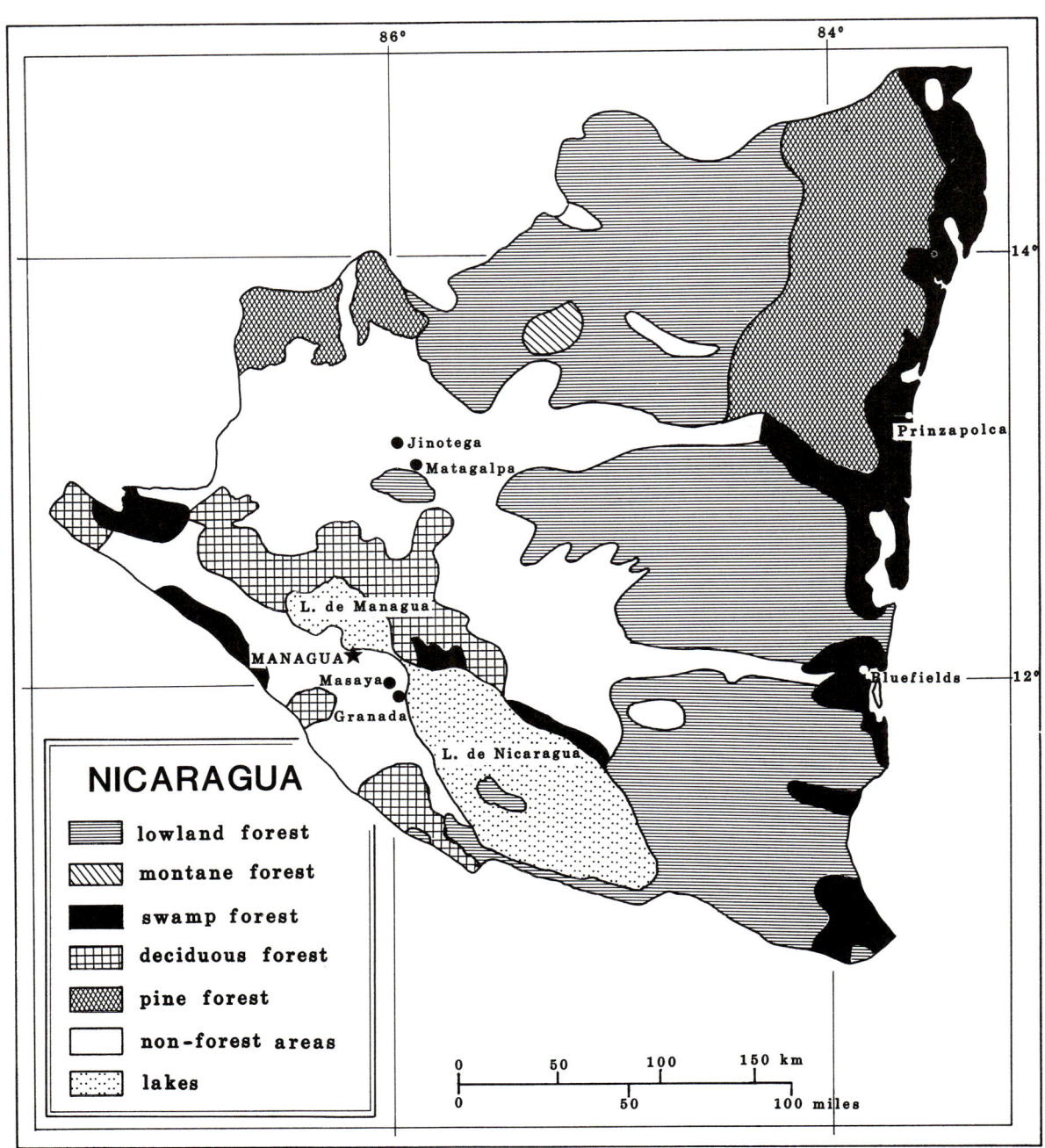

I. Description of the Region

A. GEOGRAPHICAL EXTENT AND TOPOGRAPHY

Nicaragua, with an area of 148,000 km² (41,700 km² of which are tropical forest), is the largest of the independent republics of Central America. The country can be divided into three major topographical regions. The first, a broad lowland belt about 80 km in width, runs along the Pacific coast from the Gulf of Fonseca in the north to Costa Rica. This is the most densely populated and most economically important region. In the northwest of the Pacific coastal areas there has been considerable volcanic activity and between the Volcán Momotombo and the Volcán Cosigüina near the Gulf of Fonseca, there are a string of over twenty volcanoes, some of them active. The highest is Volcán Cristobál at 1,806 m. The volcanic ash has produced a rich and fertile soil. This region is dominated by two great lakes, Lake Nicaragua and Lake Managua, lying in a depression among the mountains and drained eastward into the Caribbean by the San Juan River. Lake Nicaragua is 160 km long and 72 km wide; Lake Managua is 48 km long and ca. 20 km wide (Fig. 1).

The central mountains and the central depression together comprise the largest of the three regions, and are associated with a string of volcanoes. This area extends the length of the country with elevations between 600 and 2,150 m. The Central Highland region is dominated by three ranges, the Cordillera Segoviana, Cordillera Isabelia and the Cordillera Dariense, all running east to west.

B. CLIMATE

In the east along the Atlantic coast there is a wide belt of densely forested lowland, crossed by rivers running west to east. The climate of this area is tropical, with a rainy season of nine months and no well-defined dry season. The average annual rainfall is ca. 3,810 mm, which, together with malaria and other unhealthy conditions, has led to its being sparsely populated and economically underdeveloped. The Atlantic region consists of numerous low-lying forest-covered ranges, mostly of even height and frequently flat topped, which descend to extensive alluvial plains and swamps.

In the west the climate is tropical with mean temperatures of about 26–27°C and with a rainy season between May and November and a dry season between December and April; the average annual rainfall is 1,910 mm. The central highland area has a cooler climate and a longer, but lighter, rainy season.

C. POPULATION

The total human population in 1982 was 2,917,816. This is an average density of 24 per km². However, the distribution of the population is extremely uneven, as seen in Table I. About 85% of the people live in 26,000 km² in the Pacific

Table I
Human population densities by area for Nicaragua.

Area	Population (thousands)	Density (per Km²)
Zelaya	211,290	4
Río San Juan	30,219	4
Jinotega	131,537	14
Chontales	100,157	16
Boaco	90,286	21
Matagalpa	223,639	32
Nueva Segovia	102,262	28
Madriz	74,313	46
León	253,657	48
Chinandega	236,689	49
Rivas	112,030	51
Estelí	113,677	52
Carazo	111,619	102
Granada	115,670	117
Masaya	151,909	220
Managua	858,862	255

area. Densities range from 255 per km² in Managua, the capital, to four per km² in Zelaya, the eastern-most department of the country (comprising about one third of the whole land area). Forty percent of the population is urban and all the major towns are situated in the western lowlands. The most densely populated areas are (1) Managua, (2) Masaya, (3) Granada and (4) Carazo. The population doubled in the 20 years between 1950 and 1970.

The infant mortality rate dropped from 100/1,000 live births in 1950 to 45/1,000 live births in 1974. The life expectancy for the years 1970–1975 was 51.2 years for males and 54.6 years for females—an average of 52.9 years.

II. Vegetation Maps

The major vegetation maps for Nicaragua are Taylor (1963), US Army Corps of Engineers (1966) and Holdridge (1962).

III. Herbarium Resources

About 32,000 specimens are held in Managua (ENAG and HNMN). Gentry (1978) stated that most Nicaraguan

Table II
Nicaragua's population increase, in thousands.

Year	1800	1850	1900	1930	1950	1960	1975	1982
Number	107	257	505	638	1057	1536	2100	2917

specimens outside that country were held at MO and that the number of these is 2,000. Since the Flora de Nicaragua project started at MO in 1977, it is safe to assume that MO holds a larger collection of Nicaraguan specimens now. Nicaraguan material is also held at A, Ames, B, BM, C, DS, DUKE, EAP, ECON, F (a large orchid collection), FLAS, FSU, GH, HUH, K, L, NKH, MSC, NY, P, Clyde F. Reed Herbarium, Baltimore, S, SEL, SMU, S-PA, TENN, TEX, U, US, USF, VDB, VT, and W. Collections are usually small and no numbers have been found for this report.

The locations of important Nicaraguan collections are: A. Garnier: MICH, NY, US; P. Levy (1872): C, G, K, LG, NY, P; A. A. Orsted (1846–1848): B, C, (orig.), CGE, F, US; E. Rothshuh (1893–1895): B (650), L; B. C. Seemann: BM (338), K; R. Tate (1868–1884): K, BM (269), US.

IV. Publications

Important references on the flora of Nicaragua are IUCN (1982), Ashton (1945), Coolidge (1945), D'Arcy (1977), Denevan (1961), Gentry (1978), Seymour (1980), and Taylor (1962, 1963).

V. Present State of Inventory

The accessibility of the areas between Río Huahua and Río Coco is very good, due to logging activity and networks of logging access roads, so this area of Pine Savanna is well explored.

Presumably, the exploited areas have been well explored botanically, if not, it is certainly too late to do so. For instance, a major hardwood, *Macrohasseltia* (Flacourtiaceae) was used to supply sawmills in the Matagalpa-Jinotega area. This was only described botanically a few years ago, yet supplies are now nearly exhausted (Williams, pers. comm.).

Less-accessible areas and less-developed areas are for the most part on the Atlantic/Caribbean side. The interior of the country includes a system of valleys and ridges with unusual and rich forests. A small part of north-central Nicaragua (and a larger part of adjacent Honduras) has possibly been above sea level longer than any other parts of Central America, but have not been explored botanically in any great detail. The wet forest south from Bluefields, and especially the pluvial Serranías de Yolaina, are of great interest.

There are approximately 7,000 species of vascular plants in Nicaragua. The flora of the country is not at present well known.

A. Endemics and New Species

Endemism is possibly significant in isolated, unexplored mountains/volcanoes in the interior. However, endemics as a rule are not a strong likelihood in Nicaragua, since distribution patterns flow through the country. For instance, Nicaragua is the southern limit of the genus *Pinus* in North America. IUCN (Davis, 1986) figures state that there are 56 known endemics so far. One endemic is considered to be extinct and six are rare (IUCN).

Wet forest areas are the most likely to yield new species of any lowland area (Gentry, 1978). The Mosquitia region tropical moist forest (believed to be the largest remaining tract in Central America; about 3,600 km^2) is still largely undisturbed. In some summit areas, elfin forests, some of which are undisturbed, are also rich in new species (See also the discussion, below, of the national parks.)

B. Threatened Areas

All economically important forest species are threatened by destruction. The following species of wood are tending to disappear, "madroño" (*Calycophyllum candidissimum*), "Spanish cedar" (*Cedrela odorata*), the "indigo" (*Indigofera argentea*), the "cacao" (*Theobroma cacao*) and species of "goosefoot" (*Chenopodium anthelminticum*). Wet forests, particularly forests near the Roosevelt Highway linking Managua with the river port Rama, are all gone.

The interior around Matagalpa was once a vast forest of valuable furniture woods and pines. This is now reduced to remnants only. The growing of tree crops, such as coffee and cocoa, are also leading to deforestation and plantation clearing. Increasing acreage of cotton and vegetables also means destruction of forest. Small stands of pine savannas are left in the accessible areas north of Río Huahua and less accessible areas south of Río Huahua, particularly in upper reaches of Ríos Bambana and Kucalaya. (Pine savanna is a fire-caused disclimax, whereas true climax is in fact evergreen rainforest of tropical lowlands.)

Estimated rate of deforestation for closed broadleaved forest is 1,050 km^2/annum (Langley, 1981).

VI. Academic and Governmental Institutions

A. Native Programs

The following local institutions are active in floristic inventories, either in amassing herbarium collections from various foreign collectors or actively involved in field work themselves:

(1) Herbario, Facultad de Ciencias Agropecuarias, Universidad Nacional Autonoma de Nicaragua, Km 12½ Carretera Norte, Managua. (ENAG)
(2) Herbario Nacional de Nicaragua, Apartado 4271, Managua. (HNMN)
(3) CITES Management Authority: Departamento de Regulación y Control de Recursos Naturales, IRENA, kilómetro 14½ Carretera Norte, Apdo 5123, Managua.
(4) Herbario de Universidad Centroamericana, Apartado 69, Managua.

B. Foreign Programs

The Flora de Nicaragua project, based at the Missouri Botanical Garden in St. Louis, began in 1977 and progressed under the direction of W. D. Stevens.

The Flora of Central America-Flora Mesoamericana Project, a joint venture between the Missouri Botanical Garden, the Botany Department of The British Museum (Natural History) and Instituto de Biología, UNAM, Mexico, is also active in Nicaragua.

VII. Future of Inventory

In Nicaragua, 17,300 ha of land are already protected. Also proposed is a Natural Resources Reserve of 1,100,000 ha from Bocay-Waspuk-Huahua. On 7 March 1980 the government established the National Park Service. There are two parks with specific legislation—Volcán Masaya National Park (23 May 1979) and Saslaya NP (27 March 1971).

1) Saslaya National Park, with 11,800 ha encompasses tropical rainforest, cloud and elfin forest. An inventory is in progress (1981) and many new species are likely. As yet there is no disturbance.
2) Volcán Masaya National Park, with 5,500 ha encompasses tropical dry forest and savanna. It has very dry, poor soil and hot wind, which dwarfs trees and shrubs. There are problems with firewood collecting and animal poaching.

Presumably, the studies undertaken by the Flora de Nicaragua Project will expose areas of little botanical exploration and collection can be instigated. Exhaustive collecting is a priority in all areas.

The writing of family accounts will accentuate areas of the country with interesting species distribution patterns, although, as previously mentioned, high rates of endemism in Nicaragua seems unlikely. Coordination with local personnel in selection of floristic projects and study sites needs to be emphasized.

VIII. Resources Needed

It is important that there be constant good communication between foreign institutions involved in floristic studies of the country and those local institutions interested in conducting a floristic inventory. Local institutions should be more aware of areas threatened with destruction (for reasons stated above, i.e., beef production, tree crops, etc.). Foreign funding can assist in this. For instance, Canadian funding assisted in the protection of nearly two dozen volcanoes and migratory bird sites (Taylor, 1962). Between 1959 and 1961, the Nicaraguan Instituto de Fomento Nacional conducted a five year control program to protect 60,000 acres of pine savanna. Close cooperation with such organizations is essential to preempt situations whereby tracts of previously unexplored or important forest is to come under the ax. Species checklists are useful for land use planning and national park instigation.

It is perhaps more important to enhance existing institutions, or herbaria within institutions, rather than to develop additional ones. They will require better facilities for storage, literature retrieval, etc., and, of course, more personnel.

Floristic work has traditionally been done by visiting scientists from north temperate countries. These efforts need to become orientated to fulfilling local needs of the country. One way of accomplishing this is by training students who can carry out floristic and taxonomic work in their own country. Once trained, they can be employed in the developing herbaria.

Costs are dependent upon the manner in which local herbaria are improved and how foreign programs are to continue. It may be that students from Nicaragua can be trained in herbaria in the United States, such as MO, where expertise with the Nicaraguan flora is already developed, or specialists from abroad may stay on extended sabbaticals, developing herbaria and training students. Both methods will require grant support.

To develop herbaria, the costs will be in the region of $160,000: one specialist for two years (at $30,000 per annum) plus five assistants (at $8,000 per annum) and $20,000 for increased facilities (storage, literature, etc.). This figure represents the costs for one herbarium. Funding will be required for training of students abroad, perhaps $10,000 per student per annum.

The maintenance of national parks is also a cost to be considered with priority.

IX. Literature Cited

Ashton, J. 1945. On the plant resources and flora of Nicaragua. Pages 60–64 *in* F. Verdoorn (ed.), Plants and plant science in Latin America. Chronica Botanica. Waltham, Massachusetts.

Coolidge, H. J., Jr. 1945. Notes on conservation in the Americas. Pages 328–335 *in* F. Verdoorn (ed.), Plants and plant science in Latin America. Chronica Botanica. Waltham, Massachusetts.

D'Arcy, W. G. 1977. Endangered landscapes in Panama and Central America. Pages 89–104 *in* G. T. Prance & T. S. Elias (eds.), Extinction is forever. The New York Botanical Garden, New York.

Davis, S. D. et al. 1986. Plants in danger: What do we know? IUCN. Gland, Morges, Switzerland.

Denevan, W. M. 1961. The upland forest of Nicaragua. A study of cultural plant geography. Univ. Calif. Publ. Geogr. **12**(4): 251–320.

Gentry, A. H. 1978. Floristic knowledge and needs in Pacific Tropical America. Brittonia **30**(2): 134–153.

Holdridge, L. R. 1962. I Mapa ecológico de Nicaragua. Agencia para el Desarrollo Internacional de Gobierno de los Estados Unidos de America, Managua.

IUCN. 1982. Directory of neotropical protected areas. Morges, Switzerland.

Langley, J. P. (ed.). 1981. Tropical forest resources assessment project. Part Latin America. FAO/UNEP, Rome.

Seymour, F. C. 1980. A checklist of the vascular plants of Nicaragua. Phytologia Memoirs 1:iii-x, 1–314.

Taylor, B. W. 1962. The status and development of the Nicaraguan pine savannas. Caribbean Forester **23**: 21–26.

———. 1963. An outline of the vegetation of Nicaragua. J. Ecol. **51**: 27–54.

U.S. Army Corps of Engineers. Aid Resources Inventory Center. 1966. Inventario nacional de recursos fisicas. Washington, D.C.

Costa Rica

Luis D. Gómez

Contents

I.	Description of the Region	306
	A. Geographical Extent and Topography	306
	B. Population	306
II.	Vegetation Maps	307
III.	Herbarium Resources	307
IV.	Present State of Inventory	307
V.	Academic and Governmental Institutions	307
VI.	Future of Inventory	307
VII.	Literature Cited	308

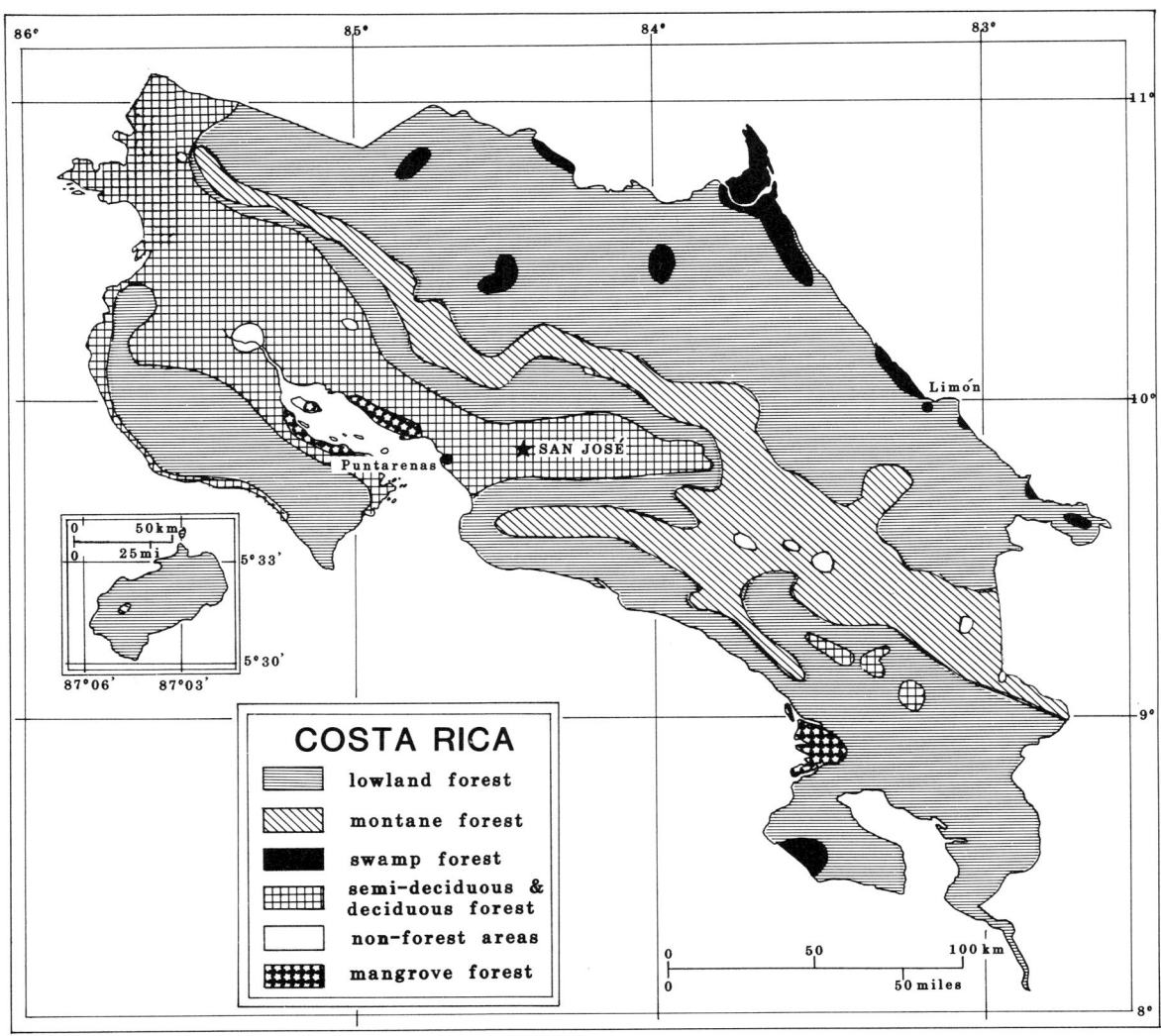

I. Description of the Region

A. Geographical Extent and Topography

Costa Rica has an area of 50,900 km². The northwest section is coastal, with irregular ridges and hill systems interspersed by narrow alluvial valleys. Inland, one finds ample alluvial valleys and low hills, ranging from flat to moderately hilly, gradually rising to the Guanacaste Volcanic Range (average of range 1,500 m).

The central section of the country is mostly mountainous, with either volcanic or tectonic ranges, such as the Central Volcanic and Talamanca ranges. The Central Valley is of volcanic origin with some alluvial deposition. Altitudes range from 1,000 m to 3,820 m (the highest point in the country).

The northeastern section of the country is mostly alluvial, flat, lowlands with isolated low ranges or rounded hill systems, becoming more abrupt towards the west and north with the Guanacaste Volcanic Range, and the Central Volcanic Range at the middle of the country. A coastal plain occurs in the south along the border with Panama.

The southwest section of the country consists of a longer valley (Río Grande de Terraba) at the foot of the Talamancas, with irregular ridges, hills and old terraces, until it opens into flatter lowlands at the Golfo Dulce. From the Osa Peninsula northward there are a series of irregular ridges, one of them the Coastal Ridge, all of tectonic and erosive origin, which, on reaching the Río Grande de Tarcoles, close the Central Valley (Fig. 1.).

B. Population

Most of the population is concentrated in the Central Valley, which contains the capitals of five of the seven provinces. The last reliable population census dates from 1983, and estimates a total population of some three million, which yields a density of 59 inhabitants per km². The population growth rate is 3.5.

II. Vegetation Maps

There are no detailed vegetation maps for Costa Rica (Gómez, 1986). Holdridge's Life Zone Ecology map of Costa Rica does reflect some of the plant formations, but it is not, nor was it intended to be, a vegetation map. Gómez (1982) depicts vegetation types of Central America and collates some vegetational descriptions from various authors. A new vegetation map for Costa Rica (scale 1:200,000) has been completed as part of the Biogeography Project of the Nature Conservancy International Program. Gómez (1986) has published a set of 21 maps covering major vegetation types and climate. Herrera and Gómez (1987–1988) have prepared a new biotic areas map and Herrera (1987) one on present natural vegetation cover (1:250,000).

III. Herbarium Resources

The National Herbarium has in excess of 100,000 Costan Rican vascular plant specimens, plus about 10,000 non-vascular plants in its collections. The herbarium at the School of Biology, University of Costa Rica, has some 15,000 specimens, the vast majority of which are non-vascular.

There may be in excess of 500,000 specimens from Costa Rica held in herbaria abroad, mainly at BM, F, K, MO, NY, US.

IV. Present State of Inventory

Two major floras exist for Costa Rica. Standley (1938) is merely an annotated checklist with descriptions for 75% of the species known at the time. The *Flora Costaricensis*, edited by William C. Burger (1971-present), of the Field Museum, is now in its fourth volume.

Areas relatively well collected in Costa Rica include the Central Valley and Volcanic Ranges. Areas poorly collected include Lower Talamanca and the northwest and northeast corners of the country.

So far we have registered over 8,000 species but expect the total to be somewhere between 10,000–12,000. The Natural Heritage program of Costa Rica has a list of 1,800 endemic vascular plants so far; with increasing research and field work this will change. The completion of *Flora Costaricensis* will render any number given here obsolete.

All areas are yielding new distributions, especially those well collected only during one of the seasons (dry, wet). At this stage of research and field work, it is impossible to estimate which families or genera are well or poorly collected.

Rates of extinction have not been determined for any of the plants species.

V. Academic and Governmental Institutions

Scholars working on the flora of Costa Rica include those of Flora Mesoamericana and Flora Neotropica and the staffs of the Field Museum and the Missouri Botanical Garden. Local scholars include Jorge Gómez Laurito, Luis Poveda, Pablo Sanchez, and Luis D. Gómez, all on the staff of the National Museum in San José. Some are part-time only, or have special interests such as medicinal plants or trees. Gómez Laurito and L. D. Gómez are the only generalists and systematists.

Some research is going on at the Universidad de Costa Rica (Department of Biology), but this is mostly concerned with education. The Technological Institute has a small collection of forest trees used for dendrology classes, but is not concerned with floristics, only timber. The only government institution actively involved in floristics is the National Herbarium at the National Museum of Costa Rica.

The Field Museum of Natural History is preparing the second *Flora Costaricensis*. Missouri Botanical Garden has had an active field program for the past five years. Institutions involved in the *Flora Mesoamericana* (University of Mexico, British Museum) have had four field expeditions to Costa Rica within the past four years (1983–1987).

VI. Future of Inventory

Deforestation in Costa Rica is as appalling as in any other Latin American country. It is in sixth place worldwide in rate of deforestation. No program can adequately deal with the developmental, demographic and political pressures. But, so far, Costa Rica has managed to set aside large tracts of land under some type of protection. Collecting efforts have been concentrated not in those areas, but in non-protected areas, as far as possible. As interesting areas are identified, every effort is made to go there, but often resources cannot compete with developmental or conversion rates.

Costa Rica needs fieldwork and needs it fast, but the cost is high. Due to the economic situation of the country, plant collecting is not one of the national priorities. Fieldwork by qualified, experienced collectors (at about US $12,000/ann. each) and equipment (jeep, camping gear, at about $14,000), plus cash to pay local guides, porters, fuel, etc. ($20,000–25,000) could complete the work on the flora and its particulars in about five to ten years time if there were well-directed efforts and administrative infrastructure. The National Herbarium is merely a housing facility.

VII. Literature Cited

Burger, W. C. (ed.). 1971-present. Flora Costaricensis. Fieldiana Bot. **35.**

Gómez, L. D. 1982. The origin of the pteridophyte flora of Central America. Ann. Missouri Bot. Gard. **69(3):** 548–556.

——— & **W. Herrera.** 1986. Vegetación y clima de Costa Rica. Vol. 1. L. D. Gómez. Vegetación de Costa Rica. 1986. Vol. 2. W. Herrera. Clima de Costa Rica. 1985. Editorial Universidad Estatal a Distancia. San José.

Standley, P. C. 1938. Flora of Costa Rica. Publ. Field Mus. Nat. Hist., Bot. Ser. **18(1–4).**

Panama

Rachel J. Hampshire

Contents

I. Description of the Region	310
A. Geographical Extent and Topography	310
B. Population	310
II. Vegetation Maps	311
III. Herbarium Resources	311
IV. Present State of Inventory	311
A. Endemics	311
B. New Taxa and Distributions	311
C. Threatened Areas and Extinction	311
V. Academic and Governmental Institutions	311
VI. Future of Floristic Inventory	312
VII. Literature Cited	312
VIII. Selected References	312

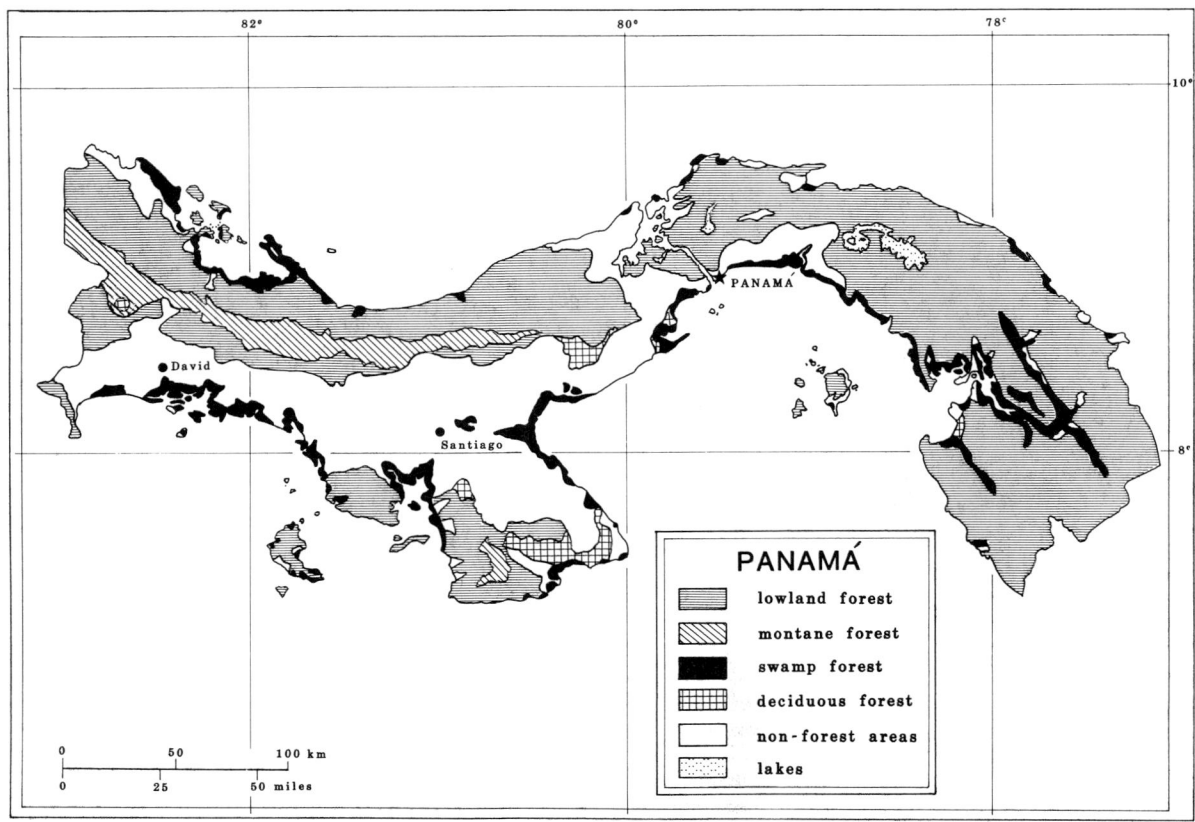

I. Description of the Region

A. Geographical Extent and Topography

Panama, with an area of 77,082 km² (of which 41,650 km² is forested) is sigmoidal in shape, approximately 480 km from east to west, and between 65 and 195 km north to south. It is divided midway along the long axis by the Panama Canal, opened in 1914. West of the Canal lies a mountain range, comprised of the Cordillera de Veragua, the Sierra de Chiriquí and the Serranía de Tabasará, with peaks reaching 1,200 m. East of the Canal, and parallel with the north coast, are the Cordillera de San Blas and Serranía del Darién, with peaks reaching 1,875 m. Two more mountain ranges, the Serranía de Cañazas, reaching 1,440 m, and the Serranía del Sapo, reaching 1,600 m, run parallel with the south coast. The mountain ranges of the north and south coast are separated by a low plateau at less than 200 m.

There are a number of islands off both the Pacific and Caribbean coasts and many islands in the Gatun Lake region of the Panama Canal (Fig. 1).

B. Population

The current (1984) population is estimated to be two million. The rate of population increase in Panama is approximately 2.9% per annum. The results of the 1980 census show that 46% of the population live in urban areas, with 21% of the total living in Panama City (Table I).

Table I
The population of Panama as of 1980.

Province	Population	Density/km²
Bocas del Toro	53,579	6.0
Coclé	140,320	27.8
Colon	166,439	0.3
Chiriquí	287,801	32.8
Darién	26,497	1.5
Herrera	81,866	33.7
Los Santos	70,200	18.1
Panama	830,278	69.0
Veraguas	173,195	15.6
TOTAL	1,830,175	23.7

II. Vegetation Maps

The principal vegetation maps for Panama include: Croat (1978), Holdridge and Budowski (1959) and Tosi (1970).

III. Herbarium Resources

A total of 100,000 botanical collections have been made in Panama (Gentry, 1978). Twenty thousand of these are held in Panama; 80,000 are held in institutions abroad. However, Dwyer (1980) estimates that 500,000 *numbers* of Panamanian plants exist worldwide. Important collections are at BM, K, MO, NY, US.

IV. Present State of Inventory

Using the more conservative estimate that 100,000 numbers have been collected in Panama, an average of more than 100 collections have been made per 100 km^2. However, some areas have been over-collected while others remain botanically unexplored. Well-collected areas of Panama include: Barro Colorado Island (well collected, but it has revealed new records in recent years), San José Island, Cerro Azul, Cerro Campana, Boquete, El Valle de Antón, Taboga Island, El Volcán de Chiriquí (from Dwyer, 1968). Localities near the Pan-American and Trans-Isthmian highways, Cerro Jefe and Llano-Cartí have been well collected.

Poorly collected areas include isolated areas such as parts of Darién Province, Bocas del Toro, and Azuero Peninsula. Areas where there is a lack of roads—the region between Boquete and El Valle and between Cerro Azul and Colombia—have also been poorly collected.

D'Arcy (1980) says that although 6,200 plant species are included in the Flora of Panama, the total number of plant species is probably in excess of 9,000 (including about 1,100 cryptogams).

A. Endemics

The number of endemics is 1,222 (Davis et al., 1986), or 19.7% of the total number of plant species. D'Arcy (1977) estimates that 50% of the flora of the Chiriquí Mountains may be endemic. Lewis (1971) estimates that 25% of the flora of the Panamanian cloud forests is unique to central and eastern Panama and Colombia.

B. New Taxa and Distributions

Most areas of Panama which were previously uncollected are yielding new species or new distributions. These include the unexplored cloud forests of Darién and San Blas (Lewis, 1971) and the Fortuna Dam region (Dwyer, 1980). A government proposal, known as the Fortuna scheme, was made to construct a power dam and farming district above Gualaca. Although the project involved the destruction of much forest, the Panamanian government authorized botanists to survey the area before felling (D'Arcy, 1977).

Most families are yielding new distributions or species, especially the Rubiaceae. Dressler (1972) estimated that for five families, in the 10–30 years since their publication in the *Flora of Panama*, there had been an increase of 85% in the number of species known from the region. All flowering plant families have been covered by authors of *Flora of Panama*, published from 1943–1980. Authors of *Flora Mesoamericana* are currently working on the flowering plant families of Central America, including Panama.

C. Threatened Areas and Extinction

The forests have been destroyed in many parts of Panama, for example, much of the cloud forest of Cerro Jefe has been cut for grazing. Pristine collecting areas which become accessible to botanists when roads and landing areas are opened also become accessible to those likely to destroy the forests for agriculture or commercial gain. The estimated rate of deforestation of closed broadleaf forest is 360 km^2 per annum. The latest IUCN figures (Davis et al., 1986) show two Panamanian endemics are now extinct (an extinction rate of .03%) and 54 species endangered or vulnerable.

The forests of Darién Province have been removed near the Pan-American Highway. As the highway has been extended, it has given peasant farmers access to cut the forest and use the land for agriculture. The forests on the Pacific side west of the Canal have almost all been destroyed, and the montane forests of Chiriquí Province are threatened. Plans to remove forest to create a dam near Gualaca have already been mentioned.

Under-collected areas, such as the mountains of Chiriquí Province have high rates of endemism. These forests are threatened with destruction; therefore they deserve highest priority for inventory.

V. Academic and Governmental Institutions

The principle Panamanian institutions conducting floristic inventory in Panama are the Universidad de Panamá, Panama City, which includes the national herbarium. Panamanian plant collecting permits are granted (by the Dirección de Recursos Naturales Renovables [RENARE]) on the condition that one set of plants are deposited there. The Smithsonian Tropical Research Institute (STRI) in Balboa, Panama gives authorization for research on Barro Colorado Island.

Foreign institutions contributing to floristic inventory in Panama incude the Missouri Botanical Garden (MO), especially since commencing the Flora of Panama and the new updated checklist (D'Arcy, pers. comm.). Panama is included in the *Flora Mesoamericana* currently being written jointly by MO, BM and UNAM. Also STRI has an active botanical research program.

Panamá has, at present, five national parks and a natural monument (Barro Colorado Island), which give protection to 6,609 km². A sixth park, the Amistad International Park, is proposed on the Panama-Costa Rica border. This would protect 2,000 km² on the Panamanian side (and 2,500 km² in Costa Rica) and envelops a vast area of Caribbean tropical rain forest. Approximately 8.6% of the area of Panama, excluding the Amistad International Park, is therefore protected.

Other parks in Panama incude the Darién National Park which covers 5,970 km² and includes montane, premontane, cloud, elfin and swamp forest; the Soberania National Park which covers 200 km² of lowland tropical wet forest; the Portobelo National Park which covers 174 km² of tropical moist, tropical wet, premontane wet and premontane rainforest; the Volcán Barú National Park, which covers 143 km² of premontane wet tropical, montane wet tropical, cloud and elfin forest with paramo nearby; the Altos de Campana National Park which covers 48 km² (less ca. 500 ha which are cultivated) of premontane humid and montane tropical forest; and Barro Colorado Island which covers about 15.6 km² of tropical moist forest. All have total legal protection (IUCN, 1982).

VI. Future of Floristic Inventory

Panamá City has good resources for maintaining a floristic inventory. However, education of people in rural areas might decrease the rates at which some forests are being cleared for farming.

VII. Literature Cited

Croat, T. B. 1978. Flora of Barro Colorado Island. Stanford University Press, Palo Alto, California.
D'Arcy, W. G. 1977. Endangered landscapes in Panama and Central America: The threat to plant species. Pages 89–104 *in* G. T. Prance & T. S. Elias (eds.), Extinction is forever. The New York Botanical Garden, New York.
———. 1980. The flora of Panama: Historical outline and selected bibliography. Ann. Missouri Bot. Gard. **67**(4): v–viii.
Davis, S. D., S. J. M. Droop, P. Gregerson, L. Henson, C. J. Leon, J. L. Villa-Lobos, H. Synge & J. Zantovska. 1986. Plants in danger: What do we know? IUCN, Gland, Morges, Switzerland.
Dressler, R. L. 1972. Terrestrial plants of Panama. Bull. Biol. Soc. Wash. **2**: 179–186.
Dwyer, J. D. 1968. A list of localities botanized in Panama. Phytologia **16**(6): 467–486.
———. 1980. The history of plant collecting in Panama (1700–1981). Ann. Missouri Bot. Gard. **67**: ix–xv.
Gentry, A. H. 1978. Floristic knowledge and needs in Pacific Tropical America. Brittonia **30**: 134–153.
Holdridge, L. R. & G. Budowski. 1956. Report of an ecological survey of the Republic of Panama. Caribbean Forest. **17**(3–4): 92–110.
Lewis, L. H. 1971. High floristic endemism in low cloud forests of Panama. Biotropica **3**(1): 78–80.
Tosi, J. A., Jr. 1970. Mapa ecológico de Panama. Programa de las Naciones Unidas para Desarollo, Rome.

VIII. Selected References

Erlanson, C. O. 1946. The vegetation of San José Island, Republic of Panama. Smithsonian Misc. Collect. **106**(2): 1–14.
Gordon, B. L. 1982. A Panama forest and shore. Boxwood Press, Pacific Grove, California.
Holdridge, L. R. 1970. Manual dendrológico para 1000 especies arboreas en la Republica de Panama. Government of Panama, Panama City.
Johnston, I. M. 1949. The botany of San Jose Island (Gulf of Panama). Sargentia **8**: 1–306.
Lamb, F. B. 1953. The forests of Darién, Panama. Caribbean Forest. **14**: 128–135.
Porter, D. M. 1973. The vegetation of Panama: A review. Pages 167–201 *in* A. Graham (ed.), Vegetation and vegetational history of North Latin America. Elsevier, London.
Schery, R. W. 1945. A few facts concerning the Flora of Panama. Pages 284–287 *in* F. Verdoorn (ed.), Plants and plant science in Latin America. Chronica Botanica Co., Waltham, Massachusetts.
Standley, P. C. 1928. Flora of the Panama Canal Zone. Contr. U.S. Natl. Herb. **27**: 1–416.
———. 1933. The Flora of Barro Colorado Island, Panama. Contr. Arnold Arbor. **5**: 5–178.
Tosi, J. A., Jr. 1971. Zonas de vida, una base ecológica para investigaciones silvicolas y inventariación forestal en la República de Panama. Organización de las Naciones Unidas para Agricultura y Alimentación, Rome.
Woodson, R. E. & R. W. Schery. 1943–1980. Flora of Panama. Ann. Missouri Bot. Gard. **30**(2)–67(4).

Regional Reports

V. Caribbean

Cuba

René Pablo Capote López
Rosalina Berazaín Iturralde
Angela Leiva Sánchez

English Translation: A. Leiva Sánchez

Contents

- I. Description of the Region (A. Leiva) 317
  - A. Geographical Extent and Topography 317
  - B. Climate 317
  - C. Soils 317
  - D. Population 317
- II. Flora and Vegetation 318
  - A. Phytogeographical Features of the Cuban Flora (R. Berazaín) 318
    1. Endemism 318
    2. Disjunctions (within Cuba) 320
    3. Vicariancy 320
    4. Relationships of the Cuban Flora with Neighboring Lands 320
    5. Phytogeographical Regionalization 321
  - B. The Present Vegetation of the Cuban Archipelago (R. P. Capote and R. Berazaín) 321
    1. Introduction 321
    2. Forest Formations 322
        - a. Rain Forests 322
        - b. Cloud Forests (Elfin Woodland) 322
        - c. Evergreen Forests 323
        - d. Semideciduous Forests 323
        - e. Pine Forests 323
        - f. Swamp Forests 324
        - g. Mangrove Forests 324
        - h. Gallery Forests 324
    3. Thicket Formations 324
        - a. Coastal and Subcoastal Xeromorphic Thicket 324
        - b. Serpentinic Thicket Formations 324
        - c. Sub-paramo Formation 325
    4. Herbaceous Formations 325
        - a. Savannas 325
        - b. Fresh-water Communities 326
        - c. Herbaceous Swamp Formation 326
        - d. Halophytic Plant Communities 326
    5. Vegetation Complexes 326
        - a. *Mogotes* Complex 326
        - b. Rocky Coasts Vegetation Complex 326
        - c. Sandy Coasts Vegetation Complex (Strand Vegetation) 326
    6. Secondary Vegetation 327
- III. Notes on the History of Botanical Exploration and Conservation of Nature in Cuba (A. Leiva) 327
  - A. Exploration and Floristic Inventories between the Eighteenth and Twentieth Centuries 327
    1. Colonial Period: Eighteenth-Nineteenth Century 327
    2. The Republican Period: First Half of the Twentieth Century 328
    3. The Modern Period: Second Half of the Twentieth Century 329
  - B. The Problem of Conservation of Nature in Cuba: Past and Present (With Emphasis on Forest Ecosystems) 333
- IV. Acknowledgments 333
- V. Literature Cited 335

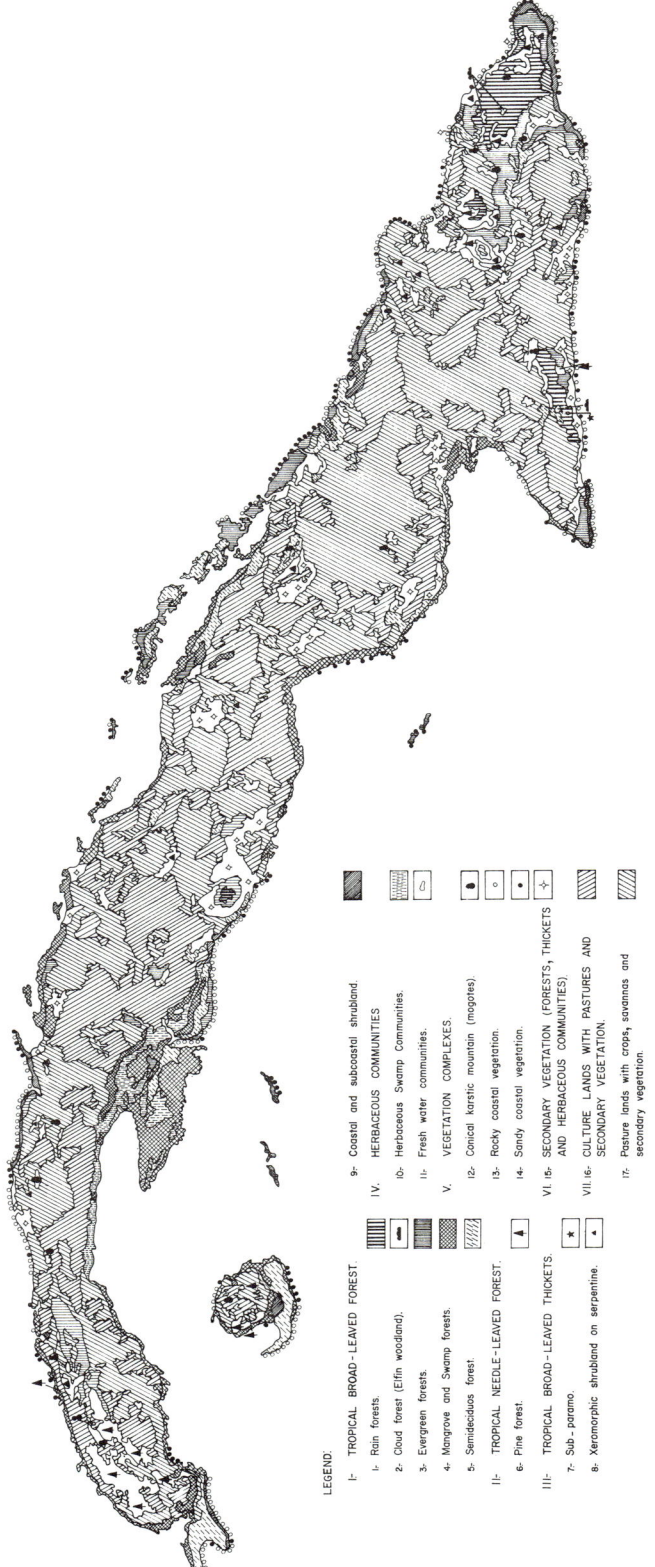

I. Description of the Region
A. Leiva

A. Geographical Extent and Topography

The Cuban Archipelago is located at the northwestern extreme of the Greater Antilles chain, in the Caribbean sea at the entrance of the Gulf of Mexico, between 19°49′–23°17′N and 74°–84°57′W.

The main island (Cuba) is 77 km from Haiti, 140 km from Jamaica, 180 km from Florida and 210 km from Mexico.

The archipelago is integrated by the island of Cuba with 105,007 km², the smaller Isla de la Juventud (formerly Isla de Pinos), with 2,200 km², and many other little islets and cays amounting to 3,715 km² more. Thus, the Republic of Cuba has a total surface area of 110,992 km² (Borhidi & Muñíz, 1980).

Two thirds of the territory are plains, interrupted by four principal mountain systems on the main island. These are, from west to east: (1) the Guaniguanico system, integrated by Sierra de los Organos at the western extreme and Sierra del Rosario at the eastern. The maximum height is Pan de Guajaibón at 692 m; (2) the Guamuhaya system, with Sierra del Escambray (Highlands of Trinidad) where the maximum point is Pico San Juan at 1,156 m; and the Highlands of Sancti-Spiritus with the maximum height at Sierra de Banao (883 m); (3) the Sierra Maestra system (including Sierra de la Gran Piedra), with the greatest heights in all the country where Pico Turquino (at 1,974 m) is the highest; and (4) the Sagua-Baracoa massif, a very extensive montane system occupying the northeastern extreme of Cuba, and with Pico Cristal at a maximal height of 1,325 m.

B. Climate

With respect to climatic conditions, Cuba possesses a tropical, warm climate heavily influenced by the East Winds and the Gulf Stream, although in some interior points a certain continentality is expressed in mesoclimatic parameters.

The annual mean temperature is about 24.5°C. The coldest month's average (January) is near 19°C and the warmest month's average temperature (July) is 29°C. The value of the minimum temperature average is 10°C and the maximum, 35°C. The altitudinal thermic gradient is between 0.6–0.9°C/100 m of height.

Rains occur mainly between May and October, and account for 1,100–1,600 mm/year as an average. Only at the northeast extreme do significant winter rains take place. Taking into account the extreme values of rainfall, figures between 300 to 3,000 mm/year can be found, distributed into two seasons: rainy and dry.

The relative air humidity is rather high, varying between 74–80% during the day and reaching more than 90% at night. In some hyperxerophytic coastal areas, such lower values as 68% can be obtained (Borhidi & Muñíz, 1980). Tropical hurricanes and cool air masses reaching northwestern coasts in winter account for annual oscillations of bulk rainfall values.

C. Soils

Soils are extremely variable; they are mostly derived from sedimentary calcareous rocks, alternating in a random "mosaic" with soils derived from serpentinite, acidic, basic, and intermediate igneous rocks; redeposited siliceous white sandy soils, hydromorphic gley soils, organic soils, and heavily lixiviated ones at the mountains. The occurrence of young, arid coastal limestone, and karstic limestone areas of different age are of particular interest. Such variability mainly accounts for the high endemism of flora and for the presence of different types of vegetation. The genetic classification of Cuban soils embraces 20 main groups and 100 genetic types (Acad. Cienc. de Cuba, 1973). Among them, soils derived from ultrabasic igneous serpentinized rocks (peridotites), with a low Ca^{++}/Mg^{++} ratio, high content of Al^{+++} and heavy metals (Ni, Cr, and Co) are responsible for the major percentage of endemism. (Borhidi & Muñíz, 1980). Deep, very ancient latosolic soils support the more endemic flora at the northeastern extreme of the Cuban island (Berazain, 1979).

Modern paleogeographical theories (Coney, 1982; Rosen, 1985) identify a Proto-Cuban archipelago as forming part of a Proto-Greater Antillean island arch that connected northern Central America (Yucatán) with northwestern South America since late Cretaceous times. During the Cenozoic (Eocene) there took place an eastward displacement of the island arch, with a subsequent reorganization (hybridization-fractionation) of islands and thus a different degree of contact between the biota established on suitable biotopes all over the island system and continental land masses. According to Iturralde-Vinent (1982) stable Cuban terrestrial biotopes have existed since upper Eocene times, and are now located in western, central, central-eastern, northeastern and southeastern mountains and highlands, which are, at the same time, the principal centers of endemism and diversification of flora. A Quaternary separation of Isla de la Juventud can explain the strong floristic relationship between its flora and the western flora of the island of Cuba.

D. Population

According to current investigations Cuba has been inhabited since ca. 6,000 years B.P., originally by prearchaic and archaic aborigines. The agroceramist cultures arrived later, from about 3,250 years B.P. till the Spanish colonization, when almost all aborigines were exterminated (Atlas Demográfico Nacional, 1985).

Urbanization started at the beginning of the sixteenth century. By 1907, 43.9% of the inhabitants lived in cities and towns, and in 1981 this figure was 69%. At a national level, 45% of the population live in 9.5% of the total territory. The overall population density is 87.7 inhabitants/km². Among Latin American and Caribbean countries, Cuba occupies ninth place regarding human population density. The total number of inhabitants now surpasses 10,100,000.

The yearly average growth is about 11.4/1,000 inhabitants. The life expectancy is 74.2 years, and infant mortality is about 13.6/1,000 successful births. (Official data from 1986).

The territory is divided into 14 provinces and a special Municipality (Isla de la Juventud).

La Habana, the capital, has a population of 1.52 million inhabitants, being the largest city of the country.

Other important cities are Santiago de Cuba, Camaguey, Santa Clara, Guantánamo, Matanzas, Cienfuegos, Bayamo and Holguín, all provincial capitals, with more than 100,000 inhabitants each (Atlas Demográfico Nacional, 1985).

II. Flora and Vegetation

A. Phytogeographical Features of the Cuban Flora (R. Berazaín)

To understand the phytogeographical features of Cuban flora, it is necessary to point up such interesting aspects as the endemism, disjunctions, vicariancy, and floristic relationships with neighboring lands, which occur in consequence of Cuba's narrow and east-west elongated configuration, its status as the largest island of the Caribbean Archipelago, and its situation at the entrance to the Gulf of Mexico, between the American continental land masses.

According to recent reports, the Cuban vascular flora embraces about 6,700 species (500 Pteridophytes and 6,200 Phanerophytes, of which nearly 20 are gymnosperms and the rest angiosperms), grouped in 1,300 genera and 181 families. Trees and shrubs make 60% of the total number of flowering plants (Alaín, 1958; Borhidi & Muñíz, 1980).

Among the best represented families are the following: Asteraceae, Poaceae, and Rubiaceae, with approximately 400 taxa each; Euphorbiaceae, Leguminosae, and Orchidaceae with nearly 300 taxa; Cyperaceae and Myrtaceae with about 200 taxa; and comprising 100 taxa each are the Bignoniaceae, Boraginaceae, Convolvulaceae, Malvaceae, Melastomataceae, Piperaceae, and Verbenaceae, among others.

The genera accounting for the major number of specific taxa are the following:

Eugenia (Myrtaceae)	ca. 120	taxa
Psychotria (Rubiaceae)	70	"
Phyllanthus (Euphorbiaceae)	65	"
Tabebuia (Bignoniaceae)	65	"
Rondeletia (Rubiaceae)	65	"
Calyptranthes (Myrtaceae)	50	"

1. Endemism

A figure of 50% endemic phanerophytes for the Cuban flora (51.3% according to Borhidi & Muñíz, 1980) has been reported by Samek (1973), thus making about 3,100 endemic species. This is the highest among the Antilles.

The endemism at the generic level amounts to 71 genera, of which 17 are Asteraceae. Other noteworthy families are the Rubiaceae, with 11 endemic genera, and subfamily Papilionoideae with seven. Numerically, *Schmidtottia* (Rubiaceae) is the largest endemic genus, with 16 species. Also *Krokia* (s.l.) (Myrtaceae) with 11 species; *Mozartia* (Myrtaceae) with eight; *Moacroton* (Euphorbiaceae) with seven; and *Platygyne* (Euphorbiaceae) with seven species, deserve special mention.

Certain genera with only a few representatives exhibit a high percentage of endemism, since their few representatives are restricted to Cuba. Nevertheless, such large genera as *Copernicia* (25 species), *Leucocroton* (19 species), *Pachyanthus* (16 species), and *Purdiaea* (11 species) are 100% endemic. Some genera show a high percentage of endemism, viz., *Tabebuia* (93%), *Rondeletia* (93%), *Coccothrinax* (95%), *Lyonia* (93%) and *Calyptranthes* (92%).

There is no endemism at the familial level. The highest percentages of endemism at the species level can be found among larger families, such as Myrtaceae (88%), Rubiaceae (68%), Euphorbiaceae (67%), and other numerically lesser families such as Arecaceae (90%), Ericaceae (92%), Cyrillaceae (80%) and Buxaceae (89%).

Some families with few taxa (even with only one species) show 100% endemism, viz., Magnoliaceae, Fagaceae, Clethraceae, and Illiciaceae. Concerning the geographical distribution of endemic taxa, the most outstanding feature is their concentration into two regions: West and East, mainly in the principal montane regions. The highest region is the eastern, where most of the endemic taxa are found; the Sagua-Baracoa montane system showing 80% endemism in its flora. In second place is the Cordillera of Guaniguanico in the western, and to a lesser extent the central altitudes of the Guamuhaya group (Escambray s.l.). In each of the above mentioned areas, some specific and more or less scattered localities can be pointed up with regard to their high number of endemics.

The actual causes for endemism remain as a matter of discussion (see Borhidi, 1985) although certain factors should not be neglected, mainly those related to soils, as it is well known that the edaphic "mosaic" of Cuba, in which fertile soils are mixed together with poor soils, cause extreme biotopes to occur (i.e., those derived from ultrabasic serpentinic rocks, the siliceous sandy soils, rocky limestone soils and karstic areas).

Among climatic factors, the amount and seasonal distribution of rainfall have a major influence, varying from very dry areas (less than 400 mm/year) to moist (more than 2,500 mm/year). Less important are the temperature fluctuations because of Cuba's insularity and situation at the limit of the Tropic of Cancer.

Topography has a noteworthy role, mainly in the eastern region, where mountain ranges above 1,000 m are common and cloud condensation causes some dryness; the altitudinal temperature gradient (0.6–0.9°C/100 m) has a mild influence.

The combination of these factors in the same area accounts for high local endemism, mainly at the northeastern extreme of Cuba, where the northern slopes of the mountains receive the direct influence of the East Winds, with rainfall

Table 1
Selected list of comprehensive works on Cuban flora and vegetation (Phanerophytes).

1. **Acuña Galé, J.** 1938. Catálogo descriptivo de las Orquídeas Cubanas. Bol. Estac. Exp. Agron. Santiago de las Vegas 60, 221 pp.
2. **Alaín (E. Liogier).** 1953. Flora de Cuba. 3. Dicotiledóneas: Malpighiaceae a Myrtaceae. Contr. Ocas. Mus. Hist. Nat. Colegio "De La Salle" **13**: 7–502.
3. _____. 1956. Flora de Cuba. 4. Dicotiledóneas: Melastomataceae a Plantaginaceae. Contr. Ocas. Mus. Hist. Nat. Colegio "De La Salle" **16**: 7–556.
4. _____. 1958. La Flora de Cuba: Sus principales características. Su origen probable. Revista Soc. Cub. Bot. **15**: 36–59, 84–96.
5. _____. 1962. Flora de Cuba 5. Rubiales–Valerianales–Cucurbitales–Campanulales–Asterales. Río Piedras, Puerto Rico.
6. _____. 1969. Flora de Cuba. Suplemento. Caracas, Venezuela.
7. **Borhidi, A.** 1985. Phytogeographic survey of Cuba I. The phytogeographic characteristics and evolution of the Flora of Cuba. Acta Bot. Acad. Sci. Hung. **31**: 3–34.
8. _____ **& O. Muñíz.** 1980. Die Vegetationskarte con Kuba. Acta Bot. Acad. Sci. Hung. **26**: 25–53.
9. _____ & _____. 1986[1987]. Phytogeographic survey of Cuba. II. Floristic relationships and phytogeographic subdivision. Acta Bot. Acad. Sci. Hung. **32(1–4)**: 3–48.
10. _____, _____ **& E. del Risco.** 1979. Clasificación fitocenológica de la Vegetación de Cuba. Acta Bot. Acad. Sci. Hung. **25**: 263–301.
11. **Capote, R. & R. Berazaín.** 1987. Clasificaciín de las formaciones vegetales de Cuba. Rev. Jard. Bot. Nac. **5(2)**: 27–75.
12. **Curtiss, A. H.** 1904. List of Curtiss Series II of West Indian Plants. Coll. 1903–1904, Isla de Pinos, Jacksonville, Florida.
13. _____. 1905. List of Series III of Curtiss' West Indian Plants. Coll. 1904–1905, Province of Havana, Cuba. Jacksonville, Florida.
14. **Eggers, H.** 1890. Botanical Exploration of Cuba. Bull. Misc. Inform. **1890**: 37–38.
15. **Gómez de la Maza Jiménez, M. & J. T. Roig Mesa.** 1916. Flora de Cuba (Datos para su estudio), ed. 2. Bol. Estac. Exp. Agron. Santiago de Las Vegas 22, ed. 2.
16. **Grisebach, A.** 1860. Plantae Wrightianae e Cuba Orientali (Polypetalae et Apetalae). Mem. Amer. Acad. Arts. n.s. **8**: 153–192.
17. _____. 1862. Plantae Wrightianae e Cuba Orientali, Part II. (Monopetalae et Monocotyledones). Mem. Amer. Acad. Arts. n.s. **8**: 503–536.
18. _____. 1866. Die Geographische Verbreitung des Pflanzen Westindiens. Abh Königl. Ges. Wiss. Göttingen **12**, Abh. Phys. Kl.: 3–80.
19. _____. 1866. Catalogus plantarum cubensium exhibens collectionem Wrightianam aliasque minores ex insula Cuba missas. Lipsiae (Leipzig).
20. **Humboldt, A. V., A. Bonpland & C. S. Kunth.** 1825. Nova Genera et Species Plantarum. 7. Paris.
21. **Hitchcock, A. S.** 1909. Catalogue of the Grasses of Cuba. Contr. U.S. Nat. Herb. **12**: 183–258.
22. _____. 1936. Manual of the Grasses of the West Indies. U.S.D.A. Misc. Publ. 243.
23. **Jacquin, N.J.** 1760. Enumeratio Systematica Plantarum, quae in Insulis caribaeis Vicinaque Americes continente detexit novas, aut jam cognitas emendavit. Leiden.
24. **León (J. S. Sauget).** 1946. Flora de Cuba 1. Gimnospermas. Monocotiledóneas. Contr. Ocas. Mus. Hist. Nat. Colegio "De La Salle" **8**.
25. _____ **& Alaín.** 1951. Flora de Cuba 2. Dicotiledóneas: Casuarináceas a Meliáceas. Contr. Ocas. Mus. Hist. Nat. Colegio "De La Salle" **10**.
26. **Marie Victorin et León.** 1942. Itinéraires Botaniques dans l'Ile de Cuba (Première Série). Contr. Inst. Bot. Univ. Montréal **41**: 1–496.
27. _____. 1944. Itinéraires Botaniques dans l'Ile de Cuba (Deuxième Série). Contr. Inst. Bot. Univ. Montréal **50**: 11–410.
28. _____. 1956. Itinéraires Botaniques dans l'Ile de Cuba (Troisième Série). Contr. Inst. Bot. Univ. Montréal **68**.
29. **Manitz, H.** 1978. Bibliographie der Verbreitungskarten cubanischer Gefässpflanzen. Beitr.. Phytotax. (Jena) **6**: 171–221.
30. _____. 1980. Bibliographie der Verbreitungskarten cubanischer Cuba 55 Gefässpflanzen, Supplement 1. Wiss. Z. Friedrich-Schiller Univ. Jena, Math. Naturwiss. **29**: 539–557.
31. _____. 1984. Bibliographie der Verbreitungskarten cubanischer Gefässpflanzen, Supplement 2. Wiss. Z. Friedrich-Schiller Univ. Jena, Math. Naturwiss. **33**: 739–757.
32. **Richard, A.** 1845. Essai d' une Flore de L'Ile de Cuba. 1. *In*: R. de La Sagra, Histoire Physique, Politique et Naturelle de l'Ile de Cuba. Paris.
33. _____. 1845–1855. Fanerogamia o Plantas Vasculares. 1–2. *In*: R. de La Sagra, Historia Física, Política y Natural de la Isla de Cuba. Parte 2. Historia Natural, 10–12. Paris.
34. **Samek, V.** 1973. Regiones Fitogeográficas de Cuba. Acad. Ci. Cuba, Ser. Forest. **15**.
35. **Samkova, Hana & V. Samek.** 1967. Bibliographia Botanica Cuba (Teorica y Aplicada) con Enfasis en la Silvicultura. Acad. Ci. Cuba, Ser. Biol. **1**.
36. **Sauvalle, F. A.** 1873. Flora Cubana. Havana.
37. **Seifriz, W.** 1940. Die Pflanzengeographie von Cuba. Bot. Jahrd. **70**: 441–462.
38. _____. 1943. The Plant Life of Cuba. Ecol. Monogr. **13**: 375–426.
39. **Smith, E. E.** 1954. The Forests of Cuba. Publ. Maria Moors Cabot Found. Bot. Res. **2**.
40. **Swartz, O.** 1794–1800. Icones Plantarum incognitarum, Fasc. 1. Erlangen.
41. _____. 1797–1806. Flora India Occidentalis aucta atque illustrata sive descriptiones plantarum in Prodromo recersitarum. 1–3. Erlangen.
42. **Urban, I.** 1893. Additamenta ad cognitionem florae Indiae Occidentalis. Particula I. Bot. Jahrb. **15**: 286–361.
43. _____. 1894–1895. Additamenta ad cognitionem florae Indiae Occidentalis. Particula II. Bot. Jahrb. **19**: 562–681.
44. _____. 1896. Additamenta ad cognitionem florae Indiae Occidentalis. Particula III. Bot. Jahrb. **21**: 514–638.
45. _____. 1897. Additamenta ad cognitionem florae Indiae Occidentalis. particula IV. Bot. Jahrb. **24**: 10–152.
46. _____ **(Ed.).** 1898–1928. Symbolae Antillaneae seu Fundamenta Florae Indiae Occidentalis. Berlin, Paris, Londres (1); Leipzig, Paris, Londres (2–7); Leipzig (8–9).
47. _____. 1914–1930. Sertum Antillanum, I–IX, XII, XIII, XV–XXX. Repert. Spec. Nov. Regni Veg. **13–22, 24, 26, 28**.

during the whole year, while on the southern slopes more dry conditions prevail, together with a higher insolation. These ecological contrasts claim for the region a higher degree of plant specialization, and are among the causes of its floristic richness.

Another interesting aspect is the great variation in geological age and its implications on endemism. The older areas coincide with the three principal mountain regions, and therefore a high degree of endemism (mainly at the generic level) is present, thus indicating the ancient evolution of the flora. A good example of this is *Microcycas calocoma* from western Cuba, a typical paleoendemic. Neoendemics are common in extreme biotopes, usually younger from a geological point of view, as the acidic siliceous sandy soils, or as in the Isla de la Juventud, in which recent (Quaternary) separation from Cuba has favored local endemism caused by geographical isolation. (*Copernicia curtissii, Paspalum insulare, Psychotria geronensis, Tabebuia geronensis, Zamia silicea*, etc.)

Concerning endemism in relation to vegetation types, it is higher in shrub formations (serpentinic thorny thickets, montane thickets and coastal and sub-coastal thorny thickets); while among forest formations, pine forests are the richest, followed by cloud forests, rainforests and microphyllous evergreen forests.

2. Disjunctions (within Cuba)

Some Cuban genera show remarkable examples of disjunct distribution, mainly those with species polarized in the western and eastern extremes of the main island, such as *Cyrilla, Moacroton, Pinus, Spathelia*, (Samek, 1973). There are good examples of western-central-eastern distribution in *Leucocroton wrightii* and *Vaccinium cubense*. And finally, some taxa bear a central-eastern disjunction: *Hedyosmum* and *Magnolia*. The above-mentioned examples demonstrate a kind of disjunction that follows the regions with a high concentration of endemic taxa. The scattered, widespread distribution of some special biotopes along the Cuban island, as with ultrabasic rock outcrops, causes some species to present highly localized ranges as in, for example, *Neobracea valenzuelana* and *Phyllanthus orbicularis* (Borhidi, 1985). In coastal and sub-coastal species, owing to the random distribution of the three types of coasts (rock, sand and marsh) such a multisectorial distribution is frequent, for example, in coastal areas is found an arborescent cactus, *Dendrocereus nudiflorus*, on the northern as well as the southern coast, or in west-east extremes.

Some external disjunctions are somewhat perplexing, as in the case of *Dracaena* from Central America, East Cuba and Africa, and the genus *Cneorum* from Cuba, the Canary Islands and the Mediterranean. In the Americas, some interesting relationships are present in *Purdiaea* (Guatemala, Colombia, Peru) and *Gaussia* (Yucatan, Cuba, Puerto Rico).

3. Vicariancy

The distribution patterns of many genera exhibit both geographical and ecological vicariancy. Geographical vicariancy can be demonstrated in species of *Pinus, Lyonia, Vaccinium*, and *Purdiaea*, from pine forests on ferritic (lateritic) soils, both in western and eastern Cuba.

In the case of the ecological vicariancy, it is clearly defined by the great soil mosaic found along the main island. Thus, for the same genus, different species are found in contrasting biotopes (e.g., limestone and serpentinic soils, geographically closely located). As good examples, such genera as *Bucida, Pseudocarpidium* and *Tabebuia* could be mentioned.

4. Relationships of the Cuban flora with neighboring lands

In order to understand better the floristic relationships of the Cuban Archipelago, it must not be taken as a whole, because the present geographical configuration shows, for western areas, a closer proximity to the northern land mass, whereas eastern regions are more related to southern lands. Moreover, other Antillean islands are in close proximity to Cuba, and as a result, the floristic affinities of Cuban plants vary according to their geographical distribution inside the Cuban Archipelago.

Western Cuba and Isla de la Juventud have somewhat close floristic relationships with southern North America (especially Florida), and such taxa as *Pinus, Salix, Fraxinus, Befaria* and *Kalmia* (mostly in pine forests) can be found in common. Semi-deciduous forests on the Guanahacabibes Peninsula (at the extreme western end of Cuba) have been reported as bearing a certain floristic affinity to the Yucatan Peninsula of Central America. Some common elements are: *Manilkara meridionalis, Diospyrus tetrasperma* and *Caesalpinia violacea* (Bisse, 1973).

Central Cuba's flora shows a lesser richness, and its main floristic relationships are with Central America, northern South America and Hispaniola.

Eastern Cuba (floristically the richest) bears an outstanding affinity with Hispaniola's flora and secondly with Jamaica, Puerto Rico, and South America.

These relationships clearly indicate a different degree of contact established in the past between the neighboring biotas and therefore both dispersal and vicariance took place (for an interesting approach on the subject, see Rosen, 1985).

As a whole, a floristic analysis demonstrates the occurrence of cosmopolitan families (Poaceae, Asteraceae, Orchidaceae), and genera (*Senecio, Euphorbia, Solanum, Drosera, Utricularia*, etc.). Some cosmopolitan species (*Plantago major, Solanum nigrum, Portulaca oleracea, Sonchus oleraceus*) are also present.

Pantropical elements, at the family level, (Annonaceae, Apocynaceae, Begoniaceae, Combretaceae, Gesneriaceae, Lauraceae, Melastomataceae, Myrtaceae, Passifloraceae, Rutaceae, Sapotaceae and Theaceae s.l.), some genera, (*Bauhinia, Eugenia, Hibiscus, Phyllanthus, Psychotria*); and a few species, (*Abrus precatorius, Caesalpinia bonduc, Dodonaea viscosa*, and *Pisonia aculeata*) can be mentioned.

Neotropical endemism is expressed by such families as Cactaceae, Bromeliaceae, Cyrillaceae, Marcgraviaceae, Cannaceae, etc. At the generic rank, *Cecropia, Croton, Jacaranda, Erythroxylum*, and *Mimosa*; and some species, e.g., *Bursera simarouba, Cedrela odorata, Cordia sebestena, Spondias purpurea* and *Tecoma stans*, exemplify the neotropical outline of the Cuban flora.

Among neotropical elements, those that mostly concern Central America and the Caribbean can be exemplified by the following genera: *Agave, Coccothrinax*, and *Thrinax*; and such species as *Lonchocarpus latifolius* and *Hippomane mancinella*. The Antillean genera *Guaiacum, Crescentia*, and species *Picrodendron macrocarpum, Hura crepitans*, and *Swietenia mahagoni* demonstrate a closer relationship with the Caribbean flora.

Holarctic influence on the Cuban flora can be demonstrated by the presence of such families as Magnoliaceae, Pinaceae, Salicaceae, Juglandaceae, Papaveraceae and Rosaceae; and among genera, *Pinguicula, Scrophularia, Lythrum, Berberis, Fraxinus, Juniperus*, and *Callicarpa*, can be mentioned as good examples of such floristic relationships.

5. Phytogeographical regionalization

The particularities of geographical distribution in the Cuban flora make it possible to attempt a phytoregionalization that, in a general way, allows us to divide the Cuban Archipelago into a system of entities.

According to Samek et al. (1986, in press) Cuba is considered to be a Province of the Antilles sub-Region. Three main Sub-Provinces can be distinguished: western Cuba, with three sectors—Western Pine Forests, Western Ridges, and Peninsulas; Central Cuba, with three sectors—Central Western, Guamuhaya massif, and Central-Eastern; and eastern Cuba, with five sectors—Mayarí-Imías, Cayo Rey-Nibujón, Sierra Maestra, Cabo Cruz-Maisí, and Central Valley. (Del Risco et al., Atlas Nacional de Cuba, in preparation).

In the Western Sub-Province pine forests on different soils can be found, with *Pinus caribaea* and *Pinus tropicalis* together with a palm, *Colpothrinax wrightii*, and *Quercus oleoides* ssp. *sagraeana* (mainly in Pinar del Río); semideciduous forests on the plains; on the hillocks shrubby vegetation occurs, with such typical elements as *Bombacopsis cubensis* and *Gaussia princeps*. On southern coasts of Isla de la Juventud, *Juniperus lucayana* is rarely encountered.

Central Cuba, the more extensive but floristically poor region, bears mostly secondary vegetation covering broad areas of plains; xeromorphic thorny thickets are found on the outcrops of serpentinized ultrabasic rocks and are places of high endemism. In the high coasts there are thorny thickets rich in microphyllous shrubs and cacti: in marshy coasts and cays, mangroves are abundant, with species typical of the Atlantic coast.

Lowland forests are mainly replaced by agricultural lands and secondary savannas rich in palms, mostly species of *Copernicia* and *Coccothrinax*, and some relicts of natural forests can be found at more or less low elevations. In the higher mountains, rain and cloud forests are present.

Eastern Cuba is floristically the richest Sub-Province, mainly in the mountain ranges. The central valley, usually devoted to agriculture, does not bear any importance from a floristic point of view. The mountains comprise two main systems: the Sierra Maestra in the south and Sagua-Baracoa in the northeast.

The Sierra Maestra mountains are the highest of Cuba, with their southern slope very close to the Caribbean sea and causing the coastal line to be a very narrow strip where xeromorphic thorny thickets are abundant and rich in cacti. Microphyllous evergreen forests are found ascending through the southern slope, followed, at higher altitudes, by rainforests rich in endemic species of *Persea, Ilex, Cleyera*, and interesting species like *Cneorum trimerum, Heliconia caribaea, Hedyosmum cubense*, etc.; above this, cloud forests with *Cyrilla racemiflora, Torralbasia cuneifolia, Weinmannia pinnata* and *Garrya fadyenii*, among others; and at the highest altitudes (Turquino group) there is a more or less meso-xeromorphic thicket with low shrubs and *Agave*, and several endemic species such as *Juniperus saxicola* and *Myrica cacuminis*. As a curious fact, isolated spots of *Pinus maestrensis* occur.

On the northern slope, natural vegetation is more destroyed by agriculture, mainly at the base of the mountains; at higher altitudes, however, rain and cloud forests, very similar to those of the southern slope, can be found.

In the Sagua-Baracoa mountain system, at the northeastern extreme of the Cuban island, extensive pine forests occur, with *Pinus cubensis* and other tree species in the understory of those pine forests. The highest endemism is present among the shrubby species of the Ericaceae and Melastomataceae. The very interesting species, *Dracaena cubensis*, is restricted to these areas. Those pine forests occupy the tablelands and the relatively dry biotopes. In more humid localities and in river valleys, rainforests typified by *Carapa guianensis* occur. Sub-xeromorphic thickets are well developed on ferritic soils bearing the highest endemic flora (about 80%).

B. THE PRESENT VEGETATION OF THE CUBAN ARCHIPELAGO (R. P. CAPOTE AND R. BERAZAÍN)

1. Introduction

In as much as Cuba and the Isla de la Juventud are the two principal islands of the Cuban Archipelago, we will refer, in describing the general outlines of Cuban vegetation, only to plant communities occurring in them, and the rest of the islets and cays are mainly occupied by mangroves (Fig. 1).

The high variability of Cuban vegetation is conditioned by climate and different geological and edaphic substrata: " As a consequence of the great edaphic richness, we can find in Cuba a large number of vegetation types edaphically conditioned, as in example: serpentinitic vegetation, riparian and coastal vegetation . . . etc." (Borhidi & Muñíz, 1980).

Human impact has favored the development of different types of secondary forests, thickets, and herbaceous communities, adding much to the complexity of Cuba's vegetation cover.

Subsequently, the different vegetation types occurring in

the Cuban Archipelago will be described with respect to their floristic, physiognomic and ecologic characters, and taking into account the criteria from Bisse (unpubl.); Berazaín, 1979, Borhidi et al., 1979; Borhidi & Muñíz, 1980, 1986; and Capote & Berazaín, 1984.

The classification of Cuban vegetation (Capote & Berazaín, 1984) groups the main vegetation types into: forests, thickets, herbaceous vegetation, complexes of vegetation and secondary vegetation. Such criteria do, in general, correspond with the principles used in the international classification of vegetation (UNESCO, 1973), making nomenclatural changes necessary owing to the particular features of our vegetation types.

According to Borhidi and Muñíz (1980), Cuban plains and low altitude mountains originally were covered by evergreen and semi-deciduous forests. The last type was particularly frequent in central and western Cuba. Among present-day formations are the following: rain, cloud, evergreen, semi-deciduous, swamp, gallery, mangrove and pine forests.

Shrubby formations are developed as different types of xeromorphic thickets that can form forests in coastal and interior areas, such as the well-known desertic coastal thickets, abundant in arborescent and columnar cacti; and also those on serpentinite and ferritic substrata, where the highest figures of endemism are reached. At Pico Turquino, in the Sierra Maestra range, a particular type of montane scrub is present, known nowadays as "subparamo."

Herbaceous communities are represented by different kinds of savannas, mostly edaphic and semianthropic ones; freshwater communities, shrubby-herbaceous communities, river and swamp grasslands, as well as halophytic herbaceous communities.

Vegetation complexes are groups of related plant communities that occur in certain areas, imparting to them a particular shape. Such vegetation complexes are present in montane limestone karst (conical karstic formations); and rocky and sandy coasts.

Secondary vegetation arises when degradation of the original one takes place, and its structural and floristic complexity is mainly related to the level achieved by successional development. Among them, forests, thickets, and herbaceous communities are encountered.

Among the herbaceous communities, anthropic savannas are considered. Ruderal and segetal vegetations associated with human activity, are developed in places constructed by man and agricultural lands respectively.

To complete the plant cover outline, there must be mentioned the different agricultural and forest plantations, the very extensive sugar cane (*Sacharum officinale*) and *Pinus caribaea* plantings, respectively.

2. Forest formations

a. Rain Forests. These are established under continuously wet conditions, with rainfall values over 2,500 mm/year, without any dry season. They are ombrophilous, evergreen forests, in which deciduous emergent trees can occur. They possess several tree and shrub strata and one herbaceous stratum; epiphytes and lianas occur abundantly.

The typical rain forest occurs at low altitude, below 400 m, in close association with river valleys, in more or less acidic soils of the north-northeast extremity of Eastern Cuba. Three tree strata are present, the height ranging between 15-35 m, with palms and tree ferns; a shrub stratum is rare or even absent; an herbaceous stratum is present, and lianas, epiphytes and epiphylls are ubiquitous.

In the first stratum tree species with compound leaves are predominant. Particularly abundant is *Carapa guianensis*, this is an unusual case, as noted by Berazaín (1979), because an abundance of a single species is not the common situation in a tropical rain forest. Other important tree species are *Calophyllum utile, Buchenavia capitata, Manilkara albescens, Micropolis polita* and *Terminalia nipensis*. In the second stratum we can find *Ochroma pyramidalis, Guarea guidonia, Oxandra laurifolia, Diospyros caribaea* and palms as *Prestoea montana* and *Calyptronoma orientalis*.

In the understory, shrubs belonging to Melastomataceae, Myrtaceae, and Ericaceae are very common. Tree ferns are abundant, and the presence of *Heliconia caribaea* is typical. Among herbaceous species, Poaceae and Cyperaceae are present. Among Cyperaceae, some species of *Scleria* occur along roads. Lianas are mostly Araceae and Cucurbitaceae. Among epiphytes, Bromeliaceae, Gesneriaceae, Orchidaceae, bryophytes, ferns and algae are present. Epiphytes are arranged into two synusiae, heliophiles in the canopy and sciophiles in the lower strata trees.

Montane rain forests are present between 800-1,600 m at Sierra Maestra and Sierra de Imías mountains in eastern Cuba, and in Sierra del Escambray in central Cuba. This forest is composed of two tree strata between 8-25 m in height, shrubs and herbaceous plants in the understory, and abundant tree ferns, mosses, liverworts, and other epiphylls. Among the trees of the first stratum the following should be mentioned: *Buchenavia capitata, Calophyllum antillanum, Alchornea latifolia, Dipholis jubilla, Didymopanax morototoni, Tabebuia* sp., *Ocotea cuneata, Zanthoxylum elephantiasis*. In the second tree stratum are found *Oxandra laurifolia, Amyris lineata, Ditta myricoides, Sapium jamaicense* and species of *Laplacea* and *Lyonia* as well as the above mentioned palms in typical rain forests. The understory is rich in species of Rubiaceae, Myrtaceae, Melastomataceae, Ericaceae, and tree ferns. Epiphytes and lianas are almost the same as cited for the preceding formation; epiphylly is also a common phenomenon here.

On purple ferritic soils (laterites) of the northeastern mountains of eastern Cuba, there is a particular type of submontane or montane rain forest (400-900 m) having two tree strata between 5-22 m in height, the understory being very rich in endemic species; one herbaceous stratum; with few epiphytes, and epiphyllous mosses and liverworts. According to Borhidi & Muñíz (1980), the canopy is typified by *Byrsonima coriacea, B. orientensis, Sloanea curatellifolia, Calophyllum utile, Podocarpus ekmanii, Hieronima nipensis*, etc. Tree ferns and sciophilous epiphytes are lacking, and the palm *Bactris cubensis* is frequent.

b. Cloud Forests (Elfin Woodland). These forests have a single tree stratum between 8-12 m in height, the trees with twisted trunks, mixed together with abundant tree ferns, rich in epiphytes mostly mosses and liverworts, ferns (Hymenophyllaceae, Polypodiaceae), Araceae, Bromeliaceae, Gesneriaceae and Orchidaceae (little ones, such as species of *Lepanthes*, *Pleurothallis*, etc.). These epiphytes almost completely cover the trunks and branches of trees, among which *Cyrilla racemiflora*, *Magnolia cubensis*, *Weinmania pinnata*, *Torralbasia cuneifolia*, *Garrya fadyenii*, *Henrietella ekmanii*, *Clethra cubensis*, and some species of the genera *Lyonia*, *Gonzalagunia* and *Miconia* can be mentioned. This forest occurs between 1,600–1,900 m.

The understory is dense, and rich in species of Ericaceae, Melastomataceae and tree ferns. An herbaceous stratum is abundant, with ferns, mosses, and terrestrial orchids.

c. Evergreen Forests. These forests have less than 30% of deciduous elements among the trees. According to leaf size it can be divided into mesophyllous (leaf size between 13–26 cm) and microphyllous (1–6 cm). As a general feature these forests have shrubs and herbaceous plants in their understory, and exhibit poor development of epiphytes and, to a lesser extent, lianas.

Mesophyllous evergreen forests usually possess a canopy of 15–25 m in height, with palms (such as *Roystonea regia*) and emergent trees up to 25–30 m, presenting epiphytes and lianas; with shrubs and herbaceous plants in the understory. These forests mainly occur, at the present time, in submontane altitudes between 300–800 m. Their floristic composition is almost the same as the semi-deciduous mesophyllous forests, but with a higher proportion of evergreen species. Among outstanding trees should be mentioned *Alchornea latifolia*, *Amaioua corymbosa*, *Antirhea radiata*, *Calophyllum antillanum*, *Dendropanax arboreus*, *Ficus* spp., *Margaritaria nobilis*, *Mastichodendron foetidissimum*, *Matayba oppositifolia*, *Oxandra lanceolata*, *O. laurifolia*, *Pseudolmedia spuria*, *Sloanea amygdalina*, *Tabebuia shaferi*, *Trophis racemosa*, *Wallenia laurifolia*, *Zanthoxylum martinicense* and *Zizyphus rhodoxylon*. The understory is poorly developed, and the herbaceous stratum is abundant in ferns (*Adiantum*, *Tectaria* and *Pteris*), and mesomorphic grasses as *Pharus glaber*, *Olyra latifolia* and *Lasiacis divaricata*. Lianas and epiphytes are scarce.

The microphyllous evergreen forest (locally called "dry forest") is located just behind the coastal and subcoastal thickets, in limestone hummocks and tablelands. The canopy is between 5–15 m in height, with evergreen and deciduous trees, with epiphytes present, as well as lianas and thorny shrubs, and some columnar or arborescent cacti among other succulents and herbaceous plants.

Several legume trees and shrubs such as *Brya ebenus*, *Belairia spinosa*, *Lysiloma latisiliqua*, *Albizia cubana*, *Peltophorum adnatum*, *Pithecellobium* spp., etc. are present. Others such as *Bucida spinosa*, *Bursera simaruba*, *Metopium brownei*, *Krugiodendrum ferreum*, *Drypetes mucronata*, *Ateramnus lucidus*, *Picrodendron macrocarpum*, *Simaruba glauca*, *Maytenus buxifolia*, *Exostema caribaeum*, *Amyris balsamifera*, *Savia sessiliflora*, *Croton lucidus*, *Adelia ricinella* and many others; arborescent succulents such as *Dendrocereus nudiflorus* as well as some species of *Leptocereus*, *Harrisia*, *Pilosocereus*, *Consolea*, and some others such as *Agave* and *Furcraea* spp. Several species of *Coccothrinax* palm are present.

d. Semideciduous Forests. These forests exhibit 40–65% deciduous trees, mostly at the canopy. Shrubs and herbaceous plants are scarce, bearing a poor development of epiphytes, in contrast with abundant lianas. Taking into account the size of leaves, two different variants are recognized: mesophyllous (leaf size between 13–26 cm) and microphyllous (1–6 cm).

The mesophyllous type bears two tree strata between 15–20 m up to 25 m in height, mostly composed of deciduous trees; emergent palms and broadleaved trees can be found, over 25 m in height. Among the low stratum trees, sclerophyllous evergreen and deciduous species are present. This type of forest is located on rendzine (red or black) and brown, carbonated soils. It is inhabited by tree species such as *Adelia ricinella*, *Alvaradoa amorphoides* ssp. *psilophylla*, *Allophyllus cominia*, *Amyris balsamifera*, *A. elemifera*, *Andira inermis*, *Bursera simaruba*, *Casearia hirsuta*, *C. spinescens*, *Cedrela odorata*, *Ceiba pentandra*, *Cordia collococca*, *C. gerascanthus*, *Drypetes* spp., *Eugenia axillaris*, *Gossypiospermum praecox*, *Ateramnus lucidus*, *Hebestigma nse*, *Hildegardia cubensis*, *Oxandra lanceolata*, *Pithecellobium cubense*, *P. lentiscifolium*, *Roystonea regia*, *Samanea saman*, *Savia sessiliflora*, *Tabebuia* spp., *Trichilia glabra*, *T. hirta*, *Zanthoxylum elephantiasis*, *Z. fagara*, etc. The herbaceous stratum bears grasses and ferns. Lianas are abundant, for example *Cydista diversifolia*, *Gouania lupuloides*, *Davilla rugosa*, and several species of *Bauhinia*, *Passiflora*, *Smilax*, *Serjania*, etc. Among epiphytes, species of *Tillandsia*, *Cattleyopsis*, and succulents such as *Rhipsalis cassuta* and *Selenicereus grandiflorus* are common.

The microphyllous semi-deciduous forest is characterized by a single stratum of deciduous and microphyllous, sometimes thorny trees as well as fan-shaped palms of 8–15 m in height.

Among the species found are *Allophyllus cominia*, *Brya ebenus*, *Belairia savannarum*, *Bursera simaruba*, *Casearia aculeata*, *Copernicia* spp., *Erythroxylum* spp., *Eugenia maleolens*, *Ficus* spp., *Gossypiospermum praecox*, *Hypelate trifoliata*, *Malpighia* spp., *Mastichodendron foetidissimum*, *Peltophorum adnatum*, *Picramnia pentandra*, *Phyllostylon brasiliensis*, *Pithecellobium arboreum*, *P. lentiscifolium*, *Piscidia havanensis*, *Randia* spp., *Spondias mombin*, *Tabebuia microphylla*, and *Zanthoxylum fagara*. This forest can be found in plains and hills at the north of the eastern provinces of Las Tunas and Holguin, as well as in the south of the Isla de la Juventud. On badly drained soils, the floristic composition slightly diverges. There can be found *Bucida subinermis*, *Cameraria retusa*, *Coccoloba armata*, *C. microphylla*, *Cipura paludosa*, *Guettarda elliptica*, *Hypoxis wrightii*, *Sabal parviflora*, *Tabebuia angustata*, and some others.

e. Pine Forests. These forests show a bipolar distribution: western Cuba (Pinar del Río province and Isla de la Juventud) with a predominance of *Pinus caribaea*, while in eastern Cuba (at Sierra Maestra and Sagua-Baracoa mountains) *P. maestrensis* and *P. cubensis* respectively occur. These pine forests are present in plains and hilly areas in western Cuba, and mainly in mountains in eastern Cuba.

The canopy is dominated by pines, with some broadleaved trees and palms. The shrubby stratum is very rich in species, and the herbaceous stratum is fairly well developed. Epiphytes and lianas are scarce. Depending upon the edaphological conditions (substrata) in which these forests occur, they have, in spite of physiognomic resemblance, noticeable floristic differences amongst them, that cannot be neglected.

Western pine forests are typified by the occurrence of *Pinus caribaea* and *P. tropicalis*. On white sandy soils at Pinar del Río and Isla de la Juventud, *P. tropicalis* is commonly associated with *Colpothrinax wrightii* (Arecaceae). Among shrubs, species of the Ericaceae genera *Lyonia* and *Kalmia*, *Byrsonima*, and palms such as *Paurotis wrightii* and *Copernicia curtissii*, are frequent. The herbaceous stratum is rich in Eriocaulaceae *Paepalanthus* and *Eriocaulon*), Lentibulariaceae (*Pinguicula* and *Utricularia*), species of *Drosera*, *Xyris*, and diverse Cyperaceae and Poaceae.

In northern Pinar del Río and Isla de la Juventud there are pine forests on slaty (oligotrophic quartz-allitic yellow) soils, with *Pinus caribaea* and *P. tropicalis*, *Byrsonima crassifolia*, *Curatella americana*, several Melastomataceae, and species of *Kalmia*, *Croton* and *Phyllanthus*. Grasses are abundant.

At Pinar del Río, those pine forests can be associated with *Quercus oleoides* spp. *sagraeana* ("encina"). This species can be dominant and constitutes a particular kind of forest known as "encinar" (Bisse, unpubl.)

On ferritic soils at Pinar del Río, *Pinus caribaea* forests have an understory very rich in endemic taxa of the genera *Mazaea* and *Rondeletia*, and such species as *Phyllanthus discolor*, *Phania cajalbanica* and several grasses in *Aristida* and *Andropogon*.

In eastern Cuba, pine forests with *Pinus cubensis* (northeastern mountains) and *P. maestrensis* (Sierra Maestra), are present. Northeastern pine forests, when occurring in low elevation tablelands, contain several taxa showing geographical vicariancy with western pine forests (*Coccothrinax*, *Ouratea*, and *Purdiaea* species among others), as well as species of *Casearia*, *Vaccinium*, *Phyllanthus*, *Lyonia*, *Andropogon* and *Rhynchospora*.

High altitude pine forests contain more broadleaved elements mostly coming from neighboring rain forests, and less endemics than the low altitude ones.

Pine forests at Sierra Maestra, with *P. maestrensis*, are the highest ones, occurring at altitudes between 800–1,800 m. Tree ferns and rain forest elements are abundant, such as *Garrya fadyenii* and *Weinmannia pinnata*; and ferns ubiquitous in the herbaceous stratum.

f. Swamp Forests. This type of forest is present in periodically or permanently inundated areas and in subcoastal marshes, on organic soils. Swamp forests are characterized by bearing a canopy between 8–15 m up to 20 m in height, with deciduous elements, helo-hydatophytes, and epiphytes. Sometimes it contains mangrove elements. Among the outstanding species are *Annona glabra*, *Bucida palustris*, *Copernicia* spp., *Dalbergia ecastophyllum*, *Fraxinus cubensis*, *Guettarda combsii*, *Hibiscus elatus*, *Ilex cassine*, *Myrsine cubana*, *Sabal parviflora*, *Salix longipes* and *Tabebuia angustata*.

g. Mangrove Forests. In marsh and low coasts, mangroves commonly occur, having a canopy of about 5–15 m in height, the shrubby stratum is lacking, and herbaceous elements and lianas are present. Among the species forming mangrove forests the following are important: *Avicennia germinans*, *Batis maritima*, *Conocarpus erecta*, *Laguncularia racemosa*, *Rhabdadenia biflora* and *Rhizophora mangle*.

h. Gallery Forests. Among plant communities associated with flowing water, gallery forests are important. Trees are disposed along rivers and small streams, and the canopy can reach about 15–20 m. This forest presents also shrubby and herbaceous strata, lianas and epiphytes. Floristic composition is highly related to heliophilous species of the surrounding vegetation, the most common ones being *Calophyllum antillanum*, *C. rivulare*, *Lonchocarpus domingensis*, *Roystonea regina* and *Tabebuia angustata*.

3. Thicket formations

a. Coastal and Sub-coastal Xeromorphic Thicket. This constitutes a shrubby formation with emergent, shrub-sized trees, and bearing deciduous, mostly sclerophyllous, micro- and nannophyllous thorny elements. It is the richest in succulents, also having palms, herbaceous elements and lianas. The abundance of arboreal and columnar succulents can be a noteworthy character, mainly in the southeast coasts of eastern provinces, reaching a "desertic" appearance, conditioned by a semidesertic climate of about 9–11 months of dry period (Borhidi & Muñíz, 1980). This kind of formation is present on coastal limestone areas (rendzines). Both soil and climate account for the relatively high endemism of its florula. The species present come from the following genera: *Acacia*, *Caesalpinia*, *Croton*, *Jaquinia*, *Plumeria*, *Eugenia*, *Picrodendron*, *Catesbaea*, *Casearia*, *Malpighia*, *Cordia*, *Cassia*, *Guettarda*, *Lantana*, *Capparis*, *Gochnatia*, *Agave*, *Coccothrinax*, *Leptocereus*, *Oplonia*, etc.

Where the abundance of succulents becomes noticeable, the following taxa are outstanding: species of *Agave*, *Consolea*, *Harrisia*, *Leptocereus*, *Melocactus*, as well as *Cylindropuntia hystrix*, *Dendrocereus nudiflorus*, *Mamillaria prolifera*, *Opuntia militaris*, *Pilosocereus robinii*, *Rhodocactus cubensis* and *Ritterocereus hystrix*.

b. Serpentinic Thicket Formations. As mentioned above, serpentine rocks have exerted an important influence on the Cuban flora and vegetation, being characterized by a strong xeromorphy that reflects even on the vertical zonation of vegetation. In mountains having such a geological substratum, the vegetation levels appear in a lower position when compared with others, and some perhumid vegetation zones are totally absent, as in the case of cloud forests (Borhidi & Muñíz, 1980).

Besides this, Bisse (unpubl.) has stated that in the highlands of the northeast montane systems in eastern Cuba, above 900 m the cloud forests are replaced by another kind of vegetation called by him as "charrascal" that can be described as a shrubby evergreen montane forest, which occurs between 900–1,100 m, and is equivalent to the evergreen shrubby subalpine forest or subalpine (sub-paramo) that is found between 1,900–1,972 m (Borhidi & Muñíz, 1980).

The plant formation accounting for the higher number of endemic taxa is the so-called "charrascal" or xeromorphic sub-thorny thicket on serpentinite (Berazaín, 1979).

The xeromorphic sub-thorny thicket on serpentinite (charrascal) shows a dense shrubby stratum of about 4–6 m in height, with emergent trees of 7–10 m, scattered herbaceous plants, lianas and epiphytes. As considered by Bisse (unpubl.), this formation bears the following characteristics: (a) palms are lacking, especially those of the genus *Coccothrinax*; (b) absence of thorny species; (c) abundance of *Arthrostylidium* spp. This formation occurs in plains, low altitude and montane areas on soils derived from serpentinite at Eastern Cuba. Among the principal species, the following can be mentioned: *Adatoa cubensis, Annona sclerophylla, Antirhea abbreviata* s.l., *Ariadne shaferi* s.l., *Ateramnus recurvus, Byrsonima biflora, B. minutifolia, Calycogonium moanum, C. rosmarinifolium* s.l., *Coccoloba* spp., *Crossopetalum ternifolium* s.l., *Erythroxylum* spp., *Euphorbia helenae, E. podocarpifolia, Galactia revoluta, Guettarda shaferi, Machaonia nipensis, Moacroton* spp., *Oplonia cubensis, Phyllanthus* spp., *Pseudocarpidium rigens, Rheedia,* spp., *Schmidtottia* spp., *Spathelia cubensis, S. splendens, Xylosma buxifolium,* etc.

The other shrubby serpentine formation is the xeromorphic thorny thicket (locally called "cuabal"), which is a rather dense thicket of 2–4 m in height, with emergent trees of 4–6 m in height, scattered palms, herbaceous plants, epiphytes and abundant lianas.

In Bisse's opinion (unpubl.), the more important physiognomic feature of this formation is the abundance of low palms of the genera *Coccothrinax* and *Copernicia*. These "cuabales" are well developed all over the country on soils derived from serpentinite, mostly skeletic ones. Among the floristic elements that inhabit this formation are several species of the genera *Bourreria, Buxus, Coccoloba, Coccothrinax, Copernicia, Eugenia, Guettarda, Leucocroton, Rondeletia, Tabebuia, Zanthoxylum* and such species as *Bonania emarginata, Bucida ophiticola, Bursera angustata, Neobracea valenzuelana, Oplonia nannophylla, Phyllanthus orbicularis, Pseudocarpidium ilicifolium, Scolosanthus crucifer* s.l., etc.

c. Sub-paramo Formation. As mentioned above, the Turquino massif at Sierra Maestra mountains, is the maximal height for the Cuban Archipelago. There, a kind of thicket is present that has received several names (see León, 1946; Berazaín, 1979; Borhidi et al., 1979; Borhidi & Muñíz, 1980; Capote & Berazaín, 1984). More recently, the term "sub-paramo" has been adopted (Atlas Nacional de Cuba, in preparation). This thicket formation is located above the cloud condensation zone, in areas exposed to wind action and with low rainfall. In the dominant stratum, shrub-sized trees, 3–8 m in height, are abundant, together with succulents, epiphytes, mosses, lianas, and occasional tree ferns. Among the outstanding species in this formation are *Eupatorium paucibracteatum, Lyonia calycosa, L. turquinii, Micromeria bucheri, Myrica cacuminis, Peratanthe cubensis, Rubus turquinensis, Vernonia praestans* var. *cacuminis*, as well as other species of *Agave, Ilex, Microlepanthes, Pleurothallis,* etc. In rocky places, *Juniperus saxicola, Haenianthus salicifolia, Viburnum* sp., *Agave* sp., *Begonia lomensis, Peperomia galioides* and *Pilea micromeriaefolia* are present, among others.

4. Herbaceous formations

These formations comprise different types of vegetation, among which savannas are important. The use of the term savanna is controversial for scientists dealing with Cuban vegetation studies.

a. Savannas. Most Cuban scientists accept the definition of savanna by Ellenberg and Mueller Dombois (1966), as those tropical ecosystems in which the dominant strata are formed by herbaceous plants (mostly Poaceae and Cyperaceae) with trees or shrubs mixed and more or less uniformly scattered here and there. These may be palms, pines, broad-leaved evergreen or deciduous, with or without thorn trees. According to Budowski (1985) those are characteristic features of savannas. This author also states, relative to present knowledge about the origin of savannas, that the general outline holds true for the particular Cuban conditions, mainly in relation to edaphical and biotical conditions.

Borhidi and Herrera (1977) established that, at the time of discovery of the Cuban island by Europeans (1492), there existed three kinds of grasslands:

1. *Open areas*
   a. Climatic savannas in alluvial valleys, very restricted in area.
   b. Edaphic savannas, in lower valleys of larger rivers.
   c. Humid grasslands related to marshes and swamps.
2. *Semianthropic savannas*, in areas where originally other forest formation occurred, potentially convertible into savannas with the intervention of meteoric effects and to a lesser extent, anthropic influence.
3. *Anthropic savannas*, originated by direct influence of aborigines, inside or at the limits of forest formations.

According to Herrera (1984), original savannas had always existed in Cuba, as scattered areas of different size, or as extensive regions having forest nuclei, and accounting for 25–28% of the total.

Capote and Berazaín (1984) recognized the existence of savannas among secondary vegetation types and called them seminatural and anthropic savannas.

As a conclusion, it can be taken as established that there are original savannas in Cuba. These types of vegetation may have mainly an edaphic and/or anthropic origin, and as a result, the edaphic ones should be considered as original herbaceous formations, and the anthropic savannas as secondary ones. The term "semianthropic" should be applied when edaphic factors impose limitations to natural regeneration of original vegetation, and human impact is not so intense; while the anthropic savannas, the origin of which is relative to intensive human management, are more close to agroecosystems.

Borhidi and Herrera (1977) described four different types of savannas in Cuba, being the following: (a) high grasses, (b) low grasses, (c) Cyperaceae-predominant, and (d) broadleaved tree savannas.

Among the outstanding tree species in high grass savannas there are *Roystonea regia, Ceiba pentandra, Copernicia* spp., and *Sabal parviflora*; among the grasses and sedges, species of *Andropogon, Rhynchospora, Paspalum, Panicum, Setaria, Scleria, Hyparrhenia*, etc. are common.

In low grasses savannas, scattered trees such as *Copernicia* spp., *Coccothrinax* spp., *Colpothrinax wrightii, Paurotis wrightii, Pinus tropicalis, P. caribaea, P. cubensis*, can be found; and low grasses or sedges of the following genera: *Andropogon, Aristida, Rhynchospora, Paspalum, Sporobolus, Cyperus, Fimbristylis*, etc.

In Cyperaceae-predominant savannas, palms such as *Paurotis wrightii* and *Sabal parviflora* are common, and herbaceous species of the genera *Eleocharis, Fimbristylis, Scirpus, Typha, Rhynchospora*, etc.

Broadleaved tree savannas contain several tree species such as *Bucida buceras, Catalpa punctata, Hibiscus elatus, Swietenia mahagoni, Prunus occidentalis*, and an herbaceous stratum characterized by Poaceae and Cyperaceae.

In sum, savanna vegetation is widespread all over the Cuban Archipelago.

b. Fresh-water Communities. In rivers, streams and ponds, there is present a kind of vegetation characterized by free-floating or rooted hydrophytes. Common free-floating species are the following: *Azolla caroliniana, Eichornia* spp., *Lemna minima, Pistia stratioides, Salvinia auriculata, Utricularia* spp., etc. Rooted species are, for example: *Brasenia schreberi, Cabomba piahuensis, Hydrocotyle umbellata, Myriophyllum verticillatum, Nymphaea* spp., *Potamogeton* spp., etc.

According to the degree of water eutrophy, different associations can be found, with the oligotrophic bodies of water being richer in species.

c. Herbaceous Swamp Formation. In swampy areas that periodically dry out, an herbaceous vegetation commonly occurs, with the following floristic composition: *Cyperus* spp., *Echinodorus* spp., *Eleocharis cellulosa, E. interstincta, Panicum aquaticum, P. lacustre, Paspalidium paludivagum, Pontederia lanceolata, Rhynchospora corniculata, R. gigantea, Sagittaria intermedia, S. lanceifolia, Scirpus olneyi, S. validus*, etc.

When water is present throughout the year, and as a consequence, peat accumulates, the floristic composition is slightly different, there being present *Centella erecta, Cladium jamaicense, Crinum oliganthum, Cyperus giganteus, Fuirena umbellata, Panicum* spp. *Paspalum giganteum, Pontederia lanceolata, Rhynchospora cyperoides, Sacciolepis striata, Solidago stricta, Thelypteris palustris, Typha domingensis*, etc.

d. Halophytic Plant Communities. This formation is present in soils with a heavy salt content, and is composed of species adapted to such extreme conditions. Among them: *Batis maritima, Chloris sagraeana, Distichlis spicata, Eragrostis salzmanii, Fimbristylis spathacea, Heliotropium curassavicum, Philoxerus vermicularis, Salicornia ambigua, S. perennis, Suaeda fruticosa, S. linearis, Spartina juncea, Sporobolus virginicus* spp. *littoralis.*

5. Vegetation complexes

In Cuba, there are three main vegetation complexes: "mogotes" (conical karstic mountains), rocky coasts and sandy coasts.

a. Mogotes Complex. Typical "mogotes" vegetation complex is found at summits and vertical hillsides of conical karstic mountains. At the summits, a thicket formation is present, with a discontinuous tree stratum of 5–10 m in height, bearing palms and deciduous trees, together with succulents, epiphytes and lianas, more or less dominant in certain areas. On hillsides there is an open vegetation with trees, shrubs, succulents, and herbaceous species, in groups or scattered.

The foothills are covered by either semi-deciduous or evergreen forests described above, conforming, together with the two other formations, the "mogotes" vegetation complex.

These "mogotes" formations are localized in western Cuba (the most representative and floristically richest ones, at Pinar del Río) though they are also found at Isla de la Juventud and central-eastern Cuba.

Among the species of western "mogotes" there are: *Agave* spp., *Ateramnus brachypodus, Anthurium* spp., *Bombacopsis cubensis, Celtis iguanea, Cuervea integrifolia, Erythrina cubensis, Ekmanianthe actinophylla, Gaussia princeps, Lantana strigosa, Leptocereus* spp., *Malpighia wigiana, Oplonia purpurascens, Plumeria sericea, Pristimera coriacea, Psidium scopulosum, Rochefortia spinosa, Tabebuia anafensis, T. calcicola, Thouinia nervosa, Thrinax morrisii, Spathelia brittonii*, etc.

Central-eastern "mogotes" are characterized by: *Anthurium* spp., *Coccothrinax* spp., *Euleria tetramera, Eupatorium carsticolum, Garrya fadyenii, Gesneria cubensis, G. heterochroa, Neobracea howardii, N. susannina, Phyllanthus epiphyllanthus* ssp. *dilatatus, Pilea* spp., *Savia erythroxyloides* var. *parviflora, Selenicereus urbanianus, Synapsis ilicifolia, Tabebuia* spp., *Thouinia* spp., *Thrinax* (sub.-gen. *Hemithrinax*) spp., *Zanthoxylum coriaceum*, etc.

b. Rocky Coasts Vegetation Complex. This vegetation complex is characterized by sparse shrubby-herbaceous communities with succulents and small shrubs, sometimes together with herbs. It is located on high limestone coasts, receiving the direct influence of the sea (salt mist) and constant winds. Among the species there are: *Borrichia arborescens, B. cubana, Chamaesyce buxifolia, Erithalis fruticosa, Flaveria linearis, Opuntia dillenii* s.l., *Pectis* spp., *Rachicallis americana, Sesuvium maritimum, S. portulacastrum, Strumphia maritima*, etc.

c. Sandy Coasts Vegetation Complex (Strand Vegetation). Cuban sandy beaches are inhabited by scattered herbaceous and suffruticose species, many of them prostrate or creeping, and well adapted to high salinity. Some tree species, such as mangroves and *Coccoloba uvifera*, can also be found. Among the species encountered in sandy coasts the following should be mentioned: *Borrichia arborescens, Canavalia maritima, Cenchrus tribuloides, Diodia maritima, Erithalis fruticosa, Ernodea littoralis, Ipomoea brasiliensis, Messerschmidtia gnaphalodes, Scaevola plumieri, Stemodia maritima, Suriana maritima, Uniola virgata*, etc.

In both types of coasts, above high tide level, is present a kind of thicket or monotypic forest (Borhidi et al., 1979) known as "uveral" (Bisse, unpubl.) dominated by *Coccoloba uvifera*, frequently together with *Thrinax radiata* and *Bursera simarouba*, among other species.

6. Secondary vegetation

Secondary vegetation has developed as a consequence of degradation of the natural vegetation, and its complexity is conditioned by the degree of succession from original to secondary forest, thickets and herbaceous communities, which have been replaced by plant communities, *Dichrostachys cinerea* thickets (marabuzal) and *Syszygium jambos* forest (pomarrosales).

Cercropia peltata, a heliophilous invader tree, contributes to the establishment of a not too dense canopy that helps the growth of forest species, allowing for the passage from initial successional stages to late or mature ones (Capote et al., 1974–1986).

Different stages of secondary vegetation are related to those described by Halle et al. (1978). The successional stages depend on the intensity and evolution of natural and anthropogenic factors causing drastic effects on natural conditions in Cuba, mainly in plains and low mountains, where the original evergreen and semi-deciduous forests have been transformed in relation to agricultural development and cattle raising since a long time ago.

III. Notes on the History of Botanical Exploration and Conservation of Nature in Cuba
A. Leiva

A. EXPLORATION AND FLORISTIC INVENTORIES BETWEEN THE EIGHTEENTH AND TWENTIETH CENTURIES

1. Colonial Period: Eighteenth-Nineteenth Century. (Mainly based on León, 1946, and Alvarez Conde, 1958).

The first European to collect plants in Cuba was the Scot William Houstoun, in the middle of the eighteenth century. His collections are conserved at the British Museum.

In the second half of the eighteenth century, two outstanding botanists, Nicolaus Joseph Jacquin and Olof Swartz, made some collections in the surrounds of the city of La Habana. Jacquin described several Cuban species in his *Selectarum Stirpium Americanarum Historia* (1781). Swartz published in *Icones Plantarum Incognitarum* and in *Flora Indiae Occidentalis* (1794–1800; 1797–1806) some of the species that he collected in Cuba.

At the end of the eighteenth century, the Frenchman, M. E. Decourtilz, spent a short time at Santiago de Cuba, where he studied the local flora. At that time, there took place the well-known expedition of the Count of Mopox and Jaruco, together with the Spanish botanist Baltasar Boldo and the illustrator José Guió. The illustrations and descriptions of the plants collected have been only recently published by the Madrid Botanical Garden.

Alexander von Humboldt, known as the second Cuban discoverer, explored, at the beginning of the nineteenth century, some interesting areas near La Habana and Trinidad cities.

In 1817, the first botanic garden in Cuba was founded under the auspices of the "Sociedad Económica de Amigos del País." José Antonio de la Ossa, a Cuban physician who was its first director, collected profusely in some places distant from La Habana city for the first time.

Ramón de La Sagra, who followed de la Ossa in the directorship of the Garden, collected actively all over the island, asking the collaboration of the clergy.

Meanwhile, several foreign botanists: W. Hamilton, G. Don, E. F. Poeppig, J. Reed, N. Funck, H. Delessert, F. E. Liebold, J. M. Despreaux, F. M. Liebmann, B. D. Greene, A. W. Lane, F. Rugel, J. J. Linden and some others, visited Cuba and made collections for European herbaria during the first half of the nineteenth century.

Between 1856 and 1866, the North American botanist Charles Wright explored Cuba, and gathered the most important collection of Cuban plants known since that time (nearly 4,000 numbers), especially from eastern and western Cuba. His very rich collections were studied principally by A. Grisebach. Francisco Adolfo Sauvalle, a North American trader established in Regla, La Habana, collaborated with Wright, and recreated an herbarium that he granted to the La Habana Academy of Sciences (more than 3,000 numbers, mostly duplicates from Wright). The Cuban José Blain (1808–1877) collected abundantly at Pinar del Río, and Isla de Pinos; his specimens anonymously improved the collections of Wright and Sauvalle.

Sebastian Alfredo de Morales (1823–1900), a Cuban naturalist, deserves mention as an explorer, and friend of Felipe Poey and Johans Gundlach (both outstanding zoologists). Morales published a *Flora de Cuba* of certain merit at that time.

In the last three decades of the nineteenth century, the exploratory activity diminished because of the difficulties caused by the first war against Spanish colonialism (1868–1878). The subsequent political and economical instability led, in 1895, to the re-starting of fights for Cuban independence. Among the very few botanists that worked in Cuba at that time, H. F. Eggers and R. Combs should be mentioned.

Table 2
Principal collections of Cuban Phanerophytes in World Herbaria (in alphabetical order of acronyms).

1. *Botanischer Garten and Botanisches museum Berlin-Dahlem* (B)
   Collections in Willdenow Herbarium (A. von Humboldt)
2. *Bereich Botanik und Arboretum des Museum für Naturkunde der Humboldt-Universität, Berlin* (BHU)
   Duplicate collections from "Proyecto Flora de Cuba" (see under HAJB)
3. *British Museum of Natural History Herbarium* (BM)
   Collections from: J. J. Linden, E. F. Poeppig, O. Swartz, N. H. Jacquin, F. Rugel, D. Don, W. Houstoun
4. *Botanical Department of the Hungarian Museum* (BP)
   Duplicates from HAC (see under HAC)
5. *Jardin Botanique National de Belgique* (BR)
   Collections from J. J. Linden
6. *Botanical Museum and Herbarium, Copenhagen* (C)
   Collections from H. F. Eggers
7. *Field Museum of Natural History, Chicago* (F)
   Collections from: C. F. Baker, G. Bucher, R. Combs, A. S. Hitchcock, H. F. Eggers
8. *Conservatoire et Jardin Botanique de la Ville de Geneve* (G)
   Collections from: R. de La Sagra, J. A. de la Ossa, C. Wright, H. Delessert, J. Fraser
9. *Gray Herbarium of Harvard University* (GH)
   Collections from: Alaín (E. Liogier), N. L. Britton, C. F. Millspaugh, R. Combs, R. A. Howard, C. Wright
10. *Systematisch-Geobotanisches Institut de Universität Göttingen* (GOET)
    Collections from: F. Rugel, C. Wright, E. L. Ekman, H. F. Eggers
11. *Herbario de la Academia de Ciencias de Cuba* (HAC)
    (incorporating HABA, LS, SV, and other minor collections)
    Collections from: J. Acuña, Alaín (E. Liogier), C. F. Baker, A. H. Curtiss, A. S. Hitchcock, E. P. Killip, León (J. S. Sauget), J. T. Roig, J. A. Shafer, C. Wright, E. L. Ekman, Hioram, Clemente, F. A. Sauvalle
    Modern collections from: A. Barreto, A. Borhidi, R. Capote, L. Catasus, M. Fernandez, I. A. Grudzinskaya, P. Herrera, L. I. Ivanina, N. Imchanitskaya, A. López, M. Moncada, O. Muñíz, R. Oviedo, E. Del Risco, G. P. Yakovlev, PFC duplicates (see under HAJB)
12. *Herbario del Jardin Botanico Nacional, Universidad de La Habana* (HAJB)
    Collections from: M. López Figueiras, and originals of "Proyecto Flora de Cuba" (PFC) (A. Alvarez, A. Areces, I. Arias, A. Bassler, C. Beurton, J. Bisse, R. Berazain, M. A. Díaz, H. Dietrich, M. E. Duharte, L. González, K. F. Gunther, J. Gutiérrez, E. Hernández, E. Köhler, G. Klotz, A. Leiva, L. Lepper, H. Lippold, M. de la Luz, H. Manitz, B. Mory,, F. K. Meyer, M. D. Ortega, C. Panfet, U. Rändel, R. Rankin, A. Rodríguez, L. Rojas, C. Sánchez, G. Stohr, A. Urquiola, M. Valentin, W. Vent)
    Minor collections from: E. Ekman, Alaín, León, Chrysogone, Clemente and others
13. *Herbarium Haussknecht University of Jena* (JE)
    Duplicate collections from "Proyecto Flora de Cuba" (PFC) (see under HAJB); Minor collections from C. Wright and E. F. Poeppig
14. *Royal Botanic Gardens, Kew (Herbarium)* (K)
    Collections from: N. L. Britton, R. Combs, A. H. Curtiss, E. L. Ekman, F. Rugel, C. Wright, H. F. Eggers
15. *Rijksherbarium, Leiden* (L)
    Collections from: R. Combs, F. S. Earle, A. S. Hitchcock, F. Rugel
16. *Herbarium of The Department of Higher Plants, V. L. Komarov Bot. Inst. of the Acad. Sci. USSR* (LE)
    Duplicates from HAC (modern collectors)
17. *University of Massachusetts* (MASS)
    Collections from W. H. Hodge
18. *Missouri Botanical Garden (Herbarium)* (MO)
    Collections from C. Wright
19. *The New York Botanical Garden Herbarium* (NY)
    Collections from: Alaín (E. Liogier), C. F. Baker, N. L. Britton, E. Britton, R. Combs, F. S. Earle, C. V. Morton, J. A. Shafer, P. Wilson, F. Rugel
20. *Museum National d'Histoire Naturelle, Laboratoire de Phanerogamie, Paris* (P)
    Collections from: C. F. Baker, A. H. Curtiss, A. Bonpland, A. von Humboldt, J. J. Linden, J. A. de la Ossa, R. de La Sagra
21. *Swedish Museum of Natural History, Stockholm* (S)
    Collections from: E. L. Ekman, C. Wright
22. *United States National Herbarium* (US)
    Collections from: G. Bucher, Clemente (A. C. Teteau), R. Combs, A. H. Curtiss, Hioram (J. F. Lagorce), A. S. Hitchcock, C. V. Norton, J. W. Seifriz, C. Wright
23. *Naturhistorisches Museum, Botanische Abteilung, Wein* (W)
    Collections from: E. F. Poeppig, N. J. Jacquin

2. The Republican Period: first half of the twentieth century

After 1902, the explorations by North American scientists were reinforced, and many such well-known botanists as A. S. Hitchcook, N. L. Britton, E. Britton, P. Wilson, J. A. Shafer, A. H. Curtiss, C. F. Baker, and others, made very important collections that improved the North American herbaria (mostly the New York Botanical Garden Herbarium). Particularly remarkable was the work done by N. L. Britton, the founder of The New York Botanical Garden, who organized, starting in 1903, a full survey of the Cuban flora together with P. Wilson, J. A. Shafer and Br. León. Britton's Cuban collections surpassed 40,000 numbers, an unprecedented figure.

In 1914 the Swede, E. L. Ekman, arrived in Cuba. He collected about 20,000 numbers and explored the highest altitudes in central and eastern Cuba. His collections, sent to Stockholm, were studied mainly by I. Urban. Several duplicates, including isotypes, are in HAC.

Two Cuban botanists stand out in the first decades of the

twentieth century: Manuel Gómez de la Maza, (1867–1916) who made important contributions as bibliographer and professor of botany at the University of La Habana; and Juan Tomás Roig Mesa (1877–1971), a clever man who, exploring all over the country, profoundly studied our flora and published important contributions on both economic and taxonomic botany.

Julian Acuña Gale (1900–1973), a Cuban botanist, collected all over the Island and collected a rather large herbarium. He was the most remarkable connoisseur of all time of the Cuban flora. His collections are conserved at HAC. The clergymen Clemente, Hioram, Chrysogone and Marie Victorin also made important collections and contributed largely to the knowledge of Cuban flora and vegetation, together with Brs. León and Alaín. These two scientists prepared and published the most outstanding work summarizing all the past knowledge on phanerophytes, in the monumental five-volume work *Flora de Cuba*, edited between 1946 and 1962. The herbarium gathered by Brs. León, Alaín, and others, amounted to more than 45,000 items (LS, now included under HAC).

Manuel López Figueiras collected abundantly in Oriente province, and his specimens were also utilized for *Flora de Cuba* from León and Alaín. Duplicate collections are deposited at HAJB, and originals at Santiago de Cuba's University.

The Cuban Society of Botany was created in 1944. This institution has contributed notably to the extension of knowledge on the Cuban flora and vegetation. Its main periodical, "Revista de la Sociedad Cubana de Botanica," edited until 1959, published very important papers from notable botanists from Cuba, and North America and Latin America. The name of Antonio Ponce de León is closely related to both the Society and the "Revista," and his participation in such very important international meetings as the VIIth and VIIIth World Botanical Congresses elevated the name of Cuban botany.

3. *The Modern Period: second half of the twentieth century*

After the victory of the Popular Revolution in 1959, deep changes were instituted in the country in order to achieve socio-economic development. As a consequence, the Government of the United States of America decided to disrupt relationships with Cuba, and decreed a total embargo that obviously reflected negatively on scientific activity. The few skilled botanists that were working in Cuba at that time, nationals or residents, emigrated (with such honorable exceptions as Dr. J. T. Roig and Ing. J. Acuña) and a sudden discontinuity arose.

Botanical research re-started in the decade of the Sixties, when some European scientists came to Cuba, not only to explore and collect, but mainly to teach and train young people. Among them, Johannes Bisse, a German botanist from Jena University (GDR), developed botanical research at the Botany Department of La Habana University, planned and founded the National Botanic Garden, taught fifteen generations of Cuban botanists, and improved the little herbarium of the University (HAJB, with ca. 4,000 numbers in 1966, and by now 60,000 numbers), until his tragic decease in 1984.

Another outstanding scientist who has extensively collaborated in advising and training young Cuban scientists is Veroslav Samek, from the Czechoslovakian Academy of Sciences, who is the author of very important phytogeographical surveys and proposals for the conservation of ecosystems. Attila Borhidi, from the Hungarian Academy of Sciences, has been actively exploring and publishing important taxonomic contributions and phytogeographical analysis during the last ten years, together with several Cuban botanists.

The Herbarium of the Cuban Academy of Sciences (HAC) has incorporated all the historical herbaria, and gathered modern collections, making nowadays a total of ca. 134,600 items (Holmgren et al., 1981).

Taking into account only the figures of herbarium specimens obtained by the principal collectors (Wright, Ekman, Britton, León, Alaín, Hioram, Clemente, Acuña, and modern collections in HAJB and HAC), we can estimate that ca. 180 collections/100 km^2 have been made, and thus say that the Cuban main islands are fairly well collected. In Figure 2 are pinpointed all the localities expored during the last two decades (data from HAJB only). The areas of highest endemism are much more explored and collected than the poor ones. If we average the number of herbarium specimens in relation to the number of native vascular plant species, the figure obtained is 27 specimens/species.

More and more of the distribution maps for Cuban plants are being completed, on the basis of well-documented herbarium specimens (Manitz, 1978, 1980, 1984) for the "Flora de la República de Cuba" project.

In 1976 computerized methods were used for the first time for some works on numerical taxonomy. In 1977, the first classification of vegetation of the Zapata Peninsula was done using numerical methods (Del Risco, pers. commun.). In 1985, the first attempts to computerize the information gathered on the Herbarium of the Cuban Academy of Sciences were done. At present a system of interdependent data-bases, including the Cuban endemic taxa, the genetic pool of cultivated plants, and phenologic observations made elsewhere in the country, is being designed, using for such purposes a group of programs in DBASE programming language (López Almirall, pers. commun.).

Today, about 155 families of Cuban vascular plants are being critically studied and prepared for the new *Flora de la República de Cuba*.

In Table III, a full list of collaborators working on the preparation of taxonomic treatments of the families of the Cuban vascular flora is provided. More than 55% are Cuban scientists.

The two main national institutions that are actively working on the Cuban flora and vegetation are the Institute of Ecology and Systematics of the Cuban Academy of Sciences, and the National Botanic Garden of the University of Havana. Minor groups are scattered all over the country, in close connection with the two centers mentioned above (universities and pedagogical institutes of Santiago de Cuba,

Table 3
Up-to-date list of collaborators in "Flora de la República de Cuba" (Vascular Plants).

Universities and Pedagogical Institutes:

A. Alvarez	Agavaceae (s.l.), Alliaceae, Smilacaceae, Amaryllidaceae
R. Berazaín	Clethraceae, Ericaceae
M. A. Diaz	Burmanniaceae (in part). Nyctaginaceae, Orchidaceae (in part).
M. E. Duarte	Cucurbitaceae, Passifloraceae
L. Gonzalez	Eriocaulaceae, Zamiaceae, Linaceae, Combretaceae, Myrtaceae (in part)
J. Guitierrez	Cactaceae, Flacourtiaceae, Sapotaceae
M. Hernández	Bromeliaceae
A. Leiva	Balanophoraceae, Loranthaceae, Olacaceae, Viscaceae, Eremolepidaceae
M. de la Luz	Amaranthaceae
A. Noa	Thymeleaceae
M. D. Ortega	Caryophyllaceae, Chenopodiaceae, Portulaceae
R. Rankin	Aristolochiaceae
A. Rodríguiz	Bombacaceae, Elaeocarpaceae, Sterculiaceae, Tiliaceae
C. Sánchez	Hymenophyllaceae, Polypodiaceae
H. Saralegui	Piperaceae
J. Sierra	Begoniaceae
A. Urquiola	Alismataceae, Haemodoraceae, Mayacaceae, Naiadaceae, Xyridaceae, Nymphaeaceae, Potamogetonaceae, Podostemonaceae, Cymodoceaceae
M. Valentín	Myricaceae
I. Arias	Araceae, Lemnaceae
C. Panfet	Droseraceae

Cuban Academy of Sciences

D. Albert	Zygophyllaceae, Bignoniaceae (in part)
A. Barreto	Caesalpinaceae, Fabaceae
L. Catusús	Cyrillaceae, Poaceae
M. Fernández	Rubiaceae (in part)
P. Herrera	Asteraceae
A. López	Podocarpaceae, Arecaceae (in part), Clusiaceae, Cupressaceae, Pinaceae
M. Moncada	Burseraceae
O. Muñíz	Arecaceae (*Coccothrinax*)
M. Vales	Capparaceae
S. Machado	Dioscoreaceae
E. Moreno	Meliaceae
C. R. Martínez	Asclepiadaceae
R. Oviedo	Erythroxylaceae
A. Cárdenas	Simaroubaceae
R. Herrera & L. Menéndez	Cecropiaceae
C. Chiappy	Goodeniaceae, Campanulaceae
L. Montes	Staphyleaceae
I. Baró	Verbenaceae (in part)

Ministry of Public Health

V. Fuentes	Goetzeaceae, Solanaceae
M. Granda	Menispermaceae

German Democratic Republic

A. Bässler	Mimosaceae
K. Beurton	Rutaceae
J. Casper	Hydrocharitaceae, Lentibulariaceae, Pontederiaceae
H. Dietrich	Acanthaceae, Burmanniaceae (in part), Orchidaceae (in part), Plantaginaceae
K. F. Günther	Cochlospermaceae, Papaveraceae, Ranunculaceae, Anacardiaceae
G. Klotz	Boraginaceae, Brunnelliaceae, Chrysobalanaceae, Connaraceae, Crassulaceae, Cunoniaceae, Rosaceae
E. Kohler	Buxaceae, Euphorbiaceae, Dichapetalaceae
L. Lepper	Myrsinaceae, Primulaceae, Theophrastaceae
D. Mai	Bonnetiaceae, Styracaceae, Symplocaceae, Theaceae (s.l.), Marcgraviaceae
H. Manitz	Convolvulaceae, Cuscutaceae, Hydrophyllaceae, Apocynaceae, Hypericaceae, Sapindaceae
F. K. Meyer	Brassicaceae, Malpighiaceae
B. Mory	Celastraceae

Table 3 (continued)

G. Natho	Aquifoliaceae, Hippocrateaceae
U. Randel	Malvaceae
H. Scharschmidt	Casuarinaceae, Fagaceae, Juglandaceae
C. Schirarend	Vitaceae
G. Storh	Polygonaceae
W. Vent	Rhamnaceae
A. Zündorf	Commelinaceae
U.S.S.R.	
I. Grundzinskaya	Celtidaceae, Moraceae, Ulmaceae, Urticaceae
N. Inchanitskaya	Annonaceae, Lauraceae, Magnoliaceae, Illiciaceae
L. Ivanina	Gesneriaceae, Scrophulariaceae
A. Vinogradova	Apiaceae, Araliaceae
G. Yakovlev	Caesalpinaceae, Fabaceae (in part)
N. Tzvelev	Typhaceae, Orobanchaceae, Juncaginaceae, Hernandiaceae, Elatinaceae, Ceratophyllaceae, Callitrichaceae, Ruppiaceae, Rhizophoraceae
Hungary	
A. Borhidi	Rubiaceae (in part)
Z. Keretzky	Verbenaceae (in part)
Poland	
K. Rostansky	Onagraceae
England	
R. M. Harley	Lamiaceae
S. A. Renvoize	Poaceae (Paniceae)
U.S.A.	
A. Gentry	Bignoniaceae (in part)
G. Proctor	Pteridophyta (in part)

Pinar del Río, Villa Clara and Camagüey; Ministry of Public Health). Research programs with foreign institutions are being conducted, both at the National Botanic Garden (*Flora de Cuba* project together with the Universities of Jena and Berlin, GDR), and the Institute of Ecology and Systematics (collaborating mainly with homologous institutions from the U.S.S.R.'s, Hungary's and Czechoslovakia's Academies of Sciences). All the programs are coordinated by the National Committee for the Flora of Cuba, integrated by scientists of Cuban institutions and led by the Cuban Academy of Sciences. The principal task is to produce a new *Flora de la República de Cuba*, embracing both non-vascular and vascular plants. This long-term program is intended to be completed in the next 15–20 years.

During the last ten years a fairly good number of floristic reports and ecological studies of important areas have been done as well as taxonomic studies on several groups of the Cuban native flora. Almost all have been published in "Revista del Jardín Botánico Nacional," "Acta Botanica Cubana," "Acta Botanica Academiae Scienciarum Hungaricae," "Acta Botanica Hungaricae," "Wissenschaftlische Zeitschrift der Friedrich-Schiller Universität Jena," "Feddes Repertorium," and "Gleditschia."

The first fascicles of the new *Flora de la República de Cuba* are by now in press, and will be available shortly. They will appear numbered, for the flowering plants, according to the Takhtajan system (Takhtajan, 1980). For the first time, the *Flora* will include Pteridophytes, Bryophytes, Fungi and Algae.

Besides this program, the cartography of vegetation on a scale of 1:250,000 is being done by specialists from the Academy of Sciences (Institute of Ecology and Systematics). Teledetection methods have been very useful tools in cartography of vegetation. Maps of different scales have been prepared using aerial and cosmic materials, both panchromatic and multizonal; and also applying methods of multizonal image synthesis. Broad ecological research programs are also being conducted by this Institution, mainly dealing with seasonal evergreen forests and savanna ecosystems. At the National Botanic Garden, minor floristic studies are being conducted on serpentinic vegetation, where endemism reaches its maximal degree.

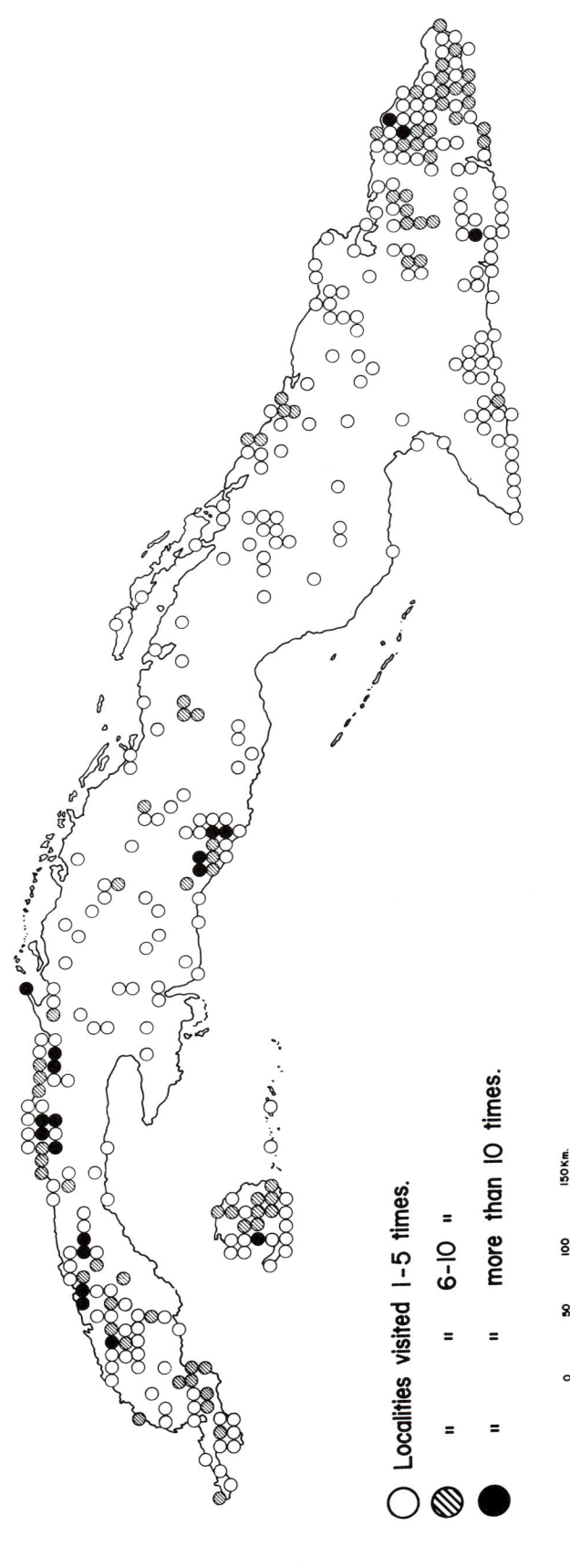

B. The Problem of Conservation of Nature in Cuba: Past and Present (With Emphasis on Forest Ecosystems)

Primitive forest ecosystems of the Cuban Archipelago before the first arrival of aborigines (about 5,000–6,000 years ago) covered more than 95% of the total area. High tropical forests probably accounted for 70–80%, and the rest was occupied by low broadleafed forests, pine forests and other types (Del Risco, pers. commun.). The simple, more or less nomadic way of life of such human activity had no significant impact on natural ecosystems.

In the sixteenth century the colonization of the main island of the Cuban Archipelago by the Spanish began. During the seventeenth and eighteenth centuries, economic activity was slight. It has been estimated that in 1812, 90% of the territory was still covered by original forests. In 1900 the area was diminished to 54%, mainly due to intensive cattle-raising, and sugar cane plantation. This destructive process was particularly reinforced during the first decades of the twentieth century. Western Cuba was the most degraded region, and since the beginning of the twentieth century, most of its forests have been irreversibly destroyed. The rise in price of cane sugar after the First World War caused such an outburst that by 1926 only 23% of the Cuban surface was still covered by forest. In 1959, 14% of Cuban forest ecosystems prevailed more or less undisturbed (figures from *Atlas de Cuba*, 1978). Nowadays, 17.6% of the total surface is covered by forests (C.I.F., 1985).

Among neotropical countries, Cuba occupies fourth place among those where the most drastic and intensive destruction of natural ecosystems has happened, topped only by Barbados, Haiti and Puerto Rico. (Muñíz, pers. commun.). About 90,000 km² of forest ecosystems have been destroyed after five centuries. Moreover, the introduction of some invader species (viz., *Dichrostachys cinerea* and *Syzygium jambos*, have displaced both low- and highland ecosystems (Del Risco, 1982). As is well known, island ecosystems are extremely vulnerable to external effects. In the case of Cuba, this vulnerability is much more pronounced because, among other reasons, its endemic flora is mostly adapted to extreme, oligotrophic biotopes and hence the competitiveness of organisms is diminished (Borhidi, 1985).

According to the opinion of A. Borhidi and O. Muñíz (1983) about 15% of the Cuban flora (Phanerophytes) are threatened (87% of them are endemics), and probably 2% are already extinct or almost lost. Before 1959, very little (or even nothing) was effectively done to protect natural ecosystems in Cuba. Moreover, the virtual absence of any technical and administrative apparatus—in spite of some legal depositions that formally existed—added much to the risk of destruction.

In 1963, the first four reserves were created under the principle of untouchability, embracing an area of ca. 25,000 ha. In the last decade, between 8–9% of the whole area of the country has been proposed for different degrees of protection. (Del Risco, pers. commun.).

In 1977 the National Committee for the Protection of the Environment and Conservation of Natural Resources (COMARNA) was created, attached to the Academy of Sciences. In every province and municipality there is a committee attached to the local government, with executive powers to decide on all the problems concerning protection of the environment and conservation/restoration of natural ecosystems. Besides this, a broad educational campaign is being developed. Since 1980, there has been in existence Law No. 33 for the protection of the environment and the rational use of natural resources. Several scientific institutions carry on important research programs in order to establish the theoretical and practical basis for the conservation of threatened ecosystems and/or species (mainly the Institute of Ecology and Systematics of the Academy of Sciences, The National Botanical Garden, and some other centers of higher education). In this context, we must mention the Reserve of the Biosphere "Sierra del Rosario," located in the western province of Pinar del Río in the Guaniguanico montane system, where since 1974 an important ecological research program is being developed by the Cuban Academy of Sciences, as a pilot project in collaboration with UNESCO's M.A.B. Program (Herrera, Menendez & Rodriguez, 1974–1986).

In Figure 3, a map showing the protected areas is provided (from *Atlas Nacional de Cuba*, in prep.). Nevertheless, the critical degree of destruction reached by Cuban ecosystems in such a short period of time, together with the actual demographic and economic growth, and to some (and not negligible) extent, the virtual absence of a conservationist tradition as a logical corollary of the past, impose on Cuban scientists, and obviously, on governmental authorities, a great responsibility for the present and the future. The basis for successful results has already been established, and future work will be a real challenge to our intelligence and skillfulness.

IV. Acknowledgments

We deeply acknowledge the help of the many colleagues who graciously assisted us in revising and criticizing the various drafts of the manuscript. Among them are: Dr. Herman Manitz, from the Herbarium Haussknecht (G.D.R.), Dr. Antonio López, Lic. Pedro Herrera and Dr. Maria Herrera, from the Institute of Ecology and Systematics of the Cuban Academy of Sciences, Dr. Enrique del Risco, from the Forest Research Institute of the Ministry of Agriculture, Dr. Miguel Rodriguez, Lic. Jorge Gutierrez and Lic. Esperanza Peña, from the National Botanic Garden. We are grateful to Nidia Palacio and Juan Carlos Santana for preparation of the maps, and to Carmen Rega for typing assistance.

We are also indebted to The New York Botanical Garden for inviting us to contribute to this book, especially to G. T. Prance, D. G. Campbell, M. L. Lebrón-Luteyn, and H. D. Hammond.

Finally, we want to honor the memory of the late Prof. Dr. Johannes Bisse, who taught us to go deeper into the knowledge of Cuban nature.

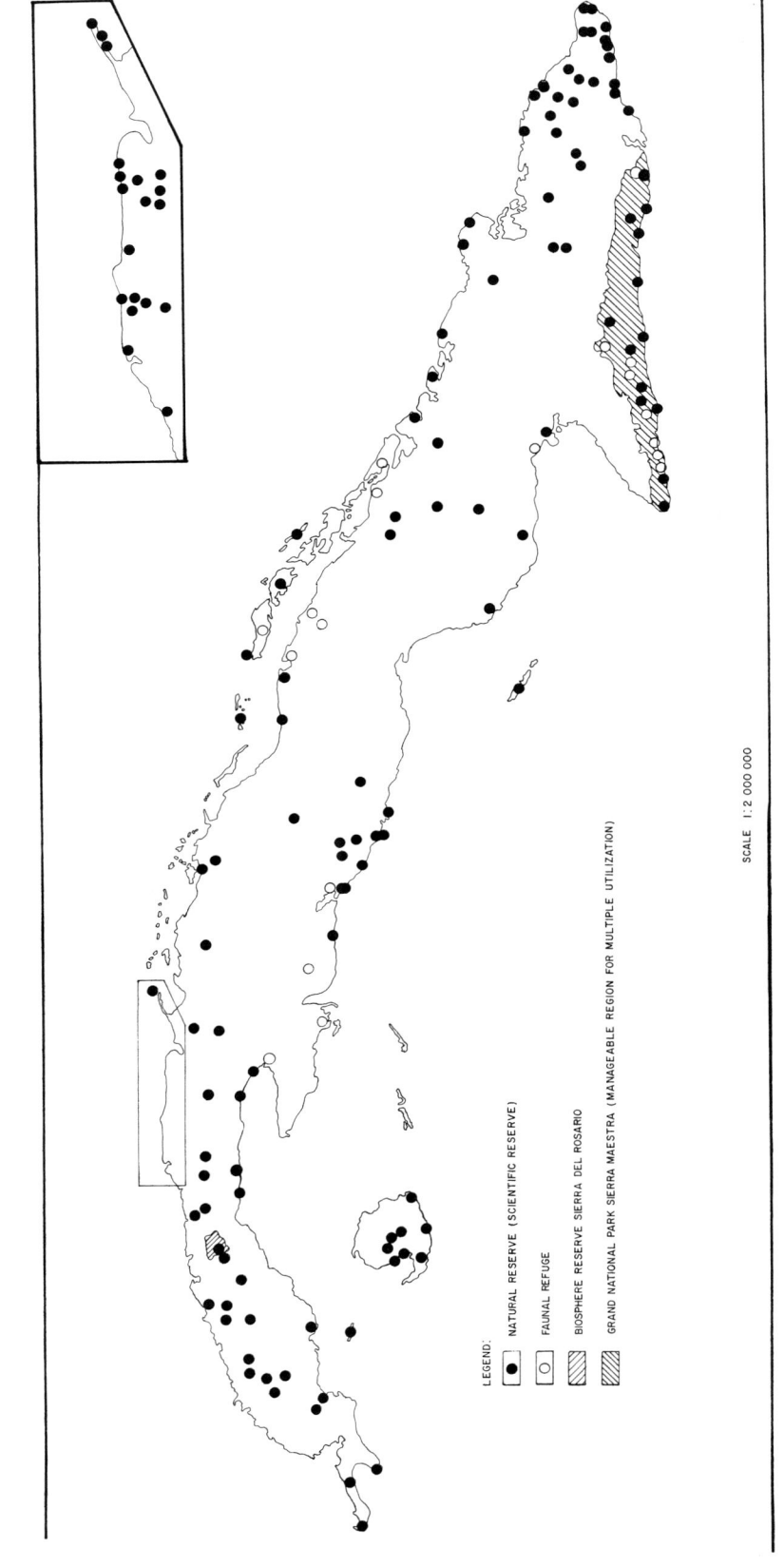

V. Literature Cited

Academia de Ciencias de Cuba. 1973. Génesis y clasificación de los suelos de Cuba. Instituto de Suelos, Havana.

Alaín (E. Liogier. 1953. Flora de Cuba, 3. Dicotiledóneas: Malpighiaceae a Myrtaceae. Contr. Ocas. Mus. Hist. Nat. Colegio "De La Salle" **13**.

———. 1956. Flora de Cuba. 4. Dicotiledóneas: Melastomataceae a Plantaginaceae. Contr. Ocas. Mus. Hist. Nat. Colegio "De La Salle" **16**.

———. 1958. La Flora de Cuba: Sus principales características. Su origen probable. Revista Soc. Cub. Bot. **15**: 36–59, 84–96.

———. 1962. Flora de Cuba 5. Rubiales–Valerianales–Cucurbitales–Campanulales–Asterales. Río Piedras, Puerto Rico. pp. 13–341.

———. 1969. Flora de Cuba. Suplemento. Caracas, Venezuela.

Alvarez Conde J. 1958. Historia de la Botánica en Cuba. Publicaciones de la Junta Nacional de Arqueología y Etnología. Havana.

C.E.E. and I.C.G.C. 1985. Atlas Demográfico Nacional, República de Cuba. Comité Estatal de Estadísticas e Instituto Cubano de Geodesia y Cartografía. Havana.

Berazaín, R. 1979. Firogeografía. Universidad de La Habana, Facultad de Biología. Havana.

Bisse, J. 1973. Guía para clasificar los diferentes tipos de montes y vegetación con especies forestales existentes en el país. Havana (unpubl.).

Borhidi, A. 1985. Phytogeographic survey of Cuba I. The phytogeographic characteristics and evolution of the Flora of Cuba. Acta Bot. Acad. Sci. Hung. **31**: 3–34.

——— & **R. Herrera.** 1977. Genesis, carácteristicas y clasificación de los ecosistemas de sabana de Cuba. Cienc. Biolog. **1**: 115–130.

——— & **O. Muñíz.** 1980. Die Vegetationskarte von Kuba. Acta Bot. Acad. Sci. Hung. **26**: 25–53.

——— & ———. 1983. Catálogo de Plantas Cubanas Amenazadas o extinguidas. Editorial Academia, Havana.

——— & ———. 1986[1987]. Phytogeographic survey of Cuba. II. Floristic relationships and phytogeographic subdivision. Acta Bot. Acad. Sci. Hung. **32**(1–4): 3–48.

Capote, R. & R. Berazaín. 1984. Clasificación de las formaciones vegetales de Cuba. Rev. Jard. Bot. Nac. **5**(2): 27–75.

Capote, R. P., L. Menéndez; E. E. García & R. A. Herrera. 1986. Sucesión vegetal, In R. A. Herrera-P., L. Menéndez & M. E. Rodríguez, Eds., Ecología de los Bosques Tropicales de la Sierra del Rosario, Cuba (1974–1986) Proyecto MAB 1. Resumenes.

———, **N. Ricardo, A. V. González, E. E. Garcia, D. Vilamajo & J. Urbino.** (In prep.) Mapa de vegetación actual 1: 100 000.

Atlas Nacional de Cuba. Academia de Ciencias de Cuba.

C.I.F. 1985. Breve caracterización de la Actividad Forestal en Cuba. Centro de Investigación Forestal, Ministerio de Agricultura, Ciudad de La Habana.

Coney, R. J. 1982. Plate Tectonic constraint on the biogeography of Middle America and the Caribbean Region. Ann. Missouri Bot. Gard. **69**: 432–443.

Del Risco, E. 1982. La Conservación de la Naturaleza y los Jardines Botánicos, Rev. Jard. Bot. Nac. **3**(1): 167–195.

———, **R. Vandama & A. Conzález.** (In prep.). Fitoregionalización, 1: 3000 000 (Atlas Nacional de Cuba).

Ellenberg, H. & D. Müeller-Dombois. 1966. Tentative physiognomic ecological classification of plant formations of Earth. Ber. Geobot. Isnt. Rübel, **37**: 21–56.

Hallé, F., R. A. A. Oldeman & P. B. Tomlinson. 1978. Tropical trees and forest. An architectural analysis. Springer Verlag, Berlin, Heidelberg, New York.

Herrera, R. 1984. El origen de las sabanas cubanas. Pages 49–97 in L. Waibe & R. Herrera, La Toponimia en el paisaje cubano. Demografia. Ed. Cienc. Soc. La Habana.

Herrera, R. A., L. Menéndez & M. E. Rodríguez (Eds.). 1986. Ecología de los Bosques Tropicales de la Sierra del Rosario, Cuba, Proyecto MAB 1 (1974–1986). Resumenes.

Holmgren, P. K., W. Keuken & E. K. Schofield. 1981. Index Herbariorum Part 1 ed. 7. Reg. Veg. 106.

I.C.G.C. 1978. Atlas de Cuba XX Aniversario del Triunfo de la Revolución. Instituto Cubano de Geodesia y Cartografía. Havana.

Iturralde-Vinent, M. A. 1982. Aspectos geológicos de la biogeografía de Cuba. Ciencias de la Tierra y el Espacio, Acad. Ci. Cuba, **5**: 85–100.

León, Hno. 1946. Flora de Cuba 1. Gimnospermas. Monocotiledóneas. Contr. Ocas. Mus. Hist. Nat. Colegio "De La Salle" **9**.

——— & **Alaín.** 1951. Flora de Cuba 2. Dicotiledóneas: Casuarináceas a Meliáceas. Contr. Ocas. Mus. Hist. Nat. Colegio "De La Salle" **10**.

Rosen, D. E. 1985. Geological Hierarchies and Biogeographic congruence in the Caribbean. Ann. Missouri Bot. Gard. **72**: 636–659.

Samek, V. 1973. Regiones Fitogeográficas de Cuba. Academia C. Cuba. Serv. Forest. **15**.

———, **E. Del Risco & R. Vandama.** (In press). Fitoregionalización del Caribe. Rev. Jard. Bot. Nac.

Takhtajan, A. L. 1980. Outline of the classification of flowering plants (Magnoliophyta). Bot. Rev. **46**(3): 349–359.

UNESCO. 1973. International classification and mapping of vegetation. Paris.

Hispaniola

T. Zanoni

Contents

- I. Description of the Region 337
  - A. Geographical Extent and Topography 337
    - 1. Dominican Republic 337
    - 2. Haiti 337
  - B. Climate 337
    - 1. Dominican Republic 337
    - 2. Haiti 337
- II. Flora of Hispaniola 337
- III. Vegetation Maps 337
- IV. Herbarium Resources 338
  - A. Herbaria in the Dominican Republic 338
  - B. Other Herbaria 338
- V. Publications 338
- VI. Magnitude of Floristic Inventory 338
  - A. Endemism 338
  - B. Extinction 338
  - C. Threatened Areas 338
  - D. Modern Collections 339
  - E. Conservation 339
- VII. Academic and Governmental Institutions 339
- VIII. Present and Future State of Inventory 340
- IX. Additional Resources 340
- X. Literature Cited 340

I. Description of the Region

A. Geographical Extent and Topography

1. Dominican Republic

Hispaniola, the island which includes Haiti and the Dominican Republic, has a total area of 77,914 km². The Dominican Republic has an area of 48,442 km².

The topography of the Dominican Republic is dominated by four principal and one minor mountain ranges: the Cordillera Septentrional, Cordillera Central, Sierra de Neiba, Sierra de Bahoruco, and the (minor) Cordillera Oriental. These ranges run northwest to southeast and parallel each other. Three valleys between the major ranges (Valle del Cibao, Valle de San Juan, Hoya de Enriquillo) are principally agricultural in use. A major low area (Llanura Costera), in the eastern part of the country, is principally in agricultural use (rice, sugar cane, grazing land for cattle). Other lowland areas (Llanura de Azua) include desert lands, but under irrigation these areas are highly productive agriculturally (root crops, melons, tomatoes, etc.). The Península Sur de Barahona is a lowland area of which about half is arid and rocky, consisting for the most part of native arid-land forest. Sorghum and cotton are cultivated where possible. The Península de Samaná is a small mostly mountainous area which has mainly coconut plantations and little native vegetation.

2. Haiti

The area of Haiti is about 27,700 km². The topography is dominated by the mountain chains of Massif de la Hotte (southwest Haiti), Massif de la Selle (southeast Haiti), Chaine des Matheux & Montagnes du Trou-d'Eau (central Haiti), Montagnes Noires (north-central Haiti), and the Massif du Nord (north Haiti). Of these the Massif de la Selle continues into the Dominican Republic as the Sierra de Bahoruco. The mountain chains in the south and north run east-west. The central chains are angled northwest-southeast. The principal valleys are the Cul-de-Sac (centrally located near Port-au-Prince), Plaine et Vallé de l'Artibonite (north-central Haiti), Plateau Central (northeast Haiti), and the Plaine du Nord (northeast Haiti). The northwestern peninsula—Presqu'île du Nord-Ouest is a low ridge with low, arid areas associated with it.

Most parts of Haiti are in agriculture or in abandoned usage after agriculture. Extremely arid areas not suited to agriculture have been harvested for firewood or charcoal, leaving few areas with much natural or even somewhat less disturbed vegetation. High mountain areas (except for two areas declared national parks) have had timber removed at least once in the last hundred years, often more frequently if the area has settlement.

B. Climate

1. Dominican Republic

The precipitation is quite varied, depending on location and relation to mountain ranges. The arid (desert) regions have 600–700 mm rain per annum (regions of northwest Cibao near Monte Cristi, near Azua, Enriquillo basin, and the Barahona Peninsula). Low elevation areas of the eastern region have 1,000–1,800 mm rainfall per annum. The mountainous areas (Cordillera Central, Sierra de Bahoruco, Cordillera Septentrional) have about 1,000–2,000 mm rainfall. The areas of highest rainfall are in the eastern Sierra de Bahoruco, Eastern Cordillera Central, Cordillera Oriental, and the eastern Cibao-Samana Peninsula, (ranging about 2,000–2,800 mm per annum). Forest development follows the rainfall patterns. "Pluvial" or heavy rainfall forest correspond to areas with 1,500 mm precipitation or more. Montane broadleaf forest and pine forests exist in areas which have about 1,000 mm rainfall or more. Lowland forests (broadleaf) existed in most areas, except the arid regions, where sclerophyll forests with trees of short stature prevailed.

2. Haiti

Arid regions such as the Presqu'île du Nord-Ouest, near Gonaives, Plaine Central, Cul-de-Sac, and smaller parts elsewhere may obtain up to 1,000 mm rainfall per annum. Low mountains, such as the areas in the central part of Haiti run 1,000–1,500 mm rainfall. The higher mountainous regions, in the Massif du Nord, Montagnes Noires, Massif de la Selle, and Massif de la Hotte, receive up to 2,000 mm or more. The areas of highest rainfall coincide with the highest summits in the last-mentioned mountain chains, receiving 2,800 mm or more. Forest patterns are similar to the rainfall regions in the Dominican Republic.

II. Flora of Hispaniola

The flora of the island of Hispaniola (Dominican Republic and Haiti) is the second most diverse for the Caribbean islands; only Cuba has a larger vascular flora. Of the estimated 5,000 species of flowering plants and conifers in Hispaniola, about 30–33% are considered endemics. This is a moderate estimate, not based on excessive splitting of taxa. An additional 500–550 species of ferns and fern allies are known. Endemism is much lower in the ferns. The mosses number 450–500 species according to recent surveys by W. R. Buck (NY). No estimates are available, however, for the hepatics, lichens, algae, etc., which are poorly inventoried.

The floristically diverse regions of the island correspond to the areas of highest rainfall, generally in the mountains. It is in these areas of varied terrain and microhabitats that the higher number of endemics can be found. The low altitude areas, particularly the arid regions, are "species poor" and have a low percentage of endemics.

III. Vegetation Maps

The principal vegetation map is the "Mapa ecológico de la República Dominicana," of the Organización de los Estados Americanos, prepared by Humberto Tasaico (1967) on a scale of 1:250,000. The sources of data for this map are aerial photographs and topographic maps, on a scale of 1:50,000, from the Series E034 and E733 of 1962, which were done by the Instituto Cartográfico Universitario of the Universidad Autónoma de Santo Domingo and the U.S. Army Map Service. This map is *potential* vegetation, based on the Holdridge system. There are no other maps of any consequence.

IV. Herbarium Resources

A. Herbaria in the Dominican Republic

The Jardin Botánico Nacional, Santo Domingo (JBSD) has over 50,000 specimens from Haiti and the Dominican Republic. The Universidad Autónoma de Santo Domingo (USD) has about 9000 specimens from the Dominican Republic. The Universidad Catolica Madre y Maestra (UCMM) has about 10,000 specimens, mainly from the Dominican Republic.

B. Other Herbaria

The Department of Agriculture, Damien, Haiti (EHH) has 10,000 specimens, mainly from Haiti, but some are from the Dominican Republic. The Smithsonian Institution (US) has approximately 30,000, about 50% from the Dominican Republic and 50% from Haiti. The New York Botanical Garden (NY) has about 30,000 specimens from the island. Stockholm (S) has about 16,000 (principally from Ekman), 60% from Haiti, 40% from the Dominican Republic. The Arnold Arboretum (A) and the Gray Herbarium (GH) have about 7000 specimens from the island. The Institute of Jamaica (IJ) has some Ekman material of an unknown quantity. They are duplicates from S. A sizeable collection of W. Buch specimens from Haiti is at IJ, making it the major collection of his specimens since the loss of the first set at Berlin.

Estimates refer only to specimens collected since about 1900. Most specimens collected before this data have inadequate or generalized data, for which localities cannot be verified. Unfortunately, some of the estimates above also include specimens with data that may be poor or of little use. These specimens are duplicated in many of the above institutions.

V. Publications

Major floras include Urban (1920–1921), Moscoso (1943), and Liogier (1981–present).

"Moscosoa," the botanical journal of the Jardín Botánico Nacional, Santo Domingo, has six published issues, one each in 1976, 1977, 1978, 1983, 1984 and 1986, and is the principal publication series on the Hispaniolan flora.

VI. Magnitude of Floristic Inventory

It is impossible to evaluate the present state of inventory in terms of specimens/100 km^2. The specimens collected here have been sent as duplicates to many institutions (without bookkeeping). Likewise, it is impossible to determine the total number of specimens *excluding* duplicates.

Areas poorly collected in the Dominican Republic are the desert areas of Cibao, Azua, Hoya de Enriquillo, and the Barahona Peninsula. The Sierra de Bahoruco and the interior of Cordillera Central are moderately collected. The Llanura Costera (present vegetation) and eastern Cordillera Central have also been collected moderately. It is now impossible to get representative specimens in some areas because of the loss of the original vegetation. These areas include the Samana Peninsula, Cordillera Septentrional, eastern Sierra de Bahoruco, Llanura Costera and the Azua Plains (now in cultivation).

The flora of Haiti and the Dominican Republic consists of approximately 5,000–5,500 species of vascular plants, of which probably 30% are endemics.

A. Endemism

There is no analysis of endemism by areas. Most endemics come from mountain ranges such as Cordillera Septentrional, Cordillera Central, Sierra Nieba, Sierra de Bahoruco, and the Samana Peninsula. There are some endemics in desert areas. The Orchidaceae is high in endemics and is the family with most new species described in the last ten years.

Areas of new and important distributions in the Dominican Republic are the Sierra Martín García, the Sierra de Neiba, the Sierra de Bahoruco (western area near Haiti) and the interior of the Cordillera Central.

B. Extinction

There is no estimate of rates of extinction, but the following areas have lost most of their original vegetation: Samana Peninsula, Cordillera Septentrional, eastern Cordillera Central, eastern Sierra de Bahoruco, Llanura Costera (which has a few, small patches of original vegetation), Valle del Cibao (also with small areas of original vegetation), and many regions of Haiti.

C. Threatened Areas

Forest areas undergoing loss due to destruction or conversion also include areas that have already lost most of their original vegetation. These are the Samana Peninsula, the Cordillera Septentrional, and the Cordillera Oriental.

Areas that are experiencing moderate disturbance are the eastern Cordillera Central and the eastern Sierra de Bahoruco. The Sierra de Neiba is experiencing slight disturbance.

Lowlands generally have drier forests than mountain areas and are greatly altered (having very little original vegetation). These are the Llanura Costera, the Valle de Cibao, the Valle de San Juan and eastern Hoya de Enriquillo near Barahona.

Most areas in Haiti suffer from severe alteration.

D. Modern Collections

Table I lists the plant families and the specialists who have collected them. There are other collaborators for the "Flora Vascular de la Isla Española," but most have not yet started studying the material. The Bromeliaceae, Cyperaceae, Orchidaceae, ferns and fern allies are among the better-collected groups. An especially poorly-collected group is the Loranthaceae.

In Haiti, in the years after Ekman (1928), collecting has been minimal. The few collectors that have visited Haiti have been concerned with monographic work and have tended to collect only in the plant groups of their special interest. The field work of Ekman in the main part of Hispaniola was exploratory, to add taxa to the known flora (Urban 1920–1921) and also to get a general idea of the distribution of the taxa. He collected over 16,000 specimens, the largest collection made on the island to that time. Recent major collections have resulted from Liogier (1960s-1970s), who collected mostly in the Dominican Republic (very few from Haiti), and Zanoni and collaborators (1980–present) who also have principally collected in the Dominican Republic. They have collected in many Haitian localities however, although not as much in terms of numbers as in the Dominican Republic. They have collected about 35,000 specimens island-wide. Walter S. Judd and collaborators have collected principally at the summit of the Massif de la Hotte and in a small area of the Massif de la Selle above Port-au-Prince, Haiti (in relation to a study to evaluate the two new national parks in Haiti). Their numbers amounted to less than 2,000 specimens, but are from areas of original, important vegetation. Yves Polynice, an agronomist at the Agronomy College, Damien, Haiti, has done collecting, but its extent is unknown.

In most of Haiti, there has been inadequate collecting in all areas. A discussion by geographic region is not useful. Areas that need attention still are the high montane regions, especially the relicts of natural vegetation in the Massifs de la Hotte, de la Selle, and du Nord. Sporadic collecting by Zanoni and his collaborators in areas of apparently complete destruction of the original vegetation has resulted in the rediscovery of endemics. The whole central region of Haiti seems to have been neglected in the last 70 years and merits attention. It should be noted that this area has a high population density and is greatly altered. Work will have to be slow and selective.

E. Conservation

The declaration of and some subsequent vigilance in the national parks of Hispaniola will provide areas of refuge for the native vegetation, particularly in the new parks—Massif de la Hotte (Pic Macaya area) and Morne La Visite in the Massif de la Selle in Haiti. These areas are somewhat small, but are of considerable importance in a country with such a low percentage of original intact vegetation.

In the Dominican Republic, about 10% of the forested land in the country is now under the charge of the national parks system. Vigilance against incursions varies from good to non-existent. The environments include a low elevation forest region, a high-rainfall karst region, montane and very high-altitude forest, arid-land forests, and a crocodile sanctuary.

Recent floristic work in the parks in Haiti and in the Dominican Republic is yielding a good representation of suites of specimens from natural habitats, which is uncommon in the last 70 years of collecting on the island.

VII. Academic and Governmental Institutions

The Jardín Botánico Nacional, Santo Domingo, is active in fieldwork throughout the year for the institution's "Flora Vascular de la Isla Española." Its herbarium now has 60,000 specimens, primarily from the Dominican Republic and Haiti. The Jardín Botánico Nacional is a "decentralized" government body with funds from the national budget and

Table I
Plant families and specialists working on collections from Hispaniola.

Family or Group	Specialists and Affiliations
Bignoniaceae	A. Gentry (MO)
Bromeliaceae	T. Zanoni and M. Mejia (JBSD)
Pitcairnia	R. W. Read (US)
Orchidaceae	D. Dod (JBSD)
Poaceae	S. Hatch (TAES)
Palmae	R. W. Read (US)
Verbenaceae	R. Sanders (FTG)
Zingiberaceae	P. J. M. Maas (U)
Aquatics	R. Lowden (UCMM)
Bryophytes	W. R. Buck (NY)
	W. C. Steere (NY)
Ferns & Fern Allies	J. T. Mickel (NY)
Lichens	R. C. Harris (NY)
General Flora	T. Zanoni, M. Mejia and Staff (JBSD)
	A. Liogier (UPR)

a fully-operating department of botany with adequate physical resources for field work and the herbarium. It is staffed by 10 people including scientists, secretaries, and technicians. The herbarium contains mostly vascular plants, with some lichens, mosses, liverworts, and a few algae.

The wildlife department of the Secretaría de Agricultura does some quick "look-see" field trips. So far there has been no active collecting or documentation of observations.

Academic institutions in the Dominican Republic include the Universidad Católica Madre y Maestra in Santiago, with one botanist working on a manual and monographic studies of aquatic plants. The herbarium has approximately 10,000 specimens, mainly from the Dominican Republic. There is no active inventory fieldwork conducted by other Dominican universities, but the Museo Nacional de Historia Natural (zoology and geology), and the Museo del Hombre Dominicano (anthropology and ethnology) have the potential to collaborate with botanical inventory in multidisciplinary studies. There are no programs conducted by foreign institutions other than by visiting collaborators of the Jardín Botánico Nacional.

The Department of Agriculture of Haiti maintains a herbarium in their main office in Damien. The herbarium is essentially inactive in field collecting. It contains a large set of Ekman specimens (maybe as many as 10,000) that were sold by him to the agricultural administration in the 1920s, but few specimens have been added recently. Yves Polynice (Agricultural College, Damien, Haiti) has expressed interest in preparing a flora of Haiti. Some fieldwork has been done in the 1980s, but no adequate facilities exist there, neither library nor herbarium. Consultation is made of the Ekman specimens in the adjacent government agricultural building. The office of the preservation of national patrimony (ISPAN), in Haiti, has coordinated the biological inventories of the two new national parks with scientists from the University of Florida, Gainesville (1980s).

VIII. Present and Future State of Inventory

To date, floristic inventory has been mostly sporadic and not planned. Exceptions are the collections of Ekman and Leonard and Abbott, which were made in areas of least representation in the 1920s–1930s, and Allard's, collections from poorly-known areas in the 1940s. Liogier collected in the late 1960s–1970s for general representation of the flora and for uncommon species.

The total number of specimens (Ekman, Leonard and Abbott, Allard, and Liogier) probably runs to 40,000 for the Dominican Republic. These same collectors collected in Haiti; however, it is unknown how many specimens came from that country.

In the 1920s–1930s collections were made in many of the areas now considered to have very little original vegetation. Surveys in the 1960s–1980s to search for and collect the same endemics have been hampered by this loss of original vegetation.

The current work at the Jardín Botánico Nacional (an inventory initiated in 1968 by Liogier and continuing under T. Zanoni in the 1980s) seeks to establish distribution patterns, as well as to find poorly-known areas and areas of endemics in the Dominican Republic and Haiti. Even though Liogier has published on the flora, renewed efforts were necessary. A revised flora is also needed.

IX. Additional Resources

There is no need to add more institutions for floristic inventory. However, for the Jardín Botánico Nacional, the following additional personnel and equipment are necessary:

More herbarium cases, more field supplies	US $50,000
Addition to herbarium building	US $50,000
Literature	US $20,000–30,000
Assist visits of collaborators for specialized collecting in coordination with present staff	US $20,000
Long term vehicle (4-wheel drive pickup replacement in 1987 or 1988)	US $25,000

X. Literature Cited

Liogier, A. H. 1981-present. Flora de Española. 4 vols. Universidad Central del Este, San Pedro de Macoris, Dominican Republic.
Moscoso, R. M. 1943. Catalogus flora domingensis.
Urban, I. 1920–1921. Flora Domingensis. Symb. Antill. Vol. 8: 1–86.

Puerto Rico

José L. Vivaldi

Contents

I.	Description of the Region	342
	A. Geographic Extent and Area	342
	B. Topography and Geology	342
	C. Population	342
II.	Vegetation Maps	342
III.	Floristic Inventory	342
IV.	Status of Floristic Inventories	343
	A. Endemics	343
	B. New Records	343
	C. Extinction	343
V.	Status of Collections	343
VI.	Theatened Areas	343
VII.	Resources for Continued Floristic Inventory	344
VIII.	Evaluation of Past, Ongoing and Planned Inventories	344
IX.	Additional Resources Needed	344
X.	Literature Cited	344
XI.	Appendix – Floristic Works for Puerto Rico and the Virgin Islands	345

I. Description of the Region

A. Geographic Extent and Area

Puerto Rico is the smallest of the Greater Antilles. Rectangular in shape, the main island of Puerto Rico is 178 km long and 58 km wide with a land area of 8,748 km². Surrounding the main island of Puerto Rico are numerous small islands and cays. The largest are Vieques and Culebra on the east, Caja de Muertos in the south, and Mona in the west. Together with them, the total land area of Puerto Rico is 8,794 km² (Picó, 1980).

B. Topography and Geology

Puerto Rico can be divided into three main physiographic units: a south-central volcanic mountainous area, a discontinuous fringe of coastal plains, and a belt of rugged karst topography in the north-central and north-western parts of the island. Mountains cover about 40% of the island, 35% by hills, and 25% by the coastal plains. At least 25% of the island has slopes greater than 45 degrees.

Puerto Rico is nearly bisected by the Cordillera Central or Central Mountain Chain that runs from east to west. The highest peaks in Puerto Rico are found in this mountain range, reaching an elevation of 1338 m at Cerro Punta. The mountains are rough, steep, and highly dissected by intermittent streams. Most of the Cordillera is Upper Cretaceous. The mountains are of volcanic origin and igneous rocks, mostly andesitic forms, cover the area. Basalt is the parent rock throughout these mountains. The Sierra de Cayey is a southeast-running extension of the Central Mountain Chain, where elevations usually are lower. To the east are the Luquillo Mountains, isolated from other mountain areas by a divide with an elevation of about 100 m (Wadsworth, 1951).

The Luquillo Mountains are of volcanic origin and consist of three topographic elements with elevations ranging from 100 to 1000 m (Wadsworth, 1951). There is a central, east to west chain from El Toro Peak to East Peak, a group of peaks to the south that are distributed irregularly but generally parallel to the central chain, and the El Yunque Peak to the north of the central chain. The topography is rough, with cliffs and rock exposures common. Waterfalls are numerous and six major river systems are born in these mountains. On the north slope the rivers drop rapidly at high elevations and more gradually at lower elevations. The reverse is true of the southern slopes.

The karst area is mostly underlain by limestone and solution is the most important geomorphologic agent. Topography varies from extremely rugged to gentle rolling hills with a relief of only a few tens of meters. Elevation varies from sea level to 530 m at the highest point near the town of Utuado (Chinea, 1980).

The karst region can be subdivided into hills, valleys, trenches, caves and minor features. The hills can be further subdivided into cone karst, mogotes, and river and coastal ramparts. Both cone karst and mogotes are sharp, pointed or oval hills. However, cone karst is surrounded by limestone whereas mogotes are surrounded by blanket deposits (Monroe, 1976). Canyons, sinkholes and subterranean rivers are common. Drainage, for the most part, is subterranean. Minor outcrops of chalk, dolomite, and volcanic rocks are also found within the karst area (Monroe, 1976).

The northern, southern, and western coastal plains cover the rest of the island. In these plains, where the pressure for development is greatest, the greater population centers are located.

C. Population

Puerto Rico is one of the most densely populated countries in the world, with 971 inhabitants per square mile. The total population of Puerto Rico in 1984 was 3.3 million, half of which are concentrated in the metropolitan area surrounding the capital city of San Juan. This large megalopolis now includes the municipalities of Carolina to the east, Bayamón and Cataño to the west, and Guaynabo to the south.

Other large population centers are the city of Caguas in the Caguas valley, south of San Juan, Ponce in the central part of the Southern coastal plain, and Mayaguez in the Western coastal plain.

Life expectancy as of 1984 was 74 years of age and the population continues to grow. Pressure for industrial, agricultural and urban development may be expected to increase, especially in the more densely populated and productive coastal plains.

II. Vegetation Maps

Numerous vegetation maps of Puerto Rico have been prepared. Most useful are those of Gleason & Cook (1927), Little & Wadsworth (1964), Dansereau (1966) and Ewel & Whitmore (1973), this last one using the Holdridge system of Life Zones. The others are based on physiognomic characteristics and are the most practical for use in the island.

III. Floristic Inventory

The number of specimens collected in Puerto Rico and housed in Institutions throughout the island probably number about 56,000. The largest herbarium is that of the Botanical Garden of the University of Puerto Rico, with about 20,000. The herbarium previously located at the Agricultural Experimental Station at Río Piedras was merged with this herbarium a few years ago. The herbaria of the Río Piedras campus and the Mayaguez campus of the University of Puerto Rico have about 10,000 each. The herbarium of the Department of Natural Resources has about 6,000. The herbarium of the Institute of Tropical Forestry has about 10,000 specimens.

Collection of flora and fauna is controlled by the Department of Natural Resources. A permit is required to collect botanical specimens in areas under the jurisdiction of the Department and a duplicate set must be deposited in a Puerto Rican Institution. This is a relatively new policy and for that reason most of the herbarium material from Puerto Rico is housed abroad.

Estimating the number of specimens from Puerto Rico that are held in institutions abroad must be a matter of guess work and educated judgement. We estimate that between 25,000 and 50,000 specimens are housed abroad, mostly in the United States. The herbaria with the largest holdings of Puerto Rican plants are New York (NY) and the U.S. National Herbarium (US). These holdings include the collections of Sintenis (NY, US), Eggers (US), Britton (NY), Woodbury (NY, US), A. A. Heller (NY), L. M. Underwood (NY), Stahl (NY, from Berlin via Urban), J. A. Shaffer (NY), and others.

Berlin used to have large and important holdings of Puerto Rican plants, foremost among them the Krug & Urban herbarium. Unfortunately, most of this collection was destroyed during World War II. Duplicates are available, in part, in G, US, NY, and GOET.

Over 50 major floristic works are known for Puerto Rico. The most important are listed in the Appendix. At least 500 additional works also deal with the flora of Puerto Rico. A partial list is housed in the card catalog of the library of the Department of Natural Resources.

IV. Status of Floristic Inventories

The best collected areas are the Luquillo Mountains, the 13 commonwealth forests, and the islands of Culebra, Vieques, and Mona. Collections in some of these areas date back to the 19th Century, prior to the massive deforestation of the Island in the early 1900's. The Central Mountain Chain and the Karst region are moderately well collected. The coastal plains are poorly collected, with the collections made after the area was converted to agriculture and pasture. Of its original vegetation only small patches remain, as a result no one knows what the original floristic composition was and what species may have been lost.

A. Endemics

The areas with the highest number of endemics are the white sands of Tortuguero and the Luquillo Mountains. Both should be considered well studied, although the Tortuguero area is poorly collected, very disturbed by man, and virtually unprotected.

B. New Records

The karst region and that area of the Central Mountain Chain at medium elevations have yielded new and important distributions. Such has been the case with the Río Abajo Forest (karst) where *Pleodendron macranthum* (Canellaceae) was located, and the Cayey area where a second population of *Banara vanderbiltii* (Flacourtiaceae) was located this year.

C. Extinction

Extinction in Puerto Rico has been surprisingly low. The areas with the highest extinction rates probably were the coastal lowlands and the Central Mountain Chain at low and middle elevations. Unfortunately very little remains of the original vegetation of the coastal plains.

V. Status of Collections

The best collected group is the ferns. This collection, housed at the Department of Natural Resources, resulted from the work carried out by Dr. George R. Proctor, who has just finished *The Ferns of Puerto Rico and the Virgin Islands*, awaiting publication by the New York Botanical Garden. Duplicates of Dr. Proctor's collection have been distributed to NY, US, and Jamaica. The grass family is well collected and the best collection is housed at US. Finally, the Orchidaceae have been collected extensively by Dr. James Ackerman, of the University of Puerto Rico at Río Piedras. Dr. Ackerman's collection is housed at the Herbarium of the University of P. R. in Rio Piedras. All other groups must be considered in need of further collection and study.

In addition to the scholars listed above, Dr. Alain Liogier, from the Botanical Garden, has devoted considerable time to the Myrtaceae.

VI. Threatened Areas

All areas of Puerto Rico, even those presumably protected by government agencies, are threatened. The U.S. Forest Service, which manages the Caribbean National Forest in the Luquillo Mountains, recently proposed a Management Plan that included the harvest of timber in several areas of the forest. One of these is the only known locality for *Calyptranthes luquillensis* (Myrtaceae) and most of the forest is considered essential habitat for the Puerto Rican Parrot. The 13 commonwealth forests, although protected by law, have been subject to increasing pressure for the construction of roads and the installation of antennae. This last use has contributed to the extirpation, and probable extinction, of species from the Elfin woodlands. A new danger is posed by plans to grow timber. Although published plans for the management of these forests indicate that native forest will not be cut to plant timber species, this policy could change in the future without strong public support from the local and international community.

In addition to these forests, the mangroves, the swamps and the wetlands are in great danger, especially those close to urban areas or of importance to the tourism industry. Examples are the mangroves along the north coast from the metropolitan area to Fajardo and those in the southwest corner of the island from Guánica to Mayaguez, the wetlands in the north and south coast, and the *Pterocarpus* swamps in Dorado and Humacao.

The white sands of Tortuguero, perhaps the area with the highest endemism in Puerto Rico, is threatened by the mining of silica sand and urban construction. This land has no formal protection.

VII. Resources for Continued Floristic Inventory

The University of Puerto Rico has plant taxonomists on the staff of the Río Piedras, the Mayaguez and the Cayey Campuses. However, within the University of Puerto Rico system only the Botanical Garden can be said to have a strong commitment to systematic botany. Other universities do not have plant taxonomists on their staff.

Of the government institutions only the U.S. Forest Service Institute of Tropical Forestry and the Puerto Rico Department of Natural Resources have shown a strong interest in floristic work. The Institute published two of the most important floristic works in the last 20 years, the *Common trees of Puerto Rico and the Virgin Islands* (Little & Wadsworth, 1964) and *Trees of Puerto Rico and the Virgin Islands* (Little, Woodbury & Wadsworth, 1974). The Department of Natural Resources has just finished *The Ferns of Puerto Rico and the Virgin Islands* by George R. Proctor. Last year the Department initiated a five year effort to produce *The Monocotyledons of Puerto Rico and the Virgin Islands* by the same author.

There are no foreign programs in Puerto Rico dealing with plant systematics at present. However, the opportunity, the infrastructure and the precedent to collaborate exists. Some of the most important floristic works of the last 150 years were done by foreigners. At present a strong cooperative program in other disciplines, such as forestry and wildlife, are carried out by the Institute of Tropical Forestry and the Department of Natural Resources.

VIII. Evaluation of Past, Ongoing and Planned Inventories

A first glance at past, ongoing and planned inventories suggests that the flora of the island is very well known. Most of Puerto Rico, including Vieques, Culebra, Mona and the other small islands around Puerto Rico, has been carefully explored. Areas of high species diversity, such as the Luquillo Mountains, the Central Mountain Chain and the karst region have been well explored. These areas, as well as others with high endemism, such as the white silica sands of Tortuguero Lagoon are, presumably, well known.

However, if the results of the work by George Proctor apply to groups other than ferns, the need for further surveys and critical floristic work is evident. After three years of work the list of known fern species for Puerto Rico increased from 320 to 404. Thus, 53 species, of which 12 are new to science, were added to the list, even though the most recent list of the ferns of Puerto Rico was published as recently as 1982 by Liogier and Martorell.

IX. Additional Resources Needed

That additional herbaria and personnel in the field of plant systematics are needed in Puerto Rico is well known. What is not understood is that, without an Institution devoted to the study of systematics and the proper care of systematic collections and libraries, many collections will disappear. Systematics is not the focal point of any of the institutions conducting floristic work in Puerto Rico, with the possible exception of the Botanical Garden. As a result, the care of the collections depends on the will-power of the systematists up against scientists and administrators who do not understand the value and fragility of the collections.

At the same time, the systematic collections and the systematists themselves, are scattered throughout the island. The lack of contact has resulted in reduced productivity. Because of the uncertainty, many collections have been deposited outside of Puerto Rico in institutions that will provide proper care. As a result, multiple trips to the mainland are needed to consult herbaria and libraries in order to conduct systematic work in Puerto Rico.

The time is ripe for a Natural History Museum in Puerto Rico. With systematics and education as primary responsibilities, this Institution would open a new era and would ensure the permanence and well being of thousands of specimens that otherwise may be lost forever.

X. Literature Cited

Chinea, J. D. 1980. The forest vegetation of limestone hills of northern Puerto Rico. Cornell University Masters Thesis, Ithaca, NY.

Dansereau, P. 1966. Studies on the vegetation of Puerto Rico: 1. Description and integration of the plant communities. Special Publ. no. 1, Univ. of Puerto Rico Institute of Caribbean Science, Mayaguez.

Ewel, J. J. & J. L. Whitmore. 1973. The ecological life zones of Puerto Rico and the Virgin Islands. U.S.D.A. Forest Service Research Paper no ITF-18, 72 pp.

Gleason, H. A. & M. T. Cook. 1927. Ecology of Porto Rico. Sci. Survey of Porto Rico and the Virgin Islands vol. 7. New York Acad. Sci.

Liogier, H. A. & L. F. Martorell. 1982. Flora of Puerto Rico and adjacent islands: A systematic synopsis. Editorial de la Universidad de Puerto Rico, 342 pp.

Little, E. J. Jr. & F. H. Wadsworth. 1964. Common trees of Puerto Rico and the Virgin Islands. U.S.D.A. Forest Service Handbook 249, 548 pp.

———, R. O. Woodbury & F. H. Wadsworth. 1974. Trees of Puerto Rico and the Virgin Islands. U.S.D.A. Forest Service Handbook 449, 1024 pp.

Monroe, W. H. 1976. The karst landforms of Puerto Rico. U.S. Geological Survey Prof. Paper **899**: 1–69.

Picó, R. 1980. Geografía de Puerto Rico. *In*: La Gran Enciclopedia de Puerto Rico, vol. 5.

Wadsworth, F. H. 1951. Forest management in the Luquillo Mountains, I. The setting. Caribbean Forest. **12**: 93–114.

XI. Appendix

Floristic Works for Puerto Rico and the Virgin Islands

Abbad y Lasierra, Fr. I. 1788. Historia geográfica, civil y natural de la isla de San Juan Bautista de Puerto Rico. Reprinted in 1966 by Editorial de la Universidad de Puerto Rico.

Acevedo-Rodríguez, P. & R. O. Woodbury. 1985. Los bejucos de Puerto Rico, Volumen 1. U.S.D.A. Forest Service General Technical Report SO-58.

Ashton, P. M. 1985. Forester's field guide to the trees and shrubs of Puerto Rico. Yale School of Forestry and Environmental Studies.

Barrett, O. W. 1925. The food plants of Porto Rico. J. Agric. **9**: 61–208.

Bello y Espinosa, D. 1881–1883. Apuntes para la flora de Puerto Rico. Anales Soc. Esp. Hist. Nat. **10**: 231–304, 1881; **12**: 103–130, 1883.

Britton, N. L. 1908. Studies of West Indian Plants—I. Bull. Torrey Bot. Club. **35**: 337–345.

———. 1909. Studies of West Indian Plants—II. Bull. Torrey Bot. Club. **35**: 561–569.

———. 1910. Studies of West Indian Plants—III. Bull. Torrey Bot. Club. **37**: 345–363.

———. 1912. Studies of West Indian Plants—IV. Bull. Torrey Bot. Club. **39**: 1–14.

———. 1914. Studies of West Indian Plants—V. Bull. Torrey Bot. Club. **41**: 1–24.

———. 1915. Studies of West Indian Plants—VI. Bull. Torrey Bot. Club. **42**: 365–392.

———. 1915. Studies of West Indian Plants—VII. Bull. Torrey Bot. Club. **42**: 487–517.

———. 1916. Studies of West Indian Plants—VIII. Bull. Torrey Bot. Club. **43**: 441–469.

———. 1917. Studies of West Indian Plants—IX. Bull. Torrey Bot. Club. **44**: 1–37.

———. 1918. The flora of the American Virgin Islands. Mem. Brooklyn Bot. Gard. **1**: 19–118.

———. 1922. Studies of West Indian Plants—X. Bull. Torrey Bot. Club. **48**: 327–343.

———. 1923. Studies of West Indian Plants—XI. Bull. Torrey Bot. Club. **50**: 35–56.

———. 1924. Studies of West Indian Plants—XII. Bull. Torrey Bot. Club. **51**: 1–12.

———. 1926. Studies of West Indian Plants—XIII. Bull. Torrey Bot. Club. **53**: 457–471.

———. 1926. Botanical Explorations of Porto Rico. J. N.Y. Bot. Gard. **27**.

——— & P. Wilson. 1923–1930. Botany of Porto Rico and the Virgin Islands. Scientific Survey of Porto Rico and the Virgin Islands, vol. 5 & 6. New York Academy of Sciences.

Chinea, J. D. 1980. The forest vegetation of limestone hills of northern Puerto Rico. Cornell University Masters Thesis, Ithaca, NY.

Claus, E. P. 1948. A study of the anemophilous plants of Puerto Rico. Bot. Gaz. **109**: 249–258.

Coll y Toste, Calletano. nd. Arboles de Puerto Rico en 1582. Bol. Hist de Puerto Rico, **1**(2): 76–77.

Cook, M. T. & H. A. Gleason. 1928. Ecological survey of the flora of Porto Rico. J. Agric. **12**: 1–139.

Cook, O. F. 1901. A synopsis of the palms of Porto Rico. Bull. Torrey Bot. Club **28**: 525–569.

——— & G. N. Collins. 1903. Economic plants of Porto Rico. Contr. U.S. Natl. Herb. **8**: 57–269.

Dansereau, P. 1966. Studies on the vegetation of Puerto Rico: 1. Description and integration of the plant communities. Special Publ. no. 1, Univ. of Puerto Rico Institute of Caribbean Science, Mayaguez.

D'Arcy, W. G. 1967. Annotated checklist of the Dicotyledons of Tortola, V.I. Rhodora **69**: 385–450.

———. 1971. The island of Anegada and its flora. Atoll Res. Bull. **139**: 1–21.

Duke, J. A. 1965. Keys for the identification of seedlings of some prominent woody species in eight forest types in Puerto Rico. Ann. Missouri Bot. Gard. **52**: 314–350.

Eggers, H. F. A. 1879. The flora of St. Croix and the Virgin Islands. Bull. U.S. Natl. Mus. **13**: 1–133.

Ewel, J. J. & J. L. Whitmore. 1973. The ecological life zones of Puerto Rico and the Virgin Islands. U.S.D.A. Forest Service Research Paper no ITF-18.

Garay, L. A. 1969. Notes on West Indian orchids, I. J. Arnold Arb. **50**: 462–468.

García-Molinari, O. 1952. Grasslands and Grasses of Puerto Rico. Bull. 102, Agric. Exp. Sta. Puerto Rico.

Gleason, H. A. & M. T. Cook. 1927. Ecology of Porto Rico. Sci. Survey of Porto Rico and the Virgin Islands vol. 7. New York Acad. Sci.

Gonzalez-Más, A. 1964. The Cyperaceae of Puerto Rico. Ph. D. Thesis, Louisiana State University.

Graham, A. & D. M. Jarzen. 1969. Studies in neotropical paleobotany, I. The oligocene communities of Puerto Rico. Ann. Missouri Bot. Gard. **56**: 308–357.

Gutiérrez del Arroyo, I. 1976. El Dr. Agustín Stahl, hombre de ciencia: perspectiva humanística. Facultad de Humanidades, Universidad de Puerto Rico, Río Piedras.

Howard, R. A. 1957. Studies in the genus *Coccoloba*, the species from Puerto Rico and the Virgin Islands and from Bahama Islands. J. Arnold Arb. **38**: 211–242.

———. 1966. Notes on some plants of Puerto Rico. J. Arnold Arb. **47**: 137–146.

———. 1970. Note on two species of *Marcgravia*. J. Arnold Arb. **51**: 41–55.

Hutchinson, J. B. 1944. The cottons of Puerto Rico. J. Agric. **28**: 35–42.

Kepler, A. K. 1975. Common ferns of Luquillo Forest, Puerto Rico. Inter American University Press.

Liogier, Brother A. 1965. Nomenclatural changes and additions to Britton and Wilson's Flora of Porto Rico and the Virgin Islands.

Rhodora **67:** 315–360.

———. 1967. Further changes and additions to the flora of Puerto Rico and the Virgin Islands. Rhodora **69:** 372–376.

Liogier, H. A. 1985. Descriptive flora of Puerto Rico and adjacent islands. Volume 1: Casuarinaceae to Connaraceae. Editorial de la Universidad de Puerto Rico.

——— **& L. F. Martorell.** 1982. Flora of Puerto Rico and adjacent islands: a systematic synopsis. Editorial de la Universidad de Puerto Rico.

Little, E. L. Jr. 1969. Trees of Jost Van Dyke (British Virgin Islands). U.S.D.A. Forest Service Research Paper ITF 9: 1–12.

——— **& F. H. Wadsworth.** 1964. Common trees of Puerto Rico and the Virgin Islands. U.S.D.A. Forest Service Handbook 249.

——— **& R. O. Woodbury.** 1976. Trees of the Caribbean National Forest, Puerto Rico. U.S.D.A. Forest Service Research Paper ITF-20.

———, ——— **& Frank H. Wadsworth.** 1974. Trees of Puertor Rico and the Virgin Islands. U.S.D.A. Forest Service Handbook 449.

———, ——— & ———. 1976. Flora of Virgen Gorda (British Virgin Islands). U.S.D.A. Forest Service Research paper ITF-21.

Nevling, L. I. 1970. A new species of *Gonocalyx* (Ericaceae). J. Arnold Arb. **51:** 221–227.

——— **& R. O. Woodbury.** 1966. Rediscovery of *Daphnopsis helleriana* Urb. (Thymelaeaceae). J. Arnold Arb. **47:** 262–265.

Otero, J. I., R. A. Toro & L. Pagán de Otero. 1945. Catálogo de los nombres vulgares y científicos de algunas plantas Puertorriqueñas. Bull. 37, Agric. Exp. Sta. Puerto Rico.

Proctor, G. R. In press. Ferns of Puerto Rico and the Virgin Islands. Mem. N.Y.B.G. **53.**

Stahl, A. 1883–1888. Estudios sobre la Flora de Puerto Rico. Reprinted by the Federal Emergency Relief Administration, San Juan, Puerto Rico, 1936.

Stimson, W. 1969. A revision of the Puerto Rican species of *Lepanthes* (Orchidaceae). Brittonia **21:** 332–345.

Urban I. 1903–1911. Flora Portoricensis. Symbolae antillanae vol. 4, 771 pp.

Vélez, I. 1939. Vegetation of the Southwestern part of Puerto Rico. Ph. D. Thesis, Louisiana State University.

———. 1950. Plantas indeseables en los cultivos tropicales. Univ. de Puerto Rico, Río Piedras.

Woodbury, R. O. & E. L. Little, Jr. 1976. Flora of Buck Island Reef National Monument (U.S. Virgin Islands). U.S.D.A. Forest Service Research Paper ITF-19.

———, **L. F. Martorell & J. C. García-Tuduri.** 1971. Flora of Desecheo Island. Agricultural Experimental Station, Río Piedras, Puerto Rico.

———, ——— & ———. 1977. Flora of Mona and Monito Islands, Puerto Rico (West Indies) Bull. 252, Agric. Exp. Sta. Puerto Rico.

The Lesser Antilles

Richard A. Howard

Contents

I. Description of the Region	348
A. Geographical Extent and Topography	348
II. Vegetation Maps	348
III. Herbarium Resources	348
A. Barbados	348
B. Guadeloupe	348
C. Montserrat	348
D. St. Lucia	348
IV. Present State of Inventory	349
V. Academic and Governmental Institutions	349
VI. Foreign Programs	349
VII. Literature Cited	349
VIII. Additional References	349

I. Description of the Region

A. Geographical Extent and Topography

The Lesser Antilles comprise the islands from Anguilla on the north, 18°N, separated from the Virgin Islands by the Anegada Passage of 100 miles, to Grenada on the south, 12°N, separated from Trinidad by 90 miles. Barbados to the east is included. The islands are of volcanic formation or of uplifted limestone. The highest mountain is Morne Diablotin on Dominica, at 4747 ft., and three mountains, Pelée on Martinique, the Soufrière on Guadeloupe, and the Soufrière on St. Vincent, have erupted during this century. Boiling lakes and other indications of suberuptic volcanic activity occur on most islands. The following figures are derived from Beard's work (1949).

Main volcanic arc	*Area in sq. miles*
Saba	5
St. Eustatius	9
St. Kitts	67
Nevis	38
Redonda	less than 1
Montserrat	33
Guadeloupe	619
Les Saints	less than 1
Dominica	304
Martinique	380
St. Lucia	233
St. Vincent	130
The Grenadines	50
Grenada	120

Limestone islands	*Area in sq. miles*
Sombrero	less than 1
Anguilla	35
St. Martin	38
St. Bartholomew	8
Barbuda	35
Antigua	108
La Désirade	10
Marie Galante	60
Barbados	166

II. Vegetation Maps

The only vegetation map of quality for the area is the Carte Ecologique de la Martinique, 1:75,000, in color, by Jacques Portecop, issued in 1978. A comparable map for Guadeloupe has been in preparation for several years.

Black and white maps, outline in nature, without scale and more general in coverage, record the vegetation of the following islands:

Anguilla. Harris 1965, p 40.
Antigua. Beard 1949, p 159; Harris 1965, p 36.
Barbuda. Beard 1949, p 159; Harris 1965, p 36.
Dominica. Beard 1949, p 108; Hodge 1954, p 16.
Grenada. Beard 1949, p 137.
Montserrat. Beard 1949, p 149.
Nevis. Beard 1949, p 96.
St. Kitts. Beard 1949, p 96.
St. Lucia. Beard 1949, p 123.
St. Vincent. Beard 1949, p 137.

III. Herbarium Resources

A. Barbados

The Department of Botany, University of the West Indies (BAR), has about 4,000 specimens, general in nature, housed in steel cases. Local botanists intend to re-issue or revise the Flora of Barbados and are assembling additional specimens and verifying records.

B. Guadeloupe

1. L'Office National des Forêts, Jardin Botanique, Basse Terre

A temporary herbarium has been established in relation to work towards a Dendrologie des Petits Antilles. About 3,000 specimens with duplicates are unmounted and housed on open shelves. The collection may eventually be housed at the local university or sent to Paris.

2. L'Institut National de la Recherche Agronomique, Petit Bourg

This herbarium, estimated at 2,000 specimens, partly mounted, and variously housed, is largely agronomic in coverage with weeds, field crops and some native vegetation represented. Its future is uncertain.

C. Montserrat

1. Department of Agriculture, Plymouth

A starter herbarium of 400 mounted specimens (collections of R. A. Howard) is stored in wooden cabinets. The local Montserrat National Trust has an interest in the biological resources of the island. They presently maintain a small museum in an old windmill base but hope to expand into larger quarters and activate a natural history museum and herbarium.

D. St. Lucia

1. Department of Forestry, but housed in the Archives Building

A young herbarium of 1,500 mounted specimens stored

in steel herbarium cases has good coverage of forest trees including vouchers for special studies. The specimens are largely the collections of a Peace Corps worker, Ms. Verna Slane, who has since left the island. A publication on the ethnobotany of St. Lucia is in preparation, as is a list of plant names in a newly written *patois* language.

IV. Present State of Inventory

In comparison with the rest of tropical America, the Lesser Antilles are well collected. Novelties are range extensions of montane species in the northern islands, particularly Saba and Montserrat. Taxa new to science are relatively few. Three new species of Dicotyledoneae have been described from St. Lucia recently. *Juniperus barbadensis*, a species long thought to be extinct in the wild in the Lesser Antilles, was found in small numbers on the top of Petit Piton, a difficult-to-climb mountain on St. Lucia. Several species known only from the original collections have been relocated in such numbers that their removal from a list of endangered species could be recommended. In general, current studies (Howard, 1974, 1977, 1979, 1988) reduce the number of endemic species by placing the names in synonymy.

With the independence of various islands as nations and the reduction of British influence in the area, the protection of the higher "crown lands," which were water catchment areas, has diminished. Poaching of trees by individuals for lumber or firewood and encroachment by cut-and-burn agriculture have increased. Commercial lumbering was started for a few years on Dominica but ceased when operations proved unprofitable.

The establishment of national parks and nature reserves is exemplary in the French islands. Dominica has also established nature reserves and studies to this end are under way in Montserrat and St. Lucia.

V. Academic and Governmental Institutions

Only the University of the West Indies has any influence in the English-speaking islands, with biology being taught on Barbados but not at the university branches on the other islands. The medical schools on several islands draw few West Indians as students and do not teach biology as a field study program.

In the French islands the Centre Universitaire Antilles/Guyane maintains branches, with France-based professors teaching biology on a visitation basis. A few French nationals serving an alternative to military service have been utilized by the park service as biologists.

There is scarcely any employment opportunity for trained people in the entire area unless in applied agriculture or forestry.

VI. Foreign Programs

Most programs for collaboration with foreign institutions require some local financial contribution, which is minimal or lacking. Scholarly meetings or seminars in the area appear to be attended by administrators or government officials, not the people doing the work, who would most benefit. From the English-speaking islands a few individuals have been selected for short courses in some aspect of biology, forestry, or park management in Jamaica and in Trinidad. West Indians rarely attend meetings in the United States. The courses in herbarium practices offered by the staff of the Royal Botanic Gardens, Kew, are *known* to some West Indians, but local funding to support their attendance is not available and fellowships usually do not cover transportation.

An exception is a still uncompleted project, Dendrologie des Petites Antilles, which was funded in part directly from France and in part through local governments on Guadeloupe and Martinique.

With outside funding collectors can be located, and with some supervision or training decent specimens and duplicates can be prepared and exchanged. Financial support is also needed for supplies and especially for postage. Only a few local people could attempt determinations. Critical botanical work could be done only in the French islands and on Barbados, where herbaria, libraries, or laboratories are available.

VII. Literature Cited

Beard, J. S. 1949. The natural vegetation of the Windward and Leeward Islands. Oxford Forestry Memoirs 21. 192 pp, 52 figs. Clarendon Press, Oxford.

Harris, D. R. 1965. Plants, animals and man in the outer Leeward Islands, West Indies. University of California Publications in Geography 18. 164 pp, 18 plates.

Hodge, W. H. 1954. Flora of Dominica. Lloydia **17**: 1–238. 1954.

VIII. Additional References

Fournet, J. 1978. Flore Illustré des Phanérogams de Guadeloupe et de Martinique. 1654 pp. Institut National de la Recherche Agronomique. Paris.

Gooding, E. R., A. R. Loveless & G. R. Proctor. 1965. Flora of Barbados. 486 pp. Ministry of Overseas Development, Publication 7. London.

Groome, J. R. 1970. A Natural History of the Island of Grenada, W.I. 115 pp. Caribbean Printers, Trinidad.

Howard, R. A. Flora of the Lesser Antilles. 1974. Vol.1. Orchidaceae, L. Garay & H. Sweet, 235 pp, 83 figs. 1977. Vol. 2. Pteridophyta, G. R. Proctor, 414 pp, 65 figs. 1979. Vol. 3. Monocotyledoneae, R. A. Howard & collaborators. 585 pp, 122 figs. 1988. Vol. 4. Dicotyledoneae, Part 1. R. A. Howard & collaborators. 500 pp, 242 figs. Cambridge, Mass.

Stoffers, A. S. Flora of the Netherlands Antilles. Vol. **1**: 1–84. 1962; Vol. **2**: 1–96. 1966, 97–209. 1980, 211–315. 1982; Vol. **3**: 61–142. 1979. With collaborators and published out of order. Utrecht.

Regional Reports

VI. South America

Colombia

Enrique Forero

Contents

I. Description of the Region	355
A. Geographical Extent and Area	355
B. Topography	355
1. Pacific Coastal Region	355
2. Atlantic Region	355
3. Andean Region	355
4. Low Altitude Region East of the Andes	355
C. Population	355
II. Review of Existing Vegetation Maps	355
III. Magnitude of Floristic Inventory in the Region	356
A. Number of Herbarium Specimens Held in Institutions in the Region	356
B. Number of Herbarium Specimens Held in Foreign Institutions	356
C. Principal Publications on the Flora of the Region	357
IV. Completeness of Floristic Inventory	357
A. Relatively Well-Collected Regions	357
B. Poorly Collected Regions	358
C. Regions High in Endemics	358
D. Regions Yielding New and Important Contributions	358
E. Extinction	358
F. Other Data	358
V. Important Areas of Tropical Forest Threatened by Destruction or Conversion	358
VI. Resources for Continued Floristic Inventory in the Region	358
A. Government Institutions	359
B. Foreign Programs	359
VII. Suitability of Past, Ongoing and Planned Floristic Inventory	359
VIII. Additional Resources Required	359
A. Additional Institutions	359
B. Additional Personnel	359
C. Training Programs	360
D. Funding	360
IX. Literature Cited	360

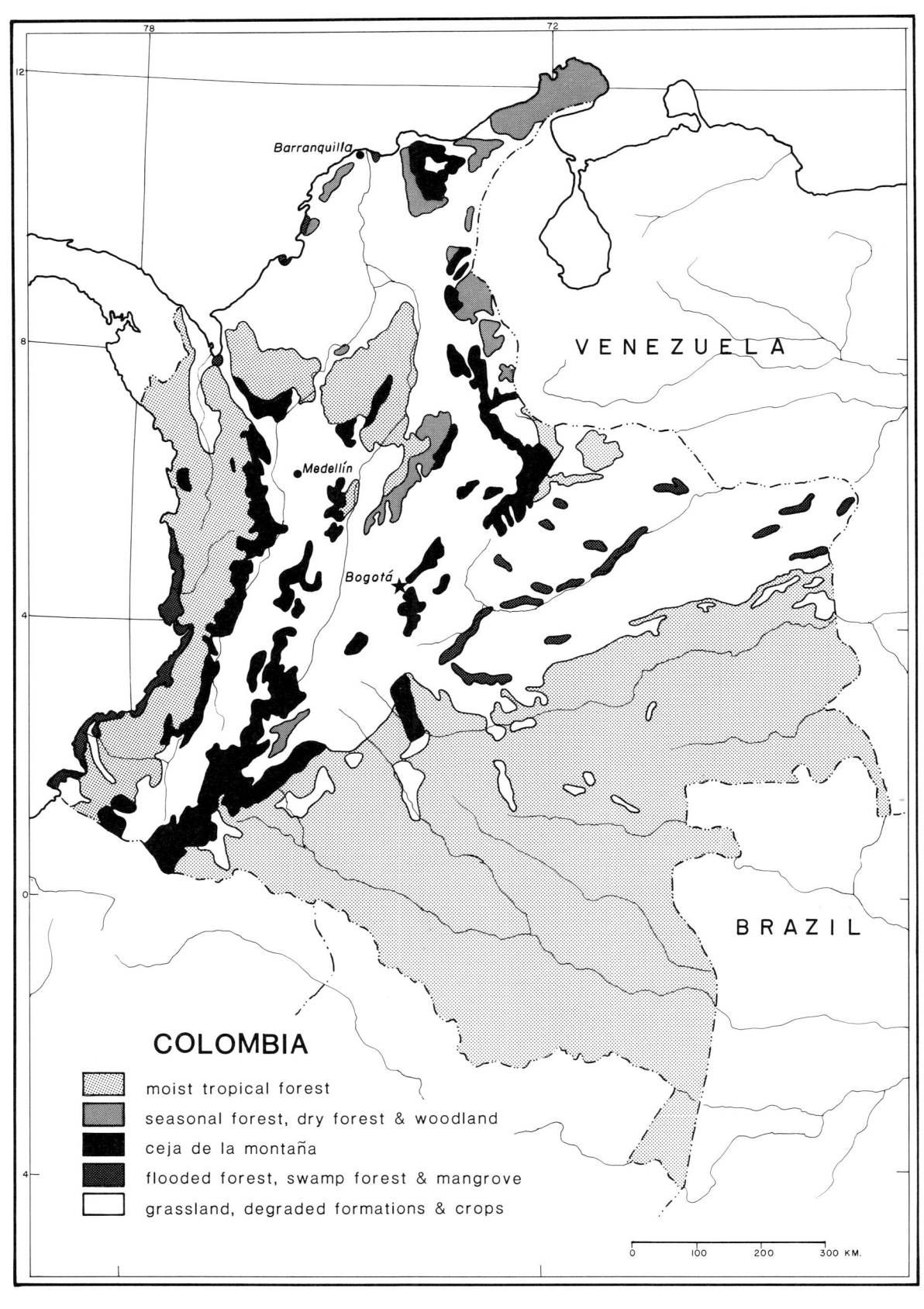

I. Description of the Region

A. Geographical Extent and Area

Colombia is located in the northwestern corner of South America and is the only country on the South American continent with coasts on two oceans: the Pacific to the west and the Atlantic or Caribbean to the north. Neighboring countries are Panama to the northwest, Venezuela to the east, Brazil to the southeast, and Peru and Ecuador to the south. Its total area is 1,138,914 sq. km.

B. Topography

The orogenic events which took place during the Tertiary and early part of the Pleistocene produced a very complicated topography in northern South America. As far as Colombia is concerned, the present topography can be conveniently divided into four main regions:

1. Pacific coastal region

The Pacific coastal region of Colombia is reduced to a narrow, irregular strip extending from Panama to Ecuador. Near the Ecuadorian border this coastland extends inland from the Pacific sometimes as much as 100 km. This apparently flat platform is interrupted by spurs of the western Cordillera (Cordillera Occidental) and by isolated peaks separated from one another by a network of rivers and creeks. The Serranía del Darién does not exceed 400 m in altitude, although some peaks to the north are higher than 1000 m. (Cerro Tacarcuna).

2. Atlantic region

The Atlantic coastal region or Caribbean Coastal Plain, stretches east from the Río Sinú to the Guajira Peninsula. It is a level or moderately hilly region, generally semi-arid and open, partly marshy along the rivers. Much of the Guajira Peninsula is arid; the Serranía de Macuira, in the north, constitutes an "island" covered by dense vegetation which is completely different from the surrounding deserts.

3. Andean region

The mountain system of the Andes forms three different branches in Colombia, separated from each other by deep valleys. These cordilleras (average altitude is estimated at 3200–3500 m), the many much higher peaks and the many inter-Andean valleys and abrupt slopes, create a great diversity of ecological conditions. Some of the volcanic peaks found in various places show perpetual snow. Of these, the Nevado del Huila is 5750 m high, the Nevado del Tolima is 5620 m, the Nevado del Cocuy in the eastern cordillera is 5453 m high. The highest peaks are found in the Sierra Nevada de Santa Marta (Picos Colón, 5800 m and Bolívar 5575 m). This Sierra appears like an island in the northern part of Colombia, without connections with any of the three cordilleras although it is considered to be part of the Central Cordillera. The Eastern Cordillera continues into Venezuela as the Cordillera de Mérida.

4. Low altitude region east of the Andes

The altitude of this region ranges from 100–250 m above sea level, but it may reach up to 1200 m in the Sierra de la Macarena (Departmento del Meta). The Llanos Orientales are located to the Northeast while the Amazon forest covers the Southeastern corner of the region. The Sierra de la Macarena and other isolated hills are considered as the western extreme of the Guayana highlands of Venezuela, Guyana and Northern Brazil.

C. Population

Most of the human population lives on the three cordilleras and in the Atlantic coastal region. The most highly populated departments are Cundinamarca, Antioquia, Valle, Santander and Boyacá, with about 55% of the total population of nearly 28,000,000. Among the least populated areas are Guajira, Chocó, Meta, the Llanos and the Amazon region. Those areas of higher relative density are Quindío, Risaralda, Atlántico, Cundinamarca and Valle. The largest cities are: Bogotá, with more than 4,000,000 inhabitants, and Medellín, Cali and Barranquilla with over 1,000,000 each.

While the government is trying to move people into "colonizing" the Llanos and the Amazon region, there is a growing number of "campesinos" moving out of the countryside and into the large cities and creating serious problems for the economy.

II. Review of Existing Vegetation Maps

The work of Cuatrecasas (1958b) was the first serious attempt at producing a classification of the natural vegetation of Colombia. His nomenclature is still widely used since it is based on altitude above sea level, precipitation and temperature, which makes it fairly easy to understand and to apply. There is no actual vegetation map, but his descriptions as well as a vegetation profile give a clear idea of his concepts.

Espinal and Montenegro (1963) produced *Formaciones Vegetales de Colombia*; this book constitutes the text for Mapa Ecológico which was printed in four sections with a scale of 1:1.000.000. The map follows Holdridge's system of classification of life zones. A new, updated edition of the map and its text was published by the Instituto Geográfico Agustín Codazzi in 1977, under the title *Zonas de vida o Formaciones Vegetales de Colombia*. The map is now printed in 21 "planchas" each with a scale of 1:500.000. In the latest version 24 life zones are recognized and several transitional stages are also shown.

The above mentioned publications constitute the basis for

any floristic, ecological or other type of work related to the natural vegetation of Colombia.

Other publications that can be mentioned are Gentry's papers on the phytogeography of northwestern South America (1978, 1981, 1982) and Cleef's work on the high Andean Vegetation (1979).

III. Magnitude of Floristic Inventory in the Region

A. Number of Herbarium Specimens Held Within Institutions in the Region

There exist in Colombia 17 herbaria which are registered in the 1981 edition of *Index Herbariorum*. While a few of these have been inactive for all practical purposes for some time now, several new ones have been created recently and are growing rapidly.

The most recent count (Forero, 1985) indicated that there are nearly 400,000 herbarium specimens available for study in Colombian herbaria.

A summary of the available information on herbarium specimens is given in Table I; also shown is the number of botanists working in each institution. The National Herbarium (COL) is the largest with more than 300,000 specimens and 12 researchers. Herbaria marked with an asterisk are not active at present.

From Table I it can also be seen that Medellín has become an active center for botanical development. Three herbaria with more than 40,000 specimens (perhaps even 50,000 now) and eight active botanists make Medellín the second most important center of botanical research in the country. There is good integration among institutions and duplication of efforts is kept to a minimum.

B. Number of Herbarium Specimens Held In Foreign Institutions

The number of herbarium specimens from the region which are held in institutions abroad is difficult to assess. The history of botanical exploration in Colombia dates back to the 1700's. It was not until 1930, however, that the National Herbarium was founded and foreign collectors began to deposit their duplicates there (although on a very irregular basis). An example of this problem is the "Cinchona Expedition," which made more than 20,000 collections in Colombia, but of which only a few are deposited in COL.

At most we can say that the largest collections of Colombian plants are deposited in the following overseas herbaria:

SMITHSONIAN INSTITUTION (many collectors, many not duplicated in Colombian herbaria; includes important collections of bryophytes)

HARVARD UNIVERSITY (R. E. Schultes, etc., many duplicated in Colombia)

Table I
Current List of Colombian Herbaria

Herbarium	No. of specimens	No. of Botanists
Nacional Colombiano, Bogotá (COL)	300,000	12
U. Nacional, Medellín (MEDEL)	25,000	2
U. de Antioquia, Medellín (HUA)	20,000	5
U. del Valle, Cali (CUCV)	15,000	3
U. Nacional, Palmira (VALLE)	12,500	2
U. Distrital, Bogotá (UBDC)*	12,000	1
U. de La Salle, Bogotá (BOG)	11,200	0
U. de Nariño, Pasto (PSO)	9,000	1
U. Pedagógica, Tunja (UPTC)	8,000	1
U. del Magdalena, Santa Marta (UTMC)	7,000	1
U. del Tolima (TOLI)	4,500	2
Jardín Botánico, Tuluá (TULV)	4,000	1
Jardín Botánico, Medellín (JAUM)	2,500	2
U. de Santander, Bucaramanga (UIS)*	2,500	0
Jardín Botánico, Cartagena (JBGP)	2,000	1
U. del Chocó, Quibdó (CHOCO)	–	1
New; Not registered		
Corp. Araracuara, Araracuara	1,000	1
U. de los Llanos, Villavicencio	–	1
U. Surcolombiana, Neiva	–	1
U. del Quindío, Armenia	–	2

* indicates that herbarium is not active at present.

FIELD MUSEUM (J. Cuatrecasas, most duplicated either in COL or VALLE)

KEW (many old collectors, most not duplicated in Colombian herbaria)

PARIS (many old collectors – Humboldt, etc. – most not duplicated in Colombia)

MISSOURI BOTANICAL GARDEN (mostly recent collectors, most duplicated in COL, HUA and/or JAUM)

VIENNA (old collections, most not duplicated in Colombia)

REAL JARDÍN BOTÁNICO DE MADRID (Mutis collections,

about 6000 numbers. A few are not duplicated in COL)

BRITISH MUSEUM (Triana's collections, most duplicated in COL)

NEW YORK BOTANICAL GARDEN (Britton & Killip, some duplicated in COL; others by recent collectors, mostly duplicated in COL, including large collections of fungi)

UTRECHT (mainly recent collectors, and mostly cryptogams)

Most holotypes of taxa described on Colombian material are deposited abroad. The National Herbarium houses about 3000–4000 types, but these are mostly isotypes or paratypes. An inventory of the types deposited in COL is being published in Mutisia at irregular intervals.

The number of herbarium specimens deposited in institutions abroad has been increased in recent years through a fairly active program of exchange with many institutions in Latin America, the U.S. and Europe.

C. Principal Publications on the Flora of the Region

1. Flora de Colombia.

This project is directed by the Instituto de Ciencias Naturales of the Universidad Nacional and financed by Fondo Colombiano de Investigaciones Científicas (COLCIENCIAS). Four monographs have been published so far: Lozano-C., G. (Magnoliaceae), 1983; Forero, E. et al. (Connaraceae), 1983; Mora-Osejo, L. E. (Haloragaceae), 1984; and Bernal, H. (*Crotalaria*), 1986.

2. Catálogo Ilustrado de las Plantas de Cundinamarca

Published by the Instituto de Ciencias Naturales and intended for the general public. Seven volumes have been published since 1966 (Camargo, 1969, 1979; García-Barriga & Forero, 1968; Huertas & Camargo, 1976; Murillo, 1966; Pinto & Mora-Osejo, 1966; Uribe-Uribe, 1972). The catalogue includes short descriptions of families, genera and, sometimes, species growing in the Departamento de Cundinamarca. Keys to genera are provided as well as very good illustrations of many species. It is widely distributed internationally.

3. Flora de Mutis

A long standing agreement between the governments of Colombia and Spain aims at publishing the so-called *Flora de Mutis* which consists of the very beautiful illustrations prepared by the Real Expedición Botánica del Nuevo Reyno de Granada between 1783 and 1816. Seven volumes have been published so far (in more than 20 years since the program started); a new impulse is being given to the preparation of additional volumes, and it is planned to have all 50 volumes published by 1992 on the occasion of the 500th anniversary of the discovery of America by Columbus.

4. Los hongos de Colombia

This is a series of mycological papers initiated in 1978. Eight contributions have been published so far, mainly in Caldasia.

5. Studies on Colombian Cryptogams

A series of papers being published by the Institute of Systematic Botany, Utrecht. Not exclusively floristic, it includes revisions and catalogues, as well as anatomical, ecological and phytogeographic studies of Lichens and Musci and Hepaticae. More than 10 contributions have been published so far.

6. Journals

There exist in Colombia several good quality scientific journals such as: "Caldasia" and "Mutisia" (Instituto de Ciencias Naturales); " Actualidades Biológicas" (Universidad de Antioquia, Medellín); "Orquideología" (Sociedad Colombiana de Orquideología); "Cespedesia" (Instituto Vallecaucano de Investigaciones Científicas, Cali); "Pérez-Arbelaezia" (Jardín Botánico de Bogotá); some of the most important papers relating to the Flora of Colombia such as Cuatrecasas' " Aspectos de la Vegetación Natural de Colombia" (1958b) and Schultes' "La riqueza de la Flora Colombiana" (1951) have been published in the "Revista de la Academia Colombiana de Ciencias Exactas, Físicas y Naturales."

7. Other publications

Cuatrecasas (1957), (1958a), (1960); García-Barriga (1974, 1975a, 1975b); Mahecha & Echeverry (1983); ICA (series of booklets); Romero-Castañeda (1961, 1965, 1966, 1969, 1971, 1985); Pérez-Arbeláez (1936); Acero (1979).

8. Special publications

In celebration of the 200th anniversary of the Real Expedición Botánica del Nuevo Reyno de Granada, a number of publications have appeared recently (1983–1984). Some, such as the four volumes of *Flora de Colombia*, have already been mentioned. Others include the new series of papers published by the Instituto de Ciencias Naturales under the general title of *Biblioteca José Jerónimo Triana* which includes several volumes, by Torres (1983a, 1983b), Glenboski (1983), Pulido (1983), and Murillo (1983).

A book on some plant species with promising economic value was published in 1983 by the Convenio Andrés Bello and COLCIENCIAS; this includes plants native to all the Andean countries and Panama.

IV. Completeness of Floristic Inventory

A. Relatively Well-Collected Regions

With 50–100 collections/100 sq. km.: Departments of Cundinamarca, Antioquia, Valle, Guajira.

B. Poorly Collected Regions

With less than 50 collections/sq. km.: Most of the rest of the country, in particular the Amazon region, Santander, Norte de Santander, Cauca, Huila, Nariño, Magdalena, the Llanos of eastern Colombia, La Macarena, Sierra Nevada de Santa Marta, Magdalena River Valley.

C. Regions High in Endemics

No. endemics/total no. spp.: La Macarena, 300 endemic species/3000 spp. Sierra Nevada de Santa Marta (no estimates available).

D. Regions Yielding New and Important Contributions

Pacific coastal region, northern Colombia, the Andes (the three cordilleras), Middle Magdalena River Valley.

E. Extinction

There are no estimates available regarding the rate of extinction (no. extinctions/total no. spp.). High rates of extinction occur in La Macarena, in the Pacific coastal region (Nariño, Cauca, Valle), the Sierra Nevada de Santa Marta (all the way from sea level to the páramo), and the slopes of the Andes and Antioquia. Extinction is not very high (so far) in the Amazon region.

F. Other Data

Well-collected families and scholars working on them, are shown in Table II.

Poorly collected taxa include all other groups, particularly: Cycadaceae, palms (and most other monocotyledons), Lecythidaceae, Chrysobalanaceae, Anacardiaceae (other genera), Myristicaceae (and most canopy trees of many families), Bignoniaceae, fungi and all bryophytes.

V. Important Areas of Tropical Forest Threatened by Destruction or Conversion

The rate of destruction is particularly serious in the Sierra de la Macarena, which is being destroyed without having been properly collected and studied. Although several attempts have been made to carry out a floristic inventory of this important isolated mountain, only a small part of it has actually been covered.

Destruction and conversion are progressing rapidly on the slopes of the Sierra Nevada de Santa Marta, a region poorly known botanically but apparently rich in endemics and in new distributional records.

The tropical forests in the departments of Nariño, Cauca and Valle are quickly being destroyed. This is true both of the inland forests and of the mangroves.

Although the rate of destruction is not very high in the Amazon region of Colombia, it must be pointed out that large areas are being cleared mainly to the north and west. No intensive floristic programs have ever been developed in the region.

Apart from the collections of Schultes (ca. 10,000 numbers) and García-Barriga, very few herbarium specimens from the Colombian portion of Amazonia are available, at least in Colombian herbaria.

The Chocó Department is obviously one of the areas most threatened by destruction.

VI. Resources for Continued Floristic Inventory in the Region

Most of the floristic work carried out in Colombia is being done by botanists attached to academic institutions, mainly government-run universities. There is now in Colombia a fairly good infrastructure from which to develop sound

Table II
Well-collected Plant Families in Colombia

Compositae: S. Díaz, Instituto de Ciencias Naturales
Leguminosae: E. Forero, Instituto de Ciencias Naturales/Missouri Botanical Garden
Orchidaceae: A. Fernández-Pérez, Popayán
Magnoliaceae: G. Lozano, Instituto de Ciencias Naturales
Gramineae: P. Pinto E., Instituto de Ciencias Naturales
Cyperaceae: L. E. Mora O., Instituto de Ciencias Naturales
Ferns: M. T. Murillo, Instituto de Ciencias Naturales; L. Atehortúa, U. de Antioquia, Medellín
Moraceae: A. Dugand, R. Jaramillo-M., Instituto de Ciencias Naturales
Piperaceae: R. Callejas, Medellín
Haloragaceae: L. E. Mora, Instituto de Ciencias Naturales
Passifloraceae: L. A. de Escobar, U. de Antioquia, Medellín
Anacardiaceae (*Mauria*): E. Rentería, Jardín Botánico, Medellín
Ericaceae: J. L. Luteyn, New York Botanical Garden
Melastomataceae: L. Uribe-Uribe, Instituto de Ciencias Naturales

floristic projects. Herbaria with basic facilities such as driers, herbarium cabinets, basic bibliography and some staff members with enough academic background exist in most capitals. There is also a growing understanding among university administrators about the importance of botanical research and about the need for systematic collections.

A. Government Institutions

Government institutions, such as the Instituto de Desarrollo de los Recursos Naturales (INDERENA) and the Instituto Colombiano Agropecuario (ICA), carry out floristic work on a very limited basis and mostly for economic reasons. INDERENA supports limited collecting as part of forestry inventories, while ICA does the collecting in connection with work on weeds, plants used or promising as fodder and so on. Their infrastructure for floristic work is very poor.

B. Foreign Programs

Due to very strict government regulations for plant collecting by foreigners, floristic programs by botanists from other countries have been kept to a minimum. In recent years only two such programs, one with the Missouri Botanical Garden and one with the University of Amsterdam (the latter mostly of an ecological nature in the Andes), have been in operation. Although permits for general collecting are hard to obtain, permits to collect one particular plant family or group are looked upon favorably. The "Flora de Colombia" program is in fact seeking the participation of botanists from other countries in the preparation of monographs. Facilities are offered for collecting particular groups under this program.

In general, floristic inventories in Colombia have been associated to some extent with disciplines such as ecology and anthropology (ethnobotany), less so to zoology. However, most of the work is done by one or two people in most cases. It is only rarely that scientists from other fields are actively involved in such inventories. Exceptions to this general rule are the "Ecoandes" project (U. of Amsterdam, mentioned above), which involves large numbers of people (15–20 or more) working on subjects such as soils, cryptogams, vascular plants, climate, and so on, and the joint efforts of phytochemists and botanists in the study of natural products.

VII. Suitability of Past, Ongoing and Planned Floristic Inventory

Floristic inventory in the tropical forests of Colombia has not followed, for the most part, a definite plan. It has been aimed at collecting particular groups of plants or at looking superficially at the vegetation of a particular region. Exceptions to this rule have been Cuatrecasas' work in Valle and the Chocó floristic survey. More recently, efforts are being made to collect intensively in the Middle Magdalena River Valley by Enrique Rentería of the Jardín Botánico de Medellín. Plans for the future include only intensive collecting in Antioquia, under the Flora of Antioquia project.

The programs just mentioned do correlate with areas that are poorly explored, have high species diversity, high rates of endemism, high rates of extinction, and are threatened by destruction. However, it is a fact that the number of floristic inventories in Colombia does not reflect the urgency of this type of work in areas where the vegetation is being destroyed very rapidly. Few Colombian botanists are actively involved in floristic projects in tropical forests. It is interesting to note that the Flora de Colombia program has invested very little money in actual field work; no provision has been made to conduct intensive plant collecting throughout the country under this program.

VIII. Additional Resources Required

A. Additional Institutions

More or less active herbaria exist at present in most Colombian capitals. It could be said that a couple of new herbaria would be enough to create a complete network of collections in the country. Therefore, more than the creation of many new herbaria, it seems more important at the present time to strengthen those already in existence, providing them with additional facilities for the adequate handling and storage of their collections. This includes in many cases better housing facilities. It is also important to reactivate the few which are inactive right now.

As far as teaching centers are concerned, these exist at the undergraduate level in several important Colombian cities as part of some of the largest universities. At the graduate level, only the National University offers at present a training program in systematics leading to a masters degree. Perhaps no additional such centers are needed right now.

B. Additional Personnel

The need for additional personnel is perhaps the most serious problem for both herbaria and teaching centers. The problem is not that there are no qualified young botanists who could carry out sound floristic inventories and careful and adequate curatorial activities in herbaria. The problem seems to be a lack of interest on the part of the institutions to hire more people to develop their botanical programs. There is, in fact, a large number of college graduates with experience in different aspects of botanical research who are out of work for that apparent reason.

Universities and other government and/or non-government institutions should be made aware of the urgent need to hire well trained botanists, even if they don't have graduate degrees (M. Sc. or Ph.D.), to carry out on a permanent basis the urgent inventory of plant resources in the country.

Finally, most herbaria suffer from a lack of technicians to carry out routine work with the collections. In many cases, the only botanist has to do collecting, drying, mounting, labeling and other chores, which take most of his or her time away from productive research.

C. Training Programs

As far as training programs are concerned, it must be pointed out that those at the undergraduate level fulfill to a large extent their goals of giving basic training in biology to their students. Some, such as the Universidad Nacional in Bogotá and the Universidad de Antioquia in Medellín are producing good botanists, some of whom are actually carrying out floristic work. Many have not had the opportunity yet. Professors at the undergraduate level don't have, in many cases, a clear understanding of the need for floristic work and, therefore, little or no interest is generated among the students for this type of research.

The graduate program in systematic botany mentioned above suffers from a lack of enough faculty with adequate academic training. Professors with masters or doctoral degrees in botany are very scarce in Colombia and no effort is being made on the part of the university to increase the graduate faculty with high level professors.

D. Funding

Several Colombian funding agencies, particularly COLCIENCIAS and Instituto Colombiano para el Fomento de la Educación Superior (ICFES) have been providing funds for the creation and development of herbaria, mainly as far as infrastructure and plant collecting are concerned. There is no doubt that much more additional money will be needed to solve, at least partially, the above mentioned problems.

Funds are, therefore, needed all over Colombia for:

1. Fieldwork.
2. More and better herbarium facilities such as cabinets, drying equipment, housing and storage, protection from insects, temperature and humidity control.
3. Many more field and herbarium botanists. A rough estimate of two botanists for each herbarium of less than 10,000 specimens, five for herbaria between 10,000 and 50,000 specimens and at least ten for larger institutions, indicates that no less than 30–35 additional people are needed.
4. At least five more professors with a Ph.D. degree and five more with a masters degree would be necessary for the only existing graduate program to achieve its objective of giving adequate training and to contribute significantly to the floristic inventory of the country.

Literature Cited

Acero, L. E. 1979. Principales plantas útiles de la Amazonia Colombiana. Proyecto Radargramétrico del Amazonas. Instituto Geográfico Agustín Codazzi. Bogotá.

Bernal M., H. Y. 1986. *Crotalaria* (Fabaceae-Faboideae). Flora de Colombia 4: 1–118.

Camargo G., L. A. 1969. Proteales, Santalales, Ranales, Rhamnales, Malvales. Catálogo Ilustrado de las Plantas de Cundinamarca IV. Publ. Instituto de Ciencias Naturales, U. Nacional, Bogotá.

———. 1979. Clethraceae, Pyrolaceae, Ericaceae. Catálogo Ilustrado de las plantas de Cundinamarca VII. Publ. Instituto de Ciencias Naturales, U. Nacional, Bogotá.

Cleef, A. M. 1979. The phytogeographical position of the Neotropical vascular Páramo flora with special reference to the Colombian Cordillera Oriental. Pages 175–184, *in* K. Larsen & L. B. Holm-Nielsen (Eds.), Tropical Botany. Academic Press.

Cuatrecasas, J. 1957. Prima Flora Colombiana. 1. Burseraceae. Webbia 12(2): 375–441.

———. 1958a. Prima Flora Colombiana. 2. Malpighiaceae. Webbia 13 (2): 343–664.

———. 1958b. Aspectos de la vegetación natural de Colombia. Revista Acad. Colomb. Ci. Exact. 10(40): 221–268.

———. 1960. Prima Flora Colombiana. 2A. Malpighiaceae, Apéndice 1. Webbia 15(2): 393–398.

Espinal, S. & E. Montenegro. 1963. Formaciones Vegetales de Colombia. Publ. Instituto Geográfico Agustín Codazzi. Bogotá.

Forero, E., E. Carbonó, C. I. Orozco, E. Ortega, J. E. Ramos, R. Ruiz, O. Salazar de Benavides, & L. A. Vidal. 1983. Connaraceae. Flora de Colombia 2: 1–83.

———. 1985. Estado actual de la investigación y la docencia en botánica en Colombia. *In* E. Forero (ed.), Memoria de la Reunión de Botánicos de los países miembros del Convenio "Andrés Bello". Ciencia y Tecnología 6: 24–41. Publ. SECAB-FUNBOTANICA, Bogotá.

García Barriga, H. 1974. Flora Medicinal de Colombia. Vol. 1. Publ. Instituto de Ciencias Naturales, Universidad Nacional, Bogotá.

———. 1975a. Flora Medicinal de Colombia. Vol. 2. Publ. Instituto de Ciencias Naturales, Universidad Nacional, Bogotá.

———. 1975b. Flora Medicinal de Colombia. Vol. 3. Publ. Instituto de Ciencias Naturales, Universidad Nacional, Bogotá.

——— & E. Forero. 1968. Mimosaceae, Caesalpiniaceae, Papilionaceae. Catálogo Ilustrado de las Plantas de Cundinamarca III. Publ. Instituto de Ciencias Naturales, Universidad Nacional, Bogotá.

Gentry, A. 1978. Floristic knowledge and needs in Pacific Tropical America. Brittonia 30: 134–153.

———. 1981. Phytogeographic patterns as evidence for a Chocó Refuge. Pages 112–136 *in* G. T. Prance (ed.) Biological diversification in the Tropics. Columbia University Press, New York.

———. 1982. Neotropical floristic diversity: Phytogeographical connections between Central and South America, Pleistocene climatic fluctuations or an accident of the Andean Orogeny? Ann. Missouri Bot. Gard. 69: 557–593.

Glenboski, L. L. 1983. The ethnobotany of the Tukuna Indians, Amazonas, Colombia. Biblioteca José Jerónimo Triana. No. 4. Publ. Instituto de Ciencias Naturales, Universidad Nacional, Bogotá.

Huertas, G. & L. A. Camargo. 1976. Verticillatae, Piperales, Salicales, Myricales, Juglandales, Fagales, Urticales, Podostemonales, Aristolochiales, Polygonales & Centrospermae. Catálogo Ilustrado de las Plantas de Cundinamarca VI. Publ. Instituto de Ciencias Naturales, Universidad Nacional, Bogotá.

Instituto Geográfico Agustín Codazzi. 1977. Zonas de Vida o Formaciones Vegetales de Colombia. Memoria Explicativa sobre el Mapa Ecológico, Bogotá.

Lozano C., G. 1983. Magnoliaceae. Flora de Colombia. 1: 1–119.

Mahecha V., G. E. & R. Echeverri R. 1983. Arboles del Valle del

Cauca. Publ. Progreso Corporación Financiera S. A. Bogotá. 208 pp.

Mora-Osejo, L. E. 1984. Haloragaceae. Flora de Colombia. **3**: 1-176.

Murillo, M. T. 1966. Pteridophyta. Catálogo Ilustrado de las Plantas de Cundinamarca II. Publ. Instituto de Ciencias Naturales, Universidad Nacional, Bogotá.

_____. 1983. Usos de los Helechos en Suramérica con especial referencia a Colombia. Biblioteca José Jerónimo Triana No. 5. Publ. Instituto de Ciencias Naturales, U. Nacional, Bogotá.

Perez-Arbeláez, E. 1936. Plantas Utiles de Colombia. Tomo I. Imprenta Nacional, Bogotá.

Pinto-E., P. & L. E. Mora-Osejo. 1966. Gramineae & Cyperaceae. Catálogo Ilustrado de las Plantas de Cundinamarca I. Publ. Instituto de Ciencias Naturales, Universidad Nacional, Bogotá.

Pulido, M. M. 1983. Los Hongos de Colombia IX. Estudios en Agaricales Colombianos. Biblioteca José Jerónimo Triana No., 7. Publ. Instituto de Ciencias Naturales, Universidad Nacional, Bogotá.

Romero-Casteñeda, R. 1961. Frutas Silvestres de Colombia. Vol. 1. Ed. San Juan Eudes, Bogotá.

_____. 1965. Flora del Centro de Bolívar. Vol. 1. Publ. Instituto de Ciencias Naturales, Universidad Nacional, Bogotá.

_____. 1966. Plantas del Magdalena I. Familia Zigofiláceas. Publ. Instituto de Ciencias Naturales, Universidad Nacional, Bogotá.

_____. 1969. Frutas Silvestres de Colombia. Vol. 2. Publ. Instituto de Ciencias Naturales, Universidad Nacional, Bogotá.

_____. 1971. Plantas del Magdalena II. (Flora de la Isla de Salamanca) Primera Parte. Publ. Instituto de Ciencias Naturales, Universidad Nacional, Bogotá.

_____. 1985. Frutas silvestres del Chocó. Publ. Instituto Colombiano de Cultura, Bogotá.

Schultes, R. E. 1951. La Riqueza de la Flora Colombiana. Revista Acad. Colomb. Ci. Exact. **8(30)**: 230-242.

Torres-Romero, J. H. 1983a. Contribución al conocimiento de las plantas Tánicas registradas en Colombia. Biblioteca José Jerónimo Triana No., 2. Publ. Instituto de Ciencias Naturales, Universidad Nacional, Bogotá.

_____. 1983b. Contribución al conocimiento de las plantas tintóreas registradas en Colombia. Biblioteca José Jerónimo Triana No. 3. Publ. Instituto de Ciencias Naturales, Universidad Nacional, Bogotá.

Uribe-Uribe, L. 1972. Passifloraceae, Begoniaceae, Melastomataceae. Catálogo Ilustrado de las plantas de Cundinamarca V. Publ. Instituto de Ciencias Naturales, Universidad Nacional, Bogotá.

Venezuela

Otto Huber and Dawn Frame

Contents

- I. Description of the Region 363
  - A. Geographical Extent and Topography 363
  - B. Geology 364
  - C. Soils 364
  - D. Climate 364
  - E. Population 364
- II. Natural Vegetation 365
  - A. Pioneer Formation 365
  - B. Herbaceous Formation 365
    - 1. Savannas 365
    - 2. Tepui campos 365
    - 3. Páramos 365
    - 4. Aquatic Herbaceous Vegetation 365
  - C. Shrub Formation 365
    - 1. Xerophytic Thorn Scrub 365
    - 2. Macro- and Mesothermic Scrub 365
    - 3. Montane Scrub 365
  - D. Forest Formation 365
    - 1. Mangrove Forests 365
    - 2. Dry Forests 365
    - 3. Semi-evergreen Forests 365
    - 4. Evergreen Rain Forests 365
    - 5. Wet Montane Forests 365
- III. Vegetation Maps 366
- IV. Herbarium Specimens 366
- V. Publications 367
- VI. Completeness of Tropical Forest Inventory 367
- VII. Endemism 367
- VIII. Rates of Extinction 367
- IX. Collections 368
- X. Threatened Forests 368
- XI. Floristic Resources 369
  - A. Venezuelan Academic Institutions 369
  - B. Venezuelan Governmental Institutions 369
  - C. Private Venezuelan Institutions 370
  - D. Foreign Programs 370
- XII. Inventories in Threatened Areas 370
- XIII. Floristically Interesting Areas 371
- XIV. Necessary Resources for Future Inventory 371
- XV. Acknowledgments 371
- XVI. Literature Cited 371
- XVII. Appendices 371

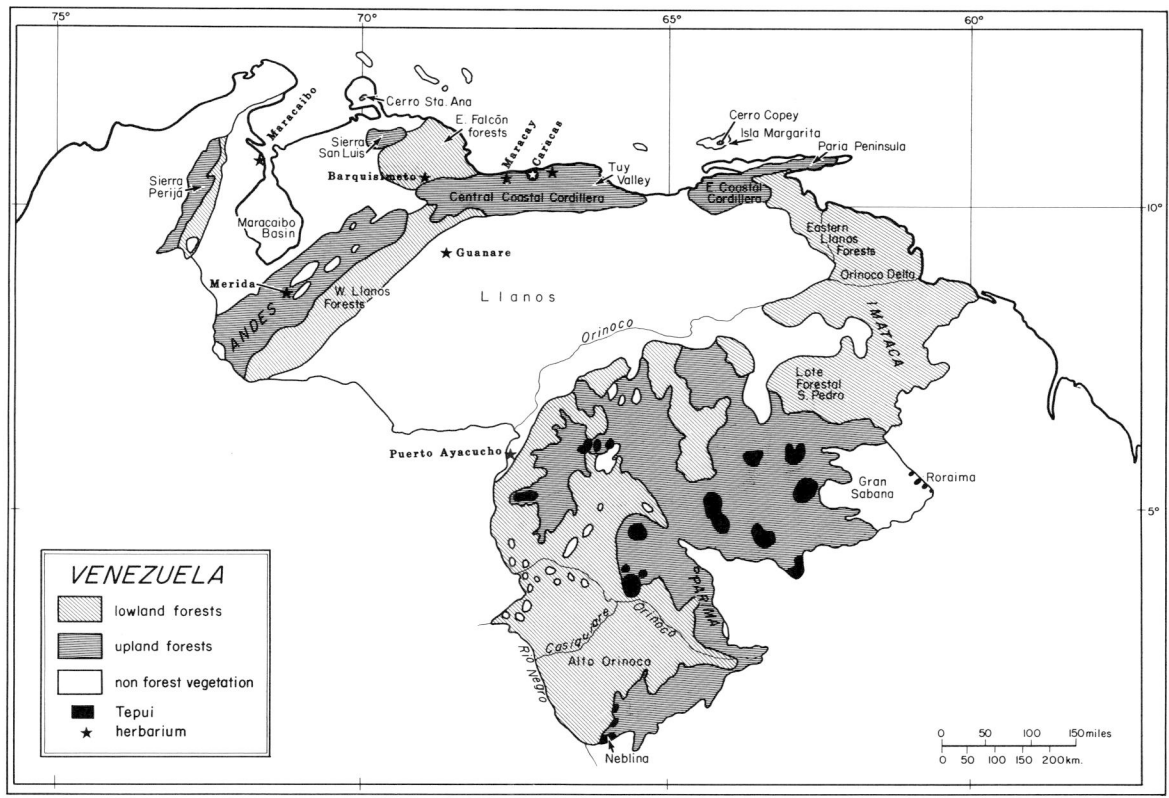

I. Description of the Region

A. Geographical Extent and Topography

Venezuela lies on the northern edge of the South American continent, extending between 1–12°N and 60–73°W. Shaped like an inverted triangle, with its apex oriented towards the south, Venezuela occupies 912,050 sq. km. The main physiographic features are:

(1). A coastline greater than 3,700 km along both the Caribbean Sea and the Atlantic Ocean.

(2). Lowland plains covering nearly one-half the country:
    (a). Depression of Maracaibo.
    (b). Llanos which cross the center of the country from east to southwest.
    (c). Delta region of the Orinoco.
    (d). Upper Orinoco lowlands (Casiquiare peneplains).

(3). Mountain systems:
    (a). Andes in the western part of the country, the highest point being Pico Bolívar (5,004 m).
    (b). Coastal Cordillera in the northern part of the country, the highest points being Pico Naiguatá (2,675 m) in the central section, and Cerro Turumiquire (2,630 m) in the eastern section.
    (c). Guayana Highland south of the Orinoco, of which the highest point (lying in Brazil) is Pico de Neblina (3,045 m).
    (d). Smaller and isolated mountains in the northwestern and northeastern sections of the country:
        (i) Sierra San Luis (approx. 1,600 m) in the state of Falcón.
        (ii) Cerro Santa Ana (approx. 980 m), on the Península de Paraguaná.
        (iii) Cerro Copey (approx. 950 m) on the island of Margarita.

The largest river is the Orinoco; it and its affluents drain about 80% of the country. Originating in the southernmost region of the Sierra Parima in the southeastern Territorio Federal (T. F.) Amazonas, the Orinoco winds its way northwest across the country until it divides (at the bifurcation near Tama tama), giving water to the Río Negro–Amazonas basin via the Casiquiare canal; from here it turns roughly to the north and then to the east, encircling the Guayana Highland and eventually emptying into the Atlantic Ocean at the northeastern edge of the continent. The Orinoco essentially divides the country into two halves, the northern and the southern, and functions as an artery. Its main affluents fall into two groups: those originating in the Guayana Highland, such as ríos Ventuari, Caura, and Caroní; and those originating in the Andes, such as ríos Meta and Apure. The second largest river system is that of Río Cuyuni basin in the eastern corner of Venezuela, which flows into Guyana. The third in importance is formed by ríos Escalante and Catatumbo in the southwestern Maracaibo basin.

B. Geology

The country presents a great variety of petrographic regions, of which the Guayana Shield region is by far the oldest (from 2,000–1,800 million years). In contrast, the Andes and the Coastal Cordilleras are of relatively more recent Tertiary origin. The prevailing rock types in the Guayana Shield are the igneous metamorphic acidic and alkaline granitic basement rocks which are overlain by one to three kilometers of sandstone belonging to the Roraima formation. These sandstones have been differentially eroded, giving rise to the characteristic table mountain landscape of the Guayana Highlands. The table mountains, known as "tepuis", rise vertically from the plains. Here, waterfalls of up to 1,000 m are known. The Coastal Cordillera consists largely of quartzitic schists, gneisses, limestones and isolated granitic and diabasic intrusions. This mountain range exhibits the character of a relatively young mountain landscape with deep valleys and dissected slopes. The Andes are of even more recent origin than the Coastal Cordillera and are of a similar geologic constitution. However, as a result of their much higher elevation, their whole uppermost region underwent marked glacial erosion during the Quaternary.

C. Soils

The predominant soil types in Venezuela north of the Orinoco river are Entisols in the lowlands, and Inceptisols in the Andean and Coastal Cordilleran mountain ranges (Comerma, 1979). Alfisols, Ultisols and Oxisols of lower fertility are widespread in the Llanos region; the latter two soil types also occur widely in the arid regions of the States of Lara and Falcón in northwestern Venezuela. Larger extensions of Histosols have accumulated so far only in the Orinoco Delta region. Recently, Spodosols (Podzols) have been described from the Southern Amazonian region, particularly the Casiquiare-Río Negro Basin (Klinge et al., 1977).

In general, Venezuelan soils are not of high fertility. A few notable exceptions are certain soils localized in portions of the Coastal Cordilleran valleys around Lake Valencia and in some Andean piedmonts. Moreover, large tracts of land in the Llanos lowlands may be heavily inundated from five to eight months of the year effectively limiting availability of arable land. In the upper Andean and parts of the Coastal Cordilleran mountain slopes, severe soil erosion has and is taking place as a result of indiscriminate deforestation practices; substantial losses of topsoil have been sustained.

D. Climate

Venezuela is influenced by two main continental climatic events: the Eastern Trade Winds and the Intertropical Convergence Zone (ITCZ). Roughly speaking, the trade winds, blowing from northeast to southwest, govern the climate during the months from November to May and are responsible for a dry season. In contrast, the ITCZ, oscillating north and south of the equator, dominates the climate during May to November and is primarily responsible for the rainy season. However, deviations from this general scheme may occur locally. A rough gradient of increasing rainfall from north to south can be discerned, but it is modified first by the Coastal Cordillera leading to orographic rain on the windward slopes, and second by the Guayana Highland, with similar results. The highest rainfall occurs at the southwestern edge of the Maracaibo Basin, where the clouds are funneled in by the prevailing winds and then trapped by the surrounding Andean mountains. Here, annual precipitation may reach 4,300 mm. Most of the coast is desertic, with 0–600 mm annual rainfall; the central Llanos are biseasonal, with 800–2,000 mm annual rainfall, and southern Amazonas is perhumid, with 2,500–4,000 mm rainfall annually.

With regard to temperature, the country is subjected to three major regimes: the macrothermic, occurring in the lowlands, with mean annual temperatures between 24–28°C and mean maximum temperatures up to 35–40°C; the mesothermic, found in the intermediate elevations (500–2,000 m), with mean annual temperatures between 15–22°C; and the microthermic, present in areas above 2,000 m, with a mean annual temperature range between 8–15°C. Frost occurs only in the uppermost Andean mountains above 4,000 m and causes solifluction in the páramos. This is the only region in Venezuela in which occasional snow falls and where small relict glaciers persist.

E. Population

The population of Venezuela, approximately 15,000,000, according to the 1981 National Census, is mainly concentrated in the northern half of the country, particularly in the Coastal Cordillera and the Andean region. The valley of Caracas has the greatest concentration of people; an estimated 4,000,000 persons reside there. The second largest urban center is the city of Maracaibo and its environs, with approx. 1,300,000. Other large urban centers are: Valencia, with approx. 650,000; Barquisimeto, with approx. 550,000; Maracay, with approx. 400,000; Puerto La Cruz and Barcelona (which are growing towards each other), with approx. 350,000; and Cumaná, with approx. 250,000 people. Whereas the most densely populated portions of the country in the north have an estimated relative density of 30 inhabitants per km^2, the entire southern region of Venezuela is believed to have an average density of scarcely two inhabitants per km^2.

Following the oil boom of recent decades, there was a mass migration of people from rural to urban areas. They came not only from the Venezuelan countryside, but from neighboring countries and other continents. As a result, the cities swelled with people. Recently there has been a decline in this trend, perhaps due in part to the economic crisis affecting most of Latin America.

II. Natural Vegetation

Venezuela offers, on a relatively small scale, a comprehensive and very illustrative view of the wealth of tropical plant life (Fig. 1).

Proceeding along a gradient of increasing biomass, one can roughly distinguish the following main vegetation types, grouped into essentially four biomes:

A. Pioneer Formation

On exposed rocks—essentially granites and sandstones of the Guayana Shield both in the lowlands and the summits—low, open vegetation types form the early colonizing stages of plant life, with algae, lichens, mosses and small herbs and forbs. Furthermore, some littoral sand dunes along the northwestern Caribbean coast harbor low herbaceous pioneer communities.

B. Herbaceous formation

1. Savannas

Extensive grasslands occupy wide areas in the macrothermic lowlands, especially in the central part of the country ('Llanos'), in northeastern Estado Bolívar, and, to a lesser extent, in T. F. Amazonas. Savannas are composed mainly of grasses and sedges, which form a relatively closed cover, in which shrubs or small trees may or may not be present, either isolated or in small groups. Venezuelan savannas show a wide variety of different types according to the prevailing climatic, edaphic and anthropogenic (mainly fire) factors.

2. Tepui campos

These are found on the summits of the flat-topped table mountains ('tepuis') of the Guayana Highland. These herbaceous communities, very rich in endemics, are dominated by many members of the family Rapateaceae, together with Xyridaceae, Eriocaulaceae, and Cyperaceae.

3. Páramos

In the upper life zone belts of the Andes (above 3,200 m), interesting herbaceous vegetation types characterized by caulirrosulate life forms, exemplified by members of the genus *Espeletia* s.l. (Asteraceae), occur. These vegetation types occupy various altitudinal zones up to 4,700 m and exhibit a high degree of morphological and physiological adaptation to harsh environmental conditions.

4. Aquatic Herbaceous Vegetation

In certain parts of the Llanos ('Llanos of Apure'), large areas are inundated for more than 5–6 months of the year and harbor dense aquatic plant communities dominated by grasses and sedges. Likewise, in the Orinoco Delta region, aquatic herbaceous communities dominated by Araceae, Marantaceae and grasses are frequent, although not so extensive.

C. Shrub formation

1. Xerophytic Thorn Scrub

This vegetation type prevails in large areas of northwestern Venezuela with low annual rainfall, mainly along the coast and on the adjacent inland hills. Dominant families include: Cactaceae, Leguminosae, and Capparidaceae. Few endemics are found here.

2. Macro- and Mesothermic Scrub

At low and medium elevations of the Guayana Shield region, extensive shrublands occupy edaphically marginal sites such as rocky slopes and white sand plains. Many endemic and peculiar taxa are known from these dense to open plant communities.

3. Montane Scrub

Occurring on tepui summits in the Guyana region, as well as on the Andes and the Coastal Cordillera in the upper montane environments, it is a low but very dense vegetation type which harbors a high number of endemic taxa.

D. Forest Formation:

1. Mangrove Forests

These occur along most of the mainland and insular coastlines, from the Orinoco Delta in the east to the mouth of the Maracaibo Lake in the northwest, and are characterized by dense, low to medium sized, evergreen trees, and low species diversity.

2. Dry Forests

This vegetation type is composed mainly of deciduous, low to medium sized, forests growing in dry parts of northern Venezuela. It is found primarily along the coast and at the base of large mountain systems.

3. Semi-evergreen Forests

These occupy a multitude of places at low to medium elevations in the northern and central part of the country. Containing medium to tall trees, and few endemic taxa, these forests are currently under heavy exploitation pressure.

4. Evergreen Rain Forests

Predominating at low and medium altitudes in southern and southeastern Venezuela, these medium to high forests with dense canopies have a great diversity of species, both in the tree layer as well as in the understory.

5. Wet Montane Forests

These cover most of the medium and upper slopes of the major mountain systems (i.e., the Andes, the Coastal Cordillera, the Guayana Shield), and contain medium to tall or very tall evergreen trees. Their physiognomy and species composition varies from place to place, culminating in the cloud forest types which may be considered the most diverse and species rich plant communities in Venezuela.

III. Vegetation Maps

The first vegetation map of Venezuela was drawn by Henri Pittier in 1920. Well known as a botanist, Pittier was also knowledgeable about the phytogeography and vegetation of the country. His map, originally drawn in different shades of brown and gray, was at an approximate scale of 1:2,000,000 and was accompanied by an explanatory text. Much later, Pittier, working in conjunction with Llewelyn Williams, developed a scheme for the vegetation of Venezuela which included a small map published in the series "Plants and Plant Science in Latin America" (1945). In 1955, Francisco Tamayo produced his "Mapa Fitogeográfico Preliminar de la República de Venezuela"; this map was on the scale of 1:2,000,000 and was the first of its type in polychromy. Explanatory notes for Tamayo's map followed in 1958. Tamayo published a modified version of this map in the 2nd edition of the Atlas de Venezuela (1979), in which 25 vegetation units were recognized. Kurt Hueck (1960) devised a color map on the same scale as Tamayo's entitled "Mapa de la Vegetación de la República de Venezuela." This was published by the Instituto Forestal Latino Americano in Mérida (Venezuela). Both Tamayo's 1955 map and Hueck's map may still be purchased from the Venezuelan Ministry of Environment (MARNR). In addition, Hueck's map can be found at a reduced scale in the "Atlas Agrícola de Venezuela" (MAC, 1960, Map No. 12) and in a modified form in "Atlas Forestal de Venezuela" (MAC, 1961). In 1968, the so-called 'Mapa Ecológico' of Venezuela elaborated by J. Ewel, A. Madriz and J. Tosi appeared. This map, on the same scale as the previous ones, in color and with a detailed explanatory text, applied the 'Life Zone' methodology of Holdridge (1947, 1967) to Venezuela. It has been well received by Venezuelan official institutions as well as by private companies interested in the exploitation of natural resources. Today this map, in its second edition published by FONAIAP-MARNR (1976, the text; 1977, the map), forms the basis for many recent studies on Venezuelan vegetation and its potential uses. Entering a new age of cartography, a map of the country at a scale of 1:250,000 entitled "Mapa de la Vegetación Actual de Venezuela" has been published (MARNR, 1982). This map consists of a set of 75 black-and-white sheets and embodies for the first time on a nationwide scale the results of remote sensing, primarily radar, as applied to the delimitation of vegetation.

There are plenty of detailed regional maps for Venezuela, the most important of which are listed in Appendix I. In addition, regional and/or local maps have been produced by MARNR as part of wider reports on soil, climate and land use, but these are either unpublished or of limited circulation. Primarily intended as land-use maps, they are generally rough maps in black and white produced from information provided by modern photographic surveillance techniques. Although fairly good for defining vegetational limits, these maps are of little use floristically because the constituents of the vegetation cover are not described in detail. Some of the most important of these recent MARNR maps are listed in Appendix II.

IV. Herbarium Specimens

When considering the magnitude of floristic inventory accomplished so far in Venezuela, estimates of the number of herbarium specimens held at home (375,500–Table I.) and abroad (225,000–Table II.) can be misleading. Lands north of the Orinoco have yielded the vast majority of specimens; in these regions inventory has been adequate and the resultant specimens have provided the primary data base for the country's flora. However, south of the Orinoco, the situation is much different. Little is known about the flora in this region and much more inventory work will be required before a realistic floristic picture may be formed.

Table I

The estimated number of herbarium specimens held in Venezuelan institutions (1985).

	Number of Specimens[1]
Caracas:	
VEN[2] (National Herbarium)	200,000
La Salle* (Private)	5,000
MYF (Universidad Central de Venezuela UCV – Faculty of Pharmacy)	2,000
Maracay:	
MY (UCV – Faculty of Agronomy)	60,000
IPMY (Instituto Pedagógico)	3,500
Mérida:	
MER (Universidad de los Andes ULA – Faculty of Forestry)	30,000
MERF (ULA – Faculty of Pharmacy)	30,000
Guanare:	
PORT (Universidad Experimental de los Llanos UNELLEZ)	25,000
Maracaibo:	
HERZU (Universidad del Zulia)	10,000
Barquisimeto:	
UCOB (Universidad Centro Occidental Barquisimeto UCO)	5,000
Puerto Ayacucho – Territorio Federal Amazonas:	
Herbario de Puerto Ayacucho* (MARNR-Zona 10)	5,000
TOTAL	375,500

[1] Note, these numbers do not include the recent Neblina collections.
[2] Acronyms according to Index Herbariorum, 7th edition (1981)*
* Not listed in above.

Table II

The major institutions holding Venezuelan collections abroad: estimated number of phanerogamic specimens.

	Number of Specimens[1]
NY[2] (New York Botanical Garden)	100,000
MO (Missouri Botanical Garden)	30,000
F (Field Museum, Chicago)	55,000
US (National Herbarium, Washington D.C.)	40,000
TOTAL	225,000

[1] Note: a) these specimens often represent duplicates and b) these numbers do not include the recent Neblina collections.
[2] Acronyms according to Index Herbariorum, 7th edition, 1981.

V. Publications

There is a wide body of literature on the flora of Venezuela, not the least of which is the ongoing "Flora de Venezuela" project. Appendix IV contains pertinent references to publications on the flora of the country as well as a list of journals in which many Venezuelan species have been described.

VI. Completeness of Tropical Forest Inventory

It is difficult to estimate the completeness of floristic inventory on the basis of how many specimens have been collected in a given area. This is because in certain instances data of this sort might suggest that a locality is well collected when in actuality it is not, perhaps due to a diversity of habitats in a small area. Other localities, although few collections have been made there, are so similar floristically to well-collected ones that they are unlikely to yield new or unusual distributions. Therefore, instead of estimating the completeness of floristic inventory in Venezuela by the aforementioned method, we have simply chosen to list those areas we believe are relatively well-collected and those poorly collected (refer to Fig. 1.).

Relatively well collected
   (in order of estimated number of specimens collected/area)
   (a) Avila
   (b) San Carlos de Río Negro
   (c) Dry areas of northern Venezuela
   (d) Andes (excepting certain places in the southeastern and northeastern ends)
   (e) Llanos
   (f) Gran Sabana
   (g) Lands around Puerto Ayacucho, T. F. Amazonas
   (h) Western base of Cerro de la Neblina
   (i) Most of the smaller tepui tops

Relatively poorly collected
   (in order of estimated number of specimens collected/area)
   (a) Sierra de Perijá
   (b) Certain cloud forests and montane regions in the central section of the Coastal Cordillera
   (c) Montane areas of the eastern section of the Coastal Cordillera
   (d) Certain Andean montane forests in the states of Apure, Táchira, Barinas, Portuguesa and Trujillo
   (e) Lands surrounding the Orinoco Delta
   (f) All forests on the slopes of the Guayana Shield mountains

VII. Endemism

Areas and biomes rich in endemics, according to present knowledge of the flora, are:
   (a) Venezuelan Andes (Cloud forests and Páramos)
   (b) Sierra San Luis (Cloud forests)
   (c) All parts of the Coastal Cordillera (Cloud forests)
   (d) All tepuis, including slope forests and summit vegetation
   (e) Granite outcrops along the Guayana Shield border

New and important data on plant distributions are coming principally from the forests, especially those located south of the Orinoco.

VIII. Rates of Extinction

Although little is known for certain, we estimate that few plant species are becoming extinct in Venezuela. Many that inhabit threatened areas have a wide distribution so that, even though they may disappear from one place, they may be found in another. Nevertheless, without further inventories to discern species richness, rates of extinction in moist forests cannot be known. Some populations of horticulturally prized orchids and bromeliads may be endangered due to predation by people especially in forests near large cities, e.g., on Avila (see Steyermark, 1977). Some rare species are or may be in danger of overcollection by zealous "naturalists". Before proper studies on the number and dynamics of populations have been made some small endemic genera and species may become endangered. As previously mentioned, north of the Orinoco we have adequate knowledge of the role of forests in the overall ecological balance. South of the Orinoco, species diversity is not well understood. However, there is no current threat of widespread deforestation here for two reasons: there is a low population density, and many places are virtually inaccessible.

IX. Collections

The most important Venezuelan plant families (in terms of numbers of species) and their recent specialists are listed in Table III.

X. Threatened Forests

In 1976, a detailed study on the status of Venezuelan tropical humid forests was published jointly by The Sierra Club and the Consejo de Bienestar Rural (C.B.R.) (Hamilton et al., 1976). In that publication, printed in both English and Spanish, experts like Julian A. Steyermark, Jean-Pierre Veillon and Edgardo Mondolfi, assessed and commented upon the state of these forests; in addition they suggested methods for future preservation.

We have arbitrarily divided the important Venezuelan forests threatened by destruction into two major headings: lowlands and uplands.

Lowlands:
- (a) The Maracaibo Basin (State of Zulia)
- (b) Llanos (States of Apure, Barinas and Portuguesa)
- (c) Lower Tuy Valley (State of Miranda)
- (d) Lote Forestal de San Pedro (State of Bolívar)

(a) Maracaibo Basin: Following the construction of the Pan American highway through the Basin in the late 1940's and early 1950's, agriculturalists moved into the southern and southeastern sections. The forests were and are being cleared to open up pasture for beef and dairy cattle. The forests are nearly gone in the southeastern section and the southwestern section is currently undergoing rapid transformation.

(b) Llanos: this land has had a long history of human intervention, dating at least as early as the eighteenth century. The Pan American highway was built through its western section in the 1930's. Squatters and farmers who settled in the eastern piedmont region of the Andes (Llanos) have had considerable impact on its semideciduous forests. Jean-Pierre Veillon (1976) published the results of his long-term study on human occupation in the western Llanos. This report, one of the few of its kind, has maps depicting the amount of forest present in 1950 and 1975. Over the space of 25 years, 33% of the forests have disappeared. It is projected that by the year 2000, forest will occupy 16% of the western Llanos (Veillon, 1976), meaning a habitat loss of 64% over 50 years. Doubtless, erosion will become increasingly serious. In connection with this, the Faculty of Forestry at the University of the Andes (ULA) is attempting to implement a plan for the "rational" usage of the remaining forests. It will be interesting to follow the results of their efforts.

(c) Tuy Valley: deforestation is occurring on a limited scale in the Valley, as parts of it are being converted to Cacao plantations and to provide for expanding industries, as the region is progressively becoming urbanized.

(d) Lote Forestal de San Pedro: this is a vast forested region in the upper Río Supamo-Río Yuruari basin south of El Manteco which is being exploited under government supervision. The past history of land management in this region has not been a happy one and it is questionable whether the future will be much different. The land to be exploited is bounded on the south by the Canaima National Park. This park, which was recently enlarged to three million ha., will limit the deforestation activities occurring further north. Specifically no transgressions into Canaima National Park are likely because it protects the watershed of the Caroní Basin, which in turn houses several important hydroelectric plants (see García, 1984).

Table III
Important Venezuelan plant families and their recent specialists.

Family	Specialist(s)
Acanthaceae	D. Wasshausen
Annonaceae	L. Aristeguieta, P. J. M. Maas
Asclepiadaceae	G. Morillo
Asteraceae (Compositae)	L. Aristeguieta*, V. Badillo, J. Cuatrecasas, J. Pruski, H. Robinson
Begoniaceae	L. B. Smith and D. Wasshausen
Bignoniaceae	A. H. Gentry*
Bromeliaceae	L. B. Smith*
Clusiaceae	B. Maguire
Chrysobalanaceae	G. T. Prance*
Cyperaceae	T. Koyama, R. Wingfield
Gesneriaceae	L. Skog, H. Wiehler
Gramineae (Poaceae)	G. Davidse, B. Garófalo, Z. Luces
Lauraceae	J. Rohwer, H. van der Werff
Leguminosae:	
Mimosoideae	R. Barneby, L. Cárdenas, R. S. Cowan, E. Forero
Caesalpinioideae	R. Barneby, R. S. Cowan, H. Irwin, R. P. Wunderlin
Papilionoideae	R. S. Cowan, J. Zarucchi
Loranthaceae	C. Rizzini
Melastomataceae	J. Wurdack*
Moraceae	C. C. Berg
Myrtaceae	R. McVaugh
Orchidaceae	G. Carnevali, G. C. K. Dunsterville, E. Foldats*, L. Garay
Palmae (Arecaceae)	M. Balick, A. Braun, A. Henderson, R. W. Read
Piperaceae	J. A. Steyermark*†
Pteridophytes	D. Lellinger, J. Mickel, B. Øllgaard, F. Ortega, A. R. Smith, R. Tryon, V. Vareschi*
Rapateaceae	B. Maguire*, J. A. Steyermark†
Rubiaceae	J. A. Steyermark*†
Solanaceae	W. G. D'Arcy, C. E. Benítez, M. Nee
Theaceae	B. Maguire, J. A. Steyermark†
Xyridaceae	R. Kral, L. B. Smith

* Have contributed treatments for the family to "Flora de Venezuela"
† Deceased

Uplands:
- (a) Sierra de Perijá
- (b) Andes
- (c) Coastal Cordillera
- (d) Guayana

(a) Sierra de Perijá: Colombian cultivators are ingressing into the upper montane forest belts on the eastern slopes of the Sierra, thus transgressing the Venezuelan border. Each year, they clear wider stretches of forests, both cloud and montane. Although reachable from the Colombian side, these lands are inaccessible from Venezuela; consequently, Colombian encroachment upon Venezuelan land goes on unhindered.

(b) Andes: throughout the Andes all montane forests (which would here include both semi-evergreen and the evergreen forests of low to mid altitudes) are being invaded with varying intensity. Most of the slopes are or have been worked by shifting cultivators. This has given the slopes a mosaic appearance.

(c) Coastal Cordillera: a similar situation may here be found at the same elevations. At present, one of the most heavily invaded areas is the Cerro Turumiquire region in the eastern section. Likewise, the lower montane forests of the Península de Paria (Cerros Humo and Patao) are being entered and settled following the completion of a road leading to Macuro.

(d) Locally in the "Gran Sabana" of Guayana, destruction of forests is occurring due to uncontrolled burning of the vegetation by the indigenous population (a process which might lead to savannization). These fires are often started without a specific agricultural purpose. A case of limited savannization has been recently observed in Sierra de Parima (see Huber et al., 1984).

In each of these cases of lands threatened by habitat destruction, the story is similar, and a basic pattern emerges. New roads are built, leading to better or increased access; human immigration follows (except for areas flooded after the construction of dams); and loss of habitat ensues, as these lands are converted to fulfill needs for food or energy production. In terms of energetics, rapid deforestation is not energy efficient, for it allows rapid expansion and immediate realization of rewards but with concomitant loss of long-term profit. It is a truism that carefully planned exploitation of resources in a manner which makes them renewable leads to long-term profit and greater economic stability.

XI. Floristic Resources

There are many agencies involved, either directly or indirectly, in the inventory and study of Venezuelan tropical forests. In this section we will name and briefly describe the most important public and private organizations concerned with this work. Acronyms followed by an asterisk are explained in Appendix III with additional notes.

A. Venezuelan Academic Institutions

UCV* has its main base in Caracas with an outlier in Maracay. It is organized the following way:

Caracas—Faculty of Sciences, School of Biology, Department of Botany.
- (a) Division of Taxonomy. Most Venezuelan taxonomists receive their training from here.
- (b) Division of General Plant Science. The different aspects of botany such as anatomy and physiology are stressed by this group.
- (c) Instituto de Zoología Tropical (IZT). The main body of plant ecological research at UCV is found here.
- Faculty of Pharmacy. They have an herbarium (MYF) which is devoted to Venezuelan plants of pharmacological, medicinal and ethnobotanical interest.

Maracay—Faculty of Agronomy. This group conducts inventories on a small scale generally around Maracay and in "Henri Pittier" National Park. Plant collecting in cultivated regions, especially the Llanos, is based here too. They have an herbarium (MY).

IVIC* is located near Caracas. Its Centro de Ecología has maintained a strong ecological research program on Venezuelan tropical forests, with particular emphasis on Amazonas, for the past 15 years.

ULA* is located in Mérida and is organized the following way:
- Faculty of Forestry. They have an herbarium (MER) and are interested specifically in Andean forests and more generally in Venezuelan forests.
- Faculty of Pharmacy. The pharmacologically and medicinally important plants of Venezuela are studied. They have an herbarium (MERF).

UNELLEZ* is located in Guanare and is organized the following way:
- Section of Earth Sciences (Agronomy). Department of Botany. They have a herbarium (PORT). The staff are actively studying the flora of Portuguesa, and in a wider sense the flora of the eastern slopes and piedmont of the Andes. They have also made general plant collecting trips to T.F. Amazonas and Apure.

UCO* is located in Barquisimeto and is organized the following way:
- School of Agronomy. They have an herbarium (UCOB) and study the flora of Lara.

Universidad del Zulia is located in Maracaibo and is organized the following way:
- Faculty of Agronomy. They have an herbarium (HERZU) and study the flora of Zulia.

B. Venezuelan Governmental Institutions (in order of floristic importance)

VEN—National Herbarium in Caracas. Since 1982, this

institution has been ascribed to INPARQUES*. The staff is responsible for the coordination and execution of the "Flora de Venezuela" project. It is the principal botanical institution in the country.

*MARNR** — The Ministry of Environment and Renewable Natural Resources has both central and regional offices. Some of the regional offices are conducting vegetational and floristic studies, the latter on a reduced scale and in connection with VEN (except in the case of Zona 10 [T.F. Amazonas] where a regional herbarium is being built up for local collections).

*CONICIT** — This institution supports scientific research in Venezuela. Under their auspices, some important floristic inventories have recently been or are in the process of being carried out. They are:

- (a) CONICIT-VEN-NSF*-MO, joint program on floristic inventory of endangered or threatened areas. Approximately 15,000 numbers have resulted from this project.
- (b) A floristic and ecological inventory of the state of Falcón. It is estimated that nearly 2,000 numbers have been collected.
- (c) A floristic and botanic inventory of savannas in T. F. Amazonas. About 6,000 numbers have been collected.
- (d) A floristic and botanic inventory of savannas of the state of Bolívar, currently in progress, has amassed some 5,000 numbers.

C. Private Venezuelan Institutions

Sociedad Venezolana de Ciencias Naturales (SVCN) has supported natural history expeditions. One such botanical example was to Jaua-Sarisariñama in 1974.

Sociedad de Ciencias Naturales La Salle has conducted many expeditions to different parts of the country. The most outstanding were the exploration of: the Sierra de Perijá, the Island of Margarita, the Cuyuni region, the T. F. Delta Amacuro.

Academia de las Ciencias Matemáticas, Físicas y Naturales, through a newly created foundation ("Fundación para el Desarrollo de las Ciencias Físicas, Matemáticas y Naturales"), working in conjunction with institutions in the United States (see below), conducted the year-long expedition to Cerro de la Neblina (1984–1985).

The Botanical Garden of Maracaibo is a private institution managed by the "Fundación Jardín Botánico de Maracaibo", and originally sponsored by the Rotary Club of Maracaibo. This garden was founded by Leandro Aristeguieta about eight years ago. Until just recently, George Bunting was a resident scientist there; he sponsored an excellent botanical collecting and research program in the northern Maracaibo Basin.

Fundación TERRAMAR, founded a few years ago, has been conducting expeditions, primarily to T. F. Amazonas and the Venezuelan islands, since its inception.

D. Foreign Programs

The aforementioned CONICIT-VEN-NSF-MO "Botanical Inventory of Endangered or Threatened Areas" is nearing completion. The field work has been done and most of the specimens have been identified.

The aforementioned Neblina Expedition received funding in the United States primarily from: NSF (Botany and Zoology), The National Geographic Society (Botany), The American Museum of Natural History (Zoology), The Smithsonian Institution (Smithsonian staff only, Botany and Zoology). Major funding in Venezuela came from the Fundación Para el Desarrollo de las Ciencias Físicas, Matemáticas y Naturales. The following herbaria participated in the expedition: NY, MO, US, VEN, F, MICH, VDB and K. The field work for this project was done from January 1984 through February 1985. The examination and study of the collected data is just beginning.

There is a NY-CONICIT-Electrificación del Caroní (EDELCA) collaborative effort conducting a botanical-ecological inventory of savanna vegetation in the state of Bolívar.

The University of Göttingen (Fed. Rep. of Germany) and EDELCA are involved in medium and long term research on a) forest-savanna ecotone dynamics and b) natural and man-made forest degradation in the Gran Sabana region.

In the past few years, MO has strengthened its ties with Venezuelan herbaria (VEN, PORT, MER, MY) and with TERRAMAR, in order to stimulate collaborative collecting programs.

Brian Boom (NY), in collaboration with MYF, PORT, and VEN conducted two, one ha. tree inventories as part of an ethno-ecological study of the Panare Indians in the Cedeño District in the state of Bolívar. As a long-term project, Dr. Boom will be studying the flora of the Guayana Highland in collaboration with various Venezuelan institutions.

Otto Solbrig (GH) and T. Givnish (WIS), are involved in a joint teaching program with the ULA Faculty of Sciences. In addition, T. Givnish conducts research on the eco-physiology of certain Venezuelan plant groups.

As is apparent from the preceding paragraphs, opportunities for foreign collaboration with Venezuelan institutions do exist, but only if they are done through the proper official channels, e.g.: Botany via the previously mentioned herbaria and/or academic institutions; Ecology, Anthropology and Limnology via IVIC; Zoology via UCV and Servicio Nacional de Fauna (MARNR); Earth Sciences (Geology, Geomorphology, Pedology) via MARNR, UCV, or Ministerio de Energía y Minas (MEM).

XII. Inventories in Threatened Areas

The CONICIT-VEN-NSF-MO joint project to inventory endangered or threatened areas focused primarily on those lands to be flooded after the construction of dams. The building of several dams was proposed during the height of the oil boom in Venezuela. With the present financial crisis, the construction of some dams has been postponed. Areas inventoried as part of the CONICIT-VEN-NSF-MO program were: parts of the Llanos, sections of Andean cloud forests, Cerro El Guapo and Cerro Bachiller in the lower Tuy Valley of the Coastal Cordilleras, the Yacambú region, the Caura region, the lower Caroní near the Guri dam, the Cataniapo area in T.F. Amazonas. The first four regions correspond to areas, or portions thereof, we previously listed as threatened. Areas where construction of dams was imminent but which have now been delayed, include: the lower Tuy Valley (Cerro El Guapo and Cerro Bachiller), and the Caura and Cataniapo regions. With the conclusion of the CONICIT-VEN-NSF-MO joint project, there is no ongoing research to inventory threatened forests in Venezuela.

XIII. Floristically Interesting Areas

It is widely accepted that the Venezuelan moist forests tend to be interesting floristically and the results of the CONICIT-VEN-NSF-MO inventories confirm this. Tepuis, which are high in endemics, are also considered to be of floristic interest; however, they are not currently threatened.

XIV. Necessary Resources for Future Inventory

Future inventory of endangered areas in Venezuela will depend upon the serious interest and major participation of Venezuelan institutions. At present, there is no need to build additional herbaria; however, improvements in existing facilities are necessary. The training of more Venezuelan botanists and field workers is also essential.

XV. Acknowledgments

We are greatly obliged to the late Dr. Julian Steyermark for his comments and recommendations on the manuscript.

XVI. Literature Cited

Atlas Agrícola de Venezuela. 1960. Ministerio de Agricultura y Cría, Dirección de Planificación Agropecuaria. Caracas.

Atlas Forestal de Venezuela. 1961. Ministerio de Agricultura y Cría, Dirección de Recursos Naturales Renovables. Caracas.

Comerma, J. 1979. Suelos. Pages 182–183 *in* Atlas de Venezuela, 2nd edition. Ministerio del Ambiente y de los Recursos Naturales Renovables, Dirección General de Información e Investigación del Ambiente, Dirección de Cartografía Nacional. Caracas.

García, J. R. 1984. Waterfalls, hydro-power, and water for industry: Contributions from Canaima National Park, Venezuela. Pages 588–591 *in* J. A. Mc. Neely & K. R. Miller (eds.), National parks, conservation and development, IUCN Commission on National Parks and Protected Areas (Proceedings of the World Congress on National Parks. Bali, Indonesia, 11–22 Oct. 1982). Smithsonian Institution Press. Washington, D. C.

Hamilton, L. S., J. A. Steyermark, J. P. Veillon & E. Mondolfi. 1976. Conservación de los bosques húmedos de Venezuela. Sierra Club–Consejo de Bienestar Rural. Caracas.

Holdridge, L. R. 1947. Determination of world plant formations from simple climatic data. Science **105**: 367–368.

———. 1967. Life zone ecology (Revised edition). Tropical Science Center. San José, Costa Rica.

Holmgren, P. K., W. Keuken & E. K. Schofield, compilers. 1981. Index Herbariorum, Part I: The Herbaria of the world, 7th edition. Bohn, Scheltema & Holkema. Utrecht/Antwerpen; Dr. W. Junk B. V., Publishers. The Hague/Boston.

Huber, O., J. A. Steyermark, G. T. Prance & C. Alès. 1984. The vegetation of the Sierra Parima, Venezuela–Brazil: Some results of recent exploration. Brittonia **36**: 104–139.

Hueck, K. 1960. Mapa de la Vegetación de la República de la Venezuela. Instituto Forestal Latinoamericano de Investigación y Capacitación. Mérida.

Klinge, H., E. Medina & R. Herrera. 1977. Studies on the ecology of Amazon Caatinga forest in southern Venezuela. I. General features. Acta Ci. Venez. **28**: 270–276.

Mapa de la Vegetación Actual de Venezuela. 75 maps, 1: 250,000. 1982. MARNR, Dirección General Sectorial de Planificación y Ordenación del Ambiente. Caracas.

Pittier, H. & Ll. Williams. 1945. A review of the flora of Venezuela. Pages 102–105 *in* Frans Verdoorn (ed.), Plants and Plant Science in Latin America. Chronica Botanica Co. Waltham, Mass., U.S.A.

Steyermark, J. A. 1977. Future outlook for threatened and endangered species in Venezuela. Pages 128–135 *in* G. T. Prance & T. S. Elias (eds.), Extinction is forever. New York Botanical Garden. New York.

Tamayo, F. 1958. Notas explicativas del ensayo del mapa fitogeográfico de Venezuela (1955). Revista Forest. Venez. **1(1)**: 7–31.

———. 1979. Vegetación. Pages 184–187 *in* Atlas de Venezuela, 2nd edition. Ministerio del Ambiente y de los Recursos Naturales Renovables, Dirección General de Información e Investigación del Ambiente, Dirección de Cartografía Nacional. Caracas.

Veillon, J.-P. 1976. Las deforestaciones en los Llanos Occidentales de Venezuela desde 1950 hasta 1975. Pages 97–110 *in* L. S. Hamilton (ed.), Conservación de de los Bosques Húmedos de Venezuela. Sierra Club–Consejo de Bienestar Rural. Caracas.

XVII. Appendices

Appendix I
Important regional vegetation maps of Venezuela.

1955 Vegetación del Edo. Yaracuy, approx. scale of 1:500,000, in color, by Aristeguieta and Vareschi.

1956 Vegetación del Guárico Occidental, scale of 1:250,000, in color, by the Consejo de Bienestar Rural (C. B. R.).

1956 Vegetación de la Cuenca de Maracaibo, approx. scale of 1:1,000,000, in color, by Lasser, Penny and Marsh (the map was published in 1957).

1957 Vegetación de los Llanos de Monagas, scale of 1: ? , by the C. B. R.

1961 Vegetación de sabanas de la zona N del Edo. Bolívar, scale of 1:500,000, in black and white, by Ramia.

1964 Vegetación de sabanas de Venezuela, scale of 1:4,000,000, in black and white, by Ramia.

1967 Tipos de sabanas en los Llanos de Venezuela, scale of 1:4,000,000, in black and white, by Ramia.

1971 Vegetación del Territorio Federal Amazonas, scale of 1:250,000, in black and white, in 19 sheets, by AERO-SERVICE-CODESUR.

1972 Vegetación actual de la Región Centro-Occidental, Falcón, Lara, Portuguesa and Yaracuy, approx. scale of 1:750,000, in color, by Smith.

1977(?) Vegetación de la Región Metropolitana, approx. scale of 1:400,000, in black and white, by Smith and Berry.

1979 Vegetación y Unidades ecológicas del Edo. Falcón, scale of 1:250,000, in black and white, by Matteucci, Colma and Pla.

1982 Vegetación del Territorio Federal Amazonas, originally drawn on a scale of 1:2,000,000, but published at an approx. scale of 1:4,000,000, in color, by Huber.

Appendix II
Important regional vegetation maps of Venezuela produced recently by MARNR.

1983 Levantamiento de información básica en vegetación y uso actual de la Cuenca Alta y Media del Río Chama, Edo. Mérida, by G. Mendoza and F. Uzcátegui (Inf. No. 825).

1983 La cobertura vegetal actual y el potencial forestal del Territorio Federal Delta Amacuro, Sector N del Río Orinoco, by H. Canales (Informe No. 902).

1983 Uso actual y cobertura vegetal de las tierras en la Región Centro Norte Costera, by R. Goitía et al. (Informe No. 919).

1983 Levantamiento de información básica en vegetación. Fase II: Cuenca media y alta del Río Escalante, Edos. Mérida y Táchira, Zona Sur del Lago, by L. Méndez et al. (Informe No. 936).

1983 Uso actual y cobertura vegetal de las tierras en la Cuenca del Lago de Valencia, by R. Goitía et al. (Informe No. 967).

In press: Conservación y manejo de los Manglares de Venezuela y Trinidad-Tobago (7 volúmenes).

(For these and other maps and reports, consult the MARNR).

Appendix III
Notes on important acronyms used in text.

Venezuelan Public Institutions

UCV = Universidad Central de Venezuela—The principal and largest university in Venezuela, with two branches, one in Caracas and another in Maracay. The institution in Maracay is composed of two faculties: Agronomy and Veterinary Science.

IVIC = Instituto Venezolano de Investigaciones Científicas—Originally founded in the 1950's and intended for Bio-medical research. It has since broadened its research activities to include such disciplines as chemistry, physics and engineering. In 1970, a center for ecological research was established. This institution remains to this day one of the most active centers for ecological field work in Venezuela. One project pursued under their auspices was the Amazonian tropical rainforest study in San Carlos de Río Negro (as part of the UNESCO International MAB program).

ULA = Universidad de los Andes—The second major Venezuelan university, located in Mérida.

UNELLEZ = Universidad Nacional Experimental de los Llanos Occidentales "Ezequiel Zamora"—An important, recently founded, regional institution. It has three branches, one in Guanare, one in Barinas and one in San Carlos. The former has a herbarium associated with it. The University is divided into three vice-rectorates. The herbarium is attached to the Agriculture Production vice-rectorate and is part of the School of Natural Resources.

LUZ = La Universidad del Zulia—An important regional institution. It is located in Maracaibo.

(continued)

Appendix III

Venezuelan Government Agencies

INPARQUES = Instituto National de Parques—An autonomous governmental institution related to MARNR. It supervises the National Herbarium (VEN) through its recently created Direccíon de Investigaciones Biológicas and is concerned with the management and protection of the Venezuelan national park system.

MARNR = Ministerio del Ambiente y de los Recursos Naturales Renovables—(1977–present): It is one of the major institutions replacing the former Ministerio de Obras Públicas (MOP). Devoted to the development and management of Venezuela's natural resources, MARNR is composed of a central and a regional organization. The central organization has four General Directions (in Caracas): Dirección General de Información e Investigación del Ambiente (DGIIA), Dirección General de Planificación y Ordenación del Ambiente (DGPOA); Dirección General de Infraestructura (DGI); Dirección General de Administración del Ambiente (DGAA). The regional organization has 14 regional offices known as "Zonas administrativas".

CONICIT = Consejo Nacional para la Investigación Científica y Tecnológica—This institution fulfills the same needs as NSF in the U.S.A. It is the governmental agency for the stimulation and financing of Venezuelan scientific research.

Foreign Governmental Agencies

NSF = National Science Foundation—The major governmental agency in the U.S.A. which proves financial backing for the stimulation and advancement of science.

Appendix IV
Pertinent Publications on the Flora of Venezuela
and
a list of Important Periodicals

FLORA DE VENEZUELA (T. Lasser ed. (1964–1981; Z. Luces de Febres with the collaboration of J. A. Steyermark ed. 1982–present).

    Vol. I, parts 1–2; 1968 (part 1), 1969 (part 2). Helechos by V. Vareschi. Merída: Inst. Bot., Minist. Agric. y Cría.

    Vol. II, part 2; 1984. Piperaceae by J. A. Steyermark. Caracas: Fund. Ed. Amb., INPARQUES.

    Vol. III, part 1; 1971. Ulmaceae; Ranunculaceae; Dichapetalaceae; Zygophyllaceae; Umbelliferae; Clethraceae; Lennoaceae; Mayacaceae; Haemodoraceae; by various contributors. Caracas: Inst. Bot., Minist. Agric. y Cría.

    Vol. IV, part 2; 1982. Loranthaceae (by C. T. Rizzini); Hernandiaceae (by K. Kubitzki); Chrysobalanaceae (by G. T. Prance). Caracas: Fund. Ed. Amb., INPARQUES.

    Vol. VIII, parts 1–2; 1973. Melastomataceae by J. Wurdack (*Mouriri* and *Votomita* by T. Morley). Caracas: Inst. Bot., Minist. Agric. y Cría.

    Vol. VIII, part 3; 1982. Hydrophyllaceae (by L. Constance); Convolvulaceae (by D. F. Austin). Caracas: Fund. Ed. Amb., INPARQUES.

    Vol. VIII, part.4; 1982. Bignoniaceae by A. H. Gentry. Caracas: Fund Ed. Amb., INPARQUES.

    Vol. IX, parts 1–3; 1974. Rubiaceae by J. A. Steyermark. Caracas: Inst. Bot., Minist. Agric. y. Cría.

    Vol. X, parts 1–2; 1964. Compositae by L. Aristeguieta. Merída: Inst. Bot., Minist. Agric. y Cría.

    Vol. XI, part 2; 1982. Podocarpaceae (by D. J. de Laubenfels); Alismataceae (by K. Rataj); Zingiberaceae (by P. J. M. Maas); Rapateaceae (B. Maguire). Caracas: Fund. Ed. Amb., INPARQUES.

    Vol. XII, part 1; 1971. Bromeliaceae by L. B. Smith. Caracas: Inst. Bot., Minist. Agric. y Cría.

    Vol. XV, parts 1–5; 1969 (part 1), 1970 (parts 2–5). Orchidaceae by E. Foldats. Caracas: Inst. Bot., Minist. Agric. y Cría.

    Vol. XV; 1970. Clave de los Géneros e Indice General by E. Foldats. Caracas: Inst. Bot., Minist. Agric. y Cría.

    No volume number, 1977. Verbenaceae by S. López-Palacios. Mérida: Consejo de Publicaciones/Facultad de Farmacia, ULA.

THE BOTANY OF THE GUAYANA HIGHLAND by Bassett Maguire and collaborators. Parts I–XII. Memoirs of the New York Botanical Garden.

    I. **8(2)**: 87–160. 1953. (Combretaceae, Compositae, Eriocaulaceae, Leguminosae, Malpighiaceae, Melastomataceae, Polygalaceae, Rutaceae)

    II. **9(3)**: 235–392. 1957. (Annonaceae, Begoniaceae, Bignoniaceae, Bromeliaceae, Compositae, Droseraceae, Eriocaulaceae, Gramineae, Haemadoraceae, Leguminosae, Musci, Piperaceae, Polygalaceae)

    III. **10(1)**: 1–156. 1958. (Apocynaceae, Bignoniaceae, Combretaceae, Guttiferae, Leguminosae, Melastomataceae, Myrtaceae, Rapateaceae, Xyridaceae)

    IV. **10(2)**: 1–37. 1960. (Musci, Xyridaceae, Marantaceae, Bromeliaceae, Annonaceae, Rutaceae)

    IV(2). **10(4)**: 1–87. 1961. (Aquifoliaceae, Ochnaceae, Guttiferae, Melastomataceae, Acanthaceae, Apocynaceae, Leguminosae—Mimosoideae, Leguminosae—Caesalpinioideae, Leguminosae-Lotoideae)

    V. **10(5)**: 1–278. 1964. (Bambusoideae, Bromeliaceae, Guttiferae, Lauraceae, Melastomataceae, Piperaceae, Rubiaceae, Xyridaceae)

    VI. **12(3)**: 1–285. 1965. (Aquifoliaceae, Cyperaceae, Euphorbiaceae, Gramineae, Lauraceae, Rapateaceae, Rubiaceae)

    VII. **17(1)**: 1–439. 1967. (Abolbodaceae, Bombacaceae, Compositae, Cyperaceae, Euphorbiaceae, Labiatae, Lentibulariaceae, Polypodiaceae, Rubiaceae)

    VIII. **18(2)**: 1–290. 1969. (Annonaceae, Apocynaceae, Bromeliaceae, Cyperaceae, Gramineae, Malpighiaceae, Mela-

(continued)

Appendix IV (continued)

stomataceae, Myrtaceae, Quiinaceae, Schizaeaceae, Velloziaceae)

IX. **23:** 1–832. 1972. (Bonnetiaceae, Caryocaraceae, Clusiaceae, Cycadaceae, Cyclanthaceae, Graminae, Meliosmaceae, Melastomataceae, Palmae, Piperaceae, Polygalaceae, Polypodiaceae, Rubiaceae, Sapotaceae, Tetrameristaceae)

X. **29:** 1–288. 1978. (Bonnetiaceae, Bignoniaceae, Cyatheaceae, Ericaceae, Icacinaceae, Saccifoliaceae, Sarraceniaceae, Styracacease, Symplocaceae, Tiliaceae)

XI. **32:** 1–391. 1981. (Tepuianthaceae, Malpighiaceae, Dipterocarpaceae, Convolvulaceae, Ebenaceae, Lissocarpaceae, Gentianaceae)

XII[a]. **38:** 1–84. 1984. (Hymenophyllopsidaceae, Hymenophyllaceae, Araliaceae)

CONTRIBUTIONS TO THE FLORA OF VENEZUELA by J. A. Steyermark and collaborators.

Numbers 1–4. Fieldiana: Botany.
Vol. **28(1):** 1–242. 1951.   Vol. **28(2):** 243–447. 1952.
Vol. **28(3):** 449–678. 1953.   Vol. **28(4):** 679–1190. 1957.

CONTRIBUCIONES A LA FLORA DE VENEZUELA by J. A. Steyermark.

Part 5 (Includes an index to parts 1–4). 1966. Published in Acta Bot. Venez. **1(3/4):** 9–256. In Spanish.

IMPORTANT FLORISTIC WORKS

Exploraciones Botánicas en la Guayana Venezolana, I. El Medio y Bajo Caura. Williams, Ll. 1942. Servicio Botánico, Minist. Agric. y Cría. Caracas.

Catálogo de la Flora Venezolana. Pittier, H., Lasser, T., Schnee, L., Luces de Febres, Z., and V. Badillo. 1945–1947 (2 Vols.). Tercera Conferencia Interamericana de Agricultura; Vargas. Caracas.

Lauráceas. Bernardi, L. 1962. ULA/Facultad de Ciencias Forestales. Mérida.

Flora de los Páramos. Vareschi, V. 1970. ULA/Ediciones del Rectorado. Mérida.

The Flora of the Meseta Del Cerro Jaua. Steyermark, J.A. & B. Maguire and collaborators. 1972. Mem New York Bot. Gard. **13:** 833–892.

Clave de las Familias de Plantas Superiores de Venezuela. Badillo, V. M. and L. Schnee. 1972. 5th edition (Revista de la Facultad de Agronomia de la UCV Maracay no. 18). 6th edition, 1983 (with C. Benitez de Rojas). Published in Ernstia **14:** 1–245.

Familias y Géneros de los Arboles de Venezuela. Aristeguieta, L. 1973. Inst. Bot., Minist. Agric. y Cría. Caracas.

Flora del Avila. Steyermark, J. A. & O. Huber. 1978. Sociedad Venezolana de Ciencias Naturales, MARNR and "Vollmer" Foundation. Caracas.

Flora de la Isla Margarita, Venezuela. Hoyos F., J. 1985. Sociedad y Fundación La Salle de Ciencias Naturales (Monogr. 34). Caracas.

IMPORTANT VENEZUELAN PERIODICALS

Acta Botanica Venezuelica. Caracas. Vol. 1+, 1965+.

Boletín de la Sociedad Venezolana de Ciencias Naturales. Caracas. Vol. 1+, 1931–1932+.

Memorias de la Sociedad de Ciencias Naturales La Salle. Caracas. Vol. 1+, 1941+.

Pittieria. Mérida. Vol. 1+, 1967+.

Ernstia. Maracay. Vol. 1+, 1981+.

IMPORTANT FOREIGN PERIODICALS

Memoirs of the New York Botanical Garden. New York, New York (U.S.A.). Vol. 1+, 1900+.

Fieldiana: Botany. Chicago, Illinois (U.S.A.). Vol. 24+, 1958+ (Preceded by: Publications of the Field Museum of Natural History. Botanical Series. Chicago, Illinois. Vols. 18–23, 1937/38–43/44. Preceded by Field Museum of Natural History. Botanical Series. Chicago, Illinois. Vols. 9–17, 1930/32–37/39, and further preceded by: Publications of the Field Columbian Museum. Botanical Series. Chicago, Illinois. Vols. 1–8, 1895/1902–30/32.).

Phytologia. New York, New York (U.S.A). Vol. 1+, 1933+.

Annals of the Missouri Botanical Garden. St. Louis, Missouri (U.S.A.). Vol. 1+, 1914+.

Kew Bulletin. Kew, England. Vol. 1+, 1946+ Volumation started with Vol. 13; preceded by Bulletin of Miscellaneous Information.

Brittonia. New York, New York (U.S.A.). Vol. 1+, 1933+.

Journal of the Arnold Arboretum. Cambridge, Mass. (U.S.A.). Vol. 1+, 1920+.

Bulletin of the Torrey Botanical Club. Lawrence, Kansas (U.S.A.). Vol. 1+, 1870+.

---

[a]XIII. **51:** 1–127. 1989. (Gentianaceae, *Ouratea* [Ochnaceae], Melastomataceae, Dioscoreaceae, Nyctaginaceae, Monimiaceae, Capparaceae, Cunoniaceae, Rhamnaceae, Rutaceae)

The Guianas

J. C. Lindeman and S. A. Mori

Contents

I.	Description of the Region	376
	A. Geographical Extent and Topography	376
	B. Geology	376
	C. Floristics	377
II.	Vegetation Types	377
	A. Seasonal Evergreen Forest	377
	B. Mangrove	378
	C. Strand Vegetation	378
	D. Marsh Forests	378
	E. Swamp Forests	379
	F. Herbaceous Swamps	379
	G. Savannas	379
	H. Montane Vegetation	379
	I. Inselbergs	380
III.	Botanical Inventory	380
	A. Guyana	381
	B. Surinam	381
	C. French Guiana	385
IV.	The Flora of the Guianas Project	387
V.	Vegetation Maps	387
VI.	Conservation	387
VII.	Conclusions	388
VIII.	Acknowledgments	388
IX.	Literature Cited	388

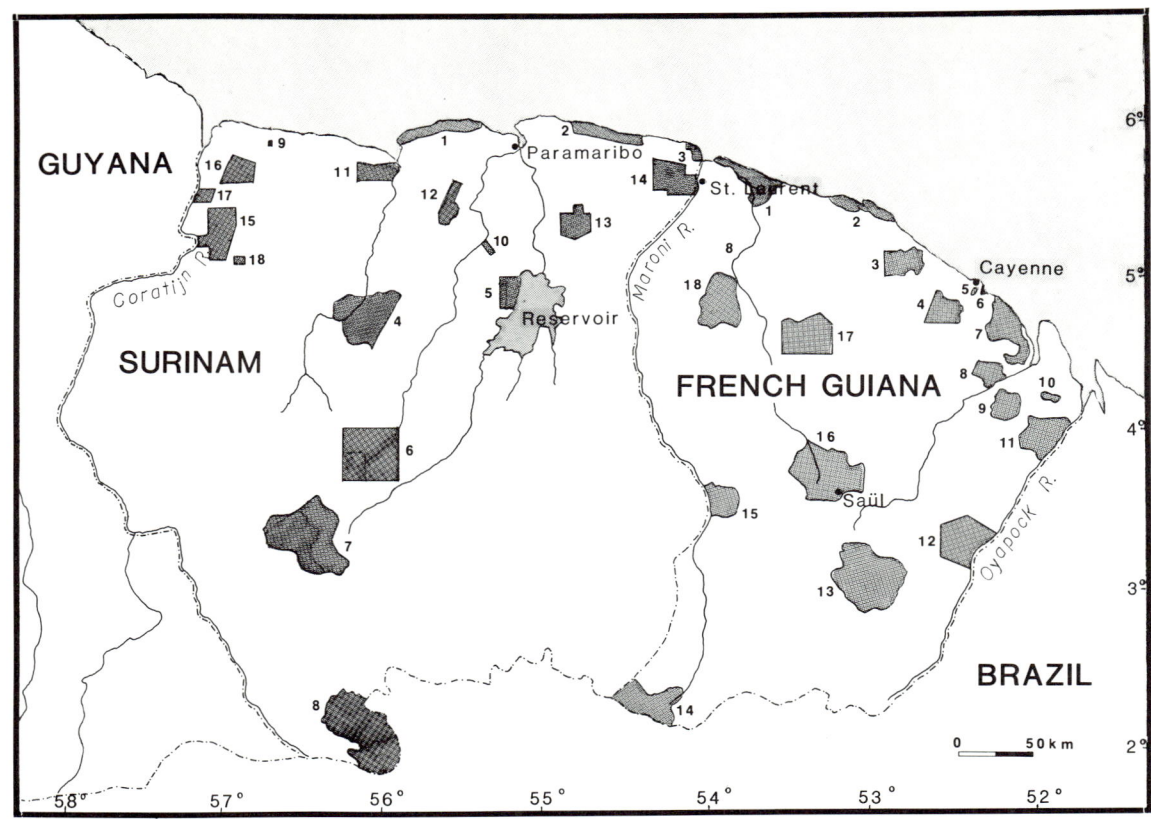

FIG. 1. Established and proposed nature and forest reserves in Surinam and proposed nature reserves in French Guiana. SURINAM. 1: Coppename (12,000 ha.), 2: Wia-Wia (36,000 ha.), 3: Galibi (4,000 ha.), 4: Voltzberg/Raleighvallen (77,000 ha.), 5: Brownsberg (8,400 ha.), 6: Tafelberg (140,000 ha.), 7: Eilerts de Haan (220,000 ha.), 8: Sipaliwini (220,000 ha.), 9: Hertenrits (100 ha.), 10: Brinckheuvel (6,000 ha.), 11: Peruvia (35,000 ha.), 12: Upper Coesewijne (27,000 ha.), 13: Copi (28,000 ha.), 14: Wanekreek (44,000 ha.), 15: Kaboericreek (proposed nature reserve of 67,000 ha.), 16: Nani (proposed nature reserve of 54,000 ha.), 17: Mac Clemen (proposed forest reserve of 6,000 ha.), 18: Snake Creek (proposed forest reserve of 4,000 ha.). FRENCH GUIANA. 1: Basse Mana (48,400 ha.), 2: Sinnamary (size not known), 3: Kourou (17,200 ha.), 4: Reserve des Cascades (5,400 ha.), 5: Reserves de La Mirande (166.4 ha.), 6: Reserve du Plateau du Mahury (800 ha.), 7: Plaine de Kaw (46,800 ha.), 8: Montagne Tortue (3,800 ha.), 9: Crique Mataroni (19,600 ha.), 10: Trois Pitons (3,600 ha.), 11: St. Georges (27,200 ha.), 12: Camopi (17,600 ha.), 13: Sommet Tabulaire (145,000 ha.), 14: Tumuc-Humac (90,800 ha.), 15: Monts Atachi Bacca (34,400 ha.), 16: Saül (133,600 ha.), 17: Montagnes de La Trinite (20,000 ha.), 18: Paul Isnard (56,400 ha.). Data from Schultz (1976), the Surinam Forest Service (pers. comm., 1986), and de Granville (1975).

I. Description of the Region

A. Geographical Extent and Topography

The political entities known as the Guianas, with the exception of Uruguay, are the smallest and least populated in South America. French Guiana, with 90,000 km² and 77,000 inhabitants, is an overseas department of France. Surinam (we will use the English spelling, without the terminal "e," in this paper), 165,942 km² in area and with 400,000 inhabitants, became an independent republic in 1975, and Guyana (215,000 km² and 865,000 inhabitants) had received its independence in 1966. The latter countries were previously known as Dutch Guiana and British Guiana respectively. Much of the population is located along the coast and is centered near the respective capital cities, Georgetown (Guyana), Paramaribo (Surinam), and Cayenne (French Guiana).

B. Geology

Guiana applies to that region of South America bounded by the Amazon, Orinoco, and Negro Rivers. Besides the three countries considered here, this region includes vast areas of Brazil and Venezuela. Included in Guiana is the crystalline Guayana Shield which, in turn, is partially covered by Roraima sandstones and the Tertiary formations of the major river basins and coastal areas. The Roraima sandstones appear as table top mountains called tepuis, which partially cover the crystalline shield. The tepuis, especially common in Venezuela, are found in the Guianas only in western Guyana (e.g., Ayanganna, Roraima, etc.) and in central Surinam (Tafelberg). A few isolated tepuis are also found in Brazil and Colombia.

The coastal region is made up of a complex of Holocene clays covered here and there by sand deposits (Janssen, 1976). Marsh forests, swamp forests, herbaceous swamps, strand vegetation, and mangroves prevail on the clays, whereas savannas dominate the sands. In Guyana, a number of low mountains in the coastal belt, now covered by seasonal evergreen forest and which harbor some endemic species, must have been islands during past sea transgressions. Numerous beach ridges present in the coastal region were deposited by alternating high and low sea levels.

C. Floristics

The Guayana Shield includes an area of about 1,000,000 square kilometers, and embraces approximately 8,000 species of vascular plants, 50% of which may be endemic (Maguire, 1970). Consequently, the Guayana Shield, including the highland elements of the tepuis and the surrounding lowland forests and savannas, is considered by most phytogeographers to be a distinct floristic province (Ducke & Black, 1954; Maguire, 1970; Mori & Prance, 1987; Steyermark, 1982).

The Guayana shield and the Brazilian shield to the south have been available to plant colonization for longer than any other part of South America (Maguire, 1970). Consequently, Guayanan plants have migrated into the more recent habitats of the Amazon Basin as the latter became free of epicontinental seas. The Guianas are a small but important part of the larger Guayanan floristic province, and together they are estimated to possess between 7,000 and 10,000 species of the flowering plants (Görts-van Rijn, pers. comm.; Smith, 1941).

Contrary to the situation in many other parts of the neotropics, the flora of the Guianas remains relatively intact. Development has taken place mainly along the coast, leaving vast areas of the interior untouched. Although only Surinam has a functioning system of national parks and biological reserves, it is still not too late to set aside representative areas of Guyana and French Guiana.

Taxonomic and ecological studies of the flora of the Guianas are also advanced in comparison with other neotropical countries. Studies began in 1775 with the publication of Aublet's "Histoire des Plantes de la Guiane Françoise" and continues today with the recently established "Flora of the Guianas" project.

Consequently, our review of the status of floristic and ecological inventories of the Guianas is more optimistic than that of other more threatened areas, such as the moist forests of the Pacific coasts of Ecuador and Colombia and those of eastern Brazil. Nevertheless, we would like to emphasize that the best time to locate, establish, and preserve natural areas is before habitats are threatened with destruction. The Guianas present a unique opportunity for rational habitat preservation before severe pressures of economic development arrive.

II. Vegetation Types

The vegetation of the Guianas is determined by the availability and the quality of water. For the most part, the quantity and availability of water throughout the year is sufficient to allow for the development of forests over most of the Guianas. Haman and Wood (1928) estimate that 86% of the total land area of Guyana is covered by forests, and estimates of forest cover for French Guiana range from 87.5% (Benoist, 1924) to 97.7% (de Granville, 1974; Sastre, 1980). Similar values are to be expected for Surinam.

Rainfall varies from 1800 mm in Guyana (Fanshawe, 1954) to 3750 mm per year in eastern French Guiana (de Granville, 1982; Smith, 1941). Throughout the Guianas, there is at least one, and often two, annual dry seasons. For example, in Surinam there are four seasons, a long dry from August through November, a short wet from December to January, a short dry from February to March, and a long wet from April through July (Schulz, 1960). These are the basic seasons, but they are subject to yearly and geographic modification. To the west in Guyana, the short dry season becomes more pronounced, and to the east, especially in eastern French Guiana, rainfall increases and the short dry season is less distinct or absent. The short wet season may be absent in some and the short dry season absent in other years, producing dry and wet years respectively. In Guyana there is a tendency for the dry years to come in 14 year cycles because of the failure of the rains of the short wet season (December to January). The resultant drought may lead to large-scale forest fires such as those known to have occurred in 1898, 1912, 1926, and 1940 (Fanshawe, 1954). However, 1954 was not dry, and a severe drought occurred over large parts of South America in 1964.

Because of the periodicity in rainfall, true tropical rain forest without dry seasons, as defined by Beard (1944), probably does not occur in the Guianas. Instead, most of the forests of the Guianas should be classified as seasonal evergreen forest (Beard, 1944). Rain forest is less widely spread in the neotropics than is generally supposed because continuously wet conditions are found only in parts of Central America, the Pacific coast of Colombia and Ecuador, western Amazonia at the foothills of the Andes, and in parts of the coastal forest of eastern Brazil.

Within this general climatic pattern, seasonal evergreen forest is replaced when soil conditions become either too dry or too wet, or when temperatures decrease with altitude. Examples are the brackish-watered mangrove swamps of the coast, the salt-sprayed strand vegetation, the periodically-inundated marsh forests, the permanently moist fresh-water swamp forests, herbaceous swamps, savannas, montane vegetation, and the vegetation of granitic outcrops (inselbergs).

A. Seasonal Evergreen Forest

Seasonal evergreen forests in the Guianas have been the subject of numerous and detailed studies. Davis (1929), Davis and Richards (1933, 1934), Fanshawe (1954), and Haman and Wood (1928) in Guyana; Lindeman (1953), Lindeman and Moolenaar (1959), Maas (1971), Sabatier (1985), and Schulz (1960) in Surinam; and Mori and Boom (1987), Oldeman (1974), Granville (1978), and the multidisciplinary team of scientists working out of ORSTOM's center in Cayenne (Anonymous, 1981, 1982) in French Guiana provide descriptions of this forest type. These studies demonstrate that structurally similar forests often differ greatly in species composition from one area to the next.

A typical seasonal evergreen forest is that of La Fumée Mountain in the vicinity of Saül, French Guiana studied by Mori and Boom (1987) and Oldeman (1974). This area consists of a relatively dissected topography ranging from 200 to 400 m alt. Here an average of 2413 mm of rain falls each year in two distinct seasons, a dry one from August through November and a wet season for the remainder of the year. The forest is characterized by a relatively high percentage of trees with: buttresses (34%); simple (58.5%), entire (94.1%), mesophyll (88.0%), mostly evergreen leaves; and latex, resin, or sap (42.7%). Vascular epiphytes are found on 51% and lianas on 42% of the trunks of all the trees. Size class distribution of the trees shows the reverse J-shape curve typical of most forests which, along with the relatively large average tree size (856.7 cm^2), suggests that this forest has not undergone major disturbance in the recent past. Three tree strata, emergents, canopy, and understory, are present. There are 619 trees ≥10 cm diam./hectare with a total basal area of 53 m^2 per hectare. The most common tree species is *Tetragastris altissima* (Aublet) Swart (Burseraceae) with an importance value of 17.9. The forest is dominated by relatively few species, as 8.8% of them account for 42.5% of the importance value. The five most important tree families in order of importance are: Leguminosae sensu lato, Burseraceae, Sapotaceae, Lecythidaceae, and Moraceae.

Within the evergreen seasonal forest environment, forest types can change over short distances. It is becoming increasingly apparent that some trees prefer ridge tops, others hillsides, and others valley bottoms (Kahn & Castro, 1985; Lescure & Boulet, 1985; Mitchell & Mori, 1987). For example, Davis and Richards (1934) have shown that associations of trees distribute themselves along a moisture gradient. In the Moraballi Creek area of Guyana, they recognized five different associations which they attributed to differences in moisture and organic content of the soil. Their Mora forest association occurs on highly organic soils that never dry out, whereas their Wallaba forest association is found on white sand soils that rapidly dry out in the dry season. The remaining associations (Morabukea, Mixed, and Greenheart) are intermediate in moisture and organic matter.

Studies of plant succession in seasonal wet forests after disturbance have been made in Surinam (Schulz, 1960) and French Guiana (Foresta et al., 1984; Prévost, 1984) and a classic study of plant-animal interactions in a secondary forest in French Guiana is provided by Charles-Dominique et al. (1981).

B. Mangrove

As in most tropical countries, a band of mangrove forest is found along the coast, especially in the estuaries of the larger rivers where they penetrate further inland. In Guyana, mangroves are well-developed along the Corentyne, Berbice, and Demerara Rivers whereas in Surinam they are widespread along the coast but do not penetrate up the Marowijne River. According to de Granville (1973), mangrove vegetation occupies 530 km^2 or only 0.6% of the surface area of French Guiana.

Because the mangrove environment is swampy and saline, it does not support a diverse flora. In Darién, Panama, Golley et al. (1975) found only 19 species in a quarter hectare of mangrove, whereas they encountered 399 species in a similar-sized adjacent moist forest.

The dominant species of mangrove in the Guianas are *Avicennia germinans* (L.) L. (Avicenniaceae), *Conocarpus erectus* L., *Laguncularia racemosa* Gaertn. (both Combretaceae), and *Rhizophora mangle* L. (Rhizophoraceae). *Rhizophora mangle* is most common along the coast and saline river banks, whereas its congener, *R. racemosa* G. F. W. Meyer, is found only along slightly saline and fresh water river banks. Where the two species meet, a probable hybrid between them, *R. harrisonii* Leechm., occurs. *Laguncularia racemosa* locally replaces species of *Rhizophora* in the estuaries, and it is there that mixed stands of mangroves develop. In other areas, mangroves tend to form pure stands. For example, in French Guiana, *A. germinans* often forms monotypic stands (de Granville, 1974). In the lower part of the tidal zone, and in front of the *Rhizophora* belt, pure stands of the spiny shrub *Machaerium lunatum* (L.f.) Ducke (Fabaceae) often develop, which are in turn replaced upstream by the giant aroid *Montrichardia arborescens* (L.) Schott (Araceae).

Structural properties of a number of neotropical mangrove forests are described by Cintron et al. (1985), and the botany of mangroves is discussed by Tomlinson (1985).

C. Strand Vegetation

The beaches of the Guianas are inhabitated by species found throughout Central and South America. Plants encountered closest to the sea are prostrate herbs adapted to high salinity and shifting sands such as *Canavalia maritima* (Aubl.) Thou. (Fabaceae), *Ipomoea pes-caprae* (L.) Sweet, *I. stolonifera* (Cyrill.) Poir. (Convolvulaceae), *Sesuvium portulacastrum* L. (Aizoaceae), *Sporobolus virginicus* (L.) Kunth (Poaceae). Above high tide level, shrubs and small trees such as *Chrysobalanus icaco* L. (Chrysobalanaceae), *Cordia macrostachya* (Jacq.) R. & S. (Boraginaceae), *Dalbergia ecastophyllum* (L.) Taub. (Fabaceae), and *Hibiscus tiliaceus* L. (Malvaceae) replace the prostrate herbs. Lindeman (1953) provides more detailed description of the strand vegetation of Surinam and Cremers (1986) provides keys, descriptions, and illustrations of the French Guianan strand species.

D. Marsh Forests

Following Beard (1944), we consider those forests of the Guianas that have soils waterlogged for part of the year and dry for the remainder of the year as marsh forests. In the Brazilian Amazon, this type of forest, associated with the rise and fall of the major rivers, is called *várzea*. In the Guianas, marsh forest may also be associated with large rivers or with the impedance of subsoil drainage due to the presence of unweathered rock or of ironpan or claypan (Beard, 1944). Marsh forests may be dominated by palms or by dicotyledonous trees. A typical palm-dominated marsh forest is that of *Mauritia flexuosa* L.f. which often borders rivers and streams penetrating savannas. In Surinam, Lindeman (1953) recognizes the following marsh forests dominated by dicotyledons: the *Triplaris surinamensis* Cham. (Polygonaceae)–*Bonafousia tetrastachya* (H.B.K.) Mgf. (Apocynaceae), the *Symphonia globulifera* L. f. (Clusiaceae), and the *Hura crepitans* L. (Euphorbiaceae) types. Profiles of marsh forest in Surinam have been prepared by Lindeman and Moolenaar (1959).

E. Swamp Forests

Following Beard (1944), we consider those forests of the Guianas that grow on soils that never completely dry out at any time of the year as swamp forests. However, we restrict our use of this term to fresh-water swamp forests and treat the saline swamp forests separately as mangroves. As Beard points out, longer periods of inundation lead to reduced stature, specialization of life form, and poverty of flora. Consequently, many swamp forests are dominated by single species. Good examples are the *Manicaria saccifera* Gaertn. palm swamps of northwestern Guyana (Davis, 1929) and the palm swamps of *Euterpe oleracea* Mart. found throughout Guayana and Amazonia. Mixed swamp wood, *Erythrina glauca* Willd. (Fabaceae) swamp wood, and *Machaerium lunatum* (L. f.) Ducke (Fabaceae) swamp wood are dicotyledonous dominated swamp forests described by Lindeman (1953). The *Mora excelsa* Benth. (Caesalpiniaceae) swamp forest of Moraballi Creek, Guyana has been studied in detail by Davis and Richards (1933). In Surinam, *Mora* swamp occurs only as far east as the Saramacca River. Profiles of swamp forest in Surinam have been prepared by Lindeman and Moolenaar (1959).

F. Herbaceous Swamps

Throughout the Guianas, vast swamps dominated by herbaceous plants are common along the coast. These swamps never, or very infrequently, dry out. They are often dominated by one or only a few species of plants. Important dominants are: *Cyperus articulatus* L., *C. giganteus* Vahl, *Eleocharis mutata* R. & S., *Lagenocarpus guianensis* Nees (all Cyperaceae), *Leersia hexandra* Sw. (Poaceae), *Rhynchospora corymbosa* (L.) Britton (Cyperaceae), and *Typha dominguensis* Pers. (Typhaceae). Lindeman (1953) points out that the herbaceous swamps of Surinam are often sharply delimited one from the other yet grow, as far as he could tell, under precisely the same ecological conditions. Some of the herbaceous swamps of western Surinam have been completely replaced by highly productive rice fields, however. De Granville (1976) has described the herbaceous swamps found in the vicinity of Sarcelle, French Guiana. His profiles give a good idea of the structure and the composition of the herbaceous swamps of the Guianas.

G. Savannas

We consider all those areas with a herbaceous stratum dominated mostly by Cyperaceae, Eriocaulaceae, Poaceae, and Xyridaceae and a woody stratum of scattered shrubs and trees as savannas. Quite frequently, islands of varying sizes, dominated by woody plants which form a closed canopy, are found scattered throughout the savannas. These islands are especially common at the transition of savanna to forest environments. In Surinam, the forest transition from savanna to seasonal evergreen forest is usually referred to as savanna forest (Lindeman, 1953). The transition from savanna to forest has been illustrated in profile by Lindeman and Moolenaar (1959).

The savannas of the Guianas may occur naturally as is the case of those caused by the alternation of waterlogged with dry soils. It is difficult for shrubs and trees to adapt to either of these conditions, let alone both of them at different times of the year. Consequently, these kinds of savannas are usually dominated by fewer species of woody plants than are found in the savannas of Central Brazil (*cerrados*). Other savannas have been formed, or at least expanded, by fires set by man. The Rupununi savannas of southwestern Guyana have been highly modified by grazing and by fires set to promote the growth of pasture grasses. These savannas have been expanded somewhat into the once forested slopes of the Kanuka Mountains.

From palynological studies, we know that savannas in the Guianas were much more extensive during the ice ages when the lowlands of South America had a much drier climate (Prance, 1982; Van der Hammen, 1974). Many of the present savannas of Surinam, if maintained free from fire, develop into closed scrub or woodland under present climatic conditions (Lindeman, unpublished data).

The savannas of Surinam have been the object of careful ecological study. Heyligers (1963) has described the white-sand savannas, Donselaar (1965, 1968) has done an ecological and phytogeographic study of the northern and southern savannas, Kramer and Donselaar (1968) have described Kappel Savanna, Donselaar-Ten Bokkel Huinink (1966) has described the structure, root systems, and periodicity of the northern savannas, and Teunissen and Wildschut (1970) have studied the savanna of the Brinckheuvel Nature Reserve. In French Guiana, Hoock (1971) described the savanna of Kourou, de Granville (1982) has mapped the location of the savannas in the Department, and Cremers (1982) has described and illustrated the plants of the Bordelaise savanna. General observations on the savannas of the Guianas have been contributed by Donselaar (1969).

H. Montane Vegetation

As Beard (1944) points out, vegetation changes progressively with higher altitudes. Decrease in temperature, increase in the amount of ultra-violet light, increase in humidity, increased exposure to wind, and less stable soils determine the structure and composition of montane vegetation. In general, in montane forests, tree stature is decreased, tree species diversity falls, leaves of trees become smaller, and cryptogams become more abundant.

The highest mountains of the Guianas are found in western Guyana with Roraima at 2772 m, Wokumung at 2134 m, and Ayanganna at 2042 m the highest peaks. These sandstone capped tepuis are among the remnants of an old peneplain that once stretched from Colombia over southern Venezuela,

Guyana, Surinam, and northern Brazil. The western Kanuku Mountains and some other peaks in southwestern Guyana reach 1000 m. In Surinam, mountains representing the Guayana crystalline shield, with peaks over 800 meters, are found in Bakkuis Gebergte (highest point 1000 m), Emma Keten (Hendrick top, 975 m), Eilerts de Haan Gebergte (884 m), Kayser Gebergte (861 m), Tafelberg with a sandstone cap (1026 m), and Wilhelmina Gebergte (Juliana Top at 1230 m). These mountains are located in western and south central Surinam. In the eastern part of the country, the lateritic capped Nassau Gebergte (564 m) and Lely Gebergte (710 m) are the highest mountains, and in the southeast the granitic Grens Gebergte (864 m) and Toemoek Hoemak Gebergte (728 m) have the highest peaks. In contrast, French Guiana has no mountain peaks above 1000 meters; the highest is Sommet Tabulaire (830 m) in the south central part of the Department. Some other peaks in west central French Guiana reach 700–800 m.

De Granville (1974) suggests that the vegetation of French Guiana begins to change from lowland to montane at about 500 m. The highest peaks of French Guiana are probably covered by montane rain forest as defined by Beard (1944). At higher elevations, especially in Surinam and Guyana, elfin woodland (Beard, 1944; Smith, 1941), often dominated by species of *Clusia*, prevails. The vegetation of the upper reaches of Tafelberg, the easternmost sandstone tepui, is a mosaic of elfin woodland, montane rain forest, savanna on rocky soils, and grass-sedge savanna (Maguire and collaborators, 1948).

I. Inselbergs

Throughout much of Surinam and the southern part of French Guiana, granitic outcrops occasionally project above the seasonal evergreen forest. A discussion of the formation and geological composition of these outcrops, or inselbergs, has been provided by Hurault (1974). Inselbergs possess a unique vegetation that differs in species composition from that of the surrounding forest. Their vegetation and that of the surrounding forests has been described in detail by de Granville and Sastre (1974) and de Granville (1978). The phytogeography of the plants inhabiting them is discussed by Sastre and de Granville (1975).

For the most part, the exposed rock possesses a xerophytic vegetation dominated by species of Bromeliaceae, Orchidaceae, and Poaceae. For example, de Granville (1978) reports that *Ischaemum guianense* Kunth (Poaceae) and *Pitcairnia geyskesii* L. B. Smith (Bromeliaceae) are dominant on the inselbergs of the Tumuc-Humac Mountains. Islands of woody plants 1.5 to 4 m tall are found in depressions where organic matter has built up enough to allow their establishment. The principal shrubs are species of *Clusia* (Clusiaceae), Melastomataceae, and Myrtaceae. The inselbergs themselves are usually surrounded by a low forest which grades into the surrounding seasonal evergreen forest.

III. Botanical Inventory

The purpose of botanical inventory is to enumerate all of the plants occurring in a given geographic area, documented by collections conserved in herbaria. Such collections, identified by the collector(s) name(s) and a unique number serve as vouchers for a given species, and provide the data for the ultimate product, a complete Flora of the region under question. Complete means the inclusion of all species present, with descriptions of all phases of their life cycle, morphological variation, ecology, uses, and distribution. Consequently, a single collection of each species is not sufficient to provide the data needed for a Flora. Because of the diversity of tropical floras, it is difficult to know when enough collections have been accumulated to justify the writing of a Flora. In fact, neotropical floristic projects are often started when not all of the species found in an area are vouchered by collections, resulting in a flora out-of-date at the time of publication. Mori (Chapter on Eastern, Extra-Amazonian Brazil) discusses this problem in more detail.

As part of the Flora of the Guianas project, discussed later, a detailed history of botanical exploration in the Guianas is being prepared by F. E. Vermeulen. The assembling of detailed itineraries and pertinent data on past and present collectors and their collections in the Guianas is intended to facilitate the preparation of revisions of plant families for the Flora of the Guianas by providing such data to the specialists preparing the treatments. The results of her work will be published as a special volume in the Flora of the Guianas, and therefore, a detailed history of botanical exploration in the Guianas will not be attempted here. Instead, we will provide a brief review of collecting in the Guianas and a discussion of the published taxonomic research based on these collections.

The first major effort in the study of the plants of the Guianas was the publication of Aublet's *Histoire des Plantes de la Guiane Françoise* (1775). Aublet's work was based on his own collections gathered from 1762 to 1764. Froidevaux (1897) provides a useful summary of Aublet's sojourn in French Guiana. The *Histoire* includes descriptions of 576 genera, 208 of which were described for the first time (Zarucchi, 1984). Howard (1983) and Zarucchi (1984) summarize the current status of Aublet's genera and species. The former points out the difficulties in typification of Aublet's species caused by many of the descriptions being based on mixed collections or due to disagreement between Aublet's text and illustrations. Nevertheless, he emphasizes that Aublet's names are valid, and therefore they must be considered when seeking the correct name for many of the plants of lowland South America, especially for those distributed in the Guianas and northeastern Amazonia. Slightly later in the same year *Plantae Surinamenses* was published by Linnaeus (1775). This work was based on the collections of Frédérique Allamand and Daniel Rolander who collected in Surinam from about 1755 to 1770 (Pennell, 1945).

Botanical collections from the Guianas continued to reach the world's herbaria through the collections of Robert and Richard Schomburgk, F. W. Hostmann, A. Kappler, P. A. Poiteau, L. C. M. Richard, P. A. Sagot, J. Martin, and others (Maguire, 1958; Pennell, 1945). A list of herbaria with significant collections from the Guianas is presented in Table I. However, it wasn't until Forest Services were founded in each of the three colonies that floristic and vegetational studies intensified. The forest services of the then British, Dutch, and French Guianas, in their efforts to obtain scientific names for the trees they wanted to exploit, provided the world's taxonomists with excellent herbarium collections.

A. GUYANA

Important early collections from Guyana are those of Robert (1835–1843 in Guyana) and Richard (1840–1844 in Guyana) Schomburgk. Their collections of nearly 2500 numbers were studied for the most part by Bentham at Kew. The collections of C. F. Appun, E. F. im Thurn, F. V. McConnell, and J. S. Quelch, mostly from the vicinity of Roraima, were also sent to Kew (Maguire, 1958).

Formal forestry began in Guiana in 1887 with Crown Lands Ordinance No. 18 which placed the administration of the Colony's forests under the Department of Lands and Mines (Welch, 1975). Collections with the Forest Department's number series (FD series) began to appear at least as early as 1909, but the Department's herbarium (FDG) was not formally established until 1926 (Holmgren et al., 1981). This was the second herbarium in the country, since G. S. Jenman (in Guyana from 1879 until his death in 1902), the government's botanist, had already started one (BRG) in 1879 (Holmgren et al., 1981) at the Department of Agriculture as part of the Georgetown Botanic Gardens. In 1937 E. Martyn, then government botanist, made an unpublished list of the Jenman herbarium. Few copies of the list were distributed, one of which is archived at U. Jenman collected more than 7000 numbers during his stay in Guyana. R. A. Alston, assistant government botanist and mycologist from 1923 to 1927, also contributed to the Jenman Herbarium. From October, 1919 to January, 1920, A. S. Hitchcock made general collections of 1250 numbers. His collections, which include many grasses, are deposited at NY and US (Maguire, 1958). A collection rich in new taxa, and now deposited at BM, was made by T. G. Tutin in 1932.

Botanical studies, especially in regard to trees and classification of vegetation, were promoted by the Forest Department. The efforts of T. A. W. Davis (1925–1939 in Guyana) and D. B. Fanshawe (1937–1952 in Guyana) were noteworthy (Welch, 1975). Not only did they provide critical botanical collections, but they also published important papers on the classification of the colony's vegetation (Davis & Richards, 1933, 1934; Fanshawe, 1954). During this time, N. Y. Sandwith of the Royal Botanic Gardens Kew, played an important role in the identification of Guyana plants, and he also described many new species from there. Although there is no flora of the country, Meyer (1818), Schomburgk (1848) and Fanshawe (1949) have published checklists. The latter includes an index to common names and provides some distributional data. Fanshawe's field books, duplicate collections, and wood duplicates are preserved in good condition at the Forest Service Headquarters in Georgetown (B. Ter Welle, pers. comm.). Therefore, even though the checklist does not give vouchers, the collections representing the species in the checklist can be obtained from these materials. A very useful flora of the Kartabo region, including keys to families, genera, and species has been published by Graham (1934). Treatments of melastomes (Gleason, 1932), palms (Im Thurn, 1884), species of Lycopodiaceae and allies (Jenman, 1886), and orchids (Rodway, 1894) of Guyana have also been published.

The New York Botanical Garden became interested in the flora of Guyana with the expedition of H.A. Gleason to the Potaro River basin where he collected 940 numbers in 1921. Gleason contracted J. S. De la Cruz, who independently gathered another 3000 numbers. There is some indication that Gleason produced an unpublished flora of Guyana but we have not yet been able to locate it (M. I. Hakki, pers. comm.). In 1937–1938, A. C. Smith gathered 1600 numbers as a member of the Terry–Holden expedition to the Kanuku and Akarai Mountains (Maguire, 1958), and in 1944 Maguire, in the company of Fanshawe, collected on the Kaieteur Plateau (Maguire and collaborators, 1948). Maguire made subsequent expeditions to Guyana in 1951–1952 with Fanshawe, in 1955 with R. S. Cowan, in 1955 with W. M. C. Bagshaw and C. K. Maguire (Maguire, 1958), and in 1960 with S. S. and L. L. Tillet, C. K. Maguire, and R. Boyan. Other collectors from the New York Botanical Garden that have visited Guyana are: Nicholas Guppy (1952, Akarai Mountains), Frère Wilson-Brown (1948, Kanuku Mountains), S. S. Tillet (1960, Kaieteur Plateau), and Scott Mori (1976, Bartica-Potaro Road and Northwest District). John Pipoly, employed by the New York Botanical Garden and the Smithsonian Institution, collected in Guyana from April 1986 to to March 1987. Brian Boom, starting in 1987, has also begun a series of botanical expeditions to Guyana.

With the onset of the Flora of the Guianas project in 1983, several recent expeditions have visited Guyana. Three took place in 1985, one led by Marion Jansen Jacobs (U) to the Kanuku Mountains, another to Mount Roraima led by Rob S. Gradstein (U), and another led by D. B. Lellinger and H. Robinson of US to Kamarang.

In spite of its rich botanical history, there is still much collecting to be done before the plants of Guyana will be known well enough to be adequately represented in the *Flora of the Guianas*. The herbaria of BRG and FDG contain only 25,000 and 8,017 collections, respectively. The latter also has a wood collection of 2,701 specimens (Holmgren et al., 1981). Moreover, in recent times both herbaria have been subject to insect damage indirectly caused by lack of funds for curation (Mori, pers. obs., 1976; Skog, pers. comm., 1985). Fortunately, many type specimens were originally deposited at K, and those of BRG were transferred to K in 1952 by Fanshawe and in 1959 by S. G. Harrison.

B. SURINAM

The botany of Surinam has been intensively studied and published upon. There is a modern flora, and numerous studies of its vegetation have been prepared. Efforts of the University of Utrecht and the Forest Service of Surinam have

Table I
Principle collections of Guianan plants in the world's herbaria. Data and acronyms from Holmgren et al., 1981 unless otherwise indicated.

BBS (Surinam Forest Service)

Now incorporated in the National Herbarium of the University of Surinam. Holds first set in BBS, LBB, and UVS series (see text). The herbarium of the Landbouwproefstation and that of CELOS were incorporated into this herbarium between 1983 and 1985. Includes duplicates of all collections made in Surinam since 1977 as well as duplicates of most post World War II collections. This herbarium has a nearly complete representation of all of the species of plants known from Surinam, including a significant collection of Orchidaceae (700 specimens).

BM (British Museum, Carruthers, 1904)

A. Anderson 1791, (Demerara); Appun (purchased 1872, Demerara); Aublet (purchased 1773-1778); Hostmann (purchased 1842-1843, Surinam); Jenman (presented 1887-1890, Guyana); Rothery (presented 1845, French Guiana); Martin (presented 1847, French Guiana in Rudge Herb.); Sagot (purchased 1868, French Guiana); R. H. Schomburgk (1836-?, Guyana); im Thurn (presented 1885, Roraima).

BR (Meise, Belgium)

First set of Wullschlaegel from Surinam (collected from 1849-1855 and numbered from 1-2052).

BRG (University of Guyana)

Contains collections of Jenman which are in part duplicated at K.

C (Copenhagen)

Holds Rolander (pupil of Linnaeus) collections and manuscripts and unknown number of duplicates from the Guianas.

CAY (ORSTOM Herb. Cayenne, Cremers, 1984, de Granville, pers. comm.)

Contains collections since 1965 from French Guiana. Has about 60,000 specimens of about 3000 taxa of phanerogams, 300 of pteridophytes, and 200 bryophytes. Most significant collections are: P. Béna, R. A. A. Oldeman (7500), J. J. de Granville (10,000), G. Cremers (5500), M.-F. Prévost (2500), C. Feuillet (3000), P. Grenand (2500), D. Sabatier (1500).

E (Edinburgh, Hedge & Lamond, 1970)

Holds duplicates from the Guianas of the collections of W. H. Campbell; Essed; Focke; Hostmann; Kappler; Leprieur; J. Martin; Mélinon; C. S. Parker; Sagot; Schomburgk; Wachenheim; and fern collections of Jenman.

F (Field Museum at Chicago)

Holds an undetermined number of duplicates from the Guianas.

FDG (Guyana Forestry Commission)

Holds collections of W. M. C. Bagshawe; J. Boyan; R. Boyan; T. A. W. Davis; D. B. Fanshawe; J. C. Fredericks; N. G. L. Guppy; K. King; B. Maguire; C. A. Persaud; N. Y. Sandwith; I. A. Welch.

GENT (Gent)

Holds an unknown number of duplicates from the Guianas.

GOET (Göttingen)

Holds first set of Kegel from Surinam, collected from 1844-1846 and numbered 1-1500; G. F. W. Meyer.

K (Kew)

Holds many important collections from the Guianas, especially from Guyana. Most important are: Abbensetts (1931, Roraima, ca. 65 collections); Abraham; Alston (including fungi, ca. 450 collections); C. W. Anderson; Appun (Ca. 900 collections); Archer (ca. 195 collections); Bartlett (duplicates from Jenman Herb., ca. 130 collections); Beckett (duplicates from Jenman Herb.); Boughton (received 1848-1856); Boz; Campbell (received 1871); Cartwright (received 1853-1855); T. A. W. Davis (in Forest Department); Fanshawe (in Forest Department); Forest Department (ca. 8,000 collections); von Goebel (1890-1891); V. Graham (ca. 500 collections); N. Guppy (in Forest Department); Hancock (1804-1810); Harrison (1924-?); Hitchcock (1919-1920); Hohenkerk (ca. 1919); Hostmann (collections from Surinam); Jenman (1879-1902); Kappler ed. Hohenacker (collections from Surinam); Leprieur (72 fern collections from French Guiana); Lockie; Martyn; McConnell & Quelch (original set from Roraima); Myers (1936); C. S. Parker (1869, original set); G. W. Parker (1904); W. Parker (1821-1824); C. A. Persaud (in Forest Department); Quelch (1895); Rothery (collections from French Guiana); Sagot (collections from French Guiana); Sandwith (1937); Richard Schomburgk; Robert Schomburgk; im Thurn; Tutin, (1908-?); Waby (in Forest Department); Ward (in Forest Department); Wilson-Browne (in Forest Department).

Table I

(continued)

L (Leiden)

Holds first sets of Samuels (1916), nos. 1–550, Splitgerber (1837–1838), nos. 1–1228 and Tulleken (1900), nos. 1–598 from Surinam and a fair representation of old collections from Surinam (e.g., Focke, Hostmann, and Kappler). Also has some duplicates of collections from Guyana and French Guiana.

MO (Missouri Botanical Garden)

Holds an undetermined number of duplicates from the Guianas.

NY (New York Botanical Garden)

Holds first set of collections of Maguire and collaborators (ca. 11,500 collections); A. C. Smith (ca. 2500 collections); H. A. Gleason (940 collections); S. A. Mori and collaborators (ca. 2769 collections); J. J. Pipoly (ca. 4,400 collections). Numerous other collections such as duplicates of those of the forest services of the Guianas and important historical collections are also deposited at NY. The total number of collections at NY from the Guianas is unknown.

OXF (Oxford; Clokey, 1964)

Holds 1840 collections of Hostmann from Surinam and about 450 collections from R. H. Schomburgk's expedition to Guyana in 1835–1839.

P (Paris)

Holds first set of many old collections from French Guiana and the collections of the Forest Service (BAFOG) and second set of all ORSTOM collections. Total number of specimens from the Guianas is estimated to be near 100,000.

U (Utrecht)

Contains over 100,000 collections from Surinam. Flora of Suriname produced at U. Holds Hermann 18th century collection from Surinam and has good sets of Focke, Hostmann, Kappler and some duplicates of other 19th century collectors. Many collections of historical interest as well as recent collections from Guyana and French Guiana are also at U. A chronological listing of significant sets of plant collections at U follows: Went (1900), nos. 1–575; H. A. Boon (not E. Boon as cited in some Flora Neotropica monographs) (1901), nos. 1001–1009, Paramaribo, 1010–1260, Coppename River, with field labels and numbering of Tulleken, see L in this list; Pulle (1902), Saramacca Exp. nos. 1–232, 401–500 (first series); G. M. Versteeg (1903–1904), Marowijne Exp. nos. 1–1004; Boschbeheer (1905–1908), nos. 1–90, some numbered by van Hall, others by van Asbeck, others by van Niel, individuals with same venacular name were sometimes given same number separated by letters if they later were found to be different species; J. Tresling (1908), Suriname River, nos. 1–502; Boschwezen (BW) (1909–1926), nos. 1–7238; I. Boldingh (1909), nos. 3801–3918, near Paramaribo in savannas; Collector Indigenous (coll. ind.) (1910), first series nos. 81–320, Apr-May in savannas, second series nos. 1–220, Oct near Paramaribo; J. F. Hulk (1910–1911), nos. 10–100, 201–420; J. Kuyper (1911–1912), first series nos. 1–114, Oct at Savanne Republiek, second series nos. 1–40, Dec near Paramaribo and nos. 65–90 (Jul 1912) at Zanderij I, nos. 502–565 with Tresling; Soeprato, also may be spelled Soeprata or Soeprapta (Jun-Aug 1913), nos. 1–50 preceded by A, B, C, D, E, F, G, H or J, and nos. 1–378 are not preceded by a letter, collected near Paramaribo and in plantations along lower Suriname and Commewijne Rivers; E. Essed (1914), nos. 1–153, 186, mostly grasses and sedges; G. Stahel & J. W. Gronggrijp, Emma Keten Exp. (Mar 1922), nos. 8–159, remaining collections numbered in the BW series between 5600 and 5950; Stahel, Wilhemina Exp. (1926), total nos. 1–625, nos. 1–283, Jan-Mar, upper Suriname River to Goddo to upper Gran Rio, nos. 284–300, Apr, upper Lucie River in mountains, nos. above 282 almost always double numbered in the BW series; van Emden 1930, unnumbered, Brownsberg; J. Lanjouw (1933), nos. 1–1355; H. E. Rombouts (1935–1938), nos. 1–50, near Paramaribo, nos. 51–192, upper Corantijne and Coeroeni Rivers, nos. 193–563, Sipaliwini Savanna, nos. 601–680, upper Tapanahoni River, nos. 700–929, Litani River; D. C. Geiskes (1939), nos. 1–99, Litani River, second set sent via NY; Stahel wood collections (1942–1945), nos. 1–380, some numbers include specimens of same species from different trees, distinguished by letters if they later were found to be different species; D. C. Geijskes (1943–1944), nos. 950–1046, upper Coppename River, second set sent via NY, continuation of Rombouts numbering; B. Maguire (1944), nos. 23738–24952, Tafelberg, second set; B. Maguire & G. Stahel (1944), nos. 22707–22806, near Paramaribo, nos. 23583–23737, nos. 24953–25072, Zanderij, all second sets; Bosbeheer Suriname (BBS) made by field men (1947–1950), nos. 1–346, 1000–1201, 1300–1302, 1400–1413, 3001–3041, special collections for research on potential paper pulp trees nos. V1–V64, US1–US14; J. Lanjouw & J. C. Lindeman (1948–1949), nos. 101–3475, coastal region, Nassau Mts. and central savannas, J. P. Schulz (to 1958), nos. to 8079; D. C. Geijskes (1950–1953), nos. 1–195, second set; P.A. Florschütz & J. Forschütz-de Waard (1950–1951), nos. 121–2293, many Bryophytes and other cryptogams; Lands Bosbeheer (LBB) began in 1950 but LBB only preceded numbers after J. P. Schulz left, starting with 8080 on to 16479, many students and visiting botanists collected under LBB series, after transfer of herbarium to University of Suriname in 1980, the code, starting with 16480, was changed to UVS; J. G. Dirven (1952–1960), nos. 223–743, second set; J. C. Lindeman (1953–1954), nos. 3501–7129, continuation of Lanjouw & Lindeman; R. S. Cowan (1954), nos. ca. 38960–39019, near Moengo, in Maguire number series, second set; R. S. Cowan & J. C. Lindeman (1954–1955), nos. 39030–ca. 39322, Nassau Mts., in Maguire number series, second set; A. M. W. Mennega (1954), nos. 1–585; B. Maguire (1955), nos. 40702–ca. 40836, second set; J. P. Schulz (1955–1957), nos. 7130–8072, continuing Lindeman nos.; F. P. Jonker & A. M. E. Jonker-Verhoef (1955–1956), nos. 1–684, including many Araceae; P. C. Heyligers (1956–1957), nos. 1–862, collections from savannas; J. van Donselaar &

Table I
(continued)

W. A. F. van Donselaar-ten Bokkel Huinink (1958–1959), nos. 35–743, C4–18; F. P. Jonker & A. G. H. Daniels (1959), nos. 711–1353, upper Saramacca River and Emma Range, continuing Jonker & Jonker-Verhoef numbering but erroneously labelled Daniels & Jonker; K. U. Kramer & W. H. A. Hekking (1960–1961), nos. 2021–3329; W. H. A. Hekking (1961–1962), nos. 750–1240; T. W. Reyenga (1962–1964), nos. 1–976, swamps of coastal plain, part of Molusca project; P. A. Florschütz & P. J. M. Maas (1964–1965), nos. 2300–3161, continuing Florschütz & Florschütz-de Waard numbering; P. J. M. Maas (1965), nos. 3162–3395, continuation of Florschütz & Maas numbering; P. J. M. Maas & J. Tawjoeran (1965), nos. LBB 10689–11068; J. van Donselaar (1964–1966, 1969), nos. 1000–3880, collected in area later flooded by Brokopondo Lake; J. G. Wessels Boer (1962–1963), nos. 151–1260, collected mostly palms; F. H. F. Oldenburger, R. Norde & J. P. Schulz (1968–1969), nos. 1–986, Sipaliwini Savanna; F. H. F. Oldenburger & J. P. Schulz (1970), nos. 987–1416; P. J. M. Maas (1971), nos. 531–566; R. W. den Outer (1974), nos. 850–987, second set; P. J. M. Maas, A. M. W. Mennega & J. Koek-Noorman (1974), nos. 2291–2365, Afobaka-Brownsberg; P. J. M. Maas & P. A. Teunissen (1974), nos. 2366–2371, collections of Musaceae; A. M. W. Mennega & J. Koek-Noorman (1974), nos. 870–913, continuing Mennega nos.; J. C. Lindeman, A. L. Stoffers, A. R. A. Görts-van Rijn, & M. Jansen-Jacobs (1975), nos. 1–842, Lely Mts.; N. M. Heyde & J. C. Lindeman (1976), nos. 1–347, many lianas; N. M. Heyde (1976–1977), nos. 349–750, many lianas; J. C. Lindeman, E. A. Mennega & students (1977), nos. 1–232; R. Tjon Lim Sang & I. H. M. van de Wiel (1975–1976), nos. 1–227, Brownsberg; P. A. Florschütz & J. Florschütz-de Waard (1975–1976), nos. 4513–4880, Bryophytes; P. A. Florschütz & J. Florschütz-de Waard (1978), nos. 4881–4938, Bryophytes; J. C. Lindeman, A. R. A. Görts-van Rijn, M. J. Jansen-Jacobs & A. M. G. Hetterscheid-Hollants (1980), nos. 1–685, area of Kabalebo dam project, special numbering system started for this project; J. C. Lindeman & A. C. de Roon (1981), nos. 686–930, Kabalebo project numbering; J. Florschütz-de Waard (1981), nos. 5000–5803, mosses for Kabalebo project; J. Bekker (1981), ca. 700 hepatics for Kabalebo project; R. Zielman (1981), ca. 120 lichens for Kabalebo project.

US (Smithsonian Institution)

Holds either first or second set of J. J. Pipoly who began collecting in Guyana in April 1986 with collection no. 7303 and ended his collections in March 1987 with no. 11792. Pipoly's collections are also deposited at NY and in the herbaria of the other members of the Flora of the Guianas Consortium.

W (Wien)

Holds an unknown number of duplicates from the Guianas.

WAG (Wageningen)

Holds modern collections of students who did graduate research in Surinam as well as duplicates of some older collections.

---

played dominant roles which will be emphasized in this review.

Important early collections in Surinam were those of C. G. Dahlberg (1746–1781, collections at LINN, S, UPS), F. L. Allamand (1755–1770, collections at LINN), D. Rolander (1755–1756, first set of collections at C), and F. W. R. Hostmann and A. Kappler who collected separately and together from 1841–1863. Rudolf Friedrich Hohenacher, German missionary, physician, and botanist who sold his and others' collections, widely distributed many of their collections. In addition, H. C. Focke (1835–1850, collections at U), H. A. H. Kegel (1844–1846, collections at BR, GOET, U), F. L. Splitgerber (1837–1838, collections at L), C. Weigelt (1827–1828, collections at BR), and H. R. Wullschlaegel (1849–1855, collections at BR) made significant early collections in Surinam (Maguire, 1958; F. W. Vermeulen, pers. comm.).

One of the most important expeditions to Surinam was A. A. Pulle's (University of Utrecht) in 1902. His interest in Surinam was stimulated by his teacher F. A. F. C. Went, the plant physiologist, who made a collecting trip there in 1900. This expedition served as the impetus for future field work and publications on the plants of Surinam. He visited again in 1920, and his visits have been followed by numerous expeditions by his students and successive generations of their students. Especially important are the expeditions of the late J. Lanjouw who collected alone in 1933 and with J. C. Lindeman in 1948–1949. The latter also visited Surinam from 1953 to 1955 and periodically from 1967 to 1981. Others, such as J. P. Schulz, P. A. Florschütz and his wife (mosses), V. F. P. Jonker and his wife, and J. van Donselaar and his wife have made significant collections of the plants of Surinam. More recently, P. J. M. Maas and his students, as well as many others from U have made important expeditions to Surinam. We estimate the number of collections from the Guianas now at U to be well over 100,000.

A very important outlet for the publication of the results of the expeditions of Utrecht botanists has been the "Medeedelingen van het Botanisch Museum en Herbarium van de Rijks Universiteit te Utrecht." From 1932 to 1984, 543 articles relating to Surinamese plants have been published in this journal. A list of these publications is available from the senior author.

The efforts of the Surinam Forest Service have been extremely important in the botanical inventory of Surinam. The herbarium of the original Forest Service ("Boschwezen") contains material collected between 1904 and 1927, which from the end of 1910 on were numbered BW 1 to BW 7238. However, this Forest Service was abolished in 1928, and after World War II the BW collections were shipped to Utrecht. In 1946, a new Forest Service ("Bosbeheer Suriname" or BBS) was established, but it's name was later changed to

"Lands Bosbeheer" (LBB). After the new Forest Service had acquired a permanent building with a separate room for a herbarium, a limited set of BW duplicates were returned to Surinam. In 1980, it was transferred to the University of Surinam and made the national Herbarium. Collections since 1981, beginning in the 16000 series are identified by UVS (the Surinamese abbreviation for the University herbarium), but the International code BBS was retained for the herbarium (Holmgren et al., 1981). Starting in 1983, the herbarium of the Agricultural Experimental Station and the herbarium of CELOS (Center for Experimental Agricultural Research in Surinam) were incorporated in the National Herbarium. The Celos herbarium includes a weed collection of 1500 numbers made by A. Everaarts. Gerold Stahel, who was the director of the Agricultural Experiment Station from 1914–1939, and the Station's entomologist, D. C. Geijskes (1938–1965 in Surinam), were largely responsible for the development of this herbarium. Stahel also made an important collection of the woods of Surinam. The National Herbarium now has a collection of 20,000 herbarium specimens and a collection of 8,000 wood samples.

A significant contribution to the botany of Surinam has been that of M. C. M. Werkhoven, formerly M. Teunissen-Werkhoven, who, since 1970, has served as curator of the herbarium of LBB and now of the National Herbarium. This herbarium, because of its careful curation and well identified collection, has served as the basis for the numerous modern studies of the taxonomy and vegetation of the country. Most of its collections are duplicated at U.

After Linnaeus' *Plantae Surinamenses* (1775), the next important contribution towards a knowledge of the flora of Surinam was Miquel's *Stirpes Surinamenses Selectae* (1850). However, it was Pulle, with his *An Enumeration of the Vascular Plants known from Suriname* (1906), *Zakflora voor Suriname* (1911), and *Flora of Suriname* (1932–present), who made the flora of Surinam among the best known of South America. The *Flora of Suriname*, under the auspices of the University of Utrecht, has been nearly completed and updated (Table II). It was its completion that led to the proposed *Flora of the Guianas* outlined below.

The *Bomenboek voor Suriname* by Lindeman et al. (1963) is extremely useful in helping to identify the trees of Surinam based on vegetative characters as well as on wood anatomy. Another useful book for plant identification in Surinam is the weed manual of Wessels Boer et al. (1976) which is published in Dutch but with a title in Sranan Tongo. Because of its many illustrations, this book is useful even to those who do not understand the language. Werkhoven (1986) has recently published a beautifully illustrated book in which 218 of the 300 known species of orchids in Surinam are treated.

The New York Botanical Garden became interested in the flora of Surinam with the expeditions of Bassett Maguire in April and from June to October, 1944. He collected in the coastal region with Gerold Stahel and independently in the interior on the Saramaca River and Tafelberg. The first set of more than 1500 collection numbers is at New York (Maguire, 1958; Maguire and collaborators, 1948). His program continued in Surinam in 1954–1955 when he collected nearly 400 numbers from the Nassau Mountains and in the vicinity of Moengo with R. S. Cowan and J. C. Lindeman. He returned to the Wilhelmina Mountains in 1963 with N. H. Holmgren, H. S. Irwin, C. K. Maguire, G. T. Prance, J. P. Schulz, T. R. Soderstrom, and G. Wessels Boer. The most recent expeditions of The New York Botanical Garden to Surinam have been those of Scott Mori and Alan Bolten in 1976 when they collected 375 numbers from several coastal localities and the Kayser and Lely Mountains of the interior.

C. French Guiana

As mentioned earlier, the study of French Guianan plants began with the collections of Aublet from 1762–1764, described in his *Histoire des Plantes de la Guiane Françoise* (Aublet, 1775). Since then, this department of France has been visited by more than 200 botanists, two-thirds of whom have been French (F. Vermeulen, pers. comm.). The collections of L. C. M. Richard (1781–1788 in French Guiana), F. R. M. Leprieur (1830–1836), J. B. Leblond (1787–1802), P. A. Poiteau (1817–1822), E. Mélinon (1840–1879), and P. A. Sagot (1854–1859), all deposited at P, are among the most important. The original collections of J. Martin, some 1172 unnumbered sheets, are at the British Museum (Maguire, 1958; Stearn & Williams, 1957). Because the Martin collections were pirated by the British, they are often mistakenly attributed as having been collected in Guyana (Stearn & Williams, 1957). Martin's collections were described by Rudge (1805–1806).

Sporadic papers on the plants of French Guiana (e.g. Benoist, 1933; Poiteau, 1822, 1825; Sagot, 1885), appeared after Aublet's work, but it was not until Lemée's *Flore de la Guyane Française* (1952–1956) that an enumeration of the Department's plant species was published, based largely on records from the *Flora of Suriname*. Well illustrated guides to the seaside plants (Cremers, 1986) and to the medicinal plants of French Guiana (Grenand et al., 1987) have recently been published.

The Department's Forest Service, first known as the "Bureau Agricole Forestier Guyanais" (BAFOG) and later as "Office National des Forêts" (ONF), played an important role in the collection and distribution of specimens from trees. Their collections are deposited at P and in the other major herbaria of the world. Several useful publications for identifying the trees of French Guiana are the direct result of the work of the Forest Service. Béna (1960) published his *Essences Forestières de Guyane* and Détienne and Mariaux (1982) have described the wood of the principle timber trees of the department.

The New York Botanical Garden first became interested in French Guiana with the 1020 numbers of plants collected there by W. E. Broadway in 1921. Subsequently, R. S. Cowan collected 310 numbers in the vicinity of Cayenne, Montagne de Kaw, and St. Laurent in 1954 (Maguire, 1958), and Scott Mori has collected 306 numbers in 1976–1977, 679 numbers in 1982, 339 numbers in 1983, 529 numbers in 1986, and 329 numbers in 1987. Most of these numbers came from the vicinity of Saül, where his student, Brian Boom, also gathered another 800 numbers with Mori in 1982.

Table II
Families treated for the Flora of Suriname. Table prepared by A. L. Stoffers.

Volume I, part 1

Contents: Gnetaceae, Loranthaceae, Amaranthaceae, Balanophoraceae, Ulmaceae, Polygonaceae, Cyperaceae Caryophyllaceae, Proteaceae, Aizoaceae, Selaginellaceae, Lycopodiaceae, Burmanniaceae, Thurniaceae, Rapateaceae, Commelinaceae, Eriocaulaceae, Xyridaceae, Mayacaceae, Typhaceae, Haemodoraceae, Lacistemaceae, Olacaceae, Gramineae, Amaryllidaceae, Iridaceae, Triuridaceae, Hydrocharitaceae, Alismataceae, Butomaceae, Portulacaceae.

Volume I, part 2

Contents: Araceae, Pontederiaceae, Batidaceae, Bromeliaceae, Marantaceae, Phytolaccaceae, Piperaceae, Chenopodiaceae.
Additions and corrections to Vol. I, part 1: Loranthaceae, Ulmaceae, Polygonaceae, Cyperaceae, Proteaceae, Burmanniaceae, Rapateaceae, Commelinaceae, Eriocaulaceae, Xyridaceae, Gramineae, Amaryllidaceaae, Iridaceae, Alismataceae.

Volume II, part 1

Contents: Euphorbiaceae, Rhamnaceae, Monimiaceae, Myristicaceae, Menispermaceae, Anacardiaceae, Malpighiaceae, Lauraceae, Hernandiaceae, Sapindaceae, Capparaceae, Polygalaceae, Rosaceae.

Volume II, part 2

Contents: Papilionaceae, Mimosaceae, Connaraceae, Annonaceae.
Additions and corrections to Vol. II, part 1: Euphorbiaceae, Monimiaceae, Myristicaceae, Menispermaceae, Anacardiaceae, Malpighiaceae, Lauraceae, Hernandiaceae, Sapindaceae, Capparaceae, Polygalaceae, Rosaceae, Chrysobalanaceae.

Volume III, part 1

Contents: Malvaceae, Bombacaceae, Sterculiaceae, Tiliaceae, Elaeocarpaceae, Guttiferae, Lecythidaceae, Punicaceae, Bixaceae, Araliaceae, Combretaceae, Melastomataceae, Flacourtiaceae, Canellaceae, Passifloraceae, Ochnaceae, Turneraceae, Quiinaceae, Caryocaraceae, Marcgraviaceae, Dilleniaceae, Linaceae, Humiriaceae, Lythraceae.

Volume III, part 2

Contents (including additions and corrections): Malvaceae, Bombacaceae, Sterculiaceae, Tiliaceae, Elaeocarpaceae, Guttiferae, Lecythidaceae, Araliaceae, Combretaceae, Melastomataceae, Flacourtiaceae, Passifloraceae, Ochnaceae, Turneraceae, Quiinaceae, Caryocaraceae, Dilleniaceae, Linaceae, Humiriaceae, Lythraceae, Erythroxylaceae, Oenotheraceae (or Onagraceae), Myrtaceae, Aquifoliaceae, Dichapetalaceae, Trigoniaceae, Vochysiaceae, Burseraceae.

Volume IV, part 1

Contents: Apocynaceae, Convolvulaceae, Loganiaceae, Pedaliaceae, Rubiaceae, Ericaceae, Campanulaceae, Boraginaceae, Labiatae, Sapotaceae, Gentianaceae, Menyanthaceae, Myrsinaceae.

Volume V, part 1 (1)

Contents: Palmae

Volume V, part 1 (2)

Contents: Moraceae, Urticaceae

Volume V, part 1 (3)

Contents: Simaroubaceae, Papaveraceae, Vitaceae, Icacinaceae, Theaceae, Theophrastaceae, Nymphaeaceae, Cabombaceae, Nelumbonaceae, Musaceae, Cannaceae, Zingiberaceae, Liliaceae.

Volume V, part 1 (4)

Contents: Cucurbitaceae, Meliaceae, Moringaceae, Crassulaceae, Begoniaceae, Caricaceae, Gesneriaceae.

Volume VI, part 1 (1)

Contents: Sphagnaceae, Fissidentaceae, Dicranaceae, Leucobryaceae, Calymperaceae, Pottiaceae, Ephemeraceae, Funariaceae, Splachnaceae, Bryaceae, Mniaceae, Drepanophyllaceae, Rhizogoniaceae, Bartramiaceae, Orthotrichaceae, Racopilaceae, Hydropogonaceae, Hedwigiaceae, Leucodontaceae, Pterobryaceae, Meteoriaceae, Phyllogoniaceae, Neckeraceae.

Volume VI, part 1 (2)

Contents: Neckeraceae, Hookeriaceae, Plagiotheciaceae

Study of the plants of French Guiana received new impetus in 1965 with the establishment of the herbarium of the "Institut Français de Recherche Scientifique pour le Developpement en Cooperation" (formerly "Office de la Recherche Scientifique et Technique Outre Mer" (ORSTOM) at Cayenne. Today this herbarium contains 60,000 collections and has provided the information needed for numerous studies on the taxonomy, ecology and ethnobotany of the Department's plants (Cremers, 1984).

IV. The Flora of the Guianas Project

Since 1983, a consortium of six major herbaria has been formed to produce a *Flora of the Guianas*. Preliminary contacts toward this goal occurred in 1982 between representatives of the University of Utrecht and ORSTOM-Cayenne and between the University of Georgetown and the University of Surinam. Discussions were held in mid-1983 among representatives of the systematic botany staffs of the University of Utrecht (U), the Botanical Garden and Museum in Berlin-Dahlem (B), and the Smithsonian Institution (US). The first formal meeting occurred in mid-October 1983 when representatives from B, US, U, ORSTOM-Cayenne (CAY), and the Muséum National d'Histoire Naturelle of Paris (P) met in Utrecht to make the initial decisions. The New York Botanical Garden (NY) was added to the consortium at the annual meeting in October, 1984 at Utrecht. One of the earliest decisions of the planners was that the Flora will be restricted to the political boundaries of Guyana, Surinam, and French Guiana rather than the natural biological boundaries which would include eastern Venezuela as well as northern and eastern Amazonian Brazil.

In principle, all groups of plants will be treated in the *Flora of the Guianas*, even those of marine offshore habitats. Treatments will be in English and the families of flowering plants will be arranged according to the Cronquist system (1981). Contributions will be published as they become available. The project is underway at one level or another in all of the participating institutions. The executive editor, A. R. A. Görts-van Rijn, is located at the administrative center for the project at the Institute of Systematic Botany of the State University of Utrecht where the Flora of the Guianas is one of the major institutional projects. Under her direction, guidelines for the *Flora* have been produced, and the first family treatments have been published (Maas, 1985; Prance, 1986).

The goals and organization of the project have recently been outlined by Görts-van Rijn (1984). It is estimated that it will take at least 20 years to complete the project. Specialists have been contacted for the preparation of treatments for many groups of plants. In the event that there is no specialist for a given group, staff members of the consortium will prepare treatments.

A series of publications entitled "Studies on the flora of the Guianas" has been initiated to accommodate the publication of new taxa and other noteworthy items too lengthy for inclusion in the Flora. These may be published in the journal of the author's choice, but it is recommended that each paper using the title be assigned a number in the series by the executive editor.

Economically important groups such as weeds, grasses, timber trees, aquatic plants, marine algae, seagrasses, etc. will be treated in separate semi-technical publications in an effort to bring botanical knowledge to students, foresters, agronomists, and the concerned public. Although English is the official language of the scientific publications, some of the semi-technical treatments may be published in the other languages of the Guianas. An example of this kind of publication is the independently produced weed flora of Surinam by Wessels Boer et al. (1976) in Dutch and van Roosmalen's (1985) book on the fruits of the Guianan flora.

V. Vegetation Maps

The vegetation maps of South America and of the Guianas are listed in Küchler (1980). Two recent maps useful for placing the Guianas and their vegetation in context with the rest of South America are Hueck and Seibert (1972) and UNESCO (1981). The UNESCO map is accompanied by explanatory text, and Hueck (1972) elaborates on the vegetation of the Guianas in his book on the forests of South America.

Numerous local maps exist for the vegetation of Guyana and Surinam (Küchler, 1980). However, the coastal area of Surinam is the only area that has had its vegetation mapped in detail. Dillewijn (1957) has aided the preparation of vegetation maps in Surinam by providing aerial photographs of the major vegetation types with accompanying photographs of the vegetation on the ground. Especially useful for the coastal region and the savanna belt of Surinam are the maps of Lindeman (1953), Lindeman and Moolenaar (1959), and Teunissen (1978).

In contrast, the vegetation of French Guiana has not been the subject of detailed vegetation mapping. The only modern vegetation map of the department is that of the *Atlas de la Guyane* (de Granville, 1979) which is accompanied by explanatory text.

VI. Conservation

Even though the natural vegetation of the Guianas has been less disturbed than that of most of Central and South America, it too has been modified in many areas by man. The impact of shifting cultivation by the Amerindian population has been so extensive in Guyana that Fanshawe (1954) questions the presence of primeval forest in much of the country. The biological consequences of shifting cultivation in French Guiana have been described by Sastre (1980) and Gély (1984). In recent times, the coastal areas of all of the Guianas have been subjected to extensive timber and firewood removal. About 7,000 hectares of plantations of *Pinus* and 2,000 hectares of plantations of other species have been established in Surinam. However, these plantations of other species have had little economic success (Jonkers & Schmidt,

1984). Cattle ranches in the Rupununi Savanna have greatly modified the original savanna by burning, which has also extended savannas into the forested areas of the adjacent Kanuku Mountains. Bauxite mining and the conversion of large areas of herbaceous swamp into rice fields have also modified the natural landscape in parts of Surinam. Nevertheless, reduced populations, confined mostly to coastal areas, have kept environmental degradation to a minimum.

An excellent system of nature preserves exists in Surinam where 18 areas have been established or are proposed as nature or forest reserves (Fig. 1). These reserves, totalling 729,000 hectares, cover all major vegetation types of the coastal and savanna zones as well as several significant areas in the interior (Held & Van der Steege, 1986; IUCN Commission on National Parks and Protected Areas, 1982; Schulz, 1976). These reserves are managed by the Surinam Forest Service, Nature Conservation Division. The Brownsberg Nature Park (8,400 hectares) is directed by the Foundation for the Preservation of Nature in Surinam ("Stichting Natuurbehoud Suriname," STINASU). STINASU organizes tours for tourists interested in nature. Because undisturbed natural areas are easy to reach and possess a rich avifauna, Surinam is known to ornithologists throughout the world as one of the best countries in which to see neotropical birds.

Guyana and French Guiana lag behind Surinam in the conservation of natural areas. In Guyana, only Kaieteur National Park (11,655 hectares) is protected and in French Guiana no reserves have been officially designated. Nevertheless, Condamin (1975) and de Granville (1975) have laid the groundwork for a series of nature preserves designed to protect French Guiana's diverse flora and fauna (Fig. 1).

VII. Conclusions

Botanical inventory in the Guianas is more complete than in most areas of South America. There is a nearly complete Flora of Surinam and at least checklists for the floras of Guyana and French Guiana. Moreover, a recently initiated Flora of the Guianas is intended to be a flora of the political entities known as the Guianas, encompassing all groups of plants, including those of marine habitats. This project is expected to require twenty years and will include considerable field work in undercollected areas. The first years of the project will concentrate on the rich flora of western Guyana. Special emphasis in collecting in the Guianas should be placed on all mountains over 500 meters altitude because they have been less well collected than the lowlands and because they possess the greatest number of endemics.

Although the Guianas are receiving a great deal of attention from foreign botanists because of the recently initiated Flora of the Guianas project, there is still much to be done by resident botanists, especially in Guyana and Surinam. The herbaria of these countries are understaffed, there are few knowledgeable tree spotters, and there are few botanists with advanced degrees to initiate and direct taxonomic and ecological research.

A positive aspect of the Guianas is the excellent condition of much of the natural vegetation. A model system of nature reserves has been established in Surinam and it is still not too late to implement similar systems in Guyana and French Guiana. The best time to set aside natural areas is before they become too pressured by development.

VIII. Acknowledgments

We are grateful to Carol Gracie for preparing Figure 1 and to Jeanne Goode for typing and editorial assistance. We thank B. Boom, G. Cremers, A. R. A. Görts-van Rijn, C. A. Gracie, J. J. de Granville, M. Jansen-Jacobs, S. A. J. Malone, M. Plotkin, L. Skog, F. E. Vermeulen and M. C. M. Werkhoven for reviewing various drafts of the manuscript.

IX. Literature Cited

Anonymous. 1981. L'ecosystème forestier guyanais. Bulletin de liaison du groupe de travail no. 3, pages 1–106. Centre de ORSTOM de Cayenne.

———. 1982. L'écosystème forestier guyanais. Bulletin de liaison du groupe de travail no. 6, pages 1–259. Centre de ORSTOM de Cayenne.

Aublet, F. Histoire des plantes de la Guiane Françoise, 4 vols. Didot, Paris.

Beard, J. S. 1944. Climax vegetation in tropical America. Ecology **25(2):** 127–158.

Béna, P. 1960. Essences forestières de Guyane. Bureau Agricole et Forestière Guyanaise, Paris.

Benoist, R. 1924. La végètation de la Guiane Française. Bull. Soc. Bot. France **71:** 1169–1177.

———. 1933. Les Bois de la Guiane Française. Arch. Bot. (Paris) 5. Mém. **1(1931):** 210–220.

Carruthers, W. 1904. The history of collections contained in the natural history museum of the British Museum, vol. I.

Charles-Dominique, P., M. Atramentowicz, M. Charles-Dominique, H. Gerard, A. Hladik, C. M. Hladik & M. F. Prévost. 1981. Les mamiferes frugivores arboricoles nocturnes d'une forêt Guyanaise: Inter-relations plantes-animaux. Rev. Ecol. (Terre et Vie) **35:** 341–435.

Cintron, G., A. E. Lugo & R. Martinez. 1985. Structural and functional properties of mangrove forests. Pages 53–68 in W. G. D'Arcy & M. D. Correa A., The Botany and Natural History of Panama. Missouri Botanical Garden, St. Louis.

Clokey, H. N. 1964. An account of the herbaria of the department of botany in the University of Oxford. Oxford University Press, Oxford, England.

Condamin, M. 1975. Projets de reserves naturelles sur le littoral Guyanais. ORSTOM, unpublished report.

Cremers, G. 1982. Végétation et flore illustrée des savanes: L'exemple de la savane Bordelaise. Office de la Recherche Scientifique et Technique Outre-Mer, Centre de Cayenne.

———. 1984. L'herbier du Centre ORSTOM de Cayenne (CAY) a 25 ans. Taxon **33(3):** 428–432.

———. 1986. Petite flore illustrée: Rivages de l'île de Cayenne. Nature Guyanaise. **1:** 1–93.

Cronquist, A. 1981. An integrated system of classification of flowering plants. Columbia University Press, New York.

Davis, T. A. W. 1929. Some observations on the forests of the North-West District. Agric. J. British Guiana. **2:** 157–166.

——— **& P. W. Richards.** 1933. The vegetation of Moraballi

Creek, British Guiana: An ecological study of a limited area of tropical rain forest. Part I. J. Ecol. **21**: 350–384.

———— & ————. 1934. The vegetation of Moraballi Creek, British Guiana: An ecological study of a limtied area of tropical rain forest. Part II. J Ecol. **22**: 106–155.

Détienne, P. J. & A. Mariaux. 1982. Manuel d'identification des bois tropicaux, tome 3, Guyane Française. Centre Technique Forestier Tropical, Paris.

Dillewijn, F. J. van. 1957. Sleutel voor de interpretatie van begroeiingsvormen uit luchtfoto's 1:40.000 van het noordelijk deel Suriname. Uitgave Dienst's Landsbosbeheer, Paramaribo, Suriname.

Donselaar, J. van. 1965. An ecological and phytogeographic study of northern Surinam savannas. Wentia **14**: 1–163.

————. 1968. Phytogeographic notes on the savanna flora of southern Surinam. Acta Bot. Neerl. **17(5)**: 393–404.

————. 1969. Observation on savanna vegetation types in the Guianas. Vegetatio **17**: 271–312.

Donselaar-Ten Bokkel Huinink, W. A. E. van. 1966. Structure of root systems and periodicity of savanna plants and vegetation in northern Surinam. Wentia **17**: 1–162.

Ducke, A. & G. A. Black. 1954. Notas sobre a fitogeografia da Amazônia Brasileira. Bol. Técn. Inst. Agron. N. **29**: 3–62.

Fanshawe, D. B. 1949. Check-list of the indigenous woody plants of British Guiana. Forest. Bull.(N. S.) **3**: 1–244.

————. 1954. Forest types of British Guiana. Caribbean Forest. **15(3/4)**: 73–111.

Foresta, H. de, P. Charles-Dominique, C. Erard & M. F. Prévost. 1984. Zoochore et premiers stades de la regeneration naturelle apres coupe en forêt Guyanaise. Rev. Ecol. (Terre et Vie) **39**: 369–400.

Froidevaux, M. H. 1897. Étude sur les recherches scientifiques de Fusée Aublet à la Guyane Française (1762–1764). Bull. Géogr. Hist. Descrip. **1897**: 425–469.

Gély, A. 1984. L'agriculture sur brulis chez quelques communautés d'amerindiens et des noirs réfugies de Guayane Française. Journ. d'Agric. Trad. et de Bota. Appl. **31**: 43–70.

Gleason, H. A. 1932. A synopsis of the Melastomataceae of British Guiana. Brittonia **1(3)**: 127–184.

Golley, F. B., J. T. McGinnis, R. G. Clements, G. I Child & M. J. Duever. 1975. Mineral cycling in a tropical moist forest ecosystem. University of Georgia Press, Athens, Georgia.

Görts-van Rijn, A. R. A. 1984. "Flora of the Guianas": An international project. Taxon **33**: 371–372.

Graham, E. 1934. Flora of the Kartabo region, British Guiana. Ann. Carnegie Mus. **22(1)**: 17–292.

Granville, J. J. de. 1973. Paysages végétaux de la Guyane. Office de la Recherche Scientifique et Technique Outre-Mer, Centre ORSTOM de Cayenne, mimeograph.

————. 1974. Paysages végétaux de la Guyane. Pages 77–78 in Regards sur la France. Editions. S.P.E.I., Paris.

————. 1975. Projets de reserves botaniques et forestières en Guyane. Office de la Recherche Scientifique et Technique Outre-Mer. Centre ORSTOM de Cayenne.

————. 1976. Un transect à travers la savane Sarcelle (Mana - Guyane Française). Cahiers ORSTOM, Sér. Biologie **11(1)**: 3–21.

————. 1978. Recherches sur la flore et la végétation guyanaises. Ph.D. Dissertation Université des Sciences et Techniques du Languedoc.

————. 1979. Carte et notice de végétation. In Atlas des departements Français d'outre-mer:IV. La Guyane. CNRS/ORSTOM, Paris.

————. 1982. Rain forest and xeric forest refuges in French Guiana. Pages 159–181 in G. T. Prance (ed.), Biological diversification in the tropics. Columbia University Press, New York.

———— & C. Sastre. 1974. Aperçu sur la végétation des inselbergs du sudouest de la Guyane Française. Compt. Rend. Séances Soc. Biogéogr. **50(439)**: 54–58.

Grenand, P., C. Moretti & H. Jacquemin. 1987. Pharmacopées traditionelles en Guyane. Éditions de l'ORSTOM, Paris.

Haman, M. & B. R. Wood. 1928. The forests of British Guiana. Trop. Woods **15**: 2–13.

Hedge, J. C. & J. M. Lamond. 1970. Index of collectors in the Edinburgh herbarium.

Held, M. M. & J. G. Van der Steege. 1986. Nature conservation and ecological research in Surinam. Pages 155–159 in SEPAN-GUY/SEPANRIT (eds.), Le littoral Guyanais. Nature Guyanaise, Cayenne.

Heyligers, P. C. 1963. Vegetation and soil of a white-sand savanna in Suriname. Verh. Kon. Ned. Akad. Wetensch., Afd. Natuurk., Tweedie Sect. **54(3)**: 1–118.

Hoock, J. 1971. Les savanes guyanaises: Kourou. Essai de phyto-écologie numérique. Mém. ORSTOM, no. 44, Paris.

Holmgren, P. K., W. Keuken & E. K. Schofield. 1981. Index Herbariorum. Part I. The herbaria of the world. Regnum Veg. **106**: 1–452.

Howard, R. A. 1983. The plates of Aublet's histoire des plantes de la Guiane Françoise. J. Arnold Arb. **64**: 255–292.

Hueck, K. 1972. As florestas da América do Sul. Editôra da Universidade de Brasília and Editôra Polígono S. A., São Paulo, Brazil.

———— & P. Seibert. 1972. Vegetationskarte von Südamerika. Gustav Fischer Verlag, Stuttgart.

Hurault, J. 1974. Les inselbergs rocheux des régions tropicales humides témoins de paléoclimats. Compt. Rend. Séances Soc. Biogéogr. **50(439)**: 49–54.

Im Thurn, E. F. 1884. Memoranda on the palms of British Guiana. Timehri **3**: 219–276.

IUCN Commission on National Parks and Protected Areas (CNPPA). 1982. IUCN directory of neotropical protected areas. Tycooly International Publishing Limited, Dublin.

Janssen, J. J. 1976. The development of the coast. Suralco Magazine, September, pp. 1–7.

Jenman, G. S. 1887. Synopsis of the Lycopodiaceae of the Guianas and their allies. Timehri **1(Ser.2)**: 35–59.

Jonkers, W. B. J. & P. Schmidt. 1984. Ecology and timber production in tropical rainforest in Suriname. Interciencia **9**: 290–297.

Kahn, F. & A. de Castro. 1985. The palm community in a forest of Central Amazonia, Brazil. Biotropica **17(3)**: 210–216.

Kramer, K. U. & J. van Donselaar. 1968. A sketch of the vegetation and flora of Kappel Savanna near Tafelberg, Suriname, I and II. Proc. Kon. Ned. Akad. Wetensch. Ser. C. **71**: 495–524.

Küchler, A. W. (ed.). 1980. International bibliography of vegetation maps, 2nd ed., section 1, South America. University of Kansas Libraries, Lawrence, Kansas.

Lemée, A. M. V. 1952–1956. Flore de la Guyane Française, Vols. 1–4. Lechevalier, Paris.

Lescure, J-P. & R. Boulet. 1985. Relationships between soil and vegetation in a tropical rain forest in French Guiana. Biotropica **17(2)**: 155–164.

Lindeman, J. C. 1953. The vegetation of the coastal region of Suriname. In J. A. Hulster, J. Lanjouw and F.W. Ostendorf (eds.), The vegetation of Surinam **1(1)**: 1–135.

————, A. M. W. Mennega & W. H. A. Hekking. 1963. Bomenboek voor Suriname. Uitgave Dienst's Lands Bosbeheer Suriname, Paramaribo, Surinam.

———— & S. P. Moolenaar. 1959. Preliminary survey of the vegetation types of northern Suriname. In. I. A. Hulster and J. Lanjouw (eds.), The vegetation of Suriname, **1(2)**: 1–145.

Linnaeus, C. 1775. Plantae surinamenses 17. Uppsala, Sweden.

Maas, P. J. M. 1971. Floristic observations on forest types in western Suriname I and II. Proc. Kon. Ned. Akad. Wetensch. Ser.

C, **74(3):** 269–284, 285–302.

———. 1985. Musaceae, Zingiberaceae and Cannaceae. Pages 1–73 *in* A. R. A. Görts-van Rijn (ed.), Flora of the Guianas. Koeltz, Konigstein, West Germany.

Maguire, B. 1958. Highlights of botanical exploration in the New World. Pages 209–246 *in* W. C. Steere, Fifty years of botany. McGraw-Hill Book Company, Inc., New York.

———. 1970. On the flora of the Guayana Highland. Biotropica **2:** 85–100.

——— & collaborators. 1948. Plant explorations in Guiana in 1944, chiefly to the Tafelberg and the Kaieteur Plateau. Bull. Torrey Bot. Club **75:** 56–115, 189–230, 286–323, 374–438, 523–580, 633–671.

Meyer, G. W. F. 1818. Primitiae florae Essequeboensis. Gottingae Sumptius Henrici Dieterich.

Miquel, F. A. W. 1850. Stirpes surinamenses selectae. Lugduni Batavorum.

Mitchell, J. D. & S. A. Mori. 1987. Chapter X. Ecology of Lecythidaceae. *In* S. A. Mori and collaborators, The Lecythidaceae of a lowland neotropical forest: La Fumée Mountain, French Guiana. Mem. New York Bot. Gard. **44:** 113–123.

Mori, S. A. & B. M. Boom. 1987. Chapter II. The forest. *In* S. A. Mori & collaborators, The Lecythidaceae of a lowland neotropical forest: La Fumée Mountain, French Guiana. Mem. New York Bot. Gard. **44:** 9–29.

——— **& G. T. Prance.** 1987. Chapter VI. Phytogeography. *In* S. A. Mori & collaborators, The Lecythidaceae of a lowland neotropical forest: La Fumée Mountain, French Guiana. Mem. New York Bot. Gard. **44:** 55–71.

Oldeman, R. A. A. 1974. L'architecture de la forêt guyanaise. Mémoires O.R.S.T.O.M. **73:** 1–204.

Pennell, F. W. 1945. Historical sketch. Pages 35–48 *in* F. Verdoorn (ed.), Plants and plant science in Latin America. Chronica Botanica.

Poiteau, M. A. 1822. Histoire des palmiers de la Guyane Française. Mém. Mus. Hist. Nat. **9:** 385–392.

———. 1825. Memoire sur les Lecythidacées. Mém. Mus. Hist. Nat. **13:** 141–165.

Prance, G. T. (ed.). 1982. Biological diversification in the tropics. Columbia University Press, New York.

———. 1986. Chrysobalanaceae. Pages 1–146 *in* A. R. A. Görts-Van Rijn (ed.), Flora of the Guianas. Goeltz, Konigstein, West Germany.

Prévost, M. F. 1983. Les fruits et les graines des espèces végétales pionnières de Guyane Française. Rev. Ecol. (Terre et Vie) **38:** 121–145.

Pulle, A. A. 1906. An enumeration of the vascular plants known from Surinam. E. J. Brill, Leiden.

———. 1911. Zakflora voor Suriname. Bulletin van het Koloniaal Museum te Haarlem No. 47.

——— (ed.). 1932–present. Flora of Suriname. Vereeniging Koloniaal Instituut, Foundation van Eedenfonds, E. J. Brill, Netherlands.

Rodway, J. 1884. The Guiana orchids. Timehri **8(Ser.2):** 1–24, 270–296.

Roosmalen, M. G. M. van. 1985. Fruits of the Guianan flora. Institute of Systematic Botany of Utrecht University and the Silvicultural Department of Wageningen Agricultural University.

Rudge, E. 1805–1806. Plantarum guianae rariorum. Richardi Taylor et Soc., London.

Sabatier, D. 1985. Saisonalité et determinisme du pic de fructification en forêt Guyanaise. Rev. Ecol. (Terre Vie) **40:** 289–320.

Sagot, P. 1885. Plantes de la Guyane Française. Ann. Sci. Nat. (Paris) VI. **20:** 198–216.

Sastre, C. 1980. Fragilité des ecosystèmes Guyanais: Quelques examples. Adansonia **19:** 435–449.

——— **& J. J. de Granville.** 1975. Observations phytogéographiques sur les inselbergs du bassin supérieur du Maroni. Compt. Rend. Séances Soc. Biogéogr. **51(444):** 7–15.

Schomburgk, R. 1848. Versuch einer fauna und flora von British-Guiana. J. J. Weber, Leipzig.

Schulz, J. P. 1960. Ecological studies on rain forest in northern Suriname. Verh. Kon. Ned. Akad. Wetensch., Afd. Natuurk., Tweede Sect. **53(1):** 1–267.

———. 1976. Nature reserves and tourism. Suralco Magazine, June, pp. 1–11.

Smith, A. C. 1941. The vegetation of the Guianas, a brief review. Chronica Botanica VI. **19/20:** 449–452.

Stearn, W. T. & L. H. J. Williams. 1957. Martin's French Guiana Plants and Rudge's plantarum Guianae rariorum icones. Bull. Jard. Bot. État **27(2):** 243–265.

Steyermark, J. A. 1982. Relationships of some Venezuelan forest refuges with tropical floras. Pages 182–220 *in* G. T. Prance (ed.), Biological diversification in the tropics. Columbia University Press, New York.

Teunissen, P. A. 1978. Suriname lowland ecosystems (coastal plain and savanna belt). Foundation for Nature Preservation in Suriname, Paramaribo, Suriname.

——— **& J. T. Wildschut.** 1970. Vegetation and flora of the savannas in the Brinckheuvel Nature Reserve, northern Suriname. Verh. Kon. Ned. Akad. Wetensch., Afd. Natuurk, Tweede Sect. **59(2):** 1–60.

Tomlinson, P. B. 1985. The botany of mangroves. Cambridge University Press, Cambridge.

UNESCO. 1981. Vegetation map of South America. United Nations Educational, Scientific and Cultural Organization, Paris.

Van der Hammen, T. 1974. The Pleistocene changes of vegetation and climate in tropical South America. J. Biogeogr. **1:** 3–26.

Welch, I. 1975. A short history of the Guyana Forest Department 1925–1975. Forest Department, Georgetown, Guyana.

Werkhoven, M. C. M. 1986. Orchids of Suriname. Vaco Uitgeversmaatschappij, Paramaribo, Suriname.

Wessels Boer, J. G., W. H. A. Hekking & J. P. Schulz. 1976. Fa joe kan tak' mi no moi, deel 1 flora (plantenbeschrijvingen), eerste helft (t/m Hypericaceae), tweede helft (vanaf Labiatae). Natuurgids Serie B no. 4, STINASU, Paramaribo, Suriname.

Zarucchi, J. L. 1984. The treatment of Aublet's generic names by his contemporaries and by present-day taxonomists. J. Arnold Arb. **65:** 215–242.

Northwest South America (Colombia, Ecuador and Peru)

Alwyn Gentry

Contents

- I. Description of the Region 393
  - A. Topography and Floristics 393
- II. Vegetation Types 393
  - A. Dry, Seasonally Deciduous Forest 393
  - B. Lowland Tropical Forests 393
  - C. Cloud Forest 394
- III. Population and Deforestation 394
  - A. Colombia 394
  - B. Ecuador 395
  - C. Peru 395
- IV. Vegetation and Other Maps 396
  - A. Colombia 396
  - B. Ecuador 396
  - C. Peru 396
- V. Herbarium Resources 397
  - A. Colombia 397
  - B. Ecuador 397
  - C. Peru 397
- VI. Floristic Publications 397
  - A. Colombia 397
  - B. Ecuador 397
  - C. Peru 398
- VII. Present State of Inventory, Institutions Involved, and the Future of Inventory ... 398
- VIII. Summary 399
- IX. Literature Cited 399

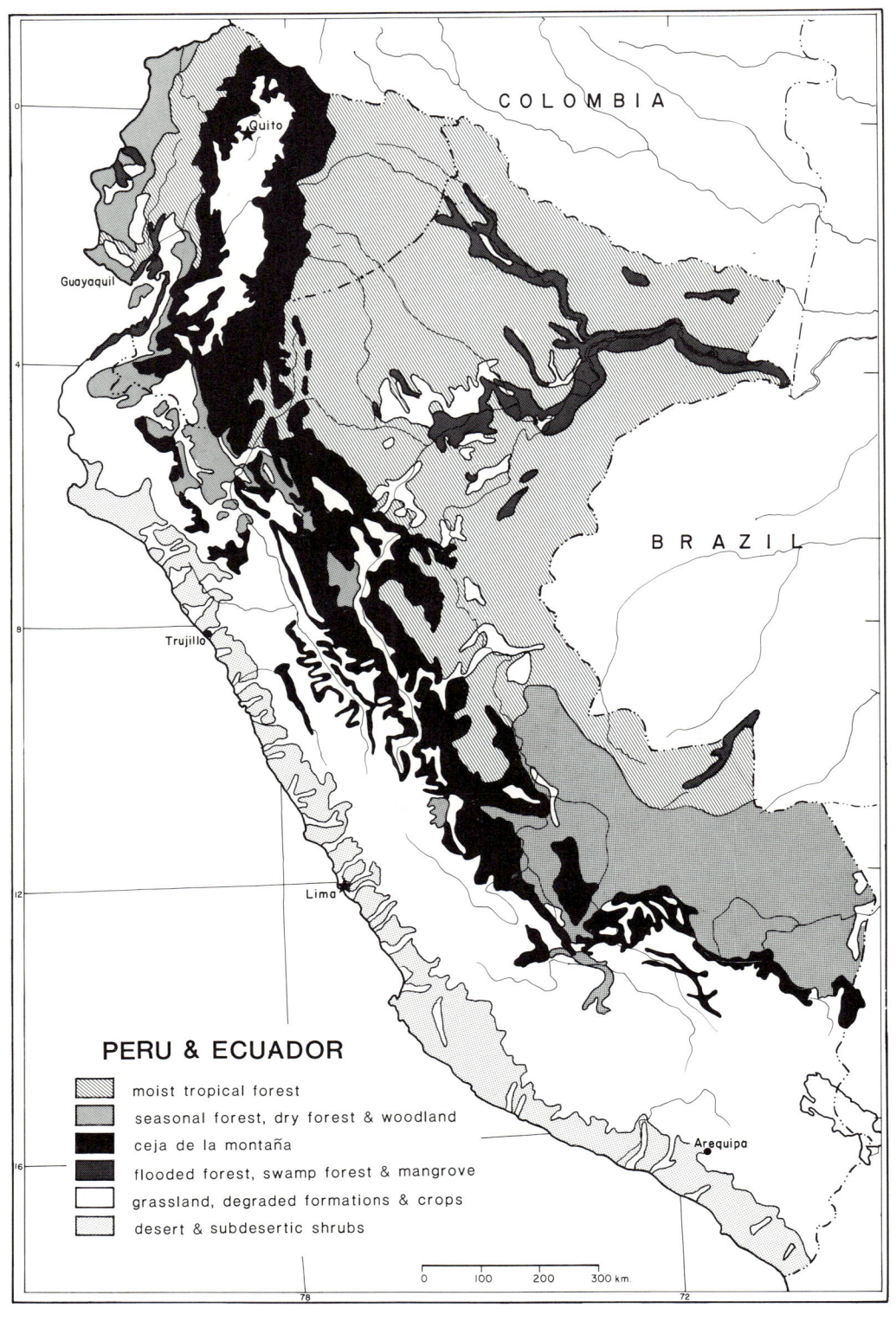

I. Description of the Region

A. Topography and Floristics

All three countries share a similar topography, with a narrow coastal plain along the Pacific, a mountainous zone parallel to the coast, and an extensive interior lowland area draining into the Amazon (in the case of Colombia the northern part of the Oriente drains into the Orinoco). In each country there are areas of naturally non-forested vegetation including (1) the high Andes (generally above 3000–3500 m); (2) the entire coastal region of Peru (desert except for a bit of seasonal dry forest in the extreme north in Tumbes, and a narrow tongue of similar forest projecting south along the western foothills above 500 m elevation into Lambayeque and Cajamarca Departments, with scattered isolated forested pockets farther south in sheltered mesic valleys between 3000 and 4000 m altitude); (3) the Llanos region of Colombia.

II. Vegetation Types

There are a number of forest types in each country (Fig. 1). The bulk of the forest is more or less evergreen lowland tropical forest. Lowland evergreen forest covers almost the entire Amazonian region of all three countries and occurs along the entire Pacific coast of Colombia into the northwest corner of Ecuador. It also occurs in the Magdalena Valley of Colombia and across the northern terminus foothills of the Colombian Central and Western Cordilleras.

A. Dry, Seasonally Deciduous Forest

This forest type occurs in the southern half of the Ecuadorian coastal region (extending into Tumbes, Peru), and along the Caribbean lowlands of northern Colombia (extending south along the Cauca Valley and in the middle to upper Magdalena Valley). There are also isolated patches in Peru (around Tarapoto, at the junction of the Ríos Ene and Perene, and near Quillabamba). In addition, there are dry areas in many rain-shadowed inter-Andean valleys (e.g., the Marañón-Huancabamba depression in northern Peru, the Huanuco area in central Peru, the upper Apurimac and Urubamba valleys in southern Peru, the Guayllambamba and Chota Valleys in Ecuador and the Patía and Dagua Valleys in Western Colombia). However, most of these inter-Andean valleys are now largely deforested, and it is generally unclear to what extent their natural vegetation was forest and what was shrubland. The old (1958) Holdridge map of Peru showed additional large dry forest areas in Amazonia, but the Gran Pajonal in Junín is almost certainly man-made savanna, and the Pucallpa region of Ucayali and the Puerto Maldonado-Iberia area on the Brazilian border in Madre de Dios are semideciduous moist forest (bosque humedo subtropical, in the "new" Holdridge system; lowland tropical evergreen and semi-deciduous forest in the system of UNESCO (1973). The Ecuadorian coastal dry forest has a fairly strong degree of endemism, even including a few distinctive endemic genera like *Macranthisiphon*. There is apparently relatively little endemism in any of the lowland Peruvian dry areas, although many remarkable disjunctions of Brazilian cerrado species are showing up. The Peruvian inter-Andean valley dry areas are generally high in endemism, but the Ecuadorian and Colombian ones are not. The semi-dry area of northeast Madre de Dios shares most of the distinctive seasonal forest species of central Amazonian Brazil, like *Bertholettia* and *Dialypetalanthus*, that have been supposed to be exclusively Brazilian; this flora is quite different from that of most of the rest of lowland Peruvian Amazonia.

B. Lowland Tropical Forests

There are a number of major local variants of this forest type. The Chocó (= Pacific coast) region of Colombia is the wettest region of the world (Gentry, 1982a) and the only true "tropical rain forest" sensu Holdridge in the entire Neotropics. It is very high in endemism (one would estimate that about one quarter of the species are strictly endemic). In general, the trees belong to Amazonian genera, in some cases with allopatric differentiation and in other cases conspecific with Amazonian taxa. The bulk of the very high local endemism is due to very active speciation on the part of epiphytes (especially Araceae, Orchidaceae, Cyclanthaceae, Bromeliaceae, etc.), palmettos (*Heliconia*, *Renealmia*, *Costus*, etc.), and shrubs (e.g., Melastomataceae, *Psychotria*). The narrow wet forest band that extends south into western Ecuador along the base of the Andes is also rich in endemics (about 20% of the Río Palenque species are endemic to western Ecuador and another 5% are endemic to western Ecuador plus extreme south-western Colombia). The currently available data suggest that such high local endemism may be characteristic of the whole Chocó region, contrary to the published opinion of this author (1982a). The northern Chocó region (centered on the Río Truando) has a distinct dry season and is phytogeographically quite distinct from the ultra-wet forest farther south. The northern Chocó region also has a good complement of endemics, as do Cerro Tacarcuna and Alto de Nique on its border with Panama.

The Magdalena Valley moist forests constitute a unique forest type that has generally not been accorded much importance in conservation planning. Much of the interest in this region centers on the occurrence of Amazonian species (and genera) that do not get farther northwest, and of Central American species (and genera) that do not get further southeast. Nevertheless, there is a distinct complement of endemic species, especially in the woody flora, and even a few genera (e.g., *Tetralocularia*, *Romeroa*, *Brachycyclix*). Also, of special interest in the Magdalena Valley, are the many endemic species that occur in the extensive swamps that extend north from the Barranca Bermeja region.

Within the large expanse of lowland Amazonian moist forest, shared by these two countries, there are many striking floristic differences. These seem generally much more closely allied to edaphic rather than climatic factors. Upper

Amazonian Peru, with its mosaic of substrates, epitomizes the strikingly different floristic units that can be represented in a small area (Gentry, 1986). In this region the kind of large scale vegetational units that, for example, differentiate the Rio Negro-Guayana Shield area from other parts of Brazilian Amazonia can be compressed into small areas with the resultant beta-diversity probably higher than in any other part of Amazonia. Around Iquitos there are different distinctive (and each intrinsically very species-rich) forest types occurring on patches of upland lateritic soil, upland white sand soil, seasonally inundated tahuampa (= varzea) forest (with very different floras in black-water and white-water inundated tahuampa forests), and relatively rich, more or less alluvial non-inundated soils (see Gentry, 1981). The tahuampa vegetation types are mostly restricted to the northern half of Peruvian Amazonia, where they seem to roughly follow the 200 m contour line, barely reaching west into Ecuador along the major Río Napo tributaries, north into Colombia's Amazonas Department and south along the Ucayali to the Pucallpa area. Most of the species are widespread in upper Amazonia and even downstream into central Brazil. Areas of distinctive campinarana-type forest on white sand soil are widespread in parts of Amazonian Colombia, but are not known at all from Ecuador, and in Peru are only known from the northern third of lowland Amazonia. These regions, heretofore extremely poorly collected in Peru, show very strong floristic affinities with the distinctive upper Rio Negro vegetation; even the endemic species are invariably more closely related to species from the Venezuela-Brazil border area than to other species from Amazonian Peru.

The relatively rich-soil lowlands along the base of the Andes (best known in the central Huallaga valley and at the Cocha Cashu Biological Station in Manu Park) are very different floristically from the rest of lowland Amazonia and have very strong floristic similarities with Central America (Foster, 1988; Gentry, 1985). There are also patches of this vegetation occurring as far out into lowland Amazonia as Iquitos (Yanamono).

C. Cloud Forest

In addition to these lowland forest vegetations there is a band of cloud forest extending the full length of the eastern side of the Andes south to northern Peru. At its upper limits (usually ca. 3000–3500 m) this Andean cloud forest merges into a shrubby subpáramo vegetation; at its lower limits (often between 500–1000 m), it gives way to lowland forest. Despite its relatively small areal extent, this cloud forest (= 'ceja de la montaña") is extremely rich in ecological variation compared with the lowlands, as a glance at the Holdridge Life Zone map of any of these countries makes obvious. The compaction of so many distinctive vegetational zones into such a small area makes the Andean slopes of special interest for conservation. Moreover, an undetermined, but surely very high, percentage of the total floras of these countries is concentrated in these Andean forests and local endemism appears to be higher here than in most lowland regions. I estimate that approaching half of all neotropical plant species are found in the minuscule region above ca. 500 m (though this includes species that also range into the lowlands) and that this high diversity is concentrated in exactly those three Andean countries treated here (Gentry, 1982b). Colombia and Peru are in a neck and neck race as to which has the most bird species in the world and all of the largest single area bird lists in the world (over 500 species in several specific localities, compared with ca. 600 in all of North America!) are concentrated along the base of the Andes (e.g., Terborgh et al., 1984; see Gentry, 1987). While alpha-diversity is not as high in the "ceja de la montaña" as it is in the lowlands at the base of the mountains, beta-diversity is probably higher due both to the habitat compression and to the apparently accelerated evolution that seems to have taken place in these regions (see Gentry, 1982b, 1986; Gentry & Dodson, 1987, for discussion and documentation).

III. Population and Deforestation

A. Colombia (See also Forero)

Colombia has by far the largest population of any country in the region, second only to Brazil and Mexico in tropical America. The population is expected to reach 38 million by 2000 A.D. (Population Reference Bureau, 1987 population data sheet).

The north coastal plain, the two major river valleys (Magdalena and Cauca) and the three Andean cordilleras are already deforested. Both the southern part and northern part of the Chocó coastal area are rapidly being deforested, as is the edge of the Amazonian forest along the Andes and abutting the Llanos in Meta, Putumayo, and Caquetá. The southeastern half of Amazonia (Amazonas, Vaupés, Guainía, part of Caquetá) has hardly been touched and much of the central Chocó region (except that part accessible from the Río San Juan or the Río Atrato) is still covered by essentially intact forest. The extensive deforestation to which Colombia has been subjected has had devastating ecological effects, especially noteworthy in loss of watersheds and electricity generating power, the shining exception is the CVC-controlled areas in the Anchicayá Valley west of Cali, a potential model for the other conservation strategies in the region.

The 1976 CONIF map (cited below) considers 35% of Colombia to be deforested (either in agriculture or abandoned), 27% naturally non-forested, and only 38% covered by natural forest. However, the great majority of the remaining natural forest is in Amazonia. Excluding Amazonia, only the Pacific coast Chocó region, part of the central Magdalena Valley, the Serrania de Perija, the north tip of the Cordillera Occidental (east of Mutata), and a tiny isolated patch between Argelia and Puerto Boyaca (Cordillera Central) are mapped as retaining natural forest. I have visited most of the areas mapped as still forested and found most of them (e.g., Serrania de Perija, Cimitarra-Carare-Opón, Mutatá, Panama border area of northern Chocó, and the Tumaco area of Nariño) undergoing exceedingly rapid deforestation. Contrary to the map, I believe that the area west of Unguía in northern Chocó, lowland Nariño south of the Tumaco road, and the eastern side of the Magdalena Valley are now more or less completely deforested. Most of this deforestation has occurred since the 1976 publication of the map. One can only describe the conservation situation in Colombia as exceedingly grim, except for the southeastern part of lowland amazonia.

B. Ecuador

In most of Ecuador the conservation picture is equally grim. Ecuador has 10 million people concentrated into an area of only 0.28 million km^2. Worse, Ecuador's population is growing at a rate of over 3% a year, the fastest population growth rate in South America, and one of the highest in the world — a mere 22-year doubling time. By 2000 A.D. Ecuador will have 14 million people and almost certainly will be completely deforested. As of a few years ago, half of Ecuador's people lived in the long-since deforested Andean uplands, and half in the recently deforested coastal plain, with almost no population in the Amazonian lowlands. Guayaquil, on the coast, is by far the largest city (814,000) despite the political dominance of Andean Quito (597,000). The third largest city, Cuenca, has a mere 104,000 people.

Exceedingly rapid colonization is taking place in Amazonia, especially the better soil areas along the base of the Andes, and much of the "Oriente" has already been deforested. This trend was already clearly visible in the map I published in 1978, and has been greatly accelerated since the 1976 FAO report from which that map was taken. Much of the coastal lowland area was virgin forest in 1960, when the first roads were pushed west from Santo Domingo de los Colorados. By 1970, in a single decade, this area had been completely deforested (see discussions in Gentry, 1978, 1980a, 1980b; and the introduction in Dodson & Gentry, 1978). We have no idea how many plant species have gone extinct here, but certainly many. (Nevertheless, it should be noted that much of coastal Ecuador has relatively rich soil and sustained agriculture is often possible, contrary to most of Amazonia, so all is not lost from the standpoint of the national economy.) The big problem associated with deforestation in western Ecuador seems to be related to the desertification associated with lesser rainfall; the dry forest of the Guayaquil area still reached west almost to Salinas about 1950, contrary to the savanna vegetation shown in this area in Harling's (1979) map. Ecuador has a well-planned National Park system on paper, but has had little success in keeping settlers from invading its parks once they become accessible from the currently actively expanding road system. The prognosis for conservation is very bad, unless there are dramatic changes in the very near future; already it is too late for most of the country except in the Oriente.

C. Peru

Peru is larger than Colombia and Ecuador but has relatively fewer people. Peru's 20.7 million citizens (estimated to reach 28 million by 2000 A.D.) inhabit an area of 1.29 million km^2. Peru's population is about equally divided between the Andean uplands and the narrow, largely desertic coastal plain, with as yet relatively few people in the Amazonian lowlands. By far the largest city is Lima, on the central coast, with over 3.25 millions (sometimes estimated as high as 4 million, including many who have only recently arrived and are living in absolute poverty). Peru's second city in size, but not importance, is Arequipa (302,000) in the southern Cordillera Occidental. Trujillo (240,000) in the northern coastal region is the second most important city, at least botanically, while Cuzco (only 120,000) in the southern uplands is widely perceived as the third most important center. The other largest cities are all in the northern coastal area (Chiclayo, Chimbote, Piura), except Huancayo in the central uplands.

Unlike Colombia and Ecuador, the natural vegetation of Peru's densely populated regions is not forest. However, there is also a very active program of Amazonian colonization in Peru. As in the other two countries, this has been largely concentrated to date in the "ceja de la montaña" and the part of lowland Amazonia adjacent to the Andes. Thus, most of the Chanchamayo, Huallaga, Mayo, and Apurimac Valleys have been deforested, the latter three mostly in the last decade. The same process has happened, and seems to be intensifying, throughout the "ceja de la montaña." There has been extensive recent deforestation also in all areas accessible from the major navigable rivers and along the new Carretera Marginal, in the Quince Mil area, and along the Tingo María-Pucallpa road. Of the relatively rich soil areas along the base of the Andes, the only one with largely intact forest is the Manu Park area, and even here a major road project is being threatened. Away from the base of the Andes and the major rivers, most of the lowland Amazonian Peru looks from the air to be covered by virgin forest. However, it appears that even in these structurally intact forests major changes have already taken place in the ecosystem, with large scale local extinction of most large game animals (including mammals, birds, reptiles, and fish) and favored tree species (*Swietenia, Cedrela, Ceiba*). Much of this apparent major extinction process has taken place during the decade I have been working in Amazonian Peru — in many ways a contemporary parallel to the famed Pleistocene extinctions in North America. There may still be time for additional major conservational efforts in Peru. Generally, the more threatened forests in Peru are in the "ceja" area and near the base of the Andes. Thus, preservation of Manu Park (see *Reporte Manu*, edited by M. Rios and published by the Nature Conservancy Centro de Datos para la Conservación), surely the most important conservational unit in the world from the standpoint of preserving species diversity, should be given the highest possible priority. Unfortunately, we still know too little about the Peruvian flora to know which areas have the highest local endemism, but a much larger area of the "ceja" (including some in the north, e.g., the Venceremos area, home of the yellow-tailed wooly monkey and an incredibly high plant species diversity for an upland area) certainly should be preserved while there is still time, especially in view of the importance of this forest in watershed management. A specific plan to preserve parts of each vegetation type in the diverse habitat mosaic around Iquitos should also be given high priority. I think that programs to develop sustained use of many of the Amazonian Peruvian forests *without cutting them* might be a feasible alternative to the cycle of cutting followed by abandonment that is now taking place (as has already taken place in most of Colombia, for example) as subsistence level highland agriculturists invade Amazonia (see Gentry, 1986b for expanded discussion). However, Peru still talks about converting Amazonia into a breadbasket for the rest of the country. Unless this attitude changes soon, to accord with ecological reality, too much of the remaining forest may have disappeared to ever make sustained use of this area possible.

IV. Vegetation and Other Maps

A. Colombia

(1.) Atlas de Colombia (Institute Geografico Agustin Codazzi). 1969. Includes a copy of the Holdridge Life Zone map plus other maps of geology, land use, etc.

(2.) Republica de Colombia Mapa Fisico-Politico. 1979. 1:1,500,000. The best political-division map.

(3.) Mapa en Relieve de la Republica de Colombia. 1965. (With the mountains actually in raised plastic.)

(4.) Mapa de las Areas de Vocación Forestal, Agropecuaria y de uso multiple de Colombia. CONIF. 1976. 1:1,000,000. Shows land use and documents the extent of deforestation.

(5.) Mapa de Bosques, Instituto Geografico "Agustin Condazzi." 1983. 1:500,000. Shows vegetation types and land use, updating 1976 CONIF map.

(6.) Republica de Colombia, Mapa Vial. 1973. 1:1,500,000. Excludes Amazonia and Llanos. Mountains indicated by shading.

(7.) Mapa de Estaciones Meterologicas. 1974. SCMH. Not all the meterological stations indicated have been established.

(8.) Departmental Maps

Antioquia	1973	1:500,000	Cundinamarca	1974	1:400,000
Arauca	1974	1:250,000	Magdalena	1975	1:250,000
Caldas	1976	1:250,000	Norte de		
Casanare	1975	1:500,000	Santander	1974	1:250,000
Chocó	1985	1:500,000	Sucre	1975	1:250,000

Some of these are out of print; there may be others published.

B. Ecuador

(1.) Atlas Geográfica de la Republica del Ecuador. (Instituto Geográfico Militar). 1977. Includes a Holdridge Life Zone map of the country plus various soil, geology, population, and land use maps.

(2.) Ecuador (IGM). 1974. 1:1,000,000. The best general map of Ecuador; includes relief.

(3.) Figure 1 (Survey of Ecuadorian vegetation types) *in* Harling (1979: 166). Good general vegetation map.

(4.) A map of deforestation (from FAO survey) is included in Gentry (1977).

C. Peru

(1.) Republica del Peru, Mapa Físico-Político. (IGM). 1973. 1:1,000,000 (4 sheets). The best general map; includes relief; there is a gazetteer of the place names included on this map, published by Llamas and Encarnación—Indice Toponimico del Mapa del Peru 1:1,000,000 del Instituto Geográfico Militar. Universidade San Marcos. 1976.

(2.) Mapa Político del Peru. 1976. Editorial "Navarrete."

(3.) Republica del Peru, Mapa Físico Político, 1970. 1:2,000,000. A single color map on a single sheet, but showing much of the same data as the 4-sheet 1973 map. The altitude is indicated with contour lines, but not color.

(4.) Mapa Ecólogico del Peru. 1976. 1:1,000,000 (ONERN, not IGM). The new Holdridge Life Zone map, which comes with an explanatory booklet. Very few were printed, however, and they are generally unavailable.

(5.) Departmental Maps (all from IGM).

Apurimac	1973	1:350,000
Arequipa	1972	1:576,000
Ancash	1976	1:400,000
Ayacucho	1973	1:520,000
Ica	1971	1:370,000
Lima	1968	1:500,000
La Libertad	1976	1:420,000
Lambayeque	1974	1:300,000
Loreto	1973	1:800,000
Moquegua	1976	1:300,000
Piura	1971	1:500,000
San Martín	1981	1:500,000
Tacna	1968	1:330,000
Tumbes	1968	1:200,000

(6.) There are also several maps of various aspects of Peruvian natural resources included in a series of publications from ONERN. For example:

Inventario, Evaluación e Integración de los Recursos Naturales de la Zona Iquitos, Nauta, Requena y Colonia Angamos. ONERN. Dec., 1976. With 6 maps in a packet (geology, forest types, forest uses, physiography, soils, potential use of natural resources. This and other similar works are generally well-executed (except for the vegetation type characterization and mapping) but are generally unavailable outside Lima.

(7.) The classical vegetational map of Peru is that of Weberbauer (most easily accessible in Macbride, 1936, Fieldiana 13(1): frontispiece). Unfortunately, this map includes only the coast and upland Andes.

V. Herbarium Resources

A. Colombia (See Forero)

B. Ecuador

(1.) Quito: QCA (15,000). The most active Ecuadorian herbarium, due in part to a collaborative program with AAU. Universidad Católica.

Q (25,300). Universidad Nacional. Mostly older collections.

QPLS (29,000). Biblioteca Ecuatoriana. This is the Sodiro collection and includes no active program nor curatorial activity.

QCNE (2000). Museo Nacional. This is the new herbarium of the new and active National Museum and likely will grow rapidly.

There are several additional Quito herbaria with ca. 2000 specimens.

(2.) Guayaquil: GUAY (formerly ECU) (8000). The only important collection in Ecuador outside Quito.

There is also a collection of 3500 specimens at the Río Palenque Science Center (RPSC), basically a documentation of the flora of the field station and surroundings. For the country there is a total of ca. 70,000 collections, with no major collection.

C. Peru

(1.) Lima: USM (250,000). Universidad de San Marcos. This is the main national plant collection. In the past badly hampered by problems of mounting, it is now actively incorporating new collections into the herbarium.

MOL (12,000). La Molina University. There are actually three separate herbaria at La Molina; one, usually locked up and inactive, has many Weberbaeuer types; the main herbarium is essentially a forestry herbarium and only recently has begun to be actively curated.

Other than USM, there are three significant herbaria in Peru:

(2.) Trujillo: HUT (20,000). Universidad de Trujillo. Currently very active in collecting in the northwestern part of Peru, with duplicates being distributed through MO.

(3.) Cuzco: CUZ (25,000). Universidad de Cuzco. Collections nearly all from the southern uplands. This herbarium has been more active recently. The figures include the herbarium of Cesar Vargas (ca. 20,000), which is now being integrated into the University's "Herbario Vargas."

(4.) Iquitos: AMAZ (35,000). Universidad de Amazonia. The most actively growing herbarium in Peru currently. Its collections are almost all from lowland Amazonia.

Smaller regional herbaria (from Index Herbariorum) include:

(5.) Huanuco: HHUA (4580).
(6.) Huancayo: HCEN (3500).
(7.) Cajamarca: CPUN (3000).

Total collections for the country are thus ca. 400,000, with far more than half at USM.

D. General Remarks

At a rough approximation, there are probably roughly equal numbers of collections from each country in foreign herbaria and native herbaria. In other words, there might be ca. 500,000 collections in all from Peru, with ca. 125,000 or more of these representing duplicates of the same numbers. For Ecuador, the extra-Ecuadorian holdings are probably relatively larger, and might be around 100,000 altogether. This is largely guesswork, but one hard datum is that there are ca. 90,000 Peruvian collections added at MO from about 1970 to 1985, including many previously-collected duplicates of Peruvian botanists as well as new collections generated by the Flora of Peru project. This may be compared with the 33,000 Peruvian collections estimated by Macbride as available to him in 1936, when the Flora of Peru was begun.

VI. Floristic Publications

A. Colombia

See Forero.

B. Ecuador

The main floristic reference for Ecuador is the *Flora of Ecuador* series, edited by G. Harling and B. Sparre, and published by the Swedish National Research Council. As of 1986, 24 families or sections of families had been published, of which the largest are Melastomataceae (J. Wurdack, 406 pp.), Orchidaceae, Part I (L. Garay, 305 pp.), Loranthaceae s.l. (J. Kuijt, 198 pp.), Scrophulariaceae (N. Holmgren & U. Molau, 189 pp.), Bignoniaceae (A. Gentry, 173 pp.), and Campanulaceae (S. Jeppesen, 170 pp.). Most of the other treatments are of small families, and altogether only a very small percent of the flora has been treated as yet. However, publication of additional families is progressing rapidly and the treatments are generally definitive ones by leading taxonomic specialists.

Ecuador also boasts several local florulas, including completely illustrated treatments of the trees of Esmeraldas Province (Little, 1969), and of the plants of two small field stations in coastal Ecuador, wet forest Río Palenque (Dodson & Gentry, 1978) and moist forest Jauneche (Dodson et al., 1985). A third, similarly completely illustrated florula of dry forest Capeira, near Guayaquil, will be published in 1987 (Dodson & Gentry, in press). An earlier work by Svenson (1945) that focuses on the Galápagos flora and its relationships with the adjacent mainland includes much floristic data on the dry part of coastal Ecuador and adjacent Peru. Flor Maria Valverde and colleagues from the University of Guayaquil have also published several books on the coastal dry forest flora (e.g., Valverde et al., 1979). Diels' important earlier work (1937, 1938–1942) on the Ecuadorian Andes hardly does more than list and describe new species. Acosta-Solis has published a number of studies of Ecuadorian plants, mostly focused on such subsets of the flora as woody plants or taxa of special interest, notably grasses and their relatives: he has also published extensively on Ecuadorian phytogeography (e.g., Acosta-Solis, 1966, 1977). Of the classic Ecuadorian floristic works, at least Ruiz and Pavon (1803), whose collections included some made by Tafalla in southern Ecuador, Humboldt and Bonpland, (HBK, 1788–1850), who collected in and described plants from all three of the countries treated here, and Bentham (1844) who described almost 50 new species from the Guayaquil region in his report on the voyage of the Sulfur, deserve mention. Harling (1979) reviews Ecuadorian vegetation types and the literature pertaining to them and Gentry (1977, 1979) summarizes the Ecuadorian plant conservation picture.

C. Peru

Even though not yet finished, the *Flora of Peru* (Macbride, 1936–1961) is by far the most significant floristic treatment of Andean or upper Amazonian plants (see Gentry, 1980, for a brief history of this project and index to the familial treatments). About ¼ of the Peruvian flora remained to be treated when Macbride ceased work on the original *Flora of Peru*. The project was re-activated in 1975 as a joint project of the Missouri Botanical Garden and the Field Museum, Chicago, in collaboration with the Universidad de San Marcos and Universidad Nacional de Amazonia Peruana, emphasizing additional collections as well as completion of publication of the remaining familial treatments for the *Flora*. A treatment of ca. ¼ of the Peruvian ferns was published separately by Tryon (1964) and work on the remainder of the Peruvian ferns by Stolze & Tryon is currently under way. A. Gentry (MO) and M. Dillon (F) are in charge of the current *Flora of Peru* Project. The "new" *Flora of Peru* (e.g. Jones, 1980; Dillon, 1981, 1982), being published in "Fieldiana, Botany," has an expanded format and relies mostly on treatments by taxonomic specialists. Many of the treatments of the original *Flora of Peru* were very incomplete, especially in Amazonia (see Gentry, 1980). Among other Peruvian floristic works are treatments of plants of the coastal lomas by Ferreyra (e.g., 1961, 1977), the Cuzco region (Herrera, 1941; Vargas, 1948), parts of Ancash (Cerrate, 1957; Smith, in prep.), and Weberbauer's (1911) classic *Pflanzenwelt der Peruanischen Anden* (translated to Spanish as *El Mundo Vegetal de los Andes Peruanas*), largely restricted to Andean and coastal regions of the country. An updated summary, in English, is included in the first volume of the *Flora of Peru* (Weberbauer, 1936). Williams' (1936) *Woods of Northeastern Peru*, although restricted to trees, remains the main published floristic reference for Amazonian Peru. There is also a recent compilation of some Peruvian tree species by Encarnación. In the last few years floristic accounts of the species that occur in the arboretum of Jenaro Herrera (a joint Swiss-Peruvian agronomic and forestry project on the Río Ucayali) have been published in "Candollea" for several woody plant families (e.g., Leguminosae (Bernardi et al., 1981; Stutz & Spichiger, 1982), Vochysiaceae (Bernardi & Spichiger, 1981), Moraceae (Spichiger 1983), Chrysobalanaceae (Spichiger & Masson, 1984), and Meliaceae (Encarnación et al., 1984).

By far the most important early floristic work in Peru is the *Flora Peruviana et Chilensis* by Ruiz & Pavon (1798–1802; see Steele, 1964, for a most interesting summary of the life and work of Ruiz and Pavon). Many of the plant species collected and described by Ruiz and Pavon have yet to be rediscovered, despite intensified modern plant collecting. Important early work on the plants of Peru was also done by Poeppig (Amazonian Peru [and Chile]) (1835–1836; Poeppig & Endlicher, 1835–1843), Humboldt & Bonpland (HBK., 1788–1850), Spruce (see 1908 travelogue), and many others.

Although not strictly floristic, two useful references for Peruvian plants are Soukup's (1970) dictionary of common and scientific names for Peruvian plants and Encarnación et al.'s (1984) bibliography of taxonomic references for Peruvian (and many other neotropical) plants.

VII. Present State of Inventory, Institutions Involved, and the Future of Inventory

One encouraging trend is an increased emphasis in all three countries on the training of new young botanists who will develop their own active collecting and research programs. The Master's degree program coordinated by E. Forero in Bogotá was an excellent example of this sort, and many of its alumni are now actively studying the floras of the regions around their home universities in all parts of the country. Several Ecuadorian botanists have recently received masters degrees from Aarhus, Denmark, and one studied in the Colombian program; all now have responsible positions and are maintaining interest in field work. From Peru, several botanists have recently received master's training at INPA in Brazil, at St. Louis, at Birmingham, U. K., at Geneva, and at Aarhus. A few of these recent graduates have been able to develop active programs of field work and plant collecting.

In all three countries additional funds for field work are urgently needed. This situation is especially acute in Peru where the country's economic plight is reflected in a dearth of funds for urgently needed botanical inventory. In recent years Colombia has made floristic inventory a high national priority, although the official rhetoric has been slow to be transformed into funds for actual botanical field work. Another favorable situation in Colombia is the growth of active new botanical institutions, well situated to study the floras of different parts of the country. Colombia is discussed more fully in the paper by Forero in this volume.

In Ecuador an infusion of money and manpower from Aarhus, Denmark has been instrumental in invigorating the systematic botany program at Universidad Católica in Quito, which has also been boosted by construction of a new biology building with a large herbarium facility. Interactions with Missouri botanists have led to collaborative floristic projects and helped stimulate interest in systematic botany at the University of Guayaquil. A new interest in systematic botany of trees has developed as part of the forestry program at the Ministerio de Agricultura with two new positions created specifically for a floristic inventory of Ecuador's trees: this program has included close interactions with the Missouri and New York Botanical Gardens. Perhaps the most significant development in floristic inventory in Ecuador has been the "discovery" by botanists of the rich Amazonian region of the country. The local florula approach (see Gentry, 1978) has rather rapidly contributed much new knowledge of the coastal region, where speedy inventory is absolutely essential due to the rapid loss of native forests. Currently a focus on forestry and useful plants seems to be providing impetus for inventory in Ecuadorian Amazonia. Although the situation is improving, many more botanists and much additional support is needed if anything approaching a thorough floristic inventory of Ecuador is to be completed before its forests are destroyed by that country's burgeoning population.

The most active Peruvian institutions in floristic inventory are the herbaria of the Universidad Nacional Mayor de San Marcos and the Universidad Nacional de Amazonia Peruana. Both of these herbaria have staff members active in floristic inventory work and both have also received many thousands of recent collections generated by the Flora of Peru project. San Marcos is in effect the National herbarium but has been most active in surveying the coastal lomas and Andean uplands. A full time collector for the Flora of Peru Project is based at San Marcos and concentrating his collecting efforts in the "ceja de la selva" and adjacent central Amazonia. The Iquitos herbarium has concentrated on the plants of the Amazonian lowlands. In tradition rich Amazonia, a logical focus of this herbarium has been on ethnobotany. A full time collector for the Flora of Peru Project is also based at the AMAZ herbarium which receives the first set of his duplicates. The University of Trujillo has also traditionally had a very active plant collecting program focused in northern Peru. Very recently two other herbaria have begun to develop into important centers for floristic inventory. The University of Cuzco herbarium is in the process of acquiring the personal herbarium of Dr. C. Vargas, which contains more than 20,000 specimens including many types. That addition, plus an enthusiastic new curator, and many incoming duplicates from a new full time Flora of Peru project collector based at the University bode well for the future of floristic work at this institution. Similarly the forestry herbarium at La Molina University in Lima has a dynamic new curator and is rapidly developing into a major center for floristic inventory. There are plans to combine several small separate herbaria at La Molina, including a collection of Weberbauer, very rich in types, with the active forestry one. La Molina is the seat of the Nature Conservancy's Peruvian Data Center, a relationship that has helped to generate new interest in biotic inventory at the University. Thus a framework for effective floristic inventory by Peruvian institutions is established, with the chief obstacle being to obtain the necessary funding both for field work and maintenance of the existing collections.

VIII. Summary

To summarize the situation: (1.) Current levels of field-oriented research and plant collecting are inadequate in all three countries; the established personnel and institutions are clearly not able to get the work done with the facilities available. (2.) The only likelihood of anything approaching an adequate inventory in the foreseeable future is through increased personnel and facilities, preferably working in a decentralized manner in the various universities in or near the lowland tropical areas of the various countries. (3.) In all three countries the needed personnel are being trained, but there are inadequate facilities, funds, and experience. (4.) Increased field work by foreign botanists is desirable, but unlikely to be forthcoming to the degree needed, due to financial and logistical constraints. (5.) A strong push to increase the ranks of the cadre of new young botanists (working in or near the tropical forests of their countries, *not* in Lima, Quito, or Bogotá) and their capabilities is needed. Support of research projects by these people, and contact and collaboration, and especially orientation, from and with outside botanists and institutions (e.g., through specimen exchange and identification, provision of needed research literature, payment of field expenses, arrangement of visits to foreign herbaria) would seem to be good places to start.

IX. Literature Cited

Acosta-Solis, M. 1966. Las divisiones fitogeograficas y las formaciones geobotanicas del Ecuador. Rev. Acad. Colomb. **12**: 401–447.

———. 1977. Ecología y Fitoecología. Casa de la Cultura Ecuatoriana. Quito.

Bentham, G. 1844. The Botany of the Voyage of H. M. S. Sulphur. London.

Bernardi, L. & R. Spichiger. 1981. Las Vochysiaceas del Arboretum Jenaro Herrera (provincia de Requena, departamento de Loreto, Peru). Candollea **36**: 131–144.

———, **F. Encarnación, & R. Spichiger**. 1981. Las Mimosoideas del Arboretum Jenaro Herrera (provincia de Requena, departamento de Loreto, Peru). Candollea **36**: 301–333.

Berry, P. 1982. The systematics and evolution of *Fuchsia* sect. *Fuchsia* (Onagraceae). Ann. Missouri Bot. Gard. **69**: 1–198.

Cerrate, E. 1957. Notas sobre la vegetación del Valle de Chiquian, Folia Biol. And. **1**: 9–39.

Diels, L. 1937. Beitrage zur kenntnis der Vegetation und Flora von Ecuador. Bibl. Bot. **29(116)**: 1–190.

———. 1938–1942. Neue Arten aus Ecuador I-V. Notizbl. **14**: 25–44; 323–241. Notizbl. **15**: 23–58; 366–393; 783–787.

Dillon, M. 1981. Family Compositae: Part II. Tribe Anthemideae. Flora of Peru. Fieldiana, Bot., n. ser. **7**: 1–21.

———. 1982. Family Compositae: Part IV. Tribe Cardueae. Flora of Peru. Fieldiana Bot., n. ser. **10**: 1–8.

Dodson, C. & A. Gentry. 1978. Flora of the Río Palenque Science Center. Selbyana **4**: 1-628.

———. 1987. Flora de Capeira. Banco Central del Ecuador, Quito. (in press).

Dodson, C., A. Gentry, & F. M. Valverde. 1985. La Flora de Jauneche, Los Rios, Ecuador. 512 pp. Banco Central del Ecuador, Quito.

Encarnación, F. 1983. Nomenclatura de las especies forestales comunes en el Peru. Proy. PNUD/FAO/PER/81/002. Lima.

Encarnación, F., L. Ramella, & R. Spichiger, 1984. Las Meliaceas del Arboretum Jenaro Herrera (provincia de Requena, departamento de Loreto, Peru). Candollea **39**: 693-713.

———, R. Spichiger, & J. Mascherpa. 1982. Bibliografia selectiva de las familias y de los generos de Fanerogamas. Boissiera **34**: 1-195.

Ferreyra, R. 1961. Las lomas costañeras del extremo sur del Peru. Bol. Soc. Arg. Bot. **9**: 87-120.

———. 1977. Endangered species and plant communities in Andean and coastal Peru. Pages 150-157 in G. Prance & T. Elias (eds.), Extinction is forever, New York Bot. Gard.

Forero, E. 1977. Index of Colombian herbaria. Taxon **26**: 488-491.

Foster, R. 1988. Floristic composition of the Manu Floodplain forest. In A. Gentry (ed.), Four Neotropical Forests. Yale University Press. Boissiera **34**: 1-195.

Gentry, A. 1977. Endangered plant species and habitats of Ecuador and Amazonian Peru. Pages 136-149 in G. Prance & T. Elias (eds.), Extinction is forever. New York Bot. Gard.

———. 1978. Floristic knowledge and needs in Pacific Tropical America. Brittonia **30**: 134-153.

———. 1979. Extinction and conservation of plant species in tropical America: a phytogeographical perspective. Pages 100-126 in I. Hedberg (ed.), Systematic botany. Plant Utilization, and Biosphere Conservation.

———. 1980. The Flora of Peru: A conspectus. Fieldiana. Bot., n.s. **5**: 1-11.

———. 1981. Distributional patterns and an additional species of the *Passiflora vitifolia* complex: Amazonian species diversity due to edaphically differentiated communities. Plant Syst. and Evol. **137**: 95-105.

———. 1982a. Phytogeographic patterns in northwest South America and southern Central America as evidence for a Chocó refugium. Pages 112-136 in G. Prance (ed.), Biological diversification in the tropics. Columbia University Press.

———. 1982b. Neotropical floristic diversity: phytogeographical connections between Central and South America: Pleistocene climatic fluctuations, or an accident of the Andean orogeny? Ann. Missouri Bot. Gard. **69**: 557-593.

———. 1985. Algunos resultados preliminares de estudios botanicos en el Parque Nacional del Manu. Pages 2/1-2/24 in M. Rios (ed.). Reporte Manu. Centro de Datos para la Conservación, La Molina University, Lima.

———. 1986a. Endemism in tropical versus temperate plant communities. Pages 153-181 in M. Soule (ed.), Conservation biology: the science of scarcity and diversity.

———. 1986b. Sumario de patrones fitogeograficos neotropicales y sus implicaciones para el desarrollo de la Amazonia. Rev. Acad. Col. Cienc. **16**: 101-116.

———. 1987. Tree species richness of upper Amazonian forests. Proc. Natl. Acad. Sci. (in prep.)

——— & C. Dodson, 1987. Diversity and Biogeography of Neotropical Vascular epiphytes. Ann. Missouri Bot. Gard. **74** (in press).

Harling, G. 1979. The vegetation types of Ecuador—a brief survey. Pages 165-174 in K. Larsen & L. Holm-Nielsen (eds.). Tropical botany. Academic Press, London.

Herrera, F. L. 1941. Sinopsis de la Flora del Cuzco. Lima.

Humboldt, F., A. Bonpland, & C. Kunth. 1788-1850. Nova Genera et Species Plantarum. Paris.

Jones, S. 1980. Family Compositae: Part I. Tribe Vernonieae. Flora of Peru. Fieldiana, Bot., n. ser. **5**: 22-73.

Little, E. 1969. Arboles Comunes de la Provincia de Esmeraldas. FAO/SF: 76:ECU 13. Rome.

Macbride, J. F. 1936-1961. The Flora of Peru. Publ. Field Mus. Nat. Hist., Bot. Ser. **13**.

Poeppig, E. F. 1835-1836. Reise in Chile, Peru und auf dem Amazonestrome wahrend der Jahre 1827-1832. Leipzig.

——— & S. Endlicher. 1835-1845. Nova genera ac species plantarum. Leipzig.

Ruiz, H. & J. Pavon. 1798-1802. Flora Peruviana et Chilensis. Madrid.

Soukup, J. 1970. Vocabulario de los Nombres Vulgares de la Flora Peruana. Colegio Salesiano, Lima.

Spichiger, R. 1983. Las Moraceas del Arboretum Jenaro Herrera (provincia de Requena, departamento de Loreto, Peru). Candollea **38**: 17-79.

——— & D. Masson. 1984. Las Crisobalanaceas del Arboretum Jenaro Herrera (provincia de Requena, departamento de Loreto, Peru). Candollea **39**: 13-43.

Spruce, R. 1908. Notes of a Botanist on the Amazon and Andes. A. R. Wallace (ed.), MacMillan, London.

Steele, A. R. 1964. Flowers for the King: The expédition of Ruiz and Pavon and the Flora of Peru. Duke Univ. Press, 1964. (Spanish Ed., Ediciones de Serbal, Barcelona, 1982).

Stutz, L.-C. & R. Spichiger. 1982. Las Cesalpinioideas y Faboideas del Arboretum Jenaro Herrera (provincia de Requena, departamento de Loreto, Peru).

Svenson, H. K. 1945. Vegetation of the coast of Ecuador and Peru and its relation to the Galapagos Islands. Amer. Jour. Bot. **33**: 394-498.

Terborgh, J. & B. Winters. 1982. Evolutionary circumstances of species with small ranges. Pages 587-600 in G. Prance (ed.), Biological diversification in the tropics. Columbia Univ. Press, N.Y.

———, J. Fitzpatrick, & L. Emmons. 1984. Annotated checklist of bird and mammal species of Cocha Cashu Biological Station, Manu National Park, Peru.

Tryon, R. 1964. The ferns of Peru. Polypodiaceae (Dennstaedtiaceae to Oleandreae). Contr. Gray Herb. **194**: 1-253.

Valverde, F. de M., G. R. de Tazan, & C. G. Rizzo. 1979. Cubierta Vegetal de la Peninsula de Santa Elena. Publ. Fac. Cienc. Nac. Univ. Guayaquil **2**: 1-236.

Vargas, C. 1941. Addenda a Sinopsis de la Flora del Cuzco de F. L. Herrera, 1941. Rev. Univ. Cuzco **94**: 1-36.

Weberbauer, A. 1911. Pflanzenwelt der Peruischen Anden. Leipzig.

———. 1936. Phytogeography of the Peruvian Andes. Publ. Field Mus. Nat. Hist., Bot. Ser. **13(1)**: 13-81.

Williams, Ll. 1936. Woods of northeastern Peru. Field Mus. Nat. Hist., Bot. Ser. **15**: 1-587.

Brazilian Amazon

Douglas C. Daly and Ghillean T. Prance

Contents

I. Description of the Region	402
A. Geographical Extent and Topography	402
B. Geology	402
C. Origin of the Flora	403
D. Geography, Climate and Soils	404
II. Occurrence of Tropical Forests	405
III. Review of Vegetation Maps and Phytogeography	408
A. Vegetation Maps	408
B. Phytogeography	410
IV. Magnitude of the Floristic Inventory	414
A. Specimens in Regional and Foreign Herbaria	414
B. Publications on the Flora	415
V. Collection Densities, Endemism	416
VI. Inventory and Study of Various Plant Groups	417
A. Taxa Well or Poorly-Collected	417
B. Specialists	417
VII. Threatened Areas	417
VIII. Resources for Continuing the Inventory	421
A. Infrastructure	421
B. Human Resources	422
C. Programming	422
D. Finances	422
IX. Suitability of Past, Present, and Planned Inventory	423
X. Additional Resources Needed	423
XI. Literature Cited	424

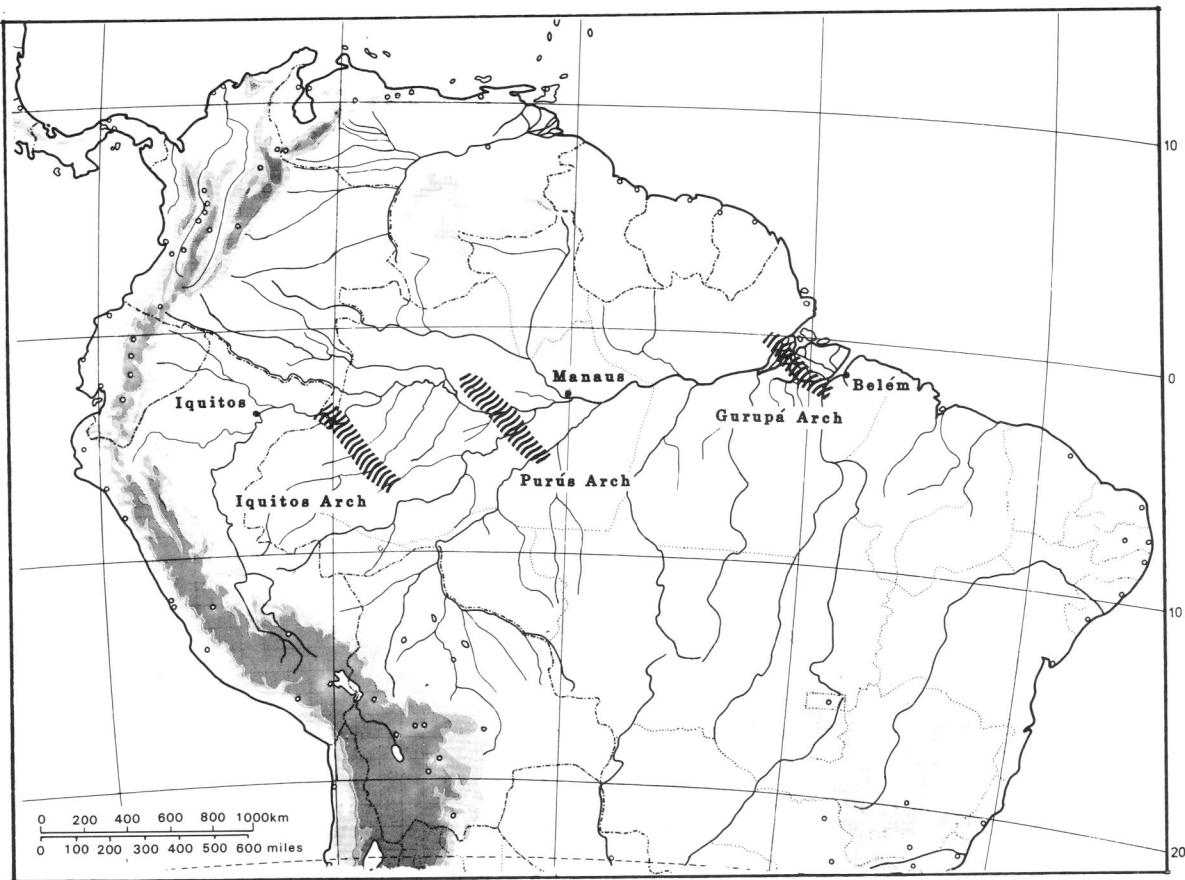

FIG. 1. Location of the three arches, or ancient zones of uplift, in Amazonia. Adapted from Putzer (1984).

I. Description of the Region

A. Geographical Extent and Topography

The Amazon basin contains the single greatest expanse of tropical forest in the world, and most of it is in Brazil. The basin is 3,500 km long and 300–1,000 km wide (Irion, 1978), and drains 7,050,000 sq. km of land. The catchment extends approximately from 79°W (the Río Chamaya, Peru) to 46°W (the Rio Palma, Goiás, Brazil), and from 5°N (Rio Cotinga, Roraima, Brazil) to 17°S (Alto Araguaia, Mato Grosso, Brazil) (Sioli, 1984).

Estimates of the size of Amazonia range from 5,000,000 sq. km (Sioli, 1984) to 6,000,000 sq. km (Pires, 1972). Pires (1973) estimated that 3,700,000 sq. km of Amazonia lie in Brazil, of which 3,374,000 sq. km is forest. This is significantly less area than the 4,975,527 sq. km decreed by the Brazilian government as the Legal Amazon (Oliveira, 1983), but the latter was meant to be an economic/political and not a natural unit.

The Amazon is also the greatest river system in the world. At 6,518 km in length, it is only the second longest river, but its discharge of 175,000 m³/sec is by far the greatest, comprising some 1/6 to 1/5 of all the discharge into the world's oceans. This enormous flow is associated with its great width. At Iquitos, 3,600 km from the Atlantic, it is 2 km wide, and below the confluence with the Rio Negro, still 1,500 km from the ocean, it is 4–5 km wide. Almost all the basin is low-lying, such that the lower Amazon shows a gradient of only 1 cm/km at low water (Sioli, 1984).

Even today it is difficult to dispel the myth of the Amazon hylaea as a single, continuous, homogeneous, repetitive mass of forest which is the same throughout. The Amazon region has a complex and incompletely known geological history. It is neither uniform nor simple geologically, topographically, or climatologically. The Amazon hylaea contains a wide range of vegetation types, both forest and non-forest, and the intricate phytogeographic patterns it presents have yet to be fully deciphered.

B. Geology

While in certain respects Amazonia is geologically very

young, its superstructure is ancient, and the history of its formation began over two billion years ago. The Amazon basin is a vast intracratonic depression which originated along a zone of weakness in the Precambrian Craton (or shield). It lies between two parts of the Precambrian Craton, the Guiana Shield to the north and the Brazilian Shield to the south. Various authors have described it as a rift valley or as a region having a graben character because it is elongate and is bordered by faults on its longer (north and south) sides. Consolidation, metamorphosis and folding of the crystalline shields occurred from 3,500 to 600 million years ago (Putzer, 1984).

The orogenies of that period produced three arches, or areas of uplift. These are sigmoid in shape but run basically NW-SE across the basin (see Figure 1). They are: 1) the Iquitos Arch near Iquitos, Peru, which crosses Acre; 2) the Purús Arch, west of Manaus, Brazil; and 3) the Gurupá Arch, east of the mouth of the Rio Xingu. These areas of uplift, formed of crystalline rock, are believed to have isolated parts of the basin from each other during marine transgressions (Putzer, 1984). The Iquitos and Gurupá arches correspond roughly to two of the phytogeographical boundaries defined by Prance (1973, 1978), and the Purús Arch corresponds more or less to the division used in discussions of the upper versus the lower Amazon. The phytogeographic significance of the arches is discussed in section III of this paper.

During the Paleozoic a series of marine transgressions from the direction of the Andean geosyncline occurred, yielding deep sediments. These shallow seas were of varying durations and extents. Some intruded as far as the lower Amazon basin; the Monte Alegre sandstones in Amapá, which are covered by savanna woodlands and deciduous forests, are derived from Paleozoic sediments. Paleozoic sediments and evaporites remain today as a syncline over the rift valley or graben and are exposed as bands along the northern and southern fringes of the Brazilian and Guiana shields, respectively, in the eastern basin. During the early Mesozoic—from the Triassic to the lower Cretaceous—there was little sedimentation and apparently no transgressions.

The upper Cretaceous to the mid-Tertiary, however, comprised a period of intense fluvial sedimentation. As South America began to separate from Africa and the Guiana and Central Brazilian floras were evolving, the upper Amazon was a catchment for westward-draining rivers, while the lower Amazon drained east into the new Atlantic sea. Marine transgressions from both directions occurred during this time: the marine sediments of Marajó Island (at the mouth of the Amazon) were deposited by a transgression of the Atlantic, and the Moa Formation of Acre was deposited by a Pacific transgression. Also, due to faults and fractures, extensive diabasic intrusions occurred between the Purús and Solimões rivers, between the Madeira and Amazonas rivers, and along the fault lines on the edges of the lower graben where the Paleozoic outcrops also occur (Putzer, 1984).

During the upper Tertiary the final and effective separation of the African and South American continents occurred. At the same time, North and South America moved toward each other, and the uplift of the Andes occurred. The latter event blocked the westward drainage of the upper Amazon, but there is some debate among geologists as to what the consequences were. One hypothesis is that almost the entire basin became a vast inland sea which finally broke through toward the Atlantic near Óbidos, an area of high ground and the narrowest point of the basin (Fittkau, 1974). Another hypothesis suggests that the Andean uplift engendered a large marsh and lake region, as well as the "junction" of the middle and upper Amazon basin (Putzer, 1984). There is also some disagreement as to what occurred during the Calabrian Transgression of the Atlantic at the Pliocene-Pleistocene interface. Putzer (1984) and Sombroek (1966) suggest that the rise in sea level caused tremendous ponding in the basin. In addition, they claimed that the Belterra clays, found in a region of the middle basin characterized by dissected plateaus, were deposited by the resulting lake(s). Irion (1978), on the contrary, asserted that there was no vast lake, and that because of their structure and chemistry the Belterra clays must have been formed in situ by weathering.

The Amazon is a cone-shaped region of extremely deep sediments just beyond the mouth of the Rio Amazonas in the Atlantic Ocean. Deposition of the sediments, now 11,000 m deep in some places, began in the Pliocene.

The glacial cycles of the Quaternary, accompanied by dramatic changes in sea level, affected the evolution of the Amazonian landscape. During periods of high sea-level, drainage of the basin was impeded, resulting in ponding, with sediments being deposited over much of the basin. During periods of low sea-level, regional erosion was accelerated (Putzer, 1984). The glacial cycles were also cycles of climatic change consisting of alternating cool-dry and warm-humid phases. The effect of the glacial cycles on Amazonian vegetation and flora is discussed below.

C. Origin of the Flora

Disagreement about the geological history of the Amazon basin consequently leaves doubts about the origins of the Amazon flora. Little has been written on the origins of the flora, and almost nothing on the evolution of Tertiary floras of northern South America. There seems to be little doubt that the Amazon flora is derived primarily from the ancient floras of the Guiana and Central Brazilian crystalline shields (Prance, 1978). Rizzini (1963) listed a number of closely related species pairs between Amazonia and the Central Brazilian Cerrado (savanna woodland). However, there has been little conjecture about other sources for the autochthonous evolution of the flora. Much of the basin was above water during the upper Cretaceous and Tertiary, so there must have been contact with other Tertiary floras.

The significance of the Iquitos, Purús, and Gurupá arches has not been examined by phytogeographers. They divide the basin into several regions in which the flora may have followed somewhat different courses of evolution of the flora. For example, these regions appear to have been affected differently by the Paleozoic marine transgressions from the Pacific; the basin drained in two directions from the appearance of the Atlantic "sea" until the uplift of the Andes; and the regions were probably affected differently by ponding during periods of high sea level.

It is also important for future phytogeographic studies to determine what areas were under water during the interglacial ponding, and what areas remained forested during the arid glacial periods.

D. Geography, Climate, and Soils

In its basic physical characters Amazonia can be divided into 1) the upper Amazon, more or less west of the Purús Arch and thus including the area west of the Iquitos Arch; 2) the lower Amazon, or middle basin; and 3) the estuary, east more or less of the Gurupá Arch (see Fig. 1). The upper Amazon is a broad region filled with sediments derived from the weathering crust of Andean bedrock. The upper reaches of the rivers are characterized by deep V-shaped valleys until they reach the lowlands, where they make *lateral* incisions in the landscape and form mazes of meanders, oxbows, and continuously changing courses (Sioli, 1984). The terrain is low-lying and topographically rather uniform, and the distinction between terra firme and floodplain (várzea) is often blurred. It has been suggested that massive ice melts in the Andes resulted in widespread catastrophic flooding and sedimentation in eastern Bolivia and Acre as recently as 5,000 BP (Campbell & Frailey, 1984), meaning that the forests there are only that old and that the terra firme of the region is a recent fluvial deposit. The rivers do not form very effective barriers to dispersal among populations of most organisms, nor are there many other readily identifiable geographical barriers to gene flow. Rainfall is abundant and well distributed throughout the year except in the extreme southwestern portion of this region.

In the upper Amazon the soils are either recent or have developed only over the past 1–2 million years from upper Cretaceous and Tertiary sediments. The parent rocks are poorly known (Irion, 1978).

The lower Amazon is a narrower basin bounded to the north and south by the crystalline shields. It is geologically, topographically and climatologically more diverse than the upper Amazon, although biologically it is not. In the western portions of the lower Amazon, the soft Tertiary sediments have been dissected into a network of small steep-sloped valleys, while in the eastern portion there are still intact sedimentary plateaus, such as that of Belterra east of the Rio Tapajós. The vast majority of the soils of lower Amazonia are part of the Barreira Formation and have been developing for 20 million years from sediments derived from Precambrian rocks of the shields. On level ground these tend to be nutrient-poor kaolinitic soils several meters thick, while on slopes they tend to be sandy podzols. Along the northern and southern periphery of the valley, the parent rocks of the strips of Paleozoic outcrops are relatively impermeable, so the resulting soils are not quite so weathered nor so infertile (Irion, 1978).

Most of the affluents of the lower Amazon share certain features. In their upper reaches they have well-defined beds, stable banks, kidney-shaped islands, and often rapids (marking the interface of the basin with the crystalline shields). However, when they reach the lowlands, the current slows, they broaden rapidly, and they form long and narrow islands—the sedimentation zone—from which the waters emerge "decanted", with little suspended matter. Some of these tributaries form huge mouth-bays, which are rich in plankton and have bottoms of many meters of soft mud. The sandy banks in the mouth-bays are products of local *lateral* erosion due to wave action during high-water storms (Sioli, 1984; see also the descriptions of Sternberg, 1975).

The lower Amazon shows a variety of topographies, especially toward the edges of the basin, where there are numerous small mountain ranges, several of them geologically unusual. Along the Guiana Shield these tend to be oriented east-west and reach significant elevations, while most of the southern ranges have a more north-south orientation.

Rainfall is well distributed throughout the year in the upper Amazon except in extreme southwestern Amazonia, but in most of the lower Amazon there is a distinct dry season. In fact, one large strip of land cutting NW-SE across the Rio Trombetas and the lower Rio Xingu receives only 1,750 mm of rain annually (Nimer, 1977). This is the so-called Aw belt; the Aw refers to Koeppen's (1901) system of climate classification. Some savannas and deciduous forests occur in this region, which is believed to have served as a corridor for exchange between the Llanos of Venezuela and Colombia and the Brazilian Planalto savannas during the dry glacial periods of the Quaternary. The Aw belt may also explain the apparent disjunct distributions of a number of plant taxa between the Guianas and central Amazonia.

The boundaries between floodplains and terra firme in the lower Amazon are generally distinct. Here the main river, some of its larger tributaries, and some of the mountain ranges provide effective barriers to dispersal of certain organisms.

The estuarine region of the Amazon is a great funnel-shaped area where the river divides into several arms. The waters of the estuary affluents move back and forth with the tides, making little forward progress. The sedimentation zone of the main river has reached the mouth. This is not the case with its clear- and black-water affluents. There is no delta because the strong Brazil Current pushes the sediment-laden effluent of the main river up the coast of Amapá, where an 80 km wide band of the coast is composed of recent Amazon sediments.

At the mouth of the Amazon is the great island of Marajó, with an area of 48,000 sq. km. The inner (western) portion of the island is composed of recent sediments while the outer portion is actually made of Tertiary sediments. This is reflected in the vegetation, with savanna dominating the former and forest the latter. Marajó and the peripheral islands are moving outward due to a tectonic shift. The true delta zone is proximal to (SW of) Marajó. Here the long and narrow inner islands are fluvial. Large areas of the estuary are flooded twice a day, and during the wetter season much of Marajó is inundated because discharge of its waterways is impeded by the main river.

Extensive Quaternary floodplains occur along the length

of the main river as well as along lower reaches of some of its major tributaries. Some of them are now terra firme, especially along the lower Amazon, because their sediments were deposited during interglacial periods of ponding when sea-level—and thus river-level—were slightly higher than now. These Pleistocene várzea or Pleistocene terrace soils have weathered to varying degrees but most remain relatively fertile (Irion, 1978).

The remaining floodplains are overlain by recent sediments which continue to be deposited annually, or diurnally in the estuary. They are almost all forested, and cover 50,000–60,000 sq. km (Sioli, 1984) or more. According to Pires (1972) there are approximately 70,000 sq. km of floodplain forests and 15,000 sq. km of floodplain grasslands. Some of the larger areas of annually inundated floodplain forests occur above and below the confluence of the rios Negro and Branco, on the middle Rio Paru in Pará, along the lower Rio Solimões/Amazonas, and especially along the Solimões at the confluence of the lower Rio Japurá.

The floodplain soils vary greatly in structure and fertility. This depends greatly on the parent material of the region being drained by each river system, and not on the adjacent terra firme soils. Consequently, Sioli (1951) defined three basic river types:

1) The white-water rivers of the upper Amazon which drain the Andes are alkaline (pH 6.2–7.2) turbid, and carry a heavy sediment load. Their floodplains have relatively fertile clay soils.

2) The clearwater rivers of the lower Amazon which drain the crystalline shields vary greatly in acidity (pH 4.5–7.8). They are virtually transparent and have a very low sediment load.

3) The acidic blackwater rivers (the Rio Negro system) drain the flat country between the Andes and the Guiana Shield. Their waters are very acidic (pH 3.8–4.9) and relatively transparent. Their color, something like weak tea, is derived from humic acids as they leach through sandy soils. These floodplains tend to be sandy and very poor in nutrients.

The distinctions between the river types is not always clear-cut geographically. For example, blackwater streams are found in sandy patches of the Barreira Formation—the region of the clearwater rivers—in the Serra dos Parecis, and in the sandy campos of the Rio Cururú near the Serra do Cachimbo. Clearwater streams are of various pH levels and are widespread, draining catchment areas of smooth, even relief (Sioli, 1984). Because of these distinctions, botanists collecting in floodplain forests should describe on their herbarium labels the type of river inundating the forests.

The floodplains present a distinctive landscape. In the upper Amazon and along the most of the main river, as the water rises and invades the land, the larger suspended particles are sedimented first, resulting in levees (barrancos) which slope away from the river. Seasonal lakes appear, some as large as 2,000 sq. km at high water. Rivers which flow independently at low water may be fused at high water, as occurs among some tributaries of the lower Rio Branco (Pires & Prance, 1985). The floodplains are especially extensive on the upper Amazon, where river levels can change 20 m between high and low water, and there is a richer flora with more large trees. There are 2,000 sq. km of floodplain on the Rio Madeira, 2,500 sq. km on the Purús, and 25,000 sq. km on the main river (Solimões). Between the Purús and Gurupá arches the floodplains are characterized by the presence of large grassy margins and "floating meadows" (Junk, 1970). Much of the estuarine floodplains are flooded twice daily by the tides. Here the majority of the floodplain grasslands occur, and palms are a strong element in the forests.

II. Occurrence of Tropical Forests

Until the 1950's, various authors proposed radically different limits for Amazonia. Then Ducke and Black (1953) discussed the limits in detail for some regions. In the same year Soares (1953), using aerial photos, field studies, overflights, and a massive bibliography, presented an entire book on the southern and eastern limits of Amazonia. He also presented maps showing the limits proposed by previous authors. Since that time most botanists have accepted Soares's limits with few modifications. Figure 2 gives an approximation of the natural boundaries of Amazonia, which must be distinguished from the broader economic/political boundaries set by the Brazilian government as "Amazonia Legal" in 1953 (Oliveira, 1983).

As Ducke and Black noted, "The only natural limits of Amazonia are the Atlantic and the Andes; on the north and south extremes the hylaea (rain forest) is gradually replaced by the flora of the drier neighboring countries" (Ducke & Black, 1953: p. 2). The boundaries of the Amazon forests—the Hylaea—are difficult to set for four reasons: 1) they do not correspond closely with geological boundaries; 2) the origin of the Amazon flora complicates attempts to set the limits using floristics; 3) purely physiognomic approaches tend to exclude regions whose woody elements are Amazonian; and 4) various kinds of transitions are found in different parts of the periphery of the region. These transitions may be very gradual and they often occur in a mosaic pattern.

A map of the geological Amazon basin will not show the limits of floristic Amazonia. The trees of Amazonia are highly adapted to weathered, nutrient-poor soils which cover most of the basin (Jordan, 1987), and in many places they occur beyond the actual basin onto the even more highly weathered soils on the parent rocks whose erosion gave rise to most of the Amazon sediments. The forests extend as far as climate and soils and other edaphic constraints allow.

Floristic delimitation of Amazonia presents several problems. Like its soils, the Amazon flora is derived primarily from the surrounding regions, especially the crystalline shields to the north and south (see section I). For that reason, only three very small families (Polygonanthaceae, Duckeodendraceae, and Dialypetalanthaceae, although the first two are not accepted by all taxonomists) are endemic to Amazonia. Of 341 genera of woody plants examined by Rizzini (1979), 188 or 55% occur in both Amazonia and the Cerrado of Central Brazil. In addition, many of these genera show

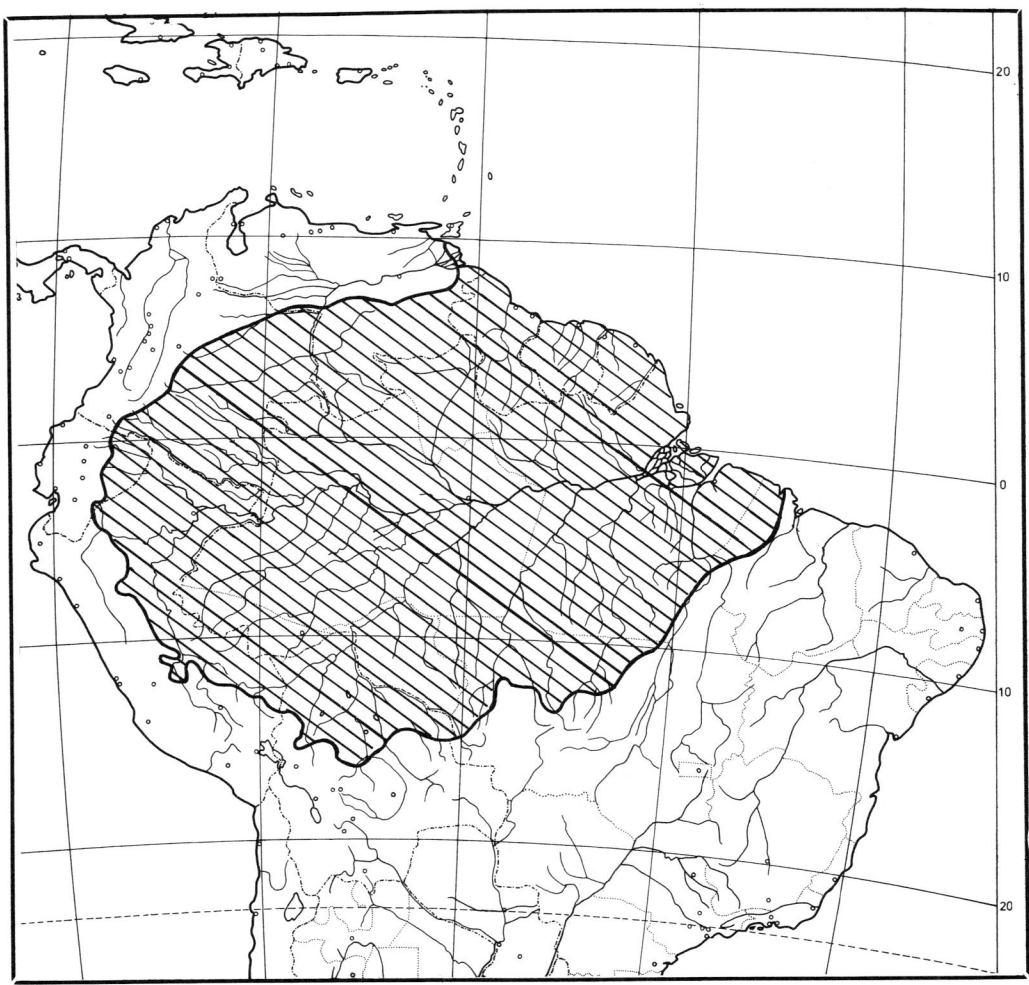

FIG. 2. The limits of Amazonian vegetation.

closely related species-pairs between the Hylaea and the Cerrado (Rizzini, 1963). Rizzini (1963) used the distributions of such a pair in *Orbignya* to mark the limits of Amazonia in the state of Maranhão, but species concepts in that palm genus have since changed (M. J. Balick, pers. comm.). Others have attempted to define the limits of Amazonia using the distribution of the genus *Hevea* (i.e., Ducke & Black, 1953).

Also in Maranhão, Fróes (1953) delineated the limits of Amazonia in that state using the distributions of some twenty taxa he regarded as being characteristic hylaean species. One must be very careful in selecting taxa for this purpose, because there are numerous genera and thousands of species endemic to Amazonia, but their distributions are discontinuous within the region. A different complement of taxa would have to be utilized in each part of the periphery of the region, and their usefulness could only be determined after the fact. One reason for this is that a number of essentially Amazonian taxa which occur on upland forest in Amazonia penetrate various distances into the Cerrado in gallery forests, in some cases reaching the Atlantic forests of Brazil via one of several river connections (Pires, 1984).

Amazonia cannot be easily defined in physiognomic terms, certainly not in terms of the distribution of tall, dense, lowland rain forest. While some locations may present unbroken horizons of such forests, the most recent vegetation map of Amazonia (Prance & Brown, 1987) indicates that rain forest on terra firme covers only about 53% of the region, and even within this category four sub-types are recognized. The remaining area is accounted for by several types of seasonal transition forests, savannas and savanna woodlands, forests on white sand soils, several types of inundated forests, and several other vegetation types (see section III).

Some non-forested areas are included because they are highly specialized formations which have unique floras, are of limited expanse, and are situated well within the limits of the classic hylaean forests. One type includes the savanna-like formations on rocky soils of the Serra do Cachimbo, the campos of the Rio Cururu, Ariramba on the Rio Trombetas, and the Canga formation on the Serra dos Carajás (see Secco & Mesquita, 1983). Floristically they are unrelated to the Cerrado of Central Brazil. Those of the Serra do Cachimbo are the most extensive, covering some 16,000 sq. km (Pires & Prance, 1985). Another type are the campinas, xeromorphic savanna-like formations on acidic coarse white sands, which occur as isolated patches of various sizes throughout Amazonia. They often grade continuously, both physiognomically and floristically, into the surrounding forest. They can also grade into sclerophyllous forest on similar soils, often called campina forests, campinaranas, or caatingas. These are discussed further below.

There are other non-forested areas which are included in Amazonia by most authors even though they occupy large areas toward its periphery. These savannas and savanna woodlands are distinguished from the Llanos of Venezuela and Colombia and from the Cerrado of Central Brazil on floristic, physiognomic, and climatic grounds (Huber, 1982; Pires, 1972). They contain only some of the more widespread elements common to most Neotropical savannas, while the remainder of their floras are quite different and in most cases highly endemic. The woody elements generally do not present such adaptations as deep root systems, xylopodia, and vegetative reproduction which are characteristic of the Cerrado, nor does the vegetation become quite so dense as that which can occur in Central Brazil. The soils are shallower, and the climate is more humid than in the Llanos or the Cerrado.

The more extensive Amazonian savannas are found on northeastern Marajó Island, subcoastal Amapá, the upper Trombetas and Paru rivers, and most of northern Roraima. The Roraima-Rupununi savannas, which extend into Guyana and cover some 54,000 sq. km, are sometimes considered separately, because they are sparsely wooded and contain many swampy areas with *Mauritia* palms (Pires & Prance, 1985).

In addition to non-forested areas, Amazonia includes several forest types which are quite different from the classic hylaean rain forest. Most notable are the caatingas, or sclerophyllous forests on white sands, covering 30,000 sq. km or more in the upper Rio Negro of Brazil and adjacent Venezuela and Colombia (Pires, 1972). They are edaphic formations not separated climatically from the rest of Amazonia. They have strong floristic affinities with the isolated campinas and campina forests elsewhere in Amazonia. Jordan (1987) and co-workers studying this vegetation in Venezuela, have found that transition of campina to campina forest is determined by the depth of the sand and of the water table. Because these white-sand formations are difficult to separate, they can be considered phases of a single vegetation type (Pires & Prance, 1985; Prance & Brown, 1987).

The several kinds of seasonal forests which occur in Amazonia are difficult to define. They can grade continuously from dense lowland forest to Cerrado, both physiognomically and climatically, and they are poorly known floristically.

Concepts about these transition forests have been changing, something which is reflected in the recent literature. The extensive semi-deciduous forests with a strong presence of *Orbignya* (babassu palm) and often *Bertholletia excelsa* (Brazil nut) which occur toward the southeastern periphery of Amazonia in Maranhão, Pará, and Goiás, were long considered to constitute a distinct formation, the "zona dos cocaes" of Sampaio (1945). Later, Pires (1972; Pires & Prance, 1985) included them as part of the "liana forests." Most recently, however, Prance and Brown (1987) again separated out the babassu palm forests on their vegetation map. These are forests with highly irregular broken canopies entangled with lianas, and containing many Amazonian elements plus some endemics and some Cerrado elements. They cover large areas between Altamira on the Rio Xingu and Cametá on the Rio Tocantins. They occur to a lesser extent also in Roraima near the savannas, such as near the Serra da Lua. While they do appear distinct, much of the babassu forests are secondary, caused by the deforestation of the last 50 years (see Pires & Prance, 1985).

Semi-deciduous forests comprise another important type of transition forest. These forests, also essentially Amazonian with some endemics and Cerrado elements, were at first considered by Pires (1972) to occur only in a discrete narrow band of territory between the Rio Araguaia and the hylaean forests west of it. He stated that the transitions to the Cerrado in Mato Grosso are more abrupt. Later, however, Prance and Brown (1987) indicated that semi-deciduous forests frequently occur in patches along the southern periphery of Amazonia, as well as in Roraima, where their floristic composition is different.

As can be seen from the preceding discussion, each portion of the periphery of Amazonia may show a different kind of transition to the extra-Amazonian floras. Some transitions have been better studied than others. In general, the northern perimeter is characterized by rather abrupt transitions to savannas, while the southern perimeter usually shows mosaic-like transitions involving semi-deciduous and/or liana forests. An exception in the north is the mosaic of campina, low caatinga, and high caatinga on the Upper Rio Negro (see section III). Proceeding counter-clockwise from the mouth of the Amazon, in subcoastal Amapá there is a large area of savanna (campo) on rolling hills, with a depauperate flora with close affinities to the Cerrado, west of the large flooded grasslands and mangroves along the coast. North of this savanna the dense lowland forest resumes and continues up to the French Guiana border.

In northwest Amapá and northern Pará, the middle and upper reaches of the rios Trombetas and Paru show extensive savannas which reach up to the Serra do Tumucumaque and the borders of French Guiana and Suriname.

As mentioned above, the Roraima campos are sparsely wooded and have many swampy areas with *Mauritia* palm (buritizais). There are some patches of little-studied semi-deciduous forest. The northern portions are on hilly terrain and continue into Guyana and southeastern Venezuela.

Brazil borders on a part of the Guayana Highland region

of sandstone plateaus (tepuis), and in places along the Venezuelan border in northwestern Roraima territory and northeastern Amazonas state the sandy Amazonian campos give way to tall forests on the lower slopes of the tepuis, which in turn may grade into montane forests on the slopes of the tepuis. These forests are humid toward the west of this area, and they extend up to approximately 2600 m elevation. Amazonian elements tend to disappear at about 900 m (J. Steyermark, pers. comm.).

The upper Rio Negro basin is characterized by the Amazonian caatingas, described above, which extend beyond the borders with southwestern Venezuela and Colombia, where a drier climate, slightly higher elevation, and other edaphic factors mark the transition to the savannas of the Llanos region (see PRORADAM, 1979).

Amazonian dense lowland forest extends past Brazil's western borders into southeastern Colombia, Ecuador, Peru, and northern Bolivia. Near the border with northeastern Bolivia, the Serra dos Parecis and the Serra Pacaás-Novos in Rondônia are long narrow strips of high ground running NW-SE which include campos of uncertain affinities. These ranges are included in Amazonia because dense lowland forest occurs south of them, primarily in river valleys.

In Mato Grosso the Amazon forest reaches south and up into many of the southern tributaries of the Amazon, gradually narrowing to gallery forests as they extend farther south into the Planalto. As mentioned, some Amazon elements penetrate deeply into the Cerrado via these gallery forests. Pará rubber (*Hevea*) trees are found as far south as Diamantino, near Cuiabá (Pires, 1972), and a number of other Amazonian species of *Enterolobium, Parkia, Simaruba*, and *Schizolobium* are found as far south as 15° S, 59°30′ W (J. M. Pires, pers. comm.) while the limits of Amazonia are considered to end at about 12° S or farther north across most of Mato Grosso. Most of the transitions to Cerrado in the state are abrupt (Pires, 1972) but stretches of semi-deciduous transition forests do occur in places.

In southeastern Pará the boundary of Amazonia angles NE, following but not including the basin of the Rio Araguaia, crossing it and cutting across the northern tip of Goiás just south of the confluence with the Rio Tocantins. The greatest occurence of liana forest and semi-deciduous forest is in Pará, as discussed above, and they extend south into Mato Grosso.

In Maranhão the transitions to Cerrado appear to be more complex than in other areas. The boundary of Amazonia crosses the Rio Tocantins near Imperatriz, then continues east to approximately the Rio Mearim and due north to and including the island of São Luís. In northwestern Maranhão, the wettest part of the state, already the climate is quite seasonal and a significant minority of the trees is deciduous. In the center of the state is a large, highly disturbed area of mostly secondary forests and anthropic savannas, both dominated by the *Orbignya* (babassu) palms, the seeds of which yield a valuable edible oil. As noted above, extensive transition forests with a strong presence of babassu also occur in Pará, but they are not as completely dominated by the palm.

Recent work on the systematics and ecology of *Orbignya* indicates that the original vegetation of central Maranhão probably was something like that of the babassu forests of Pará (A. Anderson, pers. comm.).

The limits of Amazonia in Maranhão are confused by the dissected terrain and the roles of additional vegetation types. To the east, babassu formations, seasonally flooded grasslands and hardpan savannas intergrade irregularly not into Cerrado but into the xerophytic vegetation of Northeastern Brazil, with some Cerrado elements present. In the rolling hills and small plateaus toward the southern periphery, semi-deciduous forests, liana forests and babassu forests form a mosaic with patches of Cerrado vegetation.

III. Review of Vegetation Maps and Phytogeography

A. Vegetation Maps

Brazil and Brazilian Amazonia have a long and relatively continuous history of vegetation mapping, especially since about 1910. Almost all the maps have been purely physiognomic; few have been supplemented with floristic information. Most have utilized local terminology, which is not used consistently throughout Brazil or even throughout Amazonia. During this century the map legends and subdivisions of Amazonia have become more detailed, but at present there is a plethora of classification systems from which to choose. The radar imagery of Amazonia produced by RADAMBRASIL (1973–1981) provided valuable information on the nature and extent of some vegetation types, but they may have to be reinterpreted in order to produce a truly satisfactory vegetation map of Amazonia.

When Martius published Die Physiognomie des Pflanzenreiches in Brasilien in 1824, Brazil became the first country in the Western Hemisphere to have a vegetation map (Küchler, 1982; he gave the date of Martius' work as 1858). Also, Martius' map is reproduced in Soares (1953) and in Hueck (1957). The map showed all of Amazonia as belonging to a single province, Naias. Since that time, twenty or more vegetation maps have been made for Brazil, and at least six of Amazonia. In addition, Amazonia has been treated in some detail in a few vegetation maps of South America or the world. Vegetation maps have also been published for Maranhão, Amapá, and perhaps other Amazonia regions of Brazil (Küchler, 1970).

Küchler's reviews (i.e., 1970) of vegetation maps are very useful, especially since they provide the legends for the many maps inaccessible to most researchers. Early on, mappers already distinguished swamps (pantanos), grasslands and savannas (campos), and flooded grasslands (campos inundáveis), although these were probably not mapped in much detail. By 1937 the estimated distribution of Pará rubber (*Hevea brasiliensis* Benth.) was being used to determine the limits of the Hylaea. In most maps of Amazonia since the

1940's, the region dominated by the *Orbignya* (babassú) palm in Maranhão has been mapped as a distinct formation (see section II). In the 1950's some mappers began to separate out the floodplain forests.

After the landmark works of Ducke and Black (1953) and Soares (1953), mappers of Brazil were at least aware of the array of formations which have been treated recently, usually with the exceptions of liana forest and bamboo forest. Such is the case with the *Atlas Florestal do Brasil* (Veloso, 1966). Veloso had limited field experience in Amazonia, but his text (if not the map) contains a detailed subdivision of the Amazon vegetation.

In 1972 the Amazon was treated in two important vegetation maps, those of Pires and of Hueck and Seibert. Pires (1972) produced what was probably the best overall account of Amazonian vegetation types as a whole until recently. The map, though somewhat detailed, was unfortunately presented in a very small format. As a consequence several of the formations he discussed do not appear, including "high caatinga," inundated forests, and montane forests.

The map of Hueck and Seibert (1972), when used in conjunction with Hueck's (1966, 1972) text, is one of the best vegetation maps of Amazonia overall, even though it covers all of South America. It has a relatively large scale (1:8,000,000), it is quite detailed and legible, and it was based on extensive field work and an exhaustive survey of the literature. The text is excellent, containing a great deal of floristic information. It contains one of the four phytogeographic treatments of Amazonia (discussed later in this section) that was based primarily on original work.

In 1977 Brazil's Instituto Brasileiro de Geografia e Estatística (IBGE) produced a vegetation map of Northern Brazil, which includes most of Amazonia. While the map is not very large nor highly detailed, it is useful. The text by Kuhlmann (1977) has a good deal of floristic information and good bibliography. The discussion of Amazonian phytogeography draws heavily on Hueck (1966, 1972).

The twenty volumes of Projeto RADAM (1973–1975), or Projeto RADAMBRASIL (1975–1981), are each accompanied by a set of maps based on radar imagery. The vegetation maps are extremely detailed. The legends and the accompanying texts were prepared primarily by H. P. Veloso, using his own system of vegetation classification (Veloso, 1966), which is purely physiognomic. The RADAM vegetation and phytoecological maps are based largely on remote-sensing interpretation of geomorphology. This led to designation of an excessive number of vegetation types. Projeto RADAMBRASIL allowed for wide-ranging exploration of many previously unknown parts of Amazonia by some botanists and many professional collectors (see section VIII), and involved extensive use of helicopters. These explorations helped a great deal in the targeting of later botanical expeditions. Due to the scope of the project, the amount of time spent in each location was necessarily very short and few botanical collections were made. Therefore, the amount of floristic information in the texts is limited. Because of this, and because the texts employ a system of vegetation classification that is not widely used and that is purely physiognomic without regard to available information on soils and phytogeography, few botanists consult the RADAMBRASIL maps. Moreover, they are generally not available outside Brazil.

However, the RADAM maps can still be valuable with further interpretation. Prance and Brown (1987) have done this in preparing their vegetation map. While the scale is not large, it is one of the most detailed vegetation maps of Amazonia, and the legend is perhaps the most finely subdivided to date. The vegetation map of Brazil prepared by Tosi & Vélez-Rodriguez (1983), using Holdridge's (1947) system of classification, is a preliminary effort with little detail, especially within Amazonia.

Table 1 presents parts of the systems used in the maps of Pires (1972), Pires and Prance (1985), and Prance and Brown (1987). Since Hueck divided most of the forested parts of Amazonia into 13 phytogeographic regions, most of his forest types are implicit on Hueck and Seibert's (1972) map legend but explicit in Hueck's (1966, 1972) text. Pires's "matas pesadas" and "matas de encosta" correspond roughly to Prance and Brown's "dense lowland forest" and "dense and open hill forest on richer soils," but the latter was accorded a far greater area by Prance and Brown. Rollet (1974) has reviewed the terminology of the forests referred to here as dense lowland forests. Prance and Brown also distinguished two further types of terra firme forest, "dense lowland forest mixed with open and alluvial forest types," and "periodically inundated dense lowland forest." The former corresponds to the areas with patches of "open forest and without palms" and "open forest with palms" recognized in Pires's revised system (Pires & Prance, 1985), which appeared after the Projeto RADAMBRASIL (1973–1981) maps were available. Pires as well as Prance and Brown recognized liana forest, although the latter give it a broader definition. As mentioned in section II of this paper, Pires gives dry transition forest a limited distribution along the western edge of the Rio Araguaia basin. Prance and Brown recognized babassu palm forest and the bamboo forest of Acre as separate types of transition forest. Hueck and Seibert also mapped babassu palm forests separately. Pires as well as Hueck and Siebert mapped the caatingas or campina forests of the upper Rio Negro separately from the isolated campinas which occur sporadically throughout Amazonia; Prance and Brown considered them together, along with "campo rupestre" (savanna-like vegetation on rocky soil; see section II), as a single vegetation type. They included the Serra do Cachimbo in the category, while Pires places it in his terra firme savanna (campo) type.

Prance and Brown as well as Pires include the montane forests near the Venezuelan border. Hueck and Seibert call these the 'evergreen rain forest of the higher montane area of Guayana,' and they distinguish them from some isolated 'sierra vegetation of western Amazonia.'

All these authors discuss floodplain forests and gallery forests separately, but Pires as well as Prance and Brown map them together while Hueck and Seibert map them separately.

There has been some disagreement in the past as to how to classify the various floodplain forests. Many authors have called temporarily flooded forests "várzeas" and permanently flooded formations "igapós" (i.e., Pires, 1972). Others, in

Table I

Partial comparison of map legends of Pires (1972), Pires and Prance (1985), and Prance and Brown (1987), with estimates of per cent area where given. Part of the differences in percentages may be accounted for by the fact that Prance and Brown considered all of Amazonia.

Pires	Pires & Prance	Prance & Brown
metas pesadas (90%)	dense forest	dense forest (16.5%)
matas de encosta	open forest without palms	dense and open hill forest (28.2%)
		mixed dense forest (7.9%)
		periodically inundated dense forest (0.9%)
mata de cipó (3.0%)	liana forest	open palm and liana forests, etc. (7.2%)
	open forest with palms	babassu palm forest (0.8%)
mata seca de transição (0.4%)	dry forest	semideciduous forest (3.0%)
caatingas altas (0.8%)	caatingas, etc.	white sand forests (incl. campo rupestre) (6.7%)
campinas (1.9%)	campo rupestre	
	montane forests	montane forests (1.0%)
campos de terra firme (4.4%)	savannas on terra firme	Amazonian savannas (3.0%)
floodplain forests (1.9%)	floodplain forests	floodplain and gallery forests (4.9%)

line with the terminology of limnologists, have preferred to make their primary distinctions between the floodplain forests of different river types, using "várzeas" to mean those of white-water rivers and "igapós" those of clear- and black-water rivers (Prance, 1979). The latter approach appears to have prevailed in recent publications (i.e., Pires & Prance, 1985). Using this latter terminology, the várzeas cover a far greater area than the igapós.

The "campos de terra firme" of Pires correspond to the "Amazonian savannas" of Prance and Brown, with the exception of the Serra do Cachimbo (see section II). Hueck and Seibert map the savannas of Roraima, Tumucumaque, and eastern Marajó together under the heading of "Grasssteppung, campos limpos ohne nähere Gliederung," but refer those of Amapá, Cachimbo, and the upper Rio Madeira to the Cerrado of Central Brazil.

There is still uncertainty about how to divide and classify the terra firme forests, the transition forests, and the various savanna formations of Amazonia. Still, in the last fifteen years progress in understanding the Amazonian flora has allowed the debate to reach greater precision. Further field work and more detailed interpretation of the Projeto RADAMBRASIL (1973–1981) maps should eventually resolve these disagreements.

B. Phytogeography

Thanks to increased botanical exploration and to Projeto RADAMBRASIL (1973–1981) and the field work associated with it (see section III), our understanding of Amazonia is far greater now than fifteen years ago. There are few major regions that are botanically unknown (section VI), and some of the confusion about Amazonian vegetation types has been resolved (section III). The Amazonian herbaria have grown tremendously, and there are active specialists for most of the important Amazonian plant families (sections IV and VI).

We can be encouraged by this progress, but we are still a long way from resolving the phytogeographic subdivision of Amazonia. Many tributaries and the upper stretches of most of the major rivers have not been collected at all, and to some extent, as Ducke and Black (1953) observed more than 30 years ago, "we still ignore almost completely the flora of the uplands between the navigable rivers," although extensive road construction (section VII) has allowed recent botanical exploration to focus much more on the terra firme forests of the interfluves. Many of those areas which have been collected have been visited only briefly in one season, and so are still insufficiently known.

Ducke and Black's (1953) work on the phytogeography of Brazilian Amazonia, which treated the distribution of some 500 species, is the base from which all subsequent phytogeographers have worked. Since 1953 as many as 200,000 more collections have been made in the region, hundreds of new species have been described, and many others have been put in synonymy. Rodrigues (1979) wrote an index to the taxa treated in Ducke and Black's paper, and he updated the nomenclature of some of them. But it is necessary to update the nomenclature of all the taxa treated by Ducke & Black, as well as their distributions. As for plant groups with centers of diversity in Amazonia, the upcoming revisions and monographs will provide valuable material for future phytogeographic work, but only if the specialists (or others) analyze the distributional data carefully. To date few groups with many species in Amazonia have been analyzed in this respect. Examples are Prance (1973, 1979) and Gentry (1979).

Amazon phytogeographers are hampered in the use of their primary tool, species distribution patterns. Every year, new herbarium distribution records are encountered. For example, in 1974 the records indicate that ten species of Bignoniaceae were endemic to the vicinity of Manaus, but by 1979 additional collections showed that at most two of these species were restricted even to the Manaus–Roraima phytogeographic region (Gentry, 1979).

In addition to gaps and changes in distribution data, subdividing Amazonia phytogeographically presents several other problems. First, there is little supra-specific endemism in the proposed subdivisions. Also, in addition to the species pairs occurring between Amazonia and the Cerrado which confuse the boundaries between them (see section II), there are closely-related species-pairs between adjacent phytogeographic regions within Amazonia.

Among the additional resources needed to continue the floristic inventory of Amazonia (see section X) are botanists willing to utilize the results of the inventory. Botanists must analyze the data for purposes of guiding and affecting intelligent development policies in Brazil.

Many of the earlier data on distributions of Amazonian plants came from: 1) important economic plants such as rubber (*Hevea*), the babassu palm (*Orbignya*), and timber trees; and 2) taxa of distinctive appearance, such as *Tachigali* or *Triplaris*. There were two pitfalls to using these data. First, the information may have indicated the limits beyond which it was impossible to transport the natural products, rather than the species ranges, and secondly, the taxonomy of many of these genera is only now being resolved.

The modern history of Amazonian phytogeography begins with Jacques Huber, who was director of what is now the Museu Paraense Emílio Goeldi (MG). He traveled extensively in Brazilian Amazonia, and published a number of papers on the flora and vegetation types of the lower Amazon (e.g., Huber, 1898, 1909).

Ule (e.g., 1908) worked primarily in the Peruvian and western Brazilian Amazon. His work cites relatively few species. Sampaio (1945) credited Ule with being the first to note the difference in species composition north and south of the Rio Solimões/Amazonas.

An expedition led by Schurz in 1923 (Schurz et al., 1925) traveled widely in most of navigable Amazonia under the auspices of the United States Department of Commerce in order to identify distinct vegetation zones within the region and to assess the potential of each for the cultivation of Pará rubber (*Hevea brasiliensis* Benth.). Among the Brazilian participants in the expedition was the noted botanist J. G. Kuhlmann. Their observations of physical Amazonia were primarily of topography, soils, and overall forest physiognomy, and the little floristic information they presented was based on Huber's (1909) work. Their phytogeographic subdivisions were surprisingly similar to Ducke and Black's (1953) later work: 1) Marajó; 2) Lower Amazon north; 3) Lower Amazon south; 4) Upper Amazon north; 5) Upper Amazon south; 6) Rio Madeira; 7) Acre; 8) Peru, Ecuador and Colombia. See Fig. 3a.

Sampaio (1945) included a useful review of Amazonian phytogeography up to that time, but his own subdivisions were vague and unexplained in the text: 1) Lower Amazon north; 2) Lower Amazon south; 3) Upper Amazon north; and 4) Upper Amazon south.

The important work of Soares (1953) and of Ducke and Black (1953) have been discussed. Soares's approach was purely physiognomic, while Ducke and Black's implicit criteria for defining a phytogeographic subdivision were: 1) the presence of endemic families, genera, and species; 2) physiognomy; 3) the presence of "characteristic" species; 4) topography and soils; 5) existence of a center of diversity and/or endemism for a family or genus; and 6) species reaching their limits there.

Although Aubréville's (1961) ecological study took an essentially physiognomic approach to the principal vegetation types of Brazil, he drew some interesting conclusions about regional climatic differences in Amazonia, which are discussed further below.

Rizzini (1963) based his subdivisions (Fig. 3b) on basically the same criteria as Ducke and Black (1953), and he relied heavily on their floristic information. He used Braun-Blanquet's (1932) system for classifying phytogeographic units, and unlike other Amazonian phytogeographers his nine subdivisions did not all have the same rank:

subprovinces:
1) Upper Rio Branco
2) Jari-Trombetas
3) Rio Negro
4) Tertiary Plain, sectors:
   a) Oceanic
   b) Southeast
   c) South
   d) West
   e) Southwest/Acre

Hueck's (1966, 1972) work on vegetation types is discussed in section III. In the same book he presented his own phytogeographical subdivision of Amazonia (Fig. 3c) which was also based largely on Ducke and Black's approach and data but supplemented by his own extensive field work and bibliographic research.

Prance's (1973, 1977; Fig. 3d) work likewise drew on Ducke and Black (1953), but in addition to his own wide-ranging field work he analyzed in detail the ecology and distribution patterns of the several plant families he has studied. Disregarding widespread species, gallery forest species, and taxa restricted to rare and edaphically unusual habitats, he focused instead on terra firme forest trees of limited distribution.

In addition to simply defining subdivisions, phytogeographers should also *explain* the array of distribution patterns and the regional differences in species composition found in Amazonia. Some regional differences in the geological history and in various existing edaphic factors correlate well with some of the phytogeographic subdivisions made in the four systems depicted in Figure 3, but other subdivisions cannot be explained so easily.

As noted in section I of this paper, there are three arches—ancient zones of uplift of crystalline rock—which, cutting across Amazonia (Fig. 1), isolated different regions during several periods, possibly including the Quaternary (Putzer, 1984). Although these arches are not part of the contemporary topography, their possible significance lies in the sediments deposited during past marine transgressions which have since been re-exposed. The Purús Arch corresponds roughly to a boundary separating several of Prance's (1973, 1977) subdivisions. The Gurupá Arch corresponds to

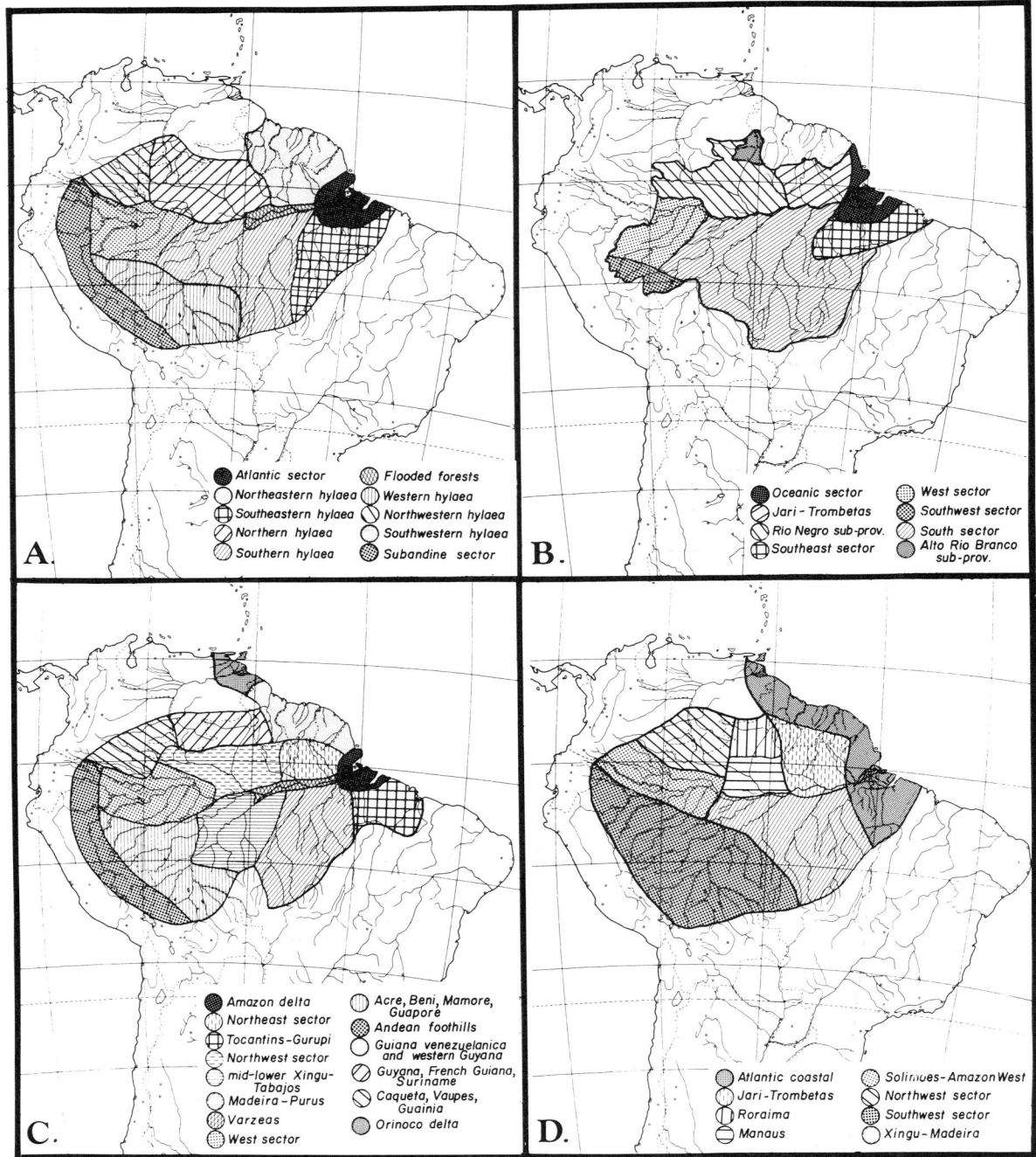

FIG. 3a–d. Four phytogeographic subdivisions of Amazonia. Adapted from: a) Ducke and Black (1953); b) Rizzini (1963); c) Hueck (1966); and Prance (1973, 1977). Note that Rizzini restricted his treatment to Brazilian Amazonia.

a division made by all four authors between the geological Amazon basin and the relatively young estuarine region.

A comparison of the geological zones of Amazonia with the four phytogeographic systems shows little correlation, with two exceptions. First, the southwestern region of Tertiary sediments corresponds roughly to Ducke and Black's (1953) Southwestern hylaea. The soils are poorly drained, low in calcium and sodium but relatively high in potassium and magnesium (Irion, 1978). Second, Prance's (1973, 1977) division of the lower and upper Rio Negro corresponds to the break between the Tertiary sediments of the basin and the ancient Guiana Shield. The only system which seems to reflect the distinction between the Brazilian Shield and the limited area of Amazonian vegetation on the Maranhão sedimentary plateau is that of Rizzini (1963).

In general, soils appear to correlate poorly with large-scale phytogeography in Amazonia but are proving extremely important on a local basis as the reason for local distribution and clustering of individual species (Lescure & Boulet, 1985). In general terms, north of the Amazon the Rio Negro and Rio Branco subdivisions constitute a large region of sandy soils between a region of relatively young alluvial clay soils to the west and a region of more highly weathered clay soils to the east.

Topography may help to characterize some phytogeographic regions and to define the boundaries of others. Ducke and Black (1953) described their Northeastern Hylaea as the "hilly country" of a particular region, and they ascribe part of the species richness in their Northern Hylaea to the different slope and summit vegetation of such mountainous areas as the Serra Curicuriari. The various low mountain ranges which run north-south in the southeastern portion of Amazonia may have served as barriers to dispersal, yielding the kind of east-west divisions reflected in all four phytogeographic systems (see Fig. 3) in that larger region.

All four systems recognize the lower Rio Solimões/Amazonas as a barrier to dispersal marking a major north-south phytogeographic division in Amazonia, while in the Upper Amazon, the river apparently ceases to be as effective a barrier. During some interglacials sea level and thus river levels were higher than at present; the significance of river valleys as barriers may have been more pronounced during those periods.

The phytogeographic subdivisions of Amazonia have been described in terms of river systems or portions of river systems. There are three reasons for this. The first is a practical reason: the floras of the interfluves are poorly known. Second, some interfluves may have served as barriers to dispersal. Third, the rivers which drain geologically distinct regions may be expected to have different floras in many cases. The different river types and their geological origins are described in section I. All four systems in Figure 3 reflect somewhat the distinctions between the regions characterized by white, clear, and blackwater rivers.

Aubréville (1961) mapped what he called the eco-climates of Brazil. He based them on the timing and duration of the dry season(s), which was defined as the number of "éco-sec" months with under 30 mm of rain. Of the six sub-climatic regions of the "Climat Amazonien," four of them correspond roughly to Ducke and Black's (1953) Atlantic Sector and Northeastern, Southeastern, and Northern Hylaea. The "Sous-climat Guyanais" corresponds to Hueck's (1966, 1972) Atlantic Guyana/French Guiana/Suriname Sector. The "Climat Haut Brancosien" corresponds to Prance's (1973, 1977) Roraima sub-sector.

Nimer (1977; also in Pires and Prance, 1977) presented a map of the "xerothermic regions" of Amazonia, which expressed length of the dry season corrected for relative humidity. The westward penetration of the 40–100 day xerothermic zone corresponds to the division of the Southern and Southwestern Hylaea of Ducke and Black (1953), and to that of the Xingu-Madeira and Southwest Amazonia sectors of Prance (1973, 1977). It also includes the Serra dos Parecis.

Forest physiognomies and vegetation types are, to a greater degree than phytogeographical subdivisions, results of the sum of edaphic and climatic factors. Most of the phytogeographic subdivisions proposed in the four systems depicted in Figure 3 can be characterized by one predominant vegetation type, but in most cases not to the exclusion of that formation from other sectors. Most sectors contain significant stretches of lowland dense forest on terra firme. The caatingas or campina forests which cover most of the Upper Rio Negro region in northwestern Amazonia also occur in patches in the isolated areas of white-sand soils which are found in various parts of the Amazon Valley. The various Amazonian savannas, which are rather unrelated floristically, are found in subcoastal Amapá, the upper rios Trombetas and Paru, the Rio Cururu in Pará, and other regions. Liana forests and semi-deciduous forests, which cover a large portion of southeastern Amazonia, have counterparts on the northern periphery of Amazonia, although the latter are restricted in area and are only somewhat related floristically. On the other hand, the large areas of diurnally inundated swamp forests with palms, which to a great degree characterize the Atlantic sector or estuary, have no real counterpart elsewhere. The bamboo forests of southwestern Amazonia in Acre and parts of Peru are also unique.

From the preceding discussion it can be seen that some of the proposed phytogeographic subdivisions of Amazonia are reasonably well correlated with one or more edaphic or historical factors, while others do not appear to be explained by these factors. Benson (1982) holds that the species diversity and centers of endemism found in northern South America can be explained by a complex of existing edaphic factors. If his arguments are valid, these factors would necessarily also serve to separate phytogeographic regions. It is hoped that these arguments will be developed.

One of the more spirited debates in biogeography in recent years is the dispute over the so-called "refuge theory." This theory holds that species diversity and centers of endemism cannot be explained by existing edaphic patterns and that these centers were islands of or refuges for rain forest biotas during cool dry climatic periods corresponding to the Pleistocene glaciations of temperate areas, when they were isolated from each other by vast areas of dry savannas. The proponents (reviewed in Prance, 1982c) believe that this repeated fragmentation and re-coalescence of rain forests generated not only much of the existing diversity but also many of the puzzling distribution patterns and taxonomic difficulties encountered in such regions as Amazonia. Some of this debate can be found in Prance (1982a).

The progress of the debate will probably not affect the phytogeographic subdivisions of Amazonia, nor will it move the centers of endemism; only new distribution records can do that. However, it can affect how the phytogeography of Amazonia is explained, and it can affect conservation policy. If the refuge theory is accepted, it will be necessary to identify and protect some of the proposed zones of secondary contact where re-coalescing forests meet. This is also true of the savannas regarded as contemporary "reverse refuges" for formerly widespread savanna biotas.

IV. Magnitude of the Floristic Inventory

A. SPECIMENS IN REGIONAL AND FOREIGN HERBARIA

However the scientific community's efforts to produce a flora of the Brazilian Amazon may be frustrated by lack of interest, inadequate programming, or lack of funds for future work, they will not suffer from a lack of collections on which to build. Compared to some other tropical regions, especially in the New World, the Brazilian Amazon is relatively well off in this respect. The region has a long and illustrious history of botanical collecting. Most of the important collections to date, including those of earlier explorers, are represented in relatively well-curated Brazilian herbaria, as well as in foreign institutions. Most of the important collectors have been Brazilians or were based in Brazil. The collections housed abroad are widely distributed and accessible for the most part. Relatively few significant collections from this region were destroyed in Berlin during World War II, with the notable exception of the first set of Ule's collections. Another unusual aspect of Brazilian Amazonia's botanical history is that there is some excellent if dated floristic literature to use for models, bases, and inspiration.

Despite this relatively strong foundation, no effort should be spared to continue floristic work in Amazonia. Of the collections made by Projeto Flora Amazônica since 1977, almost two percent have been new to science (Prance et al., 1984). Judging by the number of new species that are still being found in the region, there is still an urgent necessity to collect in the vast areas yet unexplored by botanists, in order to reach the most difficult phase of any floristic project: synthesis. Only then can all the specimens and data be put to use as tools for rational development. Meanwhile, the human population of Amazonia is growing somewhat chaotically, and the forests are being toppled faster and faster.

The itineraries and biographies of collectors in all of Brazil up to the turn of this century were listed by Urban (1906) in Martius' *Flora brasiliensis* (1840–1906). Prance (1971) compiled information on all significant collectors in Brazilian Amazonia, including dates, brief itineraries, number of collections, occasional biographical notes, and where the collections are deposited. It is very fortunate that many important collections are widely available, such as those of Spruce, Ducke, Black, Fróes, Krukoff, Pires, and the Projeto Flora Amazônica collections.

Table II indicates herbaria with major collections from Amazonian Brazil and the important collections housed there. Active collecting in Brazilian Amazonia has been sporadic but essentially continuous since the pioneering work of A. R. Ferreira in the late eighteenth century (see Fontes, 1966, and Goeldi, 1982, regarding his travels and collections). A few of the earliest Amazonian collections are scarcely represented in Brazil, such as those of Ferreira, von Martius, and Poeppig. The earliest which were deposited in Brazil are at the Museu Nacional in Rio de Janeiro (R) (a glossary of acronyms is provided at the end of this paper), which was founded in 1842, and these include partial sets of von Luetzelberg, Riedel, and Schwacke. It should be noted that many of Schwacke's, Jobert's, and other collections were later pirated by Glaziou and bear his name (see Wurdack, 1970). The Jardim Botânico do Rio de Janeiro (RB), founded in 1808, contains a number of Spruce's collections as well as some of Schwacke's.

With the founding of the Museu Paraense Emílio Goeldi (MG) in 1895, the focus of Brazilian botanical activity in Amazonia shifted north to Belém, the capital of the eastern Amazon. Housed there are the most complete sets of the collections of J. Huber, A. Goeldi, R. Siqueira Rodrigues, and Ducke's earlier collections (see Egler, 1963, for a complete account of Ducke's travels and collections), as well as some of Ule's specimens. Ducke was later based at RB. With the notable exception of Ducke's work, there was a hiatus in collecting during the period 1910–1930. The 1930's were marked by B. A. Krukoff's extensive and important expeditions to Maranhão and especially to western Amazonia, and

Table II
Partial list of herbaria with important Brazilian Amazonian collections.

Herbarium	Important Collections
BR	Martius' private herbarium
IAN	Archer, Black, Fróes, M. Oliveira, Pires (early), N. T. Silva,
INPA	D. Coêlho, L. Coêlho, C. A. Cid, Prance, W. A. Rodrigues
K	Black, Burchell, Ducke, A. R. Ferreira, Fróes, J. G. Kuhlmann, Krukoff, Riedel, Richard and Robert Schomburgk, Spruce, Ule,
LE	Riedel, Poeppig
M	von Luetzelburg, von Martius
MG	P. Cavalcante, Ducke (early), A. Goeldi, J. Huber, Pires (later), Prance, R. S. Rodrigues, Ule
NY	Egler, Fróes, Krukoff, M. Oliveira, Pires, W. A. Rodrigues, Prance, Riedel, N. T. Silva
P	A. R. Ferreira, Jobert, Richard and Robert Schomburgk
R	von Luetzelberg, Riedel, Schwacke, Ule
RB	Ducke (later), A. Goeldi, J. G. Kuhlmann, W. A. Rodrigues, Spruce
SP	Hoehne, J. G. Kuhlmann, M. Kuhlmann
US	Archer, Black, Ducke, Fróes, J. Huber, Krukoff, J. G. Kuhlmann, Pires
W	Poeppig

by Ducke's continued work in various regions. Krukoff was one of the very few foreigners whose collections unfortunately are not well represented in Brazilian herbaria, although they were widely distributed in North America and Europe.

After the mid-1940's there was a tremendous surge in activity. Again the base was Belém, but at the herbarium of the Instituto Agronômico do Norte (IAN), founded in 1945. This is the single most important herbarium of the Amazon flora of Brazil, and it houses the most complete sets of Fróes, Black, N. T. Silva, N. A. Rosa, and most of Pires.

The herbarium of the Instituto Nacional de Pesquisas da Amazônia (INPA) in Manaus was founded in 1954 and has grown rapidly since then. INPA collectors have concentrated on western Amazonia. Unfortunately, many of their earlier collections are poorly represented even in other Brazilian herbaria, including the products of W. A. Rodrigues's important expeditions.

The most active collectors during the late 1950's and early 1960's were W. A. Rodrigues and his assistants L. Coêlho and D. Coêlho at INPA and J. M. Pires and N. T. Silva at IAN. This period saw the beginning of what has become a very productive relationship between Brazilian botanists or collectors and large-scale development projects in Amazonia, a relationship which continues today. Some of the earlier projects of this nature were those of Pires on the Fazenda Pirelli rubber plantation in Pará, W. A. Rodrigues at the mines of the Serra do Navio in Amapá, and N. T. Silva at Daniel Ludwig's Jari development at Monte Dourado in Pará.

The New York Botanical Garden (NY) became involved in the early 1960's, with the work of H. Irwin in Amapá, B. Maguire in Mato Grosso, and G. T. Prance and Irwin in Mato Grosso and on the Belém-Brasília Highway. The NYBG–INPA Botanical Inventory Program began in 1965 and continued until 1977, when it was superseded by Projeto Flora Amazônica (PFA). This cooperative program resulted in the first significant collections made by foreign-based botanists since the previous century, with the exception of Krukoff. Several botanists from NY began to make regular trips in the Brazilian Amazon with colleagues at MG and INPA.

PFA, the first phase of Programa Flora, Brazil's national flora program, began in 1977 in cooperation with several U.S. herbaria. As of 1986 there have been over 30 binational expeditions to selected regions of the Brazilian Amazon (see fig. 5a, 5b). The first set of all these collections has been deposited at INPA or MG, and half of all duplicates have remained in Brazil for distribution to other Brazilian herbaria.

Table III indicates the number of Amazonian collections housed in Brazilian herbaria. For foreign herbaria, the proportion of Amazonian collections (and therefore the total number) is more difficult to estimate.

B. Publications on the Flora

The greatest floristic work in the New World to date is von Martius' massive *Flora brasiliensis* (1840–1906) for which he recruited many of the eminent botanists of the period, such as Bentham, Engler, Radlkofer, and Mez. Sixty-five specialists treated 22,767 species in forty folio volumes (Hoehne, 1939). Few angiosperm families were left unstudied, and the opus remains a useful and, for many plant groups, a fairly reliable tool today.

The botanists who participated in *Flora brasiliensis* could of course treat only the specimens they had in hand, and the amount of Amazonian material available to them was limited. Hundreds of new taxa have been described since then, and in many cases generic and family concepts have changed.

A similar handicap was faced by Hoehne and his *Flora Brasílica* (1940–1968). He intended to use *Flora brasiliensis* as the basis for producing a flora in Portuguese and with a smaller and more affordable format (Hoehne, 1939). Unfortunately, while the work was excellent, only 12 volumes of *Flora Brasílica* were produced before the project was stopped. None of the plant groups which are important in Amazonia were treated.

When Projeto Flora Amazônica was initiated in 1977,

Table III
Principal Brazilian herbaria with Amazonian collections.

Herbarium	Founded	No. Specimens	% Amazonian	Regions Emphasized
HAMAB	1978	7,000	100	Amapá
HUAM	?	2,500	100	Amazonas
IAN	1945	200,000	90	all Amazonia
INPA	1954	130,000	95	western Amazonia
MAO	?	1,200	100	western Amazonia
MG	1895	115,000	90	all Amazonia
R	1808	450,000	10	
RB	1842	210,000	15	
SP	1917	180,000	10	
UFMA	1980	4,000	50	Maranhão
UFMG	?	?	—	Mato Grosso

there were no expectations that it would produce a written flora in ten years. The project was designed to generate the collections and organize the data base needed for execution of the flora. PFA is discussed further in section VIII of this paper.

The practical way to approach a flora of Brazilian Amazonia is through intensive local and regional floras, possibly along the lines of Croat's excellent *Flora of Barro Colorado Island* (1978), but to date there are only inadequate (if useful) short papers produced by scattered expeditions or inventories. Jacques Huber wrote a series of papers under the title "Materiaes para a Flora Amazônica" (e.g., 1898) about his collecting trips to various parts of Amazônia, in which he described the vegetation in detail and enumerated the species observed and collected. The extensive bibliography in Soares (1953) contains references to many papers of this nature, although most were not so thorough as Huber. Other such publications may be found in the book *Amazônia; bibliografia* (1975). Hoehne (1951) compiled a checklist of the species collected between 1908–1921 during the Comissão Rondon project, which linked northern and southern Brazil with telegraph lines. Most of the collections were made in extra-Amazonian vegetation.

In their landmark paper on the phytogeography of Amazônia, Ducke and Black (1953) commented on the distributions of over 500 taxa. Rodrigues (1979) prepared a useful index to this work but both the distributions and names still need to be updated (see section III).

Pires (unpublished) prepared a checklist of the plants reported from Amapá. This could serve as the initial step toward a flora of Amapá, but it is already seriously outdated. W. A. Rodrigues (1959) compiled a checklist of species found in the Amazonian savannas of Roraima.

Apart from these modest efforts, floristic data can be gleaned from the texts and distribution maps of recent revisions and monographs (see section VI).

V. Collection Densities, Endemism

The state of our knowledge of floristic inventory in the Brazilian Amazon is such that statistical estimates of collection densities cannot be made. Qualitative estimates will have to suffice for the time being. There are four primary types of areas with relatively high collection densities: 1) near major cities (with herbaria); 2) near major development projects; 3) along the lower and middle stretches of the larger rivers; and 4) itineraries of Projeto Flora Amazônica (PFA) expeditions. This may seem neglectful of the wide-ranging travels of such intrepid collectors as Spruce, Huber, Krukoff, and Pires, but for the most part their collections were rather sparse from any one area. In fact, only recently have botanists begun to re-trace their predecessors' itineraries in order to collect more thoroughly those regions with unusual floras which the earlier botanists had sampled.

If only for the simple reason of accessibility, the forests around Belém and Manaus have been the most thoroughly collected. Furthermore, there have been large development projects near these cities which have provided additional opportunities for extensive collecting. During the 1950's and 1960's, these opportunities were provided through courtesy of the companies involved. Examples are the mines at the Serra do Navio, Amapá; the Fazenda Pirelli plantation near Belém; the Jari wood-pulp project at Monte Dourado, Pará.

In the 1970's such arrangements became institutionalized as well, when the World Bank began to designate part of its development loans for biological research on the areas to be affected; contracts were made with MG and INPA to carry out the research. For this reason the vicinity of the Tucuruí hydroelectric project in Pará is intensely collected, and the Serra dos Carajás, where extensive mining of iron ore and other minerals has begun, has been intensively collected. A similar arrangement has begun in the region covered by the Projeto Polonoroeste in Rondônia, where the area along the Cuiabá–Porto Velho highway is being actively collected.

Figure 5a shows the regions collected by PFA expeditions between 1977 and 1983, and figure 5b shows the foci of expeditions to be completed by the end of 1987. Most of the expeditions have been initial attempts to obtain material from poorly collected areas; therefore most of these areas should be re-visited at least once more in a different season by another major expedition before they can be said to be well collected at all. PFA has assisted at Tucuruí, the Serra dos Carajás, and the area of Projeto Polonoroeste in Rondônia.

Despite the quantity of specimens from Brazilian Amazonia in the herbaria, it is still easier to find poorly collected regions than well-collected ones, and identification of these poorly collected areas is far easier than access to them. The upper reaches of most of the tributaries of the Amazon have not been collected, nor have some entire tributaries in western Brazilian Amazonia. For those portions of the river system which have been collected, few traverses have been made across the interfluves.

While Rondônia is now being actively collected, neighboring Acre has been relatively ignored. Western and northern Amapá are largely unknown. There are relatively few collections from Roraima. Northern Mato Grosso has been receiving some attention recently, but the transitions to the savannas of Central Brazil across southern Amazônia are poorly studied. Ironically, large parts of Marajó—the giant island at the mouth of the Amazon—are completely uncollected.

The isolated peaks and small mountain ranges along the northern and southern peripheries of Amazônia have for the most part been neglected, with the exception of Carajás and the Serra do Cachimbo. In Roraima, Pico Rondón was recently collected for perhaps the first time, and the Serra do Moa in Acre was collected for the first time in several decades. The Brazilian side of the Serra do Tumucumaque in northern Amapá is almost totally unknown. The same is true of the Serra Parima and the Serra Pacaraima in Roraima on the Venezuelan border (Huber et al., 1984). Serra Aracá in Amazonas has now been collected on two recent PFA expeditions.

Prance (1977, 1982a) has portrayed centers of endemism

in South America believed to have been isolated refuges for tropical moist forest elements during the cool, dry periods of the Pleistocene glaciations. Even most of those who do not adhere to the "refuge theory" *per se* agree that these areas are centers of endemism. Of these centers, those which are situated in Brazilian Amazônia are: Imerí; São Paulo de Olivença; Tefé; Manaus; Trombetas; Belém; Tapajós; Rondônia-Aripuanã; and eastern Peru-Acre. The most controversial of these is the Manaus center/refuge; some botanists believe that the apparent endemism is an artifact of the density of collecting around the city.

Other areas of high endemism can be explained by edaphic conditions, such as the summits of some of the isolated mountain ranges in southern Amazônia where stony outcrops of various parent materials support scrubby vegetation somewhat linked to the white-sand campinas in lowland Amazônia (see section III of this paper). An example is the Serra dos Carajás, which in the past few years has yielded a number of new taxa restricted to its ironstone "canga" formations.

New distribution records are being found with almost every new expedition. The dramatic example of *Couepia longipendula* Pilg. was mapped by Prance (1978). Another example, *Corythophora rimosa* W. Rodr., was believed to be endemic to the Manaus region, but it was collected in French Guiana in 1982 and then in Amapá in 1983. Gentry (1979) has cited examples of this same phenomenon in the Bignoniaceae.

Other formerly "rare" taxa have been found to be common where they occur but range extensions have not been found. Examples are *Polygonanthus amazonicus* Ducke (Nelson et al., 1987) and *Hirtella dorvalii* Prance (Prance, 1982b).

Conclusions regarding Amazonian phytogeography should be periodically re-examined (every 5 years or so) in the light of updated distribution records and taxonomic realignments.

VI. Inventory and Study of Various Plant Groups

A. Taxa Well- or Poorly-Collected

High collection densities can be misleading. They may conceal an area that is actually poorly collected, because often inexperienced collectors make redundant collections at roadsides and airstrips and in other secondary vegetation. The number of collections does not matter if the important taxa are not collected. Kubitzki (1977) explored the problem of collecting frequencies of rare taxa. Prance (1982b) noted that in two genera of Chrysobalanaceae 28 of 32 species that were represented by a single type collection in 1972 had still not been re-collected by 1982. Some taxa may be poorly collected because they are widespread but rare where they occur, while others may have a very restricted range but are abundant where they occur. An example of the latter case is *Hirtella dorvalii* Prance, mentioned in section V, which occurs only in a small transition zone of caatinga forest near Caracaraí.

Some groups of plants are without doubt severely undercollected even in the areas with highest collection densities. By far the worst off are the hepatics, the bryophytes, and especially the fungi. Ferns have been collected (cf. Huber, 1902), but only sporadically, and some Amazonian forests have richer and more interesting fern floras than previously thought (Sperling and Daly, in prep.)

Among the flowering plants, Amazonia shares the same shortages as other parts of the tropics in collections of epiphytes, vines, and saprophytes (see Prance, 1982b). This means there are inadequate collections of Araceae, Bromeliaceae, Bignoniaceae, Dilleniaceae, Apocynaceae, some Leguminosae, some Gentianaceae, Orchidaceae, some Clusiaceae, some Solanaceae, some Sapindaceae, Menispermaceae, and others. Of the free-standing photosynthetic angiosperms, palms are certainly the least collected and understood; most of the Amazonian palm collections are incomplete (Balick et al., 1982). Taxa which are emergent trees are poorly collected relative to understory trees. Other families under-represented in proportion to their relative densities are the families of trees with inconspicuous flowers and/or fruits, such as Lauraceae, Burseraceae, Monimiaceae, Dichapetalaceae, Myristicaceae, and Euphorbiaceae. Another particular need is for greater attention to groups with dioecious flowers, such as some of the preceding as well as Clusiaceae. For many taxa in these groups only one sex is known.

B. Specialists

Most important Amazonian plant groups have specialists, or at least persons competent and willing to do identifications. Table IV presents a list of plant groups which are ecologically important and/or have centers of diversity in Amazonia, giving the specialist and institution if any. Specialists for some groups, such as Clusiaceae, Leguminosae-Caesalpinioideae, Humiriaceae, Icacinaceae, and Myristicaceae are approaching retirement, and these groups will need new specialists. Although relatively few of the important families lack specialists, the rate of monography is exceedingly slow.

Table V lists recent monographs or revisions of some plant groups with centers of diversity in Amazônia, as well as groups for which treatments are expected by the end of this decade.

VII. Threatened Areas

As in other tropical regions, the greatest threats to Amazonia's forests are large-scale development projects and colonization, and the former attracts the latter. The periodic surges of immigration which the Brazilian Amazon experienced for a hundred years have become a steady and steadily

Table IV

Flora Neotropica monographs/revisions of some plant groups with centers of diversity in Amazonia (monographs 1–14(2) published by Hafner Publishing Co., New York; all others by The New York Botanical Garden, Bronx, for The Organization for Flora Neotropica).

Completed		
Group	Author	FN #
Caryocaraceae	G. T. Prance & M. F. da Silva	12
Chrysobalanaceae	G. T. Prance	9
Flacourtiaceae	H. Sleumer	22
Lauraceae (*Aniba, Aiouea*)	K. Kubitzki & S. Renner	31
Lecythidaceae (actinomorphic)	G. T. Prance & S. A. Mori	21
Leguminosae (*Dimorphandra*)	M. F. da Silva	44
Leguminosae (*Parkia*)	H. Hopkins	43
Leguminosae (*Swartzia*)	R. S. Cowan	1
Malpighiaceae (*Banisteriopsis, Diplopterys*)	B. Gates	30
Melastomataceae (Memecycleae)	T. Morley	15
Meliaceae	T. D. Pennington & B. T. Styles	28
Moraceae (*Olmedieae, Brosimeae*)	C. C. Berg	7
Violaceae (*Rinorea* p.p., *Rinoreocarpus*)	W. H. A. Hekking	46

Expected before 1990	
Group	Author
Apocynaceae (selected genera)	J. Zarucchi
Bombacaceae (*Matisia, Quararibea*)	W. Alverson
Burseraceae (*Protium*)	D. C. Daly
Erythroxylaceae	T. Plowman
Lecythidaceae (zygomorphic)	S. A. Mori and G. T. Prance
Leguminosae (*Copaifera*)	M. F. da Silva
Myristicaceae	W. A. Rodrigues
Quiinaceae	J. M. Pires
Sapotaceae	T. D. Pennington
Sterculiaceae (*Sterculia*)	E. L. Taylor

increasing stream. In some regions the government, which heretofore had always regarded populating the Amazon a top priority, has begun to reconsider colonization programs which have worked too well for their own good. Estimates of deforestation in Amazonia have been hotly contested (cf. Fearnside, 1982), but even those who minimize the rate at which it is progressing would admit that the course and pattern of the deforestation are inadequately controlled.

Much of the historical information which follows is based on Oliveira's (1983) excellent work. A somewhat detailed account is given because this can show the nature and the foci of the forces which threaten that part of Amazonia's incomparable forests which occur in Brazil.

Mass immigration is not new to Amazonia. During the great rubber boom at the turn of the century, tens of thousands streamed into the Amazonian interior to harvest rubber. This was followed by a lesser known boom at the beginning of World War II when the U.S., facing a cutoff of rubber supplies from Asia, gave large amounts of aid to Brazil to stimulate rubber production. This was the "Acordos de Washington," and the period is called the "Batalha da Borracha," or Battle of Rubber. The Brazilian government saw a chance to implement the great development plan it had outlined in 1940, and so the Batalha was accomplished by the creation of a number of agencies charged with improving the infrastructure of the region, including colonization, transportation, and agriculture. IAN (now EMBRAPA/CPATU) and its herbarium were founded at that time. This set a precedent for the type of government involvement in Amazonian development which continues today: large federally-supervised schemes with semi-autonomous agencies to supervise each aspect of development.

In 1953 the development plan was consolidated under the SPVEA, and 3% of the gross national product was committed to the development of Amazonia. This involved legally defining the region and so that year, based on the work of Soares (1953), Law No. 1806/ Article 199 established the 4,975,527 sq. km of Amazonia Legal. SPVEA lasted until 1966. During 1955–1959 it made conventions with existing agencies to foster programs similar to those of the Batalha.

Construction of the Belém–Brasília Highway, started in 1960, prompted a new wave of immigration from the Northeast and began a change in the population structure of Amazonia from riverside to roadside colonization.

Table V

Important Amazonian plant families and their specialists.

Group	Specialist	Institution	Remarks
Fungi			
Bryophytes			
Pteridophytes			
Arecaceae	M. J. Balick	NY	
	A. B. Anderson	MG	
Apocynaceae	B. Hansen	USF	lianas
	A. Leeuwenberg	WAG	
	J. Zarucchi	MO	diverse genera
Annonaceae	P. J. Maas et al.	U	
Bignoniaceae	A. H. Gentry	MO	
Bombacaceae	W. Alverson	WIS	*Quararibea, Matisia*
Burseraceae	D. C. Daly	NY	
Chrysobalanaceae	G. T. Prance	NY	
Dichapetalaceae	G. T. Prance	NY	
Euphorbiaceae	M. J. Huft	MO	
Humiriaceae	– –		
Lauraceae	K. Kubitzki	HBG	
Leguminosae			
Mimosoideae	G. P. Lewis	K	except *Inga*
Caesalpinioideae	M. F. da Silva	INPA	*Dimorphandra, Copaifera, Peltogyne*
Papilionoideae	V. Rudd		*Dalbergia, Machaerium,* other genera
	C. Stirton	K	Sophoreae
Lecythidaceae	G. T. Prance	NY	
	S. A. Mori	NY	
Malpighiaceae	W. R. Anderson	MICH	
	B. Gates	MICH	
Melastomataceae	T. Morley	MIN	*Mouriri*
	J. J. Wurdack	US	
Meliaceae	T. D. Pennington	NY, K	
Menispermaceae	R. Barneby	NY	
Moraceae	C. C. Berg	U	
	P. Carauta	RB	
Myristicaceae	W. A. Rodrigues	INPA	
Myrtaceae	L. Landrum	ASC	
Rubiaceae	B. M. Boom	NY	
	J. Kirkbride	USDA	
Sapotaceae	T. D. Pennington	NY, K	
	J. M. Pires	IAN	
Vochysiaceae	S. Mori	NY	

Operation Amazonia was initiated in 1965–1967, and was designed to further integrate Amazonia with the nation and to increase federal influence in the region. In 1966 SPVEA became SUDAM, which exists today. In 1967 the Zona Franca de Manaus, a duty-free industrial zone, was declared in Manaus. During this period the government decided that while they needed to build the infrastructure to facilitate development, the actual process of development should be carried out by large private enterprises. About this time a group of businessmen from southern Brazil formed an association-lobby, called the Associação de Empresários da Amazônia, to steer the government's financial incentives toward development of cattle-ranching. This group helped engender the attitude—and implicit policy—of settling Amazonia 'via cow's hooves.'

The end of the 1960's saw the completion of two major highways, the Belém–Brasília (north-south) and the Cuiabá–Porto Velho (southeast-northwest). The latter stimulated a great influx of people from Southern Brazil. In 1970, as part of the Program of National Integration (PIN), even more emphasis was placed on roads, and work began on the Transamazon (east-west), Cuiabá–Santarém (north-south), and Perimetral Norte highways, the latter designed to run across the northern periphery of Amazonia. Plans were made to reserve a 10 km strip on either side of these roads for colonization.

In 1970 the PROTERRA program was instituted in order to make land acquisition easier and working conditions better, and to promote agro-industry. The early 1970's also saw the execution of the RADAM project for Amazonia, a survey and radar-mapping project meant to identify natural resources, as well as the foundation of the national institute for colonization and land reform (INCRA).

About this time there was a shift in the official conception of Amazonia from that of a problem area to that of a natural-resource frontier. This coincided with the creation of a new development strategy. The government, perceiving the problems of access, marketing, and know-how being encountered by immigrants to Amazonia, invited large companies to assume the task of developing the Amazon. What gave this policy substance was the second plan for the development of the Amazon (II PDA), which incorporated the POLAMAZONIA program. This program, like all its predecessors, included ambitious infrastructural projects, but the emphasis now was less on colonization and more on fostering large enterprises which could produce exports to defray Brazil's increasing external debt. POLAMAZONIA designated 15 development areas ("Pólos"), most of them under SUDAM's supervision, and each with a specific list of projects, both infrastructural and agro-industrial. POLAMAZONIA is in full swing today, and the Pólos will undergo intensive deforestation during the next decade.

Mining denudes the land in Amazonia as much as anywhere else. Large areas of the Serra do Navio in Amapá were deforested by the manganese mining of ICOMI/Bethlehem Steel which began in 1956, and now large-scale extraction of iron ore is under way on the Serra dos Carajás in central Pará. However, the latter project has better and more strictly controlled environmental programs than any other Amazonian development projects.

Because of the low topography of most of Amazonia, dam projects have flooded large areas and will flood many more tracts of forest if development proceeds as planned. Tucuruí, to date the largest hydroelectric project in Amazonia, flooded 2,000 sq. km when it opened in 1984. For the rest of the Tocantins River system, the government plans a total of 7 more large dams and 19 smaller ones, which would make the river a string of lakes 1900 km long (Caufield, 1982). Multiple dams for every major river system in Amazonia have been planned for years (Mitchell, 1979). Balbina, a hydroelectric project under construction on the Rio Uatumã on the lower Amazon will flood ca 2,360 sq. km, but because the reservoir will be only 6–7 m deep it will produce only 10% the energy of Tucuruí (data from unpublished Eletronorte reports; see also Caufield, 1982).

Brazilians, like North Americans, have a strong tradition of cattle-ranching, and they have tried to impose that tradition from their southern states on a reticent Amazonian ecosystem. Of all development enterprises cattle-ranching is the most destructive of forests, and it is a most wasteful and inefficient use of the land. As in the western U.S., private holdings of 20,000 ha are very common, often belonging to absentee landlords from the South who occasionally fly up north in private planes to inspect operations. Corporate holdings are much larger. However, in the western U.S. such ranches were natural grasslands. Only in 1978 did SUDAM stop giving new incentives to these huge cattle-ranching projects (Schubart, 1983). In the Paragominas region of Pará there are some indications that the rate of deforestation has gone down since SUDAM's decision (C. Uhl, pers. comm.). However, the projects which took timely advantage of the incentives are still enjoying them.

Large agricultural and agroforestry plantations are responsible for a much smaller share of deforestation than ranching, but they are becoming more common. Jari, the massive pulpwood and rice operation with 100,000 ha under cultivation, came under heavy criticism by ecologists and nationalistic Brazilians while it was personally owned by Daniel K. Ludwig; he lost almost US $1 billion on the project. Now that it is owned by a consortium of Brazilian corporations, the Brazilian press is much slower to criticize it, and the ecological and economic lessons to be learned from Jari's costly mistakes appear to have been lost. Plantations of Caribbean Pine (*Pinus caribaea*) for pulpwood have been expanded into neighboring Amapá by ICOMI/Scott Paper. Agronomists at EMBRAPA/CPATU in Belém believe that large multiple harvests of rice may be possible in the Amazon's vast floodplain forests.

Other large monocultures are in the works. Brazil's agricultural agencies have been breeding disease-resistant strains of *Hevea* rubber and large plantations are planned for the states of Amazonas and Acre. Coffee is being planted as a major crop in Rondônia, and the cocoa research institute (CEPEC) is encouraging plantations of Cacao in Rondônia and Amazonas. Other important crops in the Amazon region are black pepper in Pará and jute and malva in Pará and Amazonas (Schubart, 1983).

Large-scale development projects are responsible for much deforestation in Amazonia, but as the head of the government colonization agency (INCRA), Paulo Yokota, was recently quoted, "The basis of the frontier is the family farmer." (Ulman, 1984). For the many thousands who are fleeing the poverty of the big southern cities, the drought in the Northeast, or the high land prices of the Southeast, Amazonia is their only chance to make a better life, or in many cases simply to survive. Many are attracted by road construction or large development projects, but the majority are homesteaders.

Whatever the reasons, during the period 1960–1980 the population of Brazilian Amazonia jumped from 5,363,708 to 11,218,385 (Oliveira, 1983). The increase did not occur uniformly throughout the region, of course; the population in some regions did not increase appreciably, while in others the figures are astounding. In central Pará, between 1970–1980 the population of Conceição do Araguaia grew 311%, Altamira grew 214%, São Félix do Xingu grew 234%, and Tucuruí, the focus of the dam project, grew 517%—from 9,936 to 61,319. Near the Jari project in northern Pará, during the same period, Prainha grew 268% and Almeirim 178% (Oliveira, 1983).

The most impressive population figures, however, are from Rondônia and neighboring Mato Grosso. This region includes the Polonoroeste sector of the POLAMAZONIA development program, but much of this growth has been unplanned: by 1979 in Rondônia, 20,000 families had been placed by INCRA but 30,000 families were still seeking a home. They had been attracted not only by INCRA, but also by road construction projects and by the reputation of the relatively fertile "terra roxa" soils which occur in parts of Rondônia. During the same 1970–1980 period, Cacoal, Rondônia grew from 1,193 to 67,243, and Colides, in Mato Grosso, grew from 130 persons to a population of 34,638 (Oliveira, 1983). In a single year, the municipality of Alto Floresta, an ironic name meaning "high forest," more than doubled, from 60,000 to 140,000 (Ulman, 1983).

Clearly, Rondônia and adjacent Mato Grosso face the greatest and most immediate threat of deforestation. The population of the Polonoroeste area within this region grew from 330,000 to 980,000 during 1970–1980 and was expected to reach 2,000,000 by 1985 (Oliveira, 1983).

Another critical area is Amazonian Maranhão. In the 1960's many families immigrated to this region, drawn by construction jobs on the Belém–Brasília Highway or the land to which it gave access. During the past 10 years, tens of thousands arrived in Maranhão from the adjacent Northeast, fleeing a severe drought there. The forests of northwestern Maranhão, slightly drier than most of Amazonia, are easily burned in the dry season. Now many more settlers are being drawn by the construction of the railway from Carajás to the port of São Luis and by the colonization and silviculture schemes planned along the railway. The pre-Amazon forest has been decimated before it could begin to be studied.

All of central Pará comprises a third region in the process of being rapidly deforested, although there it has not progressed as far. The transition forests which cover a significant part of this region are poorly known. At Tucuruí, the dam has flooded a large area of forest, accompanied by extensive deforestation in a large area around the construction site. The ultimate fate of the 25,000 workers brought in for its construction is uncertain. Further dams are planned upstream on the Tocantins. Nearby, the Rio Araguari is being dredged to provide access to the deep-water port of Barcarena.

The right-of-way has been cleared and track has been laid for the Carajás–São Luís railway. Fearnside (1982) drew a dramatic comparison between Jari and Carajás. Jari was a one billion dollar investment and includes 100,000 ha of silviculture plantations, while the whole Projeto Grande Carajás, comprising Tucuruí, the mines, and the railway, will be a total investment of US $33 billion and will include 2,400,000 ha of silviculture plantations along the railway, some of which will be necessary to provide charcoal for a planned steel mill.

It should be noted that all the sectors of the POLAMAZONIA program may eventually face this scale of development and deforestation. As Prance (1977) pointed out, some of the refuges/centers of endemism in Amazonia fall entirely within these sectors—Belém, Rondônia–Aripuanã, Olivença and Tefé.

As mentioned above, there has been a great deal of debate over the actual extent and rates of deforestation in the tropical forests, including Amazonia (Fearnside, 1982). Probably the most accurate assessment of deforestation Amazonia is that of Woodwell et al. (1986). They examined a 32,000 sq. km area of Rondônia. Supplementing their laboratory work with detailed field verification, they developed a system for estimating deforestation with two types of satellite imagery, which they applied to the entire territory. They concluded that between 1976–1978 the deforestation progressed at a rate of 27,000 ha/year, and between 1978–1981 at a rate of 55,000 ha/year. They estimated that total deforestation in Rondônia up to 1982 was at least 10,338 sq. km or some 5% of its area. The deforestation continues at an increasing pace.

The Zona Bragantina, an area of 30,000 sq. km in northeastern Pará was a focus of colonization beginning in the late nineteenth century. Destruction of virtually all of its primary forests was completed years ago (cf. Fearnside, 1982). It is an area of forest destruction on such a large scale that it merited a separate entry on the legend of Hueck and Seibert's (1972) vegetation map of South America. Let us hope that the same does not become true of the POLAMAZONIA development foci.

VIII. Resources for Continuing the Inventory

Four kinds of resources are needed for pursuing floristic inventory: infrastructural, human, programming, and financial support. Brazil has an extensive botanical infrastructure of a scale that should be appropriate to the task at hand, especially if recent regional expansions continue. Brazil still maintains a rather strong botanical tradition, although the next generation of botanists is faced with difficult challenges. In recent years programming has been strong with some minor flaws, but prospects for the next decade—perhaps the most critical decade in history for tropical forests—are at best uncertain. The prospects for financial support are most disturbing. Increased support appears to be out of the question; the question is rather one of resources sufficient for continued functioning of institutions.

A. INFRASTRUCTURE

Table III lists the Brazilian herbaria with significant Amazonian holdings, and they were discussed in section IV of this paper. To a large extent these herbaria's collections are complementary. R and RB contain some of the earliest Amazonian collections, as well as some rare collections from the southeastern periphery of Amazonia. MG has much material from the first great era of Amazonia collecting IAN, from the second, and INPA and once again MG from the third which continues today. HAMAB is developing a fine regional herbarium for Amapá which should be strong enough by 1990 to serve as the base for Amazonia's first regional flora, if the program can be assembled. It is hoped that Brazil's other young regional herbaria will follow HAMAB's example.

All of these herbaria have valuable holdings which are for the most part well curated. The INPA, MG, and HAMAB herbaria are all housed in modern, air-conditioned buildings. The Amazonia herbaria try to distribute duplicates to other Brazilian herbaria, and they are cooperative about sending gifts for determination and loans to foreign institutions. Unfortunately, some major European herbaria are reluctant to send major loans, and especially type specimens, to South America, which severely hinders the monographic work of Brazilian botanists.

For all Amazonian collections accessioned at MG and INPA, the label information has been recorded on computer-readable cards, which are periodically sent down and entered on tape at the computer center at SERPRO in Brasília. The information may be retrieved in various formats, accessible through the data-basing program developed by the University of Michigan Computing Center as part of Programa Flora de Brasil. This system was recently tested with some success for the Chrysobalanaceae.

Perhaps the greatest gap in Brazil's botanical infrastructure is the difficulty of access to taxonomic literature. This problem is not quite as severe at RB and MG—the more historical institutions—but it is particularly difficult at INPA. Many recent books and journals are prohibitively expensive for all these institutions; some of them have partially circumvented this problem by exchanging duplicate specimens for literature.

Another aspect of the infrastructure which will become increasingly important in the near future is Brazil's extensive—and complicated—system of national forests, national parks, biological reserves and ecological reserves. These will be the sites of low-expense collecting and, it is hoped, of regional floristic projects, especially in those areas which have lodging. The special secretariat of the environment (SEMA), now oversees 11 ecological reserves in Amazonia, some of which already have staffs and shelters (Nogueira Neto, 1985). More are planned.

B. Human Resources

Graduate training in botany is available both at INPA and at MG (through the federal universities of Amazonas and Pará, respectively). Some of the staff at these institutions have studied in Rio de Janeiro or São Paulo for part or all of their graduate training. While graduate training at INPA and in southern Brazil is generally of good quality, undergraduate programs in biology at Amazonian universities are still extremely weak.

Since the early part of this century, a number of prodigious and productive botanists have lived and worked in Brazilian Amazonia. In addition to being competent taxonomists these scientists have traveled widely in the entire region, and they have made important contributions toward understanding the phytogeography, economic botany, and phytosociology of the Amazon forests.

However, these people are now over sixty. The next generation of Amazonian botanists consists for the most part of men and women in their late twenties and early thirties, and however enthusiastic they may be, their backgrounds are quite different. While they may have a somewhat better background in systematics than the previous generation, the breadth of their field experience is much narrower. There are several reasons for this: 1) their age; 2) some are less willing to undergo the discomfort of extended field work than the older generation; 3) some have too many administrative responsibilities in the herbarium; 4) expeditions and arrangements with large development companies are more formalized and therefore more expensive and restrictive; and 5) travel funds are difficult to obtain.

C. Programming

Two programs have been generating most of the herbarium material which will serve as the basis for future floristic work on the Brazilian Amazon. Projeto Flora Amazônica (PFA), the flora of the Amazon project, was one of five subprojects of Programa Flora, Brazil's national flora program. Begun in 1977, it was sponsored by the national research council (CNPq) and the field work was funded by the U.S. National Science Foundation (NSF) and its Brazilian counterpart, the CNPq. The program was administered by MG and INPA with the cooperation of several North American herbaria. The foreign participation was coordinated by NY. This program mounted a series of approximately 30 collecting expeditions, for a total of about 1,500 days in the field, to carefully selected areas of Amazonia that are either centers of endemism, poorly collected, and/or threatened with imminent major deforestation. Figures 5a and 5b are maps of the regions already visited by 1983 and 1987 respectively. PFA also instituted a data-basing program into which most specimens in MG and INPA were entered, as discussed above. This program terminated at the end of 1987.

The second major collecting effort does not involve funding by the government or by foreign scientific institutions. In these programs, MG and INPA have been directly or indirectly contracted by large development projects to conduct ecological studies of development sites, including floristic and quantitative forest inventory of the areas. In some cases the programs have been initiated voluntarily by the companies involved; in others the money has come ultimately from international bank loans, and a specific percentage has been earmarked by the banks for ecological surveys. Examples of the latter include projects at Tucuruí, Carajás, and Barcarena, all in Pará. The advantages and disadvantages of these programs will be discussed in section X of this paper.

D. Finances

Brazil's scientific institutions have experienced more than their share of the belt-tightening that has accompanied the country's economic straits of the past several years. CNPq's support of PFA dwindled to almost nothing, leaving a reluctant NSF to fund the remainder of the project. INPA, MG, and HAMAB have been struck with repeated hiring freezes and the enforced reduction of staff by attrition. Also, salaries at these institutions are so low as to discourage many professionals from seeking work there.

The situation is better in Brazil than in some other tropical countries, but at the present time Brazil's Amazonian herbaria are functioning on minimal budgets and staffs. They are barely coping with the task of gathering *materials* for a floristic inventory of Brazilian Amazonia at a time when a greater and better-supported effort is also urgently needed to begin the *synthesis* of the information at hand.

IX. Suitability of Past, Present and Planned Inventory

The value of a floristic inventory can be assessed by the importance of the regions selected and by the programming associated with it. The two major collecting programs, Projeto Flora Amazônica (PFA) and the arrangements with large development projects, have been described in the preceding section of this paper. Site selection by the latter is and has been inherently suitable, because these areas all undergo extensive habitat destruction. Selection by PFA had been based on further criteria. Figure 5b, which maps the PFA expeditions for the period 1985–1987, illustrates the kinds of areas chosen: those threatened with destruction (2. Transamazon Highway, 3. northern Mato Grosso, 8. Rondônia); those largely unexplored (4. Ilha do Bananal, 5. Maracá, 7. upper Rio Solimões); and those known to contain an unusual flora (1. Amapá, 6. Marajó, 9. Rio Trombetas, 10. upper Rio Negro). The areas threatened with destruction have been identified by following development of the POLAMAZONIA centers (see section VII), other largescale enterprises such as dams and mines, and colonization patterns. Selection of largely unexplored areas had been based partly on accessibility and partly on observations by Brazilian botanists involved in the exploratory field verification phase of the RADAM project (see sections II and III). Areas with unusual floras have been determined by examining the collections of Spruce and other botanical pioneers, some of whom brought back material from remote areas which had (or have) never been collected since.

Based on experience gained from the PFA field program and incoming collections, we can now suggest new emphases for future floristic efforts. In addition to particular regions, key plant groups, such as palms, must be targeted for study. As noted earlier, most important angiosperm families in Amazonia have active specialists, but Brazilian and foreign students should be encouraged to fill in the gaps.

Future efforts should encourage more participation of botanists and students from southern Brazil, in order to develop a broader base of national interest in Amazonian floristic inventory and to take advantage of the human resources in the South. PFA expeditions recently included botanists from Rio de Janeiro and São Paulo.

The agreement establishing PFA discouraged collection and study of anything other than herbarium material. Future programming should aim to stress more interdisciplinary projects. Not only are these of greater scientific value, they appear to be especially attractive to funding agencies in recent years.

Future efforts should take much greater advantage of Brazil's growing system of natural preserves (see section VIII). These preserves can be the sites for cost-efficient collection, long-term ecological projects, and local floras. Active research in these areas can also help justify their existence and protect them from encroachment and loss of government support. Selection of areas for Brazil's natural preserves focused on regions considered to be centers of endemism and Pleistocene forest refuges (Wetterberg et al., 1981); it is unfortunate that PFA was not able to begin collecting in the new preserves as they were designated. Still, PFA has carried out quantitative forest inventories on an IBDF reserve in Amapá, and a major expedition was carried out in May 1986 on the Maracá ecological reserve in Roraima.

In selecting areas for future expeditions, greater use should be made of the RADAM (see section VIII) natural resource maps, so that areas with unusual floras or which are likely to be threatened with destruction can be anticipated.

As noted in the preceding section, PFA was never intended to be a permanent program. Those involved are now considering how best to utilize the material generated by the project and considering strategies for other programs to supplement the work being carried out under the auspices of the large development enterprises. Now is the time for several independent regional floristic projects to get underway. Logical candidates for these projects are the SEMA ecological reserve and the Territory of Amapá, and well-known older reserves such as the Reserva Ducke near Manaus and the Mocambo Reserve near Belém.

If future efforts in the floristic inventory of the Brazilian Amazon become further decentralized, as it is likely they will, it will be even more important for INPA and MG to ensure that new collections and identifications continue to be entered on the computer system. Failure to do so would mean wasting a resource of inestimable value.

The research being carried out near the Tucuruí dam, the mines at the Serra dos Carajás, and other large-scale development projects is meant to assess environmental impact in addition to carrying out basic research. As these programs now function, however, the botanical results will only be known in the ecologically distant future. An effort must be made to accelerate the analysis of the results and to ensure prompt feedback from specialists in particular plant groups and from phytogeographers. The structure of these programs is ironic, of course, in that environmental impact is being often assessed after the fact, when construction—or destruction—is already under way.

Overall, recent efforts in the floristic inventory of the Amazon have been very well focused and quite effective, although some aspects of the programming currently in effect should have been initiated earlier. Future efforts are likely to be decentralized, and local and regional flora projects would be an excellent means of supplementing the programming associated with development projects and of laying foundations for an eventual comprehensive flora.

X. Additional Resources Needed

First of all, most of the additional resources needed are human resources. Botanists and local collectors are most desperately needed, so jobs must be created for those now being trained. Perhaps the best place to provide jobs is in the developing regional herbaria such as HAMAB in Amapá, which does not have any Ph.D.'s on its dedicated staff.

Second, Brazil needs to dedicate more resources to biology training at the undergraduate level especially in Belém and Manaus, so that the adequate post-graduate programs associated with MG and INPA can have a pool of students from which to train good scientists.

Third, in addition to doctoral-level staff, Amazonia's regional herbaria need additional funding for infrastructure such as laboratory facilities, vehicles, and herbarium cases. In addition, the regional herbaria in Amapá, Rondônia, Acre, and Roraima should be encouraged and supported.

Fourth, Brazil should continue and expand the multifaceted support it has accorded the SEMA ecological reserves in Amazonia. New reserves should be provided with lodging, laboratory space and facilities and a small permanent staff as soon as possible. Thus equipped, the SEMA reserves will present excellent opportunities for extensive floristic inventory on a low budget.

Brazilian institutions and those foreign institutions most closely associated with the flora of the Brazilian Amazon must commit the resources necessary to begin the synthesis of the tremendous amount of data at hand, through florulas as well as through maintenance and intensive utilization of the data bank in Brasília.

Brazilian institutions should provide funding for short-term visits to foreign herbaria by their staff. Foreign institutions should, whenever possible, contribute badly needed botanical literature. In addition, foreign institutions should support visits by Brazilians to major world herbaria.

Despite the difficulties of collecting in Amazonia, Brazilian botanists have already achieved a great deal. They have worked continuously for years through changes of governments, financial crises, and even the loss of some of their colleagues. They have left a legacy which clearly informs us about the diversity and importance of the Amazon flora. However, the inventory is not complete, and we must do all we can to continue to encourage the efforts of resident botanists and visiting foreign scientists.

XI. Literature Cited

Aubréville, A. 1961. Etude écologique des formations principales du Brésil. Centre Technique Forestier Tropicale, Nogent-sur-Marne.

Balick, M. J., A. B. Anderson, & M. F. da Silva. 1982. Palm taxonomy in Brazilian Amazonia: the state of systematic collections in regional herbaria. Brittonia 34: 463–477.

Benson, W. 1982. Alternative models for infrageneric diversification in the humid tropics: tests with passion vine butterflies. Pages 608–640 in G. T. Prance, (ed.), Biological diversification in the tropics. Columbia University Press, New York.

Braun-Blanquet, J. 1932. Plant sociology, the study of plant communities. (translation of 1928 edition). McGraw-Hill, New York.

Campbell, K. E., Jr. & D. Frailey. 1984. Holocene flooding and species diversity in southwestern Amazonia. Quaternary Res. 21: 369–375.

Caufield, C. 1982. Brazil, energy and the Amazon. New Sci. 1982: 240–243.

Croat, T. 1978. Flora of Barro Colorado Island. Stanford University Press, Palo Alto, California.

Ducke, A. & G. A. Black. 1953. Phytogeographical notes on the Brazilian Amazon. Anais Acad. Brasil. Ci. 25(1): 1–46.

Egler, W. A. 1963. Adolpho Ducke—traços biográficos, viagens e trabalhos. Bol. Mus. Paraense Hist. Nat., n.s., Botânica no. 18.

Fearnside, P. M. 1982. Deforestation in the Brazilian Amazon: how fast is it occurring? Interciência 7(2): 82–88.

―――. 1983. Development alternatives in the Brazilian Amazon: an ecological perspective. Interciência 8(2): 65–78.

―――. & J. M. Rankin. 1982. Jari and Carajás: the uncertain future of large silvicultural plantations in the Amazon. Interciência 7(6): 326.

Fittkau, E. J. 1974. Zur ökologischen Gliederung Amazoniens I. Die erdgeschichtliche Entwicklung Amazoniens. Amazoniana 5(1): 77–134.

Fontes, G. de Carvalho. 1966. Alexandre Rodrigues Ferreira. Aspectos de sua vida e obra. CNPq/INPA, Manaus.

Fróes, R. L. 1953. Estudo sobre a Amazônia Maranhense e seus limites florísticos. Revista Brasil. Geogr. 1: 96–100.

Gentry, A. H. 1979. Distribution patterns of Neotropical Bignoniaceae: some phytogeographic implications. Pages 339–354 in: K. Larsen & L. Holm-Nielsen, (eds.). Tropical Botany. Academic Press, London.

Goeldi, E. 1982 (re-issue). Alexandre Rodrigues Ferreira. Editora Univ. Brasília. 80p.

Hoehne, F. C. 1939. Flora Brasílica. Plano geral para a elaboração a publicaçao de uma obra ilustrada sobre a flora do Brasil proposto por F. C. Hoehne, S. Paulo, Brasil. (unpublished)

―――. 1940–1968. Flora Brasilica. In 12 fascicles.

―――. 1951. Indice bibliográfico e numérico das plantas colhidas pela Comissão Rondon ou Comissão de Linhas Telegráficas, Estrat. de Mato Grosso ao Amazônas, de 1908 até 1923. São Paulo, Instituto de Botânica.

Holdridge, L. R. 1947. Determination of world plant formations from simple climatic data. Science 105: 367, 368.

Huber, J. 1898. Materiaes para a flora Amazônica I. Lista das plantas colligidas na Ilha de Marajó no anno de 1896. Bol. Mus. Paraense Hist. Nat. 2: 288–321.

―――. 1902. Materiaes para a flora Amazônica III. Fetos do Amazonas inferior e de algumas regiões limitrophes, colleccionados pelo Dr. J. Huber e determinados pelo Dr. Hermann Christ de Basiléa (Suissa). Bol. Mus. Paraense Hist. Nat. 3: 60–64.

―――. 1909. Mattas e madeiras Amazônicas. Bol. Mus. Paraense Hist. Nat. 6: 91–225.

Huber, O. 1982. Significance of savanna vegetation in Amazonas Territory of Venezuela. Pages 221–224 in G. T. Prance, (ed.), Biological diversification in the tropics. Columbia University Press, New York.

―――, J. A. Steyermark, G. T. Prance, & C. Alés. 1984. The vegetation of the Sierra Parima, Venezuela–Brazil: Some results of recent exploration. Brittonia 36: 104–139.

Hueck, K. 1957. Sobre a origem dos campos cerrados do Brasil e algumas novas observações no seu limite meridional. Revista Brasil. Geogr. 19.

―――. 1966. Die Wälder Südamerikas. Okologie, Zusammensetzung und wirtschaftliche Bedeutung. Vegetationsmonographien Bd. II. Stuttgart.

―――. 1972. As florestas da América do Sul. Polígono, São Paulo (translation of Hueck, 1966).

———— & P. Seibert. 1972. Vegetationskarte von Südamerika. G. Fischer-Verlag, Stuttgart.

Instituto Brasileiro de Bibliografia e Documentação. 1975. Amazonia; bibliografia. O. Koeltz, Koenigstein.

Instituto Brasileiro de Geografia e Estatística. 1977. Geografia do Brasil I. Região Norte. IBGE, Rio de Janeiro.

Instituto Brasileiro de Geografia e Estatística. 1981. Sinopse estatística do Brasil. IBGE, Rio de Janeiro.

Irion, G. 1978. Soil infertility in the Amazon. Naturwissenschaften **65**: 515–519.

Jordan, C. F. 1987. Soils of the Amazon rainforest. Pages 83–94 *in* T. C. Whitmore & G. T. Prance, (eds.). Biogeography and Quaternary history in tropical Latin America. Oxford Biogeographical Monographs No. 3. Oxford University Press.

Junk, W. 1970. Investigations on the ecology and production-biology of the "floating meadows" (Paspalo-Echinochloetum) on the Middle Amazon. Amazoniana **2**: 449–495.

Koeppen, W. P. 1901. Versuch einer Klassifikation der Klimate vorzugweise nach ihren Beziehungen zur Pflanzenwelt. B. G. Teubner, Leipzig.

Kubitzki, K. 1977. The problem of rare and of frequent species: the monographer's view. Pages 331–336 *in* G. T. Prance and T. Elias, (eds.). Extinction is forever. New York Botanical Garden, Bronx.

Küchler, A. W. 1970. International bibliography of vegetation maps. Vol. 4. Vegetation maps of Africa, South America, and the world (general). University of Kansas Libraries, Lawrence.

————. 1982. Brazilian vegetation on maps. Vegetatio **49(1)**: 29–34.

Kuhlmann, E. 1977. Vegetação. Pages 59–90 *in* IBGE Geografia do Brasil. Região Norte. IBGE, Rio de Janeiro.

Lescure, J.-P. & R. Boulet. 1985. Relationships between soil and vegetation in a tropical rain forest in French Guiana. Biotropica **17**: 155–164.

Martius, C. F. P. von. 1824. Die Physiognomie des Pflanzenreiches in Brasilien. M. Lindauer, Munich.

————, ed. 1840–1906. Flora brasiliensis. 15 vols. Munich, Vienna, Leipzig.

Mitchell, J. G. 1979. The man who would dam the Amazon. Audubon **81(2)**: 65–81.

Nelson, B. W., C. A. Cid Ferreira, C. A. Todzia, J. L. Zarucchi, & S. R. Hill. 1987. Distributional and ecological notes on *Polygonanthus amazonicus* Ducke. Acta Amazônica, Manaus, **15(1–2)**: Supl. Mar.–Jun. **1985**: 63–70.

Nimer, E. 1977. Clima. Pages 39–58 *in* IBGE. Geografia do Brasil. Instituto Brasileiro de Geografia e Estatística, Rio de Janeiro.

Nogueira Neto, P. 1985. Getting Brazil's network of ecological stations off the ground. Annual Report 1985, Conservation Foundation.

Oliveira, A. E. de. 1983. Ocupação humana. Pages 144–327 *in* Salati, E., H. O. R. Schubart, W. Junk, & A. E. de Oliveira. Amazônia: desenvolvimento, integração, e ecologia. Editora Brasiliense/CNPq, São Paulo.

Pires, J. M. 1972. Tipos de vegetação da Amazônia. Publ. Avulsas Mus. Goeldi **20**: 179–202.

————. 1984. The Amazonian forest. Pages 581–602 *in* Sioli, H., (ed.). The Amazon. Limnology and landscape ecology of a mighty tropical river and its basin. Dr. W. Junk, Dordrecht.

———— & G. T. Prance. 1977. The Amazon forest: a natural heritage to be preserved. Pages 158–194 *in* G. T. Prance & T. Elias, (eds.). Extinction is forever. New York Botanical Garden, Bronx.

———— & ————. 1985. Vegetation types of the Brazilian Amazon. Pages 109–145 *in* G. T. Prance & T. E. Lovejoy, (eds.). Amazonia. Key Environments Series, Pergamon Press, Oxford.

Prance, G. T. 1971. An index of plant collectors in Brazilian Amazonia. Acta Amazônica **1(1)**: 25–68.

————. 1973. Phytogeographic support for the theory of Pleistocene forest refuges in the Amazon basin, based on evidence from distribution patterns in Caryocaraceae, Chrysobalanaceae, Dichapetalaceae, and Lecythidaceae. Acta Amazônica **3(3)**: 5–28.

————. 1977. The phytogeographic divisions of Amazonia and their influence on the selection of biological reserves. Pages 195–213 *in* G. T. Prance & T. Elias, (eds.). Extinction is forever. New York Botanical Garden, Bronx.

————. 1978. The origin and evolution of the Amazon flora. Interciencia **3(4)**: 207–222.

————. 1979. Notes on the vegetation of Amazonia III. The terminology of Amazon forest types subject to inundation. Brittonia **31**: 26–38.

————, ed., 1982a. Biological diversification in the tropics. Columbia University Press, New York.

————. 1982b. Forest refuges: evidence from woody angiosperms. Pages 137–158 *in* G. T. Prance, (ed.). Biological diversification in the tropics. Columbia University Press, New York.

————. 1982c. A review of the phytogeographic evidences for Pleistocene climate changes in the Neotropics. Ann. Missouri Bot. Gard. **69**: 594–624.

————. 1987. Vegetation. Pages 28–44 *in* T. C. Whitmore & G. T. Prance, (eds.). Biogeography and Quaternary history in tropical America. Clarendon Press, Oxford.

———— & K. S. Brown, Jr. 1987. The principal vegetation types of the Brazilian Amazon. Fig. 2.4, pages 30, 31 *in* T. C. Whitmore & G. T. Prance (eds.). Biogeography and Quaternary history in tropical America. Clarendon Press, Oxford.

————, B. W. Nelson, M. F. da Silva, & D. C. Daly. 1984. Projeto Flora Amazônica: Eight years of binational botanical expeditions. Acta Amazônica, Manaus **14(1–2)**, Supl.: 5–29.

Projeto RADAM. 1973–1975. Levantamento de recursos naturais. Vols. 1–7. Ministério de Minas e Energia, Rio de Janeiro.

Projeto RADAMBRASIL. 1975–1981. Levantamento de recursos naturais. Vols. 8–18. Ministério de Minas e Energia, Rio de Janeiro.

————. 1982. Fitogeografia Brasileira. Boletim Técnico. Série Vegetação. Ministério de Minas e Energia.

PRORADAM (Proyecto Radargrametrico del Amazonas). 1979. La Amazonia Colombiana y sus recursos. PRORADAM, Bogotá.

Putzer, H. 1984. The geological evolution of the Amazon basin and its mineral resources. Pages 15–46 *in* H. Sioli, (ed.). The Amazon. Limnology and landscape ecology of a mighty tropical river. Dr. W. Junk, Dordecht.

Rizzini, C. T. 1963. Nota prévia sobre a divisão fitogeografica do Brasil. Revista Brasil. Geogr. **1 (Ano XXV)**: 1–64.

————. 1979. Tratado de fitogeografia do Brasil. Vol. 2. Aspectos sociológicos e florísticos. Editora Universidade do São Paulo.

Rodrigues, W. A. 1959(?). Lista dos nomes vernáculos da flora do Território do Rio Branco. INPA/Botânica Public. no. 9.

————. 1979. Indice das espécies e assuntos contidos em "Notas sobre a fitogeografia da Amazônia Brasileira" de Ducke e Black (1954). Acta Amazônica **9**: 437–462.

Rollet, B. 1974. L'Architecture des forêts denses humides sempervirentes de plaine. Centre Technique Forestier Tropical, Nogent-sur-Marne.

Salati, E., H. O. R. Schubart, W. Junk & A. E. de Oliveira. Amazônia: Desenvolvimento, integração, e ecologia. Editora Brasiliense/CNPq, São Paulo.

Sampaio, A. 1945. Fitogeografia do Brasil (3rd ed.). Companhia Editorial Nacional, São Paulo.

Schubart, H. O. R. 1983 Ecologia e utilização das florestas. Pages 101–143 *in* E. E. Salati, H. O. R. Schubart, W. Junk, & A. E. de Oliveira. Amazônia: Desenvolvimento, integração, e ecologia. Editora Brasiliense/CNPq, São Paulo.

Schurz, W., O. D. Hargis, C. F. Marbut, & C. B. Manifold. 1925. Rubber production in the Amazon Valley. U.S. Government Printing Office (U.S. Dept. of Commerce Trade Promotion Series no. 23), Washington, D.C.

Secco, R. S. & A. L. Mesquita. 1983. Notas sobre a vegetação de canga da Serra Norte I. Bol. Mus. Paraense Hist. Nat., n.s., Botânica no. **59**: 1–13.

Simpson, B. & J. Haffer. 1978. Speciation patterns in the Amazonian forest biota. Ann. Rev. Ecol. Syst. **9**: 497–518.

Sioli, H. 1951. Alguns resultados e problemas da limnologia Amazônica. Bol. Técn. Inst. Agron. N. **24**: 3–44.

———. 1984. The Amazon and its main affluents: Hydrography, morphology of the river courses, and river types. Pages 127–165 *in* H. Sioli, (ed.). The Amazon. Limnology and landscape ecology of a mighty tropical river. Dr. W. Junk, Dordrecht.

Soares, L. de Castro. 1953. Limites meridionais e orientais da área de ocorrência da floresta Amazônica em terreno Brasileiro. Revista Brasil. Geogr. **1**: 3–96.

Sombroek, W. 1966. Amazon soils. Centre for Agricultural Publications and Documentation, Wageningen.

Sternberg, H. O'Reilly. 1975. The Amazon River of Brazil. Steiner, Wiesbaden.

Tosi, J. A., Jr. & L. L. Vélez-Rodriguez. 1983. Provisional ecological map of the Republic of Brazil. Institute for Tropical Forestry, San Juan.

Ule, E. 1908. Die Pflanzen Formation des Amazonas Gebietes II. Bot. Jahrb. Syst. **40**: 398–443.

Ulman, N. 1984. Westward ho! Brazilians are fleeing nation's economic ills to vast frontier lands. Wall Street Journal, October 11, 1984.

UNESCO. 1973. International classification and mapping of vegetation. UNESCO, Paris.

———. 1981. Vegetation map of South America. UNESCO, Paris.

Urban, I. 1906. Vitae itineraque collectorum botanicorum. *In* C. F. P. von Martius, (ed.), Flora brasiliensis **1(1)**: 1–152.

Veloso, E. 1966. Tipos de vegetação. *In* Atlas florestal do Brasil. Ministério de Agricultura, Conselho Florestal Federal, Rio de Janeiro.

Wetterberg, G., G. T. Prance, & T. E. Lovejoy. 1981. Conservation progress in Amazônia: a structural review. Parks **6(2)**: 5–10.

Woodwell, G., R. A. Houghton, & T. A. Stone. 1986. Deforestation in the Brazilian Amazon Basin: agricultural development in Rondônia based on satellite imagery. Pages 23–32 *in* G. T. Prance, (ed.), Tropical rain forests and the world atmosphere. Westview Press, Boulder Colorado. AAAS Selected Symposium 101.

Wurdack, J. J. 1970. Erroneous data in Glaziou collections of Melastomataceae. Taxon **19**: 911–913.

Glossary of Acronyms

BR	Laboratoire de Botanique Systématique, Brussels
CEPEC	Centro de Pesquisas do Cacau, Itabuna
CNPq	Centro Nacional de Desenvolvimento Científico e Político, Brasília
EMBRAPA/CPATU	Centro de Pesquisa Agropecuária do Trópico Umido, Belém
HAMAB	Herbário, Museu de Historia Natural Angêlo Moreira da Costa Lima, Macapá, Amapá
HBG	Herbarium, Institut für Allgemeine Botanik und Botanischer Garten, Hamburg
HUAM	Herbário da Universidade do Amazonas, Manaus
"HUFA"	Herbário da Universidade Federal do Acre (acronym not yet official), Rio Branco
IAN	Instituto Agronômico do Norte (now in EMBRAPA/CPATU)
IBDF	Instituto Brasileiro de Desenvolvimento Florestal
IBGE	Instituto Brasileiro de Geografia e Estatística
INCRA	Instituto Nacional de Colonização e Reforma Agrária
INPA	Instituto Nacional de Pesquisas da Amazônia, Manaus
K	Royal Botanic Gardens, Kew
LE	Herbarium of the Department of Higher Plants, V. L. Komarov Botanical Institute of the Academy of Sciences of the USSR, Leningrad
M	Herbarium, Botanische Staatssammlung, Munich
MG	Museu Paraense Emílio Goeldi, Belém
MICH	Herbarium of the University of Michigan, Ann Arbor
MIN	Herbarium, Department of Botany, University of Minnesota, St. Paul
MO	Missouri Botanical Garden, St. Louis
NSF	U.S. National Science Foundation
NY, NYBG	The New York Botanical Garden, Bronx
P	Muséum National d'Histoire Naturelle, Laboratoire de Phanérogamie, Paris
PFA	Projeto Flora Amazônica
PIN	Programa de Integração Nacional
R	Herbário, Departamento de Botânica, Museu Nacional, Rio de Janeiro
RB	Herbário, Jardim Botânico do Rio de Janeiro
SERPRO	Serviço Federal de Processamento dos Dados, Brasília
SP	Herbário do Estado "Maria Eneyda P. K. Fidalgo," Instituto do Botânica, São Paulo
SPVEA	Superintendência do Plano para a Valorização Econômica da Amazônia
SUDAM	Superintendência para o Desenvolvimento da Amazônia
U	Institute of Systematic Botany, Utrecht
US	U.S. National Herbarium, Washington, D.C.
USDA	U.S. Department of Agriculture
USF	Herbarium, Department of Biology, University of South Florida, Tampa
WIS	Herbarium, Department of Botany, University of Wisconsin, Madison

Eastern, Extra-Amazonian Brazil

Scott A. Mori

Contents

I.	Introduction	428
II.	Other Vegetation Types of Eastern Brazil	430
	A. Caatinga	430
	B. Cerrado	432
	C. Campo Rupestre	432
	D. Mangrove	434
	E. Restinga	434
	F. Granitic Outcrops	437
	G. Liana Forest	437
	H. Araucaria Zone	437
III.	Lowland Moist Forests	437
	A. Endemism	438
	B. Land Use	441
IV.	Floristic Inventory	443
V.	Ecological Studies	445
VI.	Conservation	446
VII.	Recommendations	446
VIII.	Acknowledgments	450
IX.	Literature Cited	450

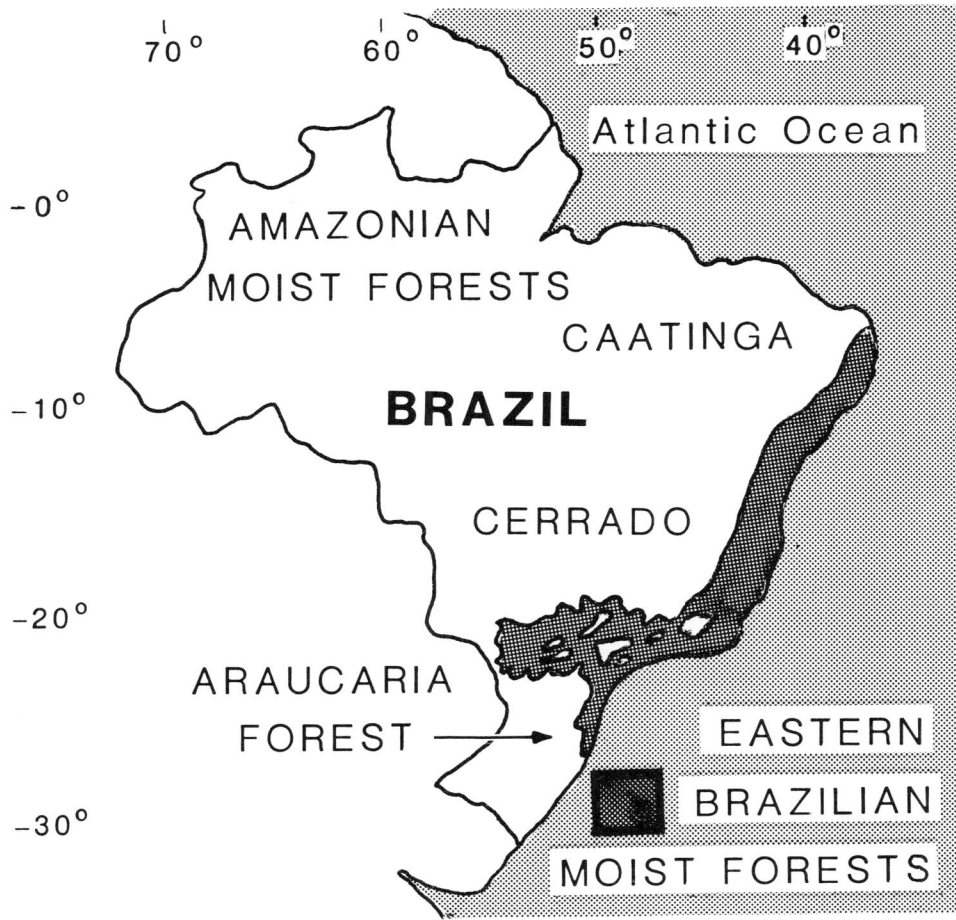

FIG. 1. The pre-colonization extent of the lowland moist forests of eastern Brazil and their relationship to major Brazilian vegetation types. Because of their reduced extent, additional vegetation types discussed in the text are not shown. Map redrawn from Fonseca, 1985.

I. Introduction

One of the biological tragedies of this century has been the accelerated destruction of the original vegetation of eastern Brazil. This area is a rich mosaic of vegetation types ranging from dry thorn-scrub *caatinga* of the northeast to wet rain forests along the coast (Fig. 1). Each vegetation type has its unique set of plants and animals, of which many are endemic to the given vegetation type in part or all of the areas of occurrence in eastern Brazil. Because Europeans first settled in eastern Brazil and because it is here that the Brazilian population is most dense (Fonseca, 1985), much of the original vegetation has been destroyed. Cocoa and sugar plantations have replaced coastal rain forests; coconut and piaçava plantations and summer homes have destroyed much of the *restinga*; coffee plantations are now found in place of liana forest; firewood gathering and grazing have drastically modified the floristic composition of *caatinga*; the search for gold and diamonds as well as for ornamental plants has taken a heavy toll on *campo rupestre*; lumbering has decimated the magnificent *Araucaria* forests; and agriculture and grazing have reduced undisturbed *cerrado* to national parks and preserves.

The importance of preserving what is left of the original vegetation of eastern Brazil has been recognized as a number one priority for all of South America. Especially critical are the moist forests (Fonseca, 1985; Mori et al., 1981; Myers, 1980; Sick & Teixeira, 1979) which are the focus of programs by the World Wildlife Fund (WWF), the International Union for the Conservation of Nature (IUCN), the Institute Brasileiro de Desenvolvimento Florestal (IBDF), the Fundação Brasileira para a Conservação de Natureza (FBCN), and the Brazilian Secretária Especial do Meio Ambiente (SEMA). These organizations have recognized the need for immediate action. It is through their efforts that areas have been set aside, and that funds for their protection have been made available.

The focus of this chapter will be the moist forests of eastern Brazil, their occurrence, their modification by man, their biological make-up, the extent to which they have been studied, the degree to which they have been preserved, and the hope for their future. To place moist forests within the

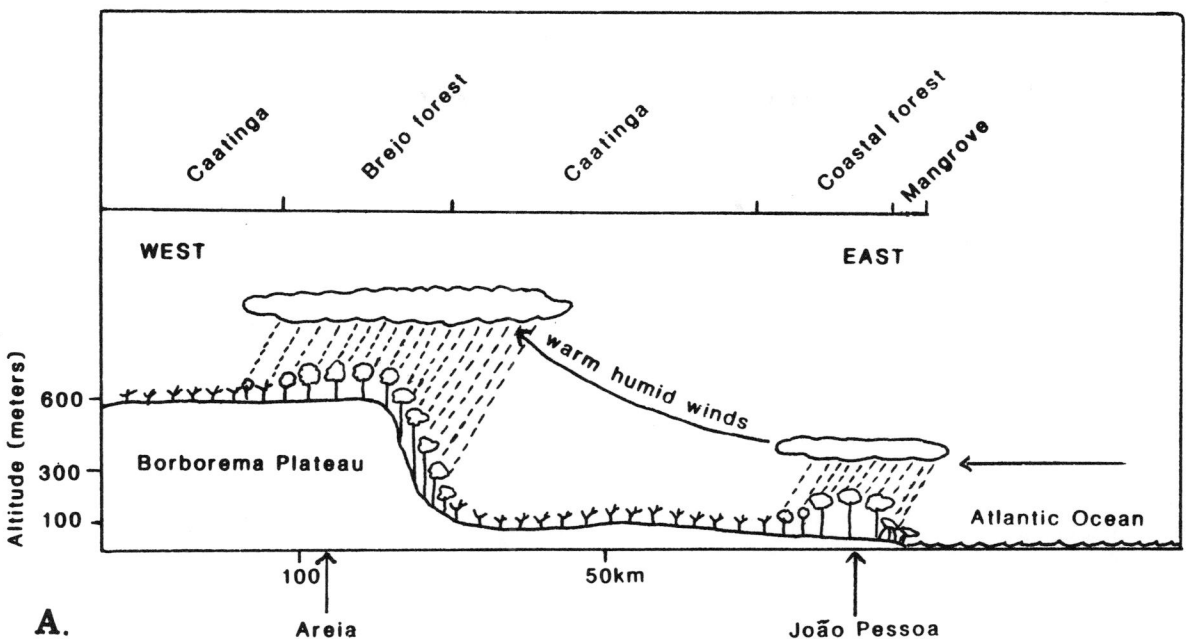

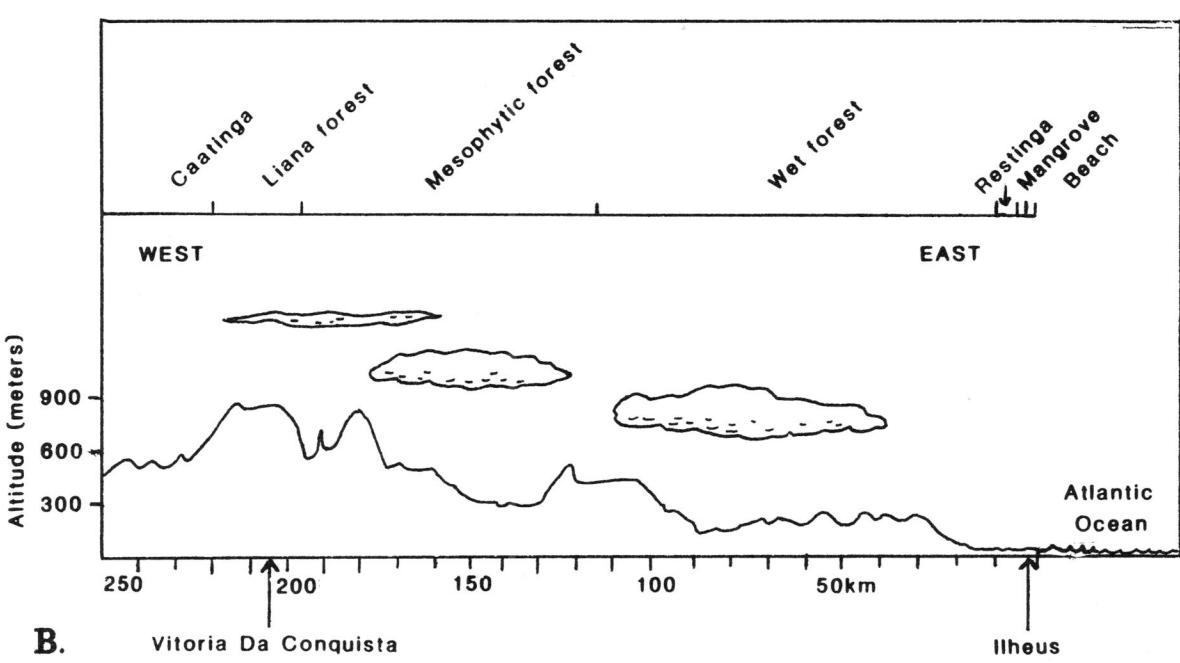

FIG. 2. Profiles from the ocean inland demonstrating the influence of moisture-laden winds on the vegetation of eastern Brazil. A. Profile from João Pessoa to the Borborema plateau in Paraíba. Moist forest occurs along the coast and at the edge of plateaus in the form of "brejo" forest further inland (from Mayo & Fevereiro, 1982). B. Profile from Ilhéus on the coast to Vitoria da Conquista in southern Bahia. Moist forest is restricted to the coastal areas (from Mori & Silva, 1979a).

context of the mosaic of eastern Brazilian vegetation types, I first will present introductions to the other types of vegetation found there. This is necessary because: 1) vegetation types often grade or interdigitate into one another, 2) animals may move freely from one vegetation type to another, 3) floras usually include more than one vegetation type, and 4) collections in herbaria of eastern Brazil are not limited to those from moist forests.

FIG. 3. Scenes from the *caatinga* of Bahia during the dry season. **A.** Cacti, such as *Cereus jamacaru* DC. shown here, are characteristic of many forms of *caatinga*. Note the single plant with evergreen leaves. **B.** The characteristic trunk of *Cavanillesia arborea* K. Schum. which serves as a water storage organ.

II. Other Vegetation Types of Eastern Brazil

Annual rainfall and temperature differences caused by differences in altitude are the important factors influencing the distribution of plant and animal species and the formation of different vegetation types in eastern Brazil (Fig. 1). The profiles of Figure 2 illustrate the influence of coastal mountains on rainfall and temperature and their subsequent influence on the vegetation. In general, moist forest is found near the coast and drier formations farther from the coast. However, inland *brejo* forests are enclaves of moist forest, surrounded by *caatinga*, which are caused by precipitation on the slopes of various plateaus. Farther to the south temperature becomes more important. For example, in Paraná, temperate forests of *Araucaria* interspersed with prairies dominate the landscape at higher elevations.

A. Caatinga

This vegetation type (Fig. 3) is characterized by much-branched, spiny trees and shrubs with tortuous trunks usually reaching only 8 to 10 m in height. Among the trees and shrubs, mimosoid legumes and Cactaceae predominate. Terrestrial Bromeliaceae with thick, fleshy leaves are frequent. The word *caatinga* comes from the Tupi Indian language and means "white forest" in allusion to its grayish-white aspect in the dry season (May-November), when most of the species lose their leaves (Alvim, 1950; Andrade-Lima, 1954). The concept of *caatinga* ranges from near desert vegetation to a closed canopy forest called *caatinga arborea*. Andrade-Lima (1981a) has attempted to classify the plant associations found within *caatinga*. The *caatinga* of eastern Brazil has no relationship with Amazonian *caatinga* (Anderson, 1981) in terms of species encountered.

Caatinga develops in areas receiving less than 800 mm rainfall per year and with sandy, often rocky soils, generally poor in organic material. Rainfall is unevenly distributed, and the dry season is pronounced and long. Modifications *caatinga* plants have evolved to withstand drought are deep root systems, deciduous leaves, leaves with reduced blades, thick, waxy cuticles, dense pubescence, and diurnally closed stomates. Some plants, such as *Spondias tuberosa* Arruda, *Sterculia striata* St. Hil., *Thiloa glaucocarpa* (Mart.) Eichl., and *Manihot* spp. store water and food in tuberous roots, while others, such as *Cavanillesia arborea* K. Schum. (Fig. 2B) and species of Cactaceae and Euphorbiaceae store water in their trunks (Alvim, 1950; Vinha et al., 1976).

The contrast between the wet and dry seasons in *caatinga*

FIG. 4. *Cerrado* vegetation. **A.** *Salvertia convallariodora* St. Hil., a common plant of Brazilian *cerrado* showing typical growth form and bark of the trees of this vegetation type. **B.** Fire in the *cerrado* of Amapá, Brazil. *Mauritia flexuosa* Linn.f., the palm in the background, is found in wetter habitats along streams.

is striking. During the rainy season the vegetation is green, crops are in full development, and livestock looks well-fed and healthy. In contrast, during the long dry season of up to six months or more, only a few plants, such as *Zizyphus cotinifolia* Reiss., retain their leaves (Fig. 3A) and a bleak landscape results. Then, the only agricultural activity is in the moister valleys, and livestock, resembling walking skeletons, are constantly on the move foraging for survival. In years of prolonged drought the *caatinga* is a region of much human suffering and the source of immigrants to the moister areas of the Amazon and southern Brazil.

Andrade-Lima (1954) provides a list of common and scientific names of the species most frequently encountered in the *caatinga* of Pernambuco, and Braga (1960) provided an alphabetical listing of the plants of northeastern Brazil, many of which are native to *caatinga*. Ducke (1959) pointed out that the flora of the *caatinga* is relatively homogeneous from the south of Piauí to the north of Minas Gerais, and that the common names are widely applied throughout this area. Luetzelburg (1922–1923) included lists of species from *caatinga*, but his names must be used with caution because they are often based on misidentifications and other errors (Ducke, 1959 p. 9). A list of economic plants native to *caatinga* is provided by Andrade-Lima (1970).

Many plants and animals are endemic to *caatinga*. Cracraft (1985) recognizes a *caatinga* center as one of his areas of endemism for birds. A close relationship between the plants and animals of the *caatinga* exists with those of the *chaco* of southwestern Brazil, southeastern Bolivia, western Paraguay, and north central Argentina (Andrade-Lima, 1977; Cracraft, 1985).

All *caatinga* has been disturbed by man in one way or the other. It is nearly always grazed by goats and exploited for timber and firewood. On moister sites sugar cane, corn, rice, beans, watermelon, and manihot are planted. In some areas large plantations of sisal, cotton, castor bean, and tobacco cover the landscape. Cattle are frequent, and in some areas the river flood plains have been converted to pasture of alien grasses such as *Cenchrus ciliaris* L. In contrast to *cerrado*, fire can not be used as a range management technique because plants of *caatinga* do not propagate vegetatively. Fire, instead of encouraging sprouting from rhizomes as it does in *cerrado*, kills the seeds of plants of the *caatinga*.

A very large ecological station, "Raso da Catarina," has recently been established in the driest part of the state of Bahia. The purpose of this station is to preserve *caatinga* dominated by shrubs and herbs. Some *caatinga* is also protected in the national parks of Serra da Capivara, Sete Cidades, Ubajara, and the biological reserve of Serra Negra (Jorge Padua, 1983). Nevertheless, there is still an urgent need to locate and preserve *caatinga* dominated by trees since most of it has already been replaced with agricultural crops because farmers believe the soils there are richer.

B. Cerrado

This complex of vegetation types is characterized by trees with contorted growth forms, thick barks, deep root systems, or underground stems (Fig. 4A). The leaves of many species often possess xeromorphic features such as pubescence, thick cuticles, sunken stomata, water-storing parenchyma, hypoderm, and mechanical tissue. Many species of the *cerrado* are tolerant of high concentrations of aluminum in the soil. Most of the trees of *cerrado* flower in the dry season and fruit at the beginning of the wet season.

Cerrado is a Portuguese word meaning "closed" or "dense" and was originally used as an adjective in the term *campo cerrado* in order to distinguish those grasslands in which trees and shrubs are present from *campo limpo*, a grassland or prairie. Later, *cerrado* came to be used as a noun to refer to a diverse array of savanna-like vegetation types, ranging from prairie to forest (Eiten, 1972). The only thing that distinguishes the *cerrados* of central Brazil from the savannas of Amapá and the *llanos* of Colombia and Venezuela is their richer species composition. Many plants (Heringer et al., 1977) and animals (Cracraft, 1985) are endemic to *cerrado*.

Cerrado develops in areas receiving 700 to 1750 mm of rain per year and with average annual temperatures from 19° to 23.5°C. This vegetation ranges from 400 to 1100 meters elevation. There is always a distinct dry season, usually from April to September, and there are often 10–25 rainless days in January or February (Eiten, 1972; Ferri, 1977; Goodland & Ferri, 1979). Rainfall occurs predictably every year in contrast to the unpredictability of precipitation in *caatinga* where several years may pass with little or no rain.

The literature on *cerrado* is extensive. Eiten (1972) and Ferri (1977) provide reviews. A detailed ecological study of *cerrado* is that of Goodland for the Triângulo Mineiro in Minas Gerais (Goodland & Ferri, 1979), and floristic lists for *cerrado* have been published by Goodland (1970), Ferri (1967), Rizzini (1963, 1971a), and Heringer et al. (1977).

Until recently there seemed to be little urgency in preserving *cerrado* because of its vast extent (1.5 million square kilometers or nearly 25% of Brazil), and because it occurs on areas once thought to be marginal for agriculture (Nogueira-Neto, 1977). This has now changed because of the Brazilian government's program to open up the *cerrado* as a new agricultural frontier. With the addition of lime and fertilizer, the level terrain of much of the *cerrado* has been converted into mechanized fields of rice, wheat, and soy bean. *Cerrado* on less desirable soils is being utilized for cattle grazing where, in order to promote the growth of native forage plants, fire is used in range management (Fig. 4B). Today there are few, if any, areas of *cerrado* outside of national parks that have not been highly modified by these activities. The Parque Nacional das Emas with 120,000 hectares and the Parque Nacional de Brasília with 30,000 hectares are two major areas set aside to protect this vast and diverse vegetation type. *Cerrado* is also protected in the national parks of Araguaia, Chapada dos Veadeiros, Pantanal Matogrossense, Serra da Canastra, and Sete Cidades (Jorge Pádua, 1983).

C. Campo Rupestre

Campo rupestre is a savanna-like vegetation that occurs at elevations over 600 m in the states of Bahia, Goiás, and

FIG. 5. *Campo rupestre* A. Velloziaceae and Bromeliaceae growing in the typical rocky habitat of this vegetation type. B. Bunches of Eriocaulaceae cut from the *campo rupestre* which will be shipped for sale as dried flowers throughout the world.

Minas Gerais. It differs from *cerrado* by occurring on rockier soils at higher elevations and by possessing a different species composition, dominated by Melastomataceae, Asteraceae, Euphorbiaceae and, in particular, by many species of Velloziaceae (Fig. 5A). It is a vegetation extremely rich in endemics as has been shown by the recent description of many new species from the *campo rupestre* of Bahia (Silva & Mori, 1981). Many species of *campo rupestre* also occur in *restinga* along the coast (Carvalho, 1981; Mori et al., 1981).

Campo rupestre lends itself poorly to economic exploitation. It is grazed by livestock in some areas but even then at low densities and, because of its poor and rocky soils, agriculture is not an important part of the landscape. A very destructive economic activity, however, has been the search for diamonds and gold. Prospectors create large holes, mounds of gravel, dammed-up ponds, and channels in search of these and other minerals. In some areas the collection of ornamental plants for landscape architecture and house plants, especially Eriocaulaceae (Fig. 5B) (Menezes et al., 1986), orchids, and bromeliads, has drastically reduced local populations of these plants. Fire, induced by man, is resulting in the destruction of *campo rupestre* in some places.

Campo rupestre is protected by the state of Minas Gerais around Serra do Cipó and in the national parks of Itatiaia (Rio de Janeiro) and Serra de Canastra (Minas Gerais). Chapada Diamantina of Bahia, a national park consisting mostly of *campo rupestre*, was established on September 17, 1985 (Jorge Pádua, pers. comm.)

D. Mangrove

Mangroves occur sporadically along the Brazilian coast from Amapá south to the mouth of the Rio Araranguá in Santa Catarina (Bascopé et al., 1959). The original distribution of mangrove has been reduced because *Rhizophora* provides excellent fuel wood and because its bark is used in the tanning industry. Mangrove vegetation is a dense tree formation found where the substrate is saline, flat, and muddy. These conditions are often met behind sand bars and at the mouths of coastal rivers. The need for warm water restricts mangroves to the tropical and subtropical regions of the world.

Mangrove trees are characterized by thick coriaceous leaves, prop roots and pneumatophores. The principal species of mangrove along the Brazilian coast are: *Avicennia schaueriana* Stap. & Lechm., *A. germinans* (L.) Stearn, *Conocarpus erectus* L., *Laguncularia racemosa* Gaertn. f., and *Rhizophora mangle* L. Because plants adapt with difficulty to the mangrove environment, species diversity is low and mangrove species are widespread.

The mangroves of the eastern Brazilian coast have been the object of numerous studies (Andrade-Lima, 1970; Bigarella, 1946; Lamberti, 1969; Luederwoldt, 1919, 1929; Luz, 1955; Rawitscher, 1944; Stellfeld, 1945; Ule, 1901; Vasconcelos, 1949). The botany of mangroves has been summarized by Chapman (1976) and Tomlinson (1985).

E. Restinga

Restinga encompasses a diverse array of vegetation (Fig. 6) found on sandy soil in a narrow band along the Brazilian coast between the high tide mark and the moist forests on lateritic soil further inland. This term has not been applied outside of Brazil, but it is also used in the Brazilian Amazon for a completely different kind of beach vegetation. In eastern Brazil, the concept of *restinga* includes beach, dunes, open shrub, closed shrub, and low forest communities. A transect from seaward inland in a typical *restinga* of southern Bahia crosses 1) low, herbaceous beach vegetation dominated by *Alternanthera maritima* (Mart.) St. Hil., *Canavalia rosea* (Sw.) DC., *Hydrocotyle bonariensis* Lam., *Ipomoea pes-caprae* (L.) R. Br., *Oxypetalum tomentosum* Wight ex Hook. & Arn., *Philoxerus vermicularis* (L.) Nutt., *Remirea maritima* Aubl., and *Sporobolus virginicus* Kunth; 2) a beach facing zone of shrubs in which *Allagoptera arenaria* Kuntze, *Chrysobalanus icaco* L., *Dalbergia ecastophyllum* (L.) Taub., and *Sophora tomentosa* L., are most common; 3) a more diversified taller zone of shrubs and small trees with *Anacardium occidentale* L., *Emmotum affine* Miers, *Humiria balsamifera* (Aubl.) St. Hil. and many species of Myrtaceae as the most frequent species; and 4) a low forest composed of a combination of species of trees from the preceding zone and from the adjacent moist forest on lateritic soil.

Restinga appears to be rich in endemics. Plowman (1987) has recently described six new species of narrow endemics from Bahian *restinga*.

Because of its easy accessibility, *restinga* has been the object of study by many botanists. Species lists have been prepared for various *restingas* along the eastern coast of Brazil (Andrade-Lima, 1951; Araújo & Henriques, 1984; Araújo & Peixoto, 1977; Bertels, 1957; Bigarella, 1946; Esteves, 1980; Lutz, 1938; Luz, 1955; Pinto et al., 1984; Rambo, 1954; Rawitscher, 1944; Sampaio, 1934; Seabra, 1949; Stellfeld, 1949; Ule, 1901). The "Flora Ecológia de Restingas do Sudeste do Brasil," published by the Museu Nacional, is extremely useful for identifying plants from the *restingas* of southeastern Brazil. Ecological studies of microclimate (Dau, 1960), regeneration after fire (Araújo & Peixoto, 1977), zonation and succession (Dansereau, 1947; Hueck, 1955; Magnanini, 1954), and community organization (Ormond, 1960) are available. A bibliography of studies of *restinga* has been prepared by Lacerda et al. (1982) and a recent symposium on the origin, structure, and processes of *restinga* gives an up-to-date summary of current research on this vegetation type (Lacerda et al., 1984).

Vast areas of *restinga* have been disturbed or destroyed, especially near villages and urban centers, by logging, grazing, and subsistence agriculture, and in some places it has been completely replaced by plantations of coconut (*Cocus nucifera* L.) and piaçava palm (*Attalea funifera* Mart.). Another threat to this vegetation is the construction of summer homes, hotels, and urban housing. For example, the dunes around Salvador, Bahia have been considerably reduced in area because of this (Fig. 6B). These dunes are of special in-

FIG. 6. *Restinga* vegetation **A.** A view of *restinga* taken from near the high tide mark looking inland. This shows the transition from the sprawling herbs of the beach to the shrubs and trees inland. The low, sprawling shrub surrounding A. M. de Carvalho is *Chrysobalanus icaco* Linn., the cocoa plum. **B.** Dunes around Lago Abaeté, Salvador, Bahia. This, and similar dune areas, are being destroyed by urban sprawl. Note "squatter" shacks in the middle of the photograph.

FIG. 7. A. Granitic outcrops surrounded by the city of Rio de Janeiro where lowland moist forest once stood. The most precipitous dome is Sugar Loaf Mountain. The flora of these outcrops harbors many endemic species. **B.** View from Corcovado Mountain, Rio de Janeiro, showing some of the extensive moist forest now surrounding the city.

terest because they are the type locality of many species collected by J. S. Blanchet between 1830 and 1856 (Mori et al., 1983).

F. Granitic Outcrops

In the southern part of the eastern Brazilian coastal forest, granitic domes appear as outcrops surrounded by moist forest at both low (e.g., Sugar Loaf Mountain of Rio de Janeiro, Fig. 7A) and high elevations (*Campos de Altitude*). The vegetation of some outcrops in the vicinity of Rio de Janeiro has been described by Carauta and Oliveira (1985) who point out the dominance of Bromeliaceae, Orchidaceae, and Velloziaceae. They add that the invasion of the African grass *Panicum maximum* Jacq. var. *maximum* tends to eliminate the natural vegetation by out-competing it and by providing easily ignited fuel that burns frequently. Martinelli (1984) and Martinelli et al. (in press) have made a detailed study of Morro du Cuca, a granitic outcrop that harbors the rare, paleoendemics *Benevidesia organensis* Sald. & Cogn. (Melastomataceae), *Glaziophyton marabile* Franchet (Poaceae), *Pitcairnia glaziovii* Baker (Bromeliaceae), *Prepusa connata* Gardner (Gentianaceae), *Tillandsia reclinata* Pereira & Martinelli (Bromeliaceae), and *Worsleya rayneri* (J. D. Hooker) Traub & Mold. They point out that destruction of the forest surrounding *Morro do Cuca* and subsequent fire threaten the existence of these species. Because these outcrops possess a unique flora, they must be given special consideration in any conservation effort.

G. Liana Forest (Mata De Cipó)

Liana forest occurs in the state of Bahia between 600 and 1000 m. It is characterized by trees of low stature (10–15 m) and small diameters, many of which lose their leaves in the dry season. Lianas are abundant and usually form a network which makes passage through this forest difficult. Some botanists argue that liana forests do not represent a distinct vegetation type, but are rather the result of disturbance. The distinction between liana forests and other forests at lower elevations is also not clear.

Liana forest is botanically the least known of all the vegetation types of eastern Brazil. Few collections have been made, and I know of no ecological studies of it. These forests have been highly modified by man, and it is doubtful if any remain undisturbed. Liana forest has been almost eliminated in the last 10 years because of its conversion to coffee plantations which grow well at these elevations on the rich soils where liana forests occurred.

H. Araucaria Zone

A vast area of 160,000 square kilometers in the states of Paraná, Santa Catarina, and Rio Grande do Sul is covered by a mosaic of temperate forests and prairies. Many of the forests were once dominated by the valuable timber tree, *Araucaria angustifolia* (Bert.) O. Kuntze, which grows at altitudes from 500 to 1000 m (Ferri, 1980; Fundação de Pesquisas Florestais do Paraná, 1978). In this region temperatures can be severe, and at times frosts and even snow cover the landscape. Consequently, the species of this region have little in common with those of the lowland moist forests of the rest of eastern Brazil. Because of the value of the timber of *A. angustifolia* and other species, as well as agricultural expansion, the only *Araucaria* forests not highly disturbed by man are those preserved in national parks (Jorge Pádua, 1983).

III. Lowland Moist Forests

The moist forests of eastern Brazil (Figs. 1, 7B, 8) once occupied about one million square kilometers and extended from Rio Grande do Norte to Rio Grande do Sul in a strip ranging from several to 120–160 km wide (Andrade-Lima, 1977; Bigarella et al., 1975; Rizzini, 1979; Smith, 1962). Scattered patches of forest related to the coastal forest are found further inland, especially in Minas Gerais and São Paulo. Dean (1983) estimates that about a third of Minas Gerais and four-fifths of São Paulo were covered with forest before man began to destroy it. Troppmair (in Angely, 1970) calculates that 190,000 km² or 77% of the state of São Paulo was originally covered by forest. In the northernmost states, moist forests with Amazonian and coastal forest elements appear on the slopes of plateaus as well as along the coastal plain (Fig. 2). In northeastern Brazil these forests are called *brejos*, a term which is also applied to marshes and swamps throughout Brazil. Hueck (1972) estimates original forest cover in the northernmost states as 25% in Rio Grande do Norte, 35% in Alagôas and Pernambuco, and 40% in Paraíba and Sergipe. The altitudinal limits of coastal moist forests are not clear because forests ascend the coastal mountains to nearly 1800 m in southeastern Brazil. Although there are no clear discontinuities in forest types, species composition of lowland and higher altitude forests differ, and, therefore, any conservation efforts must preserve forests at all altitudinal levels.

Typical coastal forest climate is found in southern Bahia (Mori et al, 1983; Roeder, 1975). Here the moist forests fall within the Af climatic region of Köppen (1936), which is that of hot, humid tropical forests, without a dry season and with rainfall greater than 1300 mm/year. In this region it rains more than 150 days each year. However, short, unpredictable dry seasons of one to three months may occur. In dry years the rainfall averages 1200–1350 mm, and in wet years it averages 1650–1800 mm. The maximum precipitation recorded for a 24 hour period is 200 mm. The evapotranspiration potential is 1200–1300 mm and the annual average relative humidity is 80 to 90 percent. The temperature averages between 24° and 25°C annually, with the warmest years between 28° and 29° C. The maximum temperature recorded is 38° C. The warmest months are October through April. The mean annual minimum temperature is between 18° and 20° C, and in the coolest months (June to August) of the coolest years, it can drop to 7.4° C. The average annual fluctuation is 7–8° C.

FIG. 8. Moist forest in the process of being destroyed near the Pau-Brazil Ecological Station in southern Bahia. A. Newly-opened logging road in virgin forest. B. Logging road shortly after most of the valuable timber had been removed.

Lowland moist forests of eastern Brazil are related both structurally and in species composition to those of Amazonia. Disjunction of species ranges between Amazonia and the moist forests of eastern Brazil (Fig. 9) has been well documented by Andrade-Lima (1953, 1969, 1977, 1981b). These two forests were probably once continuous in the past, but, as aridity increased in the Tertiary, they became separated by the xeromorphic *cerrado* and *caatinga* (Fig. 1) (Bigarella et al., 1975). Mori et al. (1981) have calculated that 7.8% of the forest tree species of eastern Brazilian moist forests are also found only in the Amazon, and they point out that many other widespread species are distributed in these as well as other areas. A botanist, placed in the midst of either of these forests without knowing where he was and without a detailed species list, would not be able to say if he were in Amazonia or eastern Brazil because of the structural similarities of the two forests. Buttresses, latex, and mesophyll leaves with entire margins characterize many of the trees, and epiphytes and lianas are common. These forests, as well as those of Amazonia, are extremely rich in species. For example, Mori et al. (1983) found 178 species of trees with diameters equal to or greater than 10 cm in a 600 tree sample of a southern Bahian forest.

Considerable variation is found within my concept of moist forest. For example, there are gallery forests in regions of *cerrado*, periodically inundated forests called *várzea* along some of the larger rivers, swamp forests along the coast in which the soil never dries out, more mesophytic forests with a greater percentage of deciduous species further inland (Gouvêa et al., 1976), and cloud forests at the higher elevations. The differences among these types of moist forests are still in need of study, and because data to separate them are not available, I group them all as moist forest. To further complicate the landscape, moist forests are often dissected by other vegetation types, especially *cerrado* and the numerous secondary growths and plantations created by man.

A. Endemism

The moist forests of eastern Brazil are of special interest to biologists because of their high endemism in all groups of organisms. Studies of birds (Cracraft, 1985; Haffer, 1974), reptiles (Jackson, 1978a; Müller, 1973), primates (Fonseca, 1985; Kinzey, 1982), butterflies (Brown, 1979), and plants (Calderón & Soderstrom, 1980; Mori et al, 1981, 1983) all indicate that these forests possess many species unique to them. Mori et al. (1981) analyzed the distributions of 1245 species of trees treated in Flora Neotropica monographs and calculated that 53.5% of those found in the moist

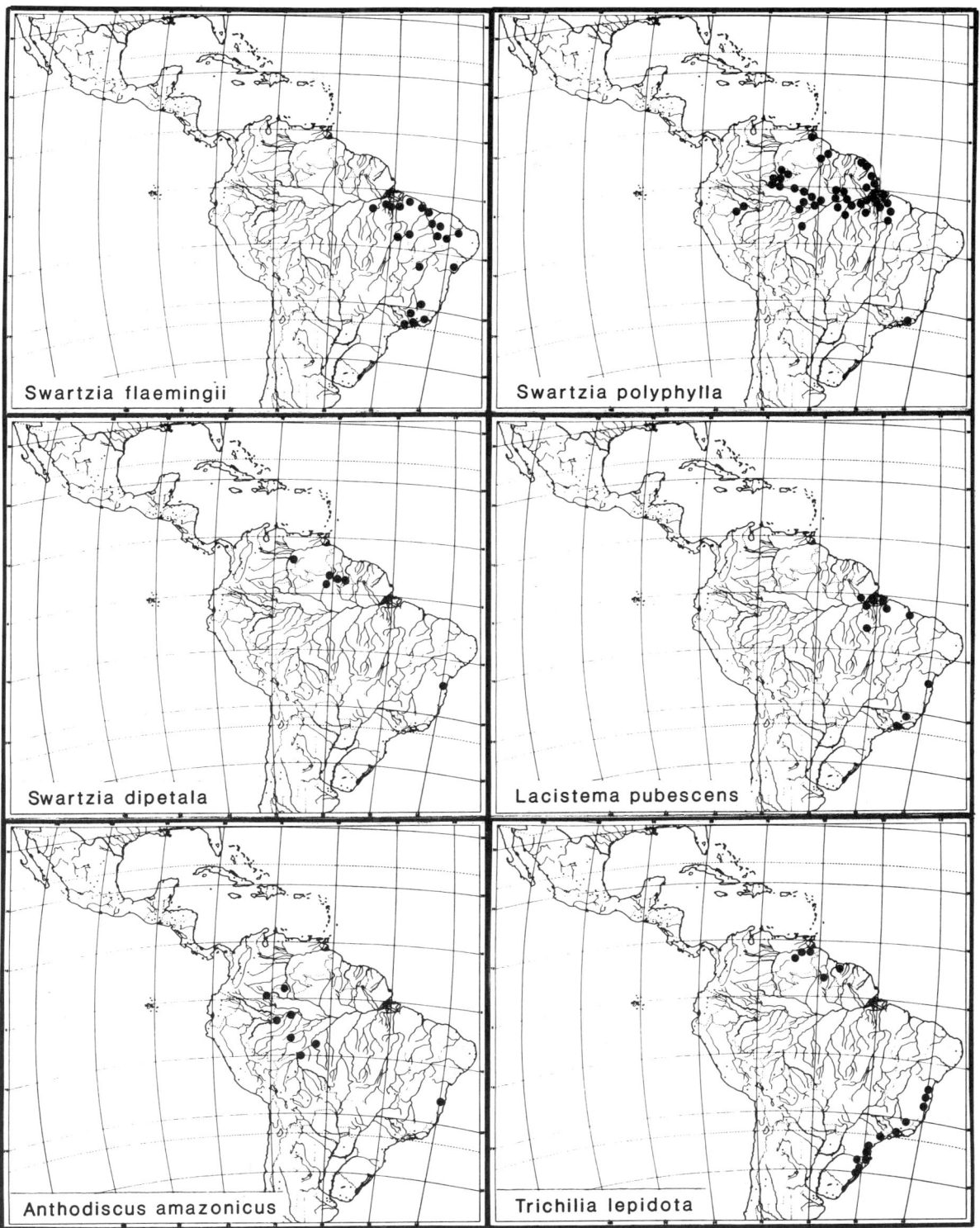

FIG. 9. Disjunctions of representative tree species between the moist forests of eastern Brazil and those of Amazonia (from Mori et al., 1981).

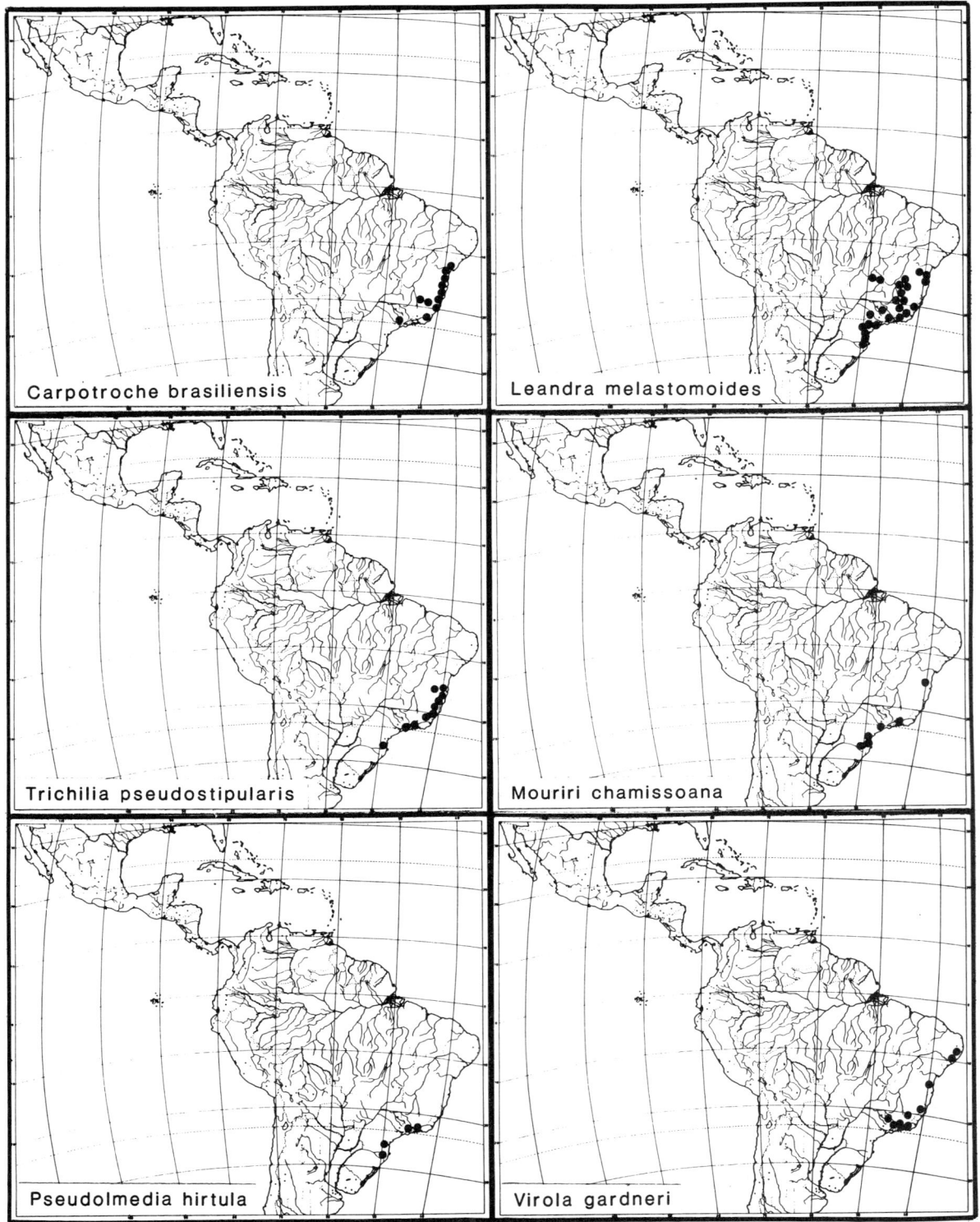

FIG. 10. Selected tree species endemic to the moist forests of eastern Brazil (from Mori et al., 1981).

forests of eastern Brazil only occur there (Fig. 10). Some animals, such as the woolly spider monkey (*Brachyteles arachnoides*) (Mittermeier, 1987), the lion tamarins (*Leontopithecus*), an endemic sloth (*Bradypus torquatus*), the recently rediscovered thin-spined porcupine (*Chaetomys subspinosus*) (Anonymous, 1987), the eastern Brazilian races of the razor-billed curassow (*Crax mitu*), the red-billed curassow (*Crax blumenbachi*) (Collar, 1986) and the bearded bellbird (*Procnias averano*) are nearing extinction because their forest habitats in eastern Brazil have been nearly eliminated. Even the once common Brazil-wood tree (*Caesalpinia echinata* Lam.), after which Brazil was named, is now difficult to find in native habitats. Continued destruction of the moist forests of eastern Brazil will mean the loss of these and many other species of plants and animals.

Within the moist forests of eastern Brazil there are separate centers of endemism. Brown (1979), using distributions of butterflies, hypothesized four centers of endemism while Jackson's study of reptiles (1978a) led him to recognize 11 centers in eastern Brazil. Mori et al. (1981) found that the distribution of tree species supported two of the centers of endemism recognized by Brown (1979). Nearly all studies indicate that the areas around Rio de Janeiro, southern Bahia/ northern Espírito Santo, and coastal Pernambuco are centers of endemism.

Teixeira (1985) has pointed out differences between the avifaunas of the uplands (500–600 m) and lowlands of the northeastern part of the coastal forests of eastern Brazil. He emphasizes that each forest type harbors its own endemic taxa. This must also be true for all other groups of organisms. Consequently, in order to preserve all taxa, both upland and lowland forests must be identified and protected.

B. Land Use

Amerindians reached the forests of eastern Brazil about ten thousand years ago. According to Dean (1983), they were hunters and gatherers whose use of fire extended the range of *cerrado* at the expense of forest (Fig. 4B). About 1500 years ago, the Tupi-Guaraní culture drove the hunter-gatherers to the less productive highland environments of the interior. The Tupi-Guaraní practiced swidden or slash-and-burn agriculture, which involves felling the forest in the dry season and burning the accumulated, dried-out litter just before the onset of the rainy season. Nutrients released from the burned vegetation are taken up by the crops which are planted at the onset of the rainy season. The staple crop of the Tupi-Guaraní was manioc (*Manihot esculenta* Crantz), both bitter and sweet, grown in association with maize, beans, peppers, squash, peanuts, and other minor crops (Dean, 1983). It is estimated that the Tupi-Guaraní cleared one hectare of forest per family per year. The first Portuguese and French navigators reached Brazil around 1500, and these early colonists learned swidden cultivation from the Tupi-Guaraní. Nineteenth century discussions of swidden agriculture estimate that three hectares per family per year were cleared to make way for agricultural fields (Dean,

1983). Consequently, swidden farming has been responsible for the destruction of much of the moist forest of eastern coastal Brazil.

Colonization of Brazil began in the coastal forests of Bahia, and, because of the relatively rich soils upon which these forests grow throughout their extension, they were first exploited for forest products and then destroyed and converted into agricultural plantations. The first economically important product of the region was the dye extracted from the wood of the Brazil-wood tree (*Caesalpinia echinata*), used in the textile industry. As early as 1809, Brazil-wood had been so decimated in the state of Alagôas that a plea was made for its rational exploitation (Fraga, 1960).

The removal of native trees for their timber has been, and continues to be, a major factor in the destruction of eastern Brazilian moist forests. Although Brazil-wood is no longer used as a dye, its lumber is favored for making the bows of musical instruments, and today it is difficult to find wild individuals of the tree for which Brazil was named. Other trees such as *Cariniana estrellensis* (Raddi) Kuntze, *C. legalis* (Mart.) Kuntze, *Dalbergia nigra* Allem. ex Benth., *Tabebuia* spp., *Aspidosperma* spp., *Cedrela odorata* L., *Plathymenia foliolosa* Benth., *Astronium concinnum* Schott, and *Centrolobium* sp., are sought for their timber used in construction and furniture making (Vinha et al., 1976). Forestry practices are so destructive in eastern Brazil that, even though only a few species are removed, much of the rest of the forest is destroyed (Fig. 8). The completion of the last stretch of the Brazilian coastal highway (BR-101) in southern Bahia in 1973 opened this final enclave of forest to logging. Forest destruction since then has occurred at such an accelerated rate that the fame of southern Bahian expanses of moist forest is now mostly a memory (Mori & Silva, 1979a). Although timber extraction itself usually does not completely destroy the forest, access to new land via logging roads opens it to colonists and the subsequent swidden agriculture practiced by them. Subsequently, once-forested areas are often turned into cattle pasture.

Extraction of firewood and the production of charcoal contribute to forest destruction. According to Dean (1985), household firewood consumption in southeastern Brazil at the turn of the century amounted to about 2 m^3 per capita. Adding industrial uses to this raises the figure to 2.3 m^3 per person. Consequently, it was necessary to clear about 57 km^2 of forest per year to supply Rio de Janeiro, where 10,000 persons made their living by woodcutting and charcoal preparation. Particularly destructive was the use of firewood as locomotive fuel. By 1900 there were 6,000 km of railroad in the forested regions of southeastern Brazil. In addition to fuel, these railroads needed 1500 crossties per kilometer which had to be replaced every six years (Dean, 1983). A growing concern for the dwindling firewood supply at the turn of the twentieth century resulted in the Paulista Railroad contracting Edmundo Navarro de Andrade to plant forests to guarantee a steady supply of firewood. After a series of experiments, he selected the Australian genus *Eucalyptus* because it was the fastest growing (Dean, 1985). Consequently, vast areas of southeastern Brazil are now covered by *Eucalyptus* plantations instead of *cerrado* and moist forests.

Although bottled gas is used for cooking in most urban homes, wood is still the most common cooking fuel in rural areas. In addition, firewood is used widely to dry cocoa beans in southern Bahia and for baking in many bakeries throughout eastern Brazil. Charcoal is employed to barbecue meat in the widespread and popular restaurants called "churrascarias." More importantly, charcoal is still extensively used in the steel industry. Domed ovens used in charcoal production and trucks carrying vast loads of charcoal to the market are still very common sights throughout eastern Brazil. Fonseca (1985) has shown the impact that 56 steel companies established in the coastal zone of Minas Gerais have had on deforestation in this state.

Dean (1983) has pointed out the impact that gold prospecting and mining have had on the forests of Minas Gerais and São Paulo. Wilderness was transformed into a heavily populated district of 350,000 people at the peak of mining. In order to facilitate mining, forest and *cerrado* were burned, and, in order to feed the population, agriculture supplanted many forests. Dean estimates that 20,000 km², most of it forested land, was burned for mining and that another 25,000 km² of forest was cleared for the planting of manioc, corn, and rice. This mining boom, which began in the 1690's and lasted for about 100 years, is no longer an important factor in forest destruction in eastern Brazil. Most mining activity is on a smaller scale and is restricted to the higher altitude *campo rupestre*.

Fonseca (1985) has pointed out that wars with Amerindians have taken their toll on forests. The Military used fire as a means of driving Indians from their refuges.

The first plantation crop responsible for massive deforestation in eastern Brazil was sugar cane, introduced to the area of Santos, São Paulo in 1531 or possibly even earlier (Dean, 1983). It was already an important crop in the first decades of the 16th century, with centers of production in the coastal areas of northern Bahia, Pernambuco, the São Vicente area of São Paulo, and Rio de Janeiro (Sick & Teixeira, 1979). By 1709, Rio de Janeiro was exporting about 5,000 tons a year, about 20 percent of the colony's total (Dean, 1983).

In order to make room for sugar cane plantations, the original forests are cut and burned. In addition, forests are exploited for firewood to provide the fuel necessary for processing the cane. Today, the moist forests of Sergipe, Alagôas, Pernambuco, and eastern Rio de Janeiro are mostly replaced by sugar cane plantations. With the increased emphasis the Brazilian government has placed on alcohol as an alternative energy source, sugar plantations have been expanded at the expense of forest and other crops (Fonseca, 1985).

Coffee has played an important role in the Brazilian economy and in the destruction of its forests. By the 1830's, planters in southeastern Brazil were turning from sugar cane to coffee (Dean, 1983), and by the middle of the 19th century, coffee was the principal crop in many areas. Initially, coffee cultivation centered in the moist forest areas of the Paraíba do Sul valley around Rio de Janeiro. Subsequently, it pushed westward to higher elevations. Coffee cultivation is extremely destructive of forests because it is widely believed that coffee does best in areas once forested. In southern Bahia, coffee plantations have almost completely replaced liana forest at elevations between 700 and 1000 m. This replacement has been accelerated in the last several decades due to the occurrence of frosts in the traditional coffee producing areas further to the south.

Cocoa is the most important crop in the moist forest areas of southern Bahia. *Theobroma cacao* L. was introduced into Bahia in 1746 from the Amazon (Mori et al., 1983). It is an understory tree of Amazonian forests which, like so many other crops grown outside its native range, thrives under the similar conditions of eastern Brazilian moist forests. Its success in eastern Brazil may be because its coevolved predators and diseases were left behind.

Traditionally, cocoa has been planted under a system known as *cabruca* in which native trees are left as shade. The understory is removed and the canopy is thinned to provide the proper shade of about 25 trees/hectare. This system has the advantage of preserving part of the native flora and fauna as well as structurally mimicking the original forest and thus minimizing any changes in the hydrological cycle and climate. The native epiphytic flora may still provide breeding sites for the midges essential for the pollination of cocoa (Fish & Soria, 1978). In addition, the high diversity of the native forest may also help in controlling insect pests. However, with time, the native shade trees die out because natural regeneration of trees does not occur due to the constant weeding of the understory. Therefore, continuation of the *cabruca* system depends on replacement of the native shade trees.

A system of clear-cut and burn with the subsequent replacement of native trees by monotypic stands of *Erythrina fusca* Loureiro has been developed. In this system banana trees are used to shade the young cocoa plants and to provide income until the cocoa trees produce fruit. Fertilizers, herbicides, and pesticides are used in both systems but more so in the cut-and-burn system.

Cocoa cultivation is dependent on the forest for shade trees, firewood to dry cocoa beans, poles for supporting branches heavily laden with fruit, insects that control pests, midges that pollinate its flowers, and perhaps even for the regional climate that allows cocoa plants to grow. Therefore, preservation of moist forest has direct economic benefits for this region of eastern Brazil.

Other plantation crops that account for the destruction of moist forest in eastern Brazil are: bananas (*Musa* spp.), rubber (*Hevea brasiliensis* Muell. Arg.); "dendê" palm (*Elaeis guineensis* Jacq.) cultivated for its seed oil; "piaçava" palm (*Attalea funifera* Mart.) cultivated for its fibers used in making brooms, brushes, mats, etc.; "guaraná" (*Paullinia cupana* HBK var. *sorbilis* (Mart.) Ducke), the source of a stimulating soft drink; black pepper (*Piper nigrum* L.); and cloves (*Eugenia caryophyllata* Thunb.). Manioc (*Manihot esculenta*) is cultivated throughout the region, and was initially responsible for the deforestation of large areas, as pointed out in my discussion of swidden agriculture.

Cattle raising is also responsible for the destruction of

large areas of moist forest in eastern Brazil. Pastures are mostly used to raise beef cattle. In southern Bahia, an average of only 0.9 head/ha is supported (Ettinger et al., 1977). The few native trees left in pasture are occasional shade trees and palms of economic value, neither of which are replaced as the individual trees die. The predominant pasture plants are species of imported grasses, and weeds are quick to invade if the pastures are poorly managed. This results in pasture of reduced productivity dominated by a low, dense cover of subshrubs belonging mainly to Melastomataceae, Moraceae, Myrtaceae, Rubiaceae, Leguminosae sensu lato, Verbenaceae, and Clusiaceae, none of which are edible by cattle.

Minor disturbance to moist forests has been caused by the collection of orchids and bromeliads. Dean (1983) reports that large horticultural firms in the nineteenth century exported as many as 100,000 orchids annually to Europe. Bromeliads have not only been removed because of their ornamental value but also because many of the species with water accumulating tanks are breeding sites for malaria carrying mosquitos. Consequently, they have sometimes been completely removed from the vicinity of population centers (Smith, 1955).

IV. Floristic Inventory

Brazil has a long history of botanical exploration. The first descriptions of Brazilian plants are based on studies of the eastern Brazilian coastal flora (Christovão, 1968, (based on work done between 1624 and 1635); Piso & Marcgrave, 1648). Many of the early botanical explorers of Brazil began their expeditions in Rio de Janeiro and proceeded along the coast from where they penetrated the interior. Accounts of these explorers are found in Hoehne et al. (1941), Mori and Silva (1979a), Mori et al. (1983), Prance (1971), and Urban (1906).

The data base of much taxonomic research is derived from dried botanical specimens housed in herbaria. The first step in any floristic inventory is to make sure that all species are represented by at least one collection. However, a single collection is not sufficient for a flora or a monograph to be complete because each species must be represented by enough specimens to show: 1) geographic distribution, 2) ecological preferences, 3) abundance, 4) all stages of the plant's life cycle, and 5) morphological variation. Considerable botanical exploration is consequently needed to gather the necessary specimens.

Failure to have a sufficient number of collections on hand at the beginning of a flora often means that the flora is out-of-date even at the time of its publication. For example, the recently completed *Flora of Panama*, even though covering 6,200 species of vascular plants, is missing 2800 to 3899 species (D'Arcy, 1980). An extreme case is that of the Araceae which included only 68 species when it was treated for the *Flora of Panama* by Standley in 1944. Croat (1985) now estimates that there are 302 species, an increase of 344% in the last 41 years! A similar situation has occurred with the *Flora of Suriname*, where it has been necessary to update most of the original treatments. When the Euphorbiaceae was first treated for this flora by Lanjouw in 1932, it contained 82 species, a figure which increased to 107 species (including four additional genera) when it was revised by Görts-van Rijn in 1976.

The present extent of botanical exploration of the moist forests of eastern Brazil is not sufficient for the preparation of a regional flora. In Table I, I present a list compiled, from *Index Herbariorum* (Holmgren et al., 1981), of those Brazilian herbaria with the possibility of having at least some col-

Table I

Major herbaria of eastern Brazil with at least some collections from lowland moist forests. The total number of specimens is given in parenthesis. Data from Holmgren et. al, 1981 unless otherwise noted.

Alagoas
No herbaria listed.

Bahia
ALCB (15,000, Faria, 1985)
BAH (2763)
CEPEC (20,000)
HRB (5,000)
IAL (8,000)

Ceará
EAC (8,000)

Espírito Santo
CVRD (2,000)

Minas Gerais
BHCB (3,000)
BHMH (6,000)
CESJ (18,000)
EM (5,000)
ESAL (5,000)
HXBH (4,600)
OUPR (26,000)
PAMG (15,000)
PMG (6,500)
VIC (10,000)

Paraíba
EAN (2,000)
JPB (3,133)

Paraná
EFC (?)
IPB (6,300)
MBM (65,000)
PKDC (20,800)
UPCB (10,800)

Pernambuco
IPA (24,920)
UFP (8,000)
URM (72,800 fungi)

Rio de Janeiro
FCAB (3,000)
GUA (15,000)

Rio de Janeiro (con.)
HB (71,000)
NIT (1,400)
R (450,000)
RB (200,000)
RBE (5,000)
RBR (12,000)
RFA (38,000)
RIZ (3,900)
UFRJ (7,000 fungi)

Rio Grande do Norte
No herbaria listed

Rio Grande do Sul
BLA (13,700)
HAS (14,180)
HURG (5,000)
ICN (50,000)
MPUC (3,500)
PACA (15,000)
PEL (12,000)
SFPA (4,478 fungi)
SMDB (1,570)

Santa Catarina
HBR (66,000)
FLOR (13,000)

São Paulo
BOTU (10,000)
ESA (2,500)
HRCB (5,000)
IAC (28,000)
IACM (8,000)
IBA (13,528, mostly fungi)
SP (170,000)
SPF (42,000)
SPSF (6,120)
UEC (25,000)

Sergipe
ASE (1,100)

Total Collections   1,689,592

lections from the moist forests of eastern Brazil. These 61 herbaria contain 1,689,592 specimens of which I judge 25% to be vascular plants from the moist forests of eastern Brazil. Assuming that moist forests occupy 1,000,000 km², this gives a collecting density of 0.4 collections/km². It should be noted that this is a very rough estimate because it does not include collections in foreign herbaria not duplicated in Brazilian herbaria, and because it probably overestimates the number of specimens from moist forest. Most collections in herbaria come from the easier to collect *cerrado*, *campo rupestre*, and *restinga* habitats. Moreover, large trees of moist forests are under represented in all herbaria.

When the *Flora Ilustrada Catarinense* was initiated, the herbarium upon which it is based had 65,354 collections or 0.7 collections/km² (Mori & Silva, 1979b). The highest figures reported for Central and South America are those of 1.38 and 1.32 collections/km² for Costa Rica and Panama, respectively (Prance, 1977). These figures clearly indicate that much additional collecting is needed before the floristic inventory of the moist forests of eastern Brazil is complete.

A commonly practiced way of obtaining names for species of neotropical plants is to make duplicate collections of the species in question. All duplicates are given the same number in the series of the collector, and one of the duplicates is sent in exchange for determination to a specialist on the family to which the species belongs or to a botanist familiar with the flora where the plant was collected. Specialists on the families of plants of eastern Brazil are essentially those found on the specialist lists of the The New York Botanical Garden and in the recently updated list of specialists preparing treatments for *Flora Neotropica*. Local experts on the floras of eastern Brazil are found in the Brazilian herbaria listed in Table I. The addresses for these herbaria are in *Index Herbariorum* (Holmgren et al., 1981). Specialists should be consulted before sending gifts for determination.

A useful way of obtaining scientific names for unknown specimens is through use of the local name. Several lists of common names with corresponding scientific names are applicable to the plants of eastern Brazil. The most important is Pio Corrêa's six volume Dicionário das Plantas Utéis do Brasil (1926, 1931, 1952, 1969, 1978a, 1978b) in which much useful information is given about the taxonomy and economic botany of Brazilian plants. Other checklists of common names for the plants of eastern Brazil are those of Braga (1960) and Silva et al. (1982). However, caution should be employed in converting common names to scientific names. Because of the regional variation in local names, names for specimens of an unknown plant should be confirmed by comparison with properly identified material in a herbarium.

Angely (1969, 1970) has compiled a checklist, based on herbarium collections and the literature, of the 219 families, 1694 genera, and 7747 species known by him to occur in the state of São Paulo. A checklist of the plants of the moist forests of southern Bahia, without corresponding common names, is provided by Mori et al. (1983). Harley and Mayo (1980) and Lewis (1979) have published checklists for the plants of Bahia.

The only flora encompassing all of Brazil is Martius's *Flora Brasiliensis* which first appeared in 1840 and was completed in 1906 under the consecutive editorships of Martius, A. W. Eichler, and I. Urban. This flora includes some 23,000 species, 5939 of which were new to science (Hoehne et al., 1941). It is still very useful in identifying unknown specimens from the moist forests of eastern Brazil, but, as might be expected, it is out-of-date because of the discovery of many new species and because of changes in nomenclature. Hoehne's *Flora Brasilica* was an attempt to update *Flora Brasiliensis* but only 12 fascicles treating parts of the Aristolochiaceae, Lamiaceae, Leguminoseae, Lycopodiaceae, Onagraceae, and Orchidaceae were published between 1940 and 1968.

The first attempt at a local flora for a part of eastern Brazil is Vellozo's *Flora Fluminensis* 1825, which treats the plants around Rio de Janeiro (Carauta, 1973). Many of its species were new to science, and, therefore, it is necessary to consult this work in any study of the plants of eastern Brazilian moist forests. An outstanding modern flora is that prepared for the state of Santa Catarina. This flora is based on a well conceived collection program which placed 180 collection stations in the six phytogeographic regions of the state. A square kilometer with a grid system of trails was botanically explored during all seasons of the year at many of these stations, and the collections were sent to specialists throughout the world for determination. By 1979, the *Flora Ilustrada Catarinense* contained treatments of 96 families with 1958 species, 314 of which were new to science. It is estimated that there are 4709 species of vascular plants in the state, an 8-fold increase from the 597 listed in *Flora Brasiliensis* (Reitz, 1980). An especially useful part of this series for other parts of Brazil is the key to families of seed plants by Goldberg and Smith (1975). *Flora Ilustrada Catarinense* is the standard upon which other Brazilian state floras should be modeled.

Publications on the plants of eastern Brazil are voluminous, and to list them would be beyond the scope of this review. Many revisions and monographs of plants have been published in the Brazilian journals "Arquivos do Jardim Botânico do Rio de Janeiro" and "Rodriguésia." For example, Vattimo (1956, 1957, 1959) has published extensively on the Lauraceae, and Barroso (1974) has revised the Dioscoreaceae for southeastern Brazil in these journals. A relatively complete taxonomic bibliography for the Brazilian literature up to 1977 has been published (Abreu et al., 1974, 1976; Ferreira et al., 1979; Ribeiro et al. 1979a, 1979b; Silva et al, 1978; Valente et al., 1984). Frodin (1984) provides further information on Brazilian floristics and botanical bibliographics.

Brazil's longstanding interest in natural history is also reflected in local treatments of flora and fauna. For example, Brade (1956), Castellanos (1965), and Mello (1950) have published on the flora, the ferns, and the trees of Itatiaia National Park. On a larger scale, Rizzini's *Arvores e Madeiras Utéis do Brasil* (1971b) and Rizzini and Mors' *Botânica Econômica Brasileira* (1976) include much information about the plants of the moist forests of eastern Brazil.

One of the foreign publications most useful in identifying plants from eastern Brazil is *Flora Neotropica*. This journal is dedicated to monographic treatments of plants from the New World tropics. To date, 45 monographs have been published in this series. However, at the current rate of publication, Prance (1977) has estimated that it will take 300 more years to complete. It is also unfortunate that, because of their high cost, these monographs are often not available in Brazilian herbaria.

An extremely useful type of publication is the revision of plant families for various parts of eastern Brazil. An example of this is Renvoize's treatment of the grasses of Bahia (1984). This book contains keys to all taxa, detailed descriptions, and excellent illustrations of all genera of grasses known from Bahia. Consequently, it has a much wider application than just for Bahia. Other specialists should be encouraged to contribute similar treatments for parts or all of eastern Brazil. Floras of individual national parks and reserves should be prepared in the same way that Venezuelan botanists are preparing them. Steyermark and Hubers' *Flora del Avila* (1978) is the first in a series of local floras to be published in Venezuela. Once the local floras are known it will lead to many other biological studies which were previously not possible. A good example of this is the *Flora of Barro Colorado Island* (Croat, 1978) which has recently been followed by a classic study of the forest ecology of the island (Leigh et al., 1982). Studies of the interaction of animals and plants are especially critical in eastern Brazil because many of its animals, such as the woolly spider monkey, are threatened with extinction. Without knowing what plants these animals depend on, it may not be possible to save them.

Most of the foregoing discussion has dealt with monographic and floristic work on vascular plants. Less has been written about the non-vascular plants of eastern Brazil. Important references are Yano's (1981) check list of Brazilian mosses and Yano's checklist of Brazilian liverworts and hornworts (1984). Comparable checklists for lichens and fungi are unknown to me.

Brazilian botanists are capable of producing the floras and monographs needed. There is a vast network of herbaria throughout eastern Brazil, and there are many botanists trained at least at the undergraduate level. As an example of Brazilian interest in botany, the annual Brazilian Botanical Congress usually attracts 500 to 1200 participants. Nevertheless, there is still a need for training Brazilian botanists with advanced degrees. The Brazilian Conselho Nacional de Pesquisas (CNPq) has recognized this need by establishing the Programa Flora of the CNPq. The goals of this program are to 1) produce the most complete inventory possible of the plants of Brazil, 2) produce a computerized data retrieval system so that information about all aspects of Brazilian plants will be available, 3) establish centers throughout the country capable of carrying out botanical inventory as well as producing ecological and environmental impact studies, and 4) train Brazilian botanists in systematic botany and in processing data. This program has been divided into five projects: Projeto Flora Amazônica, Projeto Flora Centro-Oeste, Projeto Flora Nordeste, Projeto Flora Sudeste, and Projeto Flora Sul. The program began in 1977 and by 1980, the first three projects were in full operation (Teixeira, 1980). Collections of the major herbaria in these regions have been entered on computer forms. However, because of recent financial problems in the country, the latter two projects have not been started, and activity in the initial three projects has slackened. Moreover, Brazilian taxonomists associated with herbaria are generally underpaid, poorly equipped for modern systematic work, and lack adequate support staff.

V. Ecological Studies

Küchler's bibliography of vegetation maps (1980) shows that Brazil leads all other South American countries in the number of vegetation maps. This review supplies a chronological listing of all maps completed in Brazil up to 1980. Each entry includes the date of mapping, the title, the color and scale, the complete legend as it appears on the original, the author, and the bibliographic reference.

For a general overview of the distribution of vegetation types of eastern Brazil, the maps of Hueck and Seibert (1972) and UNESCO (1981) should be consulted. Detailed vegetation maps based on radar images are being prepared for all of Brazil by RADAMBRASIL (Av. Antônio Carlos Magalhães 1131, Edificio Mal. Ademar de Queiroz, Pituba, 40.000-Salvador, Bahia, Brazil). Each RADAMBRASIL study includes maps of the geology, geomorphology, soils, vegetation, and economic capacity which are accompanied by detailed explanations. The first volume for eastern Brazil, Volume 24 for Salvador, Bahia, has been completed (RADAMBRASIL, 1981) and others are under way. Satellite images, and maps based on them, are under the auspices of the Instituto de Pesquisas Espaciais (INPE, Av. dos Astronautas, 1758, Jardim de Granja, Caixa Postal 515, 12.200-São José dos Campos-São Paulo, Brazil). The efforts of INPE are particularly important in monitoring deforestation throughout Brazil. An excellent detailed vegetation map of a limited part of the moist forest of eastern Brazil is that of southern Bahia by Gouvêa et al. (1976). Tosi (1983) has prepared a provisional vegetation map of Brazil based on the life zone system of ecological classification of L. R. Holdridge.

Martius (1906) provided one of the first general descriptions of the vegetation of Brazil. Recent descriptions and illustrations have been contributed by Aubreville (1961), Eiten (1970, 1983), Ferri (1980), Hueck (1972), Rizzini (1976), and Romariz (1974). Discussions of the vegetation of the southern (Alonso, 1977a; Lindman & Ferri, 1974), southeastern (Alonso, 1977b), and northeastern (E. Kuhlmann, 1977) parts of eastern Brazil appear in the *Geografia do Brasil*, published by the Instituto de Geografia e Estatística (IBGE).

Quantitative ecological studies of the structure and composition of moist forests of eastern Brazil began with Veloso (1946a, 1946b, 1946c) and have continued with Gibbs et al. (1980), Mayo and Fevereiro (1982), Mori et al. (1983), Silva

(1980), and Soares and Ascoly (1970). These authors provide data on the frequency, density, dominance, and diversity of eastern Brazilian forest trees which show that these forests are similar to Amazonian forests in diversity and structure. However, because of differences in sampling methods, it is difficult to compare one study with another. Further ecological studies are needed to establish the differences between forests at different altitudes and to distinguish between the lowland wet forests near the coast and the more mesophytic ones in the interior.

Phenological studies of the moist forests of eastern Brazil have been published by Alvim and Alvim (1978), Andrade-Lima (1957), Jackson (1978b), Mori et al. (1980c, 1982), and Veloso (1945). In these studies, spring and summer, regardless of whether they are wet or dry, appear to be the periods of greatest vegetative and reproductive activity. In southern Bahia, where there are no reliable wet and dry seasons, leaf fall, leaf flush, and flowering peak in the spring, nevertheless, there is considerable activity throughout the year (Alvim & Alvim, 1978; Mori et al., 1982).

The initial stages of secondary succession have been studied by Mori et al. (1980b). They sampled an area 105 days after it had been completely cleared of vegetation but not burned and found 63 species in 24 plant families with a biomass of 477.7 g/m^2 dried weight. This represents a net primary productivity of 16,425 kg/hectare/year. Veloso (1946a, 1946c) studied forest succession in southern Bahia and found eight associations which represent seral stages in the regeneration of climax forest after abandonment of sugar and cocoa plantations. Continued studies on forest succession are needed in order to provide the data needed for the recuperation of the vast areas of disturbed moist forests in eastern Brazil.

An important aspect of tropical ecology, that of coevolution of animals and plants (Gilbert & Raven, 1973), has been neglected in eastern Brazil. An example of the interdependence of tropical plants and animals in eastern Brazil is that of the "sapucaia" (*Lecythis pisonis* Cambess.), the carpenter bees (*Xylocopa frontalis* (Olivier)) that pollinate its flowers (Mori & Orchard, 1979, Mori et al., 1980a), and the bats that disperse its fruits (Greenhall, 1965; Prance & Mori, 1983). These three organisms are so dependent on one another that environmental changes that affect one will also affect the others. Moreover, because the carpenter bee is known also to pollinate some crops, such as the passion fruit (*Passiflora spp.*), changes in environment may have economic implications.

VI. Conservation

Brazil is an immense country that covers nearly one-half of South America. With 126,000,000 inhabitants, it has the world's sixth largest population, and it is growing at about 2.2% per year (IUCN/World Wildlife Fund, 1982). Conservation of eastern Brazilian moist forests is difficult because it was here that Brazil was first colonized, and it is here that about one-third of the Brazilian population is located on only six percent of the land surface (Dean, 1983).

The dense human population of eastern Brazil has had a severe impact on the moist forests. Mori et al. (1981) state that between 65.8 and 93% of the original forest has been destroyed. Others have estimated that only 1–2% of the coastal forest is still in a condition worth saving (IUCN/World Wildlife Fund, 1982). All that is left of these magnificent forests are isolated patches separated by vast areas of pasture and plantation. In addition, most of the remaining forest patches have been disturbed in one way or the other by man (Fonseca, 1985). Consequently, it is an entire floristic province, rather than individual forest areas, that is threatened. The severity of this problem has been recognized by the World Wildlife Fund which regards the forests of eastern coastal Brazil, along with Madagascar, as the most endangered areas on earth. Air pollution, spreading of soil erosion, and landslides, all of which accompany the loss of plant and animal species, have drawn Brazil's attention to the human cost of deforestation. Consequently, in recent years, many new areas have been set aside as national parks and biological reserves (Jorge Pádua, 1983). In Table II, I list those parks and reserves with at least some eastern Brazilian moist forest. Pressure from within Brazil to preserve these forests is so great that the entire 750 kilometer strip of moist forest in the state of São Paulo has recently been declared a national landmark (Lamlein Villa-Lobos, 1985).

Especially critical is the preservation of moist forest from Espírito Santo northward. The southern portion of the coastal forest is somewhat protected because of the very steep slopes of the Serra do Mar where much of the forest is located. Further to the north, the more level terrain is easier to log and more readily converted to agriculture.

The apparently untouched forests behind Rio de Janeiro and north of the city of São Paulo are testimonies to the importance of forests to local water supplies. In the 1860's the government of Rio de Janeiro was forced to invest in water supply lines because streams supplying the city were drying up. Consequently, reforestation of the Tijuca massif was carried out as a means of preserving water supply (Dean, 1985). The presence of coffee (*Coffea arabica* L.), jackfruit (*Artocarpus integrifolia* L.), and eucalyptus (*Eucalyptus* spp.) within the forests of Tijuca National Park is evidence that it was once covered by plantations (Mori et al., 1981). Recently, some of these areas have been invaded by squatters, and a secondary cycle of destruction of these reserves has begun to alarm conservationists (W. Dean, pers. comm.).

VII. Recommendations

1. The moist forests of eastern Brazil contain a high number of endemic organisms. Further depletion of these forests will result in the continued extinction of species of plants and animals. Consequently, continued efforts to educate the public about these forests, and to locate and preserve remaining forest patches must be made. These efforts have recently been highly successful, but much remains to be done.

Table II
National parks, reserves, and private areas known to possess some hectarage of eastern Brazilian moist forest.

Alagoas(AL)
Pedra Branca Massif—7,000 hectares, 1,500 to 2,000 of which are in good shape, privately owned? (Teixeira, 1985).

Bahia(BA)
CEPEC Botanical Reserve—43 hectares of old cocoa plantation previously shaded with native trees, left to regenerate since 1971; 6 km E of Itabuna on grounds of Cocoa Research Center (CEPEC) (Mori & Silva, 1979a).

Gregorio Bondar Agricultural Experimental Station—710 hectares of which 440 are forested, near Barrolândia. Owned by CEPEC (Mori & Silva, 1979a).

Lemos Maia Agricultural Experimental Station—495 hectares of which 393 are forested; near Una. owned by CEPEC (Mori & Silva, 1979a).

Monte Pascoal National Park—14,700 hectares, most of which is lowland moist forest; in southern Bahia (Jorge Pádua, 1983).

Pau-brasil Ecological Station—1145 hectares of which 800 are old secondary forest and perhaps some virgin forest. Silvicultural experiment station owned by CEPEC (Mori & Silva, 1979a).

Una Biological Reserve—11,400 hectares, one-third owned by the federal government. Mostly lowland moist forest, much of which has been, and continues to be, degraded. Important for protection of golden-headed lion tamarin (*Leontopithecus rosalia* (Liannaeus)). Located near Una (Jorge Pádua, 1983).

Ceará(CE)
Araripe Forest Park—with moist forest called "brejos" on higher hills (Andrade-Lima, 1977).

Ubajara National Park—563 hectares with some "brejo" forests on moist slopes; Serra Ibiapaba, in NW part of state (Andrade-Lima, 1977; Jorge Pádua, 1983).

Espírito Santo(ES)
Córrego do Veado Biological Reserve—2,400 hectares, entirely of lowland moist forest; near Linhares. Owned by federal government (Jorge Pádua, 1983).

Duas Bocas State Forest Reserve—2,200 hectares of coastal moist forest between 300 and 738 m; 90% covered with forest; in Municipality of Cariacica. Administered by the Secretaria de Agricultura of ES (Strang et al., 1982).

Fazenda Klabin—originally 4,000 hectares, now reduced to 1,240 hectares (Collar, 1987).

Forno Grande State Forest Reserve—5,000 hectares between 1,100 and 2,039 m; mostly occupied by colonists; in Municipality of Castelo. Administered by the Secretaria de Agricultura of ES (Strang et al., 1982).

Ilha de Comboios Biological Park of the Eastern Region—9,960 hectares between 30 and 80 m; 1,878 hectares in private ownership, also occupied by the Tupinquins Indians; in Municipalities of Linhares and Aracruz. Administered by the Secretaria de Agricultura of ES (Strang et al., 1982).

Mestre Álvar State Biological Reserve—2,217 hectares between 100 and 833 m; originally covered by lowland moist forest, now mostly covered by agricultural fields, pasture, and secondary forest; in Municipality of Serra, District of Pitanga. Administered by the Secretaria de Agricultura of ES (Strang et al., 1982).

Nova Lombardia Biological Reserve—4,450 hectares of moist forest located between 500 and 1,200 m in hills near Santa Tereza. Owned by federal government (Jorge Pádua, 1983).

Pedra Azul State Forest Reserve—1,119 hectares between 1,100 and 1,909 m; has subdeciduous forest; in Municipality of Domingos Martins, District of Venda Nova. Administered by the Secretaria of Agricultura of ES (Strang, et al., 1982).

Santa Lucia Reserve—6,000 hectares of moist forest. Field station of the Museu Nacional (Mittermeier, pers. comm. 1981).

Sooretama Biological Reserve—25,232 hectares of moist forest below 200 m in Municipality of Linhares. A 19,000 hectare private reserve belonging to the Companhia do Vale do Rio Doce is adjacent to this reserve (Collar, 1986, Jorge Pádua, 1983).

Minas Gerais(MG)
Colônia 31 de Março Biological Reserve—5,000 hectares; invaded by colonists, forests disturbed by extraction of wood for charcoal production; in Municipality of Felixlândia near Três Marias Dam. Administered by the Instituto Estadual de Florestas of MG (Strang et al., 1982).

Caparaó National Park—an undetermined portion of this park at its lowest elevations has moist forest. Park extends into Espírito Santo (Jorge Pádua, 1983).

Fazenda de Lapinha Biological Reserve—345 hectares; partially occupied by experimental agricultural fields; in Municipality of Leopoldina. Administered by the Instituto Estadual de Florestas of MG (Strang et al., 1982).

Fazenda Montes Claros—800 hectares of state owned forest near Caratinga, partially protected for the last 40 years. Important for protecting one of the last large populations of woolly spider monkey (*Brachyteles arachnoides*) (IUCN/WWF, 1982).

Mar de Espanha Biological Reserve—220 hectares; in Municipality of Mar de Espanha. Administered by the Instituto Estadual de Florestas of MG (Strang et al., 1982).

Mata dos Ausentes Biological Reserve—750 hectares at 80 m; secondary forest, *cerrado*, and forest; 50% occupied by colonists; in Municipality of Senador Modestino Gonçalves. Administered by the Instituto Estadual de Florestas of MG (Strang et al., 1982).

Mata do Acauá Biological Reserve—4,000 hectares; partially invaded by colonists and charcoal producers; in Municipality of Turmalina. Administered by the Instituto Estadual de Florestas of MG (Strang et al., 1982).

Table II
(continued)

Mata do Jambreiro—912 hectares of subdeciduous forest; in Municipality of Nova Lima. Administered by the Instituto Estadual de Floresta of MG (Strang et al., 1982).

Rio Doce State Forestry Park—36,000 hectares at 300–400 m; in municipalities of Marliéria, Timóteo and Dionísio. Protected by the Instituto Estadual de Florestas of Minas Gerais (Strang et al., 1982).

Paraíba(PB)

Mata de Pau Ferro—"brejo" forest protected by the state to preserve the watershed of Campina Grande (Mayo & Fevereiro, 1982).

Paraná(PR)

Amaporã State Reserve—204 hectares of which 180 are forested; tropical sub-deciduous forest; in Municipality of Amaporã. Administered by the Instituto de Terras e Cartografia of PR (Strang et al., 1982).

Figueira Forest Reserve—100 hectares of tropical subdeciduous forest; in Municipality of Eugenheiro Beltrão. Administered by the Instituto de Terras e Cartografia of PR (Strang et al., 1982).

Ibicatu Forest Park—57 hectares of tropical subdeciduous forest; in Municipality of Centenário do Sul (Strang et al., 1982).

Marumbi Park—6,657 hectares of coastal lowland moist forest; in Municipalities of Paranaguá and Matinhos (Strang et al., 1982).

Vila Rica—353 hectares of tropical subdeciduous forest; in Municipality of Fenix. Administered by the Instituto de Terras e Cartografia of PR (Strang et al., 1982).

Pernambuco(PE)

Curados Reserve—50 hectares of moist forest (Mori & Boom, 1981).

Dois Irmãos Reserve—132 hectares of lowland moist forest owned by the Federal University of Pernambuco (Mori & Boom, 1981).

São Bento Reserve—100 hectares near Recife (Mori & Boom, 1981).

Serra Negra Biological Reserve—1,100 hectares with some "brejo" forest (Jorge Pádua, 1983).

Tapacurá Ecological Research Station—100 hectares of subdeciduous forest in Municipality of São Lourenço de Mata about 40 km from Recife. Owned by the Federal Rural University of Pernambuco (Andrade-Lima, 1977).

Rio de Janeiro(RJ)

Araras Biological Reserve—2,068 hectares of coastal moist forest; mostly in Municipality of Petrópolis with a small part in the Municipality of Miguel Pereira. Administered by the Departamento Geral de Recursos Renováveis of RJ (Strang et al., 1982).

Cairucu Environmental Protection Area—peninsula on southern coast of state, including areas to 1,000 m alt. Extends up to Bocaina National Park (IUCN/WWF, 1982).

Crubixás Forest Reserve—710 hectares of coastal moist forest; in Municipality of Macaé (Strang et al., 1982).

Desengano State Park—22,000 hectares between 200 and 1788 m; harbors several species of endangered animals; in Municipalities of Santa Madalena, São Fidelis, and Campos. Administered by the Departamento Geral de Recursos Renováveis of RJ (Strang et al., 1982).

Pedra Branca State Park—12,577 hectares between 10 and 1024 m, 30% in forest; in Municipality of Rio de Janeiro. Administered by the Departamento Geral de Recursos Renováveis of RJ (Strang et al., 1982).

Poço das Antas Biological Reserve—5,138 hectares (of which 10% is mature forest). Owned by the federal government (Jorge Pádua, 1983).

Praia do Sul na Ilha Grande State Biological Reserve—3,600 hectares, with *restinga*, mangrove, moist forest etc.; in Municipality of Angra do Reis. Administered by the Fundação Estadual de Engenharia do Meio Ambiente of RJ (FEEMA) (Strang et al., 1982).

Serra da Bocaina National Park—120,000 hectares (of which less than 25% is owned by federal government), ranging from sea level to 2132 m, forest extends to about 1,000 m. Part of park is in São Paulo (Jorge Pádua, 1983).

Serra dos Orgãos National Park—5,000 hectares, ranging from 400 to 2263 m, while mostly secondary, some virgin forest still remains; near Teresópolis (Jorge Pádua, 1983).

Tamoios—90 hectare ecological station of the Secretaria Especial do Meio Ambiente (SEMA). Comprised of several islands and a tract of mainland with second-growth forest (Nogueira-Neto, 1984).

Tijuca National Park—3,300 hectares surrounded by the city of Rio de Janeiro, highest point is Pico da Tijuca at 1,021 m, entirely secondary as area was once covered by sugar and coffee plantations (Jorge Pádua, 1983).

Santa Catarina(SC)

Canela-Preta State Biological Reserve—1,844 hectares of coastal moist forest; in Municipalities of Botuverá, Novo Trento, and Vidal Ramos. Administered by the Fundação de Amparo à Tecnologia e Meio Ambiente of SC (FATMA) (Strang et al., 1982).

Sassafrás State Biological Reserve—5,068 hectares of coastal moist forest between 500 and 1,000 m; in Municipality of Benedito Novo. Administered by FATMA (Strang et al., 1982).

Table II

(continued)

Serra do Tabuleiro State Park—90,000 hectares, 28% under government ownership, with many vegetation types but including coastal moist forest; in Municipalities of Florianópolis, Palhoça, Santo Amaro da Imperatriz, Águas Mornas, São Bonifacio, São Martinho, Imaruí, Garopaba, and Paulo Lopes. Administered by FATMA (Strang et al., 1982).

Serra Furada State Park—1,329 hectares between 500 and 1400 m of coastal moist forest; in Municipalities of Grão-Pará and Orleans. Administered by FATMA (Strang et al., 1982).

São Paulo(SP)

Unless otherwise stated, the data comes from Strang et al., 1982 and the reserve is administered by the Coordenadoria da Pesquisa de Recursos Naturais da Secretaria de Agricultura e Abastecimento of São Paulo.

Águas da Prata State Reserve—48 hectares of subdeciduous tropical forest with transition to *cerrado*; in Municipality of Águas da Prata. Administered by the Instituto Florestal of SP.

Alto da Serra de Paranapiacaba Biological Reserve—336 hectares of sub-deciduous tropical forest; in Municipalities of Campo Grande and Paranapiacaba. Administered by the Instituto de Botânica of SP.

Alto Ribeiro State Park—35,712 hectares of coastal moist forest; in Municipality of Iporange. Administered by the Instituto Florestal of SP.

Ara State Park—40 hectares of subdeciduous tropical forest in transition to *cerrado*; in Municipality of Campinas. Administered by the Instituto Florestal of SP.

Bananal State Reserve—884 hectares of subdeciduous tropical forest; in Municipality of Bananal.

Caetetus State Park—2,187 hectares of subdeciduous tropical forest; in Municipalities of Garça and Gália. Administered by the Instituto Florestal of SP.

Cantareira State Reserve—5,647 hectares of subdeciduous tropical forest; in Municipalities of Franco da Rocha, São Paulo, Guarulhos, and Mairiporã.

Capão Bonito State Reserve—6,534 hectares of coastal moist forest; in Municipality of Capão Bonito.

Capital State Park—5 hectares, originally of subdeciduous tropical forest; in Municipality of São Paulo.

Caraguatatuba State Park—13,769 hectares of coastal moist forest; in Municipality of Caraguatatuba.

Carlos Botelho State Reserve—7,189 hectares of coastal moist forest; in Municipality of São Miguel Arcanjo.

Cunha State Reserve—2,854 hectares of coastal moist forest; in Municipality of Cunha.

Fazenda Barreiro Rico—2,198 hectares; privately owned (Mittermeier, pers. comm. 1981).

Fontes do Ipiranga State Park—549 hectares between 770 and 825 m, originally of subdeciduous tropical forest. Administered by the Instituto de Botânica of SP.

Ilha Anchieta State Park—1,000 hectares of coastal moist forest; in Municipality of Ubatuba.

Ilhabela State Park—27,025 hectares of coastal moist forest; in Municipality of Ilhabela.

Ilha do Cardoso State Park—22,500 hectares, with some coastal moist forest; in Municipality of Cananéia.

Itaberá State Reserve—180 hectares of coastal moist forest; in Municipality of Itaberá.

Ibicatu State Reserve—76 hectares of subdeciduous tropical forest in transition to *cerrado*; in Municipality of Piracicaba.

Itapeti State Reserve—89 hectares of subdeciduous tropical forest; in Municipality of Mogi das Cruzes.

Itatins State Reserve—12,058 hectares of coastal moist forest; in Municipalities of Iguapé, Iatiri, Pedro de Saledo, and Miracatu.

Jacupiranga State Park—150,000 hectares of coastal moist forest; in Municipalities of Jacupinga, Eldorado Paulista, Barra do Turvo, and Cananéia.

Jaraguá State Park—488 hectares, originally of subdeciduous tropical forest; in Municipality of São Paulo.

Jureia Ecological Reserve—30,000 hectares, between sea level and 600 m, mostly of coastal moist forest. Owned by federal government in order to protect site of future nuclear energy plant (Por & Imparatriz-Fonseca, 1984).

Lagoa São Paulo State Reserve—13,343 hectares of which 7,850 are owned by the government, subdeciduous tropical forest; in Municipality of Presidente Epitácio.

Mogi-Guaçu Biological Reserve—469 hectares of subdeciduous tropical forest; in Municipalities of Mogi-Guaçu and Bairro de Martinho Prado (Gibbs et al., 1980; Strang et al., 1982).

Morro do Diabo State Reserve—subdeciduous tropical forest; in Municipality of Teodoro Sampaio.

Perimetro do São Roque (2°) State Forest Reserve—23,900 hectares of coastal moist forest; in Municipalities of Ibiuna and Piedade.

São Carlos State Reserve—75 hectares of subdeciduous tropical forest; in Municipality of São Carlos.

Serra do Mar State Park—315,423 hectares, with moist forest on the coast and subdeciduous forest inland; in many different municipalities.

Sete Barras State Reserve—15,646 hectares of coastal moist forest; in Municipality of Sete Barras.

Travessão State Reserve—8,273 hectares of coastal moist forest; in Municipalities of Barras and Tapirai.

Valinhos State Reserve—16 hectares of subdeciduous tropical forest in transition to *cerrado*. Administered by the Instituto Florestal of SP.

Vassununga State Park—1,675 hectares; in Municipality of Santa Rita do Passo Quatro.

Xitué State Reserve—2,397 hectares of subdeciduous tropical forest; in Municipality of Capão Bonito.

2. It is especially important to locate and preserve forest reserves in the states of Bahia, Sergipe, Alagôas, Pernambuco, Paraíba, and Rio Grande do Norte where no significant areas have been set aside.

3. The available plant collections in Brazilian herbaria are not sufficient for producing a complete botanical inventory of the moist forests of eastern Brazil. Continued expeditions are therefore needed, especially to gather specimens of the more difficult-to-collect trees.

4. Local floras of national parks and biological reserves, as well as for entire states, should be encouraged. A knowledge of plants is the first step in any biological study of forests, and it is essential for developing an appreciation of nature.

5. Ecological studies comparing forests at higher and lower elevations and contrasting the wetter forests of the coast with the more mesophytic ones of the interior are needed. Because of species differences that occur among them, it is essential that we be able to recognize, and then preserve, representative samples of all forest types.

6. Studies of regeneration of moist forest after natural and man-made disturbances are needed in order to understand how moist forest can be restored.

7. Studies of plant/animal interactions should be encouraged. Especially important are relationships of plants and primates. Because of the highly endangered status of many species of the latter, we need to know about the plants necessary for their survival. In addition, efforts should be made to point out ecological relationships that have economic impact. Studies of the pollinators of economically important plants should be especially encouraged.

8. The role of forests in preserving local watersheds, and their influence on local and global climatic patterns needs further study. The public should be made much more aware of the manifold adverse effects of deforestation on watersheds and climate.

9. There is a continued need to train Brazilian botanists at the graduate level in plant taxonomy and ecology. More importantly, jobs must be made available for them upon graduation. Brazilian biologists trained abroad should be encouraged to study Brazilian problems rather than problems of the countries in which they are studying. SEMA's system of ecological stations (Nogueira-Neto, 1984) offers ideal conditions for the study of the taxonomy and ecology of Brazilian plants. Their use by Brazilian as well as foreign students of the environment should be encouraged

10. Brazilian botanical institutions need more reliable funding to operate effectively. It is of little use to have an active collecting program if funds are not available for the processing, distribution, preservation, and study of the collections. It is also important that competitive salaries are provided in order to attract and retain Brazilian botanists to work on taxonomic and ecological problems.

VIII. Acknowledgments

I am grateful to B. M. Boom, D. Daly, W. Dean, C. Gracie, R. M. Harley, M. T. Jorge Pádua, G. P. Lewis, G. Martinelli, S. J. Mayo, J. D. Mitchell, F. Okamoto-Nishida, T. Plowman, and B. L. Stannard for reviewing the manuscript, and to C. Gracie for her help in preparing the illustrations. I thank Jeanne Goode for her typing and editorial assistance with the numerous drafts of the manuscript. I am especially thankful to the "Centro de Pesquisas do Cacau" at Itabuna, Brazil for providing me with the opportunity to work as the curator of its herbarium from 1978 to 1980, and to the World Wildlife Fund-US for supporting a field study of southern Bahian moist forests in 1981. Many of the ideas expressed in this paper were developed during these stays in Brazil.

IX. Literature Cited

Abreu, C. L. B. de, N. M. Ferreira da Silva & P. C. A. Fevereiro. 1974. Bibliografia de botânica. I. Taxonomia de Angiospermae Dicotyledoneae. Rodriguésia **27 (39-anexo):** 1–79.

_____, _____, _____, & A. L. Peixoto. 1976. Bibliografia de botânica. II. Taxonomia de Angiospermae Dicotyledoneae. Rodriguésia **28 (40-anexo):** 1–60.

Alonso, M. T. A. 1977a. Vegetação. Pages 81–109 *in* Fundação Instituto Brasileiro de Geografia e Estatística, Diretoria Técnica (ed.), Geografia do Brasil, Região Sul. Fundação Instituto Brasileiro de Geografia e Estatística, Diretoria de Divulgação, Rio de Janeiro, Brazil.

_____. 1977b. Vegetação. Pages 91–118 *in* Fundação Instituto Brasileiro de Geografia e Estatística, Diretoria Técnica (ed.), Geografia do Brasil, Região Sudeste. Fundação Instituto Brasileiro de Geografia e Estatística, Diretoria de Divulgação, Rio de Janeiro, Brazil.

Alvim, P. de T. 1950. Observações ecológicos sobre a flora da região semiárida do nordeste. Bol. Geogr. **7:** 75–82.

_____ & R. Alvim. 1978. Relation of climate to growth periodicity in tropical trees. Pages 445–463 *in* P. B. Tomlinson & M. H. Zimmerman (eds.). Tropical trees as living systems. Cambridge University Press, New York.

Anderson, A. B. 1981. White-sand vegetation of Brazilian Amazonia. Biotropica **13(3):** 199–210.

Andrade-Lima, D. 1951. A flora da praia de Boa Viagem. Bol. Secr. Agric. (Recife) **18(1–2):** 121–125.

_____. 1953. Notas sobre a dispersão de algumas espécies vegetais no Brasil. An. Soc. Biol. Pernambuco **11(1):** 25–49.

_____. 1954. Contribution to the study of the flora of Pernambuco, Brazil. Bol. Técn. Inst. Pesqu. Agron., Recife **41:** 1–32.

_____. 1957. Notas para a fenologia de zona da mata de Pernambuco. Rev. de Biologia (Lisboa) **1(2):** 125–135.

_____. 1969. Pteridófitas que ocorrem nas floras extra amazônica e amazônica do Brasil e proximidades. An. Soc. Bot. Bras. (Goiânia) **20:** 33–40.

_____. 1970. Recursos vegetais de Pernambuco. Bol. Téc. Inst. Pesquisas Agron., Recife **41:** 1–32.

_____. 1977. Preservation of the flora of northeastern Brazil. Pages 234–239 *in* G. T. Prance & T. S. Elias (eds.), Extinction is Forever. The New York Botanical Garden, New York.

_____. 1981a. The caatingas dominium. Rev. Bras. Bot. **4(2):** 149–163.

_____. 1981b. Present-day forest refuges in northeastern Brazil. Pages 245–251 *in* G. T. Prance (ed.), Biological diversification in the tropics. Columbia Univ. Press, New York.

Angely, J. 1969. Flora analítica e fitogeográfica do estado de São Paulo, vol. 1. Edições Phyton, São Paulo, Brazil.
_____. 1970. Flora analítica e fitogeográfica do estado de São Paulo, vols. 2–6, Edições Phyton, São Paulo, Brazil.
Anonymous. 1987. Unusual porcupine, feared extinct, is rediscovered in Brazil. World Wildlife Fund Focus 9(2): 1,7.
Araújo, D. S. Dunn de & R. P. B. Henriques. 1984. Análise florística das restingas do estado do Rio de Janeiro. Pages 159–193 in L. D. de Lacerda, D. S. Dunn de Araújo, R. Cerqueira & B. Turcq (eds.), Restingas: Origem, estrutura, processos. CEUFF, Niterói, Brazil.
_____ & A. L. Peixoto. 1977. Renovação da communidade vegetal de restinga após queimada. Anais do Congresso Nacional da Sociedade Botânica do Brasil 26: 1–17.
Aubreville, A. 1961. Étude écologique des principales formations végétales du Brésil et contribution à la connaissance des forêts de l'Amazonie brésilienne. Nogent-sur-Marne (Seine) Centre Technique Forestier Tropical.
Barroso, G. M. 1974. Dioscoreaceae em flora de Guanabara. Sellowia 25: 9–256.
Bascopé, F., A. L. Bernardi, R. N. Jorgensen, K. Hueck & H. Lamprecht. 1959. Los manglares en América. Descripciones de árboles florestales no. 5. Inst. For. Lat. Amér. Invest. Capacit. Mérida, Venezuela.
Bertels, M. A. 1957. Monocotiledôneas psamofiticas do litoral do Rio Grande do Sul. Bot. Técn. Inst. Agron. S. 17: 29–34.
Bigarella, J. J. 1946. Contribuição ao estudo da planície litorânea do estado do Paraná. Arch. Biol. Técn. Paraná 1: 75–111.
_____, D. de Andrade-Lima & P. J. Riehs. 1975. Considerações a respeito das mudanças paleoambientais na distribuição de algumas espécies vegetais e animais no Brasil. Anais. Acad. Brasil. Ci. 47(supl.): 411–464.
Brade, A. C. 1956. A flora do Parque Nacional do Itatiaia. Parque Nacional do Itatiaia Boletim 5: 1–85.
Braga, R. 1960. Plantas do Nordeste, especialmente do Ceará, ed. 2. Imprensa Oficial, Fortaleza, Ceará, Brazil.
Brown, K. S. 1979. Ecologia, geografia e evolução nas florestas neotropicais. Universidade Estadual de Campinas, São Paulo, Brazil.
Calderón, C. & T. Soderstrom. 1980. The genera of Bambusoideae (Poaceae) of the American continent. Keys and comments. Smithsonian Contr. Bot. 44: 1–27.
Carvalho, A. M. V. de. 1981. Aspectos fitogeográficos de Bahia. Page 64 in Resumos do XXXII Congresso Nacional de Botânicas, Piauí, Brazil.
Castellanos, A. 1965. Catálogo dos Pteridófitas Pteridophyta Fasc. 1. Parque Nacional do Itatiaia Boletim 8: 1–45.
Carauta, J. P. P. 1973. The text of Vellozo's Flora Fluminensis and its effective date of publication. Taxon 22(2/3): 281–284.
_____ & R. Ribeiro de Oliveira. 1984. Plantas vasculares dos morros da Urca, Pão de Açúcar e Cara de Cão. Rodriguésia 36(59): 13–24.
Chapman, V. J. 1976. Mangrove vegetation. Vaduz, J. Cramer.
Christovão, Frei. 1968. Historia dos animães e árvores do Maranhão. Universidade Federal do Paraná, Conselho de Pesquisas, Curitiba, Paraná, Brazil.
Collar, N. J. 1986. The best-kept secret in Brazil. World Birdwatch 8(2): 14–15.
_____. 1987. Hook-billed hermit. World Birdwatch 9(1): 5.
Cracraft, J. 1985. Historical biogeography and patterns of differentiation within the South American avifauna: Areas of endemism. Pages 49–84 in P. A. Buckley, M. S. Foster, E. S. Morton, R. S. Ridgely & F. G. Buckley (eds.), Neotropical ornithology. Ornithological Monographs No. 36, The American Ornithologists' Union, Washington, D. C.
Croat, T. B. 1978. Flora of Barro Colorado Island. Stanford University Press, Stanford, California.
_____. 1985. The large monocots of Panama. Pages 5–12 in W. G. D'Arcy & M. D. Correa A. (eds.)., The botany and natural history of Panama. Missouri Botanical Garden, St. Louis.
Dansereau, P. 1947. Zonation et succession sur la restinga de Rio de Janeiro. I. Halosere. Rev. Canad. Biol. 6(3): 448–477.
D'Arcy, W. G. 1980. The flora of Panama: Historical outline and selected bibliography. Ann. Missouri Bot. Gard. 67: v–viii.
Dau, L. 1960. Microclima das restingas do sudeste do Brasil. I-Restinga interna de Cabo Frio. Arq. Mus. Nac. Rio de Janeiro 50: 79–133.
Dean, W. 1983. 4. Deforestation in southeastern Brazil. Pages 50–67 in R. Tucker & J. F. Richards (eds.), Global deforestation and the nineteenth-century world economy. Chapel Hill, North Carolina.
_____. 1985. Forest conservation in southeastern Brazil, 1900 to 1955. Environ. Rev. 9(1): 54–69.
Ducke, A. 1959. Estudos Botânicos no Ceará. Anais Acad. Bras. Ci. 31(2): 211–308.
Eiten, G. 1970. A vegetação do estado de São Paulo. Bol. Inst. Bot. 7: 1–147.
_____. 1972. The cerrado vegetation of Brazil. Bot. Rev. (Lancaster) 38: 201–341.
_____. 1983. Classificação de vegetação do Brasil. CNPq/Coordenação Editorial, Brasília, Brazil.
Esteves, G. Lopes. 1980. Contribuição ao conhecimento da vegetação da restinga de Maceió. Bot. Técn. Coord. Meio Amb. Maceió 1: 1–36.
Ettinger, A. E. de M., A. M. F. de Carvalho, M. H. Alencar, C. P. Pereira & S. R. Asmar. 1977. Diagnóstico-econômico do desenvolvimento da pecuária das regiões sul e sudoeste da Bahia. CEPLAC and SUDENE, Ilhéus, Bahia, Brazil.
Faria, L. S. S. 1985. Herbário Alexandre Leal Costa (ALCB) Salvador, Bahia. Bol. Bot. Latinoamer. 16: 8.
Ferreira, V. F. M. da Conceição Valente, C. L. B. de Abreu, L. Mautone, A. de Souza Leão, A. L. Peixoto, H. P. Bautista, F. P. L. Salles, A. F. R. de Souza, H. H. Correa, V. M. Schettino, A. M. C. Studart de Fonseca, V. M. Conti, & J. F. A. Baumgratz. 1979. Bibliografia de botânica-IV. Taxonomia de Angiospermae Dicotyledoneae. Rodriguésia 31 (50-anexo): 1–25.
Ferri, M. G. 1967. Plantas do Brasil, espécies do cerrado. Editora Edgard Blücher Ltda. and Editora Universidade de São Paulo, São Paulo, Brazil.
_____. 1977. Ecologia dos cerrados. Pages 15–36 in M. G. Ferri (ed.). IV. Simpósio sobre o cerrado. Editora Itatiaia, Belo Horizonte, Brazil.
_____. 1980. Vegetação brasileira. Editora Itatiaia Limitada, Belo Horizonte, Brazil and Editora da Universidade de São Paulo, Brazil.
Fish, D. & S. Soria. 1978. Water-holding plants (phytotelmata) as larval habitats for Ceratopogonid pollinators of cacao in Bahia, Brazil. Revis. Theobroma 8: 133–146.
Fonseca, G. A. B. da. 1985. The vanishing Brazilian Atlantic forest. Biol. Conserv. 34: 17–34.
Fraga, M. V. G. 1960. A questão florestal ao tempo do Brasil-colônial. Anuário Brasil. Econ. Florest. 3: 7–96.
Frodin, D. G. 1984. Guide to standard floras of the world. Cambridge University Press, Cambridge.
Fundação de Pesquisas Florestais do Paraná. 1978. Inventário

florestal do pinheiro no sul do Brasil. F.U.P.E.F.-Fundação de Pesquisas Florestais do Paraná, Curitiba, Paraná.

Gibbs, P. E., H. F. Leitão Filho & R. J. Abbot. 1980. Application of the point-centered quarter method in a floristic survey of an area of gallery forest at Mogi-Guaçu, SP, Brazil. Rev. Bras. Bot. 3(1/2): 17–22.

Gilbert, E. & P. H. Raven. 1973. Coevolution of animals and plants. University of Texas Press, Austin and London.

Goldberg, A. & L. B. Smith. 1975. Chave para as famílias espermatofíticas do Brasil. Separata de Flora Ilustrada Catarinense. Itajaí, Santa Catarina, Brazil.

Goodland, R. 1970. Plants of the cerrado vegetation of Brazil. Phytologia 20: 57–78.

―――― & M. G. Ferri. 1979. Ecologia do cerrado. Editora Itatiaia, Belo Horizonte, Brazil.

Görts-van Rijn, A. R. A. 1976. Euphorbiaceae, additions and corrections. In J. Lanjouw & A. Stoffers (eds.), Flora of Suriname. E. J. Brill, Leiden.

Gouvêa, J. B. S., L. A. M. Silva & M. Hori. 1976. I. Fitogeografia. Pages 1–7 in Diagnóstico socioeconômico da região cacaueira, recursos florestais, vol. 7. Comissão Executiva do Plano da Lavoura Cacaueira and Instituto Interamericano de Ciências Agrícolas-OEA. Ilhéus, Bahia, Brazil.

Greenhall, A. M. 1965. Sapucaia nut dispersal by greater spearnosed bats in Trinidad. Caribbean J. Sci. 5: 167–171.

Haffer, J. 1974. Avian speciation in tropical South America. Nuttall Ornithological Club, Cambridge, Massachusetts.

Harley, R. M. & S. J. Mayo. 1980. Towards a checklist of the flora of Bahia. Royal Botanic Gardens, Kew, Richmond, Surrey, England.

Heringer, E. P., G. M. Barrosso, J. A. Rizzo & C. T. Rizzini. 1977. A flora do cerrado. Pages 211–232 in M. G. Ferri (ed.), IV Simpósio sobre o cerrado. Editora Itatiaia, Belo Horizonte, Brazil.

Hoehne, F. C., M. Kuhlmann & O. Handro. 1941. O jardim botânico de São Paulo. Secretária da Agricultura, Indústria e Comércio de São Paulo, Brazil.

Holmgren, P. K., W. Keuken & E. K. Schofield. 1981. Index herbariorum, part I, the herbaria of the world. Bohn, Scheltema & Holkema, Utrecht/Antwerpen. Dr. W. Junk B. V., Publishers, The Hague/Boston.

Hueck, K. 1955. Plantas e formação organogênica das dunas no litoral Paulista. Secretrária da Agricultura do Estado de São Paulo, Instituto de Botânica, São Paulo.

――――. 1972. As florestas da América do Sul. Editôra da Universidade de Brasília and Editôra Polígono, S.A., São Paulo, Brazil.

―――― & P. Seibert. 1972. Vegetationskarte von Südamerika. Gustav Fischer Verlag, Stuttgart.

IUCN/World Wildlife Fund. 1982. Tropical forest campaign fact sheet no. 14–Brazil.

Jackson, J. F. 1978a. Differentation in the genera *Enyalius* and *Stobilarus* (Iguanidae): Implications for Pleistocene climatic changes in eastern Brazil. Arq. Zool. 30: 1–79.

――――. 1978b. Seasonality of flowering and leaf-fall in a Brazilian subtropical lower montane moist forest. Biotropica. 10: 38–42.

Jorge Pádua, M. T. 1983. Os parques nacionais e reservas biológicas do Brasil. Instituto Brasileiro de Desenvolvimento Florestal, Brasília, Brazil.

Kinzey, W. G. 1982. Distributions of primates and forest refuges. Pages 455–482 in G. T. Prance (ed.), Biological diversification in the tropics. Columbia University Press, New York.

Köppen, W. 1936. Das geographisches system der klimate. Chapter 3 in W. Köppen & W. Geiger (eds.), Handbuch der klimatologie vol. I Teil C. Ebr. Bornträger, Berlin.

Küchler, A. W. (ed.). 1980. International bibliography of vegetation maps, 2nd ed., section 1, South America. University of Kansas Libraries, Lawrence, Kansas.

Kuhlmann, E. 1977. Vegetação. Pages 85–110 in Fundação Instituto Brasileiro de Geografia e Estatística, Diretoria Técnica (ed.), Geografia do Brasil, região nordeste. Fund. Inst. Brasil. Geogr. Estatís., Directoria de Divulgaçao, Rio de Janeiro, Brazil.

Lacerda, L. D. de, D. S. Dunn de Araújo & N. C. Maciel. 1982. Restingas brasileiras uma bibliografia. CNPq, Proc. no. 107410/79.

――――, ――――, R. Cerqueira & B. Turcq. (eds.). 1984. Restingas: Origem, estrutura, processos. Universidade Federal Fluminense-CEUFF, Niterói, Brazil.

Lamberti, A. 1969. Contribuição ao conhecimento da ecologia das plantas do manguezal de Itanhaém. Univ. São Paulo, Fac. Filos. Ciên. e Letras Bol. 317, Bot. 23: 7–218.

Lamlein Villa-Lobos, J. 1985. Atlantic forest protected. Threatened Plants Newslett. 15: 16–17.

Lanjouw, J. 1932. Euphorbiaceae. Pages 1–101 in A. Pulle (ed.), Flora of Suriname. E. J. Brill, Leiden.

Leigh, E. G. Jr., A. S. Rand & D. M. Windsor. 1982. The ecology of a tropical forest: Seasonal rhythms and long-term changes. Smithsonian Institution Press, Washington, D.C.

Lewis, G. P. 1979. A preliminary checklist of the Leguminosae of Bahia. Royal Botanic Gardens, Kew, Richmond, Surrey, England.

Lindman, C. A. M. & M. G. Ferri. 1974. A vegetação no Rio Grande do Sul. Editora da Universidade de São Paulo and Livraria Itatiaia Editora, Belo Horizonte, Brazil.

Luederwoldt, H. 1919. Os manquesães de Santos. Revista Mus. Paul. Univ. São Paulo 11: 101 pp.

――――. 1929. Resultados de uma excursão ciêntifica a Ilha de São Sebastião no litoral de estado de São Paulo. Rev. Mus. Paul. 16: 1–79.

Leutzelburg, P. von. 1922–1923. Estudo botânico do nordeste, vols. 1, 2, 3. Ministério da Viação e Obras Publicas, Inspectoria Federal de Obras contra as Secas, Publ. no. 57, Serie I, A.

Lutz, B. 1938. Flora fluminense litoral. Apontamentos decorrentes do herbário do Museu Nacional e de observaçaões feitas no litoral, mimeograph, 55 pp.

Luz, A. A. da. 1955. Aspetos fisiograficos e biológicos da orla marítima de Ararangua (geografica, geologia, flora, fauna e ecologia de um trecho da costa de Santa Catarina. Florianópolis II, 54 pp.

Magnanini, A. 1954. Contribuição ao estudo das zonas de vegetação da praia Sernambetiba, D. F., Brasil. Arq. Serv. Florest. 8: 147–232.

Martinelli, G. 1984. Nota sobre *Worsleya rayneri* (J. D. Hooker) Taub & Moldenke, espécie ameaçada de extinção. Rodriguésia 36(58): 65–72.

――――, H. C. de Lima, T. S. Pereira, J. F. Baumgratz & L. A. P. Gonzaga. In press. Campos de altitude da floresta costeira do Brasil I-Morro do Cuca, Estado do Rio de Janeiro. Rodriguésia.

Martius, C. F. P. von. 1906. Tabulae physiognomicae. Brasiliae regiones iconibus expressas descripsit deque vegetatione illius terra uberius. Fl. bras. 1(1): I–CX.

Mayo, S. J. & V. P. B. Fevereiro. 1982. Mata de Pau Ferro–A pilot study of the brejo forest of Paraíba, Brazil. Royal Botanic Gardens, Kew, England.

Mello, E. C. 1950. Estudo dendrológico de essências florestais do Parque Nacional de Itatiaia e os caracteres anatômicas de seus lenhos. Parque Nac. Itatiaia Bol. 2: 1–272.

Menezes, N. L. de, A. M. Giulietti, R. Harley & S. Mayo. 1986. Campo rupestre. Threatened Plants Newslett. **16**: 11–12.

Mittermeier, R. A. 1987. Rescuing Brazil's muriqui: Monkey in Peril. Natl. Geogr. Mag. **171**: 387–395.

Mori, S. A. & B. M. Boom. 1981. Final report to the World Wildlife Fund-US on the botanical survey of the endangered moist forests of eastern Brazil. The New York Botanical Garden, New York.

———, ———, **A. M. de Carvalho & T. S. dos Santos.** 1983. Southern Bahian moist forests. Bot. Rev. (Lancaster) **49(2)**: 155–232.

———, ——— **& G. T. Prance.** 1981. Distribution patterns and conservation of eastern Brazilian coastal forest tree species. Brittonia **33**: 233–245.

———, **G. Lisboa & J. A. Kallunki.** 1982. Fenologia de uma mata higrófila sul-baiana. Revis. Theobroma **12(4)**: 217–230.

——— **& J. E. Orchard.** 1979. Fenologia, biologia floral e evidência sobre dimorfismo fisiológico de pólen de *Lecythis pisonis* Cambess. (Lecythidaceae). Anais Soc. Bot. Brasil **30**: 109–116.

———, ——— **& G. T. Prance.** 1980a. Intrafloral pollen differentiation in the new World Lecythidaceae, subfamily Lecythidoideae. Science **209**: 400–403.

——— **& L. A. M. Silva.** 1979a. The herbarium of the "Centro de Pesquisas do Cacau" at Itabuna, Brazil. Brittonia **31**: 177–196.

——— **& ———.** 1979b. Flora da região cacaueira da Bahia—Plano geral para sua elaboração. Anais Soc. Bot. Brasil **30**: 101–106.

———, ———, **G. Lisboa, R. C. Pereira & T. S. dos Santos.** 1980b. Subsídios para estudos de plantas invasoras no sul da Bahia I. Produtividade e fenologia. Centro de Pesquisas do Cacau, Bol. Técn. **73**: 1–18.

———, ——— **& T. S. dos Santos.** 1980c. Observações sobre a fenologia e biologia floral de *Lecythis pisonis* Cambess. (Lecythidaceae). Revis. Theobroma **10(3)**: 103–111.

Müller, P. 1973. The dispersal centres of terrestrial vertebrates in the neotropical realm. W. Junk, The Hague.

Myers, N. 1980. Conversion of tropical moist forests. National Research Council. Committee on Research Priorities in Tropical Biology.

Nogueira-Neto, P. N. 1977. Conservação da natureza no cerrado. Pages 349–352 *in* M. G. Ferri (ed.), IV. Simpósio sobre o cerrado. Editora Itatiaia, Belo Horizonte, Brazil.

———. 1984. Getting Brazil's network of ecological stations on the ground. Conservation Foundation Annual Report.

Ormond, W. T. 1960. Ecologia das restingas do sudeste do Brasil. Comunidades vegetais das praias arenosas. Parte I. Arq. Mus. Nac. Rio de Janeiro **50**: 185–236.

Pinto, G. C. P., H. P. Bautista & J. D. C. A. Ferreira. 1984. A restinga do litoral nordeste do estado da Bahia. Pages 195–216 *in* L. D. de Lacerda, D. S. Dunn de Araujo, R. Cerqueira & B. Turcq, Restingas: Origem, estrutura, processos. Universidade Federal Fluminense-CEUFF, Niterói, Brazil.

Pio Corrêa, M. 1926. Dicionário das plantas úteis do Brasil e das exóticas cultivadas, Vol. I. Ministério da Agricultura, Rio de Janeiro, Brazil.

———. 1931. Dicionário das plantas úteis do Brasil e das exóticas cultivadas, Vol. II. Ministério da Agricultura, Rio de Janeiro, Brazil.

———. 1952. Dicionário das plantas úteis do Brasil e das exóticas cultivadas, Vol. III. Ministério da Agricultura, Serviço de Informação Agrícola, Rio de Janeiro, Brazil.

———. 1969. Dicionário das plantas úteis do Brasil e das exóticas cultivadas, Vol. IV. Ministério da Agricultura, Instituto Brasileiro de Desenvolvimento Florestal, Brazil.

———. 1978a. Dicionário das plantas úteis do Brasil e das exóticas cultivadas, Vol. V. Ministério da Agricultura, Instituto Brasileiro de Desenvolvimento Florestal, Rio de Janeiro, Brazil.

———. 1978b. Dicionário das plantas úteis do Brasil e das exóticas cultivadas, Vol. VI. Ministério da Agricultura, Instituto Brasileiro de Desenvolvimento Florestal, Rio de Janeiro, Brazil.

Piso, G. & G. Marcgrave. 1648. Historia naturalis brasiliae, auspicio et beneficio illustris I. Mauritii Com. Nassau, Leiden.

Polowman, T. 1987. Ten new species of *Erythroxylum* (Erythroxylaceae) from Bahia, Brazil. Fieldiana, Bot. n.s. **19**: 1–41.

Por, F. Dov & V. L. Imperatriz-Fonseca. 1984. The Jureia ecological reserve, São Paulo, Brazil—Facts and plans. Environ. Conserv. **11(1)**: 67–70.

Prance, G. T. 1971. An index of plant collectors in Brazilian Amazonia. Acta Amazonica **1**: 25–65.

———. 1977. Floristic inventory of the tropics: Where do we stand? Ann. Missouri Bot. Gard. **64**: 659–684.

——— **& S. A. Mori.** 1983. Dispersal and distribution of Lecythidaceae and Chrysobalanaceae. Sonderbd. Natur. Wiss. Ver. Hamburg **7**: 163–186.

RADAMBRASIL. 1981. Projeto RADAMBRASIL, levantamento de recursos naturais vol. 24, folha SD.24 Salvador. Ministério das Minas e Energia Secretaria-Geral, Projeto RADAM-BRASIL, Rio de Janeiro, Brazil.

Rambo, B. 1954. História da flora do litoral riograndense. Sellowia **6(6)**: 113–172.

Rawitscher, F. K. 1944. Algumas noções sôbre a végetação do litoral brasileiro. Bol. Asoc. Geogr. Bras. **5**: 13–128.

Reitz, R. 1980. Preparação de uma flora regional tendo a "Flora Ilustrada Catarinense" como exemplo. Unpublished paper presented at the XXXI Congresso Nacional de Botânica da Sociedade Botânica do Brasil, Itabuna, Bahia, Brazil.

Renvoize, S. A. 1984. The grasses of Bahia. Royal Botanic Gardens, Kew, England.

Ribeiro, V. M. L., L. Mautone, M. da Conceição Valente, V. F. Ferreira, C. L. B. de Abreu, H. P. Bautista, A. de Souza Leão, A. L. Peixoto, E. F. Guimarães & J. Fontella Pereira. 1979a. Bibliografia de botânica V. Taxonomia de Angiospermae Dicotyledoneae. Rodriguésia **31 (50 anexo-I)**: 1–80.

———, ———, **A. de Souza Leão, V. F. Ferreira, M. da Conceição Valente, C. L. B. de Abreu, V. L. Gomes & G. Martinelli.** 1979b. Bibliografia de botânica VI. Taxonomia de Angiospermae Dicotyledoneae. Rodriguésia **31 (51-anexo-II)**: 1–44.

Rizzini, C. T. 1963. A flora do cerrado. Pages 125–177 *in* Simpósio sobre cerrado. Editora Universidade de São Paulo, São Paulo, Brazil.

———. 1971a. Arvores e arbustos do cerrado. Rodriguésia **26**: 63–77.

———. 1971b. Arvores e madeiras úteis do Brasil. Editora E. Blücher, São Paulo, Brazil.

———. 1976. Tratado de fitogeografia do Brasil, aspectos ecológicos, vol. I. Editora de Humanismo, Ciência e Técnologia "Hucitec" Ltd. and Editora da Universidade de São Paulo, São Paulo, Brazil.

———. 1979. Tratado de fitogeografia do Brasil. Editora da Universidade de São Paulo, São Paulo, Brazil.

——— **& W. B. Mors.** 1976. Botânica econômica brasileira. São Paulo, EPU, Editora da Universidade de São Paulo, São Paulo, Brazil.

Roeder, M. 1975. Reconhecimento climatológico. Pages 1–89 *in* Diagnóstico socioeconômico da região cacaueira, vol. 4. Comissão Executiva do Plano da Lavoura Cacaueira and Instituto Interamericano de Ciências Agrícolas-OEA. Ilhéus, Bahia, Brazil.

Romariz, D. A. 1974. Aspectos da vegetação do Brasil. Instituto Brasileiro de Desenvolvimento Florestal, Rio de Janeiro, Brazil.

Sampaio, A. J. 1934. Phytogeografia do Brasil. Brasiliana no. 35, Companhia Editorial Nacional, São Paulo, Brazil.

Seabra, J. J. A. 1949. Flora das dunas (apontamentos sobre a flora psamófila das dunas de Itapoan-Bahia). Lilloa 20: 187–192.

Sick, H. & D. M. Teixeira. 1979. Notas sobre aves brasileiras raras ou ameaçadas de extinção. Publ. Avulsas Mus. Nac.-Univ. Fed. Rio de Janeiro 62: 1–39.

Silva, A. F. da. 1980. Composição florística e estrutura de um trecho da mata atlântica de encosta no município de Ubatuba-SP. Disertação apresentada ao Instituto de Biologia da Universidade Estadual de Campinas para a obtenção do titulo de mestre em biologia (Ecologia), Campinas, São Paulo, Brazil.

Silva, L. A. M., G. Lisboa & T. S. dos Santos. 1982. Nomenclatura vulgar e científica de plantas encontradas na região cacaueira da Bahia. Centro de Pesquisas do Cacau, Bol. Técn. 95: 1–79.

―――― & S. A. Mori. 1981. Os tipos do herbário Centro de Pesquisas do Cacau. Centro de Pesquisas do Cacau, Bol. Técn. 89: 1–72.

Silva, N. M. Ferreira da, C. L. B. de Abreu, M. da Conceição Valente, A. L. Peixoto, M. do Carmen Mendes Marques, A. F. R. de Souza & V. F. Ferreira. 1978. Bibliografia de botânica III. Taxonomia de Angiospermae Dicotyledoneae. Rodriguésia 29(44–anexo): 1–92.

Smith, L. B. 1955. The Bromeliaceae of Brazil. Smithsonian Misc. Collect. 126(1): 1–290.

――――. 1962. Origins of the flora of southern Brazil. Contr. U. S. Natl. Herb. 35(3): 215–249.

Soares, R. O. & R. B. Ascoly. 1979. Florestas costeiras do littoral leste (inventário florestal de reconhecimento). Brasil Florest. 1(2): 9–20.

Standley, P. C. 1944. Araceae. Pages 1–60 in R. E. Woodson & R. W. Schery (eds.). Flora of Panama. Ann. Missouri Bot. Gard. 31: 405–464.

Stellfeld, C. 1945. Contribuição para o estudo da flora marítima do Paraná. " As iriuba dos manguezais." Arq. Mus. Paraná 4 (Art. X): 237–248.

――――. 1949. Fitogeografia geral do estado do Paraná. Arq. Mus. Paraense 7(3): 309–350.

Steyermark, J. A. & O. Huber. 1978. Flora del Avila. Publicación Especial de la Sociedad Venezolana de Ciencias Naturales bajo los auspicios de "Vollmer Foundation" y Ministerio del Ambiente y de los Recursos Naturales Renovables, Caracas, Venezuela.

Strang, H. E., J. de P. Lanna Sobrinho & L. D. Tosetti. 1982. Parques estaduais do Brasil, sua caracterização e essências nativas mais importantes. Silvic. São Paulo 16A(3): 1583–1609.

Teixeira, A. 1980. O programa flora do CNPq: Estágio atual. Unpublished paper presented at the XXXI Congresso Nacional de Botânica da Sociedade Botânica do Brasil. Itabuna, Bahia, Brazil.

Teixeira, D. L. 1985. Endangered forest birds of northeastern Brazil. World Wildlife Fund Monthly Report, January, 1985.

Tomlinson, P. B. 1985. The botany of mangroves. Cambridge University Press, Cambridge.

Tosi, J. 1983. Provisional ecological map of the republic of Brazil. Institute of Tropical Forestry, Río Piedras, Puerto Rico.

Ule, E. 1901. Die vegetation von Cabo Frio an der küste von Brasilien. Bot. Jarhrb. Syst. 28: 511–528.

UNESCO. 1981. Vegetation map of South America. United Nations Educational, Scientific and Cultural Organization, Paris.

Urban, I. 1906. Vitae itineraque collectarum botanicorum. In Martius (ed.), Fl. bras. 1(1): 1–154.

Valente, M. da C., E. de Araujo Schwarz, V. Flechtmann Ferreira, A. de Souza Leão, & A. Luna Peixoto. 1984. Bibliografia de botânica. VIII. Taxonomia de Angiospermae Dicotyledoneae. Boletim do Museu Botânico Kuhlmann 7(4): 1–130.

Vasconcelos, S. 1949. As regiões naturais do Pernambuco, o meio e a civilização. Inst. Est. Agr. Secr. de Agr. Ind. e Com. do Est. de Pernambuco 2, 219 pp.

Vattimo, I. de. 1956. Lauraceae do Itatiaia. Rodriguesia 18/19 (30/31): 39–86.

――――. 1957. "Lauraceae" do estado do Rio de Janeiro. Arch. Jard. Bot. Rio de Janeiro 15: 113–144.

――――. 1959. Flora da cidade do Rio de Janeiro-Lauraceae. Rodriguesia 21/22 (33/34): 157–176.

Vellozo, J. M. da C. 1825. Flora fluminensis. Flumine Januario.

Veloso, H. P. 1945. As comunidades e as estações botânicas de Teresópolis, estado do Rio de Janeiro. Bol. Mus. Nac. Rio de Janeiro 3: 1–95.

――――. 1946a. A vegetação no município de Ilhéus, Estado da Bahia I—Estudo sinecológico das areas de pesquisas sôbre a febre amarela silvestre realizado pelo S.E.P.F.A. Mem. Inst. O. Cruz 44: 13–103.

――――. 1946b. A vegetação no município de Ilhéus. Estado da Bahia II—Observação e ligeiras considerações acêrca de espécies que ocorrem na região. Chave analítica das espécies árboreas. Mem. Inst. O. Cruz 44: 221–293.

――――. 1946c. A vegetação no município de Ilhéus, Estado da Bahia III—Caracterização de vegetação pelo valôr dos ídices das espécies. Mem. Inst. O. Cruz 44: 323–341.

Vinha, S. G. da, T. de Jesus Soares Ramos & M. Hori. 1976. 2. Inventário Florestal. Pages 20–212 in Anonymous (ed.), Diagnóstico Socioeconômico da Região Cacaueira, Recursos Florestais, vol. 7. Comissão Executiva do Plano da Lavoura Cacaueira and Instituto Intermericano de Ciências Agrícolas—OEA. Ilhéus, Bahia, Brazil.

World Wildlife Fund. 1984. The cry of the muriqui. World Wildlife Fund Monthly Report, March, 1984.

Yano, O. 1981. A checklist of Brazilian mosses. J. Hattori Bot. Lab. 50: 279–456.

――――. 1984. Checklist of Brazilian hornworts and liverworts. J. Hattori Bot. Lab. 56: 481–548.

Bolivia

James C. Solomon

Contents

- I. Description of the Region ... 457
  - A. Geographical Extent and Topography ... 457
    - 1. The Altiplano ... 457
    - 2. The Western Cordillera ... 457
    - 3. The Eastern Cordillera and Sub-Andean Ranges ... 457
      - a. The Yungas ... 458
      - b. The Valles ... 458
    - 4. The Chaco ... 458
    - 5. The Beni Plain ... 459
    - 6. The Brazilian Shield ... 459
- II. Population ... 459
- III. Vegetation Maps ... 459
- IV. Present State of Floristic Inventory ... 460
  - A. General ... 460
  - B. History of Collections ... 460
- V. Publications ... 460
- VI. Resources for Inventory ... 460
  - A. Herbario Nacional de Bolivia ... 461
  - B. Herbario Florestal Nacional ... 461
  - C. Museo de la Salle ... 461
  - D. Instituto de Botánica ... 461
- VII. Current Trends ... 461
- VIII. Literature Cited ... 462
- IX. Appendix I ... 463

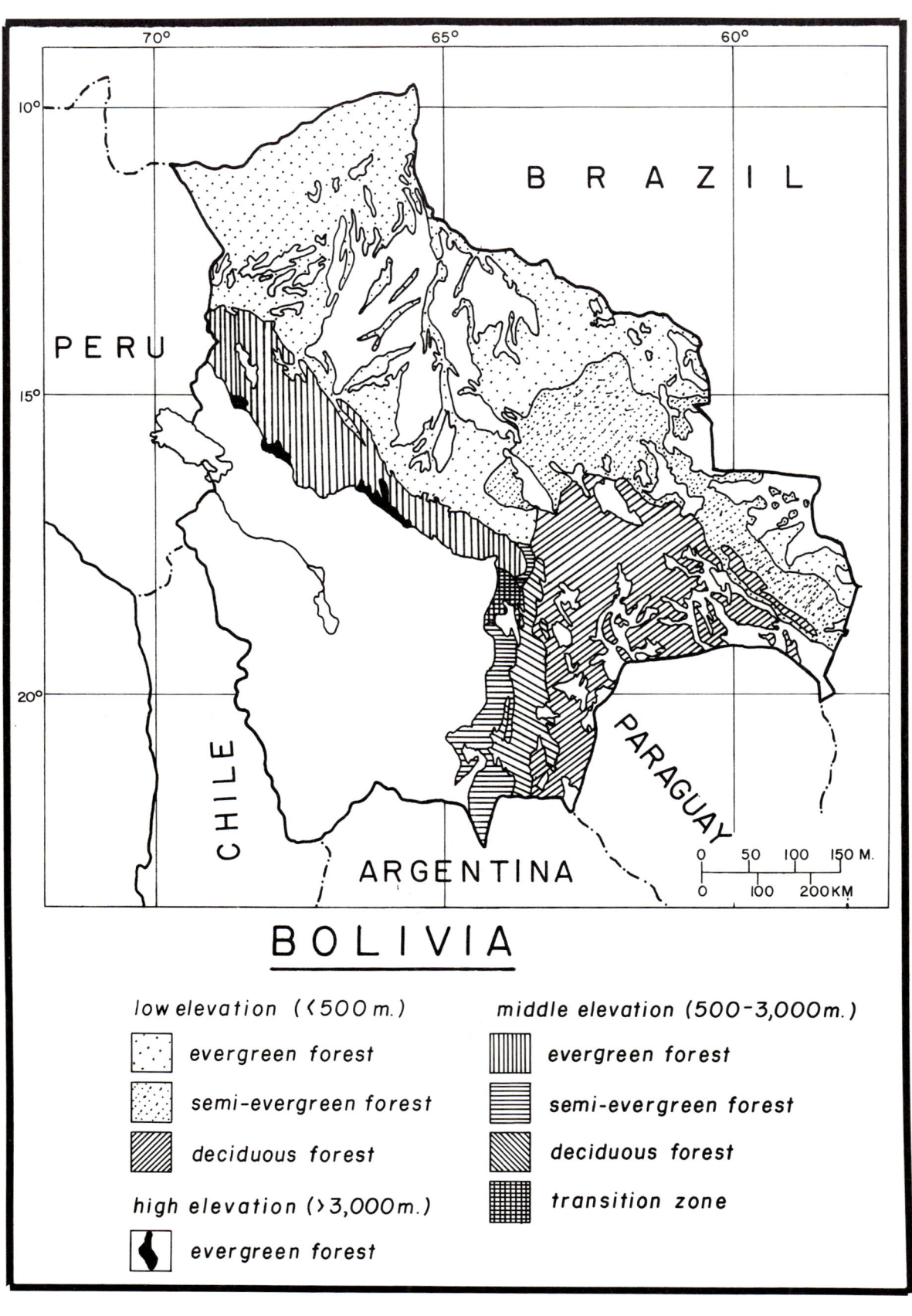

I. Description of the Region

The land-locked Republic of Bolivia covers an area of 1,098,581 km² (Fig. 1[1], Table 1), somewhat less than three times the size of California, lying between 10° and 23°S latitude. Within this large area a great diversity of ecological zones can be found, ranging from permanent snow on mountain peaks which extend to over 6500 m in elevation to Amazonian lowland forest at 150 m, and from pluvial zones receiving up to 6 m of rain per year to arid regions receiving less than 100 mm. The physiographic features of the country, its location, and the variety of geological formations encountered, provide Bolivia with an exceedingly rich and varied flora. By far the most biologically diverse zones are those which contain tropical and subtropical moist and wet forests.

A. Geographical Extent and Topography

Geographically, the country can be divided in a number of different ways; the broadest interpretation recognizing seven more or less distinct zones. Mountainous areas comprise about one third of the country, with representation of four major vegetation types: puna, moist to wet montane tropical forest, semi-arid mesothermic thorn woodland, and moist montane sub-tropical forest. The other two-thirds is relatively level or with low relief, mostly less than 500 m in elevation and sloping gently from south to north. A few isolated mountains occur in the eastern portion of Santa Cruz Department, formed from the ancient Brazilian Shield.

Floristically, these lowlands can be divided into four major types: chaco, seasonally inundated savannas, Amazonian forest and cerrado. These floristic zones mostly correspond to the geographic divisions outlined below, but with some degree of overlap and finer partitioning. Brockmann (1978) indicates that forests cover approximately 564,684 km² of Bolivia, or slightly more than 51% of its land area (Table II).

1. The Altiplano

The Altiplano is a broad expanse of interior basins and valleys with extensive alluvial and lacustrine sediments, having a mean elevation of 3800 m and punctuated by a few volcanic peaks and short mountain ranges. With the western and eastern Cordilleras as boundaries, the major portion of the Altiplano forms a watershed without external drainage, with two lakes, Titicaca and Poopo, and two saltflats, Uyuni and Coipasa, as the major interconnected catchments. The vegetation, usually termed puna, is composed of semi-arid or seasonal grass- or shrublands, with such genera as *Adesmia, Baccharis, Parastrephia*, and *Tetraglochin*, and seasonally green bunch grasses (*Festuca, Stipa*, and *Poa*). The northern portions, particularly in the vicinity of Lake Titicaca, are the most equable in terms of temperature and precipitation, with a decrease in both from the northeast to southwest. The puna extends northwards into Peru and southwards into Argentina and adjacent Chile. Along the northeastern margin of the Altiplano, where there is increased precipitation, the dry puna gives way to a moist puna of different species composition, and above this is a low, cushion-forming high Andean vegetation type which is contiguous with that of Peru.

2. The Western Cordillera

The Western Cordillera forms the border between Bolivia and Chile and is composed almost entirely of material laid down by volcanic activity during the Miocene and Pliocene (Ahlfeld, 1972). Not a continuous mountain chain, but a series of discrete volcanic cones trending primarily north-south, it forms the western boundary of the closed interior basins of the Altiplano and contains the highest peak in Bolivia, Volcan Sajama, at 6,520 m. The vegetation of this region is little different from that of the Altiplano, being principally affected most strongly by the low temperatures and low precipitation, which become more pronounced to the south and west.

3. The Eastern Cordillera and Sub-Andean Ranges

The Eastern Cordillera and Sub-Andean ranges which form the eastern boundary of the Altiplano are exceedingly complex geologically, but can be divided into two basic units based on their floristic affinities.

Table I
Political subdivisions of Bolivia.

Departments	Area (km²)	Population[a] (1980 est.)	Capital
1. Pando	63,827	34,493	Cobija
2. La Paz	133,985	1,778,230	La Paz
3. Beni	213,564	204,355	Trinidad
4. Oruro	53,588	376,757	Oruro
5. Cochabamba	55,631	875,051	Cochabamba
6. Santa Cruz	370,621	862,637	Santa Cruz
7. Potosí	118,218	798,331	Potosí
8. Chuquisaca	51,524	435,147	Sucre
9. Tarija	37,623	227,217	Tarija

[a] Source: Montes de Oca, 1982.
Scale = 100 km/cm.

Table II
Forested areas of Bolivia by department.

Department	Forest Cover (km²)	Area of Department (km²)	% in Forest
Beni	105,083	213,564	49
Cochabamba	26,664	55,631	48
Chuquisaca	17,798	51,524	36
La Paz	61,381	133,985	46
Pando	60,816	63,827	95
Santa Cruz	266,478	370,621	72
Tarija	26,464	37,623	70
Oruro	— —	53,588	0
Potosí	— —	118,218	0
TOTAL	564,684	1,098,581	51

Source: Brockmann, 1978.

---

[1] Modified from Brockman, 1978.

The Yungas. This is the name commonly given to the eastern slopes of the northern portion, which runs east-southeast from the Peruvian border, through La Paz, Cochabamba and Santa Cruz Departments, to just northwest of Santa Cruz at about 18°S. It consists of four mountain ranges extending for nearly 600 km and containing a substantial number of peaks in excess of 6,000 m. The Chapare, the zone of moist Andean slopes north of Cochabamba, is part of the general formation indicated here as the Yungas, although historically and politically it has usually been considered distinct from the area to the northeast of La Paz. The eastern slopes of these mountains descend precipitously through nearly 5,000 m elevation before meeting the Beni plains, crossing a wide variety of vegetation and precipitation zones.

Climatically this area is dominated by northeasterly and northerly winds, which produce orographic precipitation throughout the year, especially above 2,000 m. elevation. Chimore, at 300 m in the Chapare northeast of Cochabamba, receives about 4,000 mm of precipitation annually, making it the wettest recording station in Bolivia. Undoubtedly, the mountain slopes above this area receive more, although data are completely absent.

Precipitation patterns and temperature have a strong effect on the vegetation of the Yungas, producing three roughly defined zones: the Ceja de la Montana, the medio Yungas and Yungas verdadero. The Ceja ranges from open shrublands at 4,000 to 3,800 m to low cloud forest at about 3,000 m. This zone is cool and moist with frequent fog and mists, although relatively drier during the dry season. The vegetation is quite diverse, but dominated by shrubby Compositae, Ericaceae, Melastomaceae, *Myrica,* and *Fuchsia*, with trees of *Brunellia, Clusia, Weinmannia,* and bambusoid grasses at the lower elevations. In some restricted areas there exist low forests of *Polylepis* at the upper limit with the moist puna grasslands. The medio Yungas, between 3,000 and 2,000 m, is moist to wet cloud forest with a rich flora of trees, shrubs, and tree ferns, covered with an abundance of mosses, other bryophytes, and epiphytic vascular plants, principally ferns, orchids, and bromeliads. However, the species richness of the epiphytic flora is much less than in comparable areas in Ecuador or Peru, for example. The Yungas verdadero (true Yungas) lies below the 2,000 m level, and is warmer and, in some places, somewhat drier than the higher zones, with a marked increase in the diversity of vascular plants. The lack of fogs and mists below 2,000 m is especially noticeable during the dry season, although there may be variation between one valley and the next depending on slope and aspect. The large tree ferns disappear and palms make their appearance. This zone continues across the sub-Andean ranges, which are nearly 300 km wide opposite La Paz, to approximately 500 m elevation, where the mountains meet the Beni plains. The Yungas are continuous with the same formation on the eastern slopes of the Peruvian Andes. At the southern boundary of this formation, the majority of the typically montane tropical species, including entire genera, reach their southernmost limit along the Andes.

The Valles. The second portion of the eastern Cordillera is strikingly different from the northern Yungas. South of about 18°S latitude (vicinity of Santa Cruz) the mountains run directly north-south and are not nearly as elevated as those to the north. Because of the change in orientation and shifting weather patterns, orographic precipitation is much less evident, resulting in generally drier vegetation types, and the valleys are in rain-shadows. The tablelands and mountains of the eastern Altiplano and sub-Andean ranges are cut by numerous arid to semi-arid, mesothermic valleys, occupied by xerophytic thorn woodland, the northernmost in Bolivia being the realtively isolated valley of the Río Sorata, followed by the upper Río La Paz. From the Cochabamba valley southwards, this vegetation type becomes increasingly common. These valleys form the sources of the Rios Pilcomayo and Gaupay (Grande). Elevated portions of this zone continue to be covered with the dry puna grasslands typical of the Altiplano, while middle elevations in the valleys (2000 to 3000 m) are occupied by woodlands containing a variety of genera, among others, *Acacia, Aloysia, Caesalpinia, Ephedra, Prosopis, Schinus,* and many cacti. These valleys have been centers of human settlement for thousands of years, and are thus greatly modified. The floristic affinities of the mesothermic valleys lie with the semi-arid regions of Argentina and with the dry, interior valleys of Peru. In the southernmost portions of the Eastern Cordillera and Sub-Andean ranges, principally in Chuquisaca and Tarija Departments, increased precipitation from southerly winds produces a diverse, sub-tropical, evergreen forest, usually termed the Tucumano-boliviana forest. It is less diverse than more tropical forests under the same precipitation regime, but does contain, among other genera, *Cedrela, Juglans, Pategonula, Tabebuia, Terminalia,* some Lauraceae and many Myrtacae, with *Alnus, Podocarpus* and *Polylepsis* at higher elevations. This vegetation continues southwards into Argentina to the vicinity of Tucuman.

The mountainous zones just outlined comprise roughly the southwestern one-third of Bolivia. The other two-thirds, to the north and east are relatively level, with a few isolated mountains in the east, these formed from the ancient Brazilian Shield.

4. The Chaco

The Chaco covers the large, relatively level area south and east of the city of Santa Cruz, including the eastern parts of Chuquisaca and Tarija Departments, with floristic elements penetrating into the eastern ranges of the mountains. The vegetation is continuous with the zerophytic Chaco formation of western Paraguay and northern Argentina, with such common genera as *Aspidosperma, Bulnesia, Celtis, Prosopis, Schinus,* and *Schinopsis.* Forests are mostly deciduous, or semi-evergreen. Large parts of the Chaco, such as the Banados de Izozog, are swampy or seasonally inundated. In the portions where the water table is very near the surface, palm savannas of *Copernicia* are formed. Other savannas are found scattered throughout the Chaco. To the North, the

Chaco gradually merges with the more mesic, tropical vegetation of the Beni plains. Of particular interest are the isolated Chacoan sandstone mountains, principally the Serranias de Santiago and San Jose. The highest peak in these ancient mountains reaches nearly 1,300 m and is separated by more than 400 km from the nearest area of similar elevation.

5. The Beni Plain

The Beni Plain occupies the northern third of Bolivia and is drained primarily by the Rios Itenez (Guapore), Mamore, Beni and Madre de Dios. Approximately 40% of this area is seasonally inundated, edaphic grassland, mostly centered in Beni Department but also in the northern parts of La Paz and Santa Cruz Departments. In the savanna areas, woody vegetation is limited to the margins of rivers and streams, where it forms gallery forest, and to better drained sites. The forests of the savanna zone are rich in species, although not as diverse as the Amazonian forest farther north and west.

The northern and western portion of the Beni plain is covered with lowland, Amazonian forest. Geographically this area is comprised of the Department of Pando, the northernmost portion of Beni and La Paz Departments, and a narrow band extending southeast along the base of the Andes from central La Paz Department into Santa Cruz. Most of the Amazonian forest is evergreen, but merges gradually with semi-evergreen forest along its southern limit. This is probably the most species-rich formation in Bolivia. An exceedingly large percentage of the species in the Amazonian forest flora reach their southernmost distributional limits in the vicinity of the city of Santa Cruz.

6. The Brazilian Shield

The Brazilian Shield appears as a series of short mountain chains or elevated land in eastern Santa Cruz Department. The Serrania de Huanchaca and the area south of there along the Brazilian border has the characteristic vegetation of the Cerrado, the dominant vegetation type further east in Brazil. Otherwise, the vegetation is a mixture of Chacoan and tropical forest elements, plus scattered open savannas. Many typical Cerrado species penetrate west into the savannas of the Beni plain and as far north as the northernmost savannas in the vicinity of Riberalta.

Phytogeographically then, Bolivia is in a unique position, containing the southernmost extensions of Amazonian and moist montane vegetation in South America, the western limit of the Cerrado, the northern limit of the Chaco, plus Andean and savanna floras. As a result, the Bolivian flora is one of the most diverse for its area in all of South America.

II. Population

Bolivia has a population currently estimated at nearly 6 million people (Population Reference Bureau, 1984). At the present time about 85% of the population lives in the Altiplano and Valles portions of the country, where traditional agricultural practices have reached their maximum productivity under present circumstances, and nearly all pasture lands are heavily overgrazed. Many centuries of human occupation have resulted in little, if any, undisturbed vegetation, except in the driest or most inaccessible parts. In addition, the population is increasing at about 2.7% per year, i.e., will double in 26 years (Population Reference Bureau, 1984), so settlement of lowland forest areas is being actively encouraged. Spontaneous and government sponsored colonization is opening up many areas which were previously undisturbed. These areas, such as the foothills of the Yungas verdadero, the southwestern edge of the Beni plain, central Pando, and the territory along the Paraguayan and Brazilian borders, are seen as an attractive frontier for agricultural colonization to absorb a steadily increasing population.

The three largest cities are La Paz, Cochabamba and Santa Cruz. All are growing rapidly, with a tremendous influx of people from the countryside in recent years due to adverse climatic and economic conditions. There are no data for rates of increase, but Santa Cruz (450 m elevation), at the base of the mountains and on the margin of the evergreen and semi-evergreen Amazonian forest, is certainly increasing most rapidly, followed by La Paz and then the other major cities. In general, Bolivians see their country as being underpopulated, considering that the density is only 5.5 persons per km^2, making it one of the least populated countries in South America.

III. Vegetation Maps

Küchler (1980) has recently reviewed the existing vegetation maps of South America. Of the sixteen maps he lists, only three are really recent (since 1974), and two of these are indispensable for an overview of the Bolivian vegetation and the planning of floristic studies. The first is an ecological map, with a scale of 1:1,000,000, based on the Life Zone system of Holdridge (Tosi et al., 1975) with an expanatory analysis of each zone existing in Bolivia by Unzueta (1975). The second consists of two complementary maps of present land cover and use, developed by the ERTS program in Bolivia (Brockmann, 1978; ERTS, 1978). The larger scale map (1:1,000,000) by Brockmann shows grasslands-shrublands, forests (evergreen, semi-evergreen, deciduous, secondary), cultivated lands, and swamps at low elevation (<500 m), middle elevation (500 to 3,000 m) and high elevation (>3,000 m), based on satellite images taken by LANDSAT 1 and 2, with a comparative analysis of the area covered by each formation within Bolivia, as a whole, and by department (Figure 1, Table 1). The smaller scale map (1:4,000,000) is a summary of the larger scale one, indicating only the gross categories given above.

IV. Present State of Floristic Inventory

A. GENERAL

Considering its varied and interesting flora, Bolivia is sadly undercollected and little explored. In a summary of the state of tropical floristics, Prance (1977, 1979) indicates that "Bolivia is probably the least collected of all South American countries." As a result, relatively little is known about the number of species of vascular plants and even less about endemism, the distribution of individual species in flora or the composition of the vegetation. Almost nothing can be said about the degree of endemism, although it is likely that the eastern Andean slopes (Yungas) and the interior valleys will have the highest percentages. Based on the rate at which new records are being added to the flora, and an analysis of recent monographic studies, the best current estimate of the size of the Bolivian flora is on the order of 18,000 species, of which somewhat more than 10,000 have actually been recorded from the country. It is estimated that no more than 90,000 collections of vascular plants have been made in all of Bolivia, giving an overall density of 8 collections per 100 km^2. There are at least 10 times as many collections from Central America, which has only one-half the area of Bolivia (Gentry, 1979; Prance, 1977). Certainly there are areas with a fair number of collections, mostly from the vicinity of La Paz, the upper Yungas valleys near La Paz, the region around Lake Titicaca, the Cochabamba valley, and, to a lesser extent, Santa Cruz and Tarija. No area of the country, however, can be said to have been well collected, and most parts are completely unexplored. Despite the fact that the Yungas region near La Paz is one of the better known areas in Bolivia from previous collections, many new records, some new taxa, and second and third collections of poorly known but often common taxa, continue to come from this area. With steadily increasing human pressures, the need to salvage botanical material from threatened localities becomes increasingly important. There is a great need for collections from everywhere on the eastern slopes of the eastern Cordillera and the lowlands, especially the Amazonian forest region and the Brazilian Shield.

B. HISTORY OF COLLECTIONS

Beginning in the late 18th century with Thaddeus Haenke, a succession of naturalists and botanists have visited Bolivia, but most have collected relatively few specimens, or confined their collecting to the more accessible Altiplano or high montane areas, especially in the vicinity of La Paz and Cochabamba. The most often cited collections are those of E. Asplund, M. Bang, W. M. A. Brooke, O. Buchtien, M. Cardenas, K. Fiebrig, T. Herzog, B. Krukoff, G. Mandon, H. Rusby, J. Steinbach, G. Tate and R. S. Williams. A few of the early collectors reached the last ranges of the Andes or worked in the lowlands, most notably Herzog, Rusby, Steinbach and Williams. Much more recently S. Beck, J. Solomon, M. Nee and a number of others have begun collecting in forested lowland areas. Up to the present, Bolivian specimens collected by 200 individual collectors or groups have been seen or, at least, referenced in the literature, with a marked increase in the number of collectors since 1980. While this count is not yet to be considered exhaustive, it certainly includes the majority if collectors.

V. Publications

Publications dealing specifically with the systematics or floristics of Bolivian plants are relatively few and scattered. The earliest of these was Weddell's *Chloris Andina* (1855–1857), which treated the entire Andean region. The collections of Bang, Rusby and Williams are detailed in a long series of articles (E. Britton, 1896; N. Britton, 1889–1892; Rusby, 1888, 1893–1896, 1898–1902, 1907, 1910–1912, 1927), while Buchtein (1910) published a short florula based on some of his collections. Herzog (1913–1921, 1945), along with numerous collaborators, also published an enumeration of the collections from his two trips to Bolivia. The most important general floristic studies are those of Karl Fiebrig (1911), who discussed the vegetation of Tarija Department, and Theodor Herzog (1910, 1923), who summarized all of the previous work on Bolivia and, in particular, dealt with the extensive area covering southeastern Santa Cruz Department through the city of Santa Cruz and along the mountains to Cochabamba and La Paz. A preliminary checklist of the vascular plants was published in 1958 (Foster, 1958), containing about 10,000 names representing 1,678 genera. In addition to the references above, which are cited in the bibliography, a selected bibliography dealing with Bolivian plants is included as Appendix I. There also exists a very large number of systematic treatments covering specific families, genera or groups of plants having one or more species represented in the flora of Bolivia.

VI. Resources for Inventory

There are presently four established herbaria in Bolivia, with combined holdings of about 35,000 herbarium collections. Only two are listed in *Index Herbariorum* (Holmgren et al., 1981), namely BOLV and LPB. As is so frequently the case, the balance of the herbarium collections are held outside of Bolivia. The largest holdings of Bolivian material are at the New York Botanical Garden (NY), the U.S. National Herbarium (US), Missouri Botanical Garden (MO), and Kew Gardens (K). There are also large collections at the Instituto Darwinion (SI), Universidad del Nordeste (Corrientes, CTES) and Instituto Lillo (LIL) in Argentina. For more than a century many of the staff members of these Argentine institutions have made numerous collections in Bolivia, which are little consulted.

A. Herbario Nacional de Bolivia

With regards to botanical institutions within Bolivia, the largest, and certainly the most active, is the Herbario Nacional de Bolivia (LPB). This herbarium was formed in 1978 as a part of the Instituto de Ecología, a joint project between the University of Gottingen and the Universidad Mayor de San Andres (La Paz). In 1981, a second herbarium was created at the Museo Nacional de Historía Natural in La Paz by a Cooperative agreement between the Missouri Botanical Garden and the Academia Nacional de Ciencias de Bolivia. In January 1984, a new agreement was signed between the Academia de Ciencias and the Universidad Mayor de San Andres to combine these separate herbaria into a single Bolivian National Herbarium, which is administered jointly by the Museo de Historia Natural and the Instituto de Ecología. The herbarium has somewhat more than 30,000 specimens from all parts of the country. The creation of this herbarium and its current activities are a major step towards the compiling of information on plant resources in Bolivia. The Herbario Nacional de Bolivia is actively involved in botanical inventories and general collecting throughout the country. Since its formation, staff and students have made major collections in the Altiplano (both moist and dry puna), the Yungas, the mesothermic valleys, the savannas of the Beni, the Amazonian forest near Riberalta and the moist, subtropical forests of Tarija. There has recently been an increase in the number of foreign scientists visiting Bolivia, no doubt due to the presence of an active institution like the Herbario Nacional, which serves as a focal point and source of logistical help.

B. Herbario Florestal Nacional

In Cochabamba, the Herbario Forestal Nacional 'Martin Cardenas' (BOLV) was formed in 1976 with the aid and support of FAO (Proyecto Forestal FAO/CDF, Bol/74/031). It is administered by the Centro de Dessarollo Forestal (CDF) of the Ministerio de Asuntos Campesinos y Agropecuarios. They presently have about 3000 collections, concentrating principally on forest trees.

C. Museo de la Salle

There also exists a second herbarium in Cochabamba, housed at the Museo de la Salle (Colegio La Salle), and consisting of about 1,500 specimens. Brother Adolfo Jiménez has been principally responsible for the formation of this collection. He has produced a number of publications on the flora of Cochabamba based on this material (Jiménez, undated, 1962, 1976, 1984).

D. Instituto de Botánica

A fourth herbarium, consisting of slightly more than 1000 collections is to be found at the Instituto de Botánica of the Universidad Autonoma 'Juan Misael Saracho' in Tarija. These collections are primarily oriented towards important forest trees and forage plants, but with a good representation of the flora of the Tarija valley. They were utilized in the production of a recent summary of the vegetation of the Tarija valley (Coro, 1982).

VII. Current Trends

It has become readily apparent that human activity is rapidly depleting or removing the natural vegetation of tropical zones throughout the world (Myers, 1980). Large parts of Bolivia are undergoing increasingly aggressive development, especially in the moist tropical and sub-tropical zones where, until recently, the lack of good communications has retarded extensive development. The greatest impact at the present time is coming in the form of spontaneous and planned colonization, extensive logging operations in the vicinity of Santa Cruz, selective logging throughout the Beni, and conversion of large areas to annual crops and pasture. In the Llanos de Mojos, savanna grazing lands are being continuously enlarged by clearing gallery forests and forest islands, in addition to expansion into the more continuous forests towards the base of the Andes. The forests of the lower Yungas and the Beni are probably undergoing the most rapid alteration of any zone, but its magnitude is not known and there still remain substantial undisturbed areas (Freeman et al., 1980). Portions of the Yungas are fairly accessible, with a long history of human settlement due to their relatively fertile soils. Currently, colonization efforts are intensifying, brought about, in part, by adverse climatic and economic conditions. The agricultural methods of the colonists result in repeated clearing and burning of the vegetation cover, beginning in the lowest parts of a valley and progressively working up the steep slopes. These practices are leading to severe erosion problems in many areas. The Chaco still has a high proportion of undisturbed forest, while the Amazonian forests of Pando Department remain largely untouched.

Bolivia, like most of Latin America, is currently undergoing real and severe economic dislocation, in part provoked by the incredible foreign debts they have accumulated. As a result, internal social strife is rising, and the government is directing most of its efforts at controlling immediate social problems. Thus, minimal resources are available for basic floristic and vegetational research. The Ministerio de Asuntos Campesinos y Agropecuarios has an interest in any research dealing with the economic potential of plant resources and land use, but also has other major social priorities.

At present, the most viable institutions within Bolivia are the Museo de Historia Natural, universities and such foreign financed programs as those of the Missouri Botanical Garden, the Instituto de Ecología and ORSTOM. The prospects for continued foreign support of floristic work appear quite good, although at modest levels. There already exist a number of herbaria, and an increasing interest and understanding among students and professionals of the need for information on native plants. There is an absolutely essential and pressing need for the formation of cadres of professional Bolivian botanists and technicians of all types, and the development of an institutional infrastructure to support them in their studies. When one considers the size and diversity of the Bolivian flora, and the relatively few people involved in its study, present resources are simply not adequate to complete a floristic inventory within a reasonable length of time, or before many threatened areas have been substantially modified.

VIII. Literature Cited

Ahlfeld, F. 1972. Geologia de Bolivia. Editorial Los Amigos del Libro, La Paz. 190 pp.

Britton, E. G. 1896. An enumeration of the plants collected by H. H. Rusby in Bolivia, 1885–1886. Part II. Bull. Torrey Bot. Club **23**: 471–499.

Britton, N. L. 1889. An enumeration of the plants collected by Dr. H. H. Rusby in South America, 1885–1886. Part IV. Bull. Torrey Bot. Club **16**: 13–20; Part V. **16**: 61–64. 1889; Part VI. **16**: 153–160. 1889; Part VII. **16**: 189–192. 1889; Part VIII. **16**: 259–262. 1889; Part IX. **16**: 324–327. 1889; Part X. **17**: 9–12. 1890; part XI **17**:53–60. 1890; Part XII. **17**: 91–94. 1890; Part XIII. **17**: 211–214. 1890; Part XIV. **17**: 281–284. 1890; Part XV. **18**: 35–38. 1891; Part XVI. **18**: 107–110. 1891; Part XVII. **18**: 261–264. 1891; Part XVIII. **18**: 331–334. 1891; Part XIX. **19**: 1–4. 1892; Part XX. **19**: 148–151. 1892; Part XXI. **19**: 263–366. 1892; Part XXII. **19**: 371–374. 1892; Part XXIII. **20**: 137–140. 1893.

Brockman, C. E. (ed.). 1978. Mapa de cobertura y uso actual de la tierra, Serie sensores remotos 2. GEOBOL.

———. 1978. Mapa de cobertura y uso actual de la tierra, Bolivia. 1:1,000,000. Memoria explicativa. 116 p. La Paz: Servicio Geologico de Bolivia (GEOBOL), Programa ERTS.

Buchtien, O. 1910. Contribuciones a la flora d Bolivia, I Parte. Direccion General de Estadistica y Estudios Geograifcos, Seccion Museo Nacional. J. M. Gamarra, La Paz. 197 pp.

Coro, M. 1982. El algorrobo y la vegetación del valle de Tarija. Revista Co. Tec. (Tarija) **3(4)**: 29–107.

ERTS. 1978. Mapa de cubertura y uso actual de la tierra, Bolivia. 1:4,000,000. Servicio Geologico de Bolivia (GEOBOL) Programa ERTS, La Paz.

Fiebrig, K. 1911. Ein Beitrag zur Pflanzengeographie Boliviens. Bot. Jahrb. Syst. **45**: 1–68.

Foster, R. C. 1958. A catalogue of the ferns and flowering plants of Bolivia. Contrib. Gray Herb. **184**:1–223.

Freeman, P. H., Cross, R. D. Flannery, D. A. Harcharik, G. S. Hartshorn, G. Simmonds & J. D. Williams. 1980. Bolivia: State of the environment and natural resources, a field study. AID Contract PDC-C-Q247. (unpaged).

Gentry, A. 1979. Extinction and conservation of plant species in tropical America. Pages 110–126 *in* I. Hedberg, ed. Systematic botany, plant utilization and Biosphere conservation. Almquist & Wiksell, Stockholm.

Herzog, T. 1910. Beitrage zur Kenntnis von Ostbolivien. Petermann's Mitth. Justus Prthes' Geogr. Anst. **56**: 136–138, *pl.* 25; 194–200, *pl.* 34–35.

———. 1913. die von Dr. Th. Herzog auf seiner zweitne Reise durch Bolivien in den Jahren 1910 und 1911 gesammelten pflanzen. Meded. Rijks-Herb. Teil 1. **19**: 1–84; Teil 2. **27**: 1–89. 1915; Teil 3. **29**: 1–94. 1916; Teil 4. **33**: 1–19. 918; Teil 5. **40**: 1–77. 1921; Teil 6. **46**: 1–31. 1922.

———. 1923. Die Pflanzenwelt der bolivischen Anden und ihres Ostlichen Vorlandes. *In* A. Engler & O. Drude, eds. Die Vegetation der Erde **15**: 1–258.

———. 1945. Plantae a Th. Herzog in itinere eius boliviensi altero annis 1910 et 1911 collectae pars VIII. Blumea **5**: 641–685.

Holmgren, P. K., W. Keuken, and E. K. Schofield. 1981. Index Herbariorum. Ed. 7. Regnum Vegetabile vol. 106. 452 pp.

Jiménez, A. (undated). Juncáceas de la región de Cochabamba. Editorial Bruño, La Paz. 52. pp.

———. 1962. Plantas del valle de Cochabamba. Fasc. I. Editorial Canelas, Cochabamba. (unpaged). Fasc. II. 90 pp. 1966.

———. 1976. Gramineas Bolivianas. Imprenta Visión, Cochabamba. 286 pp.

———. 1984. Flora de Cochabamba. Imprenta Los Huérfanos, Santa Cruz. 390 pp.

Küchler, A. W. 1980. Bolivia. Pages 116–123 *in* International bibliography of vegetation maps. 2nd ed. Sec. I: South America. Univ Kansas Public Library Ser. 45.

Montes de Oca, I. 1982(1983). Geografia y recursos naturales de Bolivia. Imprenta Superel, Ltda., La Paz. 628 pp.

Myers, N. 1980. Conversion of tropical moist forests. Washington, National Academy of Sciences, D.C. 205 pp.

Population Reference Bureau. 1984. World Population Data Sheet. Population Reference Bureau, Washington, D.C.

Prance, G. T. 1977. Floristic inventory of the tropics: Where do we stand? Ann. Missouri Bot. Gard. **64**: 659–684.

———. 1979. South America. Pages 57–70 *in* Hedberg, I., ed. Systematic botany, plant utilization and Biosphere conservation. Almquist & Wiksell, Stockholm.

Rusby, H. H. 1888. An enumeration of the plants collected by Dr. H. H. Rusby in South America, 1885–1886. Part I. Bull. Torrey Bot. Club **15**: 177–184; Part III. **15**: 247–253.

———. 1893. On the collections of Mr. Miguel Bang in Bolivia. Part I. Mem. Torrey Bot. Club **3**: 1–67; Part II. **4**: 203–274. 1895; Part III. **6**: 1–130. 1896.

———. 1898. An enumeration of the plants collected by Dr. H. H. Rusby in South America, 1885–1886. Part XXIV. Bull. Torrey Bot. Club **25**: 495–500; Part XXV. **25**: 542–545. 1898; Part XXVI. **26**: 145–152. 1899; Part XXVII. **26**: 189–200. 1899; Part XXVIII. **27**: 22–31. 1900; Part XXIX. **27**: 69–84. 1900; Part XXX. **27**: 124–137. 1900; Part XXXI. **28**: 301–313. 1901; Part XXXII. **29**: 694–704. 1902.

———. 1907. An enumeration of the plants collected in Bolivia by Miguel Bang, with descriptions of new genera and species. Part 4. Bull. New York Bot. Gard. **4**: 309–470.

———. 1910. New species from Bolivia collected by R. S. Williams. Part 1. Bull. New York Bot. Gard. **6**: 487–517; Part 2. **8**: 89–135. 1912.

———. 1927. Descriptions of new genera and species of plants collected on the Mulford Biological Exploration of the Amazon valley, 1921–1922. Mem. New York Bot. Gard **7**: 205–387.

Tosi, J., O. Unzueta, L. Holdridge & A. Gonzalez. 1975. Mapa ecologico de Bolivia. Ministerio de Asuntos Campesinos y Agropecuarios, La Paz.

Unzueta, O. 1975. Memoria explicativa: Mapa ecologico de Bolivia. Ministerio de Asuntos Campesinos y Agropecuarios, La Paz. 312 pp.

Weddell, H. A. 1855. Chloris andina. Vol. I. 232 pp., *pl.* 1–42; Vol. II. 316 pp., *pl.* 43–90. 1857.

Appendix I

A Selected Bibliography of Bolivian Botany Bibliographies

Aliaga de Vizcarra, I. 1978. Bibliografia boliviana de recursos vegetales. Academia Nacional de Ciencias de Bolivia, La Paz. 14 pp.

Cardozo, A. 1974. Bibliografia de bibliografias agricolas bolivianas. Ed. 2 Bogotá. 18 pp.

Flick, F. J. 1952. Bolivia. Pages 42–44 *in* The forests of continental Latin America; a bibliography of selected literature, 1920–1950. U.S.D.A. Biblio. Bull. 18.

Huber, O. 1974. Bolivia. Pages 756–766 *in* Las sabanas neotropicales. Instituto Italo-latinoamericano, Roma. Serie Tecnico-Cientifico 3.

Meneces, E. 1975. Bibliografia de las palmeras de Bolivia. Revista Soc. Hist. Nat. (Cochabamba) **2(2):** 18–21.

Muñoz, J. 1967. Bibliografia geografica de Bolivia. Publ. 16. 171 pp.

Floras/Checklists/Mixed Treatments

Asplund, E. 1926. Contributions to the flora of the Bolivian Andes, I. Pteridophyta, Gymnospermae, Helobiae. Ark. Bot. **20:** 1–38.

Cardenas, M. 1932. Contribución al estudio de la flora del sud de Bolivia. Plantae Potosinae, Catalago I. Potosi. 26 pp.

Cogniaux, A., A. Lingelsheim, F. Pax & H. Winkler. 1910. Plantae novae bolivianae, IV. Repert. Spec. Nov. Regni Veg. **8:** 1–6.

Cuatrecasas, J. 1935. Plantae Isernianae. Anales Ci. Univ. Madrid **3:** 206–265.

Diaz Romero, B. 1919. Florula pacensis: Descripcion de las plantas indigenas y exoticas que existen en la hoya de la Paz. Bol. Direc. Nac. Estad Estud. Geogr. Epoca 2. **25,26,27:** 40–87; **28,29,30:** 30–76; **31,32,33,:** 27–76, 1920.

Fries, R. E. 1906. Zur kenntnis der Phanerogamenflora der grenzgebiete zwischen Bolivia und Argentinien. II. Malvales. Ark. Bot. **6(2):** 1–16, *pl.* 1–2; III. Einige gamopetale Familien. Ark. Bot. **6(11):** 1–32, *pl.* 1–4, 1097; IV. Einige Choripetale und Monokotyledone Familien. Ark. Bot. **8(8):** 1–51, *pl.* 1–2, 1909.

Hemsley, W. B. & H. Pearson. 1901. On a small collection of dried plants obtained by Sir Martin Conway in the Bolivian Andes. J. Linn. Soc., Bot. **35:** 78–90.

Herzog, T. 1909a. Siphonogamae novae Bolivienses in itinere per boliviam orientalem ab auctore lectae. Repert. Spec. Nov. Regni Veg. **7:** 49–69.

———. 1909b. Nachtrage zu Siphonogamae Novae Bolivienses. Repert. Spec. Nov. Regni Veg. **7:** 354–359.

Hieronymus, G. 1896. Plantae steubelianae novae, quas descripsit adjuvantibus aliis auctoribus. Bot. Jahrb. Syst. **21:** 306–378.

Kempff, N. 1976. Flora Amazonica Boliviana. Academia Nacional de Ciencias de Bolivia, La Paz. 71 pp.

Kranzlin, F. 1915. Novitiae quaedam Bolivianae. Repert. Spec Nov. Regni Veg. **13:** 117–120.

Kuntze, O. 1898. Revisio generum plantarum, III(2). **33:** 1–384.

Lingelsheim, A., F. Pax & H. Winkler. 1909. Plantae novae bolivianae. Part II. Repert. Spec. Nov. Regni Veg. **7:** 107–114; Part III. **7:** 241–251.

Pax, F. 1908. Einige neue Pflanzen der bolivianischen Flora. Repert. Spec. Nov. Regni Veg. **5:** 225–227.

Pena, R. 1901. Flora Curcena. Imprenta Bolivar de M. Pizarro. Sucre. 287 pp.; Ed. 2. Ministerio de Educación, Bellas Artes y Asuntos Indigenas. Biblioteca Boliviana 3a, Ser. Editorial del Estado. La Paz. 475 pp. 1944; Ed. 3. Universidad 'Gabriel Rene Moreno'. La Paz. 371 pp. 1976.

Perkins, J. R. 1913. Beitrage zur Flora von Bolivia. Bot. Jahrb. Syst. **49:** 170–233.

Presl, K. B. 1825–1835. Relique Haenkeanae. Vol. 1, 356 pp. 1825–1830; Vol. 2, 152 pp. 1831–1835.

Remy, J. 1846. Analecta boliviana: pars I. Ann. Sci. Nat. Bot. Ser. 3, **6:** 345–357, *pl.* 20; Pars II. **8:** 224–240, 1847.

Rusby, H. H. 1893. New genera of plants from Bolivia. Bull. Torrey Bot. Club **20:** 429–434, *pl.* 167–170.

———. 1894. Two new genera of plants from Bolivia. Bull. Torrey Bot. Club **21:** 487–489, *pl.* 223–224.

———. 1920. Descriptions of three hundred new species of South American plants. (Privately printed by author). 170 pp.

———. 1922. New species of trees of medical interest from Bolivia. Bull. Torrey Bot. Club **49:** 259–264.

———. 1934. New species of plants of the Ladew Expedition to Bolivia. Phytologia **1:** 49–80.

Schultz-Bipontinus, C. 1865. Premiere liste des plantes des Andes Boliviennes recueilles et distribuées. Bull. Soc. Bot. France **12:** 79–82.

Smith, A. C. 1950. Studies of South American plants, part 12. Contrib. U.S. Natl. Herb. **29:** 317–393.

Solms-Laubach, H. G. 1907. Uber eine kliene Suite hochandiner Pflanzen aus Bolivien de Prof. Steinmann von seiner Reise in Jahre 1903 mitgebracht hat. Bot. Zeitung **7:** 119–138, *pl.* II.

Urban, I. 1906. Plantae novae andinae imprimis weberbauerianae. Part I. Bot. Jahrb. Syst. **37:** 373–462; Part II. **37:** 503–646; Part III. **40:** 225–395; Part IV. **42:** 49–117. 1908.

Miscellaneous

Cardenas, M. 1969. Manual de plantas econimicas de Bolivia. Cochabamba: Imprenta Icthus. 421 pp.

Organización de Estados Americanos (O.E.A.). 1965. Lista de especies de fauna y flora en vias de extincion en los estados miembros de Argentina, Bolivia y Ecuador. *In* Conferencia expecializada latinoamericana sobre problemas de conservacion de recursos naturales en el continente. Doc. 16. 8 pp.

Ravenna, P. 1978. Neotropical species threatened and endangered by human activity in the Iridaceae, Amaryllidaceae and allied bulbous families. Pages 257–266 *in* G. T. Prance, & T. S. Elias, eds. Extinction is forever. New York Botanical Garden, New York.

Section Three: Appendices— Collection Methods and Historical Reviews

Collection and Preparation of Bark and Wood Samples

Ben J. H. ter Welle

I. Introduction

Inventories of tropical vegetations are usually carried out by botanists working on systematic studies, or ecologists dealing with the composition of a given vegetation. They both use the material collected to increase the knowledge in their particular field. Generally, the proper names for the taxa collected is the first goal of an inventory. The value of the material collected is directly linked to the use that is made of the information gathered through a proper identification and the systematic position of the taxa. This can be achieved with the information provided by the description of the morphology of the flowers, fruits, leaves and habit, and the information from the labels for the herbarium vouchers on location, ecology, etc.

However, much more valuable information can be obtained if other disciplines can be involved, such as pollen morphology, wood and bark anatomy, karyology, seedlings, etc. Unfortunately, for these disciplines specific material has to be collected to provide the desired data. Although all these disciplines have proven their value in the past, sampling proper specific material is often still neglected.

There are two main reasons why this neglect exists. Collectors either are insufficiently aware of the value of such material, or, if they realize the value, they often think that making wood, bark or other collections is very difficult or time consuming. Knowing that, botanists from institutes where, for instance, wood anatomists or karyologists are also employed, always collect such samples, proving that neither reason holds in every case.

In view of the fact that the conservation of botanical diversity in the world should be attempted, at least through museum collections, it is strongly recommended that only complete collections, including the specific samples mentioned here, be made.

II. Bark Samples

The value of bark anatomy, especially in systematic studies, is striking. A wonderful example is the recent publication by Richter (1981) on the Lauraceae. The family is infamous for the difficulty in delimiting the individual genera. However, bark characteristics are very useful in this respect. As a matter of fact, Richter's work has led to the establishment of a new genus, and to a much better understanding of the classification of the genera.

A. Method

Making bark samples is very simple. A sample of approximately 1.5 × 1.5 × 1.5 cm is sufficient, unless the bark is very thick. A sample should include a small piece of wood, as the cambial area might provide some extra information. The samples should be collected in 70% alcohol or FAA.

III. Wood Samples

Since Solereder (1899) published his work on the anatomy of the dicotyledons, wood anatomy has been used for several goals. The accumulated knowledge in this field was compiled by Metcalfe and Chalk (1950), and updated since then (1983). Wood anatomical characters can be used in systematic studies; phylogenetic studies (using the phylogenetic trends established in 1957 by Bailey and his students); in ecological studies, e.g., dealing with the adaptations of the secondary xylem under various circumstances (Baas, 1973; Carlquist, 1975); and for identification purposes.

Finally, wood anatomy is used as an important tool for the interpretation of the performance of the various timber species in industry, furniture, house construction, etc.

A. Methods

Wood samples should contain adult wood. This is the most important factor for the selection of a wood sample. As species variously grow into big trees, shrubs, lianas, treelets, or herbs with only a very limited woody base, the diameter or size of wood sample is less important. In general, a wood sample of about 12 cm long × 8 cm wide × 4 cm thick is sufficient. This applies especially for trees. However, such a sample can be taken from a branch or the main trunk. The anatomy of a branch deviates from that of the tree trunk. Therefore, a sample from the trunk is much more valuable. In order to avoid unnecessary destruction of trees, a sampling such as the one illustrated in Figure 1 is advised.

This technique can be carried out using an axe or machete, following the sequence of cuts 1, 1a, 2, 2a, and 3, as seen in Figure 1. This sequence is essential, since collapsing or splitting of the wood sample occurs if one starts with 1a or 2a, for example. The thickness of about 4 cm includes the bark.

For treelets, shrubs and herbs with a woody base, a sample of about 12 cm long from the main trunk is sufficient.

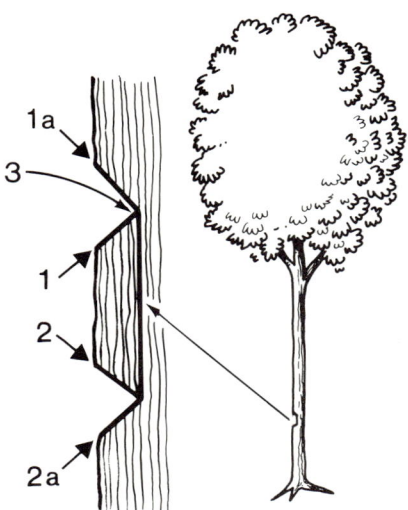

If the diameter of the trunk is over 6 cm, the sample can be split in the axial direction through the pith. Finally, for lianas the same length of sample is needed. If possible, both a sample from the trunk near ground level, and one from the canopy should be collected.

Wood samples can be preserved without any specific treatment. Solar heat is all right, as well as various types of stoves. Do not dry them too fast. Fungal attack is not a problem at all, since the anatomy of the wood is hardly affected. Fungi occur only when the sample is wet or green.

For a restricted number of families, preservation of a small sample of the same size as the bark samples, in 70% alcohol or Formalin-Acetic Acid-Alcohol (FAA), is useful. These families, like some taxa of the Leguminosae, Loganiaceae, Nyctaginaceae, Menispermaceae, Solanaceae and Urticaceae have internal or included phloem. Included phloem (= interxylary phloem) is phloem embedded in the secondary xylem of a stem or root. It is completely destroyed during drying.

Sometimes botanists are willing to collect wood samples, but they do not know what to do with them afterwards. Wood collections are found all over the world. An Index Xylariorum was published by Stern (1978). Copies are available through him (Department of Botany, University of Florida, Gainesville, Florida 32611), or the International Association of Wood Anatomists (office at the Institute of Systematic Botany, Utrecht, The Netherlands). The I.A.W.A., founded in 1931 to advance the knowledge of wood anatomy in all its aspects, is also helpful in suggesting museums to store material and to find wood anatomists that might be interested in collaborating in special fields.

Requests for collaboration with wood anatomists in the fields of taxonomy, systematics and/or phylogeny in a certain group or family can be published in the I.A.W.A. Bulletin. Through the Bulletin over 500 wood anatomists in more than 60 countries can be reached. These requests can be sent to the editor (Dr. P. Baas, Rijksherharium, P.O. Box 9514, 2300 RA Leiden, The Netherlands, or to the Executive Secretary of the I.A.W.A., Ben J. H. ter Welle, Inst. Syst. Botany, Heidelberglaan 2, 3584 CS Utrecht, The Netherlands).

IV. Literature Cited

Baas, P. 1973. The wood anatomical range in *Ilex* (Aquifoliaceae) and its ecological and phylogenetic significance. Blumea 21: 193–258.

Bailey, I. W. 1957. The potentials and limitations of wood anatomy in the study of phylogeny and classification of angiosperms. J. Arnold Arbor. 38: 243–254.

Carlquist, S. 1975. Ecological strategies of xylem evolution. University of California Press, Berkeley.

Metcalfe, C. R. & L. Chalk. 1950. Anatomy of the dicotyledons. Clarendon Press, Oxford.

———— & ————. 1983. Anatomy of the dicotyledons. 2nd ed. Vol. II. Wood structure and conclusions of the general introduction. Clarendon Press, Oxford.

Richter, H. G. 1981. Anatomie des sekundären Xylems und der Rinde der Lauraceae. Sonderbände des Naturwissenschaftlichen Vereins Hamburg 5. Verlag Paul Parey, Hamburg.

Solereder, H. 1899. Systematische Anatomie der Dicotyledonen. Verlag Ferdinand Enke, Stuttgart.

Stern, W. L. 1978. Index Xylariorum. Institutional wood collections of the world. 2. Taxon 27(2/3): 233–269.

Collection and Preparation of Karyological Samples

W. Morawetz

I. Introduction

One of the most important tools for plant systematics is karyology. Chromosome numbers are the basic information. In many cases, however, they allow only the interpretation of generic or subfamilial relationships (e.g., Lecythidaceae: Kowal et al., 1977). Further karyological parameters such as karyotype architecture, condensing behavior, interphase nuclei, etc. are characters which may indicate relationships between families and sometimes keep stability at the ordinal level (Ehrendorfer, 1970; Goldblatt & Briggs, 1979; Morawetz, 1981b; Okada, 1975). Doubtless, the best results are obtained with cultivated material. Nevertheless, in most cases field fixations allow accurate chromosome counts (Annonaceae: Morawetz, 1984a, 1984c; several plant families: Fig 1a, b from Morawetz, 1985) or even the application of more refined methods like C-banding (*Porcelia*: Morawetz, 1984b). Our experience indicates that 80% of field fixations allow accurate counts, while 20–50% are useful for interpretation of interphase nuclei or banding patterns. The fixation and preparation method was selected from a wide range of different methods, and allows one to obtain well-contrasting staining of very small chromosomes with low DNA content. Fixations can be stored up to 20 (!) years without loss of quality. The resultant permanent slides can be used for purposes of comparison over many years. The methods were tested in about 600 taxa of 50 families of tropical (mostly woody) plants by the author.

Every botanical institute or herbarium, especially those in tropical countries, is welcome to communicate with our research group concerning collaboration in karyosystematic studies. Monographers in particular should find such collaboration valuable. The basis for a fruitful collaboration is only the receipt of well-fixed material (as described below) and/or viable seeds or seedlings from tropical woody plants growing wild, along with the corresponding vouchers. In most cases results can be published quickly, with the collector as co-author.

II. Methods

A. Fixation Liquid

Specimens should be fixed and stored with Carnoy's solution (96% alcohol: glacial acetic acid, 3:1). The fixation fluids should always be mixed immediately before use.

B. Selection of Materials

1. Choose flower buds (3 to 8 units) of a maximum size half as large as the flower in anthesis; the petals can be removed, and the gynoecium should be cut in slices (1 to 3 mm) with a razor blade before fixation.
2. Sprouting leaf buds (10 to 15), at a maximum 1 cm long, should be cut only once longitudinally; young leaves (maximally 1.5 cm long) should be cut vertically to venation.
3. Fruits and fruitlets (3 to 5) can be fixed as they can be easily cut with a razor blade into slices.
4. Root tips, if available, can be fixed from seedlings and young plants; they must not be cut.
5. If possible, flower buds, leaf buds and fruitlets should be fixed simultaneously.

C. Fixation and Storing

The ratio of material to fixation liquid should be at least 1:5, i.e., enough fixation liquid should be used. The fixed material should be sent as soon as possible by airmail to ones institute where it should be kept in a freezer. If it is not possible to do that, it should be stored until it can be transported in a freezer or refrigerator. Usually, the fixations remain useful for 2 to 8 weeks at room temperature (18°–25°C).

D. General Remarks

If easy to do, one individual should be fixed on different days and/or different times. Fixations from one species at different localities (e.g., Amazonas and Surinam) are of interest because polyploidy may occur. Every fixation should represent only one individual and should have an individual number and voucher.

E. Preparation

Best results come from root tips, young ovules, leaf primordia and young anthers for meiosis. For mitotic counts, the plates must be selected carefully because of frequently occurring endopolyploidy. The preparation method follows that of Guerra (1983; see also Morawetz, 1981a): the material is washed in distilled water for 10 to 30 minutes then incubated in 5 M HCl for 20 min. at room temperature and afterwards washed in distilled H_2O for some minutes; very small pieces of meristematic tissue are selected (maximum 1 mm^2, and less material is better) and squashed in 45% ace-

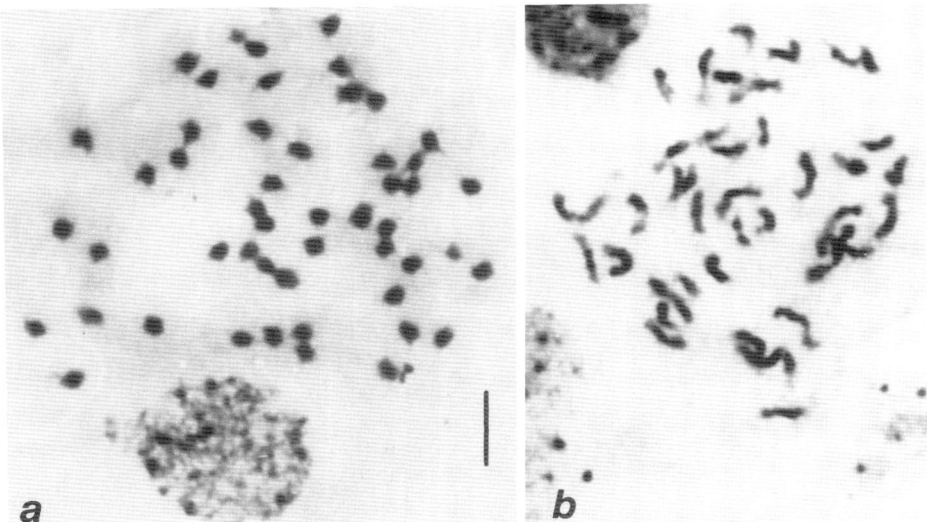

FIG. 1. Chromosome preparations from field fixations (HCl/Giemsa method). a. *Virola elongata* (Benth.) Warb. (Morawetz et al. 31-7983, INPA, WU); fully contracted metaphasic chromosomes, 2n=52; material from leaf primordium. b. *Hortonia angustifolia* Trim. (Morawetz 41-9284, WU); prometaphasic chromosomes, 2n=38; note the proximally condensed chromatin and the ± light colored distal ends; material from minute flower buds. Bar (for a, b): 5 µm.

tic acid; the preparation can be controlled under a phase contrast microscope if well-spread metaphasic plates (or meiotic chromosomes) are present; if the preparation is satisfactory, it can be put on dry ice and the cover-slip prized off (following Conger & Fairchild, 1953). Air-dried preparations are stained with 2-4% Giemsa solution in distilled water, air dried, and mounted in Euparal. If the cytoplasm stains too dark, the preparations can be incubated before staining in 45% acetic acid at 60°C for 10 to 15 minutes. *Note*: in many tropical field fixations, one well-spread metaphasic plate per slide is a good result!

III. Literature Cited

Conger, A. D. & L. M. Fairchild. 1953. A quick freeze method for making smear slides permanent. Stain Technol. 28: 281–283.

Ehrendorfer, F. 1970. Chromosomen, Verwandtschaft und Evolution tropischer Holzpflanzen, I. Allgemeine Hinweise. Oesterr. Bot. Z. 118: 30–37.

Goldblatt, P. & B. G. Briggs. 1979. Chromosome number in two primitive dicots, *Xymalos monospora* (Monimiaceae) and *Piptocalyx moorei* (Trimeniaceae). Ann. Missouri Bot. Gard. 66: 898, 899.

Guerra, M. 1983. O uso de Giemsa na citogenetica vegetal – comparação entre a coloração simples e o bandeamento. Ci. & Cult. 35: 190–193.

Kowal, R. R., S. A. Mori & J. A. Kallunki. 1977. Chromosome numbers of Panamanian Lecythidaceae and their use in subfamilial classification. Brittonia 29: 399–410.

Morawetz, W. 1981a. Karyologie and morphologisch ökologische. Variation von *Peumus boldus* (Monimiaceae, Laurales). Pl. Syst. Evol. 138: 157–173.

———. 1981b. Zur systematischen Stellung der Gattung *Prockia*: Karyologie und Epidermisskulptur im Vergleich zu *Flacourtia* (Flacourtiaceae), *Grewia* (Tiliaceae) und verwandten Gattungen. Pl. Syst. Evol. 139: 57–76.

———. 1984a. Karyological races and ecology of *Duguetia furfuracea* as compared with *Xylopia aromatica* (Annonaceae). Flora 175: 195–209.

———. 1984b. How stable are genomes of tropical woody plants? Heterozygosity of C-banded karyotypes of *Porcelia* as compared with *Annona* (Annonaceae) and *Drimys* (Winteraceae). Pl. Syst. Evol. 145: 29–39.

———. 1984c. Karyologie, Ökologie und Evolution der Gattung *Annona* in Pernambuco, Brasilien. Flora 175: 435–447.

———. 1985. Remarks on karyological differentiation patterns in tropical woody plants. Pl. Syst. Evol. 152: 49–100.

Okada, H. 1975. Karyomorphological studies of woody Polycarpicae. J. Sci. Hiroshima Univ., Ser. B., Div. 2, Bot 15(2): 115–200.

Collecting and Germinating Fern Spores

M. C. Roos and G. P. Verduyn

I. Introduction

Ferns are peculiar amongst the vascular plants because of their life-cycle consisting of two free-living generations. The sporophyte, the 'fern' everybody is familiar with, is the asexual, diploid generation. Sporangia are situated on the lower side of mature fronds (in most ferns) which, after meiosis, form halpoid spores. These spores germinate, forming the sexual haploid generation (viz., the gametophytes). On the lower side of mature gametophytes, sex organs (the antheridia and/or archegonia) are present. In the antheridia, antheridiozoids (sperm) are formed which fertilize the single egg present in a ripe archegonium. The zygote first grows parasitically into a young sporophyte which soon forms a new, free-living generation. See, for example, Sporne (1979) for a general survey on ferns, and Tryon and Tryon (1982) for a detailed account on American ferns.

Living ferns for cultivation can be obtained in two different ways. The first one is, not surprisingly, to collect the living plants themselves. This can be done quite easily in the case of (tropical) ferns with creeping rhizomes (Fig. 1; e.g., Polypodiaceae, Davalliaceae). A piece of rhizome of ca. 1 dm, including the rhizome tip and a few full-grown fronds, will do. Young circinate fronds (the fiddleheads), as well as dead and broken fronds, must be removed. Kept in shady places in moist air, such rhizomes can survive for several weeks. For transportation they only have to be wrapped in moist peat moss to prevent them from drying out completely. It is important to avoid rotting, which can be caused by keeping them in plastic bags.

There is another very easy, and often overlooked, way to obtain living ferns, viz., to collect spores. Spores are formed in sporangia (Fig. 2), as noted above, which are usually located in clusters (sori, soral patches) on the lower frond surface. These clusters are occasionally situated on specialized (fertile) fronds (e.g., Deer Fern, *Blechnum spicant*) or modified frond parts (e.g., Pine Fern, *Anemia adiantifolia*) and may be covered with a membrane (indusium). Compare the Male Fern (*Dryopteris filix-mas*) to the Common Polypody (*Polypodium vulgare*), for example. The sori are variably sized and shaped. They are often round or oblong, as in the Common Polypody, Male Fern, or Lady Fern (*Athyrium filix-femina*), or, less commonly, linear, as in the Cretan Brake (*Pteris cretica*). Occasionally, sporangia cover large parts of the lower surface of the fronds, as with the soral patches of Staghorn Ferns (*Platycerium*).

Using a 10x hand lens, it is possible to examine the maturity of the sporangia and thus of the spores. Young sporangia are white, green, or light brown. If present, the indusium is green. When the sporangia ripen, they are swollen and turn yellow, orange, or shiny brown. The indusia are still intact, firm, and variously colored (not green). Sporangia that have already shed their spores are dull brown, open, and frayed-looking. The indusium is either shed or shriveled.

Fertile fronds are suitable for collecting spores when the sporangia are mature (or nearly so), especially when a number of sporangia have already opened and the laminae are covered with yellow, green or dark-colored spore-dust. It is necessary to collect a complete fertile frond, or at least, in the case of large-sized plants, a large part of it in order to get a sufficient amount of spores. Furthermore, it is recommended that one collect a complete herbarium specimen (rhizome, sterile fronds, fertile fronds) for identification and voucher purposes. Otherwise, the plant must be identified using the collected fragments only, which may not be sufficient.

II. Spore Preservation

To get good results, place the collected fronds (or parts) in a paper pouch in an envelope and keep it in a dry condition, preferably in air currents. The cover will prevent the spores from blowing away. After one or two days most of the spores will have been shed.

Do not forget to note the collector's name, collecting number, date of the specimen, locality, and other important or striking information (e.g., "growing on rocks in gallery forest," "epiphytic on *Quercus*").

Green spores, such as those present in a relatively small number of species (e.g., Royal Fern, *Osmunda*, and Cinnamon Fern, *Grammitis*), must be sown very soon after having been collected. These spores remain viable only for a period of several hours or days. This period can be extended by keeping them in the refrigerator. Yellow or dark spores, which are present in the greater number of fern species, remain viable for much longer periods, even up to several years. However, in all cases, the older the spores are, the more the percentage of germination decreases.

The collecting of fern spores can be done easily simultaneously with the collecting of herbarium specimens. Spores take up very little space and can be easily transported. Moreover, most botanical gardens have suitable supplies and conditions for the cultivation of ferns from spores.

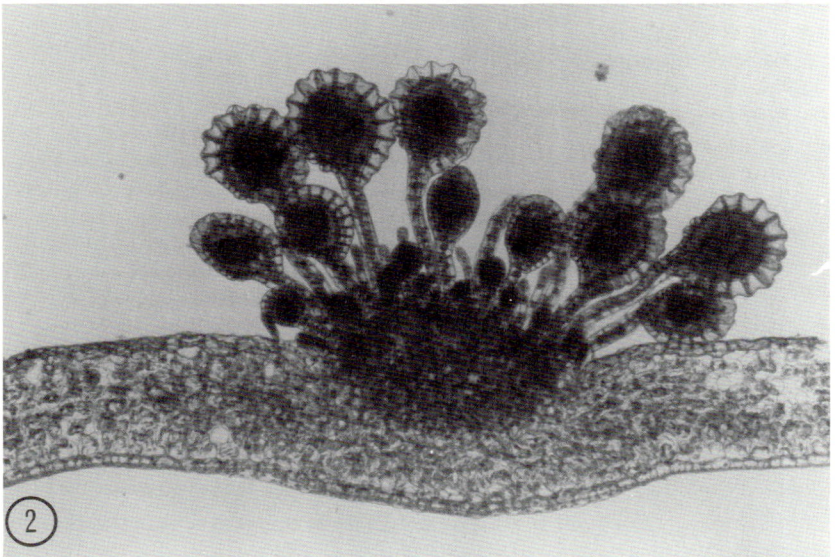
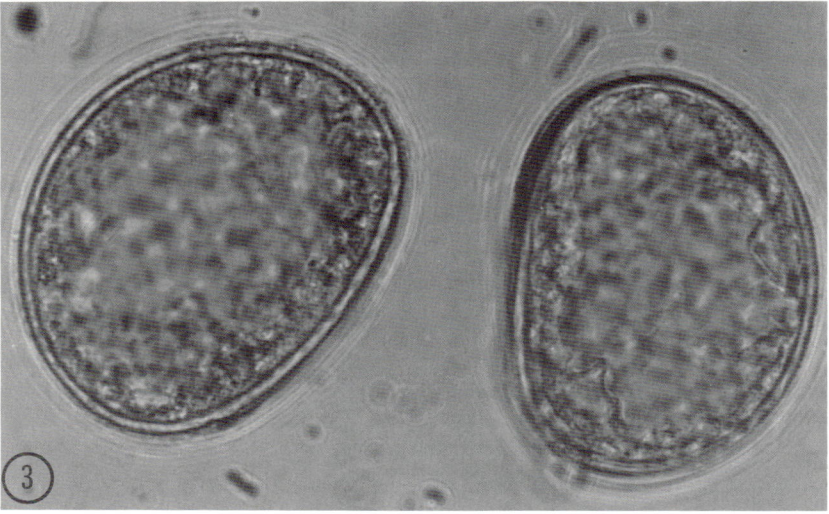

FIGS. 1–3. 1. *Aglaomorpha cornucopia*; creeping rhizome with young circinnate frond (x2). 2. *Crypsinus taeniatus*; sorus in cross-section (x60). 3. *Pessopteris crassifolia*; spores (x1000).

III. Spore Germination and Cultivation

The spores (Fig. 3) which have been shed from the sporangia are visible on the paper pouch by use of a hand lens or binocular dissecting microscope. In our laboratory, they are sown on sterile nutrient agar plats after removing the sporangia stalks and heads by a gentle tapping on the paper. The spores stick to the paper, whereas the often contaminated bigger structures fall off. The agar dishes are placed in a climate room with a constant temperature of about 20°C and sufficient illumination (e.g., by fluorescent tube).

The spores will germinate between about five days to one month after sowing. The first step is the formation of a rhizoid which grows into the agar, followed by the formation of a small thread of cells which develops into an often heart-shaped prothallium (the gametophyte). On the underside of the prothallium the sex organs are formed: the antheridia, which produce the sperm, and the archegonia, which produce an egg each. If a sufficient amount of water is present at this stage, the swimming sperm may reach the eggs and fertilization can take place.

When the gametophytes are three to eight mm in cross-section, and the sex organs are present, they can be removed from the agar dishes to sterilized fern soil which consists mostly of leafmold, peatmoss, and inorganic material. The gametophytes are covered by translucent plastic to keep the humidity at a sufficient level for fertilization. After fertilization, the zygote develops into a juveline sporophyte which produces a series of juvenile fronds, the first ones often quite differently shaped from the later, mature fronds. The sporophytes can be removed to pots, baskets, or blocks made from the bark of tree ferns. For young sporophytes, a plastic coverage might be useful to prevent them from dessication.

For further information on culturing gametophytes and growing ferns, see Dyer (1979) and Hoshizaki (1979).

IV. Literature Cited

Dyer, A. F. 1979. The culture of fern gametophytes for experimental investigation. Pages 253–305 *in* A. D. Dyer (ed.), The experimental biology of ferns. Academic Press, New York.

Hoshizaki, B. J. 1979. Fern growers manual. Alfred Knopf, New York.

Sporne, K. R. 1979. The morphology of Pteridophytes. Hutchinson & Co., London.

Tryon, R. M. & A. F. Tryon. 1982. Ferns and allied plants: With special reference to tropical America. Springer Verlag, New York.

Collection and Preparation of Pollen Samples

Annick Le Thomas

I. Introduction

The study of morphology of pollen grains is as old as most other branches of botanical science, but it has been a self-standing science for only about forty years. The term palynology was proposed for the first time by Hyde and Williams (1944).

Since then, palynology has developed quickly and extensively in many ways, in both a geological and biological context. The sample collections therefore have two aspects: the preparation of samples (of sediment, hay-fever pollen or honey-pollen extracts, and atmospheric pollens) for analysis, on the one hand, and the preparation of recent material for comparative or morphological studies, on the other. For pollen analysis, studies of the technique for preparation of samples have been published repeatedly, and textbooks dealing with these technical aspects have been published by Brown (1960) and Faegri and Iversen (1975).

In botanical studies, since Wodehouse (1935) and Erdtman (1943, 1952) published their works on pollen grains, the characters of pollen have been used to advance several aims: systematics and phylogeny, ecology, biology, etc. Many techniques have been developed, especially for electron microscopy, but most often the slide collections are prepared by the classical method of Erdtman: acetolysis.

II. Methods

If naturally-shed pollen, or that gathered and fixed in the field, is not available, satisfactory material can generally be obtained from herbarium specimens. It is better to include flowers at various stages of development in the material collected. Often, it is only necessary to tap the dry flowers or crush a few anthers to collect the polliniferous material, which is transferred into a centrifuge tube and washed down with a mixture of acetic anhydride: concentrated sulfuric acid::9:1. The centrifuge tubes are then filled with the acetolysis mixture and transferred to a water bath. When the temperature has reached the boiling point (or a little less for delicate material), heating is stopped or prolonged not more than one minute. The tubes are then transferred to an electric centrifuge for five minutes at 1200 rpm. After centrifuging, the reaction maixture is decanted; glacial acetic acid then added. After centrifuging again at 1200 rpm and decanting, the pollen is washed with distilled water three times; then a mixture of glycerine: water :: 1:1 is added to the tubes for one-half hour. After centrifuging at 1200 rpm and decanting, the tubes are placed upside down on filter paper to drain.

To prepare the microscopic slides, a minute piece of glycerine jelly is fixed on a platinum needle and allowed to touch the pollen. The jelly and the polliniferous material sticking to it is then transferred to a slide, and the pollen grains, after a gradual heating, can be examined at low magnification.

For delicate pollen grains, this method is often much too drastic, and many adaptations have been described, especially for scanning electron microscopy: e.g., critical point drying, which is a method of drying tissue without collapsing or deforming the structure.

The acetolysis method allows one to observe only the exine, i.e., the external part of the sporoderm, with most of the morphological characters, since the inner part, or intine, and the cytoplasm are destroyed by the chemical treatment. To observe the intine, the Wodehouse method is used: the pollen grains are placed on the center of a microscopic slide, and one or two drops of alcohol are added and allowed to partly evaporate. A small drop of glycerine jelly is added and heated by passing the slide over a small flame. A cover-slip is then placed over the specimen and the slide gently heated.

All palynology laboratories keep slide collections of their studies or slide references for comparison, and many exchanges have been made amongst them. Nowadays, they often also have a picture collection from electron microscopy.

Twenty-two national societies of palynologists exist in the world and are federated in the International Federation of Palynological Societies (IFPS). The office address is that of the elected president at the general assembly during the last International Congress, in Brisbane (1988).

III. Literature Cited

Brown, C. A. 1960. Palynological techniques. C. A. Brown, Baton Rouge, Louisiana.
Erdtman, G. 1943. An introduction to pollen analysis. Chronica Botanica, Waltham, Massachusetts.
———. 1952. Pollen morphology and plant taxonomy. Almqvist & Wiksell, Stockholm
Faegri, K. & J. Iversen. 1975. Textbook of pollen analysis, Ed. 3. Blackwell Scientific Publications, London.
Hyde, H. A. & D. A. Williams. 1944. The right word (letter to Paul B. Sears, dated July 15, 1944). Pollen Analysis Circular no. 8, p. 6, Oct. 28, 1944. Mimeograph. Oberlin, Ohio.
Wodehouse, R. P. 1935. Pollen grains. McGraw-Hill Book Company, New York.

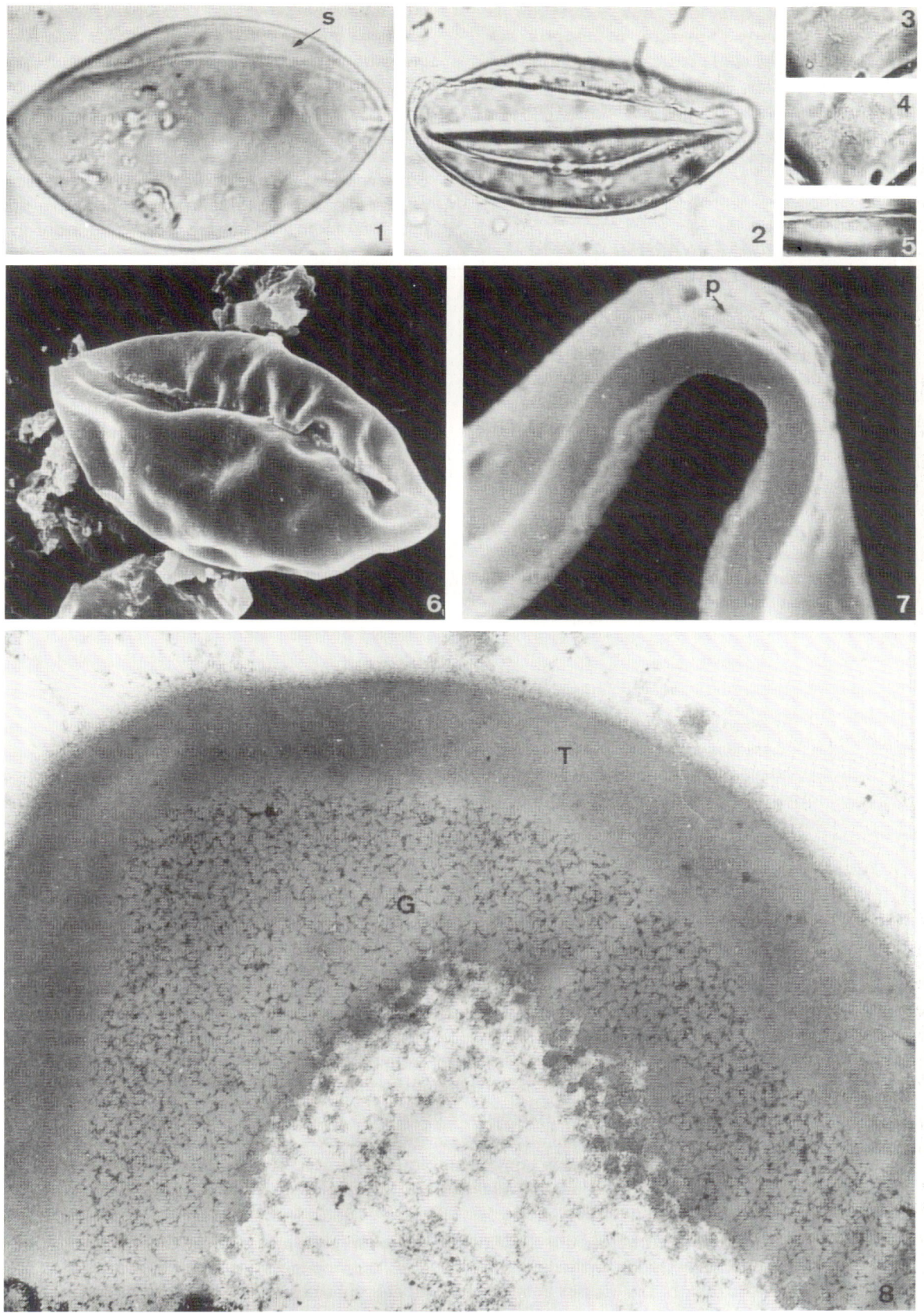

FIG. 1. An example of the three microscopic techniques used in palynology. *Piptophyllum calophyllum* (Annonaceae). 1–5. Mph; 6–7. MeB; 8. Ultrasection in MeT. (Extrait des Annales de Mines de Belgiques, 1978.)

Collecting Tropical Plant Germplasm

Michael J. Balick

I. Introduction

Along with the diversity in numbers of plant species in the tropical rainforest, there is also a vast diversity of uses and potential uses for many plants. A number of economically important cultivated crops have wild relatives found only in rain forests, and these related species possess many valuable genes, such as for disease and insect resistance, improved nutrition, better agronomic habit, drought tolerance, etc. Many other species are used by local people for a variety of subsistence and commercial purposes. It is possible to say that, from the standpoint of utilization, the flora in many regions of the Tropics is more unknown than known to botanical scientists, and that discoveries of potentially great consequence to all of humankind still await us in the future.

Given the heavy rate of deforestation mentioned elsewhere in this volume, many of these potential contributions will never be possible, as the plants, and the human cultures which utilize them, will be destroyed. Thus, it is essential to incorporate into our inventory programs the collection of genetic material (germplasm) of as many useful or potentially useful tropical plants as possible, and to safeguard this genetic material in a viable form for the future. Unfortunately, even with the best facilities, germplasm is often lost after being collected. Ultimately, then, the only sure way of preserving the diversity of tropical forest plant life for future utilization is to preserve the forest itself.

II. Methods

Ayad (1980) in his glossary of plant genetic resources terms, defined plant germplasm as the "sum total of the genetic material in a plant." For the purposes of this contribution, however, germplasm can be defined as a living physical unit containing the genetic composition of a particular organism with the ability to be reproduced. Thus, germplasm can be collected and stored as seeds, plants, cuttings, pollen, and/or tissue culture, depending on the species.

In recent years the International Board for Plant Genetic Resources (IBPGR) has taken global responsibility for the coordination of germplasm collection and repository efforts, and for resolving the problem of the loss of genetic diversity of economic plants. Plants have been preserved on a priority basis, and by 1981, fifty crops were assigned "first priority" status (IBPGR, 1981). Most of the priority species are the major and minor cultivated crops, and a smaller proportion of the organization's resources have been devoted to the so-called underexploited plants (National Academy of Sciences, 1975). Work with these species has proceeded at a slower pace, and has been the research responsibility of universities, botanical gardens and governmental institutions in the Tropics.

The usual procedure is to send out germplasm expeditions with a specific mission, led by knowledgeable specialists in the particular group under study. During the expedition, other germplasm can be collected and brought back, assuming that proper storage or growth facilities are available.

While the germplasm centers for the major crops are relatively well funded, those for lesser-known species often are not. A germplasm bank has a wide range of responsibilities, which require physical infrastructure, a skilled staff and stable financial support. Ellis (1985) divided these responsibilities into four main tasks:

1. Banking: storing the accessions and keeping a current inventory
2. Monitoring: taking care of the accessions according to proper procedures
3. Regeneration: reproducing fresh material from all the accessions in the collection
4. Distribution: sending out living material to those who request it.

Problems in germplasm banking still exist. While the collecting expeditions are relatively well funded, material that might be transferred back to an agronomic station or tissue culture laboratory does not always survive due to problems in shipping, handling upon receipt or with long-term maintenance of the material itself. A simple occurrence such as a power failure in a tropical country might, within a few hours' time, result in the destruction of important plant cultures. Or, an unpredicted drought or labor problem could result in the loss of a complete nursery or recently planted field or important genetic material. However, in the past decade the situation has improved, at least for the major crops, and funding agencies have provided resources to the international germplasm centers that work with these agencies.

The IBPGR has published directories of germplasm collections, such as the one on tropical and subtropical fruits and tree nuts by Gulick and van Sloten (1984). Using the species index, researchers can locate germplasm collections for hundreds of fruits and nuts around the world.

The tropical biologist is therefore presented with an opportunity to collect and distribute germplasm of plants seen

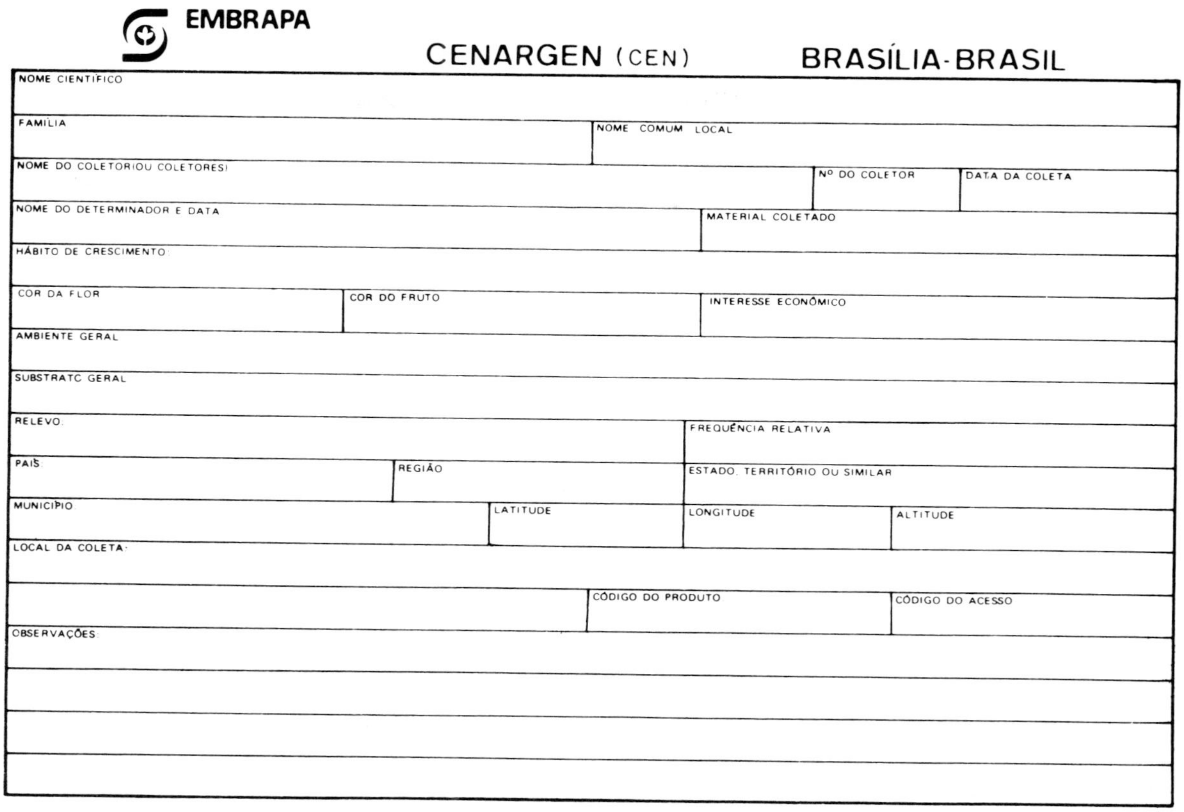
EMBRAPA

CENARGEN (CEN) BRASÍLIA-BRASIL

NOME CIENTÍFICO

FAMÍLIA | NOME COMUM LOCAL

NOME DO COLETOR(OU COLETORES) | Nº DO COLETOR | DATA DA COLETA

NOME DO DETERMINADOR E DATA | MATERIAL COLETADO

HÁBITO DE CRESCIMENTO

COR DA FLOR | COR DO FRUTO | INTERESSE ECONÔMICO

AMBIENTE GERAL

SUBSTRATC GERAL

RELEVO | FREQUÊNCIA RELATIVA

PAÍS | REGIÃO | ESTADO, TERRITÓRIO OU SIMILAR

MUNICÍPIO | LATITUDE | LONGITUDE | ALTITUDE

LOCAL DA COLETA

CÓDIGO DO PRODUTO | CÓDIGO DO ACESSO

OBSERVAÇÕES

FIG. 1. Germplasm collection format, reproduced from the field books used by CENARGEN/EMBRAPA in Brazil.

during fieldwork, plants which otherwise might not be preserved in collections. This is an especially attractive opportunity when working in areas that are soon to be destroyed, such as dam sites, or amongst indigenous groups who are losing their knowledge of the local flora. When appropriate, fieldwork should include a concern for, and awareness of, local useful plants.

The ideal situation would be to negotiate in advance with researchers utilizing germplasm, and obtain specific instructions for the collection, storage, and shipment of that material. Alternatively, one can also collect interesting material as it appears in the field and hope to transmit this to qualified researchers upon return to the home base. While the techniques for collection and storage of germplasm of commercially important tropical and temperate crops such as rice, wheat, corn, etc. are relatively advanced, methodology for the collection of tropical germplasm is now in an embryonic stage. At the present time there is no overall protocol for handling the quantity and variety of material of interesting and useful plants that one could possibly observe during an expedition. Within a single family, such as the palms, the seeds of some species must be germinated within a few days' time and handled quite carefully, while others will tolerate a year of storage, even treatment with fire, while still remaining viable.

Germplasm collection requires careful documentation of the location and material (accessions). For most of the major crops, descriptor lists have been written; for many of the underexploited crops, no such information exists. Basically, the descriptor list involves passport information, data on characterization and preliminary evaluation. The passport data includes accession and collection data and this is, for the most part, obtained in the field. Characterization and preliminary evaluation are done in the germplasm bank. Further characterization and evaluation can be carried out in the germplasm bank or at other sites where plant breeders are working.

Figure 1 is a page reproduced from the field books used by the Centro Nacional de Recursos Genéticos/Empresa Brasileira de Pesquisa Agropecuária (CENARGEN/EMBRAPA) in Brazil in their very extensive program of germplasm collection, characterization and evaluation. The preformatted type of field book leaves little room for error or omission when collecting data on germplasm. Booklets, such as those published on sunflower descriptors (IBPGR, 1985), coconut (IBPGR, 1978), *Capsicum* (IBPGR, 1983), or the increasing number of descriptor lists being produced for other crops, provide an excellent format for the description of germplasm, and can be modified for other crops. Probably the most detailed list of descriptors for a neotropical palm has been produced by Clement (1986) as his

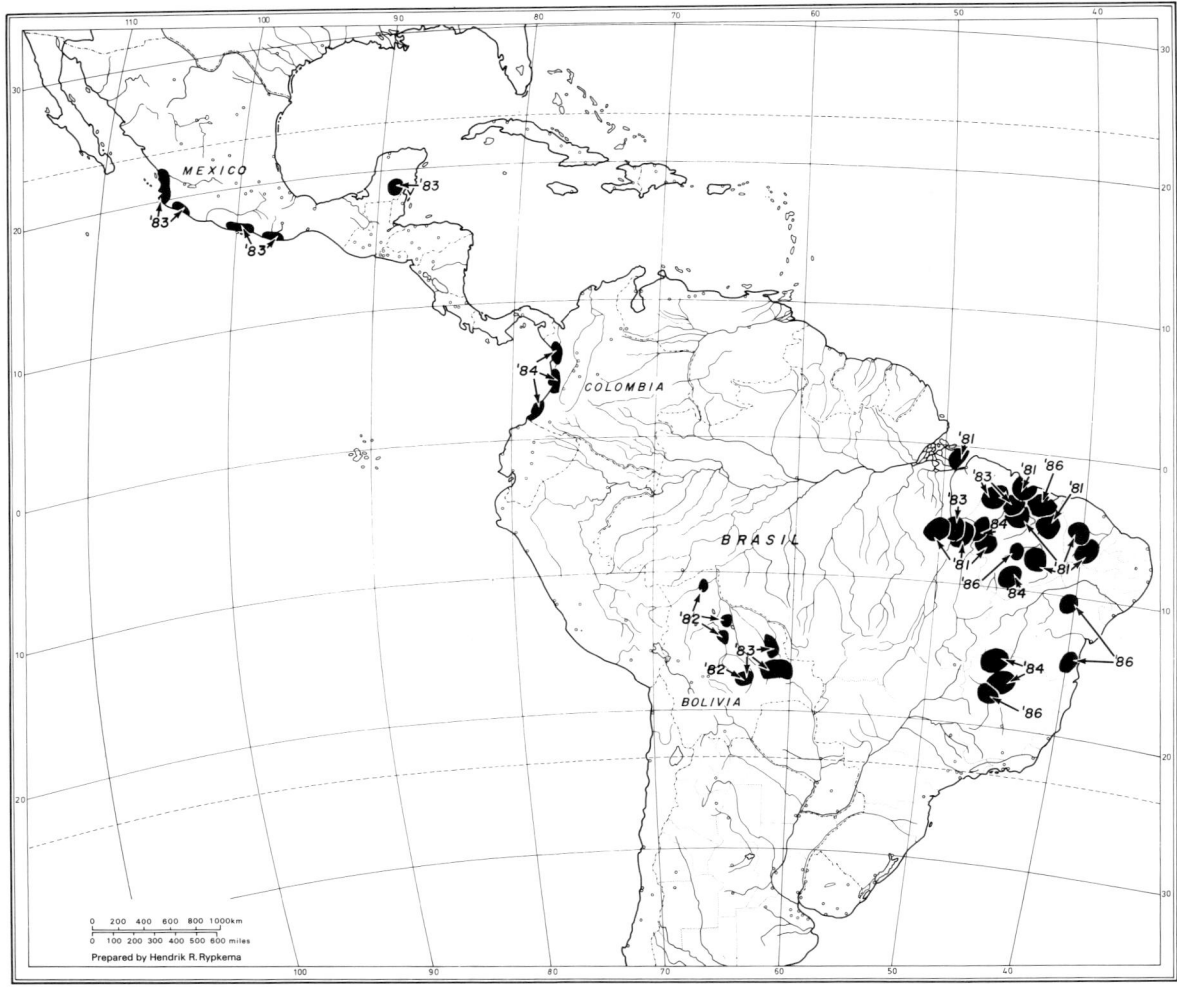

FIG. 2. Collecting sites for the babassu domestication program during 1981–1986.

masters degree research on the peach palm, *Bactris gasipaes* H.B.K.

It may be useful to discuss a specific example of germplasm collection and conservation to illustrate some of the functions of the collector and the role of the germplasm bank. The New York Botanical Garden Institute of Economic Botany has been involved in a program of domestication of the babassu palm complex (*Orbignya* spp.) since 1980. This program was initiated with the full collaboration of CENARGEN in Brazil, as well as a state institution in Piauí, Brazil, Unidada de Execução de Pesquisa de Ambito Estadual de Teresina (UEPAE de Teresina).

Under this program, one germplasm expedition is undertaken each year with joint funding from U.S./Brazilian sources. Babassu palms are a source of edible oil, charcoal, fiber, and a host of subsistence products, and comprise the largest oilseed industry in the world dependent on a wild species for total production. These palms were selected because of their commercial value, over US$ 150,000,000 annually (Pick et al., 1985), and because the entire industry is based on the harvest of wild palms.

There are two germplasm banks for babassu in Brazil, in Bacabal, Maranhão and Teresina, Piauí. The objective of the domestication program is to assemble, grow and select superior palms for utilization, either as agroforestry crops for smallholder production, or for industrialization. Since 1980, expeditions have been carried out in a number of countries (Mexico, Bolivia, Brazil and Colombia) and, as of March 1986, 369 individuals from 46 populations had been sampled, for a total of 19,360 fruits collected. These have been deposited in germplasm banks in Brazil, distributed elsewhere for field trials, and used in laboratory analyses to further domestication studies. Figure 2 is a map of the areas that have been collected during this multinational effort. Figure 3 is a flowchart of activities involved in the collection and characterization of babassu germplasm.

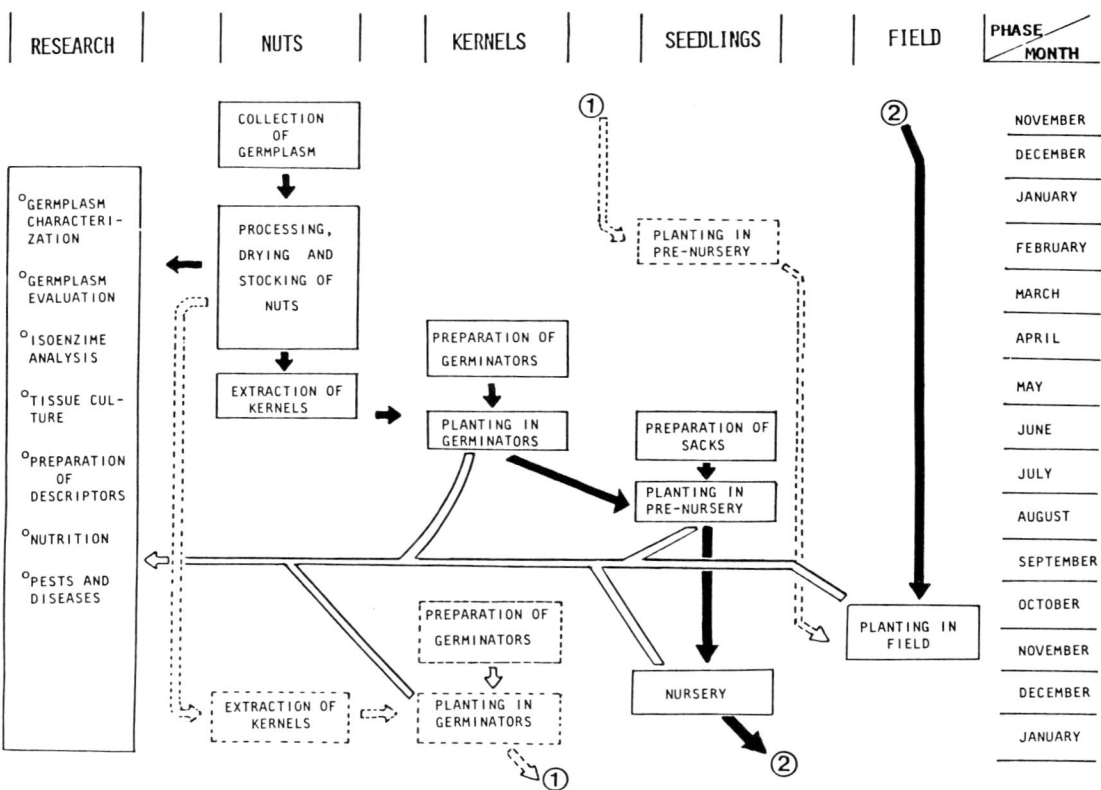

FIG. 3. Flowchart of activities involved in the collection and conservation of babassu germplasm. Black arrows: flow of material; white arrows: flow of research data; dotted arrows: alternate flow of material. Reproduced from Frazão and Pinheiro, 1986.

During these expeditions, an attempt is made to collect a minimum of 20 individual trees per population, and ca. 70 fruits per individual. This is not always possible. For example, because *Orbignya cuatrecasana* is an endangered species and difficult to locate, fewer collections were made. Additionally, as each panicle of this species contains only 10–15 fruits, it is difficult to harvest 70 fruits per tree. In general, a minimum conserved population size of 50 individuals is necessary to maintain genetic variability for genetic adaptation (Frankel, 1983).

Further details on the babassu domestication program, along with information on another effort to domesticate *Bactris gasipaes*, the peach palm, is presented in Balick (1984). Future fieldwork will be designed to collect the diversity in critical areas of palm distribution outlined in Lleras et al. (1983). This effort to domesticate useful palms is but a small part of the overall germplasm collection/conservation effort being undertaken by the Institute of Economic Botany in collaboration with local institutions at sites around the world (Fig. 4), but it is a useful example of how an international project can be developed that unites systematic biological studies with ecology, agronomic utilization and economics, with the aim of developing a better strategy for utilization of the tropical ecosystem.

III. Local Permits

One problem facing the tropical collector in many countries is that the movement of germplasm is tightly controlled both inside and outside the national boundaries. Because tropical germplasm is now recognized as an important natural resource by many governments, local laws in many countries limit or control its transmittal outside of the individual country. Other laws have been developed to limit the transfer of living material for fear of spreading disease.

Strict legislation has been passed in some countries allowing the collection of living material, but only for use within that particular country. The collector should be aware of local germplasm facilities and supply these with material whenever possible. As germplasm is usually traded by scientific organizations working on specific crops, it is often more appropriate to leave germplasm in a country where it can be conserved, utilized and distributed, than to take it outside the national boundaries for use elsewhere. It is essential that one become familiar with any local laws dealing with the exchange of plant material and adhere to them rigorously. The collection and deposition of germplasm in an institution in its country of origin is a worthy goal in itself; one does

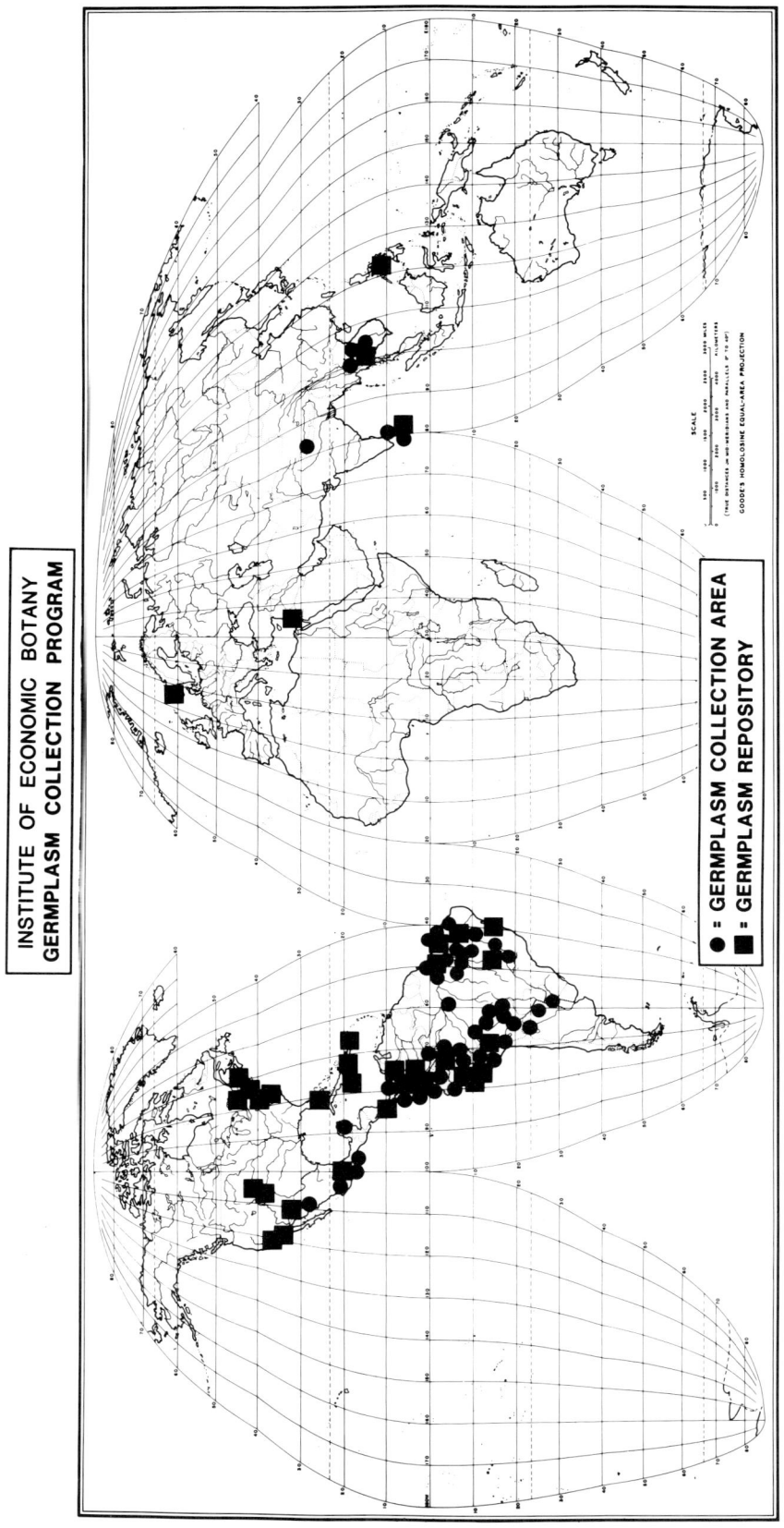

not always have to distribute it elsewhere for successful conservation to be achieved. For specific information on the collection, documentation and curation of germplasm, the reader is referred to Hawkes (1980), who discusses some of the general philosophy and methodology for crop plant germplasm collection.

IV. Conclusion

It is impossible to address all of the problems of working with tropical plant germplasm in the few pages allotted. While this is not considered the main responsibility of the tropical biologist at present, it is critical for researchers to develop a better understanding of the magnitude of the problem of the loss of tropical plant germplasm, as well as the reduction in the diversity of all tropical life. While most tropical biologists do not wish to assume the role of a full time germplasm collector, niches exist where contributions to this kind of work can and are being made. Working with local governmental agencies responsible for germplasm collection and banking is a good way to strengthen both one's own research program as well as overall institutional relations. Scientists and policymakers in tropical countries are increasingly expressing the opinion that, given the increasing rate of forest destruction, the collection of herbarium specimens and the preparation of floras and monographs is not enough. These countries, and indeed the world economy, do and will continue to require useful plants from the tropical forests for many purposes. Efforts by biologists toward meeting this need are very important and are a vital and lasting contribution to biology.

V. Literature Cited

Ayad, W.G. 1980. A glossary of plant genetic resources terms. International Board for Plant Genetic Resources (IBPGR) Secretariat, Rome.

Balick, M. J. 1984. Palms, people and progress. Horizons **3(4)**: 32–37.

Clement, C. R. 1986. Descriptores minimos para el pejibaye (*Bactris gasipaes* H.B.K.) y sus implicaciones filogenéticas. Masters thesis. Ciudad Universitaria "Rodrigo Facio," Costa Rica.

Ellis, R. H. 1985. Information required within genetic resources centres to maintain and distribute seed accessions. Pages 11–20 *in* J. Konopka & J. Hanson (eds.), Documentation of genetic resources: information handling systems for genebank management. International Board for Plant Genetic Resources (IBPGR) Secretariat, Rome.

Frankel, O. H. 1983. The place of management in conservation. Pages 1–14 *in* C. M. Schonewald-Cox, S. M. Chambers, B. MacBryde & L. Thomas (eds.), Genetics and conservation. Benjamin/Cummings Publishing Co., Inc., London.

Frazão, J. M. F. & C. U. B. Pinheiro. 1986. Conservation of babassu germplasm. Typewritten report.

Gulick, P. & D. H. van Sloten. 1984. Directory of germplasm collections. 6.I. Tropical and subtropical fruits and tree nuts. International Board for Plant Genetic Resources (IBPGR) Secretariat, Rome.

Hawkes, J. G. 1980. Crop genetic resources field collection manual. International Board for Plant Genetic Resources (IBPGR) and European Association for Research on Plant Breeding, Wageningen.

IBPGR. 1978. IBPGR Consultation on coconut genetic resources: Report. International Board for Plant Genetic Resources (IBPGR) Secretariat, Rome.

———. 1981. Crop genetic resources. International Board for Plant Genetic Resources (IBPGR) Secretariat, Rome.

———. 1983. Genetic resources of *Capsicum*: A global plan of action. International Board for Plant Genetic Resources (IBPGR) Secretariat, Rome.

———. 1985. Descriptors for cultivated and wild sunflower. International Board for Plant Genetic Resources (IBPGR) Secretariat, Rome.

Lleras, E., D. C. Giacometti & L. Coradin. 1983. Areas criticas de distribución de palmas en las Américas para coleta, evaluación y conservación. Pages 67–102 *in* FAO, Informe de la reunión de consulta sobre palmeras poco utilizadas de América Tropical. Organizada por CATIE/FAT, Turrialba, Costa Rica.

National Academy of Sciences. 1975. Underexploited tropical plants with promising economic value. NAS, Washington, D.C.

Pick, P. J. et al. 1985. Babassu (*Orbignya* species): Gradual disappearance vs. slow metamorphosis to integrated agribusiness. A report prepared for The New York Botanical Garden Institute of Economic Botany.

FIG. 4. Germplasm collecting sites and repositories for material collected during 1981–1985 by the staff of the Institute of Economic Botany (NYBG) in collaboration with local institutions.

Collection and Preparation of Palm Specimens

Michael J. Balick

I. Introduction

Palms are distributed throughout much of the Tropics, including the Neotropics. In some areas they occur scattered throughout the vegetation, while in other zones palms are the dominant forest trees. In fact, natural monocultures of palms are quite common, especially in habitats that are not well tolerated by most dicotyledonous species, such as swamps and areas subject to seasonal inundation. Despite their relative abundance and importance as a part of the overall tropical rainforest biota, palms are very poorly represented in the study collections used by biologists. For example, in Brazilian Amazonia where palms often dominate the landscape, this family is very poorly represented in the local herbaria. As a result, local studies of ecology, population biology, floristics, pollination, biology, ethnobotany, etc., are hindered by not having adequate and complete reference collections with which to compare material being brought in from the field. The lack of good palm collections from one important region was documented by Balick et al. (1982) who found that only 37.5% of the 232 recognized palm species in Brazilian Amazonia were present in any of the three local herbaria within that zone. Of the species present, many lacked crucial information on their labels and others were in an unreasonable state of preservation.

With the vigorous collection programs underwritten by agencies such as the U.S. National Science Foundation as well as similar scientific agencies throughout the world, access to the rainforest has been greatly facilitated, and increasing numbers of plants are being distributed to systematists worldwide. Given that (1) adequate drying facilities now exist in may areas of the Tropics, that (2) reasonable transportation systems are in place, and that (3) improved methods of drying and preservation have been developed, we can expect to see more and more palm collections being made over the next few decades. It is important to emphasize that good collections of these plants must be made. These collections require the devotion of a great deal of field time in order to be made properly, and if done correctly will serve future generations of biologists in their quest to understand the tropical ecosystem. A poor palm collection is worse than no collection at all, in that it tends to give a misleading impression of the plant or creates too may unanswerable questions, thus adding only confusion to what could be a clear systematic picture.

II. Collection Techniques and Methods

In this section I would like to summarize some of the techniques and methods to follow for making a good palm collection in the hope of encouraging field workers to devote more effort towards working with these important organisms.

This information is presented in greater detail in Balick et al. (1982), and the interested reader is also directed to Fosberg and Sachet (1965) for additional information on this topic.

Because of the rather large size of many palm parts, it is difficult, to say the least, to try and collect entire organs such as leaves, which may be up to 8 meters in length and weigh 20 kilos or more when fresh. Similarly, a fruiting panicle with 30 kilos of cocosoid fruits would be extremely difficult to collect, dry, ship and curate in its entirety. Because of this, it is advisable to make collections with representative parts of each of the major organs, rather than making collections which encompass the complete organ. Through the use of photographs and very detailed measurements, one can subsequently reconstruct in the herbarium the general nature and habit of the palm. Ten elements of a good palm collection can be outlined:

1. *The leaf apex in pinnat-leaved palms; leaf hastula in palmate- or costapalmate-leaved palms.* The apical pinnae can be fused in a distinctive way or have dimensions altogether different from the middle pinnae. The hastula or ligula, as it is sometimes known, often varies in shape and size and can provide some element of distinction at the specific level.
2. *Leaf pinnae and at least a portion of the rachis in pinnate-leaved palms; leaf segments in palmate-leaved palms.* The pinnae are often crucial for making generic and specific determinations. For example, the presence or absence of peltate trichomes on the pinna can be used to separate the closely related genera, *Jessenia* and *Oenocarpus*. The pinna size and insertion (clustered versus regular) is very useful for separating out Cocosoid palms.
3. *Flowers or fruits, or both.* Flowers, and to a lesser degree fruits, usually contain the most essential criteria for systematic identification in palms, both at the generic and specific level. It is essential that col-

lections include such fertile material. It is often possible to locate an individual plant in flower, while at the same time locating a similar individual of the same species having fruit in some degree of maturation. While it is preferable to obtain flowering and fruiting material from the same tree, it is also permissible to collect flowering and fruiting material from different individuals, if great caution is taken to collect from the same species. The herbarium label should note that sexual material was taken from different trees.

4. *Flowering or fruiting axes, or both.* The morphology of the flowering or fruiting axes often provides important criteria for taxonomic identification. Smaller axes should be collected in their entirety, while larger ones should be photographed and measured quite carefully.

5. *Supplementary material: bracts, sheaths, spines, wood samples, stems, and seedlings.* Such materials may be helpful in identification, depending on the taxa in question.

6. *Quantitative and qualitative information on vegetative, reproductive and morphological characters not apparent on the specimen.* As previously mentioned, it is essential to spend as much time as needed to properly document the presence, distribution and dimensions of vegetative, reproductive and morphological characters. In herbarium collections there are many important bits of information that cannot be fully and accurately preserved or discerned. For example, some palm fruits change color when dry. Another example would be the dimensions of a large leaf, which might be represented in the herbarium collection by only a sample of the apical, middle and basal pinnae and rachis. It is very important that proper measurements be made in the field, and faithfully transcribed on the herbarium label.

7. *Information on habitat.* These data should include comments on altitude, substrate, vegetation type and degree of habitat disturbance. Such data are especially useful in palms, as many species are limited to specific habitats and serve as ecological indicators.

8. *Reasonably precise locale.* As is the case in all herbarium collections, information on locale is essential, especially if the specimen represents a newly discovered, rare or endemic species.

9. *Vernacular names and uses.* Many palm species are referred to by vernacular names and many have local uses. Vernacular names are extremely helpful in relocating species populations in the field; and ethnobotanical uses may be indicative of species of potential economic value worth developing further. It is surprising how many palm collections are lacking such information.

10. *Good quality black-and-white or color photographs.* Even if they are not contained in the herbarium collection, it is extremely important that the collector take adequate pictures of the palm before and after it is collected. For example, habitat photographs, a shot of the individual specimen(s) collected for harvest as well as closeups of the reproductive, leaf, stem and other subsidiary organs. These photographs are often kept by the collector, and their availability indicated on the label. While useful, it is admittedly rather expensive to include original photographs in each herbarium collection.

Because the proper collection of a palm often results in sacrifice of the tree, it is important to make as many specimens as possible from that particular individual. In general, I find that six specimens are sufficient—from one to three deposited in the country of origin as requested, and three duplicates arriving to the home herbarium. More duplicates can be made, although drying these in the field is somewhat difficult, especially if more than one or two good collections are made each day and the collecting activity lasts for several weeks or more. It is not uncommon for palm collectors to find that this material has filled up the field dryers and takes many days to process, resulting in a potential backlog of both palm and non-palm collections. Figure 1 illustrates a complete herbarium collection of the Cocosoid palm *Maximiliana maripa* from Brazil.

While seemingly difficult, palm collection becomes an interesting and rewarding endeavor once the hurdle of the first few collections has been passed. The collector should be armed with a good ax, machete, clippers, measuring tape, 35 mm camera and copious supplies of color slide and black-and-white print film. Strong cord and woven plastic sacks are important for carrying the collected material back to the location where it is to be pressed and otherwise processed. The fact that these plants have been somewhat ignored by collectors in the past gives an opportunity to provide great treasures in the form of new hybrids and species, country records, uses, distribution, and other information for today's generation of tropical botanists.

III. Literature Cited

Balick, M. J., A. B. Anderson & M. F. da Silva. 1982. Palm taxonomy in Brazilian Amazônia: the state of systematic collections in regional herbaria. Brittonia **34**(4): 463–477.

Fosberg, F. R. & H. M. Sachet. 1965. Manual for tropical herbaria. Regnum Veg. **39**: 1–132.

Review of Bryological Studies in the Tropics

William R. Buck and Barbara M. Thiers

Contents

I. Introduction	485
II. Collecting Bryophytes	485
III. Synopsis of Tropical Bryophyte Literature	486
A. Mosses	486
1. American Tropics	487
2. Africa	487
3. Asia	488
4. Australia and Polynesia	488
B. Hepatics	488
1. General Remarks	488
2. American Tropics	489
3. Africa	489
4. Asia and Polynesia	490
5. Australia	490
IV. Conclusions	490
V. Literature Cited	490

I. Introduction

Bryophytes are frequently overlooked in the tropics by plant collectors, ecologists and conservationists alike. However, these small plants, in mass, are often conspicuous components of the ecosystem and perform vital ecological roles. Indeed, they are even more sensitive to habitat modification than most vascular plants and can therefore serve as bioindicators (Low et al., 1985). They have been generally neglected because they are perceived as taxonomically difficult and are not included in the education of most botanists. However, like in the flowering plants, there are difficult genera and families but most can be named with relative certainty. Bryophytes have an advantage over higher plants: most can be identified from sterile material. This is due to the fact that most early bryophyte collectors were not themselves bryologists, therefore, bryologists examining exotic dried specimens were forced to look carefully at microscopic leaf details. Today's bryologists perpetuate this heritage.

Unlike many tropical phanerogams, there are very few individual species of bryophytes with economic value or potential. Rather, their value is to the ecosystem since as a community they are influential, if not critical, in water relations and mineral cycling. Bryophyte mats on trees can retain water and thus serve as seed beds for vascular epiphytes. Often times bryophytes are among the few colonizers on cleared land. After rains they can serve to slow or prevent erosion through soil stabilization by their rhizoids. Even in undisturbed habits bryophytes perform this role and help retain a layer of organic material at the soil surface, thereby providing a habitat for micro-organisms involved in decomposition.

Bryophytes also help maintain the humidity levels in the forest. Due to their growth forms, bryophytes cover relatively large surface areas, thereby moderating evaporation and preventing wild fluctuations in humidity.

A bryophyte colony may harbor a variety of other organisms such as fungi, algae, bacteria, cyanobacteria, protozoans and invertebrates (especially Rotifera, Nematoda and Tardigrada) which live in the moist, protected environment of the spaces among leaves and stems. The association of bryophytes with microorganisms may be important for the tropical rainforest as a whole in the case of the cyanobacteria that are sheltered by epiphyllous hepatics (mostly of the family Lejeuneaceae). The nitrogen-fixing cyanobacteria (usually *Scytonema* or *Nostoc*) are an ecologically significant but inconspicuous component of the epiphyllous community. Recent studies (e.g., Bentley & Carpenter, 1984) have demonstrated that N-fixation occurs on leaves in proportion to the quantity of hepatic cover on the leaves. Although the N-fixers are not truly symbiotic with the epiphyllous hepatics, the colonization and growth of the cyanobacteria is promoted by the moist, protected environment the hepatics provide. Most of the nitrogen fixed by these cyanobacteria is washed off the leaf by the rain into the soil where it is absorbed by the roots of vascular plants. The input to the soil of nitrogen that has been fixed on the surfaces of living leaves may be large enough to be of significance in the nutrient cycling of an area where epiphyllous growth is abundant.

Bryophytes are among the most environmentally sensitive plants in a forest. Once a forest is disturbed, the bryophytes are likely to be among the first organisms adversely affected. The humidity levels which bryophytes help maintain are also requisite for their survival. Although a single tree, if all others around it are felled, can often survive, its epiphytic bryophytes can rarely tolerate the greatly altered environment. Bryophyte species, though, are usually more widespread than most vascular plants due to their numerous, wind-borne spores. Therefore, species extinction may be less of a problem than with higher plants unless all of a particular type of habitat is destroyed.

Despite the sensitivity of some bryophytes, others are among the most resilient plants on earth, being able to withstand desiccation for long periods of time. On parched, lateritic soils cleared of vegetation, pioneer species can cover large areas with their protonemata and, eventually, leafy plants. This can be one of the first stages in succession. Amelioration of the environment by bryophytes thus provides other plants a more favorable microenvironment for establishment.

In naturally harsh environments (e.g., windy, high elevation, rocky habitats) bryophytes and lichens often colonize bare rock. These small tufts, acting as a focal point for soil accumulation and a less exposed situation for seeds, allow higher plants access to habitats which, without bryophytes, they could not enter.

Bryophytes are the second largest group of green land plants after the flowering plants. There are about 20,000 species of bryophytes, more than pteridophytes and gymnosperms combined. Since they are often thought to "all look alike" by the uninitiated, it is understandable that they are often only accessible to the specialist who works full-time on the group. Indeed, even workers on mosses or hepatics rarely have more than a superficial knowledge of the other group. Also, the literature for the two groups is often not together since muscology and hepaticology have been separate fields since the early 19th century. However, both mosses and hepatics grow in similar environments and similar collecting techniques can be used to sample them.

II. Collecting Bryophytes

Bryophytes are among the easiest plants to collect. Only with extreme abuse do specimens lose their value. Although different workers have different field techniques, they are usually just variations on a theme. Bryophytes have many growth forms and actual collection will depend on the plant at hand. Loose plants in tufts or mats can simply be picked up by hand, as can pendent forms. However, at times plants are small or tightly adherent to the substrate. In such circumstances it is best to collect a small amount of the substrate with the plant so as to keep the plant from being lost or from breaking into many fragments. A knife or wood chisel are the best implements. Soil often stays around colonies of small, terrestrial plants, indeed it may be the plants which bind the soil, and should not be broken up although some soil may be scraped from the bottom of the clod. Small corticolous species are best collected with a shallow strip of bark so that slender stolons or stems which may be present will not be broken or lost. Clippers may be used to collect branch or twig sections with small bryophytes on them. As a general rule, for average-sized bryophytes, a piece the size of the palm of a hand is a good size for a single herbarium specimen. Additional material should be collected, when available, for sending to specialists, depositing in host country herbaria, and other functions. There are small species, though, which even when common, do not make large patches. Some bryophytes only occur as scattered plants and should not be ignored just because of scanty, but well developed material. General collectors should be careful when dividing material to send to specialists: bryophyte species, like those of higher plants, have distinctive aspects, but at times these may be too subtle for the untrained eye to discern.

Bryophytes are best collected into individual paper bags, with a single collection per bag. Although some collectors place large numbers of species into a single bag, individual species are often seen more easily in nature and separation in the laboratory can be dirty and tedious. Plastic bags should never be used unless one intends to examine the specimens within a few days. Bryophytes, especially hepatics, are susceptible to becoming moldy if not allowed to dry or have air circulation. Also some mosses will continue to grow, but in a very odd, etiolated fashion, if left in plastic bags. We have generally found 2 lb, flat-bottomed, brown paper bags to be the most convenient type. They are inexpensive and come bound in groups of 500 thus allowing easy transport and dispensability. One should purchase these before going to the tropics since they are often either not available there or are of unacceptable quality (with V-bottoms and glue which does not hold the bag together under tropical humidity). If paper bags are not available, packets can be folded from any available paper or bryophytes can be put in a press with higher plants. Although pressing does not alter critical morphological features, it does give the plant, especially mosses, a less than natural aspect when dry. In the case of leafy hepatics, the ideal method of preparation is to remove excess soil and other debris from the still moist (or rewetted) colony, and place the plants between papers (or in a folded packet) in a plant press for about 24 hours (without heat). Tension on the press should be light.

Individual bags can also be used to record any desired field data. Although ecologists may be interested in complex habitat parameters, bryologists most often only need basic habitat information. Substrate is often recorded (e.g., tree trunk, moist soil, submerged on rock in stream) as is general habitat type (e.g., tropical rainforest, páramo, thorn scrub). Otherwise, label data should be as for other plants, i.e., locality (including latitude and longitude when possible), elevation, date and collector. There is no need to give a descriptive account of the plant, as one often does for higher plants, since the whole plant is in the collection and almost never is altered on drying.

Although in very humid habitats it can be helpful to use a plant drier, in most cases the bags can be left to air dry. If in a vehicle, the paper bags can be put in a net or burlap bag and tied to the top of the car so that air is forced over them, facilitating drying. Very wet mosses, such as *Sphagnum* or aquatic species, should be squeezed out, not wrung, to remove excess water. For both mosses and hepatics air drying is preferred, but limited exposure to heat in a mechanical drier does not seriously diminish the value of the specimen.

The value of leafy hepatic specimens will be greatly enhanced if oil-body data are gathered from fresh specimens. Oil-bodies are lipid-containing, membrane-bound organelles found only in this group of organisms. The number of these structures per cell, as well as their size, shape and color, are becoming increasingly important in the taxonomy of hepatics at the generic and specific levels. In order to obtain these data a compound microscope with a resolving power of 400x is required, a luxury rarely afforded in tropical collecting. The easiest way to observe oil-bodies is to place a leaf or portion thereof on a microscope slide in a drop of water and put a coverslip over it. No stains are needed. Unfortunately, oil-bodies are ephemeral: they will disappear rapidly if the leaf is mounted in a medium other than water (such as Hoyer's solution or lactic acid) and they begin to disintegrate as the plant dies. They may remain intact for several days to several weeks (in air-dried plants), but for best results oil-body data should be taken immediately after collecting and before pressing or drying. Oil-body data should be added to the label, and will be very helpful for specialists. A brief note is sufficient to convey the information, e.g., oil-bodies large, 2–3 per cell, botryoidal, brownish.

Generally bryophytes are not excessively brittle and dried plants do not need special care. An added advantage of individual paper bags is that they act as packing and insulation for the specimens in shipping. Some hepatics and *Sphagnum* (and a few other mosses) can be quite brittle on drying so specimens should not be abused. Usually just rolling up the bags and putting them in cardboard boxes is ample protection.

III. Synopsis of Tropical Bryophyte Literature

A. Mosses

The diversity of mosses, as a rule, increases with elevation. Low elevations (below 500 m) often do not offer as wide a selection of microhabitats as do higher elevations, which may account for this relative paucity of species. However, because of this, low elevations near mountains are often less well known than the mountains themselves. A similar increase in diversity is often correlated with a moisture gradient. Each of the major land masses in the tropics has its own distinctive bryoflora and there are a few families of mosses restricted to one continent or another. Widespread families are often more successful on certain continents thus comprising a more conspicuous part of the flora. For example, the Hookeriaceae s.l. have been particularly successful in Latin America. For absolute diversity the Australasian tropics are the richest and it is here that mosses have reached their most spectacular statures.

1. American Tropics

The American tropics have the second richest tropical moss flora. This diversity was cataloged over 100 years ago with keys and descriptions (Mitten, 1869). *Musci Austroamericani*, although now out of date, is still the only work for many parts of its range. Central America itself is not exceptionally rich, but there are notable exceptions. Mexico, due to its large size, elevational gradient and habitat diversity, has a very rich bryoflora. Although this diversity has been documented (Crum, 1951), there is no guide to allow identification of specimens. This should soon be remedied with the appearance of a multi-authored moss flora organized by A. J. Sharp with, recently, the assistance of Howard Crum. The mosses from Guatemala to Nicaragua are relatively uninteresting due to mostly low elevations and disturbance since antiquity. However, unchallenged endemics have been reported. Although somewhat out of date, Bartram's (1949) *Mosses of Guatemala* is still the most useful work for all Central America. The most recent reference for Honduras, with cross references to previous contributions, is Crum's 1968 treatment. Costa Rica and Panama, due to the extensive mountains, are scarcely more than a poor Andean flora and collections from their high elevations are best named by using Andean treatments. The most recent checklist of Costa Rican mosses (Bowers, 1974) provides access to the earlier literature. A key to the moss genera of Costa Rica has been recently published (Griffin & Morales, 1983). Bryophytes have been ignored by the "Flora of Panama" project however, and only a checklist is available (Crosby, 1969, 1971; Salazar A. & Crosby, 1985).

Although the West Indies have been the subject of bryological interest for over 200 years, there is as yet no comprehensive guide to its flora. However, a moss flora project is currently underway for the entire area by William R. Buck and William C. Steere. Until that is completed, the only treatments available are keys to the mosses of Jamaica (Crum & Bartram, 1958) and Guadeloupe (Foucault, 1977). A moss flora of Puerto Rico, with keys and descriptions (Crum & Steere, 1957) was recently supplemented with a key to the genera (Miller & Russell, 1975).

Although the Andes have the richest bryoflora in tropical America, there are no keys that allow identification of general collections (Griffin & Gradstein, 1982). Griffin's (1982) key to the genera of mosses of the state of Mérida, Venezuela, is the only available work. It is particularly useful in conjunction with the checklist by López F. (1976). Dana Griffin, though, is preparing a moss flora of the Venezuelan páramos. Robinson's (1967) preliminary work on the mosses of Colombia—providing keys to species of selected genera—is also of use. Only checklists are available for Venezuela (Pursell, 1973) and Colombia (Florschütz-de Waard & Florschütz, 1979). Only checklists are available for Ecuador (Steere, 1948) and Peru (Hegewald & Hegewald, 1975), but a more comprehensive account of Ecuador's mosses is under way by William C. Steere. Even though Bolivia remains one of the least known areas for vascular plants, it is probably one of the best known areas for mosses. Although quite old, Herzog's (1916, 1921) account of his bryophyte collecting in Bolivia remains very useful, not only for Bolivia but for the entire Andean region. Although Herzog's works lack keys and descriptions (except to the numerous new taxa), they are a remarkably complete account of the Bolivian bryoflora with numerous illustrations. Hermann's (1976) checklist provides more recent literature additions.

The tepuis of the Guayana Highland, although highly acclaimed for their vascular plant endemism, have proved disappointing for mosses. The only published account (Bartram, 1960) suggested a low diversity but this was suspect due to lack of critical collecting. However, several recent expeditions to Cerro de la Neblina, as well as more eastern tepuis, have verified the low diversity of mosses (with the notable exception of *Sphagnum*) on tepui summits. This phenomenon may be explained by the fact that prevailing winds (i.e., the carrier of spores) blow from lowland areas, where species would not be adapted to high elevation existence.

The Guianas, due to their low elevational relief, have a relatively poor, widespread flora. The first two volumes of a moss flora of Surinam has been published (Florschütz, 1964; Florschütz-de Waard, 1986) and additional volumes are in preparation by Jeanne Florschütz-de Waard.

The lack of species diversity of lowland South America is also in evidence throughout Brazilian Amazonia. Although only a small moss flora is present, only one recent work (Griffin, 1979) provides keys and illustrations. Four moss families dominate Amazonia: Sematophyllaceae, Callicostaceae, Leucobryaceae and Calymperaceae. The latter two have received considerable attention by Yano (1975) and Reese (1961, 1977, 1978), respectively. Southern Brazil, with its more variable climate and landscape, has a more diverse flora. An unfinished, marginally usable flora for this region was published by Sehnem (1969–1980) before his death. A checklist is available for all of Brazil (Yano, 1981). Only a recent checklist is available for Paraguay (Buck, 1985).

2. Africa

Africa, due to its relative aridity, has the poorest moss flora of any continent with tropical regions. Yet, its moss flora is poorly known. Recent research (e.g., Buck & Griffin, 1984) has provided evidence that much of the tropical African moss flora is conspecific with that of South America, but has gone under different names. This will prove to be a major challenge to the conceived but not-yet-started *Bryologia Africana*. Two out-of-date but still usable moss floras by Potier de la Varde are available for wet, tropical Africa. In 1927 he wrote a moss flora of Oubangui (Central African Republic) with species descriptions but few keys or illustrations. His 1936 moss flora of Gabon, in addition to descriptions, also provides keys. Although East Africa harbors one of Africa's best known endemic moss families, the Rutenbergiaceae, it lacks a floristic treatment of the mosses and only a recent checklist (Kis, 1985) for the area exists. Otherwise, only expedition reports, scarcely usable for collection identification, are available. Madagascar has (or had) a very rich flora. The most useful guide for this region is the long since outdated catalog of the mosses prepared by Renauld (1897, 1909), although there is a recent checklist (Crosby et al., 1983). The most usable guide to the mosses of the Mascarenes remains Bescherelle's (1878, 1880) treatment—with many descriptions but few keys. Southern Africa, presumably due to prolonged European presence, has fared better than the rest of Africa in cataloging its mosses. In 1926 Sim produced a complete bryoflora with keys, descriptions and (poor) illustrations. The current *Flora of Southern Africa* has published the first two moss fascicles (Magill, 1981, 1987), and more are in progress.

3. Asia

Asia, with its numerous intrusions into the tropics, has the richest bryoflora. India itself has a remarkably rich flora. Gangulee's (1969–1980) moss flora of eastern India is indispensable. Also of use are a preliminary treatment of all Indian mosses (Chopra, 1975), the first fascicle of a moss flora of the western Himalayas and adjacent plains (Chopra & Kumar, 1981), and Vohra's (1983) treatment of most of the Himalayan pleurocarps. Sri Lanka has benefited from the publications of Abeywickrama (1960) and Abeywickrama and Jansen (1978b). Although continental Southeast Asia has a very rich bryoflora, regional conflicts have prevented adequate modern study. Only checklists for the bryophytes of Vietnam (Pócs, 1965; Tixier, 1966), Mount Bokor in southern Cambodia (Tixier, 1979), Thailand (Horikawa & Ando, 1964; Touw, 1968) and western Indochina (including Burma) (Tixier, 1973) are worthy of note. Peninsular Malaysia has fared somewhat better with a generic moss flora (Manuel, 1981) and a beginner's guide to the mosses (Johnson, 1980).

The Philippines have long been one of the best known tropical moss floras. Numerous collectors, both native and Western, have gathered mosses extensively. These resulted in Bartram's (1939b) very useful *Mosses of the Philippines*. However, like in all tropical areas, much remains to be discovered, as evidenced by the additions in the recent checklist by Iwatsuki and Tan (1979). Only a literature-based checklist exists for Taiwan (Kuo & Chiang, 1987).

The East Indies have been an area of interest for many years, presumably a result of the early spice trade. The *Bryologia Javanica* (Dozy & Molkenboer, 1855–1861) is incomplete and out of date, but its illustrations are so elegant it deserves mention. Of much use is the masterful treatment of Javan mosses by Fleischer (1904–1923), probably the first moss flora in a modern sense. New Guinea, though, with its incredibly rich moss flora, lacks even a checklist. There are two current moss flora projects focusing on New Guinea. One project by Timo Koponen, Daniel H. Norris and collaborators, serialized in "Annales Botanici Fennici," is concentrating on the Huon Peninsula. The other project, headed by Alan Eddy, will include all of peninsular Malaysia, the East Indies, New Guinea and the Philippines. Only the first volume is published to date (Eddy, 1988).

4. Australia and Polynesia

Only the northern part of Australia is in the tropics and much of that is desert. A recent, albeit preliminary, flora of the mosses of southern Australia (Scott & Stone, 1976) nevertheless lists reports from throughout the country. There is no source for identification of Queensland mosses.

Many of the Pacific islands have long been studied, but few have more than recently published checklists. A checklist for the whole Pacific basin with a key to the genera (Miller et al., 1978) is useful, as are lists of mosses for Fiji (Whittier, 1975) and New Caledonia (Pursell & Reese, 1982). The Society Islands (including Tahiti) have a modern flora (Whittier, 1976) with keys, descriptions and some illustrations. Hawaii's fully illustrated moss flora (Bartram, 1933, 1939a) has been updated by Hoe's (1974) annotated checklist.

B. Hepatics

1. General remarks

A humid tropical climate supports a much more diverse hepatic flora than anywhere else in the world. In lowland and montane forests in the tropics, hepatics carpet tree trunks, living and dead leaves of vascular plants, boulders, soil banks, and to a lesser extent, fallen logs. Diversity is correlated with rainfall, elevation and light levels. In the tropics of Australasia, at least, a good rule of thumb is that hepatic diversity will be poor in areas with less than 900 mm of rain per year. Generally, lowland areas are relatively low in diversity and often populated by pantropical ("weedy") species. Floristic deversity appears to be greatest at elevations of 1000–2000 m. Above 2000 m hepatic biomass continues to increase, but species diversity begins to decline. Although many hepatics are able to tolerate the very low light levels found within a tropical rainforest, the number of epiphytic hepatic species is often greater on trees along a stream, trail or some other type of break in the forest, suggesting that higher light regimes may promote greater diversity.

The tropical hepatic flora has three main components: neotropical, African and Asian. In addition to their unique elements, each component has some pantropical species, which are usually dominant in lowland rainforests, especially in coastal and disturbed areas. Because revisionary works are so few in the Hepaticae, it is very difficult to make accurate comparisons of species numbers and floristic makeup in the tropics. As a result of monographic studies, each year many species fall into synonymy, sometimes with species described from other continents, a trend which suggests that the tropical flora is smaller and less distinct geographically than previously thought. However, this decline in diversity is offset by the number of new species described each year as a result of increased collecting activity in previously ignored areas.

Unlike the mosses—in which some families predominate over others in different continents—the families Lejeuneaceae, Plagiochilaceae, Jubulaceae, Lepidoziaceae, Jungermanniaceae and Geocalycaceae dominate the hepatic floras of all tropical regions in approximately equal proportions. Endemism occurs mostly at the generic or specific level. Two centers of distribution have been proposed for the Hepaticae—Asia (especially the East Indies) and northern South America (Amazonia anad the Guiana Shield)—because of the richness of these floras and the proportion of putatively primitive taxa found there. The flora of Africa is derivative, and rather depauperate by comparison.

The identification of hepatics from any area of the tropics can be a difficult and frustrating endeavor, especially for the novice, because of the paucity of literature. Modern floristic treatments, with keys, descriptions and illustrations are rare. Much of the literature is in the form of annotated species lists of areas, sometimes with thorough descriptions or illustrations, and sometimes with synonymies, but usually without keys. The authors often state in the introductions that their contributions are only preliminary surveys, or words to that effect. Independently, such publications will usually not allow a positive determination to be made, but may at least give the student a list of possible names, descriptions and illustrations of which may then be sought in other publications. The single most comprehensive reference for hepatic species worldwide is still Stephani's *Species Hepaticarum* (1898–1924), although it is outdated and marred by incomplete and inaccurate descriptions. The standard nomenclatural guide is *Index Hepaticarum* (Bonner, 1962–1976; Bischler & Lamy, 1978; Geissler & Bischler, 1985), a work still in progress (currently complete only through the letter M). Although mostly concerned with an extratropical flora, Schuster's *Hepaticae and Anthocerotae of North America* (1966–1980) is an excellent reference for literature, descriptions of morphological features, supergeneric classifications (including keys to families), and species descriptions, discussions and illustrations. The publications discussed on the following pages are limited to floristic accounts (concerning more than one genus or family) because these may be the most useful to the non-specialist. Works of a monographic nature will become more important as familiarity with the hepatics increases, and ample reference to such works may be found in the bibliographies of the papers cited herein.

2. American Tropics

The two most essential references for the identification of hepatic species from the American tropics are Fulford (1963–1976) and Spruce (1885–1886). Although incomplete, Fulford's series is geographically the most comprehensive, covering all Latin America. Spruce's book is a monumental treatment of the hepatics of the Amazon basin and the Andes, and is usually the best source for descriptions of species not covered in Fulford's manual. Within South America three floristic regions seem to be somewhat distinct: the Guayana Highlands, the Amazon Basin, and the Andes. The only floristic survey of any of the sandstone tepuis of the Guayana region is that of Fulford (1967), although it is clear from her other works (Fulford 1963–1976) that these mountains have an interesting and distinctive hepatic flora, especially rich in taxa considered primitive. Very little has been published on the lowland flora of the Guianas, but S. R. Gradstein and colleagues at Utrecht are currently conducting intensive research on this area as part of the "Flora of the Guianas" project. The Andean flora is very rich, and has received more attention. Studies by Gradstein and collaborators have resulted in a checklist of Colombian hepatics (Gradstein & Hekking, 1979). Publications on the hepatics of Peru (Evans, 1914; Herzog, 1955) and Bolivia (Herzog, 1916, 1921) may also prove useful for indentification of Andean hepatics. The rather uniform flora of the vast area of Amazonia is treated in detail only by Spruce. Griffin (1979) treated the common species in the vicinity of Manaus and Schiffner (1893), Herzog (1924) and Hell (1969) treated some species from southern Brazil. A checklist for all Brazil was published by Yano (1984).

The hepatics of Central America and Mexico have received little study. Cole (1983, 1984) provided keys to the genera of Costa Rican hepatics, with discussion of some species, and excellent illustrations. Some thalloid hepatics of the drier areas of the region are treated by Howe (1934). Perhaps the best overall view of the Central American flora can be obtained from Herzog (1938, 1951). Review of the species he included indicates that the lowland flora is not particularly rich, and is quite similar to that found elsewhere in the neotropics, and the flora at higher elevations contains some novelties, but has strong affinities to that of the Andes.

Previous research on the hepatics of the West Indies has resulted in checklist-type publications on the flora of the Virgin Islands (Evans, 1918), the Bahamas (Evans, 1911), Bermuda (Evans, 1906), Puerto Rico (Pagán, 1939), and the French Antilles (Bescherelle, 1893; Stephani, 1901–1902; Pagán, 1942; Jovet-Ast, 1947–1950). Based on these publications it appears that the hepatic flora shows similar trends to that in Central America—a rather depauperate flora in lowland areas, with a more diverse flora, including numerous endemics, at higher elevations.

3. Africa

The most comprehensive publications on the African hepatic flora cover species of southern Africa (Sim, 1926; Arnell, 1963; Vanden Berghen, 1972, 1977, 1978). Some species occurring on Mauritius and Réunion have been enumerated by Arnell (1965), and a preliminary account of the hepatics of Madagascar was provided by Jovet-Ast (1949). The wet, equatorial regions of Africa appear to be the richest on the continent in species of hepatics, and these have been enumerated in a variety of catalog-like publications (Mitten, 1887; Vanden Berghen, 1953, 1965, 1973; Bizot & Pócs, 1974, 1979, 1982; Bizot et al., 1976, 1978; Váňa et al., 1979; Aké-Assi & Pócs, 1983). From this literature it appears that the African hepatic flora is most similar to that of Asia. Although quite a few species have been reported only from Africa, the flora contains very few endemics at the generic or familial level.

4. Asia and Polynesia

In terms of the hepatic flora, Asia covers an area from India to Hawaii. Although some regions within this area seem to have developed rather distinctive floras (e.g., the Himalayas, New Caledonia), the differences among the taxa occurring in the major geographic subdivisions, such as India, Southeast Asia, Malaysia, Melanesia and Polynesia, are too few to allow each to be considered distinct floristic provinces. Although many areas are still very poorly known, the Asian flora has received more attention in recent years than other areas of the tropics and, as work continues, an increasing number of species once thought to be regional endemics fall into synonymy.

The first major contribution on the hepatics of India and Southeast Asia was that of Mitten (1861), based on the collections made by early botanical explorers such as Hooker, Wallich and Gardner. Two more modern works on the flora of India are those of Chopra (1944) and Kashyap (1926). The hepatics of the Himalayas have been fairly well covered in the publications of Grolle (1966), Hattori (1966) and Long (1979). The generic flora of Sri Lanka by Abeywickrama (1959) is the only recent guide to the species of this area, updated by a later checklist (Abeywickrama & Jansen, 1978a), For Indochina the publications by Pócs (1965) and Tixier (1962) on the hepatics of Vietnam may prove useful for an overview of the flora.

Unfortunately, very few keys are available for the hepatics of what may be generally called the East Indies (including Malaya, Indonesia, New Guinea and the Philippines), and this is probably the richest in the world in hepatic species. A recent publication by Miller et al. (1983) is centered on the Polynesian flora, but includes a key to genera which may be helpful for the East Indies as well. Grolle and Piippo's (1984) annotated guide to the hepatics and anthocerotes of western Melanesia is an indispensable reference for the hepatics of New Guinea, and an excellent source of literature on species occurring throughout the East Indies. Although outdated, the publications of Schiffner (1893, 1898, 1900) and Goebel (1928) can still be used for species descriptions. Herzog treated some species occurring in the Philippines (1931) and Borneo (1930, 1950).

Miller et al. (1983) give a thorough review of the literature on Polynesian hepatics. Especially useful publications are those on the floras of Samoa and Fiji by Mitten (1873), Jack and Stephani (1894), and Grolle and Schulze-Motel (1972); of New Caledonia by Pearson (1922); of Micronesia by Miller (1968), and Miller and Bonner (1963); and of Hawaii by Evans (1891, 1900).

5. Australia

There are no published accounts of the hepatics of tropical Australia, although a hepatic flora of Australia (part of the *Flora of Australia* series), coordinated by G. A. M. Scott, will cover this region. Much of the area within tropical Australia is arid, and thus supports only very few hepatic species. However, in the rainforests of coastal northeast Queensland, hepatic diversity is substantial. Most of the species found there are widespread in Asia, but endemics do occur, as well as a few species from temperate Australia. Publications by Gottsche (1857) and Stephani (1889, 1895) described some new species from tropical Australia, and Henderson and Prentice (1971) give a preliminary account of the species found on Norfolk Island. Some temperate species that extend their ranges into tropical Australia are treated by Watts (1901, 1902, 1904), Stephani and Watts (1914), and Scott (1985).

IV. Conclusions

Certainly we have not included reference to every checklist. Nevertheless, as can be seen, much remains to be done. Few of the references cited above are modern floras in the sense that specimens can be identified with them through the use of keys, descriptions and illustrations. Although much exotic material continues to enter herbaria, and the need for floras is greater than ever, fewer and fewer bryologists have a broad enough base of knowledge to produce such floras. Of some consolation is the modern trend for coordinated, multi-authored floras. Certainly there are advantages in such projects since authors working on individual taxonomic groups over extended periods of time are more likely to understand the groups better than generalists. However, these patchwork floras are difficult for the user due to the varying quality of work from family to family, and the greatly disparate species concepts. Perhaps when floristic information is more readily available from published monographs, the generalist will once again come into his own. However, even today, monographs contain repeated examples of species still only known from type specimens. This serves to emphasize the need for additional field work in bryology. Certainly considering their ease of collection, bryophytes should be gathered by all collectors while in the tropics.

V. Literature Cited

Abeywickrama, B.A. 1959. The genera of the liverworts of Ceylon. Ceylon J. Sci., Biol. Sci. **2**: 33–81.
―――. 1960. The genera of the mosses of Ceylon. Ceylon J. Sci., Biol. Sci. **3**: 41–122.
――― & **M. A. B. Jansen.** 1978a. A check list of the liverworts of Sri Lanka. UNESCO: Man & Biosphere Natl. Committee for Sri Lanka Publ. **1**: 1–10.
――― & ―――. 1978b. A check list of the mosses of Sri Lanka. UNESCO: Man & Biosphere Natl. Committee for Sri Lanka Publ. **2**: 1–25.
Aké-Assi, L. & T. Pócs. 1983. Hépatiques de Côte d'Ivoire, I. Cryptog. Bryol. Lichénol. **4**: 65–70.
Arnell, S. 1963. Hepaticae of South Africa. Swedish Natl. Sci. Res. Council, Stockholm.
―――. 1965. Hepaticae collected by Mr. Gillis Een in Mauritius and Réunion in 1962. Svensk Bot. Tidskr. **59**: 65–84.
Bartram, E. B. 1933. Manual of Hawaiian mosses. Bernice P.

Bishop Mus. Bull. **101:** 1–275.

———. 1939a. Supplement of the Manual of Hawaiian mosses. Occ. Pap. Bernice P. Bishop Mus. **15:** 93–108.

———. 1939b. Mosses of the Philippines. Philipp. J. Sci. **68:** 1–437.

———. 1949. Mosses of Guatemala. Fieldiana Bot. **25:** 1–442.

———. 1960. Musci. *In*: B. Maguire, J. J. Wurdack & collab., The botany of the Guayana Highland—part IV. Mem. New York Bot. Gard. **10(2):** 2–10.

Bentley, B. L. & E. J. Carpenter. 1984. Direct transfer of newly-fixed nitrogen from free-living epiphyllous microorganisms to their host plant. Oecologia **63:** 52–56.

Bescherelle, É. 1878, 1880. Florule bryologique de la Réunion et des autres îles austro-africaines de l'Océan Indien. Ann. Sci. Nat. Bot. VI, **9:** 291–380. 1878; **10:** 233–332. 1880.

———. 1893. Énumeration des hépatiques connues jusqui'ici aux Antilles françaises (Guadeloupe et Martinique). J. Bot. (Morot) **7:** 174–180; 183–194.

Bischler, H. & D. Lamy. 1978. Index Hepaticarum. Pars IX. *Jungermanniopsis* to *Lejeunites*. J. Cramer Verlag, Vaduz.

Bizot, M. & T. Pócs. 1974. East African bryophytes. I. Acta Acad. Paed. Agriensis II, **12:** 383–449.

——— & ———. 1979. East African bryophytes. III. Acta Bot. Acad. Sci Hungaricae **25:** 223–261.

——— & ———. 1982. East African bryophytes, V. Acta Bot. Acad. Sci. Hungaricae **28:** 15–64.

———, M. N. Dury & T. Pócs. 1976. East African bryophytes, II. Collections made by L. Ryvarden in Malawi, SE Africa. Acta Bot. Acad. Sci. Hungaricae **22:** 1–8

———, I. Friis, J. Lewinsky & T. Pócs. 1978. East African bryophytes IV. Danish collections. Lindbergia **4:** 259–284.

Bonner, C. E. B. 1962–1976. Index Heparticarum. Parts 1–9. J. Cramer Verlag, Vaduz.

Bowers, F. D. 1974. The mosses reported from Costa Rica. Bryologist **77:** 150–171.

Buck, W. R. 1985. A preliminary list of the mosses of Paraguay. Candollea **40:** 201–209.

——— & D. Griffin, III. 1984. *Trachyphyllum*, a moss genus new to South America with notes on African–South American bryogeography. J. Nat. Hist. **18:** 63–69.

Chopra, R. S. 1943 (1944). A census of Indian hepatics. J. Indian Bot. Soc. **22:** 237–259.

———. 1975. Taxonomy of Indian mosses (an introduction). Bot. Monogr. **10:** 1–631.

——— & S. S. Kumar. 1981. Mosses of the western Himalayas and adjacent plains. Ann. Cryptog. Phytopath. **5:** 1–142.

Cole, M. 1983. An illustraed guide to the genera of Costa Rican Hepaticae. I. Brenesia **21:** 137–201.

———. 1984–(1985). Thallose liverworts and hornworts of Costa Rica. Brenesia **22:** 319–348.

Crosby, M. R. 1969. The mosses reported from Panamá. Bryologist **72:** 513–521.

———. 1971. Additional Panamanian bryophytes. Ann. Missouri Bot. Gard. **58:** 258–260.

———, U. Schultze-Motel & W. Schultze-Motel. 1983. Katalog der Laubmoose von Madagaskar und den umliegenden Inseln. Willdenowia **13:** 187–255.

Crum, H. 1951. The Appalachian–Ozarkian element in the moss flora of Mexico with a check-list of all known Mexican mosses. Dissertation, University of Michigan, Ann Arbor.

———. 1968. Mosses from Honduras. III. Adv. Frontiers Pl. Sci. **21:** 189–193.

——— & E. B. Bartram. 1958. A survey of the moss flora of Jamaica. Bull. Inst. Jamaica Sci. Ser. **8:** 1–90.

——— & W. C. Steere. 1957. The mosses of Porto Rico and the Virgin Islands. Sci. Surv. Porto Rico & Virgin Isl. **7(4):** 399–599.

Dozy, F. & J. H. Molkenboer. 1855–1861. Bryologia Javanica seu descriptio muscorum frondosorum archipelagi indici iconibus illustrata. E. J. Brill, Leiden.

Eddy, A. 1988. A handbook of Malesian mosses. Vol. 1. Sphagnales to Dicranales. British Museum (Natural History). 204 pp.

Evans, A. W. 1891. A provisional list of the Hepaticae of the Hawaiian Islands. Trans. Connecticut Acad. Sci. **8:** 253–261.

———. 1900. Hawaiian Hepaticae of the tribe Jubuloideae. Trans. Connecticut Acad. Sci. **10:** 387–462.

———. 1906. The Hepaticae of Bermuda. Bull. Torrey Bot. Club **33:** 129–135.

———. 1911. The Hepaticae of the Bahama Islands. Bull. Torrey Bot. Club **38:** 205–222.

———. 1914. Hepaticae: Yale Peruvian Expedition of 1911. Trans. Connecticut Acad. Sci. **18:** 291–345.

———. 1918. Hepaticae of St. Croix, St. Jan, St. Thomas and Tortola. *In*: N. L. Britton, The flora of the American Virgin Islands. Mem. Brooklyn Bot. Gard. **1:** 104–109.

Fleischer, M. 1904–1923. Die Musci der Flora von Buitenzorg (zugleich Laubmoosflora von Java). Vols. 1–4. E. J. Brill, Leiden.

Florschütz, P. A. 1964. Musci [part 1]. Flora of Suriname **4(1):** 1–271. E. J. Brill, Leiden.

Florschütz-de Waard, J. 1986. Musci [part II]. Flora of Suriname **4(1):** i–x, 273–361. E. J. Brill, Leiden.

——— & P. A. Florschütz. 1979. Estudios sobre criptógamas colombianas III. Lista comentada de los musgos de Colombia. Bryologist **82:** 215–259.

Foucault, B. de. 1977. Flore des bryophytes de Guadeloupe. Parc Naturel de Guadeloupe, Basse-Terre.

Fulford, M. 1963–1976. Manual of the leafy Hepaticae of Latin America—parts I–IV. Mem. New York Bot. Gard. **11:** 1–172, 1963; 173–276, 1966; 277–392, 1968; 393–535, 1976.

———. 1967. Hepaticae. *In*: J. Steyermark (ed.), Flora del Auyántepui. Acta Bot. Venez. **2:** 72–99.

Gangulee, H. C. 1969–1980. Mosses of eastern India and adjacent regions. **1:** 1–170, 1969; **2:** 171–566, 1971; **3:** 567–830, 1972; **4:** 831–1134, 1974; **5:** 1135–1462, 1976; **6:** 1463–1546, 1977; **7:** 1547–1752, 1978; **8:** 1753–2145, 1980. Calcutta.

Geissler, P. & H. Bischler. 1985. Index Hepaticarum. Vol. 10. *Lembidium* to *Mytilopsis*. J. Cramer Verlag, Vaduz.

Goebel, K. 1928. Malesische Lebermoose. Ann. Jard. Bot. Buitenzorg **39:** 1–116.

Gottsche, C. 1857. Hepaticae australalasiae a D^{re}· Ferd. Mueller lectae. Linnaea **28:** 545–561.

Gradstein, S. R. & W. A. Hekking. 1979. Studies on Colombian cryptogams. IV. A catalogue of the Hepaticae of Colombia. J. Hattori Bot. Lab. **45:** 93–144.

Griffin, D., III. 1979. Guia preliminar para as briófitas freqüentes em Manaus e adjacências. Acta Amazonica **9(3, Supl.):** 1–67.

———. 1982. Los musgos del estado Mérida. Clave para los géneros. Pittieria **11:** 6–56.

——— & S. R. Gradstein. 1982. Bryological exploration of the tropical Andes: Current status. Beih. Nova Hedwigia **71:** 513–518.

——— & M. I. Morales. 1983. Keys to the genera of mosses from Costa Rica. Brenesia **21:** 299–323.

Grolle, R. 1966. Die Lebermoose Nepals. *In*: W. Hellmich (ed.), Ergebnisse Forschung-Nepal-Himalaya **1:** 262–298. Springer-Verlag, Berlin.

——— & S. Piippo. 1984. Annotated catalogue of western Melanesian bryophytes. I. Hepaticae and Anthocerotae. Acta Bot. Fenn. **125:** 1–86.

———— & W. Schultze-Motel. 1972. Vorlaufiges Verzeichnis der Lebermoose von Samoa. J. Hattori Bot. Lab **36:** 75–89.

Hattori, S. 1966. Anthocerotae and Hepaticae. Pages 501–536 *in*: H. Hara (ed.), The flora of eastern Himalaya. University of Tokyo Press, Tokyo.

Hegewald, P. & E. Hegewald. 1975. Verzeichnis der Laubmoose von Peru nach Literaturangaben. J. Hattori Bot. Lab. **39:** 39–66.

Hell, K. G. 1969. Briófitas talosas dos arredores da cidade de São Paulo (Brazil). Bol. Fac. Filos. Ci. Letr. Univ. São Paulo **335(Bot. 25):** 1–190.

Henderson, D. M. & H. T. Prentice. 1971. On the bryophyte flora of Norfolk Island. Rev. Bryol. Lichénol. **37:** 657–661.

Hermann, F. J. 1976. Recopilación de los musgos de Bolivia. Bryologist **79:** 125–171.

Herzog, T. 1916. Die Bryophyten meiner zweiten Reise durch Bolivia. Biblioth. Bot. **87:** 1–347.

————. 1921. Die Bryophyten meiner zweiten Reise durch Bolivia. Nachtrag. Biblioth. Bot. **88:** 1–31.

————. 1924. Beiträge zur Kenntnis der Moosflora von Brasilien. Arch. Bot. São Paulo **1:** 23–105.

————. 1930. Hepaticae. *In*: E. Irmscher, Beiträge zur Flora von Borneo. I. Mitt. Inst. Allgem. Bot. Hamburg **7:** 182–216.

————. 1931. Hepaticae philippinensis a cl. C. J. Baker lectae. Ann. Bryol. **4:** 79–94.

————. 1938. Hepaticae Standleyanae costaricensis et hondurensis. I. Rev. Bryol. Lichénol. **11:** 5–30.

————. 1950. Hepaticae borneensis (Oxford University Expedition to Sarawak, 1932). Trans. Brit. Bryol. Soc. **1:** 275–326.

————. 1951. Hepaticae Standleyanae costaricensis et hondurenses. II. Rev. Bryol. Lichénol. **20:** 126–175.

————. 1955. Hepaticae aus Colombia und Peru. Feddes Repert. **57:** 156–203.

Hoe, W. J. 1974. Annotated checklist of Hawaiian mosses. Lyonia **1:** 1–45.

Horikawa, Y. & H. Ando. 1964. Contributions to the moss flora of Thailand. *In*: T. Kira & T. Umesao (eds.), Nature and Life in Southeast Asia **3:** 1–44.

Howe, M. A. 1934. The Hepaticae (chiefly *Riccia* and *Anthoceros*) of the Galapagos Islands and the coast and islands of Central America. Proc. Calif. Acad. Sci. IV, **21:** 199–210.

Iwatsuki, A. & B. C. Tan. 1979. Checklist of Philippine mosses. Kalikasan **8:** 179–210.

Jack, J. B. & F. Stephani. 1894. Hepaticae in insulis vitiensibus et samoanis a D^re. Ed. Graeffe anno 1864 lectae. Bot. Centralbl. **60:** 97–109.

Johnson, A. 1980. Mosses of Singapore and Malaysia. Singapore University Press, Singapore.

Jovet-Ast, S. 1947–1950. Hépatiques des Antilles françaises récoltées par P. & V. Allorge en 1936. I. Rev. Bryol. Lichénol. **16:** 17–46, 1947; II. **17:** 24–34, 1948; III. **18:** 35–42, 1949; IV. **19:** 24–31, 1950.

————. 1949. Les hépatiques de Madagascar. Mém. Inst. Sci. Madagascar, sér. B, **1:** 39–42.

Kashyap, S. R. 1929. Liverworts of the western Himalaya and the Punjab plain. Part 1. University of the Punjab, Lahore.

Kis, G. 1985. Mosses of south-east tropical Africa: An annotated list with distributional data. Inst. Ecol. & Bot., Hungarian Acad. Sci., Vácrátót.

Kuo, C.-M. & T.-Y. Chiang. 1987. Index of Taiwan mosses. Taiwania **32:** 119–207.

Long, D. G. 1979. Hepaticae from Bhutan, east Himalaya. Lindbergia **5:** 54–62.

López F., M. 1976. Lista preliminar de musgos del estado Mérida. Revista Fac. Farm. Univ. Los Andes **18:** 31–39.

Low, K. S., C. K. Lee, S. T. Loi & A. Phoon. 1985. The use of the moss, *Calymperes delessertii* Besch., as a bioindicator to airborne heavy metals. Pertanika **8:** 109–114.

Magill, R. E. 1981. Mosses. Fascicle 1. Sphagnaceae–Grimmiaceae. Flora of Southern Africa. Bryophyta **1:** 1–291.

————. 1987. Mosses. Fascicle 2. Gigaspermaceae–Bartramiaceae. Flora of Southern Africa. Bryophyta **1(2):** 293–443.

Manuel, M. G. 1981. A generic moss flora of peninsular Malaysia and Singapore. Fed. Mus. J. **26(2):** 1–163.

Miller, H. A. 1968. Bryophyta of Guam and northern Micronesia. Micronesica **4:** 49–83.

———— & C. E. B. Bonner. 1963. Hepaticae. *In* H. A. Miller et al., Bryoflora of the atolls of Micronesia. Beih. Nova Hedwigia **11:** 42–72.

———— & K. W. Russell. 1975. Key to the mosses of Puerto Rico. Florida Sci. **38:** 175–182.

————, H. O. Whittier & B. A. Whittier. 1978. Prodromus florae muscorum Polynesiae with a key to genera. Bryophyt. Biblioth. **16:** 1–334.

————, ———— & ————. 1983. Prodromus florae hepaticarum Polynesiae with a key to genera. Bryophyt. Biblioth. **25:** 1–423.

Mitten, W. 1861. Hepaticae Indiae orientalis: an enumeration of the Hepaticae of the East Indies. J. Proc. Linn. Soc. Bot. **5:** 89–128.

————. 1869. Musci Austro-americani. J. Linn. Soc. Bot. **12:** 1–659.

————. 1873. Muscineae. Pages 378–419 *in*: B. Seeman, Flora Vitiensis. L. Reeve & Co., London.

————. 1887. The mosses and Hepaticae collected in central Africa by the late Right Rev. James Hannington. J. Linn. Soc. Bot. **22:** 298–329.

Pagán, F. M. 1939. A preliminary list of the Hepaticae of Puerto Rico, including Vieques and Mona Island. Bryologist **42:** 1, 2, 37–50, 71–82.

————. 1942. Catalogue of the Hepaticae of Guadeloupe. Bryologist **45:** 76–110.

Pearson, W. H. 1922. Hepaticae. *In*: A systematic account of the plants collected in New Caledonia and the Isle of Pines by Mr. R. H. Compton, M.A., in 1914. Part III. Cryptogams. J. Linn. Soc. Bot. **46:** 13–44.

Pócs, T. Prodrome de la bryoflore du Vietnam. Acta Acad. Paed. Eger II, **3:** 453–495.

Potier de la Varde, R. 1927. Mousses de l'Oubangui, Arch. Bot. Mém. **1(3):** 1–152.

————. 1936. Mousses du Gabon. Mém. Soc. Natl. Sci. Nat. Math. Cherbourg **42:** 1–271.

Pursell, R. A. 1973. Un censo de los musgos de Venezuela. Bryologist **76:** 473–500.

———— & W. D. Reese. 1982. The mosses reported from New Caledonia. J. Hattori Bot. Lab. **53:** 449–482.

Reese, W. D. 1961. The genus *Calymperes* in the Americas. Bryologist **64:** 89–140.

————. 1977. The genus *Syrrhopodon* in the Americas I. The elimbate species. Bryologist **80:** 2–31.

————. 1978. The genus *Syrrhopodon* in the Americas II. The limbate species. Bryologist **81:** 189–225.

Renauld, F. 1897. Prodrome de la flore bryologique de Madagascar, des Mascareignes et des Comores. Monaco.

————. 1909. Essai sur les *Leucoloma* et supplément au prodrome de la flore bryologique de Madagascar, des Mascareignes et des Comores. Monaco.

Robinson, H. 1967. Preliminary studies on the bryophytes of Colombia. Bryologist **70:** 1–61.

Salazar A., N. & M. R. Crosby. 1985. Distribución y diversidad de la flora de musgos de Panamá. *In*: The botany and natural history of Panama. Monogr. Syst. Bot. Missouri Bot. Gard. **10**: 49–52.

Schiffner, V. 1893. Über exotische Hepaticae, hauptsächlich aus Java, Amboina und Brasilien, nebst einigen morphologischen und kritischen Bermerkungen über *Marchantia*. Nova Acta Acad. Caes. Leop.-Carol. **60**: 217–316.

———. 1898. Conspectus Hepaticarum Archipelagi Indici. Staatendruckeri, Batavia.

———. 1900. Die Hepaticae der Flora von Buitenzorg. Fl. Buitenzorg **4**: 1–220. E. J. Brill, Leiden.

Schuster, R. M. 1966–1980. Hepaticae and Anthocerotae of North America east of the hundredth meridian. **1**: 1–802, 1966; **2**: 1–1062, 1969; **3**: 1–880, 1974; **4**: 1–1334, 1980. Columbia University Press, New York.

Scott, G. A. M. 1985. Australian liverworts. Australian Flora and Fauna Series number 2. Australian Government Publishing Service, Canberra.

——— & I. G. Stone. 1976. The mosses of southern Australia. Academic Press, London.

Sehnem, A. 1969–1980. Musgos sul-brasileiros. Pesquisas Bot. **27**: 1–36, pls. I–V, 1969; II. **28**: 1–117, 1970; III. **29**: 1–70, 1972; IV. **30**: 1–79, 1976; V. **32**: 1–170, 1978; VI. **33**: 1–149, 1979; VII. **34**: 1–121, 1980.

Sim, T. R. 1926. The Bryophyta of South Africa. Trans. Roy. Soc. S. Afr. **15**: 1–475.

Spruce, R. 1885–1886. Hepaticae Amazonicae et Andinae. Trans. Proc. Bot. Soc. Edinburgh **15**: 1–588. [Reprinted in: Contr. New York Bot. Gard. **15**. 1985.]

Steere, W. C. 1948. Contribution to the bryogeography of Ecuador. I. A review of the species of Musci previously reported. Bryologist **51**: 65–167.

Stephani, F. 1889. Hepaticae Australiae. I–III. Hedwigia **28**: 128–135; 155–175; 257–278.

———. 1895. Hepaticarum species novae VII–VIII. Hedwigia **34**: 43–65; 232–253.

———. 1898–1924. Species Hepaticarum. **1**: 1–113, 1898–1900; **2**: 1–615, 1901–1905; **3**: 1–693, 1906–1909; **4**: 1–824, 1909–1912; **5**: 1–1044, 1912–1917; **6**: 1–763. 1917–1924. Georg., Geneva.

———. 1901–1902. Hepaticae novae Dussiana, I and II. *In*: I. Urban, Symbollae Antill. **2**: 469–472, 1901; **3**: 275–279, 1902.

——— & W. W. Watts. 1914. Hepaticae australes. J. Proc. Roy. Soc. New South Wales **48**: 94–134.

Tixier, P. 1962. Bryophytes du Vietnam. Premières récoltes dans le massif de Back-Ma. Rev. Bryol. Lichénol. **31**: 190–203.

———. 1966. Bryophytes du Vietnam. Récoltes de A. Petelot et V. Demange au Nord Vietnam (relictae Henryanae). Rev. Bryol. Lichénol. **34**: 127–181.

———. 1973. Bryophytae Indosinicae. Lists of western Indochina mosses (Assam, Chittagon, Burma). Nat. Hist. Bull. Siam Soc. **25**: 67–132.

———. 1979. Bryogéographie du Mont Bokor (Cambodge). Bryophyt. Biblioth. **18**: 1–121.

Touw, A. 1968. Miscellaneous notes on Thai mosses. Nat. Hist. Bull. Siam Soc. **22**: 217–244.

Váňa, J., T. Pócs & J. L. De Sloover. 1979. Hépatiques d'Afrique tropicale. Lejeunia II, **98**: 1–23.

Vanden Berghen, C. 1953. Quelques hépatiques récoltées par O. Hedberg sur les montagnes de l'Afrique oriental. Svensk Bot. Tidskr. **47**: 263–283.

———. 1965. Hépatiques récoltées par le Dr. J. J. Symoens dans le region peri-Tanganyikaise. Bull. Soc. Roy. Bot. Belg. **98**: 129–179.

———. 1972. Hépatiques et anthocerotées du Shaba. *In*: J. J. Symoens, Exploration hydrobiologique du Bassin et du Lac Bangweolo et du Luapula 7.

———. 1973. Quelques hépatiques récoltées au Gabon par G. Le Testu. Rev. Bryol. Lichénol. **34**: 365–385.

———. 1977. Hépatiques épiphylles récoltées par J. L. De Sloover au Kivu (Zaire) au Rwanda et au Burundi. Bull. Jard. Bot. Nat. Belg. **47**: 199–246.

———. 1978. Hépatiques épiphylles récoltées en Rhodesie. Rev. Bryol. Lichénol. **44**: 443–452.

Vohra, J. N. 1983. Hypnobryales suborder Leskeineae (Musci) of the Himalayas. Rec. Bot. Surv. India 23: 1–336.

Watts, W. W. 1901. Notes on some Richmond River hepatics. Proc. Linn. Soc. New South Wales **26**: 215, 216, 633, 634.

———. 1902. Some New South Wales hepatics. Proc. Linn. Soc. New South Wales **27**: 493, 494.

———. 1904. Further notes on Australian hepatics. Proc. Linn. Soc. New South Wales **29**: 558–560.

Whittier, H. O. 1975. A preliminary list of Fijian mosses. Florida Sci. **38**: 85–106.

———. 1976. Mosses of the Society Islands. University Presses of Florida, Gainesville.

Yano, O. 1975. Leucobryaceae (Musci) do estado de São Paulo. Dissertation, Escola Paulista de Medicina.

———. 1981. A checklist of Brazilian mosses. J. Hattori Bot. Lab. **50**: 279–456.

———. 1984. Checklist of Brazilian liverworts and hornworts. J. Hattori Bot. Lab. **56**: 481–548.

Review of Mycological Studies in the Neotropics

Florence H. Nishida

Contents

I.	Introduction	495
II.	Historical Collections	495
III.	Exsiccati	497
IV.	Early Twentieth Century Examples	497
V.	Recent Activity	498
VI.	Rust Fungi	498
VII.	Myxomycetes	499
VIII.	Fungus-Insect Relationships (Ectoparasitic Fungi)	500
IX.	Soil Fungi	500
X.	Endomycorrhizal Fungi	501
XI.	Coprophilous Fungi	501
XII.	Host-Specific Fungi	502
XIII.	Aquatic Hyphomycetes	502
XIV.	Marine Fungi	502
XV.	Lichenized Fungi	503
XVI.	Ascomycetes	504
	A. Discomycetes	504
	B. Pyrenomycetes	505
	C. Powdery Mildews	506
	D. Foliicolous Fungi	506
XVII.	Basidiomycetes	507
	A. Polypores	507
	B. Corticiaceae and Stereaceae	508
	C. Jelly Fungi	508
	D. Mycoparasitism	509
	E. Agarics	509
	F. Rolf Singer	510
	G. Fungal Ecology; Ectomycorrhizae	510
	H. Mexico	511
	I. Brazil	512
	J. Other Areas Studied	512
	K. Gasteromycetes	512
XVIII.	Remarks on Fungal Distribution	512
XIX.	Conclusions and Recommendations	512
XX.	Acknowledgments	514
XXI.	Literature Cited	514

I. Introduction

Neotropical mycological activity historically parallels that of other botanical groups, although knowledge of it is not as widespread. Circumstances of political and economic developments in the New World, as well as the training and nature of fungal interests of early mycologists influenced the beginnings of neotropical mycology. The earliest botanical and zoological explorations of the neotropical regions were accomplished by such legendary naturalists as Condamine, who in 1735 led a ten year expedition in South America, including descending the Amazon River; Humboldt and Bonpland, who traveled extensively in Colombia, Venezuela, Ecuador, Peru, Mexico, and Cuba from 1799 to 1804; Martius, who traveled in Brazil, Colombia, and Peru from 1817 to 1820; Darwin, on the Beagle from 1831 to 1836; Bates, entomologist and naturalist, who collected on the Amazon between 1848 and 1859, accompanied by Wallace for two years; and Spruce, botanist specializing in bryology, who traveled and collected on the Amazon and its tributaries from 1849 to 1864. An interesting account of these first explorations is given by B. Lowy in his introduction to the neotropical Tremellales (1971b). However, despite the abundant plant material collected, forming the foundation for neotropical systematic botany, fungi were overlooked except for some specimens collected by Humboldt, Spruce, and Darwin. Mycology, especially neotropical mycology, received scant attention by botanists until the publication of Fries' *Systema Mycologicum* in 1821.

Neotropical activity relating to fungi passed through four different periods. The eighteenth and early part of the nineteenth century was highly exploratory, with all manner of fungi collected by those eager to visit a new, "exotic" region being sent back to Europe for study by French, German, and English mycologists, who described a wide range of fungi. The late nineteenth century was dominated by interest in plant pathogenic fungi, due to the importance of export crops. Many "leaf spot" specimens were studied by Europeans, but the early involvement of American mycologists, mostly trained or influenced by European mycology, can be seen. In the early decades of the twentieth century, even into the late 1950s in some cases, mycological work was influenced by political and economic events which stimulated a greater U.S. interest in Latin America, e.g., the annexation of Puerto Rico, the completion of the Panama Canal, Neotropics-based U.S. agriculture (bananas, coffee, cacao), and attendant U.S. Department of Agriculture interest, such as surveys of natural resources. These developments involved increasing numbers of mycologists trained in the United States. Finally, recent years (since 1960), have witnessed an emphasis on systematics, the support of basic research, greater involvement of in-country mycologists, and more international cooperative programs, such as The Organization for Flora Neotropica, Projeto Flora Amazônica (PFA), as well as support of field work by research agencies of various countries.

In the last 15 years a widening of fungal research in general, has contributed to the understanding of natural taxonomic groups. The fields of cytology, genetics, cultural characters, mycorrhizal relationships, fungus-insect relationships, the study of specialized habitats, such as mangroves, the marine environment, the phyllosphere (leaf surface habitat), and of course, the investigation of the mycologically under-explored neotropical regions have all become important. Mycorrhizae have become a subject of intense research among temperate and tropical mycologists but have, so far, not received much attention from tropical botanists. Endomycorrhizal relationships are estimated to occur among 90% of the plants in tropical ecosystems (J. Menge, pers. comm.). Since fungi play a significant role in the recycling of nutrients through litter decomposition (of extreme importance in nutrient-poor tropical soils), as well as a vital symbiotic role in plant acquisition of mineral nutrients through the soil solution, discussions of tropical forest systems need to include mycological studies.

II. Historical Collections

The first records of fungi from Central and South America and the Caribbean are found in citations from 1774 (Lindau & Sydow, 1908–1917). Jean Baptiste Fusée Aublet, French explorer and botanist known as the founder of the knowledge of the flora of Guyana, is recorded as the first person publishing on neotropical fungi. The earliest reports, however, of fungi in the New World, are shared by Brazil and Mexico. In 1560 José de Anchiéta, a Jesuit living among the indigenous people in São Paulo, writing to his superior in Portugal, described a stone which was "flexible in several planes." This is probably the first report of the sclerotium (vegetative portion) of a stipitate polypore, *Polyporus saprema* Möller. Known from southern Brazil as "pão do indio," or Indian bread (Viégas, 1959), the dense mycelial mass can attain a weight of 28–60 kgs. Since only the sclerotium had been collected in nature, its connection to the genus *Polyporus* was unknown until it was induced to fruit by Möller in the laboratory at the Botanic Garden of Berlin-Dahlem in 1897, producing a mushroom out of a sclerotium weighing 20 kg (Gonçalves, 1937). Recently, a new species of "pão do indio" (sclerotium only) was reported from Amazonia (Araujo & Sousa, 1981) and was induced to fruit in the laboratory. Since then a "pão do indio" fruiting in nature, probably *P. saprema* and not *P. indigenus*, has been collected by the author in Acre, Brazil. Sahagun's reference in the 1500s to fungi used by the indigenous people in the Valley of Mexico appeared in Kingsborough's *Antiquities of Mexico* in 1837.

Trends in the way mycological knowledge was acquired in the Neotropics is evident from the titles, dates, and annotations in literature sources. Lindau and Sydow's thesaurus of mycological literature, the first fifteen volumes of P. A. Saccardo's *Sylloge Fungorum* (1882–1901) and J. Stevenson's (1971) account of exsiccati issued from the Neotropics substantiate considerable activity by European and American botanists. In the eighteenth and especially the nineteenth centuries, over 200 people were associated with neotropical collecting or studies of fungi. Still, by 1895, only about one percent of the total number of fungal species thus far recorded had come from tropical regions (Lowy, 1971b).

The West Indies and South America yielded the first collections, dating from 1774 to 1800. Fungus collecting gradually increased from 1800 to 1850 and more so in the latter half of the nineteenth century, accompanied by a change in the nature of mycological activity. The earliest workers were general collectors who sent their material to Europe for identification. Neotropical fungi were sent out of the region and curated by French, English, and German authors, who named and described them without ever having seen them in their natural habitat or their natural variability. By the beginning of the twentieth century, the trend had changed. Mycologists who were collector-authors (still mainly Europeans) were replacing the botanist-collectors.

Most of the long-distance tropical mycologists had an impressive ability to deal with a wide range of fungi; nevertheless, the not uncommon phenomenon of a tropical fungus being described over and over, sometimes by the same mycologist, is attributable to the lack of konwledge of the fungus in the field, and in the case of macrofungi, sometimes due to reliance upon macroscopical characters alone. *Xerulina asprata* (Berk.) Pegler, a very distinctive, pantropical agaric has 26 synonyms in 14 genera (Redhead & Ginns, 1980; Pegler, 1983a). Polypores, among the most frequently collected tropical fungi (easy to spot and easy to carry back), have been very susceptible to such multiple naming due to environmentally produced morphological variations.

Accounts of early mycological activity in the Caribbean region are given by Pegler (1983a, 1983b) and Stevenson (1975). A few examples illustrate the nature of nineteenth century tropical mycology in which temperate-region workers, often non-specialists, described fungi collected by others. The Reverend M. J. Berkeley (England), whose broad range of interests included molluscs, mosses, lichens, and algae, "gave accounts of fungi from the Colonies, [and] America," describing 6,000 new species in all (Hawksworth et al., 1983). He began publishing on neotropical fungi in 1839 from specimens sent to him from Brazil, British Guiana, Surinam, Puerto Rico, Mexico, Santo Domingo, and Cuba (e.g., Berkeley, 1867; Berkeley & Cooke, 1876). Sources included fungi sent to the Hookers, successive directors of the Royal Botanic Gardens, Kew, by various collectors, including Darwin. With M. A. Curtis (American minister and student of American fungi), Berkeley published in 1851, on fungi collected by the United States Exploring Expedition. With Curtis and J. P. F. C. Montagne (a former medical man in Napoleon's army and student of French fungi), Berkeley published a series of articles entitled "Decades of fungi," including data on neotropical fungi. The Berkeley herbarium is now located at Kew. Recent type studies (Ryvarden, 1975, 1984) reveal the extent to which identification of material from a completely different region, without knowledge of fresh field characters, but with heavy reliance on gross morphological characters can lead one astray. Of Berkeley's 336 described polypores, 219 are synonyms.

J. P. F. C. Montagne also began working with neotropical cryptogams in 1839 with a treatise on collections made in Brazil by A. de Saint-Hilaire. An early plant specialist who also concentrated on fungi was F. R. Leprieur in French Guiana. Montagne described many new taxa between the years 1837 to 1857 from Leprieur's 40 years' collecting, as well as fungi sent to him by others from Cuba, Mexico, Brazil, and Venezuela. Patouillard (in Paris) published several new taxa between 1889 and 1903 from collections sent to him by the Rev. R. P. A. Düss from Martinique, where he resided. C. Weigelt was sent to Surinam by the government of Saxony in 1827 to make botanical collections. He made an enormous collection of fungi as well, before he died a year later. A colleague provided the Rev. von Schweinitz, the first important American mycologist, with the collections (Hawksworth et al., 1983). He worked up the specimens and wrote brief diagnoses, but did not publish them. At his death, some of the collections were published by Berkeley and Curtis in 1853. From the former Danish West Indies, now the American Virgin Islands, fungi collected by Benzon, Oersted, and other Danish travelers were described by E. Fries in Sweden (1851). Another enthusiastic plant collector, C. Wright, an American botanist who had participated in surveys of the U.S. Southwest and the Mexican boundary, traveled around the world on the Ringgold and Rodgers North Pacific Exploring Expedition, all in the 1850s, and finally spent nearly eleven years in botanical collecting in Cuba. His accumulated fungal collection of 1,600 specimens were sent to Berkeley and M. A. Curtis, who described over 600 new species (Berkeley & Curtis, 1869). Collections were ultimately deposited both at Kew and at the Farlow Herbarium. Spegazzini (1899) published on Argentine fungi.

As Pegler (1983a) has pointed out, many of these new generic and specific descriptions were based on specimens which had been collected by non-specialists and sent to European mycologists. The material was often poorly preserved and annotated; consequently, many of the early described species are no longer recognized. Nevertheless, the importance of tropical collections being available to temperate region mycologists cannot be over-emphasized. Though Patouillard's collecting experience of "exotics" was limited to field work in North Africa, he took a keen interest in them and received numerous fungi from South America as well as Asia by the 1880s. More material became available as the French extended their protectorates through the tropical regions, including the Caribbean. His careful work, descriptions in copious detail, and consistent use of microscopic characters made him an important consultant for mycologists throughout the world (Pfister, 1977). He described 1,100 new species and 111 new genera in all, mostly from tropical and subtropical regions. The Farlow Herbarium acquired his herbarium, which contained 50,000 specimens, in 1927 after his death. His pioneering investigations of the basidium resulted in his recognition of two principal groups of Basidiomycetes, the Heterobasidiomycetes and Homobasidiomycetes.

III. Exsiccati

Popular devices for the distribution of neotropical fungal collections were the *exsiccati*, defined as the issuance of uniform series of dried specimens (Stevenson, 1971). These sets of specimens, intended to be replicates, were distributed by mycologically interested people to other fungal workers and to herbaria. This practice probably grew out of a need for specimens for comparative study, and those from the tropical regions were especially sought. Until the 1953 prohibition by the International Code of Botanical Nomenclature, the exsiccati label was used to describe new taxa or make nomenclature changes. The practice provided a small but useful income for collectors.

Fungi from the American Tropics were distributed in this form in the latter part of the nineteenth century and into the twentieth. Culberson (1959) and Stevenson (1971) discussed 34 exsiccati issued from 1856 to 1964 of lichens and other fungi that included neotropical material. Europeans distributed exsiccati from Brazil, British Guyana, Guatemala, Colombia, Costa Rica, Cuba, the Dominican Republic, Martinique, Mexico, and Surinam. Hans Sydow collected fungi in Costa Rica, Ecuador, and Venezuela. In a preliminary announcement of his *Fungi Exotici Exsiccati*, issued in 1912, he promised to issue fungus material that was then poorly represented in mycological herbaria, in an effort to lay the ground work for usable fungus floras of the non-European countries (Stevenson, 1975). Other exsiccati prepared and published in the United States contained specimens from the Dominican Republic, Guatemala, Mexico, Nicaragua, Puerto Rico, and South America (Argentina, Bolivia, Brazil, Ecuador).

Neotropical collectors or mycologists who have issued geographically focused exsiccati are: J. Rick, R. Spruce, F. Theissen, C. Torrend, E. Ule, and E. A. Vainio (Brazil); C. Wright (Cuba); R. Ciferri (the Dominican Republic); W. A. Kellerman (Guatemala); C. G. Pringle (Mexico); C. L. Smith (Mexico and Nicaragua); and the A. A. Hellers (Puerto Rico). The Hellers' collections, including 25 new species, were described by F. S. Earle (1901, 1904), a mycologist at the New York Botanical Garden, and ultimately distributed to large herbaria in Europe and America. Rick's exsiccati were sent to Bresadola (Italy), C. G. Lloyd (U.S.A.), H. Rehm and H. & P. Sydow (Germany). Ule sent material to Berlin where it was studied by P. Hennings and Rehm, but much of this was destroyed during World War II. Fortunately some Ule collections were also sent to Hamburg (Friederichsen, 1973). The recent find of A. Möller's material by Oberwinkler (Friederichsen, 1977) in Hamburg has allowed their study by current mycologists. While Rick collected in a generally accessible area, Rio Grande do Sul, in southern Brazil, Ule was the rare mycological collector to visit the Amazon valley in the early period. He made an extended expedition there, even into Amazonian Peru, from June 1900 to March 1903.

IV. Early Twentieth Century Examples

Neotropical mycology near the end of the nineteenth century became the domain of the specialist rather than the generalist identifier of species. And the trend for mycologists to do their own field work, as well as identification, grew. Systematic treatment of fungal orders, families, genera, and not simply description of new species became the focus of studies. A review of the mycological history of Puerto Rico as related by Stevenson (1975) shows the characteristic pattern: fungi were collected by botanists from Italy and Germany in the early 1800s; collections were sent to Europeans, e.g., Bresadola, P. Hennings, and P. Magnus, for identification in the late nineteenth century; American botanists collected considerable numbers of fungi and distributed them in exsiccati form; American mycologists visited the island to investigate numerous pathogenic fungi (F. S. Earle, 1901, 1904; E. W. D. Holway, 1911; H. H. Whetzel and E. W. Olive, 1915–1916). J. C. Arthur (1925) published on the rust fungi collected by several mycologists.

F. L. Stevens became a neotropical resident from 1912 to 1914 while a dean at the University of Puerto Rico and collected several thousand fungi from practically every part of the island and the American Virgin Islands. He contributed greatly to the taxonomic mycology of Puerto Rico and other neotropical regions. The first native-born mycologist, Carlos Chardón, who trained at Cornell University with H. H. Whetzel (Uredinales) and H. M. Fitzpatrick (Ascomycetes), returned to Puerto Rico in 1921 and thereafter collected in Venezuela, Colombia, and the Dominican Republic and had a productive career in taxonomic mycology and plant pathology. He established a scientific survey project for Puerto Rico and the Virgin Islands, carried out under the auspices of the New York Academy of Sciences with the cooperation of the Puerto Rican government, which resulted in numerous publications besides that of the scientific survey *per se* (Seaver & Chardón, 1926; Seaver, Chardón et al., 1932). This brought numerous specialists to the region and contributed to the growing "modernism" of mycology—the emphasis on specialization. Although plant pathogens were still considered important for collecting and study, other American specialists were invited to participate (R. Hagelstein, Myxomycetes; F. Seaver, Discomycetes; R. A. Toro, pyrenomycetes; L. O. Overholts, polypores (e.g., 1934); and B. Fink, lichenized fungi), making the mycoflora of Puerto Rico better known to date than any other of comparable size (Fitzpatrick, 1927).

William A. Murrill (1869–1957), an active and avid collector (more than 70,000 specimens of agarics and polypores), fails to fit the pattern above, having visited the "exotic" places (tropical Mexico, Cuba, and Jamaica) earlier (1904–1909) than most American mycologists, to collect non-pathogenic fungi. As assistant director of the New York Botanical Garden (1908–1919) and editor of "Mycologia," he also led an active publishing career as author of more than 500 papers and articles, including " Agaricaceae of tropical North America" (1911-1918). His descriptions were occasionally hasty, and one suspects, sometimes made without resort to a microscope. His collections and descriptions nevertheless provide a valuable storehouse of fungi from regional vegetations which are rapidly disappearing. Murrill's localities of "virgin tropical forest" in southern Mexico, for example, have become poor agricultural or meadow lands (Guzmán, 1983a).

Otero & Cook's *Bibliography of mycology and phytopathology of Central and South America, Mexico and the West Indies* (1937) demonstrates the emphasis in mycological activity of the 1920s, the 1930s, and continued into the 1940s, upon foliicolous and plant pathogenic fungi. Due to the importance of food crops such as banana, cacao, sugar cane, sweet potato, and manioc, research on fungi from the days of colonization to the present has concentrated on phytopathology. A general work on the ascomycetes of Brazil is Viégas (1944). Even now, among the numerous Brazilian mycologists listed (Fidalgo & Bononi, 1979), 143 out of 169 are phytopathologists, and 106 others are medical mycologists. Only a scant 20 are systematists.

A very active mycologist of that period and interest was R. Ciferri (1897-1964), born in Italy, but living in Santo Domingo from 1925 to 1932. He worked on a variety of fungi from the Dominican Republic, Cuba, and Haiti, including phytopathogens, publishing on Uredinales (rusts) (Kern & Ciferri, 1930), Ustilaginales (smuts), and Fungi Imperfecti (Baldacci, 1965). His efforts increased the known mycoflora of the Dominican Republic to 2000 species, including 400 new species (Ciferri, 1961). After his return to Italy, he collaborated with A. Chavez Batista, Recife, Brazil, in work on foliicolous, ascomycetous fungi (Batista & Ciferri, 1963). John A. Stevenson's professional life spanned this period, beginning with his employment as a plant pathologist for the Puerto Rico Department of Agriculture (1914-1918), where he acquired a fluency in Spanish and a life-long communication and involvement with neotropical mycologists. He published in collaborations on fungal pathogens in Puerto Rico, Peru, El Salvador, Chile, Bolivia, Nicaragua, on *Hevea* (Brazilian rubber tree) blight, and coffee rust, and concluded his career with two extremely useful references for neotropical mycology (1971, 1975).

The historical pattern of mycological activity in the Neotropics was from generalist-collector to mycologist-collector and from European generalist mycologists identifying a variety of fungi sent to them, to European collector-authors and specialists, and finally to American mycological specialists. Brazil and Mexico had the most nineteenth century mycological attention and this trend continues, followed by Puerto Rico and Cuba. Between 1850 and the 1920s, Central America received the most attention due to the importance of agriculture.

V. Recent Activity

Recent mycological activity reflects a broad range of fungal groups and methods. While emphasis is upon systematics, extremely large geographical areas are still unsampled so that extensive monographic work must await the results of greater efforts in field work. Some fungal groups have received greater regional attention; others have had the attention of uniquely active specialists. The growing knowledge for most fungal groups is being enhanced by culture work, cytological studies, cooperative collecting efforts with resident workers, and some expedition efforts (although the emphasis is still far greater upon plant rather than fungus collection). The author apologizes for the omission of numerous resident mycological workers. Lack of sufficient time has made a thorough perusal of literature published in the Neotropics impossible. Information on current activity was based primarily on direct responses from surveyed mycologists, including requests to them for names of other neotropical mycologists, and a review of publications which have appeared in major mycological journals over the last 25 years.

It is difficult to say exactly how many fungi there are, and especially so for the largely uncollected regions of the Neotropics. G. W. Martin, in 1951, thought Bisby and Ainsworth's 1943 estimate of total fungi at 100,000 to be extremely conservative, based on his survey of lists of parasitic fungi and their plant hosts, which showed a 3:1 ratio (Ainsworth & Sussman, 1968). He concluded that the number of fungal species probably was of the same order of magnitude as that of the vascular plants, ca. 250,000. If fungus-plant specialization has multiplied fungal species, and with new substrates being investigated such as invertebrates, soil, dung, and marine environment, it can be suggested that the potential number of undiscovered fungal taxa for a region in which more than half of the world's species of plants and insects are said to live, would be truly enormous.

VI. Rust Fungi

Systematic mycologists of today are more concerned with clarifying natural taxonomic groups, than with taking a phytopathological approach to fungal studies. Yet, an important order of obligate plant parasites, the Uredinales (rust fungi), is still in need of basic treatment in the Neotropics, i.e., finding and naming. Estimated to be a very large group (6,000 species) with a cosmopolitan distribution (Hawksworth et al., 1983), its variety of life cycles, spore states, and sometimes hosts, have been well studied in regional treatments in the temperate zones, but extensive collecting in the Neotropics is still needed for a complete flora and distribution record. Two important early rust collectors stand out. E. W. D. Holway made the most extensive collections from the Neotropics (Mexico, Guatemala, Costa Rica, Cuba, Puerto Rico, Peru, Bolivia, Brazil, and Ecuador). Both J. C.

Arthur (1925) and H. S. Jackson (1926–1932) published much on Holway's fungi and other neotropical collections. E. Ule's extensive collections from Brazil were determined mainly by P. Hennings and the Sydows (Stevenson, 1975). Hennen et al. (1982) cited most of Hennings' and the Sydows' publications on neotropical rusts. Rusts from Venezuela and the West Indies were studied by F. D. Kern and H. W. Thurston (Thurston & Kern, 1933; Kern & Thurston, 1944) and H. Sydow (1930). Colombia's first surveys by Chardón and Toro (1930) were done in areas near large cities on the coast, such as Medellín, and in the cordilleras, encouraging Colombian workers to continue work on parasitic fungi. G. B. Cummins (1940) identified collections sent to him from Mexico and Venezuela; I. Jorstad, 1959, identified South American rusts collected by Swedish botanists; Benjamin and Slot (1969) included rusts in their survey of the fungi of Haiti.

In recent years (Hennen & Cummins, 1967, 1973a, 1973b, 1973c; Carrión & Galván, 1984), the rusts of Mexico, Central America, and the Caribbean have been studied. Rusts have been collected in Colombia, Ecuador, and Peru since 1967 by Buritica (1978), who has discussed the use of teliospore ontogeny as a criterion for rust phylogeny (Buritica, 1980). In Brazil, the presence of numerous phytopathologists would suggest the availability of abundant collections, added to the historical collections of the nineteenth century. The last rust treatments for that region cited in recent literature (Buritica & Hennen, 1980), however, are the works of Thurston in 1940 for Minas Gerais, and Viégas in 1945. Systematic treatments of floras of rusts are still lacking for Bolivia and Ecuador.

Despite these scattered studies, tropical regions are still poorly known for rusts. Because natural vegetation there is being destroyed so rapidly, the highest research priority needs to be given to basic field and herbarium inventory studies (J. Hennen, pers. comm.). In a modern treatment of rusts (Buritica & Hennen, 1980), developmental studies are used as a basis for the taxonomy of the tribe Pucciniosireae (Pucciniaceae). The Flora Neotropica monograph (#24) includes type studies, host and distribution information. Buritica and Hennen's recent field work strongly suggested to them the potential for rust diversity in the Tropics. During the first visits to the site (a 100 acre tract), 25 species were found; at the end of the period, 100 species had been found, 20% of which were new taxa. About one third of the plant species in the area had their own rust species. If this held true in other regions of the Neotropics, "rust diversity would be truly amazing" (Hennen, pers. comm.).

VII. Myxomycetes

As with rusts, collecting of Myxomycetes has had a long history, although the motivation was completely different, the latter having no known pathogenic activity upon economic plants. Rather, their aesthetic appeal and curious nature have likely prompted their nearly universal collecting by botanists and mycologists, though they are not related to true fungi. A surprising number of regional floras (an exception, with polypores, to the general rule) are listed by Farr (1976, Flora Neotropica #16) in her thorough discussion of their collecting history and distribution: Argentina (Spegazzini, Fries, Deschamps); Bolivia (Fries, Stevenson & Cárdenas); Brazil (Torrend, Hertel, Farr, Gottsberg, Calvacanti); Colombia (Muenscher, Martin); Costa Rica (Welden); Cuba (Berkeley); the Dominican Republic (Ciferri, Toro); the Galapagos Islands (Bonar, Martin, Eliasson); Guyana and Surinam (Gilbert); Jamaica (Farr); the Lesser Antilles (Lister, Farr); Mexico (Emoto); Nicaragua (Macbride); Puerto Rico (Seaver & Chardón, Hagelstein); Trinidad and Tobago (Baker & Dale, Barnes, Dennis); and Venezuela (Patouillard & Gaillard, Rodriguez, Dennis). Recent additions to the knowledge of Myxomycetes in the Neotropics include: 47 species reported from Ecuador (Farr et al., 1979); Myxomycetes of numerous regions of Mexico (Braun & Keller, 1976, 1986; Keller & Braun, 1977), and from Veracruz (Villarreal, 1983). Places which appear to be lacking in studies altogether are Belize, El Salvador, French Guiana, Guatemala, Honduras, and Peru.

While the great majority of the known Myxomycete species (ca. 430) occur in the temperate zones, about 280 are reported from humid and lowland Tropics (Farr, 1976). More taxa would be expected if one considered the variety of habitats in the Neotropics, from arid zones like the *caatinga* in northeastern Brazil, seasonally deciduous forests, montane, alpine and cloud forests, coastal strands and islands. Because of the ease of travel of Mycomycete spores, their saprophytic lifestyle, and adaptability to a wide range of environmental conditions, distribution of Myxomycetes in the Neotropics may be, even more than for fungal groups, a reflection of the distribution of mycologists working there. About 40 species of Myxomycetes are known from deserts of North America, northern Africa, and Israel, but they are known because specialists have induced their appearance and growth in moist chambers from dung or plant material collected in deserts. Survival of plasmodial sclerotia at 60°C for up to 32 days was shown (Blackwell et al., 1984). Searches for Myxomycetes in desert regions of the Neotropics ought to be an interesting project.

Farr reports some particulars of slime mold occurrence. Unexpectedly, they are rare inside dark, moist, tropical forests with abundant substrates, and generally rare on decomposing wood. Also, while temperate species exhibit seasonal activity, being less active during the winter, tropical and subtropical species are more inclined to respond primarily to the amount and distribution of rainfall. Ecological considerations were suggested by Farr, and Alexopoulos and Saenz (1975), such as competition by filamentous fungi due to the high humidity, lack of air movement for spore entry and travel within the deep forest, action of torrential rain in mechanical wear, acid conditions of substrates in the tropical forest, and invertebrate competition or predation. These are all questions worthy of investigation not only for Myxomycetes, but for true fungal groups as well. Careful and repeated collecting reveals patterns not observed by the occasional or one-time collector. Under certain conditions, especially of adequate moisture, some Myxomycetes fail to produce sporangia. Because that form of the life cycle is the most easily observed, the difficulty of finding large numbers of Myxomycetes in the wet Tropics may be due to the low frequency of sporangia formation there (Blackwell, 1984).

In a cross-disciplinary study of neotropical beetles feeding on a slime mold, Wheeler (1980) reported the discovery in montane Panama of a beetle, *Anisotoma*, which was previously known only from montane oak and pine forests of Mexico and Guatemala; no species are known from South America. Additional collecting of *Anisotoma* is thought to be extremely useful for analysis of biogeographic relationships and evolutionary hypotheses relating to this monophyletic insect group. Information on insect feeding on Myxomycete sporangia and spores suggests that besides benefit to the insect, the Myomycete may benefit by direct spore dispersal onto suitable substrates, facilitation of spore germination, and elimination of competition from larger, more active colonies (Blackwell, 1984).

VIII. Fungus-Insect Relationships (Ectoparasitic Fungi)

Research in certain insect-fungus relationships (entomogenous fungi) has grown since Lichtwardt's MSA presidential address (1973) in which he raised and answered the question of whether Trichomycetes (fungi living as commensals within the guts of arthropods) are really fungi. Physiological and cytological studies confirm that they are indeed. Collections of Trichomycetes have been made in Panama, Trinidad, and Brazil; further collecting by Lichtwardt is planned in Costa Rica.

Tropical ectoparasitic Zygomycetes and Ascomycetes were reported in nineteenth century literature, but most commonly from the Paleotropics. The monographs of Thaxter (1896-1931) cover 160 species of Laboulbeniales, ascomycetous parasites of the chitin on insect bodies, mostly from Trinidad and Venezuela. Termites are a large component of the insect population in the tropical forest. Most termite parasites, except *Termitaria*, were considered to have separate distributions in the Eastern and Western Hemispheres (Blackwell & Kimbrough, 1978), until the discovery of a second species in Mexico of *Cordycepioideus* (described from Tanzania) suggested a wider occurrence (Blackwell & Gilbertson, 1981). The Neotropics are vastly understudied for ectoparasitic fungi. Since the original discovery in 1929 of a species of *Mattirolella*, parasitizing the integument of a termite in British Guiana, Khan and Kimbrough (1974) have re-determined a species of *Termitaria* collected in Panama to be a second species of *Mattirolella*. Termite colonies from Barro Colorado Island, Panama, provided information on developmental stages of *Mattirolella* (Kimbrough & Thorne, 1982). Because the fungus was collected from *Rhinotermes*, a member of the dominant subfamily of Termitidae to which belong 75% of the termites of the Neotropics, it is likely that extensive collecting and examination of these termites will increase the number of species and enhance the present biogeographical knowledge of these ectoparasitic fungi (see also Blackwell & Rossi, 1986). Future studies are needed on the impact of these obligate fungus-insect relationships.

IX. Soil Fungi

Since Waksman asked in 1917, half tongue-in-cheek, "Is there any fungus flora of the soil?" the investigation of soil fungi, since the first isolation by Adametz in 1886, has confirmed early observations. Much of the soil flora is cosmopolitan (on the generic level at least), but Penicillia seem to be more largely represented in the cooler climates, while Aspergilli seem to be more abundant in the soils of warmer climates. The more fertile soils contain more fungi, both in number and species; but there are some ecological preferences, e.g., acid and water-logged soils are richer than normal agricultural soils in numbers and species of Trichodermae. Most of the studies on the distribution of soil fungi have dealt with those in agricultural soils and less is known about their occurrence under natural soil conditions (Warcup, 1951). And it should be added, even less is known about their occurrence in natural soil conditions in the Neotropics. *Penicillium* and other soil fungi have been reported from the northeast of Brazil (Batista et al., 1965, 1967; Upadhyay, 1967, 1970); and from the Amazon in Pará (Dunn et al., 1985, 1986). Other observations have been made from isolations from soils of the Galapagos Islands (Mahoney, 1971, 1976); Honduras (Goos, 1963); Costa Rica and Panama (Farrow, 1954; Goos, 1960); and Colombia (Veerkamp & Gams, 1983).

In an overview of *Aspergillus* reported in 62 recent soil surveys, Christensen and Tuthill (1986) conclude that there are the "highest absolute and relative numbers of species in the Subtropics, particularly in cultivated soils, saline soils and subtropical deciduous forest soils." But of these soil surveys, only Peru was listed from the Neotropics. The total number of *Aspergillus* species in tropical lists was 76, of which 58 species appear restricted to tropical or subtropical soils. According to Christensen and Tuthill, an average soil survey will have 18 species of *Aspergillus*. A recent survey of soil fungi isolated from Amazonian Brazil yielded 17 species (Dunn et al., 1985, 1986). Gochenaur (1970) observes that Aspergilli account for the highest proportion of species in dry tropical soils generally, and in subtropical soils in a tree crop area. The ascigerous state, now known for 65 species, has been studied only by temperate region mycologists, mostly on temperate species. A high proportion of Aspergilli (40% of named species) are known from single localities and single habitats, a fact which increases the necessity of neotropical soil collection for the revelation of patterns of distribution. Perhaps tropical botanists and mycologists surveying the botanical resources of a region can be encouraged to collect samples of the flora beneath trees.

The main role of soil bacteria and fungi is to decompose organic residues. Fungi are important in the early stages of the breakdown of plant material, which is composed of cellulose and lignin, and in fixing large amounts of nitrogen as 'microbial' protein (Hawksworth, et al., 1983). Some of the fungi may be widely distributed in soil while others are limited to certain habitats. Their distribution is influenced by the abundance and nature of organic matter, as well as by soil and climatic conditions, and surface vegetation. Bacteria may be outnumbered by fungi when the soil is acid. Succession of some fungi in leaf litter has been well studied in temperate regions, but insufficiently in the Neotropics. Maia (1983) studied the decomposition of leaves of two species of *Licania* and of *Hortia* in a mesophytic forest (Recife, Brazil), over a period of seven to nine months, or until the leaves were indistinguishable from litter. Seventy-five fungal taxa were present, of which 70% were imperfect fungi, mainly Hyphomycetes, and 30% Zygomycetes and Ascomycetes. It was concluded that the species of host leaves did not significantly influence the fungal population. The prevailing factor in fungal succession in decomposing leaves was the stage of decomposition of such leaves, since environmental conditions appeared to be adequate for year round good development of fungi. Studies of fungal succession in litter of the various ecological habitats of the Neotropics are needed. Rates of decomposition, hence of nutrient recycling, are still not well known for many ecosystems in the Tropics.

X. Endomycorrhizal Fungi

A recent and important area of research for the Neotropics is that of fungi forming endomycorrhizae, i.e, VA mycorrhizae-forming fungi (Endogonaceae, Zygomycetes). These fungi form symbiotic relationships with most plants. New species and genera are being discovered and recognized rapidly. It is hoped that cytological investigations and culture work will reveal useful taxonomic characters. These fungi are difficult to differentiate, and are rarely obvious, since they do not usually produce fruit bodies, unlike ectomycorrhizal fungi, which also occupy the rhizosphere and hypogeous habitat. An interesting hollow, spherical fungus from Cuba was described by Berkeley and Curtis in 1869 as a *Xylaria* (Ascomycete), and subsequently collected in Brazil, and the West Indies. It was later ascribed to the Endogonaceae because its large, often empty spores resembled chlamydospores (asexual spores of that family). But ultrastructure evidence points to its belonging to the Ascomycetes instead, while its peculiar set of characters place *Glaziella aurantiaca* in its own family and order (Gibson et al., 1985). The first reported neotropical sporocarpic species of *Acaulospora* (Endogonaceae) has been described by Schenck et al. (1986) from a collection made in Colombia in 1981. The fungus occupies cryptic habitats such as soil crevices and roots, empty seed teguments, cast insect exoskeletons, and empty spores of other Endogonaceae. It is reported presently as restricted to Colombia and Peru by Schenck et al., who also advise researchers that small, hyaline-spored species of Endogonaceae, can be very easily overlooked if very fine sieves are not used in surveys for mycorrhizal fungi. Improved collecting, pot culture technique, and attention to wall structure, as proposed by Walker (1983), has enabled the description of several new species from Colombia (Schenck et al., 1984). Janos and Trappe (1982) also report new species of *Acaulospora* from Costa Rica and Panama. Collecting hindrances (ease of escape from usual collection techniques and cryptic habitats), added to the low numbers of neotropical mycologists, have influenced the acquisition of knowlege of fungi in the region far more than occurs for vascular plants. Reporting of successful techniques for recovering and identifying cryptic fungi may stimulate others to look into these habitats, which will certainly rapidly increase our knowledge of their numbers and distribution.

Mycorrhizae may influence plant succession in the Tropics (Janos, 1980). Dependence on mycorrhizae is variable among plants. Generally, by improving mineral nutrition, especially phosphorus uptake, mycorrhizae enable plants to grow better in poor soils. Non-mycorrhizae-forming plants (weedy types) are the only ones likely to grow in soils lacking an appropriate fungus. In later stages of succession, facultatively mycotrophic tree species, which can grow in fertile soils without the fungi, will compete well with the introduction of the fungus in the soil. The mycorrhizal content of a soil interacting with the varying degrees of plant dependence upon mycorrhizae can influence the competitive abilities of plants and thereby affect plant succession. For example, obligate mycotrophic species are likely to dominate plant communities on poor soils and many mature forest species are probably obligately mycotrophic (Janos, 1975a, 1975b). Janos (1980) suggested that there was apparently a lack of host specificity among endomycorrhizal fungi, and with only about 50 species described thus far (Gerdemann & Trappe, 1974), it indicated that the great plant-species richness of tropical forests, in which endomycorrhizae predominate, was probably not due to niche differentiation related to mineral uptake. But recent evidence points to some specific host-fungus reactions, despite the broad host range of the fungal symbiont, and the number of described species of VA (vesicular-arbuscular) mycorrhizae-forming fungi has risen to over 100 (Schenck, 1985).

XI. Coprophilous Fungi

Coprophilous fungi, distributed among a wide range of fungal groups (Myxomycetes, Mucorales, Ascomycetes such as Pezizales, Helotiales, Sphaeriales, Sordariales, and Basidiomycetes such as *Coprinus, Panaeolus*, and *Psilocybe*) have been collected in the Neotropics, usually on cow, burro, horse, or rodent dung. Most of the systematic work and description of new species is found in the temperate regions; next best known are India and other paleotropic areas. Among the few records of coprophilous microfungi in the Neotropics (Jeng & Cain, 1976; Jeng et al., 1977), the Flora Neotropica collecting trip to Venezuela (1972) has yielded many. Much of the neotropical forest is being changed into pastureland. Because the introduction of domesticated herbivores is relatively recent, there may be opportunities to investigate patterns of speciation from the distribution of these fungi. Dungs of many native herbivores in neotropical rainforests have not been examined for fungi. The difficulty of the collector in finding a substance (dung) which is of considerable value for creatures there and is thus quickly recycled may be the reason behind this, rather than any paucity of mycological interest. A delightful account in Forsyth and Miyata's book (1984) on the ecological aspects of life in the rainforest describes the immediate scramble by various foraging and opportunistic organisms to obtain a share of the precious resources which dung offers to the inhabitants of a nutrient poor environment. Reports of fungi isolated from lizard dung of Mexico, and bat dung from Panama (Benny & Benjamin, 1976), and coprophilous fungi collected on the volcano Popocatepetl in Mexico (Udagawa & Kobayashi, 1979) give some idea of the potential for study in a largely unexplored area, geographically as well as mycologically.

XII. Host-Specific Fungi

Some mycologists have been interested in examining fungi that occur on certain plant substrates. Fungi brought into the United States on imported orchids were brought to the attention of Edith K. Cash by the Plant Quarantine Branch of the USDA. The new species of various Ascomycetes and Deuteromycetes, several of pantropical distribution, were described from these orchids from places such as Puerto Rico, Venezuela, Mexico, Brazil and Central America (Cash & Watson, 1955). Petrini and Dreyfuss (1981) recovered 36 taxa of endophytic fungi (30 Deuteromycetes and 6 Ascomycetes) by culture work with 22 tropical epiphytes belonging to Araceae, Bromeliaceae, and Orchidaceae. The preferred substrate was Araceae (29 species), the least preferred, Bromeliaceae (16 species).

Studies of Hyphomycetes on rotting wood and specific substrates such as palm petioles and bamboo have resulted in the description of new species in Cuba (Holubova-Jechova, 1982; Holubova-Jechova & Mercado-Sierra, 1982), Venezuela (Crane & Dumont, 1975, 1978), Jamaica and Puerto Rico (Crane & Dumont, 1975). A Japanese expedition into the Peruvian Amazon also recorded Hyphomycetes found on wood, leaf litter, and soil (Matsushima, 1981, 1983). Morris (1978) described Belizean Hyphomycetes collected in 1969-1978 and briefly reviewed mycological activity in Belize through the 1930s. Since the Neotropics are rich in palm species occupying a variety of climatic and ecological areas, the opportunities are abundant for investigating substrate specificity.

XIII. Aquatic Hyphomycetes

Despite the abundant aquatic habitats which exist in the tropical region, collection of aquatic Hyphomycetes has been reported only from Cuba, Jamaica, Venezuela, and Puerto Rico (Betancourt & Caballero, 1983). Uneven and limited sampling is a problem: the area most intensively studied in Puerto Rico shows the greatest number of species (29), only 22 are listed for Venezuela, and of 42 species reported from the Caribbean, 24 are from one location. Milanez, reporting on 'water molds' (Oomycetes) from the *cerrado* region in southern Brazil (Milanez, 1968), points to the need of gathering ecological information and to the importance of technique. Many river systems ought to be studied, using a variety of baits (to maximize the variety of fungi recovered). The recording of the amount of organic material, pH, rainfall pattern (which influences the composition at a given place), level of water, temperature, and characteristics of the region, such as vegetation and soil are also important. Chytridiomycetes, parasites of plant and animal inhabitants of aquatic habitats, have been collected from Central America and Brazil (Johnson, pers. comm.).

XIV. Marine Fungi

In the marine environment, filamentous higher fungi are especially important as decomposers of dead organic material, while parasitic fungi appear to play a lesser role (Kohlmeyer & Kohlmeyer, 1979). Tropical marine fungi are much less known than those of well-studied temperate regions. Early investigations in the 1950s on tropical marine fungi were carried out in the Pacific Ocean off Australia; more recent research has been conducted on the western Atlantic Ocean, but the geographical distribution throughout the tropical areas of the world is still not well known. Collection of 54 species from the Bahamas, Tobago, Trinidad, and the U.S. Virgin Islands were represented by 34 Ascomycetes, 18 Deuteromycetes, and 2 Basidiomycetes, which were collected from a variety of substrates (submerged parts of mangroves, driftwood, seagrasses, in calcium carbonate animal shells, or inside calcified green macroalgae, e.g., *Halimeda*) (Kohlmeyer, 1984). Spores were also collected in seafoam. Deutromycetes apparently can produce sclerotia (resting bodies) attached to sand grains, which can survive unfavorable periods in the interstitial habitats of sandy beaches (Kohlmeyer, 1980). Further collecting in waters off Martinique yielded 20 species (Kohlmeyer, 1981) and off Belize and Cozumel, Mexico, 35 new records, including 4 new species (Kohlmeyer, 1984). A total of 73 taxa of marine fungi are known (Kohlmeyer, 1984) at the present time from the Tropics and Subtropics, making up about one third of all described marine fungi. Exclusively tropical and subtropical fungi comprise a greater share (77%) than cosmopolitan species in this region. Cosmopolitan species are expected to have a wider temperature tolerance than tropical species, which show optimal growths at 35°-40° C. Temperature is the most important parameter of the environmental factors controlling the distribution of marine fungi; otherwise, there appears to be no barrier to their latitudinal distribution (Kohlmeyer, 1983).

Booth (1979) reviews the current knowledge of marine fungi in Brazilian and South American waters and stresses the need for further work particularly near Brazil, which has the longest continuous coastline washed by tropical waters. Out of 43 South American localities previously surveyed for marine fungi, only 21 reported more than one fungus, and out of 62 marine taxa reported from South American waters, more than half (38) were actually collected from temperate localities. A large part of the tropical regions remains yet unsampled. Booth gives detailed methods for more successful recovery of marine fungi. Sampling a wide variety of microhabitats and substrates (wood, leaves, roots, algae, and soil), from 12 localities off southern Brazil, he collected 50 lignicolous fungi, 21 foliicolous fungi, 10 rhizosphere fungi, 17 algicolous fungi, 18 chytrids and thraustochytrids, and 2 nematode trapping fungi. In another study of lignicolous substrates in the waters off São Paulo, Booth (1983) found the ratio of fungi among 32 species to be: 24 Ascomycetes, 7 Deuteromycetes, and 1 Basidiomycete.

The coastal waters of São Paulo are a region where the mixing of temperate and tropical waters occurs and are thus highly variable in salinity and temperature, so "rigorous description of various inhabiting mycological taxa may indicate evidence of morphological and physiological races" (Booth, 1983). The tropical-subtropical-temperate interface of coastal Ecuador also has yet to be studied. Other marine studies are by J. W. Fell and I. Master (1975) and Fell et al., (1977) in red mangrove detritus and in mangrove swamps near São Paulo and Rio de Janeiro, Brazil by A. Ulken (1970, 1972).

XV. Lichenized Fungi

About one fifth of all fungi are lichenized. While they are a biological rather than a systematic group (polyphyletic in origin), the mycobiont, or fungal component of this symbiotic association of a fungus and alga is very often an Ascomycete. Of the 37 orders of Ascomycetes, 16 include lichenized taxa totalling ca. 13,500 species (Hawksworth et al., 1983). Mason Hale, an active collector and monographer of lichenized fungi (1959, 1960, 1965, 1975, 1976a, 1976b, 1976c, 1976d), echoes the sentiments of most tropical mycologists: that herbarium holdings of neotropical groups are notoriously poor. It is particularly true in the case of crustose lichens such as Thelotremataceae, which are rarely collected in abundance by phanerogamic botanists or naturalists, the main source of lichen specimens in the past. But determined and organized collecting is productive; e.g., after good collecting of this family in Dominica (Hale, 1974), three visits to Panama yielded twice as many species. With only four specimens from Panama on file previously at the herbarium at the Smithsonian Institution, and no record of species, the final collection of 700 specimens, representing 99 species from 11 localities, was far different from predicted results (Hale, 1978). The tropical rainforest seems rich in thelotremes, with 100 species already added, and another 100 likely. Panama apparently still has areas of virgin rainforest, difficult of access, which provide a diverse lichen flora. Dominica was studied also for Parmeliaceae (Hale, 1971) and Graphidaceae (Wirth & Hale, 1978). Collections were made by separate collectors, working in higher elevation rainforests (that were logged in 1968-1972), as well as in lower, disturbed forests. About 30% of the species were new to science. Collection of Thelotremataceae in the Lesser Antilles (Hale, 1980), islands of volcanic origin, revealed a high level of endemism. In lichen distribution, as with other fungi, collecting in the Neotropics is so small in comparison to knowledge of the mycoflora in the temperate region, that one is always reluctant to call something endemic. However, on all islands in the Lesser Antilles, the thelotremes, an almost exclusively tropical family, are almost totally confined to undisturbed virgin rain forest. If the Lesser Antilles might be considered a point of origin for some of these species, the prevailing easterly trade winds could easily carry spores westward to the western Caribbean, northern South America and Panama.

Foliicolous lichens, common in tropical rainforests, were the focus of studies by Santesson (1952). Trips to South America in 1939-1941 for field work in Tierra del Fuego and Patagonia caused him to pass through tropical regions of Brazil and Venezuela and to see the rich and tantalizing lichen flora on living leaves there. As everyone investigating neotropical species has discovered, upon his return and among his collections, he found that it was necessary to examine collections described from other tropical regions, and to search for very scattered literature and authentic material of species previously described. Botanizing in herbaria turned out to be useful, examining collected plant specimens for foliicolous fungi and bryophytes. In the history of neotropical collecting of lichenized fungi, as with other fungi, Elias Fries' name appears as an early author, associated with collections made in the West Indies. Montagne, between 1838 and 1856, published upon species collected by de la Sagra in Cuba and by Leprieur in French Guiana. A large collection of foliicolous lichens was made by R. Spruce along the Amazon River (1849-1855) and published on by Müller Arg. in 1892. The latter also received specimens from other tropical regions: Cuba (C. Wright), Brazil (J. Puiggari, A. Glaziou, E. Ule), Venezuela (A. Ernst), Paraguay (B. Balansa), and Costa Rica (H. Pittier, A. Tonduz). Vainio published several papers between 1890 and 1926 on Brazilian foliicolous lichens (Santesson, 1952). Müller Arg. and Vainio each account for about a quarter of the 700-800 species described to 1952; Santesson (1952) recognized 236 species (including 44 new taxa) of obligately foliicolous species, distributed in ten families.

All obligately foliicolous lichens (those that are confined to living leaves) are crustaceous. It is interesting that none of the genera that are important in the tropical, corticolous lichen flora are also represented among the foliicolous lichens. A living leaf, termed the "phyllosphere" by Ruinen, is a very particular habitat (Santesson, 1952), characterized by its usually very smooth surface (particularly true in tropical rainforests), and its relatively short life span as a fungal substrate. The ecology of the phyllosphere in tropical rainforests with abundant, persistent, coriaceous leaves, is distinct from that in temperate forests, which frequently have deciduous leaves, or in arid or seasonally dry regions, where leaves are often covered with hairs for drought resistance. Obligately foliicolous lichens occur most commonly on durable leaves of the humid forest, although occasional species are found on deciduous or very thin leaves. However, they are scarce upon the durable leaves of Agave, Bromeliaceae, of Cactaceae. This may be due to climatic conditions rather than to substrate preference; generally, foliicolous lichens grow on a variety of taxonomic hosts and apparently have a greater pantropical distribution than do other fungi. These plant families are frequently found in open, sunny, exposed areas, suffering periods of drought, in rocky or sandy soils with little humus, conditions to which they are adapted to survive and even compete successfully. But these same conditions may not favor the establishment of foliicolous lichens.

The region of the phyllosphere has been characterized (Reynolds, 1972): in a lowland, primary forest (Osa Peninsula, Costa Rica), epiphyllous plants such as lichens, bryophytes, algae, and fungi were most abundant in the lower 5 m of the understory. Above this height, leaves had very few epiphylls growing on them. Bryophytes grew best on leaves from ground level up to 10 m, lichens were present at all levels and peaked at 22 m. Collections from trees adjacent to the trunk of a felled 32 m tree showed a drastic reduction in epiphylls, and exposed leaves from cut-open areas of the forest showed very few lichens.

Collections have been made of foliose and fruticose lichens in the páramos of Colombia, a cool montane environment (3000 m) quite distinct from tropical lowlands, which contain more crustose lichens (Sipman & Aguirre, 1982). Hekking and Sipman (1988) have produced a list of all lichens reported from the Guianas before 1987. In a more general work, Sipman (1983, 1986) discusses the lichen family Megalosporaceae. M. Plunkte (1984) has produced a bibliography of the Cuban lichen flora. Brako et al. (1984) report on lichens from the Amazonian part of Brazil. New species of the family Parmeliaceae have been described from the páramos of Venezuela, from Brazil, Argentina, Chile, and one from montane Peru, which may be endemic (Hale, 1978, 1985). Harris (1984) reports on the Trypetheliaceae from Amazonian Brazil. Kalb (1987) reports on the Brazilian species of *Pyxine*. Lichens have also been collected in Haiti, the Dominican Republic, Puerto Rico and Costa Rica (Wetmore, pers. comm.).

XVI. Ascomycetes

Ascomycetes constitute the largest known group of fungi (2720 genera and 28,650 species) worldwide, and occur in all ecosystems and geographical areas as saprobes, parasites (especially of plants), or as the mycobiont in a lichen association (Hawksworth et al., 1983). A recent systematic approach to this large group of frequently microscopic fungi has been to limit or omit higher ranks, e.g., 228 families in 39 orders are recognized by Eriksson (1984). Teleomorphic, i.e., ascomycetous states of *Aspergillus* and *Penicillium* of the soil fungi and lichenized fungi have been referred to above. The great majority of Ascomycetes are not obvious to the untrained eye, occurring as small, often dark-colored, buried or scarcely erumpent or pulvinate inhabitants of plant parts, living and dead. General collectors usually overlook all but species with brightly colored or macroscopic fruit bodies, such as the Discomycetes (cup fungi).

A. Discomycetes

One's search for literature on Discomycetes of the Americas at one time would certainly have begun with Fred J. Seaver's various papers (1925, 1928; Seaver & Chardón, 1926; Seaver et al., 1932; Seaver & Waterson, 1940, 1942) covering mycological explorations in Bermuda and Puerto Rico, from which he described several new species. Chardón and Toro (1934) described fungi in Venezuela and provided valuable comments on their distribution. Since then, the *Fungus flora of Venezuela and adjacent countries* (R. W. G. Dennis, 1970) has been the starting place for many workers in search of descriptions fitting their neotropical collections. The lack of monographic treatments for neotropical species has hampered many mycologists until quite recently. Thus, Dennis' flora is still the only comprehensive flora for a neotropical region, covering as it does, five major categories (Basidiomycetes, Ascomycetes, Fungi Imperfecti, Phycomycetes, and Myxomycetes). Therefore, the number of species must necessarily be small in comparison to what exists in nature, and recent taxonomic treatments have resulted in some name changes; nevertheless, it is of great use for anyone entering neotropical mycology. Also, with the exception of obligate parasites, fungi sometimes have a wider distribution than most terrestrial green plants, so a good flora of one region is often useful for similar regions. Dennis (1970) refers to a Basidiomycete, *Marasmius cladophyllus*, a brightly colored mushroom, as an easily recognized index species for the lowland tropical forest, occurring from the Rio Negro in Brazil to Belize.

The cup fungi are among the rare colorful elements (yellow, orange, to scarlet) in an otherwise very green and dark tropical forest. Since they are also among the largest of the Ascomycetes, they are likely to be seen and collected. Some members are specific in their ecological preferences: dung inhabiting species can be found near animal water holes; some prefer soggy wood; while others only grow on charred wood or scorched soil. The operculate Discomycete order Pezizales, containing members with macroscopic bodies, is well-represented in the temperate zone, and some occur in the Tropics as well (50% of the Humariaceae and 80% of the Ascobolaceae of Venezuela are European species), possibly due to their coprophilous habit (Dennis, 1970). Thirty-two species of operculate cup fungi were reported from Costa Rica (W. C. Denison, 1963). Central American species of *Cookeina*, the most conspicuous and frequently collected cup fungi were discussed (Denison, 1967); later studies added new species of *Phillipsia* and two genera, *Geodina* and *Nanoscypha* (1969, 1972). Examination of collections of Seaver and Chardón among others, from Bermuda, Mexico, and Venezuela were used for monographic treatments of genera in Pezizales (Kimbrough et al., 1969, 1972). An interesting series of spore germination and cultural studies (Paden, 1974, 1975, 1977) on species of Pezizales (*Cookeina, Phillipsia*) and Sphaeriales (*Poronia*) collected in Costa Rica, have revealed the production of conidium-like structures, the rapid germination of ascospores, and the anamorphic connection to several teleomorphs, resulting in the description of a new Hyphomycete, *Molliardiomyces* (Paden, 1983).

Mycological explorations led by R. Korf (1970–1971) involving several mycologists (J. R. Dixon, K. P. Dumont, R. W. Erb, D. H. Pfister, D. R. Reynolds, A. Y. Rossman, G. J. Samuels) to the Caribbean (Puerto Rico, Dominica, and Jamaica) became the basis for a number of publications of a systematic nature. Material from Jamaica as well as from Mexico and Brazil was used for the Flora Neotropica monograph on Sarcosomataceae (Pezizales), a family of darkish colored, leathery or gelatinous textured cup fungi (Paden, 1983). Culture studies by Pfister (1973) on a collection of *Nanoscypha* from Puerto Rico appeared to support the taxonomic separation of members of Sarcoscyphaceae and the Sarcosomataceae (Korf, 1970). Tiny, usually colored inoperculate Discomycetes (Helotiales), occurring on rotten wood, were collected and studied from Puerto Rico, Dominica and Jamaica by Dumont (1981) and Korf (1977, 1978a, 1978b).

The several explorations of K. P. Dumont (1971–1976) and other Ascomycete mycologists under the Flora Neotropica project collected abundant material from Venezuela, Colombia, Ecuador, Panama, and Peru, which was used by primary investigators in numerous monographic treatments published in a variety of journals: P. Buritica (1978; Buritica & Hennen, 1980); S. Carpenter (1975; Carpenter & Dumont, 1975, 1978); K. P. Dumont (1974, 1980; Dumont & Carpenter, 1978, 1982; Dumont & Umana, 1978; Dumont et al., 1978); J. H. Haines (1980, 1984; Haines & Dumont, 1983); D. R. Reynolds (1977, 1978, 1979, 1982, 1983, 1985); A. Y. Rossman (1977, 1979a, 1979b); G. J. Samuels (1985; Samuels & Dumont, 1982; Samuels & Müller, 1978–1979); and M. Sherwood-Pike (1974, 1977a, 1977b, 1977c). These collections, made by specialists, are a valuable storehouse of material which can be used for comparison with historical material. Recent studies have enabled evaluations of generic limits, life-history studies through culture work and investigation of teleomorph-anamorph connections (Samuels; Paden, 1984, 1986), and ultrastructure studies (Benny et al., 1985a, 1985b). New genera are proposed as more careful examination of characters, the assessment of features previously ignored, and more material for comparison allow for better evaluation of what might be morphological variation due to environmental variation. For example, a group of common lignicolous Discomycetes from the Neotropics which has a characteristic tendency to form multiple discs on a common stipe was first described as *Peziza* by Berkeley and Curtis from a C. Wright collection from Cuba. Later, it was transferred to *Dasyscyphus*. A unique combination of characters places it in Hyaloscyphaceae but as it is unlike any existing genus there, it was redescribed as *Proliferodiscus* (Haines & Dumont, 1983). Study of the prevalent and brightly colored inoperculate Discomycete, *Dasyscyphus* (Hyaloscyphaceae) collected from large tropical ferns revealed that most specimens were related to one species, *D. varians*, but with wide variation in spore, ascus, and apothecial characters (Haines, 1980).

Dumont (1971), studying Sclerotineaceae in Central and South America, needed to find distinguishing characters other than spore color, conidial states, host relationships and stromatal morphology, the characters generally used. The sterile tissue of the apothecium, (cup of the fungus) was considered useful (Korf & Dumont, 1968). The hyphal arrangement there was found to be a taxonomic character. An example of the value of extensive collecting is found in his ability to synonymize several taxa with *Helotium rufo-corneum*, first described by Berkeley and Broome in 1873 (Dumont, 1980). From several hundred collections of inoperculate Discomycetes growing in leaf and on wood he was able to see that it was an extremely variable fungus, widely distributed in all tropical regions. Its spores can vary from 25 μm long in the apical part of the ascus to 41 μm in the basal part. Dumont suggested that a wide range of anatomical and morphological variation occurs in tropical species, more than in temperate species, and that the amount of variation seen in each species is dependent on the number of collections being studied.

Monographs of Ostropales and Phacidiales (Sherwood-Pike, 1977a, 1977b, 1977c, 1980), predominantly saprophytic discoid fungi usually immersed in the plant substrate, have been based upon collections made in Guadeloupe, Colombia, Ecuador, and Peru, as well as extensive examination of neotropical material in herbaria. The lack of conspicuous size or color would make these fungi more easily overlooked by general collectors or botanists. But their long-lived and buried fruit bodies may permit their discovery on dead plant material collected for other purposes. Mycologists might encourage botanists to recognize some of these less obvious fungi. Sherwood cautions against using host substrate as a criterion for identifying a fungus in this group. Only two genera of Ostropales appear to be restricted to tropical America. A conspicuous feature of this group, though not unique to fungi, is the production of calcium oxalate crystals (Sherwood-Pike, 1977a).

B. Pyrenomycetes

A specific plant substrate, Podocarpaceae, with a greater distribution in temperate regions, was used for the investigation of the parasitic fungi of the Coryneliales from various southern hemisphere regions. Benny et al. (1985a, 1985b) have shown the usefulness of SEM for revealing minute taxonomic characters. *Corynelia* seems to have evolved in the Old and New Worlds into two separate groups. Three out of seven species are confined entirely to the Neotropics while two are pantropical. Fitzpatrick's collections from the Neotropics as well as other herbarium material provided the basis for the study (Benny, 1985a).

Many tropical pyrenomycetous fungi are redescribed in von Arx and Müller (1954) and Müller and von Arx (1962).

Hypocreales, brightly colored, perithecial Ascomycetes characterized by Clark Rogerson (1970), have been the focus of Samuels and Rossman, who both have had field experience in northern South America, and most recently on the Neblina, ancient highland outcrops of Venezuela. See Samuels and Rogerson (1984) for new ascomycetes (Hypocreales and Discomycetes) from Amazonia. Comprehensive treatments of common pyrenomycetes are badly needed. Such a treatment of *Nectria* (Samuels & Dumont, 1982), based on Panamanian collections of G. W. Martin, Martin and Welden, Dumont et al. in 1976, the abundant collections at the New York Botanical Garden, as well as the collections made by the authors from Venezuela, Colombia, Ecuador, Peru, and Brazil provide useful descriptions and a key, permitting the identification of most neotropical species. Culture studies of *Nectria*, as well as species of Sphaeriales and Dothideales have demonstrated anamorphic connections for many species and have elucidated relationships among taxa (Samuels, 1985; Samuels & Müller, 1978, 1979, 1980; Samuels & Rossman, 1979; Samuels, Rossman & Müller, 1978). Monographic studies by Rossman (1977, 1979b) have clarified the status of *Ophionectria* and reviewed many taxa described in *Calonectria* through type studies. Such life history studies and monographs are essential for the identification of species by mycological workers, particularly those in the region, for whom the scarcity of literature has been a serious problem. Another series of papers revising the Hypocreales, with cultural observations, have contributed further knowledge of these fungi from Colombia, Bolivia, Peru, and Brazil (Doi, 1975–1979).

The common Ascomycete order, Sphaeriales, contains conspicuous fungi, likely to be collected by non-specialists and botanists because of their frequently carbonaceous fruiting bodies, common on tree trunks. Cytological studies have contributed to the understanding of Hypoxylon species occurring in the Neotropics (Rogers, 1978). Several new species have been described from Colombia (Rogers & Dumont, 1979) and Brazil (Rogers, 1981; Rogers & Samuels, 1985). Two new genera of the Sphaeriaceae (*Porosphaeria* and *Striatosphaeria*) were described with their anamorphs, from collections made in Roraima, Brazil (Samuels & Müller, 1978). See also the series of papers on the life histories of Brazilian ascomycetes by Samuels and Müller (1980). Although coprophilous species of Sordariaceae had been well studied, lignicolous members were poorly known. Collections from cloud forests in Costa Rica resulted in the description of a new genus, *Mycomediospora*, characterized by vermiform, fragmenting ascospores, as well as eight new species of other genera (Carroll & Munk, 1964).

The new genus *Javaria* J. R. Boise was described from Serra Aracá (Boise, 1984).

C. POWDERY MILDEWS

Erysiphales (powdery mildew), which are obligate plant parasites on leaves, with a cosmopolitan distribution, are well studied in the temperate regions, but poorly known in the Neotropics. In temperate zones, at least, they are parasitic on some 7187 species in 44 orders of angiosperms (Yarwood, 1973). This wide host range is exceeded only by the Uredinales (rusts), and the greatest abundance of the powdery mildews appears to be in areas of rain-free summers and intensive agriculture such as California and Israel. Two papers appeared in one year on the genus *Brasiliomyces*, originally described by Viegas in 1944 from *Malvastrum*. Re-interpretation by Zheng (1984) of its "lack of appendages," which separated the genus from *Erysiphe*, resulted in placing *Erysiphe trina* Harkness collected from a California oak, (later transferred to *Californiomyces* by Braun in 1981), finally into *Brasiliomyces trina* (Harkness) Zheng. Another species, *Erysiphe malachrae* Seaver from Puerto Rico is found to be conspecific with *Brasiliomyces malvastri* Viegas. *Brasiliomyces* seems far-flung in distribution (Peru, Puerto Rico, Brazil, California), though relatively rare, and is "interesting ecologically in that it is the only genus of powdery mildews whose species form abundant ascocarps in a tropical climate" (Hanlin & Tortolero, 1984).

D. FOLIICOLOUS FUNGI

The cosmopolitan, parasitic leaf-dwelling Ascomycete, Meliolineae had been monographed (Hansford, 1961, 1963) using herbarium specimens collected worldwide and those collected by the author from Jamaica. Species were based on host relationships. And the warmer regions of the world were least represented. A modern treatment and intensive collecting in the Neotropics would probably revise the group considerably.

Foliicolous, ascomycetous fungi (sooty molds) forming dense, black mats on living leaves and stems, are particularly abundant in the Tropics. Many of these species are redescribed by von Arx and Müller (1954) and Müller and von Arx (1962). A. Chavez Batista had an influential role in Brazilian mycology by building an active mycological group, the Instituto de Micologia, at the Universidade do Recife (now Univ. Federal de Pernambuco) as well as publishing more than 600 papers on a variety of fungal topics. He also contributed a monographic treatment of the sooty molds in Capnodiaceae (Batista & Ciferri, 1963). These fungi, as well as other foliicolous fungi, have been collected from all parts of the Neotropics, Central as well as South America, and monographed by D. R. Reynolds (1971, 1977, 1978, 1979, 1982, 1983, 1985). Reproduction in the sooty mold colony shows periodicity related to food availability and a direct association of the fungi with leaf inhabiting, honey dew secreting insects such as scale. Further, there are differential growth forms apparently related to geographical or ecological distributions (Reynolds, 1975). A deciduous form, characterized as a thin layer peeling away in drier cooler months, is found predominantly in lowland tropical regions, but, interestingly, absent inside a rainforest. Instead, it is present in intermittent clearings where the climax vegetation has been disturbed, or where air moves more easily. This is also the common form found along the coast of southeastern United States, throughout Florida, and in a disjunct population occurring on the Pacific coast of the U.S.A. The permanent growth form, inhabiting some stems and twigs, forms pulvinate, thick black mats, and retains its reproductive units in the phyllosphere. This form is predominant in neotropical highlands, temperate lowlands (Pacific northwest of the U.S.A.), and possibly montane regions of neotropical islands. Ascocarpic and hyphal characters of neotropical taxa are analyzed, supporting inferences leading to recognition of the monophyly of Capnodiaceae with subfamilies Antennulariellodieae, Capnoideae, and Limacinioideae (Reynolds, 1986).

New areas of collection were opened by the Projeto Flora Amazônica expeditions (1977–present) in the Brazilian Amazon, in which a total of 12 U.S.A. and 5 Brazilian mycologists participated (compared to over 50 plant collectors). Of the U.S.A. mycologists, slightly more than half were Ascomycete specialists: K. P. Dumont, D. R. Reynolds, G. J. Samuels (twice), M. Dibben and L. Brako (lichenized fungi). Flora Neotropica and Projeto Flora Amazônica contributed a great resource of neotropical fungal material to add to historical collections for systematic studies (see, e.g., Farr, 1985).

XVII. Basidiomycetes

Basidiomycetes (ca. 16,000 species) comprise the majority of macrofungi. Agaricales (mushrooms) are terrestrial, humicolous or lignicolous Tremellales (jelly fungi) and Aphyllophorales (poroid, hydnoid, cantharelloid, clavarioid, and thelephoroid fungi) are largely lignicolous. The latter two orders are well represented in tropical regions. In herbarium holdings of neotropical Basidiomycetes, the greatest amount of space is probably taken up by the polypores, which are seen by even the casual visitor to a tropical forest. These fungi are large, robust, persistent, frequently abundant, often attractive, and easy to carry back, compared to ephemeral and fragile agarics. Taxonomists working in Polyporaceae are blessed with abundant material for study, but people trying to identify specimens have been cursed by a wealth of names, scattered literature, and constantly changing taxonomic systems.

A. Polypores

Teixeira (1962) reviewed the taxonomy of the polypores; the early systems were based upon gross morphology, which resulted in confusion and conflict. Numerous taxa were described in the nineteenth century, based on morphological variations. Even though Patouillard as early as 1887 had described microstructures (hyphal septation, clamp-connections, hyphal contents, crystals, lactiferous hyphae, structure of the pileus, basidia, cystidia, as well as spores), showing their importance in classification, he had not considered them for the segreatation of genera. Since Ellis and Everhart in 1889 and Donk in 1933 began basing generic separation on special microstructures, about 125 genera have been described. Hyphal characters were considered important taxonomic characters (Pinto-Lopes, 1952), reaction of spores and other structures to a chemical reagent (Melzer's reagent) was applied by Singer (1944, 1951) for generic separation, and the use of cultural characters, by Nobles in 1958.

The earliest polypores described from the Neotropics were those collected by Humboldt in Colombia, published obscurely by W. J. Hooker in Kunth (1822), but which remained relatively unnoticed until E. Fries included them in his "Epicrisis systematici mycologici" in 1838. Type studies by Ryvarden (1975) of these and other collections, e.g., Gaudichaud's fungi, determined by C. H. Persoon in 1827 (Ryvarden, 1973) and Léveillé's tropical fungi (Ryvarden, 1981) have placed most of the species into synonymy. More important contributions to the knowledge of neotropical polypores were made by J. F. C. Montagne between the years 1835 to 1857 on fungi from French Guiana, Cuba, and Brazil (Ryvarden, 1982a); J. M. Berkeley from 1856 to 1868 on fungi from Brazil, Mexico, and Cuba (Ryvarden, 1984); N. Patouillard from 1876 to 1924 on fungi from Venezuela and Guadeloupe (Ryvarden, 1983a); and W. A. Murrill from 1904 to 1943 on his own collections from Cuba, Jamaica, and Mexico, as well as others' from Nicaragua, Costa Rica, Guatamela, Puerto Rico, and British Honduras (now Belize) (Ryvarden, 1985a). Type studies by Ryvarden on the above collections show that the more enthusiastic the describer, the greater the tendency to describe species already known to science: of Montagne's 93 species, 57% are synonyms; of Berkeley's 336 species, 65%; of Patouillard's 235, 69%; of Murrill's 343, 78%. Much of the synonymy is due to the gradual awareness of the importance of special microstructures for defining natural groups, and thus the few large genera have been broken into 125 smaller ones. In fact, Murrill's generic concepts, which were ignored in his lifetime have, in recent decades, found wide acceptance (Ryvarden, 1985a). But errors in not recognizing previously described taxa were also due to undue reliance upon morphological variation resulting from environmental effects or natural variation.

A few, generally brief, regional studies of polypores are: the pore fungi of the French Antilles and Guiana (David & Rajchenberg, 1985); Colombian Aphyllophores, which are known from Overholts' descriptions (Chardón & Toro, 1930), and recently from the 1974 and 1976 collections of K. P. Dumont et al. (Setliff & Ryvarden, 1983); wood-rooting fungi of Costa Rica (Carranza-Morse, 1982; Carranza-Morse & Sáenz, 1984); Cuban polypores (Kotlaba, Pouzar & Ryvarden, 1984) showing several taxa (41% out of 234 species) in common with those in the eastern pine forests of the United States and Canada (Herrera & Bondartseva, 1975). Polypores of Venezuela, Trinidad and Tobago were described by Overholts (Chardón & Toro, 1934), Fidalgo and Fidalgo (1966, 1967, 1968), Dennis (1970), whose fungal flora is still the most comprehensive for this group, and Setliff (1984) recently. Mexico was treated by Murrill (1912) and Guzmán (1979). Brazilian polypores of Mato Grosso were described by M. E. P. K. Fidalgo (1968b). Most of Argentina is extra-limital, except for the 'Misiones' region near Paraguay, and therefore exhibits a fungus flora in many respects similar to that of the northern temperate regions. But an active mycological group (Rajchenberg & Wright, 1982; Wright & Blumenfeld, 1984; Rajchenberg, 1986) reports new polypores of Argentina, in their "gallery forests" containing "tropical" elements from the Amazonian basin in spite of the latitude (Wright, pers. comm.). Yet the absence of a consolidated treatment or complete flora of the neotropical polypores still renders it difficult for in-country mycologists to identify species encountered there, despite good type studies.

A synopsis of *Poria* (resupinate polypores) (Lowe, 1963) from the tropical regions brings together scattered lists and provides a useful key with microscopic features. Lowe also points out that the higher elevations in the Tropics, likely to have widely distributed temperate plants, have been less investigated than the lower. Other treatments of common neotropical genera were provided by the Brazilians: M. E. P. K. Fidalgo on *Hexagonia* (1968a), and on *Osmoporus* (1962); Furtado on *Ganoderma* (1967), and on *Amauroderma* (1981).

Some recently described polypore genera are interesting. *Fuscocerrena*, monotypic, is known only from the U.S.A. (Iowa and Minnesota) and Brazil and on deciduous wood. It has a farinose layer, which may cause it to escape detection. *Navisporus* has two species, one in tropical Africa and the other known only from Brazil, also on deciduous wood (Ryvarden, 1982b, 1983b). *Aporpium*, collected on deciduous wood, and first described by Bondartseva and Singer, was later discovered by Teixeira and Rogers (1955) to have longitudinally septate basidia and thus, transferred to the Tremellaceae (jelly fungi) as a polyporoid genus. Of four species ascribed to *Aporpium*, two are reported in the Neotropics (Panama, Guadeloupe, Venezuela). Mycoparasitism was suggested by David (1974) for *A. dimidiatum*, and rejected after a fragment of the trama was cultured and produced the characteristic tremellaceous basidia. Still, the possibility may exist of one or more of these fungi being a polypore with a systemic Heterobasidiomycete parasite (Setliff & Ryvarden, 1982). The monotypic genus *Henningsia*, described by Möller as developing from a smooth, telephoraceous layer to a poroid hymenium, known only from Brazil, has been re-characterized and compared to *Rigidoporous* (Ginns, 1979).

Hymenochaetaceae, widespread lignicolous fungi, causing white rot, is frequently seen in herbarium collections under *Phellinus* and *Hymenochaete*. They are reported from Cuba (Herrera & Bondarceva [sic], 1982), the West Indies (Reeves & Welden, 1967), Costa Rica (Gomez & Ryvarden, 1985), the Brazilian Amazon (Sousa, 1980), as well as in the floras of Venezuela (Dennis, 1970) and Mexico (Guzmán, 1979). The tropical genus *Hydnochaete* (8 species), whose type species was collected in Brazil by Moller, is monographed (Ryvarden, 1982c). The species have distinctly separate Old World and New World distributions. A new stipitate genus, *Stipitochaete*, known only from the Neotropics, is distinctive by its occurrence on the ground, actually as a parasite on roots of both deciduous trees and *Bambusa* (Ryvarden, 1985b).

B. CORTICIACEAE AND STEREACEAE

Two wide-spread non-poroid families of Aphyllophorales are the Corticiaceae and Stereaceae. Corticiaceae (on wood, stems, or litter) are easily overlooked by the general collector, because of their smooth hymenia and effuse habits, but can be discovered by rolling logs over and searching for what appear to be "paint spatters." Neotropical corticiums are being studied from Brazilian collections (Hjortstam & Ryvarden, 1980, 1982). By contrast, Stereaceae, with effused-reflexed to pileate habits are rather noticeable, being often vase-shaped, or resembling ruffles on logs. They are of wide distribution and have been monographed (Reid, 1959, 1965) and described from collections made by A. Welden and G. W. Martin in Panama and Costa Rica (Welden, 1954), and by Welden (1958, 1960) in Jamaica and Cuba of *Cotylidia* and *Cymatoderma*.

C. JELLY FUNGI

Jelly fungi (Heterobasidiomycetes in several families: Auriculariaceae, i.e., "tree's ear," Tremellaceae, Dacrymycetaceae) are well represented in tropical regions. Tremellaceous fungi in C. G. Lloyd's herbarium, which included neotropical specimens collected by Martin (Panama), J. Rick, Möller, and Torrend (Brazil), were reviewed and synonymized (Bandoni, 1958). Descriptions and keys for neotropical Tremellales are available in a series of papers, including a very useful Flora Neotropica monograph (#6) by Lowy (1970, 1971b, 1975, 1976a, 1976b, 1980), enabling the identification of most neotropical species. Additional new species have been described from Mexico, Brazil, Colombia, Panama, Guatemala, Ecuador, and recently from Amazonian Brazil from PFA expeditions (Lowy, 1981, 1982, 1986).

Several curious heterobasidiomycetous fungi have been recently described or surveyed, including material from the Neotropics. A new order, Atractiellales, was described (Oberwinkler & Bandoni, 1982) for fungi having tiny gastroid basidiocarps resembling synnemata-type deuteromycetous fungi in their general morphology, but having basidia which are mostly cylindric and transversely septate as in the Auriculariales, Uredinales, and Ustilaginaceae. Spores bud out in yeast-like fashion. *Agaricostilbum* was collected by Wright in Argentina on palm spathes and described as a Deuteromycete (Wright, 1970), later emended by Wright, Bandoni, and Oberwinkler (1981) as an auriculoid Basidiomycete. *Agaricostilbum* and *Atractiella*, a tiny gelatinous fungus found on recent wounds on palms could possibly be insect dispersed (Oberwinkler & Bandoni, 1982). No doubt, awareness of these fungi and further collecting will increase the record.

D. MYCOPARASITISM

Mycoparasitism has been reported in *Tremella* species on pyrenomycetes and Basidiomycetes (Bezerra & Kimbrough, 1978). The genus *Naematelia* was established by Fries in 1816 based on a *Stereum* parasitized by *Tremella encephala*. Cultural studies established the presence of two distinct fungi, despite the appearance, due to a changed morphology, of a single fungus having dual characteristics. It was therefore disposed of as a *nomen confusum* (Bandoni, 1961). Cultural studies have established a new species, *Tremella rhytidhysterii*, of a mycoparasitic fungus on an Ascomycete reported from the southeast U.S.A., Puerto Rico, the Dominican Republic, Colombia, as well as Southeast Asia (Bezerra & Kimbrough, 1978). This type of mycoparasitism, in which internal haustoria or hyphae are not formed and the host is unharmed, is classified as a biotrophic (balanced) contact parasitism (Barnett & Binder, 1973). Among the Brazilian *Tremella* collections of A. Möller re-discovered by Oberwinkler (Friedrichsen, 1977) in Hamburg, was a collection of a common neotropical species, *T. fibulifera* Möller, which seemed to be parasitized by an unknown fungus which possessed basidium-like structures (Bandoni & Oberwinkler, 1983). Another interesting fungus recognized as a mycoparasite of the *Tremella* type is *Syzygospora alba*, described by Martin (1937) as an auriculariaceous Basidiomycete, through the interpretation of fusing conidia as a heterobasidium. A closer examination (Oberwinkler & Lowy, 1981) of this fungus has shown that it has a teleomorph (holobasidia) and anamorph (conidiophores) present in the same fruiting body, as well as basidiospores capable of budding and producing yeast-like cells. Thus far, the species is known only from six collections from Panama and one from Mexico. Awareness of these mycoparasitic relationships will undoubtedly result in other discoveries in time.

E. AGARICS

Agarics of the Neotropics are enormously under-represented in herbarium collections though they form a larger group than the Aphyllophorales (4,000 species vs. 1,200 species worldwide). Ectomycorrhizal agarics, which are in the majority in the temperate regions, are limited in lowland neotropical forests (Singer, 1983). Also, their fleshy and fragile fruit bodies decompose readily in warm tropical climates. A mushroom collector there must gather his fungi early in the morning, take his notes, and dry the specimens in a timely fashion. Transporting dried, easily-shattered mushrooms requires special care; they cannot be flattened in a plant press as foliicolous Ascomycetes can, or stuffed in a backpack as lichens, wood-decomposing polypores and pyrenomycetes can. Thus *Lentinus*, a pantropical coriaceous fungus is probably the most commonly represented gilled fungus genus in herbarium collections of neotropical fungi. In his recent monograph, Pegler (1983b) placed *Lentinus* in Polyporaceae (Aphyllophorales) on the basis of its hyphal constitution. Unlike most agarics, it has a firm, durable body, is slow-growing or long-lived (hence, likely to be seen), and dries easily. Like the polypores, *Lentinus* is also well represented in herbaria because of its great morphological variability; it tended to be collected and described anew. Even Murrill (1911) spoke of the tendency to describe certain common species and the resulting confusion and synonyms: "Variations in these strong, prevailing types, which easily run together when examined in hundreds of specimens in the field, loom to specific proportions in a herbarium several thousand miles distant."

The islands of the Caribbean are among the most extensively studied regions. A good review of the work done there has been given by Pegler (1983a). A few mushrooms were among the earliest fungi collected by the early general collectors, usually plant specialists, who sent them to European mycologists. Fries (1851) named several macrofungi from the then Danish West Indies. Leprieur's collections from French Guiana were described by Montagne between 1837 and 1857. Patouillard between 1889 and 1902 described some mushroom species from the Lesser Antilles collected by Düss, who lived in Martinique. Many of these early species have since not been recognized, as they were based on material which was poorly preserved. Early in the twentieth century, Murrill described numerous species (1911, 1913, 1918) based on his own collections in Jamaica (mostly from elevations of 5,000 feet) and Mexico, and on Earle's collections from Cuba.

Studies by Pegler and Fiard (1978, 1979) and Pegler and Singer (1980) on fungi of the Lesser Antilles were based on collections sent to Kew by Fiard, an active collector, as well as personal collecting by Pegler in 1977. A comprehensive and monographic agaric flora of the Lesser Antilles (Pegler, 1983a) makes this region the best known for agarics in the Neotropics. It treats 106 genera and 457 species of neotropical distribution, including 70 new species and required a review of the collections of Murrill, Patouillard, Berkeley and Curtis, Massee, and Dennis. Pegler describes the agaric flora of the humid forests as rich and extensive in the Lesser Antilles, whose extremely mountainous terrain has discouraged large scale cultivation and destruction of the primary forest. Pegler notes that in the Lesser Antilles the characteristically tropical families of Entolomataceae, Hygrophoraceae, Tricholomataceae, and Agaricaceae dominate. Tricholomataceae is by far the largest family with many species of *Collybia, Marasmiellus*, and *Marasmius* present, whereas the genera *Clitocybe, Laccaria*, and *Lepista* are rarely encountered. The genus *Inopilus* appears to be an ancient group among the entolomatoid genera. Among well-known ectomycorrhizae-forming fungi of the temperate regions, Russulales is well represented, the boletes are mostly represented by non-mycorrhizal or facultatively mycorrhizal genera such as *Chaciporus, Pulveroboletus,* and *Xerocomus,* while obligate mycorrhizal genera such as *Boletus, Leccinum,* and *Suillus* are absent. Cortinariaceae, an important temperate region family, is limited to the lignicolous *Gymnopilus* and some *Inocybe* (ectomycorrhizae-formers) in dry or in disturbed areas. Frequently, casual visitors to neotropical rainforests form an impression that the agaric flora is depauperate, but the contributions of Singer and Pegler would belie that.

Caribbean and Central American agarics were treated by Singer (1970, 1973, 1974) mostly through monographic studies of important and widespread neotropical fungi, which included examination of collections of Murrill and Dennis. Other important Singer papers treat Russulaceae of Trinidad and Venezuela (1952), *Lactarius* in the Gulf area (1976b), Basidiomycetes of Costa Rica (Singer & Gomez, 1982, 1984), and Agaricales from the French Antilles (Singer & Fiard, 1976). Agarics of Cuba were described by Kreisel (1970, 1971a, 1971b, 1971c [the last a key]) and Berkeley & Curtis (Pegler, 1987); of Puerto Rico and the Virgin Islands by Stevenson (1975).

The Guyanas are relatively unstudied, except for Heim's (1967) description of some agarics. Trinidad and Tobago, off the coast of Venezuela, are well known, from Baker and Dale's (1951) list of fungi and Dennis' publications (1951a, 1951b, 1951c, 1952a, 1952b, 1953, 1960, 1961). After visits to Trinidad and Venezuela in 1949 and again in 1958, Dennis provided modern, detailed descriptions and type revisions of many neotropical agarics. The comprehensive "Fungus Flora of Venezuela and adjacent countries" (Dennis, 1970) which incorporated these earlier studies, included 601 species of Agaricales, 322 Aphyllophorales, and 1185 species of pyrenomycetes (Ascomycetes) collected from various altitudinal and vegetation zones of Venezuela, excepting the Guyana Highlands and Amazonas, i.e., the region east of the Andes. Recently, Halling (Redhead & Halling, 1987) has investigated the agaric flora of the Guyana highlands and has current plans for more field work there, in Roraima, and in French Guiana (pers. comm.). The larger fungi of the lowland forest appeared to have pantropical affinities or ranges, especially with the fungi of tropical Africa (*Lepiota* and *Marasmius* species) while those of the cloud forest and Andean regions have affinities with north temperate species (Dennis, 1970). The occurrence of temperate coprophilous species of larger Ascomycetes and Agaricales species is understandable, but Dennis reports several noncoprophilous fungi mixed in with a number of apparently endemic species. While there may have been a migration of temperate fungi from Mexico on high mountains (over 3000 m) in Costa Rica and Guatamela, the agaric flora of such a migration path is still not well known and needs investigation.

F. Rolf Singer

Rolf Singer's contribution to the study of agarics is unique: establishing the use of modern concepts (1951, 1975), i.e., greater investigation and use of microscopic structures for generic delimitation; consideration of fungal ecology (1963a, 1984a; Singer & Araujo, 1979; Singer & Morello, 1960; Singer et al., 1965); phylogenetic relationships (1971), biogeography (1953); and finally, for his enormous contribution to the knowledge of neotropical agarics. He has collected extensively during repeated trips to South Florida, Mexico, Colombia, the Amazonian forests of Bolivia, Ecuador, and Brazil, the Andean chain in Colombia, Ecuador, Peru, Bolivia, Argentina and Chile, and the lowland tropical and subtropical Atlantic coastal forests of Brazil, Paraguay, Uruguay and Argentina in the last 40 years (from ca. 1948 to the present) (see, e.g., Singer, 1963b, 1965b). His Flora Neotropica and other monographs of important neotropical fungi—*Crinipellis* and *Chaetocalanthus,* (1942); *Pluteus* (1958b), *Marasmius* (1965a), *Crepidotus, Simocybe* (1973), *Favolaschia* (Aphyllophorales) (1974), *Campanella* and *Aphyllotus* (1976a), Marasmieae (1976c), Russulaceae (*Lactarius*) (1976b, 1984b); and *Psilocybe* (Singer & Smith, 1958a, 1958b)—have contribute innumerable new species and vastly increased the knowledge of their distribution and ecology. The monographs and hundreds of papers, including type studies, have provided a modern taxonomic base for agaric studies in general. The first major keys to South American mushrooms were provided by Singer & Digilio (1953).

G. Fungal Ecology; Ectomycorrhizae

Singer's wide collecting experience in South America, which included the historical collecting places of Spegazzini and Rick around Buenos Aires and Rio Grande do Sul, subtropical, temperate regions, including the *Nothofagus* forests, montane regions, down to Patagonia and Tierra del Fuego, as well as tropical lowland forests enabled him to make useful observations on the ecology and biogeography of the fleshy fungi (Singer, 1953; Singer et al., 1965). The predominance of European and American species in earlier floristics lists of South Brazil, Uruguay, and Argentina are partly due to introductions of those fungi with pine seedlings or with domestic animal stock, as well as misdeterminations by mycologists of native fungal species. A factor in the distribution of some fungi, e.g., *Marasmiellus* species, is their adaptability to man-made environments, such as plantations of sugar cane, banana, and coffee. Certain pantropical or "American" species of South America are also found in the eastern North American flora, especially that of Florida. There seems to be little similarity in the agaric flora of the important ectomycorrhizal *Alnus* forests of the Selva Boliviano-Tucumán and of the *Nothofagus* forests of southern Argentina. Rather, there is a definite relationship between the fungi of the *Nothofagus* area in South America and the fungi of New Zealand and Australia. Therefore, the Alneta of the Selva Boliviano-Tucumán is not considered a route for the southward migration of agarics from the *Alnus* forests of Mexico.

Singer (1976c) pointed out that ecologists had generally omitted the fungi in descriptions of tropical vegetation types and associations, due to lack of collaboration with appropriate mycologists. He preferred to use the term mycosociology rather than phytosociology in his study of fungal associations. He investigated the distribution of the Marasmieae (a taxon with 85.9% of its species in the Neotropics) among five localities (three in terra firma vegetation of the Amazonian lowlands and two in inundated or inundable forests). The lack of commonality of species among the five localities was interesting. He interpreted that as due to the extreme sensitivity of the fungi to minor differences in micro-climate, host distribution, and geographical isolation. It must be kept in mind that fungi are heterotrophic organisms with varying degrees of dependence upon their hosts or substrates. In other words, some fungi, at least, may be just scattered in the forest as many phanerogamic species and even frequent collecting will not register every population in the forest. Further, collecting in second growth or disturbed forests can lead to erroneous conclusions about the distribution of fungi in vegetation or ecological regions (terra firme, várzea, etc.) based on such stands, which tend to shelter a large number of species otherwise rare or absent, including "cosmopolitan" species and those forming cicatrizing mycorrhizae (Singer et al., 1965).

Studies of plant-fungal interactions of 'ectotrophic' forest tree mycorrhizae have shed light on such phenomena as litter accumulation and ectomycorrhizal fungi distribution (Singer, 1963a, 1984a; Singer & Araujo, 1979; Singer et al., 1983; Singer & Morello, 1960). In white-sand podzol 'campinarana' type forests of lowland Central Amazonia, the dominant trees are obligatorily ectotrophically mycorrhizal and litter is accumulataed as a consequence of ectotroph dominance. In primary terra firme forests, the trees are almost all non-ectomycorrhizal and leaf litter does not accumulate as a deep humus layer because of the higher number of leaf litter decomposing fungi. Ectotrophic plants may be more widespread than believed earlier (Singer et al., 1983). Ectotroph forests are dominated by Gnetaceae, Leguminosae, and Sapotaceae in the primary forest of Amazonia, while in secondary forests, Fagaceae, Nyctaginaceae, Pinaceae, Polygonaceae, Rubiaceae, and Sapindaceae are involved in ectomycorrhizal associations. The most important ectomycorrhizal fungi belong to the same genera as in the temperate regions, but the smaller number of species (ca. 114) are mostly strictly neotropical, excepting introductions. In comparison, the most important non-ectotrophically mycorrhizal family, Marasmieae, has 274 species (85.9% of the family) occurring exclusively in the American Subtropics or Tropics (Singer, 1976c). Boletaceae, together with Russulaceae, represent the families richest in species in the ectotrophically mycorrhizal fungi in tropical lowlands. Cortinariaceae, a large and important temperate region mycorrhizal family, is represented by *Inocybe* but not by *Cortinarius* in the Neotropics.

H. MEXICO

More than half of Mexico lies below the Tropic of Cancer (south of Mazatlán and Tampico). Its variable topography provides a diversity of interesting habitats, and a rich plant flora of ectotrophic plants: forests of *Pinus* at between 1000 and 4000 m, *Abies* between 3000 and 4000 m in the sierras, isolated volcanic peaks with conifers on the upper slopes and diversified pine-oak, mesophytic to subtropical forests between 1000 and 2500 m. Sadly, these upland forests are rapidly disappearing due to the planting of corn, coffee, and recently, potatoes. Despite the efforts of several active Mexican mycologists, much of the mycota has yet to be well sampled and is in danger of loss due to destruction of habitats. There are also distinct alpine and subalpine forests at elevations of 3500–4000 m where a diversified bryophyte and lichen flora is found. The formerly extensive tropical coastal forests along the Gulf of Mexico are now greatly reduced by agriculture. Arid to semi-arid regions contain drought adapted fungi, such as desert puffballs (*Battarea*, *Tulostoma*, and *Podaxis*) (Guzmán & Herrera, 1969).

Mexico is unique among neotropical regions for the dual interest in macrofungi, of floristics and systematics on the one hand, and ethnobotany and food interest on the other. The latter interest is reflected in Guzmán's flora (1979) in the lists provided of edible, poisonous, and hallucinogenic mushrooms. The long-standing folk knowledge of mushrooms is also indicated by the long list of common names of macrofungi. Mushrooms were collected in market places and forests by mycologists for the description of the knowledge and use of mushrooms by the Purépecha in Pátzcuaro, Michoacán (Mapes et al., 1981). Ethnomycological observations for Mexico and Guatemala were also made by Lowy (1971a).

The history of folk and mycological interest in Mexican hallucinogenic mushrooms (81 species) is given in the world monograph of *Psilocybe* (Guzmán, 1983a). Of 467 taxa previously described, 144 are recognized as valid, 52 species being described by the author. *Psilocybe* is a genus of saprophytic fungi on diverse substrate, but exhibiting some ecological preferences: some species are distinctly fimicolous, others lignicolous; those growing in muddy or swampy soil are from deciduous or subtropical forests; all hallucinogenic species grow in acid media; the majority prefer deciduous forests, then conifer forests, and lastly, lowland tropical forests. The rapidly diminishing *Quercus* forests inside tropical rainforests of southeastern Mexico may contain endemic species.

Mexican macrofungi have been regionally surveyed, but not collected by Maury. An interest in hallucinogenic mushrooms brought Heim and Singer to Mexico; a number of publications appeared around 1958 (Heim & Wasson, 1958; Singer & Smith, 1958a, 1958b). Perez-Silva (1967) published on *Inocybe* collected by Heim in 1956, 1959, and 1961, and treated only 43 species, indicating a need to collect more widely in Mexico's pine-oak forests. Besides the Gasteromycetes, Guzmán and Herrera reviewed knowledge of the agarics and Aphyllophorales from the nineteenth century to date (T. Herrera & Guzmán, 1972, Guzmán & Herrera, 1972). These surveys were compiled for the flora, "Identificación de los hongos" (Guzmán, 1979).

Floristic studies of macrofungi have been undertaken by several Mexican mycologists, mostly in the pine-oak forests: Chiapas (Guzmán & Chacón, 1985); Durango (Rodriguez-

Scherzer & Guzmán-Dávalos, 1984); Hidalgo (Frutis & Guzmán, 1983; Varela & Cifuentes, 1979); Jalisco (Guzmán-Dávalos, 1984; Guzmán & García-Saucedo, 1973; Guzmán-Dávalos et al., 1983); Morelos (Mora & Guzmán, 1983); *Lepiota* of Quintana Roo (Guzmán-Dávalos & Guzmán, 1982); Veracruz (Guzmán & Villarreal, 1984; Guzmán-Dávalos & Guzmán, 1984); the Yucatan Peninsula (Guzmán, 1982, 1983b). Tropical agarics have also been collected and studied by Cifuentes (1981) in Chiapas, Guerrero, Hidalgo, Michoacán, and Veracruz. Publications on Mexican fungi (1968–1978) can be located by checking the index to the Boletín de la Sociedád de Micologia (Vol. 12, 1978).

I. Brazil

The knowlege of the agaric flora in Brazil is enhanced by current efforts, such as by the Instituto de Botânica in São Paulo (Bononi, 1979, 1981, 1984a, 1984b, Bononi et al., 1984c; Guzmán et al., 1984); the participation of several Basidiomycete specialists in Projeto Flora Amazônica expeditions (Prance, pers. comm.): Araujo, 1977, 1978; Bononi, 1980; Hosford, 1977; Peterson, 1979 (clavarias); and Nishida, 1985. However, since an agaric flora for Brazil is nonexistent and literature is still scattered or difficult to obtain by resident mycologists, the floras of Dennis (1970) and Pegler (1983a) are used to a great extent by resident mycologists. Pegler (pers. comm.) is planning a study of Brazilian agarics (1986 to 1989).

Unlike Mexico, indigenous knowledge in Brazil of mushrooms as food appears limited or lost, and reports are scarce (Torrend, 1951; Fidalgo, 1965, 1968; Fidalgo & Prance, 1976; Prance, 1972, 1973). On the other hand, the humid tropical regions are ideal locations for the production of edible mushrooms for its inhabitants. Waste byproducts of tropical agricultural grops (sugar cane bagasse, corn stalks, coffee husks, etc.) are usable by fungi to produce economical food, containing vegetable protein. The region naturally provides good growing conditions—high humidity and even temperatures. Investigation by INIREB in Veracruz, Mexico appears promising for using coffee waste (Martinez et al., 1984), for example. Interest by the public in populated Brazil is substantial. After a television broadcast (April 1985) by the Instituto de Botânica in São Paulo, Brazil on mushroom cultivation, the Institute received 2000 letters of enquiry per week for over five months! (Bononi, 1987).

J. Other Areas Studied

Other recent areas of concentration on agarics have been: Costa Rica (Singer & Gomez, 1982) in tropical dry forest and cloud forest, boletes in tropical montane forest (Singer & Gomez, 1984); Colombia, visited by Guzmán in 1964 and 1971 (Guzmán & Varela, 1978), revealed species shared with Mexico, particularly in disturbed areas of the tropical forest, and also of the subtropical deciduous forest, (but at higher elevations in Colombia). Some endemism was reported by Singer (1963a) in the *Quercus* forests of Colombia, which reach their southern limit in South America there. Additional studies of agarics in the Cordillera are by Boekhout (1983) and Pulido (1983). Cuban agarics are presently being revised by Pegler (1987). Areas lacking in extensive investigation in agarics are southern and northern Mexico, much of Central America, except Costa Rica, especially in the pine regions, the Guyanas, areas away from coastal and southern Brazil, e.g., Mato Grosso, and montane islands of northeastern Brazil. These are the same areas which generally lack study of fungi. In addition, tropical high mountains except for some efforts in Colombia, Peru (Yokoyama, pers. comm.) and in the Venezuelan highlands, and dry scrub vegetation of desert regions have had little investigation.

K. Gasteromycetes

There are very few studies on Gasteromycetes in the Neotropics, probably because, although they are cosmopolitan in general distribution, they tend to occur more commonly in dry regions. However, some useful treatments are: from Mexico, (Guzmán & Herrera, 1969; Wright et al., 1972), on *Scleroderma* (Guzmán, 1970); South American Gasteromycetes (though mostly from temperate regions there) (Homrich & Wright, 1973); phalloid fungi (Sáenz, 1980; Sáenz et al., 1983; Sáenz & Nassar-C., 1982; Wright, 1960); and hypogeous Gasteromycetes (Trappe & Guzmán, 1971; Trappe et al., 1979; Hosford & Trappe, 1980).

XVIII. Remarks on Fungal Distribution

Some general differences in fungal distribution between temperate and tropical regions are: a preponderance of foliicolous and parasitic leaf Ascomycetes in the Tropics as opposed to the temperate regions (Dennis, 1970); a preponderance of fleshy fungi in general, and of ectomycorrhizal fungi particularly, (mainly Basidiomycetes), in the temperate regions. There are occasional dimorphisms in taxa shared by the two regions, such as smaller spores in tropical representatives and larger ones in temperate counterparts (Guzmán, 1983a; Horak, 1977).

XIX. Conclusions and Recommendations

Early mycological activity in the Neotropics involved European non-specialists or plant collectors, who sent fungal material to European workers for description and naming. The colonial period encouraged interest in fungi as pathogens on agricultural crops, and general collecting provided a resource for comparison with later and specialized collecting. Gradual involvement of Western Hemisphere mycologists saw the beginning of specialization in fungal collecting and systematics. The last 25 years have seen increasing specialization, monographic and type studies, the use of cultural and cytological work, and the consequent revision of generic limits, and lately, ecological and biogeographical observations. Expedition type field work within the last 15 years and NSF grants to individual reserarchers are gradually building a storehouse of collections from previously unexplored areas (particularly the lowland forest), which are used in systematics studies. A well-defined project, e.g., Projeto Flora Amazônica, concentrates on a particular region (albeit, relatively large), and puts mycologists of varying specializations into the field, who can make good specialist collections as well as general fungal collections to be directed to specialists. Requirements of splitting collections for deposit in both U.S.A. and host country herbaria, as well as encouragement to publish some results of the collecting in a timely fashion, in a neotropical periodical provide advantages of information and resources to the host country.

Some areas of fungal systematics have been adequately treated: the Aphyllophorales, Tremellales, discoid Ascomycetes and Myxomycetes. Most large geographical regions are represented by collections from these groups, but many other fungal groups are greatly under-collected and under-represented in herbaria and published studies. More seriously, despite the appearance recently of several good monographs, needed floras and keys are non-existent for many fungal groups and most geographic areas. Basic identification aids, such as keys, ought to be constructed for non-specialist workers in the region (G. Samuels, pers. comm.). Hanlin and Tortolero's key (1985) to the Ascomycetes associated with cultivated plants is an example of a useful collaborative effort, inexpensively produced and easily distributed to people who can use it.

Additionally, within neotropical regions themselves, the number of workers with access to good libraries and laboratories is very small compared to the vast area in need of study. Even a comprehensive literature guide to neotropical fungi is practically non-existent. The Watlings' (1980) guide to literature for mushrooms, extremely useful for temperate region mycologists, has a tiny fraction of the available literature for the Neotropics, including the list of Neotropics-based periodicals.

The fact that the greater number of peer-review journals are published in the temperate regions, together with the unfavorable monetary exchange rates, makes it difficult for resident mycologists in the Neotropics to obtain important systematics literature. In the other direction, non-resident mycologists usually do not publish in regional journals and often do not receive and read them on a regular basis. The inability to read Spanish or Portuguese contributes to this lack of communication. The recommendations of D. Pearson for ways in which U.S.A. systematists can avoid being a kind of "ugly American" (ASC newsletter, February 1986) are especially appropriate. Exchanges of literature even when one is not visiting are important.

Because of the long distances and consequent expense of travel for field work in the Neotropics and difficulties of access to collecting areas, cooperation with resident mycologists is an essential and provides benefit for both. Cooperative publications and exchange or donation of useful literature would help built a "network" for the sharing of information and specimens. Communication with other neotropical researchers, particularly with plant collectors, would greatly contribute to the effectiveness of a mycological expedition. Mycologists ought to communicate more and directly with plant specialists, encouraging them to include fungal collections. For example, botanists collecting to make sets of plants for herbarium sheets tend to discard branches of dirty-looking leaves which the mycologist specializing in foliicolous fungi would be happy to have. Botanical expeditions generally are equipped to collect from the upper reaches or from the inaccessible heights at which many tropical trees have their lowest branches. Collections of soil can be part of any forest inventory, broadening the collecting range for fungi inhabiting that zone.

Specialist collecting is the ideal, since the collections would tend to be qualitatively superior for that fungal group, as weedy sorts of fungi would be left in the forest. A systematist using only herbarium material can be misled: a fungus of seeming rare occurrence, based on herbarium representation, turns out to be nearly weedy when a specialist is actually searching for it in the field (J. Boise, pers. comm.). However, because of the vastness of areas still unstudied and the great unlikelihood of return visits, general fungal collections by a mycologist participating in most botanical expeditions would contribute much to further fungal knowledge and, consequently, to information on ecological interactions and biogeography for the Neotropics. Mycological collections ought standardly to be accompanied with minimal ecological data, such as vegetation type, plant community, soil type, rain and flooding patterns, etc.

While classical collections in herbaria are known and accessible, herbaria of neotropical specimens are also represented by the backlog of collections that is in the possession of every neotropical worker. The author has compiled a list of current workers (Nishida, in press) which could be used to encourage communication and subsequent exchange of material among those contributing to neotropical mycological knowledge. Interested parties could communicate with the author. Otherwise, publications citing actual collections and their deposition are the sole source of information concerning existing and accessible fungal specimens from the Neotropics. The deposition of authenticated material in at least one major herbarium of the neotropical country would aid resident workers in their own floristic and monographic studies.

Finally, the same imperative for increased and effective collecting of fungi exists as for plants. Primary rainforests are rapidly disappearing in many parts of the Neotropics. *Quercus* stands, which represent naturally occurring ectomycorrhizal communities, inside primary rainforests in Mexico and Central America are vanishing. Thus, it is becoming more difficult to see possible dispersal patterns. On the other hand, some fungi apparently can quickly invade and colonize new habitats. This is evident in the frequent presence of certain pantropical to "cosmopolitan" species in disturbed places and in forest clearings. Man's activities such as farming and cattle raising may even introduce fungi from distant areas through infection of feed. The opportunity is rapidly diminishing to see fungal distribution and interactions with other fungi and plants under "natural" conditions, i.e., without the disruptive changes which man's activities frequently encourage. In view of the rapid degradation of the natural vegetation, the most urgent task is thorough, carefully planned collecting—of all fungal groups in representative areas—what Fosberg (1979) has referred to as "salvage botany." These activities need to result in revisions of genera represented in the Neotropics, up-dated floristics lists based on accurate identification and current taxonomic concepts, and modern comprehensive floras usable by the non-specialist as well, with drawings and/or photographs, and current ecological and distribution information.

XX. Acknowledgments

The author wishes to express appreciation to Ghillean Prance of The New York Botanical Garden and to Don R. Reynolds of the Natural History Museum of Los Angeles for providing opportunities to do mycological field work in the Neotropics through the Projeto Flora Amazônica and an NSF grant to D. R. Reynolds. Appreciation is also expressed to D. R. Reynolds for useful comments and suggestions regarding this contribution.

XXI. Literature Cited

Ainsworth, G. C., F. K. Sparrow & A. S. Sussman. 1973. The Fungi. Vol. IVA. Academic Press, New York.

─── & A. S. Sussman. 1968. The Fungi. Vol. III. Academic Press, New York.

Alexopoulos, C. J. & J. A. Sáenz. 1975. The Myxomycetes of Costa Rica. Mycotaxon 2(2): 223-271.

Araujo, I. & M. A. Sousa. 1981. *Polyporus indigenus*, nova especie da Amazonia. Acta Amazonica 11(3): 449-455.

Arthur, J. C. 1925. The grass rusts of South America; based on the Holway collections. Proc. Amer. Phil. Soc. 64: 131-223.

Arx, J. A. von & E. Müller. 1954. Die Gattungen der amerosporen Pyrenomyceten. Beitrage zur Kryptogamenflora der Schweiz. Band II, heft 1.

Baker, R. E. D. & W. T. Dale. 1951. Fungi of Trinidad and Tobago. Mycol. Papers. 33: 1-123.

Baldacci, R. 1965. Raffaele Ciferri. Mycologia 57: 198-201.

Bandoni, R. J. 1958. Some tremellaceous fungi in the C. G. Lloyd collection. Lloydia 21(3): 137-151.

───. 1961. The genus *Naematelia*. Amer. Midl. Naturalist. 66(2): 319-328.

─── & F. Oberwinkler. 1983. On some species of *Tremella* described by Alfred Möller. Mycologia 75(5): 854-863.

Barnett, H. L. & F. L. Binder. 1973. The fungal host-parasite relationship. Annual Rev. Phytopathol. 11: 273-292.

Batista, A. C., F. A. C. Barros, J. O. da Silva, A. L. Castrillon, M. J. P. Maciel. 1965. Especies fungicas dos solos do Estado do Maranhão. Atlas Inst. Micologia 2: 309-317.

─── & R. Ciferri. 1963. Capnodiales. Saccardoa 2: 1-298.

───, M. M. J. Peres, J. A. de Lima. 1967. Curiosas especies de Penicillia dos solos florestais do noroeste do Estado do Maranhão. Atlas Inst. Micologia 4: 191-201.

Benjamin, C. R. & A. Slot. 1969. Fungi of Haiti. Sydowia, Ser. II, 23: 125-163.

Benny, G. L. & R. K. Benjamin. 1976. Observations on Thamnidiaceae (Mucorales) II. *Chaetocladium, Cokeromyces, Mycotypha*, and *Phascolomyces*. Aliso 8: 391-424.

───, D. A. Samuelson & J. W. Kimbrough. 1985a. Studies on the Coryneliales II. Taxa parasitic on Podocarpaceae: *Corynelia*. Bot. Gaz. 146: 238-251.

───, ───, & ───. 1985b. Studies on the Coryneliales III. Taxa parasitic on Podocarpaceae: *Lagenulopsis* and *Trispora*. Bot. Gaz. 146: 431-436.

Berkeley, M. J. 1867. On some new fungi from Mexico. J. Linn. Soc., Bot. : 423-425.

─── & M. C. Cooke. 1876. The fungi of Brazil including those collected by J. W. H. Trail, Esq. in 1874. J. Linn. Soc. Bot. 15: 363-398.

─── & M. A. Curtis. 1851. Descriptions of new species of fungi collected by the U.S. Exploring Expedition under C. Wilkes, U.S.N. Commander. Amer. J. Sci. Arts, 2 ser., XI, p. 93.

─── & ───. 1853. Exotic fungi from the Schweinitzian herbarium, principally from Surinam. J. Acad. Nat. Sci. Philadelphia, New Ser., II, p. 277.

─── & ───. 1868. Fungi Cubenses (Hymenomycetes). J. Linn. Soc., Bot. 10: 280-341.

─── & C. Montagne. 1844. Decades of fungi. Hooker J. Bot. III: 239.

Betancourt, C. & M. Caballero. 1983. Aquatic hyphomycetes (Deuteromycotina) from Los Chorros, Utuado, Puerto Rico, Caribbean J. Sci. 19(3-4): 41-42.

Bezerra, J. L. & J. W. Kimbrough. 1978. A new species of *Tremella* parasitic on *Rhytidhysterium rufulum*. Canad. J. Bot. 56(24): 3021-3033.

Blackwell, M. 1984. Myxomycete life cycles and arthropod spore dispersers. Pages 67-90 *in* Q. Wheeler & M. Blackwell (eds.), Fungus-insect relationships: Perspectives in ecology and evolution. Columbia University Press, New York.

─── & R. L. Gilbertson. 1981. *Cordycepioideus octosporus* sp. nov., a termite suspected pathogen from Jalisco, Mexico. Mycologia 73: 358-362.

─── & J. W. Kimbrough. 1978. *Hormiscioideus filamentosus* gen. and sp. nov., a termite-infecting fungus from Brazil. Mycologia 70: 1273-1280.

─── & W. Rossi. 1986. Biogeography of fungal ectoparasites of termites. Mycotaxon 25: 581-601.

───, J. Vandewaa & M. Reynolds. 1984. Survival of myxomycete sclerotia after exposure to high temperatures. Mycologia 76: 752-754.

Boekhout, T. 1983. Distribución y ecología de hongos macroscopicos (datos iniciales). Pages 210-215 *in* T. van der Hammen, A. P. Preciado, P. Pinto (eds.), Studies in tropical Andean ecosystems I, La Cordillera Central Colombiana. J. Cramer, Vaduz.

Boise, J. 1984. New and interesting fungi (Loculoascomycetes) from the Amazon. Acta Amazonica 15(1/2) PFA Supl., 1: 49-53.

Bononi, V. L. R. 1979. Basidiomycetes da Ilha do Cardoso. I–III. Rickia 8: 63-74; 85-99, 105-121.

───. 1981. Alguns basidiomicetos hidnoides da região Amazonia. Rickia **9**: 17–30.

———. 1984a. Basidiomycetes da Reserva Biologica de Mogi-Guaçu, SP. Rickia **11**: 1–24.

———. 1984b. Basidiomycetes da Ilha do Cardoso. IV. Rickia **11**: 43–52.

———, G. Guzmán & M. Capelari. 1984c. Basidiomycetes da Ilha do Cardoso. V. Rickia **11**: 91–97.

Booth, T. 1979. Strategies for study of fungi in marine influenced ecosystems. Rev. Microbiol. (São Paulo) **10**: 123–138.

———. 1983. Lignicolous marine fungi from São Paulo, Brazil. Canad. J. Bot. **61**(2): 488–506.

Brako, L., M. J. Dibben & I. do Amaral. 1984. Preliminary notes on the macrolichens of Serra do Cachimbo, north-central Brazil. Acta Amazonica **15**(1/2) PFA Supl., 3: 123–135.

Braun, K. L. & H. W. Keller. 1976. Myxomycetes of Mexico I. Mycotaxon **3**: 297–317.

——— & ———. 1986. Myxomycetes of Mexico III. Biotica (in press).

Braun, U. 1981. Taxonomic studies in the genus *Erysiphe*. Nova Hedwigia **34**: 679–719.

Buritica, P. 1978. Los hongos de Colombia. II. Nuevas especies de Uredinales. Caldasia **12**(57): 170.

———. 1980. Teliospore ontogeny as a criterion for rust phylogeny. Rep. Tottori Mycol. Inst. **18**: 296.

——— & J. F. Hennen. 1980. Pucciniosireae (Uredinales, Pucciniaceae). Flora Neotropica, Monogr. **24**: 1–51. The New York Botanical Garden, New York.

Carpenter, S. 1975. *Bisporella discedens* and its *Cystodendron* state. Mycotaxon **2**(1): 123–126.

——— & K. P. Dumont. 1975. Leoticeae I. Nannfeldt's Phialeoideae: the genera *Belonioscypha, Cyathicula*, and *Phialea*. Mycologia **70**: 1223–1238.

——— & ———. 1978. Los hongos de Colombia-IV. *Bisporella tri-septata* and its allies in Colombia. Caldasia **12**(58): 339–348.

Carranza-Morse, J. 1982. Polypores new to Costa Rica. Mycotaxon **15**(1): 405–408.

——— & J. A. Sáenz. 1984. Wood decay fungi of Costa Rica. Mycotaxon **19**: 151–166.

Carrión, G. & M. Galván. 1984. Plant pathogen fungi from the state of Veracruz. III. Bol. Soc. Mex. Micol. **19**: 15–64.

Carroll, G. C. & A. Munk. 1964. Studies on lignicolous Sordariaceae. Mycologia **61**(1): 77–98.

Cash, E. & A. J. Watson. 1955. Some fungi on Orchidaceae. Mycologia **47**: 729–747.

Chardón, E. C. & R. A. Toro. 1930. Mycological explorations of Colombia. J. Dept. Agric. Porto Rico **14**: 195–369.

——— & ———. 1934. Mycological explorations of Venezuela. Monogr. Univ. Puerto Rico. Ser. B, **2**: 85–90.

Christensen, M. & D. E. Tuthill. 1986. *Aspergillus*: An overview. *In* R. A. Samson and J. I. Pitt (eds.), Advances in the systematics of *Penicillium* and *Aspergillus*. Plenum Press, London (in press).

Ciferri, R. 1961. Mycoflora Domingensis integrata. 1st. Bot. Univ. Pavia, Quaderno 19. Pavia, Italy.

Cifuentes, J. 1981. Descripción y distribución de hongos tropicales (Agaricales) no conocidos previamente en Mexico. Bol. Soc. Mex. Micol. **16**: 35–61.

Crane, J. L. & K. P. Dumont. 1975. Hyphomycetes from the West Indies and Venezuela. Canad. J. Bot. **53**(9): 843–851.

——— & ———. 1978. Two new Hyphomycetes from Venezuela. Canad. J. Bot. **56**(20): 2613–2616.

Culberson, W. L. 1959. Lichenes exsiccati in herbariis Americae septentrionalis asservati. Bryologist **62**: 48–52.

Cummins, G. B. 1940. Descriptions of tropical rusts–III. Bull. Torrey Bot. Club **67**: 607–613.

David, A. 1974. *Aporpium dimidiatum*, nouvelle Tremellale porée. Bull. Trimest. Soc. Mycol. France **90**: 179–185.

——— & M. Rajchenberg. 1985. Pore fungi from French Antilles and Guiana. Mycotaxon **22**(2): 285–325.

Denison, W. C. 1963. A preliminary study of the operculate cup-fungi of Costa Rica. Revista Biol. Trop. **11**(1): 99–129.

———. 1967. Central American Pezizales. II. The genus *Cookeina*. Mycologia **59**: 306–317.

———. 1969. Central American Pezizales. III. The genus *Phillipsia*. Mycologia **61**: 289–304..

———. 1972. Central American Pezizales. IV. The genera *Sarcoscypha, Pithya*, and *Nanoscypha*. Mycologia **64**(3): 609–623.

Dennis, R. W. G. 1951a. Some Agaricaceae of Trinidad and Venezuela. Leucosporae: Part 1. Trans. Brit. Mycol. Soc. **34**: 411–480.

———. 1951b. *Lentinus* in Trinidad, B. W. I. Kew Bull. **5**: 320–333.

———. 1951c. Agaricaceae referred by Berkeley and Montagne to *Marasmius, Collybia* and *Heliomyces*. Kew Bull. **6**: 387–410.

———. 1952a. The *Laschia* complex in Trinidad and Venezuela. Kew Bull. **7**: 325–332.

———. 1952b. *Lepiota* and allied genera in Trinidad, West Indies. Kew Bull. **7**: 459–499.

———. 1953. Some West Indian collections referred to *Hygrophorus* Fr. Kew Bull. **8**: 253–268.

———. 1960. Fungi Venezuelani: III. Kew Bull. **14**: 418–458.

———. 1961. Fungi Venezuelani: IV. Agaricales. Kew Bull. **15**: 67–156.

———. 1970. Fungus flora of Venezuela and adjacent countries. Kew Bull. Addit. Ser. **3**: i–xviv, 1–531. HMSO, London.

Doi, Y. 1975a. Revision of the Hypocreales with cultural observations VIII. The genus *Hypocrea* and its allied genera in South America (1). Bull. Natl. Sci. Mus. Tokyo, ser. B. **1**: 1–33.

———. 1975b. Revision of the Hypocreales VIII. *Hypocrea peltata* (Jungh.) Berk. and its allies. Bull. Natl. Sci. Mus. Tokyo. ser. B. **1**: 121–134.

———. 1976a. Revision of the Hypocreales IX. The genus *Hypocrea* (2). Bull. Nat. Sci. Mus. Tokyo, ser. B. **2**(4): 119–131.

———. 1976b. *Protocreopsis*, a new genus of the Hypocreales. Kew Bull **31**: 551–555.

———. 1977. Revision of the Hypocreales X. Two new species of the genus *Hypocreopsis*. Bull. Nat. Sci. Mus. Tokyo, ser. B. **3**(3): 99–104.

———. 1978. A revision of the genus *Protocreopsis*. Bull. Nat. Sci. Mus. Tokyo, ser. B. **4**(3): 113–121.

———. 1979. Revision of the Hypocreales XII. Bull. Nat. Sci. Mus. Tokyo, ser. B. **5**(2): 37–49.

Donk, M. A. 1983. Revision der niederländischen Homobasidiomycetae-Aphyllophoraceae II. Ned. Mycol. Ver. Medel. **22**: 1–278.

Dumont, K. P. 1971. Sclerotiniaceae II. *Lambertella*. Mem. New York Bot. Gard. **22**(1): 1–178.

———. 1974. Sclerotiniaceae V. On some tropical *Lambertella* species. Mycologia **66**(2): 341–346.

———. 1980. Sclerotiniaceae XVI. On *Helotium rufo-corneum* and *Helotium fraternum*. Mycotaxon **12**(1): 255–277.

———. 1981. Leotiaceae II. A preliminary survey of the neotropical species referred to *Helotium* and *Hymenoscyphus*. Mycotaxon **12**: 313–371.

——— & S. Carpenter. 1978. Sclerotiniaceae XIV. *Asterocalyx*. Mycologia **70**: 68–75.

———— & ————. 1982. Leotiaceae IV: Los hongos de Colombia-VII. *Hymenoscyphus caudatus* and related species from Colombia and adjacent regions. Caldasia **13(64)**: 567–602.

———— & M. I. Umana. 1978. Los hongos de Colombia. V. *Laternea triscapa* y *Calostoma cinnabarina* en Colombia. Caldasia **12(58)**: 349–352.

————, P. Buritica & E. Forero. 1978. Los hongos de Colombia—I. Caldasia **12(57)**: 159–164.

Dunn, P., D. R. Reynolds, F. H. Nishida & S. Barro. 1985. *Penicillium* in Brazil. Acta Amazonica **15(1/2) PFA Supl. 2**: 137–143.

————, ————, ————, & ————. 1986 (in press). *Aspergillus* in Brazil. Acta Amazonica, PFA Supl., 4.

Earle, F. S. 1901. Some fungi from Puerto Rico. Muhlenbergia **1**: 10–17.

————. 1904. Report on observations in Puerto Rico. Rep. P. R. Agric. Exp. Sta. **1903**: 454–468.

Ellis, J. B., B. M. Everhart. 1889. A new genus of Polyporaceae. J. Mycol. **5**: 28–29.

Eriksson, O. 1984. Outline of the Ascomycetes. Systema Ascomycetum **3**: 1–72.

Escobar, G. A. & J. D. Toledo. 1977. El "Tenquique," hongo comestible de El Salvador. Communicaciones **IV(1)**: 15–22.

Farr, M. A. 1969. Myxomycetes from Dominica. Contr. U.S. Natl. Mus. **37(3)**: 397–440.

————. 1976. Myxomycetes. Flora Neotrop. Monogr. **16**: 1–304.

————. 1985. Amazonian foliicolous fungi III. A preliminary list of Ascomycotina, mostly Dothidiales, s.l. Acta Amazonica **15(1/2) PFA Supl. 2**: 29–53.

————, U. Eliasson & K. P. Dumont. 1979. Myxomycetes from Ecuador. Mycotaxon **8(1)**: 127–134.

Farrow, W. M. 1954. Tropical soil fungi. Mycologia **46**: 632–646.

Fell, J. W. & I. M. Master. 1975. Phycomycetes (*Phytophthora* spp. nov. and *Pythium* sp. nov.) associated with degrading mangrove (*Rhizophora mangle*) leaves. Canad. J. Bot. **53**: 2908–2922.

————, ————, S. Y. Newell & A. S. Tallman. 1977. The role of fungi in the red mangrove (*Rhizophora mangle*) detrital system. Abst. Sec. Intl. Mycol. Congr., p. 192.

Fidalgo, M. E. P. K. 1962. The genus *Osmoporus* Sing. Rickia **1**: 95–138.

————. 1968a. The genus *Hexagonia* Mem. New York. Bot. Gard. **17(2)**: 35–108.

————. 1968b. Contributions to the fungi of Matto Grosso, Brazil. Rickia **3**: 171–220.

Fidalgo, O. 1965. Conhecimento micologico dos indios brasileiros. Rickia **2**: 1–10.

————. 1968. Conhecimento micologico dos indios brasileiros. Rev. Antropol. **15–16**: 27–34 (1967–1968).

———— & V. L. Bononi. 1979. Mycology in Brazil: Available infrastructure for its development. Taxon **28(4)**: 435–464.

———— & M. E. P. K. Fidalgo. 1966. Polyporaceae from Trinidad and Tobago I. Mycologia **58**: 862–904.

———— & ————. 1967. Polyporaceae from Trinidad and Tobago II. Mycologia **59**: 833–869.

———— & ————. 1968. Polyporaceae from Venezuela. I. Mem. New York Bot. Gard. **17(2)**: 1–34.

———— & G. T. Prance. 1976. The ethnomycology of the Sanama Indians. Mycologia **68(1)**: 201–210.

Fitzpatrick, H. M. 1927. A mycological survey of Puerto Rico and the Virgin Islands (Review). Mycologia **19**: 144–149.

Forsyth, A. & K. Miyata. 1984. Tropical nature. Charles Scribner's Sons, New York.

Fosberg, F. R. 1979. Tropical floristic botany—concepts and status—with special attention to tropical islands. Pages 89–105 *in* K. Larsen & L. B. Holm-Nielsen (eds.), Tropical botany. Academic Press, London.

Friederichsen, I. 1973. Liste der Pilze der Kollektion E. Ule aus Brasilien (1883–1903) im Herbarium Hamburgense. Mitt. Staatsinst. Allg. Bot. Hamburg **14**: 95–134.

————. 1977. Das Schicksal der von A. Möller in Brasilien (1890–1895) gesammelten Pilze sowie eine Liste der noch vorhandenen Sammlungsstucke (The fate of the fungi collected by A. Möller in Brazil (1890–1895) and a list of specimens of the collections still existing). Mitt. Inst. Allg. Bot. Hamburg, **15**: 99–104.

Fries, E. 1821. Systema mycologicum I. Lundae.

————. 1851. Novae symbolae mycologicae, in peregrinis terris a botanicis Danicis collectae. Acta R. Soc. Sci, Upsala ser. 3, **1**: 17–136.

Frutis, I. & G. Guzmán. 1983. Contribución al conocimiento de los hongos del estado de Hidalgo. Bol. Soc. Mex. Micol. **18**: 219–265.

Furtado, J. S. 1967. Some tropical species of *Ganoderma* (Polyporaceae) with pale context. Persoonia **4**: 379–389.

————. 1981. Taxonomy of *Amauroderma* (Basidiomycetes, Polyporaceae). Mem. New York Bot. Gard. **34**: 1–109.

Gerdemann, J. W. & J. M. Trappe. 1974. The Endogonaceae in the Pacific Northwest. Mycol. Mem. No. **5**: 1–76.

Gibson, J. L., J. W. Kimbrough & G. L. Benny. 1985. Ultrastructural evidence for the Ascomycete-like nature of *Glaziella aurantiaca*. Page 431 *in* R. Molina (ed.), Proceedings of the 6th North American Conference on Mycorrhizae. Forest Research Laboratory, Corvallis, Oregon.

Ginns, J. 1979. *Henningsia* (Polyporaceae) and a description of the type species. Mycologia **71**: 305–309.

Gochenaur, S. E. 1970. Soil mycoflora of Peru. Mycopathol. Mycol. Appl. **42**: 259–272.

Gomez, L. D. & L. Ryvarden. 1985. *Inonotus fimbriatus*, nov. sp. (Hymenochaetaceae, Basidiomycetes). Mycotaxon **23**: 291–292.

Gonçalves, R. D. 1937. Saporema. O Biologico **III(10)**: 302–306.

Goos, R. D. 1960. Soil fungi from Costa Rica and Panama. Mycologia **52**: 877–883.

————. 1963. Further observations on soil fungi in Honduras. Mycologia **55**: 142–150.

Guzmán, G. 1970. El genero *Scleroderma*. Darwiniana **16**: 233–407.

————. 1979. Identificación de los hongos. Edit. Limusa, Mexico City.

————. 1982. New species of fungi from the Yucatan peninsula. Mycotaxon **16(1)**: 249–261.

————. 1983a. The genus *Psilocybe*. Nova Hedwigia **74**: 1–439. J. Cramer.

————. 1983b. Los hongos de la peninsula de Yucatan. II. Nuevas exploraciones y adiciones miciologicas. Biotica **8(1)**: 71–100.

————, V. L. Bononi & R. A. Piccolo. 1984. New species, new varieties, and a new record of *Psilocybe* from Brazil. Mycotaxon **19**: 343–350.

———— & S. Chacón. 1985. Nuevas observaciones sobre los hongos, liquenes, y mixomicetos de Chiapas. Bol. Soc. Mex. Micol. **19**: 245–251.

———— & D. A. García-Saucedo. 1973. Macromicetos del Estado de Jalisco, I. Bol. Soc. Mex. Micol. **7**: 129–143.

———— & T. Herrera. 1969. Macromicetos de las zonas áridas de México, II. Gasteromicetos. An. Inst. Biol. Univ. Nac. Aut. Mexico **40**: 1–92.

———— & ————. 1972. Especies de macromicetos citados de México, II. Fistulinaceae, Merulinaceae y Polyporaceae. Bol. Soc. Mex. Micol. **5**: 57–77.

———— & L. Varela. 1978. Los hongos de Colombia—III. Observaciones sobre los hongos, liquenes y mixomicetos de Colombia. Caldasia 12(58): 309–338.

———— & L. Villarreal. 1984. Estudio sobre los hongos, liquenes y mixomicetos del Cofre de Perote, Veracruz I: Introducción a la microflora de la región. Bol. Soc. Mex. Micol. 19: 107–124.

Guzmán-Dávalos, L. 1984. Hongos del estado de Jalisco, III. Bol. Inst. Bot. Univ. Guadalajara 5(10): 21–34.

———— & G. Guzmán. 1982. Contribución al conocimiento de los Lepiotaceos (Fungi, Agaricales) de Quintana Roo. Bol. Soc. Mex. Micol. 17: 43–54.

———— & ————. 1984. Nuevas registros de hongos en el estado de Veracruz. Bol. Soc. Mex. Micol. 19: 221–244.

————, G. Nieves & G. Guzmán. 1983. Hongos del estado de Jalisco, II. Bol. Soc. Mex. Micol. 18: 165–181.

Haines, J. H. 1980. Studies in the Hyaloscyphaceae I: Some species of *Dasycyphus* on tropical ferns. Mycotaxon 11(1): 189–216.

————. 1984. Studies in the Hyaloscyphaceae III: The long-spored, lignicolous species of *Lachnum*. Mycotaxan 14: 1–39.

———— & K. P. Dumont. 1983. Studies in the Hyaloscyphaceae II: *Proliferodiscus*, a new genus of Arachnopezizoideae. Mycologia 75(3): 535–543.

Hale, M. E., Jr. 1959. New or interesting Parmelias from North and Tropical America. Bryologist 62: 123–132.

————. 1960. A revision of the South American species of *Parmelia* determined by Lynge. Contr. U.S. Natl. Herb. 36: 1–41.

————. 1965. A monograph of *Parmelia* subgenus *Amphigymnia*. Contr. U.S. Natl. Herb. 6: 193–358.

————. 1971. *Parmelia affluens*, a new species of lichen in subgenus *Amphigymnia* with a yellow medulla. Phytologia 22: 141–142.

————. 1974. Morden-Smithsonian Expedition to Dominica: The lichens (Thelotremataceae). Smithsonian Contr. Bot. 16: 1–46.

————. 1975. A revision of the lichen genus *Hypotrachyna* (Parmeliaceae) in tropical America. Smithsonian Contr. Bot. 25: 1–73.

————. 1976a. Synopsis of a new lichen genus, *Everniastrum* Hale (Parmeliaceae). Mycotaxon 3: 345–353.

————. 1976b. A monograph of the lichen genus *Pseudoparmelia* Lynge (Parmeliaceae). Smithsonian Contr. Bot. 31: 1–62.

————. 1976c. A monograph of the lichen genus *Bulbothrix* Hale (Parmeliaceae). Smithsonian Contr. Bot. 32: 1–29.

————. 1976d. A monograph of the lichen genus *Parmelina* Hale (Parmeliaceae). Smithsonian Contr. Bot. 33: 1–60.

————. 1978. A revision of the lichen family Thelotremataceae in Panama. Smithsonian Contr. Bot. 38: 1–60.

————. 1980. Systematics and evolution of the family Thelotremataceae (Lichens) in the Lesser Antilles. Natl. Geogr. Soc. Res. Rep. 12: 305–308.

————. 1985. New species of *Xanthoparmelia* (Vain.) Hale (Ascomycotina: Parmeliaceae). Mycotaxon 20: 73–79.

Hanlin, R. T. & O. Tortolero. 1984. An unusual tropical powdery mildew. Mycologia 76(3): 439–442.

———— & ————. 1985. Clave preliminar para géneros de Ascomicetos asociados con plantas cultivadas in Venezuela. Bol. Soc. Venezolana Fitopathol. 5: 1–28.

Hansford, C. G. 1961. The Meliolineae, a monograph. Sydowia, Ser. II, Beih. 2: 1–806. Verlag von Ferdinand Berger, Horn, Austria.

————. 1963. Iconographia Meliolinearum. Sydowia, Ser. II, Beih. 5: 1–285. Verlag von Ferdinand Berger, Horn, Austria.

Harris, R. C. 1984. The family Trypetheliaceae (Loculoascomycetes: Lichenized Melanommatales) in Amazonian Brazil. Acta Amazonica 14(1/2) PFA Suppl., 1: 55–80.

Hawksworth, D. L., B. C. Sutton & G. C. Ainsworth. 1983. Ainsworth's & Bisby's Dictionary of the Fungi. CMI, Kew.

Heim, R. 1967. Hygrophores tropicaux recuielles par Roger Heim 1: Espèces de Guyane française et de Nouvelle-Guinée australienne. Rev. Mycol. 32: 16–27.

———— & R. G. Wasson. 1958. Les champignons hallucinogènes du Mexique, L'études ethnologiques, taxonomiques, biologiques, physiologiques et chimiques. Arch. Mus. Nat. Hist. Nat. ser. 7, 6: 1–332.

Hekking, W. H. A. & H. J. M. Sipman. 1988. The lichens reported from the Guianas before 1987. Willdenowia 17: 193–228.

Hennen, J. F. & G. B. Cummins. 1967. The Mexican species of *Uromyces* Southw. Naturalist 12: 146–155.

———— & ————. 1973a. Additions to *Uromyces* (Uredinales) from Mexico. Southw. Naturalist 18: 73–77.

———— & ————. 1973b. The Mexican species of *Puccinia*. Bol. Soc. Mex. Micol. 7: 59–88.

———— & ————. 1973c. New taxa of Mexican rust fungi. Rep. Tottori Mycol. Inst. 10: 169–182.

————, M. M. Hennen & M. B. Figueiredo. 1982. Indice das Ferrugems (Uredinales) do Brasil. State of São Paulo Ministry of Agriculture, Brazil, Arq. Inst. Biol. 49 (suppl. 1): 1–201.

Herrera, S. & M. A. Bondartseva. 1975. Particularities of the Cuban flora of polypores (Polyporaceae *s. lato*). Abstr. First International Mycological Congress, Leningrad.

———— & ————. 1982. Especies del género *Phellinus* (Basidiomycetes: Hymenochaetaceae) nuevas o poco conocidas en Cuba. Acta Bot. Cubana 8: 1–17.

Herrera, T & G. Guzmán. 1972. Especies de macromicetos citadas de México, III. Agaricales. Bol. Soc. Mex. Micol. 6: 61–91.

Hjortstam, K. & L. Ryvarden. 1980. Studies in tropical Corticiaceae (Basidiomycetes) I. Mycotaxon 10: 269–287.

———— & ————. 1982. Studies in tropical Corticiaceae (Basidiomycetes) IV. Type studies of taxa described by J. Rick. Mycotaxon 15: 261–276.

Holubova-Jechova, V. 1982. Some new or interesting micro-fungi from Cuba. Mycotaxon 14(1): 309–315.

———— & A. Mercado-Sierra. 1982. Some new or interesting micro-fungi from Cuba. Mycotaxon 14(1): 309–315.

Homrich, M. H. & J. E. Wright. 1973. South American Gasteromycetes. I. The genera *Gastropila*, *Lanopila*, and *Mycenastrum*. Mycologia 65(4): 779–794.

Horak, E. 1977. *Entoloma* in South America. I. Sydowia 30: 40–111.

Hosford, D. R. & J. M. Trappe. 1980. Taxonomic studies on the genus *Rhizopogon* II. Notes and new records of species from Mexico and the Caribbean countries. Bol. Soc. Mex. Micol. 14: 3–15.

Jackson, H. S. 1926–1931. The rusts of South America based on the Holway collections. I. Mycologia 18: 139–162. 1926; II. Mycologia 19: 51–63. 1927; III. Mycologia 23: 96–116. 1931a; IV. Mycologia 23: 332–364. 1931b; V. Mycologia 23: 463–503. 1931c; VI. Mycologia 24: 62–186. 1932.

Janos, D. P. 1975a. Vesicular-arbuscular mycorrhizal fungi and plant growth in a Costa Rican lowland rainforest. Doctoral dissertation, The University of Michigan, Ann Arbor.

————. 1975b. Effects of vesicular-arbuscular mycorrhizae on lowland tropical rainforest trees. Pages 437–446 *in* F. E. Sanders, B. Mosse & P. B. Tinker (eds.), Endomycorrhizas. Academic Press, London.

————. 1980. Mycorrhizae influence tropical succession. Biotropica 12 (Suppl.): 56–64.

―――― & J. M. Trappe. 1982. Two new *Acaulospora* species from tropical America. Mycotaxon 15: 515–522.

Jeng, R. S. & R. F. Cain. 1976. *Collematospora*, a new genus of the Trichosphaericeae. Canad. J. Bot. 54(21): 2429–2433.

――――, E. R. Luck-Allen & R. F. Cain. 1972. New species and new records of *Delitschia* from Venezuela. Canad. J. Bot. 55(4): 383–392.

Jorstad, I. 1959. Uredinales from South America and tropical North America, chiefly collected by Swedish botanists. II. Ark. f. Bot., Ser. 2, 4(5): 59–103.

Kalb, K. 1987. Brasilianische Flechten. 1. Die Gattung *Pyxine*. Biblioth. Lichenol. 24: 1–89.

Keller, H. W. & K. L. Braun. 1977. Myxomycetes of Mexico II. Bol. Soc. Mex. Micol. 11: 167–180.

Kern, F. D. & R. Ciferri. 1930. Fungi of Santo Domingo. III. Uredinales. Mycologia 22: 111–117.

―――― & H. W. Thurston, Jr. 1944. Additions to the Uredinales of Venezuela. III. Mycologia 36: 54–64.

Khan, S. R. & J. W. Kimbrough. 1974. Taxonomic position of *Termitaria* and *Mattirolella* (entomogenous Deuteromycetes). Amer. J. Bot. 61: 395–399.

Kimbrough, J. W., E. R. Luck-Allen & R. F. Cain. 1969. (*Iodophanus*, the Pezizaceae segregate of *Ascophanus* (Pezizales). Amer. J. Bot. 56: 1187–1202.

――――, ―――― & ――――. 1972. North American species of *Coprotus* (Thelebolaceae: Pezizales). Canad. J. Bot. 50: 957–971.

―――― & B. L. Thorne. 1982. Structure and development of *Mattirolella crustosa* (Termitariales, Deuteromycetes) on Panamanian termites. Mycologia 74: 201–209.

Kohlmeyer, J. 1980. Tropical and subtropical filamentous fungi of the Western Atlantic Ocean. Bot. Mar. 23: 529–544.

――――. 1981. Marine fungi from Martinique. Canad. J. Bot. 59: 1314–1321.

――――. 1983. Geography of marine fungi. Austral. J. Bot., Suppl. Ser. 10: 67–76.

――――. 1984. Tropical marine fungi. Marine Ecol. (P.S.Z.N.I.) 5(4): 329–378.

―――― & E. Kohlmeyer. 1979. Marine mycology. The higher fungi. Academic Press, New York.

Korf, R. P. 1970. Nomenclatural notes. VII. Family and tribe names in the Sarcoscyphineae (Discomycetes), and a new taxonomic disposition of the genus. Taxon 19(5): 782–786.

――――. 1977. A new species of *Lasiobelonium* (Helotiales, Hyaloscyphaceae, Arachnopezizeae) from the neotropics. Trans. Mycol. Soc. Japan 17: 206–208. '1976.'

――――. 1978a. Nomenclature and taxonomic notes on *Lasiobelonium*, *Erioscypha* and *Erioscyphella*. Mycotaxon 7: 399–406.

――――. 1978b. Revisionary studies in the Arachnopezizoideae: A monograph of the Polydesmiae. Mycotaxon 7: 457–492.

―――― & K. P. Dumont. 1968. The case of *Lambertella brunneola*: An object lesson in taxonomy of the higher fungi. J. Elisha Mitchell Sci. Soc. 84: 242–247.

Kotlaba, F., Z. Pouzar & L. Ryvarden. 1984. Some polypores rare or new to Cuba. Česká Mykol. 38: 137–145.

Kriesel, H. 1970. El papél de los hongos en la vegetación forestál de Cuba. Bol. Soc. Mex. Micol. 4: 39–43.

――――. 1971a. Charakeristika der Pilzflora Kubas. Biol. Rundschau, Jena 9: 65–73.

――――. 1971b. Ektotrophe Mykorrhiza bei *Coccoloba uvifera* in Kuba. Biol. Rundschau 9: 97–98.

――――. 1971c. Clave para la identificación de los macromicetos de Cuba. Ciéncias, Habana, ser. 4, Cien. Biol. 16: 1–101.

Kunth, C. S., ed. 1822. Synopsis plantarum quas in itinere ad plagam ascuinoctialem orbis novi collegerunt A. de Humboldt et A. Bonpland. Vol. 1.

Lichtwardt, R. W. 1973. The Trichomycetes: What are their relationships? Mycologia 65: 1–20.

Lindau, G. & P., Sydow. 1908–1917. Thesaurus litteraturae mycologicae et lichenologicae. Vol. 1, vii + 903 pp., 1908; 2, 808 pp. 1909; 3, iv + 766 pp., 1912; 4, xiii + 609 pp., 1915 (exsiccati pages 31–39); 5, vii + 527 pp., 1917. Leipzig.

Lowe, J. L. 1963. A synopsis of *Poria* and similar fungi from the tropical regions of the world. Mycologia 55: 453–486.

Lowy, B. 1970. Keys to neotropical Tremellales. Nova Hedwigia 19: 407–438.

――――. 1971a. Some observations on ethnomycology in Mexico and Guatemala. Revista/Review Interamericana 1: 39–49.

――――. 1971b. Tremellales. Flora Neotrop. Monogr. 6: 1–153.

――――. 1975. Additional neotropical Tremellales. Mycologia 67: 991–1000.

――――. 1976a. A new *Tremella* from Ecuador. Mycotaxon 4: 163–165.

――――. 1976b. New Tremellales from Panama. Mycologia 68: 1103–1108.

――――. 1980. Supplement to Neotropical Tremellales. Flora Neotrop. Monogr. 6: 1–18.

――――. 1981. A new species of *Dacryopinax* from Brazil. Mycotaxon 13: 428–430.

――――. 1982. New Tremellales from West Brazilian Amazon. Mycotaxon 15: 95–102.

――――. 1985. Some Phragmobasidiomycetes from Acre and Amazonas. Acta Amazonica 15(1/2) PFA Supl., 2: 35–42.

Mahoney, D. P. 1971. Soil and litter microfungi of the Galapagos Islands. Ph.D. Thesis, Univ. of Wisconsin, Madison.

――――. 1976. A new *Neocosmospora* from Galapagos Island soil. Mycologia 68(5): 1111–1116.

Maia, L. C. 1983. Sucessão de fungos em folhedo de floresta tropical umida. Thesis, Univ. Fed. Pernambuco, Recife, Brazil.

Mapes, C., G. Guzmán & J. Caballero. 1981. Etnomicología Purépecha. Soc. Mex. Micol., Inst. Biol., UNAM, Mexico.

Martin, G. W. 1937. A new type of heterobasidiomycete. J. Wash. Acad. Sci. 27: 112–114.

Martinez, D., M. Quirarte, C. Soto, D. Salomones & G. Guzmán. 1984. Prospects of Mexican edible mushroom cultivation on agricultural wastes. Bol. Soc. Mex. Micol. 19: 207–219.

Matsushima, T. 1981. Matsushima Mycological Memoirs No. 2.

――――. 1983. Matsushima Mycological Memoirs No. 3.

Milanez, A. I., 1968. Aquatic fungi of the "Cerrado" region of São Paulo state. I. First results. Rickia 3: 97–109.

Montagne, C. D. M. 1837. Centurie de plantes cellulaires exotiques nouvelles. Ann. Sci. Nat. Bot. ser. 2, 8: 345–370.

――――. 1839. Cryptogamae Brasilienses seu Plantae quas in itinere per Brasilien a cl. A. de Saint-Hilaire collectas observationibusque nonnullis illustravitt. Ann. Sci. Nat. 2. Ser., XII: 42–55.

――――. 1840. Seconde centurie de plantes cellulaires exotiques. Ann. Sci. Nat., Bot. ser. 2, 13: 193–207.

――――. 1842. Troisième centurie des plantes cellulaires exotiques nouvelles. Ann. Sci. Nat., Bot. ser. 2, 18: 241–282.

――――. 1843. Quatrième centurie de plantes cellulaires exotiques nouvelles. Ann. Sci. Nat., Bot. ser. 2, 20: 352–379.

――――. 1854. Crytogamia Guyanensis seu plantarum cellularum in Guyana gallica annis 1835–1849 a cl. Leprieur collectarum enumeratio universalis. Ann. Sci. Nat., Bot. ser. 4, 1: 91–144.

――――. 1857. Note sur deux champignons Bull. Soc. Bot. France 4: 444.

Mora, V. & G. Guzmán. 1983. Agaricales poco conocidos en el Estado de Morelos. Bol. Soc. Mex. Micol. **18**: 115–139.

Morris, E. F. 1978. Belizean hyphomycetes. Mycotaxon **7(2)**: 265–274.

Müller, E. & J. A. von Arx. 1962. Die gattungen der didymosporen Pyrenomyceten. Beitrage zur Kryptogamenflora der Schweiz, Band II, Heft 2. Wabern, Bern.

Müller Arg., J. 1892. Lichenes epiphylli Spruceani. J. Linn. Soc., Bot. **29**.

Murrill, W. A. 1911–1918. The Agaricaceae of tropical North America. I. Mycologia **3**: 23–36. 1911a. II. Mycologia **3**: 79–91. 1911b. III. Mycologia **3**: 189–199. 1911c. IV. Mycologia **3**: 271–282. 1911d. VI. Mycologia **5**: 18–36. 1913a. VIII. Mycologia **10**: 62–85. 1918.

_____. 1912. The Polyporaceae of Mexico. Bull. N.Y. Bot. Gard. **8(28)**: 137–153.

Nishida, F. H. 1989. Current and active neotropical mycological workers. Mycotaxon (in press).

Nobles, M. K. 1958. Cultural characters as a guide to the taxonomy and phylogeny of the Polyporaceae. Canad. J. Bot. **36**: 883–926.

Oberwinkler, F. & R. J. Bandoni. 1982. A taxonomic survey of the gastroid auriculoid Heterobasidiomycetes. Canad. J. Bot. **60**: 1726–1750.

_____ & B. Lowy. 1981. *Syzygospora alba*, a mycoparasitic heterobasidiomycete. Mycologia **73(6)**: 1108–1115.

Otero, J. & M. T. Cook. 1937. Bibliography of mycology and phytopathology of Central and South America, Mexico and the West Indies. J. Agric. Univ. Puerto Rico **21(3)**: 249–486.

Overholts, L. O. 1934. Hymenomycetes. Pages 304–316. *in* C. E. Chardón & R. A. Toro, Mycological exploration of Venezuela. Monogr. Univ. Puerto Rico.

Paden, J. W. 1974. Ascospore germination in *Phillipsia crispata* and *P. lutea*. Mycologia **66(1)**: 25–31.

_____. 1975. Ascospore germination, growth in culture and imperfect spore formation in *Cookeina sulcipes* and *Phillipsia crispata*. Canad. J. Bot. **53(1)**: 56–61.

_____. 1977. Two new species of *Phillipsia* from Central America. Canad. J. Bot. **55(21)**: 2685–2692.

_____. 1983. Sarcosomataceae (Pezizales, Sarcoscyphineae). Flora Neotrop. Monogr. **37**: 1–16. New York Botanical Garden.

_____. 1984. A new genus of Hyphomycetes with teleomorphs in the Sarcoscyphaceae (Pezizales, Sarcoscyphineae). Canad. J. Bot. **62(2)**: 211–218.

_____. 1986. On the anamorph of *Phillipsia crispata*. Mycotaxon **25(1)**: 165–174

Patouillard, N. 1876–1924. Collected mycological papers. Asher, Amsterdam.

_____. 1889. Fragments mycologiques. Notes sur quelques champignons de la Martinique. J. Bot. (Morot) **3**: 335–343.

_____. 1899. Champignons de la Guadeloupe. Bull. Soc. Mycol. France **15**: 191–209.

_____. 1900. Champignons de la Guadeloupe recueillis par le R. P. Düss (2e Serie). Bull. Soc. Mycol. France **16**: 175–188.

_____. 1902. Champignons de la Guadeloupe recueillis par le R. P. Düss (3e Serie). Bull. Soc. Mycol. France **18**: 171–186.

Pearson, D. 1986. United States systematics in foreign countries: The new Ugly Americans? ASC Newsletter, February, **14(1)**: 4–6.

Pegler, D. N. 1983a. Agaric flora of the Lesser Antilles. Kew Bull. Addit. Ser. 9. HMSO, London.

_____. 1983b. The genus *Lentinus*: A world monograph. Kew Bull. Addit. Ser. 10. HMSO, London.

_____. 1987. A revision of the agaricales of Cuba 1. Species described by Berkeley & Curtis. Kew Bul **42(3)**: 501–585.

_____ & J. P. Fiard. 1978. *Hygrocybe* sect. *Firmae* (Agaricales) in tropical America. Kew Bull. **32**: 297–312.

_____ & _____. 1979. Taxonomy and ecology of *Lactarius* (Agaricales) in the Lesser Antilles. Kew Bull. **33**: 601–628.

_____ & R. Singer. 1980. New taxa of *Russula* from the Lesser Antilles. Mycotaxon **12**: 92–96.

Perez-Silva, E. 1967. Les inocybes du Mexique. Anales Inst. Biol. Univ. Nac. Mexico **38 (Ser. Bot.1)**: 1–60.

Petrini, O. & M. Dreyfuss. 1981. Endophytische Pilze in Epiphytischen Araceae, Bromeliaceae und Orchidaceae. Sydowia **34**: 135–148.

Pfister, D. H. 1973. Notes on Caribbean Discomycetes. III. Ascospore germination and growth in culture in *Nanoscypha tetraspora* (Pezizales, Sarcoscyphineae). Mycologia **65**: 952–956.

_____. 1977. Annotated index to fungi described by N. Patouillard. Contr. Reed. Herb. **25**: 1–211.

Pinto-Lopes, J. 1952. Polyporaceae – Contribuição para a sua biotaxonomia. Mem. Soc. Broteriana **8**: 5–214. Coimbra.

Plunkte, M. 1984. Die Flechtenflora Kubas (Flora Lichenum Cubensis): Bibliographie. Terrestrial Ökol., Sonderh. **4**: 1–157.

Prance, G. T. 1972. An ethnobotanical comparison of four tribes of Amazonian Indians. Acta Amazonica **2(2)**: 7–27.

_____. 1973. The mycological diet of the Yanomam Indians. Mycologia **65**: 248–250.

Pulido, M. M. 1983. Estudios en Agaricales Colombianos. Biblioteca Jose Jeronimo Triana No. 7. Museo de Historia Natural, Inst. Ciencias Naturales, Bogotá.

Rajchenberg, M. 1986. On *Trametes aethalodes* Mont. and other species of *Daedalea* (Polyporaceae). Canad. J. Bot. **64**: 2130–2135.

_____ & J. Wright. 1982. Two new South American species of *Perenniporia*. Mycotaxon **15**: 306–310.

Redhead, S. A. & J. Ginns. 1980. *Cyptotrama asprata* (Agaricales) from North America and notes on the five other species of *Cyptotrama* sect. *Xerulina*. Canad. J. Bot. **58**: 731–740.

_____ & R. E. Halling. 1987. *Xeromphalina nubium* sp. nov. (Basidiomycetes) from Cerro de la neblina, Venezuela. Mycologia **79(3)**: 383–386.

Reeves, F., Jr. & A. L. Welden. 1967. West Indian species of *Hymenochaete*. Mycologia **59**: 1034–1049.

Reid, D. A. 1959. The genus *Cymatoderma* Jungh. (*Cladoderris*). Kew Bull. **13**: 518–530.

_____. 1965. A monograph of the stipitate stereoid fungi. Beih. Nova. Hedwigia **18**: 1–382.

Reynolds, D. R. 1971. The sooty mold ascomycete genus *Limacinula*. Mycologia **63**: 1173–1209.

_____. 1972. Stratification of tropical epiphytes. Kalikasan, Philippine J. Biol. **1**: 1–10.

_____. 1975. Observations on growth forms of sooty mold fungi. Nova Hedwigia **26**: 179–193.

_____. 1977. Foliicolous ascomycetes 1: The capnodiaceous genus *Scorias*, reproduction. Nat. Hist. Mus. Los Angeles County Contr. in Sci. **288**: 1–16.

_____. 1978–1986. Foliicolous ascomycetes 2: *Capnodium salicinum* Mont. emend. Mycotaxon **7**: 501–507. 1978. 3: The stipitate capnodiaceous species. Mycotaxon **8**: 417–445. 1979. 4: The capnodiaceous genus *Trichomerium*. Mycotaxon **14**: 198–220. 1982. 5: The capnodiaceous clypeate genus *Treubiomyces*. Mycotaxon **17**: 349–360. 1983. 6: The capnodiaceous genus *Limacinula*. Mycotaxon **23**: 141–152. 1985.

_____. 1986. Foliicolous ascomycetes 7: Phylogenetic systematics of the Capnodiaceae. Mycotaxon 27: 377–403.

Rodriguez-Scherzer, G. & L. Guzmán-D. 1984. Los hongos (Macromicetos) de las Reservas de la Biosfera de la Michilia y Mapimi, Durango. Bol. Soc. Mex. Micol. **19:** 159–168.

Rogers, J. D. 1978. *Hypoxylon conostomum*: Cytology of the ascus. Canad. J. Bot. **56:** 1946–1948.

———. 1981. *Camarops rickii* sp. nov. from Brazil and comments on *C. peltata*. Canad. J. Bot. **59:** 2539–2542.

——— & K. P. Dumont. 1979. Los hongos de Colombia VI. Two new applanate species of *Hypoxylon*. Mycologia **71:** 807–810.

——— & G. J. Samuels. 1985. New taxa of *Hypoxylon*. Mycotaxon **22:** 367–373.

Rogerson, C. T. 1970. The hypocrealean fungi (Ascomycetes, Hypocreales) Mycologia **62(5):** 865–910.

Rossman, A. Y. 1977. The genus *Ophionectria* (Euascomycetes, Hypocreales). Mycologia **69:** 355–391.

———. 1979a. *Calonectria* and its type species, *C. daldiniana*, a later synonym of *C. pyrochroa*. Mycotaxon **8(2):** 321–328.

———. 1979b. A preliminary account of the taxa described in *Calonectria*. Mycotaxon **8(2):** 485–558.

Ryvarden, L. 1973. Type studies in the Polyporaceae—I. Tropical species described by C. H. Persoon. Persoonia **7(2):** 305–312.

———. 1975. Type studies in the Polyporaceae—6. Species described by W. J. Hooker. Norweg. J. Bot. **22:** 285–287.

———. 1981. Type studies in the Polyporaceae—13. Species described by J. H. Léveillé. Mycotaxon **13(1):** 175–186.

———. 1982a. Type studies in the Polyporaceae—11. Species described by J. F. C. Montagne, either alone or with others. Nord. J. Bot. **2:** 75–84.

———. 1982b. *Fuscocerrena*, a new genus in the Polyporaceae. Trans. Brit. Mycol. Soc. **79(2):** 279–281.

———. 1982c. The genus *Hydnochaete* (Hymenochaetaceae). Mycotaxon **15:** 425–447.

———. 1983a. Type studies in the Polyporaceae—14. Species described by N. Patouillard, either alone or with other mycologists. Occ. Pap. Farlow Herb. **18:** 1–39.

———. 1983b. The genus *Navisporus* (Polyporaceae). Nord. J. Bot. **3:** 411–413.

———. 1984. Type studies in the Polyporaceae—16. Species described by J. M. Berkeley, either alone or with other mycologists from 1856 to 1886. Mycotaxon **22(2):** 329–363.

———. 1985a. Type studies in the Polyporaceae—17. Species described by W. A. Murrill. Mycotaxon **23:** 169–198.

———. 1985b. *Stipitochaete*, gen. nov. Hymenochaetaceae, Basidiomycotina). Trans. Brit. Mycol. Soc. **85:** 535–539.

Saccardo, P. A. 1881–1901. Sylloge fungorum, Vol. I–XV. Patavii. [reprint by Edwards Brothers, Inc., Ann Arbor, Michigan, 1944.]

Sáenz, J. A. 1980. *Ligiella*, a new genus for the Clathraceae. Mycologia **72(2):** 338–349.

———, A. V. Macaya-L. & M. Nassar-C. 1983. Hongos comestibles, venenosos y aluccinatorios de Costa Rica. Revista Biol. Trop. **31(2):** 201–207.

——— & M. Nassar-C. 1982. Hongos de Costa Rica: familias Phallaceae y Clathraceae. Revista Biol. Trop. **30(1):** 41–52.

Samuels, G. J. 1985. Four new species of *Nectria* and their *Chaetopsinia* anamorphs. Mycotaxon **22(1):** 13–32.

——— & K. P. Dumont. 1982. The genus *Nectria* (Hypocreaceae) in Panama. Caldasia **13(63):** 379–423.

——— & E. Müller. 1978. Life-history studies of Brazilian Ascomycetes, (1–4): 1. Two new genera of Sphaericeae (*Porosphaeria* and *Striatosphaeria*) having, respectively *Sporoschisma*-like and *Codinaea* anamorphs. Sydowia **31:** 126–136; 2. A new species of *Thaxteriella* and its helicosporous anamorph. Sydowia **31:** 137–141; 3. *Melanomma radicans* sp. nov. and its *Aposphaeria* anamorph, *Trematosphaeria perrumpens* sp. nov. and *Berlesiella fungicola* sp. nov. and its *Ramichloridium* anamorph. Sydowia **31:** 142–156; 4. Three species of *Herpotrichia* and their *Pyrenochaeta*-like anamorphs. Sydowia **31:** 157–168.

——— & ———. 1979. Life-history studies of Brazilian Ascomycetes. 7. *Rhytidhysteron rufulum* and the genus *Eutryblidiella*. Sydowia **32:** 277–292.

——— & ———. 1980. Life-history studies of Brazilian Ascomycetes, (8–9). 8. *Thamnomyces chordalis* (anam.: *Nodulisporium*) and *Camillea bacillum* (anam.: *Geniculosporium*) with notes on taxonomy of the Xylariaceae. Sydowia **33:** 274–281; 9. *Fluviostroma wrightii* gen. et sp. nov. (Syn.: *Sphaerostilbe wrightii* nom. illegit.) and its synnematous anamorph (*Stromatostilbella* gen nov.). Sydowia **33:** 282–288.

——— & C. T. Rogerson. 1984. New Ascomycetes from Amazonia. Acta Amazonica **14(1/2) PFA Supl. 1.** 81–93.

——— & A. Y. Rossman. 1979. Conidia and classification of the nectrioid fungi. Pages 167–182 *in* B. Kendrick (ed.), The whole fungus, Kananaskis II, Vol. I.

——— & ——— & E. Müller. 1978. Life-history studies of Brazilian Ascomycetes 6. Three species of *Tubeufia* with, respectively, dictyosporous/pycnidial and helicosporous anamorphs. Sydowia **31:** 180–193.

Santesson, R. 1952. Foliicolous lichens I. (A revision of the taxonomy of the obligately foliicolous, lichenized fungi). Symb. Bot. Ups. **12:** 1–590.

Schenck, N. 1985. VA mycorrhizal fungi: 1950 to the present—the era of enlightenment. Pages 56–60 *in* R. Molina (ed.), Proc. 6th North Amer. Conf. Mycorrhizae. Forest Research Lab., Corvallis, Oregon.

———, J. L. Spain & E. Sieverding. 1986. A new sporocarpic species of *Acaulospora* (Endogonaceae). Mycotaxon **25(1):** 111–117.

———, ———, ——— & R. H. Howeler. 1984. Several new and unreported vesicular-arbuscular mycorrhizal fungi (Endogonaceae) from Colombia. Mycologia **76:** 685–699.

Seaver, F. J. 1925. Studies in tropical Ascomycetes—III. Porto Rican cap fungi. Mycologia **17:** 45–50.

———. 1928. Studies in tropical Ascomycetes. V. Species of *Phyllachora*. Mycologia **20:** 214–225.

——— & C. E. Chardón. 1926. Mycology. Sci. Surv. Porto Rico and the Virgin Isl. N.Y. Acad. Sci. **8(1):** 1–208.

———, ———, R. A. Toro & F. D. Kern. 1932. Supplement to Mycology. Sci. Survey P. R. and the Virgin Isl. N.Y. Acad Sci. **8:** 209–240.

——— & J. M. Waterston. 1940. Contributions to the mycoflora of Bermuda—I. Mycologia **32:** 388–407.

——— & ———. 1942. Contributions to the mycoflora of Bermuda—III. Mycologia **34:** 515–524.

Setliff, E. C. 1984. Flora neotropica I. Some lignicolous polypores from Venezuela. Mycotaxon **19:** 213–217.

——— & L. Ryvarden. 1982. The genus *Aporpium* and two additional poroid fungi. Canad. J. Bot. **60:** 1004–1011.

——— & ———. 1983. Los hongos de Colombia. VII. Some Aphyllophoraceous wood-inhabiting fungi. Mycotaxon **18:** 509–526.

Sherwood-Pike, M. A. 1974. New Hyphomycetes from Guadeloupe, F. W. I. *Albosynnema filicicola*, *Tetracrium musicola*, and *Thozetellopsis calicioides*. Mycotaxon **1:** 117–120.

———. 1977a. The Ostropalean fungi. Mycotaxon **5:** 1–277.

———. 1977b. Taxonomic studies in the Phacidiales: *Propolis* and

Propolomyces. Mycotaxon 5: 320–330.

———. 1977c. The Ostropalean fungi II. *Schizoxylon*, with notes on *Stictis*, *Acarosporina*, *Coccopeziza*, and *Carestiella*. Mycotaxon 6: 215–260.

———. 1980. Taxonomic studies in the Phacidiales: The genus *Coccomyces* (Rhytismataceae). Occ. Pap. Farlow Herbarium 15: 1–120.

Singer, R. 1942. A monographic study of the genera *Crinipellis* and *Chaetocalathus*. Lilloa 8: 441–534.

———. 1944. Notes on taxonomy and nomenclature of the polypores. Mycologia 36: 65–69.

———. 1951. The 'Agaricales' (Mushrooms) in modern taxonomy. Lilloa 22: 1–832.

———. 1952. Russulaceae of Trinidad and Venezuela. Kew Bull. (1952): 295–302.

———. 1953. Four years of mycological work in southern South America. Mycologia 45: 865–891.

———. 1958a. Mycological investigations on *teonanacatl*, the Mexican hallucinogenic mushroom. I. Mycologia 50: 239–261. II. (with A. Smith). Mycologia 50: 262–303.

———. 1958b. Monographs of South American Basidiomycetes, especially those of the east slope of the Andes and Brazil. I. The genus *Pluteus* in South America. Lloydia 21: 195–299.

———. 1963a. Oak mycorrhiza fungi in Colombia. Mycopathol. Mycol. Appl. 20: 239–252.

———. 1963b. Un nuevo hongo comestible de Sudamerica. Bol. Soc. Arg. Bot. 10: 207–208.

———. 1965a. Monographic studies on South American Basidiomycetes, especially those of the east slope of the Andes and Brazil. II. The genus *Marasmius* in South America. Sydowia 18: 106–358.

———. 1965b. Interesting and new Agaricales from Brazil. IMUR Acta 2: 15–59.

———. 1970. Flora Neotropica Monographs: 3. Omphalinae (Clitocybeae-Tricholomataceae Basidiomycetes); 4. *Phaeocollybia* (Cortinariaceae, Basidiomycetes); 5. Strobilomycetaceae (Basidiomycetes). OFN Hafner, 84 + 13 + 34 pp.

———. 1971. A revision of the genus *Melanomphalia* as a basis for the phylogeny of the Crepidotaceae. In R. H. Petersen (ed.), Evolution of the Higher Basidiomycetes. Knoxville.

———. 1973. The genera *Marasmiellus*, *Crepidotus* and *Simocybe* in the Neotropics. Nova Hedwigia, Beih. 44: 1–517.

———. 1974. A monograph of *Favolaschia*. Nova Hedwigia, Beih. 50: 1–108.

———. 1975. The Agaricales in modern taxonomy, Third edition. Verlag Cramer, Vaduz.

———. 1976a. The neotropical species of *Campanella* and *Aphyllotus*. Nova Hedwigia 26: 847–896.

———. 1976b. Tropical Russulaceae; *Lactarius* Sect. Polysphaerophori in the Gulf area. Nova Hedwigia 26: 897–901.

———. 1976c. Marasmieae (Basidiomycetes, Tricholomataceae). Flora Neotrop. Monogr. 17: 1–347.

———. 1983. Acanthocytes in *Amparoina* and *Mycena*. Cryptogamie, Mycol. 4: 111–115.

———. 1984a. Adaptation of higher fungi to várzea conditions. Amazoniana 8: 311–319.

———. 1984b. Tropical Russulaceae 2: *Lactarius* sect. Panuoideae. Nova Hedwigia 40: 435–447.

——— & **I. Araujo.** 1979. Litter decomposition and ectomycorrhiza in Amazonian forests. Acta Amazonica 9: 25–41.

———, ——— & **M. H. Ivory.** 1983. The ectotrophically mycorrhizal fungi of the neotropical lowlands, especially Central Amazonia. Nova Hedwigia, Beih. 77: 1–352.

——— & **A. P. L. Digilio.** 1953. Prodromo de la Flora Agaricina Argentina. Lilloa 25: 5–462.

——— & **J. P. Fiard.** 1976. Agaricales nouvelles des Antilles françaises. Bull. Soc. Mycol. France 92: 445–447.

——— & **L. Gomez.** 1982. Basidiomycetes of Costa Rica. I. Brenesia 19/20: 31–47.

——— & ———. 1984. Basidiomycetes of Costa Rica. III. The genus *Phylloporus* (Boletaceae). Brenesia 22: 163–181.

——— & **H. S. Morello.** 1960. Ectotrophic forest tree mycorrhizae and forest communities. Ecology 4: 549–551.

——— & **M. Moser, I. Gamundi, E. R. de la Sota & G. Sarmiento.** 1965. Forest mycology and forest communities in South America I. Mycopathol. Mycol. Appl. 26: 129–191.

——— & **A. H. Smith.** 1958a. Mycological investigations on teonanacatl, the Mexican hallucinogenic mushroom. II. A taxonomic monograph of *Psilocybe*, section *Caerulescentes*. Mycologia 50: 262–303.

——— & ———. 1958b. New species of *Psilocybe*, Section *Caerulescentes*. Mycologia 50: 141–142.

Sipman, H. J. M. 1983. A monograph of the lichen family Megalosporaceae. Bibliot. Lichen. 18: 1–241.

———. 1986. Additional notes on the lichen family Megalosporaceae. Willdenowia 15: 557–564.

——— & **J. Aguirre-C.** 1982. Contribución al conocimiento de los liquenes de Colombia-I. Clave genérica para los liquenes foliosos y fruticosos de los paramos Colombianos. Caldasia 13(64): 603–634.

Sousa, M. 1980. O genero *Phellinus* Quelet (Hymenochaetaceae) na Amazonia Brasileira. Tesis. INPA. Manaus, Brazil.

Spegazzini, C. 1899. Fungi Argentini novi vel critici. An. Mus. Nac. Buenos Aires. 6: 81–367.

Sydow, H. 1930. Fungi venezuelani. Ann. Mycol. 28: 29–224.

Stevenson, J. A. 1971. An account of fungus exsiccati containing material from the Americas. Nova Hedwigia 36: 1–563.

———. 1975. The fungi of Puerto Rico and the American Virgin Islands. Contr. Reed Herbarium 13: 1–743.

Teixeira, A. R. 1962. The taxonomy of the Polyporaceae. Bio. Rev. 37: 51–81.

——— & **D. P. Rogers.** 1955. *Aporpium*, a polyporoid genus of the Tremellaceae. Mycologia 47: 408–415.

Thaxter, R. 1986-1931. Contributions towards a monograph of the Laboulbeniales I-V. Mem. Amer. Acad. Arts. Sci. 12–16.

Thurston, H. W., Jr. 1940. The rusts of Minas Gerais, Brazil, based on collections by A. S. Müller. Mycologia 32: 290–309.

——— & **F. D. Kern.** 1933. Distribution of the West Indian rusts. Mycologia 25: 58–64.

Torrend, C. 1951. Os cogumelos na sua alimentação e sua cultura. 2nd. ed. São Paulo: Edit. Chacaras e Quinteis.

Trappe, J. & G. Guzmán. 1971. Notes on some hypogeous fungi from Mexico. Mycologia 63: 317–332.

———, ——— & **C. Vazquez-S.** 1979. Observaciones sobre la identificación, distribución y usos de los hongos del género. *Elaphomyces* en Mexico. Bol. Soc. Mex. Micol. 13: 145–150.

Udagawa, S. & Y. Kobayashi. !979. Coprophilous fungi of Mexican volcano Popocatepetl. J. Jap. Bot. 54: 161–168.

Ulken, A. 1970. Phycomyceten aus der Mangrove bei Cananeia (São Paulo, Brasilien). Veroff. Inst. Meeresforsch. Bremerh. 12: 313–319.

———. 1972. Physiological studies on a phycomycete from a mangrove swamp at Cananeia, São Paulo, Brazil. Veroff. Inst. Meeresforsch. Bremerh. 13: 217–230.

Upadhyay, H. P. 1967. Soil fungi from North-East Brazil. III. Phycomycetes. Mycopathol. Mycol. Appl. 31(1): 49–62.

_____. 1970. Soil fungi from north-east and north Brazil. Persoonia **6**: 111–117.

Varela, L. & J. Cifuentes. 1979. Distribución de algunos macromicetos en el norte del estado de Hidalgo. Bol. Soc. Mex. Micol. **13**: 75–88.

Veerkamp, J. & W. Gams. 1983. Los hongos de Colombia—VIII. Some new species of soil fungi from Colombia. Caldasia **13(65)**: 709–715.

Viégas, A. P. 1944. Alguns fungos do Brasil. II. Ascomicetos. Bragantia **4**: 5–392.

_____. 1945. Alguns fungos do Brasil. IV. Uredinales. Bragantia **5**: 1–144.

_____. 1959. A pedra flexivel descrita por Anchiéta. Bragantia **18**: 31–37.

Villarreal, L. 1983. Algunas especies de Myxomycetes no registradas del estado de Veracruz. Bol. Soc. Mex. Micol. **18**: 153–164.

Waksman, S. A. 1917. Is there any fungus flora in the soil? Soil Sci. **3**: 565–589.

Walker, C. 1983. Taxonomic concepts in the Endogonaceae: Spore wall characteristics in species descriptions. Mycotaxon **18**: 443–455.

Warcup, J. H. 1951. The ecology of soil fungi. Trans. Brit. Mycol. Soc. **34**: 376–399.

Watling, R. & A. E. Watling. 1980. A literature guide for identifying mushrooms. Mad River Press, Eureka, California.

Welden, A. 1954. Some tropical American stipitate stereums. Bull. Torrey Bot. Club **81**: 422–439.

_____. 1958. A contribution toward a monograph of *Cotylidia* (Thelephoraceae). Lloydia **21(1)**: 38–44.

_____. 1960. The genus *Cymatoderma* (Thelephoraceae) in the Americas. Mycologia **52(6)**: 856–876.

Wheeler, Q. 1980. Studies on neotropical slime mold/beetle relationships, Part I: Natural history and description of a new species of *Anisotoma* from Panama. Proc. Entomol. Soc. Wash. **82(3)**: 493–498.

Wirth, M. & M. E. Hale, Jr. 1978. Morden-Smithsonian Expedition to Dominica: The lichens (Graphidaceae). Smithsonian Contr. Bot. **40**: 1–69.

Wright, J. E. 1960. Notas sobre Faloideas Sud y Centroamericanas. Lilloa **30**: 339–359.

_____. 1970. *Agaricostilbum*, a new genus of Deuteromycetes on palm spathes from Argentina. Mycologia **62**: 679–682.

_____, R. J. Bandoni & F. Oberwinkler. 1981. *Agaricostilbum*, a Basidiomycete. Mycologia **73**: 880–886.

_____ & S. Blumenfeld. 1984. New South American species of *Phellinus* (Hymnochaetaceae). Mycotaxon **21**: 413–425.

_____, T. Herrera & G. Guzmán. 1972. Estudios sobre el género *Tulostoma* en México. Ciéncia (Mexico) **27(4/5)**: 109–122.

Yarwood, C. E. 1973. Pyrenomycetes: Erysiphales. Pages 71–86 *in* G. C. Ainsworth, F. K. Sparrow & A. S. Sussman (eds.). The Fungi IVA. Academic Press, N.Y.

Zheng, R. 1984. The genus *Brasiliomyces* (Erysiphaceae). Mycotaxon **19**: 281–289.

Quantitative Inventory of Tropical Forests

David G. Campbell

Contents

I. Introduction	524
II. Field Methods of Quantitative Inventory	524
A. Voucher Specimens	524
B. Diameter at Breast Height	525
C. Measurement of Height	526
D. Choice and Size of Study Site	527
E. Design of Sampling System	527
1. Quadrat or Circular Plot	527
2. Belt Transect	528
3. Point-Centered Quarter Method	528
F. Measurement for Other Information	528
1. Light Gaps	528
2. Soil Characteristics	528
3. Terrain and Slope	528
III. The Interpretation of Phytosociological Data from Tropical Forests	528
A. Forestry	529
B. Ecology and Conservation	529
1. Patterns of Distribution	529
2. Patterns of Abundance	529
3. Structure of the Forest	530
4. Life Histories and Life Strategies	530
5. Phenology of Forest	530
6. Long-Term Dynamics of Forest	530
7. Human Ecology and Anthropology	531
IV. Recommendations	531
V. Acknowledgments	531
VI. Literature Cited	531

I. Introduction

". . . These gigantic trees are in a constant struggle for their own preservation, and impede one another's growth . . . even the stems which have grown to a considerable height, and require a large sample of nutriment, feel the influence of their more powerful neighbors. . . ". With this observation, Johann Baptist von Spix and Karl Friedrich Philipp von Martius (1824), made one of the earliest contributions toward understanding the relationships of tropical forest trees in the Amazon River Basin. Since that time, the study of the organization of plants in tropical forests has generated a considerable literature from Asia, Australia, Africa, Central America and South America, and has been the testing ground for important hypotheses concerning ecology and evolution, as well as being integral to a number of pragmatic concerns, such as conservation and the yield of economically important woods.

The methods employed in these studies are those of quantitative ecological inventory, which is defined for the purposes of this paper as the enumeration of individuals and species of trees in a small patch of forest, the measurement of several important parameters of those individuals, and the analysis of the abundance and distribution of those individuals as functions of their physical and biotic environments. This field is therefore quite distinct from, and should not be confused with, floristic inventory, which is the stock-taking and description of plant species, and which is the subject of most of the rest of this volume. (Floristic inventory is, of course, requisite to quantitative inventory, since the latter depends on the accurate identification of species.) The methods of quantitative inventory have been highly diverse and therefore have not always yielded information of comparable format and dimension that can be used to contrast one study site with another.

The purposes of this paper are A) to review the field methods of data collection employed in the quantitative ecological inventory of tropical forests, particularly as they have been applied to the Neotropics, B) to evaluate some of the strengths and weaknesses of each method, C) to review and evaluate some methods of data analysis, D) to review some of the uses of inventory data, with particular emphasis on ecology and conservation, and E) to recommend methods that insure comparability of future quantitative inventories in tropical forests.

II. Field Methods of Quantitative Inventory

Quantitative forest inventory employs a modest array of tools and measurements, all the more remarkable in light of the diversity of standards, methods and results that are generated through their use. The field tools of the researcher are tapes and measures, an instrument to measure heights, instruments to measure meteorological variables, devices to sample soil, and voucher specimens for the exact identification of plants to species. The parameters of analysis are likewise few: diameter at breast height, height of tree (both bole and total height) or height of canopy, crown diameter, location of each tree on some sort of grid or coordinate system and environmental, meteorological and soil information.

A. Voucher Specimens

The high species richness of trees in tropical forests is at once the joy and bane of the botanical ecologist. It is a joy because of the vast array of life forms that the researcher observes. It is a bane because of the difficulty of identification of trees, particularly if they are neither in flower nor fruit, and because of the morphological similarities of the sterile parts of many tropical tree species. This problem is exacerbated by the lack of comprehensive floras for many areas of tropical forests, particularly in the Neotropics. Notable exceptions are the florulas of Río Palenque (Dodson & Gentry, 1978), Avila (Steyermark & Huber, 1978) and Barro Colorado Island (Croat, 1979). Without such florulas, field identification of plants is often misleading and is unacceptable for rigorous studies. Therefore, voucher specimens, even sterile ones (and at any one particular sampling time, the majority of individual plants in a forest will be sterile), must be collected for each individual in the sample. The only exception is when one already has a series of vouchers for a particularly common species in a study site, and one is certain that the species can be reliably identified in the field. Then it is sometimes permissible, for reasons of economy, not to collect further specimens.

In tall tropical forest, the collection of voucher specimens is often the most time-consuming, expensive and dangerous activity of all. The methods of collection have been diverse. Human climbers, using spiked griffes or a foot strap (called a "peconha" in Brazil), are the most commonly employed climbing tools, at least in the Neotropics. Griffes, however, injure trees and probably increase mortality rates, and are therefore unacceptable for long-term studies. The peconha is not harmful to the tree, but its use is limited to smaller girths. Professional climbers have been widely used in the Brazilian Amazon, where they are valued permanent employees of museums and research organizations, and in eastern Zaire. However, in many parts of the world finding such skilled and courageous people is next to impossible. In Malesia, Corner used trained monkeys to climb into the canopy of tropical forest trees and break off branches. Others have used projectiles to break off specimens. Shotguns often yield small but adequate samples, even from very tall trees. In Suriname, van Roosmalen (1980) used a slingshot to fire weights attached to monofilament over boughs, and then rigged wire saws to cut off branches. Often trees are cut down in order to be collected, an obviously unacceptable activity for phytosociological studies on permanent study sites (however, one which is convenient on survey areas that are doomed to be flooded by a dam or destroyed by a road).

During certain times of the year, particularly in seasonal forests, some species will be deciduous and foliar voucher specimens are unobtainable. Under these circumstances it is advisable to collect twigs, fallen leaves, or small wood and bark samples as vouchers.

Many tropical forest inventories, particularly those conducted by foresters, have been limited to living, boled trees and have excluded dead trees, acaulate palms and lianas, even though they may have DBH's large enough for inclusion into the sample, which has thereby considerably diminished the value of the studies to ecology. Consequently, I recommend that palms and lianas, as well as dead trees (particularly if the mortality is to be monitored by resurveys in the future), be included in inventory data.

B. Diameter at Breast Height

Diameter at breast height (DBH), is a fundamental measurement in quantitative inventory of forests. Is is also the easiest measurement to make, since it is made directly, without use of distance estimates or optical tools. DBH is the primary measured (as opposed to enumerated) parameter for both the purposes of ecology and forestry, being used in static studies to compute cross-sectional area, dominance, ground cover, as a component in the calculation of biomass, and in dynamic, long-term studies to measure growth.

Breast height above ground level is obviously a subjective term. Some authors have rigorously defined it to two decimal places (for example, Mori et al., 1983: 1.37 m); most define it to one decimal place ranging from 1.3 m to 1.5 m. In forestry practice, an exact definition is of little importance, since the taper of a tree trunk over the few cm variation in breast height among observers is negligible and since irregularities of stem always exceed this variance. However, regardless of the breast height chosen, for long-term dynamic studies the diameter of the bole must be remeasured in exactly the same spot. The most accurate method of insuring replicability of breast height is to paint a strip at the location of the original measurement (fig. 1). Rankin and Pires, working in Manaus and the Mocambo Reserve (Belém) respectively, both used paint to permanently mark the site of DBH measurement. By contrast, one of the least accurate methods of permanently marking height of measurement is to use nails as a reference point. Nails tend to induce the growth of scar tissue, which may exaggerate measures of growth.

The presence of buttresses, which often extend several meters above breast height, often complicates the measurement of DBH in tropical forests, where many of the trees are buttressed (see discussion in Richards, 1952). For purposes of measureing biomass and related indices, measurement of DBH at the level of the buttresses is necessary. But for purposes of ecology and dominance, which are concerned with the amount of ground surface area controlled by an individual, regardless of whether it physically occupies that area, the measurement of DBH at the level of the buttresses may be warranted.

Indeed, not only buttresses, but also chancres, bulges and other distortions of tree trunks, as well as inclined terrain, will affect DBH measurements. Rankin (1985) proposed

Table I

The relationship between minimum diameter at breast height (DBH) and the number of individuals, families, genera and species included in an inventory. The data are from a three hectare inventory of *terra firme* tropical forest at O Desierto on the Rio Xingu in the Brazilian Amazon (Campbell et al., 1986).

Minimum DBH (cm.)	10	15	20	25	30	35	40
Number of:							
Individuals	1904	1345	1033	900	805	731	665
Families	39	35	32	31	30	26	21
Genera	127	109	84	72	62	53	42
Species	265	202	143	112	84	68	51

measuring DBH 10 cm above the distortion, and in the case of inclined terrain, the 1.30 m mark was measured from the highest intersection of ground level with the trunk (fig. 1).

The minimum DBH for inclusion in an inventory defines the sample size and therefore the completeness of the survey. Most commercial forest inventories have used a minimum 60 cm or 40 cm DBH (Heinsdijk & de Bastos, 1965), which has made them relatively useless to the ecologist or the investigator interested in species diversity. Only a few, for the most part recent, inventories have employed a 20 cm or a 10 cm minimum (for a general review, see Rollet, 1978, and for a review of the New World inventories, see Campbell et al., 1986; Gentry, 1982). A few exceptionally valuable inventories have included trees as small as one cm DBH (Hubbell, 1979; Knight, 1975).

Table I shows the relationship between minimum DBH and sample size in terms of number of individuals, families, genera and species for three hectares of tropical forest on the Rio Xingu, the Brazilian Amazon (data from Campbell et al., 1986). From these data, it is clear that sample size increases rapidly as minimun DBH decreases. In general, the increase in the number of trees sampled as a function of decrease in DBH is nearly linear for trees greater than 20 cm DBH and is nearly exponential for smaller stems (for a review of mathematical models of this function, see Rollet. 1978).

The amount of effort required to conduct an inventory is a function of minimum DBH. There is a tradeoff here. Smaller DBH's yield more information per unit area of forest, but usually limit the overall geographical sample size. Larger DBH's sample fewer juvenile trees, but enable a larger area to be sampled. Many ecologists have compromised by using a minimum DBH of 10 cm. Whitmore (1985) called this the "industrial standard" of forest phytosociology; Prance (1984) also advocated that it become the standard DBH for quantitative inventory. One can reasonably assume that by using a minimum DBH of 10 cm, most of the species of trees in the forest are sampled and that juveniles less than 10 cm DBH which are not sampled are represented as adults in the sample. Indeed, as Webb et al.

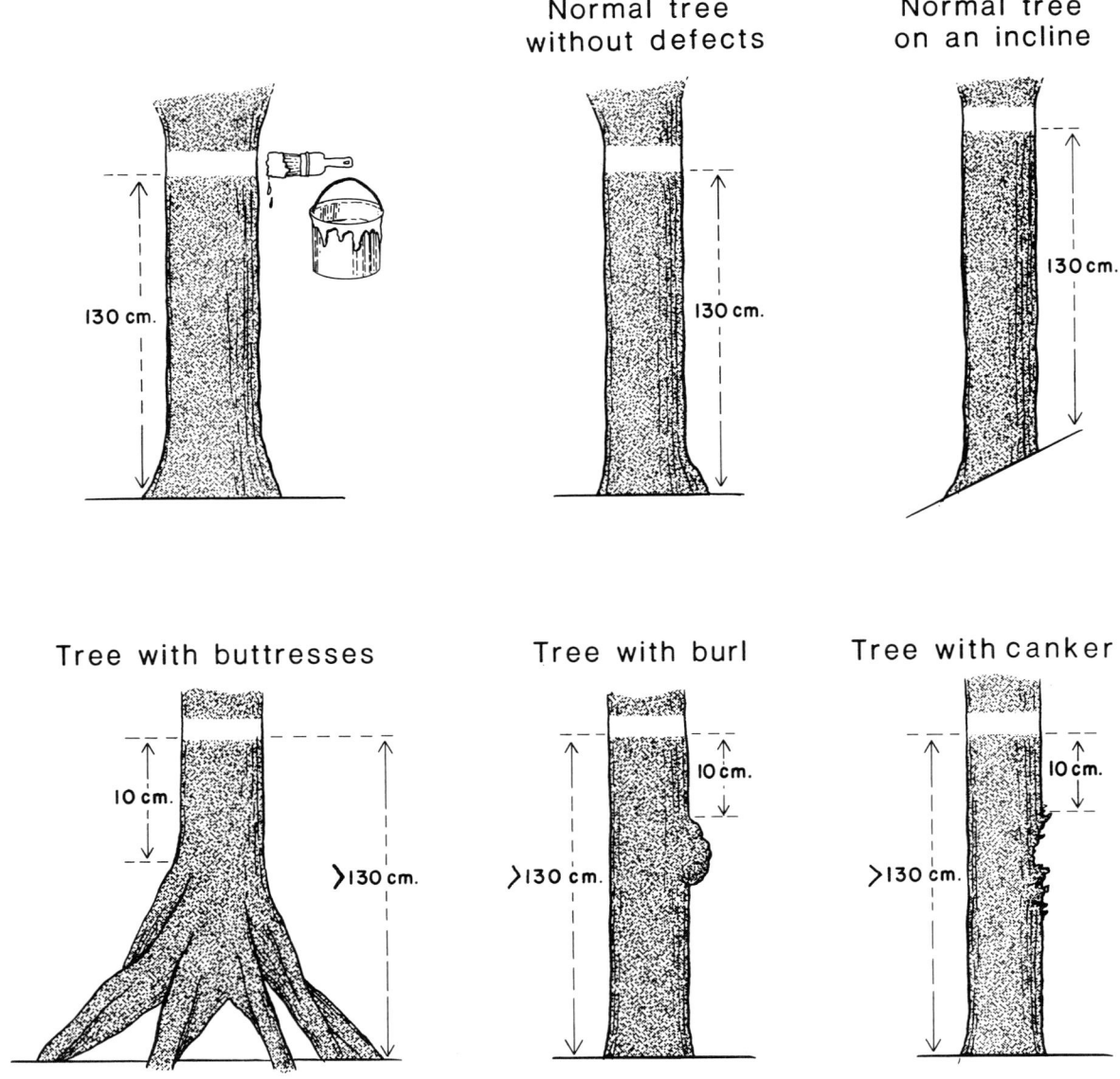

FIG. 1. Location of measurement of DBH on trees with irregular trunks or on inclined terrain (after Rankin, 1985).

(1967) pointed out, large trees contain most of the classificatory information in a forest. Finally, the choice of this arbitrary cutoff is reinforced by the difficulty of keying to species many juvenile tropical forest trees, which are often reproductively immature and which have leaf morphologies and habits that are different from those of adults.

It is therefore recommended that a minimum DBH of 10 cm be employed in quantitative ecological inventory of tropical forests. This will insure the comparability of separate surveys and, given a sample of one or several hectares, will embrace most of the species of trees found in the forest (see discussion of sampling design, below). This recommendation in no way implies that, when resources permit, a smaller minimum DBH should not be employed or, in fact, that epiphytes and herbaceous understory plants should not also be included. Indeed, studies with these smaller thresholds will best answer questions regarding spatial distribution patterns, dispersal strategies and reproductive strategies of tropical forest trees. It is recommended, however, that when a DBH of less than 10 cm is employed, the results also be presented in terms of a 10 cm minimum DBH, so that they may be directly compared with other surveys.

C. Measurement of Height

The measurement of bole height—that is, trunk height to the first major branch—is a parameter inherited from forestry, since the highest quality, knotless timber-bearing wood occurs in the bole of a tree. However, bole height is a useful parameter for the ecologist as well, since by subtracting it from total height, one approximates the vertical depth of the canopy, or the sun-soaked (euphotic) zone of the tree. The plane described by bole heights over the whole forest is therefore equivalent to the level of the "morphological inversion surface" (Oldeman, 1974) of the forest, above which the forest is fully exposed to sunlight and air currents, below which it is dark and the air is still (Richards, 1983).

Tropical forest trees commonly grow to heights of 20 to 30 meters and some individuals may exceed 60 meters. Tall trees will predominate, of course, if one's sample has a ten cm DBH minimum size, which selects for large trees that are almost invariably taller than a human. Therefore, the measurement of height, being something that can seldom be done directly, is potentially fraught with inaccuracy. Various methods have been used to measure height in tropical forests. Heinsdijk (1960) and Heinsdijk and de Bastos (1965), working in the Amazon, visually estimated tree height and claimed, probably with justification, that their method was as accurate as any other. Other workers have directly measured tree height by dropping metered lines from the tops of trees, or taken the opposite approach and used tethered balloons (Prance et al., 1977). These methods are probably highly accurate, but impractical on a large scale. Many workers, particularly foresters, use a clinometer and trigonometric conversions to calculate tree height. This method requires a measure of the distance of the observer to the tree (accuracy improves the greater the ratio of the distance to the tree height) and, if the observer has an unobstructed view, has an error of less than five percent for trees up to 50 meters tall. Unfortunately, in dense tropical forest, an observer seldom has an unobstructed view of the canopy, and therefore clinometric measurement in tropical forest should always be considered to have an error of at least ten percent.

D. Choice and Size of Study Site

Study sites for phytosociological inventory may be specifically selected for biological features, such as species complexes, or physical features, such as soil (Dantas & Muller, 1980), or—using a flexible sampling scheme—for as much heterogeneity as possible (Smartt, 1978).

More commonly, study sites are chosen to be "representative" of a more general forest type (such as lowland tropical forest). However, there is tremendous variation in species composition within any seemingly homogeneous sample of lowland forest, due to the clumping of species and the patchy distribution of rare species. This was demonstrated by Hubbell and Foster (1983) using spatial autocorrelation of 50 one-hectare quadrats on Barro Colorado Island and by Campbell et al. (1986), who found low coefficients of similarity between three hectares of tropical forest on a three-kilometer long belt transect on the Rio Xingu in Brazil. It is therefore probably impossible to ever find a "representative" study site, short of sampling the total forest, in most areas of lowland tropical forest. (Exceptions, of course, are monospecific stands adapted to unusual conditions, such as the *buritizais* forests, which are almost pure stands of *Mauritia flexuosa* on flooded soils of Amazonia, described by Pires, 1984). Furthermore, in light of the inherent variation among small samples of lowland forest, it is often dangerous to extrapolate one's data from a small sample to generate conclusions as to the forest as a whole.

The variation in species obviously directly relates to one's choice of size of study site (which of course also relates to minimum DBH, discussed above, and type sampling system, discussed below). The simplest quantitative method of determining whether one's sample encompasses all, or most, of the species in an area is by determining whether one's sample approaches asymptote on a species-area curve, which is one of the oldest and most powerful tools of ecology. (For a review of the assumptions, strengths, weaknesses and common misinterpretations of the species-area curve, see McGuiness, 1984.) Using this technique in Amazonian terra firme forest, Campbell et al. (1986) showed that the species-area curve approached asymptote between two and three hectares, for trees greater than ten cm DBH on a linear transect line 10 meters wide. A three-hectare study site is, unfortunately, considerably larger than most phytosociological study sites in tropical forests. Yet, even using a 31 hectare sample, Rankin et al. (1984), working near Manaus, found that the number of species of Chrysobalanaceae and Sapotaceae was still increasing linearly. It should be noted, however, that the 31 hectares were not all contiguous.

Some researchers employ rigorous criteria for determining when a species-area curve or a species-number of individuals curve approaches asymptote. For example, Rankin (pers. comm.) has suggested that a sample be considered complete when the number of new species per one hundred new individuals in a sample drops to less than five.

E. Design of Sampling System

There are a number of study site designs commonly used in quantitative inventory of forests; however, this paper will only deal with the three most common systems: the quadrat or circular plot, the belt transect and point-centered quarter method. For a review of the methodologies of each, one should consult a basic forestry or ecology text (such as Smith, 1980). For use in quantitative inventory of tropical forests, each method has its own limitations and strengths and the choice of one is really a function of the researcher's goals. Ashton (1965) and Cain and Castro (1959) have discussed the advantages and disadvantages of rectangular vs. linear plots in quantitative inventory; in 1983 Daly et al. (in prep.) field tested and compared the three techniques in tropical forest in Amapá, Brazil. The strengths and weaknesses of the three techniques are reviewed below.

1. Quadrat or circular plot

Quadrats and circular plots produce the most circumscribed study sites. Their advantages are 1) the study site is not spread out, and therefore the least amount of energy is expended in surveying it, and 2) certain spatial patterns, particularly clumping of individuals of the same species, are easily detected when trees are plotted on a system of Cartesian coordinates.

The main disadvantages are 1) that the sample, being limited to a small area, does not include as great a variety of microhabitats nor encompass as large a number of species as other sampling methods and 2) in the process of laying out the study site and taking measurements, much of the understory may be inadvertently trampled, and therefore the value of the study site for long-term studies is diminished.

2. Belt transect

The belt transect is probably the most popular sampling format. Advantages are 1) it samples a broad spectrum of microhabitats and species in comparison to a quadrat or circular plot of the same area, and is therefore useful for detecting internal patterns of distribution as functions of terrain, soil type, inundation, etc., and 2) the chance of accidental destruction of undertory is less than in a quadrat of the same area, particularly if workers avoid using the transect line as a highway to and from the current work sites.

The disadvantage is that, being a long and narrow sample that encompasses many microhabitats, it is inefficient as a method to characterize a particular community type, except in very homogeneous terrain.

3. Point-centered quarter method

The main advantages of the point-centered quarter method are 1) that it permits the broadest sampling of microhabitats and species per number of individuals sampled (of course, this advantage is greatest when the sampling pattern is organized linearly), 2) by its very design, it is ideal for analysis of nearest neighbors and therefore has utility in detecting patterns of species distribution, and 3) it minimizes damage to the understory.

The principal disadvantages of the method are that 1) in contrast to the quadrat circular and belt transect formats, which permit exact measurement of sampling area, it is a "dimensionless" sampling format and cannot yield species richness as a function of unit area and 2) it is the most labor-intensive method, since only four trees are sampled per sampling interval, and therefore to get a sufficiently large sample, the sampling line may have to be spread out over several kilometers of forest.

F. Measurement for Other Information

Environmental factors, such as soil characteristics, terrain, slope, disturbance and light gaps are integral to forest ecology, yet many quantitative inventories of tropical forests have failed to record these data.

1. Light gaps

The dissection of a forest by tree falls, or "light gaps" is of course related to its age and successional state. The most commonly employed method of measuring disturbance is the simple qualitative noting of a light gap (and sometimes its dimensions) in one's field notebook. However, such a subjective measure is of negligible value in quantitative analysis. More quantitative techniques employ the direct measurement of canopy levels above points in the sample (Hubbell & Foster, 1983), the measurement of canopy cover with wide-angle photography, or photometric measures using quantum light meters integrated over the course of the day or one-time measures an hour before and an hour after midday.

2. Soil characteristics

Nutrient levels in tropical forest soils, although characteristically very low, are often highly variable, even on the small scale of several square meters, and this variability affects vegetation. Much of this variation is secondary to the terrain, since, all other variables being equal, nutrients in soils leach faster on ridgetops than in valley bottoms, and valley bottoms often accumulate litter faster than steeper terrain. Likewise, the particle size and porosity of soils (and their concomitant effect on soil hydrology) is also often a function of the variable sorting rates of particles due to terrain and slope.

The relationship between vegetation and edaphic factors has been excellently demonstrated by (among others) Ashton (1976), Austin et al. (1972), working in Brunei and Malaysia, Hubbell and Foster (1983), working in Panama, and Lescure and Boulet (1985), working in French Guiana.

One can never assume uniformity of nutrient levels or porosity in tropical forest soils. It is recommended that soils be characterized at regular intervals in quantitative inventory, particularly if the sampling design employs an elongated belt transect or the point-centered quarter method. When feasible, measurements should include levels of important nutrients, such as available nitrogen, magnesium, phosphorus, potassium, iron and calcium, as well as pH and particle size. Trees at different life stages may have roots that invade different soil horizons; therefore it is important to sample all soil horizons, at least down to one meter.

3. Terrain and slope

Not only do terrain and slope affect the characteristics of soil, as described above, but they also determine the access of a tree to the euphotic zone, since the lateral insolation of steep slopes often illuminates relatively short trees that otherwise would be growing in the understory. Therefore, slope on which a tree is growing, and the extent of that slope, are important values to record during quantitative inventory. Slope may be conveniently, but crudely, measured by a clinometer. For accurate measurement of contours, surveyor's tools are necessary.

III. The Interpretation of Phytosociological Data from Tropical Forests

Quantitative inventory of tropical forests is elemental to the understanding of tropical forest ecology and the rational management of tropical forests. These uses may be divided into the general categories of forestry, ecology and conservation. The category of forestry is really beyond the purview of this paper, but since most methods in quantitative inventory were developed by foresters, it will be discussed briefly in this context.

A. FORESTRY

Early methods for quantitative inventory of forests were developed by commercial foresters and academicians in forestry science working in temperate regions, as a means for evaluating their resource. (For reviews of these methods, see Beleya, 1931; Bruce & Schumacher, 1942; Chapman & Demeritt, 1936; Graves, 1906; Spurr, 1952). These methods had limited utility for or application to ecology or conservation. More general methods with applications in ecology and conservation, but still principally targeted at forestry, appeared later (Loetsch & Haller, 1973; Prodan, 1968).

Foresters are interested in the yield of economically valuable wood in a portion of forest plot. Therefore, the forester often concentrates only on large trees and often only on certain species, consequently generating data of only limited utility to other disciplines. Nevertheless, several papers useful to ecologists and conservationists have resulted from forestry studies in tropical forests. For example, de Carvalho (1983) conducted a study of the population biology of rosewood (*Aniba duckei*) in Pará. Burgess (1975) conducted an excellent autecological study of the commercially important Dipterocarp, *Shorea curtisii*, in Malaysia. By contrast, Heinsdijk (1960) sampled 918 hectares of forest in the Brazilian Amazon for commercially important species, but failed to use Latin nomenclature, which invalidated his extensive data for ecology.

B. ECOLOGY AND CONSERVATION

The questions addressed by ecologists and conservationists striving to preserve ecological diversity are the patterns of distribution and abundance of the species in the forest. These questions can be divided into seven subcategories.

1. Patterns of distribution

The ecologist seeks to discover which factors determine the distributions of species in a tropical forest; whether species are randomly or nonrandomly distributed, and if nonrandomly distributed, whether their distributions conform to the intrinsic biological properties of the species, patterns of terrain, soil, disturbance (such as light gaps and the changing courses of rivers), or the presence or absence of conspecifics.

Disturbance, and particularly light gaps, have been invoked as a source of heterogeneity and therefore a cause of high species diversity of tropical forests. Grieg-Smith et al. (1967) showed a correlation between forest type and physiognomy and the prevalence of storms in the Solomon Islands. Certain tropical forest species have long been recognized by ecologists as being favored by light gaps, at least during certain stages of their lives (Pires, 1984; Pires & Prance, 1977), however attempts to detect consistent successional patterns as functions of gap size and age have had mixed success (Brokaw, 1985; Denslow, 1980; Williams et al., 1969).

One of the foci of ecological research in tropical forests has been the investigation of non-random distributions of trees and, more specifically, clumping. The impetus for this research derived from papers by Janzen (1970) and Connell (1970), in which they hypothesized that seed predation, herbivory and pathogens would select against clumping of tropical forest trees, and favor their wide dispersion. A number of experiments employing quantitative inventory have been designed to directly or indirectly test this very testable hypothesis. For example, Clark and Clark (1984) showed that survival of *Dipteryx panamensis* seedlings increased as distance from parent tree increased. Hubbell and Foster (1983), using a quadrat one-half square kilometer in area on Barro Colorado Island, showed three patterns of distribution for canopy trees: random or near-random distribution, clumping as a function of topographic and edaphic features and clumping that was uncorrelated with topography. Connell et al (1984) showed decreased mortality among rare species of trees in tropical forest in Queensland.

2. Patterns of abundance

Species richness is one of the two components of species diversity. The simplest index of species richness is the number of species per unit area. In the parlance of forest ecology, which is divorced from the familiar and logical terminology of statistics, species richness is addressed by relative "density", which is defined as the number of trees of species A divided by the total number of trees in the sample.

The evenness of the distribution of the species is the other component of diversity; for a discussion of indices of species richness, evenness and diversity in tropical forest, see Stocker et al. (1985). In the standard terminology of forest ecology, a measure of evenness is obtained from the likewise illogically-named relative "frequency," defined as the number of sampling units of containing species A divided by the total number of sampling units for all species.

The basal area for individual trees (or for a particular species) is a key index of its ecological dominance and is an indirect measure of biomass. Indeed, this is the definition of "dominance" in the jargon of forest ecology; relative dominance is defined as the cross-sectional area for species A divided by the basal area for all species.

The sum for a species of the three relative values described above—relative density, relative frequency and relative dominance —equals the "importance value" for that species, as defined by Curtis and Cottam (1962). Importance value has become a standard ranking index for forest tree species. Goodall (1970) validly cricitized the worth of relative density, relative frequency and therefore importance value, arguing that they are not commensurable between species. Furthermore, comparisons between community types based on composite measures are intrinsically difficult. Regardless, these values have been widely used in the literature and workers in the field of quantitative inventory should be familiar with them.

3. Structure of the forest

Forest structure has been traditionally depicted by sketched profiles of eight by four meter transects of forests (Richards, 1952). In spite of the obvious qualitative nature of this method, and the unrepresentativeness of the small sample size, profiles have proven to be a useful tool for comparing forest physiognomies. Because of the small sample size, and the simplicity of profiles, stratification of forest into vertical layers seems most apparent using this method. It is less clear, however, in larger samples and frequency distributions which aggregate individual trees into classes of height (see below and Whitmore, 1984). For this reason, the concept of stratification has been criticized by, among others, one of its originators (Richards, 1983).

In a quantitative inventory, the sample population of each species will vary in size as a function of the extent of the study site and the commonness of the species. The advantage of large study sites, and concomitant large populations, to the population biologist is obvious, yet in tropical forests even relatively large study sites have vanishingly small populations of many species. For example, Campbell et al. (1986) found that 99 species (37% of the total number of species) on a three-hectare sample of trees of DBH ten cm or greater were represented by only one individual.

When a sample is sufficiently large to permit analysis, one may construct frequency histograms of the distribution by DBH size class (or by height class) of all of the sampled trees in the forest, or of a particular species of tree. This method is highly diagnostic of the condition of the forest and type of stand structure, ranging from a bell-shaped curve for an evenly-distributed small stand of one species, to a bimodal distribution indicating a growing cohort of trees that may have experienced a past disturbance, to a "reverse J-shaped" curve for most large forest samples of many species. De Carvalho (1981) used this technique to analyse the size structure of populations of six species of trees in Amazonian tropical forests.

For reverse J-shaped distributions, the depletion rate through the DBH size classes can be derived as a constant in a negative exponential function. This constant can be readily compared between stands (Veblen et al., 1981).

However, it is important to note that DBH and height do not necessarily correlate with age of trees in tropical forests, where an individual tree may wait years, with little growth, for a light gap to appear. Indeed, age data are usually unobtainable for tropical trees, most of which do not have distinct, interpretable annual growth rings (Bormann & Merlyn, 1981). (An exception to this may be trees that grow in regimes of annual inundation, such as in the *várzea* forests of the Brazilian Amazon.) In order to overcome this problem, Lieberman and Lieberman (1985) simulated growth curves as a function of age from periodic increment data in tropical forest trees of La Selva, Costa Rica.

4. Life histories and life strategies

In spite of the difficulty and inaccuracy of their measurement, bole height and total height are useful parameters in characterizing the light requirements and growth strategies of individual species. For example by regressing height against DBH, Pires and Prance (1977) were able to discern the adaptive strategies regarding light for certain tree species on the Mocambo Reserve in Brazil. Campbell et al. (1986) used this technique to categorize 33 species of trees on the Rio Xingu into life strategy classes.

The standard analytic techniques of population biology may be applied to inventory data. For example, Hartshorn (1975) used a projection matrix model to analyse the population dynamics of *Pentaclethra macroloba* on four hectares of tropical forest in Costa Rica. By monitoring the mortality and recruitment within 15 life stages of *P. macroloba* over a year, he was able to construct a life table for the species.

5. Phenology of forest.

The cycling of foliation, flowering and fruiting may be monitored by the frequent sampling of the same study site over the course of at least a year. Since some tropical forest trees take longer than a year to complete their reproductive cycles (and some, such as *Tachigalia myrmecophila* flower only once in their lives), monitoring over several years is often necessary. For example, Mori et al. (1982), working in a Bahian forest, monitored phenology for six years, and although they detected a peak in foliation and flowering during the southern hemisphere spring, they detected no pattern of fruiting.

6. Long-term dynamics of forest

There have been few studies of the long-term dynamics of primary tropical forests: the measurement of mortality recruitment and growth rates for entire forests, species or size classes of individuals. This is especially true in the Neotropics.

Discussion of forest dynamics is confused by the variety of ways used to quantify mortality. Some authors use linear mortality models, from which can be derived stand turn-over rate, which is the inverse proportion of trees dying per year (Brokaw, 1982). Others define turn-over as the inverse of the rate of light gap formation per year (Hartshorn, 1978). It is important to note, however, that overall mortality rates do not correlate with the rate of light gap formation (Brokaw, 1982), since tree falls do not always open up the canopy. Still others (Swaine & Lieberman, 1987; Campbell & Maciel, in prep.) use a log model mortality rate and stand half-life which, because it is a negative exponential function, is equivalent to slightly more than half of the stand turn-over rate.

Measures of mortality and growth rates increase in accuracy with longer time intervals and larger sample sizes. For example, Weaver (1979) measured growth in two Puerto Rican forests in an important study spanning 30 years. The fifty hectare study site containing ca. 250,000 trees greater than 2.5 cm DBH, which has been monitored on Barro Colorado Island (Panama) since 1980 by Hubbell and Foster (1983), because of its magnitude and resolution, will probably become a benchmark study for long-term change in tropical forests.

Long-term forest dynamics also affect the choice of minimum park and reserve size for the incorporation, and equally importantly the maintenance, of all species, which is a priority for conservation. These considerations, of course, relate to the definition of community and species-area relationships, described above. However, the maintenance of diversity within reserves is likewise a function of the size of a reserve, which is a virtual island and subject to island population dynamics (MacArthur & Wilson, 1967). The periodic monitoring of reserves over time is essential to understanding trends of extinction in these finite areas. The WWF-US/INPA minimum critical size program in Brazil (Lovejoy et al., 1983) is using inventoried plots of tropical forest, ranging in size from one hectare to 10,000 hectares, to determine these trends.

7. Human ecology and anthropology

Quantification of human impact on tropical forest has been discussed above. Inversely, the impact of tropical forest species on humans is also quantifiable by means of inventory. For example, using data from a one-hectare inventory of tropical forest, Boom (1985) showed that the Chácobo Indians of Bolivia utilized, for one purpose or another, 82% of the tree species on the plot. This finding is one of the most evocative proofs of the generally unappreciated economic value of tropical forest species to humans.

IV. Recommendations

To insure the quality and comparability of future quantitative inventory of tropical forests, I recommend the following standard procedures.

1. The study site should be in an area that will be protected in perpetuity, enabling future teams to return to the site to monitor phenology and long-term dynamics.

2. Vouchers should always be collected, even if they are sterile and for common species, and entered into a permanent herbarium collection.

3. Each tree in the sample should be permanently labelled with an aluminum tag attached with an aluminum nail. The nail should not be placed in proximity to the level where DBH is measured.

4. One should employ a maximum of 10 cm DBH cutoff for inclusion of trees into the sample. Cutoffs smaller than 10 cm DBH are, of course, better; however, if a smaller cutoff is employed, the summary data should be presented for both the smaller DBH and the 10 cm DBH, so that comparisons may be made with other inventories employing the "industrial standard" of 10 cm DBH.

5. All plants with stems larger than the selected cutoff should be measured, including lianas, multiple-stemmed palms (of which each rosette should be considered one individual) and acaulate palms.

6. The layout of one's study site should serve the goals of one's research. In tropical forests a belt transect or point-centered quarter method sample will encompass more microhabitats and differences in terrain and therefore will sample more species than a quadrat or circular plot. However, a transect is inappropriate if one wishes to define community type or to conduct an inventory within a community type. If one is interested in the distributions of species within the sample, a quadrat or the point-centered quarter method (enabling nearest neighbor analysis) are preferred. However, the point-centered quarter method is dimensionless, and therefore cannot be used for the comparison of exact areas.

7. A minimum of one hectare is necessary to sample the majority of tree species in species-rich tropical forest where the species-area curve won't begin to approach asymptote until at least two to five hectares are sampled. In certain low-diversity forest types, such as mangrove or peat forests, a smaller sample may be adequate.

8. The basic summary data for each sampled tree should be entered in a data bank, so that future researchers can proceed with further analysis of the data. This should be done even if the analysis only deals with several principal components of the forest. At a minimum, the following basic data should be included for each tree: A) the collection number or a voucher of the herbarium specimen, B) the DBH, C) the number of its sampling unit or representation, and D) its location on a grid of Cartesian coordinates (this is useful for analysis of distribution even on linear transects). Fundamental summary statistics of forest ecology (such as relative density, relative frequency, relative dominance, importance value and many others not mentioned in this paper) can be derived from these basic data.

9. If time and resources permit, one should collect the following additional information from one's sample: A) height and bole height for each tree, B) two crown diameters at right angles to each other and always in the same compass directions C) details on buttresses, latex and supported lianas and other epiphytes for each tree, D) notes on terrain and gradient for the location of every tree, E) profiles of soil nutrients and hydrology, F) locations of dead trees and G) profiles of light gaps, using photometric methods, photographic images of canopy or direct measurement of canopy height.

V. Acknowledgments

The author is grateful to P. S. Ashton, D. C. Daly, S. M. Mori, B. W. Nelson, G. T. Prance and J. M. Rankin for valuable comments on this paper.

VI. Literature Cited

Ashton, P. S. 1986. Some problems arising in the sampling of mixed rain forest communities for floristic studies. Pages 235–239 *In* Symposium on ecological research in humid tropics vegetation. Proceedings of the Kuching Symposium. UNESCO.

———. 1976. Mixed Dipterocarp forest and its variation with habitat in the Malayan lowlands: A re-evaluation at Pasoh. Malaysian

Forester **39**(2): 56–72

Austin, M. P., P. S. Ashton & P. Grieg-Smith. 1972. The application of quantitative methods for vegetation survey III. A re-examination of rain forest data from Brunei. J Ecol. **60**: 305–324.

Belyea, H. C. 1931. Forest measurement. John Wiley & Sons. New York.

Boom, B. 1985. Amazonian Indians and the forest environment. Nature 314: 324.

Bormann, F. H. & G. Berlyn, eds. 1981. Age and growth rates of tropical trees: New directions for research. Yale Univ. School of For. & Env. Stud. Bull. No. 94. New Haven, CT.

Brokaw, N. V. 1982. Treefalls: Frequency, timing and consequences. Pages 101–108. *In* E. G. Leigh, Jr., A. S. Rand & D. M. Windsor, eds. The ecology of a tropical forest: Seasonal rhythms and long-term changes. Smithsonian Institution Press, Washington, D. C.

———. 1985. Treefalls, regrowth, and community structure in tropical forests. Pages 53–71 *in* S. T. A. Pickett & P. S. White (eds), The ecology of natural disturbance and patch dynamics. Academic Press. New York.

Bruce, D. & F. X. Schumacher. 1942. Forest mensuration. McGraw-Hill Book Company. New York.

Burgess, P. F. 1975. Silviculture in the hill forests of the Malay Peninsula. Malaysian Forestry Dept. Research Pamphlet 66.

Cain, S. A. & G. M. De O. Castro. 1959. Manual of vegetation analysis. Harper & Bros. New York.

Campbell, D. G., D. C. Daly, G. T. Prance & U. N. Maciel. 1986. Quantitative ecological inventory on of terra firme and várzea tropical forest on the Rio Xingu, Brazilian Amazon. Brittonia. **38**(4): 369–393.

——— & U. N. Maciel. In prep. Five-year dynamics of three hectares of tropical forest, Rio Xingu, Brazilian Amazon.

Carvalho, J. O. P. de. 1981. Distribução diametrica de especias comerciais e potencias em floresta tropical umida natural na Amazônia. Boletim de Pesquisa No. 23. EMBRAPA. Belem.

———. 1983. Abundancia, frequencia e grau de agregação do pau-rosa (*Aniba duckei* Kostermans) na Floresta Nacional do Tapajos. Boletim de Pesquisa No. 53. EMBRAPA. Belem.

Chapman, H. C. & D. B. Demeritt. 1936. Elements of forest mensuration. J. B. Lyon Company. Albany.

Clark, D. A. & D. B. Clark. 1984. Spacing dynamics of a tropical rain forest tree: evaluation of the Janzen-Connell model. Amer. Naturalist **124**(6): 769–788.

Connell, J. H. 1970. On the role of natural enemies in preventing competitive exclusion in some marine animals and in rain forest trees. Proc. Adv. Study. Inst. Dynamics Numbers Population. 298–312.

———, J. G. Tracey & L. J. Webb. 1984. Compensatory recruitment, growth, and mortality as factors maintaining rain forest tree diversity. Ecol. Monogr. **54**(2): 141–164.

Croat, T. 1979. Flora of Barro Colorado Island. Stanford Univ. Press. Palo Alto, Calif.

Curtis, J. T. & G. Cottam. 1962. Plant ecology workbook. Burgess Publishing Company. Minneapolis.

Daly, D. C., D. G. Campbell & S. M. Mori. In prep. Comparison of three sampling methods in Amazonian tropical forest.

Dantas, M. & N. R. Muller. 1980. Estudos fito-ecologicos do tropico umido Brasileiro: I aspectos fitosociologicos de mata sobre terra roxa na região de Altamira. Anais do Congresso Nacional de Botanica. São Paulo.

Denslow, J. S. 1980. Gap partitioning among tropical rainforest trees. Biotropica. **12**. suppl.: 47–55.

Dodson, C. & A. H. Gentry. 1978. Flora of the Río Palenque Science Center. Selbyana **4**: 1–628.

Gentry, A. H. 1982. Patterns in neotropical plant species diversity. Pages 1–80 *in* M. K. Hecht, B. Wallace & G. Prance (eds), Evolutionary Biology, Vol 15. Plenum Press. New York.

Goodall, D. W. 1970. Statistical plant ecology. Annual Rev. Ecol. Syst. **1**: 99–124.

Graves, H. S. 1906. Forest mensuration. John Wiley & Sons. New York.

Grieg-Smith, P., M. P. Austin & T. C. Whitmore. 1967. The application of quantitative methods to vegetation study I, association-analysis and principal components ordination of rain forest. J. Ecol. **55**(2): 483–503.

Hartshorn, G. S. 1975. A matrix model of tree population dynamics. Pages 41–52 *in* F. B. Golley & E. Medina (eds) Tropical ecological systems. Springer-Verlag. New York.

———. 1978. Tree falls and tropical forest dynamics. Pages 617–638 *in* P. B. Tomlinson & M. H. Zimmerman, eds. Tropical trees as living systems. Cambridge University Press, Cambridge, U. K.

Heinsdijk, D. 1960. Interim report to the Government of Brazil on the dry land forests of the Tertiary and Quaternary south of the Amazon River. FAO Report No. 1284. Rome.

——— & M. A. de Bastos. 1965. Forest inventory on the Amazon. FAO Report No. 2080. Rome.

Hubbell, S. P. 1979. Tree dispersion, abundance, and diversity in a tropical dry forest. Science 213: 1299–1309.

——— & R. B. Foster. 1983. Diversity of canopy trees in a neotropical forest and implications for conservation. Pages 25–41 *in* S. L. Sutton, T. C. Whitmore, & A. C. Chadwick (eds) Tropical rain forest: ecology and management. Blackwell. Oxford.

Janzen, D. H. 1970. Herbivores and the number of tree species in tropical forest. Amer. Naturalist **104**(940): 501–528.

Knight, D. H. 1975. A phytosociological analysis of a species-rich tropical forest on Barro Colorado Island, Panama. Ecol. Monogr. **45**: 259–284.

Lescure, J-P. & R. Boulet. 1985. Relationships between soil and vegetation in a tropical rain forest in French Guiana. Biotropica **17**(2): 155–164.

Lieberman, M. & D. Lieberman. 1985. Simulation of growth curves from periodic increment data. Ecology 66: 632–635.

Loetsch, F. & K. E. Haller. 1973. Forest Inventory, Volumes I and II. BLV Verlagsgesellschaft. Munich.

Lovejoy, T., R. O. Bierregaard, J. Rankin & H. O. R. Schubart. 1983. Ecological dynamics of tropical forest fragments. Pages 377–384 *in* S. L. Sutton, T. C. Whitmore & A. C. Chadwick (eds). Tropical rain forest: ecology and management. Blackwell. Oxford.

MacArthur, R. H. & E. O. Wilson. 1967. The theory of island biogeography. Princeton University Press. Princeton, N. J.

McGuiness, K. A. 1984. Equations and explanations in the study of species-area curves. Biol. Rev. **59**: 423–440.

Mori, S. A., G. Lisboa & J. A. Kallunki. 1982. Fenologia de uma mata higrofila sul-bahiana. Revista Theobroma. CPC. Ilheus. **12**: 217–230.

———, B. M. Boom, A. M. de Carvalho & T. S. dos Santos. 1983. Southern Bahian Moist Forests. Bot. Rev. **49**(2): 155–232.

Oldeman, R. A. A. 1974. L'Architecture de la forêt guyanaise. Memoires ORSTOM No. 73. Paris.

Pires, J. M. 1984. The Amazonian Forest. Pages 581–602 *in* H. Sioli (ed.). The Amazon. Dr. W. Junk Publishers. Dordrecht.

——— & G. T. Prance. 1977. The Amazon Forest: a natural heritage to be preserved. Pages 158–194 *in* G. T. Prance & T. S. Elias (eds.). Extinction is forever. New York Botanical Garden. Bronx, N. Y.

Prance, G. T. 1984. Completing the inventory. Pages 365–397 *in* V. H. Heywood & D. M. Moore (eds.). Current concepts in plant taxonomy. Academic Press. London.

Prodan, M. 1968. Forest biometrics. Pergamon Press. Oxford.

Rankin, M. 1985. Instruction sheet given to workers on WWF-US/INPA "Minimum critical size of ecosystems" Project. Manaus, Brazil.

———, **G. T. Prance, M. F. da Silva, W. A. Rodrigues & M. Ueling.** 1984. Resultados preliminarios de um levantamento florestal de 31 ha. de terra firme na Amazônia central: descrição geral de vegetação e dados taxonomicos. Submitted to Acta Amazonica.

Richards, P. W. 1952. The tropical rain forest. Cambridge Univ. Press. Cambridge.

——— 1983. The three-dimensional structure of tropical rain forest. Pages 3–10 *in* S. L. Sutton, T. C. Whitmore & A. C. Chadwick (eds) Tropical rain forest ecology and management. Blackwell Scientific Publications. Oxford.

Rollet, B. 1978. Organization. Pages 112–143 *in* Tropical forest ecosystems. UNESCO. Paris.

Roosmalen, M. G. M. van. 1980. Habitat preferences, feeding strategy and social organization of the black spider monkey (*Ateles paniscus* Linnaeus 1758) in Suriname. Ph.D. Thesis. Agricultural University of Wageningen.

Smartt, P. F. M. 1978. Sampling for vegetation survey: A flexible systematic model for sample location. J. Biogeogr. **5**: 43–56.

Smith, R. L. 1980. Ecology and field biology. Third Edition. Harper & Row. New York.

Spix, J. B. von & K. F. Von Martius. 1824. Travels in Brazil in the Years 1817–1920. Longman, Hurst, Rees. London.

Spurr, S. H. 1952. Forest inventory. The Ronald Press Company. New York.

Steyermark, J. & O. Huber. 1978. The flora of Avila. Ministerio del Ambiente y de los Recursos Naturales Renovables. Caracas, Venezuela.

Stocker, G. C., G. L. Unwin & P. W. West. 1985. Measures of richness, evenness and diversity in tropical rainforest. Austral. J. Bot. **33(2)**: 131–137.

Swaine, M. D. & D. Lieberman. 1987. The dynamics of tree populations in tropical forest. J. Trop. Ecol. **3(Pt. 4). Spec. Iss.**: 289–375.

Veblen, T. T., Z. C. Donoso, F. M. Schlegel & R. B. Escobar. 1981. Forest dynamics in south-central Chile. J. Biogeogr. **8**: 211–247.

Weaver, P. L. 1979. Tree growth in several tropical forests of Puerto Rico. U. S. Forest Serv. Res. Paper SO-152.

Webb, L. J., J. G. Tracey, W. T. Williams & G. N. Lance. 1967. Studies in the numerical analysis of complex rain-forest communities II. The problem of species sampling. J. Ecol. **55**: 525–538.

Whitmore, T. C. 1984. Tropical rain forests of the Far East. Second edition. Clarendon Press. Oxford.

———. 1985. Particular plant aspects of species diversity. Paper presented to Third International Congress of Systematic and Evolutionary Biology. University of Sussex.

Williams, W. T., G. N. Lance, L. J. Webb, J. G. Tracey & J. H. Connell. 1969. Studies in the numerical analysis of complex rain-forest communities. IV. a method for the elucidation of small-scale forest pattern. J. Ecol. **57**: 635–652.

Index

Bold face numbers indicate tables, asterisked numbers indicate maps, *italicized* numbers indicate graphs and illustrations.

Abundance, interpretation of, 529
Academic institutions (*see* Herbaria)
Africa (*see also* Regional Reports, 189–250)
    hepatics, 489–490
    number of herbarium specimens, 8, **9**
    priorities for inventory, **21, 22**, 27
    rates of botanical inventory, **9**, *11*
Afromontane region
    priorities for inventory, **21, 28**
Amapá, 403, 416
Amazon R. Basin (*see also* Regional Reports, 353–426, 455–463)
    Colombia, 358, 392*, 394
    Ecuador, 392*
    geology (*see also* Guiana Shield and Brazilian Shield), 402*
    Peru, 392*
    phytogeographic regions, 410–414, 412*
    priorities for inventory, **25**, 28
    quantitative inventory, 524
Anambas I., 104*
Andaman I., 135
Andean forests
    Colombia, 355, 394
    priorities for inventory, **25, 28**
    Venezuela, 363–364
Andes, mosses of, 487
Angola
    number of herbarium specimens, **9**
    rates of botanical inventory, **9**, *11*
Aquatic vegetation (*see also* *Igapó*, Mangrove and *Várzea*)
    Venezuela, 365
Arauacaria forest, 437, 428*
Aru I., 121*
Asia (*see also* Regional Reports, 35–146)
    number of herbarium specimens, 8, **9**
    priorities for inventory, **18, 19**, 27
    rates for botanical inventory, **9**, *10*
Australia (*see also* Kershaw & Whiffin, 149–165)
    adequacy of inventory, 161, 163–164
    hepatics, 490
    mosses, 488
    number of herbarium specimens, **9**, 158, 159
    priorities for inventory, **20**
    rates for botanical inventory, **9**, *10*
Babassu palm
    forest, 408
    germplasm, collection of, *478, 479*
Bacan Is., 116
Baguio District, 53
Banda Is., 119
Banggai I., 114*
Barahona Peninsula, 337
Bark, collection of, 467–468
Belize (*see also* Hampshire, 286–289)
    number of herbarium specimens, **9**
    priorities for inventory, **23**
    rates of botanical inventory, **9**, *12*
Benin (*see also* Hepper, 189–197)
    number of herbarium specimens, **9**, 288
    rates of botanical inventory, **9**, *11*
Bight of Benin forest mosaic (*see also* Hepper, 189–197, and Kendrick, 203–216)
    priorities for inventory, **21, 28**
Bismarck Archipelago, (*see also* New Guinea), 121*

Bohol I., 45*
Bolivia (*see also* Solomon, 455–463)
    no. of herbarium specimens, **9**, 460
    priorities for inventory, **26, 28**
    rates of botanical inventory, **9**, *12*
Borneo (*see also* Ashton, 91–100)
    priorities for inventory, **18**, 27
Borneo, 93*
Botanical inventory, world rates of, 8, **9**, *10, 11, 12,* 13
Bragantina, Zona, 421
Brazil (*see also* Daly & Prance, 401–426, and Mori, 427–454)
    number of herbarium specimens, **9**, **415**, **443**
    priorities for inventory, **25–26**, 28
    rates of botanical inventory, 7, **9**, *12*
Brazilian Shield, 377, 403–404
Brunei (*see also* Ashton, 91–99), 93*
    number of herbarium specimens, **9**, **94**
    rates of botanicial inventory, **9**, *10*
Bryophytes (*see* Buck & Thiers, 484–493)
Burma
    number of herbarium specimens, **9**
    rates of botanical inventory, **9**, *10*
Buru I., 114*, 118
Caatinga formations, 407, 413, 438
    E. Brazil, 428*, *429*, 430–432
Calamian I., 45*
Cambodia (*see also* Schmid, 83–90)
    mosses, 488
Cameroun (*see also* Hepper, 189–197 and Kendrick, 203–216)
    number of herbarium specimens, **9**
    priorities for inventory, **21**, 27
    rates of botanical inventory, **9**, *11*
Campinas vegetation, 394, 413
Campo rupestre, E. Brazil, 428, 432–434, *433*
Campos (*see* Savannas)
Canga formations, ironstone, 417
Canopy measurement, 527, 531
Carajás-São Luís railway, 421
Caribbean Islands (*see also* Regional Reports, 315–346)
    priorities for inventory, **24**, 28
Casiquiare Canal, 363, 364
Cebu I., 45*
Ceja de la montaña (*see also* Cloud forests and *Yungas*)
    Colombia, 354*, 394
    Ecuador, 392*, 394
    Peru, 392*, 394, 395
Celebes I. (*see* Sulawesi)
Centinela Mt., 13
Central African Republic (*see also* Kendrick, 203–216)
    number of herbarium specimens, **9**
    rates of botanical inventory, **9**, *11*
    mosses, 488
Central America (*see also* Regional Reports, 253–312)
    number of herbarium specimens, 8, **9**
    priorities for inventory, **23, 24**
    rates of botanical inventory, **9**, *12*
Ceram I., 114*, 118, 212*
Cerrado vegetation, Brazil
    Amapá, *431*
    Eastern Brazil, 428*, 432, *431*, 438
    transition from lowland forest, 408
Cerro Bachiller, 371
Cerro El Guapo, 371
Chaco, 458–459, 461

Chao Phraya R., 65
China (see also Wang, Chen & Wang, 35–43)
    number of herbarium specimens, 9
    priorities for inventory, **18**
    rates of inventory, 9, *10*
Chocó Dept., 358, 394–395
Chromosome number, in taxonomy, 469, **470**
Circular plots, 527–528
Cloud forests (see also *Ceja de la montaña* and *Yungas*)
    Colombia, 394
    Cuba, 316*, 322–323
    Ecuador, 394
    Oaxaca, 256–257
    Perú, 394
    priorities for inventory, **28**
    Venezuela, 367, 369
Coastal vegetation (see also Mangroves)
    Africa, West, 190*
    Brazil
        Eastern, priorities for inventory, **26, 28**
    Caribbean S. America
        Colombia, 355, 393
        priority areas for inventory, **25, 28**
    China, coral islands, 39–40
    Cuba, 324, 326–327
    Ecuador, 393–394
    Fiji, 167
    Guianas, 377–379
    Java, 101
    Kenya, 222
    Moluccas, 116
    New Guinea, 123
    Pacific Central America
        priority areas for inventory, **24, 28**
    Pacific, Colombia, 355, 393
    Philippines, 47
    Puerto Rico, 342
    Venezuela, 363
Collecting expeditions
    Africa (see also by Country)
        Central, 210–211
        East, 220–227
        West, 191–192, 194
    Amazon Basin, **7, 29**, 410
    Australia, 158–160
    Bolivia, 460
    Borneo, 97–98
    Brazil, Eastern, 443–445
    Cuba
        colonial period, 327
            modern period, 329, 332*
        republican period, 328
    Ecuador, **29**, 397
    Fiji, 169–170
    French Guiana, 385–387
    fungi, 495–498, 505
    Gabon, 200
    Guianas, 380–387
    Guyana, 381
    Hawaii, 183–184
    India, 137–138
    Java, 102
    Madagascar, 241–246
    México
        Campeche, 273
        Chiapas, 274
        Oaxaca, 261–262
        Quintana Roo, 275
    Moluccas, 116–119
    New Guinea, 125–126, 130
    Perú, 397

Philippines
    American colonial period, 51–52
    post-war period, 52–54
    pre-Linnean period, 51
    Spanish colonial period, 51
Sri Lanka, 143
Sulawesi, 111
Sumatra, 106
Surinam, 381–385
Thailand, 71–73, 77
Yucatán, 275
Collection density (see also Inventory, adequacy of)
    Australia, 160
    Bolivia, 460
    Brazil, Amazon, 416–417, 422
    Colombia, 357
    Fiji, 170
    Hawaiian I., 183–184
    Moluccas, 116
    New Guinea, 127
    Sulawesi, 111
    Sundaland, **94**
    Venezuela, 367
Collection density index (see also Collection density)
    definition, 7–8
Colombia (see also Forero, 353–361 and Gentry, 391–400)
    hepatics, 489
    number of herbarium specimens, 9, 356
    priorities for inventory, **25–26, 28**
    rates of botanical inventory, 9, *12*
Computers
    at INPA herbarium, 422
    in inventory data, 13
    in taxonomy, x, 15
Congo (see also Kendrick, 203–216)
    number of herbarium specimens, 9, 210
    priorities for inventory, **21, 28**
    rates of botanical inventory, 9, *11*
Congo River (see Zaire River Basin)
Coniferous forest (see Pine forest)
Conservation (see also Protected areas)
    importance of inventory to, 15
Convergence zone, intertropical 364
Conversion of tropical forest
    Africa (see also by Country)
        West, 190–193, **191**
    Angola, 209
    Australia, 151, 162–163
    Belize, 288, *288*
    Bolivia, 457, 459, 461–462
    Borneo, 92*
    Brazil
        Amazon, 417–421
        cerrado, 432
        E. *caatinga*, 432
        E. rainforests, 428, 441–443, 446
        Maranhão, 421
        Mato Grosso, 420–421
        restinga vegetation, 434–437
        Rondônia, 420–421
    Burundi, 209
    Cameroun, 208–209
    Cauca, 358
    China
        Qionglei, 36
        Xishuan Banna, 39
    Chocó, 358, 394
    Colombia, 355, 394–395
        Amazon, 358
        Chocó, 394
        Cauca, 358

Magdalena, 394–395
Nariño, 358, 394
Sierra de Macarena, 358
Valle, 358
Cuba, 316*, 317–318, 321, 327
anthropogenic origin of savannas, 325
Dominican Republic, 338–339
Ecuador, 392*, 395
Pacific coastal forest, 395
El Salvador, 296*, 297
Fiji, 167–169, 171–173
Gabon, 199–200
Guatemala, 282*, 283
Guianas, 377, 387
Guinée Republic, 193
Haiti, 339
Hawaiian I., 182
Honduras, 291
India, 134–136
Indochina, 86, 88
Ivory Coast, 193
Java, 92*
Kenya, 219
Liberia, 193
Madagascar, 238, 240, 245–246
Malay Peninsula, 104*
México
Oaxaca, 255–261, 264–265
transisthmic, 271
Moluccas, *114*, 115–119
New Caledonia, 178
New Guinea, 121*, 128
Nicaragua, 300*, **301**, 301–302
Panamá, 310*, **310**, 311
People's Rep. Congo, 206, 209, 212
Perú, 392*
Amazon, 393–394
Philippines, 45*, 47, 49, 55–56
Negros I., 55
priorities for inventory, 6, 17
Puerto Rico, 342–344
Rwanda, 209
Sierra Leone, 193
Sri Lanka, 142
Sulawesi, 109*, 110
Sumatra, 104*, 106
Sundaland, 94, 96
Tanzania, 219, 233
Thailand, 67
Uganda, 219
Venezuela, 363*, 364
Bolívar State, 368
Llanos, 368
Maracaibo Basin, 368
Tuy Valley, 368–369
Zaire, 213
Costa Rica (*see also* Gómez, 305–308)
mosses, 487
number of herbarium specimens, **9**, 307
priority areas for inventory, **23–24**, **28**
rates of botanical inventory, **9**, *12*
Costs of inventory, 20, 29 (*see also* Inventory, adequacy of)
Cryptograms (*see also* Buck & Thiers, 484–493 and Nishida, 494–522)
Philippines, 54
Sundaland, 96
Dams (*see* Hydroelectric projects)
Deciduous forest (*see* Seasonal forest)
Deforestation (*see* Conversion)
Desert vegetation, Oaxaca, 261
Diameter at breast height, **525**, **526**, 531

Dipterocarp forest
Indochina, 85, 86–87
Philippines, 48, **50**
Sumatra, 105
Sundaland, 94
Thailand, 64*, 67–68, 70
Discovery rates, new species (*see also* Inventory, adequacy of), **7**
Disjunctions, botanical
Brazil, 439
Cuba, 320
Distribution
interpretation of, 529
of plants, 6, 16
Diversity
Andean cloud forest, 394
Australia, **162**
bryophytes, 485
Cameroun, 205–206
ceja de la montaña, 394
China, 40
Costa Rica, 307
Cuba, 320–321
Isla de la Juventud, 321
Gabon, 200, **201**
Guayanan Floristic Province, 377
hepatics, 488
Hispaniola, 337
Java, 101
Moluccas, 116
Oaxaca, 262
Perú, Amazon, 395–396
Philippines, 54
Rwanda, 211
Sulawesi, 111
Sumatra, 106
Sundaland, 94
world, *ix*, 6
Diwata Range, 53
Dominican Republic (*see also* Zanoni, 336–340)
priority areas for inventory, **24**, **28**
Dongsha, 39–40
Drought-deciduous forest (*see* Seasonal forest)
Dynamics, forest, 530–531
Ecological studies
Eastern Brazil, 445–446, 450
Ecuador (*see also* Gentry, 391–400)
number of herbarium specimens, **9**, 397
priorities for inventory, **25–26**, **28**
rates of botanical inventory, **7**, **9**, *12*
El Salvador (*see also* Hampshire, 295–398)
number of herbarium specimens, **9**, 297
rates of botanical inventory, **9**, *12*
Endangered species (*see also* Conversion)
China, 41
Hawaiian Is., 184
India, 138
Indochina, 88
Philippines, 55
Sri Lanka, 144
Thailand, 74
Zaire, 213
Endemism, 16
Amazon Basin, 405, 413–414, 416–417
Andean valleys, 393–394
Angola, 212
Australia, 160
Bolivia, 460
Brazil
Acre, 417
caatinga, 432

campinas vegetation, 394
E. forests, 438–441, 440*
Rondônia-Apipuanã, 417
São Paulo de Olivença, 417
Tapajós, 417
Tefé, 417
Trombetas R., 417
Burundi, 212
Cameroun, 193
China, 41
Colombia, 358, 359, 393
Chocó, 393
Cuba, 318, 320–321
Dominican Republic, 338
Ecuador
Centinela Mt., 13
Río Palenque, 13, 15, 393
Fiji, 170
Gabon, 199–200
Ghana, 193
Hawaiian I., 184
hepatics, 489
India, 138
Indochina, 88
Ivory Coast, 193
Java, 96
Liberia, 193
Madagascar, 238, 244
Mayombe Escarpment, 205
México
Chiapas, 276
Oaxaca, 263
mosses, East Africa, 488
New Caledonia, 179
New Guinea, 128
Nicaragua, 302
Nigeria, 193
Panamá, 311
People's Rep. Congo, 212
Peruvian Amazon, 394, 417
Philippines, 46, 54
Puerto Rico, 343
Rwanda, 212
Sri Lanka, 143
Sulawesi, 111
Sundaland, 95, 98
Tabasco, 276
Tanzania, 234
Thailand, 65, 74
Venezuela, 367
Yucatán, 276
Zaire, 212
Enggano I., 104*
Entomogenous fungi (*see* Fungi, ectoparasitic)
Epiphytes, 97
quantitative inventory, 531
Equatorial Guinea (*see also* Kendrick, 203–216)
priorities for inventory, **21**
Ethnobotany, Oaxaca, 265
Extinction (*see also* Conversion and Endangered species), 16
"anonymous", 13
Antilles, Lesser, 349
Brazil, Eastern forests, 441
Chiapas, 276–277
Colombia, 358–359
Dominican Republic, 338
Ecuador, Pacific coastal forest, 395
Fiji, 170, 173
Gabon, 200
Ghana, 193

India, 138
Ivory Coast, 193
Liberia, 193
Madagascar, 244
New Caledonia, 179
New Guinea, 128
Nigeria, 193
Oaxaca, 264
Panamá, 311
Puerto Rico, 343
Sri Lanka, 143
Sulawesi, 111
Tabasco, 276
Venezuela, 367–368
Yucatán, 276
Extraction, sustainable in rainforest, 396
Fiji (*see also* Ash & Vodonaivalu, 166–176)
hepatics, 490
priority areas for inventory, **20**
Floodplain vegetation (*see* *Igapó*, Mangrove and *Várzea*)
Floras (*see also* Literature Cited and Selected References, for each Regional Report)
Africa (*see also* by Country)
East, 225, 233
Australia, 159–160
Bolivia, 460, **463**
Brazil
Amazon, 415–416
Eastern forests, 443–445
Burundi, 212
Cameroun, 211
China, 40–41
Colombia, 357
Cuba, **319, 330–331**
Dominican Republic, 338
Ecuador, 397–398
Fiji, 169–170
French Guiana, 385–387
Gambia, 196
Ghana, 197
Guianas, 380–387
Guinée Republic, 196
Guyana, 381–384
Hawaiian Is., 183
India, 137–138
Ivory Coast, 196
Liberia, 196
Madagascar, 241–243
Mayombe, 211
México
Chiapas, 276
Oaxaca, 262–263
mosses, 486–488
New Caledonia, 178–179
New Guinea, 124–125
Nicaragua, 302
Nigeria, 197
People's Rep. Congo, 211
Perú, 398
Philippines, 50–54
Puerto Rico, 344, **345–346**
Quintana Roo, 276
Rwanda, 212
Senegal, 196
Sierra Leone, 196
Sri Lanka, 143
Surinam, 384–385
Tabasco, 276
Thailand, 73–74
Venezuela, 367, **373–374**
Yucatán, 275

Zaire, 212
French Guiana (*see* Guianas)
Fungi (*see also* Nishida, 494-522)
    Agarics, 509
    coprophilous, 501-502
    Corticiaceae, 508
    Discomycetes, 504-505
    ecological significance of, 495, 501, 510-511
    ectoparasitic, 500
    endomycorrhizal, 501
    foliicolous, 506-507
    freshwater, 502
    jelly, 508-509
    lichenized, 503-504
    marine, 502-503
    parasitic, 502, 506, 509
    Polyporaceae, 507-508
    powdery mildew, 506
    Pyrenomycetes, 505-506
    rust, 498-499
    soil, 500-501
    Stereaceae, 508
Gabon (*see also* Breteler, 198-202)
    number of herbarium specimens, **9**, 199
    priorities for inventory, **21**
    rates of botanical inventory, **9**, *11*
Germplasm
    banks, 476, *480*
    collection of, 476-481
Ghana (*see also* Hepper, 189-197)
    number of herbarium specimens, **9**, **192**
    priorities for inventory, **21**, **27**
    rates of botanical inventory, **9**, *11*
Ghats, 138
Glaciers, Venezuela, 364
Gondwanaland, 93, 142, 178
Gradient, see Slope
Gran Sabana (*see also* Llanos, and Savannas), 369
Granitic outcrops, E. Brazil, *436*, 437
Guangxi, 37
Guaraná, 442
Guatemala (*see also* Tebbs, 281-285)
    mosses of, 487
    number of herbarium specimens, **9**, 283
    priority areas for inventory, **23-24**, **28**
    rates of botanical inventory, **9**, *12*
Guiana Highlands
    hepatics, 489
    Venezuela, 363, 364
Guiana Shield, 364-365, 376, 377, 394, 403-404
    priority areas for inventory, **25**, **28**
Guianas (*see also* Lindeman & Mori, 375-390)
    mosses of, 487
    number of herbarium specimens, **9**, **382-384**
    priorities for inventory, 25-26, 28
    rates of botanical inventory, **9**, *12*
Guineo-Congolian forest mosaic, 199, 207-208, 220-221
    priority areas for inventory, **21**, **28**
Guyana (*see* Guianas)
Hainan I., 36-37, 41, 46
Halmahera I., 114*, 116-117
Hawaiian Is. (*see also* Frame *et al.*, 181-186)
    completeness of inventory, 184-185
    herbaria, 183-185
    mosses of, 488
    number of collections, **183**
    priority areas for inventory, **20**
Height, measurement of, 526-527
Herbaria, 6-8
    Africa
        East, 225
        West, 191, **192**
    Angola, 210
    Australia, 158-159, 163-164
    Barbados, 348
    Belize, 288
    Bolivia, 461
    Brazil
        Amazon, **414**, **415**, 421-424, **426**
        E. Brazil, **443**
    Burundi, 211
    China, **40-42**
    Colombia, 356, **356**, 359
    Costa Rica, 307
    Cuba, **328**, 329
    Dominican Republic, 338-340
    Ecuador, 397
    El Salvador, 296-297
    Gabon, 200
    Ghana, **192**, **194**
    Guadelope I., 348
    Guatemala, 283-284
    Guianas, **382-384**
    Haiti, 338, 340
    Honduras, 291-292
    Indochina, **86**, 88
    Ivory Coast, **192**, 193
    Liberia, **192**
    Madagascar, 241
    Madagascar, 246
    México, 275
        Campeche, 277
        Chiapas, 277
        Oaxaca, 262
        Quintana Roo, 277
        Yucatán, 277
    Montserrat, 348
    New Guinea, 128
    Nicaragua, 301-302
    Nigeria, **192**, 194
    Panamá, 311-312
    People's Republic of Congo, 210
    Perú, 397
    Philippines, 52-53
    Puerto Rico, 342-344
    Rwanda, 211
    Sierra Leone, **192**, 193
    St. Lucia I., 348
    Sumatra, 106
    Sundaland, 97, 98
    Tabasco, 277
    Thailand, 75
    Venezuela, 363*, 367, 369-370, **372-373**
    Zaire, 210-211
Herbarium labels, 13
Herbarium resources (*see* Herbaria, and Inventory, adequacy of)
Herbarium specimens (*see* Herbaria)
    number of (*see* by Country, and Inventory, adequacy of)
Honduras (*see also* Nelson, 290-294)
    number of herbarium specimens, **9**
    priority areas for inventory, **23-24**, **28**
    rates of botanical inventory, **9**, *12*
Human population (*see also* Conversion), 15
Hydroelectric projects
    Tucuruí, 416
    Venezuela, 371
    Balbina, 420
Igapó forest, 405
Immigration (*see* Transmigration projects)
Importance value, 529
India (*see also* Kendrick, 133-140)

adequacy of inventory, 138
herbaria, 136, 139
number of herbarium specimens, 9, 136–137
priorities for inventory, 18, 27
rates of botanical inventory, 9, *10*
Indo-Malayan floristic region, 38
Indochina (*see also* Schmid, 83–90)
number of herbarium specimens, 87, **87**
number of species, 87
priorities for inventory, 18, 27
Indonesia (*see also* Ashton, 91–99; van Balgooy, 100–102; de Wilde, 103–107; de Vogel, 108–112, and Stevens, 120–132)
number of herbarium specimens, 9, 102, 106, 110, 115, **116**
priorities for inventory, 18–19, 27
rates of botanical inventory, 9, *10*
Inselbergs, 380
Inventory adequacy of (*see also* Density index), 6–8
Africa (*see also* by Country)
East, 225–228, 234
West, 192–193
Angola, 210
Belize, 288–289
Bolivia, 460–462
Brazil
Amazon, 417, 422
Eastern, 443–445
Colombia, 357–360, 398
Costa Rica, 307
Ecuador, 397
El Salvador, 297
Fiji, 170–173
fungi, 498–510
fungi, 512–514
Gabon, 199–201
general, 16, 17
Guatemala, 283–284
Guianas, 377
Hispaniola, 340
Honduras, 292
Lesser Antilles, 349
Madagascar, 243–248
México, 277
Campeche, 275
Chiapas, 275
Quintana Roo, 275
Yucatán, 275
mosses, 486–490
Nicaragua, 302–303
Oaxaca, 263–266, 269
palms, 482
Panamá, 311
Perú, 398
Puerto Rico, 344
Sangha R. Interval, 210
Sundaland, 98, 99
Tanzania, 234
Venezuela, 367, 370–371
Zaire R. Basin, 210–211
Inventory, completeness of (*see* Inventory, adequacy of)
Inventory, priorities of (*see* Inventory, adequacy of)
Inventory, rates of (*see* Inventory, adequacy of)
Irian Jaya (*see* New Guinea)
Ivory Coast (*see also* Hepper, 189–197)
number of herbarium specimens, 9, **192**
priorities for inventory, 21, 27
rates of botanical inventory, 9, *11*
Japanese Current, 46
Jari Project, 420
Java (*see also* van Balgooy, 100–102), 93*

priorities for inventory, **19, 27**
journals, *see* publications on flora
Kampuchea (*see* Indochina)
Kenya (*see also* Polhill, 217–231)
number of herbarium specimens, 9, 225
priorities for inventory, **21–22, 28**
rates of botanical inventory, 9, *11*
Kuri-Siwo (*see* Japanese Current)
Lake Victoria regional mosaic, 220–221
Land use (*see* Conversion)
Laos (*see* Indochina)
Laurasian Plate, 93
Leizhou Peninsula, 36
Lesser Antilles (*see also* Howard, 347–350)
priority areas for inventory, **24**
Leyte I., 45*
Liana forest
Amazon, 408
E. Brazil, 437
Lianas, quantitative inventory of, 525, 531
Liberia (*see also* Hepper, 189–197)
number of herbarium specimens, **9**
priorities for inventory, **21, 27**
rates of botanical inventory, 9, *11*
Lichens, 503–504
Life history, interpretation of, 530
Light gaps, measurement of, 528–529, 531
Limestone vegetation
China
Guangxi, 37–38
Yunnan, 38
Cuban *mogotes* karst formation, 326
Fiji, 171
Indochina, 85
karst formation, Puerto Rico, 342
Mexico, 271
Tabasco, 276
Moluccas, 117–118
New Guinea, 122, 127
Sulawesi, 109*
Sundaland, 95
Thailand, 65
Lingga I., 104*
Literature (*see* Floras)
Llanos (*see also* Gran Sabana, and Savannas)
Venezuela, 363, 371
Lowland rainforest
Africa (*see also* by Country)
Central, 204*, 205–208
West, 190*
Australia, 150*, 151
Belize, 287*
Bolivia, 456*, **457**, 457, 459
Borneo, 92*
Brazil,
Amazon Basin, 405–408, 406*
Eastern Brazil, *429*, 437–438, **438**
disjunctions with Amazon, 439*
Colombia, 354*, 392*, 393
Costa Rica, 306*
Cuba, 316*, 322
Ecuador, 392*
El Salvador, 296*
Guatemala, 282*
Honduras, 291*
Java, 92*
Madagascar, 237*, 238–240
Malay Peninsula, 104*
México
Chiapas, 273
Oaxaca, 256–257, 264–265

       Tabasco, 273
       Moluccas, *114*
       New Guinea, 121*
       Nicaragua, 300*
       Panamá, 310*
       Perú, 392*
       Sri Lanka, 142
       Sulawesi, 109*
       Sumatra, 104*
       Thailand, 67
       Venezuela, 363*
Luzon I., 45*, 53
Madagascar (*see also* Dorr *et al.*, 236–250)
    number of herbarium specimens, **9**, 243
    priority areas for inventory, **21, 28**
    rates of botanical inventory, **9**, *11*
Magdalena R. Valley, 393
    priority areas for inventory, **25, 28**
Makiling Mt., 53
Malay Peninsula, 104*
Malaysia (*see also* Ashton, 91–99)
    number of herbarium specimens, **9**
    priorities for inventory, **18, 27**
    rates of botanical inventory, **9**, *10*
Malesia (*see also* Tan & Rojo, 44–62; Ashton, 91–99; van Balgooy, 100–102; de Wilde, 103–107; de Vogel, 108–112; de Vogel, 113–119; and Stevens, 120–132)
    rates of botanical inventory, *10*
Mangrove forest
    Africa (*see also* by Country)
        East, 225
    Belize, 287*
    Brazil, Eastern, *429*, 434
    China
        Hainan I., 37
        Qionglei, 36
    Costa Rica, 306*
    Cuba, 316*
    Ecuador, 392*
    El Salvador, 296*
    Fuji, 167–168
    Guianas, 378
    India, Sundarbans, 134
    México
        Campeche, 274
        Oaxaca, 261
        Tabasco, 274
    Nicaragua, 300*
    Nigeria, 193
    Philippines, 47, **50**
    Puerto Rico, 343
    Thailand, 67, 69
    Venezuela, 365–366
    Yucatán, 274
    Zaire, 207
Maps, vegetation, 15–16
    Africa (*see also* by Country)
        Central, 204*, 209–210
        East, 218*, 219
        West, 190*, 190–191
    Amazon Basin, 408–410, **410**
    Angola, 210
    Australia, 150*, 152, **153**, 154*, 155*, 156*, 157*, 158*
    Belize, 287*, 288
    Bolivia, 456*, 459
    Borneo, 92*
    Brazil, Eastern, 428*, 445
    Cameroun, 209
    China, 40
    Colombia, 354*, 355, 396
    Costa Rica, 306*, 307

    Cuba, 316*, 331
    Dominican Republic, 337–338
    Ecuador, 392*, 396
    El Salvador, 296, 296*
    Fiji, 168
    Gabon, 199
    Ghana, 191
    Guatemala, 282*, 283
    Guianas, 387
    Hawaiian I., 172–173
    Honduras, 291, 291*
    India, 136
    Indochina, 84*, 87
    Ivory Coast, 191
    Java, 92*
    Lesser Antilles, 348
    Liberia, 191
    Madagascar, 237*, 240–241
    Martinique, 348
    México
        Campeche, 272
        Chiapas, 272
        Oaxaca, 255–256, 269
        Yucatán Peninsula, 272
    Moluccas, *114*
    New Caledonia, 178
    New Guinea, 121*, 123
    Nicaragua, 300*, 301
    Nigeria, 191
    Panamá, 310*, 311
    Perú, 392*, **396**, 396
    Philippines, 45*, 49
    Puerto Rico, 342
    S. America, 421
    Sierra Leone, 191
    Sri Lanka, 143
    Sulawesi, 109*
    Sumatra, 104*, 105
    Sundaland, 92*, 94
    Thailand, 64*, 71
    Venezuela, 363*, 366, **372**
    Zaire R. Basin, 209–210, 215–216
Mariana Trench, 46
Mato Grosso, 408
Mayan Empire, decline of, 271
Mentawai I., 104*
Meranti forest (*see* Dipterocarp forest)
Mexico (*see also* Lorence & Mendoza, 253–269, and Rzedowski & Rzedowski, 270–280)
    hepatics, 489
    number of herbarium specimens, **9**
    priority areas for inventory, **23, 28**
    rates of botanical inventory, **9**, *12*
Mindanao I., 45*, 53
Mindoro I., 45*, 46
Misool I., 114*
Mixtecan Shield, 254
Molave forest, 48
Moluccas (*see also* de Vogel, 120–312)
    priorities for inventory, **19, 27**
    number of herbarium specimens, 115, **116**
Monsoonal forest (*see* Seasonal forests)
Montane forest (*see also* Andean vegetation, and *Ceja de la montaña*)
    Africa (*see also* by Country)
        Central, 204*
        East, 218*
        West, 190*
    Afromontane region, 223–225
    Bolivia, 456*, **457**, 457–458
    Borneo, 92*

Brazil, Eastern, *429, 436*, 437
    ecotone with lowland, 441
Burundi, 205, 207, 215–216
Central America
    priority areas for inventory, **23**
China, Hainan I., 37
Colombia, 394
Costa Rica, 306*
Cuba, 316*, 317, 322, 324
Dominican Republic, 337
Ecuador, 392*, 394
    ecotone with lowland forest, 394
Gabon, 199
Guatemala, 282*
Guianas, 379–380
Himalayas, 134
Honduras, 291*
India, Western Ghats, 135
Indochina, 85, 86
Java, 92*, 101
Kenya, 222
Madagascar, 238–240
Malay Peninsula, 104*
México
    Chiapas, 271, 273–274
    Oaxaca, 254, 256–257
Moluccas, 114*, 115, 117–119
New Caledonia, 178
New Guinea, 121*, 123, 126
Nicaragua, 300*
Panamá, 310*
Perú, 392*–393
Philippines, 49
Puerto Rico, 342
Rwanda, 205, 207, 215–216
Sulawesi, 109*
Sumatra, 96, 104*, 105
Sundaland, 95
Taiwan, 36
Tanzania, 219, 222, 233
Thailand, evergreen, 68
Venezuela, 363*, 364–365
Zaire, 205
Morotai I., 116
Mycorrhizae (*see* Fungi, endomycorrhizal)
Myxomycetes, 499–500
    relationship with beetles, 500
Nariño, 358
New Caledonia (*see also* Kendrick, 177–180)
    herbaria, 179
    mosses, 488
    number of herbarium specimens, 178
    priority areas for inventory, **20**
New Guinea (*see also* Stevens, 120–132)
    completeness of inventory, 125–128, *127*, 130–131
    hepatics, 490
    herbaria, 124
    number of herbarium specimens, 9, 124
    priorities for inventory, **19, 27**
New distributions (*see* Inventory, adequacy of)
Nias I., 104*
Nicaragua (*see also* Sutton, 299–304)
    mosses of, 487
    number of herbarium specimens, 9, 301
    priority areas for inventory, **23–24, 28**
    rates of botanical inventory, 9, *12*
Nicobar I., 135
Nigeria (*see also* Hepper, 189–197)
    number of herbarium specimens, 9, **192**
    priorities for inventory, **21, 27**
    rates of botanical inventory, 9, *11*

Number of species (*see* Diversity)
Number of vascular plants (*see* Diversity)
Oak forest
    Campeche, 274
    Chiapas, 274
Obi I., 114*, 117
Oil-body, bryophytes, 486
Orinoco R. Basin, 363, 367
    priority areas for inventory, **25**
Pacific Islands (*see also* Ash & Vodonaivalu, 166–176; Kendrick, 177–180; and Frame *et al.*, 181–186)
    number of herbarium specimens, 9, 178, **183**
    priority areas for inventory, **20**
    rates of botanical inventory, 9
Palawan I., 45*, 46, 93, 96, 97
Paleotropic floristic province, 36
Palm forest (*see also* Babassu forest)
    Campeche, 273
Palms
    collection of, 482–483
    quantitative inventory of, 525, 531
Palynology, 474, *475*
Panamá (*see also* Hampshire, 309–312)
    number of herbarium specimens, 9, 311
    priority areas for inventory, **23–24, 28**
    rates of botanical inventory, 9, *12*
Panay I., 45*
Papua New Guinea (*see also* Stevens, 120–132)
    number of herbarium specimens, 9, 124
    priorities for inventory, **19, 27**
    rates of botanical inventory, 9, *10*
Parks (*see* Protected areas)
Páramos (*see also* Andean forests, and Montane vegetation)
    Cuba, 325
    Venezuela, 365
Perú (*see also* Gentry, 391–400)
    number of herbarium specimens, 9, 397
    priorities for inventory, **26, 28**
    rates of botanical inventory, 9, *12*
Phenology, 530
Phetchabun Mt., 66
Philippines (*see also* Tan & Rojo, 44–62)
    mosses, 488
    number of herbarium specimens, 9, 53
    priorities for inventory, **18, 27**
    rates of botanical inventory, 9, *10*
Pico Rondón, 416
Pine forest
    Belize, 287*
    Cuba, 316*, 320, 323–324
    El Salvador, 296*
    Guatemala, 282*
    Honduras, 291*
    Indochina, 85, 86
    Philippines, 49, **50**
    Sumatra, 105
    Thailand, 64*, 69
Plantations (*see also* Conversion)
    Africa, West, 190
    Brazil
        Amapá, pine, 420
        Amazon, black pepper, 420
        Eastern, cattle, 442–443
            coffee, 437, 442
            sugar cane, 442
        Eastern, cacao, 442
        Rondônia, cacao, 420
            coffee, 420
    China
        Guangxi, 37
        Qionglei, 36

Index

South, 40
Yunnan, rubber, 38, 41
Dende, (*see* Plantations, oil palm)
Fiji, 171
Malaysia, rubber, 93
oil palm, Eastern Brazil, 193, 442
Philippines, 50
dendrothermal, 47
Pyrethrum, 213
Surinam, pine, 388
Thailand, rubber, 67
Pleistocene refuge theory (*see also* Endemism, Amazon Basin), 413, 417
Point-centered quarter method, 528
Pollen samples, 474, *475*
Poorly-collected areas (*see* Inventory, adequacy of)
population, human (*see* Conversion)
Priority areas for inventory (*see* by Country, and Region), **20-28**
Projeto Flora Amazônica, 7, 442-423, 507
Protected areas
   Africa (*see also* by Country)
     East, 226
     West, 190
   Angola, 213
   Bolivia, *puna,* 457
   Brazil
     Eastern, 432, 446, **447-449**
     Amazon, 422-423
   Cameroun, 212
   China, 38, 41
     Hainan I., 37
     Qionglei, 36
     Yunnan, 38-39
   Cuba, 333, 334*
   Dominican Republic, 338
   Ecuador, 395
   Fiji, 171-172
   French Guiana, 388
     (proposed), 376*
   Ghana, **192**
   Guatemala, 284
   Guyana, 388
   Indochina, 88
   Kenya, 221-222, 224
   New Guinea, 129
   Nicaragua, 303
   Oaxaca, 265
   Panamá, 312
   Philippines, **50**
   Rwanda, 213
   Sulawesi, 111
   Sundaland, 97
   Surinam, 376*, 377, 388
   Taiwan, 36
   Tanzania, 221-222, 224-225
   Thailand, 75
   Uganda, 220-221, 224
   Zaire, 213
Quadrats, 527-528
Quantitative ecological inventory, 524-533
   French Guiana, 378
   Philippines, 57
Rare species (*see also* Conversion, and Endangered species), 7
Refuge theory (*see* Pleistocene refuge theory)
Reserves (*see* Protected areas)
Resources for inventory, 17
Restinga forest (*see also* Coastal forests, E. Brazil)
   E. Brazil, 434-437, *435*
Riau I., 104*

Rift Lake forests (*see also* Polhill, 217-231)
   priority areas for inventory, **21, 28**
Riparian forest
   Chiapas, 274
   Tabasco, 274
Río Palenque, 7, 13, 397
Roraima, Mt., 380
Salt pan substratum, Thailand, 66
Samana Peninsula, 337
Samar I., 45*
Sandstone vegetation (*see also* Tepuis)
   Amazon Basin, 407-408
   Brazil, Amapá, 403
   Indochina, 85
   Sundaland, 93
Satellite monitoring of forests
   Philippines, 50
   Rondônia, 421
   Sundaland, 96
Savanna vegetation (see also *Chaco,* Gran Sabana, and *Llanos*)
   Amazon Basin, 406-407
   Bolivia, 457
     inundated, 459
   Colombia, 354*
   Cuba, 325
     anthropogenic origin of, 325
   Guianas, 379-380
   Perú, 393*
   Thailand, 71
   Venezuela (*see also* Llanos), 365
Seasonal forest
   Belize, 287*
   Bolivia, 456*
   Brazil
     Amapá, 403
     Amazon Basin (*see also* Liana forest) 407-408
   China, 36-39
   Colombia, 354*
   Costa Rica, 306*
   Cuba, 323
   Ecuador, 392*, 393, 395
   Guatemala, 282*
   Guianas, 377-378
   India, 134-135
   Java, 101-102
   Madagascar, 237*
   México,
     Campeche, 273
     Oaxaca, 257, 259-260, 264-265
     Quintana Roo, 273
     Yucatán, 273
   Nicaragua, 300*
   Panamá, 310*
   Perú, 392*-393
   Philippines, 46
   quantitative inventory of, 524-525
   Sri Lanka, 142
   Taiwan, 36
   Thailand, 64*, 67, 70
   Tibet, 39
   Venezuela, 365
Secondary vegetation (*see* Conversion)
Semi-deciduous forest (*see* Seasonal forest)
Seram Is., 118
Serpentinic formations, Cuba, 324
Serra do Cachimbo, 406-407
Serra do Tumucumaque, 416
Serra dos Carajás, 407, 417, 420
Serra Pacaraima, 416
Serra Parima, 416

Sierra Atrevesada, 254
Sierra de Banao, 317
Sierra de Juárez, 263
Sierra de Macarena, 358
Sierra Maestra, Cuba, 321
Sierra de Perijá, 367, 369
Sierra Leone (see also Hepper 189–197)
    number of hebarium specimens, 9, **192**
    priorities for inventory, **21**, 27
    rates of botanical inventory, 9, *11*
Sierra Madre de Chiapas, 255
Sierra Madre de Oaxaca, 254
Sierra Madre del Sur (Oaxaca), 254
Sierra Madre Range, Philippines, 53
Sierra Nevada de Santa Marta, 355
Simeulue I., 104*
Singapore (see also Ashton, 91–99)
    number of herbarium specimens, **9**
    rates of botanical inventory, 9, *10*
Slash and burn (see Conversion)
Slime molds (see Myxomycetes)
Slope, measurement of, 528
Soil characteristics, measurement of, 528
Solomon Islands (see New Guinea)
Somali-Masai region
    priority areas of inventory, **21**
South America (see also Regional Reports, 353–462)
    number of herbarium specimens, 8, **9**
    priorities for inventory, **25, 26,** 28
    rates for botanical inventory, 9, *12*
Specialists, 17
    Australia, **162**
    Brazil
        Amazon, 417, **419**
        Eastern, 444, 450
    China, **42**
    Colombia, **358**
    Cuba, **330–331**
    Dominican Republic, **339**
    fungi, **497**
    Guatemala, **283**
    Guianas, **382–384**
    Hawaiian I., **185**
    Honduras, **292**
    Indochina, **89**
    Madagascar, **245**
    México
        Campeche, 274
        Oaxaca, 268
    New Guinea, 128, **129**
    Philippines, **52, 56, 57**
    Thailand, 71–73, **75–76**
    Venezuela, 368
Species/area curves, 527
Spores, fern
    collection of, 471, *472*, 473
    germination of, 473
Sri Lanka (see also Kendrick, 141–146)
    hepatics, 490
    number of herbarium specimens, 9, **143**
    priorities for inventory, **19**
    rates of botanical inventory, 9, *10*
Strand vegetation (see Coastal vegetation)
Structure, forest, measurement of, 530–531
Sula Is., 114*, 117–118
Sulawesi I. (see also de Vogel, 113–119), 114*
    priorities for inventory, **19,** 27
    number of herbarium specimens, 110
Sulu Archipelago, 45*
Sumatra (see also de Wilde, 103–107), 104*
    priorities for inventory, **18,** 27
    number of herbarium specimens, 106
Sundaland, 91–99
Surinam (see Guianas)
Swamp forest
    Africa, Central, 204*
    Cuba, freshwater, 326
    Fiji, freshwater, 167
    Guatemala, 282*
    Guianas, 379
    New Guinea, 126
    Panamá, 310*
    saltwater (see Mangrove forest)
    Sulawesi, freshwater, 111
    Sumatra, 105
    Thailand, freshwater, 69
Tahuampa forest (see also *Várzea* forest), 394
Taiwan (see also Wang, Chen & Wang, 35–43)
    number of herbarium specimens, **9**
    priorities for inventory, **18**
    rates of botanical inventory, 9, *10*
Tanzania (see also Polhill, 217–231, and Lovett, 232–235)
    number of herbarium specimens, 9, **225**
    priorities for inventory, **21–22,** 28
    rates of botanical inventory, 9, *11*
Tehuantepec, Isthmus of, 254, 271
Tepuis, 376
    mosses of, 487
    Venezuela, 363*, 364–365
Thailand (see also Smitinand, 63–82)
    number of herbarium specimens, **9**
    priorities for inventory, **18,** 27
    rates of botanical inventory, 9, *10*
Thorn scrub forest
    Campeche & Quintana Roo, 272
    Oaxaca, 259
    Venezuela, 365
Threatened areas (see Conversion)
Tibet, 39
Tijuca Massif, 446
Togo (see also Hepper, 189–197)
    number of herbarium specimens, **9**
    rates of botanical inventory, 9, *11*
Transamazon highway system, 416, 418–421
Transects, 528
Transmigration projects (see also Transamazon highway system)
    Brazilian Amazon, 416–421
        rubber boom, 418
    Fiji, 167
    New Guinea (Irian Jaya), 123
    Perú
        Amazon, 395
        ceja de la montaña, 395
        Tingo María-Pucallpa Road, 395
    Sulawesi, 110
    Sundaland, 93
Uganda (see also Polhill, 217–231)
    number of herbarium specimens, 9, **225**
    priorities for inventory, **21–22,** 28
    rates of botanical inventory, 9, *11*
Ultrabasic vegetation
    Moluccas, 117, 118
    New Caledonia, 178–179
    New Guinea, 127
Universities (see Herbaria)
Upland forest (see Montane forest)
Valle, Colombia, 358
Valles, Bolivia, 458
Várzea forest, 404–405
Vegetation maps (see Maps, vegetation)
Venezuela (see also Huber & Frame, 362–374)

number of herbarium specimens, **9**, 366
   priorities for inventory, **25–26**, **28**
   rates of botanical inventory, **9**, *12*
Vicariancy, Cuba, 320
Vietnam (*see* Indochina)
Visayan Islands, 53
Volcanic soils
   Malaysia, 93
   Sundaland, 97
Voucher specimens, *ix*, 524–525, 531
Waiego I., 121*, 127
well-collected areas (*see* Inventory, adequacy of)
West African forest mosaic (*see also* Hepper, 189–197)
   priority areas for inventory, **21**, **27**
West Indies (*see also* Caribbean regional reports, 315–347)
   hepatics, 489
   mosses of, 487
   number of herbarium specimens, **9**
   priorities for inventory, **24**, **28**
   rates of botanical inventory, **9**, *12*
Wetlands (*see* Mangrove forest, and Swamp forest)
Wilkes' U.S. Exploring Expedition, 51
Windbreak forests Qionglei, 36
Wood samples, 467–*468*
Xisha, 39
Xishaun Banna, 38
Yungas (see also *Ceja de la montaña*)
   Bolivia, 458, 460
Yunnan, 38
Zaire (*see also* Kendrick, 203–216)
   number of herbarium specimens, **9**, 211
   priorities for inventory, **21**, **28**
   rates of botanical inventory, **9**, *11*
Zaire River Basin, 205
Zanzibar-Inhambane regional mosaic, 221–223
   priority areas for inventory, **21**, **28**
Zhongsha, 39